Lecture Notes in Artificial Intelligence 9799

Subseries of Lecture Notes in Computer Science

LNAI Series Editors

Randy Goebel
University of Alberta, Edmonton, Canada
Yuzuru Tanaka
Hokkaido University, Sapporo, Japan
Wolfgang Wahlster
DFKI and Saarland University, Saarbrücken, Germany

LNAI Founding Series Editor

Joerg Siekmann
DFKI and Saarland University, Saarbrücken, Germany

More information about this series at http://www.springer.com/series/1244

Hamido Fujita · Moonis Ali
Ali Selamat · Jun Sasaki
Masaki Kurematsu (Eds.)

Trends in Applied Knowledge-Based Systems and Data Science

29th International Conference
on Industrial Engineering and Other Applications
of Applied Intelligent Systems, IEA/AIE 2016
Morioka, Japan, August 2–4, 2016
Proceedings

 Springer

Editors
Hamido Fujita
Iwate Prefectural University
Iwate
Japan

Jun Sasaki
Iwate Prefectural University
Iwate
Japan

Moonis Ali
Department of Computer Science
Texas State University
San Marcos, Texas
USA

Masaki Kurematsu
Iwate Prefectural University
Iwate
Japan

Ali Selamat
Universiti Teknologi Malaysis (UTM)
Bahru
Malaysia

ISSN 0302-9743   ISSN 1611-3349   (electronic)
Lecture Notes in Artificial Intelligence
ISBN 978-3-319-42006-6   ISBN 978-3-319-42007-3   (eBook)
DOI 10.1007/978-3-319-42007-3

Library of Congress Control Number: 2016943422

LNCS Sublibrary: SL7 – Artificial Intelligence

Printed on acid-free paper

This Springer imprint is published by Springer Nature
The registered company is Springer International Publishing AG Switzerland

Preface

The International Conference on Industrial, Engineering and Other Applications of Applied Intelligent Systems (IEA/AIE), sponsored by The International Society of Applied Intelligence (ISAI), has provided a thriving and outstanding series of conferences for almost three decades. This is done through its yearly worldwide activities that gather practitioners, scientists, businessmen, and students from different cultural backgrounds to meet and discuss hot topics in computational and applied intelligence fields. The IEA/AIE series is one of the distinguished active events bringing theory into practice for real problems in intelligent systems and problem solving. These prestigious conferences are held every year, and the 29th round of IEA/AIE in 2016 was held in Iwate, Japan. It was approved to be held in Japan during the IEA/AIE 2011 meeting in Syracuse (USA) just a few months after the March 11 earthquake that destroyed much of the infrastructure in the North-East of the Main Island of Japan, known as the "Tohoku" region. After five years of rebuilding, we believe that technology and computational intelligence are on the scientific frontier that has provided a substantial and sustainable lead to accelerate scientific practices toward a better estimation of risk analysis and prediction of catastrophic phenomena. The president of ISAI predicted in 2011 that IEA/AIE 2016 will see Japan as it was before, if not better. This prediction was right, and we were able to welcome everyone to the 29th event of AIE/AIE in Japan.

We received a variety of submissions related to several topics in applied intelligence, and this year there was an emphasis on applied knowledge-based systems and data sciences. Selected papers also cover classification of data science, an essential field in big data analysis and social networking used for variety of predictions, such as risk analysis among others. Because of the underlying trend of the submissions, we titled the conference proceedings: "Trends in Applied Knowledge-Based Systems and Data Science."

We received submissions from authors worldwide; each one was reviewed by three to four reviewers selected from the conference Program Committee. The selections of the 85 accepted articles presented in this book were based on the quality, relevance to the conference, and technical outcomes of the reported research and state of the art. These accepted papers were resubmitted as revised articles after incorporating the reviewers' required changes and modifications. In this book there are also papers presented at two special sessions: one on applied "Neural Networks," and the other on "Innovations in Intelligent Systems and Applications," as well as the following 13 chapters: 1, Data Science; 2, Knowledge-Based Systems; 3, Natural Language Processing and Sentiment Analysis; 4, Semantic Web and Social Networks; 5, Computer Vision; 6, Medical Analysis and Bio-Informatics; 7, Applied Neural Networks; 8, Innovations in Intelligent Systems and Applications; 9, Decision Support Systems; 10,

Adaptive Control; 11, Soft Computing and Multi-agent Systems; 12, Evolutionary Algorithms and Heuristic Search; and 13, System Integrations for Real-Life Applications.

We also had four well-known keynote speakers at the conference: Dr. Rose Alinda Alias, Deputy Vice-Chancellor (Academic & International) Universiti Teknologi Malaysia (UTM); Johor Malaysia, expert in learning systems and intelligent tutoring; Prof. Enrique Herrera-Viedma, Vice-President for Research and Knowledge Transfer, University of Granada Spain, expert in recommender systems; Dr. Jie Lu from the University of Technology Sydney Australia, expert in decision support systems; and Dr. Hiroshi Okuno from Waseda University and Emeritus Professor of Kyoto University, Japan, expert in voice sound analysis. These four speakers were selected based on their overwhelming experience in research and education as well as their advanced achievements that were shared with the conference participants and audience.

We would like to thank all the authors who participated by providing the quality work presented in these proceedings. Much appreciation goes to the conference Program Committee members, who devoted a great deal of time to review all submissions with the quality and technical feedback that assisted the authors in providing good-quality revisions of their papers for inclusion in this book. We also appreciate the financial and logistical support of the President of Iwate Prefectural University, Prof. Atsuto Suzuki, and the Foundation of Tateishi Science and Technology, as well as MITCA of Morioka City.

We also wish to express our appreciation to Microsoft Support making available the Conference Management Tool (CMT) used in managing the conference submission system.

May 2016   H. Fujita
                                                                                          M. Ali
                                                                                       J. Sasaki
                                                                                     A. Selamat
                                                                                  M. Kurematsu

Organization

General Chairs

Moonis Ali, USA
Hamido Fujita, Japan

Organizing Chairs

Jun Sasaki, Japan
Masanori Takagi, Japan

Program Chairs

Ali Selamat, Malaysia
Masaki Kurematsu, Japan

Web Chairs

Keizo Yamada, Japan
Issei Komatsu, Japan

Publication Chairs

Love Ekenberg, Sweden
Gajo Petrovic, Japan

Special Session Chair

Jun Hakura, Japan
Prima Oky Dicky A, Japan

Program Committee

Abdul Syukor	Universiti Teknikal Melaka, Malaysia
Adel Ben Zina	University of Carthage, Tunisia
Ahmed El-Serafy	Ain Shams University, Egypt
Aida De Haro	University of Cordoba, Spain
Akram Zeki	International Islamic University Malaysia, Malaysia
Alex Syaekhoni	Dongguk University, Korea
Alexander Vazhenin	University of Aizu, Japan

Ali Selamat	Universiti Teknologi Malaysia, Malaysia
Amruth Kumar	Ramapo College of New Jersey, USA
Ana Funes	Universidad Nacional de San Luis, Argentina
Antonio Bahamonde Rionda	Universidad de Oviedo, Spain
Ariel Monteserin	Universidad Nacional del Centro de la Provincia de Buenos Aires, Argentina
Aristides Dasso	Universidad Nacional de San Luis, Argentina
Aymen Gammoudi	University of Carthage, Tunisia
Azurah A. Samah	Universiti Teknologi Malaysia, Malaysia
Balsam Abdul Jabbar Mustafa	Universiti Malaysia Pahang, Malaysia
Beata Czarnacka-Chrobot	Warsaw School of Economics, Poland
Cheng-Fa Cheng	National Taiwan Ocean University, Taiwan
Chen Heng-chou	Chienkuo Technology University, Taiwan
Chen-Chiung Hsieh	Tatung University, Taiwan
Chi-Yo Huang	National Taiwan Normal University, Taiwan
Chidchanok Choksuchat	Silpakorn University, Thailand
Chien-Chung Chan	University of Akron, USA
Chiou-Shann Fuh	National Taiwan University, Taiwan
Chung-Hsien Kuo	National Taiwan University, Taiwan
Daniela D'Auria	University of Naples Federico II, Italy
Darryl Charles	University of Ulster, UK
Domenico Pisanelli	Institute of Cognitive Sciences and Technologies, Italy
Don Potter	University of Georgia, USA
Don-Lin Yang	Feng Chia University, Taiwan
Duco Ferro	Almende, The Netherlands
Edurne Barrenechea	Universidad Pública de Navarra, Spain
Elke Pulvermüller	University of Osnabrueck, Germany
Erik Cambria	Nanyang Technological University, Singapore
Eugene Ko	Chung Hua University, Taiwan
Farid Adaili	Conservatoire National des Arts et Métiers, France
Doutor Fernando Sérgio	Instituto Politécnico de Castelo Branco, Portugal
Fevzi Belli	University of Paderborn, Germany
Francesco Marcelloni	University of Pisa, Italy
Francisco Chiclana	De Montfort University, UK
Francisco Javier Cabrerizo	University of Granada, Spain
Frank Klawonn	Ostfalia University, Germany
Gabriella Cortellessa	Institute of Cognitive Sciences and Technologies, Italy
Gajo Petrovic	Iwate Prefectural University, Japan
Georgios Dounias	University of the Aegean, Greece
Hakan Altincay	Eastern Mediterranean University, Cyprus
Hamido Fujita	Iwate Prefectural University, Japan
Hatam Ali	Syria, Universiti Teknologi Malaysia, Malaysia
He Jiang	Dalian University of Technology, China
Hector Perez-Morago	National University of Distance Education, Spain

Hitoaki Yoshida	Iwate University, Japan
Hoshang Kolivand	Universiti Teknologi Malaysia, Malaysia
Hung-Yuan Chung	National Central University, Taiwan
José Valente de Oliveira	Universidade do Algarve , Portugal
Jae C. Oh	Syracuse University, USA
Jean-Charles Lamirel	Loria 2016, France
Jinsiang Shaw	National Taipei University of Technology, Taiwan
Jiunn-Lin Wu	National Chung Hsing University, Taiwan
Joao M. Sousa	Universidade de Lisboa, Portugal
Joao Paulo Carvalho	Instituto de Engenharia de Sistemas e Computadores, Portugal
Jooyoung Lee	Innopolis University, Canada
Jun Hakura	Iwate Prefectural University, Japan
Jun Sasaki	Iwate Prefectural University, Japan
Jyh Horng Chou	National Kaohsiung University of Applied Sciences, Taiwan
Kazuhiko Suzuki	Okayama University, Japan
Keizo Yamada	Iwate Prefectural University, Japan
Kensuke Onishi	Tokai University, Japan
Kishan Mehrotra	Syracuse University, USA
Lei Zhang	University of Illinois at Chicago, USA
Lorena Baigorria	Universidad Nacional de San Luis, Argentina
Maroua Gasmi	LISI Lab Insat, Tunis, Tunisia
Martijn Warnier	Delft University of Technology, The Netherlands
Masaki Kurematsu	Iwate Prefectural University, Japan
Masanori Takagi	Iwate Prefectural University, Japan
Nazri Kama	Universiti Teknologi Malaysia, Malaysia
Noorfa Haszlinna	Universiti Teknologi Malaysia, Malaysia
Ondrej Krejcar	University of Hradec Kralove, Czech Republic
Pak Wong	University of Macau, China
Patrick Brezillon	Accueil LIP6, France
Peter Breuer	Birmingham City University, UK
Philippe Fournier-Viger	Harbin Institute of Technology Shenzhen Graduate School, Canada
Pi-Chung Wang	National Chung Hsing University, Taiwan
Prima Oky Dicky	Iwate Prefectural University, Japan
Riccardo De Benedictis	Institute of Cognitive Sciences and Technologies, Italy
Riichiro Mizoguchi	Japan Advanced Institute of Science and Technology, Japan
Roliana Ibrahim	Universiti Teknologi Malaysia, Malaysia
Roselina Sallehuddin	Universiti Teknologi Malaysia, Malaysia
Ruben Heradio	Universidad Nacional de Educacion a Distancia, Spain
Rudolf Keller	PMOD Technologies LLC, Canada
Samir Ouchani	University of Luxembourg, Luxembourg
Satoshi Kawamura	Iwate University, Japan

Contents

Knowledge Based Systems

Natural Language Processing and Sentiment Analysis

Semantic Web and Social Networks

Computer Vision

Medical Diagnosis System and Bio-informatics

Applied Neural Networks

Innovations in Intelligent Systems and Applications

Evolutionary Algorithms and Heuristic Search

System Integration for Real-Life Applications

Data Science

Intelligent Systems in Modeling Phase of Information Mining Development Process

Sebastian Martins[1,3], Patricia Pesado[2],
and Ramón García-Martínez[3(✉)]

[1] PhD Program on Computer Science,
National University of La Plata, La Plata, Argentina
smartins089@gmail.com
[2] III-LIDI. Computer Sc School,
National University of La Plata – CIC Bs as, La Plata, Argentina
ppesado@lidi.info.unlp.edu.ar
[3] Information Systems Research Group,
National University of Lanus, Lanús, Argentina
rgml960@yahoo.com

Abstract. The Information Mining Engineering (IME) understands in processes, methodologies, tasks and techniques used to: organize, control and manage the task of finding knowledge patterns in information bases. A relevant task is selecting the data mining algorithms to use, which it is left to the expertise of the information mining engineer, developing it in a non-structured way. In this paper we propose an Information Mining Project Development Process Model (D-MoProPEI) which provides an integrated view in the selection of Information Mining Processes Based on Intelligent Systems (IMPbIS) within the Modeling Phase of the proposed Process Model through a Systematic Deriving Methodology.

1 Introduction

Information Mining is defined as the sub-discipline of information systems which provides to the Business Intelligence [1, 2] the tools to transform information into knowledge [3, 4]. Information mining based on intelligent systems [5] refers especially to the application of intelligent systems-based methods to discover and enumerate existing patterns in information. Intelligent systems-based methods allow retrieving results about the analysis of information bases that conventional methods fail to achieve [6], such as: TDIDT algorithms (Top Down Induction Decision Trees), Self-Organizing Maps (SOM) and Bayesian networks. TDIDT algorithms allow the development of symbolic descriptions of the data to distinguish between different classes [7]. Self-Organizing Maps can be applied in the construction of information clusters. They have the advantage of being tolerant to noise and the ability to extend the generalization when needing the manipulation of new data [8]. Bayesian networks can be applied to identify discriminative attributes in large information bases and detect behavior patterns in the analysis of temporal series [9].

© Springer International Publishing Switzerland 2016
H. Fujita et al. (Eds.): IEA/AIE 2016, LNAI 9799, pp. 3–15, 2016.
DOI: 10.1007/978-3-319-42007-3_1

An Information Mining Process is defined as a group of logically related tasks [10], which from a set of information with a degree of value for the organization obtains knowledge pieces that generalize the previous information. The Information Mining Engineering understands in processes, methodologies, tasks and techniques used to: organize, control and manage the task of finding knowledge patterns in information bases [11]. A Process Model for Information Mining Engineering is defined as set of phases, where each phase comprehends a set of tasks and each task has well defined inputs and outputs. Each phase of the Process Model is oriented to obtain a partial product. The whole Process Model is oriented to obtain a final product. This final product is derived from the partial products of phases. In case of Information Mining Engineering, the final product is a set of knowledge pieces [12]. There are several process models for information mining engineering: KDD [13], CRISP-DM [14], KDD + Std. IEEE 1074 [16]. Failure rate of this kind of projects is over 60 % [15].

In previous work [12] we have pointed out that the process models mentioned above are deficient in structuring the general tasks, leading to unnecessary iterations that cause delays and increasing costs. Applying data preparation tasks before determining the tools and algorithms, leads to a possible need of looping back to the stage of data preparation, when it is possible to identify those elements in advance. Furthermore, selection of data mining algorithms to use is left to the expertise of the information mining engineers.

In this context, this paper proposes a possible solution to reduce the need of human expertise in the selection of information mining algorithms. To deal with this problem, we present in Sect. 2 an Information Mining Project Development Process Model (D-MoProPEI). This Process Model has a phase called "Modeling" in which the Information Mining Processes to be used are selected. The Information Mining Processes based on intelligent systems used in the "Modeling" phase are presented in Sect. 3. A process to derive the Information Mining Processes in "Modeling" phase is presented in Sect. 4. Section 5 provides a concept proof and Sect. 5 presents the conclusions of our work.

2 Information Mining Project Development Process Model (D-MoProPEI)

MoProPEI is composed by two sub-processes: *Development*, focus on the technical activities and *Management* which covers those activities oriented to control and organize the process. Each sub-process is integrated by a set of phases, which group several activities according to their goals. Management sub-process contains five phases: Initiation, Project Planning, Support, Quality and Control and Closure. Development is integrated by six phases: Business Understanding, Data Understanding, Modeling, Data Preparation, Implementation and Assessment and Presentation.

The execution of both sub-process is not sequential, they are applied in parallel. The activities of the Management sub-process provide support to those activities associated with the construction of the final product (pieces of knowledge).

Development sub-process covers those activities associated with the identification of relevant and novel patterns, as well as the analysis and comprehension of the result obtained to generate interesting and valuable pieces of knowledge that bring added value to the organization. Figure 1 shows the internal dependencies between the development activities. Phases and their activities, inputs and outputs (on the left and right side respectively) are presented in the same horizontal line of the picture.

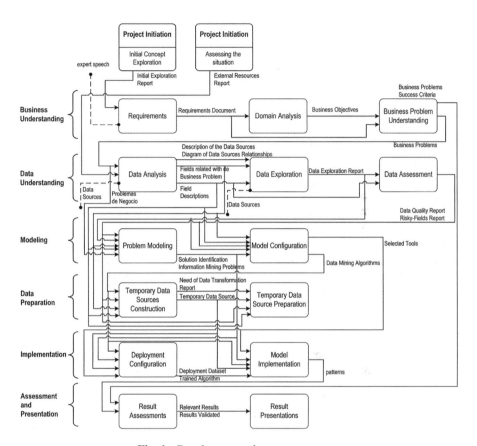

Fig. 1. Development sub-process structure

Phase 1 - Business Understanding, consist of 3 activities: *Requirements* whose inputs are expert speech (external) and Initial Exploration Report created as output in the Initial Concept Exploration phase from the Management sub-process and produces as output the Requirements Document; *Domain Analysis* whose input is the output of the previous activity and identifies the Business Objectives; and *Business Problem Understanding* whose inputs are the outputs of the two previous activities and produces the Business Problems and their Success Criteria.

Phase 2 - Data Understanding, compound by 3 activities: *Data Analysis*, whose inputs are Business Problem, External Resources Report (from the Management sub-process) and Data Sources and its outputs are Description of the Data Sources, Diagram of Data Sources Relationships, Fields related with de Business Problem, and Field Descriptions; *Data Exploration* whose inputs are the elements generated in the first activity of the phase and Data Sources and generates as output Data Exploration Report; and *Data Assessment* whose inputs are Fields related with de Business Problem and Data Exploration Report and produces as outputs Data Quality Report and Risky-Fields Report.

Phase 3 – Modeling, integrated by 2 activities: *Problem Modeling,* whose inputs are the outputs of Data Assessment, Data Exploration Report, Fields related with de Business Problem, Field Descriptions and Business Problem, and generates as output Solution Identification and Information Mining Problems; and *Model Configuration* whose inputs are Data Quality Report, Risky-Fields Report, Data Exploration Report, Fields related with de Business Problem, Field Descriptions and the outputs of the previous activity, and its outputs are Selected Tools and Data Mining Algorithms.

Phase 4 – Data Preparation, conformed by 2 activities: *Temporary Data Sources Construction* whose inputs are Data Mining Algorithms, Data Quality and Risky-Fields Reports, Fields related with de Business Problem, Field Descriptions and Business Problem, and generates as output Temporary Data Source and Need of Data Transformation Report; and *Temporary Data Source Preparation* whose inputs are the outputs of the previous activity, Data Quality and Risky-Fields Reports and transforms the Temporary Data Source.

Phase 5 – Implementation, integrated by 2 activities: *Deployment Configuration* whose inputs are Selected Tools and Data Mining Algorithms, Solution Identification and Information Mining Problems and Temporary Data Source and its outputs are Deployment Dataset and Trained Algorithm; and *Model Implementation* whose inputs are the outputs of the previous activity, Solution Identification and Information Mining Problems, Selected Tools, Data Mining Algorithms and Temporary Data Source and it generates as output the knowledge patterns.

Phase 6 – Assessment and Presentation, conformed by 2 activities: *Result Assessments* whose inputs are the results produced from the implementation model and Business Problems Success Criteria and its outputs are Relevant Results and Results Validated; and *Result Presentations* whose inputs are the outputs of the previous activity and produces as output the Final Project Report and the results are presented to the client. As result of implementing the assessment activity, belonging to the current phase, new questions about the considered problem may raise as well as the need of analyzing it in depth, being necessary to iterate over the previous stages.

Please note that activities shown Fig. 1 provide an abstraction layer which groups sets of tasks according to their goals, presenting only the results that are dependents of other activities.

3   Information Mining Processes Based on Intelligent Systems (IMPbIS)

In this section, the information-mining processes based on Intelligent Systems proposed in [12] are presented: simple processes (Sect. 3.1) and compound processes (Sect. 3.2).

3.1 Simple Processes

There have been defined three types of simple processes: [a] the **process for discovery of behavioral rules** which applies when it is necessary to identify which are the conditions to get a specific outcome in the problem domain. The following problems are examples among others that require this process: identification of the characteristics for the most visited commercial office by customers, identification of the factors that increase the sales of a specific product, definition of the characteristics or traits of customers with high degree of brand loyalty, definition of demographic and psycho-graphic attributes that distinguish the visitors to a website. For the discovery of behavioral rules from classes attributes in a problem domain that represents the available information base, it is proposed the usage of TDIDT induction algorithms [16] to discover the rules of behavior for each class attribute; [b] the **process of discovery of groups** which applies when it is necessary to identify a partition on the available information base of the problem domain. The following problems are examples among others that require this process: identification of the customers seg-ments for banks and financial institutions, identification of type of calls of customer in telecommunications companies, identification of social groups with the same charac-teristics, identification of students groups with homogeneous characteristics. For the discovery of groups [17, 18] in information bases of the problem domain for which there is no available "a priori" criteria for grouping, it is proposed the usage of Kohonen's Self-Organizing Maps or SOM [19–21]. The use of this technology intends to find if there is any group that allows the generation of a representative partition for the problem domain which can be defined from available information bases; and [c] the **Process of Discovery of Significant Attributes** which applies when it is necessary to identify which are the factors with the highest incidence (or occurrence frequency) for a certain outcome of the problem. The following problems are examples among others that require this process: factors with incidence on the sales, distinctive features of customers with high degree of brand loyalty, key-attributes that characterize a product as marketable, key-features of visitors to a website. Bayesian Networks [22] allows seeing how variations in the values of attributes, impact on the variations in the value of class attribute. The use of this process seeks to identify whether there is any interde-pendence among the attributes that model the problem domain which is represented by the available information base.

3.2 Compound Processes

There have been also defined two complex processes: [a] the **process of discovery of group membership rules** applies when it is necessary to identify which are the conditions of membership to each of the classes of an unknown partition "a priori", but existing in the available information bases of the problem domain. The following problems are examples among others that require this process: types of customer's profiles and the characterization of each type, distribution and structure of data of a web site, segmentation by age of students and the behavior of each segment, classes of telephone calls in a region and the characterization of each class. For running the process of discovery of group-membership rules it is proposed to use of self-organizing maps (SOM) for finding groups and once the groups are identified, the usage of induction algorithms (TDIDT) for defining each group behavior rules [21, 23, 24]; and [b] the **Process of Weighting of Behavior or Group-membership Rules** that first required that all sources of information (databases, files, others) are identified, and then they are integrated together as a single source of information which will be called integrated data base. Based on the integrated data base, the Self-Organizing Maps (SOM) is applied. As a result of the application of SOM, a partition of the set of records in different groups is achieved which is called identified groups. The associated files for each identified group are generated. This set of files is called "ordered groups". The "group" attribute of each ordered group is identified as the class attribute of that group, establishing it in a file with the identified class attribute (GR). Then TDIDT is applied to the class attribute of each "GR group" and the set of rules that define the behavior of each group is achieved.

4 Deriving IMPbIS in Modeling Phase of Proposed Information Mining Development Process

We have developed a methodology [25] (shown in Fig. 2) to derive the processes of information mining from frames and semantic nets. The methodology has three phases: "Analysis of Business Domain", "Analysis of the Problem of Information Mining", and "Analysis of the Process of Information Mining".

The phase "Analysis of Business Domain" develops three tasks: "Identification of the Elements and Structure of the Business Domain", "Identification of Relationships

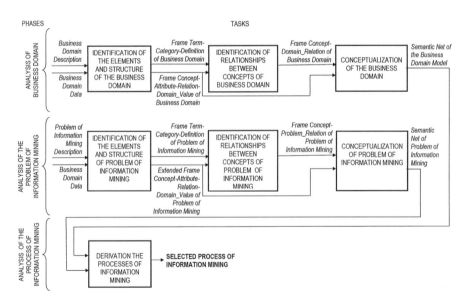

Fig. 2. Methodology to derive the processes of information mining

between Concepts of Business Domain", and "Conceptualization of the Business Domain". The task "Identification of the Elements and Structure of the Business Domain" has as input the "Business Domain Description" and the "Business Domain Data"; and produces as output the "Frame Term-Category-Definition of Business Domain" and the "Frame Concept-Attribute-Relation-Domain_Value of Business Domain". The task "Identification of Relationships between Concepts of Business Domain" has as input the "Frame Term-Category-Definition of Business Domain" and the "Frame Concept-Attribute-Relation-Domain_Value of Business Domain"; and produces as output the "Frame Concept-Domain_Relation of Business Domain". The task "Conceptualization of the Business Domain" has as input "Frame Concept-Domain_Relation of Business Domain" and the "Frame Concept-Attribute-Relation-Domain_Value of Business Domain"; and produces as output the "Semantic Net of the Business Domain Model".

The phase "Analysis of the Problem of Information Mining" develops three tasks: "Identification of the Elements and Structure of Problem of Information Mining", "Identification of Relationships between Concepts of Problem of Information Mining" and "Conceptualization of Problem of Information Mining". The task "Identification of the Elements and Structure of Problem of Information Mining" has as input the "Problem of Information Mining Description" and the "Business Domain Data"; and produces as output the "Frame Term-Category-Definition of Problem of Information Mining" and the "Extended Frame Concept-Attribute-Relation-Domain_Value of Problem of Information Mining". The task "Identification of Relationships between Concepts of Problem of Information Mining" has as input the "Frame Term-Category-Definition of Problem of Information Mining" and the "Extended Frame Concept-Attribute-Relation-Domain_Value of Problem of Information Mining"; and produces as output the "Frame Concept- Problem_Relation of Problem of Information Mining". The task "Conceptualization of Problem of Information Mining" has as input the "Extended Frame Concept-Attribute-Relation-Domain_Value of Problem of Information Mining" and the "Frame Concept- Problem_Relation of Problem of Information Mining"; and produces as output the "Semantic Net of Problem of Information Mining".

The phase "Analysis of the Process of Information Mining" develops one task: "Derivation the Processes of Information Mining" which has as input the "Semantic Net of the Business Domain Model" and the "Semantic Net of Problem of Information Mining"; and produces the "Selected Process of Information Mining".

5   Concept Proof

In this section, it presents a case of study solved by applying MoProPEI. In the following paragraphs, we present some of the documents generated by implementing the tasks that compound each phase of the Development sub-process.

Phase 1 - Business Understanding. Interactions with the experts of the different business areas are performed to comprehend the characteristics, goals and business problems of the organization. Identifying requirements, resources (either information documents/sources, as person experts in the various areas of interest), describing the domain terminology, goals/problems, risks/contingencies and success criteria. In the following paragraph the most relevant contents obtained by interviews with the experts are presented:

> "...*The purpose of the project is to facilitate the appropriation of knowledge in college/university education in massive contexts, in this particular case, focusing on the subject informatics. Providing information for proper design of public policies in college/university education, contributing to a better ownership of knowledge. In this direction, a relevant dimension is associated with student characteristics, the main actor in this complex scenario. It is desired to understand the features that provide clues about the difficulties of a student in the fulfillment of the career curricula, being able to identify and act, in an early stage, providing to the student with tools which allow him/her to overcome as far as possible obstacles that may arise. It is highlighted the interest of comprehend how the socioeconomics features are related with the academic performance of the students in massive contexts...*"

In Table 1, the business problem is presented which is related to a Business Objective previously identified. Then, experts, risk/contingencies and success criteria related with that problem are identified.

Table 1. Business problem

Superior Educational Project – Cordoba National University			Project ID:	ES.UNC
Business Problem			Document ID:	D.1.3.1
			Version:	1.0
ID	ID# Business Objectives	Problem Description	Comments	
PN.1	ON.1	Is it possible to find aspects from students to comprehend and identify in an early stage those who could present difficulties to fulfill the career curricula in the expected way?		
General Comments:				

Phase 2 - Data Understanding. From the resources previously identified, a detailed analysis of the organization's information sources (digital and non-digital) is performed and then it proceeds to analyze the characteristics of the data in relation to their applicability to business problems. To accomplished that a detailed description of the data sources and existence fields is performed (making use of existent documents such as ER model), selecting in conjunction with the expert those relevant fields (existent or to be generated from existing fields) to solve the business problems. Furthermore, a quality and risk analysis is performed pointing out possible risky fields. In Table 2, we present the data dictionary (showing only the fields related with the problem) and list the set of relevant data, explaining how to create them (Table 3).

Phase 3 – Modeling. The aim of the phase is to identify the set of tools, techniques and data mining algorithms to find pieces of knowledge to support the decision making process associated with the business problem. To reach that goal, we define the information mining problem (a technical and detailed description of the business process) and apply the methodology to derive the processes of information mining (describe in Sect. 4) identifying the set of algorithms to implement. In Table 4, it is

Table 2. Fields description

Superior Educational Project – Cordoba National University			Project ID:	ES.UNC
Fields Description			ID:	D.2.1.3
			Version:	1.0
Data Source:	RM.1			
ID	Field	Type	Description	
4	courses Informatics	Integer	Year in which the student courses Informatics	
5	Year entered	Integer	Year that the student entered to the career	
6	Sex	Boolean	sex	
8	Country	String	Country of origin	
9	Province	String	Province of origin	
10	Department	String	Department of origin	
...	...	...	...	
17	Pay studies	String	The way the student pays his/her studies	
19	Last studies father group	String	Last studies reach by the father	
22	Last studies Mother group	String	Last studies reach by the mother	
24	Quantity of subjects coursed	Integer	How many subjects course the student the first half	
25	Approved informatics Date	Date	Approved informatics	
27	subjects approved	Integer	How many subjects approved	
General Comments:				

pointed out the Information Mining Problem (IMP), and in Fig. 3 we present the conceptualization model derived from IMP getting as result the **Process of Discovery of Group-Membership Rules**.

Phase 4 – Data Preparation. After identifying the problem, tools and algorithms to use, we have all the necessary information to understand the format requirements of each of the relevant fields. At this stage the database to be used to implement the algorithms is generated and the fields are formatted and cleaned, ready to knowledge extraction.

Phase 5 – Implementation. In this phase the registers that would be used in the information mining process are defined (training and/or testing dataset). Additionally, we determine the configuration and optimization strategy that will be adopted to obtain the best results (results' understandability and success rate), according to the client's objectives. At the end of the phase, the knowledge patterns are discovered.

Phase 6 – Assessment and Presentation. The aim of this phase is identifying and filtering those patterns which provide novel and interesting knowledge through verifying the results against the success criteria and then validating them against the area expert's opinion.

Once analyzed and covered the various aspects required by the customer, it proceeds to carry out and presents the report in which the results are transferred to the customer. For example, rules that describe students who present worse academic performance are identified: [a] one group that work and are delayed to attend informatics (experts explaining that the fact of working is a strong causal of delay, being able of correcting the situation through scholarships) and [b] one group which do not work (his income comes from their families) and are delayed to attend informatics

Table 3. Fields related with the business problem

Superior Educational Project – Cordoba National University				Project ID:	ES.UNC
Fields Related with the Business Problem				Document ID:	D.2.1.4
				Version:	1.0
Business Problem		PN.1			
#IDBD.#ID Field	**To build**	**Name**	**Comments**		
RT.1.1	x	work	Generated from the **RM.1.17** variable, indicating 1 if one of the value of the multivalued field is "with their own work," and 0 otherwise.		
RT.1.2	x	family	Generated from the **RM.1.17** variable, indicating 1 if one of the value of the multivalued field is "with family's support" and 0 otherwise.		
RT.1.3	x	scholarship	Generated from the **RM.1.17** variable, indicating 1 if one of the value of the multivalued field is "with scholarship" and 0 otherwise.		
RM.1.19		Last studies father group			
RM.1.22		Last studies Mother group			
RT.1.4	x	Approved Informatics before year	Indicates whether the student passed the subject until a year of having attended it (1) or not (0), from the **RM.1.25** variable.		

RT.1.5	x	Fulfill career curricula	If **RS.1.1**> 5 => 0,
			else:
			  If **RS.1.1** < 5:
			    If (**RS.1.1** * 9 – 3 > **RM.1.27**) => 3,
			    else => 4
			    If ((**RS.1.1** – 1) * 9 +7 > **RM.1.27**) => 1
			    else => 2		

RT.1.6	x	Delay in Studying	Scale indicating the time it takes to study informatics, from the difference between the **RM.1.5** and **RM.1.4** variables: If attend the same year => 0, If attends within 2 years later => 1, else 2.		
RM.1.6		Sex			
RT.1.7	x	Initial rhythm	Scale indicating the student's progress in the first half of the career, taking into account the **RM.1.24** and **RT.1.6** variable: If assist to 5 subjects (including informatics) => 3, If assist to 3 or more subjects (including informatics) => 2, If assist to 2 or 1 subjects (including informatics) => 1, else 0.		
RT.1.8	x	Argentine	if **RM.1.8** is "Argentine" => 1,else 0		
RT.1.9	x	Cordoba	if **RM.1.9** is "Cordoba" => 1,else 0		
RT.1.10	x	Main Town	if **RM.1.10** is "Main Town" => 1, else 0		
General Comments: Fields RT.1.[8,9,10] are dependent variables (**location**)					

Table 4. Information Mining Problem

Superior Educational Project – Cordoba National University			Project ID:	ES.UNC
Business Problem			Document ID:	D.3.1.1
			Version:	1.0
ID	**#ID PN**	**Problem Description**	**Comments**	
PEI.1	PN.1	How academic student response is characterized, in relation to their similarity among socio-economic variables?		
General Comments:				

(experts pointed out that their failure of the career curricula may not being related with socioeconomics aspects, but with their personal interest, being harder to intervene).

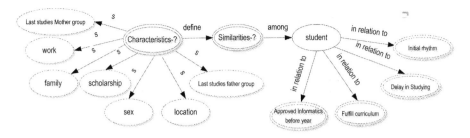

Fig. 3. Information mining problem conceptualization

6   Conclusions

Intelligent Systems has been the corner stone of data mining from the early stages. Data mining has evolved to Information Mining, and in the last decade academic movements towards an information mining engineering are increasing.

There has been pointed a weakness of current process models for information mining engineering related to the selection of information mining algorithm. This weakness lies in dependency on expertise of information mining engineer to select de information mining algorithm. Additionally, an unnecessary iteration that cause delays and increases the project' costs is highlighted.

In order to cope with this problem we propose an Information Mining Project Development Process Model (D-MoProPEI). This Process Model has a phase called "Modeling" in which the Information Mining Processes based on Intelligent Systems (IMPbIS) to be used are selected by a systematic deriving methodology and restructure the sub-process, switching the Modeling and Data Preparation phases.

We are working in an Information Mining Project Management Process Model (M-MoProPEI). Next step is to integrate both process Models in a single one to study its effectiveness to carry out information mining projects.

Besides, there are in the literature many papers and results about the convenience of the usage of certain data mining algorithms compared to others, but it is rarely raised the information mining process associated to these algorithms or the convenience of the usage of one algorithm compared to other for that process. In this context, it is an interesting open problem the identification of the relationship between the data mining algorithm and the process of information mining.

Acknowledgments. The research reported in this paper was partially funded by Project ME-SPU-PROMINF-UNLa-2015-2017 of the Argentinean Ministry of Education and Project UNLa-33A205 of the Secretary of Science and Technology of National University of Lanus (Argentina).

References

1. Thomsen, E.: BI's promised land. Intell. Enterp. **6**(4), 21–25 (2003)
2. Negash, S., Gray, P.: Business intelligence. In: Bursteiny, F., Holsapple, C. (eds.) Handbook on Decision Support Systems 2. IHIS, pp. 175–193. Springer, Heidelberg (2008)
3. Langseth, J., Vivatrat, N.: Why proactive business intelligence is a hallmark of the real-time enterprise: outward bound. Intell. Enterp. **5**(18), 34–41 (2003)
4. Grigori, D., Casati, F., Castellanos, M., Dayal, U., Sayal, M., Shan, M.: Business process intelligence. Comput. Ind. **53**(3), 321–343 (2004)
5. Michalski, R., Bratko, I., Kubat, M.: Machine Learning and Data Mining, Methods and Applications. Wiley, New York (1998)
6. Michalski, R.: A theory and methodology of inductive learning. Artif. Intell. **20**, 111–161 (1983)
7. Quinlan, J.: Learning logic definitions from relations. Mach. Learn. **5**, 239–266 (1990)
8. Kohonen, T.: Self-Organizing Maps. Springer, Heidelberg (1995)
9. Heckerman, D., Chickering, M., Geiger, D.: Learning bayesian networks, the combination of knowledge and statistical data. Mach. Learn. **20**, 197–243 (1995)
10. García-Martínez, R., Britos, P., Rodríguez, D.: Information Mining Processes Based on Intelligent Systems. In: Ali, M., Bosse, T., Hindriks, K.V., Hoogendoorn, M., Jonker, C.M., Treur, Jan (eds.) IEA/AIE 2013. LNCS, vol. 7906, pp. 402–410. Springer, Heidelberg (2013)
11. García-Martínez, R., Britos, P., Pesado, P., Bertone, R., Pollo-Cattaneo, F., Rodríguez, D., Pytel, P., Vanrell. J.: Towards an information mining engineering. In: Software Engineering, Methods, Modeling and Teaching, pp. 83–99. Medellín University Press. ISBN: 978-958-8692-32-6 (2011)
12. Martins, S., Pesado, P., García-Martínez, R. (2014). Process Mining Proposal for Information Mining Engineering: MoProPEI (in spanish). Latin-American Journal of Software Engineering, 2(5): 313–332. http://dx.doi.org/10.18294/relais.2014.313-332. ISSN: 2314-2642
13. Fayyad, U., Piatetsky-Shapiro, G., Smyth, P.: From data mining to knowledge discovery in databases. AI Mag. **17**(3), 37–54 (1996)
14. Chapman, P., Clinton, J., Keber, R., Khabaza, T., Reinartz, T., Shearer, C., Wirth, R.: CRISP-DM 1.0 Step by step BI guide. Edited by SPSS (2000)
15. Marbán, Ó., Mariscal, G., Menasalvas, E., Segovia, J.: An engineering approach to data mining projects. In: Yin, H., Tino, P., Corchado, E., Byrne, W., Yao, X. (eds.) IDEAL 2007. LNCS, vol. 4881, pp. 578–588. Springer, Heidelberg (2007)
16. Britos, P., Jiménez Rey, E., García-Martínez, E.: Work in progress: programming misunderstandings discovering process based on intelligent data mining tools. In: Proceedings 38th ASEE/IEEE Frontiers in Education Conference (2008)
17. Kaufmann, L., Rousseeuw, P.: Finding Groups in Data: An Introduction to Cluster Analysis. Wiley, New York (1990)
18. Grabmeier, J., Rudolph, A.: Techniques of Cluster Algorithms in Data Mining. Data Min. Knowl. Disc. **6**(4), 303–360 (2002)
19. Ferrero, P., Britos, P., García-Martínez, R.: Detection of Breast Lesions in Medical Digital Imaging Using Neural Networks. In: Debenham, J. (ed.) IEA/AIE 2008. IFIP, vol. 218, pp. 1–10. Springer, Boston (2008)
20. Britos, P., Cataldi, Z., Sierra, E., García-Martínez, R.: Pedagogical protocols selection automatic assistance. In: Nguyen, N.T., Borzemski, L., Grzech, A., Ali, M. (eds.) IEA/AIE 2008. LNCS (LNAI), vol. 5027, pp. 331–336. Springer, Heidelberg (2008)

21. Britos, P., Grosser, H., Rodríguez, D., García-Martínez, R.: Detecting unusual changes of users consumption. In: Bramer, M. (ed.) IEA/AIE 2008. IFIP, vol. 276, pp. 297–306. Springer, Boston (2008)

22. Britos, P., Felgaer, P., García-Martínez, R.: Bayesian networks optimization based on induction learning techniques. In: Bramer, M. (ed.) IEA/AIE 2008. IFIP, vol. 276, pp. 439–443. Springer, Boston (2008)

23. Britos, P., Abasolo, M., García-Martínez, R., Perales, F.: Identification of MPEG-4 patterns in human faces using data mining techniques. In: Proceedings 13th International Conference in Central Europe on Computer Graphics, Visualization and Computer Vision 2005, pp. 9–10 (2005)

24. Cogliati, M., Britos, P., García-Martínez, R.: Patterns in temporal series of meteorological variables using SOM & TDIDT. In: Bramer, M. (ed.) AITP. IFIP, vol. 217, pp. 305–314. Springer, Boston (2006)

25. Martins, S., Rodríguez, D., García-Martínez, R.: Deriving processes of information mining based on semantic nets and frames. In: Ali, M., Pan, J.-S., Chen, S.-M., Horng, M.-F. (eds.) IEA/AIE 2014, Part II. LNCS, vol. 8482, pp. 150–159. Springer, Heidelberg (2014)

Performance Evaluation of Knowledge Extraction Methods

Juan M. Rodríguez[1,2,3], Hernán D. Merlino[2,3], Patricia Pesado[4],
and Ramón García-Martínez[3(✉)]

[1] PhD Program on Computer Science,
National University of La Plata, La Plata, Argentina
jmrodriguez1982@gmail.com
[2] Intelligent Systems Group, University of Buenos Aires,
Buenos Aires, Argentina
jmrodriguez1982@gmail.com, hmerlino@gmail.com
[3] Information Systems Research Group,
National University of Lanús, Lanús, Argentina
rgml960@yahoo.com
[4] III-LIDI. Computer Science School,
National University of La Plata – CIC Bs As, La Plata, Argentina
ppesado@lidi.info.unlp.edu.ar

Abstract. This paper shows the precision, the recall and the F-measure for the knowledge extraction methods (under Open Information Extraction paradigm): ReVerb, OLLIE and ClausIE. For obtaining these three measures a subset of 55 newswires corpus was used. This subset was taken from the Reuters-21578 text categorization and test collection database. A handmade relation extraction was applied for each one of these newswires.

1 Introduction

The goal of this research is to decide which knowledge extraction method (for semantic relations) is the more accurate one for a given database. In this case, the chosen was Reuters-21578, a text categorization and test collection database (Lewis 1997). This collection was widely used in natural language process research projects; more specifically in text classification works (Joachims 1998). As each newswire has a quite short text and being Reuters-21578 a well known database, a subset of it has been chosen for this research. The selected extraction methods were those that, according with the state of the art research made in (Rodriguez et al. 2015), proved to be among the top three in terms of quantity and quality of the extracted knowledge pieces.

Knowledge extraction is any technique which allows the analysis of unstructured sources of information, for instance: text in natural language, using an automated process to extract the embedded knowledge in order to show it in a structured form, capable of being manipulated for an automated reasoning process, for instance: a production rule or a sub graph in a semantic network. The output information for this kind of process is called: piece of knowledge (Rancan et al. 2007). If knowledge extraction is presented as an algebraic transformation, the formula could be formulated as follows:

© Springer International Publishing Switzerland 2016
H. Fujita et al. (Eds.): IEA/AIE 2016, LNAI 9799, pp. 16–22, 2016.
DOI: 10.1007/978-3-319-42007-3_2

$$knowledge\_extraction(information\_structures) = piece\_of\_knowledge. \quad (1)$$

Since (Banko et al. 2007) presented a method of knowledge extraction for the Web, many other knowledge extraction methods for the Web have been presented. The paradigm that encompasses this type of self-supervised methods is called Open Information Extraction.

Open Information Extraction (OIE) is a paradigm that facilitates domain independent discovery of relations extracted from text and readily scales to the diversity and size of the Web corpus. The sole input to an OIE system is a corpus, and its output is a set of extracted relations. An OIE system makes a single pass over its corpus guaranteeing scalability with the size of the corpus (Banko et al. 2007).

2 Related Work

A state of the art research was made (Rodríguez et al. 2015) over a set of eight relevant semantic relation extraction methods; methods which work according Open Information Extraction paradigm. In our previous work the quality of each method's output has been compared, trying to understand which one performs a better extraction than other. The analyzed methods were: KnowItAll (Etzioni et al. 2005), TEXTRUNNER (Banko et al. 2007), WOE (Wu and Weld 2010), SRL-Lund (Christensen et al. 2011), ReVerb (Fader et al. 2011), OLLIE (Schmitz et al. 2012), ClausIE (Del Corro and Gemulla 2013), ReNoun (Yahya et al. 2014) y TRIPLEX (Mirrezaei et al. 2015). Our comparison work is summarized in Table 1, which is a double entry table, where each cell must be understood as a comparison made between two methods. The method indicated in the column against the method indicated in the row. The intersection cell shows the method that has achieved a higher quality and quantity of extracted pieces of knowledge (approximated), regardless of the measure used in the article. References from where comparison was taken, are also given.

Table 1. Summary of comparisons between methods

Methods	TextRunner	WOE	SRL-Lund	ReVerb	OLLIE	ClausIE	ReNoun	TRIPLEX
KnowItAll	TextRunner[a]							
TextRunner		WOE [b,e,i]	SRL-Lund [d]	ReVerb [e,i]		ClausIE [i]		
WOE				ReVerb [e,i]	OLLIE [f,i]	ClausIE [i]		
SRL-Lund					SRL-Lund [f]			
ReVerb					OLLIE [f,h]	ClausIE [i]		TRIPLEX, TRIPLEX + ReVerb [h]
OLLIE					ReVerb [i]	ClausIE [i]		OLLIE, TRIPLEX + OLLIE [h]
ClausIE								
ReNoun								
TRIPLEX								

References: a. (Banko et al. 2007), b. (Wu and Weld 2010), c. (Mesquita et al. 2010), d. (Christensen et al. 2011), e. (Fader et al. 2011), f. (Schmitz et al. 2012), g. (Yahya et al. 2014), h. (Mirrezaei et al. 2015), i. (Del Corro and Gemulla 2013)

We can draw the following preliminary conclusions:

[i] The best studied method, in terms of quantity and quality of knowledge pieces extracted, is ClausIE.
[ii] Since TRIPLEX in combination with OLLIE is only slightly better than OLLIE alone, we would expect that ClausIE exceeds TRIPLEX+OLLIE in precision.
[iii] If we consider again quantity and quality of knowledge pieces extracted, after ClausIE, the next methods are: OLLIE, ReVerb and WOE, in that order.

3   Experiment

The goal of this experiment is to obtain a reliable estimation about which of these three methods: ReVerb, OLLIE o ClausIE (the top three methods according to our state of the art research), has the better precision, the better recall and the better F-measure for a given database. The precision, recall, and F-measure will be calculated using the following formulas:

$$\text{precision} = \frac{\text{amount of relevant extracted knowledge pieces}}{\text{amount of extracted knowledge pieces}} \tag{2}$$

$$\text{recall} = \frac{\text{amount of relevant extracted knowledge pieces}}{\left(\text{amount of handmade relation extraction} \; + \; \text{new extracted pieces}\right)} \tag{3}$$

$$F_\beta = \frac{\left(1 + \beta^2\right) \cdot \text{precision} \cdot \text{recall}}{\left(\beta^2 \cdot \text{precision}\right) + \text{recall}} \tag{4}$$

The "new extracted pieces" in formula (3), are the relevant extracted knowledge pieces which are not in the handmade set. In formula (4), the selected value for β is 1, for simplicity the F-measure will be called F1-measure or just F1.

To calculate the confidence level and the associated margin of error for a given number of samples, the following formula (see Hamburg 1979) for sample size determination will be used:

$$n = \frac{N \cdot Z^2 \cdot p \cdot (1 - p)}{(N - 1) \cdot e^2 \; + \; Z^2 \cdot p \cdot (1 - p)} \tag{5}$$

Where:
N: total number newswire articles in Reuters-21578 (21578)
Z: is the deviation from the mean accepted to achieve the desired level of confidence
p: is the ratio we hope to find (for an unknown sample 50 % is usually taken)
e: is the maximum permissible margin error

The research goal is to get the highest confidence level with a maximum margin error of 10 %. According to the formula (5), the current confidence level will be 86 %, to know which of the three evaluated method would be the preferred one to extract semantic relations of the Reuters-21578 database, with the established error margin.

3.1 Manual Extraction

The first part of our experiment was to develop a semantic relation extraction manually, for each selected newswire of the selected subset. During this part of the experiment we were helped with several senior students of Information Engineering Bachelor Degree level. The semantic relation extraction procedure was explained to them. Finally, the authors made a revision of the students extraction work. To show an example of these handmade extractions, let's see the newswire with id 44:

> "...McLean Industries Inc's United States Lines Inc subsidiary said it has agreed in principle to transfer its South American service by arranging for the transfer of certain charters and assets to Crowley Mariotime Corp's American Transport Lines Inc subsidiary. U.S. Lines said negotiations on the contract are expected to be completed within the next week. Terms and conditions of the contract would be subject to approval of various regulatory bodies, including the U.S. Bankruptcy Court..."

The following semantic relations were obtained manually:

- (McLean Industries Inc; is subsidiary of; United States Lines Inc.)
- (McLean Industries Inc; said; it has agreed in principle to transfer its South American service by arranging for the transfer of certain charters and assets to Crowley Mariotime Corp's American Transport Lines Inc. subsidiary)
- (McLean Industries Inc; has agreed to transfer; its South American service by arranging for the transfer of certain charters and assets to Crowley Mariotime Corp's American Transport Lines Inc. subsidiary)
- (U.S. Lines; said; negotiations on the contract are expected to be completed within the next week)
- (negotiations on the contract; are expected to be completed; within the next week)
- (Terms and conditions of the contract; would be; subject to approval of various regulatory bodies)

3.2 Verification

The next step was to run the methods over the same 55 Reuter's articles and made a validation by hand for each automatic extraction. A category of three values was used: right, invalid and more-or-less-right. This last value was used for extractions in a limit, when was difficult to see if the extraction was right or not. An extraction marked as more-or-less-right was not taken into consideration for obtaining the precision and recall, in this way a penalization for do a more-or-less-right extraction was avoided, or a double penalization if we think in the F1-measure. This value (more-or-less-right) was also used to avoid compute twice two right extractions very similar each other, extractions where the only difference was in the second entity scope (typically in ClausIE). For the automatic extraction made by ClausIE over newswire with Id 44, the following two are of our interest:

- *(it; has agreed; to transfer its South American service)*
- *(it; has agreed; in principle to transfer its South American service)*

Both extractions were correct, and both made reference to the same sentence. In this particularly case the first was marked as right and the second was marked as more-or-less-right. A second consideration we had made, before mark an automatic extraction as right, was to identify if there was a manual extraction to match with the automatic one, in other words we check that both extractions that made reference to the same sentence and to the same relation, regardless of minor details. Continuing with the same example, the following manual extraction:

(McLean Industries Inc; has agreed to transfer; its South American service by arranging for the transfer of certain charters and assets to Crowley Mariotime Corp's American Transport Lines Inc subsidiary)

was considered equivalent to the following automatic extraction:

(it; has agreed; to transfer its South American service)

Even though, in this case, there were differences within the two entities and differences with the relation too; the manual version and the automatic version were considered semantically equals. Then, for cases where a valid automatic extraction was identified, but there was not matching with any handmade extraction, it was marked as "new". So at the moment of calculate the recall, the amount of valid relations was computed as all the handmade extractions plus all the automatic extraction marked as "new" (for a given method).

4 Results

The Table 2 shows (for each method) a summary of the total amount of automatic extractions made, the right ones and the total of the valid semantic relations, calculated in the way just described. The precision, recall and F1-measure were calculated using values in Table 2, with the formulas (2), (3) and (4). The obtained results are shown in Table 3.

Table 2. Summary of relations and extractions for each method.

Method	Total relations	Right extractions	Total extractions
ClausIE	650	327	638
ReVerb	569	202	301
OLLIE	633	266	545

Table 3. Precision, Recall and F1-measure for each method.

Method	Precision	Recall	F1-measure
ClausIE	0.513	0.503	0.508
ReVerb	0.671	0.355	0.464
OLLIE	0.488	0.420	0.451

5   Conclusions and Future Research

According with results summarized in the Table 1, ClausIE should have obtained the higher precision, followed by OLLIE and then by ReVerb but the obtained results contradict these assumptions. What we see is that ReVerb is the method with a higher precision, followed by ClausIE and finally by OLLIE. But if we see the obtained recall, this value is consistent with the expected results. ClausIE has a better recall than OLLIE, and OLLIE a better recall than ReVerb. ReVerb extracts less semantic relations than ClausIE or OLLIE (see in Table 2, 301 against 638 and 545), but the valid extraction percentage is bigger. To conclude, the F1-measure shows that ClausIE has a better F1-measure than ReVerb and OLLIE. The ReVerb F1-measure is a little better than OLLIE but they have almost the same value, the difference is only 0.013.

The next step in our research is to increase the evaluated newswires to 96 in order to get a confidence level of 95 % to establish which one of the three methods is the best to extract the semantic relations in Reuters-21578 database.

Acknowledgments. The research reported in this paper was partially funded by Projects UNLa-33A205 and UNLa-33B177 of National University of Lanus (Argentina). Authors wish to thank to senior students in our courses within Information Engineering Bachelor Degree at Engineering School - University of Buenos Aires for their help during the experiment.

References

Banko, M., Cafarella, M. J., Soderland, S., Broadhead, M., Etzioni, O.: Open information extraction for the web. In: IJCAI, vol. 7, pp. 2670–2676, January 2007

Christensen, J., Soderland, S., Etzioni, O.: An analysis of open information extraction based on semantic role labeling. In: Proceedings of the Sixth International Conference on Knowledge Capture, pp. 113–120. ACM (2011)

Del Corro, L., Gemulla, R.: ClausIE: clause-based open information extraction. In: Proceedings of the 22nd International Conference on World Wide Web, pp. 355–366. International World Wide Web Conferences Steering Committee, May 2013

Etzioni, O., Cafarella, M., Downey, D., Popescu, A. M., Shaked, T., Soderland, S., Weld, D.S., Yates, A.: Unsupervised named-entity extraction from the web: an experimental study. Artif. Intell. **165**(1), 91–134 (2005)

Fader, A., Soderland, S., Etzioni, O.: Identifying relations for open information extraction. In: Proceedings of the Conference on Empirical Methods in Natural Language Processing, pp. 1535–1545. Association for Computational Linguistics, July 2011

Hamburg, M.: Basic Statistics: A Modern Approach. Jovanovich, New York (1979)

Joachims, T.: Text categorization with support vector machines. In: Nédellec, C., Rouveirol, C. (eds.) Learning with many relevant features, pp. 137–142. Springer, Heidelberg (1998)

Lewis, D.D.: Reuters-21578 text categorization test collection, distribution 1.0. http://www.research.att.com/ ~ lewis/reuters21578.html

Mesquita, F., Merhav, Y., Barbosa, D.: Extracting information networks from the blogosphere: State-of-the-art and challenges. In: Proceedings of the Fourth AAAI Conference on Weblogs and Social Media (ICWSM), Data Challenge Workshop (2010)

Mirrezaei, S.I., Martins, B., Cruz, I.F.: The triplex approach for recognizing semantic relations from noun phrases, appositions, and adjectives. In: The Workshop on Knowledge Discovery and Data Mining Meets Linked Open Data (Know@LOD) Co-located with Extended Semantic Web Conference (ESWC), Portoroz, Slovenia (2015)

Rancan, C., Kogan, A., Pesado, P., García-Martínez, R.: Knowledge discovery for knowledge based systems. Some experimental results. Res. Comput. Sci. J. **27**, 3–13 (2007)

Rodríguez, J.M., García-Martínez, R., Merlino, H.D.: Revisión Sistemática Comparativa de Evolución de Métodos de Extracción de Conocimiento para la Web. XXI Congreso Argentino de Ciencias de la Computación (CACIC 2015), Buenos Aires, Argentina (2015)

Schmitz, M., Bart, R., Soderland, S., Etzioni, O.: Open language learning for information extraction. In: Proceedings of the 2012 Joint Conference on Empirical Methods in Natural Language Processing and Computational Natural Language Learning, pp. 523–534, July 2012

Wu, F., Weld, D.S.: Open information extraction using Wikipedia. In: Proceedings of the 48th Annual Meeting of the Association for Computational Linguistics, pp. 118–127. Association for Computational Linguistics, July 2010

Yahya, M., Whang, S.E., Gupta, R., Halevy, A.: Renoun: fact extraction for nominal attributes. In: Proceedings 2014 Conference on Empirical Methods in Natural Language Processing (EMNLP), Doha, Qatar, October 2014

Various Classifiers to Investigate the Relationship Between CSR Activities and Corporate Value

Ratna Hidayati[✉], Katsutoshi Kanamori, Ling Feng, and Hayato Ohwada

Department of Industrial Administration,
Faculty of Science and Technology, Tokyo University of Science,
2641 Yamazaki, Noda-shi, Chiba-ken 278-8510, Japan
7415623@ed.tus.ac.jp, {katsu,ohwada}@rs.tus.ac.jp,
fengl@rs.noda.tus.ac.jp

Abstract. The relationship between corporate social responsibility (CSR) and financial performance is complex and nuanced. Many studies have reported positive, negative, and neutral impacts of CSR on financial performance. This inconsistency is due to differences in methodologies, approaches, and selection of variables. Rather than focusing on specific variables, the present study aims to classify as many variables as possible in CSR if they contribute to shaping corporate value. In this study, we calculate corporate value using the Ohlson model based on income, since many previous studies focus on only a market-based approach. We chose some common classifiers that were appropriate for the nature of our data. After evaluating the performance of each classifier, we found that the Decision Tree is the best classifier to analyze the relationship between CSR activities and corporate value. Based on the tree, companies with high or medium corporate values seek to enhance their CSR activities or to empower secondary stakeholders (e.g., communities, societies), as indicated by cooperation with NPO/NGO. In contrast, companies with low corporate values still focus their CSR activities on primary stakeholders (e.g., customers, employees).

Keywords: Corporate social responsibility · Corporate value · Ohlson model · Income approach · Classifiers

1 Introduction

As consumer awareness of global social issues continues to grow, so does the importance of corporate social responsibility (CSR) for companies. CSR activities are an important way to increase competitive advantage, protect or raise brand awareness, and build trust with customers and employees. These benefits can be achieved through various CSR activities in which the business chooses to engage for the benefit of its stakeholders (e.g., employees, suppliers, shareholders, government, community/society, and customers). Though these activities certainly result in societal benefits, they can also enhance financial performance.

© Springer International Publishing Switzerland 2016
H. Fujita et al. (Eds.): IEA/AIE 2016, LNAI 9799, pp. 23–30, 2016.
DOI: 10.1007/978-3-319-42007-3_3

Many empirical studies have been conducted to investigate the relationship between corporate social performance (CSP) and corporate financial performance (CFP). Some researchers have found inconsistency in results, due to differences in methodologies, approaches and selection of variables [1]. In addition, few studies have based their estimations on corporate value. In fact, it is important to focus on corporate value, rather than stock returns, which have been at the center of most market-based financial studies. Many studies also examined only specific factors in CSR. In fact, there are many indicators of CSR performance. This study aims to investigate the relationship between CSR and corporate value calculated using the Ohlson model and as many factors as possible in CSR. Some machine learning techniques have also been chosen.

This paper proceeds as follows. The following section provides an overview of literature pertaining to CSR and financial performance. Section 3 explains the data collection and methodology used in this study. Section 4 discusses the study's findings. The paper ends with the limitations of the study and recommendations for further research.

2 Related Works

Correlating corporate social performance and corporate financial performance is not an easy task. Many studies have found positive, negative, or neutral impacts of CSR on financial performance. Most studies also indicate a direct or indirect relationship between CSR and financial performance. For instance, companies know that CSR is linked to their reputation and brand identity. Using structural equation modeling (SEM), one study found that CSR positively affects customer satisfaction and loyalty. The importance order of CSR factors is as follows: consumer protection, philanthropic responsibility, legal responsibility, ethical responsibility, economic responsibility, and environmental contribution [2]. CSR has become a useful tool to gain customer loyalty in today's markets. Customers who identify more strongly with a company tend to purchase more and recommend both the company and its products more often [3]. Retaining loyal customers can certainly increase profits. It is plausible that the more loyal customers companies have, the higher their profits.

Although CSR is important to win customer loyalty, not all companies have a CSR department in their organizations. According to [4], the size of corporate community involvement (CCI) programs is related to the allocation of responsibility in the company's organizational structure. Companies with small CCI programs manage their community involvement through their central administrative functions rather than a marketing/HR or CSR department. The allocation of responsibility for CCI is also related to the industry in which the company operates. Central administration of CCI was found to be disproportionately associated with service sector companies; marketing/PR departments were positively correlated with utility companies; and CSR departments were significantly more likely among financial services companies.

Regarding profitability, one study indicated no relationship between the level of directors' social responsiveness and corporate profitability, perhaps because of their limited involvement in the organization. However, the study reported that both profits and CSR are possible [5]. Another study concluded that CSR levels and their relationship with profit vary by industry. Moreover, stock market measures and accounting measures also respond differently to CSR measures, where stock market measures are better than accounting measures [6]. A study using a time series fixed effects approach found the relationship between CSR and financial performance is weaker than traditional statistical results indicated. Focusing on stakeholder management, the study also found little evidence of causality between financial performance and social performance. In detail, strong stock market performance simply leads to greater company investment in aspects of CSR addressed to employee relations. Hence, CSR activities do not affect financial performance [7].

3   Data and Methodology

For the present study, CSR data was obtained from Toyo Keizai, and financial data was obtained from the NikkeiNEEDS_CD-ROM (Nikkei Economic Electronic Databank System) in 2015. According to the Toyo Keizai database, CSR data are grouped into three categories: (1) employment and human resources (HR), (2) CSR in general, and (3) environment. Data on employment and HR include such information as number of employees, re-employment system, health and safety management system, and flextime system. Data on CSR in general include CSR department, CSR officers, stakeholder engagement, and ISO26000. Data on environment include the environmental department, climate change initiatives, environmental policy, and green purchasing. In this study, after extracting some attributes, we used a total of 37 attributes. All the data are categorical.

The next step is calculating corporate values using the Ohlson model, which is good for policy recommendations [8]. Since not all companies provide their financial data in the NikkeiNEEDS_CD-ROM, we have only 260 corporate values. The highest value is 1766976, and the lowest value is −51160.2. We cluster these values using simple k-means and the elbow method to determine the optimal number of clusters for k-means clustering [9]. We found the k in k-means clustering is 3. The k-means provide results from 260 samples, with a low value of 205, a medium value of 39, and a high value of 16. Since our objective is to classify corporate values based on CSR activities and our data are categorical, we selected some common classifiers in machine learning to address the problem. These common classifiers are Gradient Boosting, Decision Tree, Random Forest, Support Vector Machine, and k-Nearest Neighbors.

4   Results

4.1   Classifier Accuracy, Classification Report, and Confusion Matrix

In this section, we evaluate all classifiers to determine which classifier would best classify corporate values based on CSR activities. Since the data are imbalanced, we used a stratified k-fold to evaluate classifier performance. The sample is randomly partitioned into 10 k equal-sized subsamples (a common method). The accuracy of each classifier is presented in Table 1.

Table 1. Accuracy of classifiers

No.	Classifier	Accuracy
1	Gradient Boosting Classifier	0.74
2	Decision Tree Classifier	0.73
3	Support Vector Machine (SVM)	0.79
4	Random Forest Classifier	0.79
5	K-Nearest Neighbor Classifier	0.75

Table 1 indicates that the lowest accuracy (0.73) is obtained from the Decision Tree. The highest accuracy is obtained from Support Vector Machine or Random Forest, k-Nearest Neighbor, and Gradient Boosting classifier, in that order. However, Table 2 indicates that the classification report of Decision Tree is better than others.

Table 2. Classification report

class	precision					recall					f1 score					support
	GB	SVM	RF	KNN	DT	GB	SVM	RF	KNN	DT	GB	SVM	RF	KNN	DT	
0	0.81	0.79	0.8	0.83	0.85	0.93	1	0.94	0.92	0.85	0.9	0.88	0.87	0.87	0.9	250
1	0.29	0	0.3	0.27	0.36	0.15	0	0.15	0.23	0.44	0.2	0	0.2	0.25	0.4	39
2	0	0	0	0	0.11	0	0	0	0	0.06	0	0	0	0	0.1	16
avg./total	0.68	0.62	0.7	0.69	0.73	0.76	0.79	0.76	0.76	0.74	0.7	0.7	0.71	0.72	0.7	260

In the classification report, the f1-score indicates the harmonic mean of precision and recall. The results indicate that Decision Tree is sufficient to classify the data from all classes. After a classification system has been trained to distinguish between clusters 0, 1, and 2, we compute the confusion matrix of all classifiers. A confusion matrix tabulates the results of a predictive algorithm. Each row of the matrix represents an actual class, and each column represents a predicted class. Correct predictions are tallied in the table's diagonal (yellow block). Non-zero values outside the diagonal indicate incorrect predictions.

Even though its accuracy is the lowest, Decision Tree is the only classifier that can predict high corporate value or cluster 2. Table 3 indicates how frequently the predictive Decision Tree algorithm correctly classifies each type of value. The sample contains 260 companies: 205 in cluster 0, 39 in cluster 1, and 16 in cluster 2. Of the 205 in cluster 0, Decision Tree predicted that 166 are low, 31 are medium, and 8 are high. The Decision Tree confusion matrix indicates that the algorithm is not very successful in distinguishing between clusters 1 and 2. Only fourteen of the 39 companies in cluster

Table 3. Confusion matrix

classifier/label	GB			DT			SVM			RF			KNN		
0	189	11	5	166	31	8	205	0	0	192	12	1	187	18	0
1	29	8	2	23	14	2	39	0	0	29	9	1	32	7	0
2	9	7	0	7	7	2	16	0	0	14	2	0	9	7	0

1 predicted correctly as medium corporate value, and only two in cluster 2 predicted as high corporate value. However, only Decision Tree can predict clusters 1 and 2 better than other classifiers. In the Decision Tree confusion matrix, 182 samples are correctly classified, and 78 are incorrectly classified.

4.2   K-Values

Many studies that use k-means clustering usually do not offer any explanation for selecting particular values for k. Some researchers use only one or two values for k, while others utilize relatively large k values compared with the number of objects [10, 11]. In this study, for the first time to select which classifiers can address the problem, we divided corporate value into three clusters using the elbow method. In the process of building the tree, we also tried several k values to know whether the more k-values can give more information or not.

Table 4 indicates that the best accuracy is obtained for two values of k. However, the most informative attribute for this k-value differ from that for the others. The more k values, the lower the accuracy obtained. If the number of k is 3 or more, the most informative attribute is cooperation with NPO/NGO, and the best accuracy (0.804) is achieved with 3 values for k. Therefore, we keep using 3 clusters to build the tree for this study.

Table 4. Comparing k values on the tree

Number of K	Accuracy model	Most informative attribute
2	0.862	Environmental department
3	0.804	Cooperation with NPO/NGO
4	0.688	Cooperation with NPO/NGO
5	0.662	Cooperation with NPO/NGO
6	0.650	Cooperation with NPO/NGO
7	0.562	Cooperation with NPO/NGO

4.3   Building the Tree

Since our data are categorical, we used C4.5 algorithm to build the tree. As indicated in Fig. 1, cooperation with NPO/NGO is the most relevant attribute. Of 260 samples, 169 are cooperating with NPO/NGO and 91 are not. The 169 cooperating samples consist of 116 companies with low value, 38 companies with medium value, and 15 companies with high value. In the total data here, almost all companies with medium and high values are cooperating with NPO/NGOs. This result indicates many Japanese companies engage in much prudent dialogue with various NPOs/NGOs to help them verify which organization of society they should reach for, or the problems that need to be solved.

Moving on to the next leaf or the right leaf, although companies that have low value are not cooperating with NPOs or NGOs, they have a CSR officer within their

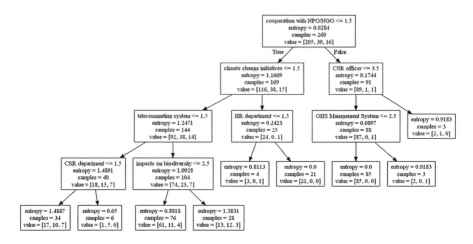

Fig. 1. Tree for corporate value and CSR activities

organizations. In addition, they realize that the risks for occupational accidents and diseases at work must be managed. To ensure effective action, they establish occupational health and safety (OHS) management systems at their workplaces. After cooperation with NPOs and NGOs, many Japanese companies are undertaking efforts in the environmental area, more correctly the tree shown about addressing climate change issues. As an integral part of sustainable development, the impact of climate change is a critical challenge for most Japanese companies. We can see most companies with medium values and high values consider with this issue. Companies with low value that are not concerned about climate changes issues, also do not have an HR department. This department usually helps manage diversity of HR in the organization.

The most prominent result involves a telecommuting system in the organization or company. Most companies with low value do not implement a telecommuting system in their organization. However, nearly half of the companies that have medium and high values implement a telecommuting system in their organization. In contrast, the other half (i.e., companies that do not implement a telecommuting system) consider with the impacts on biodiversity by business activities. The last leaf includes having a CSR department in the organization. As Fig. 1 indicates, 5 of the 15 companies with medium value do not have a CSR department in their organization. This is consistent with the previous result indicating that not all companies have a CSR department. Having a CSR department in an organization is still debatable. Recently, many companies argue in favor of eliminating the CSR department.

4.4   Variable Importance

Usually, only a few variables appear explicitly as splitters when building a tree for predicting data. The other variables might be interpreted as not important in understanding or predicting the dependent variable. However, a variable can be considered highly important even if it never appears as a node splitter.

Only 10 attributes from the 37 attributes tested are included in Table 5 for variable importance, because they have relative importance above 40. The rest are below 40. Based on Table 5, some environmental variables have high importance, indicating that many Japanese companies still include environmental issues on their agendas. We also have some CSR variables that involve with consumers or employees.

Table 5. Variable importance

No.	Variable/attribute	Relative importance
1	Consumer Support Department	80–100
2	Pro Bono Support	60–80
3	Department for managing diversity of HR	60–80
4	Green Purchasing Efforts	60–80
5	Impacts on biodiversity by business activities	60–80
6	ISO26000	60–80
7	Discretionary Labor System	40–60
8	Human Rights Efforts	40–60
9	Satellite Office	40–60
10	Re-employment system of employees	40–60

5   Conclusions and Further Works

The relationship between corporate social performance and financial performance has not been fully established, and many studies draw blurred conclusions. A number of factors influence their relationship, for instance, the relationship between CSR and corporate financial performance is researched in a certain country and in different industries. In cross-industry studies, the type of industry usually plays a key role as a control variable, which confirms that the CSP-CFP relationship differs in different industries [12].

In this study, we analyzed the relationship between CSR and corporate value with some classifiers. We then found the Decision Tree to be the best approach to this issue. Based on the tree, we conclude that companies with higher values tend to engage in CSR activities that impact their community or society more widely (secondary stakeholders). In contrast, companies with low corporate values still focus their CSR activities directly, on to their primary stakeholders, such as customers or employees. Decision Trees are simple to use and easy to understand, and offer many more advantages than other decision-making tools. However, the reliability of the information in the Decision Tree depends on using precise internal and external information at the onset. Changing variables, excluding duplicate information, or altering the sequence midway can lead to major changes and might possibly require redrawing the tree. Hence, the effect of instability result produced by the tree and also the complexity in information as a result from large tree with many branches should be considered.

For further research, we hope to improve our study by allowing the tree to become bigger in an effort to know more details about the characteristics of Japanese companies engaged in CSR activities based on their corporate values. Moreover, we would like to conduct cross-industry study.

References

1. Girerd-Potin, I., Jimenez-Graces, S., Louvent, P.: Which dimensions of social responsibility concern financial investors? J. Bus. Ethics **10**(1), 559–576 (2013). Univ. Grenoble Aples
2. Chung, K.H., Yu, J.E., Choi, M.G., Shin, J.I.: The effects of CSR on customer satisfaction and loyalty in China: the moderating role of corporate image. J. Econ. Bus. Manag. **3**(5) (2015). doi:10.7763/JOEBM.2015.V3.243
3. Ahearne, M., Bhattacharya, C.B., Gruen, T.: Antecedents and consequences of customer – company identification: expanding the role of relationship marketing. J. Appl. Psychol. **90**(3), 574–585 (2010)
4. Brammer, S., Millington, A.: The effect of stakeholder preferences, organizational structure and industry type on corporate community involvement. J. Bus. Ethics **45**, 213–226 (2003). Kluwer Academic Publishers, Printed in the Netherlands
5. O'Neil, H.M., Saunders, C.B., McCarthy, A.D.: Board members, corporate social responsiveness and profitability: are tradeoffs necessary? J. Bus. Ethics **8**, 353–357 (1989). Kluwer Academic Publishers, Printed in the Netherlands
6. Beliveau, B., Cottril, M., O'Neil, H.M.: Predicting corporate social responsiveness: a model drawn from three perspectives. J. Bus. Ethics **13**, 731–738 (1994). Kluwer Academic Publishers, Printed in the Netherlands
7. Nelling, E., Elizabeth, A.E.: Corporate social responsibility and financial performance: the "virtuous circle" revisited. Rev. Quant. Finan. Acc. **32**, 197–209 (2009). doi:10.1007/s11156-008-0090-y. Accessed 14 May 2008. Springer Science+Business Media, LLC 2008
8. Hand, J.R.M., Landsman, W.R.: Testing the Ohlson Model: v or not v, That is the Question, Working Paper, University of North Carolina at Chapel Hill (1999)
9. Madhulatha, T.S.: An overview on clustering methods. IOSR J. Eng. **2**(4), 719–725 (2012)
10. Han, J., Kamber, M.: Data Mining: Concepts and Techniques. Morgan Kaufmann, San Francisco (2000)
11. Castro, V.E., Yang, J.: A fast and robust general purpose clustering algorithm. In: Proceedings of the Fourth European Workshop on Principles of Knowledge Discovery in Databases and Data Mining (PKDD 2000), Lyon, France, pp. 208–218 (2000)
12. Brammer, S., Millington, A.: Does it pay to be different? An analysis of the relationship between corporate social and financial performance. Strateg. Manag. J. **29**, 1325–1343 (2008)

Matching Rule Discovery Using Classification for Product-Service Design

A.F. Zakaria and S.C.J. Lim[(✉)]

Department of Engineering Education,
Faculty of Technical and Vocational Education,
Universiti Tun Hussein Onn Malaysia (UTHM), Johor, Malaysia
aniesfaziehan12@gmail.com, scjohnson.lim@gmail.com

Abstract. Product-service design plays an important role in offering an optimal mix of product and accompanied service for the best customer experience and satisfaction. Previously, there are a number of design methods in literature that are focused at proposing the best combination of product and service package pertaining to customer requirements. However, the relationship between product and service elements from the perspective of customer demographics is less emphasized. In this study, we proposed a methodology to discover the matching relationship between product and service from the perspective of customer demographics. We detailed how a survey can be designed and conducted using openly available product and service information. A classification algorithm, C4.5, is applied to discover the possible product-service relationships. In order to showcase our approach, a case study of mobile phone choices and telecommunication services is presented. We have also discussed our results with some indication for future works.

Keywords: Product-service design · Rule discovery · Classification rules

1 Introduction

Product-service design is an increasingly important design area that aimed at fulfilling customer satisfaction through offering the best combination of product and service offers. It combines the tangible product and intangible service in order to optimize the product life cycle performances [1]. The main challenge for product-service design is to understand customer needs on product and associated service needs to design a product-service package that satisfy these needs [2, 3]. In this design discipline, the benefits of designing a successful product-service package can be viewed from two perspectives, which is service can complement the functional product experience; and services can ensure the value provided by product and create additional values through the entire lifecycle of product.

In the literature, early studies of product-service design methodology were conducted mainly in the viewpoint of marketing perspectives [4, 5]. Later, design engineering techniques are being proposed in literatures to offer systematic approaches towards product-service design. From the literatures, a number of studies have proposed methods such as quality function deployment (QFD) [6], fuzzy-QFD [7],

© Springer International Publishing Switzerland 2016
H. Fujita et al. (Eds.): IEA/AIE 2016, LNAI 9799, pp. 31–42, 2016.
DOI: 10.1007/978-3-319-42007-3_4

analytical hierarchy process (AHP) [8] and analytic network process (ANP) in QFD [6] that can be very useful in quantifying both product and service needs. Despite the progresses, not many of these works have studied how the design of product-service elements can be considered with regards to certain targeted customer segment. On the other perspective, there are a number of studies that have suggested several approaches for matching customer needs towards product-service attributes, such as artificial neural network [9], association rules mining [10, 11] and rough set theory [12]. Nevertheless, these matching are usually considered simultaneously. The association between product features and service features from the perspective of customer demographics are less explored in literature, which we think can be useful to determine the optimal combination of product-service package for market success.

In this paper, we proposed a methodology to discover the relationship between customer demography, product features and service through processing of customer, product and service information. Specifically, we design a survey using extracted market information in order to obtain customer preferences on different product-service packages and the survey information is processed to discover possible matching rules. The rest of this paper is organized as follows: Sect. 2 reviews the related literatures on product service design methodologies and product-service relationship discovery. We detailed our overall methodology in Sect. 3, and showcase our approach in Sect. 4. Last but not the least; we conclude our findings in Sect. 5 with some discussion on future work.

2 Literature Review

2.1 Product Service Design Methodologies

There are a number of methodologies in product-service design. From the literature, Shimomura and Arai [5] proposed a design process model for service–oriented product to increase the level of customer satisfaction. They have proposed a design process model for service-oriented product. QFD is applied to determine the effects of a service on customers. Geng et al. [6] focused on mapping customer requirements (CRs) to engineering characteristics (ECs) which includes the product related and service related ECs. Their requirement evaluation process starts with a fuzzy pair-wise comparison using analytic network process (ANP) in QFD, followed by a data envelopment analysis approach to identify initial and final weight of ECs respectively, and then a categorization of ECs into five different Kano attribute classes such as 'like', 'must-like', 'neutral' using two types fuzzy Kano's questionnaires (i.e. functional and dysfunctional). Peruzzini et al. [13] proposed a structured methodology to identify the product-service functionalities using QFD. They applied QFD to correlate customer needs with new design functionalities by selecting tangible assets (e.g., machine, material, devices and sensor) and intangible assets (e.g., competences, skills, knowledge) that are necessary to realize the new product-service from company ecosystem. The selection of assets is optimized according to the global sustainability performance. In addition, Sung [14] studied the differentiation among customers through customer segmentation discovery and customer shift tracking to predict the customer behavior patterns using recency, frequency and monetary analysis. Song et al. [8] proposed the rough analytic hierarchy process (AHP) set approach for prioritizing the vague customer PSS requirements at the earliest stage. Their work has taken into account the merging of varying opinion of different stakeholders including experts. Wang et al. [15] suggested modular development of product-service system through functional, product and service modularization using QFD approach. Functional modularization serves to discover the function decomposed of product-service system, and product or service modularization is aimed at translating the functional modules into physical or service modules.

2.2 Product-Service Relationship Discovery

The aim of discovering product-service relationship is to match a product to the right service package. From the conventional product design perspective, there are already a number of literatures that address the approach on how to perform matching from product to services. For instance, Yu et al. [9] proposed a knowledge-based artificial neural network with CART decision tree to realize the translation of customer needs towards product specification. Shao et al. [12] combined data mining and rough set to find the relationship among customer and product specification and configuration alternative. Yang [16] proposed a classification-based engineering system for modeling customers' perception responses. However, the aforementioned approaches are product-centric and may not directly consider the service elements. From the perspective of service, Jiao et al. [10] have applied ARM in the context of affective design to discover guidelines that map customer needs into the product or service. They discovered association rules using conjoint analysis for a given set of customer needs and suggested product or service specification derived from the rules. Geng et al. [11] proposed an association rule mining and maintaining a dynamic database for aiding product-service system conceptual design. An association rules mining algorithm, *Apriori*, is proposed to translate CRs into design requirements and then into module characteristics of product and services. Long et al. [17] connects customer perception needs and product-service system using multi-class support vector machines.

2.3 Summary of Literatures

From the literature, service efficiency is usually obtained based on customer expectations [4]. Therefore, service design need to consider the underlying human factors before it can be implemented with selected product. Since the level of a service success is usually a result of how customers perceive it and how the service delivers satisfaction, one of the major challenges is that important service features that matters are difficult to be identified. Compared to product specifications which are relatively easier to quantify, integrating a product with the right service elements is a challenging one. Current literatures in product-service design are mainly focused in matching customer needs to product and services which are considered simultaneously. It is noticed that the association between product features and service features given customer background are less emphasized in literature. In this study, we wish to focus on best matching of a product to corresponding service, i.e. to discover the possible relationship or matching rules between product and service, with customer demographics perspective considered. In order to reduce the service perception uncertainty that may arise, we think that it would be useful to include customer demographic information, e.g. some general customer's background information to aid in determining the right service choice. The customer demographic attributes shall be useful in determining possible differentiation in matching product with the right service. In this study, we propose a methodology of determining matching rules or relationship between product, customer demography and service using a machine learning algorithm on perceived service preference, which is detailed in the following sections.

3   Methodology

Figure 1 shows an overview of our proposed methodology toward product-service matching. As shown in Fig. 1, there are three main tasks involved: Task 1 collects market information, extracting product feature values and service feature values to generate product-service packages; Task 2 deploys a survey to gather and analyze customer preferences on packages; and Task 3 performs matching that links between product features, customer demographics and service features. In this section, we use mobile phones and the accompanied telecommunication services as an illustrative example.

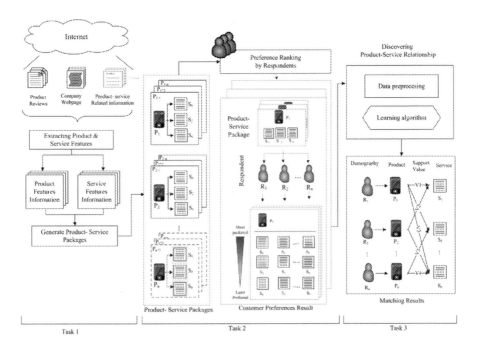

Fig. 1. Overview of methodology for product-service matching

From the figure, Task 1 starts with collecting market information from the internet. Products and services related feature information, such as package reviews and product information pages are downloaded. Generally, there are a finite number of products, N_p and services offering, N_s that can be obtained, such that there are a set of P products, $P = \{P_i | i = 1, 2, 3, \ldots N_P\}$ and a set of related service offerings, $S = \{S_j | j = 1, 2, 3, \ldots N_s\}$. The selections of products are based on their technical features such as capacity, weight, etc. while for services these features may differ depending on the type of product considered, such as the amount of call minutes for a mobile phone service, the subscription period for a streaming video service, etc. The corresponding feature values for each product and service are extracted from collected market information. Upon completing the extraction process, typical product and service feature values are identified. These values are used to categorize products and services into categories (such as high-end, mid-end or low-end). Matching product and service feature values in each product and service categories are then determined to generate typical product-service packages.

Task 2 deploys a survey with the aim to gather customer preference on available or given package choices. For this purpose, there are a number of ways where this information can be gathered, such as preference survey, wish list, actual purchasing behavior, etc. This study proposes a survey design that utilizes the generated product-service package. A survey can be designed using rating-based, ranking-based, or choice-based responses [18]. In the context of this study, we propose a ranking-based survey to measure choice importance at the individual level for product and service offered in market. An indispensable part of this survey is the demography section where customers' background is obtained. Given the survey, a number of N_r respondents are required to rank their preference on available product-service packages. Specifically, given a certain product from a product category, the respondents are asked to rank the available service packages in that product category from the 'most preferred' to the 'least preferred'. These ranking results are then aggregated to indicate service preferences for each product in their respective category.

In Task 3, available ranking results are used to perform matching that links between product and demography with service package. All obtained survey data are treated as transactional instances that undergo the pre-processing stage, where raw data are converted into suitable format for further processing. A machine algorithm is then used to perform matching between product, demography and service package. While there are a number of different algorithm that can be used, we have explored classification as our choice of learning algorithm in the context of this study to discover possible matching rules among different attributes.

For limited instances collected or small training set, the C.45 algorithm that has low bias or high variance is applied in this study [19]. Pseudo-code of the algorithm are as shown in Table 1 [20, 21]. The important parameters involved are such as confidence factor, C_f (for pruning), minimum numbers per leaf, l and number of folds, f (the amount of data used reduced-error pruning). In order to determine the suitable confidence factor in experiment, the percentage of the performance accuracy from selected classifier is obtained from cross-validation testing. Once the suitable C_f value is indicated, service categories are set as class attributes and basic steps in classification approach are used in order to calculate the potential of information gain of each services class output. Such a splitting criterion is used to determine which feature is the best split for the portion of data that reaches the particular node. The highest information gain of each branch for product features and demography information for services class will be selected as the criteria for selecting best matching rules in this study.

Table 1. C4.5 classification algorithm pseudo-code [20, 21]

C4.5 algorithm
// C4.5 algorithm is used to build a decision tree, given a set of non categorical attributes, $C_1, C_2, ..., C_n$, the categorical attributes, C and training set T of record.

function :(R a set of non-categorical attributes,
        C: the categorical attribute,
        S: a training set) returns a decision tree;

begin
        If S is empty, return a single node with value Failure;
        If S consists of records all with the same value for
          the categorical attribute,
          return a single node with that value;
        If R is empty, then return a single node with as value
          the most frequent of the values of the categorical attribute
          that are found in records of S; [note that then there
          will be errors, that is, records that will be improperly
          classified];
        Let D be the attribute with largest Gain (D, S)
          among attributes in R;
     Let $\{d_j | j = 1, 2, ..., m\}$ be the values of attribute D;
     Let $\{S_j | j = 1, 2, ..., m\}$ be the subsets of D consisting
     respectively of records with value d_j for attribute D;
     Return a tree with root labelled D and arcs labelled
     $d_1, d_2, .., d_m$ going respectively to the trees

        ID3(R-{ D }, C, S_1), ID3(R -{ D }, C, S_2), .., ID3(R -{ D }, C, S_m); end ID3;

4   An Illustrative Example

In order to showcase our methodology, a case study of matching product features and customer demography with service packages is performed using mobile phones and the accompanied telecommunication services. Corresponding to the methodology presented in Fig. 1, we have collected product features and service features information of mobile phone and telecommunication services package that are openly available[1]. First, all the selected mobile phone are sorted based on their selling prices. A total of six mobile phones are selected and sorted into three different categories: high-end, middle-end, and lower-end, with two mobile phones in each category. Similar to

---

[1] www.cnet.com, www.newdigi.com.my, www.maxis.com.my

mobile phone categorization, price is also used to determine service categories. For service offerings, there are three different service packages that are categorized: basic, standard, and premium service packages.

In determining product and service features, we have chosen a few features that are typically considered for mass consumers. The selected features and their typical values are as shown in Table 2. Common or typical product feature values are determined for each product category based on collected information. For instance, the most commonly offered feature value for phone camera resolution is 18 megapixel (MP) for high-end products. Thus, 18 MP camera resolution is considered as the typical features values for this study.

Table 2. Product and service features and features values

Product category			
Features (units)	High-end	Middle-end	Lower-end
Camera (Megapixel)	18/20	10/12	4/8
Display (Inch)	4.0/4.5	5.0/5.5	6/7
Weight (Grams)	120/130	140/170	190/200
Battery (mAH)	2800/3000	2300/2500	1400/1800
RAM (Gigabytes)	2	1.5	1

Service category			
Features (units)	Basic	Standard	Premium
Calls (Minutes)	100/200	350/450	500/1500
SMS (Number)	350/500	250/300	500/1000
Internet (Gigabytes)	1/2	4/6	8/10
Monthly Fee (RM)	50/70	100/120	150/200

RM = Malaysian Ringgit

Next, an online survey is developed based on the information in Table 1. In this survey, for each item there is a combination one product from each product category with three different service offerings. Table 3 shows a typical item of the developed preference survey. Respondents are required to state their preference by indicating their most preferred package given a mobile phone of choice with ranking scales from 1 (most preferred) to 3 (least preferred) (with non-repeating rank values). The survey also includes demographic items (to obtain respondent's background) and a completion time of 10 to 15 min is expected. The survey is distributed at our university and posted online at an online forum. The online community members were invited to complete the survey. Respondents are primarily assumed to have experience in purchasing mobile phones and subscribing mobile services. The results indicate that 60 respondents have participated in the survey. After the customer preference survey completed, descriptive statistical analysis was performed to analyze and interpret the respondent's profile information. Table 4 shows demographic profile of respondents. Generally, the results showed that majority of the respondents are female, aged between 26–32, and are pursuing postgraduate courses.

Table 3. Sample item for product-service preference survey

Phone 1	Service Package	Basic	Standard	Premium
	SMS	100	600	1000
	Call per minutes	200	500	1500
	Internet	2 GB	6 GB	8GB
Retail Price- ^{RM}2700 **Product Specification** 4.55 oz (129g) 4.7' inch 8MP Rear Camera 1820mAH 1GB RAM	Device Price	^{RM}2389	^{RM}1839	^{RM}1559
	Monthly Commitment	^{RM}50	^{RM}155	^{RM}250

Preference
1 = Most preferred; 2= moderately preferred; 3 = Least preferred

Table 4. Summary of respondents' demographic information

Demography profile									
Gender		Age range		Education		Salary range (RM)		Occupation	
Male	38.4 %	19–25	18.4 %	Dip.	5.0 %	<1000	21.7 %	Study	45.0 %
Female	61.6 %	26–32	61.6 %	B.Deg.	35.0 %	1001 – 2000	16.7 %	Work	55.0 %
		33–49	20.0 %	Post	60.0 %	2001 – 3000	23.3 %		
						3001 – 4000	11.6 %		
						>4000	26.7 %		

*Dip. – Diploma, B.Deg. – Bachelor Degree, Post – Post Graduate Degree

All the raw survey data are collected and formatted into a suitable transactional format for learning matching rules. In this study, we have applied the C4.5 classification algorithm [19]. Figure 2 the shows the comparison of classification performance accuracy of C4.5 algorithm across a range of confident factor values from 0.1 to 0.9 under different cross validation testing. Cross validation is an evaluation technique used to assess how the results of a statistical analysis will generalize to an independent test set [22]. In this study, we use 10-fold and 5-fold cross validation to test classification accuracy, where the overall data are segmented into 10 equally-sized and 5 equally-sized of test sets, respectively. For instance, in a 10-fold cross validation, full dataset are randomly divided into 10 sets, where 9 sets are randomly chosen for training to perform classification on the remaining one set. This process is repeated a few times to obtain an averaged accuracy value. In our 10-fold cross validation test, the highest percentage of performance accuracy C_f values is 0.6 (accuracy of 54.37 %), with no changes in performance increments with incremental C_f value. For 5-fold cross validation test, the highest percentage of performance accuracy is 51.39 % at $C_f = 0.4$. Given these two results, we choose $C_f = 0.6$ as our parameter to generate the best

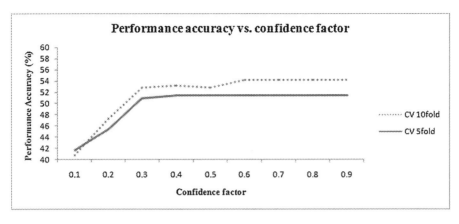

Fig. 2. Performance accuracy versus confidence factor for C4.5 algorithm

possible classification rules. With this, associated parameters chosen is $f = 2$ (number of folds) and $l = 3$ (minimum number per leaf).

Next, based on the aforementioned parameter values, we selected the best matching rules based on information gain as our criteria. Table 5 lists some of the best and interesting rules that are generated from our survey results. From the table, RAM is determined as the primary branch in the decision tree of C4.5. For the first product category, RAM = 1 GB belongs to low-end product category. It was found that under this product category, the most preferred service offering is premium service packages. Such a combination is found evident for working male respondents with age between 33–49 years old. This result shows that respondents from this age group may require the highest services offering for work related purpose with less emphasize on the phone features. On the other hand, there is another group of respondents age 26–32 years old with income below than RM 1000 also preferred the premium service packages, which may indicate heavy internet usage for premium entertainment purposes. The second category of RAM = 1.5 GB belongs to the middle-end product category. For this product category, standard and basic service offering are the most preferred service packages offering but not premium packages. The matching middle-end product features and standard service offering are deemed affordable package for all range of respondents whether their income is high or low. However, for male respondents of age group such as 33–49 years old, premium service package is preferable. Then, for high-end product category (RAM = 2 GB), the most preferred service offering are basic and standard packages. Our survey results also showed that young customers in the age group of 19–26 years old who preferred high-end mobile phones have considerably less commitment in terms of service offering fees. In overall, our study has shown that interesting and valuable insights can be discovered using our proposal, where service offerings can be tied not only to product features but also customer demographics. These can be very useful to aid product-service designers in customizing or optimizing their package offerings.

Table 5. Generated rules for product-service matching

Product	Demography	Service	Info. gain (log)
RAM = 1 GB Weight = 190 g	A = 33–49, O = W, G = M	Premium	16
	A = 26–32, I = Below 1000	Premium	9
	A = 33–49, O = W, G = Male	Premium	8
	A = 26–32, G = M, O = S, I = 2001–3000	Basic	8
	A = 19–25, E = BacDg, G = F	Standard	8
	A = 33–49, O = S, G = Male	Basic	6
RAM = 1.5 GB	A = 19–25, E = BacDg, G = F	Standard	8
	I = 3001–4000, O = W, G = M	Standard	8
	I = Below 1000, E = BacDg	Basic	6
	A = 26–32, G = F, I = 1001–2000	Standard	7
	Age–26–32, G = M, I = 2001–3000, O = W	Standard	6
	A = 33–49, I = 4000 and above, O = W, G = M	Premium	6
	A = 26–32, G = F, O = S, I = 1001–2000	Standard	5
	A = 33–49, O = W, G = F, E = PostGg	Basic	5
RAM = 2 GB Weight = 120 g	I = 4001 and above, O = W, G = M	Standard	16
	I = 2001–3000, G = F, E = PostDg	Standard	14
	I = 4001 and above, O = W, G = M	Standard	11
	I = 3001–4000, O = W, G = M	Standard	8
	I = Below 1000, A = 19–25	Basic	7
	I = 1001–2000, G = F, A = 26–32, O = S	Standard	5
	I = 2001–3000, G = M	Basic	5
	A = 26–32, G = F, O = W	Basic	4

A – Age, I – Income, G – Gender, O – Occupation, E – Education, Bac.Dg – Bachelor's Degree, PostGd – Postgraduate, M – Male, F – Female, W – Working, S – Study

5   Conclusion and Future Works

Product-service design focused on maximizing customers' satisfaction through an optimal blend of product and service package offerings. In this study, we have proposed a methodology to learn the best matching relationship between customer demography, product features and service using classification. A case study of matching user demography, mobile phone choices and telecommunication services was conducted to showcase our approach. The case study involves a small group of 60 respondents from various backgrounds. The results of the case study have indicated the usefulness of our proposal in discovering the important connections or relationship among the three factors considered. Given that, we do realize the limitations of a small-scaled preliminary study of which the classification accuracy can always benefit with a larger survey data size. For future works, we wish like to explore the possible difference between actual customer purchasing behavior and their perceptive preference of product-service package and to include such a consideration in our analysis. Particularly, we wish to investigate whether a mismatch is present between the two behaviors and how can we best minimize the gap to ensure customer satisfaction if it occurs. While we have only experimented with C4.5 classification algorithm in this study, we are currently exploring other algorithms that may be best suited for our purpose of relationship discovery. We intend to apply our findings towards proposing an optimal product-service design methodology that emphasizes on maximizing customer satisfaction. All these are research areas that await our further exploration.

Acknowledgment. The work described in this paper was partially supported by a research grant by Ministry of Higher Education, Malaysia (Grant Ref: RAGS R029).

References

1. Cook, M.B., Bhamra, T.A., Lemon, M.: The transfer and application of product service systems: from academia to UK manufacturing firm. J. Cleaner Prod. **14**(7), 1455–1465 (2006)
2. Baines, T.S., et al.: State-of-art in product service systems. Proc. Inst. Mech. Eng. Part B: J. Eng. Manuf. **221**(10), 1543–1552 (2007)
3. Goedkoop, M., et al.: Product service system, ecological and economic basic. In: Report for Dutch Ministers of Enviroment (VROM) and Economi Afffair (EZ), Netherland (1999)
4. Sakao, T., Shimomura, Y.: Service engineering: a novel engineering discipline for procedures to increase value combing service and product. J. Cleaner Prod. **15**(6), 590–604 (2007)
5. Shimomura, Y., Arai, T.: Service engineering-methods and tools for effective PSS development. In: Sakao, T., Lindahl, M. (eds.) Introduction to Product/Service-System Design, pp. 113–135. Springer, London (2009)
6. Geng, X., et al.: An integrated approach for rating engineering characteristics' final importance in product service system development. Comput. Ind. Eng. **59**, 585–594 (2010)
7. Shen, J., Wang, L.: A methodology based on fuzzy extended quality function deployment for determining optimal engineering characteristics in product-service system design. In: IEEE International Conference on Service Operations and Logistics, and Informatics, IEEE/SOLI 2008, pp. 331–336. IEEE (2008)
8. Song, W., et al.: A rough set approach for evaluating vague customer requirement of industrial product-service system. Int. J. Prod. Res. **51**(22), 6681–6701 (2013)
9. Yu, L., Wang, L., Yu, J.: Identification of product definition patterns in mass customization using a learning-based hybrid approach. Int. J. Adv. Manuf. Technol. **38**(11), 1061–1074 (2008)
10. Jiao, J., Zhang, Y., Helander, M.: A Kansei mining system for affective design. Expert Syst. Appl. **30**(4), 658–673 (2006)
11. Geng, X., Chu, X., Zhang, Z.: An association rule mining and maintaining approach in dynamic database for aiding product–service system conceptual design. Int. J. Adv. Manuf. Technol. **62**(1–4), 1–13 (2012)
12. Shao, X.Y., et al.: Integrating data mining and rough set for customer group-based discovery of product configuration rules. Int. J. Prod. Res. **44**(14), 2789–2811 (2006)

13. Peruzzini M., Marilungo E., Germani, M.: A QFD- based methodology to support product - service design in manufacturing industry. In: 2014 International ICE Conference on Engineering, Technology and Innovation (ICE), Bergamo (2014)
14. Sung, H.H.: Applying knowledge engineering techniques to customer analysis in the service industry. Adv. Eng. Inf. **21**, 293–301 (2007)
15. Wang, P.P., et al.: Modular development of product service systems. Concurrent Eng. **19**(1), 85–96 (2011)
16. Yang, C.C.: A classification-based Kansei engineering system for modeling consumers' affective responses and analyzing product form features. Expert Syst. Appl. **38**(9), 11382–11393 (2011)
17. Long, H.J., et al.: Product service system configuration based on support vector machine considering customer perception. Int. J. Prod. Res. **51**(18), 5450–5468 (2013)
18. Green, P.E., Srinivasan, V.: Conjoint analysis in consumer research: issue and outlook. J. Consum. Res. **5**, 110–123 (1978)
19. Salzberg, S.L.: C4.5: programs for machine learning by J. Ross Quinlan. Morgan Kaufmann Publishers Inc., 1993. Mach. Learn. **16**(3), 235–240 (1994)
20. Quinlan, J.R.: Induction of decision trees. Mach. Learn. **1**(1), 81–106 (1986)
21. Quinlan, J.R.: C4. 5: Programs For Machine Learning. Elsevier, Amsterdam (2004)
22. Witten, I.H., Frank, E., Hall, M.A.: Data Mining: Practical Machine Learning Tools and Techniques. Sabre Foundation, United States (2011)

Rare Event-Prediction with a Hybrid Algorithm Under Power-Law Assumption

Mina Jung$^{(\boxtimes)}$ and Jae C. Oh

Department of Electrical Engineering and Computer Science,
Syracuse University, Syracuse, NY 13210, USA
{mijung,jcoh}@syr.edu

Abstract. We present an algorithm for predicting both common and rare events. Statistics show that occurrences of rare events are usually associated with common events. Therefore, we argue that predicting common events correctly is an important step toward correctly predicting rare events. The new algorithm assumes that frequencies of events exhibit a power-law distribution. The algorithm consists of components for detecting rare event types and common event types, while minimizing computational overhead. For experiments, we attempt to predict various fault types that can occur in distributed systems. The simulation study driven by the system failure data collected at the Pacific Northwest National Laboratory (PNNL) shows that fault-mitigation based on the new prediction mechanism provides 15 % better system availability than the existing prediction methods. Furthermore, it allows only 10 % of all possible system loss caused by rare faults in the simulation data.

Keywords: Event prediction · Rare event · Power-law distribution · Logistic regression · Bayesian inference · Online learning · Fault mitigation

1 Introduction

Prediction of events is an important area of research with a varieties of application areas. Due to the rapid advances in computing power, accurate models can be devised to predict significant phenomena in nature such as earthquakes and tsunamis. Predicting natural disasters benefits public safety by informing citizens and authorities about what to expect and allowing them an adequate amount of time to prepare when emergency measures are required.

Various events occur with various frequencies and severities. In general, frequent events often cause relatively less severe consequences [1]. On the other hand, rare ones are extremely difficult to predict yet the consequences can be catastrophic. A challenge in predicting rare event is that available data for rare events are often insufficient; this obviously is true otherwise these events are not called as rare events. With some statistical analysis, we observed that a combination of multiple common events repeatedly happens before a certain type of rare

© Springer International Publishing Switzerland 2016
H. Fujita et al. (Eds.): IEA/AIE 2016, LNAI 9799, pp. 43–55, 2016.
DOI: 10.1007/978-3-319-42007-3_5

events occur. We hypothesize, therefore, in order to predict rare events, common events also must be observed and characterized. We combine the principle of power-law distribution in event frequencies, logistic regression, and Bayesian inference to design a new adaptive event prediction model that is capable of predicting events with various frequencies including rare ones.

In our experiments, we use the PNNL (Pacific Northwest National Laboratory) system failure data collected from about 1,000 computing nodes over 4 years [2]. The data follows the power-law distribution in events occurrence in terms of their severities and frequencies. We analyze the data, and classify events into 39 types, then we group the 39 types into 10 categories based on the severity which is magnitude of fault effects on the system and its locality. (See Table 1.) Simulation studies are conducted by using the data as if events in the data are occurring in real-time. Our prediction algorithm is deployed to learn and predict as events are occurring. Therefore, this is an online learning. The simulation results show that the new prediction mechanism provides 15 % better system availability than the probability-based prediction with incremental updates and it allows only 10 % of all possible system loss caused by rare faults in the simulation data.

This paper is organized as follows. Section 2 presents a background on common and rare events and the PNNL failure data to test our prediction mechanism. Section 3 presents the proposed event prediction mechanism combining power-law, logistic regression, and Bayesian inference. In Sect. 4, we conduct experiments to predict and detect fault events and mitigate them in the simulated system based on the PNNL system and evaluate results. Finally, Sect. 5 concludes the paper and discuss the future work.

2   Motivation and Background

The motivating application is to predict faults in distributed systems as in the PNNL system. In such a setting, the algorithm starts with no knowledge about the characteristics of the environment or event frequencies. As events start to occur, the algorithm accumulates information about each type of event. This is an online reinforcement learning to predict certain event types [3]. Key characteristics of the event are updated after the prediction is made and the actual event instance is discovered. Information from actual instances is used to refine the hypothesis [4].

2.1   Common and Rare Events

In earthquake research, seismic researchers have measured everyday tremors and earthquakes on the Richter scale over a number of years [5]. Many foreshocks are observed before a larger seismic event and this suggests that some small events trigger a larger one. Earthquake magnitudes are empirically observed to follow the Gutenberg-Richter law which exhibits a power-law behavior in its frequency/magnitude distribution [1]. Not all distributions follow power-laws, however, power-law distributions seems to appear quite often naturally.

In general, rare events are often assumed to be high-magnitude (i.e., severe) and drawn from the tail of a distribution while common events – i.e., frequently observed events–have low-magnitudes. It is important to observe that rare events and common events share significant causal similarity [1]. Studying common events from sufficient observations of the *symptom-event causalities* can help to predict rare events. For example, some degree of tremors and low Richter scale earthquakes had been observed over time in a certain region before Richter 5 earthquakes occurred and therefore, the probability of the Richter 5 earthquake could be estimated. When the Richter 5 or 7 earthquakes often occurred in the region, the big one (> Richter 8) hasn't happened yet and it will hit someday with a predicted probability corresponding to the probabilities of common ones.

2.2   Test Case: PNNL (Pacific Northwest National Laboratory) System Failure Data

We introduce the failure data from the aforementioned PNNL dataset. We use this data to test our new algorithm. The failure data was collected on the High Performance Computing System-2 (MPP2) of the Molecular Science Computing Facility (MSCF) at the Pacific Northwest National Laboratory [2]. According to the data, the system has the following configuration: 924 computing nodes, 15 servers, 27 switches, 38 storages and 1 tape driver. Computing nodes have computing capabilities of 10 GB RAM and 10 to 430 GB local HDD to run large-scale scientific simulations or visualization applications. Servers and computing nodes are computational nodes, and switches, storages and tape driver are non-computational nodes.

The recorded failures were collected from November 2003 through September 2007. Total number of events are 4650. Table 1 classifies the PNNL data into 39 types of events, and groups 39 types into 10 categories. The ten categories of severity are decided with magnitude of fault effects on the system and locality of faults. For example, f_4 (memory) faults happened on node components only for two times. In terms of frequency, f_4 type is one of the extremely rare events, but it just affects a local node and has similar level of effect on the system as f_3 (dimm) faults which are also related to memory chips. There are some associations between faults. Hence, faults belonging to severity s_i are a collection of multiple types of faults with similar effect and locality.

3   Hybrid Event Prediction Algorithm

3.1   Probability-Based Prediction

We start with a question: *how often and when* an event occur? *How often* refers to the estimated frequency of the event and *when* refers to the estimated time

Table 1. Severity table of PNNL failure data

Severity	Component	Fault type		♯ of events	Total	Per node	%
s_1	node	f_1	disk	2036	2038	2.206	43.82
		f_2	hsv	2			
s_2	node	f_3	dimm	838	1373	1.486	29.63
		f_4	memory	2			
		f_5	cpu	327			
		f_6	platform	106			
		f_7	cable	86			
		f_8	cntlr	59			
		f_9	scsi_bp	31			
		f_{10}	fan	17			
		f_{11}	console	7			
s_3	storage	f_{21}	hsv	469	472	12.42	10.15
		f_{22}	disk	1			
		f_{23}	console	1			
		f_{24}	cntlr	1			
s_4	node	f_{12}	os	271	328	0.355	7.05
		f_{13}	misc	46			
		f_{14}	appl	6			
		f_{15}	mlb	2			
		f_{16}	select hw/sw	1			
		f_{17}	core	2			
s_5	node	f_{18}	unknown	147	147	0.159	3.16
s_6	node	f_{19}	elan	77	77	0.083	1.66
s_7	switch	f_{25}	elan	51	65	2.407	1.4
		f_{26}	misc	4			
		f_{27}	core	4			
		f_{28}	fc	3			
		f_{29}	qsw	2			
		f_{30}	cntlr	1			
s_8	node	f_{20}	ps	25	25	0.027	0.54
s_9	server	f_{31}	misc	11	24	1.6	0.52
		f_{32}	platform	3			
		f_{33}	cntlr	3			
		f_{34}	cpu	2			
		f_{35}	dimm	2			
		f_{36}	disk	1			
		f_{37}	unknown	1			
		f_{38}	console	1			
s_{10}	tape	f_{39}	library	1	1	1	0.02

of event occurrence. First, we start using a simple probability-based prediction about a particular type of event with sigmoid model.

$$p = 2 \times \left(\frac{1}{1 + e^{-dt}} - \frac{1}{2}\right) \tag{1}$$

In the sigmoid function of Eq. (1), we use two pieces of information: *sensitivity*, d, to events ($d > 0$) and *silence time of event*, t, which is a length of estimated time since the last prediction. Figure 1 shows the probability distribution of sigmoid function taking any real value between negative and positive infinity. Since d and t values are always positive, the output of the sigmoid function $\frac{1}{1+e^{-dt}}$ is more than 0.5 and presented in the gray section of Fig. 1.

Equation (1) adjusts the output range between 0 to 1. As shown in Fig. 1, with a greater d value as t increases we can more frequently expect the associated event to happen. On the other hand, a smaller d value means the associated event would happen less by the model. Regardless of the value of d, the longer the event has not been observed, the higher its probability that will happen or already happened [6]. Note that d is initialized so that the rate of prediction is slightly

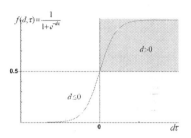

Fig. 1. Sigmoid curve of $\frac{1}{1+e^{-dt}}$

higher than actual event frequency. After each prediction, the sensitivity value d will be updated for next prediction.

3.2 Power-Law Distribution

In a power-law distribution, the magnitude x events are drawn from a probability distribution:

$$p(x) = c \cdot x^{-\alpha}, \tag{2}$$

where c is a fitted parameter and α is a constant scaling parameter [1,7].

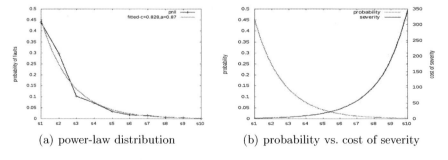

(a) power-law distribution   (b) probability vs. cost of severity

Fig. 2. Probability of event severity in power-law distribution of the PNNL data. (Color figure online)

We observed a power-law distribution in the PNNL failure data. Figure 2(a) shows a fitted power-law distribution of the PNNL data, $p(x) = 0.828 \cdot x^{-0.87}$. Less severe events occur more frequently with higher probability. More severe events occur rarely with smaller probabilities appearing at the tail area. Cost of a failure event depends on its severity; in this study, it means the degree of degradation on the system performance and reliability caused by the event. Figure 2(b) presents cost of an event based on severity, exponentially increasing. Rare events scarcely happen, however, when they happen, they may have disastrous effects. The rarest events in the PNNL data is a fault on the tape device usually used for backup. A Tape drive hardware replacement is easy and not much expensive, but the system may not be available until the hardware recovery completes, so the event of tape driver failure should be prevented, made rarer, or its effect should be at least reduced.

3.3   Bayesian Inference

Bayesian inference is a well known model for uncertainty to calculate a probability by combining an estimate and observations [8]. In this paper, this method is used to update subjective beliefs about rare events based upon observations on common events, i.e., given a prior state of knowledge. The fundamental equation in Bayesian inference is from the Bayes' rule, which is a learning model to calculate a posterior probability from a prior probability using the definition of conditional probability:

$$
\begin{aligned}
P(XD) &= \frac{P(DX)P(X)}{P(D)} \\
&\propto P(DX)P(X),
\end{aligned} \tag{3}
$$

where X is an estimate of event x, D is observed data, $P(X)$ is a prior probability of even t x, $P(DX)$ is a likelihood, and $P(D)$ is a constant normalization factor for a fixed data [9].

The observation data consists of a set of event variables. If states of some variables are True when the estimate (also a variable of event) is True, there are conditional dependencies; otherwise, they are independent. Initially, we assume that there is no prior knowledge about dependencies among variables. In general, many Bayesian models give initial values based on either common-sense or expert knowledge [10,11]. In this paper, the initial prior probabilities are given by the power-law distribution.

In order to calculate conditional probabilities from observations, we use a simple table containing evidence of event occurrence. In this study, the table is updated with a time interval to help finding conditional relations between events. Table 2 is an example of Bayes table containing evidence of the PNNL data during observed time interval. The collected data (# of events) is translated to the observed evidence represented with either True or False, for example, "an event of s_i level occurs" is interpreted as the evidence variable of s_i is True and represented with 1 in the binary table. Bayesian inference updates the probability of an estimate of event s_i after new evidence [8,9].

Table 2. Example of binary Bayes table

s_1	s_2	s_3	s_4	s_5	s_6	s_7	s_8	s_9	s_{10}

collected events during observed time interval									
14	16	7	1	0	1	1	0	0	0

$\Downarrow$

binary observational evidence									
1	1	1	1	0	1	1	0	0	0

3.4 Logistic Regression

Logistic regression is generally used for learning to predict the probability of an event. We are interested in an on-line parametric learning function with boolean classification. The variables of vector $X = <X_1, X_2, \ldots, X_n>$ are independent features of the event. Then an approximate estimation of the event is explained by a logistic function:

$$f(z) = \frac{1}{1 + e^{-z}}, \tag{4}$$

where z is a total contribution of independent variables with individual weights to estimate the probability, and $z = w_0 + w_1 x_1 + w_2 x_2 + \ldots + w_n x_n$, x_i is a boolean or continuous value of X_i, and w_i is a weight and $W = <w_0, w_1, w_2, \ldots, w_n>$ is a vector of weights. z is rewritten as $z = w_0 + \sum_{i=1}^{n} w_i x_i$ and when all evidence factors are zero, the event occurs independently with the fixed probability due to a given w_0 value [12].

In logistic regression learning, the weight parameters are updated with the sum of the gradients caused by each prediction:

$$w_i = w_i + \eta \sum_{l} X_i^l (Y^l - \hat{P}) - \eta \lambda w_i, \tag{5}$$

w_i is a weight parameter of X_i feature, X^l is a lth evidence factor, $\hat{P}$ is a True estimation with lth evidence and weight vector, η is a learning rate and λ is a penalty term [13]. Y^l is an actual state and $Y^l - \hat{P}$ is the true gradient of lth prediction called as prediction error [12,14]. By fitting observed data to a logistic function, logistic regression finds weights of prediction variables and predicts the probability of occurrence [12–14].

3.5 Proposed Hybrid Prediction Mechanism

Figure 3 presents our new event prediction mechanism incorporating *power-law* distribution and *online learning* algorithms. From the simple probability-based prediction of Eq. (1), we build up a new probability-based prediction based on observed knowledge including cumulated characteristics of target events. We develop a strategy for dealing with both common and rare events to achieve more accurate prediction. The estimated probability of a target event is calculated separately either when it is common, i.e., it belongs to events of Severity 1 to

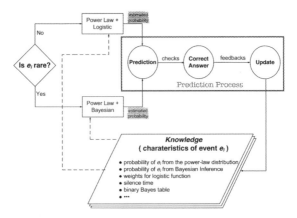

Fig. 3. Proposed event prediction mechanism

6 or when it is a rare event of Severity 7 to 10. The prediction mechanism makes a prediction when the estimated probability is more than standard, then it checks whether the target event occurs, and collects information and updates its knowledge about the target.

For rare events, the mechanism calculates estimated probability of rare events using Bayesian inference and silence time t since the last prediction. As mentioned, there is no prior knowledge about which common events influence which rare events until rare events are detected. Therefore, initial values are based on the power-law principle. Bayesian inference updates the estimated probability based on the relationships and the evidence in the Bayes table. The estimated Bayesian probability is denoted by bp. The probability-based prediction for a particular type of rare event is

$$p = 2 \times \left(\frac{1}{1 + e^{-(w_0 + w_1 \cdot bp + w_2 \cdot t)}} - \frac{1}{2}\right) \tag{6}$$

We calculate an estimated probability of non-rare (common) events using power-law probability pl and silence time t with two individual weights. The probability-based prediction for a particular type of common event is

$$p = 2 \times \left(\frac{1}{1 + e^{-(w_0 + w_1 \cdot pl + w_2 \cdot t)}} - \frac{1}{2}\right) \tag{7}$$

For both Eqs. (6) and (7), there are weight values for variables; w_0 is a constant value to guarantee a minimum probability of event occurrence, w_1 is a weight value for either bp or pl, and w_2 is a weight value for t. w_1 is a "big knob" to control learning and w_2 is a "small knob" because t is often oscillating. Both weight values are updated by Eq. (5). The big knob, w_1 is updated when w_2 is not within the defined boundary.

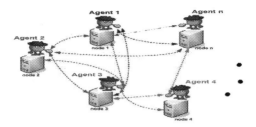

Fig. 4. Distributed problem solving agents. An agent checks some nodes that would have high fault potential through the proposed prediction mechanism, then mitigates detected faults. The direction of arrows means which agents check which nodes through message passing.

4   Experiments and Results

We developed a simulator for the PNNL system that exhibits the failure behavior as in the data described in Sect. 2.2 and a distributed multi-agent approach to fault-tolerance. In this simulation study, the system consists of 50 computing nodes, 5 servers, 5 switches, 8 storages and 1 tape, and the period of simulation is 3 years. We modeled the simulated system with independent computing nodes that communicate with each other mimicking the same software framework as in PNNL. We consider events of Severity 7 to 10 as rare events as in Table 1. These events belong to the long tail part of the power-law distribution.

Our prediction mechanism is deployed as a part of distributed agent embedded in each computational nodes and the agent predicts faults on other nodes and mitigate them as presented in Fig. 4. We will refer these prediction mechanism as *distributed problem solvers* as they are distributed over the computing nodes. There are 55 agents and 69 components. Power-law probability is fitted and updated every 5 months, and Bayesian probability and internal Bayes' tables of agents are updated every 3 days. We perform experiments and evaluate the correctness of predictions in terms of the performance of the simulated PNNL system in order to investigate the efficacy and limits of our event prediction mechanism for fault-mitigation.

4.1   Bayesian, Power-Law and Logistic Regression

The proposed event prediction mechanism presented in Fig. 3 integrates the power-law function from the PNNL data. We test the following two methods of computing the probability of fault and compare these methods.

(I) **BPIL** (Bayesian and Power-law Inside Logistic function)

$$p = \begin{cases} 2 \times (\frac{1}{1+e^{-(w_1 \cdot pl+w_2 \cdot t)}} - \frac{1}{2}), & \text{\textit{for common events}}; \\ 2 \times (\frac{1}{1+e^{-(w_1 \cdot bp+w_2 \cdot t)}} - \frac{1}{2}), & \text{\textit{for rare events}} \end{cases}$$

(II) **BPOL** (Bayesian and Power-law Outside Logistic function)

$$p = \begin{cases} 2 \times (w_1 \cdot pl + \frac{1}{1+e^{-w_2 \cdot t}} - \frac{1}{2}), & \text{for common events;} \\ 2 \times (w_1 \cdot bp + \frac{1}{1+e^{-w_2 \cdot t}} - \frac{1}{2}), & \text{for rare events} \end{cases}$$

Each distributed problem solver maintains a binary Bayes table representing the occurrence of each fault event and calculate conditional probabilities from observations. The table is updated with an interval to help finding conditional relations between events. The initial bp is given by the power-law distribution, which is also fitted with real-time observations. bp and pl are regularly updated with different periods.

Figure 5 presents the average system availability using Bayesian and power-law embedded methods. These methods achieve very high availability ($>98\%$) and problem solvers start with high number of prediction of potential of fault and the number converges to a single digit number near $4 \cdot 10^5$ time step. BPOL starts with three times of more prediction than BPIL.

As shown in Table 3, we compare possible system loss and communication overhead in several experiments: centralized heartbeat (HB), simple probability-based prediction (Step), logistic regression (Logistic), Power-law embedded methods (PL*Ls), and Bayesian and Power-law embedded methods (BP*Ls). We observe that the BP*Ls have less system loss but with more prediction of high potential of fault than Step and Logistic methods. For common events, PL*Ls and BP*Ls have similar amount of damage and cost because of the same mechanism. By positioning pl and bp inside the logistic function, we could reduce the overhead by 30% with 10% increment of system loss: 31 K (O) < 34 K (I) in system loss and 10 K (O) > 7 K (I) in overhead, where the O means "outside the logistic function" and the I means "inside the logistic function".

For rare events, PL*Ls and BP*Ls achieve a dramatic decrease in system loss with more overhead than the step and logistic methods. We found some meaningful difference between PL*Ls and BP*Ls in both system loss and overhead.

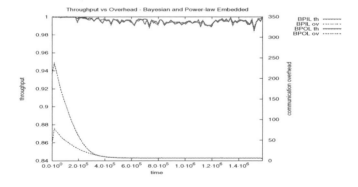

Fig. 5. System Performance: throughput vs communication overhead of Bayesian and power-law embedded methods (Color figure online)

Table 3. System loss vs. communication overhead comparison

	Centralized		Distributed											
	Heartbeat		Step		Logistic		PLIL		PLOL		BPIL		BPOL	
	L	C	L	C	L	C	L	C	L	C	L	C	L	C
Common (Σ_1^6)	8.7 G	0.8 K	1.1 G	1.2 K	0.6 G	1.2 K	34.4 K	7.9 K	31.4 K	10.4 K	34.4 K	7.7 K	31.4 K	10.4 K
Rare (Σ_7^{10})	15.6 G	1.4 K	9.9 G	5.8 G	5.4 G	3.2 K	357 K	11.7 K	375 K	12.6 K	154 K	31.4 K	150.9 K	35.4 K

L = Possible System Loss, C = Communication Overhead, $K = 10^3$, $G = 10^6$

Unlike common events, the position of *pl* and *bp* doesn't make a big difference. However, our proposed methods (BP*Ls) accomplish half the system loss of PL*Ls with twice the overhead: 357 K (PLIL) > 154 K (BPIL) in system loss and 11.7 K (PLIL) < 31.4 K (BPIL) in overhead. Therefore, the BPIL method achieved the best result with less damage and cost for both common and rare event prediction.

We also compare our experimental results with the centralized heartbeat checking. The single centralized problem solver periodically checks target nodes with a fixed rate, and our distributed problem solvers predict a fault on a target node with their own prediction mechanism. The results show huge difference between centralized heartbeat (HB) and BPIL: 24.3 G (HB) > 188.4 K (BPIL) in system loss and 2.2 K (HB) < 39.1 K (BPIL) in overhead for all events.

Figure 6 presents performance comparison of the distributed six experiments. As shown, the Step method shows the worst performance of about 80 % system availability. The Logistic approach shows about 10 % improvement compared to the step approach. We can see the communication overhead of all methods is adapting to a single digit number after learning event occurrence. The Step and Logistic approaches have lower overhead but their system performance is

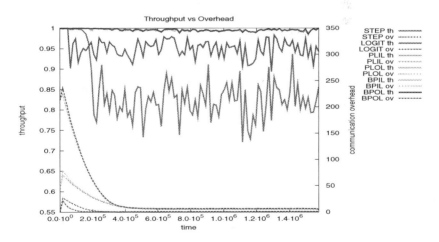

Fig. 6. Performance Comparison : throughput vs communication overhead (Color figure online)

considerably lower. Our proposed prediction mechanism makes the system available by predicting and mitigating faults with about 98 % throughput and applying power-law and Bayesian probability inside the logistic function has less overhead than outside.

5   Conclusion

This paper proposed a new prediction algorithm to accurately predict both rare and common events. We combined the principle of a power-law distribution in events, logistic regression, and Bayesian inference to design an adaptive event prediction mechanism that is capable of learning and predicting various events online.

We conducted various experiments to show the efficacy of our proposed mechanism using the PNNL system data. For common events, our distributed prediction algorithm reached a good degree of convergence in its learning. The algorithm shows some divergence for rare event learning. Our prediction mechanism improves the prediction of rare events and reduces the possible damage of the system overall. As a consequence, it provides about 98 % throughput.

There are many possible directions for further work, especially in rare event prediction when the amount of event data is extremely small. Future research should focus on minimizing the prediction error and enhancing decision-making to predict extremely rare events as well as associated common events in other event data such as earth quakes data.

References

1. JASON: The MIRTE Corporation: Rare Events. Technical report, JSR 09–108 (2009)
2. USENIX - Computer Failure Data Repository. http://cfdr.usenix.org/data.html# pnl
3. Kaelbling, L., Littman, M., Moore, A.: Reinforcement learning: a survey. J. Artif. Intell. Res. **4**, 237–285 (1996)
4. Kivinen, J., Smola, A., Williamson, R.: Online Learning with Kernels (2003)
5. Stanley, H., Gabaix, X., Gopikrishnan, P., Plerou, V.: Economic fluctuations and statistical physics: quantifying extremely rare and less rare events in finance. Physica A **382**(1), 286–301 (2007)
6. Kuznar, L., Frederick, W.: Environmental constraints and sigmoid utility: implications for value, risk sensitivity, and social status. Ecol. Econ. **46**(2), 293–306 (2003)
7. Clauset, A., Shalizi, C., Newman, M.: Power-law distributions in empirical data. J. SIAM Rev. **51**, 661–703 (2009)
8. Tenenbaum, J., Griffiths, T.: Generalization, similarity, and bayesian inference. J. Behav. Brain Sci. **24**(4), 629–640 (2001)
9. Oakley, J., O'Hagan, A.: Bayesian inference for the uncertainty distribution of computer model outputs. J. Biometrika **89**(4), 769–784 (2002). http://biomet.oupjournals.org/cgi/doi/10.1093/biomet/89.4.769

10. Silva, W., Milidi, R.: Algorithms for combining belief functions. Int. J. Approximate Reasoning **7**(1–2), 73–94 (1992)
11. Smets, P.: Belief functions: the disjunctive rule of combination and the generalized Bayesian theorem. Int. J. Approximate Reasoning **9**(1), 1–35 (1993)
12. Schein, A., Ungar, L.: Active learning for logistic regression: an evaluation. J. Mach. Learn. **68**(3), 235–265 (2007). http://www.springerlink.com/index/10.1007/s10994-007-5019-5
13. Mitchell, T.: Machine Learning. McGraw Hill Series in Computer Science. McGraw-Hill, New York (1997)
14. King, G., Zeng, L.: Logistic regression in rare events data. Polit. Anal. **9**(2), 137–163 (2001)

"Anti-Bayesian" Flat and Hierarchical Clustering Using Symmetric Quantiloids

Anis Yazidi[1], Hugo Lewi Hammer[1(✉)], and B. John Oommen[2]

[1] Department of Computer Science,
Oslo and Akershus University College of Applied Sciences, Oslo, Norway
hugo.hammer@hioa.no
[2] School of Computer Science, Carleton University, Ottawa, Canada

Abstract. A myriad of works has been published for achieving data clustering based on the Bayesian paradigm, where the clustering sometimes resorts to Naïve-Bayes *decisions*. Within the domain of clustering, the Bayesian principle corresponds to assigning the unlabelled samples to the cluster whose mean (or centroid) is the closest. Recently, Oommen and his co-authors have proposed a novel, counter-intuitive and pioneering PR scheme that is radically opposed to the Bayesian principle. The rational for this paradigm, referred to as the "Anti-Bayesian" (AB) paradigm, involves classification based on the non-central *quantiles* of the distributions. The first-reported work to achieve clustering using the AB paradigm was in [1], where we proposed a flat clustering method which assigned unlabelled points to clusters based on the AB paradigm, and where the distances to the respective learned clusters was based on their quantiles rather than the clusters' centroids for uni-dimensional and two-dimensional data. This paper, extends the results of [1] in many directions. Firstly, we generalize our previous AB clustering [1], initially proposed for handling uni-dimensional and two-dimensional spaces, to arbitrary d-dimensional spaces using their so-called *"quantiloids"*. Secondly, we extend the AB paradigm to consider how the clustering can be achieved in hierarchical ways, where we analyze both the Top-Down and the Bottom-Up clustering options. Extensive experimentation demonstrates that our clustering achieves results competitive to the state-of-the-art flat, Top-Down and Bottom-Up clustering approaches, demonstrating the power of the AB paradigm.

1 Introduction

Clustering is the task of grouping data points in a way that elements that exhibit some similarity, or that inherently belong to the same class, end up in the same group. It is a fundamental task in data analysis and inference, and it is, arguably, among the most popular machine learning and data mining techniques [2,3]. A range of different clustering methods have been proposed and each of them vary

---

B.J. Oommen—*Chancellor's Professor; Fellow: IEEE* and *Fellow: IAPR*. This author is also an *Adjunct Professor* with the University of Agder in Grimstad, Norway.

© Springer International Publishing Switzerland 2016
H. Fujita et al. (Eds.): IEA/AIE 2016, LNAI 9799, pp. 56–67, 2016.
DOI: 10.1007/978-3-319-42007-3_6

with the understanding of what a cluster, actually, is. For instance, density models, such as OPTICS [4] and DBSCAN [5], coalesce most dense regions in the space into a single cluster. As opposed to this, in hierarchical clustering [6,7], the aim is to arrange the data points into an underlying hierarchy which then determines the various clusters. A third group of clustering algorithms constitute the so-called "centroid" methods where all the points within a computed cluster are represented by a single point, for example the cluster's centroid. The most prominent example of a scheme within this family is the acclaimed k-means clustering algorithm where a centroid is represented by the mean value of the points in the cluster. The central strategy motivating *these* clustering schemes involves classifying *unassigned* data points to the different clusters based on the distances to the means (or centroids) of the clusters. From the above, one can informally see that any specific pattern classification algorithm can be conceptually expanded to yield a clustering scheme. Thus, if we have k previously-determined clusters, an unknown unlabelled sample can be assigned to any one of the k classes by the corresponding classification algorithm, whence the specific cluster can be grown to include this specific sample. Almost all the well-known classifiers involved in pattern classification are based on a Bayesian principle which aims to maximize the *a posteriori* probability. Quite recently, Oommen and his co-authors proposed a completely counter-intuitive paradigm, known as CMQS, the Classification by Moments of Quantile Statistics. CMQS works with a counter-intuitive philosophy, and essentially compares the testing sample with points from each class which are distant from the mean – as opposed to the Bayesian principle which essentially compares it to the clusters' means or the centroids

The question that begged investigation and that was considered open was that of invoking these "Anti-Bayesian" (AB) PR algorithms to design the corresponding clustering algorithms. This is the avenue of research undertaken here. The pioneering steps taken in this direction were reported in [1], where we introduced a novel alternative to the k-means clustering algorithm. The algorithm presented in [1] follows the same steps dictated by a typical k-means clustering algorithm. The main difference, however, is the manner by which it assigns the data points to the already-formed clusters. Indeed, rather than follow a Bayesian classification methodology, it traverses one of the AB-based PR CMQS-based schemes reported earlier. In fact, unlike the k-means clustering strategies that rely on centroid-based criteria, we resort to *quantiles positions distant from the cluster means* [8–10], which is a strategy just as counter-intuitive and non-obvious as the CMQS schemes themselves. Central to the development of such CMQS clustering algorithms is the concept of a "Quantiloid" We will elaborate on the phenomenon of Quantiloids in the next section. It is pertinent to mention that by working with Quantiloids, we will have effectively extended our previous work [1]. However, apart from doing it in the "vanilla" manner, we shall accomplish it by also invoking hierarchical clustering approaches.

1.1 Structure of the Paper

In Sect. 2, we present the fundamental principles of AB clustering. In Sect. 3, we demonstrate the development of AB flat clustering in d-dimensional spaces. The principles of hierarchical AB clustering are given in Sect. 4. In Sect. 5, we report our experimental results which compare our AB flat and hierarchical clustering schemes to their Bayesian counterparts. Section 6 concludes the paper.

2   The "Anti-Bayesian" Clustering Solution

2.1   Quantiloids

As alluded to earlier, the solution we propose is based on the concept of "Quantiloids". What then is a Quantiloid? The quantiloid associated with the real number, θ, is, quite simply, for a uni-dimensional distribution, the unique point where the Cumulative Distribution Function (CDF) has the value θ. This is the unique point where the probability mass (i.e., the integral of the Probability Density Function (PDF)) attains the value of θ. While this is an elementary concept for uni-dimensional variables, the concept can be extended for multi-dimensional vectors to be the hyper-surface under which the CDF has the value θ. The goal of this paper is to develop quantiloid-based clustering algorithms that work in an AB paradigm just as the centroid-based clustering algorithms worked within the "Bayesian" paradigm. Indeed, rather than characterizing a cluster by its centroid, we shall attempt to characterize it by its quantiloids, which will then lead to the various AB clustering algorithms.

Although the concept of quantiloids is valid for multi-dimensional vectors, the question of *how* they can be computed and represented is still open. We shall thus restrict ourselves to uni-dimensional quantiloids by processing the multi-dimensional distribution in terms of its uni-dimensional marginals.

2.2   "Anti-Bayesian" Classification Rules

We first summarize the AB classification rules designed and proven in [8–10] for uni-dimensional features. To do this, we use the notation that for the jth dimension of the feature vector of class ω_i, $q_p^{i,j}$ is the quantiloid for the value p, i.e., the position where the feature's CDF has a value of p. In the case when both the classes are characterized by only a *single* feature X, q_p^i is ω_i's quantiloid for the value p, i.e., more formally $q_p^i = Pr(X < p | X \in \omega_i)$. Observe that we encounter the cases when the quantiloids overlap (i.e., $q_{1-p}^1 < q_p^2$) or when they do not overlap (i.e., $q_{1-p}^1 > q_p^2$). Using this notation, the uni-dimensional AB classification rules for the testing sample x^* are:

Case 1: When the quantiloids are non-overlapping (see Fig. 1 on the left):

$$
\begin{aligned}
&\text{If } x^* < q_{1-p}^1 && \Rightarrow x^* \in \omega_1; \\
&\text{If } x^* > q_p^2 && \Rightarrow x^* \in \omega_2; \\
&\text{If } (q_{1-p}^1 < x^* < q_p^2) \ \wedge \ (\|x^* - q_{1-p}^1\| < \|x^* - q_p^2\|) \Rightarrow x^* \in \omega_1; \\
&\text{If } (q_{1-p}^1 < x^* < q_p^2) \ \wedge \ (\|x^* - q_{1-p}^1\| > \|x^* - q_p^2\|) \Rightarrow x^* \in \omega_2.
\end{aligned}
\tag{1}
$$

The reader will observe that the cases are mutually exclusive and that the classification border is: $\frac{q_{1-p}^1 + q_p^2}{2}$.

Case 2: When the quantiloids are overlapping (see Fig. 1 on the right):

$$
\begin{aligned}
&\text{If } x^* < q_p^2 && \Rightarrow x^* \in \omega_1;\\
&\text{If } x^* > q_{1-p}^1 && \Rightarrow x^* \in \omega_2;\\
&\text{If } (q_p^2 < x^* < q_{1-p}^1) \wedge (\|x^* - q_p^1\| < \|x^* - q_{1-p}^2\|) \Rightarrow x^* \in \omega_1;\\
&\text{If } (q_p^2 < x^* < q_{1-p}^1) \wedge (\|x^* - q_p^1\| > \|x^* - q_{1-p}^2\|) \Rightarrow x^* \in \omega_2.
\end{aligned}
\tag{2}
$$

In this case, the comparison is based on the distant quantiloids and so the classification border is: $\frac{q_p^1 + q_{1-p}^2}{2}$.

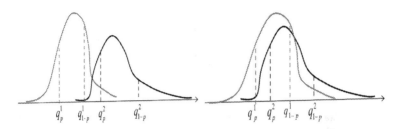

Fig. 1. The AB scheme: (a) When the quantiloids are non-overlapping on the left, and (b) When the quantiloids are overlapping on the right.

The reader will observe that the latter case (Case 2) is the one that uses the so-called "Dual" scenario (please see [8–10]), and where the extreme quantiloids are used for the classification as opposed to the quantiloids that are close to the discriminant. In the symmetric cases analyzed in [8–10], it is easy to see that the assignments in the so-called "Dual" scenario reduce to those involving comparisons to the quantiloids that are *close to the discriminant*, but where the assignment is to the class *that is the more distant one*. The decision rule for this is given below.

Case 2 (Revised): When the quantiloids are overlapping (again see Fig. 1 on the right):

$$
\begin{aligned}
&\text{If } x^* < q_p^2 && \Rightarrow x^* \in \omega_1;\\
&\text{If } x^* > q_{1-p}^1 && \Rightarrow x^* \in \omega_2;\\
&\text{If } (q_p^2 < x^* < q_{1-p}^1) \wedge (\|x^* - q_p^2\| < \|x^* - q_{1-p}^1\|) \Rightarrow x^* \in \omega_2;\\
&\text{If } (q_p^2 < x^* < q_{1-p}^1) \wedge (\|x^* - q_p^2\| > \|x^* - q_{1-p}^1\|) \Rightarrow x^* \in \omega_1.
\end{aligned}
\tag{3}
$$

The difference between the two versions of Case 2 (Eqs. (2) and (3)) lies in the assignments made in the last two statements, where they, however, are *done to the non-adjacent classes*. In this case, the comparison is based on the closer quantiloids and so the classification border is: $\frac{q_p^2 + q_{1-p}^1}{2}$. To distinguish between these two scenarios, we shall refer to this version of the "Dual" scenario as the "Swapped Border" scenario.

The cases when the second distribution (for ω_2) is to the left of the first (for ω_1), is shown in Fig. 2. Observe that this is identical to the case of the figure on the left of Fig. 1, except that the identities of the classes is interchanged.

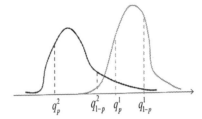

Fig. 2. This figure depicts the case of when the quantiloids do not overlap but when second distribution (for w_2) is to the left of the first (for w_1).

There is one additional scenario, and that occurs when there is a huge overlap between the distributions (See Fig. 3). The classification decision rule to be used is not that obvious because the classes are highly overlapping. Apart from this, the classification of an unknown sample itself is not just non-obvious, it is actually "meaningless". This case never occurred in our experiments.

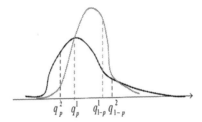

Fig. 3. This figure depicts the scenario when there is a huge overlap between the distributions.

3   The AB Multi-dimensional Clustering

We now consider the extensions of the results in [1] to the multi-dimensional scenario. To explain this, we state that in [1], as explained above, we used the concept of the closest quantile corners in two dimensions. For the multi-dimensional scenario, instead of measuring the distances between the centroids as as done in the Bayesian paradigm, we measure the distances between the quantiloids.

3.1   The Quantiloids Used

In the d-dimensional feature space, let $Q^1 = [Q_1^1, Q_2^1, ..., Q_d^1]$ and $Q^2 = [Q_1^2, Q_2^2, ..., Q_d^2]$ denote the quantiloids of the distributions (clusters) of $f_1(X)$ and $f_2(X)$ respectively. Q^1 and Q^2 are computed as follows:

– In each dimension, we decide which distribution (cluster) is to the left and which is to the right. To decide this, we exactly follow the principles explained in [1].
– For each of the three cases defined in [1] the elements in the ith dimension of the quantiloid vectors is computed as follows:
  • Case 1: Here $f_1()$ is to the left of $f_2()$. Here we set $Q_i^1 = q_{i,p}^1$ and $Q_i^2 = q_{i,1-p}^2$. In this case, we also have to consider the case when an exception occurs, i.e., when there is a degree of overlap between them. Indeed, if the $f_1()$ and $f_2()$ are close in the ith dimension such that the quantiles overlap, i.e., that $q_{i,p}^1$ is to the right of $q_{i,1-p}^2$, then the point should be classified to ω_1 if it is closer to $q_{i,1-p}^2$ and to ω_2 if it is closer to $q_{i,p}^1$. This corresponds to "Swapped Border" scenario (Case 2 (Revised)) in Sect. 2.2. With such overlapping quantiles we therefore set $Q_i^1 = q_{i,1-p}^2$ and $Q_i^2 = q_{i,p}^1$.
  • Case 2: Here $f_1()$ is to the right of $f_2()$ and, if the quantiles do not overlap, we set $Q_i^1 = q_{i,1-p}^1$ and $Q_i^2 = q_{i,p}^2$. If the quantiles overlap, we switch the quantiles, as described above, to account for the "Swapped Border" scenario as in Case 1.
  • Case 3: Here we set $Q_i^1 = \frac{q_{i,p}^1 + q_{i,1-p}^1}{2}$ and $Q_i^2 = \frac{q_{i,p}^2 + q_{i,1-p}^2}{2}$. This is the case when the overlap is significant and the classification can be considered to be "meaningless". As mentioned earlier, this case occurs very rarely in the domain of clustering.

The first two scenarios encountered above can be explained in the following figures drawn in two dimensions. In each case, we have plotted the hyper-rectangles defined by the quantiloids. If the overlap is small, the distances are measured from the nearest quantiloids, as seen in Fig. 4.

If the overlap is significant, the distances are measured from the farthest quantiloids (Case 2 of Sect. 2.2), or equivalently from the "Swapped Border" (Case 2 (Revised)) quantiloids explained in Sect. 2.2 and shown in Fig. 5.

3.2 The Distance Measures Used

Based on the definition of the quantiloids, we are now ready to define two types of distances used in the framework of our AB clustering paradigm. The two types of distance metrics we use are listed below:

– *Data Point to Cluster (DPC) Distance:* Once the quantiloids have been computed following the procedure above, the points Z is classified to ω_1 if Z is closer, in terms of its Euclidean distance to Q^1 than to Q^2. Otherwise, Z is classified to ω_2. The DPC Distance has been used for flat clustering as well as for Top-Down clustering.
– *Cluster to Cluster (CC) Distance* The same notion can be used to characterize the distance between two clusters. The CC distance between two clusters C_1 and C_2 is the Euclidean distance between their corresponding quantiloids Q_1 and Q_2. The notion of the CC Distance is usually used for Bottom-Up clustering techniques.

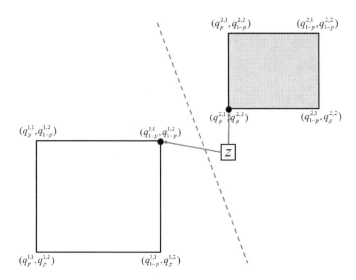

Fig. 4. The case when the multi-dimensional distributions have little overlap. The rectangles representing the quantiloids.

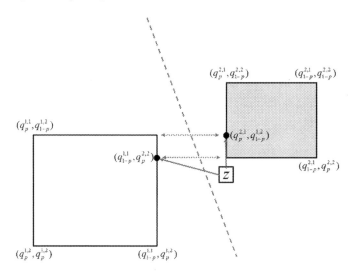

Fig. 5. The case when the multi-dimensional distributions have a large overlap. The rectangles representing the quantiloids. Observe that in this case, we have utilized the "Swapped Border" scenario to compute the quantiloids.

4 Principles of Our Hierarchical Clustering

It is well-known that clustering can also be achieved hierarchically, where the scheme is either of a Bottom-Up paradigm or of a Top-Down paradigm. These traditional paradigms can be extended to our AB paradigm by *merely modifying the concept of the distances invoked*, where in the AB scheme, the distance is based on the concept of quantiloids. Thus, in essence, our algorithms follow the classical hierarchical clustering philosophy [11] in all the relevant steps, except that we consider the distances to the qunatiloids rather than the distances to the centroids of the clusters. To explain these, we present the hierarchical AB clustering methods. These are, precisely, the counter-parts of the classical hierarchical clustering methods [11]. The only difference is the way by which we specify the distances, i.e., whether we invoke the DPC or CC distance measures based on the principles of the quantiloids rather than centroids.

4.1 Bottom-Up AB Clustering

A Bottom-Up clustering works with the principle that all the points are individually specified in the d-dimensional space. The points and then gathered together to form clusters, to which the unclassified points are then subsequently added. Thus, the steps of a Bottom-Up AB clustering are described below:

– Compute all pair-wise similarity distances between the different clusters and populate the proximity matrix. The distance between the clusters is merely the Euclidean distance between their corresponding quantiloids.
– Identify the closest clusters in terms of their similarity and merge them into a single cluster. This results in updating the proximity matrix and decreasing its order by unity.
– Repeat the above steps until we obtain the desired (pre-specified) number of clusters.

4.2 Top-Down AB Clustering

A Top-Down clustering works with the principle that all the points are collectively grouped into a single cluster in the d-dimensional space. The most distant points are then separated to be the nuclei of two distinct clusters, and the points closest to these are then included into their respective clusters. Again, in an AB paradigm, the distances are measured in terms of the quantiloids rather than the centroids. Thus, the steps involved in Top-Down AB clustering are described below:

– Start at the top level with all the data points coalesced in a single cluster.
– Use a flat clustering scheme in order to split the cluster.
– Apply the procedure recursively until a termination condition on the depth of the tree is reached or until each data point (singleton) ends up as its own cluster (maximum depth). Usually, one invokes a termination condition which involves the desired (pre-specified) number of clusters.

5 Experimental Results: "Anti-Bayesian" Clustering

In order to test the validity of the concepts proposed in this paper, we conducted numerous experiments on synthetic data. In the interest of space and brevity, we report the salient ones here.

In all our experiments, we used $K = 3$ clusters. All the synthetic data were from multivariate Normal distributions, where we fixed d, the dimension of the space, to be 4.

5.1 Data Generation

We shall first explain how the data points were generated for Normally-distributed distributions. Let $N(\mu_k, \Sigma_k)$ denote a multivariate Normal distribution with an expectation vector μ_k and a covariance matrix Σ_k, where $k = 1, \ldots, K$ (where we are dealing with K clusters). To generate the K distributions, it is crucial that we determine how μ_k and Σ_k ($k = 1, \ldots, K$) are set.

In our experiments, the expectations, $\{\mu_k\}$ were uniformly spread on the d–dimensional cube $[0, D]^d$ where D was chosen such that the clusters were reasonably spread in space to their inter-class overlaps to be minimal. In the experiments for which we report the results, we used $D = 6$.

The covariance matrices Σ_k for each cluster was generated by the following procedure:

- Set the diagonal element to be equal to 1, i.e., the marginal variance was equal to unity in all the dimensions;
- The correlation between each of the variables in the ith and j^{th} dimensions for $i = 1, \ldots, j < i$ was drawn uniformly from the interval $[-\rho_{max}, \rho_{max}]$, where $\rho_{max} < 1$. In the experiments we used $\rho_{max} = 0.8$;
- We checked if the generated covariance matrix was positive definite. If it was not, we repeated the previous step, above[1].

In what follows, we let n denote the number of samples generated from each underlying cluster in the synthetic data. Thus, the total number of data points generated were $n \cdot K$. In the experiments that we report, we used two values of n, i.e., $n = 20$ and $n = 100$.

5.2 Quantile and "Distance" Estimations

Since we constantly need to estimate the quantiloids, we opted to achieve this using the corresponding quantiles in each of the projected dimensions. This was done non-parametrically and parametrically as below:

- Non-parametrically (referred to in the columns titled "AB Non-parametric" in the tables below): This was achieved in a manner similar to the work presented in [1].

---

[1] With $\rho_{max} = 0.8$ and for $d = 3$ the matrix were almost always positive definite on the very first attempt. For $d = 5$, on the average, about every third matrix that was generated was positive definite.

– Parametrically: This was achieved by assuming normality Here we estimated the corresponding μ and σ and computed the respective quantiles from the Normal distributions (referred to in the columns titled "AB Parametric" in the tables).

The corresponding "distance" estimations for the experiments done were achieved as below:

– Row titled "Bottom up": These represent the classical version where all the inter-point distances are computed. The distances were computed between the centroids in case of the Bayesian clustering, and between the corresponding quantiloids in the case of the AB clustering.
– Row titled "Bottom-Up Distance UD (uni-dimensional)": In this case, we sorted the data by the first dimension and repeatedly merged the closest points in this dimension. This approach was a simplification of the general approach where we should have considered all the dimensions. The present approach required less computations. We expected that such a simplification would result in a reduced accuracy as we only relied on the first dimension of the data for executing the clustering, and this was, indeed, our experience.
– Row titled "Top-Down": In this case, the points were repeatedly split in two using k-means and the AB analog [1].
– Row titled "Top-Down Distance UD (uni-dimensional)": In this case, the data was sorted by the first dimension and repeatedly split in such a way that the distance (Bayes or AB) between the clusters was as large as possible. Again, this approach required less computations than the "Top-Down" one. As before, it is reasonable to expect such a simplification to result in a reduced accuracy. This was, indeed, the case.

5.3   Evaluation of Clustering Performance

The better a clustering algorithm performs, the better we expect that the following is satisfied: If two points are in the same clusters for the true clusters described by the "state of nature", they should ideally also be in the same cluster in the results obtained from the clustering algorithm. Conversely, if two points are *not* in the same clusters for the true clusters, they should not be in the same cluster in the results of the clustering algorithm either. As a measure of clustering performance, we measured the portion of pair of points satisfying this agreement between the true clusters and the clusters from the algorithm.

5.4   Clustering Performance

Tables 1 and 2 show results for the different clustering algorithms when $n = 20$ and $n = 100$, respectively.

Comparing the non-parametric and parametric AB approaches, we do not observe any (significant) differences in the results showing that both approaches perform about equally well. Comparing the AB approaches to the Bayesian approach we observe that the two methods perform about equally well for all the

Table 1. The clustering errors of the various methods with $n = 20$.

	Bayes	AB Non-parametric	AB Parametric
Flat clustering	0.098 (0.094, 0.103)	0.098 (0.094, 0.102)	0.099 (0.094, 0.103)
Bottom-Up	0.21 (0.201, 0.219)	0.277 (0.266, 0.288)	0.275 (0.265, 0.286)
Bottom-up Distance UD	0.423 (0.415, 0.43)	0.434 (0.427, 0.442)	0.443 (0.435, 0.45)
Top down	0.13 (0.125, 0.135)	0.132 (0.127, 0.137)	0.13 (0.125, 0.135)
Top-Down Distance UD	0.335 (0.329, 0.341)	0.369 (0.363, 0.375)	0.358 (0.352, 0.365)

Table 2. The clustering errors of the various methods $n = 100$.

	Bayes	AB Non-parametric	AB Parametric
Flat clustering	0.079 (0.064, 0.094)	0.089 (0.072, 0.107)	0.082 (0.065, 0.099)
Bottom-Up	0.247 (0.203, 0.291)	0.322 (0.268, 0.375)	0.369 (0.311, 0.428)
Bottom-up Distance UD	0.479 (0.451, 0.507)	0.481 (0.451, 0.512)	0.504 (0.476, 0.532)
Top-Down	0.108 (0.087, 0.129)	0.097 (0.08, 0.114)	0.126 (0.104, 0.148)
Top-Down Distance UD	0.346 (0.32, 0.372)	0.364 (0.342, 0.385)	0.334 (0.311, 0.358)

methods except for the Bottom-Up approach in which the Bayes perform a little better. We also observe that the uni-dimensional approaches (rows three and five) perform far poorer than the other methods documenting that such a uni-dimensional approaches are not satisfactory. Overall we see that the Anti-Bayesian paradigm performs very well in clustering data and competitive to the Bayesian paradigm.

6   Conclusion

In this paper, we have considered an "Anti-Bayesian" (AB) paradigm for clustering. All of the reported clustering algorithms (except the one reported in [1]) operate on Bayesian principles, (where the Bayesian principle corresponds to assigning the unlabelled samples to the cluster whose mean (or centroid) is the closest). Our aim here has been to see if the "Anti-Bayesian" (AB) classification philosophy, introduced recently by Oommen and his co-authors, can be extended into the domain of clustering. The AB principle involves classification based on the non-central *quantiles* of the distributions, which involves utilizing the information resident in the outlier samples.

In this paper, we have extended the first-reported AB clustering methods proposed in [1]. This paper has extended the results of [1] in many directions. Firstly, we have generalized our previous AB clustering [1], initially proposed for handling uni-dimensional and two-dimensional spaces, to arbitrary d-dimensional spaces using their so-called *"quantiloids"*. Secondly, we have extended the AB paradigm to consider how the clustering can be achieved in hierarchical ways, where we have analyzed both the Top-Down and the Bottom-Up clustering options. The AB paradigm can also use an anti-Naïve-Bayesian *computational* mechanisms. The paper contains the results of extensive experimentation that demonstrate that our clustering achieves results competitive to the state-of-the-art flat, Top-Down and Bottom-Up clustering approaches.

In the future, we envisage an ambitious goal of devising an AB clustering method based on applying majority voting on the decision made in each dimension of the quantile vector.

References

1. Hammer, H.L., Yazidi, A., Oommen, B.J.: A novel clustering algorithm based on a non-parametric "Anti-Bayesian" paradigm. In: Ali, M., Kwon, Y.S., Lee, C.-H., Kim, J., Kim, Y. (eds.) IEA/AIE 2015. LNCS, vol. 9101, pp. 536–545. Springer, Heidelberg (2015)
2. Jain, A.K., Dubes, R.C.: Algorithms for Clustering Dats. Prentice Hall, Englewood Cliffs (1988)
3. Xu, R., Wunsch, D.: Survey of clustering algorithms. Trans. Neur. Netw. **16**(3), 645–678 (2005)
4. Ankerst, M., Breunig, M.M., Kriegel, H.P., Sander, J.: Optics: ordering points to identify the clustering structure, pp. 49–60. ACM Press (1999)
5. Ester, M., Kriegel, H.P., S, J., Xu, X.: A density-based algorithm for discovering clusters in large spatialdatabases with noise, pp. 226–231. AAAI Press (1996)
6. Murtagh, F., Contreras, P.: Methods of hierarchical clustering. CoRR abs/1105.0121 (2011)
7. Sibson, R.: SLINK: an optimally efficient algorithm for the single-link cluster method. Comput. J. **16**(1), 30–34 (1973)
8. Thomas, A., Oommen, B.J.: The fundamental theory of optimal "Anti-Bayesian" parametric pattern classification using order statistics criteria. Pattern Recogn. **46**(1), 376–388 (2013)
9. Thomas, A., Oommen, B.J.: Order statistics-based parametric classification for multi-dimensional distributions. Pattern Recogn. **46**(12), 3472–3482 (2013)
10. Oommen, B.J., Thomas, A.: Anti-Bayesian parametric pattern classification using order statistics criteria for some members of the exponential family. Pattern Recogn. **47**(1), 40–55 (2014)
11. Jain, A.K., Murty, M.N., Flynn, P.J.: Data clustering: a review. ACM Comput. Surv. (CSUR) **31**(3), 264–323 (1999)

On the Online Classification of Data Streams Using Weak Estimators

Hanane Tavasoli[1], B. John Oommen[1], and Anis Yazidi[2(✉)]

[1] Department of Computer Science,
Oslo and Akershus University College of Applied Sciences, Oslo, Norway
[2] School of Computer Science, Carleton University, Ottawa, Canada
anis.yazidi@hioa.no

Abstract. In this paper, we propose a novel *online* classifier for complex data streams which are generated from non-stationary stochastic properties. Instead of using a single training model and counters to keep important data statistics, the introduced online classifier scheme provides a real-time self-adjusting learning model. The learning model utilizes the multiplication-based update algorithm of the Stochastic Learning Weak Estimator (SLWE) at each time instant as a new labeled instance arrives. In this way, the data statistics are updated every time a new element is inserted, without requiring that we have to rebuild its model when changes occur in the data distributions. Finally, and most importantly, the model operates with the understanding that the correct classes of previously-classified patterns become available at a later juncture subsequent to some time instances, thus requiring us to update the training set and the training model.

The results obtained from rigorous empirical analysis on multinomial distributions, is remarkable. Indeed, it demonstrates the applicability of our method on synthetic datasets, and proves the advantages of the introduced scheme.

Keywords: Weak estimators · Learning automata · Non-stationary environments · Classification in data streams

1 Introduction

In the past few years, due to the advances in computer hardware technology, large amounts of data have been generated and collected and are stored permanently from different sources. Some the applications that generate data streams are financial tickers, log records or click-streams in web tracking and personalization, data feeds from sensor applications and call detail records in telecommunications. Analyzing these huge amounts of data has been one of the most important challenges in the field of Machine Learning (ML) and Pattern Recognition (PR). Traditionally, ML methods are assumed to deal with static data

B.J. Oommen—*Chancellor's Professor; Fellow: IEEE* and *Fellow: IAPR.* This author is also an *Adjunct Professor* with the University of Agder in Grimstad, Norway.

© Springer International Publishing Switzerland 2016
H. Fujita et al. (Eds.): IEA/AIE 2016, LNAI 9799, pp. 68–79, 2016.
DOI: 10.1007/978-3-319-42007-3_7

stored in memory, which can be read several times. On the contrary, streaming data grows at an unlimited rate and arrives continuously in a single-pass manner that can be read only once. Further, there are space and time restrictions in analyzing streaming data. Consequently, one needs methods that are "automatically adapted" to update the training models based on the information gathered over the past observations whenever a change in the data is detected.

Mining streaming data is constrained by limited resources of time and memory. Since the source of data generates a potentially unlimited amount of information, loading all the generated items into the memory and achieving off-line mining is no longer possible. Besides, in non-stationary environments, the source of data may change over time, which leads to variations in the underlying data distributions. Thus, with respect to this dynamic nature of the data, the previous data model discovered from the past data items may become irrelevant or even have a negative impact on the modeling of the new data streams that become available to the system.

A vast body of research has been performed on the mining of data streams to develop techniques for computing fundamental functions with limited time and memory, and it has usually involved the sliding-window approaches or incremental methods. In most cases, these approaches require some *a priori* assumption about the data distribution or need to invoke hypothesis testing strategies to detect the changes in the properties of data.

In this article we will study classification problems in non-stationary environments (NSE), where sequential patterns are arriving and being processed in the form of a data stream that was potentially generated from different sources with different statistical distributions. The classification of the data streams is closely related to the estimation of the parameters of the time varying distribution, and the associated algorithms must be able to detect the source changes and to estimate the new parameters whenever a *switch* occurs in the incoming data stream.

Apart from the "traditional" classification problem involving *unique and distinct* training and testing phases, this paper pioneers the concept when these phases are not so clearly well-defined. Rather, we consider the fascinating phenomenon in which the testing patterns can subsequently be considered as training patterns, once their true class identities are known. Thus, the model operates with the understanding that the correct classes of previously-classified patterns become available at subsequent time instances (after some time has lapsed), thus requiring us to update the training set and the training model. This renders the whole PR problem intriguing.

Finally, to render the problem more complex, we consider the case where the classes' stochastic properties potentially vary with time as more instances become available. In this perspective, with regard to the training, we will argue that using "strong" estimators that converge with probability of 1 is inefficient for tracking the statistics of the data distributions in non-stationary environments. However, "weak" estimator approaches are able to rapidly unlearn what they have learned and adapt the learning model to new observations.

This feature of "weak" estimators makes these approaches the most effective methods for estimation in non-stationary environments. In this work, we will employ a particular family of weak estimators, referred to as Stochastic Learning Weak Estimation (SLWE) methods [5], for classification in non-stationary environments. The SLWE has been successfully used to solve *two*-class classification problems by Oommen and Rueda [5] by applying it on non-stationary *one*-dimensional datasets. In this article we will study the performance of the SLWE with more complex classification schemes.

1.1 Contributions of the Paper

The main contributions of the paper are the following:

- We have pioneered an *online* classification scheme that is composed of three phases. In the first phase, the model learns from the available labeled samples. In the second phase, the learned model predicts the class label of the unlabeled instances currently observed. In the third phase, after knowing the true class label of these recently-classified instances, the classification model is adjusted in an *online* manner.
- Most of the data stream mining approaches have involved building an initial model from a sliding window of recently-observed instances and thereafter, refining the learning model periodically or whenever its performance degrades based on the current window of observed data. We present a novel framework to deal with concept and distribution drift over data streams in non-stationary environments, which is more efficient and provides more accurate results. We emphasize that this non-stationarity could even be *abrupt*.
- Our classifier scheme provides a real-time self-adjusting learning model, utilizing the multiplication-based update algorithm of the SLWE at each time instance, as new labeled instances arrive. Instead of using a single training model and maintaining counters to keep important data statistics, we have used a technique to replace these frequency counters by data estimators. In this way, the data statistics are updated every time a new element is observed, without needing to rebuild its model when a change in the distributions is detected.
- Extensive experimental results that we have obtained, for multi-dimensional distributions, demonstrate the efficiency of the proposed classification schemes in achieving a good performance for data streams involving non-stationary distributions under different scenarios of concept drift, and the new model of computation in which the training and testing samples are not completely dichotomized.

1.2 Organization of the Paper

In Sect. 2, we proceed with discussing the issues and challenges encountered when one learns from data streams and provide a brief explanation about the theoretical properties of the SLWE. In Sect. 3, we present the details of the design and implementation of the online classifier where the samples that were testing samples at any given time instant can, at a subsequent juncture, be considered as training data. We then explain how this solution can be used to perform online classification, and present the new experimental framework for concept drift in Sect. 4. This section also contains the experimental results we have obtained from rigorous testing. Section 5 concludes the paper.

2   Literature Review

Estimation theory is a fundamental subject that is central to the fields of Pattern Recognition (PR) and data mining. The majority of problems in PR require the estimation of the unknown parameters that characterize the underlying data distributions.

2.1   Learning Methods for Data Streams in NSE

In general, most algorithms in the data stream mining literature have one or more of the following modules: a Memory module, an Estimator module, and a Change Detector module [1]. The Memory module is a component that stores summaries of all the sample data and attempts to characterize the *current* data distribution. Data in non-stationary environments can be handled by three different approaches, namely, by using partial memory, by window-based approaches and by instance-based methods. The term "partial memory" refers to the case when only a part of the information pertaining to the training samples are stored and used regularly in the training. In window-based approaches, data is presented as "chunks", and finally, in instance-based methods, the data is processed upon its arrival. In fact, the Memory module determines the forgetting strategy used by the mining algorithm operating in the dynamic environments.

The Estimator module uses the information contained in the Memory or only the observed information to estimate the desired statistics of the time varying streamed data.

The Change Detector module involves the techniques or mechanisms utilized for detecting explicit drifts and changes, and provides an "alarm" signal whenever a change is detected based on the estimator's outputs.

Apart from the above schemes, many other incremental approaches have been proposed that infer change points during estimation, and use the new data to adapt the learning model trained from historical streaming data. The learning model in incremental approaches is adapted to the most recently received instances of the streaming data. Let $X = \{x_1, x_2, \ldots, x_n\}$ be the set of training examples available at time $t = 1 \ldots n$. An incremental approach produces a sequence of hypothesis $\{\ldots, H_{i-1}, H_i, \ldots\}$ from the training sequence, where each hypothesis, H_i, is derived from the previous hypothesis, H_{i-1}, and the example x_i. In general, in order to detect concept changes in these types of approaches, some characteristics of the data stream (e.g., performance measures, data distribution, properties of data, or an appropriate statistical function) are monitored over time. When the parameters switch during the monitoring process, the algorithm should be able to adapt the model to these changes.

We now briefly review some *other* schemes used for learning in non-stationary environments. The review here will not be exhaustive because the methods explained can be considered to be the basis for other modified approaches.

FLORA. Widmer and Kubat [6], presented the FLORA family of algorithms as one of the first supervised incremental learning systems for a data stream. The initial FLORA algorithm used a fixed-size sliding window scheme. At each time step, the elements in the training window were used to incrementally update the learning model. The updating of the model involved two processes: an incremental *learning* process that updated the concept description based on the new data, and an incremental *forgetting* process that discarded the out-of-date (or stale) data.

The initial FLORA system did not perform well on large and complex data domains. Thus, FLORA2 was developed to solve the problem of working with a fixed window size, by using a heuristic approach to adjust the window size dynamically. Further improvements of the FLORA were presented to deal with recurring concepts (FLORA3) and noisy data (FLORA4).

Statistical Process Control (SPC). The SPC was presented by Gama *et al.* [4] for change detection in the context of data streams. The principle motivating the detection of concept drift using the SPC is to trace the probability of the error rate for the streamed observations. While monitoring the errors, the SPC provides three possible states, namely, "in control", "out of control" and "warning" to define a state when a warning has to be given, and when levels of changes appear in the stream. When the error rate is lower than the first (lower) defined threshold, the system is said to be in an "in control" state, and the current model is updated considering the arriving data. When the error exceeds that threshold, the system enters the "warning" state. In the "warning" state, the system stores the corresponding time as the warning time, t_w, and buffers the incoming data that appears subsequent to t_w. In the "warning" mode, if the error rate drops below the lower threshold, the "warning" mode is canceled and the warning time is reset. However, in case of an increasing error rate that reaches the second threshold, a concept change is declared and the learning model is retrained from the buffered data that appeared after t_w.

ADWIN. Bifet and Gavalda [2,3] proposed an adaptive sliding window scheme named ADWIN for change detection and for estimating statistics from the data stream. It was shown that the ADWIN algorithm outperforms the SPC approach and that it has the ability to provide rigorous guarantees on false positive and false negative rates. The initial version of ADWIN keeps a variable-length sliding window, W, of the most recent instances by considering the hypothesis that there is no change in the average value inside the window. To achieve this, the distributions of the sub-windows of the W window are compared using the Hoeffding bound, and whenever there is a significant difference, the algorithm removes all instances of the older sub-windows and only keeps the new concepts for the next step. Thus, a change is reliably detected whenever the window shrinks, and the average over the existing window can be considered as an estimate of the current average in the data stream.

2.2 Stochastic Learning Weak Estimator (SLWE)

Using the principles of stochastic learning, Oommen and Reuda [5] proposed a strategy to solve the problem of estimating the parameters of a binomial or multinomial distribution efficiently in non-stationary environments. This method is referred to as the SLWE, where the convergence of the estimate is "weak", i.e., with respect to the first and second moments. Unlike the traditional MLE and the Bayesian estimators, which demonstrate strong convergence, the SLWE converges fairly quickly to the true value, and it is able to just as quickly "unlearn" the learning model trained from the historical data in order to adapt to the new data.

The SLWE is an estimator method that estimates the parameters of a binomial/multinomial distribution when the underlying distribution is non-stationary. In non-stationary environments, the SLWE updates the estimate of the distribution's probabilities at each time-instant based on the new observations. The updating, however, is achieved by a *multiplicative* rule. To formally introduce the SLWE, let X be a random variable of a multinomial[1] distribution, which can take the values from the set $\{`1`,\dots,`r`\}$ with the probability of S, where $S = [s_1,\dots,s_r]^T$ and $\sum_{i=1}^{r} s_i = 1$. In the other words: $X = `i`$ with probability s_i.

Consider $x(n)$ as a concrete realization of X at time 'n'. In order to estimate the vector S, the SLWE maintains a running estimate $P(n) = [p_1(n), p_2(n),\dots,p_r(n)]^T$ of vector S, where $p_i(n)$ is the estimation of s_i at time 'n', for $i = 1,\dots,r$. The value of $p_i(n)$ is updated with respect to the coming data at each time instance, where Eqs. (1) and (2) show the updating rules:

$$p_i(n+1) \leftarrow p_i + (1-\lambda)\sum_{j\neq i} p_j \quad \text{when } x(n) = i \tag{1}$$

$$\leftarrow \lambda p_i \quad \text{when } x(n) \neq i. \tag{2}$$

Similar to the binomial case, the authors of [5] explicitly derived the dependence of $\mathrm{E}[P(n+1)]$ on $\mathrm{E}[P(n)]$, demonstrating the ergodic nature of the Markov matrix. The paper[2] also derived two explicit results concerning the convergence of the expected vector $P(.)$ to S, and the rate of convergence based on the learning parameter, λ.

---

[1] The case of estimating binomial distributions is a particular case of multinomial distributions where $r = 2$.

[2] The proofs of the theorems are omitted in the interest of brevity.

Theorem 1. *Consider $P(n)$, the estimate of the multinomial distribution S at time 'n', which is obtained by Eqs. (1) and (2). Then, $E[P(\infty)] = S$.* □

Theorem 2. *Consider $P(n)$, the estimate of the multinomial distribution S at time 'n', which is obtained by Eqs. (1) and (2). The expected value of P at time 'n + 1' is related to the expectation of $P(n)$ as $E[P(n+1)] = \mathbf{M}^T E[P(n)]$, where $\mathbf{M}$ is a Markov matrix. Further, every off-diagonal term of the stochastic matrix, $\mathbf{M}$, has the same multiplicative factor, $(1 - \lambda)$, and the final solution of this vector difference equation is independent of λ.* □

Theorem 3. *Consider $P(n)$, the estimate of the multinomial distribution S at time 'n', which is obtained by Eqs. (1) and (2). Then, all the non-unity eigenvalues of $\mathbf{M}$ are exactly λ, and therefore the convergence rate of P is fully determined by λ.* □

Theoretically, since the derived results are asymptotic, they are valid only as $n \to \infty$. However, in practice, by choosing λ from the interval $[0.9, 0.99]$, the convergence happens after a relatively small value of 'n'. Indeed, if λ is as "small" as 0.9, the variation from the asymptotic value will be in the order of 10^{-50} after 50 iterations. In other words, the SLWE will provide good results even if the distribution parameters change after 50 steps. The experimental results in [5] demonstrated a good performance achieved by using the SLWE in dynamic environments.

3   Online Classification Using SLWE

In traditional ML learning and particularly supervised learning, the training phase is performed in an *offline* manner, i.e., the training set is used to learn the stochastic properties of each class. Subsequently, the learned model is deployed and used to classify unlabeled data instances that appeared in the form of data streams.

In many real life applications, it is not possible to analyze the stochastic model of the classes in an *offline* manner because of their dynamic natures. In fact, *offline* classifiers assume that the entire set of training samples can be accessed. However, in many real life applications, the entire training set is not available either because it arrives gradually or because it is not feasible to store it so as to infer the model of each class. Consequently, one is forced to constantly make the classifier update the learning model using the newly-arriving training samples.

We present a novel *online* classifier scheme, that is able to update the learned model using a single instance at a time. Our goal is to predict the source of the arriving instances as accurately as possible, with the added complexity that the testing patterns can subsequently be considered as training patterns. To achieve this, we first define the general structure of the *Online* classifier, and then provide some experimental results on synthetic multinomial datasets in the next section.

Online classifiers deal with data streams, in which the labeled and unlabeled samples are mixed. Therefore, the training, testing and deploying phases of the *online* classifiers are interleaved as they are applied to these types of data streams. This fascinating avenue is our domain, and we have investigated the performance of SLWE-based classifiers to this new scheme.

Devising a classifier that deals with the data streams generated from non-stationary sources poses new challenges, since the probability distribution of each class might change even as new instances arrive. An important characteristic of our model for *online* learning is that the actual source of the data is discovered shortly after the prediction is made, which can then be used to update the learned model. In other words, our *online* algorithm includes three steps, which are described in Algorithm 1. First, the algorithm receives a data element. Using it and the currently-learned model, the classifier predicts the source of that element. Finally, the algorithm receives the true class of the data, which is then used to update and refine the classification model.

In order to perform the *online* classification of the instances, we need to obtain the *a posteriori* probability of each class. Analogous to the previous classification models, we assign a label to the new unlabeled data element by comparing the obtained *a posteriori* probabilities and the estimated probability from the unlabeled test stream. Finally, after receiving the true label of the instance, the *a posteriori* probabilities are updated using the algorithm explained in Eqs. (1) and (2).

In this classification model the training phase and the testing phase were performed simultaneously, and so the problem can be described as follows. We are given the stream of unlabeled samples generated from different sources arriving in the form of a (Periodically Switching Environment) PSE, in which, after every T time instances, the data distribution and the source of the data might change. In this case, in addition to the switching of the source of the data elements, the probability distribution of each source also possibly changes at random time instances. The aim of the classification is to predict the source of the elements arriving at each time step by using the information in the detected data distribution, and also the information of current model of each class. In the *online* classification model, shortly after the prediction is made, the actual class label of the instance is discovered, which can be utilized to update the classification model to be used by the SLWE updating algorithm.

The process is formalized in Algorithm 1.

4   Experimental Results

In this section, we present the results of this classifier on synthetic data. To assess the efficiency of the SLWE-based *online* classifier, we applied it for multinomial randomly generated data streams. Our results demonstrate the applicability of our method on synthetic datasets, and proves the advantages of the introduced scheme. We also classified and compared the data streams' elements by following the traditional MLE with a sliding window, whose size is also selected randomly.

Algorithm 1. Online Classification Algorithm

1: $X \leftarrow$ data stream for classification
2: $\hat{S} \leftarrow$ initialize *posterior* probabilities for each class
3: **while** there exists an instance $x \in X$ **do**
         Step 1. Receiving data:
4:     The model receives the unlabeled sample
5:     **for** all dimensions d of x **do**
6:        $p_i(n) \leftarrow$ Estimate the probability p_i using the SLWE
7:     **end for**
         Step 2. Prediction:
8:     $P(n) \leftarrow \{p_1(n), p_2(n), \ldots, p_d(n)\}$
9:     $\hat{\omega} \leftarrow \arg_i \min KL(\hat{S}_i \| P(n))$
         Step 3. Updating the model:
10:    After some delay, t_d, the true category of the instance x is received
11:    $\omega \leftarrow$ true class of x
12:    Update *posterior* probabilities $\hat{S}$ using ω and the SLWE
13: **end while**

4.1 Multinomial Data Stream

In this section, we report the results for simulations performed for *multinomial* data streams with *two* different non-stationary categories. Here, the classification problem was defined as follows. We are given a stream of unlabeled multinomially distributed random d-dimensional vectors, which take on the values from the set $\{1, \ldots, r\}$, and which are generated from *two* different periodically switching sources (classes), say, S_1 and S_2. Each class was characterized with probability values, S_{i1}, S_{i2}, which demonstrate the probability of the value 'i', where $i \in \{1, \ldots, r\}$.

The multinomial data stream classification started with the estimation of the *a priori* probability of each possible value of 'i', in all the 'd' dimensions, for each class 'j' from the available labeled instances, which we refer to as $\hat{S}_{ij}$. To assign a label to the newly arriving unlabeled element, the SLWE estimated the probabilities of each possible value 'i', in all the 'd' dimensions, from the unlabeled instances, which we refer to as $P_i(n)$. Thereafter, these probabilities were used to predict the class label that a new instance belonged to class 'j', with the probability vector of the $\hat{S}_j = \{\hat{S}_{1j}, \hat{S}_{2j}, \ldots, \hat{S}_r\}$, that had the minimum distance to the estimated probability of $P = \{P_1(n), P_2(n), \ldots, P_r(n)\}$, was chosen as the label of the observed element. The distances between the learned SLWE probabilities, $P_i(n)$, and the SLWE estimation during training, $\hat{S}_{ij}$ were again computed using the KL divergence measure, using Eq. (3).

$$KL(U\|V) = \sum_i u_i \log_2 \frac{u_i}{v_i}. \tag{3}$$

Thereafter, after some delay, t_d, at time $n+t_d$ the algorithm received the true class of the nth instance and used it to refine and update the true class probabilities. The true value of the category for the nth instance was read and added to the previously trained model by updating the probability of the corresponding class based on the updating algorithm in Eqs. (1) and (2).

The classification procedure explained above was performed on multinomial data streams generated from two different classes where the probability of the distributions of each class switched four times. For the results which we report, each element of the data stream could take any of the four different values, namely 1, 2, 3 or 4. The specific values of S_{i1} and S_{i2}, were changed and set to random values four times at random time instances, which were assumed to be unknown to the classifiers. The results are shown in Table 1, and again, the uniform superiority of the SLWE over the MLEW is noticeable. For example, when T=100, the MLEW-based classifier yielded an accuracy of only 0.7443, but the corresponding accuracy of the SLWE-based classifier was 0.8012. We also notice that the results of the classification in periodic environments with a varying T chosen randomly from $[50, 150]$ were also similar to the fixed $T = 100$ case, as the classifier achieved the accuracy of 0.8092 and 0.8012 in the first and second environments, respectively. The results also show that the SLWE-based algorithm handles the concept drift and provides satisfactory performance.

Table 1. The ensemble results for 100 simulations obtained from testing multinomial classifiers which used the SLWE (with $\lambda = 0.9$) and the MLEW for classifying one-dimensional data streams generated by two non-stationary different sources.

T	MLEW	SLWE
50	0.6786	0.7582
100	0.7443	0.8012
150	0.7476	0.8153
200	0.7595	0.8197
250	0.7514	0.8282
300	0.7509	0.8367
350	0.7587	0.8322
400	0.7598	0.8354
450	0.7635	0.8344
500	0.7574	0.8387
Random $T \in (50, 150)$	0.7523	0.8092

The experiment explained above was repeated on different 2-class multinomial datasets with different dimensionalities. These sets were generated randomly based on random vectors with different random distribution probabilities involving 2, 3 and 4 dimensions and each element could take on four different values. The results obtained are shown in Tables 2 and 3, from which we see that classification using the SLWE was uniformly superior to classification using the MLEW. For example, for the 2-dimensional data, when $T = 250$, the MLE-based classifier resulted in the accuracy of 0.7544 and the SLWE achieved significantly

Table 2. The ensemble results for 100 simulations obtained from testing multinomial classifiers which used the SLWE (with $\lambda = 0.9$) and the MLEW for classifying 4-dimensional data streams generated by two non-stationary different sources.

T	MLEW	SLWE
50	0.6906	0.8847
100	0.7535	0.9402
150	0.7557	0.9592
200	0.7532	0.9669
250	0.7550	0.9730
300	0.7531	0.9760
350	0.7522	0.9785
400	0.7460	0.9818
450	0.7572	0.9817
500	0.7562	0.9839
Random $T \in (50, 150)$	0.7526	0.9512

better results with the accuracy of 0.9706. Here the accuracy of the classifier, increased with the dimensionality of the datasets as the classifiers could process the data more efficiently. For example, in the case of $T = 150$ the SLWE-based *online* classifier resulted in the average accuracy of 0.9181 over several different two-dimensional datasets, while with more useful information in 4-dimensional datasets, it yielded better results with the accuracy of 0.9569.

Table 3. The ensemble results for 100 simulations obtained from testing multinomial classifiers which used the SLWE (with $\lambda = 0.9$) and the MLEW for classifying 2-dimensional data streams generated by two non-stationary different sources.

T	MLEW	SLWE
50	0.6913	0.8560
100	0.7402	0.9082
150	0.7463	0.9333
200	0.7480	0.9393
250	0.7478	0.9430
300	0.7397	0.9474
350	0.7486	0.9463
400	0.7525	0.9535
450	0.7554	0.9505
500	0.7518	0.9559
Random $T \in (50, 150)$	0.7436	0.9241

5   Conclusion

In this paper we have considered the problem of classification when these phases of training and testing are not so clearly well-defined, i.e., where the testing patterns can subsequently be considered as training patterns. This paradigm of classification is further complicated because we have assumed that the class-conditional distributions of the classes are time-varying or non-stationary. Here, we consider the scenario when the patterns arrive sequentially in the form of a data stream with potentially time-varying probabilities that change over time for each class. The proposed *online* classification algorithm was used to perform the training and the testing simultaneously in three phases. In the first phase, the algorithm received a new unlabeled instance. After this, the scheme assigned a label to it based on the distributions' estimated probabilities using the SLWE. Finally, after a few time instances, the algorithm received the actual class of the instance and used it to update the training model by invoking the SLWE updating algorithm. In this way the training and testing phases are almost inter-twined. Thereafter, the classification model was adjusted to the new available instances in an *online* manner.

References

1. Bifet, A.: Adaptive learning and mining for data streams and frequent patterns. Ph.D. thesis, Departament de Llenguatges i Sistemes Informatics, Universitat Politcnica de Catalunya, Barcelona Area, Spain (2009)
2. Bifet, A., Gavaldá, R.: Kalman filters and adaptive windows for learning in data streams. In: Todorovski, L., Lavrač, N., Jantke, K.P. (eds.) DS 2006. LNCS (LNAI), vol. 4265, pp. 29–40. Springer, Heidelberg (2006)
3. Bifet, A., Gavalda, R.: Learning from time-changing data with adaptive windowing. In: Proceedings SIAM International Conference on Data Mining, vol. 8, pp. 443–448 (2007)
4. Gama, J., Medas, P., Castillo, G., Rodrigues, P.: Learning with drift detection. In: Bazzan, A.L.C., Labidi, S. (eds.) SBIA 2004. LNCS (LNAI), vol. 3171, pp. 286–295. Springer, Heidelberg (2004)
5. Oommen, B.J., Rueda, L.: Stochastic learning-based weak estimation of multino-mial random variables and its applications to pattern recognition in non-stationary environments. Pattern Recogn. **39**(3), 328–341 (2006)
6. Widmer, G., Kubat, M.: Learning in the presence of concept drift and hidden contexts. Mach. Learn. **23**, 69–101 (1996)

Explicit Contrast Patterns Versus Minimal Jumping Emerging Patterns for Lazy Classification in High Dimensional Data

Marzena Kryszkiewicz[(✉)] and Przemyslaw Podsiadly

Institute of Computer Science, Warsaw University of Technology,
Nowowiejska 15/19, 00-665 Warsaw, Poland
{mkr, P.Podsiadly}@ii.pw.edu.pl

Abstract. Minimal jumping emerging patterns have been proved very useful for classification purposes. Nevertheless, the determination of minimal jumping emerging patterns may require evaluation of candidate patterns, the number of which might be exponential with respect to the dimensionality of a data set. This property may disallow classification by means of minimal jumping emerging patterns in the case of high dimensional data. In this paper, we derive an upper bound on the lengths of minimal jumping emerging patterns and an upper bound on their number. We also propose an alternative approach to lazy classification which uses explicit contrast patterns instead of minimal jumping emerging patterns, but produces the same classification quality as a lazy classifier based on minimal jumping emerging patterns. We argue that our approach, unlike the approach based on minimal jumping emerging patterns, can be applied in the case of high dimensional data.

1 Introduction

In [3, 4, 9], the authors proposed to use *jumping emerging patterns* for classification purposes. A pattern was regarded as a jumping emerging pattern if it occurred in only one decision class of a given decision table. In a lazy approach presented there, only jumping emerging patterns consisting of attribute value pairs occurring in an evaluated object T were searched and used to identify objects in a decision table that were useful for classifying object T. However, the number of jumping emerging patterns may be huge. To avoid this drawback, the authors of [8, 9] proposed to use only minimal jumping emerging patterns for classification instead of all jumping emerging patterns. Nevertheless, their number, which might be exponential with respect to the dimensionality of a data set, may be still huge. In fact, it may be so large that classification by means of minimal jumping emerging patterns becomes impossible. This happens in the case of high dimensional data such as e.g. DNA data with hundred thousands of attributes (genes), even if their number is reduced by an order or two orders of magnitude in a preprocessing step consisting in selecting attributes most relevant (and, eventually, irredundant) for classification purposes.

Our main contribution in this paper is the derivation of an upper bound on the lengths of minimal jumping emerging patterns and an upper bound on their number, as

© Springer International Publishing Switzerland 2016
H. Fujita et al. (Eds.): IEA/AIE 2016, LNAI 9799, pp. 80–94, 2016.
DOI: 10.1007/978-3-319-42007-3_8

well as the proposal of an alternative approach to lazy classification which does not generate minimal jumping emerging patterns at all, but produces the same classification quality as the lazy classifier from [9]. We argue that our approach, which is based on explicit contrast patterns, unlike the approach based on minimal jumping emerging patterns, can be applied in the case of high dimensional data.

Our paper has the following layout. In Sect. 2, we present basic notions and formulate properties, a lemma and corollaries useful in the remaining part of the paper. In Sect. 3, we recall an earlier approach to lazy classification based on minimal jumping emerging patterns [9] and show that it classifies in the same way irrespective of all minimal jumping emerging patterns are applied or all jumping emerging patterns are applied. In Sect. 4, we derive upper bounds on lengths and number of minimal jumping emerging patterns and argue that direct classification by means of minimal jumping emerging patterns is infeasible in the case of high dimensional data. In Sect. 5, we introduce the notions of explicit patterns, contrast objects and explicit contrast patterns. We also examine relations between them and minimal jumping emerging patterns in order to propose an algorithm, which will not depend in an exponential way on the number of attributes. Based on these findings, in Sect. 6, we offer an ECPC algorithm for finding explicit contrast patterns and using them for classification, which is capable of classifying in high dimensional data. Some implementation considerations and additional remarks are made in Sect. 7. Section 8 summarizes our work.

2 Decision Table, Decision Classes, Jumping Emerging Patterns and Minimal Jumping Emerging Patterns

Let AT be a non-empty set of attributes. In the paper, we assume that each *object R* is characterized by a *unique identifier Id(R)* and a *descriptor Desc(R)* being a set of at most $|AT|$ attribute-value pairs (a, v), where a is an attribute from AT and v is a value from a domain of attribute a.

In the remainder of the paper, an attribute-value pair will be also called an *item*, and a set of items whose attributes are distinct will be called an *itemset*. Clearly, a descriptor of an object is a maximal itemset characterizing the object in terms of attributes from AT. The number of items in an itemset is called its *length*.

In the paper, we assume that DT is a *decision table* consisting of a non-empty set of n objects, for which their *decision values* are known. The decision value for an object R will be denoted by $d(R)$.

A *decision class* is defined as a non-empty set of objects in DT that have the same decision value. We assume that DT consists of r, where $r > 1$, mutually exclusive decision classes $D_1, ..., D_r$ such that $DT = D_1 \cup ... \cup D_r$. By D_i' we denote the complement of D_i in DT; that is, $D_i' = DT \backslash D_i$.

Definition 1. *An object R supports an itemset X if the descriptor of object R contains itemset X; that is, if Desc(R) $\supseteq$ X.*

Property 1. If an object R supports an itemset X, then R supports all subsets of X.

Proof. Let X, Y be itemsets such that $X \supseteq Y$ and R be an object. If object R supports itemset X, then $Desc(R) \supseteq X \supseteq Y$. Hence, R supports Y.   □

Definition 2. *Support of an itemset X in an object set* $S \subseteq DT$ *is denoted by* $sup_S(X)$ *and is equal to the number of objects supporting X in* S.

Property 2. Let X, Y be itemsets such that $Y \subseteq X$ and $S \subseteq DT$. Then $sup_S(Y) \geq sup_S(X)$.

In the paper, we will be interested in calculating supports of itemsets in the whole decision table as well as in its parts: in a decision class and in the complement of a decision class.

Definition 3. An itemset X is a *jumping emerging pattern of a decision class* D_i if it has non-zero support in D_i ($sup_{D_i}(X) > 0$) and zero support in its complement D_i' ($sup_{D_{i'}}(X) = 0$). *The set of all jumping emerging patterns of decision class* D_i *will be denoted by* JEPs(D_i).

Definition 4. An itemset X is a *jumping emerging pattern* if there is a decision class in DT for which X is its jumping emerging pattern. *The set of all jumping emerging patterns in* DT *will be denoted by* JEPs; *that is,* JEPs $= \cup_{i \in \{1,...r\}}$ JEPs(D_i).

Property 3. An itemset X is a *jumping emerging pattern of a decision class* D_i if and only if $sup_{D_i}(X) > 0$ and $\forall j \in \{1,..., r\} \setminus \{i\}$ ($sup_{D_j}(X) = 0$).

Property 4.

(a) An itemset can be a jumping emerging pattern of at most one decision class.
(b) $\varnothing$ is not a jumping emerging pattern.

Proof. Ad (a) By Property 3.

Ad (b) Since each of decision classes in DT is assumed to be non-empty and DT is assumed to have at least two decision classes, then $\varnothing$ occurs in more than one decision class. Hence, by Property 3, $\varnothing$ is not a jumping emerging pattern.   □

Lemma 1. Let X be a jumping emerging pattern of D_i and Y be a proper superset of X. Then:

(c) $sup_{D_{i'}}(Y) = 0$.
(d) Y is not a jumping emerging pattern of D_i'.
(e) If Y has a non-zero support in D_i, then Y is a jumping emerging pattern of D_i.
(f) If Y has a zero support in D_i, then Y is not a jumping emerging pattern of D_i.
(g) If Y is a jumping emerging pattern of D_i, then each pattern Z such that $Y \supseteq Z \supset X$ is a jumping emerging pattern of D_i.

Proof. Ad (a) By definition of a jumping emerging pattern of D_i, $sup_{D_{i'}}(X) = 0$. Since $Y \supset X$, then $sup_{D_{i'}}(Y) = 0$ (by Property 2).

Ad (b) Follows from Lemma 1a.

Ad (c) Let $sup_{D_i}(Y) > 0$. By Lemma 1a, $sup_{D_{i'}}(Y) = 0$. Hence, Y is a jumping emerging pattern of D_i.

Ad (d) Let $sup_{Di}(Y) = 0$. Then, by definition, Y is not a jumping emerging pattern of D_i.

Ad (e) Let Y be a jumping emerging pattern of D_i and Z be an itemset such that $Y \supseteq Z \supset X$. Then, $sup_{Di}(Y) > 0$, so $sup_{Di}(Z) > 0$ (by Property 2). Since X is a jumping emerging pattern of D_i, then $sup_{Di'}(X) = 0$, so $sup_{Di'}(Z) = 0$ (by Property 2). As $sup_{Di}(Z) > 0$ and $sup_{Di'}(Z) = 0$, then Z is a jumping emerging pattern of D_i. □

Property 5. Let X be not a jumping emerging pattern and have non-zero supports in at least two distinct decision classes D_i and D_j (that is, $sup_{Di}(X) > 0$, $sup_{Dj}(X) > 0$) and $i \neq j$). Then:

(a) Supports of all subsets of X in both decision classes D_i and D_j are non-zero ($\forall Y \subseteq X$ ($sup_{Di}(Y) > 0$ and $sup_{Dj}(Y) > 0$)).
(b) Subsets of X are not jumping emerging patterns ($\forall Y \subseteq X$ ($Y \notin$ JEPs)).

Proof. Ad (a) By Property 2.

Ad (b) By Properties 3 and 5a.   □

Definition 5. An itemset X is a *minimal jumping emerging pattern* if X is a jumping emerging pattern and all proper subsets of X are not jumping emerging patterns.

Property 6. Let X be a minimal jumping emerging pattern of D_i, $Y \subset X$, $j \neq i$, and the support of Y in D_j is non-zero.

(a) Y has non-zero supports in at least two decision classes D_i and D_j, and is not a jumping emerging pattern.
(b) Each subset of Y has non-zero supports in at least two decision classes D_i and D_j, and is not a jumping emerging pattern.

Proof. Let X be a minimal jumping emerging pattern of D_i, $Y \subset X$, $j \neq i$ and $sup_{Dj}(Y) > 0$ (*).

Ad (a) Since X is a jumping emerging pattern of D_i, then $sup_{Di}(X) > 0$. So, by Property 2, all subsets of X, including Y, have non-zero supports in D_i. Hence and by (*), $sup_{Di}(Y) > 0$ and $sup_{Dj}(Y) > 0$. Thus, by Property 3, Y is not a jumping emerging pattern.   □

Ad (b) By Property 6a, $sup_{Di}(Y) > 0$, $sup_{Dj}(Y) > 0$ and $Y \notin$ JEPs. Hence and by Property 5, each subset of Y has non-zero supports in at least two decision classes D_i and D_j, and is not a jumping emerging pattern.   □

Corollary 1. Let X be a jumping emerging pattern of D_i, Y be a minimal jumping emerging pattern of D_i and $X \supseteq Y$. Then the maximal set of objects in D_i that support X constitutes a subset of the maximal set of objects in D_i that support Y; that is,

$$\{R \in D_i | Desc(R) \supseteq X\} \subseteq \{R \in D_i | Desc(R) \supseteq Y\}.$$

Proof. By Property 1.   □

Corollary 2. The maximal set of objects in D_i each of which supports at least one jumping emerging pattern of D_i equals the maximal set of objects in D_i each of which supports at least one minimal jumping emerging pattern of D_i:

$$\bigcup{}_{X \text{ is a jumping emerging pattern of } D_i} \{R \in D_i | Desc(R) \supseteq X\} =$$
$$\bigcup{}_{Y \text{ is a minimal jumping emerging pattern of } D_i} \{R \in D_i | Desc(R) \supseteq Y\} \cup$$
$$\bigcup{}_{X \text{ is a non-minimal jumping emerging pattern of } D_i} \{R \in D_i | Desc(R) \supseteq X\} =$$
$$\bigcup{}_{Y \text{ is a minimal jumping emerging pattern of } D_i} \{R \in D_i | Desc(R) \supseteq Y\}.$$

Proof. By Corollary 1.   □

Corollary 3. For each jumping emerging pattern X of D_i, there is a subset of X being a minimal jumping emerging pattern of D_i and there is a superset of X being a maximal jumping emerging pattern of D_i.

Finally, we would like to note that the Rough Set community widely applies the concept of a (minimal) jumping emerging pattern, however, in the context of certain decision rules; namely, a (minimal) jumping emerging pattern X of a decision class D_i is known by the Rough Set researchers under name of *an antecedent of an (optimal) certain decision* rule $X \Rightarrow d_i$, where d_i is a common decision value of the objects in decision class D_i [10]. A rule $X \Rightarrow d_i$ is called *certain* if X occurs only in D_i (which means that X is a jumping emerging pattern of D_i). $X \Rightarrow d_i$ is called *optimal certain* if X is a minimal itemset which occurs only in D_i (which means that X is a minimal jumping emerging pattern of D_i).

3   Lazy Classification Based on Minimal Jumping Emerging Patterns

In [9], the authors described the following approach to lazy classification of an object, say T, with respect to a decision table DT by means of minimal jumping emerging patterns:

Step 1.   Create a *T-reduced decision table* DT^T by intersecting $Desc(T)$ with descriptor of each object R in DT:

$$DT^T = \{(Id(R), Desc(T) \cap Desc(R), d(R)) | R \in DT\}$$
$$= D_1^T \cup \ldots \cup D_r^T, \text{ where}$$
$$D_i^T = \{(Id(R), Desc(T) \cap Desc(R), d(R)) | R \in D_i\}, i \in \{1, \ldots, r\}.$$

Step 2.   Determine all minimal jumping emerging patterns in DT^T.

Step 3.   For each decision class D_i^T in DT^T, identify the maximal set of objects each of which supports at least one minimal jumping emerging pattern of that decision class.

Step 4. For each decision class D_i^T in DT^T, determine its *voting strength* as the ratio of the cardinality of the set of objects identified in Step 3 for D_i^T to the number of all objects in D_i:

$$VotingStrength_{Di}(T) = \frac{\left| \bigcup_{X \text{ is a minimal jumping emerging pattern of } D_i^T} \left\{ R \in D_i^T | Desc(R) \supseteq X \right\} \right|}{|D_i|}.$$

Step 5. Assign object T to the decision class with the maximal value of the voting strength $VotingStrength_{Di}(T)$.

Please note that a descriptor of an object in DT^T may have less items than its descriptor in DT. In fact, a descriptor of an object in DT^T may become even an empty itemset, nevertheless, the cardinality of a decision class D_i^T is the same as the cardinality of D_i.

Corollary 2 allows us to infer that the voting strength will be the same if one uses all jumping emerging patterns instead of all minimal jumping emerging patterns (see Corollary 4).

Corollary 4. $VotingStrength_{Di}(T) = \dfrac{\left| \bigcup_{X \text{ is a jumping emerging pattern of } D_i^T} \left\{ R \in D_i^T | Desc(R) \supseteq X \right\} \right|}{|D_i|}.$

Proof. By Corollary 2.   □

Hence, the presented algorithm would produce the same classification results irrespective of all jumping emerging patterns were applied or all minimal jumping emerging patterns were applied. Clearly, using minimal jumping emerging patterns is more efficient as their number is less than the number of jumping emerging patterns. Moreover, minimal jumping emerging patterns (or, in the Rough Set terminology, antecedents of optimal certain rules) are so called *generators* (see the definition beneath) in DT^T [2, 7], which allows optimizing their discovery [1, 6, 7, 11].

Definition 6. An itemset X is a *generator* in an object set S if supports of all proper subsets of X in S are greater than the support of X in S.

Property 7 [5, 6, 11]. Let S be an object set. If X is a *generator* in object set S, then all subsets of X are generators in S. If X is not a generator in S, then all supersets of X are not generators in S.

Property 7 allows inferring that if a pattern X is not a generator, then neither X nor its supersets are minimal jumping emerging patterns. In the next section, we examine minimal jumping emerging patterns in more detail and discuss their limitations in classifying.

4 Upper Bounds on Lengths and Number of Minimal Jumping Emerging Patterns

In this section, we derive an upper bound on the length of a minimal jumping emerging pattern and an upper bound on the number of minimal jumping emerging patterns in a T-reduced decision table DT^T. We will formulate these bounds in terms of the number m, where $m \leq |AT|$, of items in the evaluated object T and the number n of objects in DT.

Proposition 1. A minimal jumping emerging pattern in DT^T has no more than $n - 1$ items.

Proof. Let X be a minimal jumping emerging pattern in DT^T and consist of k items $\{x_1, ..., x_k\}$, where $k > 0$. Let us consider the following subsets of X: $X_0 = \emptyset$, $X_1 = \{x_1\}$, $X_2 = \{x_1, x_2\}, ..., X_k = \{x_1, ..., x_k\} = X$. Since X is a minimal jumping emerging pattern in DT^T, then X is a generator in DT^T. Hence, by Property 7, all its subsets are generators in DT^T, too. Thus, for $i = 0..k-1$, $sup_{DT^T}(X_i) > sup_{DT^T}(X_{i+1})$; that is, $sup_{DT^T}(X_i) - sup_{DT^T}(X_{i+1}) \geq 1$. Therefore, $\sum_{i=0..k-1}(sup_{DT^T}(X_i) - sup_{DT^T}(X_{i+1})) \geq k$ (*). On the other hand, $\sum_{i=0..k-1}(sup_{DT^T}(X_i) - sup_{DT^T}(X_{i+1})) = sup_{DT^T}(X_0) - sup_{DT^T}(X_k) = sup_{DT^T}(\emptyset) - sup_{DT^T}(X) = n - sup_{DT^T}(X)$ (**). Hence, by (*) and (**), $n - sup_{DT^T}(X) \geq k$. Since X is a jumping emerging pattern in DT^T, then it is supported by at least one object in DT^T, so $sup_{DT^T}(X) \geq 1$. The value of k will be maximal for minimal value of $sup_{DT^T}(X)$; that is, for $sup_{DT^T}(X) = 1$. So, maximal value of k, and by this of $|X|$, is equal to $n - 1$.   □

Now, we will focus on deriving an upper bound on the number of minimal jumping emerging patterns.

Proposition 2. The number of all minimal jumping emerging patterns in DT^T is not greater than $\binom{m}{\lfloor m/2 \rfloor}$.

Proof. Minimal jumping emerging patterns found in DT^T are subsets of descriptor $Desc(T)$ of length m. We also note, that any set of minimal jumping emerging patterns is an anti-chain; that is, for each pair of two distinct patterns X and Y in this set, neither X is a subset of Y nor Y is a subset of X. The upper bound on the cardinality of an anti-chain whose itemsets are subsets of an itemset of length m is equal to $\binom{m}{\lfloor m/2 \rfloor}$. Hence, the number of all minimal jumping emerging patterns is not greater than $\binom{m}{\lfloor m/2 \rfloor}$.   □

By Proposition 2, the number of all minimal jumping emerging patterns of DT^T is not greater than the number of all $\lfloor m/2 \rfloor$ item subsets of itemset $Desc(T)$ of length m. Proposition 2, nevertheless, does not take into account the derived upper bound on the length of a minimal jumping emerging pattern given in Proposition 1. Beneath we derive an improved upper bound on the number of minimal jumping emerging patterns based on both Propositions 1 and 2.

Proposition 3. If $n - 1 < \lfloor m/2 \rfloor$, then the number of all minimal jumping emerging patterns in DT^T is not greater than $\binom{m}{n-1}$, Otherwise, the number of all minimal jumping emerging patterns in DT^T is not greater $\binom{m}{\lfloor m/2 \rfloor}$.

Proof.

Case $n - 1 < \lfloor m/2 \rfloor$: Then lengths of all minimal jumping emerging patterns are at most $n - 1$ (by Proposition 1). In addition, all minimal jumping emerging patterns constitute an anti-chain and are subsets of $Desc(T)$ of length m. Hence, their number cannot be greater than $\binom{m}{n-1}$, which is less than $\binom{m}{\lfloor m/2 \rfloor}$.

Case $n - 1 \geq \lfloor m/2 \rfloor$: Proposition 3 follows immediately from Proposition 2. $\square$

As follows from Proposition 3, the number of minimal jumping emerging patterns found in a T-reduced decision table DT^T strongly depends on the number of items in the descriptor of evaluated object T. If the number of items is small, then it is feasible to both calculate and store discovered minimal jumping emerging patterns. However, even in the case when $m = n = 50$, the upper bound on the number of minimal jumping emerging patterns is very large; namely, 1.26×10^{14}. If $m = n = 100$, the upper bound on the number of minimal jumping emerging patterns is 1.01×10^{29}. In the case of a bio-informatics data, where m is equal to, for example, 10000 and $n = 100$, the upper bound on the number of minimal jumping emerging patterns is 6.98×10^{105}. In this case, classification with minimal jumping emerging patterns is quite unlikely. Actually, bio-informatics data can easily have around a few hundred thousand of attributes (and by this, items in descriptors of objects) and a few thousand of objects, which makes classification by means of minimal jumping emerging patterns infeasible.

5   Explicit Patterns, Contrast Objects, Explicit Contrast Patterns and Minimal Jumping Emerging Patterns

In this section, we investigate properties of minimal jumping emerging patterns and objects supporting them as well as introduce notions of an *explicit pattern*, a *contrast object* and an *explicit contrast pattern*. The last notion will play an important role in the *ECPC* classification algorithm we will offer in Sect. 6.

Definition 7. *An explicit pattern of* D_i^T *is a descriptor of an object in* D_i^T.

Definition 8. *A contrast object of* D_i^T *is an object supporting a jumping emerging pattern of* D_i^T. *An explicit contrast pattern of* D_i^T *is a descriptor of a contrast object in* D_i^T.

Proposition 4. *Let X be an explicit pattern of* D_i^T. *X is an explicit contrast pattern of* D_i^T *if and only if X is a jumping emerging pattern of* D_i^T.

Proof. ($\Rightarrow$) Let X be an explicit contrast pattern of D_i^T. Then, there is a contrast object, say R, in D_i^T such that $X = Desc(R)$. Hence, the support of X is non-zero in D_i^T (*). Since R is a contrast object, then there is a jumping emerging pattern, say Y, of D_i^T, such that $X = Desc(R) \supseteq Y$ (**). By (*), (**) and Lemma 1c, X is a jumping emerging pattern of D_i^T.

($\Leftarrow$) Let X be an explicit contrast pattern of D_i^T and a jumping emerging pattern of D_i^T. Then, there is an object, say R, in D_i^T such that $X = Desc(R)$ and R supports jumping emerging pattern X of D_i^T. Hence, R is a contrast object D_i^T. So, its descriptor $Desc(R)$, and by this, itemset X, is an explicit contrast pattern of D_i^T. $\square$

Proposition 5. *The maximal set of objects each of which supports at least one jumping emerging pattern of* D_i^T *is equal to the set of all contrast objects in* D_i^T:

$$\cup_{X \text{ is a jumping emerging pattern of } D_i^T} \left\{ R \in D_i^T | Desc(R) \supseteq X \right\}$$
$$= \left\{ R \in D_i^T | R \text{ is a contrast object in } D_i^T \right\}$$

Proof. By definition of contrast objects.   □

We will use Proposition 6 to express voting strength first in terms of contrast objects, and then in terms of explicit contrast patterns being their descriptors.

Proposition 6.

(a) $VotingStrength_{Di}(T) = \dfrac{|\{R \in D_i^T | R \text{ is a contrast object of } D_i^T|}{|D_i|}.$

(b) $VotingStrength_{Di}(T) = \dfrac{\sum_{X \text{ is an explicit contrast pattern of } D_i^T} |\{R \in D_i^T | Desc(R) = X|}{|D_i|}.$

Proof Ad (a) By Corollary 4 and Proposition 5.

Ad (b) A descriptor of each contrast object in D_i^T is an explicit contrast pattern of D_i^T. It may happen that two or more contrast objects share the same descriptor. Let us split all contrast objects in D_i^T into object clusters with respect to descriptors of these contrast objects. Then, the number of all contrast objects in $D_i^T (|\{R \in D_i^T | R \text{ is a contrast object of } D_i^T|)$ will be equal to the sum of the cardinalities of the object clusters: $\sum_{X \text{ is an explicit contrast pattern of } D_i^T} |\{R \in D_i^T | Desc(R) = X|$. Hence and by Proposition 6a, we obtain Proposition 6b.   □

Proposition 6b allows us to calculate voting strength of D_i^T based on the number of objects in D_i^T the descriptors of which are explicit contrast patterns. Please note that the number of explicit contrast patterns does not exceed the number of objects in DT^T with non-empty descriptors. In the next section, we will offer the ECPC algorithm, which first finds explicit contrast patterns in accordance with Proposition 4 and numbers of objects supporting them in D_i^T, and then calculates voting strengths by means of Proposition 6b.

6 Algorithm for Finding Explicit Contrast Patterns and Using Them for Classification

Let DT be a decision table with decision classes D_1, ..., D_r and T be an object to be classified based on DT. Beneath we offer an algorithm *ECPC* (Explicit Contrast Patterns Classifier) for finding explicit contrast patterns in DT^T, which are needed for determining voting strengths of decision classes, and for classifying based on them. The ECPC algorithm we propose consists of three steps:

Step 1.   Calculation of all *non-empty explicit patterns* EPs of DT^T.

Step 2.   Identification of all explicit contrast patterns ECPs of DT^T as those EPs of DT^T that are jumping emerging patterns (based on Proposition 4 and Properties 1 and 5).

Step 3.   Calculation of voting strengths based on the numbers of objects supporting explicit contrast patterns ECPs (in accordance with Proposition 6b).

Notation applied in the *ECPC* algorithm:

- X – an itemset
- $X.d$ – takes -1 value if X occurs in many decision classes; otherwise, it is an identifier (a value from $\{1, ..., r\}$) of a decision class in which X occurs;
- $X.\#$ – the number of objects having X as a descriptor in DT^T;
- VS – an array storing voting strengths for respective decision classes $D_1, ..., D_r$;
- EPs – non-empty explicit patterns of DT^T;
- ECPs – explicit contrast patterns of DT^T.

function *ECPC*(DT, *T*);
{* Calculation of non-empty explicit patterns EPs of DT^T *}

EPs = $\varnothing$;
foreach object R in DT **do begin**
   $X = Desc(R) \cap Desc(T)$;
   if $X \neq \varnothing$ **then begin**
     $X.d = R.d$;
     if there is an itemset Y in EPs such that $Y = X$ **then**
       if $Y.d = X.d$ **then**
         $Y.\# = Y.\# + 1$;   // update the number of duplicates of Y
       else
         $Y.d = -1$;   // Y occurs in more than one decision class
       endif
     else begin
       $X.\# = 1$;
       insert X to EPs;   // insert new non-empty explicit pattern X to EPs
     endif;
   endif;
endfor;

{* Identification of explicit contrast patterns ECPs of DT^T among non-empty explicit patterns EPs. *}
{* By Proposition 4, an explicit pattern is an explicit contrast pattern if and only if it is a JEP.   *}
ECPs = $\varnothing$;
while EPs $\neq \varnothing$ **do begin**
   X = a longest itemset in EPs;
   if $X.d = -1$ **then**   // X is not a JEP (by Property 3)
     delete X and all its proper subsets from EPs;   // subsets of X are not JEPs either (by Property 5)
   else begin
     move X from EPs to ECPs;   // X is a JEP of decision class $D^T_{X.d}$
     while there is Y in EPs such that $Y \subset X$ and $Y.d \neq X.d$ **do**
       {Since support of X in $D^T_{X.d}$ is non-zero, then support of Y in $D^T_{X.d}$ is non-zero (by Property 2). }
       {Y is not a JEP as it has non-zero supports in $D^T_{Y.d}$ and $D^T_{X.d}$, which are distinct decision classes.}
       delete Y and all its proper subsets from EPs   // subsets of Y are not JEPs (by Property 5)
     endwhile;
   endif;
endwhile;

{* Calculating voting strengths based on the numbers of objects supporting explicit contrast patterns ECPs according to Proposition 6b *}
initialize all elements of the array VS with 0;
forall X in ECPs **do**
   $VS[X.d] = VS[X.d] + X.\#$;
endfor;
for $i = 1..r$ **do**
   $VS[i] = VS[i] / | D_i |$;   // $| D_i |$ is the cardinality of D_i in non-reduced DT
endfor;
$maxVS = max\{VS[i]| i = 1..r\}$ // determine the maximal voting strength
return i for which $VS[i] = maxVS$;   // return the identifier of D_i with maximal voting strength

Our algorithm identifies explicit contrast patterns for all decision classes simultaneously in contrast to approaches presented in [8], where minimal jumping emerging patterns were determined separately for each decision class in the case of a decision table with more than two classes. The complexity of our proposed ECPC algorithm is $O(m \times n^2)$.

Example 1. Let T be an object to be classified and its descriptor $Desc(T) = \{abcde\text{-}fghijk\}$. Items $a, b, c, d, e, f, g, h, i, j, k$ stand for respective attribute-value pairs (for example, a may stand for (apple, Mackintosh)). Let us assume that Table 1 presents all non-empty explicit patterns EPs of DT^T together with the information about their lengths, numbers of objects in DT^T having them as descriptors and decision classes to which they belong.

Table 1. Non-empty explicit patterns EPs of some T-reduced decision system DT^T

Length	X in EPs	X.d	X.#
9	$\{cdefghijk\}$	2	4
7	$\{abcdefg\}$	1	2
7	$\{adeghjk\}$	−1	3
7	$\{bcdehij\}$	1	5
5	$\{bcdef\}$	2	2
5	$\{cdegh\}$	1	2
5	$\{eghjk\}$	1	1
4	$\{bcde\}$	1	3
3	$\{cde\}$	2	2

Table 2 presents results of analyzing the longest itemset ($\{cdefghijk\}$) in the set of all non-empty explicit patterns EPs. $\{cdefghijk\}$ belongs to the decision class D_2^T. Since $\{cdefghijk\}$ is a longest itemset, it is clear that there are no proper supersets of $\{cdefghijk\}$ in other decision classes. Hence, $\{cdefghijk\}$ is a JEP, and by this, an explicit contrast pattern. Its proper subsets $\{cdegh\}$ and $\{eghijk\}$ belong to another decision class (D_1^T). Hence, $\{cdegh\}$, $\{eghijk\}$ and all their proper subsets (here: $\{cde\}$) are not jumping emerging patterns. As a result: $ECPs(D_1^T) = \emptyset$; $ECPs(D_2^T) = \{\{cdefghijk\}\}$.

Illustration of an analysis of the next longest itemset ($\{abcdefg\}$) in the modified set of EPs is presented in Table 3. As a result: $ECPs(D_1^T) = \{\{abcdefg\}\}$; $ECPs$ $(D_2^T) = \{\{cdefghijk\}\}$.

Illustration of an analysis of the third longest itemset ($\{adeghjk\}$) in the modified set of EPs is presented in Table 4. Since $\{adeghjk\}$ occurs in more than one decision class (which is indicated by $\{adeghjk\}.d = -1$), it is not a jumping emerging pattern. Nor its proper subsets are (here: no itemset in the current EPs is a proper subset of $\{adeghjk\}$). As a result the sets of explicit jumping emerging patterns do not change: $ECPs(D_1^T) = \{\{abcdefg\}\}$; $ECPs(D_2^T) = \{\{cdefghijk\}\}$.

Table 2. Results of analyzing the longest explicit pattern ($\{cdefghijk\}$)

Length	X in EPs	X.d	X.#	Comment
9	{cdefghijk}	2	4	• {cdefghijk} is moved to ECPs; • {cdegh}.d ≠ {cdefghijk}.d, so {cdegh} and all its proper subsets (here: {cde}) are removed from EPs; • {eghjk}.d ≠ {cdefghijk}.d, so {eghjk} and all its proper subsets (here: no proper subset) are removed from EPs.
7	{abcdefg}	1	2	
7	{adeghjk}	-1	3	
7	{bcdehij}	1	5	
5	{bcdef}	2	2	
5	{cdegh}	1	2	
5	{eghjk}	1	1	
4	{bcde}	1	3	
3	{cde}	2	2	

Table 3. Results of analyzing the second longest explicit pattern ($\{abcdefg\}$)

Length	X in EPs	X.d	X.#	Comment
7	{abcdefg}	1	2	• {abcdefg} is moved to ECPs; • {bcdef}.d ≠ {abcdefg}.d, so {bcdef} and all its proper subsets (here: {bcde}) are removed from EPs.
7	{adeghjk}	-1	3	
7	{bcdehij}	1	5	
5	{bcdef}	2	2	
4	{bcde}	1	3	

Table 4. Results of analyzing the third longest explicit pattern ($\{adeghjk\}$)

Length	X in EPs	X.d	X.#	Comment
7	{adeghjk}	-1	3	• {adeghjk} and all its proper subsets are not JEPs, so {adeghjk} and all its proper subsets (here: there is no proper subset of {adeghjk} in EPs) are removed from EPs.
7	{bcdehij}	1	5	

Table 5. Results of analyzing the last explicit pattern ($\{bcdehij\}$)

Length	X in EPs'	X.d	X.#	Comment
7	{bcdehij}	1	5	• {bcdehij} is moved to ECPs; • There are no proper subsets of {bcdehij} in EPs (belonging do a decision class different from $D_1{}^T$).

Illustration of an analysis of the last itemset ($\{bcdehij\}$) in the current set of EPs is presented in Table 5. $\{bcdehij\}$ belongs to D_1^T. If there was previously a proper superset of $\{bcdehij\}$ belonging to a decision class different from D_1^T, then $\{bcdehij\}$ would be found a non-jumping emerging pattern and would be deleted. As this did not happen, $\{bcdehij\}$ is found a jumping emerging pattern of D_1^T, and by this, an explicit contrast pattern of D_1^T.

As a result, the final sets of explicit contrast patterns are determined as follows: $ECPs(D_1^T) = \{\{abcdefg\}, \{bcdehij\}\}$; $ECPs(D_2^T) = \{\{cdefghijk\}\}$. These sets are used to calculate voting strengths:

- $VotingStrength_{D1}(T) = \frac{\{abcdefg\}.\# + \{bcdehij\}.\#}{|D_1|} = \frac{2+5}{|D_1|}$.

- $VotingStrength_{D2}(T) = \frac{\{cdefghijk\}.\#}{|D_2|} = \frac{4}{|D_2|}$.

T will be assigned to the class with higher voting strength.   □

7   Implementation Considerations and Additional Remarks

In order to improve the efficiency of the presented algorithm, certain optimizations were performed. One optimization is based on the observation that all EPs and ECPs used in the ECPC algorithm may be represented exclusively using (identifiers of) attributes instead of attribute-value pairs. It is possible because for each ECP or EP, a value of each attribute is exactly the same as the value of that attribute for object T, or otherwise the attribute is not present in a pattern. This way the creation, multiplication and comparison is simplified, and memory footprint decreased.

In order to further improve efficiency, a hash function was used to quickly compare objects. An inexpensive but still well performing hash function was chosen (Fowler-Noll-Vo FNV-1A) and its calculation was performed during itemset creation to minimize overhead. Hash values were stored together with the object descriptor and were used in a hash table lookup to check if an itemset is already present in a set of EPs.

Let us also note that the ECPC algorithm may be used to analyze transaction data, where a transaction is a list of elements such as, for example, products purchased during a single visit in a shop. Such a transaction can be treated as an object the descriptor of which consists of items being attribute-value pairs (a,v), where a corresponds to a product and $v = 1$ informs that this product was purchased. The information about products that were not purchased could be also stored if useful. Then item (a,v), where $v = 0$ would inform that product a was not purchased.

Furthermore, in our implementation of the ECPC algorithm, we allow numerical data to be treated as indiscernible if their values differ only slightly. Our solution is analogous to that presented in [9]. In ECPC, numerical attributes are treated in a special way only when calculating explicit patterns EPs of DT^T; that is, when intersecting descriptors of objects in DT with a descriptor of a classified object T. In the case of DT with nominal attributes only, an EP is the result of the set-theoretical intersection of $Desc(R)$, where R is an object in DT, with $Desc(T)$, and may be represented using only attributes, values of which are equal for $Desc(R)$ and $Desc(T)$. If DT contains (also) numerical attributes, a numerical attribute is included into the resulting EP if the values of that attribute for R and T are sufficiently similar. To achieve this, all the numerical values can be normalized and two values can be treated as sufficiently similar if they belong to their respective ε-neighborhoods, where ε is a threshold value.

8 Summary

In the paper, we derived upper bounds on lengths and number of minimal jumping emerging patterns. We argued that direct classification by means of minimal jumping emerging patterns is infeasible in the case of high dimensional data. That motivated us to propose a new algorithm that, without using minimal jumping emerging patterns, would provide the same classification results as the successful lazy classifier from [9], which applies minimal jumping emerging patterns. We started by introducing the notions of explicit patterns, contrast objects and explicit contrast patterns, as well as examining relations between them and minimal jumping emerging patterns in order to devise a new algorithm. Eventually, we devised and offered the ECPC algorithm, which finds explicit contrast patterns and uses them for classification. The complexity of the ECPC algorithm is $O(m \times n^2)$, where m is the number of items in the descriptor of a classified object and n is the number of objects in the decision table DT. Our algorithm is capable of classifying in high dimensional data.

A side-effect of our main contribution in this paper is the formulation of a few properties, a lemma and corollaries related to jumping emerging patterns. In particular, we formally proved that the lazy classification based on minimal jumping emerging patterns, as presented in [9], is equivalent to the lazy classification based on jumping emerging patterns.

References

1. Bastide, Y., Pasquier, N., Taouil, R., Stumme, G., Lakhal, L.: Mining minimal non-redundant association rules using frequent closed itemsets. In: Palamidessi, C., Moniz Pereira, L., Lloyd, J.W., Dahl, V., Furbach, U., Kerber, M., Lau, K.-K., Sagiv, Y., Stuckey, P.J. (eds.) CL 2000. LNCS (LNAI), vol. 1861, pp. 972–986. Springer, Heidelberg (2000)
2. Bailey, J., Ramamohanarao, K.: Mining emerging patterns using tree structures or tree based searches. In: Contrast Data Mining: Concepts, Algorithms, and Applications. Taylor & Francis Group LLC, Boca Raton (2013)
3. Dong, G., Li, J., Zhang, X.: Discovering jumping emerging patterns and experiments on real datasets. In: IDC 1999 (1999)
4. Fan, H., Kotagiri, R.: An efficient single-scan algorithm for mining essential jumping emerging patterns for classification. In: Chen, M.-S., Yu, P.S., Liu, B. (eds.) PAKDD 2002. LNCS (LNAI), vol. 2336, pp. 456–462. Springer, Heidelberg (2002)
5. Kryszkiewicz, M.: Concise representation of frequent patterns based on disjunction-free generators. In: Proceedings of ICDM 2001, pp. 305–312. IEEE Computer Society (2001)
6. Kryszkiewicz, M.: Concise representations of frequent patterns and association rules. Prace Naukowe Politechniki Warszawskiej. Elektronika **142** (2002)
7. Kryszkiewicz M.: Using generators for discovering certain and generalized decision. In: HIS 2005, pp. 181–186 (2005)
8. Li, J., Dong, G., Ramamohanarao, K.: Making use of the most expressive jumping emerging patterns for classification. In: PAKDD 2000, pp. 220–232 (2000)
9. Jinyan, L., Guozhu, D., Kotagiri, R., Limsoon, W.: DeEPs: a new instance-based lazy discovery and classification system. ML **54**(2), 99–124 (2004)

10. Skowron A.: Boolean reasoning for decision rules generation. In: ISMIS 1993, pp. 295–305 (1993)
11. Stumme, G., Taouil, R., Bastide, Y., Pasquier, N., Lakhal, L.: Fast computation of concept lattices using data mining techniques. Proc. KRDB **2000**, 129–139 (2000)

An Evaluation on KNN-SVM Algorithm for Detection and Prediction of DDoS Attack

Ahmad Riza'ain Yusof[1,2], Nur Izura Udzir[1], and Ali Selamat[2(✉)]

[1] UTM-IRDA Digital Media Centre, Universiti Teknologi Malaysia,
81310 Johor Bahru, Johor, Malaysia
rizaain@utm.my, izura@upm.edu.my
[2] Faculty of Computing, Universiti Teknologi Malaysia,
81310 Johor Bahru, Johor, Malaysia
aselamat@utm.my

Abstract. Recently, damage caused by DDoS attacks increases year by year. Along with the advancement of communication technology, this kind of attack also evolves and it has become more complicated and hard to detect using flash crowd agent, slow rate attack and also amplification attack that exploits a vulnerability in DNS server. Fast detection of the DDoS attack, quick response mechanisms and proper mitigation are a must for an organization. An investigation has been performed on DDoS attack and it analyzes the details of its phase using machine learning technique to classify the network status. In this paper, we propose a hybrid KNN-SVM method on classifying, detecting and predicting the DDoS attack. The simulation result showed that each phase of the attack scenario is partitioned well and we can detect precursors of DDoS attack as well as the attack itself.

Keywords: Distributed Denial of Services (DDoS) · Machine learning classifiers · Security · Intrusion detection · Prediction · Support Vector Machine (SVM) · k-nearest neighbor (KNN) · KNN-SVM

1 Introduction

Three aspects usually involve in computer related issues such as integrity, confidentiality and availability. Security threats fall into three categories such as breach of confidentiality, failure of authenticity and unauthorized denial of services [1]. Distributed Denial of Services (DDoS) become the major problem and it gives the latest threat to the users, organizations and infrastructures of the internet. This type of intrusion (DDoS) attacker attempts to disrupt a target, by flooding it with illegitimate packets, exhausting its resource and overtaking it to prevent legitimate inquiries from getting through. According to the security report of Arbor 2005–2010 [5].

This paper analyzes current research challenges in DDoS by evaluating machine learning algorithms for detecting and predicting DDoS attack, which includes feature extraction, classification, and clustering. Besides, various hybrid approaches have been employed. It is illustrated that these evaluation results of research challenges are mainly suitable for machine learning technique.

© Springer International Publishing Switzerland 2016
H. Fujita et al. (Eds.): IEA/AIE 2016, LNAI 9799, pp. 95–102, 2016.
DOI: 10.1007/978-3-319-42007-3_9

This paper is organized as follows. Section 2 provides a related study on an overview of machine learning techniques and briefly describes a number of related techniques for intrusion detection. Section 3 compares related work based on the types of classifier design, the chosen baselines, datasets used for experiments, etc. Conclusion and discussion for future research are given in Sect. 4.

2 Related Study

Lately, there are many reports that show the involvement of DDoS attack on commercial or government website [3]. Along with the advancement technique of DDoS attack, the studies on detection also evolve and as a result, various methods have been suggested to counter DDoS attack. As we know, DDoS attack can be classified into anomaly-based, congestion-based and others [2]. A network traffic controller using machine learning (ML) techniques was proposed in 1990, aiming to maximize call completion in a circuit-switched telecommunications network [1]. This was one of the works that marked the point at which ML techniques expanded their application space into the telecommunications networking field. In 1994, ML was first utilized for Internet flow classification in the context of intrusion detection. It is the starting point for much of the work using ML techniques in Internet traffic classification that follows.

Gavrilis et al. [3] utilized RBF-NN detector which is a two-layer neural network. It uses nine packet parameters and the frequencies of these parameters are estimated. Based on the frequencies, RBF-NN classifies traffic into attack or normal class. In this study, the IP spoofing characteristic which is one of the most definite DDoS attack evidences is not considered for a correct attack detection. Regarding UDP type attacks, the detection efficiency is lower than that of TCP type attacks and is apparently low in the beginning period of attacks. Defining k-means center which minimizes the quantization error is also a difficult task.

The hybrid technique proposed by Ming-Yang Su et al. [4] is a method to weigh features of DDoS attacks and it analyzed the relationship between detection performance and number of features. The study proposed a genetic algorithm combined with KNN (k-nearest-neighbor) for feature selection and weighting. All initial 35 features in the training phase were weighted, and the top ones were selected to implement Network Intrusion Detection System (NIDS) for testing. A fast mechanism to detect DDoS attack is by extracting features from the network traffic, so that all these features come from the headers, including IP, TCP, UDP, ICMP, ARP and IGMP. According to the framework of Genetic Algorithm (GA), the proposed NIDS is described by three parts in the section. The first subsection will present all features that are considered in the study; the second subsection will state the encoding of a chromosome and the fitness function; the third subsection will provide details on the selection, crossover, and mutation in the GA. There is also an evaluation on machine learning technique on DDoS attack, proposed by Suresh [7] which indicates that Fuzzy c-means clustering gives better classification and it is fast compared to the other algorithms.

3   Propose Work

In this section, we discuss the details of methods that have been utilized in this work for detection and prediction of DDoS attack. There are k-nearest neighbor (KNN) and support vector machine (SVM) or known as KNN-SVM.

3.1   Support Vector Machine (SVM)

In classification and regression, Support Vector Machines are the most common and popular method for machine learning tasks [12]. In this method, a set of training examples is given with which each example is marked belonging to one of the two categories. Then, by using the SVM algorithm, a model that can predict whether a new example falls into one category or the other is built.

3.2   K-Nearest Neighbor (K-NN)

A k-NN algorithm has been shown to be very effective for a variety of problem domains including text categorizing [13]. It determines the class label of a test example based on its k neighbor that is close to it. The similarity score of each neighbor document to test document is used as the weight of categories of the neighbor document. Referring to Fig. 1, it has been effectively used to calculate the distance among neighbors.

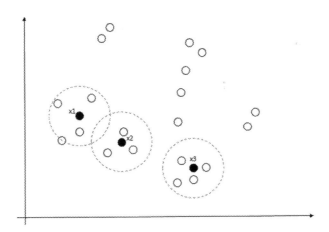

Fig. 1.   A k-nearest neighbor (KNN) classifier [12]

3.3   Features Extraction

Various types of DDoS attacks are studied to select the traffic parameters that change unusually during such attacks. There are eight features extracted from both datasets using information gain rank. Then we rank all the features to identify which one is more relevant. Many machine learning problems can actually enhance their accuracy by applying features selection and extraction. This situation intensively indicates that feature selection is also important for ranking [7]. Information gain is applied to measure the importance of each feature. The information gain of a given attribute X with respect to the class Y is the reduction in uncertainty about the value of Y, after observing values of X. The uncertainty about the value of Y is measured by its entropy defined as

$$H(Y) = -\sum_i P(Y_i) \log_2((P(Y_i)) \tag{1}$$

where P(Yi) is the prior probabilities for all values of Y. The uncertainty about the value of Y after observing values of X is given by the conditional entropy of Y given X defined as

$$H(Y|X) = -\sum_i P(x_j \sum_i P(y_i|x_j) \log_2(P(y_i|x_j) \log_2(P(y_i|x_j)) \tag{2}$$

where $P(y_i|x_j)$ is the posterior probabilities of Y given the values of X. The information gain is thus defined as

$$IG(Y|X) = H(Y) - H(Y|X) \tag{3}$$

By calculating information gain, the correlations of each attribute can be ranked to the class using the dataset from [13]. The most important attributes can then be selected based on the ranking. Based on the result, the following eight feature vectors are selected for detection of DDoS attacks as shown in Table 1.

Table 1. The feature vectors that are selected for detection of DDoS attacks from KDD datasets [13] based on Information Gain feature ranking.

Info. Gain Rank	Features no	Features	Description
1	5	Src bytes	Number of data bytes from source to destination
2	23	Count	Number of connections to the same host as the current connection in the past two seconds
3	3	Service	Network service on the destination, e.g., http, telnet, etc.
4	24	Srv count	Number of connections to the same service as the current connection in the past two seconds
5	36	Dst host same src port rate	Percentage of connections to the current host having the same source port
6	2	Protocol Type	Connection protocol (TCP, UDP, ICMP)
7	33	Dst host srv count	Count of connections having the same destination host and using the same service
8	35	Dst host diff srv rate	Percentage of different services on the current host

3.4   Machine Learning Algorithms

In this part, we briefly describe machine learning algorithm which is used in our experiment.

3.4.1   Naive Bayes

The Naïve Bayes is a simple probabilistic classifier. According to Livadas et al. [8], a widely used framework for classification is provided by a simple theorem of probability known as Bayes' rule, Bayes' theorem, or Bayes' formula:

3.4.2   C4.5

Among classification algorithms, the C4.5 system of Quinlan [9], shows the result of research in machine learning that traces back to the $ID3^2$ [10] system that tries to locate small decision tree.

3.4.3   K-Mean Clustering

K-means or hard c means clustering is basically a partitioning method applied to analyze data and treat observations of the data as objects based on locations and distance between various input data points. Partitioning the objects into mutually exclusive clusters (K) is done by it in such a fashion that objects within each cluster remain as close as possible to each other but as far as possible from objects in other clusters [11].

3.4.4   K-NN Classifier

The k-NN algorithm is a similarity-based learning algorithm and is known to be highly effective in various problem domains, including classification problems. Given a test element dt, the k-NN algorithm finds its k-nearest neighbors among the training elements, which form the neighborhood of dt. Majority voting among the elements in the neighborhood is used to decide the class for dt.

3.4.5   FCM Clustering

Fuzzy c-means (FCM) is a method of clustering which allows one piece of data to belong to two or more clusters. This method (developed by Dunn in 1973 and improved by Bezdek in 1981) is frequently used in pattern recognition. It is based on minimization of the following objective function:

$$J_m(U, \mathrm{v}) \; = \; \sum\nolimits_{k=1}^{n} \sum\nolimits_{i=1}^{c} (u_{ik}) m \|y_k - v_i\|_A^2 \qquad (4)$$

Where m is any number greater than 1, u_{ij} is the degree of membership of x_i in the cluster j, x_i is the ith of the d-dimensional measured data, c_j is the d-dimension center of the cluster, and $\|*\|$ is any norm expressing the similarity between any measured data and the center.

4   Experimental Result

The KDD99 dataset [13] is used in the experiments as the attack component. Classification of attack and normal traffic is done using WEKA. Table 2 shows the dataset and the normal traffic. Table 3 shows the correct classification and the attack detection time. Table 4 shows the F-measure details and Fig. 2 shows the evaluation results using ROC curves for the selected machine learning techniques.

Table 2. Sample collected

Network data	Data type	Total number of record
Trained	Full set data	494,021
	Normal	97,277
	DDoS Attack	391,458

Table 3. Classification results

Method used	Correct classification %	Detection time (In Second)
SVM	96.4	0.23
KNN	96.6	0.26
Decision Tree	95.6	0.25
K-Mean	96.7	0.20
Naive Bayesian	92.9	0.52
Fuzzy C Mean	98.7	0.15

Table 4. F-Measure details of classifiers

Method	TP	FP	TN	FN	F-Measure
SVM	281	18	253	20	0.96
KNN	280	20	243	30	0.97
Decision Tree	277	22	218	55	0.96
K-Mean	285	15	273	0	0.97
Naive Bayesian	292	10	256	17	0.97
Fuzzy C Mean	298	2	270	3	0.99

4.1   Performance Evaluation Criteria

Two criteria are chosen for evaluating performance of the classifier: True Positive Rate (TPR) and False Positive Rate (FPR).

$$\text{TPR} = \frac{TP}{TP + FN}, \ \text{FPR} = \frac{FP}{TN + FP} \tag{5}$$

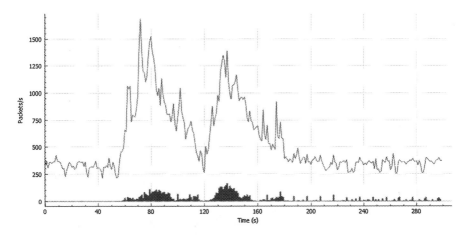

Fig. 2. Attack traffic trace at 11.30 a.m [13]

In formula (3), TP(True Positive), FN(False Negative), FP(False Positive) and TN (True Negative) are defined in [6]. TPR describes the sensitivity of our classifier while FPR shows the rate of false alarms. According to TPR and FPR, a Receiver Operating Characteristic (ROC) curve can be drawn, which is from signal detection theory.

5   Conclusion

The dataset is evaluated by using machine learning algorithms for effectively detecting the DDoS attacks. KDD99 dataset is used as the attack data and based on information gain ranking, relevant features have been selected. Experimental results show that Fuzzy c-means clustering gives better classification and it is fast compared to other algorithms.

Acknowledgement. The authors would like to thank anonymous reviewers for their constructive comments and valuable suggestions. The authors wish to thank Universiti Teknologi Malaysia (UTM) under Research University Grant Vot-02G31 and Ministry of Higher Education Malaysia (MOHE) under the Fundamental Research Grant Scheme (FRGS Vot-4F551) for completion of the research.

References

1. Silver, B.: Netman: A learning network traffic controller. In: Proceedings of the Third International Conference on Industrial and Engineering Applications of Artificial Intelligence and Expert Systems. Association for Computing Machinery (1990)
2. Ferguson, P., Senie, D.: Network ingress filtering: Defeating denial of service attacks which employ IP source address spoofing. In: RFC 2267, January 1998

3. Gavrilis, D., Dermatas, E.: Real-time detection of distributed denial-of-service attacks using RBF networks and statistical features. Comput. Netw. **48**(2), 235–245 (2005). doi:10.1016/j.comnet.2004.08.014

4. Lee, K., Kim, J., Kwon, K.H., Han, Y., Kim, S.: DDoS attack detection method using cluster analysis. Expert Syst. Appl. **34**, 1659–1665 (2008)

5. Geng, X., Liu, T., Qin, T., Li, H.: Feature selection for ranking 2. Learning **49**, 407–414 (2007)

6. Suresh, M., Anitha, R.: Evaluating machine learning algorithms for detecting DDoS attacks. In: Wyld, D.C., Wozniak, M., Chaki, N., Meghanathan, N., Nagamalai, D. (eds.) CNSA 2011. CCIS, vol. 196, pp. 441–452. Springer, Heidelberg (2011)

7. Livadas, C., Walsh, R., Lapsley, D., Strayer, W.T.: Using machine learning techniques to identify botnet traffic. In: Proceedings 2006 31st IEEE Conference on Local Computer Networks, pp. 967–974 (2006). doi:10.1109/LCN.2006.322210

8. Quinlan, J.R.: Induction of decision trees. Mach. Learn. **1**(1), 81–106 (1996)

9. Ghosh, S., Dubey, S.: Comparative analysis of K-Means and fuzzy C-Means algorithms. IJACSA **4**(4), 35–39 (2013). doi:10.14569/IJACSA.2013.040406

10. Vapnik, V.: The Nature of Statitical Learning Theory. Springer, Heidelberg (1995)

11. Guo, G., Wang, H., Bell, D., Bi, Y., Greer, K.: Using kNN model-based approach for automatic text categorization. Soft. Comput. **10**(5), 423–430 (2006)

12. Tavallaee, M., Bagheri, E., Lu, W., Ghorbani, A.A.: A detailed analysis of the KDD CUP 99 data set. In: IEEE Symposium on Computational Intelligence for Security and Defense Applications CISDA 2009, pp. 1–6 (2009)

13. The CAIDA UCSD 'DDoS Attack 2007' Dataset (2013). http://www.caida.org/data/passive/ddos-20070804_dataset.xml

Reliable Clustering Indexes

Jean-Charles Lamirel[(✉)]

SYNALP Team, LORIA, Bâtiment B, 54506 Vandoeuvre Cedex, France
lamirel@loria.fr

Abstract. This paper deals with a major challenge in clustering that is optimal model selection. It presents new efficient clustering quality indexes relying on feature maximization, which is an alternative measure to usual distributional measures relying on entropy or on Chi-square metric or vector-based measures such as Euclidean distance or correlation distance. Experiments compare the behavior of these new indexes with usual cluster quality indexes based on Euclidean distance on different kinds of test datasets for which ground truth is available. This comparison clearly highlights altogether the superior accuracy and stability of the new method, its efficiency from low to high dimensional range and its tolerance to noise.

1 Introduction

Unsupervised classification or clustering is a data analysis technique which is increasingly widely-used in different areas of application. If the datasets to be analyzed have growing size, it is clearly unfeasible to get ground truth that permits to work on them in a supervised fashion. The main problem which then arises in clustering is to qualify the obtained results in terms of quality. A quality index is a criterion which makes possible to decide which clustering method to use, to fix an optimal number of clusters and also to evaluate or develop a new method. Many approaches have been developed for that purpose as it has been pointed out in [1, 19, 20, 23]. However, even if recent alternative approaches do exist [3, 10, 11], the usual quality indexes are mostly based on the concepts of dispersion of a cluster and dissimilarity between clusters. Computation of the latter criteria themselves relies on Euclidean distance. Most popular such indexes are the Dunn index [7], the Davis-Bouldin index [5], the Silhouette index [21], the Calinski-Harabasz index [4] and the Xie-Beni index [24]. They implement the afore mentioned concepts in slightly different ways.

The Dunn index (DU) identifies clusters which are well separated and compact. It combines dissimilarity between clusters and their diameters to estimate the most reliable number of clusters. The Davies-Bouldin index (DB) is similar to the Dunn index and identifies clusters which are far from each other and compact. The Silhouette index (SI) computes a width depending on the membership of a data point in any cluster. A negative silhouette value for a given point means that the point is most suited to belong to a different cluster from the one it is allocated. The Calinski-Harabasz index (CH) computes a weighted ratio

© Springer International Publishing Switzerland 2016
H. Fujita et al. (Eds.): IEA/AIE 2016, LNAI 9799, pp. 103–114, 2016.
DOI: 10.1007/978-3-319-42007-3_10

between the within-group scatter and the between group scatter. Well separated and compact clusters should maximize this ratio. The Xie-Beni index (XI) is a compromise between the approaches provided by the Dunn index and by the Calinski-Harabasz index.

As stated in [9, 23] usual indexes have the defect to be sensitive to the noisy data and outliers. In [16], we also observed that the proposed indexes are not suitable to analyze clustering results in highly multidimensional space as well as they are unable to detect degenerated clustering results. Also these indexes are not independent of the clustering method with which they are used. As an example, a clustering method which tends to optimize WGSS, like k-means [18], will also tend to naturally produce low value for that criteria which optimizes indexes output, but does not necessarily guarantee coherent results, as it was also demonstrated in [16]. Last but not least, as Hamerly et al. pointed out in [12], the experiments on these indexes in the literature are often performed on unrealistic test corpora made up of low dimensional data with a small number of "well-shaped" (mostly hyperspheric) embedded virtual clusters. As an example, in their reference paper, Milligan and Cooper [19] compared 30 different methods for estimating the number of clusters. They classified CH and DB in the top 10, with CH the best but their experiments only used simulated data described in a low dimensional Euclidean space. The same remark can be made about the comparison performed in [23] or in [6]. However, Kassab et al. [13] used the Reuters test collection to show that the aforementioned indexes are often unable to identify an optimal clustering model whenever the dataset is constituted by complex data which need to be represented in both high-dimensional and sparse description space, obviously with embedded non-Gaussian clusters, as is often the case with textual data. The silhouette index is considered one of the more reliable indexes among those mentioned above especially in the case of multidimensional data, mainly because it is not a diameter-based index optimized for Gaussian context. However, like the Dunn and Xie-Beni indexes, its main defect is that it is computationally expensive, which could represent a major drawback for use with large datasets constituted by high-dimensional data.

There are also other alternatives to the usual indexes. For example, in 2009 Lago-Fernãndez et al. [14] proposed a method using negentropy which evaluates the gap between the cluster entropy and entropy of the normal distribution with the same covariance matrix, but again their experiments were only conducted on two-dimensional data. Also other recent indexes attempts were limited by the researchers' choice of complex parameters [23].

Our aim was to get rid of the method-index dependency problem and the issue of sensitivity to noise while also avoiding computation complexity, parameter settings and dealing with a high-dimensional context. To achieve goals, we exploited features of the data points attached to clusters instead of information carried by cluster centroids and replaced Euclidean distance with a more reliable quality estimator based on the feature maximization measure. This measure has been already successfully used by Lamirel et al. to solve complex high-dimensional classification problems with highly imbalanced and noisy data gathered in similar classes thanks to its very efficient feature selection and data resampling capabilities [17]. As a complement to this information, we shall show in the upcoming experimental section that cluster quality indexes relying on this measure do not possess any of the defects of usual approaches including computational complexity.

Section 2 presents a feature maximization measure and our proposed new indexes. Section 3 presents our first experimental context based on reference datasets. Section 4 details our first results. Section 5 draws our conclusion and ideas for future work.

2   Feature Maximization for Feature Selection

Feature maximization is an unbiased measure which can be used to estimate the quality of a classification whether it be supervised or unsupervised. In unsupervised classification (i.e. clustering), this measure exploits the properties (i.e. the features) of data points that can be attached to their nearest cluster after analysis without prior examination of the generated cluster profiles, like centroids. Its principal advantage is thus to be totally independent of the clustering method and of its operating mode.

Consider a partition C which results from a clustering method applied to a dataset D represented by a group of features F. The feature maximization measure favours clusters with a maximal feature F-measure. The feature F-measure $FF_c(f)$ of a feature f associated with a cluster c is defined as the harmonic mean of the feature recall $FR_c(f)$ and of the feature predominance $FP_c(f)$, which are themselves defined as follows:

$$FR_c(f) = \frac{\Sigma_{d \in c} W_d^f}{\Sigma_{c \in C} \Sigma_{d \in c} W_d^f}, \quad FP_c(f) = \frac{\Sigma_{d \in c} W_d^f}{\Sigma_{f' \in F_c, d \in c} W_d^{f'}} \tag{1}$$

with

$$FF_c(f) = 2 \left(\frac{FR_c(f) \times FP_c(f)}{FR_c(f) + FP_c(f)} \right) \tag{2}$$

where W_d^f represents the weight of the feature f for the data d and F_c represents all the features present in the dataset associated with the cluster c.

There is some important similarities between Recall and Predominance used in the proposed approach and Recall and Precision used in information retrieval. We have already exploited this analogy more thoroughly in some of our former works, like in [15], but the measures proposed here must be considered as generalizations of such information retrieval measures which are no more based on agreement but on influence of a feature materialized by a weight. Weight represents the importance of a feature for a data and furthermore for a cluster. The choice of the weighting scheme is not really constrained by the approach instead of producing positive values. Such scheme is supposed to figure out the significance (i.e. semantic and importance) of the feature for the data.

Feature recall is a scale independent measure but feature predominance is not. We have however shown experimentally in [17] that the F-measure which is a combination of these two measures is only weakly influenced by feature scaling. Nevertheless, to guaranty full scale independent behavior for this measure, data must be standardized.

Feature maximization measure can be exploited to generate a powerfull feature selection process [17]. In the clustering context, this kind of selection process can be defined as non-parametrized process based on clusters content in which a cluster feature is characterized using both its capacity to discriminate between clusters ($FP_c(f)$ index) and its ability to faithfully represent the cluster data ($FR_c(f)$ index). The set S_c of features that are characteristic of a given cluster c belonging to a partition C is translated by:

$$S_c = \{f \in F_c \mid FF_c(f) > \overline{FF}(f) \text{ and } FF_c(f) > \overline{FF}_D\} \tag{3}$$

where

$$\overline{FF}(f) = \Sigma_{c' \in C} \frac{FF_{c'}(f)}{|C_{/f}|} \text{ and } \overline{FF}_D = \Sigma_{f \in F} \frac{\overline{FF}(f)}{|F|} \tag{4}$$

where $C_{/f}$ represents the subset of C in which the feature f occurs.

Finally, the set of all selected features S_C is the subset of F defined by:

$$S_C = \cup_{c \in C} S_c. \tag{5}$$

In other words, the features judged relevant for a given cluster are those whose representations are better than average in this cluster, and better than the average representation of all the features in the partition, in terms of feature F-measure. Features which never respect the second condition in any cluster were discarded.

A specific concept of contrast $G_c(f)$ can be defined to calculate the performance of a retained feature f for a given cluster c. It is an indicator value which is proportional to the ratio between the F-measure $FF_c(f)$ of a feature in the cluster c and the average F-measure $\overline{FF}$ of this feature for the whole partition[1]. It can be expressed as:

$$G_c(f) = FF_c(f)/\overline{FF}(f) \tag{6}$$

The active features of a cluster are those for which the contrast is greater than 1. Moreover, the higher the contrast of a feature for one cluster, the better its performance in describing the cluster content. Conversely, the passive features in a cluster are selected features present in the cluster's data for which contrast

---

[1] Using p-value highlighting the significance of a feature for a cluster by comparing its contrast to unity contrast would be a potential alternative to the proposed approach. However, this method would introduce unexpected Gaussian smoothing in the process.

is less than unity[2]. A simple way to exploit the features obtained is to use active selected features and their associated contrast for cluster labelling as we proposed in [17]. A more sophisticated method (as we shall propose hereafter) is to exploit information related to the activity and passivity of selected features in clusters to define clustering quality indexes identifying an optimal partition. This kind of partition is expected to maximize the contrast described by Eq. 6. This approach leads to the definition of two different indexes:

The PC index, whose principle corresponds by analogy to that of intra-cluster inertia in the usual models, is a macro-measure based on the maximization of the average weighted contrast of active features for optimal partition. For a partition comprising k clusters, it can be expressed as:

$$PC_k = \frac{1}{k} \sum_{i=1}^{k} \frac{1}{n_i} \sum_{f \in S_i} G_i(f) \tag{7}$$

The EC index, whose principle corresponds by analogy to that of the combination between intra-cluster inertia and inter-cluster inertia in the usual models, is based on the maximization of the average weighted compromise between the contrast of active features and the inverted contrast of passive features for optimal partition:

$$EC_k = \frac{1}{k} \sum_{i=1}^{k} \left(\frac{\frac{|s_i|}{n_i} \sum_{f \in S_i} G_i(f) + \frac{|\overline{s_i}|}{n_i} \sum_{h \in S_i} \frac{1}{G_i(h)}}{|s_i| + |\overline{s_i}|} \right) \tag{8}$$

where n_i is the number of data associated with the cluster i, $|s_i|$ represents the number of active features in i, and $|\overline{s_i}|$, the number of passive features in the same cluster.

3   Experimental Data and Process

To objectively calculate the accuracy of our new indexes, we used several different datasets of varying dimensionality and size for which the optimal number of clusters (i.e. ground truth) is known in advance.

A part of the datasets came from the UCI machine learning repository [2] and is more usually exploited for classification tasks. The 4 selected UCI datasets represent mostly low to middle dimensional datasets and small datasets (except for PEN dataset which is large). The ZOO and SOY datasets which includes variables with modalities are transformed into binary files. IRIS is exploited both in standard and in binarized version to obtain clearer insight into the behavior of quality index on binary data.

The VERBF dataset is a dataset of French verbs which are described both by semantic features and by subcategorization frames. The ground truth of this

---

[2] As regards the principle of the method, this type of selected features inevitably have a contrast greater than 1 in some other cluster(s) (see Eq. 3 for details).

dataset has been established both by linguists who studied different clustering results and by a gold standard based on the VerbNet classification, as in [22]. This binary dataset contains verbs described in a space of 231 Boolean features. It can be considered a typical middle size and middle dimensional dataset.

The R8 and R52 corpora were obtained by Cardoso Cachopo from the R10 and R90 datasets, which are derived from the Reuters 21578 collection[3]. The aim of these adjustments was to only retain data that had a single label. Considering only monothematic documents and classes that still had at least one example of training and one of test, R8 is a reduction of the R10 corpus (the 10 most frequent classes) to 8 classes and R52 is a reduction of the R90 corpus (90 classes) to 52 classes. The R8 and R52 are large and multidimensional datasets with respective size of 7674 and 9100 and associated bag of words description spaces of 1187 and 2618 words. This datasets can be considered as large and high dimensional datasets.

The summary of datasets overall characteristics is provided in Table 1. We

Table 1. Datasets overall characteristics (Binarization of IRIS dataset results in 12 binary features out of 4 real-valued features).

	IRIS	IRIS-b	WINE	PEN	SOY	ZOO	VRBF	R8	R52
Nbr class	3	3	3	10	16	7	12–16	8	52
Nbr data	150	150	178	10992	292	101	2183	7674	9100
Nbr feat	4	12	13	16	84	114	231	3497	7369

exploited 2 different usual clustering methods, namely k-means [18], a winner-take-all method, and GNG [8], a winner-take-most method with Hebbian learning. For text and/or binary datasets we also used the IGNGF neural clustering method [16] which has already been proven to outperform other clustering methods, including spectral methods [22], on this kind of data. We have reported on the method that produced the best results in the following experiments.

As class labels were provided in all datasets and considering that the clustering method could only produce approximate results as compared to reference categorization, we also used purity measures to estimate the quality of the partition generated by the method as regards to category ground truth. Following [22], we use modified purity (mPUR) to evaluate the clusterings produced and this was computed as follows:

$$mPUR = \frac{|P|}{|D|} \tag{9}$$

where $P = \{d \in D \mid \text{prec}(c(d)) = g(d) \wedge |c(d)| > 1\}$ with D being the set of exploited data points, $c(d)$ a function that provides the cluster associated to data d and $g(d)$ a function that provides the gold class associated to data d.

---

[3] http://www.research.att.com/lewis/reuters21578.html.

Clusters for which the prevalent class has only one element are considered as marginal and are thus ignored.

For the same reason, we also varied the number of clusters in a range up to 3 times that determined by the ground truth. An index which gave no indication of optimum in the expected range was considered to be out-of-range or diverging index (- out-). We finally obtained a process which consists of generating disturbance in the clustering results by randomly exchanging data between clusters to different fixed extents (10 %, 20 %, 30 %) whilst maintaining the original size of the clusters. This process simulated increasingly noisy clustering results and the aim was to estimate the robustness of the proposed estimators.

4   Results

The results are presented in Tables 2, 3 and 4. Some complementary information is required regarding the validation process. In the tables, MaxP represents the number of clusters of the partition with highest mPUR value (Eq. 9), or in some cases, the interval of partition sizes with highest stable mPUR value. When a quality index identified an optimal model with MaxP clusters and MaxP differed from the number of categories established by ground truth, its estimation was still considered valid. This approach took into account the fact that clustering would quite systematically produce sub-optimal results as compared to ground truth. The partitions with the highest purity values were thus studied to deal with this kind of situation. For similar reason, all estimations in the interval range between the optimal k (ground truth) and MaxP values were also considered valid. When indexes were still increasing and decreasing (depending on whether they were maximizers or minimizers) when the number of clusters was more than 3 times the number of expected classes, there were considered out-of-range (-out- symbol in Tables 2, 3 and 4). The Fig. 1 draws the trends of evolution of EC and PC indexes in the case of the R52 dataset. It highlights what is a

Table 2. Overview of the indexes estimation results on low dimensional data (Bold numbers represent valid estimations).

	IRIS	IRIS-b	WINE	PEN	SOY	Number of correct matches
DB	2	5	**5**	7	**19**	**2/5**
CH	2	**3**	6	8	5	**1/5**
DU	1	1	8	17	8	**0/5**
SI	4	2	7	14	14	**1/5**
XB	2	7	-out-	19	24	**0/5**
PC	**3**	**3**	**4**	9	**16**	**4/5**
EC	**3**	**3**	**4**	9	**16**	**4/5**
MaxP	3	3	5	11	19	
Method	K-means	K-means	GNG	GNG	GNG	

Table 3. Overview of the indexes estimation results on average to high dimensional data (Bold numbers represent valid estimations).

	ZOO	VRBF	R8	R52	Number of correct matches
DB	**8**	-out-	5	58	**1/4**
CH	4	7	**6**	-out-	**1/4**
DU	**8**	2	-out-	-out-	**1/4**
SI	4	-out-	-out-	**54**	**1/4**
XB	-out-	23	-out-	-out-	**0/4**
PC	7	18	-out-	-out-	**1/4**
EC	**7**	**15**	**6**	**52**	**4/4**
MaxP	10	12–16	6	50–55	
Method	IGNGF	IGNGF	IGNGF	IGNGF	

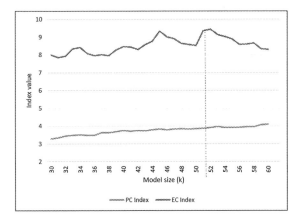

Fig. 1. Trends of PC and EC indexes on Reuters R52 dataset. (Color figure online)

suitable index behaviour (EC index) and in a parallel way what represents the out-of-range index behaviour we mentioned before (PC index).

When considering the results presented in Tables 2 and 3, it should first be noted that one of our tested indexes, the Xie-Beni (XB) index never provides any correct answers. These were either out of range (i.e. diverging) or answers (i.e. minimum value when this index was a minimizer) in the range of the variation of k, but too far from ground truth or even too far from optimal purity among the set of generated clustering models. Some indexes were in the low mid-range of correctness and provide unstable answers. This was the cases with the Davis-Bouldin (DB), Calinski-Harabasz (CH), Dunn (DU) and Silhouette (SI) indexes. When there was dimension growth, these indexes were found to become generally unable to provide any correct estimation. This phenomenon has already been observed in previous experiments with Davis-

Table 4. Indexes estimation results in the presence of noise (UCI ZOO dataset).

	ZOO	ZOO Noise 10%	ZOO Noise 20%	ZOO Noise 30%	Number of correct matches
DB	**8**	4	3	3	**1**/4
CH	4	5	3	3	**0**/4
DU	**8**	2	2	2	**1**/4
SI	14	-out-	-out-	-out-	**0**/4
XB	-out-	-out-	-out-	-out-	**0**/4
PC	6	4	11	**9**	**1**/4
EC	**7**	5	6	**9**	**2**/4
MaxP	10	7	10	10	
Method	IGNGF	IGNGF	IGNGF	IGNGF	

Bouldin (DB) and Calinski-Harabasz (CH) indexes [13]. Davis-Bouldin (DB) performed slightly better than average on low dimensional data. Our PC index was found to perform significantly better than average on low dimensional data but obviously remains a better low dimensional problem estimator than a high dimensional one. Help from passive features somehow seems mandatory to estimate an optimal model in the case of high dimensional problems. Hence, the EC index which exploited both active and passive features was found to have from far the best performance, whatever it faced with low (Table 2) or high dimensional estimation problem (Table 3). According to our evaluation criteria, this index only do wrong in the case of the PEN dataset. However, even in this case its estimation (model of size 9) is still in the close neighbourood of the optimal one (model of size 10). Additionally, both the EC and PC indexes, were both found to be capable of dealing with binarized data in a transparent manner which is not the case of some of the usual indexes namely the Xie-Beni (XI) index, and to a lesser extent, Calinski-Harabasz (CH) and Silhouette (SI) indexes.

Interestingly, on the UCI ZOO dataset, the results of noise sensitivity analysis presented in Table 4 underline the fact that noise has a relatively limited effect on the operation of PC and EC indexes. The EC index was again found to have the most stable behavior in that context. The Fig. 2 presents a parallel view of the different trends of EC value on non noisy and noisy clustering environment, respectively. It shows that noise tends to lower the index value in an overall way and to soften the trends related to its behaviour relatively to changes in k value. However, the index is still able to estimate, either the optimal model in the best case, or a neighbour model in the worst case. The usual indexes do not work as well at all in the same context. For example, the Silhouette index firstly delivered the wrong optimal k values on this dataset before getting out of range when the noise reached 20% on clustering results. The Davis-Bouldin (DB) and

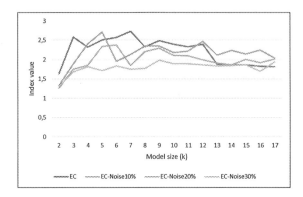

Fig. 2. Trends of EC indexes on UCI ZOO dataset with and without noise. (Color figure online)

Dunn (DU) indexes were found to shift from a correct to a wrong estimation as soon as noise began to appear.

In all our experiments, we observed that the quality estimation depends little on the clustering method. Morever, we noted that the computation time of the index was one of the lowest among the indexes studied. As an example, for the R52 dataset, the EC index computation time was 125 s as compared to 43000 s for the Silhouette index using a standard laptop with 2,2 GHz quadricore processor and 8 GB of memory.

5   Conclusion

We have proposed a new set of indexes for clustering quality evaluation relying on feature maximization measurement. This method exploits the information derived from features which could be associated to clusters by means of their associated data. Our experiments showed that most of the usual quality estimators do not produce satisfactory results in a realistic data context and that they are additionally sensitive to noise and perform poorly with high dimensional data. Unlike the usual quality estimators, one of the main advantages of our proposed indexes is that they produce stable results in cases ranging from a low dimensional to high dimensional context and also require low computation time while easily dealing with binarized data. Their stable operating mode with clustering methods which could produce both different and imperfect results also constitutes an essential advantage.

However, further experiments are required using both an extended set of clustering methods and a larger panel of high dimensional datasets to confirm this promising behavior.

Additionally, we plan to test the ability of our indexes to discriminate between correct and degenerated clustering results in the context of large and heterogeneous datasets.

References

1. Angel Latha Mary, S., Sivagami, A.N., Usha Rani, M.: Cluster validity measures dynamic clustering algorithms. ARPN J. Eng. Appl. Sci. **10**(9), 4009–4012 (2015)
2. Bache, K. and Lichman, M.: UCI machine learning repository. University of California, School of Information and Computer Science. Irvine, CA, USA (2013). http://archive.ics.uci.edu/ml
3. Bock, H.-H.: Probability model and hypothese testing in partitionning cluster analysis. In: Arabie, P., Hubert, L.J., De Soete, G. (eds.) Clustering and Classification, pp. 377–453. World Scientific, Singapore (1996)
4. Calinsky, T., Harabasz, J.: A dendrite method for cluster analysis. Commun. Stat. **3**(1), 1–27 (1974)
5. Davies, D.L., Bouldin, D.W.: A cluster separation measure. IEEE Trans. Pattern Anal. Mach. Intell. **PAMI–1**(2), 224–227 (1979)
6. Dimitriadou, E., Dolnicar, S., Weingessel, A.: An examination of indexes for determining the number of clusters in binary data sets. Psychometrika **67**(1), 137–159 (2002)
7. Dunn, J.: Well separated clusters and optimal fuzzy partitions. J. Cybern. **4**, 95–104 (1974)
8. Fritzke, B.: A growing neural gas network learns topologies. In: Tesauro, G., Touretzky, D.S., Leen, T.K. (eds.) Advances in Neural Information Processing Systems 7, pp. 625–632 (1995)
9. Guerra, L., Robles, V., Bielza, C., Larrañaga, P.: A comparison of clustering quality indices using outliers and noise. Intell. Data Anal. **16**, 703–715 (2012)
10. Gordon, A.D.: External validation in cluster analysis. Bull. Int. Stat. Inst. **51**(2), 353–356 (1997). Response to comments. Bull. Int. Stat. Inst. **51**(3), 414–415 (1998)
11. Halkidi, M., Batistakis, Y., Vazirgiannis, M.: On clustering validation techniques. J. Intell. Inf. Syst. **17**(2/3), 147–155 (2001)
12. Hamerly, G., Elkan, C.: Learning the K in K-Means. In: Neural Information Processing Systems (2003)
13. Kassab, R., Lamirel, J.-C.: Feature based cluster validation for high dimensional data. In: IASTED International Conference on Artificial Intelligence and Applications (AIA), Innsbruck, Austria, pp. 97–103, February 2008
14. Lago-Fernández, L.F., Corbacho, F.: Using the negentropy increment to determine the number of clusters. In: Cabestany, J., Sandoval, F., Prieto, A., Corchado, J.M. (eds.) IWANN 2009, Part I. LNCS, vol. 5517, pp. 448–455. Springer, Heidelberg (2009)
15. Lamirel, J.-C., Francois, C., Al Shehabi, S., Hoffmann, M.: New classification quality estimators for analysis of documentary information: application topatent analysis and web mapping. Scientometrics **60**(3), 445–462 (2004)
16. Lamirel, J.-C., Mall, R., Cuxac, P., Safi, G.: Variations to incremental growing neural gas algorithm based on label maximization. In: Proceedings of IJCNN 2011, San Jose, CA, USA, pp. 956–965 (2011)
17. Lamirel, J.-C., Cuxac, P., Chivukula, A.S., Hajlaoui, K.: Optimizing text classification through efficient feature selection based on quality metric. J. Intell. Inf. Syst. **45**(3), 379–396 (2014). Special Issue on PAKDD-QIMIE 2013, 1–18
18. MacQueen, J.B.: Some methods for classification and analysis of multivariate observations. In: Proceedings of 5th Berkeley Symposium on Mathematical Statistics and Probability, vol. 1, pp. 281–297. University of California Press (1967)

19. Milligan, G.W., Cooper, M.C.: An examination of procedures for determining the number of clusters in a dataset. Psychometrika **50**(2), 159–179 (1985)
20. Rendón, E., Abundez, I., Arizmendi, A., Quiroz, E.M.: Internal versus external cluster validation indexes. Intern. J. Comput. Commun. **5**(1), 27–34 (2011)
21. Rousseeuw, P.J.: Silhouettes: a graphical aid to the interpretation and validation of cluster analysis. J. Comput. Appl. Math. **20**, 53–65 (1987)
22. Sun, L., Korhonen, A., Poibeau, T., Messiant, C.: Investigating the cross-linguistic potential of VerbNet-style classification. In: Proceedings of ACL, Beijing, China, pp. 1056–1064 (2010)
23. Yanchi, L., Zhongmou, L., Xiong, H., Gao, X., Wu, J.: Understanding of internal clustering validation measures. In: Proceedings of the 2010 IEEE International Conference on Data Mining, ICDM 2010, pp. 911–916 (2010)
24. Xie, X.L., Beni, G.: A validity measure for fuzzy clustering. IEEE Trans. Pattern Anal. Mach. Intell. **13**(8), 841–847 (1991)

FHM+: Faster High-Utility Itemset Mining Using Length Upper-Bound Reduction

Philippe Fournier-Viger[1(⊠)], Jerry Chun-Wei Lin[2],
Quang-Huy Duong[3], and Thu-Lan Dam[3,4]

[1] School of Natural Sciences and Humanities,
Harbin Institute of Technology Shenzhen Graduate School, Shenzhen, China
philfv@hitsz.edu.cn
[2] School of Computer Science and Technology,
Harbin Institute of Technology Shenzhen Graduate School, Shenzhen, China
jerrylin@ieee.org
[3] College of Computer Science and Electronic Engineering,
Hunan University, Changsha, China
huydqyb@gmail.com
[4] Faculty of Information Technology,
Hanoi University of Industry, Hanoi, Vietnam
lanfict@gmail.com

Abstract. High-utility itemset (HUI) mining is a popular data mining task, consisting of enumerating all groups of items that yield a high profit in a customer transaction database. However, an important issue with traditional HUI mining algorithms is that they tend to find itemsets having many items. But those itemsets are often rare, and thus may be less interesting than smaller itemsets for users. In this paper, we address this issue by presenting a novel algorithm named FHM+ for mining HUIs, while considering length constraints. To discover HUIs efficiently with length constraints, FHM+ introduces the concept of Length Upper-Bound Reduction (LUR), and two novel upper-bounds on the utility of itemsets. An extensive experimental evaluation shows that length constraints are effective at reducing the number of patterns, and the novel upper-bounds can greatly decrease the execution time, and memory usage for HUI mining.

Keywords: Pattern mining · High-utility itemsets · Length constraints

1 Introduction

High-Utility Itemset Mining (HUIM) [2,5–11,13] is a popular data mining task. It consists of enumerating all high-utility itemsets (HUIs), i.e. groups of items (itemsets) having a high utility (e.g. yielding a high profit) in customer transaction databases. HUIM is a generalization of the problem of Frequent Itemset Mining (FIM) [1], where items can appear more than once in each transaction and where each item has a weight (e.g. unit profit). HUIM is widely viewed as more difficult

© Springer International Publishing Switzerland 2016
H. Fujita et al. (Eds.): IEA/AIE 2016, LNAI 9799, pp. 115–127, 2016.
DOI: 10.1007/978-3-319-42007-3_11

than FIM because the utility measure used in HUIM is neither anti-monotonic nor monotonic, i.e. a high utility itemset may have supersets or subsets having lower, equal or higher utilities [2]. HUIM has a wide range of applications [9,11]. However, an important issue of traditional HUIM algorithms is that they tend to find itemsets containing many items, as they are more likely to have a high utility. This is an issue because itemsets containing many items are generally less useful than itemsets containing fewer items. The reason is that itemsets containing many items generally represent situations that are more specific, and thus rare. For example, consider a retail manager that has found two high-utility itemsets {mapleSyrup, pancake}, and {mapleSyrup, pancake,orange,cheese,cereal} in a customer transaction database. If the retail manager wants to increase the overall profit of his retail store, it will be more effective to promote the first itemset than the second one, as the former contains only two items and may be quite common, while the second contains five items and is rarer. To provide HUIs that are more useful to users while filtering those that may be less useful, it is thus desirable to incorporate the concept of length constraints in HUIM. To our knowledge, no HUIM algorithms offer this feature. To incorporate length constraints in HUIM, a naive approach would be to discover all HUIs using a traditional HUIM algorithm, and then to apply the constraints as a post-processing step. But this approach would be inefficient, as the algorithm would not take advantage of the constraints to prune the search space. Hence, it is desirable to push the constraints as deep as possible in the mining process to improve the performance of the mining task. In frequent pattern mining, length constraints have been previously used such as the maximum length constraint [4]. The key idea of algorithms using a maximum length constraint is that since itemsets are generated by recursively appending items to itemsets, no item should be appended to an itemset containing the maximum number of items. Although this approach can prune the search space using length constraints, there is a need to find novel ways of reducing the search space using length constraints, to further improve the performance of algorithms. In this paper, we address these issues by presenting a novel algorithm named FHM+ (Fast High-utility itemset Mining+) for discovering HUIs with length constraints. It extends the state-of-the-art FHM algorithm for HUIM with a novel concept named Length Upper-bound Reduction (LUR), to reduce the upper-bounds on the utility of itemsets using length constraints, and thus prune the search space. An extensive experimental evaluation shows that the proposed algorithm can be much faster than the state-of-the-art FHM algorithm, and greatly reduce the number of patterns presented to the user. Moreover, results show that the LUR concept greatly improves the algorithm's efficiency. This is an interesting result as the LUR concept introduced in this paper is quite general, and thus could be integrated in other utility pattern mining algorithms. The rest of this paper is organized as follows. Sections 2, 3, 4, and 5 respectively present the problem of HUIM and related work, the proposed FHM+ algorithm, the experimental evaluation, and the conclusion.

2   Problem Definition and Related Work

The problem of high-utility itemset mining is defined as follows. Let there be a set of items (symbols) I. A *transaction database* is a set of transactions $D = \{T_1, T_2, ..., T_n\}$ such that for each transaction T_c, $T_c \subseteq I$ and T_c has a unique identifier c called its Tid. Each item $i \in I$ is associated with a positive number $p(i)$, called its external utility (e.g. representing the unit profit of this item). For each transaction T_c such that $i \in T_c$, a positive number $q(i, T_c)$ is called the internal utility of i (e.g. representing the purchase quantity of item i in transaction T_c). For instance, consider the database of Table 1, which will be used as running example. It contains five transactions $(T_1, T_2...T_5)$. For example, transaction T_4 indicates that items a, c, e and g appear in this transaction with an internal utility of respectively 2, 6, 2 and 5. Table 2 indicates that the external utilities of these items are respectively 5, 1, 3 and 1. The utility of an item i in a transaction T_c is defined as $u(i, T_c) = p(i) \times q(i, T_c)$. The utility of an itemset X (a group of items $X \subseteq I$) in a transaction T_c is defined as $u(X, T_c) = \sum_{i \in X} u(i, T_c)$. The utility of an itemset X is denoted as $u(X)$ and defined as $u(X) = \sum_{T_c \in g(X)} u(X, T_c)$, where $g(X)$ is the set of transactions containing X. The *problem of high-utility itemset mining* is to discover all high-utility itemsets. An itemset X is a *high-utility itemset* if its utility $u(X)$ is no less than a user-specified minimum utility threshold *minutil* given by the user. For example, the utility of item a in T_4 is $u(a, T_4) = 5 \times 2 = 10$. The utility of the itemset $\{a, c\}$ in T_4 is $u(\{a, c\}, T_4) = u(a, T_4) + u(c, T_4) = 5 \times 2 + 1 \times 6 = 16$. The utility of the itemset $\{a, c\}$ is $u(\{a, c\}) = u(a) + u(c) = u(a, T_1) + u(a, T_3) + u(a, T_4) + u(c, T_1) + u(c, T_3) + u(c, T_4) = 5 + 5 + 10 + 1 + 1 + 6 = 28$. If *minutil* = 30, the complete set of HUIs is $\{a, c, e\}$: 31, $\{a, b, c, d, e, f\}$: 30, $\{b, c, d\}$: 34, $\{b, c, d, e\}$: 40, $\{b, c, e\}$: 37, $\{b, d\}$: 30, $\{b, d, e\}$: 36, and $\{b, e\}$: 31, where each HUI is annotated with its utility. Several HUIM algorithms have been proposed. They can generally be categorized as one-phase or two-phase algorithms [2]. Two-phase algorithms such as Two-Phase [10], BAHUI [8], PB [5], and UPGrowth+ [11] operate in two phases. In the first phase, they identify itemsets that may be high-utility itemsets by considering an upper-bound on the utility of itemsets called the *Transaction-Weighted Utilization (TWU)* [10]. Then, in the second phase, they scan the database to calculate the exact utility of all candidates found in the first phase and filter those having a low utility. The TWU measure and its pruning property are defined as follows. The *transaction utility (TU)* of a transaction T_c is the sum of the utility of all the items in T_c. i.e. $TU(T_c) = \sum_{x \in T_c} u(x, T_c)$. The *transaction-weighted utilization (TWU)* of an itemset X is defined as the sum of the transaction utility of transactions containing X, i.e. $TWU(X) = \sum_{T_c \in g(X)} TU(T_c)$. For example, the TUs of T_1, T_2, T_3, T_4 and T_5 are respectively 30, 20, 8, 27 and 11. The TWU of single items a, b, c, d, e, f and g are respectively 65, 61, 96, 58, 88, 30 and 38. $TWU(\{c, d\}) = TU(T_1) + TU(T_2) + TU(T_3) = 30 + 20 + 8 = 58$. Because the TWU measure is anti-monotonic, it can be used to prune the search space.

Table 1. A transaction database

TID	Transaction
T_1	$(a,1),(b,5),(c,1),(d,3),(e,1),(f,5)$
T_2	$(b,4),(c,3),(d,3),(e,1)$
T_3	$(a,1),(c,1),(d,1)$
T_4	$(a,2),(c,6),(e,2),(g,5)$
T_5	$(b,2),(c,2),(e,1),(g,2)$

Table 2. External utility values

Item	a	b	c	d	e	f	g
Unit profit	5	2	1	2	3	1	1

Property 1 (Pruning search space using the TWU). Let X be an itemset, if $TWU(X) < minutil$, then X and its supersets are low utility. [10]

A drawback of two-phase algorithms is that they generate a huge number of candidates in the first phase. To address this issue, one-phase algorithms were proposed such as FHM [2], HUI-Miner [9], and EFIM [13], which discover HUIs directly using a single phase. To our knowledge, the fastest HUIM algorithm is EFIM, which was shown to outperform FHM, which was shown to be up to 6 times faster than HUI-Miner [2]. The FHM and HUI-Miner algorithms use the concept of remaining utility upper-bound to prune the search space, which is defined as follows. Let $\succ$ be any total order on items from I (e.g. lexicographical order). The *remaining utility of an itemset X in a transaction T_c* is defined as $ru(X) = \sum_{i \in T_c \wedge i \succ x \forall x \in X} u(i, T_c)$. The *remaining utility of an itemset X in a database* is defined as $reu(X) = \sum_{T_c \in g(X)} ru(X, T_c)$. For example, assume that $\succ$ is the alphabetical order. The remaining utility of itemset $\{a, d\}$ in the database is 3, when assuming the alphabetical order. FHM and HUI-Miner are depth-first search algorithms. The FHM algorithm associates a structure named *utility-list* to each itemset [2,9]. Utility-lists allow calculating the utility of any itemset by making join operations with utility-lists of shorter patterns. Utility-lists are defined as follows. The *utility-list $ul(X)$* of an itemset X in a database D is a set of tuples such that there is a tuple $(tid, iutil, rutil)$ for each transaction T_{tid} containing X. The *iutil* element of a tuple is the utility of X in T_{tid}. i.e., $u(X, T_{tid})$. The *rutil* element of a tuple is defined as $\sum_{i \in T_{tid} \wedge i \succ x \forall x \in X} u(i, T_{tid})$. The utility-list of $\{a\}$ is $\{(T_1, 5, 25), (T_3, 5, 3), (T_4, 10, 17)\}$. The utility-list of $\{d\}$ is $\{(T_1, 6, 3), (T_2, 6, 3), (T_3, 2, 0)\}$. The utility-list of $\{a, d\}$ is $\{(T_1, 11, 3), (T_3, 7, 0)\}$. The FHM algorithm scans the database once to create the utility-lists of itemsets containing a single item. Then, the utility-lists of larger itemsets are constructed by joining the utility-lists of smaller itemsets. The join operation for single items is performed as follows. Consider two items x, y such that $x \succ y$, and their utility-lists $ul(\{x\})$ and $ul(\{y\})$. The utility-list of $\{x, y\}$ is obtained by creating a tuple $(ex.tid, ex.iutil + ey.iutil, ey.rutil)$ for each pair of tuples $ex \in ul(\{x\})$ and $ey \in ul(\{y\})$ such that $ex.tid = ey.tid$. The join operation for two itemsets $P \cup \{x\}$ and $P \cup \{y\}$ such that $x \succ y$ is performed as follows. Let $ul(P)$, $ul(\{x\})$ and $ul(\{y\})$ be the utility-lists of P, $\{x\}$ and $\{y\}$. The utility-list of $P \cup \{x, y\}$ is obtained by creating a tuple $(ex.tid, ex.iutil + ey.iutil - ep.iutil, ey.rutil)$ for each set of tuples $ex \in ul(\{x\})$, $ey \in ul(\{y\})$, $ep \in ul(P)$ such that $ex.tid = ey.tid = ep.tid$. The utility-list structure allows to calculate the utility-list of itemsets and prune the search space as follows.

Property 2 (Calculating the utility of an itemset using its utility-list). The utility of an itemset is the sum of *iutil* values in its utility-list [9].

Property 3 (Pruning search space using a utility-list). Let X be an itemset. Let the *extensions* of X be the itemsets that can be obtained by appending an item y to X such that $y \succ i$, $\forall i \in X$. If the sum of *iutil* and *rutil* values in $ul(X)$ is less than *minutil*, X and its extensions are low utility [9].

Although much work has been done on HUIM, a key problem of current HUIM algorithms is that they tend to find a huge amount of itemsets containing many items. As explained in the introduction, these itemsets may be less useful for users, as they generally represent specific and rare cases. To let users find itemsets that are more useful, we define the problem of mining high-utility itemsets with length constraints as follows.

Definition 1 (High-utility itemset mining with length constraints). *Let minutil, minlength, and maxlength be parameters set by the user. The problem of mining high-utility itemsets with length constraints is to find all itemsets having a utility no less than minutil and containing at least minlength items, and at most maxlength items.*

Example 1. If $minutil = 30$, $minlength = 1$ and $maxlength = 3$, the set of HUIs is: $\{a, c, e\}$, $\{b, c, d\}$, $\{b, c, e\}$, $\{b, d\}$, $\{b, d, e\}$, and $\{b, e\}$.

3   The FHM+ Algorithm

This section presents the proposed FHM+ algorithm for efficiently mining HUIs with length constraints. FHM+ extends the state-of-the-art FHM [2] algorithm with novel techniques for pruning the search space using length constraints.

3.1   Length Upper-Bound Reduction

As previously mentioned, FHM performs a depth-first search to discover HUIs. To enforce the length constraints in FHM, a simple solution is to modify FHM to not extend an itemset with an item if its number of items is equal to *maxlength*, and to check if the *minlength* constraints is respected for any HUI found by FHM. This approach would find all HUIs when considering length constraints. However, a drawback of this solution is that it does not reduce upper-bounds on the utilities of itemsets to prune the search space. But having tight upper-bounds is crucial in HUIM for pruning the search space efficiently [2,9,11]. To address this issue, we next propose a novel concept of *length upper-bound reduction*. It consists of a set of techniques for reducing upper-bounds on the utilities of itemsets using length constraints. This results in two novel tighter upper-bounds on the utility of itemsets called the revised TWU and revised remaining utility. The proposed *revised TWU upper-bound* is defined as follows:

Definition 2 (largest utilities in a transaction). *Let there be a transaction* $T_c = \{i_1, i_2, \ldots i_k\}$. *The largest utilities in* T_c *is the set of the maxlength largest values in the set* $\{u(i_1, T_c), u(i_2, T_c), \ldots, u(i_k, T_c)\}$, *and is denoted as* $L(T_c)$.

Definition 3 (revised Transaction-Weighted Utilization). *Let there be a transaction* $T_c = \{i_1, i_2, \ldots i_k\}$. *The revised transaction utility of* T_c *is defined as* $RTU(T_c) = \sum L(T_c)$, *and represents the maximum utility that an itemset respecting the maxlength constraint could have in* T_c. *The revised TWU of an itemset* X *is defined as the sum of the revised transaction utilities of transactions where* X *appears, i.e.* $RTWU(X) = \sum_{T_c \in g(X)} RTU(T_c)$.

For example, the RTU of transactions T_1, T_2, T_3, T_4 and T_5 are respectively 21, 17, 8, 22, and 9. Hence, $RTWU(\{c, d\}) = RTU(T_1) + RTU(T_2) + RTU(T_3) = 21 + 17 + 8 = 48$, which is a tighter upper-bound on the utility of {c,d} and its supersets than the original TWU, which was calculated as 58. The proposed RTWU has the two following important properties.

Property 4 (The revised TWU is a tighter upper-bound than the TWU). Let there be an itemset X. The relationship $RTWU(X) \leq TWU(X)$ holds.

Proof. By definition, $RTU(X) \leq TU(X)$, for any itemset X. Hence, $RTWU(X) = \sum_{T_c \in g(X)} RTU(T_c) \leq TWU(X) = \sum_{T_c \in g(X)} TU(T_c)$.

Property 5 (Pruning the search space using the revised TWU). Let X be an itemset, if $RTWU(X) < minutil$, then X and its supersets are not high-utility itemsets respecting the *maxlength* constraint.

Proof. For any transaction T_c, $RTU(T_c)$ represents the maximum utility that an itemset respecting the *maxlength* constraint could have in T_c. Thus, $RTWU(X)$ is an upper-bound on $u(X)$. Furthermore, it is also a upper-bound on the utilities of supersets of X since those cannot appear in more transactions than X.

The second tighter upper-bound introduced in this paper is called the *revised remaining utility upper-bound*.

Definition 4 (largest utilities in a transaction w.r.t. an itemset). *Let there be a transaction* T_c *and an itemset* X. *Let* $V(T_c, X) = \{v_1, v_2, \ldots v_k\}$ *be the set of items occurring in* T_c *that can extend* X, *i.e.* $V(T_c, X) = \{v \in T_c | v \succ x, \forall x \in X\}$. *The maximum number of items that can be appended to* X *so that the resulting itemset would respect the maxlength constraint is defined as* $maxExtend(X) = maxlengh - |X|$, *where* $|X|$ *is the cardinality of* X. *The largest utilities in transaction* T_c *with respect to itemset* X *is the set of the* $maxExtend(X)$ *largest values in* $\{u(v_1, T_c), u(v_2, T_c), \ldots, u(v_k, T_c)\}$ *and is denoted as* $L(T_c, X)$.

Definition 5 (revised remaining utility). *Let there be a transaction T_c and an itemset X. The revised remaining utility of an itemset X in a transaction T_c is defined as $rru(X, T_c) = \sum L(T_c, X)$, and represents the maximum utility that an extension of X respecting the maxlength constraint could have. The revised remaining utility of an itemset X in a database is defined as $rreu(X) = \sum_{T_c \in g(X)} rru(X, T_c)$.*

For example, the revised remaining utility of itemset $\{a\}$ in the running example is 31, while the remaining utility of $\{a\}$ is 45. This illustrates that the proposed revised remaining utility can be a much tighter upper-bound than the remaining utility upper-bound used in previous work. The proposed revised remaining utility upper-bound has the two following important properties.

Property 6 (The revised remaining utility is a tighter upper-bound than the remaining utility). Let there be an itemset X. The relationship $rreu(X) \leq reu(X)$ holds.

Proof. It can be easily shown that $rru(X) \leq ru(X)$, for any itemset X and transaction T_c. Thus, $rreu(X) = \sum_{T_c \in g(X)} rru(X, T_c) \leq reu(X) = \sum_{T_c \in g(X)} ru(X, T_c)$.

Property 7 (Pruning search space using the revised remaining utility). Let X be an itemset. If the sum of $u(X) + rreu(X)$ is less than *minutil*, X and its extensions are not HUIs respecting the *maxlength* constraint.

Proof. Since $u(X)$ represents the utility of X, and rreu(X) represents the highest utilities of items that could be appended to X while respecting the *maxlength* constraint, in transactions where X appears, it follows that the property holds.

We have so far introduced two novel tighter upper-bounds on the utility of itemsets, with the goal of using the *maxlength* constraint for reducing the search space. We next present a novel structure called *revised utility-list* to calculate the revised remaining utility of any itemset efficiently. This structure is a variation of the utility-list structure used in FHM.

Definition 6 (revised utility-list structure). *The revised utility-list $rul(X)$ of an itemset X in a database D is a set of tuples such that there is a tuple $(tid, iutil, llist)$ for each transaction T_{tid} containing X. The difference with the utility-lists used in FHM is that the rutil element is replaced by the llist element, which stores the set $L(T_c, X)$.*

For example, consider the running example, with *maxlength* = 3. The revised utility-list of $\{a\}$ is $\{(T_1, 5, \{10, 6\}), (T_3, 5, \{2, 1\}), (T_4, 10, \{6, 6\})\}$, and the revised utility-list of $\{b\}$ is $\{(T_1, 10, \{6, 3\}), (T_2, 8, \{6, 3\}), (T_5, 4, \{3, 2\})\}$. The proposed revised utility-list structure stores the necessary information for pruning an itemset X and its extensions using Property 7.

Property 8 (Pruning search space using the revised utility-list structure). Let X be an itemset. If the sum of the *iutil* values and the *llist* elements in $rul(X)$, is less than *minutil*, X and its extensions are not high-utility itemsets respecting the length constraints.

Although this property is useful, an important question is how to construct the revised utility-list of of any itemset encountered in the search space. This is done as follows. FHM+ initially builds the revised utility-lists of each item by scanning the database. Then, the revised utility-lists of each larger itemset X is obtained by performing a join operation using smaller itemsets. This operation is the same as the join operation of FHM, except that *llist* elements are calculated instead of *rutil* elements, for each tuple in $rul(X)$. Consider two itemsets $P \cup \{x\}$ and $P \cup \{y\}$ such that $x \succ y$. Let there be a tuple *epxy* in the revised utility-list $rul(P \cup \{x, y\})$. Let *ey* be the tuple in $rul(P \cup \{y\})$ such that $ey.tid = epxy.tid$. Recall that $maxExtend(P \cup \{x, y\})$ is the number of items that can be appended to $P \cup \{x, y\}$ to generate an itemset that respects the *maxlength* constraint. The *llist* element of the tuple *epxy* is calculated as the $maxExtend(P \cup \{x, y\})$ largest values in the *llist* element of the tuple *ey*. For example, consider the join of the revised utility-lists of items $\{a\}$ and $\{b\}$ to generate the revised utility-list of itemset $\{a, b\}$, when $maxlength = 3$. The revised utility-list of $\{a, b\}$ contains a single tuple for the transaction T_1. Since $maxExtend(\{a, b\}) = maxlength - |\{a, b\}| = 1$, the *llist* element in the tuple corresponding to T_1 in $rul(\{a, b\})$ is set to the largest utility value in the corresponding *llist* element of $rul(\{b\})$, that is 6. The revised utility-list of $\{a, b\}$ is thus $\{(T_1, 15, \{6\})\}$.

3.2   The Proposed Algorithm

We next describe the proposed FHM+ algorithm in detail, which relies on the two novel upper-bounds, and revised utility-list structure introduced in the previous subsection. The main procedure of FHM+ (Algorithm 1) takes a transaction database with utility values as input, and the *minutil*, *minlength* and *maxlength* parameters. The algorithm first scans the database once to calculate the RTWU of each item. Then, the algorithm identifies the set I^* of all items having a RTWU no less than *minutil* (other items are ignored since they cannot be part of a high-utility itemset by Property 5). The RTWU values of items are then used to establish a total order $\succ$ on items, which is the order of ascending RTWU values (similarly to the TWU ascending order used in FHM). A database scan is then performed. During this database scan, items in transactions are reordered according to the total order $\succ$, the revised utility-list of each item $i \in I^*$ is built and a structure named EUCS (Estimated Utility Co-Occurrence Structure) is built [2]. This latter structure is defined as a set of triples of the form $(a, b, c) \in I^* \times I^* \times \mathbb{R}$. A triple (a,b,c) indicates that $RTWU(\{a, b\}) = c$. The EUCS can be implemented as a triangular matrix that stores these triples for all pairs of items. The EUCS is very useful as it stores the RTWU of all pairs of items, an information that will be later used for pruning the search space. Building the EUCS is very fast (it is performed with a single database scan) and occupies a small amount of memory, bounded by $|I^*| \times |I^*|$. The reader is referred to the paper about FHM [2] for more details about the construction of this structure and how it can be implemented efficiently using hash maps. Then, if $minlength \geq 1$, each item having a utility no less than $minutil$ according to its revised utility list, is output as a high-utility itemset. After the construction of the EUCS, the depth-first search exploration of itemsets starts by calling the recursive procedure $Search$ with the empty itemset $\emptyset$, the set of single items I^*, $minutil$, $minlength$, $maxlength$, and the EUCS structure. This exploration is performed only if the user wants to find itemsets containing more than one item.

Algorithm 1. The FHM+ algorithm

---

   input : D: a transaction database, $minutil, minlength, maxlength$:
              user-specified parameters
   output: the set of high-utility itemsets

1 Scan D once to calculate the RTWU of single items;
2 $I^* \leftarrow$ each item i such that RTWU$(i) \geq minutil$;
3 Let $\succ$ be the total order of RTWU ascending values on I^*;
4 Scan D to build the revised utility-list of each item $i \in I^*$ and build the $EUCS$ structure;
5 **if** $minlength \leq 1$ **then** output each item $i \in I^*$ such that SUM($\{i\}.utilitylist.iutils) \geq minutil$ **if** $maxlength > 1$ **then** Search ($\emptyset$, I^*, $minutil$, $minlength$, $maxlength$, $EUCS$)

The $Search$ procedure (Algorithm 2) takes as input (1) an itemset P, (2) extensions of P having the form Pz meaning that Pz was previously obtained by appending an item z to P, (3) $minutil$, $minlength$, $maxlength$, and (4) the EUCS. The search procedure operates as follows. For each extension Px of P, if the sum of $iutil$ and $llist$ values in the revised utility-list of Px are no less than $minutil$, it means that Px and its extensions should be explored (Property 7). This is performed by merging Px with all extensions Py of P such that $y \succ x$ to form extensions of the form Pxy containing $|Px| + 1$ items. The revised utility-list of Pxy is then constructed by calling the $Construct$ procedure to join the utility-lists of P, Px and Py. This latter procedure performs the steps described in the previous subsection for constructing a revised utility-list. Then, if Pxy respects the length constraints, and the sum of the $iutil$ values of the utility-list of Px is no less than $minutil$, then Pxy is a high-utility itemset, and it is output (cf. Property 2). Then, if the length of Pxy is less than $maxlength$, a recursive call to the $Search$ procedure with Pxy is done to calculate its utility and explore its extension(s). Since the $Search$ procedure starts from single items, it recursively explores the search space of itemsets by appending single items and it only prunes the search space based on Properties 5 and 7, it can be easily seen based on Properties 1, 2 and 3 that this procedure is correct and complete to discover all high-utility itemsets, while considering the length constraints.

Algorithm 2. The *Search* procedure

input : P: an itemset, *ExtensionsOfP*: a set of extensions of P, *minutil*,
                     minlength, maxlength: user-specified parameters, *EUCS*: the *EUCS*
output: the set of high-utility itemsets

1 **foreach** *itemset* $Px \in$ *ExtensionsOfP* **do**
2 | **if** *SUM(Px.utilitylist.iutils)+SUM(Px.utilitylist.llist)* $\geq$ *minutil* **then**
3 | | *ExtensionsOfPx* $\leftarrow \emptyset$;
4 | | **foreach** *itemset* $Py \in$ *ExtensionsOfP such that* $y \succ x$ **do**
5 | | | **if** $\exists(x,y,c) \in$ *EUCS such that* $c \geq$ *minutil* **then**
6 | | | | $Pxy \leftarrow Px \cup Py$;
7 | | | | $Pxy.utilitylist \leftarrow$ Construct (P, Px, Py);
8 | | | | *ExtensionsOfPx* $\leftarrow$ *ExtensionsOfPx* $\cup$ *Pxy*;
9 | | | | **if** *SUM(Pxy.utilitylist.iutils)* $\geq$ *minutil and*
        | | | | *minlength* $\leq |Pxy| \leq$ *maxlength* **then** output Px
10 | | | **end**
11 | | **end**
12 | | **if** $|Pxy| <$ *maxlength* **then** Search $(Px,$ *ExtensionsOfPx, minutil*$)$
13 | **end**
14 **end**

---

4   Experimental Study

We performed an experiment to assess the performance of FHM+. The experiment was performed on a computer with a third generation 64 bit Core i5 processor running Windows 7 and 5 GB of free RAM. We compared the performance of the proposed FHM+ algorithm with the state-of-the-art FHM algorithm for mining HUIs. All memory measurements were done using the Java API. The experiment was carried on three real-life datasets commonly used in the HUIM literature: *chainstore, retail*, and *mushroom*. These datasets have varied characteristics and represent the main types of data typically encountered in real-life scenarios (dense, sparse, and long transactions). Let $|I|$, $|D|$ and A represents the number of transactions, distinct items and average transaction length. *chainstore* is a sparse dataset ($|I| = 46{,}086$ $|D| = 1{,}112{,}949$, $A = 7.2$). *retail* is a sparse dataset with many different items ($|I| = 16{,}470$, $|D| = 88{,}162$, $A = 10{,}30$). *mushroom* is a dense dataset ($|I| = 119$, $|D| = 88{,}162$, $A = 23$). *chainstore* contains real external and internal utility values. For the other datasets, external utilities for items are generated between 1 and 1,000 by using a log-normal distribution and quantities of items are generated randomly between 1 and 5, as the settings of [2,9,11]. The source code of all algorithms and datasets can be downloaded at http://goo.gl/Qr8diZ. In the experiment, FHM+ was run with five different *maxlength* threshold values (1, 2, 3, 4, 5), and the *minlength* threshold was set to 1 as it has no influence on efficiency. Algorithms were first run on each dataset, while decreasing the *minutil* threshold until they became too long to execute, ran out of memory or a clear trend was observed. Figure 1 compares the performance of the algorithms in terms of execution time and pattern count.

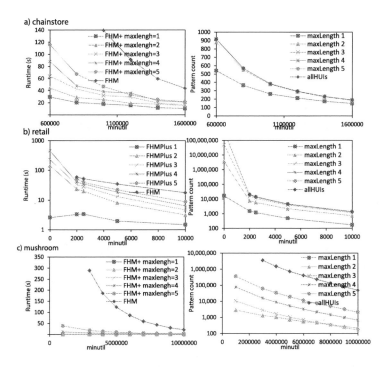

Fig. 1. Execution times and pattern count comparison

It can be first observed that using the *maxlength* constraint can greatly speed up the discovery of HUIs. Depending on how the *maxlength* parameter is set, on the *chainstore*, *retail*, and *mushroom* datasets, FHM+ is respectively from 3 to 10 times, 2 to 17 times, and from 15 to 1400 times faster than FHM. It can also be observed that the number of patterns can be greatly reduced by using length constraints. On the *chainstore*, *retail*, and *mushroom* datasets, the number of patterns found by FHM+ was up to 0.5 times, 13 times, and 2,700 times smaller than the number found by FHM. But note that on the *retail* datasets, no result was obtained for FHM when *minutil = 1*, because it had to be stopped after generating more than 20 GB of patterns, while FHM+ was still able to run, even with *maxlength = 5*. Thus, a benefit of using constraints is that it allows the algorithms to run for smaller *minutil* values. Memory consumption was also compared (detailed results are not shown due to space limitations). On the *chainstore*, *retail*, and *mushroom* datasets, FHM+ used from 5 % to 50 %, 5 % to 50 %, and 25 % to 50 % less memory than FHM. This is due to the ability to prune a larger part of the search space using the proposed upper-bounds. Lastly, the efficiency of the proposed LUR concept introduced in this paper was evaluated (detailed results not shown due to space limitations). It was found that on the *chainstore*, *retail*, and *mushroom* datasets, the proposed upper-bound reduced the execution time by up to 4 times, 2 times, and 2 times.

5   Conclusion

This paper presented an efficient algorithm named FHM+ to efficiently discover high-utility itemsets while considering length constraints. The proposed algorithm integrates a novel concept called Length Upper-Bound Reduction (LUR) to reduce the search space using length constraints. In particular, two novel upper-bounds named revised TWU and revised remaining utility were presented, and a novel data structure called revised utility-list. An extensive experimental evaluation shows that LUR is effective at reducing the number of patterns, and can greatly decrease the execution time and memory requirements for HUI mining. This is very interesting for the end user since a huge amount of long HUIs, representing rare cases are filtered and not presented to the user. The source code of algorithms and datasets can be downloaded as part of the SPMF open source data mining library [3] at http://www.philippe-fournier-viger.com/ spmf/. In future work, the concept of LUR could be incorporated in other utility mining problems such as high-utility sequential rule mining [12]. Moreover, we intend to apply the concept of LUR in the EFIM algorithm [13], which was recently shown to outperform FHM for high-utility itemset mining.

References

1. Agrawal, R., Srikant, R.: Fast algorithms for mining association rules in large databases. In: Proceedings of the International Conference on Very Large Databases, pp. 487–499 (1994)
2. Fournier-Viger, P., Wu, C.-W., Zida, S., Tseng, V.S.: FHM: faster high-utility itemset mining using estimated utility co-occurrence pruning. In: Andreasen, T., Christiansen, H., Cubero, J.-C., Raś, Z.W. (eds.) ISMIS 2014. LNCS, vol. 8502, pp. 83–92. Springer, Heidelberg (2014)
3. Fournier-Viger, P., Gomariz, A., Gueniche, T., Soltani, A., Wu, C., Tseng, V.S.: SPMF: a Java open-source pattern mining library. J. Mach. Learn. Res. (JMLR) **15**, 3389–3393 (2014)
4. Pei, J., Han, J.: Constrained frequent pattern mining: a pattern-growth view. ACM SIGKDD Explor. Newsl. **4**(1), 31–39 (2012)
5. Lan, G.C., Hong, T.P., Tseng, V.S.: An efficient projection-based indexing approach for mining high utility itemsets. Knowl. Inf. Syst. **38**(1), 85–107 (2014)
6. Krishnamoorthy, S.: Pruning strategies for mining high utility itemsets. Expert Syst. Appl. **42**(5), 2371–2381 (2015)
7. Lin, J.C.-W., Gan, W., Hong, T.-P., Pan, J.-S.: Incrementally updating high-utility itemsets with transaction insertion. In: Luo, X., Yu, J.X., Li, Z. (eds.) ADMA 2014. LNCS, vol. 8933, pp. 44–56. Springer, Heidelberg (2014)
8. Song, W., Liu, Y., Li, J.: BAHUI: fast and memory efficient mining of high utility itemsets based on bitmap. Int. J. Data Warehous. Min. **10**(1), 1–15 (2014)
9. Liu, M., Qu, J.: Mining high utility itemsets without candidate generation. In: Proceedings of the 22nd ACM International Conference on Information and Knowledge Management, pp. 55–64 (2012)
10. Liu, Y., Liao, W., Choudhary, A.K.: A two-phase algorithm for fast discovery of high utility itemsets. In: Ho, T.-B., Cheung, D., Liu, H. (eds.) PAKDD 2005. LNCS (LNAI), vol. 3518, pp. 689–695. Springer, Heidelberg (2005)

11. Tseng, V.S., Shie, B.-E., Wu, C.-W., Yu, P.S.: Efficient algorithms for mining high utility itemsets from transactional databases. IEEE Trans. Knowl. Data Eng. **25**(8), 1772–1786 (2013)
12. Zida, S., Fournier-Viger, P., Wu, C.-W., Lin, J.C.W., Tseng, V.S.: Efficient mining of high utility sequential rules. In: Proceedings of the 11th International Conference on Machine Learning and Data Mining, pp. 1–15 (2015)
13. Zida, S., Fournier-Viger, P., Lin, J.C.-W., Wu, C.-W., Tseng, V.S.: EFIM: a highly efficient algorithm for high-utility itemset mining. In: Sidorov, G., Galicia-Haro, S.N. (eds.) MICAI 2015. LNCS, vol. 9413, pp. 530–546. Springer, Heidelberg (2015). doi:10.1007/978-3-319-27060-9_44

MC^2: An Integrated Toolbox for Change, Causality and Motif Discovery

Yasser Mohammad[1,2(✉)] and Toyoaki Nishida[1,2]

[1] Assiut University, Assiut 71516, Egypt
yasserm@aun.edu.eg
[2] Kyoto University, Kyoto, Japan
nishida@i.kyoto-u.ac.jp

Abstract. Time series are being generated continuously from all kinds of human endeavors. The ubiquity of time-series data generates a need for data mining and pattern discovery algorithms targeting this data format which is becoming of ever increasing importance. Three basic problems in mining time-series data are change point discovery, causality discovery and motif discovery. This paper presents an integrated toolbox that can be used to perform any of these tasks on multidimensional real-valued time-series using state of the art algorithms. The proposed toolbox provides practitioners in time-series analysis and data mining with several tools useful for data generation, preprocessing, modeling evaluation and mining of long sequences. The paper also reports real world applications that uses the toolbox in HRI, physiological signal processing, and human behavior modeling and understanding.

Keywords: Change point discovery · Motif discovery · Causality analysis

1 Introduction

Machine learning have focused on supervised, unsupervised and reinforcement learning algorithms that are mostly based on vector representation of data. Discovery problems on the other hand are less represented in the literature due to their ill specification [24]. Discovery problems are problems in which the target is to find *interesting* patterns or relations in data with little or no knowledge of their location, size, and shape. Discovery problems find their natural application in time-series analysis in which we can define three major problems: motif, change point and causality influence discovery. Motif Discovery (MD) is the problem of finding approximately recurring patterns. Change Point Discovery (CPD) is the problem of discovering points at which the underlying dynamics of the time-series generation process are changing. Causality Influence Discovery (CID) is the problem of discovering causal influences between different time-series.

This paper presents a modular toolbox that implements several state-of-the-art approaches to these three core problems in time-series analysis. MD,

© Springer International Publishing Switzerland 2016
H. Fujita et al. (Eds.): IEA/AIE 2016, LNAI 9799, pp. 128–141, 2016.
DOI: 10.1007/978-3-319-42007-3_12

CPD, and CID can be used as building blocks for several applications in human behavior understanding [26], physiological signal processing [19], forecasting [8], analysis of climate data [10], cognitive modeling [33], etc. The proposed toolbox is the successor of CPMD [30] which focused on CPD and MD alone and implemented a single approach to each of them. MC^2 (for Motif Change and Causality discovery), on the other hand, implements multiple approaches to each of the three main problems of discovery in time-series and provides evaluation routines to compare the adequacy to different mining tasks. Like its predecessor, MC^2 is available from the authors as open source.

One main contribution of the paper is to provide an integrated view of major approaches to the three discovery problems implemented by the toolbox. The paper also reports a study of three variations of the GEMODA [11] algorithm to real-valued time-series data.

The rest of this paper is organized as follows. Section 2 gives an overview of the toolbox and its design. Section 3 reports the three core problems of the toolbox. Section 4 reports briefly some of the most important generic time-series analysis routines in MX^2, and Sect. 5 gives some showcase real-world applications.

2 Toolbox Design

Figure 1 shows an overview of the main building blocks of MC^2. There are three major areas covered by the algorithm: MD, CPD and CID. Section 3 provides details about these problems and the main algorithms implemented in the toolbox for each of them. Other than the three main discovery problems implemented in the toolbox, MC^2 implements a host of supporting routines that can be used for pre-processing, post-processing, evaluation, signal generation, etc. These will be discussed in Sect. 4.

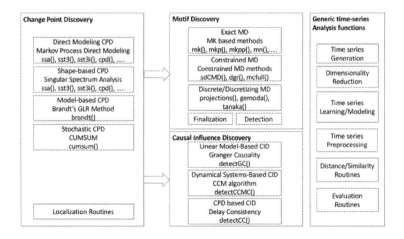

Fig. 1. The general structure of the MC^2 toolbox showing the key algorithms.

An important design decision was that related algorithms were implemented using a single general routine. For example, a single routine can be used to run 48 different related algorithm variants for solving CPD. The rationale for this decision, is that this common implementation reveals the relation between all of these algorithms even on the expense of slight performance degradation. Another design decision was to use an elaborate optional parameter set for each one of the key routines implemented. The routines provide default values for these parameters that are selected based on either theoretical or experimental evaluation to balance execution time and accuracy (with more focus on accuracy). Another important feature of the proposed toolbox is the availability of objective evaluation routines for comparing different algorithms used to solve the same problem. This, combined with a common interface to all of these algorithms in the toolbox, enables easy comparison of different approach in any new dataset the practitioner is interested in.

3   Core Algorithms

MC^2 implements several algorithms for solving three discovery problems: CPD, MD, and CID. These problems are deeply connected. Discovering change in time-series is the most basic of these problems as it underlies some of the algorithms for the other two as indicated by the arrows in Fig. 1. Knowing where a time-series changes can improve the performance of any motif discovery algorithm because, in most cases, a motif occurrence is marked by a change at its beginning and another at its end. This improvement was shown to be super-linear [17]. Moreover, as will be shown in Sect. 3.3, causality can be ascertained by measuring the effect of *change* in one signal on the *change* in another which again links change point and causality discovery. This was shown to be more effective in discovering causal connection between a robot's motor command and its motion behavior and in relating user gestures to actions executed by an operator watching them [18].

Consider the behavior of two players in a game like table-tennis. The behaviors of the two players are linked causally as the behavior of one of them causes the other to respond in the appropriate way according to game rules. This may not though appear immediately by a CID algorithm applied to motion-capture (mocap) data collected from the two players. The reason is that the instantaneous behavior of one player is not affected by the instantaneous behavior of the other at an earlier time (say like prices are affected by demand) but a pattern of behavior of one player (e.g. shooting the ball to the right) will affect the probability of execution of a pattern of behavior of the other (e.g. moving orienting the racket to the right to intercept the ball). To detect such casual relations, MD can be used to discover recurrent patterns and their occurrence probability can then be used as the input to the CID algorithm.

3.1   Change Point Discovery

CPD is the problem of discovering points in time-series data at which the generation dynamics change abruptly. The exact definition of what counts as a *change* and *generating dynamics* is application dependent and MC^2 implements methods with minimal reliance on time-series models to provide a general tool for the practitioner. As shown in Fig. 1.

CPD has a long history in time-series analysis literature, with survey papers in 1976 [40]. This resulted in several algorithms [1,2,6,9,20,35]. Four main approaches can be found in this wide literature and MC^2 implements at least one example of each of them.

The first approach models the distribution of change points themselves directly usually using a Markov process and uses information about the past locations of change points to estimate the next change point location. We call this the direct modeling approach in Fig. 1. An example of this approach can be found in the work of Fearnhead and Liu in which the probability of a change at time s given a change at time t is assumed to depend only on the difference $s - t$ [6]. This approach is implemented in the function *mpcpd*() in the toolbox.

The second approach is called the two-models approach (see Fig. 1). In this case, instead of modeling the probability of change at one point given a change in another, we fit two generating models using data before and after each point in the time-series (we call them λ_b and λ_a hereafter) and announce a change (or calculate a change score) based on the difference between these two models. One of the earliest approaches for CPD was based on comparing λ_b and λ_a and using an autoregressive model for both $(AR(p))$ with additive Gaussian noise. The difference between the predictions of the two models is measured using Generalized Likelihood Ratio (GLR) test and this difference gave an estimate of the change point score at every point. [1] applied this method to speech segmentation as early as 1988. This approach is implemented in the function *brandt*() in the MC^2. The implementation in the toolbox extends the original algorithm by allowing the two models to use Autoregressive Moving Average (ARMA) processes instead of only autoregressive processes. This gives more flexibility in modeling time-series when the noise cannot be considered Gaussian and independent at every point.

The third approach is one of the oldest and finds change in some stochastic parameter estimated from the sequences before and after a candidate change point. The simplest such stochastic parameter to deal with is the mean. Mean change can be found using the CUMSUM algorithm [35] (also known as Page-Hinkley's algorithm). This algorithm uses a likelihood ratio test (similar to Brandt's GLR) that is continuously calculated based on estimates of the mean before and after each point in the time-series. MC^2 implements this algorithm in the function $page_h inkley()$.

The final approach to CPD implemented by MC^2 is based on Singular Spectrum Analysis. This is a special case of shape-based CPD (Fig. 1) in which we focus on the shape of the signal with no apriori given generation model. In this case also, subsequences before and after each point in the time-series are used to estimate the change score at that point. SSA is used to find the subspace spanned by the two subsequences and some estimate of the difference between these two subspaces is used as the change score. This difference can be measured as the angle between the two subspaces, the relative length difference between the first Eigen-vector of the future subsequence relative to its projection on the past subspace, weighted summation of future eigen-vectors' projections on the past subspace, etc. In previous work, we compared 64 different variants of this SSA based approach to CPD [20]. The function *cpd()* implements this approach and can be used to experiment with many variants using optional parameters.

3.2   Motif Discovery

The history of motif discovery is rich and dates back to the 1980s. The problem started in the bioinformatics research community focusing on discovering recurrent patterns in RNA, DNA and protein sequences leading to several algorithms that we will discuss later in this chapter. Since the 1990s, data mining researchers started to shift their attention to motif discovery in real-valued time-series and several formulations of the problems and solutions to these formulations were and are still being proposed. The motif discovery problem in real-valued time-series is harder to pose compared to the discrete case due to the complications arising from the fact that noise and inaccurate productions never allow the pattern to be repeated *exactly* and this forces the practitioner to decide upon allowed distortions to consider two time-series as realizations of the same pattern. Even in the discrete case, noise may cause repetition, deletion or letter-change of a part of the motif occurrences rendering the problem again more difficult and more interesting. The toolbox provides routines for solving both discrete and real-valued MD problems but we will focus on the more challenging real-valued case.

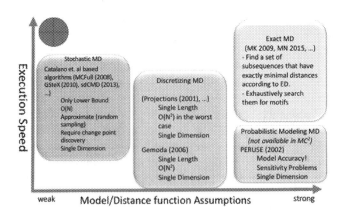

Fig. 2. A unified view of different motif discovery approaches in MC^2.

Figure 2 shows a unified view of the most widely used approaches to real-valued MD in literature and distributes them according to two main criteria: execution time and strong modeling or distance evaluation assumptions. Four main approaches can be distinguished.

The first approach is stochastic MD in which we sample subsequences from the time-series and compare them using some predefined distance function (that may use modeling or shape) searching for recurrent patterns with small distances. The simplest of these algorithms was proposed by Catalano et al. [4] and it samples subsequences randomly (i.e. from a uniform distribution over subsequence start points) then uses the distances between short overlapping parts of these subsequences to discover if two full occurrences of the same motif exist in these sampled subsequences. This approach is only viable when motifs appear frequently in the time-series that random sampling can have a chance of sampling two compete occurrences. Constrained MD is a special case of Stochastic MD which utilizes some constraint to bias the sampling process toward parts of the time-series in which there is higher probability of finding a motif occurrence. This can be achieved using domain knowledge but, more interestingly, it can be achieved by utilizing a CPD algorithm. This use of CPD for seeding MD is based on that when we assume that motif occurrences to not overlap or immediately follow each other, a change in the generation process must – by definition – happen at the beginning and end of each occurrence and CPD can be used to find these locations and bias the search process of MD. The toolbox has three Constrained MD functions. The function $mcfull()$ implements MCFull and MCInc [17]. These algorithms extend Catalano et al.'s algorithm by sampling from the output of CPD as described before. The function $dgr()$ implement variants of the Distance Graph Relaxation [15] which first samples short subsequences using the same approach as MCFull and then builds a graph representing their distances and uses the distribution of distances to find short stems of recurrent motifs. These short stems can then be extended until a stopping criterion determined by a statistical test is met using GSteX [16]. The final and most robust function for constrained MD is $sdCMD()$ which implements the shift-density based approach proposed by the authors [27] which converts the problem of MD into a search in one dimensional shift space and has the advantage of being capable of discovering motifs even if partial occurrences of them appeared in the sampled subsequences.

The second approach to MD implemented in MC^2 is to discretize the time-series (usually but not necessarily using SAX [5,13]) then apply a discrete motif discovery algorithm from the bioinformatics literature. The most widely used such algorithm is Projections implemented in the function $projections()$. Projections is a seeding step that aims at discovering probable candidates for the hidden motif M. The main idea of the algorithm is to select several random hash functions $f_j(s_i^t)$ and use them to hash the input sequences. Occurrences of the hidden motif are expected to hash frequently to the same value (called *bucket*) with a small proportion of background noise. Noise sequences on the other hand are not expected to hash to similar values. If the hash functions are different enough (and complex enough), we can use the buckets with largest hits as representing occurrences of the motif. This is an initialization step that can then be refined using the EM algorithm in order to recover the full motif. An extension of Projections-based MD was proposed by Tanaka [38] that uses minimum description length (MDL)and PCA to handle multidimensional time-series and find only statistically significant ones. This algorithm is implemented in the function $tanaka()$.

Another important discrete MD discovery algorithm is GEMODA [11]. Gemoda runs in three stages (comparison, clustering, and convolution). During the comparison stage, a similarity matrix $\mathbb{M}_{T-L+1 \times T-L+1}$ is built by scanning the input time-series and calculating all pairwise distances between all subsequences of length L which is binary-thresholded as $\mathbb{B}$. The clustering stage, clusters the subsequences in the time-series using any clustering algorithm like K-means. Jensen et al. [11] proposed two simple clustering algorithms based on graph operations that are implemented by the toolbox. The final convolution stage is the heart of the Gemoda algorithm and it tries to extend the motif stems found in the clustering stage. In our implementation (the function $gemoda()$), it is possible to restrict the allowable overlap between different motifs and different motif occurrences. It is also possible to build the similarity matrix from a predefined set of locations in the time-series rather than using exhaustive search. These options render Gemoda an approximate algorithm but they can significantly increase its speed. Moreover, the algorithm is extended to handle discrete, real-valued and multidimensional time-series.

The final approach implemented by MC^2 is exact motif discovery. The base of this approach is the MK algorithm [32] which finds the efficiently most similar pair of subsequences of a given length in a single-dimensional time-series using extensively the triangular inequality and pruning of unneeded calculations. Other than the standard MK algorithm implemented in $mk()$, the toolbox implements several extensions that can discover multiple pairs ($mkp()$ [29]), multiple motif lengths ($mkpp()$ [22]) and full motif enumeration of scale invariant motifs ($mkpp()$ [23], $mn()$ [25] and $moen()$ [31]).

A fourth approach that is not implemented in MC^2 is probabilistic modeling MD in which a model of motif generation is used to find the occurrences. An important example of this approach is the PERUSE algorithm [34] in which a motif is defined by a coupled set of Gaussians with known transition probabilities. Due to the utilization of a specific model, this approach suffers from stability problems with the time-series is not generated from a similar model even though impressive results in learning words from spoken stories was shown using this approach. In the future, an implementation of PERUSE will be added to MC^2.

3.3   Causality Influence Discovery

The central problem in CID can be stated as follows: Given a set of n_t time-series of length T characterizing the performance of n processes (where $n_t \geq n$), build a n-nodes directed graph representing the causal relations between these processes.

Each edge in the DAG should be labeled by the expected delay between events in the source processes and their causal reflection in the destination node. MC^2 implements three approaches for solving CID problems.

The first approach assumes linear influence between time-series and uses a model of time-series generation (e.g. AR or ARMA models). The most widely used example of this approach is the Granger-causality test [7] which uses and AR model of the candidate influenced process. Two models are fit to the time-series one without any components of the tested influencing process and the other using components from both processes. If the accuracy of predicting the candidate influenced process is shown to be higher when using the candidate influencing process in a statistically significant sense, then a causal edge is drawn from the influencing to the influenced process. This approach is implemented in the function $detectGC()$.

Another, recently proposed, approach to CID is to announce a causal connection between two variables if it can be shown that they are part of a common dynamical system. A state-of-the-art algorithm for deciding whether such dynamical system exists is called Convergent Cross Mapping (CCM) [37]. The idea behind CCM is to study the prediction of each variable based on the manifold of the other variable (i.e. $\hat{Y}|M_x$ and $\hat{X}|M_y$). If the two time-series belong to the same dynamical system then the longer the time-series used for prediction the more accurate will these prediction be compared with the original time-series. By studying the convergence of this cross mapping, we can quantify the causal relation between the two time-series and decide its direction. This approach is implemented in the function $detectCCMC()$.

A third approach to CID is to base the decision of causal influence of whether or not a change in one process predicts a change in the other. This connects CID to CPD (the same way as constrained MD connected MD to CPD). The toolbox implements three methods for quantifying these predictability that can all be conducted using the function $detectCC()$: Granger causality between changes cores, consistency of delays between changes around the median delay and a normality test of delay distribution between change points.

4   Generic Time-Series Analysis Functions

MC^2 does not only provide the core algorithms discussed in the previous section but also implements several generic time-series analysis functions that are commonly used for either pre-processing the time-series before applying a discovery algorithm to it (e.g. dimensionality reduction, noise attenuation, etc.) or post-processing the results of a discovery algorithm (e.g. modeling a motif from its occurrences). This section briefly describes some of the most important generic time-series analysis functions in the toolbox.

4.1 Evaluation Functions

Objective evaluation functions for different approaches to the same problem are essential for practical usability. The toolbox provides a rich set of such function for each of the three discovery problems targeted.

MC^2 provides three approaches to compare and evaluate CPD algorithms. The first approach uses the traditional confusion matrix based statistics including the F measure, MCC, Precision, Recall, etc. and is implemented by the function *cpquality()*. The second approach is using information theoretic measures like the Kullback-Leibler divergence or Jennsen-Shannon Divergence between the true change point locations and estimated locations. These are implemented in the function *jsdiv()* and *kldiv()*. The third approach is to use Equal Sampling Rate (ESR) which is defined for any two probability distributions as the probability that two samples taken from the distributions will coincide. In CPD comparison case, the two distributions are simply normalized versions of the change scores found by two algorithms (or one algorithm and ground truth). This approach was shown in [20] to provide more intuitive comparison results relative to the other two approaches mentioned above. ESR is implemented in the function *esr()* in the toolbox.

MC^2 provides two approaches to compare MD algorithms. The first uses a multi-dimensional criterion assuming that ground-truth information about the motifs and their occurrences is available.

For each of the discovered motifs, four quantities are calculated:

- Correct Motifs: The fraction of discovered motifs that completely cover at least some occurrences of a single ground truth motif.
- Covering None: The fraction of discovered motifs that cover no parts of any ground truth motif occurrence.
- Covering Partially: The fraction of discovered motifs that cover only parts of some occurrences of a single ground truth motif.
- Covering Multiple: The fraction of discovered motifs that cover occurrences from multiple ground truth motifs.

These four criteria are calculated from the view-point of the discovered motifs. We also calculate two criteria from the ground-truth motifs view point: *Fraction Covered* is the fraction of occurrences of each one of the ground-truth motifs that is covered by discovered motifs. *Extra Fraction* is the fraction of the discovered motif occurrences used in calculating the fraction covered that are not covering a part of the ground truth occurrence. This approach to comparison of motifs is implemented in the function *mdquality()*.

The other approach is to collect the output motif locations in a single list and compare it to ground truth using the same techniques suggested for comparing change point discovery algorithms. This is done in the MC^2 toolbox using the function *mdq2cpq()*.

MC^2 provides a single evaluation function for comparing causality models called *cmpModels()* which takes as input the graph representing the learned model from some algorithm and a graph representing the ground-truth causal structure and finds a list of confusion matrix based metrics. It also generates a set of time related statistics that measure how accurate are the delays estimated between causes and effects in the graph including its mean and standard deviation in the whole graph.

4.2 Signal Generation, Conditioning and Modeling

Generating signals with predefined change points, motif occurrences, or causality structure is essential to test any discovery algorithm. The toolbox provides several routines for generating time-series using a variety of models (model-based generation routines) and another set of routines (problem-specific generation routines) that utilize the first set to generate appropriate inputs to the three discovery problems considered in Sect. 3. Model-based generation routines can be used to generate multidimensional time-series from moving average models AR models, ARMA models, ARIMA models, Markov chains, Gaussian Processes, Gaussian Mixture Models, Hidden Markov Models, and random walks.

Problem specific generation routines can generate time-series targeting any of the three core problems discussed in this paper: $generatePatterns()$ for both CPD and MD problems and $generateCausalGraph()$ in combination with the set of functions $produce * ()$ for the CID problem.

The toolbox provides several routines to simplify common preprocessing and postprocessing operations that can be needed in some applications before or after calling the core algorithms for discovery. These include dimensionality reduction using PCA ($tspca()$), detrending using Empirical Mode Decomposition ($detrendEMD()$), Singular Spectrum Analysis ($detrendSSA()$) and simple linear detrending ($detrendLinear()$), thinning for detecting local maxima ($thin()$), among many others.

5   Showcase Real-World Application

This section gives few real world applications of the discovery problems implemented in MC^2 from different research areas that can be achieved in few lines of code using the proposed toolbox.

5.1   Evaluating the Proposed Extension of GEMODA

Using MC^2, it is possible to apply GEMODA in at least three different ways depending on whether the data is discretized and whether CPD is used to focus the search for motifs as in constrained MD. This leads to three versions of the algorithm that we call GEMODA-D for the discretization version that uses SAX for symbolization and the Hamming distance in GEMODA. The version that applies the Euclidean distance directly to the time-series is called hereafter GEMODA-E while the version in which SSA based change point discovery is used to limit the search space of the algorithm is called GEMODA-C.

The first evaluation experiment compared these three alternatives using a synthetic dataset generated using the toolbox in order to have accurate ground-truth information. A hundred time-series with lengths ranging from 2000 to 10000 points were generated using $generatePatterns()$ with known motif occurrence locations. Motif lengths ranged from 50 to 100 points with five to seven occurrences each. The three GEMODA alternatives and shift-density constrained motif discovery were applied (see Sect. 3.2) with $cpd()$ used as the CPD function. Evaluation was done using $mdquality()$ function.

Shift-density based CMD achieves the highest correct motif discovery rate of 82 %. The GEMODA-D comes second with correct motif discovery rate of 81 % but with an improved speed of 234 %. GEMODA-E achieves the worst correct discovery rate of 76 % but with another improvement in speed of 32.4 % compared with GEMODA-D. The figure shows that GEMODA-D was the algorithm with minimal partial motif discovery yet with highest rate of false alarms (i.e. motifs that cover no ground-truth motifs).

5.2 Gesture Discovery from Accelerometer Data

Given accelerometer data collected while a person is doing some gestures in a continuous stream, it is possible to find individual gestures and learn a model of them in 3 lines of code using MC^2. The data for this experiment was sampled 100 times/second leading to a 78000 points 3D time-series. The time-series was converted into a single space time series by calling $tspca()$. SSA based CPD is then used to find a constraint on motif occurrence locations by calling $cpd()$. Finally, sdCMD as well as DGR with GStex for extension were applied to this projected time-series by calls to $sdCMD()$ and $dgr()$. sdCMD discovered 9 gestures, the top seven of them corresponded to the true gestures (with a discovery rate of 100 %) while DGR discovered 16 gestures and the longest six of them corresponded to six of the seven gestures embedded in the data (with a discovery rate of 85.7 %) and five of them corresponded to partial and multiple coverings of these gestures. Finally a call to $detectMotif()$ can be used to enhance discovered motifs by adding all of their other occurrences in the time-series to them. If a model of a gesture is needed, a simple call to $LearnHMM()$ can be used to learn it as a HMM, $learnGMM()$ can be used to learn it as a GMM, and $learnARMA()$ can be used to learn it as an ARMA process.

5.3 Fluid Imitation in HRI

Fluid imitation is the process of learning new skills by just watching them conducted by experienced people without during their day to day activities. This kind of imitation is very common in humans but challenging for robots because the motions to be learned are embedded in a stream of undifferentiated sensed time-series. The fluid imitation engine [21] was proposed to solve this problem. The engine consists of a battery of CPD processes that find change locations in both the states of objects in the environment (O), their perceived state from the perspective of the model from which the learner is learning (P), and model's actions (A). The outputs of these change processes are passed through a CID process to determine the significance of each point of the time-series A. The significance level as well as the CPD score for each dimension of A are passed to a constrained MD algorithm that finds recurrent patterns representing important motions executed by the model. These patterns are then modeled by a standard learning from demonstration system (e.g. GMM/GMR [3]). The GMM learning routine in MC^2 can be used for the modeling. These models are then aggregated in a library of primitive motions that can be used by the learner later. In MC^2, fluid imitation can be implemented by less than 20 lines of code.

A simple extension to the fluid imitation engine is to use CIP again to find the relation between learned motion patterns and changes in the state of the environment (as found in the stream O) and other learned models. This allows the robot not only to learn the basic actions but their sequencing and relation to object state manipulations. This extension will be evaluated in future work.

6 Conclusions

The paper describes a toolbox for solving three discovery problems in time-series mining: change point discovery, motif discovery and causality influence discovery. The toolbox implements examples of the major approaches for each of these three problems and provides a consistent interface for all of these approaches. This allows easy prototyping of solutions to more complex problems in which these discovery problems are building blocks. Moreover, the paper presents a simple extension of the GEMODA motif discovery algorithm that increases its efficiency compared with its standard form while having approximately the same correct discovery rate. Four showcase real world applications that are simplified by the proposed toolbox are also presented.

Acknowledgments. The work reported in this paper was financially supported by JSPS KAKENHI Grant Number 15K12098, and AFOSR/AOARD Grant No. FA2386-14-1-0005.

References

1. Andre-Obrecht, R.: A new statistical approach for the automatic segmentation of continuous speech signals. IEEE Trans. Acoust. Speech Signal Process. **36**(1), 29–40 (1988)
2. Basseville, M., Kikiforov, I.: Detection of Abrupt Changes. Printice Hall, Englewood Cliffs (1993)
3. Calinon, S., Billard, A.: Incremental learning of gestures by imitation in a humanoid robot. In: Proceedings of the ACM/IEEE International Conference on Human-Robot Interaction, pp. 255–262. ACM (2007)
4. Catalano, J., Armstrong, T., Oates, T.: Discovering patterns in real-valued time series. In: Fürnkranz, J., Scheffer, T., Spiliopoulou, M. (eds.) PKDD 2006. LNCS (LNAI), vol. 4213, pp. 462–469. Springer, Heidelberg (2006)

5. Chiu, B., Keogh, E., Lonardi, S.: Probabilistic discovery of time series motifs. In: KDD 2003: The 9th ACM SIGKDD International Conference on Knowledge Discovery and Data Mining, pp. 493–498. ACM, New York (2003)
6. Fearnhead, P., Liu, Z.: On-line inference for multiple changepoint problems. J. Roy. Stat. Soc. Ser. B (Stat. Methodol.) **69**(4), 589–605 (2007)
7. Granger, C.W.: Investigating causal relations by econometric models and cross-spectral methods. Econometrica J. Econometric Soc. **37**, 424–438 (1969)
8. Hassani, H., Heravi, S., Zhigljavsky, A.: Forecasting European industrial production with singular spectrum analysis. Int. J. Forecast. **25**(1), 103–118 (2009)
9. Ide, T., Inoue, K.: Knowledge discovery from heterogeneous dynamic systems using change-point correlations. In: SDM 2005: SIAM International Conference on Data Mining, pp. 571–575 (2005)
10. Itoh, N., Kurths, J.: Change-point detection of climate time series by nonparametric method. In: Proceedings of the World Congress on Engineering and Computer Science, Citeseer, vol. 1, pp. 445–448 (2010)
11. Jensen, K.L., Styczynxki, M.P., Rigoutsos, I., Stephanopoulos, G.N.: A generic motif discovery algorithm for sequential data. Bioinformatics **22**(1), 21–28 (2006)
12. Kulic, D., Nakamura, Y.: Incremental learning and memory consolidation of whole body motion patterns. In: International Conference on Epigenetic Robotics, pp. 61–68 (2008)
13. Lin, J., Keogh, E., Wei, L., Lonardi, S.: Experiencing SAX: a novel symbolic representation of time series. Data Min. Knowl. Disc. **15**(2), 107–144 (2007)
14. Minnen, D., Starner, T., Essa, I.A., Isbell Jr., C.L.: Improving activity discovery with automatic neighborhood estimation. In: IJCAI 2007: 16th International Joint Conference on Artificial Intelligence, vol. 7, pp. 2814–2819 (2007)
15. Mohammad, Y., Nishida, T.: Learning interaction protocols using augmented Baysian networks applied to guided navigation. In: IROS 2010: IEEE/RSJ International Conference on Intelligent Robots and Systems, pp. 4119–4126. IEEE (2010)
16. Mohammad, Y., Ohmoto, Y., Nishida, T.: GSteX: greedy stem extension for free-length constrained motif discovery. In: IEA/AIE 2012: The International Conference on Industrial, Engineering, and Other Applications of Applied Intelligence, pp. 417–426 (2012)
17. Mohammad, Y., Nishida, T.: Constrained motif discovery in time series. New Gener. Comput. **27**(4), 319–346 (2009)
18. Mohammad, Y., Nishida, T.: Mining causal relationships in multidimensional time series. In: Szczerbicki, E., Nguyen, N.T. (eds.) Smart Information and Knowledge Management. SCI, vol. 260, pp. 309–338. Springer, Heidelberg (2010)
19. Mohammad, Y., Nishida, T.: Using physiological signals to detect natural interactive behavior. Appl. Intell. **33**, 79–92 (2010)
20. Mohammad, Y., Nishida, T.: On comparing SSA-based change point discovery algorithms. In: SII 2011: IEEE/SICE International Symposium on System Integration, pp. 938–945 (2011)
21. Mohammad, Y., Nishida, T.: Fluid imitation. Int. J. Soc. Robot. **4**(4), 369–382 (2012)
22. Mohammad, Y., Nishida, T.: Exact discovery of length-range motifs. In: Nguyen, N.T., Attachoo, B., Trawiński, B., Somboonviwat, K. (eds.) ACIIDS 2014, Part II. LNCS, vol. 8398, pp. 23–32. Springer, Heidelberg (2014)
23. Mohammad, Y., Nishida, T.: Scale invariant multi-length motif discovery. In: Ali, M., Pan, J.-S., Chen, S.-M., Horng, M.-F. (eds.) IEA/AIE 2014, Part II. LNCS, vol. 8482, pp. 417–426. Springer, Heidelberg (2014)

24. Mohammad, Y., Nishida, T.: Data Mining for Social Robotics. Springer, Cham (2015)
25. Mohammad, Y., Nishida, T.: Exact multi-length scale and mean invariant motif discovery. Appl. Intell. **44**(2), 322–339 (2015)
26. Mohammad, Y., Nishida, T.: Learning interaction protocols by mimicking: understanding and reproducing human interactive behavior. Pattern Recogn. Lett. **66**(15), 62–70 (2015)
27. Mohammad, Y., Nishida, T.: Shift density estimation based approximately recurring motif discovery. Appl. Intell. **42**(1), 112–134 (2015)
28. Mohammad, Y., Nishida, T., Okada, S.: Unsupervised simultaneous learning of gestures, actions and their associations for human-robot interaction. In: IROS 2009: IEEE/RSJ International Conference on Intelligent Robots and Systems, IROS 2009, pp. 2537–2544. IEEE Press, Piscataway (2009)
29. Mohammad, Y., Nishida, T.: Unsupervised discovery of basic human actions from activity recording datasets. In: SII 2012: IEEE/SICE International Symposium on System Integration, pp. 402–409. IEEE (2012)
30. Mohammad, Y., Ohmoto, Y., Nishida, T.: CPMD: a matlab toolbox for change point and constrained motif discovery. In: Jiang, H., Ding, W., Ali, M., Wu, X. (eds.) IEA/AIE 2012. LNCS, vol. 7345, pp. 114–123. Springer, Heidelberg (2012)
31. Mueen, A.: Enumeration of time series motifs of all lengths. In: 2013 IEEE 13th International Conference on Data Mining (ICDM). IEEE (2013)
32. Mueen, A., Keogh, E., Zhu, Q., Cash, S., Westover, B.: Exact discovery of time series motifs. In: SDM 2009: SIAM International Conference on Data Mining, pp. 473–484 (2009)
33. Nagai, Y., Nakatani, A., Qin, S., Fukuyama, H., Myowa-Yamakoshi, M., Asada, M.: Co-development of information transfer within and between infant and caregiver. In: 2012 IEEE International Conference on Development and Learning and Epigenetic Robotics (ICDL), pp. 1–6. IEEE (2012)
34. Oates, T.: PERUSE: an unsupervised algorithm for finding recurring patterns in time series. In: International Conference on Data Mining, pp. 330–337 (2002)
35. Page, E.: Continuous inspection schemes. Biometrika **41**, 100–115 (1954)
36. Pantic, M., Pentland, A., Nijholt, A., Huang, T.: Machine understanding of human behavior. In: AI4HC 2007: IJCAI 2007 Workshop on Artificail Intelligence for Human Computing, pp. 13–24. University of Twente, Centre for Telematics and Information Technology (CTIT), January 2007
37. Sugihara, G., May, R., Ye, H., Hsieh, C., Deyle, E., Fogarty, M., Munch, S.: Detecting causality in complex ecosystems. Science **338**(6106), 496–500 (2012)
38. Tanaka, Y., Iwamoto, K., Uehara, K.: Discovery of time-series motif from multidimensional data based on MDL principle. Mach. Learn. **58**(2/3), 269–300 (2005)
39. Vahdatpour, A., Amini, N., Sarrafzadeh, M.: Toward unsupervised activity discovery using multi-dimensional motif detection in time series. In: IJCAI 2009: The International Joint Conference on Artificial Intelligence, pp. 1261–1266 (2009)
40. Willsky, A.S.: A survey of design methods for failure detection in dynamic systems. Automatica **12**(6), 601–611 (1976)

Knowledge Based Systems

Hidden Frequency Feature
in Electronic Signatures

Orcan Alpar and Ondrej Krejcar[✉]

Faculty of Informatics and Management,
Center for Basic and Applied Research, University of Hradec Kralove,
Rokitanskeho 62, Hradec Kralove 500 03, Czech Republic
orcanalpar@hotmail.com, ondrej@krejcar.org

Abstract. Forensics is a science discipline that deals with collecting evidence in crime scene investigation. However if we're dealing with signatures, the crime scene is the signed paper itself. Therefore, for any kind of investigation, there should be a sample and a master signature to benchmark the similarities and differences. The characteristics of a master signature could easily be identified by forensics techniques, yet it is still infeasible for electronic signatures due to ease of copy-pasting. Through the emerging touchscreen technologies, the features of the signatures could be stealthily extracted and stored while the user is signing. Given these facts, the novelty we put forward in this paper is a feature extraction method using short time Fourier transformations to identify frequencies of a simple master signature. We subsequently presented the spectrogram analysis revealing the differences between the original and fake signatures. Finally a validation method for the analysis of the spectrograms is introduced which resulted in a significant gap between real and fraud signatures for various window sizes.

Keywords: Forensics · Touchscreen · Biometrics · Electronic signature · Frequency: fourier transforms

1 Introduction

Electronic signature term fundamentally represents the real signature to approve an electronic document, in a digital framework or an interface, consisting of letters that generally form a name or a surname or both. Although it seems very easy to implement, these kind of systems are not widely accepted due to security deficits. Above all, most of the document types are either an image file type (.jpg) or a portable document format (.pdf) or a document format (.doc), which all enable the imitation of the signatures. When the instance of a signature is somehow stolen, it can be used in the documents without any consent of the owner, by an image processing software. In this case, the fake signature would have the same characteristics of the real one that disables all forensics methods for discriminations.

On the other hand, if there is an interface to be signed, it is still possible to digitally imitate the signatures. However, as long as the user is signing on an electronic interface just like on a paper, the characteristics could be extracted through the technological

© Springer International Publishing Switzerland 2016
H. Fujita et al. (Eds.): IEA/AIE 2016, LNAI 9799, pp. 145–156, 2016.
DOI: 10.1007/978-3-319-42007-3_13

developments of touchscreen devices. There are three types of features to be extracted such as: global, local, and geometrical [1, 2]. Global features include very superficial features of the signatures like height-width ratio, area, angles and such [3–5]. Local features are fundamentally similar however they are extracted from a portion of the signature [6, 7]. Geometric features are the combination of these two with geometrical calculations added [8, 9].

Regarding very recent studies, there are many research on feature extraction of offline signature identification such as direction based [10], geometric based [11] and moment based [12]. As the prominent papers: Tolasana et al. [13] dealt with five major features to be extracted: Time, Kinematic, Direction, Geometry and Pressure, for written signatures. They used the Mahalanobis distance to compare the similarity of original signatures with the attempted ones.

Although the methods mentioned above are used in forensics by investigating the signed hard-copy documents, they could also be beneficial when dealing with the extraction of biometric features of electronic signatures. Since we deal with online feature extraction while the user is signing, there a few of relevant papers in the literature. López-García et al. [14] presented an input device for signing process with shape comparison using dynamic time warping. They captured many signatures to identify the real ones using Gaussian Mixture model. Forhad et al. [15] introduces an interface that collects the signing points to generate a string in 2D plane. They discriminated the real signatures from the fake ones using approximate string matching algorithms.

Moreover, Kumar et al. [16] proposed an touchscreen system where the signatures are acquired using a digitizing tablet which captures both dynamic and spatial information. They also dealt with strings and the similarities are searched by defining a threshold of the original signature and the signed one. Bhateja et al. [17] divided the entire signatures into slices in time domain and the compared the similarities using artificial neural network based classifier. They extracted six major features from signing process: average pressure, average speed along x-axis, average speed along y-axis, average speed, average azimuth and average altitude.

Considering these papers, we propose a very novel user interface for touchscreens to collect biometric features of electronic signatures using hidden frequency characteristics. Initially we decided a simple and a basic signature "Amber" to be the main template on a 600×355 plane, with a unique style of signing therefore some portion of the signature are signed very gently while the rest a little harsh. Afterwards we wrote an interface in Matlab to emulate the collection of touch data in x and y coordinates.

In every 0.01 s, the interface collects the x-y data and subsequently calculates the distance covered in scalar format and turns the data into an interpolated signal. The hidden frequency feature is extracted from the signal by short time Fourier transforms and eventually the spectrograms are analyzed within various window sizes. The workflow is shown in Fig. 1 below which could also be considered as the graphical abstract of our work.

The fundamentals of Fourier transforms are introduced in the following section prior to the presentation of the kernel of our paper: how the spectrograms change as the signature style alters.

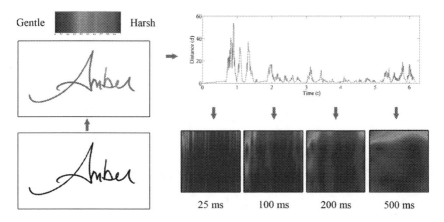

Fig. 1. Workflow of frequency feature extraction interface. (Color figure online)

2 Preliminaries

Since we turned the signatures into signals to identify hidden frequency feature, all aspects of Fourier transforms and spectrogram analysis are presented in following subsections.

2.1 Signal Classification

Signals are generally classified into three major categories depending on the characteristics of x-y data and the gathering method. First type of the signals is analog or continuous signals that have y values for $\forall x \in R$ which mostly obtained by an equation. If a continuous signal is sampled by a rate the discrete signals are achieved so that $\forall x, y \in R$ yet there is no information of some x-y couples anymore. In addition, the discrete signal could be achieved by measurement as well. If the discrete signal is quantized a digital signal is obtained where $\forall x, y \in R$ are quantized values. The y-values could be integers of binary vale depending on quantizing method yet most of the x-y couple data is lost.

On the other, it is still possible to turn discrete and digital signals into continuous by interpolation, when necessary. If the sample rate is small enough, it is feasible to interpolate the discrete signal to estimate the continuous signal yet it won't be identical. The interpolation is made by finding the shortest distance between two succeeding discrete values and drawing a line. The examples of a sinusoidal wave could be seen in Fig. 2.

The common way to analyze the signals in time-domain is using Fourier transforms.

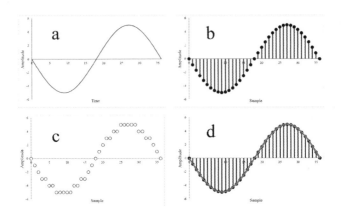

Fig. 2. Classification of the signals (a: continuous signal b: discretized signal c: digitized signal d: interpolated signal)

2.2 Fourier Transforms

Fourier transforms are widely used to identify all frequency components of a signal in time domain and it is essential to reveal how the signal is behaving in frequency domain. The Fourier transform of an analog signal is calculated by:

$$F(u) = \frac{1}{N} \int_0^{N-1} f(x)e^{-j2\pi ux} dx, u = 0, 1, \ldots, N - 1 \tag{1}$$

and for the discrete signals, discrete Fourier transform, converted from continuous Fourier transforms, is computed by;

$$F(u) = \frac{1}{N} \sum_{x=0}^{N-1} f(x)e^{\frac{-j2\pi ux}{N}}, \quad u = 0, 1, \ldots, N - 1 \tag{2}$$

where; $F(u)$ is the frequency domain representation of signal time-series signal $f(x)$ for every u^{th} frequency component; $u = 0, 1, \ldots N - 1$, for every time series signal x; $x = 0, 1, \ldots N - 1$ while N is the total number of samples in a signal and j is the imaginary unit. Both transformations are very useful to identify the frequency components of steady signals however not effective to analyze time-variant and/or unsteady signals. For instance, if we try to turn a continuous and time-variant sinusoidal signal into frequency domain, the results will be inadequate and thus vague. In Fig. 3, an infinite sinusoidal signal with 5 Hz and 10 Hz sections are analyzed with Fourier transforms and achieved the identical results. However in first signal, the 10 Hz section is before 5 Hz and in the second signal it is the opposite. Even if the frequency components are correctly computed, there is no information about when the corresponding frequency takes place.

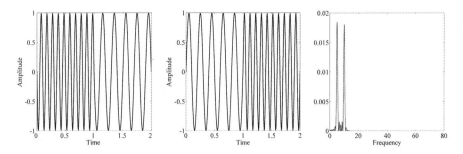

Fig. 3. Time-variant unsteady infinite sinusoidal signal starting with 10 Hz and turns into 5 Hz after 1 s (on the left), starting with 5 Hz and turns into 10 Hz after 1 s (in the middle), and the Fourier transforms in frequency domain (on the right).

2.3 Spectrogram Analysis

The most common method to reveal the time and corresponding frequencies together is the short-time Fourier transformations. Despite the continuous and discrete Fourier transforms, the signal is divided into desired width of windows and the signals inside are separately transformed. The short-time Fourier transform is calculated for continuous signals as

$$STFT_f^u(t', u) = \int_0^{N-1} [f(x)W(t-t')]e^{-j2\pi ux}dx, \quad u = 0, 1, \ldots, N-11 \qquad (3)$$

and for discrete signals as;

$$STDFT_f^u(t', u) = \sum_{x=0}^{N-1}[f(x)W(t-t')]e^{-j2\pi ux}, \quad u = 0, 1, \ldots, N-1 \qquad (4)$$

where the short time Fourier is computed for each window centered at $t = t'$, W is windowing function, $f(x)$ is signal, u is frequency parameter and t is the time parameter. The determination of the window size here is very crucial since if it is selected too small, the transform will reveal only the time data. On the contrary, if the window size is exceedingly large, there will be only frequency component just like Fourier transforms. The main benefit of short time Fourier is the layered constitution of frequencies since all Fourier transforms of separate windows are finally placed together. Therefore the results of short time transforms could be three dimensional graphics nonetheless the easiest way to exhibit the time vs frequency relation is the spectrograms with color identifiers. Spectrograms which are computed for each window combined by $\left|STFT_f^u(t', u)\right|^2$ are shown in Fig. 4 for various window sizes.

Considering these facts, we present an extraction and validation methodology for electronic signatures in following section.

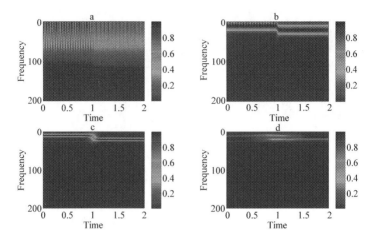

Fig. 4. Spectrogram analysis of time-variant infinite sinusoidal signal. Window sizes: a.25 ms b.125 ms c.375 ms, d.1000 ms (Color figure online)

3   Extraction of Frequencies from Signatures

The code of written interface starts with reading the image of the signature as the template that will be imitated while signing on the frame. The plane is emulated with 600 × 355 resolution JPG file for better precision even though it would be smaller when used in touchscreens. The style of the signature "Amber" is designated for this research as in Fig. 1 top-left, so that it starts with rapid motion, then slows down and finally faster again. While handling the signing process, the x and y coordinates are extracted in every 0.01 s, and the distance is calculated starting from $x_0 = 0$ and $y_0 = 0$. The data extracted includes $y \in R$ values for $\forall x \in R$ however still discrete due to sample time $\Delta x = 0.01$ s. Since in short time transformation, it is necessary to resample the signal, the discrete data is interpolated to form a continuous signal.

After signing with the style mentioned above, the matrix consisting of distance covered and corresponding time is saved for further analysis. Given the total time of signing process, the interval is divided into equal parts by several with sizes and the signal is sampled at 400 Hz sample size. The spectrograms are generated by "Hamming" window function: $W_n(t - t') = 0.54 - 0.46\cos\left(2\pi\frac{n}{N}\right)$, where $0 \leq n \leq N$. Given the original signing style, some sections are faster than the other, however it doesn't always mean that their frequencies are higher. The spectrograms revealed by the original signing style could be seen in Fig. 5.

In the figures, the red areas represent the higher frequencies regardless of the signing speed. It is see that smaller window sizes reveal the signal in time domain itself where the larger window size only shows the frequency component. The spectrogram analysis of the real signature exhibits a slowing down behavior since the higher frequencies tend to left. Three more experiments are conducted by imitating the signature by three rigid styles: gentle, harsh and normal. The gentle style, most probably how the copycat will imitate the original signature, doesn't have any major frequency peaks

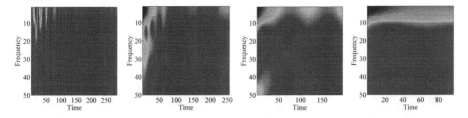

Fig. 5. Spectrogram analysis of real signature signal Window sizes: 25 ms, 100 ms, 250 ms, 500 ms from left to right. (Color figure online)

since it is signed very slowly. On the other hand, some red spikes could be seen in spectrograms with 25 ms window size, since they possibly represent the sliding from one letter to another. The example of gentle signing style is shown in Fig. 6.

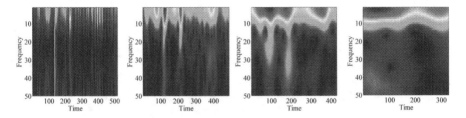

Fig. 6. Spectrogram analysis of fake signature signal example with very gentle signing style. Window sizes: 25 ms, 100 ms, 250 ms, 500 ms from left to right. (Color figure online)

The other experiment is done by signing very harshly from the beginning to the end. The interface collects the distance covered vs time therefore the frequencies remain similar as long as the signature stays same however the values of the frequencies are higher and more spread as in Fig. 7.

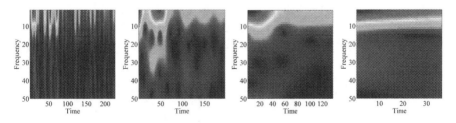

Fig. 7. Spectrogram analysis of fake signature signal example with very harsh signing style. Window sizes: 25 ms, 100 ms, 250 ms, 500 ms from left to right. (Color figure online)

The final signing style is "normal" that the signature is imitated only by copying without any speed effect. Therefore and as expected the higher frequencies have spread along time axis which seems more homogenous. On the other hand it has some similarities with the gentle and the harsh style depending on the window sizes however still hugely different than the original style. The spectrograms could be seen in Fig. 8.

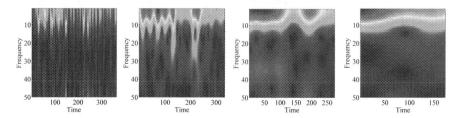

Fig. 8. Spectrogram analysis of fake signature signal example with normal signing style. Window sizes: 25 ms, 100 ms, 250 ms, 500 ms from left to right. (Color figure online)

4 Validation Through Closeness Analysis of Spectrograms

The spectrograms revealed by the short time Fourier transforms actually are matrices consisting of numbers from 0 to 1. It is very hard to compare the outcomes without the spectrograms since the durations of signing process are various and thus the x-axis limits. This fact could be observed from the sample spectrograms presented above with the same window sizes. For instance the rapid signature has significantly smaller x-axis interval than the gentle signing if the same window-sized spectrograms are compared, since their matrix sizes are different. However, this difference wiped out through the spectrograms since the matrices are turned into square images with the same resolution but various x and y axis values. Therefore the spectrograms revealed by short time Fourier transforms are totally comparable by image closeness methods. Although there are numerous methods in the literature on identification of the similarities of the images, the most appropriate technique for our research is closeness of red pixels regardless of others.

Given the most common methods, the entire images or patches are compared by several algorithms mostly focused on shifting along both axes and rotating the images. On the contrary in this research these processes are impracticable since the x-axis of the spectrograms are strictly dependent upon time thus cannot be rotated. In addition, it is not possible to shift the image along the y-axis, since it is dependent upon the frequency vs time correlation. Even if the spectrograms only consist of the signing process, we start to record the coordinate vs time data after the first click, therefore there could be a minor gap between the first click and process initiation. However the results show that the shifting is unnecessary for comparing the signatures, for now.

Based on these facts, spectrogram images consisting of $735 \times 735 \times 3$ RGB structure are briefly stated as; $I_{i,j,c}(i \in [1:735], j \in [1:735], c \in [1:3])$ where I is the 3d image space, i and j are the column and row indices, c is the color channel and all are integers. Since we focus on higher frequencies, we firstly obtain the red channel of the images. The RGB images are formed by $i \times j \times 3$ arrays therefore if we delete the second green and third blue matrices, the image will turn a grayscale image representing the intensity of red color, namely; $R_{i,j} = I_{i,j,1}(i \in [1:735], j \in [1:735])$. Each element $s_{i,j}$ in $R_{i,j}$ are between 0 and 255 since the red intense pixels in $R(k)_{i,j,c}$ will have a value so close to 255 while if there is no red color, the value will be zero. The red channel separation $I = 1 * red + 0 * green + 0 * blue$ is different than grayscaling which has a universal equation of $I = 0.2989 * red + 0.5870 * green + 0.1140 * blue$. Afterwards, the red channels of the spectrograms are tuned into black & white images by 50 % binarization so that $0 \leq \dot{s}_{i,j} \leq 1$ will be provided in the real and testing spectrograms $P_{i,j}$ and $Q_{i,j}$ (Fig. 9).

Fig. 9. Several spectrograms with window size = 100 ms. From left to right: RGB image, Red Channel, Grayscale (on the right), Black & white of red channel (Color figure online)

Based on the two binarized red channels, square $n \times n$ matrices $P_{i,j}$ and $Q_{i,j}$ of the images will be compared, therefore we propose a closeness algorithm which perfectly fits our requirements namely;

$$c_0 = \left[\left(\sum_{i=1}^{n} \sum_{j=1}^{n} P_{i,j} \cdot Q_{i,j} \right) \bigg/ \left(\sum_{i=1}^{n} \sum_{j=1}^{n} P_{i,j} \right) \right] - \left[\left(\sum_{i=1}^{n} \sum_{j=1}^{n} |P_{i,j} - Q_{i,j}| \right) \bigg/ n^2 \right] \quad (5)$$

where $0 \leq c_0 \in R \leq 1$ is the closeness of the images as a closeness ratio, i is the place in x-axis and j is the place in y-axis, and the multiplication is dot-product. The results are as follows in Table 1.

Table 1. The validation experiment for window size = 100 ms

Attempt	Original	Real #1	Real #2	Real #3	Real #4
Image					
c	100%	64.62%	70.33%	72.33%	86.23%
Attempt	Original	Fraud #1	Fraud #2	Fraud #3	Fraud #4
Image					
c	100%	27.00%	0.273%	0%	0%

As seen in Table 1, the results are highly promising for the window size = 100 ms with our validation algorithm. The real set is formed by the signing style mentioned in Fig. 1, while the fraud set includes various copying styles without changing the signing speed or frequency. However, for window size = 250 ms all closeness indicators increased and one fraud is too close to the real attempts in Table 2.

Table 2. The validation experiment for window size = 250 ms

Attempt	Original	Real #1	Real #2	Real #3	Real #4
Image					
c	100%	93.42%	97.75%	89.05%	91.06%
Attempt	Original	Fraud #1	Fraud #2	Fraud #3	Fraud #4
Image					
c	100%	40.15%	84.47%	49.03%	0%

Given these facts, the significant separation of real and fraud closeness values are revealed in window size = 100 ms experiment.

5   Results

Briefly in this paper an interface is written to extract the hidden frequency component while signing electronically. The theory beneath the kernel of the research is that everyone has unique signatures with unique signing style as usual, however this theory is implemented to touchscreens. Results of the various experiments prove that as long as a user has a unique signing style, the signature has a unique spectrogram. While copying the original signature, the copycat would try to follow the signature very slowly with a pen, therefore the frequency component of the original signature significantly changes. Moreover if the copycat tries to sign in normal speed or even faster, it is very hard to copy the frequency of the original one.

Furthermore, several window sizes are investigated and the most feasible one we decided is 100 ms, yet we made two experiments for 100 ms and 250 ms separately. The lower window sizes could be used when the frequency component is not so important since they look like the results of time-domain graphs. On the contrary, when the window size increase, the spectrograms start to look very similar since the time component disappears. Even though the results seem like images, they are formed by short time Fourier transforms and thus they actually are matrices which can be compared with the original signatures. Therefore we compared the spectrograms by image closeness technique we created for this research. Experiments proved that the original signatures have very high closeness ratios, significantly greater than frauds.

6   Conclusion and Discussion

Forensics deals with the post-analysis of the signatures however thanks to touchscreen technology, the signatures of the users could be saved with their features. In this paper, we dealt with extraction of frequency feature while a user is signing the interface we wrote. It is seen that every signature style has unique spectrograms no matter the signature itself is copied. We conducted two experiments to show the performance of our algorithms with image closeness techniques. Although both of the experiments resulted in high separation intervals, 100 ms still seems better and plausible.

As the future research ideas, some more experiments could be conducted for each window sizes with a real and fraud group to compute the system performance. For the classifier, sequential bins could be most proper technique to quantize the images. Afterwards any intelligent classifier could be utilized to discriminate fraud from real signatures and to conclude in "reject" or "accept" decisions, as in our previous researches [18–21] such as Gauss-Newton or Levenberg-Marquardt based neural network classifiers or adaptive neuro-fuzzy based ones.

Acknowledgment. This work and the contribution were supported by project "SP/2016/2102 - Smart Solutions for Ubiquitous Computing Environments" from FIM, University of Hradec Kralove.

References

1. Saikia, H., Sarma, K.C.: Approaches and issues in offline signature verification system. Int. J. Comput. Appl. **42**(16), 45–52 (2012)
2. Kumar, S.: A survey on handwritten signature verification techniques. Int. J. Adv. Res. Comput. Sci. Manage. Stud. **3**, 182–186 (2015)
3. Baltzakis, H., Papamarkos, N.: A new signature verification technique based on a two-staged neural network classifier. Eng. Appl. Artif. Intell. **14**, 95–103 (2001)
4. Biswas, S., Kim, T., Bhattacharyya, D.: Features extraction and verification of signature image using clustering technique. Int. J. Smart Home **4**(3), 43–56 (2010)
5. Nguyen, V., Blumenstein, M., Leedham, G.: Global features for the off-line signature verification problem. In: 10th International Conference on Document Analysis and Recognition, 2009. ICDAR 2009. IEEE (2009)
6. Madasu, V.K., Lovell, B.C.: An automatic off-line signature verification and forgery detection system. In: Pattern Recognition Technologies and Applications, Recent Advances, pp. 63–89 (2008)
7. Qiao, Y., Liu, J., Tang, X.: Offline signature verification using online handwriting registration. In: IEEE Conference on Computer Vision and Pattern Recognition CVPR 2007, Minneapolis (2007)

8. Al-Omari, Y.M., Abdullah, S.N.H.S., Omar, K.: State-of-the-art in offline signature verification system. In: 2011 International Conference on Pattern Analysis and Intelligent Robotics (ICPAIR). IEEE (2011)
9. Arya, M.S., Inamdar, V.S.: A preliminary study on various off-line hand written signature verification approaches. Int. J. Comput. Appl. 1(9), 50–56 (2010)
10. Pal, S., Alireza, A., Pal, U., Blumenstein, M.: Off-line signature identification using background and foreground information. In: International Conference on Digital Image Computing Techniques and Applications (DICTA) (2011)
11. Wang, N.: Signature identification based on pixel distribution probability and mean similarity measure with concentric circle segmentation. In: Fourth International Conference on Computer Sciences and Convergence Information Technology, 2009, ICCIT 2009 (2009)
12. Pal, S., Pal, U., Blumenstein, M.: Off-line English and Chinese signature identification using foreground and background features. In: The 2012 International Joint Conference on Neural Networks (IJCNN), pp. 1–7. IEEE (2012)
13. Tolosana, R., Vera-Rodriguez, R., Fierrez, J., Ortega-Garcia, J.: Feature-based dynamic signature verification under forensic scenarios. In: 2015 International Workshop on Biometrics and Forensics (IWBF), IEEE (2015)
14. Lopez-Garcia, M., Ramos-Lara, R., Miguel-Hurtado, O., Canto-Navarro, E.: Embedded system for biometric online signature verification. IEEE Trans. Ind. Inform. 10(1), 491–501 (2014)
15. Forhad, N., Poon, B., Amin, M.A., Yan, H.: Online signature verification for multi-modal authentication using smart phone. In: Proceedings of the International MultiConference of Engineers and Computer Scientists (2015)
16. Kumar, S., BH, V.P.: Embedded Platform For Online Signature Verification. IJSEAT 3(4), 126–131 (2015)
17. Bhateja, A.K., Chaudhury, S., Saxena, P.K.: A robust online signature based cryptosystem. In: 2014 14th International Conference on IEEE Frontiers in Handwriting Recognition (ICFHR) (2014)
18. Alpar, O.: Keystroke recognition in user authentication using ANN based RGB histogram technique. Eng. Appl. Artif. Intell. 32, 213–217 (2014)
19. Alpar, O.: Intelligent biometric pattern password authentication systems for touchscreens. Expert Syst. Appl. 42(17), 6286–6294 (2015)
20. Alpar, O., Krejcar, O.: Biometric swiping on touchscreens. In: Computer Information Systems and Industrial Management (2015)
21. Alpar, O., Krejcar, O.: Pattern password authentication based on touching location. In: Intelligent Data Engineering and Automated Learning–IDEAL 2015 (2015)

A Multimodal Approach to Relevance and Pertinence of Documents

Matteo Cristani(⊠) and Claudio Tomazzoli(⊠)

University of Verona, Verona, Italy
{matteo.cristani,claudio.tomazzoli}@univr.it

Abstract. Automated document classification process extracts information with a systematical analysis of the content of documents. This is an active research field of growing importance due to the large amount of electronic documents produced in the world wide web and made readily available thanks to diffused technologies including mobile ones. Several application areas benefit from automated document classification, including document archiving, invoice processing in business environments, press releases and search engines. Current tools classify or "tag" either text or images separately. In this paper we show how, by linking image and text-based contents together, a technology improves fundamental document management tasks like retrieving information from a database or automatically routing documents. We present a formal definition of pertinence and relevance concepts, that apply to those documents types we name "multimodal". These are based on a model of conceptual spaces we believe compulsory for document investigation while using joint information sources coming from text and images forming complex documents.

1 Introduction

Nowadays the wide availability of electronic documents through the Internet or private business networks has changed the way people search for information. We deal with a huge quantity of knowledge which has to be organized and searchable to be utilized. Also for this reason in information technology research community there is an always growing interest in the field of automatic document classification. Although several innovative studies are produced every year, some topics are still to be deeply investigated. Among these, the problem of efficient classification and retrieval of documents containing both text and images has been treated in a non multidisciplinary approach. There are several publications of efficient information retrieval from text. There are also publications about information extraction from images and even text contained in images [1], but the joint analysis of text and image information from a complex document still lacks a well documented solution. For example, if a brochure from an isolated hotel in the Dolomites describes the hotel's features and includes maps and pictures of mountainous surroundings, the categorizer will automatically discover the content and link the text and the images together. Then someone searching

© Springer International Publishing Switzerland 2016
H. Fujita et al. (Eds.): IEA/AIE 2016, LNAI 9799, pp. 157–168, 2016.
DOI: 10.1007/978-3-319-42007-3_14

for an isolated mountain lodge within a certain price range would retrieve the brochure even if "isolated lodge in the mountains" were never mentioned in the actual text. The paper is organized as follows: Sect. 2 presents the areas in which automatic document classification is relevant; Sect. 3 summaries the main approaches existing in current literature; Sect. 4 presents the model and the approach of extracting joint textual and image information; finally in Sect. 5 we give the formal definition of the model, of *Pertinence* and *Relevance* and make some conclusions.

2   Motivations

Automatic document classification is an interesting process for a wide variety of application areas, due to the huge amount of electronic documents in which is stored the information a user can search for. Among these there are Web Mining, Press Survey, Scientific Research, Image Indexing.

Press Survey

Press Survey is the task of retrieving what has been "printed" and diffused on the mass media, usually newspapers, about a particular topic.

Politicians are interested in knowing who is writing about them or about a particular subject they are interested in. Firms are interested in knowing how the Press responded to a particular marketing event or a new product release. Most of this work is performed by humans, who scan the several sources for relevant information. As the Press are going to deploy on the Internet their former printed daily or weekly magazines, we can consider the sources of information to be digital, thus eligible for automatic elaboration. Due to the visual impact of images, articles are very often equipped with pictures which add informative content to the article itself. Articles are an example of documents in which textual and visual information are related and concur to form the meaning of the work. Therefore, a classifier able to use the joint information of both text and images can build a good tool in constructing collection of articles related to a specific topic, leading to more accurate and efficient surveys (Fig. 1).

3   State of the Art

During this research we develop a model that had significant previous references. In particular we employed techniques used in Text and Image Mining.

3.1   Text Mining

Text mining is about inferring structure from sequences representing natural language text, and may be defined as the process of analyzing text to extract information that is useful for particular purposes, such as extraction of hierarchical phrase structures from text, identification of keyphrases in documents,

Fig. 1. Magazine information is contained in both text and image

locating proper names and quantities of interest in a piece of text, text categorization, word segmentation, acronym extraction, and structure recognition. There are several text mining task; among the most frequently used are *Supervised Learning* (or Classification), *Unsupervised Learning* (or Clustering) and *Probabilistic Latent Semantic Analysis*.

3.2 Image Mining

Image search is traditionally obtained mainly through relational database search of caption or keywords [2]; the automatic classification is often achieved using content-based image retrieval (CBIR) systems [3]; in this topic research focus is divided between low-level (or visual) feature extraction algorithms and high-level (or textual) feature extraction, the latter used to reduce the so called 'semantic gap' between the visual features and the richness of human semantics. We identify five major categories of the state-of-the-art techniques in narrowing down the 'semantic gap' [4]:

(1) using object ontology to define high-level concepts;
(2) using machine learning methods to associate low-level features with query concepts;
(3) using relevance feedback to learn users' intention;
(4) generating semantic template to support high-level image retrieval;
(5) fusing the evidences from HTML text and the visual content of images for WWW image retrieval.

There are low-level features extraction algorithms which make use of text mining techniques above explained. Features like color, texture, shape, spatial relationship among entities of an image and also their combination are used for the computation of multidimensional feature vector [5]; *color*, *texture* and *shape* are known as primitive features. *Color* and *texture* are used as a base for image detection and classification using a support vector machine (SVM), where color is represented using HSV (hue, saturation, value) color model because this model is closely related to human visual perception and texture is computed using the entropy of rectangular regions of the image in [6]. According to [7] *shapes* can be described textually using parts, junction line and disjunction line using XML language for writing descriptors of outline shapes. Thus, we can build a method for shape comparison and similarity measure which is computed directly from the textual descriptor.

There are high-level features extraction algorithms which make use of text close to the image. Text-based image retrieval (TBIR) first labels the images in the database according to text close to the image and then uses the database management system to perform image retrieval based on those labels [8], sometimes taking into account the extent to which a word can be perceived visually in images [9] exploiting a self-organizing neural network architecture [10] to extract labels or combining high and low level features [11], or using an ontology model that integrates both these information [12]. Other techniques make use of the 'bags of visual words' model, having images as documents, and categories as topics (for example, grass and houses) so that an image containing instances of several objects is modeled as a mixture of topics [13] or define a scene categorization method based on contextual visual words, and introducing contextual information from the coarser scale and neighborhood regions to the local region of interest based on unsupervised learning [14]. Images are classified through the surrounding text also with statistical methods, such as $TFIDF$ [15]: For a single piece of text, a word's *term frequency (TF)*, is the number of times that this word occurs in that text. For a category (such as all indoor images), the TF assigned to a word is the number of times that word occurs in all documents of that category. A word's *inverse document frequency (IDF)*, is the logarithm of the ratio of the total number of documents to the word's *document frequency (DF)*, which is the number of documents that contain that word; this measure remains constant independently of the particular document or category examined. There is also a wide documentation about the task of *Text Extraction from Images* in which images containing text are analyzed to automatically extract the included text [16], having to deskews the image, extracts text regions, segments text regions into text lines [17] or differentiating between region of text, graphics and background, using a neuro-fuzzy methodology [18], finally using local energy analysis for segmenting text [19] or Support Vector Machine [20]. The visual appearance of a document can be used as a feature to achieve clustering [21] where a statistical approach is used to characterize typical texture patterns in document images.

3.3   Text and Document Joint Information Retrieval

In [22] is proposed a method to learn the relationships between images and the surrounding text. For an image, a term in the description may relate to a portion of an image. If we divide an image into several smaller parts, called blocks or regions, we could link the terms to the corresponding parts. This is analogous to word alignment in a sentence aligned parallel corpus. Here the word alignment is replaced with the textual-term/visual-term alignment. If we treat the visual representation of an image as a language, the textual description and visual parts of an image are an aligned sentence. The correlations between the vocabularies of two languages can be learned from the aligned sentences. First, images are segmented into regions using a segmentation algorithm (in [22] "Blobword" is used).

Finally, in [23] we face a paper which deals with document similarity extracting both textual and visual information, which are called "mode" of a document, so that the authors refer to them as "multimodal" documents. Image similarity is computed using a "bag of visual word" representation (Fisher vector) in which the visual vocabulary is obtained with a Gaussian mixture model (GMM) which approximates the distribution of the low-level features in images. The similarity measure between two images is then defined as the L1-norm of the difference between the normalized Fisher Vectors of the two images. Text similarity is computed with text being pre-processed including tokenization, lemmatization, word decompounding and standard stop-word removal. The authors in [23] define a global similarity measure between two multi-modal objects d and d_q using, for instance, a linear combination:

$$sim_{glob}(d, d_q) = \lambda_1 sim_{TT}(d, d_q) + \lambda_2 sim_{TV}(d, d_q) + \lambda_3 sim_{VT}(d, d_q) + \lambda_4 sim_{VV}(d, d_q)$$

4   The Model

We found the model described in [23] as a valuable starting point for our model; we will use accordingly the term "mode" of a document for both text and image and we will use the "bag of word" representation for features set of both modes, but we define those contributes in a more general sense than in [23]; we showed in [24] that the model has solid experimental ground truth and leads to computable algorithms. Then we apply "noise" and we define our general model, which will be used later in the framework.

4.1   Latent Semantic Analisys

The problem of classification can be considered the problem of properly attach tags (class names) to documents. Suppose we have n *documents* $\mathcal{D} = \{d_1, d_2, ..., d_n\}$ and m *tags* $\mathcal{T} = \{t_1, t_2, ..., t_m\}$.

The links between these n *documents* and the m *tags* are denoted by a $n \times m$ matrix A. The elements $A_{i,j} \in \mathbb{R}^{n \times m}$ of this matrix represent the *weight* of link, e.g., $A_{i,j} = 1$ if jth tag is assigned to ith document, or $A_{i,j} = 0$ otherwise. The goal is to construct a set of feature vectors $\{X_1, X_2, \ldots, X_n\}$ in a latent semantic space $\mathbb{R}^k$ to represent these multimedia objects in the form $A = U\Sigma V^T$. Here, U and V are orthogonal matrices such that $UU^T = VV^T = I$, and the diagonal matrix Σ has the singular values as its diagonal elements. By retaining the largest k singular values in Σ and approximating others to be zero, we can create an approximated diagonal matrix Σ^k with fewer singular values.

This diagonal matrix is used to approximate Σ as $A \cong U\Sigma^k V^T$. Then the matrix $X = U\Sigma^k$, $X \in \mathbb{R}^{n \times k}$ yields a new feature representation, each row of which is a k-dimensional feature vector of a document, $X = [X_1, X_2, \ldots, X_n]^T$.

4.2   The Model for Multimodal Documents

We are considering both textual and visual contributions to the meaning of a document. Details of this model and its motivations can be found in [24].

Suppose we are given a matrix Q of content links, where $Q_{i,j}$ can represent the similarity measurement between the ith document and the jth document. Recalling the works in latest literature [23] we have that documents can be described as *multimodal* when made of both text and visual content, each defined as "mode"; a repository that contains a set of multimodal documents is then $D = \{d_1, d_2, \ldots, d_n\}$ (Fig. 2).

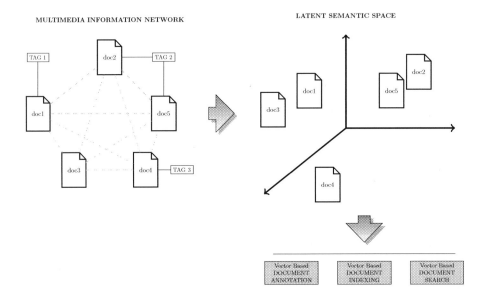

Fig. 2. Model for multimodal documents

We can define a global similarity measure between two multi-modal objects d and d_q using, for instance, a linear combination as in [23]:

$$sim_{glob}(d_i, d_j) = \lambda_1 sim_{TT}(d_i, d_j) + \lambda_2 sim_{TV}(d_i, d_j) + \lambda_3 sim_{VT}(d_i, d_j) + \lambda_4 sim_{VV}(d_i, d_j)$$

so we have that the elements in our matrix Q of similarity of multimodal content of documents can be

$$Q_{i,j} = \lambda_1 sim_{TT}(d_i, d_j) + \lambda_2 sim_{TV}(d_i, d_j) + \lambda_3 sim_{VT}(d_i, d_j) + \lambda_4 sim_{VV}(d_i, d_j) \ (1)$$

We assume that the documents with stronger links ought to be closer to each other in the latent semantic space. Based on this assumption, we introduce the quantity Ω to measure the smoothness of documents in the underlying latent space.

$$\Omega(X) = \frac{1}{2} \sum_{i,j=1}^{n} Q_{i,j} \|X_i - X_j\|_2^2 = \frac{1}{2} \sum_{i,j=1}^{n} Q_{i,j}(X_i - X_j)(X_i - X_j)^T \quad (2)$$

where, $\|M\|_2^2$ is the l_2 norm of matrix M, and X_i and X_j are the ith and jth row of X. It is easy to see that by minimizing the above regularization term, a pair of documents with larger $Q_{i,j}$ will have closer feature vectors X_i and X_j in the latent space (Fig. 3).

Given D as the diagonal matrix with its elements as the sum of each row of Q and $L = D - Q$, with some matrix operations we obtain

$$\Omega(X) = trace(X^T L X) \tag{3}$$

using the factorization $X = U\Sigma^k$, defining H as $H = U\Sigma_k V^T = XV^T$ and knowing that $VV^T = I$ we have

$$\Omega(X) = trace(H^T L H) \tag{4}$$

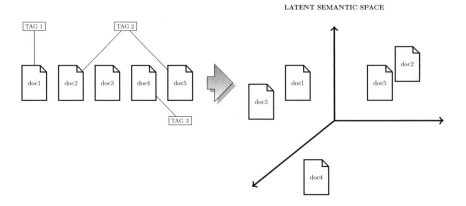

Fig. 3. Documents with stronger links will be closer

4.3   The Noisy Model

Due to the fact that we consider both textual and visual contribution to the meaning of a document, we have to consider the existence of noise in process so a noise term ε exist on the matrix Q such that $Q = H + \varepsilon$ where H is the matrix which denotes the noise-free tag links, after the noise ε has been removed. The goal is to make a correctly representative H of 'minimal rank'. The problem, as shown in [24] can be solved using the *nuclear norm* of a matrix M ($\|M\|_*$)

$$min\|Q - H\|_F + \gamma\|H\|_* \tag{5}$$

where $\|M\|_F$ is the squared summation of all elements in a matrix M (the Frobenius norm) and γ is a balancing parameter. Always in [24], a consistent solution to the problem is found to be

$$min\|Q - H\|_F + \gamma\|H\|_* = H_\gamma = U\Sigma_+^\gamma V^T \tag{6}$$

The difference with normal *Latent Semantic Indexing* is that it directly selects the largest k singular values of A where this Formulation subtracts something $\left(\frac{\gamma}{2}\right)$ from each singular value and thresholds them by 0. Suppose the resulting noise free matrix H is of rank k, then the Support Vector Machine of H has form as $H = U\Sigma_k V^T$ where Σ_k is a $k \times k$ diagonal matrix. Similar with *Latent Semantic Indexing*, the row vectors of $X = U\Sigma_k$ can be used as the latent vector representations of documents in latent space. It is also worth noting that minimizing the rank of H gives a smaller k so that the obtained latent vector space can have lower dimensionality, and then the storage and computation in this space could be more efficient.

4.4   The Global Model for Multimodal Documents

Considering both contribution to the model we can make use of both Eqs. 4 and 5 so our problem can be completely described as finding

$$min\|Q - H\|_F + \gamma\|H\|_* + \lambda trace(H^T LH) \tag{7}$$

Here λ is another balancing parameter. In contrast to Formulation (4), Formulation (7) does not have an closed-form solution. Fortunately, this problem can be solved by the *Proximal Gradient method* known from literature which uses a sequence of quadratic approximations of the objective function in order to derive the optimal solution.

5   The Framework

5.1   The Matrix Q of Similarity for Multimodal Content

We have considered in [24] both textual and visual contributions to the meaning of a document. We defined matrix Q_t of content links, where $Q_t(i,j)$ can represent the similarity measurement between the text of the ith document and the text of the jth document. We defined matrix Q_v of content links, where $Q_v(i,j)$ can represent the similarity measurement between the image of the ith document and the image of the jth document. Following PLSA approach as above specified we have, respectively for textual and visual mode

$$Q_t \cong U_t\Sigma_t^k V_t^T \qquad Q_v \cong U_v\Sigma_v^k V_v^T \tag{8}$$

We have similarly, the dual representation of the visual and textual part as:

$$S_T = U_t\Sigma_t^k \qquad S_V = U_v\Sigma_v^k \tag{9}$$

We denote the textual part of d_j by $S_T(d_j)$ and its visual part $S_V(d_j)$ which are the jth rows of matrix S_T and S_V. Recalling that in our model in Eq. 1

$$Q_{i,j} = \lambda_1 sim_{TT}(d_i, d_j) + \lambda_2 sim_{TV}(d_i, d_j) + \lambda_3 sim_{VT}(d_i, d_j) + \lambda_4 sim_{VV}(d_i, d_j)$$

and *assuming that the both text and image part of a document shall define the same meaning for the document in the meaning space* we will use these partial latent semantic representations to define the single components of the equation above

$$sim_{TV}(d_i, d_j) = \|S_T(d_i) - S_V(d_j)\|_F \tag{10}$$

$$sim_{VT}(d_i, d_j) = \|S_V(d_i) - S_T(d_j)\|_F \tag{11}$$

$$sim_{TT}(d_i, d_j) = \|S_T(d_i) - S_T(d_j)\|_F \tag{12}$$

$$sim_{VV}(d_i, d_j) = \|S_V(d_i) - S_V(d_j)\|_F \tag{13}$$

This model benefits from two major aspects: it is simple to understand and it is simple to implement, both because it involves only measure of distance in a vector space. The main assumption is that there is *one* meaning space so that features in text and features in images all refers to a set of concepts or meanings which are the same but are expressed with words and with images.

When these meanings are expressed with words the dimensionality of the feature space is different than the dimensionality of the feature space coming from the images, but using a dimensionality reduction algorithm we can reduce these different dimensions to be the same, so that we could compute a distance. Experiments performed with a knowledge base of almost a million newspaper articles shows [24] that model and framework holds.

5.2   Pertinence and Relevance

We have that $H = U_H \Sigma_H^k V_H^T$ and $X = U_H \Sigma_H^k$ will be the our full latent vector representations of documents in latent space.

Definition 1. *We define the Pertinence of the text in a document informally as how near is the meaning of the text to the meaning of the whole document. This leads to the definition of a distance which in our vector space is*

$$P_T(d_i) = \|X(d_i) - S_T(d_i)\|_F \tag{14}$$

This definition can be used for other modes of a document, so for the image the *Pertinence* of the image in a document is how near is the meaning of the image to the meaning of the whole document, so we have

$$P_V(d_i) = \|X(d_i) - S_V(d_i)\|_F \tag{15}$$

Definition 2. *We define the Relevance of the text in a document informally as how important is the meaning of the text in defining the meaning of the whole document. This also leads to the definition of a distance in our vector space as*

$$R_T(d_i) = \|\frac{S_T(d_i)}{X(d_i)}\|_F \tag{16}$$

Fig. 4. Example of pertinent and not relevant

and as above for the image the *Relevance* of the image in a document is how important is the meaning of the image in defining the meaning of the whole document, so we have

$$R_V(d_i) = \| \frac{S_V(d_i)}{X(d_i)} \|_F \tag{17}$$

These definitions are sound with the fact that a meaning might be pertinent but not relevant or not pertinent and not relevant to a document but the same meaning can not be relevant and not pertinent, which reflects everyday life experience. These definitions are simple to understand and to implement, mainly because they follow the model above which is both simple and computable. In Fig. 4 we have and example with the concept of 'panda': the image of the car *Fiat Panda* is not pertinent and not relevant while *WWF* contribute is pertinent but not relevant whereas the article of the family of bears is both pertinent and relevant in defining the meaning of the document.

6   Conclusions and Further Work

The first part of this work was dedicated to point out the overview of the research and the problems and choices we got through during the path of this research. Then we focused on the model we would use to determine different contribution to classification of the text and image information of a document; we've given the details of the definition of a meaning space using Latent Semantics for multimodal documents including consideration and modeling of the possible noise that shall be considered in this process and how to deal with it. Then we focused on the definitions of the distances in the meaning space and we've given the formal definition of *Persistence* and *Relevance* which will lead to a computable algorithm for our model, which will then enable a better understanding of semantic gap between the different parts, or "modes" of a document. This can be extended also to other kind of multimodal documents, such as videos, which have a spoken (i.e. text) and visual parts and the correlation with time can be explored as further research.

References

1. Ye, Q., Huang, Q., Gao, W., Zhao, D.: Fast and robust text detection in images and video frames. Image Vis. Comput. **23**(6), 565–576 (2005)
2. Kahn, C.: Dynamic inline images: context-sensitive retrieval and integration of images into web documents. J. Digit. Imaging **21**(3), 274–279 (2008)
3. Park, G., Baek, Y., Lee, H.-K.: Web image retrieval using majority-based ranking approach. Multimed. Tools Appl. **31**(2), 195–219 (2006)
4. Liu, Y., Zhang, D., Guojun, L., Ma, W.-Y.: A survey of content-based image retrieval with high-level semantics. Pattern Recogn. **40**(1), 262–282 (2007)
5. Schettini, R., Brambilla, C., Ciocca, G., Valsasna, A., De Ponti, M.: A hierarchical classification strategy for digital documents. Pattern Recogn. **35**(8), 1759–1769 (2002)
6. Seo, K.-K.: An application of one-class support vector machines in content-based image retrieval. Expert Syst. Appl. **33**(2), 491–498 (2007)
7. Larabi, S.: Textual description of shapes. J. Vis. Commun. Image Represent. **20**(8), 563–584 (2009)
8. Sagara, N., Sunayama, W., Yachida, M.: Image labeling using key sentences of HTML. Electron. Commun. Jpn. (Part III Fundam. Electron. Sci.) **89**(7), 31–41 (2006)
9. Fei, W., Han, Y.-H., Zhuang, Y.-T.: Multiple hypergraph clustering of web images by MiningWord2Image correlations. J. Comput. Sci. Technol. **25**(4), 750–760 (2010)
10. de Mello, R.F., Bueno, J.M., Senger, L.J., Yang, L.T.: Image indexing and retrieval using an ART-2A neural network architecture. Int. J. Imaging Syst. Technol. **18**(2–3), 202–208 (2008)
11. Shen, H.T., Zhou, X., Cui, B.: Indexing and integrating multiple features for www images. World Wide Web **9**(3), 343–364 (2006)
12. Wang, H., Liu, S., Chia, L.-T.: Image retrieval with a multi-modality ontology. Multimed. Syst. **13**(5), 379–390 (2008)
13. Bosch, A., Zisserman, A., Munoz, X.: Scene classification using a hybrid generative/discriminative approach. IEEE Trans. Pattern Anal. Mach. Intell. **30**(4), 712–727 (2008)
14. Qin, J., Yung, N.H.C.: Scene categorization via contextual visual words. Pattern Recogn. **43**(5), 1874–1888 (2010)
15. Sable, C.L., Hatzivassiloglou, V.: Text-based approaches for non-topical image categorization. Int. J. Digit. Libr. **3**(3), 261–275 (2000)
16. Zhao, M., Li, S., Kwok, J.: Text detection in images using sparse representation with discriminative dictionaries. Image Vis. Comput. **28**(12), 1590–1599 (2010)
17. Srihari, S.N., Tao, H., Geetha, S.: Machine-printed Japanese document recognition. Pattern Recogn. **30**(8), 1301–1313 (1997)

18. Caponetti, L., Castiello, C., Gorecki, P.: Document page segmentation using neuro-fuzzy approach. Appl. Soft Comput. J. **8**(1), 118–126 (2008)
19. Chan, W., Coghill, G.: Text analysis using local energy. Pattern Recogn. **34**(12), 2523–2532 (2001)
20. Chang, Y., Chen, D., Zhang, Y., Yang, J.: An image-based automatic arabic translation system. Pattern Recogn. **42**(9), 2127–2134 (2009)
21. Wen, D., Ding, X.-Q.: Visual similarity based document layout analysis. J. Comput. Sci. Technol. **21**(3), 459–465 (2006)
22. Lin, W.-C., Chang, Y.-C., Chen, H.-H.: Integrating textual and visual information for cross-language image retrieval: a trans-media dictionary approach. Inf. Process. Manage. **43**(2), 488–502 (2007)
23. Ah-Pine, J., Bressan, M., Clinchant, S., Csurka, G., Hoppenot, Y., Renders, J.-M.: Crossing textual and visual content in different application scenarios. Multimed. Tools Appl. **42**(1), 31–56 (2009)
24. Cristani, M., Tomazzoli, C.: A multimodal approach to exploit similarity in documents. In: Ali, M., Pan, J.-S., Chen, S.-M., Horng, M.-F. (eds.) IEA/AIE 2014, Part I. LNCS, vol. 8481, pp. 490–499. Springer, Heidelberg (2014)

Fuzzy-Syllogistic Systems: A Generic Model for Approximate Reasoning

Bora İ. Kumova[(⌂)]

Department of Computer Engineering, İzmir Institute of Technology,
Urla 35430, Turkey
borakumova@iyte.edu.tr

Abstract. The well known Aristotelian syllogistic system $\mathbb{S}$ consists of 256 moods. We have found earlier that 136 moods are distinct in terms of equal truth ratios that range in $\tau = [0,1]$. The truth ratio of a particular mood is calculated by relating the number of true and false syllogistic cases that the mood matches. The introduction of $(n - 1)$ fuzzy existential quantifiers, extends the system to fuzzy-syllogistic systems $^n\mathbb{S}$, $1 < n$, of which every fuzzy-syllogistic mood can be interpreted as a vague inference with a generic truth ratio, which is determined by its syllogistic structure. Here we introduce two new concepts, the relative truth ratio $^r\tau = [0,1]$ that is calculated from the cardinalities of the syllogistic cases of the mood and fuzzy-syllogistic ontology (FSO). We experimentally apply the fuzzy-syllogistic systems $^2\mathbb{S}$ and $^6\mathbb{S}$ as underlying logic of a FSO reasoner (FSR) and discuss sample cases for approximate reasoning.

Keywords: Syllogistic reasoning · Fuzzy logic · Approximate reasoning

1 Introduction

Multi-valued logics were initially introduced by Łukasiewicz [13], as an extension to propositional logic, which was then generalised by Zadeh using fuzzy sets [22] to fuzzy logic. After he had introduced approximate reasoning [23], he proposed fuzzy-syllogistic reasoning as a theory of common sense [24] and discussed fuzzy quantifiers again in the context of fuzzy logic [25]. However, these initial fuzzifications of syllogistic moods were experimentally applied to only a few true moods and did not cover all moods systematically, in terms of the four syllogistic figures. Only fuzzy quantifications based on interval arithmetic [6] comply to some extend with traditional figures [12]. The first systematic application of multi-valued logics on syllogisms were intermediate quantifiers and their reflection on the square of opposition [17]. However only set-theoretic representation of moods as syllogistic cases allow analysing the fuzzy-syllogistic systems $^n\mathbb{S}$ mathematically exactly, such as by calculating truth ratios of moods [8] and their algorithmic usage in fuzzy inferencing [9]. Here we present a sample application of $^n\mathbb{S}$ for fuzzy-syllogistic ontology reasoning.

Learning from scratch can be modelled probabilistically, as objects and their relationships need to be first synthesised from a statistically significant number of perceived instances of similar objects. This leads to probabilistic ontologies [4, 14, 18], in which attributes of objects may be synthesised also as objects.

© Springer International Publishing Switzerland 2016
H. Fujita et al. (Eds.): IEA/AIE 2016, LNAI 9799, pp. 169–181, 2016.
DOI: 10.1007/978-3-319-42007-3_15

There are more probabilistic ontology reasoners than fuzzy or possibilistic ones and most of them reason with probabilist ontologies [10]. Several ontology reasoners employ possibilistic logic and reason with fuzzy ontologies. The most popular reasoning logic being hyper-tableau, for instance in HermiT [15]. Other experimental reasoning logics are also interesting to analyse, such as fuzzy rough sets and Łukasiewicz logic [3] in FuzzyDL [1], Zadeh and Gödel fuzzy operators in DeLorean [2], Mamdani inference in HyFOM [21] or possibilistic logic in KAON [18]. Fuzzy-syllogistic reasoning (FSR) can be seen as a generalisation of both, fuzzy-logical and possibilistic reasoners.

A fuzzy-syllogistic ontology (FSO) extends the concept of ontology with the quantities that led to the ontological concepts. A FSO is usually generated probabilistically, but does not preserve any probabilities like probabilistic ontologies [14] or probabilistic logic networks [7] do. A FSO can be a fully connected and bidirectional graph.

Several generic reasoning logics are discussed in the literature, like probabilistic, non-monotonic or non-axiomatic reasoning [20]. Fuzzy-syllogistic reasoning in its basic form [26] is possibilistic, monotonic and axiomatic.

Syllogistic reasoning reduced to the proportional inference rules deduction, induction and abduction are employed in the Non-Axiomatic Reasoning System (NARS) [19]. Whereas FSR uses the original syllogistic moods and their fuzzified extensions [27].

There is one implementation mentioned in the literature that is close to the concept of syllogistic cases: Syllogistic Epistemic REAsoner (SEREA) implements poly-syllogisms and generalised quantifiers that are associated with combinations of distinct spaces, which are mapped onto some interval arithmetic. Reasoning is then performed with concrete quantities, determined with the interval arithmetic [16].

First the fuzzy-syllogistic systems $^n\mathbb{S}$ are discussed, thereafter fuzzy-syllogistic reasoning is introduced, followed by its sample application on a fuzzy-syllogistic ontology and the introduction of relative truth ratios $^r\tau$.

2   Fuzzy-Syllogistic Systems

The fuzzy-syllogistic systems $^n\mathbb{S}$, with $1 < n$ fuzzy quantifiers, extend the well known Aristotelian syllogisms with fuzzy-logical concepts, like truth ratio for every mood and fuzzy quantifiers or in general fuzzy sets. We discuss first the systems $^n\mathbb{S}$ and introduce them further below as the basic reasoning logic of FSR.

2.1   Aristotelian Syllogistic System $\mathbb{S}$

The Aristotelian syllogistic system $\mathbb{S}$ consists of inclusive existential quantifiers ψ, i.e. I includes A and O includes E as one possible case:

Universal affirmative $\psi = A$ **: All S are P :** $\{x|\ S - P = \varnothing \wedge x \in S \cap P\}$

Universal negative $\psi = E$ **: All S are not P :** $\{x|\ x \in S - P \wedge P \cap S = \varnothing\}$

Inclusive existential affirmative $\psi = I$ **: Some S are P :**

$A \cup \{x|\ x \in S \cap P \vee (x \in S \cap P \wedge P - S = \varnothing)\} \Leftrightarrow A \cup \{x|\ x \in S \cap P \vee P - S = \varnothing\}$

Inclusive existential negative $\psi = O$ **: Some S are not P :**

$E \cup \{x|\ x \in S - P \vee (x \in S - P \wedge P - S = \varnothing)\} \Leftrightarrow E \cup \{x|\ x \in S - P \vee P - S = \varnothing\}$

A categorical syllogism $\psi_1\psi_2\psi_3 F$ is an inference schema that concludes a quantified proposition $\Phi_3 = S\psi_3 P$ from the transitive relationship of two given quantified proportions $\Phi_1 = \{M\psi_1 P, P\psi_1 M\}$ and $\Phi_2 = \{S\psi_2 M, M\psi_2 S\}$:

$$\psi_1\psi_2\psi_3 F = (\Phi_1 = \{M\psi_1 P, P\psi_1 M\}, \Phi_2 = \{S\psi_2 M, M\psi_2 S\}, \Phi_3 = S\psi_3 P)$$

where $F = \{1, 2, 3, 4\}$ identifies the four possible combinations of Φ_1 with Φ_2, namely syllogistic figures. Every figure produces $4^3 = 64$ moods and the whole syllogistic system $\mathbb{S}$ has $4 \times 64 = 256$ moods.

2.2 Syllogistic-Cases

Syllogistic cases are an elementary concept of the fuzzy-syllogistic systems $^n\mathbb{S}$, for calculating truth ratios [8] of the moods algorithmically [9].

For three sets, 7 distinct spaces δ_i, $i = [1, 7]$ are possible, which can be easily identified in a Venn diagram (Table 1). There are in total $j = 96$ distinct combinations of the spaces $\Delta_j = \delta_1\delta_2\delta_3\delta_4\delta_5\delta_6\delta_7$, $j = [1,96]$ [27], which constitute the universal set of syllogistic moods. Within that universe, we determine for every mood true and false matching space combinations (Fig. 1).

Table 1. Binary coding of the 7 possible distinct spaces for three sets.

Sample Syllogistic Case $\Delta_j = \delta_1\delta_2\delta_3\delta_4\delta_5\delta_6\delta_7^*$; $\Delta_{95} = 1111110^\#$							
Venn Diagram	Space Diagram[+]						
	S-M-P	P-S-M	M-S-P	$M \cap S$-P	$M \cap P$-S	$S \cap P$-M	$M \cap S \cap P$
	δ_1	δ_2	δ_3	δ_4	δ_5	δ_6	δ_7

*Binary coding of all possible distinct space combinations Δ_j, $j=[1,96]$ that can be generated for three sets.
#$\delta_i=0$: space i is empty; $\delta_i=1$: space i is not empty; $i=[1,7]$.
+Every circle of a space diagram represents exactly one distinct sub-set of $M \cup P \cup S$.

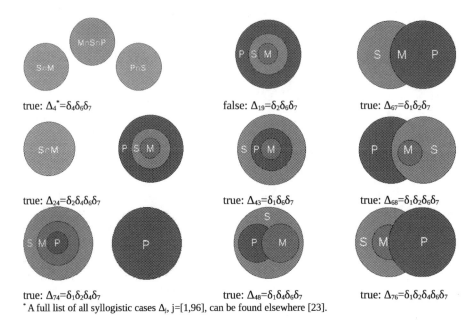

true: $\Delta_4^* = \delta_4\delta_6\delta_7$ false: $\Delta_{19} = \delta_2\delta_6\delta_7$ true: $\Delta_{67} = \delta_1\delta_2\delta_7$

true: $\Delta_{24} = \delta_2\delta_4\delta_6\delta_7$ true: $\Delta_{43} = \delta_1\delta_6\delta_7$ true: $\Delta_{68} = \delta_1\delta_2\delta_6\delta_7$

true: $\Delta_{74} = \delta_1\delta_2\delta_4\delta_7$ true: $\Delta_{48} = \delta_1\delta_4\delta_6\delta_7$ true: $\Delta_{76} = \delta_1\delta_2\delta_4\delta_6\delta_7$

* A full list of all syllogistic cases Δ_j, j=[1,96], can be found elsewhere [23].

Fig. 1. 9 syllogistic cases Δ_j of the mood $^{2/1}IA^1I4$ of the fuzzy-syllogistic systems $^2\mathbb{S}$.

2.3   Fuzzy-Syllogistic Moods

We extend the ancient binary truth classification of moods, to a fuzzy classification with truth values in [0,1]. For this purpose, first the above set-theoretical definitions of the quantifiers of a particular mood are compared against the set of all syllogistic cases Δ_j, j = [1,96], in order to identify true and false matching cases:

$$\textbf{True syllogistic cases}: \Lambda^t = {}_{j=1}\cup{}^{96}\Delta_j \in \left(\Phi_1^\Delta \cap \Phi_2^\Delta\right) \rightarrow \Delta_j \in \Phi_3^\Delta$$

$$\textbf{False syllogistic cases}: \Lambda^f = {}_{j=1}\cup{}^{96}\Delta_j \in \left(\Phi_1^\Delta \cap \Phi_2^\Delta\right) \rightarrow \Delta_j \notin \Phi_3^\Delta$$

where Λ^t and Λ^f is the set of all true and false matching cases of a particular mood, respectively (Fig. 1) and Φ^Δ is a proposition in terms of its true and false matching syllogistic cases. For instance, the two premises Φ_1 and Φ_2 of the mood IAI4 of the syllogistic system $\mathbb{S}$, match the 10 syllogistic cases $\Lambda^t = \{\Delta_4, \Delta_{19}, \Delta_{67}, \Delta_{24}, \Delta_{43}, \Delta_{46}, \Delta_{68}, \Delta_{74}, \Delta_{48}, \Delta_{76}\}$, which are all true for the conclusion Φ_3 as well. Thus the mood has no false cases $\Lambda^f = \varnothing$.

The truth ratio of a mood is then calculated by relating the amounts of the two sets Λ^t and Λ^f with each other. Consequently the truth ratio becomes either more true or more false $\tau \in \{\tau^f, \tau^t\}$:

More true $: \tau^t = 1 - |\Lambda^f|/(|\Lambda^t| + |\Lambda^f|) = [0.545, 1]$ **for** $|\Lambda^f| < |\Lambda^t|$

More false $: \tau^f = |\Lambda^t|/(|\Lambda^t| + |\Lambda^f|) = [0, 0.454]$ **for** $|\Lambda^t| < |\Lambda^f|$

where $|\Lambda^t|$ and $|\Lambda^f|$ are the numbers of true and false syllogistic cases, respectively. A fuzzy-syllogistic mood is then defined by assigning an Aristotelian mood $\psi_1\psi_2\psi_3 F$ the structurally fixed truth ratio τ:

$$\text{Fuzzy-syllogistic mood} : (\psi_1\psi_2\psi_3 F, \tau)$$

The truth ratio identifies the degree of truth of a particular mood, which we will associate further below in fuzzy-syllogistic reasoning with generic vagueness of inferencing with that mood.

The analysis of the Aristotelian syllogistic system $\mathbb{S}$ with these concepts reveals several interesting properties, like $\mathbb{S}$ has 136 distinct moods, 25 true moods $\tau = 1$, of which 11 are distinct, and 25 false moods $\tau = 0$, of which 11 are distinct, and that $\mathbb{S}$ is almost point-symmetric on syllogistic cases and truth ratios of the moods [11, 27].

2.4 Fuzzy-Syllogistic System $^2\mathbb{S}$

In the fuzzy-syllogistic system (FSS) $^2\mathbb{S}$, the universal cases A and E are excluded from the existential quantifiers I and O, respectively:

Exclusive existential affirmative : Some S are P $: \psi = I : \{x \mid x \in S \cap P \vee P - S = \varnothing\}$

Exclusive existential negative : Some S are not P $: \psi = O : \{x \mid x \in S - P \vee P - S = \varnothing\}$

For instance the mood IAI4 of $\mathbb{S}$, becomes $^{2/1}IA^1I4$ in $^2\mathbb{S}$. Because of the exclusive existential quantifier $^{2/1}I$, the case Δ_{46} is no more matched by the first premiss Φ_1 and the conclusion Φ_3 becomes false for the case Δ_{19} (Fig. 1).

The analysis of the FSS $^2\mathbb{S}$ shows that $^2\mathbb{S}$ has 70 distinct moods, 11 true moods $\tau = 1$, of which 5 are distinct, and 40 false moods $\tau = 0$, of which 13 are distinct, and that $^2\mathbb{S}$ is not point-symmetric [11, 27].

2.5 Fuzzy-Syllogistic System $^n\mathbb{S}$

By using $(n - 1)$ fuzzy-existential quantifiers, the total number of fuzzy-syllogistic moods of the FSS $^n\mathbb{S}$ increases to $(2n)^3$. For instance the mood IAI4 of $\mathbb{S}$ can be generalised in $^n\mathbb{S}$ to $^{n/k_1}IA^{k_2}I4$, k_1, $k_2 = [2, n]$. $^{n/k_1}IA^{k_2}I4$ consists of $(n - 1)^2$ fuzzy-moods, all having the very same 9 syllogistic cases (Fig. 1).

Same linguistic terms used in different FSSs do not necessarily equal each other. For instance, "most" may have different value ranges in the FSSs $^3\mathbb{S}$, $^4\mathbb{S}$, $^5\mathbb{S}$, $^6\mathbb{S}$ and therefore are in general not equal $^{3/2}I \neq {}^{4/3}I \neq {}^{5/3}I \neq {}^{6/4}I$, respectively. Likewise for "half" in ${}^{4}\mathbb{S}$ and ${}^{6}\mathbb{S}$ the quantifiers may not exactly equal ${}^{4/2}I \neq {}^{6/3}I$, respectively (Table 2).

Table 2. Value ranges of affirmative fuzzy quantifiers[#] of n fuzzy-syllogistic systems ${}^{n}\mathbb{S}$

Syllogistic System			Fuzzy Quantifier ψ*					
Aristotelian	$\mathbb{S}$	A=all	I=some (including A)					
Fuzzy	${}^{2}\mathbb{S}$	A=all	${}^{2/1}I$=some (excluding A)					
	${}^{3}\mathbb{S}$	A=all	${}^{3/2}I$=most			${}^{3/1}I$=several		
	${}^{4}\mathbb{S}$	A=all	${}^{4/3}I$=most		${}^{4/2}I$=half		${}^{4/1}I$=several	
	${}^{5}\mathbb{S}$	A=all	${}^{5/4}I$=many		${}^{5/3}I$=most	${}^{5/2}I$=several		${}^{5/1}I$=few
	${}^{6}\mathbb{S}$	A=all	${}^{6/5}I$=many	${}^{6/4}I$=most	${}^{6/3}I$=half	${}^{6/2}I$=several		${}^{6/1}I$=few
	${}^{n}\mathbb{S}$	A=all	${}^{n/n-1}I$		...			${}^{n/1}I$

[#] Negative quantifiers are arranged analogously.
* Column breadths are not drawn proportional to the overall value range or to other quantifiers.

3   Fuzzy-Syllogistic Ontology

A fuzzy-syllogistic ontology (FSO) is an extended semantic network, whose concept relationships consist of possibly multiple bi-directional fuzzy quantifiers ψ. A FSO may be obtained in two ways, by extending an existing crisp ontology or by extending a probabilistically learned ontology. Here we provide a definition for FSO and discuss learning FSOs, along with some distinguishing properties of FSOs.

3.1   Definition

A FSO consists of concepts, their relationships and assertions on them, whereby all quantities are given with fuzzy quantifications:

$$\text{Fuzzy-syllogistic ontology}: \text{FSO} = {}^{k}(\mathbf{C}, \mathbf{R}, \mathbf{A})$$

where C is the set of all concepts, R is the set of all binary relationships between the concepts, A is the set of all assertions and k is a particular FSS ${}^{k}\mathbb{S}$, k = [2, n]. A FSO is in compliance with a particular FSS ${}^{k}\mathbb{S}$, if all quantifiers ψ of the FSO comply with ${}^{k}\mathbb{S}$ (Fig. 2).

3.2   Learning Fuzzy Quantifiers

Although existing learning approaches generate ontological concepts and their relationships through probabilistic analysis of domain data [4, 14, 18], the quantities that

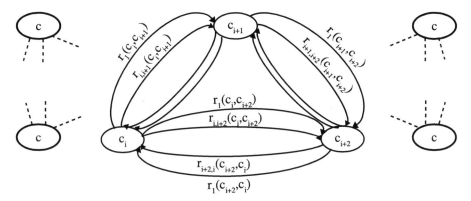

Fig. 2. Poly-bi-directional fuzzy-quantified binary concept relationships of a fuzzy-syllogistic ontology (FSO).

actually imply the concepts and relationships, are not preserved in probabilistic ontologies [10]. Exactly this data, ie the quantities of the samples, needs to be preserved in the learning phase of a FSO, as they imply the fuzzy quantifiers ψ of the fuzzy moods $(\psi_1\psi_2\psi_3 F, \tau)$.

For any two concepts c_i and c_{i+1} of a FSO, all binary relationships have to be stored with the FSO. Hence, all possible bi-directional binary relationships between all concepts of a FSO constitute a poly-bi-directional graph (Fig. 2):

Poly-uni-directional binary relationships :
$$\mathbf{R}_{ci,ci+1} = \{\mathbf{r}_1(\mathbf{c}_i, \mathbf{c}_{i+1}), \ldots, \mathbf{r}_{i,i+1}(\mathbf{c}_i, \mathbf{c}_{i+1})\}; \ 2 < i \leq o$$

Poly-bi-directional binary relationships : $\mathbf{R}_{ci,ci+1} \cup \mathbf{R}_{ci+1,ci}; \ 2 < i \leq o$

Since the data may imply for any three concept multiple bi-directional relationships, multiple ternary relationships may be generated for those concepts. Every ternary relationship of a FSO $= {}^k(C, R, A)$ may be interpreted as a fuzzy-syllogistic mood:

Ternary relationships/fuzzy-syllogistic moods :
$$(\psi_1\psi_2\psi_3 F, \tau) = (\psi_1 \in \{\mathbf{r}_{i,i+1}(\mathbf{c}_i, \mathbf{c}_{i+1}), \mathbf{r}_{i+1,i}(\mathbf{c}_{i+1}, \mathbf{c}_i)\}$$
$$\psi_2 \in \{\mathbf{r}_{i+1,i+2}(\mathbf{c}_{i+1}, \mathbf{c}_{i+2}), \mathbf{r}_{i+2,i+1}(\mathbf{c}_{i+2}, \mathbf{c}_{i+1})\} \psi_3 = \mathbf{r}_{i,i+2}(\mathbf{c}_i, \mathbf{c}_{i+2}), \tau)$$

Ones concepts and relationships of a FSO $= {}^k(C, R, A)$ are learned, the final step is to determine the most appropriate fuzzy quantifier system, i.e. FSS, ${}^k\mathbb{S}$. This is achieved by matching the average quantity distributions between all concepts of the FSO to the closest FSS ${}^k\mathbb{S}$.

Since fuzzy quantifiers are calculated by accumulating samples, new samples can continuously be learned by cumulatively updating the quantifiers. The most appropriate FSS ${}^k\mathbb{S}$ out of ${}^n\mathbb{S}$, $k = [2, n]$ can be re-calculating, if necessary. For instance, in case of significant amounts of quantifier updates, which change the quantity distributions.

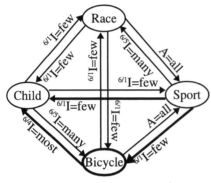

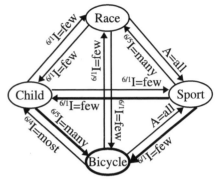

^{6}**S**: ($^{6/4}$I^{1}I^{1}I1, 40/47=0.851)
Φ_1: Most **bicycles** are good for children
Φ_2: Few sports are good for **bicycles**
Φ_3: Few **sports** are good for **children**

^{6}**S**: ($^{6/5}$I^{1}I^{1}I2, 40/48=0.833)
Φ_1: Many children have **bicycles**
Φ_2: Few sports are good for **bicycles**
Φ_3: Few **sports** are good for **children**

Fig. 3. Sample fuzzy-syllogistic ontology with affirmative relationships and the best matching fuzzy-syllogistic moods from the syllogistic Figs. 1 and 2.

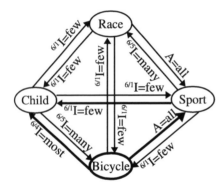

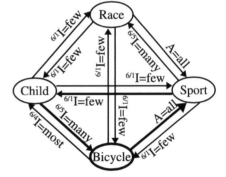

^{6}**S**: ($^{6/4}$IA^{1}I3, 6/6=1.0)
Φ_1: Most **bicycles** are good for children
Φ_2: All **bicycles** are good for sports
Φ_3: Few **sports** are good for **children**

^{6}**S**: ($^{6/5}$IA^{1}I4, 8/9=0.888)
Φ_1: Many children have **bicycles**
Φ_2: All **bicycles** are good for sports
Φ_3: Few **sports** are good for **children**

Fig. 4. Sample fuzzy-syllogistic ontology with affirmative relationships and the best matching fuzzy-syllogistic moods from the syllogistic Figs. 3 and 4.

3.3   Relative Truth Ratio

Relative truth ratios are calculated from the exact quantities of all syllogistic cases of a particular mood, rather than from just the amount of the cases:

$$\textbf{Relative true :}\, ^r\tau^t = \lambda^f/(\lambda^f + \lambda^t)\ \textbf{for}\ \lambda^f < \lambda^t$$
$$\textbf{Relative false :}\, ^r\tau^f = \lambda^t/(\lambda^t + \lambda^f)\ \textbf{for}\ \lambda^t < \lambda^f$$

where $\lambda^t =_{j=1} \sum^{|\Lambda t|} |\Delta_j^t|$ and $\lambda^f =_{j=1} \sum^{|\Lambda f|} |\Delta_j^f|$ is the total number of elements accumulated over all true and false syllogistic cases, respectively. Where $|\Lambda^t|$ and $|\Lambda^f|$ is the number of true and false cases of the mood, respectively. Accordingly, we re-define the truth of a fuzzy-syllogistic mood in terms of relative truth ratio $^r\tau$:

$$\text{Fuzzy-syllogistic mood with relative truth ratio}: (\psi_1\psi_2\psi_3F,^r\tau)$$

The structural truth ratio τ of a particular mood represents the generic vagueness of the mood and is constant, whereas the relative truth ratio $^r\tau$ adjusts τ by weighting every case of the mood with its actual quantity.

The concept of a relative truth ratio is a set-theoretic representation of a weighted logic, which is not new in the literature [5].

3.4 Mood Semantics

Truth ratios τ, $^r\tau$ of a mood provide solely structural evaluations for any propositions loaded on the mood. The semantics of a mood for given sample propositions can be determined with following two principle approaches:

- Top-down: Specifying from existing knowledge; possibilistic.
- Bottom-up: Learning from data sources; probabilistic.

In either approach the objective is to determine for reasonable concepts {M, P, S}, reasonable propositions $\Phi_1 \in \{M\psi_1P, P\psi_1M\}$, $\Phi_2 \in \{S\psi_2M, M\psi_2S\}$ with reasonable fuzzy quantifiers ψ_1, ψ_2, ψ_3 and to find the most reasonable concluding proposition $\Phi_3 = (S, P)$, which is the one with the highest truth ratio (Fig. 5).

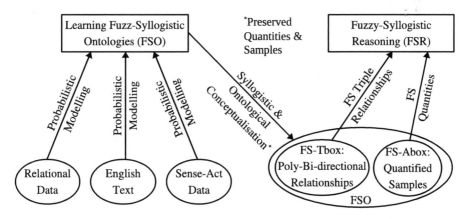

Fig. 5. Learning fuzzy syllogistic ontologies (FSO) from various sources and fuzzy-syllogistic reasoning with FSOs.

The specification of a sample FSO is sketched for the below discussed reasoning examples (Figs. 3 and 4). The quantifier distributions of the FSO represent personal perceptions from that domain. Therefore relative truth ratios cannot be calculated for the moods.

In bottom-up approaches, concepts and their relationships are synthesised from probability distributions that are calculated from source data. Such a process is typically associated with learning ontologies [10, 14]. In case of statistically sufficient numbers of samples found in the source data for ever concept and every concept relationship, we can assume that a learned FSO sufficiently represents the domain. and that uncertainties become increasingly more tolerable and eventually neglectable. Since source data is available in this approach, relative truth ratios $^r\tau$ can be calculated for the moods.

Thus mood semantics of FSOs are learned from the source data. The semantic of a sample syllogism is determined by the FSR, by searching for the fuzzy mood with the highest truth ratio τ.

4 Fuzzy-Syllogistic Reasoning

The fuzzy-syllogistic systems $\mathbb{S}$, $^2\mathbb{S}$ and $^6\mathbb{S}$ are currently implemented experimentally as the reasoning logic of the fuzzy-syllogistic reasoner (FSR), for reasoning over FSOs [26]. Our objective is to generalise the logic of the reasoner to $^n\mathbb{S}$ and to use it as a cognitive primitive for modelling other cognitive concepts within a cognitive architecture. We now sketch the algorithmic design of the FSR.

4.1 Sample Reasoning Processes

For any directly connected ternary concept relationship of the FSO, seven distinct relationships are possible (Table 1). FSR is concerned with identifying for any given concept $c \in C$, all possible ternary concept relationships $r \in R$, $r = \{M,P,S\}$, of a given FSO $= {}^k(C, R, A)$ and to reason with the most appropriate fuzzy-syllogistic moods of the FSS $^k\mathbb{S}$, $k = [2, n]$. Whereby associated assertions $a \in A$ may be used for exemplifying a particular reasoning.

For instance, for the concept $c = $ Bicycle, multiple ternary relationships $r = \{$Bicycle, Child, Sports$\}$ exist in the sample FSO $= {}^6(C, R, A)$ (Fig. 3). The reasoner iterates for all moods of the FSS $^6\mathbb{S}$ and matches the moods with the closest fuzzy-syllogistic quantities of relationships r. As best match the mood $^{6/k1}IA^{k2}I4$, $0 < k_1, k_2 < 6$ is found for this particular relationship r.

In the below example with $\mathbb{S}$, I in Φ_3 may include A and therefore is less true. Whereas in $^3\mathbb{S}$, $^{3/1}I$ in Φ_3 is still too general. The best matching quantifiers are found in $^6\mathbb{S}$ (Fig. 4).

4.2   Reasoning Algorithm

For a FSO = ^{k}(C, R, A) and any given directly connected three concepts c_1, c_2, $c_3 \in C$, the FSR searches all fuzzy moods for the highest matching truth ratio τ. That mood is determined as the most reasonable syllogism for the given ternary concepts.

The steps of the reasoning algorithm are as follows:

```
For given c₁, c₂, c₃∈C
For all fuzzy moods in ᵏS
Match the fuzzy mood having maximum truth ratio τ or ʳτ
```

4.3   Less Reasonable FSOs

For the sample domain (Figs. 3 and 4), the FSSs $\mathbb{S}$ and $^3\mathbb{S}$ are less reasonable than $^6\mathbb{S}$, because their quantifiers cover a too broad range or the coverage of the quantifiers of $^6\mathbb{S}$ are closer to the domain data.

In the Aristotelian system $\mathbb{S}$, the existential quantifier I may include A, thus in the below example, the proposition Φ_1 is wrong for the A case. Φ_1 is further wrong for all bicycles having larger wheel sizes than those sizes more suitable for children.

```
S: (IAI3, 10/10=1.0)
Φ₁: Some bicycles are good for children
Φ₂: All bicycles are good for sports
Φ₃: Some sports are good for children
```

The below example is based on $^3\mathbb{S}$. The quantifier $^{3/2}I$ = Most of the proposition Φ_1 expresses a closer quantity representation of the domain data than I = Some, since it excludes the above mentioned quantity ranges. Although, the quantity $^{3/2}I$ = Several in Φ_3, covers the domain quantity closer than I = Some, it matches weaker than $^{6/4}I$ = Few of $^6\mathbb{S}$. Here we assume that in reality only few sports are actually suitable for children, ie most spots become suitable for children only in simplified versions.

```
³S: (³ᐟ²IA¹I3, 6/6=1.0)
Φ₁: Most bicycles are good for children
Φ₂: All bicycles are good for sports
Φ₃: Several sports are good for children
```

In general, the closer a particular FSS $^k\mathbb{S}$ to the quantity distributions of the domain data, the more realistic conclusions can be expected from the FSR.

4.4   Cognitive Primitive

The FSS $^n\mathbb{S}$ serves as the underlying logic of the FSO and the FSR. The current implementation of the components comprises further learning FSOs, which is also based on the same logic (Fig. 4).

Learning FSOs is the emergent component of the symbolic FSR. Therefore, the depicted architecture is a hybrid cognitive architecture, in which the FSS serves as a cognitive primitive logic.

5 Conclusion

The FSS $^n\mathbb{S}$ was introduced as the fundamental logic of the FSR and its application to approximate reasoning on FSOs was shown on the sample FSSs $^2\mathbb{S}$ and $^6\mathbb{S}$. The relative truth ratio $^r\tau$ of a mood was introduced, which adapts the structural truth ratio τ of a mood to the cardinalities of the syllogistic cases of the mood. FSR with FSOs has been proposed as a generic possibilistic reasoning approach, since the underlying logic $^n\mathbb{S}$ is structurally generic. We have further proposed learning mood semantics statistically.

Currently we are generalising the FSS analysis tool, such that all reasonable systems $^n\mathbb{S}$ can be exploited algorithmically for realistic numbers of quantifiers [2, n]. That will enable us to implement a comprehensive FSR. Currently we are further developing applications for learning FSOs from English text sources and from robotic sense-act data relationships. Our ultimate goal is to employ the components learning FSOs and FSR.

References

1. Bobillo, F., Straccia, U.: fuzzyDL: an expressive fuzzy description logic reasoner. In: Fuzzy Systems (FUZZ-IEEE). IEEE Computer Society (2008)
2. Bobillo, F., Delgado, M., Gomez-Romero, J.: DeLorean: a reasoner for fuzzy OWL 1.1. In: Uncertainty Reasoning for the Semantic Web (URSW), LNAI. Springer (2008). http://ceur-ws.org/Vol-423
3. Bobillo, F., Straccia, U.: Reasoning with the finitely many-valued Łukasiewicz fuzzy description logic SROIQ. Inf. Sci. **181**, 758–778 (2011)
4. Cimiano, P., Völker, J.: Text2Onto: a framework for ontology learning and data-driven change discovery. In: Montoyo, A., Muñoz, R., Métais, E. (eds.) NLDB 2005. LNCS, vol. 3513, pp. 227–238. Springer, Heidelberg (2005)
5. Dubois, D., Godo, L., Prade, H.: Weighted logics for artificial intelligence – an introductory discussion. Int. J. Approx. Reason. **55**, 1819–1829 (2014)
6. Dubois, D., Godo, L., Prade, H.: On fuzzy syllogisms. Comput. Intell. **4**, 171–179 (1988)
7. Goertzel, B., Iklé, M., Goertzel, I.L.F., Heljakka, A.: Probabilistic Logic Networks: A Comprehensive Conceptual, Mathematical and Computational Framework for Uncertain Inference. Springer, New York (2008)
8. Kumova, Bİ., Çakir, H.: The fuzzy syllogistic system. In: Hernández Aguirre, A., Reyes García, C.A., Sidorov, G. (eds.) MICAI 2010, Part II. LNCS (LNAI), vol. 6438, pp. 418–427. Springer, Heidelberg (2010)
9. Kumova, Bİ., Çakır, H.: Algorithmic decision of syllogisms. In: García-Pedrajas, N., Herrera, F., Fyfe, C., Benítez, J.M., Ali, M. (eds.) IEA/AIE 2010, Part II. LNCS, vol. 6097, pp. 28–38. Springer, Heidelberg (2010)

10. Kumova, Bİ.: Generating ontologies from relational data with fuzzy-syllogistic reasoning. In: Kozielski, S., Mrozek, D., Kasprowski, P., Małysiak-Mrozek, B., Kostrzewa, D. (eds.) BDAS 2015. CCIS, vol. 521. Springer, Switzerland (2015)

11. Kumova, B.İ.: Properties of the syllogistic system. J. Rev. (2016)

12. Lui, Y., Kerre, E.E.: An overview of fuzzy quantifiers. (II). Reasoning and applications. Fuzzy Sets Syst. **95**, 135–146 (1998)

13. Łukasiewicz, J.: O logice trójwartościowej (translation from Polish: On three-valued logic). Ruch filozoficzny 5 (1920)

14. Lukasiewicz, T., Straccia, U.: Managing uncertainty and vagueness in description logics for the SemanticWeb. Web Semant.: Sci. Serv. Agents WWW **6**, 291–308 (2008). Elsevier

15. Motik, B., Shearer, R., Horrocks, I.: Hypertableau reasoning for description logics. J. Artif. Intell. Res. **36**, 165–228 (2009). AAAI

16. Pereira-Fariña, M., Vidal, J.C., Díaz-Hermida, F., Bugarín, A.: A fuzzy syllogistic reasoning schema for generalized quantifiers. Fuzzy Sets Syst. **234**, 79–96 (2014)

17. Peterson, P.: Intermediate Quantifiers: Logic, Linguistics, and Aristotelian Semantics. Ashgate, Farnham (2000)

18. Qi, G., Pan, J.Z., Ji, Q. A possibilistic extension of description logics. In: Description Logics (DL). Sun SITE Central Europe (CEUR) (2007)

19. Wang, P.: From inheritance relation to non-axiomatic logic. Int. J. Approx. Reason. **7**, 281–319 (1994)

20. Wang, P.: Reference classes and multiple inheritances. Int. J. Uncertain. Fuzziness Knowl. Based Syst. **3**(1), 79–91 (1995)

21. Yaguinuma, C.A., Magalhães Jr., W.C., Santos, M.T., Camargo, H.A., Reformat, M.: Combining fuzzy ontology reasoning and mamdani fuzzy inference system with HyFOM reasoner. In: Hammoudi, S., Cordeiro, J., Maciaszek, L.A., Filipe, J. (eds.) ICEIS 2013. LNBIP, vol. 190, pp. 174–189. Springer, Heidelberg (2014)

22. Zadeh, L.A.: Fuzzy Sets. Inf. Control **8**, 338–353 (1965)

23. Zadeh, L.A.: Fuzzy logic and approximate reasoning. Syntheses **30**, 407–428 (1975)

24. Zadeh, L.A.: A theory of commonsense knowledge. In: Skala, H.J., Trillas, E., Termini, S. (eds.) Aspects of Vagueness. Reidel, Dordrecht (1984)

25. Zadeh, L.A.: Syllogistic reasoning in fuzzy logic and its application to usuality and reasoning with dispositions. IEEE Trans. Syst. Man Cybern. **15**(6), 754–763 (1985)

26. Zarechnev, M., Kumova, Bİ.: Ontology-based fuzzy-syllogistic reasoning. In: Ali, M., Kwon, Y.S., Lee, C.-H., Kim, J., Kim, Y. (eds.) IEA/AIE 2015. LNCS, vol. 9101. Springer, Switzerland (2015)

27. Zarechnev, M., Kumova, B.İ.: Truth ratios of syllogistic moods. In: IEEE International Conference on Fuzzy Systems (FUZZ-IEEE). IEEE Xplore (2015)

Towards a Knowledge Based Environment for the Cognitive Understanding and Creation of Immersive Visualization of Expressive Human Movement Data

Christopher Bowman[1(✉)], Hamido Fujita[2], and Gavin Perin[1]

[1] Faculty of Design, Architecture and Building,
University of Technology Sydney, Ultimo, Australia
{chris.bowman,Gavin.Perin}@uts.edu.au
[2] Iwate Prefectural University, Takizawa, Iwate 020-0693, Japan
issam@iwate-pu.ac.jp

Abstract. How the classification of expressive movement data and use of such classifications may be used to inform the development of a knowledge-based system of movement analysis. In this paper, we present an example that adopts a practice-led design approach to the creation of immersive animation environments that features a method of indexing dance movement used by artists and designers to characterize this movement using "narrative grammar" based on pathemic, kinesthetic, cinematographic and aesthetic criteria referred to as a "movement index". The utility of the system described in this experiment was tested principally in art and design but this use can be generalized, to inform the development of cognitive medical applications for analyzing irregular movement patterns or the unusual motion behavior often characteristic of the ill or the elderly.

Keywords: Motion capture · Data visualization · Knowledge-based systems · Similarity · Taxonomy · Graphic user interface · Virtual reality

1 Introduction

Recent progress in motion capture and gesture recognition technologies have provided the foundation for the development of sophisticated motion analysis frameworks and expanded research in the classification and visualization of complex human movement. The ability to record, classify and visualize human movement or gesture is desirable when attempting to create consumer devices, therapeutic agents or, in this case, specialist tools for designers and artists. The current market for gesture recognition, interpretation and visualization devices includes automotive, biomechanical, entertainment, gaming and smartphone industries. Concurrent with these developments is the bespoke development of specialist applications used by artists and designers, in particular choreographers, architects and animators who wish to visualize expressive human movement and gesture data for analysis, visualization and exhibition. More specifically, we refer to the visualization of dance movement in the form of abstract animation for immersive and emotional exhibition experiences that in turn may have further application beyond the arts as a therapeutic device.

© Springer International Publishing Switzerland 2016
H. Fujita et al. (Eds.): IEA/AIE 2016, LNAI 9799, pp. 182–192, 2016.
DOI: 10.1007/978-3-319-42007-3_16

The envisaged application for this research is the further development of interoperability interaction with a knowledge-based system that enables analysis, selection and visualization of data from large labeled data sets of complex human movement pre-labeled and sampled based on their features extracted through supervised learning related to multiclass feature extraction. We expect that this classification technique can be used to sort the features of movement data so that we can identify the similarity calculation and understand these movements.

The authors have previously experimented with this [1] and describe in this paper how we are expanding this experiment by adding other modal feature related to motion for dancers movement. The indexing is based on a practice led research methodology used by the authors in the creation of immersive animation assets. Practice led research are approaches used by artists in order to gain new knowledge and the outcomes of that practice are the finished artwork or, in this case, animation assets [2].

The system features:

1. a library of expressive movement data created by dancers and choreographers (Movement Library);
2. an intuitive interface based on an index of movement that supports selection of all or part of the movement data (each "movement data") in response to the criteria (Movement Index);
3. capability to pre-visualize the movement data stored in the Movement Library to be viewed as an abstract animation assets (Pre-visualization);
4. an intuitive interface based on "narrative grammar" that supports rapid prototyping and visualization of assets in response to animation design pre-sets (color, style, speed, pattern) in Maya, Houdini, MotionBuilder or others (Visualization); and
5. capacity to capture information that is used to add new criteria to the Movement Index in order to develop a taxonomy of movement.

Central to the process is the development of a "movement index" that has its foundations in the language of movement that emerged in the early 20thC from the work of artists, dancers, filmmakers and photographers who sought to codify movement in a range of visually expressive and informative ways.[1] In the context of the process described in Fig. 1 the authors propose that such a Movement Index would inform the initial curation of data into sets of labeled movements that are meaningful for the artist.

The creation of a knowledge-based system that incorporates the Movement Index would create an efficient and effective means for the workflow and production of abstract animation assets for testing in 2D and 3D immersive cinematic environments. These environments range in scale from large cinematic facilities like the 360° Data Arena located at the University of Technology, Sydney (UTS) to small intimate devices like the new eye wear used for virtual and augmented reality immersion experiences (Oculus Rift, Microsoft Holo or Samsung's Gear VR).

---

[1] It is beyond the scope of this paper to expand on this large body of knowledge that includes key thinkers in the area of dance movement such as Rudolf Laban's *Labanotation* published in 1928 and works of the early pioneers of abstract animation such as Viking Eggling's *Symphonie Diagonale* (1924) or Osckar Fishinger's *Motion Painting No.1* (1947).

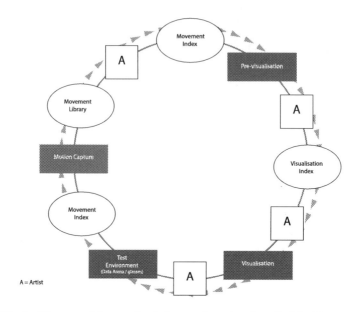

Fig. 1. Diagram of the practice-led research system described in this paper.

This is also an opportunity in the context of machine recognition using pre-sampled learning based features to inform the development of the rules and behaviors that may be applied to interoperability interaction with knowledge based system that classifies complex sets of labeled features related to human movement data. In so doing, we hope to better understand the physical and psychological schemas and condition of the user and the augmentation of senses activated by visual stimulae such as immersive animation that may be used for therapeutic purposes. This also, is built on the language of film to inform and/or to develop an annotated experience in the context of therapeutic purposes that can be used for further semantical analysis as stored features in the knowledge-based systems.

For example the Virtual Doctor System designed by the authors could recognize changes both facial and bodily movements of human users in a subjective context like that of a 'patient' and conclude from the knowledge-based system using Bayesian analysis the degree of change including stress, distress, spasms and other physical anomalies. The data used in this instance are image frames and sound data [1]. A question arises as to whether or not this system could be enhanced through the application of new ways of indexing movement used by artists to identify data movement traits that are sought after for creative development.

This paper describes the practice-led research undertaken to date and includes a case study that sets out the processes used by artists for the creation of immersive animation assets.[2] In so doing, the authors will illustrate areas within the design process

_____

[2] Practice led research are approaches used by artists in order to gain new knowledge and the outcomes of that practice are the finished artwork or, in this case, animation assets [2].

that may be enhanced by a knowledge based system and how such a system may be used in the creation of immersive animation for exhibition and potentially, therapeutic purposes. We further expect that the movement index developed by artists may have further application in the development of the Virtual Doctor System and other similar systems that rely on the analysis of human movement, expression and gesture.

2 The Practice Led Design Process

In 2013, a group of designers, architects, animators and programmers ("artists") in the Faculty of Design Building and Architecture at the University of Technology Sydney commenced development of a system for the classification and visualization of dance movement data recorded in the UTS Motion Data Laboratory. The data is sourced from an optical-passive motion capture system that uses approximately 40 retro-reflective markers placed on the dancer's body that are then tracked by infrared cameras and recorded as date comprising x, y and z coordinates. The data is then visualized in a custom application referred to as MxCAP.01.

3 MxCap.01

MxCAP.01 was developed by the authors to overcome a technical issue that existed at that time. Generally, to enable real-time visualization of movement captured in the movement capture systems. The omnitrack system used to capture movement data was only able to render movement data into a vector based human figure (stick figure or other). This kind of figurative rendering is generally not useful for artists and chore-ographers due to the figures inability to embody the expressive qualities of the movement while it is being performed. MxCAP.01 sought to overcome this by enabling near real-time visualization of human movement data in the form of abstract ribbons of light (see Fig. 2).

Fig. 2. MxCAP.01- developed by Daniel Scott. MxCAP.01 rendered in near real-time movement data as "ribbons" of movement. On the right, reflective markers placed on Anton Lock, dancer in the UTS Motion Data Laboratory.

"Central to this is work is the development of a new digital software interface, MxCAP.01, which is an interdisciplinary design visualisation tool that non-figuratively mediates and transcribes movement data. Working in combination with an optical motion capture suite and the three-dimensional software Maya™, MxCAP.01 captures gesture data and converts it into abstract representations so as to facilitate an understanding of movement by furnishing visualisations that populate 'movement' taxonomies. Built on the work of others, this extraction and abstraction interface processes and configures emotional and perceptive movement responses that may be used to inform performance design and foster better appreciation of three-dimensional spatial logic" [3].

MxCAP.01 enables artists and choreographers to make decisions 'in-situ' allowing immediate steps to be taken to adjust movements to better suit the needs of the project. It also presented an opportunity for the dancer to consider the visualized movement and, in particular, the "instrumentalisation" of the movement inherent in the technology. The importance of this pre-visualization cannot be overstated. It deepens the artist's understanding of the movement and enables an iterative process of experimentation and refinement to occur.

In addition, MxCAP.01 had a customized menu that enabled artists to select one or more of the 40 optical markers on the dancer's body to be viewed in isolation from each other. For example, the artist may wish to view hand movements only. This ability to select points is particularly important as it enables artists to explore the language of the movement data discussed below (Fig. 3).

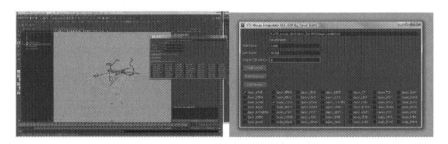

Fig. 3. The graphic user interface of MxCAP.01 illustrating the menu devised for the selection of optical markers and the replay of the selection in the form of lines.

Today, MxCAP.01 has been superseded by the features that exist in the UTS Data Arena, a state of the art 360-degree data visualization facility equipped with an omnitrack, optical-passive motion capture system consisting of 12 infrared movement cameras. The software configured for the Data Arena allows for real-time visualization of movement data in either figurative or abstract form using MotionBuilder as the rendering program.

The Data Arena terminal allows for the live visualization of data and extrapolation of selected data for further manipulation and rendering in Maya, Houdini or other 3D animation software programs in order to achieve finished animation assets as shown in Fig. 4 above (Fig. 5).

Fig. 4. The UTS Data Arena real-time data visualization facility and movable terminal display.

Fig. 5. Immersive animation created from the motion data in the Movement Library. The image on the right is a still from "Heroic" created by Daniel Scott and Rachel Walls. The image on the left is from "Estuary" by students in the Data Arena.

4 "Territory" and "Intruder" a Design Case Study

The case study described below describes a practice-led approach used by a group of artists to explore the language of movement in the creation of immersive animation assets. In 2014 creative work began on a 15 min animated film entitled "*Heroic*". Artists from tranSTURM collective and the University of Technology, Sydney were commissioned by Sydney Olympic Park Authority to create an animated film for the Pulman Façade outdoor screen space. The artists were asked to respond to the theme "nature". The artists chose to examine the site of the *Heroic*, a metal-hulled barge that rests in the tidal mudflat and mangrove of Homebush Bay in Sydney, Australia. This projection work explores the "ecology" of animated movement that reflects the bio-diversity of the site interpreted in part through dance movement. Working with one of Australia's leading choreographers Dean Walsh (dancer and aquatic environmentalist) the artists drew from Dean's own taxonomy of movement to choreograph aquatic animals resident to the site of the *Heroic*.

Underpinning, the language of movement developed by the artists for *Heroic* is Dean's taxonomy of movement based on his own "scoring" system that, in part, examines and interprets the weight and buoyancy of aquatic animals through the study of movement of marine species referred to as "modalities". For *Heroic* the artists elected to work with only those modalities that are inhabitants of Homebush Bay including siphonophores, arthropods and cephalopods. In his choreographic interpretation of these modalities Dean draws upon the distinct movement patterns for each modality. For example, siphonophores (such as jellyfish) are invertebrates that live in colonies and give the appearance of a single organism. Their collective movement is generally fluid, floating and buoyant. In sharp contrast, the bodies of arthropods such crustaceans (lobsters, crabs, shrimps) are encased in an exoskeleton so that most of the animal's movement is limited to its limbs that are typically both strong and flexible. Cephalopods (squid and octopus) have an outer muscular covering over its internal organs. However, in between these two is a cavity filled with water that enables the cephalopod to propel itself quickly through water. It tentacles (8–10 in number) are particularly tactile capable of great dexterity and capable of rapidly extending and contracting.

The creative design brief for *Heroic* asked artists to consider the narrative drivers of "Territory" and "Intruder" as key concepts used to inform the selection of movement data from the Movement Library for development as a set of immersive animation assets. These opposing concepts were chosen for their metaphorical, symbolic or literal meanings. For example, "territory" may be understood as a spatial constraint that can be expressed in political, physical or confinement terms.

In 2015 and 2016 this research was further extended through an Interdisciplinary Design Lab that consisted of students in the UTS School of Design participating in a project entitled "Estuary". *Estuary* built on the work undertaken in *Heroic* and extended the visual exploration of Dean's taxonomy of movement for the purpose of creating immersive animation experiences for testing in the UTS Data Arena. In the time available the students produced a range of animation assets that successfully demonstrated to an audience of artists, animators, and programmers the potential for these animation assets to be immersive, engaging and to create an environment that embodied the viewer in the cinematic experience as illustrated below (Fig. 6).

Fig. 6. Student work from the "Estuary" project demonstrated in the UTS Data Arena.

Working with movement data from the Movement Library the artists used an established language of movement to identify particular movement data sets that had the potential to exemplify aspects of "territory" or "intrusion". For artists working in abstract film and animation, the language of movement based on "Absolute Film" [4] presents an approach to abstraction that uses the characteristics of film (such as duration, motion, frame, and montage) to produce pure visual effect that are only achievable using cinematic processes. Using these characteristics the artists were able to make informed decisions regarding which data movement they wanted to explore as part of the design process.

In order to better understand how the language of movement and narrative drivers could be re-framed in a knowledge-based system the authors have turned to the work of A.J. Greimas and his analytical model known as "narrative grammar". "For Greimas", fundamental grammar is composed of a constitutional or taxonomic model and its narrativization via operations of syntax, logic, and ordering. This taxonomic model may equally be named "the elementary structure of signification" or the "semiotic square". The achronic or static constitutional model may be defined by as the juxta-position of pairs of contradictory terms" [5].

The authors have considered how Griemas's model of "narrative grammar" could be used to inform the indexation of movement data based on the concepts "Territory" and "Intruder". To achieve this the authors identified four grammatical layers (sub-categories) by which movement may be characterized. These layers are consid-ered critical to the creation of immersive abstract animation and include a pathemic layer (somatics, pathos or emotions), an aesthetic layer (beauty or poetics), a kines-thetic layer (the study of body motion) and a cinematographic layer (the principles of the moving image). Together these layers enable a highly sophisticated assessment and characterization of movement data as demonstrated in Table 1.

Table 1. The proposed narrative grammar of *Heroic*.

Grammatical layers	Territory	Intruder
Pathemic (Somatic)	Comforting, desirable, protective	Disturbing, overpowering
Aesthetic	Soft, fluid, repetitive	Rough, linear, erratic
Kinesthetic	Confident, harmonious, repetitive	Aggressive, clashing, penetrating
Cinematographic	Fluid montage, movement contained within the frame, long transitions	Hard edits, movement penetrates the frame, short transitions

In practice, the authors propose that narrative grammar could inform the devel-opment of an intuitive interface to allow artists to work through a series of iterative design processes making decisions about movement data along the way. Continued application of the principles of "narrative grammar" will increase the author's own

Fig. 7. Dean Walsh choreographer and dancer performing in the UTS Motion Data Laboratory showing the 3D animations created by Jason Benedek, Daniel Scott and Holger Deuter. The first four images (top) feature performance of siphonopore and territory by Dean Walsh and two animation assets of siphonophores. The bottom four images feature performance of cephalopod, an animation asset exploring intruder, performance of arthropod and animation asset exploring cephalopod.

language of movement [6], which in turn may be used to enhance the software and algorithms that are used in the system to categorize expressive human movement (Fig. 7).

For example, through the interface artists will be able to:

1. search the Movement Library for movement data that expresses the pathemic, aesthetic and kinesthetic qualities sought by the artist in response to a set of narrative drivers (usually opposites);
2. select and isolate parts of the movement data for visualization;
3. select the means by which the movement data will be visualized (using an palette as illustrated in Fig. 8 below);
4. assemble the visualized data on a time-line that enables the artist to manipulate cinematographic properties such as montage, timing, framing and editing as a movie; and
5. view the movie in a test environment such as the Data Arena, terminal display or VR/AR wearable devices.

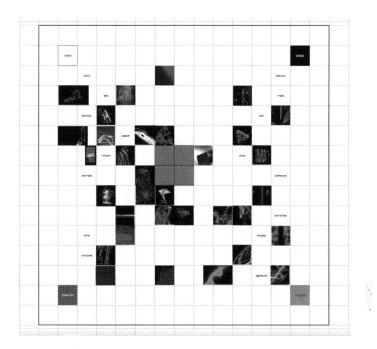

Fig. 8. Model of the aesthetic pallet of the proposed graphic user interface.

5   Conclusion

This case study is part of an on-going program of research conducted by the artists into an index of movement and development of a graphic user interface for rapid proto-typing of immersive animation assets. We expect that the benefit this system will have over existing scientific and dance based indexical systems, will include the ability to inform the selection of movement data from a large Movement Library that meet a set of sophisticated criteria posed by the artist.

We further propose that this process has potential to be adapted into labeled fea-tures categorized for interoperability on a supervised learning knowledge-based system that may be used by artists in the creation of immersive animations for exhibition purposes. However, we believe the utility of such a system and the resulting abstract animations extend beyond use by artists into other disciplines such as biomedical or psychological applications.

Firstly, the system itself presents opportunities for software designers to engage with artists to better understand how the creative process informs decisions made by artists. In this case, how artists use a set of criteria to select movement data from the Movement Library for use in animation. In particular, the ways in which artists con-sider, classify and curate movement in response to a design problem may be useful to establish a system that can develop a taxonomy of movement that enhances the ability of intelligent systems to recognize and understand expressive aspects of human movement.

Secondly, the abstract animations produced by artists using this system are designed to be highly immersive experiences when viewed in any format including large cinematic display systems or intimate display systems such as the current virtual/augmented reality technologies. The authors expect that such abstract animation may be effective if used as part of a therapeutic device or process. This is evidenced by the reported effectiveness of immersive sound, audio and object devices used in the treatment of anxiety disorders, dementia, Alzheimer's disease and autism [7, 8] which may in the future include cutting edge technologies like Microsoft Holo, Samsung's Gear VR or Oculus Rift.

We expect that a knowledge-based system developed from this practice led processes would create opportunities for other disciplines to engage with the movement data and create immersive visualizations for a range of purposes including biomedical, psychological and architectural uses.

Acknowledgements. We acknowledge the University of Technology Faculty of Design Building and Architecture, the Sydney Olympic Park Authority, Dean Walsh, the Hochschule Kaiserslautern College and University, Ben Simons and Darren Lee of the UTS Data Arena and the tranSTURM artist collective.

References

1. Fujita, H., Herrera-Viedma, E.: Virtual doctor systems for medical practices. In: Eren, H., Webster, J.G. (eds.) Examples of Integrating Technologies: Virtual Systems, Image Processing, BioKinematics, Measurements, and VLSI, pp. 435–463. CRC Press (2015)
2. Edmonds, E., Candy, L.: Explorations in Art and Technology. Springer, London (2002)
3. Perin, G.J., Bowman, C.P.: Thresholds: observations on motion capture. In: Proceedings of the 19th International Symposium on Electronic Art, ISEA2103, pp. 1–4. ISEA International, the Australian Network for Art and Technology and the University of Sydney. Sydney (2013)
4. Herozogenrath, W.: Light-play and kinetic theatre as parallels to absolute film. In: Film as Film, Formal Experiments in Film 1910–1975, p. 22. Hayward Gallery, London (1979)
5. Grabócz, M.: A.J. Greimas's narrative grammar and the analysis of sonata form. In: Intégral, vol. 12, pp. 1–23 (1998)
6. Bowman, C.P.: Explorations of visual representation: towards a language of movement, international symposium on electronic art. In: Funke, J., Riekeles, S., Broeckmann, A. (eds.) Proceedings of ISEA 2010 RUHR Hartware, pp. 171–173. MedienKunstVerein, Expanded Visual Spaces (2010)
7. Powers, B.M., Emmelkamp, M.G.: Virtual reality exposure therapy for anxiety disorders: a meta-analysis. J. Anxiety Disord. **22**(3), 561–569 (2008)
8. Gregg, L., Tarrier, N.: Virtual reality in mental health – a review of the literature. Soc. Psychiatry Psychiatr. Epidemiol. **42**(5), 343–354 (2007)

Bibliometric Tools for Discovering Information in Database

Enrique Herrera-Viedma[1(✉)], M. Angeles Martinez[3], and Manuel Herrera[2]

[1] Department of Computer Science and A.I.,
University of Granada, 18071 Granada, Spain
viedma@decsai.ugr.es
[2] UNIR Research, Universidad Internacional de la Rioja, Madrid, Spain
[3] Department of Social Work and Social Services,
University of Granada, Granada, Spain

Abstract. In bibliometrics, there are two main procedures to explore a research field: performance analysis and science mapping. Performance analysis aims at evaluating groups of scientific actors (countries, universities, departments, researchers) and the impact of their activity on the basis of bibliographic data. Science mapping aims at displaying the structural and dynamic aspects of scientific research, delimiting a research field, and quantifying and visualizing the detected sub-fields by means of co-word analysis or documents co-citation analysis. In this paper we present two bibliometric tools that we have developed in our research laboratory SECABA: (i) H-Classics to develop performance analysis based on Highly Cited Papers and (ii) SciMAT to develop science mapping guided by performance bibliometric indicators.

Keywords: Bibliometrics · H-index · Science mapping · Citations

1 Introduction

Bibliometrics is an academic science whose aim is to evaluate the research developed by any scientific community in any field. Concretely, Bibliometrics is a set of methods used to study or measure the research through the scientific publications stored or indexed in big databases. Many scientific communities use bibliometric methods to explore the impact of their field, the impact of a set of researchers, or the impact of a particular paper [46]. Bibliometrics contributes to the progress of science because it allows us to discover information in many different ways: allowing assessing progress made, identifying the most reliable sources of scientific publication, laying the academic foundation for the evaluation of new developments, identifying major scientific actors, developing bibliometric indices to assess academic output, and so on. Therefore, bibliometrics has become an essential tool in most scientific areas that aims to progress (medicine, mathematics, economics, computer science, physics, sociology, psychology, etc.) [29].

© Springer International Publishing Switzerland 2016
H. Fujita et al. (Eds.): IEA/AIE 2016, LNAI 9799, pp. 193–203, 2016.
DOI: 10.1007/978-3-319-42007-3_17

There are two main bibliometric methods for exploring a research field: performance analysis and science mapping [32, 46]. While performance analysis aims to evaluate the citation impact of the scientific production of different scientific actors, science mapping aims to display the conceptual, social or intellectual structure of scientific research and its evolution and dynamical aspects. The former is focused on the citation-based impact of the scientific production. For example, some popular performance metrics are the Journal Impact Factor [20] and Hirsch index [25]. The second approach is focused on the discovering of the conceptual structure of the scientific production by means of science maps. More particularly, it is focused on monitoring a scientific field and delimiting research areas to determine its conceptual structure and scientific evolution [9].

In this contribution, we present two bibliometric tools that our research laboratory SECABA has developed to analyze science: H-Classics and SciMAT. We analyze their performance, drawbacks, advantages and we show some examples of use.

The rest of the paper is set out as follows: In Sect. 2, we introduce the concept of H-Classics. Section 3, presents the software tool SciMAT. Finally Sect. 4 closures this work pointing our conclusions.

2   H-Classics

A *"citation classic"* or also called, *"classic article"* or *"literary classic"*, is a bibliometric concept introduced by Eugene Garfield [21] to designate those highly cited papers of a scientific discipline. Citation classics are as the *"gold bullion of science"* [40] and they could help us to discover potentially important information for the development of a discipline and understand the past, present and future of its scientific structure [30]. For example, it is possible to recognize the major advances in the discipline, to identify emergent or hot topics, to identify also the main intellectual markers of the research field, which could be journals or researchers or countries or research groups or institutions [21, 40]. Therefore, the development of studies on citation classics is becoming one of the most popular strategies to analyze scientific disciplines. Some examples are "Integrative & Complementary Medicine" [43], "Parkinson" [35], "Deviant Behavior" [41], "Epilepsy" [27], "Dentistry" [18], etc.,.

A common characteristic of studies on highly cited papers is to fix a selection criterion based on a threshold value following Garfield's recommendations [21, 22]. There exist two approaches to do it [30]: (i) Setting the threshold values on the citations received [27, 35]; or (ii) Setting the threshold values on the number of highly cited papers to be retrieved [18, 41, 43]. Both approaches do not take into account the citation patterns and the scientific evolution of the research areas. Therefore, the identification parameters are set according to the traditional recommendations provided by Garfield [21, 22], without considering a rigorous scientific argument and neither the circumstances of the research area when the study is done, which could introduce a bias in the choice of the highly cited papers. To overcome those problems we introduced in [30] the concept of

H-Classics based on the popular H-index [25], which provides us an unbiased and fair criterion to construct a systematic search procedure for citation classics for any field of research.

H-index was originally introduced by [25] to measure the scientific performance of a researcher through his/her publications:

"A scientist has index h if h of his or her N_p papers have at least h citations each, and the other $(N_p - h)$ papers have $\leq h$ citations each."

[4] points out that the H-index identifies the most productive core of an author's output in terms of the most cited papers. For this core, consisting of the first h papers, [38] introduced the term *Hirsch core* (*H*-core), which can be considered as a group of high-performance publications with respect to the scientist's career [28].

Then, if we have retrieved N articles and their respective citations subject scientific category of A, we could also calculate the H-index of category A as we calculate the H-index of a researcher [30], i.e.,

a paper P of scientific category A is considered an H-Classic of A if and only if P is inside of the H-core of A.

In such a way,

H-Classics of a research area A could be defined as the H-core of A that is composed of the H highly cited papers with more than H citations received.

Therefore, the identification process of highly cited papers of a research area through the concept of H-Classics could be carried out in the following steps [30]:

1. *Choosing the bibliographic database to locate the scientific production and citations.* For example, Google Scholar, Scopus and WoS could be used. The latter is used in this study.
2. *Set the research area under study.* This is done by identifying those core journals that are traditionally used to disseminate scientific advances made in the area and by using two types of papers, "article" and "review". Sometimes, it is necessary to configure more complex queries in order to delimit the research area [30].
3. *Compute the H-index of the research area.* The computation of H-index of research area is done by establishing a ranking of the papers according to their citations. If WoS is used to retrieve the scientific production, it provides us filtering tools to compute easily the H-index of the research area.
4. *Compute the H-core of the research area.* This step consists in recovering the H highly cited papers that are included in the H-core of the research area, i.e., H-index = #(H-Classics). Again, we should point out that using WoS this operation is easy to be carried out.

Some advantages of H-Classics to characterize the most influential papers of a research category are the following [30]:

1. It comprises in a single procedure the number of papers published in the field and the impact of those publications.
2. It provides a scientific and transparent criterion to identify the most influential papers in the scientific literature.
3. It is very simple to compute.
4. It is a criterion sensitive to the dimension of the research area.
5. And it is a criterion sensitive to the citation pattern of each research area.

3   Science Mapping Analysis: SciMAT

Science mapping or bibliometric mapping is a spatial representation of how disciplines, fields, specialities, and documents or authors are related to one another [39]. It has been widely used to show and uncover the hidden key elements (documents, authors, institutions, topics, etc.) in different research fields [8–10,14,19,26,31,34,36,37,44]. The general workflow in a science mapping analysis has a number of different steps [3,12]: data retrieval, preprocessing, network extraction, normalization, mapping, analysis and visualization. At the end of this process the analyst has to interpret and obtain conclusions from the results.

There are several possible on-line bibliographic databases to retrieve data. The most important ones are the ISI Web of Science, Scopus, and Google Scholar. These databases do not cover the scientific fields and journals in the same way and have their respective advantages and limitations, which are somewhat discipline dependent [2,17].

Science mapping analysis cannot be usually applied directly to the data retrieved from the bibliographic sources because they could contain errors. Thus, to improve the quality of the data, a preprocessing step is needed. Different preprocessing methods can be applied, among which it is worth mentioning those that detect duplicate and misspelled items, time slicing, and data reduction.

Once the data has been preprocessed a network is built using a unit of analysis, such as journals, documents, cited references, authors, author's affiliation, and descriptive terms or words [3]. Usually, words are the most common units and can be selected from the title, abstract, author's keywords or body of the documents, or from combinations of them. Furthermore, we can select the indexing terms provided by the bibliographic data sources (e.g. ISI Keywords Plus) as words to analyze. Several relations among the units of analysis can be established, such as co-occurrence, coupling or direct linkage. A co-occurrence relation is established between two units (authors, terms or references) when they appear together in a set of documents, i.e., when they co-occur throughout the corpus. A coupling relation is established between the documents when they have a set of units in common. A direct linkage establishes a relation between documents and references, particularly a citation relation. In addition, different aspects of a research field can be analyzed depending on the selected units of analysis. For example, when using words, a co-word analysis can be performed to obtain the conceptual structure of a discipline and the main topics researched in that knowledge field.

When the network of relationships between the selected units of analysis has been built, a normalization process is needed to correct the data for differences in the number of occurrences of units of analysis [45]. In Bibliometrics, the normalization process is carried out by using a similarity measure [45], such as *Salton's cosine, Jaccard's index,* or *equivalence index* [13].

Once the normalization process of the network is completed, different techniques could be applied to build science maps, such as principal component analysis or clustering algorithms [3].

Science mapping analysis methods allow the data to yield useful knowledge [12]. For example, a network analysis [15] makes it possible to perform a statistical study in order to show different measures of the relationship or overlapping of the different detected clusters, while a temporal or longitudinal analysis [23] aims to show the conceptual, intellectual or social evolution of a research field, discovering patterns, trends, seasonality and outliers.

Visualization techniques are used to represent both science maps and the results of the different analyses applied. The visualization technique employed is very important in order to achieve a good understanding and better interpretation of the output. For example, the networks resulting from the mapping step can be represented with thematic networks; the clusters detected in a network can be categorized using a strategic diagram; the evolution of detected clusters in successive time periods (temporal or longitudinal analysis) can be represented by means of thematic areas. Moreover, visualization can be improved using the results of a performance analysis, which allows us to add a third dimension to the visualized elements. For example, a strategic diagram could show spheres of a volume proportional to the citations achieved by the documents of cluster.

Finally, when the science mapping analysis is completed, the analysts have to interpret the results and maps using their experience and knowledge. In this interpretation step, the analyst examines the information to discover and extract useful knowledge that could be used to make decisions.

SciMAT (Science Mapping Analysis software Tool) was presented in [13] as a powerful science mapping software tool that integrates the majority of the advantages of available science mapping software tools [12]. It was designed according to the workflow shown above and also using the science mapping analysis approach presented in [11]. SciMAT can be freely downloaded, modified and redistributed according to the terms of the GPLv3 license. The executable file, user-guide and source code can be downloaded via the following website (http://sci2s.ugr.es/scimat).

In [11] it was defined a bibliometric approach that combines both, performance analysis tools and science mapping tools, to analyse a research field and to detect and visualize its conceptual subdomains (particular topics/themes or general thematic areas) and its thematic evolution. It is based on a co-word analysis [6] and the h-index [25].

The construction of maps using co-word analysis in a longitudinal framework provides information on the themes or topics of a research field and make it possible to analyse and track the evolution of a research field throughout consecutive periods of time [23]. Additionally, the h-index is used to measure the impact of the different identified themes and thematic areas.

This approach establishes four stages to analyse a research field in a longitudinal framework:

1. *Detection of the research themes.* This phase summarizes the first five steps of the workflow of science mapping analysis. In each period studied, the corresponding research themes are detected by applying a co-word analysis [6] to raw data of all the published documents in the research field, followed by a clustering of keywords to topics/themes using the simple centres algorithm [16]. Formally, the methodological foundation of co-word analysis is based on the idea that the co-occurrence of keywords describes the content of the documents in a corpus [7]. These co-occurrences of keywords can be used to build co-word networks [6] and these networks can be associated with research themes using clustering tools. The co-occurrence frequency of two keywords is extracted from the corpus by counting the number of documents in which the two keywords appear together. Once the co-word network is built, each arc/edge will have in its weight the co-occurrence value of the linked terms. Next, the weight of each edge is transformed in order to normalize it (extract the similarity relations between terms) using their keyword and co-occurrence frequencies [45]. The similarity between the keywords is assessed using the equivalence index [7]: $e_{ij} = c_{ij}^2/c_i c_j$, where c_{ij} is the number of documents in which two keywords i and j co-occur and c_i and c_j represent the number of documents in which each one appears. Note that when two keywords always appear together, the equivalence index equals unity; but it is zero when they are never associated. At the end of this phase, the keywords are clustered into topics/themes by the simple centre algorithm [16]. The clustering process locates keyword networks that are strongly linked to each other and that correspond to centres of interest or to research problems that are the subject of significant interest among researchers.

2. *Visualizing research themes and thematic network.* In this phase the detected themes are visualized by means of two different visualization instruments: strategic diagram [5,24,33,47] and thematic network. Each theme can be characterized by two measures [7]: *centrality* and *density*. Centrality measures the degree of interaction of a network with other networks and can be defined as $c = 10 * \sum e_{kh}$, with k being a keyword that belongs to the theme and h a keyword that belongs to other themes. Centrality measures the strength of external ties to other themes. This value can be taken as the measure of the importance of a theme in the development of the entire research field analysed. The density measures the internal strength of the network and can be defined as $d = 100(\sum e_{ij}/w)$, where i and j are keywords belonging to the theme and w the number of keywords in the theme. Density measures the strength of internal ties among all the keywords that describe the research theme. This value can be understood as a measure of the theme's development. Once the centrality and density rankings have been calculated, the themes can be laid out in a strategic diagram. Given both measurements, a research field can be

visualised as a set of research themes, mapped in a two-dimensional strategic diagram (Fig. 1) and classified into four groups:

(a) Themes in the upper-right quadrant are both well developed and important for the structuring of a research field. They are known as the *motor-themes* of the speciality, given that they present strong centrality and high density.

(b) Themes in the upper-left quadrant have well-developed internal ties but unimportant external ties and as they are of only marginal importance for the field. These themes are very *specialized and peripheral*.

(c) Themes in the lower-left quadrant are both weakly developed and marginal. The themes in this quadrant have low density and low centrality and mainly represent either *emerging or disappearing* themes.

(d) Themes in the lower-right quadrant are important for a research field but are not developed. This quadrant contains *transversal and general*, basic themes.

Note that the addition of a third dimension can enrich the strategic diagrams as this will allow the representation of further informative data [11]. For example, the themes could be represented using spheres with volume proportional to another alternative measure, such as the number of documents associated with the theme or the total number of citations achieved.

3. *Discovery of thematic areas.* In this phase, the evolution of the research themes over a set of periods of time is first detected and then analysed to identify the main general areas of evolution in the research filed, their origins and their interrelationships. Their evolution over the whole period is then measured as the overlapping of clusters from two consecutive periods. For

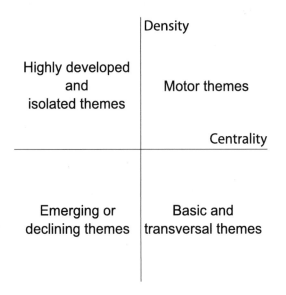

Fig. 1. The strategic diagram.

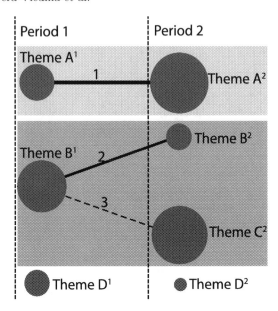

Fig. 2. Example of thematic evolution. (Color figure online)

this purpose, the inclusion index [42] is used to detect conceptual nexuses between research themes in different periods and, in this way, to identify the thematic areas in a research field. A thematic area is defined as a set of themes that have evolved over several periods of time. It is worth noting that interrelationships between research themes could indicate that a particular research theme belongs to a unique thematic area or to more than one thematic area. It could also be that a particular research theme cannot be associated with any of the thematic areas identified and, therefore, it could be interpreted as the origin of a new thematic area in the research field. For example, Fig. 2 shows a bibliometric map of thematic evolution over two time periods. The solid lines (lines 1 and 2) mean that the linked themes share the same name: either the themes are labelled with the same keywords, or the label of one theme is part of the other theme (name of theme $\in$ {thematic nexuses}). A dotted line (line 3) means that the themes share elements that are not the names of the themes (name of theme $\notin$ {thematic nexuses}). The thickness of the lines is proportional to the inclusion index and the volume of the spheres is proportional to the number of published documents associated with each theme. Hence, two different thematic areas can be observed, shaded in different colours, while $Theme D^1$ is discontinued, and $Theme D^2$ is considered a new theme. As each theme is associated with a set of documents, each thematic area could also have an associated collection of documents, obtained by combining the documents associated with its set of themes.

4. *Performance analysis*. In this phase, the relative contribution of research themes and thematic areas to the whole research field is measured (quantitatively and qualitatively) and used to establish the most prominent, most productive and

highest-impact subfields. This performance analysis is developed as a complement to the analysis step of the science mapping workflow. Some bibliometric indicators used: the number of published documents, the number of citations, and, or the different types of h-index [1, 25, 30].

4 Conclusions

In this paper, we have presented two important bibliometric tools that allow us to analyze the research output and also discover important information to evaluate the research developed in a scientific field: H-Classics and SciMAT. Former provides performance analysis of scientific field by means of the identification of the highly cited papers published in databases. Later is a software tool that provides content analysis of papers published.

Nowadays, many institutions related with research (centers, universities, department, hospitals, etc.) need such bibliometric tools to support their research policies. The publications in research are growing exponentially and we have to work to develop bibliometric tools that allow us to deal with this new big data framework that is appearing in science.

Acknowledgments. The authors would like to acknowledge FEDER financial support from the Project TIN2013-40658-P, and also the financial support from the Andalusian Excellence Project TIC-5991.

References

1. Alonso, S., Cabrerizo, F.J., Herrera-Viedma, E., Herrera, F.: H-index: a review focused in its variants, computation and standardization for different scientific fields. J. Inf. **3**(4), 273–289 (2009)
2. Bar-Ilan, J.: Citations to the "introduction to informetrics" indexed by WOS, Scopus and Google Scholar. Scientometrics **82**(3), 495–506 (2010)
3. Börner, K., Chen, C., Boyack, K.W.: Visualizing knowledge domains. Ann. Rev. Inf. Sci. Technol. **37**, 179–255 (2003)
4. Burrell, Q.L.: On the h-index, the size of the Hirsch core and Jin's A-index. J. Inf. **1**(2), 170–177 (2007)
5. Cahlik, T.: Comparison of the maps of science. Scientometrics **49**(3), 373–387 (2000)
6. Callon, M., Courtial, J.P., Turner, W.A., Bauin, S.: From translations to problematic networks: an introduction to co-word analysis. Soc. Sci. Inf. **22**(2), 191–235 (1983)
7. Callon, M., Courtial, J.P., Laville, F.: Co-word analysis as a tool for describing the network of interactions between basic and technological research - the case of polymer chemistry. Scientometrics **22**(1), 155–205 (1991)
8. Cartes-Velásquez, R., Manterola-Delgado, C.: Bibliometric analysis of articles published in ISI dental journals, 2007–2011. Scientometrics **98**(3), 2223–2233 (2014)
9. Cobo, M.J., Chiclana, F., Collop, A., de Oña, J., Herrera-Viedma, E.: A bibliometric analysis of the intelligent transportation systems research based on science mapping. IEEE Trans. Intell. Transp. Syst. **11**(2), 901–908 (2014)

10. Cobo, M.J., López-Herrera, A.G., Herrera, F., Herrera-Viedma, E.: A note on the ITS topic evolution in the period 2000–2009 at T-ITS. IEEE Trans. Intell. Transp. Syst. **13**(1), 413–420 (2012)

11. Cobo, M.J., López-Herrera, A.G., Herrera-Viedma, E., Herrera, F.: An approach for detecting, quantifying, and visualizing the evolution of a research field: a practical application to the fuzzy sets theory field. J. Inf. **5**(1), 146–166 (2011)

12. Cobo, M.J., López-Herrera, A.G., Herrera-Viedma, E., Herrera, F.: Science mapping software tools: review, analysis and cooperative study among tools. J. Am. Soc. Inform. Sci. Technol. **62**(7), 1382–1402 (2011)

13. Cobo, M.J., López-Herrera, A.G., Herrera-Viedma, E., Herrera, F.: Scimat: a new science mapping analysis software tool. J. Am. Soc. Inform. Sci. Technol. **63**(8), 1609–1630 (2012)

14. Cobo, M.J., Mártinez, M.A., Gutiérrez-Salcedo, M., Fujita, E., Herrera-Viedma, H.: 25 years at knowledge-based systems: a bibliometric analysis. Knowl.-Based Syst. **80**, 3–13 (2015)

15. Cook, D.J., Holder, L.B.: Mining Graph Data. Wiley-Interscience, Hoboken (2006)

16. Coulter, N., Monarch, I., Konda, S.: Software engineering as seen through its research literature: a study in co-word analysis. J. Am. Soc. Inform. Sci. Technol. **49**(13), 1206–1223 (1998)

17. Falagas, M.E., Pitsouni, E.I., Malietzis, G.A., Pappas, G.: Comparison of PubMed, Scopus, Web of Science, and Google Scholar: strengths and weaknesses. FASEB J. **22**(2), 338–342 (2008)

18. Feijoo, J.F., Limeres, J., Fernández-Varela, M., Ramos, I., Diz, P.: The 100 most cited articles in dentistry. Clin. Oral Investig. **18**(3), 699–706 (2013)

19. Gao-Yong, L., Ji-Ming, H., Hui-Ling, W.: A co-word analysis of digital library field in China. Scientometrics **91**(1), 203–217 (2012)

20. Garfield, E.: Citation analysis as a tool in journal evaluation. Science **178**(60), 417–479 (1972)

21. Garfield, E.: Introducing citation classics the human side of scientific reports. Curr. Comments **1**, 5–7 (1977)

22. Garfield, E.: 100 citation classics from the journal of the American medical association. J. Am. Med. Assoc. **257**, 52–59 (1987)

23. Garfield, E.: Scientography: mapping the tracks of science. Curr. Contents Soc. Behav. Sci. **7**(45), 5–10 (1994)

24. He, Q.: Knowledge discovery through co-word analysis. Libr. Trends **48**(1), 133–159 (1999)

25. Hirsch, J.E.: An index to quantify an individual's scientific research output. Proc. Nat. Acad. Sci. **102**, 16569–16572 (2005)

26. Huang, M.-H., Chang, C.-P.: Detecting research fronts in oled field using bibliographic coupling with sliding window. Scientometrics **98**(3), 1721 (2014)

27. Ibrahim, G.M., Snead, O.C., Rutka, J.T., Lozano, A.M.: The most cited works in epilepsy: trends in the "citation classics". Epilepsia **53**(5), 765–770 (2012)

28. Jin, B.H., Liang, L.M., Rousseau, R., Egghe, L.: The R- and AR-indices: complementing the h-index. Chin. Sci. Bull. **52**(6), 855–863 (2007)

29. Martínez, M.A., Cobo, M.J., Herrera, M., Herrera-Viedma, E.: Analyzing the scientific evolution of social work using science mapping. Res. Soc. Work Pract. **5**(2), 257–277 (2015)

30. Martínez, M.A., Herrera, M., López-Gijón, J., Herrera-Viedma, E.: H-classics: characterizing the concept of citation classics through h-index. Scientometrics **98**, 1971–1983 (2014)

31. Murgado-Armenteros, E., Gutiérrez-Salcedo, M., Torres-Ruiz, F.J., Cobo, M.J.: Analysing the conceptual evolution of qualitative marketing research through science mapping analysis. Scientometrics **102**(1), 519–557 (2014)

32. Noyons, E.C.M., Moed, H.F., Luwel, M.: Combining mapping and citation analysis for evaluative bibliometric purposes: a bibliometric study. J. Am. Soc. Inform. Sci. **50**(2), 115–131 (1999)

33. Ozel, B.: Individual cognitive structures and collaboration patterns in academia. Scientometrics **91**(2), 539–555 (2012)

34. Peters, H.P.F., van Raan, A.F.J.: Co-word-based science maps of chemical engineering. part i: representations by direct multidimensional scaling. Res. Policy **22**(1), 23–45 (1993)

35. Ponce, F.A., Lozano, A.M.: The most cited works in Parkinson's disease. Mov. Disord. **26**(3), 380–390 (2011)

36. Porter, A.L., Youtie, J.: How interdisciplinary is nanotechnology? J. Nanopart. Res. **11**(5), 1023–1041 (2009)

37. Rodriguez-Ledesma, A., Cobo, M.J., Lopez-Pujalte, C., Herrera-Viedma, E.: An overview of animal science research 1945–2011 through science mapping analysis. J. Anim. Breed. Genet. **132**(6), 475–497 (2014)

38. Rousseau, R.: New developments related to the Hirsch index. Sci. Focus **1**(4), 23–25 (2006). (in Chinese) An English translation can be found at: http://eprints.rclis.org/7616/

39. Small, H.: Visualizing science by citation mapping. J. Am. Soc. Inform. Sci. **50**(9), 799–813 (1999)

40. Smith, D.R.: Ten citation classics from the New Zealand medical journal. N. Z. Med. J. **120**, 2871–2875 (2007)

41. Stack, S.: Citation classics in deviant behavior: a research note. Deviant Behav. **34**(2), 85–96 (2013)

42. Sternitzke, C., Bergmann, I.: Similarity measures for document mapping: a comparative study on the level of an individual scientist. Scientometrics **78**(1), 113–130 (2009)

43. Tam, W.W., Wong, E.L., Wong, F.C., Cheung, A.W.L.: Citation classics in the integrative and complementary medicine literature: 50 frequently cited articles. Eur. J. Integr. Med. **4**, e77–e83 (2012)

44. Tang, L., Shapira, P.: China-US scientific collaboration in nanotechnology: patterns and dynamics. Scientometrics **88**(1), 1–16 (2011)

45. van Eck, N.J., Waltman, L.: How to normalize cooccurrence data? an analysis of some well-known similarity measures. J. Am. Soc. Inform. Sci. Technol. **60**(8), 1635–1651 (2009)

46. van Raan, A.F.J.: Measuring science. In: Moed, H.F., Glanzel, W., Schmoch, U. (eds.) Handbook of Quantitative Science and Technology Research, pp. 19–50. Springer, Netherlands (2005)

47. Zong, Q.-J., Shen, H.-Z., Yuan, Q.-J., Xiao-Wei, H., Hou, Z.-P., Deng, S.-G.: Doctoral dissertations of library and information science in china: a co-word analysis. Scientometrics **94**(2), 781–799 (2013)

Natural Language Processing
and Sentiment Analysis

The Statistical Approach to Biological Event Extraction Using Markov's Method

Wen-Juan Hou[(✉)] and Bamfa Ceesay

Department of Computer Science and Information Engineering,
National Taiwan Normal University,
No. 88, Section 4, Ting Chou Road, Taipei 116, Taiwan, ROC
{emilyhou, bmfceesay}@csie.ntnu.edu.tw

Abstract. Gene Regulation Network (GRN) is a graphical representation of the relationship for a collection of regulators that interact with each other and with other substances in the cell to govern the gene expression levels of mRNA and proteins. In this study, we examine the extraction of GRN from literatures using a statistical method. Markovian logic has been used in the natural language processing domain extensively such as in the field of speech recognition. This paper presents an event extraction approach using the Markov's method and the logical predicates. An event extraction task is modeled into a Markov's model using the logical predicates and a set of weighted first ordered formulae that defines a distribution of events over a set of ground atoms of the predicates that is specified using the training and development data. The experimental results has a state-of-the-art F-score comparable 2013 BioNLP shared task and gets 81 % precision in forming the gene regulation network. It shows we have a good performance in solving this problem.

Keywords: Biological entity · Biological event · First order logic · Gene regulation network · Bayesian network · Markov's model

1 Introduction

Automatically mining knowledge from the dramatically growing biomedical texts is a hot research topic and direction of bioinformatics. To extract the relevant events from given texts, several information extraction systems have been proposed with the natural language processing (NLP) techniques. However, this poses a number of research challenges due to variations and complex properties of texts. Automatic event extraction has a broad range of applications in systems biology, ranging from support for the creation and annotation of pathways to automatic population or enrichment of data. Systems biology recognizes in particular the importance of the interactions between biological components and the consequences of these interactions. Such interactions and their downstream effects are known as events. A trigger is any word in a sequence that serves as an indicator to an event. To computationally mine the events and triggers from biological literatures using text mining methods is a challenge to interactions and relationships between lexical terms.

A summarized review of event extraction for system biology by text mining the literature in [1] describes several approaches such bases as incorporated linguistic

© Springer International Publishing Switzerland 2016
H. Fujita et al. (Eds.): IEA/AIE 2016, LNAI 9799, pp. 207–216, 2016.
DOI: 10.1007/978-3-319-42007-3_18

representation (pattern matching versus full parsing), the use of lexical or ontological resources, text analysis (rule-based versus machine learning) and domain specificity. The event extraction approaches can also be viewed as statistical and non-statistical approaches. Bui and Sloot [2] proposed a system to extract biological events from texts using simple syntactic patterns and they adopt a rule-based approach to derived event candidates from syntactic patterns in a parse tree. A non-statistical approach to event extraction for gene regulation network in [3] presented the semantic and syntactic methods to extract events triggers from biological literatures leveraging on the linguistic representation incorporated within the text. A similar non statistical event extraction approach by McGrath *et al.* [4] used the signatures of linguistic and semantic features to extract complex biological events from texts. This specific work used traditional linguistic features and semantic knowledge to predict event triggers and their candidate arguments.

Several supervised machine learning approaches adopted statistical classifiers in training event extraction systems. To overcome the limitations and complexity of biological texts, Kim *et al.* [5] proposed a two-phase learning for biological event extraction and verification using the maximum entropy (ME) classification method. The work exploited a machine learning method to generate rules for event extraction from texts and utilizes a statistical classifier to verify event components.

Several statistical models are employed in the extraction of events from texts. Markov models [6] have been extensively applied in event extraction systems. It is due to the ability of the Markov model to represent repetitive events and the time dependence of both probabilities and utilities which allows for more accurate representation. A maximum entropy Markov models for information and segmentation [7] adopted a Markovian sequence model to present observations as arbitrary overlapping features. This approach was able to combine the advantages of both the Markov model and the maximum entropy model into a general model that allowed state transitions to depend on non-independent features of the sequences under analysis. Hidden Markov Models (HMMs) have been applied extensively to the problem of statistical modeling, database searching and sequence alignment etc. In Natural Language Processing, HMMs have been used in speech recognition, where observations are sounds forming a word model through a hidden random process that generates the sound with high probability [8]. A good model would therefore assign the high probability to all sound sequences that are likely the utterances of the word it models, and give the low probability to any other sequence. An approach that integrated speech recognition and natural language capabilities [9] took advantages of both syntactic and semantic constraints provided by NLP assigning tractable search space tasks.

In text mining, representing biological literatures with the regulatory network requires the extraction of biological events, relations, and entities. To illustrate this, consider the biological expression:

FAS ligation induces the activation of the src family kinase of jun-kinase.

The above expression mentions the relationship between two biological entities, *FAS* and *jun-kinase*. This paper presents an extension of Markov model for extraction of events and event arguments from biological texts with the following primary contributions.

- Integrating the advantages of both Markov logic model [10] and Bayesian Network [11] to a single event extraction model.
- Using logical inferences to extract events from our proposed model.

Unifying statistical probability and logical inferences allow us to handle both uncertainty and complexity in the real world.

The rest of this paper is organized as follows. Section 2 presents the overview of our system architecture. We describe network models and event extraction methods in details in Sects. 3 and 4. The experimental results are shown and discussed in Sect. 5. Finally, we express our main conclusions and the future research directions.

2 Architecture Overview

2.1 System Architecture

Figure 1 shows the flowchart of the statistical approach for biological event extraction using Markov's method. The extraction system takes as an input, a sentence describing a biological process from pubMed abstracts[1]. The components of the system network model are composed of the Bayesian Network model and the Markov Network model as mentioned in Sects. 3.1 and 3.2 respectively. That is, the model integrates the two network components to formulate a logical network for event extraction using the logical formulae as mentioned in Sect. 4. The system network model explores the information in the annotated training data to enrich the extraction process. The details are described in the following sections.

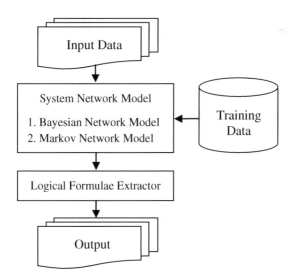

Fig. 1. The main system architecture

---

[1] https://www.nlm.nih.gov/bsd/policy/structured_abstracts.html.

2.2   Experimental Data

The experimental data is obtained from the BioNLP-2013 Shared Task's GRN task [12]. The purpose of the GRN task is to use information extraction techniques to retrieve all the genetic interaction of reference networks at least one occurrence per interaction independent of where it is mentioned in the given literature. The results are then represented as a GRN network. The experimental data consists of three parts: the training data, the development data, and the test data. The file formats contain the text file, the entity annotation file and the event annotation file. The text files in the training and development data have both entity and event annotation files. However, the test data only have entity annotation files. Therefore, the task in the shared task is simplified to find the event annotations for the test data and to generate a GRN network.

3   System Network and Model for Event Extraction

Probabilistic logical models are models combining aspects of probability theory with aspects of logical programming, first-order logic, or relational languages. Bayesian networks (BNs) and Markov networks (MNs) [13] are languages for representing independencies and each can represent independent constraints that the other cannot. A BN can be transformed to an MN from two perspectives:

1. Given a BN, B, how to represent its distribution as a parameter of an MN or
2. Given a graph G, how to represent independencies in G using an undirected graph H.

Considering a distribution P_B where B is a parameterized BN over a graph G, the parameters of B can be viewed as parameters for a Gibbs distribution[2] by taking each conditional probability distribution and viewing it as a factor of scope. The BN with evidence e is a Gibbs with $Z = P(e)$, that is:

If B is a BN over distribution X, and $E = e$, an observation, letting $W = X$-E then $P_B(W \mid e)$ is a Gibbs distribution with factor $\Phi = \{\phi_{Xi}\}$, $X_i \in X$ where $\phi_{Xi} = P_B(X_i \mid Pa_{Xi})$ $[E = e]$ where Pa_{Xi} is the parent of X_i in the graph. Thus any BN conditioned on evidence can be regarded as a Markov network. Hence given a BN of relationships between lexical items of a text, we can incorporate independencies deduced from an MN.

3.1   Bayesian Network

Bayesian networks were developed and were used extensively to model distributed processing by combining both semantical expectations and perceptual evidences to form a coherent interpretation. Bayesian networks are directed acyclic graphs (DAGs) in which the nodes represent variables of interest (e.g., the lexical items of a text in our study case) and the links represent informational or causal dependencies among the variables – arguments in this study. The strength of a dependency is represented by conditional probabilities that are attached to each cluster of parent-child nodes in the

---

[2] https://en.wikipedia.org/wiki/Gibbs_measure.

network. In this study, we represent the causal relationship among the nodes using the probability distribution in the given annotated training and development data mentioned in Sect. 2.2[3]. To determine the candidate of events and event triggers from the input lexical items, we use the conditional probability distribution over the annotated lexical items in the training data. If "A" is the set of annotated lexical items in the training data and "y" is the lexical item from test data, the distribution can be computed as $P(y \mid A)$. This computation uses the features of the parse structure for the lexical items. The features include the parts of speech of lexical items, the subject and object properties within the expression. This enables us to find candidate triggers and we can use the triggers and the annotated name entities to represent variables of nodes in the Bayesian network. Finally, we represent relationships between nodes in a Bayesian network as in Fig. 2.

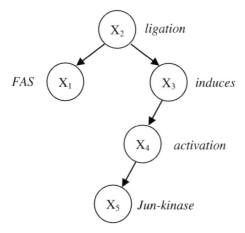

Fig. 2. Bayesian network of representing relationships among variables

Figure 2 above represents the Bayesian network for the expression *"FAS ligation induces the activation of the src family kinase of jun-kinase."* In this figure, variables X_i is a lexical term in the expression where $i = \{1, 2 \ldots N\}$ and the number of terms is N. The position and relationship between lexical terms are by the probability distribution and dependency within the parse structure of the expression. However, this structure mainly presents the relationship between candidate triggers and terminal nodes that are the target entities of the triggers. In Fig. 2, the root node X_2 with the lexical term "ligation" as the key word determining through the probability distribution has two related triggers (induces and activation) with the target object, "Jun-kinase" and a subject "FAS," determined by the term dependency in the parse structure.

3.2   Markov Logic Network

Markov logic [10] is a statistical relational learning language model based on the first order logic [14] and Markov networks [13]. This study considers an event extraction task as a relational structure over tokens. We consider the Markov logic network [15] as an extension of the first order logic to instantiate an MN for a repetitive or nested case of events and arguments. Since logical predicates can be used as bases to model the MN, this study uses a set of logical predicates such as *eventType (event_token, type), entity (entity_name), argument (argument_type, role)* and *function (argument (), role)*. A set of weighted first order formulae is used to define distributions over sets of ground atoms of these predicates. An event *e* is described by a tuple (*event_index, event_name, function()*), where *event_index, event_name* and *function()* are the index position of the event token, the event token and the functional role and argument of the event respectively. In Markov logic, a set of the weighted first order formulae forms a Markov logic network. To assign weights to the logic predicates, the annotated training data is used as the ground atom or the observed truth, thus simplifying the event extraction task to finding the hidden roles of lexical items.

Regard events in a given sentence as a sequence of occurrences with the specific features such as the part of speech, index, and grammatical role etc. Therefore, for each lexical item, there is a probability distribution over the training data that measures its likelihood as an event trigger. Feature function $f(i, t)$ is used to assign the weight to lexical items in the input with respect to the training data, where i and t are features with respect to the input term and the training data respectively.

$$f(i,t) = \begin{cases} 1 & for \ \ i = t \\ 0 & otherwise \end{cases} \tag{1}$$

The goal is to find a model that accurately represents a sequence of events in a given text by assigning a large probability to a more possible event sequence. From the training sequence, one can see how well a model fits by calculating the likelihood of the model. If F is a set of features for a training sequence with size n, the likelihood of lexical item i with respect to F can be calculated as Eq. (2).

$$prob(i|\mathrm{F}) = \frac{\sum_j^n f(i,j)}{n} \tag{2}$$

Subsequently, the posterior probability of a sequence of events given a model is as follows.

$$prob(sequence|model) = prob(model|sequence)prob(sequence) \tag{3}$$

In Eq. (3), the normalizing factor *prob(model)* can be ignored since it is independent of the sequence. This allows for the assignment of the weight to the logical predicates for the Markov logic network. Using a normalizing factor, Z, and feature element, k, we can make the distribution sum to one across all the possible cases including the next event or event argument with the following constraints:

$$prob(i|j) = \frac{1}{Z}e^{\sum_{k}^{n} w_k f(i,j)} \tag{4}$$

where w_k is a feature weight of a given feature calculated as the ration of a given feature and its distribution in the training data.

4 Extraction of Events Using Logical Formulae

In order to extract events effectively, we can take advantage of both Bayesian logic and Markov logic analysis. The Bayesian logic analysis in Sect. 3.1 presents one-way relationship in a Bayesian network missing the nested relationship that might be present among events. However, Markov logic has the ability in capturing this lost information in a logic network. Combining the two networks enables to fetch the benefit of both logics. As mentioned in Sect. 3, using the training evidence, the Bayesian network is transformed to a Markov network for the logical inference. The logical predicate over tokens and token pairs used in this is similar to the work by Riedel *et al.* [16].

The set of logical predicates used in this study can be put into predicates with no observed information from the training data (hidden predicates) and those with observed information such as annotated entities (observed predicates). With this predicates, events and arguments are predicted using the logical formulae, also called logical expressions, shown in Table 1.

Table 1. Logical formulae for event extraction

Logical expression	Description
$\exists i_t$ (baground_data.etity(i_t) $\rightarrow$ $\exists r$ bacground_data.role (i_t, r))	If there is any test entity i_t that occurs in the background data, then there is a role r taken by i_t
$\exists r$ (role(r) $\rightarrow$ $\exists i$ (event(i) $\lor$ eventTrigger(i)) $\land$ fill_role (i, r))	If there exists a role, then there must exist an event or event trigger filing the role
$\exists i$ (event(i) $\rightarrow$ $\exists t$ eventType(i, t))	If there exists event i, then there is an event type t describing event i
$\exists i_t$ (entity(i_t) $\land$ weight(i_t) $\geq$ 0) $\land \exists e_t$ $\exists t$ (event(e_t) $\in$ background_data.event() $\land$ eventType(e_t,t)) $\rightarrow$ add(background. annotation, test.annotation)	If an entity i_t with weight $w \geq 0$ has event e_t found in the background data, add the background annotation to the test annotation
$\forall e$ $\exists t$ $\forall x$ (eventType(e, t) $\land$ ($t \neq x$) $\rightarrow$ $\neg$eventType(e, x))	Every event e must have only one event type
$\forall i$ $\forall j$ $\forall k$ $\forall r$ ((index(i) < index(j)) $\land$ (index (j) < index(k)) $\land$ role(i, j, r)) $\rightarrow$ $\neg$role(i, k, r)	We cannot have roles within roles
$\forall i$ $\forall j$ $\forall r$ $\forall s$ (role(i, j, r) $\land$ ($r \neq s$) $\rightarrow$ $\neg$role(i, j, s))	There can only be a single role assignment

5   Experimental Results and Evaluation Matrix

This work uses the BioNLP-2013 shared task evaluation service for the Gene Regulation Network to test the experimental result [17]. The evaluation service gives recall, precision, F-score and Slot Error Rate (SER) [17, 18]. In addition, it also gives the number of prediction for substitutions (number of slots in system results that are aligned to slots in the reference gold data but are scored as incorrect), deletions (missing slots or false rejections), and insertions (spurious slots or false acceptances). The table below shows the system's result (Table 2).

Table 2. System score for the BioNLP-2013 GRN task

Substitution	Deletion	Insertion	Recall	Precision	F-score	SER
0	56	6	0.31	0.81	0.45	0.77

We compare the result of this approach to the results of the following works: University of Ljubljana [17], K.U.Leuven [19], TEES-2.1 [20], IRISA-TexMex [21], EVEX [22] which are the systems participating in the BioNLP GRN shared task[4] and NTNU [3]. The results are shown in Table 3.

Table 3. Results of the systems on BioNLP-2013 GRN shared task

Participant	SER	Recall	Precision	F-score
University of Ljubljana	0.73	0.34	0.68	0.45
K.U.Leuven	0.83	0.23	0.50	0.31
TEES-2.1	0.86	0.23	0.54	0.32
IRISA-TexMex	0.91	0.41	0.40	0.40
EVEX	0.92	0.13	0.44	0.19
NTNU	0.89	0.23	0.51	0.32

Comparing with Table 3, we show that our use of Markov's logic approach has a state-of-art F-score comparable to participants with 81 % precision in forming the gene regulation network. It shows we have a good performance in solving this problem.

6   Conclusion

This study investigates the application of the statistical probability and logical inferences to extract biological events from texts. We argue that advantages can be driven from both Bayesian and Markovian analysis to improve extraction of events. This has not come with a surprise as the mathematical modeling plays a significant role in many

---

[4] http://2013.bionlp-st.org/tasks/gene-regulation-network/test-results.

research areas. Markov logic is one of the statistical modeling methods that is prominent in related research areas such as speech recognition.

In future works, we shall focus on the capabilities of the statistical modeling approaches in relation information retrieval problems. We believe a careful study in this area will be a good research interest area and will be very significant in understanding the statistical roles of lexical items in the literatures.

References

1. Ananiadou, S., Pyysalo, S., Tsujii, J.I., Kell, D.B.: Event extraction for systems biology by text mining the literature. Trends Biotechnol. **28**(7), 381–390 (2010)
2. Bui, Q. C., Sloot, P.: Extracting biological events from text using simple syntactic patterns. In: Proceedings of the BioNLP Shared Task 2011 Workshop, pp. 143–146 (2011)
3. Hou, W.J., Ceesay, B.: Event extraction for gene regulation network using syntactic and semantic approaches. In: Ali, M., Kwon, Y.S., Lee, C.-H., Kim, J., Kim, Y. (eds.) IEA/AIE 2015. LNCS, vol. 9101, pp. 559–570. Springer, Heidelberg (2015)
4. McGrath, L.R., Domico, K., Corley, C.D., Webb-Robertson, B.J.: Complex biological event extraction from full text using signatures of linguistic and semantic features. In: Proceedings of the BioNLP Shared Task 2011 Workshop, pp. 130–137 (2011)
5. Kim, E., Song, Y., Lee, C., Kim, K., Lee, G.G., Yi, B.K., Cha, J.: Two-phase learning for biological event extraction and verification. ACM Trans. Asian Lang. Inf. Process. (TALIP) **5**(1), 61–73 (2006)
6. Ghahramani, Z.: An introduction to hidden Markov Mmodels and Bayesian networks. Int. J. Pattern Recogn. **15**(1), 9–42 (2001)
7. McCallum, A., Freitag, D., Pereira, F. C.: Maximum entropy Markov models for information extraction and segmentation. In: Proceedings of the Seventeenth International Conference on Machine Learning (ICML 2000), vol. 17, pp. 591–598 (2000)
8. Kita, K., Kawabata, T., Saito, H.: HMM continuous speech recognition using predictive LR parsing. In: Proceedings of 1989 International Conference on Acoustics, Speech, and Signal Processing (ICASSP 1889), vol. 2, pp. 703–706 (1989)
9. Murveit, H., Moore, R.: Integrating natural language constraints into HMM-based speech recognition. In: Proceedings of 1990 International Conference on Acoustics, Speech, and Signal Processing (ICASSP 1990), vol. 1, pp. 573–576 (1990)
10. Domingos, P., Kok, S., Lowd, D., Poon, H., Richardson, M., Singla, P.: Markov logic. In: Raedt, L., Frasconi, P., Kersting, K., Muggleton, S.H. (eds.) Probabilistic ILP 2007. LNCS (LNAI), vol. 4911, pp. 92–117. Springer, Heidelberg (2008)
11. Koller, D., Friedman, N., Getoor, L., Taskar, B.: 2 graphical models in a nutshell. In: Statistical Relational Learning, p. 13 (2007)
12. Nédellec, C., Bossy, R., Kim, J.D., Kim, J.J., Ohta, T., Pyysalo, S., Zweigenbaum, P.: Overview of BioNLP shared task 2013. In: Proceedings of the BioNLP Shared Task 2013 Workshop, pp. 1–7 (2013)
13. Bromberg, F., Margaritis, D., Honavar, V.: Efficient Markov network structure discovery using independence tests. J. Artif. Intell. Res. **35**, 449–484 (2009)
14. Smullyan, R.M.: First-Order Logic. Dover Publications, New York (1995)
15. Richardson, M., Domingos, P.: Markov logic networks. Mach. Learn. **62**(1–2), 107–136 (2006)

16. Riedel, S., Chun, H. W., Takagi, T., Tsujii, J.: A Markov logic approach to bio-molecular event extraction. In: BioNLP 2009 Proceedings of the Workshop on Current Trends in Biomedical Natural Language Processing: Shared Task, pp. 41–49 (2009)
17. Bossy, R., Bessières, P., Nédellec, C.: BioNLP shared task 2013–an overview of the genic regulation network task. In: Proceedings of the BioNLP Shared Task 2013 Workshop: The Genia Event Extraction Shared Task, pp. 153–160 (2013)
18. Makhoul, J., Kubala, F., Schwartz, R., Weischedel, R.: Performance measures for information extraction. In: Proceedings of DARPA Broadcast News Workshop, pp. 249–252 (1999)
19. Provoost, T., Moens, M. F.: Detecting relations in the gene regulation network. In: Proceedings of the BioNLP Shared Task 2013 Workshop: the Genia Event Extraction Shared Task, pp. 135–138 (2013)
20. Björne, J., Salakoski, T.: TEES 2.1: automated annotation scheme learning in the BioNLP 2013 shared task. In: Proceedings of the BioNLP Shared Task 2013 Workshop: The Genia Event Extraction Shared Task, pp. 16–25 (2013)
21. Claveau, V.: IRISA participation to BioNLP-ST 2013: lazy-learning and information. In: Proceedings of the BioNLP Shared Task 2013 Workshop: The Genia Event Extraction Shared Task, pp. 188–196 (2013)
22. Hakala, K., Van Landeghem, S., Salakoski, T., Van de Peer, Y., Ginter, F.: Application of the EVEX resource to event extraction and network construction: shared task entry and result analysis. BMC Bioinform. **16**, S3 (2015)

Citation-Based Extraction of Core Contents from Biomedical Articles

Rey-Long Liu$^{(\boxtimes)}$

Department of Medical Informatics, Tzu Chi University, Hualien, Taiwan
rlliutcu@mail.tcu.edu.tw

Abstract. Retrieval of biomedical articles about specific research issues (e.g., gene-disease associations) is an essential and routine job for biomedical researchers. An article *a* can be said to be about a research issue *r* only if its *core content* (goal, background, and conclusion of *a*) focuses on *r*. In this paper, we present a technique CoreCE (Core Content Extractor) that, given a biomedical article *a*, extracts the textual core content of *a*. The core contents extracted from biomedical articles can be used to index the articles so that articles about specific research issues can be retrieved by search engines more properly. Development of CoreCE is challenging, because the core content of an article *a* may be expressed in different ways and scattered in *a*. We tackle the challenge by considering titles of the references cited by *a*, as well as the passages (in *a*) used to explain why the references are cited (i.e., the citation passages). Empirical evaluation shows that, by representing biomedical articles with the core contents extracted by CoreCE, retrieval of those articles that are judged (by biomedical experts) to be about specific gene-disease associations can be significantly improved. CoreCE can thus be a front-end processor for search engines to preprocess biomedical scholarly articles for subsequent indexing and retrieval. The contribution is of technical significance to the retrieval and mining of the evidence already published in biomedical literature.

Keywords: Biomedical article · Core content · Citations · Content extraction

1 Introduction

Biomedical researchers often need to retrieve articles about specific research issues (e.g., associations among biomedical entities, such as genes, diseases, chemicals, and proteins). For example, to construct and maintain databases of the gene-disease associations that were published in biomedical literature, several websites (e.g., Genetic Home Reference and Online Mendelian Inheritance in Human) recruit a large number of experts that carefully and frequently retrieve and check biomedical articles.[1] Therefore, search engines (e.g., Google Scholar[2] and PubMed[3]) often provide the

---

[1] The ways of database update by Genetic Home Reference and Online Mendelian Inheritance in Human can be found at http://ghr.nlm.nih.gov/ExpertReviewers and http://www.omim.org/about, respectively.

[2] Google Scholar is available at https://scholar.google.com.

[3] PubMed is available at http://www.ncbi.nlm.nih.gov/pubmed.

© Springer International Publishing Switzerland 2016
H. Fujita et al. (Eds.): IEA/AIE 2016, LNAI 9799, pp. 217–228, 2016.
DOI: 10.1007/978-3-319-42007-3_19

service of retrieving those scholarly articles that are related to a given article or a textual query consisting of keywords.

In this paper, we present a novel technique CoreCE (Core Content Extractor) that, given a biomedical article a, extracts the *core content* of a. The core content of a consists of the text (in a) describing the goal, background, and conclusion of a. Therefore, a can be said to be highly related to a research issue r only if its core content focuses on r. The core contents extracted from biomedical articles can thus be used to index the articles so that articles about specific research issues can be retrieved by search engines more properly. CoreCE can thus be a front-end processor for the search engines to preprocess the articles for subsequent indexing and retrieval. To our knowledge (ref. related work discussed in Sect. 2), CoreCE is the first technique that aims at improving the indexing and retrieval functions of search engines by extracting textual core contents from biomedical scholarly articles.

Extraction of the core content from an article a is challenging, due to three reasons: (1) textual parts about the core content may be scattered in the title, the abstract, and the main body of a, (2) each textual part may have different degrees of relatedness to the core content, and (3) the core content may be expressed in different ways. As illustrated in Fig. 1, we tackle the challenge by considering (1) titles of the references cited by a, and (2) the passages (in a) used to explain why the references are cited (i.e., the *citation passages*). Titles of the references cited by a can be a resource to enrich the core content of a, because these reference should be closely related to the core content of a, and their titles can indicate their main goals. Moreover, the reasons why the references are cited by a are described in the citation passages of the references. These citation passages can thus be helpful for the extraction of the core content of a. They tend to appear immediately before the place where the references are cited in a, because authors tend to discuss a reference followed by a citation to the reference (as illustrated in Fig. 1).

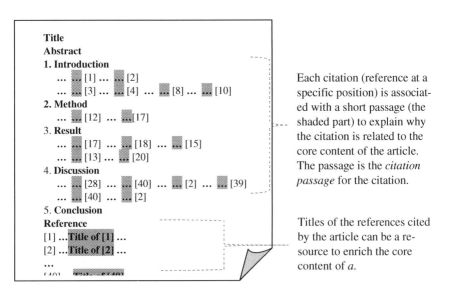

Fig. 1. Both the *citation passages* and the *title* of a reference cited by the article can be helpful for the extraction of the core content of the article.

We also design and conduct a series of experiments to evaluate the contribution of CoreCE. The results show that, by representing biomedical articles with the core contents extracted by CoreCE, retrieval of those articles that are judged (by biomedical experts) to be about specific gene-disease associations can be significantly improved. The textual contents extracted by CoreCE for these articles can thus be used to represent the core contents of these articles. CoreCE can thus be a front-end processor for search engines to preprocess biomedical scholarly articles for better indexing and retrieval. The contribution is of technical significance to the retrieval and mining of the evidence already published in biomedical literature.

2 Related Work

As we aim at extracting textual contents of biomedical scholarly articles for better indexing and retrieval, we survey different kinds of information that was employed by previous techniques to retrieve scholarly articles. There are two typical kinds of information: *citation links* among scholarly articles and *textual contents* of the articles. For an article a, the citation links include those references that are cited by a (i.e., *out-link* references of a) as well as those articles that cite a (i.e., *in-link* citations of a). The textual content of a was typically from the title, the abstract, and/or the main body of a. Both kinds of information can provide certain evidence for the previous techniques to retrieve related articles.

2.1 Using Citation Links (Relationships) Among Articles

Out-link references were employed by those techniques that were based on bibliographic coupling [14], which expects that two articles a_1 and a_2 may be related to each other if they co-cite some articles. Out-link references were useful in classifying scientific articles [8, 16], clustering scientific articles [6], retrieving legal judgments [13], detecting plagiarism [9], and measuring the professional similarity between authors of articles [12]. On the other hand, in-link citations were employed by those techniques that were based on co-citation [21], which expects that two articles a_1 and a_2 may be related to each other if they are co-cited by other articles. In-link citations were found to be useful in classifying webpages [7, 8] and measuring the professional similarity between authors of articles [23]. Proximity of the in-link citations may be helpful as well, as two articles may be related to each other if they are cited in nearby areas in many articles that cite them [4, 10]. Moreover, out-link references and in-link citations can also be simultaneously considered [2, 25, 26].

Instead of considering the citation links (relationships) among articles, we aim at extracting *textual* contents from biomedical articles (based on the titles and citation passages of out-link references, ref. Fig. 1). Our motivation lies on the fact that search engines (e.g., PubMed) often receive *textual* queries (e.g., keyword-based queries) and employ *textual* contents of articles to index and retrieve related articles.[4]

---

[4] The way PubMed employs to retrieve related articles can be found at http://www.ncbi.nlm.nih.gov/books/NBK3827/#pubmedhelp.Computation_of_Similar_Articl.

CoreCE can thus be a front-end processor for these search engines to index and retrieve articles more intelligently.

2.2   Using Textual Contents of Articles

Many previous techniques considered textual contents of articles as well. Given a scholarly article a, one typical way was to employ the citation passages around the citations to a in other articles. Such citations are actually *in-link* citations of a, and these citation passages are the textual contents that authors of the other articles employ to comment a. These passages can thus indicate the main contents of a [18], although the citing articles may focus on different topical parts of a [13]. These citation passages may be used for inter-article similarity estimation [1], topic-based article retrieval [17], and disambiguation of named entities [19]. However, application of these techniques is restricted, because many scholarly articles have very few (or even no) in-link citations. CoreCE presented in this paper extracts the core textual content from each article per se, rather than from other articles.

Textual contents of scholarly articles were typically from titles, abstracts, and main bodies of the articles. As the essential content of an article a is often scattered in a, several previous techniques considered certain topic-indicative parts of a (e.g., the title and the abstract of a [6, 8]), while many other techniques employed *term weighting* to improve their performance (those terms that were more important would get higher weights). A typical term weighting technique is the *tf-idf* technique, which prefers those terms that frequently appear in a (i.e., the term frequency, *tf*, is large) but rarely in other articles (i.e., the inverse document frequency, *idf*, is large). Based on the term weighting, various techniques were developed to retrieve those articles that share similar textual contents. The vector space model (VSM) is a typical technique. Similarity between two articles is the cosine similarity on their vectors of term weights. However the cosine similarity did not always perform well [5, 22]. Latent Semantic Analysis (LSA) was a typical technique to improve the vector representation of scholarly articles [11, 15], however it did not perform well for scholarly articles either [5]. Among many other techniques, OK was one of the best techniques to identify related articles [22]. It was developed based on BM25 [20], which was one of the best techniques in finding related scholarly articles [5]. OK employs Eq. 1 to estimate the similarity between two articles a_1 and a_2, where k_1 and b are two parameters, $|a|$ is the number of terms in article a (i.e., length of a), *avgal* is the average number of terms in an article (i.e., average length of articles), $tf(t, a)$ is the term frequency of a term t in an article a, and $idf(t)$ is the inverse document frequency of a term t.

$$Similarity_{OK}(a_1, a_2) = \sum_{t \in a_1 \cap a_2} \frac{tf(t, a_1)(k_1 + 1)}{tf(t, a_1) + k_1(1 - b + b\frac{|a_1|}{avgal})} \frac{tf(t, a_2)(k_1 + 1)}{tf(t, a_2) + k_1(1 - b + b\frac{|a_2|}{avgal})} Log_2 idf(t)$$

(1)

However, all these techniques did not aim at extracting the core contents of biomedical scholarly articles, and hence they did not retrieve related articles based on the core contents. CoreCE presented in this paper aims at extracting the core contents.

It can be a front-end processor for these previous techniques to achieve better performance in estimating inter-article similarity based on the core contents extracted. We will show that (see Sect. 4), with the core contents extracted by CoreCE, the state-of-the-art technique OK (ref. Eq. 1) can be significantly improved when retrieving those biomedical articles that were judged (by domain experts) to be about the same research issues.

3   CoreCE: A Core Content Extractor

Extraction of the core content from an article a is challenging, because the core content may be scattered in a and expressed with different terms. CoreCE tackles the challenge by a novel citation-based extraction strategy. It employs two types of citation-based information from the out-link references cited by a: (1) titles of the references, and (2) citation passages that authors of a used to explain why the references are cited by a. The out-link references are selected by authors of a to highlight the main contributions of a, and hence their titles and citation passages can indicate the core content of a. The titles can "enrich" the core content of a with different terms, because they were written by other authors to describe the main goals of the references, which are closely related to the core content of a. The citation passages can indicate the core content of a as well, because they aim at describing the critical differences between the references and a.

More specifically, as defined in Eq. 2, the core content of an article a consists of two concatenated parts: (1) basic core content (i.e., *BasicCore*), and (2) citation passages of the out-link references in a (i.e., *CitationP*), where '+' is a string concatenation operator. The first part is composed of three concatenated subparts (see Eq. 3): the title (i.e., *Title*), the abstract (i.e., *Abstract*), and the titles of the out-link references (i.e., *RefTitles*). The first part should have contained the core content of a, however it may still contain many words not quite related to the core content.

$$CoreContent(a) = BasicCore(a) + CitationP(a) \qquad (2)$$

$$BasicCore(a) = Title(a) + Abstract(a) + RefTitles(a) \qquad (3)$$

Therefore, the second part (i.e., the citation passages, *CitationP*) is employed. As defined in Eq. 4, CoreCE considers α words that appear immediately before the place where the references are cited in an article a,[5] where α is a parameter governing the window size for the citation passages.[6] As not all out-link references may be closely related to the core content of a, the words in a citation passage window may not always be related to the core content of a. Therefore, CoreCE only extracts the words that appear in *both* the citation passage windows *and* the basic core content (i.e., *BasicCore* defined above). Therefore, term frequencies of the words in *BasicCore* will be

---

[5] When extracting the α words, stopwords are excluded and hence not counted.

[6] We expect that α should be in the range of [5, 15], which is the typical number of words employed to comment a reference. The expectation will be justified in the experiments reported in Sect. 4.

increased if these words are employed (by authors of a) to explain why a reference is cited by a. In that case, these words should be more likely to be related to the core content of a.

$$CitationP(a) = \sum_{c\in\{references\ in\ a\}} \sum_{\substack{p\ \in\ \{places\ where \\ c\ appears\ in\ d\}}} \{\alpha\ words\ before\ p\} \cap BasicCore(a)$$

(4)

Therefore, CoreCE employs citation-based information from out-link references to extract the core content of a scholarly article. The extraction strategy has several interesting features:

(1) Both titles and citation passages of the out-link references are considered as the resource employed by CoreCE.
(2) By simultaneously considering the two parts (the basic core content and the citation passages), the effects of those words that appear in both parts can be "amplified" (by increasing their term frequencies) as these words should be more related to the core content of the article.
(3) The core content extracted by CoreCE is represented by plain text, which is a common input format for search engines and text rankers, making CoreCE able to serve as a front-end processor for them.

4 Empirical Evaluation

Experiments are conducted to evaluate the contribution of CoreCE in retrieving those biomedical articles whose core contents focus on the same gene-disease associations.

4.1 Basic Experimental Settings

We employ the articles that have been judged (by biomedical experts) to be dedicated to specific gene-disease associations [16]. Articles that are dedicated to a gene-disease association should have the core contents focusing on the association, and hence they are highly related to each other. They can thus be used to investigate whether the textual contents extracted by CoreCE are really the core contents of biomedical articles. Given an article a about a gene-disease association p, the state-of-the-art article ranker OK (ref. Eq. 1 noted in Sect. 2) is invoked to rank articles so that those articles that are highly related to a (dedicated to p) can be ranked higher. Both the original content of a and the content extracted by CoreCE are input to OK for performance comparison. The contribution of CoreCE can be verified if OK can perform better (rank highly related articles higher) by employing the content extracted by CoreCE.

More specifically, the articles were collected from DisGeNET,[7] which maintains a database of gene-disease associations. We focus on those gene-disease associations for

---

[7] DisGeNET is available at http://www.disgenet.org/web/DisGeNET/menu/home.

which related articles were selected by biomedical experts of Genetic Association Database (GAD)[8] or Comparative Toxicogenomics Database (CTD)[9] for human. Biomedical experts were recruited to select articles to annotate specific gene-disease pairs [3, 24]. For CTD, doctoral-level curators achieved good curation accuracy with a high degree of inter-curator agreement in an experiment [24].

There are 53 gene-disease associations in our experiments, and an association corresponds to a test in the experiment (we thus have 53 tests in the experiment). We experiment on those articles that have full text in PubMed Central (PMC).[10] For each gene-disease association $<g, d>$, we designate one article as the *target*, while the others as the *highly related candidates*. Given the target article, the article ranker (OK) aims at identifying the highly related candidates, among other non-relevant candidate articles that are *not* dedicated to $<g, d>$. These non-relevant candidates were collected by sending two queries to PMC: "*g* NOT *d*" and "*d* NOT *g*" (at most 200 articles were collected). They mention *g* or *d* but not both, and hence are not dedicated to $<g, d>$. We thus totally have 9,876 articles in which there are 435,786 out-link references.

4.2   The Baselines

To have the baselines for performance comparison with CoreCE, we implement several typical ways to extract the core content of an article: (1) using the tile of the article (i.e., *Title Only*), (2) using the abstract of the article (i.e., *Abstract Only*), (3) using both the title and the abstract (i.e., *Title+Abstract*), (4) using the title, the abstract, and the titles of the out-link references of the article (i.e., *Title+Abstract+RefTitles*), and (5) using the title, the abstract, and the main body of the article (i.e., *Whole Article*). The first three baselines can represent the typical ways of identifying the core content, because the title and the abstract of an article should express the main goal, background, and findings of the article. With the three baselines, we can investigate whether CoreCE can further improve the extraction of the core content. The fourth baseline aims at investigating the contribution of employing *citation passages* to extract the core content, because CoreCE employs both the titles and the citation passages of out-link references of the article. The fifth baseline is implemented to show that using the whole article as the core content is not a good way, because many parts of the article are not quite related to the core content of the article.

4.3   Evaluation Criteria

Two evaluation criteria are used to measure the performance of CoreCE and the baselines. They are *Mean average precision* (MAP) and average P@X. MAP is defined in Eq. 5, where $|T|$ is the number of tests, and $AvgP(i)$ is the average precision in the ith test. MAP is simply the average of the *AvgP* values in all the testes.

---

[8]   GAD is available at http://geneticassociationdb.nih.gov.

[9]   CTD is available at http://ctdbase.org.

[10]   PMC provides full-text biomedical articles at http://www.ncbi.nlm.nih.gov/pmc. All articles that are not included in PMC are excluded in the experiments.

$$MAP = \frac{\sum_{i=1}^{|T|} AvgP(i)}{|T|} \qquad (5)$$

$$AvgP(i) = \frac{\sum_{j=1}^{h_i} \frac{j}{Seen_i(j)}}{h_i} \qquad (6)$$

$AvgP(i)$ is defined in Eq. 6, where h_i is the number of articles that are judged (by domain experts) to be highly related to the target article in the ith test, and $Seen_i(j)$ is the number of articles that readers have seen when the jth highly related article in the ith test is shown. Therefore, given a target article a in the ith test, if a system can rank higher those articles that are highly related to a, $AvgP(i)$ will be higher.

On the other hand, average P@X works on those articles that are ranked at top-X positions. Average P@X is defined in Eq. 7. It is the average of the P@X values in all the 53 tests. Equation 8 defines P@X, which is the precision when top-X articles are shown to the readers. To measure how a system ranks highly related articles at top positions, we set X to 1, 3, and 5.

$$Average\ P@X = \frac{\sum_{i=1}^{|T|} P@X(i)}{|T|} \qquad (7)$$

$$P@X(i) = \frac{Number\ of\ top - X\ articles\ that\ are\ highly\ related\ to\ the\ target\ in\ the\ i^{th}\ test}{X}$$

$$(8)$$

4.4   Results

Figure 2 shows MAP and average P@X achieved by CoreCE and all the baselines (i.e., performance of OK when using the contents extracted by CoreCE and the baselines). To investigate whether the performance difference is *statistically significant*, we conducted two-sided and paired t-test with 95 % confidence level on the *AvgP* (ref. Eq. 6) and P@X (ref. Eq. 8) values of CoreCE and each baseline in the 53 tests. The results show that various combinations of the original parts of an article (i.e., the contents extracted by the four baselines: *Title Only*, *Abstract Only*, *Title+Abstract*, and *Whole Article*) lead to significantly poorer performance than CoreCE. As noted in Sect. 2, many previous techniques (e.g., [6, 8]) and search engines (e.g., PubMed) retrieved related articles based on titles and abstracts of the articles, our results show that CoreCE can be used as a front-end processor for them to further improve performance in retrieving highly related articles.

Figure 3 shows whether CoreCE and each baseline can rank highly related articles at top positions for a large percentage of the 53 tests (recall that a test corresponds to a

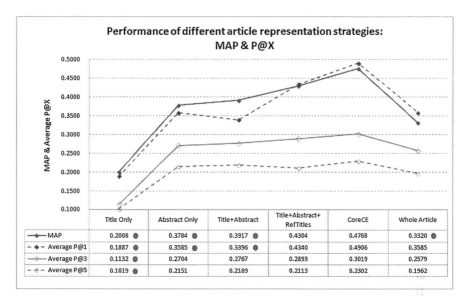

Fig. 2. MAP and average P@X: Various combinations of the textual parts of an article (i.e., title, abstract, and main body of the article) lead to significantly poorer performance than CoreCE (a dot on a performance result *r* indicates that the difference between *r* and the performance of CoreCE is statistically significant). (Color figure online)

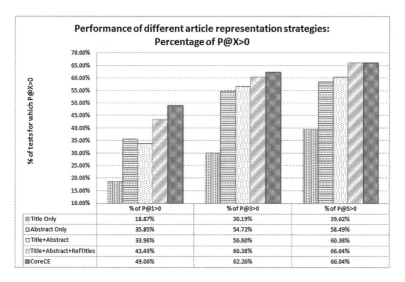

Fig. 3. Percentage of the tests for which P@X > 0: CoreCE helps to rank highly related articles at top positions (top-1 and top-3) for a higher percentage of the testes. (Color figure online)

gene-disease association). CoreCE successfully helps to rank highly related articles at the top positions (top-1 and top-3) for a higher percentage of the 53 testes, indicating that with the core contents extracted by CoreCE, biomedical researchers can read highly related articles at top positions for more gene-disease associations.

Figure 4 shows the effects of the size of the citation passage window. CoreCE performs better when the size is set to 5, however the performance differences are *not* statistically significant, indicating that the window size can be easily set in [5, 15], which is the typical number of words used to comment an out-link reference.

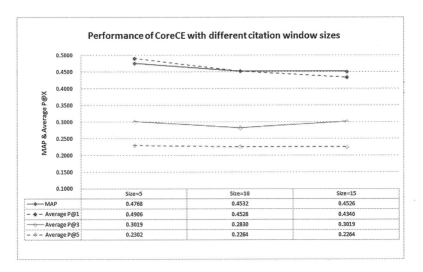

Fig. 4. Effects of the size of the window for citation passages: CoreCE performs better when the size is set to 5, however the performance differences are *not* statistically significant. (Color figure online)

5   Conclusion and Future Extension

We have presented a novel technique CoreCE that, given a biomedical article *a*, extracts the core content of *a*. The core contents extracted from biomedical articles can be used to index the articles so that articles about specific research issues can be retrieved by text rankers and search engines more properly. As the core content of an article *a* may be scattered in *a* and expressed with different terms, CoreCE extracts the core content based on titles and citation passages of the references cited by *a*. Empirical evaluation shows that, by representing biomedical articles with the core contents extracted by CoreCE, retrieval of those articles that are judged (by biomedical experts) to be dedicated to specific gene-disease associations can be significantly improved. CoreCE can thus be a front-end processor for search engines to preprocess biomedical scholarly articles for subsequent indexing and retrieval. The contribution is of significance to the retrieval and mining of the evidence already published in biomedical literature. We expect that CoreCE may be applied to domains other than biomedicine.

We are also investigating the possible contribution of considering the *importance* of each reference, based on the expectation that *not* all references cited by an article can be quite related to the core content of the article. Estimation and application of the degrees of importance are interesting research tasks to further improve CoreCE.

Acknowledgment. This research was supported by the Ministry of Science and Technology of Taiwan under the grant MOST 104-2221-E-320-005.

References

1. Aljaber, B., Stokes, N., Bailey, J., Pei, J.: Document clustering of scientific texts using citation contexts. Inf. Retrieval **13**(2), 101–131 (2010)
2. Amsler R.A.: Application of citation-based automatic classification. Technical report, Linguistics Research Center, University of Texas at Austin (1972)
3. Becker, K.G., Barnes, K.C., Bright, T.J., Wang, S.A.: The genetic association database. Nat. Genet. **36**(5), 431–432 (2004)
4. Boyack, K.W., Small, H., Klavans, R.: Improving the accuracy of co-citation clustering using full text. J. Am. Soc. Inform. Sci. Technol. **64**(9), 1759–1767 (2013)
5. Boyack, K.W., Newman, D., Duhon, R.J., Klavans, R., Patek, M., Biberstine, J.R., et al.: Clustering more than two million biomedical publications: comparing the accuracies of nine text-based similarity approaches. PLoS One **6**(3), e18029 (2011)
6. Boyack, K.W., Klavans, R.: Co-citation analysis, bibliographic coupling, and direct citation: which citation approach represents the research front most accurately? J. Am. Soc. Inform. Sci. Technol. **61**(12), 2389–2404 (2010)
7. Calado, P., Cristo, M., Moura, E., Ziviani, N., Ribeiro-Neto, B., Goncalves, M.A.: Combining link-based and content-based methods for web document classification. In: Proceedings of the 2003 ACM CIKM International Conference on Information and Knowledge Management, New Orleans, Louisiana, USA (2003)
8. Couto, T., Cristo, M., Gonçalves, M.A., Calado, P., Nivio Ziviani, N., Moura, E., Ribeiro-Neto, B.: A comparative study of citations and links in document classification. In: Proceedings of the 6th ACM/IEEE-CS Joint Conference on Digital Libraries, pp. 75–84 (2006)
9. Gipp, B., Meuschke, N.: Citation pattern matching algorithms for citation-based plagiarism detection: greedy citation tiling, citation chunking and longest common citation sequence. In: Proceedings of 11th ACM Symposium on Document Engineering, Mountain View, CA, USA (2011)
10. Gipp, B., Beel, J.: Citation proximity analysis (CPA) – a new approach for identifying related work based on co-citation analysis. In: Proceedings of the 12th International Conference on Scientometrics and Informetrics, vol. 2, pp. 571–575 (2009)
11. Glenisson, P., Glanzel, W., Janssens, F., De Moor, B.: Combining full text and bibliometric information in mapping scientific disciplines. Inf. Process. Manag. **41**, 1548–1572 (2005)
12. Heck, T.: Combining social information for academic networking. In: Proceedings of the 16th ACM Conference on Computer Supported Cooperative Work and Social Computing (CSCW), San Antonio, Texas, USA (2013)
13. Kumar, S., Reddy, P.K., Reddy, V.B., Singh, A.: Similarity analysis of legal judgments. In: Proceedings of the Fourth Annual ACM Bangalore Conference (COMPUTE), Bangalore, Karnataka, India (2011)

14. Kessler, M.M.: Bibliographic coupling between scientific papers. Am. Documentation **14**(1), 10–25 (1963)
15. Landauer, T.K., Laham, D., Derr, M.: From paragraph to graph: latent semantic analysis for information visualization. Proc. Natl. Acad. Sci. U.S.A. **101**(Suppl 1), 5214–5219 (2004)
16. Liu, R.-L.: Passage-based bibliographic coupling: an inter-article similarity measure for biomedical articles. PLoS One **10**(10), e0139245 (2015)
17. Liu, S., Chen, C., Ding, K., Wang, B., Xu, K., Lin, Y.: Literature retrieval based on citation context. Scientometrics **101**(2), 1293–1307 (2014)
18. Liu, X., Zhang, J., Guo, C.: Full-text citation analysis: a new method to enhance scholarly networks. J. Am. Soc. Inform. Sci. Technol. **64**(9), 1852–1863 (2013)
19. Nakov, P.I., Schwartz, A.S., Hearst, M.: Citances: citation sentences for semantic analysis of bioscience text. In: Proceedings of the SIGIR 2004 Workshop on Search and Discovery in Bioinformatics, pp. 81–88 (2004)
20. Robertson, S.E., Walker, S., Beaulieu, M.: Okapi at TREC-7: automatic ad hoc, filtering, VLC and interactive. In: Proceedings of the 7th Text REtrieval Conference (TREC 7), Gaithersburg, USA, pp. 253–264 (1998)
21. Small, H.G.: Co-citation in the scientific literature: a new measure of relationship between two documents. J. Am. Soc. Inform. Sci. Technol. **24**(4), 265–269 (1973)
22. Whissell, J.S., Clarke, C.L.A.: Effective measures for inter-document similarity. In: Proceedings of the 22nd ACM International Conference on Information and Knowledge Management, pp. 1361–1370 (2013)
23. White, H.D., Griffith, B.C.: Author cocitation: a literature measure of intellectual structure. J. Am. Soc. Inform. Sci. Technol. **32**(3), 163–171 (1981)
24. Wiegers, T.C., Davis, A.P., Cohen, K.B., Hirschman, L., Mattingly, C.J.: Text mining and manual curation of chemical-gene-disease networks for the Comparative Toxicogenomics Database (CTD). BMC Bioinformatics **10**, 326 (2009)
25. Yoon, S.-H., Kim, S.-W., Park, S.: A link-based similarity measure for scientific literature. In: Proceedings of the 19th International World Wide Web Conference (WWW), North Carolina, USA (2010)
26. Zhao, P., Han, J., Sun, Y.: P-Rank: a comprehensive structural similarity measure over information networks. In: Proceedings of the International Conference on Information and Knowledge Management, pp. 553–562 (2009)

Event Extraction and Classification by Neural Network Model

Bamfa Ceesay and Wen-Juan Hou[(⊠)]

Department of Computer Science and Information Engineering,
National Taiwan Normal University,
No. 88, Section 4, Ting Chou Road, Taipei 116, Taiwan, ROC
{bmfceesay, emilyhou}@csie.ntnu.edu.tw

Abstract. To understand and automatically extract information about events presented in a text, semantically meaningful units expressing these events are important. Extracting events and classifying them into event types and subtypes using Natural Language Processing techniques poses a challenging research problem. There is no clear-cut definitions to what an event from a text is and what the optimal representation of semantic units within a given text is. In addition, events in a text can be classified into types and subtypes of events; and a single event can have multiple mentions in a given sentence. In this paper, we propose a model to determine events within a given text and classify them into event types or subtypes and REALIS by the distributional semantic role labeling and neural embedding techniques. For the task of the event nugget detection, we trained a three-layer network to determine the event mentions from texts achieving F1-score of 77.37 % for macro average and 71.10 % for micro average, respectively.

Keywords: Event extraction · Event nuggets · Event classification · Neural embedding · Semantic role labeling

1 Introduction

With an increasing availability of a large volume of unstructured data, it is difficult to fully process these data manually. This creates the necessity to build and apply automatic systems to extract important information for data analysis. One important trend towards achieving this goal is the event extraction from text problem in the natural language processing (NLP) domain. Events are frequently discussed in text, for example, a criminal activity such as theft in a police report, changes in the futures on goods and the stock market in a business report, and a binding of proteins in a biological process etc. A trigger is any word in a sequence that serves as an indicator to an. An event extraction process automatically identifies triggers and arguments that constitute the textual mention of an event in the real world. Event detection, a step towards event extraction, is the task of identifying the main lexical items or semantically meaningful units (event nuggets) that indicate an event from texts. A traditional process of event detection first determines the event triggers and classifies events into event types. In this work, we also annotate the realis values (i.e., ACTUAL, GENERIC,

© Springer International Publishing Switzerland 2016
H. Fujita et al. (Eds.): IEA/AIE 2016, LNAI 9799, pp. 229–241, 2016.
DOI: 10.1007/978-3-319-42007-3_20

and OTHER) of events. An event mention will be referred to as ACTUAL if the event actually occurred, GENERIC when it is with no specific or unknown time and/or place, and OTHER if the event failed. An OTHER event is a future event or non-generic variation. Consider the following sentence (a):

Several militants are shot death during clashes near Kabul. (a)

Here the trigger "*shot death*" (i.e., Conflict.Attack in Table 1) predicates the interaction between the argument, shooter, and "militant", the target argument of the event. This single trigger would be used to identify two events, "shot (Conflict.Attack) and "dead" (Life.Die). The study considers lexical items or short phrases in a text as triggers, expressing the occurrence of an event. Feature extraction and processing will be applied to classify a trigger as an event or non-event. Therefore, every event mention is considered a trigger but not every trigger is considered an event. An event nugget is an event with semantically meaningful units such as event types, event subtypes and event realis.

In this study, we use the predefined event types and subtypes in TAC KBP 2015 shared task[1] [1]. The task defines 9 event types (Business, Contact, Conflict, Justice, Life, Manufacture, Movement, Personal, Transaction) and 38 subtypes such as Life. Die, Conflict.Attack, Contact.Meet etc. Each event mention is classified into an event type and a subtype of that event type. Table 1 depicts the distribution of event types in the training and test data.

An event mention can be tagged multiple times for different event types/subtypes. For example, when an event nugget instantiates different events as illustrated below.

*Suicide explosion **killed** 1 and **injured** 2 others.* (b)

In Sentence (b), the bold faced letters represent our events and will be tagged as Life.Die and Life.Injure.

Previous studies categorize the adaptation of techniques for event extraction into the rule-based approach, machine learning approach and hybrid approach [2]. It is also argued that the use of the rule-based and machine learning methods varies across industries and academia. The rule-based information extraction approach has enjoyed wider adaptation in industries, due to its explainable ability and the rapid development features. However, due to lack of the state-of-the-art approach to formulating the rules, the machine learning approach gains wider adaptation in academia [3].

In this paper, we adapt the machine learning approach to event extraction. This is a challenging problem because there is no clear-cut definition to what an event is from a set of lexical items. We adopt the definition of an event as an explicit occurrence involving participants or a change of the state in place and time [4]. This work presents the following contribution to event extraction:

- Presenting a single model for multiple tasks in event detection and classification into types and subtypes.
- Extracting semantic features using semantic role labeling.
- Predicting event types and subtypes using neural embedding.

---

[1] http://cairo.lti.cs.cmu.edu/kbp/2015/event/index.

Table 1. Distribution of event types in the provided training and test datasets

Type	Subtype	#Train	#Test
Business	Declare Bankruptcy	33	44
Business	End Org	13	6
Business	Merge Org	28	33
Business	Start Org	18	35
Conflict	Attack	800	591
Conflict	Demonstrate	200	149
Contact	Broadcast	417	510
Contact	Contact	337	587
Contact	Correspondence	95	110
Contact	Meet	244	272
Justice	Acquit	30	31
Justice	Arrest-Jail	37	69
Justice	Appeal	287	348
Justice	Charge-Indict	190	155
Justice	Convict	222	96
Justice	Execute	66	97
Justice	Extradite	63	60
Justice	Fine	55	45
Justice	Pardon	239	51
Justice	Release-Parole	73	124
Justice	Sentence	144	158
Justice	Sue	55	72
Justice	Trial-Hearing	196	155
Life	Be Born	19	17
Life	Die	514	408
Life	Divorce	45	49
Life	Injure	133	87
Life	Marry	76	83
Manufacture	Artifact	22	90
Movement	Transport-Artifact	70	66
Movement	Transport-Person	517	439
Personnel	Elect	97	71
Personnel	End Position	209	291
Personnel	Nominate	35	63
Personnel	Start Position	77	94
Transaction	Transaction	51	63
Transaction	Transfer-Money	551	554
Transaction	Ownership	280	265
Total:		6538	6438

2   System Framework

Figure 1 illustrates a schematic representation of our approach. Our system follows a pipeline paradigm as shown in the framework in Fig. 1. First, the input raw text is preprocessed through sentence splitting, tokenization, tagging and parsing. The pre-processed data are passed to feature extractors and event trigger identifier. The Wiki-pedia Word2Vec is used for semantic embedding to enhance features. The rest of the stages are discussed in the sections below.

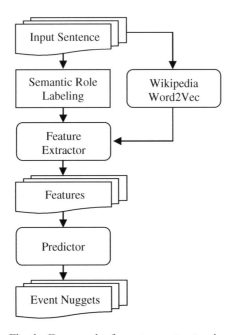

Fig. 1.   Framework of event nugget extraction

3   Semantic Role Labeling and Feature Extraction

Semantic Role Labeling (SRL) is the task of detecting and labeling semantic arguments of a predicate or verb of a sentence and classifying them to specific roles. A typical semantics includes *Agents, Patient, Theme,* etc. It can also include adjunctive argu-ments indicating time, location, manner, etc. Consider the expression (c):

Mary sold the book to John. (c)

The SRL task would recognize the semantic role "to sell" for the lexical item "sold" as representing a predicate, the seller (*Agent*) as "Mary", the goods being sold (*Theme*) as "the book", and the recipient or the buyer represented as "John". It is important towards making sense of the meaning of a sentence.

Many semantic representation resources such as FrameNet [5], VerbNet [6], PropBank [7], etc. are rich with human annotations. This has enhanced their popular use in the supervised machine learning approach. There is an extensive use of these corpora in different approaches to semantic role labeling. Recent studies show that information in the syntactic structure of lexical items in a text can be exploited further for more meaningful feature analysis. A common syntactic information use to extract features from texts is the parse tree. A recent work by Xue and Palmer [8] proposes the feature extraction for semantic role labeling with an overview of the PropBank corpus.

Other studies in the semantic role tagging consider the domain-specific semantic roles such as SPEAK, MESSAGE, and TOPIC or abstract semantic roles such as AGENT or PATIENT. Gildea and Jurafsky [9] use the statistical classifiers trained on sentences from the FrameNet semantic corpus for extraction of lexical and syntactic features.

Semantic Role Labeling is also used in event extraction. A semantic role labeling approach to extracting events from Wikipedia [10] uses the semantic roles for the event argument identification and property extraction. This work also uses external resources for disambiguation and linking before mapping the predicate structures to event models. The research on domain-independent detection, extraction and labeling of atomic events [11] has good success in developing a domain independent event extraction system for texts using atomic level items (i.e., sentences and predicates).

In this study, a probabilistic distribution of features across parse structures is built to define semantic roles for a given sentence. An external corpus is used for disambiguation between roles. Parsing a text provides rich syntactic relations between lexical terms. These relationships can further be processed for semantic relations. We employ several features to assign semantic roles, and use the statistical probability of semantic roles across these features. The following features are used to assign scores for semantic role determination.

Phrase Type. Phrases are important indicators of the semantic roles of lexical items. Phrase types can be obtained from the parse tree by parsing the text. The phrase type used in this study includes the noun phrase (NP), verb phrase (VP) and clause from the parse tree.

Grammatical Function. The relationships among lexical items in a given text are important features to determine the semantic roles of lexical items. We only consider the subject and object relations and we apply this idea to only NPs. NPs have more effect on the subject and object relations. NPs with an S (sentence) ancestors are assigned the subject roles and NPs with VP ancestors are assigned the object roles.

Position. This feature considers if a constituent is before or after a predicate when defining a frame. Generally subjects occur before verbs and objects occur after verbs in an active voice.

Voice. The voice feature refers to the active or passive nature of predicates to capture the connection between the semantic role and the grammatical function.

FrameNet and VerbNet Features. The FrameNet corpus is a lexical database of English that is both human and machine readable, based on annotating examples of how words are used in actual texts. FrameNet is based on a theory of meaning called frame semantics [12]. A frame is a script-like conceptual structure that describes a particular type of situation, object, or event along with the participants and props that are needed for that frame. For example, the "Apply_Heat" frame describes a common situation involving a "Cook", some "Food" and a "Heating_Instrument", and is evoked by words such as bake, blanch, boil, broil, steam, etc. The roles of frames are called "frame elements" (FEs) and the frame-evoking words are called "lexical units" (LUs). FrameNet also includes relations between frames. Several types of relations are defined, of which the most important are:

- Inheritance: An IS-A relation. The child frame is a subtype of the parent frame, and each FE in the parent is bound to a corresponding FE in the child. An example is the "Revenge" frame which inherits from the "Rewards_and_punishments" frame.
- Using: The child frame presupposes the parent frame as background. For example, the "Speed" frame uses or presupposes the "Motion" frame. However not all parent FEs are bound to child FEs.
- Subframe: The child frame is a subevent of a complex event represented by the parent. For instance, the "Criminal_process" frame has the subframes of "Arrest", Arraignment", "Trial", and "sentencing".
- Perspective_on: The child frame provides a particular perspective on an un-perspectivized parent frame. A pair of examples consists of the "Hiring" and "Get_a_job" frames, which perspectivize the "Employment_start" frame from the employer's and the employee's point of view, respectively.

The main idea of using FrameNet is that there are variations in semantic role types available in a particular event. Hence we can constraint the identification of important frames relevant to a particular sentence or a predicate to the role of searching problem. A generative model for semantic role labeling uses the FrameNet corpus for semantic role and frame identification [13].

Similarly, VerbNet also provides network structure, revealing relationships such as the sense of application. VerbNet verb classes are constructed based on syntactic frames, with verbs in the same class sharing a similar syntactic behavior. Thematic roles are provided for each argument in a syntactic frame, together with the selectional restrictions. The major difference between the VerbNet thematic roles and FrameNet semantic roles is that the thematic roles are generic and global with respect to language, while the semantic roles from FrameNet are local and specific only to their frame. VerbNet only focuses on lexical terms that are verbs, thus limiting its overall contribution to our model. However, this limitation does not overrun its significance since verbs are sufficiently important in determining the event nugget.

In this study, FrameNet and VerbNet features are used in training our model. The target of our system is to identify event nuggets in a text. Therefore, given a constituent from a sentence, we have to decide what the semantic type is with respect to FrameNet. This can be determined since Frame Elements and their associated Lexical Units in FrameNet reside in the semantic space via frame-to-frame relations and semantic types. To incorporate features from both corpus, we reduce the problem to linking the FrameNet to VerbNet using the following two steps:

- Associate VerbNet verb entries with FrameNet semantic frames.
- Associate VerbNet syntactic frame arguments to semantic roles.

In order to link a VerbNet verb entry to a FrameNet semantic frame, we need to identify the corresponding verb meanings in two lexical resources. First, VerbNet entries are divided into two sets, depending on whether they have a direct match in FrameNet. To create a division of these two sets, we recognize the identical word forms among VerbNet verb entries and FrameNet lexical units, and find the intersection of their corresponding sense lists. The first verb set is composed of VerbNet entries that have a direct counterpart in FrameNet. We label the entries with the corresponding frames. Note that multiple frames can be assigned to a single VerbNet verb entry. For example, the VerbNet sense numbers 1, 2, and 3 of "admonish" belong to the verb class "advise-31.9-1" while in FrameNet sense numbers 1 and 2 of "admonish" are defined under the frame "Attempt_suasion" and sense number 3 under "Judgement_direct_address".

For the second set, where no corresponding lexical unit can be identified in FrameNet, the study uses Word2Vec to find synonym related words, and uses such words to identify frames for those verb entries. Only the most similar word for which a frame is defined in FrameNet is used. Once each VerbNet verb entry is mapped to a semantic frame, we can map FrameNet semantic roles and VerbNet arguments. To find these correspondences between a VerbNet argument and a FrameNet semantic role, we compare features against each other. The features used includes Grammatical Function (GF) (e.g. subject, object), Phrase Type (PT) (e.g. noun phrase NP, prepositional phrase PP), Voice (active and passive) and Selectional Restriction (SR). If an argument and a semantic role have equivalent GF, PT and Voice features, a correspondence is established. If the phrase is a prepositional phrase, then their corresponding preposition should also agree and the head word of the semantic role needs to meet the selectional restrictions of the argument. Probability distribution of these features in the annotated training data is used for training purposes.

For training the model, the study uses probabilities calculated from features mentioned above. Details about description of probability distribution of features and illustration of semantic roles computation are shown in Sect. 4.

4 Probability Distribution of Features

This study uses the data from TAC 2015 event nugget track[2]. For training, statistical probabilities are determined across the training data for features mentioned in Sect. 3 above to train the model in Sect. 5. As an illustration, considering a lexical item, l, and a phrase type, pt, we can define a distribution for semantic role sr, as in (1).

$$P(sr|l, pt, X) = \frac{\#(sr, l, pt, X)}{\#(l, pt, X)} \tag{1}$$

---

[2] http://www.nist.gov/tac/2015/KBP/data.html.

The probability is calculated as the ratio of the count of each role, $\#(sr, l, pt, X)$, to the number of observations for each event nugget, $\#(l, pt, X)$. In (1), X is a feature such as grammatical function (gf), Voice, or Position. It is worth to mention that this method significantly works for simple event nuggets with only one lexical item. For complex events with two or more lexical items, we find the joint probability for the two items forming the event nugget. As an illustration, consider Sentence (a) "*Several militants are shot death during clashes near Kabul.*" in Sect. 1.

Considering the lexical item "Several", the model obtains the probability distribution shown in Table 2 below.

Table 2. An example probability distribution over features

$P(sr\|l, pt, gf)$	Scores
$P(sr = AGT\|l = Several, pt = NP, gf = Subj)$	0.145
$P(sr = THM\|l = Several, pt = NP, gf = Subj)$	0.131
$P(sr = THM\|l = Several, pt = NP, gf = Obj)$	0.120
$P(sr = AGT\|l = Several, pt = JJ)$	0.547
$P(sr = THM\|l = Several, pt = JJ)$	0.348

In Table 2, sr means the semantic role such as agent (AGT) and theme (THM) as mentioned in Sect. 3. The distribution is over three key features: (1) the phrase type (pt), e.g., noun phrase (NP), adjective (JJ), (2) lexical item (l), and (3) grammatical function (gf) such as subject ($Subj$) and object (Obj).

A similar distribution is defined for FrameNet semantic types and relations across FrameNet-VerbNet Mapping[3] from SemLink project[4]. Using the lexical units (lu) within a frame and frame relations (fr), the probability estimates are computed below. The value $\#(l \in lu)$ and $\#(l \in fr)$ represent the count of a lexical item, l, in lexical units and frame relations respectively.

$$P(st|l, lu, fr) = p(st|l, lu) + p(st|l, fr) \tag{2}$$

where

$$P(st|l, lu) = \frac{\#(l \in lu)}{\#(lu)} \tag{3}$$

$$p(st|l, fr) = \frac{\#(l \in fr)}{\#fr} \tag{4}$$

It can be noted that the framework for our probability distribution of features emphasizes the probability distribution and the joint probability distribution of semantic roles of lexical items.

---

[3] https://verbs.colorado.edu/semlink/semlink1.1/vn-fn/.

[4] https://verbs.colorado.edu/semlink/.

5   Event Nugget Extraction Model

After extensive extraction of features as discussed in Sect. 3 above, the study utilizes these features to present lexical terms with values through the neural embedding.

Take Sentence (a) as an example. Using the feature values, each lexical item can be numerically represented by a vector. The lexical term "Several", for example, can be represented as a vector V as follows.

$$V = [0.145, 0.131, 0.12, 0.547, 0.348, 0.478]$$

The first five elements of V are $P(r|pt, l, gf)$ shown in Table 2 and the last element is $P(st|l, lu, fr)$ value calculated with (2). Vectors such as V are computed for each lexical term and bigram in the input sentence. The study uses the bigram to expand the coverage of the complex event nuggets that are the combination of two lexical terms. Each sentence is represented by a pair set of vectors for single lexical terms and for the bigram of lexical terms respectively. Using the feature vectors for sentences in the training data, a skip-gram model [14] is used to learn feature patterns of the annotated event nuggets.

The training objective of the skip-gram model is to find word representations that are useful for predicting the surrounding words in a sentence [15]. Unlike most of the other neural network architectures for learning word vectors, the training of the skip-gram model does not involve dense matrix multiplications. Recently an extension of the original skip-gram model proposed by Mikolov *et al.* [15] has shown their speed-up and accuracy in training and prediction. Figure 2 below illustrates the skip-gram model architecture in our system. In our model, for a given set of words in a sentence in the training data, the context for a word is in relation to the annotated event nuggets. The annotations in the training data those not only provide the event type annotations but also event subtype and realis. Our proposed approach provides a single model that learns features and determines the event mention, event types, event subtypes and realis.

In Fig. 2, the input of the model is a feature vector of a single lexical term or a bigram V; and the output is the event nuggets in context of a lexical term, $\{y_1, y_2, y_N\}$, defined in the window size N. The hidden layer h_i is composed of vector W of $1 \times N$ dimension. To parameterize the model, the study follows the neural-network language model literature, and models the conditional probability using soft-max [16] as shown in (5). Here v_c and w_a in R^d are vector representations for c and w respectively. Set C is the set of all available contexts; c is a member of C; and w is a feature vector. The parameter θ represents the feature and context vectors respectively (v_c and v_w).

$$p(c|w; \theta) = \frac{e^{v_c \cdot w_a}}{\sum_{c' \in C} e^{v_{c'} \cdot w_a}} \tag{5}$$

For the final determination of event nuggets from this model, only those terms that are assigned output values in relations to the training model are considered as the event nuggets. Therefore, the model performance will be negatively affected by unseen training samples. In handling this problem, the Wikipedia corpus is used for the larger

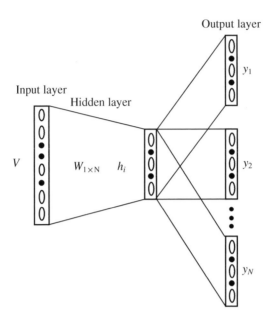

Fig. 2. Skip-gram model architecture for learning and predicting event nuggets

scale training in reference to the annotated training data. However, this strategy does not make a significant difference, and thus recall is affected.

6 Evaluation and Results

The evaluation system is the tool provided for TAC 2015 Event track[5] evaluation. The evaluation metrics are recall (R), precision (P), and F1 score (F) both at micro and macro averages as shown below.

$$R = \frac{TP}{(TP + FN)} \tag{6}$$

$$P = \frac{TP}{(TP + FP)} \tag{7}$$

$$F = \frac{2PR}{(P + P)} \tag{8}$$

where *TP*, *FP* and *FN* are the true positive, false positive and false negative with respective to the gold standard annotation of the test data respectively. These values are

---

[5] http://cairo.lti.cs.cmu.edu/kbp/2015/event/scoring.

used to calculate micro and macro values for precision and recall as shown below where c is the total number of event nuggets the system proposes.

$$R_{micro} = \frac{\sum_{i=1}^{|c|} TP}{\sum_{i=1}^{|c|} (TP + FN)} \tag{9}$$

$$P_{micro} = \frac{\sum_{i=1}^{|c|} TP}{\sum_{i=1}^{|c|} (TP + FP)} \tag{10}$$

$$R_{macro} = \frac{1}{C} \frac{\sum_{i=1}^{|c|} TP}{\sum_{i=1}^{|c|} (TP + FN)} \tag{11}$$

$$P_{macro} = \frac{1}{C} \frac{\sum_{i=1}^{|c|} TP}{\sum_{i=1}^{|c|} (TP + FP)} \tag{12}$$

The evaluation considers four attributes: Event Mention, Types, REALIS Status, and Types and REALIS Status combined as shown in Table 3.

Table 3. Final mention detection results for event nuggets extractions

Macro average			
Attributes	Precision	Recall	F1-score
Plain	98.51 %	63.70 %	77.37 %
Type	98.51 %	63.70 %	77.37 %
Realis	98.51 %	63.70 %	77.37 %
Type + Realis	98.51 %	63.70%	77.37%
Micro average			
Attributes	Precision	Recall	F1-score
Plain	100 %	55.16 %	71.10 %
Type	100 %	55.16 %	71.10 %
Realis	100 %	55.16 %	71.10 %
Type + Realis	100 %	55.16 %	71.10 %

From Table 3, for both macro and micro averages, our system has compromised efficiency for recall. This is due to the weak handling of unseen data during testing. The result also shows the same values in precision, recall and F1-Scores for plain annotation, type annotations and realis. This is because features to classify event mentions are learned from the same source.

Various related works such as [17] that is about the disease-related nugget extraction and [18] that investigates the study of event nugget extraction in 2014 use similar measures of evaluation on the event nugget extraction task, on categories of events nuggets and a plot study. They show the best F1-score of 61.0 % and 70 %

respectively. Comparison shows that our system's performance is competitive with outstanding precision and F1-Scores.

7 Conclusion

In this study, we present a single model for extracting and classifying events into types, subtype and realis. It also shows the effectiveness of feature extraction via the semantic role labeling and from the reference corpora. The study also demonstrates a probability distribution across features to translate lexical terms into feature vectors. Neural embedding techniques such as the skip-gram model in distributional semantics have an extensive usage in speech recognition. While the system reaches a good precision, 98.51 % in macro average and 100 % in micro average, there is still a need to improve its recall. There is a big margin of difference between the recall and the precision.

In future, there is a plan to continue to improve this work by exploiting more lexical resources for the model's weak point. Extending and diversifying features for individual sub-tasks, maybe give different results and/or much better result.

References

1. Song, Z., Bies, A., Strassel, S., Riese, T., Mott, J., Ellis, J., Wright, J., Kulick, S., Ryant, N., Ma, X.: From light to rich ERE: annotation of entities, relations, and events. In: Proceedings of the 3rd Workshop on EVENTS at the NAACL-HLT, pp. 89–98 (2015)
2. Aguilar, J., Beller, C., McNamee, P., Durme, B.V., Strassel, S., Song, Z., Ellis, J.: A comparison of the events and relations across ACE, ERE, RAC-KBP, and Framenet Annotation Standards. In: Proceedings of the 2nd Workshop on EVENTS: Definition, Detection, Coreference, and Representation, pp. 45–53 (2014)
3. Chiticariu, L., Krishnamurthy, R., Li, Y., Reiss, F., Vaithyanathan, S.: Domain adaptation of rule-based annotators for named-entity recognition rasks. In: Proceedings of the 2010 Conference on Empirical Methods in Natural Language Processing, pp. 1002–1012 (2010)
4. Mitamura, T., Yamakawa, Y., Holm, S., Song, Z., Bies, A., Kulick, S., Strassel, S.: Event nugget annotation: processes and issues. In: Proceedings of the 3rd Workshop on EVENTS at the NAACL-HLT, pp. 66–76 (2015)
5. Baker, C.F., Fillmore, C.J., Lowe, J.B.: The Berkeley Framenet Project. In: Proceedings of the 17th International Conference on Computational Linguistics, vol. 1, pp. 86–90 (1998)
6. Schuler, K.K.: VerbNet: A Broad-Coverage, Comprehensive Verb Lexicon. University of Pennsylvania, Philadelphia (2005)
7. Palmer, M., Gildea, D., Kingsbury, P.: The proposition bank: an annotated corpus of semantic roles. Comput. Ling. 31(1), 71–105 (2005)
8. Xue, N., Palmer, M.: Calibrating features for semantic role labeling. In: Proceedings of the 2004 Conference on Empirical Methods in Natural Language Processing, pp. 88–94 (2004)
9. Gildea, D., Jurafsky, D.: Automatic labeling of semantic roles. Comput. Ling. 28(3), 245–288 (2002)
10. Exner, P., Nugues, P.: Using semantic role labeling to extract events from wikipedia. In: Proceedings of the First Workshop on Detection, Representation, and Exploitation of Events in the Semantic Web (DeRiVE 2011), Workshop in Conjunction with the 10th International Semantic Web Conference, pp. 38–47 (2011)

11. Filatova, E., Hatzivassiloglou, V.: Domain-independent detection, extraction, and labeling of atomic events. Ubiquity: ACM IT Mag. Forum **4**(4), 18–24 (2003)
12. Fillmore, C.J.: Frame semantics and the nature of language. In: Annals of the New York Academy of Sciences: Conference on the Origin and Development of Language and Speech, pp. 21–32 (1976)
13. Thompson, C.A., Levy, R., Manning, C.D.: A generative model for semantic role labeling. In: Lavrač, N., Gamberger, D., Todorovski, L., Blockeel, H. (eds.) ECML 2003. LNCS (LNAI), vol. 2837, pp. 397–408. Springer, Heidelberg (2003)
14. Guthrie, D., Allison, B., Liu, W., Guthrie, L., Wilks, Y.: A closer look at skip-gram modelling. In: Proceedings of the 5th International Conference on Language Resources and Evaluation (LREC-2006), pp. 1–4 (2006)
15. Mikolov, T., Sutskever, I., Chen, K. Corrado, G.S., Dean, J.: Distributed representations of words and phrases and their compositionality. Adv. Neural Inf. Process. Syst., 3111–3119 (2013)
16. Monti, S., Cooper, G.F.: Learning hybrid Bayesian networks from data. In: Jordan, M.I. (ed.) Learning in Graphical Models, pp. 521–540. Springer, Dordrecht (1998)
17. Imran, M., Elbassuoni, S. M., Castillo, C., Diaz, F., Meier, P.: Extracting information nuggets from disaster-related messages in social media. In: Proceedings of the 10th International Conference on Information Systems for Crisis Response and Management (ISCRAM 2013), pp. 791–801 (2013)
18. Liu, Z., Mitamura, T., Hovy, E.: Evaluation algorithms for event nugget detection: a pilot study. In: Proceedings of the 3rd Workshop on EVENTS at the NAACL-HLT, pp. 53–57 (2015)

A Hybrid Approach to Sentiment Analysis with Benchmarking Results

Orestes Appel[1(✉)], Francisco Chiclana[1], Jenny Carter[1], and Hamido Fujita[2]

[1] Centre for Computational Intelligence (CCI),
De Montfort University, Leicester, UK
orestes.appel@email.dmu.ac.uk, {chiclana,jennyc}@dmu.ac.uk
[2] Iwate Prefectural University (IPU), Takizawa, Iwate, Japan
issam@iwate-pu.ac.jp

Abstract. The objective of this article is two-fold. Firstly, a *hybrid approach* to Sentiment Analysis encompassing the use of Semantic Rules, Fuzzy Sets and an enriched Sentiment Lexicon, improved with the support of SentiWordNet is described. Secondly, the proposed hybrid method is compared against two well established Supervised Learning techniques, Naïve Bayes and Maximum Entropy. Using the well known and publicly available *Movie Review Dataset*, the proposed hybrid system achieved higher *accuracy* and *precision* than Naïve Bayes (NB) and Maximum Entropy (ME).

Keywords: Sentiment Analysis · Fuzzy sets · Semantic Rules · Natural language processing · Computational linguistic · SentiWordNet

1 Introduction

In this section we will cover the basics of Sentiment Analysis (SA), or Opinion Mining (OM) as it is frequently called as well, and the motivation that led us to explore the solution and results we are discussing in this article.

1.1 Sentiment Analysis – Basics

Sentiment Analysis (SA) has been at the front of research efforts for the last few years. The data volumes generated through multiple channels and media are too bulky and complex for human digesting, hence the need for a computer-aided process capable of telling the end-user (a product consumer, a researcher, a teacher, a political analyst, etc.) whether a document, a sentence or a tweet are carrying an opinion or factual information. If it is the former, the users will be keen on telling positive opinions from negatives. Furthermore, there is even room for understanding the degree of positiveness or negativeness of a given piece of information. Typically, SA is performed at specific levels, such as feature/aspect level, sentence level, document level, etc. In this research, we are focusing at carrying SA at the sentence level. For a complete review of the evolution of the Sentiment Analysis field, please refer to the work of [1,2].

© Springer International Publishing Switzerland 2016
H. Fujita et al. (Eds.): IEA/AIE 2016, LNAI 9799, pp. 242–254, 2016.
DOI: 10.1007/978-3-319-42007-3_21

1.2 Motivation

Most of approaches to address the SA problem belong either to the category of Supervised or Unsupervised Machine Learning. However, it seems to the author that fuzzy sets, considering their mathematical properties and their ability to deal with vagueness and uncertainty, are well equipped as well to model sentiment-related problems. It can be hypothesised that a combination of different techniques could be more effective at succeeding at addressing the SA challenges than specific techniques used in isolation. In the next few paragraphs we will address our motivation to explore this realm of possibilities.

Dzogang et al. stated in [4] that usually authors refer mainly to psychological models when addressing the SA problem. However, other models may be successful as well. As per Dzogang et al. "...it must be underlined that some appraisal based approaches make use of graduality through fuzzy inference and fuzzy aggregation for processing affective mechanisms ambiguity and imprecision...". On the other hand, Bing Liu [10], one of the main world experts in SA, says that "...we probably relied too much on Machine Learning" when dealing with SA. Hence, the following arguments combined together have sparked the research here reported: (i) the concept of graduality expressed through fuzzy sets; (ii) the idea that other alternatives, besides Supervised Machine Learning, may be viable as well when extracting sentiment from text; (iii) the positive contribution that semantic rules and a solid opinion lexicon can have in identifying polarity; and (iv) the success brought in by the use of effective NLP techniques, like parsing and smart-tokenisation.

During the rest of this article we will cover: (I) the Research Methodology used; (II) the Proposed Hybrid Classification Method for SA; (III) a comparison of the experimental results obtained; and (IV) our Conclusions.

2 Research Methodology

The research methodology used is discussed from two different perspectives: the process to follow and the data to use for measuring the performance of the proposed SA solution.

2.1 The Process

In order to measure success, any proposed solution should perform *same or better* than today's most accepted solutions. In the specific case of the SA problem, the proposed hybrid solution is compared against two Supervised Learning methods that enjoy a high level of acceptance and credibility in the classification research community and that are relatively easy to implement: Naïve Bayes (NB) and Maximum Entropy (ME) [10,13–15]. The comparison will focus on two aspects at the sentence level: (1) *Subjectivity Determination* (being able to tell an opinion from a fact), and (2) *Opinion polarity/graduality Classification* (assigning a value to an opinion inside a given range [positive, negative, neutral, etc.]).

Our approach will consist of comparing our results with those attained by the Supervised Learning methods described above when identifying subjectivity and estimating opinion polarity on the subjective content at the sentence level. The focus will be on *Accuracy* and *Precision*.

2.2   The Data

The main dataset used in our research is the *Movie Review Dataset*, published and utilised by Pang and Lee [14]. The results obtained by using the aforementioned dataset are addressed in [13,15]. The fact that many articles in SA discuss this dataset and have used it to validate their own methods and approaches makes it an ideal candidate from the benchmarking angle. The dataset contains 5,331 positive snippets and an equal number of negative ones. Each line corresponds to a single snippet that could contain more than one sentence. As a results-validation dataset, we will use the data presented in *Sentiment140*, which is available at http://help.sentiment140.com/for-students.

2.3   Most Commonly Used Measurements in the Evaluation of SA

It has become customary to evaluate the performance of sentiment classification systems utilising the following four indexes, as defined in [18] (refer to the so-called *confusion matrix* given in Table 1):

- *Accuracy*: the portion of all true predicted instances against all predicted instances $\equiv \dfrac{(TP + TN)}{(TP + TN + FP + FN)}$
- *Precision*: the portion of true positive predicted instances against all positive predicted instances $\equiv \dfrac{TP}{(TP + FP)}$
- *Recall*: the portion of true positive predicted instances against all actual positive instances $\equiv \dfrac{TP}{(TP + FN)}$
- *F1-score*:   a   harmonic   average   of   precision   and   recall $\equiv \dfrac{(2 \times Precision \times Recall)}{(Precision + Recall)}$

Table 1. Confusion matrix

	Predicted positives	Predicted negatives
Actual positive instances	# of true positive instances (TP)	# of false negative instances (FN)
Actual negative instances	# of false positive instances (FP)	# of true negative instances (TN)

3 A Hybrid Approach to Sentiment Analysis - The Proposed Method

Let us discuss a little further what exactly we mean by utilising a 'Hybrid Approach', a concept that is key to our proposed solution. Our intention is to manage *hybrid concepts* at two different levels: (i) the methods employed by the sentiment classifiers (Naïve Bayes, Maximum Entropy, Decision Tree, Fuzzy Sets/Logic, and others), and (ii) the techniques utilised to build key components of our approach, like the creation and population of the Sentiment/Opinion Lexicon. Our study will focus on addressing the SA problem at the *sentence level*. The following paragraphs will present the components and processes that encompass our proposed Hybrid Solution.

3.1 The Sentiment Lexicon

Dr. Liu compiled an Opinion Lexicon a few years ago, as mentioned at http://www.cs.uic.edu/~liub/FBS/sentiment-analysis.html#lexicon: "it does include a list of positive and negative opinion words or sentiment words for English (around 6800 words) [...] compiled over many years starting from our first paper (Hu and Liu, KDD-2004)" [8]. it was decided to use Liu's Lexicon to re-use data resulting from a quality effort in words compilation. In generating our Opinion Lexicon, we have taken the following approach:

1. We have utilised the *opinion-conveying-words* that are part of the Opinion Lexicon used by Prof. Bing Liu et al. in [8] and other pieces of research work. They correspond to lists containing 'positive meaning words' and 'negative meaning words'. They include *only* nouns, verbs, adjectives and adverbs. These four elements of Part-of-Speech (PoS) have been proven to be capable of delivering opinions [6,7,9,20].
2. We have used SentiWordNet [5] to extract *polarity or valence scores* for words carrying opinion sense.
3. We have combined both elements in (1) and (2) above. As such, we have substituted the words in the original Liu's opinion lexicons for their Synset-equivalent in SentiWordNet (at least partially). This way, we have added a *positive score* and a *negative score* to the existing words in Liu's lexicon, enriching the Lexicon. It is important to keep in mind as well that

$$0 \leq PositiveScore, NegativeScore, ObjectivityScore \leq 1$$

$$0 \leq (PositiveScore + NegativeScore) \leq 1$$

$$ObjectivityScore = 1 - (PositiveScore + NegativeScore)$$

As such, when the sum of *PositiveScore* and *NegativeScore* is equal to 1 for a given word $Word_k$, then the term $Word_k$ is fully opinionated, as opposed to the case when the addition of these two scores is zero, in which case the term $Word_k$ is fully Neutral or Objective.

4. The above results in an *improved* Sentiment Lexicon containing lists of Positive and Negative words which have as attributes polarity/valence scores. Here is the description of the attributes of our proposed lexicon:

   Word: word in the lexicon (entries).

   PoS: part of speech (n=noun; v=verb; a=adjective; r=adverb; s=adjective satellite).

   PSC: Positive Score as taken from SentiWordNet [5].

   NSC: Negative Score as taken from SentiWordNet [5].

   COBJ: Calculated Objectivity Score [5].

   VDX: Versioning index for identifying/managing synonyms (future use).

   UPDC: Update Counter to keep track of every time a given entry in the lexicon is updated.

   PL: Polarity Label (either **pos** for positive or **neg** for negative)

   A typical occurrence of an item in the sentiment lexicon is represented as:

$$(\#(word_k \text{ PoS PSC NSC COBJ VDX UPDC}) \text{ PL}).$$

3.2 Semantic Rules (SR)

In this section we address those cases for which new rules need to be defined in order to model the problem of SA in a more accurate fashion. Indeed, a number of authors, among them [12, 19, 21], have pointed out the fact that negation and the use of specific part-of-speech particles, like 'but, despite, unless, ...' could affect the final outcome of a classification exercise. Thus, some rule strategies are needed to be put in place as the order of the different part-of-speech play a role in the semantic of a sentence. Researchers have been, through time, improving the quality of these semantic rules so that they are more encompassing of the possible cases that must be managed. These research efforts are summarised by Xie et al. in a very well organised, easy to read semantic rule tables in [21]. Despite the apparent completeness of existing Semantic Rules, *two new rules* are incorporated for managing particular part-of-speech particles: the particle **while** and the particle **however**, which were not included in the original set of rules provided in [21], resulting in the following set of Semantic Rules (Table 2).

3.3 Negation Handling

According to Dr. Christopher Potts from Stanford University, Linguistics Department http://sentiment.christopherpotts.net/lingstruc.html, "Sentiment words behave very differently when under the semantic scope of negation" [17]. The complex nature of *negation* suggests that it would be difficult to have a general *a priori* rule for how to handle negation. The technique that Dr. Potts favours for approximating the effects of negation is due to Das and Chen [3] and Pang, Lee, and Vaithyanathan [15]. This method utilises *Regular Expressions* and has been implemented in our research effort as an extension to the publicly available Tokenizer that we have used [16]. Notice that even long-distance negation effects can be managed using this technique.

Table 2. Semantic Rules for proposed hybrid system

Rules	Semantic Rules	Example
R1	Polarity (not var_k) = −Polarity (var_k)	'not *bad*.'
R2	Polarity (NP_1 of NP_2) = Compose (NP_1, NP_2)	'*Lack* of crime in rural areas.'
R3	Polarity (NP_1 VP_1) = Compose (NP_1, VP_1)	'*Crime* has decreased.'
R4	Polarity (NP_1 be ADJ) = Compose (ADJ, NP_1)	'*Damage* is minimal.'
R5	Polarity (NP_1 of VP_1) = Compose (NP_1, VP_1)	'*Lack* of killing in rural areas.'
R6	Polarity (ADJ to VP_1) = Compose (ADJ, VP_1)	'*Unlikely* to destroy the planet.'
R7	Polarity (VP_1 NP_1) = Compose (VP_1, NP_1)	'*Destroyed* terrorism.'
R8	Polarity (VP_1 to VP_2) = Compose (VP_1, VP_2)	'*Refused* to deceive the man.'
R9	Polarity (ADJ as NP) = $1_{(Polarity(NP=0))}$ · Polarity(ADJ) + $1_{(Polarity(NP\neq0))}$ · Polarity(NP)	'As *ugly* as a rock.'
R10	Polarity (not as ADJ as NP) = −Polarity (ADJ)	'That wasn't as *bad* as the original.'
R11	If sentence contains "but", disregard all previous sentiment and only take the sentiment of the part after "but"	'And I've never liked that director, *but* I loved this movie.'
R12	If sentence contains "despite", only take the sentiment of the part before "despite"	'I love the movie, *despite* the fact that I hate that director.'
R13	If sentence contains "unless" followed by a negative clause, disregard the "unless" clause	'Everyone likes the video *unless* he is a sociopath.'
R14 (New)	If sentence contains "while", disregard the sentence following the 'while' and take the sentiment only of the sentence that follows the one after the 'while'	'*While* they did their best, the team played a horrible game.'
R15 (New)	If sentence contains "however", disregard the sentence *before* 'however' and take only the sentiment of sentence *after* 'however'	'The film counted with good actors. *However*, the plot was very poor.'

Table 3. Compose functions referenced in Table 2

Compose functions	Algorithms
Compose1 (arg1, arg2)	1. Return −Polarity(*arg2*) if *arg1* is negation.
	2. Return Polarity(*arg1*) if (Polarity(*arg1*) = Polarity(*arg2*).
	3. Otherwise, return the majority term polarity in *arg1* and *arg2*.
Compose2 (arg1, arg2)	1. Return Polarity(*arg2*) if *arg1* is negative and *arg2* is not neutral.
	2. Return −1 if *arg1* is negative and *arg2* is neutral.
	3. Return Polarity(arg2) if *arg1* is positive and *arg2* is not neutral.
	4. Return 2*Polarity(arg1) if Polarity(*arg1*) = Polarity(*arg2*).
	5. Return Polarity(*arg1*) + Polarity(*arg2*) if *arg1* is positive and *arg2* is neutral.
	6. Return Polarity(*arg1*) + Polarity(*arg2*) if *arg2* is positive and *arg1* is neutral.
	7. Otherwise, return. **0**

3.4 Linguistic Variables and Fuzzy Sets

According to Miller [11], *7 plus or minus 2*, is the effective number of categories that a subject (individual or person) can maintain. In our case, we have chosen a conservative approach and have devised 5 labels (7 minus 2), symmetrically distributed in the interval $[0 \ldots 1]$. Our choice of trapezoidal function obeys to the fact that the latter generalises a triangular function and we have aimed for more generality and for more than one value at the top of every category. We have opted for linguistic modifiers that have the ability to change the level of

granularity. In our research, the classification labels for the **intensity** of semantic orientation and/or polarity of a given sentence are identified as: (a) Poor,
(b) Slight, (c) Moderate, (d) Very and (e) Most. In essence, we either classify a
sentence as *Objective* or *Subjective*. If it is considered as *Subjective*, then it could
either be *Negative* or *Positive*, with an intensity as qualified by the modifiers
belonging in the set {*Poor, Slight, Moderate, Very, Most*}. A generic trapezoidal membership function would take the following form as shown in Fig. 1
and described in Eq. 1 (Table 3).

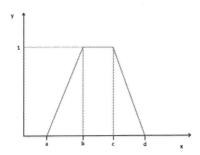

Fig. 1. General form of the trapezoidal family of curves

$$\mu_{\tilde{A}}(x) = \begin{cases} \frac{x-a}{b-a} & \text{if} \quad a \le x \le b; \\ 1 & \text{if} \quad b \le x \le c; \\ \frac{d-x}{d-c} & \text{if} \quad c \le x \le d; \\ 0 & \text{otherwise.} \end{cases} \tag{1}$$

A trapezoidal membership function (MF) can be represented as well using
the following 4-tuple (a, b, c, d). Our MFs in 4-tuple format are as follows:

– MF (Poor): (0, 0, 0.050, 0.150)
– MF (Slightly): (0.050, 0.150, 0.250, 0.350)
– MF (Moderate): (0.250, 0.350, 0.650, 0.750)
– MF (Very): (0.650, 0.750, 0.850, 0.950)
– MF (Most): 0.850, 0.950, 1, 1)

Let us refresh some fuzzy sets theory concepts.

– T-norm ($\min(a, b)$): it is a binary function $T : [0, 1] \times [0, 1] \rightarrow [0, 1]$ that is
  commutative, associative, monotonic, and $T(a, 1) = a$.
– T-conorm ($\max(a, b)$): it is a binary function $S : [0, 1] \times [0, 1] \rightarrow [0, 1]$, that is
  commutative, associative, monotonic, and $S(a, 0) = a$.
– Any t-conorm, S, can be generated by a t-norm, T, and viceversa: $S(a, b) = 1 - T(1 - a, 1 - b)$. When this is the case (T, S) is said to be a dual pair of
  t-norm and t-conorm, as it is the pair $(\min, \max)$.

3.5   Calculating Polarity Scores for Sentences

In order to illustrate the mechanism used, let us calculate the polarity of a given sentence S_k. As an example, let us assume that out of the 11 words included in sentence S_k, 4 are in our lexicon and will be used to determine the polarity of the sentence. *Sentence (S_k)*: "Many good actors. However, the film was simplistic, silly, unrealistic and tedious". The *sentiment-conveying words* in S_k, are:

– Good: (#(good s 0.75 0.0 0.25 0 0) pos)
– Simplistic: (#(simplistic s 0.0 0.0 1.0 0 0) neg)
– Tedious: (#(tedious no-info no-info no-info no-info 0 0) neg)
– Unrealistic: not in our Opinion Lexicon
– Silly: (#(silly n 0.125 0.0 0.875 0 0) neg)

Our approach to determining the qualified semantic orientation of the sentences belonging in our test dataset, involves a two-step approach, as described below. Please notice that we will present a summarised version of the processes behind the proposed Hybrid Method.

Step I: Hybrid Standard Classification (HSC): This process is performed by using a number of techniques, as follows. Every intermediate step has as an outcome a list of features that is passed to the next sub-step in the process chain.

1. Takes as input the output from the Tokeniser, PoS tagging and Smart Parsing (negation is partially handled here)
2. Applies the Semantic Rules presented in Table 2.
3. Extracts essential particles that convey sentiment/opinion (adjective, nouns, verbs and adverbs). It generates a list with those key PoS particles
4. Searches the Sentiment Lexicon and substitutes the words in the list-expression with the associated matches in the lexicon
5. Generates an exception list for those key-words that are not in the Opinion Lexicon
6. Calculates the semantic orientation of each sentence by taking into consideration: (a) the word-label semantic orientation present in the Lexicon (POS, NEG or OBJ), and (b) the Positive and/or Negative Scores of the words in the sentence that appear in the lexicon. The calculations are performed as per the Semantic Rules and a classification label belonging in the set {Positive, Negative, Objective, No-Semantic-Orientation} is assigned to each sentence.
7. Re-scans the resulting list in search for words still labelled as OBJ that could be now converted into either POS or NEG labels. This situation is resolved by using the services of a dictionary previously generated for all sentences being processed. The structure of the Dictionary is [Word, Sentence-numbers-where-Word-appears, Frequency-of-Appearances].
8. Generates a new list with the classification POS/NEG of all sentences in the test dataset (particles initially marked as OBJ, are substituted with their final POS/NEG labels as per resolved in the previous step).

Step II: Hybrid Advanced Classification (HAC): This approach enhances the standard classification process by incorporating:

1. Determination of the degree with which a given sentence leans towards being *positive* or *negative* (a fuzzy approach rather than a crisp method).
   Let us assume that IP corresponds to the Intensity of Polarity of a given word, $Word_i$

   $$\min\left(IP(Word_1) \ \ldots \ IP(Word_n)\right) = \delta,$$

   where $Word_i$ is the i^{sm} sentiment-carrying word in a given sentence, and $i = 1 \ldots numwords$, where $numwords$ corresponds to the total number of sentiment-conveying words, n, found in the Opinion Lexicon for a given sentence S_k; $\delta \in [0 \ldots 1]$.

   $$\max\left(\mu_{poor}(\delta), \ \mu_{slight}(\delta), \ \mu_{moderate}(\delta), \ \mu_{very}(\delta), \ \mu_{most}(\delta)\right) = \beta,$$

   where, μ_j, with $j \in \{poor, slightly, moderate, very, most\}$, corresponds to the evaluation of the membership function μ_j for the sub-index j; $\beta \in [0 \ldots 1]$. In essence, we calculate the T-conorm represented by the value β.
2. Diagnosing when a given sentence could be considered rather *objective/neutral* as opposed to either *positive* or *negative*. Not all sentences have been created equal, and even in the test dataset that has been carefully chosen, there are some sentences that one could argue are rather neutral (not leaning towards negative or positive). With the Advance Classification system we will ponder which sentences could be borderline neutral/objective.

4   Experimental Results

In this section we will look at the experimental results, starting with the outcome of using two ML supervised learning methods (Naïve Bayes and Maximum Entropy), and then we will show the results obtained applying our hybrid method.

4.1   Naïve Bayes Classifier (NB)

We trained the NB classifier using some of the recommendations presented by Perkins in [16]. The classifier uses the concept of 'bag of words'. Using this representation model of a sentence, the classifier creates 'feature vectors' exhibiting the main traits of such a sentence. The NB classifier is a binary classifier. It will classify a sentence either as 'negative' or 'positive', and these categories are exclusive. The classifier returns a probability value that represents the *probability* that the sentence belongs with a specific label (negative or positive). The probability value has to be *0.5* or higher for a sentence to belong in a specific category. The results of applying the NB classification algorithm to the data set created by Pang and Lee in [13] are summarised in Table 4.

Table 4. NB results

Accuracy	0.67
Precision	0.63
Recall	0.85
F1-score	0.72

Table 5. ME results

Accuracy	0.68
Precision	0.63
Recall	0.86
F1-score	0.73

Table 6. HSC/HAC results

Accuracy	0.76
Precision	0.73
Recall	0.83
F1-score	0.77

Table 7. HAC classifier results for **POS** dataset - *Increased Granularity*

False negatives	929
No semantic orientation	35
True positives	4,402
Poor	577
Slight	1,106
Moderate	1,041
Very	1,365
Most	313
Number of snippets	5,331

Table 8. HAC classifier results for **NEG** dataset - *Increased Granularity*

False positives	1,646
No semantic orientation	76
True negatives	3,685
Poor	770
Slight	1,089
Moderate	789
Very	864
Most	173
Number of snippets	5,331

4.2 Maximum Entropy (ME)

We trained the ME classifier using Perkins's recommendations as presented in [16]. We used the Generalized Iterative Scaling (GIS) learning method to train the ME classifier. As in the case of the Naïve Bayes classifier, the ME classifier returns a probability value that represents the *probability* that the sentence belong with a specific label (negative or positive). The probability value has to be of *0.5* or higher for a sentence to belong in a specific category. Once we applied the trained classifier to our test dataset we obtained the results presented in Table 5.

4.3 Our Hybrid Method: HSC/HAC

Our Hybrid Method utilises a number of different components, coming from different disciplines, to achieve the results presented in Table 6, which are very encouraging when compared to the outputs obtained by applying the NB and ME techniques. In addition, when we incorporate the fuzzy set approach (HAC) we can provide a much better granularity level in the classification process. See Tables 7 and 8 for details.

As an example, the intensity of sentence 'The movie was simplistic, silly and tedious' (classified as *Most Negative*) is certainly stronger than the one exhibited by the phrase 'An alternately frustrating and rewarding experience' (labelled by HAC as *Slightly Negative*).

4.4   Comparison of Results

In this subsection we will take a closer look at the results that so far we have obtained. We present the metrics compiled for two other methods and our two hybrid approaches. The results are very encouraging, especially with respect to Accuracy, Precision and F1-score. However, the Recall indicator -known as well as Sensitivity- is better for the results shown by the NB/ME methods. Experiments show that the NB/ME method does slightly better than HSC/HAC for identifying positive sentences, but it does much worse than HSC/HAC when classifying negative snippets. As Recall represents the portion of true positive predicted instances against all actual positive instances, it seems reasonable that NB/ME carries a better Recall than HSC/HAC. In general, our method HSC/HAC shows good improvements when compared to the results achieved by the NB/ME method. The main difference between the results achieved with HSC and HAC, is that the latter adds a fuzzy approach that provides not only a classification result (POS or NEG), but in addition, supplies important data in establishing the intensity or strength with which a given sentence is positive or negative. This outcome could be used in the future to determine additional properties and characteristics, like sentences that are borderline between being subjective or objective.

5   Conclusions

In general, our proposed hybrid system works very well with a high level of *accuracy* and *precision*. Indeed, the fact that our hybrid system improved the results obtained when we applied Naïve Bayes (NB) and Maximum Entropy (ME) to the same dataset *satisfies* one of our initial hypotheses, that a hybrid method using natural language processing techniques, semantic rules and fuzzy sets should be able to perform well. Additionally, by the utilisation of fuzzy sets we can determine when a given sentence has a stronger/weaker intensity in terms of polarity. In closing, there are some lessons learned and observations that we would like to share: (a) using efficient NLP techniques (like tokenising, parsing, negation handling, etc.), contribute positively to the application of our Hybrid Method; (b) the creation of an *improved* Sentiment Lexicon was decisive in obtaining good experimental results; (c) SentiWordNet became an important component of our proposed solution and certainly *enriched* dramatically the quality of our Lexicon. Our expectation is that the quality of the content of SentiWordNet would continue improving with time, reflecting positively in the performance of our Hybrid Method; (d) work should continue on improving the completeness and quality of Semantic Rules. In essence, hybrid techniques can play an important role in the advancement of the SA discipline by combining together the elements we described in our research contribution.

References

1. Anbananthen, K.S.M., Elyasir, A.M.H.: Evolution of opinion mining. Aust. J. Basic Appl. Sci. **7**(6), 359–370 (2013)
2. Appel, O., Chiclana, F., Carter, J.: Main concepts, state of the art and future research questions in sentiment analysis. Acta Polytech. Hung. J. Appl. Sci. **12**(3), 87–108 (2015)
3. Das, S.R., Chen, M.Y., Agarwal, T.V., Brooks, C., Chan, Y.S., Gibson, D., Leinweber, D., Martinez-Jerez, A., Raghubir, P., Rajagopalan, S., Ranade, A., Rubinstein, M., Tufano, P.: Yahoo! for Amazon: sentiment extraction from small talk on the web. In: 8th Asia Pacific Finance Association Annual Conference (2001)
4. Dzogang, F., Lesot, M.-J., Rifqi, M., Bouchon-Meunier, B.: Expressions of graduality for sentiments analysis - a survey. In: 2010 IEEE International Conference on Fuzzy Systems (FUZZ), pp. 1–7 (2010)
5. Esuli, A., Sebastiani, F.: Senti Word Net - a publicly available lexical resource for opinion mining. In: Proceedings of the 5th Conference on Language Resources and Evaluation (LREC06), pp. 417–422 (2006)
6. Hatzivassiloglou, V., McKeown, K.: Towards the automatic identification of adjectival scales: clustering adjectives according to meaning. In: Schubert, L.K. (ed.) ACL: Proceedings of the 31st Annual Meeting of the Association for Computational Linguistics, pp. 22–26, Ohio State University, Columbus, Ohio, USA, pp. 172–182. ACL, June 1993
7. Hatzivassiloglou, V., McKeown, K.: Predicting the semantic orientation of adjectives. In: Proceedings of the 35th Annual Meeting of the ACL and the 8th Conference of the European Chapter of the ACL, New Brunswick, NJ, USA. ACL, pp. 174–181 (1997)
8. Hu, M., Liu, B.: Mining and summarizing customer reviews. In: Proceedings - ACM SIGKDD International Conference on Knowledge Discovery and Data Mining (KDD-2004 full paper), Seattle, Washington, USA, 22–25 August 2004
9. Kamps, J., Marx, M., Mokken, R.J., de Rijke, M.: Using Word Net to measure semantic orientations of adjectives. In: Proceedings of LREC-04, 4th International Conference on Language Resources and Evaluation, LREC 2004, vol. IV, pp. 1115–1118 (2004)
10. Liu, B.: Sentiment Analysis and Opinion Mining, Synthesis Lectures on Human Language Technologies, 1st edn. Morgan and Claypool Publishers, San Rafael (2012)
11. Miller, G.: The magical number seven, plus or minus two: some limits on our capacity for processing information. Psychol. Rev. **63**, 81–97 (1956)
12. Nadali, S., Murad, M., Kadir, R.: Sentiment classification of customer reviews based on fuzzy logic. In: 2010 International Symposium in Information Technology (ITSim), vol. 2, pp. 1037–1040, Kuala Lumpur, Malaysia, June 2010
13. Pang, B., Lee, L.: Seeing stars: exploiting class relationships for sentiment categorization with respect to rating scales. In: Proceedings of the 43rd Annual Meeting on Association for Computational Linguistics (ACL 2005), ACL2005, pp. 115–124 (2005)
14. Pang, B., Lee, L.: Opinion mining and sentiment analysis. NOW Essence Knowl. Found. Trends Inf. Retrieval **2**(1–2), 1–135 (2008)
15. Pang, B., Lee, L., Vaithyanathan, S.: Thumbs up? sentiment classification using machine learning techniques. In: Proceedings of the ACL-02 Conference on Empirical Methods in Natural Language Processing (EMNLP), vol. 10, pp. 79–86 (2002)

16. Perkins, J.: Python Text Processing with NLTK 2.0 Cookbook. Packt Publishing, Birmingham (2010)
17. Potts, C.: Sentiment Symposium Tutorial: Linguistic structure (part of the Sentiment Analysis Symposium held, San Francisco, 8–9 November 2011. Stanford Department of Linguistics, Stanford University (2011). Accessed Dec. 2011
18. Sadegh, M., Othman, R.I.Z.A.: Combining lexicon-based and learning-based methods for twitter sentiment analysis. Int. J. Comput. Technol. **2**(3), 171–178 (2012)
19. Subasic, P., Huettner, A.: Affect analysis of text using fuzzy semantic typing. IEEE Trans. Fuzzy Syst. **9**(4), 483–496 (2001)
20. Wiebe, J.: Learning subjective adjectives from corpora. In: Proceedings of the Seventeenth National Conference on Artificial Intelligence and Twelfth Conference on Innovative Applications of Artificial Intelligence, pp. 735–740. AAAI Press (2000)
21. Xie, Y., Chen, Z., Zhang, K., Cheng, Y., Honbo, D.K., Agrawal, A., Choudhary, A.N.: Mu SES: a multilingual sentiment elicitation system for Social Media Data. IEEE Intell. Syst. **29**(4), 34–42 (2014)

Mixture of Language Models Utilization in Score-Based Sentiment Classification on Clinical Narratives

Tran-Thai Dang[1(✉)] and Tu-Bao Ho[1,2]

[1] Japan Advanced Institute of Science and Technology,
1-1 Asahidai, Nomi, Ishikawa, Japan
{dangtranthai,bao}@jaist.ac.jp
[2] John Von Neumann Institute, VNU-HCM, Ho Chi Minh City, Vietnam

Abstract. Sentiment classification on clinical narratives has been a groundwork to analyze patient's health status, medical condition and treatment. The work posed challenges due to the shortness, and implicit sentiment of the clinical text. The paper shows that a sentiment score of a sentence simultaneously depends on scores of its terms including words, phrases, sequences of non-adjacent words, thus we propose to use a linear combination which can incorporate the scores of the terms extracted by various language models with the corresponding coefficients for estimating the sentence's score. Through utilizing the linear combination, we derive a novel vector representation of a sentence called language-model-based representation that is based on average scores of kinds of term in the sentence to help supervised classifiers work more effectively on the clinical narratives.

Keywords: Sentiment shifters · Language-model-based representation · Linear combination

1 Introduction

The clinical narratives reflect the patient's health status through observations of symptoms, progress in treatment, and physician's assessments. Therefore, determining such observations and assessments as positive or negative or neutral towards a disease plays an important role in therapeutic assistance and abnormality recognition.

The text in clinical narratives has several particular characteristics that pose some challenges for sentiment classification on such text. Beside lack of domain-specific resources, implicit sentiment mentioned in [1], we have to face with two main challenges as the following:

- The diversity of sentiment shifters used in clinical text.
- The shortness of clinical text.

H. Fujita et al. (Eds.): IEA/AIE 2016, LNAI 9799, pp. 255–268, 2016.
DOI: 10.1007/978-3-319-42007-3_22

Sentiment shifters are known as expressions used to change the sentiment orientation of a sentence such as negation words. The clinical text contains descriptions of patient's health status, to express the improvement of patient status, nurses or doctors often use the negation of symptoms and negative observations. However, the negation is in various variants not only negation words. For instance, we consider the following sentences/clauses:

- "There has significant improvement in pleural effusion." (positive)
- "There is no evidence of pleural effusion." (positive)
- "There has been marked decrease in right pleural effusion." (positive)
- "less nauseous than previous." (positive)

The example shows that the sentiment shifters are not only strong positive/negation words such as "improvement", "no" but also phrases like "less nauseous", or sequences of non-adjacent words as "decrease ... pleural effusion".

The problem of sentiment shifters was mentioned and solved by several methods on product-review domain. Such methods follow one of two main approaches, one is negation words and scope of the negation detection, the other is simple voting for overall sentence's sentiment score by word/phrase scores. The first approach often gives a better performance than the second one due to the intensive analysis of word contexts while the second one is more flexible because of the specific language independence. For our case, the first approach seems to be not effective because it is difficult to exactly capture all variants of sentiment shifters. Therefore, the second one is more appropriate, but it requires some modifications to enhance word's contexts considering instead of individually aggregating scores at word-level or phrase-level. For example, the word "improvement" is a strong positive word, so its score can dominate the other and rule the sentence's score while the word "less" may not due to a weaker positive sense. However, the phrase "less nauseous" with more positive purity volume can make a bigger influence on the sentence's score. It helps us raise an idea that the sentence score does not separately depends on word or phrase score. Thus, we simultaneously sum up word and phrase scores by a linear combination in which the coefficients characterize how words and phrases affect the sentence sentiment orientation. Besides, sequences of non-adjacent words are also used to capture more contexts of words. All words, phrases, sequences of non-adjacent words (terms) are extracted by using different language models.

The shortness of text requires a particular representation method instead of popular methods such as bag-of-words, bag-of-n-grams because the short length of text does not provide enough word co-occurrence or shared context for good similarity measures [11]. Through the idea of using the linear combination of different kinds of term extracted by their corresponding language models in estimating the sentence's score, it is clear to see that the sentence score depends on the score of each kind of term. Thus, that raises an idea of a novel vector representation for a sentence based on the average scores of such kinds of term called language-model-based representation to deal with the problem of representing short text. Different from the strategy of using topic model to enhance the co-occurrence of words in sentences for improving the similarity measures that are based on the appearance of common words, language-model-based representation measures the similarity between two sentences by comparing the scores of such sentences according to each kind of term.

In this paper, we present two our contributions for sentiment classification on clinical narratives in case of lack of sentiment resources for medical domain:

- Effectively using a linear combination of different kinds of term extracted by various language models to estimate the sentence's score.
- Deriving a novel vector representation of a sentence called language-model-based representation to deal with the problem of short text representation.

2   Related Work

The techniques for sentiment classification on clinical text are mainly based on available techniques used in product review domain. In [21,24], the authors made a review of sentiment classification techniques that follow three approaches: machine learning-based approach, lexicon-based approach, hybrid approach. In machine learning-based approach, the feature set is determined through part-of-speech, n-gram, or sentiment words [23] before applying classification methods such as Naive Bayes, Support Vector Machine [22]. Besides, with a simpler way, in [2,3], the authors just summed up sentiment scores of words, phrases to estimate a sentence score for decision making.

In product review domain, sentiment shifters are mainly indicated via negation terms, so several works attempt to detect such terms, and the scope of negation in the sentence. In [4], Polanyi et al. described how the base attitudinal valence of lexical item can be modified by context and proposed a simple "proof of concept" implication for some context shifters. In other work, Li et al. [5] presented a shallow semantic parsing approach to learn the scope of negation. Ikeda et al. [6] proposed a method that models polarity shifters better than simple voting by sentiment word method. The effect of valence shifters on classification was examined in [7]. The parser and some heuristic rules was used to identify the scope of negation [8]. In [9], Li et al. proposed a feature selection method to generate scale polarity shifting training data, and a combination of classifiers to improve the performance. In [14], Kiritchenko et al. determined the sentiment of words in the presence of negation by detecting negation context via computing two scores of term in two parts: affirmative context, and negated context.

To improve similarity measures for short text, the probabilistic topic model is commonly utilized. LSA, pLSA, LDA have been widely applied to discover the latent topics for short text representation [10,11,15]. In addition, PMM-based classifier based on conditional probabilities of upcoming symbol given several previous symbols was applied for topic and non-topic classification [12]. Dai et al. [13] proposed cluster-based representation method named CREST to deal with the shortness and sparsity of text.

Several works also try using sentiment classification on clinical text, but on nurse/doctor narratives, the ontained results are not good enough. Ali et al. [16]

applied the methods such as Naive Bayes, SVM, Logistic-R to classify the posts in medical forums. Additionally, SVM and Naive Bayes were also used in [20] to determine the watchlist of drugs as positive or negative in drug surveillance. In [17], Deng et al. applied dictionary-based method to classify nurse letters, radiology reports in the MIMIC II database, they also presented some difficulties when doing classification on such data set. Besides, Na et al. [18] did clause-level sentiment classification using pure linguistic approach.

3   Mixture of Language Models Utilization in Sentiment Classification on Clinical Narratives

3.1   Our Approach

The core of our solution for sentiment classification on clinical text is to simultaneously sum up the score of words, phrases, sequences of non-adjacent words extracted by different language models by a linear combination. The linear combination is a simple and efficient model for voting sentiment score of the sentence with low computational cost that characterizes the importance of its components via the corresponding coefficients. Moreover, relying on such linear combination, we are able to derive a novel vector representation of a sentence called language-model-based representation. The proposed idea is formulated as the following.

Assume that $\mathbf{L} = \{L_1, L_2, ..., L_m\}$ is a set of m language models used to extract terms. $\mathbf{T} = \{L_1(s), L_2(s), ..., L_m(s)\}$ where $L_i(s), i = 1, 2..., m$, is a set of terms extracted from the sentence s according to the language model L_i. For each term $t \in L_i(s)$ compute $Score(t)$. An average score over all terms belonging to $L_i(s)$ is computed by the following equation:

$$Score(L_i(s)) = \frac{\sum_{t \in L_i(s)} Score(t)}{N_i} \tag{1}$$

where N_i is the number of terms in $L_i(s)$.

The sentiment score of the sentence s is defined as a linear combination over $Score(L_i(s))$ as the following:

$$Score(s) = \sum_{i=1}^{m} w_i \times Score(L_i(s)) \tag{2}$$

$$\begin{cases} Score(s) > 0 \Rightarrow positive \\ Score(s) < 0 \Rightarrow negative \end{cases}$$

In the linear combination, the coefficients $(w_1, w_2, ..., w_m)$ characterize how the sentence's score depends on each $Score(L_i(s))$. If the sentence's score is strongly related to a kind of term, its coefficient is larger, that means there is a bias for such kind of term. Besides, some kinds of term contribute to sentence's score identification with equal roles. Therefore, we pose three assumptions regarding the coefficient's values:

- Assumption 1: The value of coefficients $(w_1, w_2, ..., w_m)$ are different. That means there is a bias in the voting process.
- Assumption 2: The value of coefficients are equal, and set as 1.
- Assumption 3: That incorporates assumption 1 and assumption 2. There exists a subset of language models following assumption 1, and the rest is appropriate with assumption 2. In this case, the sentence's score is computed as the following:

$$Score(s) = \sum_{i=1}^{k} w_i \times Score(L_i(s)) + \sum_{i=k+1}^{m} Score(L_i(s)) \qquad (3)$$

where k, $m - k$ are the number of language models following assumption 1, assumption 2 respectively.

Through the experiments and interpretations, we assess that if the components $Score(L_i(s))$ have a weak linear relationship, assumption 1 is more appropriate to obtain a better performance because in this case, there will has a conflict when aggregating such components, so we need to adjust the aggregation by a priority setting via adding the different weights for the components. Otherwise, in case such components have a strong linear relation that means we can use one of them to make the aggregation to make the decision, and we do not need to adjust them, thus assumption 2 is more appropriate. The detail and explanation are presented in Subsect. 4.2.

Equation 2 gives an idea of a vector representation for a sentence that is different from most of previous works using topics of words. In this equation, the sentence's score depends on the concurrent contribution of the components $Score(L_i(s))$, thus the set $\mathbf{S} = \{Score(L_1(s)), Score(L_2(s)), ..., Score(L_m(s))\}$ could be considered a feature set to represent the sentence that is called language-model-based representation. By such method, the similarity measure of two sentences is based on the comparison between the sentence's scores which are decomposed into the components $Score(L_i(s))$ instead of enhancing the co-occurrence of common words like using the topic model.

3.2   Proposed Method

Relying on the proposed approach mentioned above, we propose a method that includes three main steps: language-model-based terms extraction, sentiment score measure, and feature derivation and linear combination coefficients estimation.

Language-Model-Based Terms Extraction. In our work, language models including n-gram and skip-gram models play a role as templates in terms extraction. More general than n-gram models that help to extract sequences of adjacent words, skip-gram [19] models can capture not only sequences of adjacent words but also sequences of non-adjacent words. For example, we consider the following sentence:

"There is no evidence of pleural effusion."

Various language models such as unigram, bigram, trigram, 1-skip-bigram, 2-skip-bigram, 3-skip-bigram, 4-skip-bigram, 1-skip-trigram, 2-skip-trigram are used in this step. Table 1 shows an example of language model utilization for term extraction.

Table 1. Terms extraction by language models

Language model	Extracted terms
unigram	there, is, no, evidence, of, pleural, effusion
bigram	there is, is no, no evidence, evidence of, of pleural, pleural effusion
1-skip-bigram	there is, there no, is no, is evidence, no evidence, no of, evidence of, evidence pleural, etc.
2-skip-bigram	there is, there no, there evidence, is no, is evidence, is of, no evidence, no of, no pleural, etc.
trigram	there is no, is no evidence, no evidence of, evidence of pleural, of pleural effusion.
1-skip-trigram	there is no, there is evidence, there no evidence, is no evidence, is evidence of, is no of, etc.

As the definition in [19], k-skip-n-grams consider k or less skips to construct n-gram. For example, 3-skip-bigram includes 3 skips, 2 skips, 1 skip, 0 skips (bigram). Relying on number of tokens in terms, the language models are divided into three groups as the following:

- Group 1: occurrence of words individually (unigram)
- Group 2: co-occurrence of two words (bigram, 1-skip-bigram, 2-skip-bigram, 3-skip-bigram, 4-skip-bigram).
- Group 3: co-occurrence of three words (trigram, 1-skip-trigram, 2-skip-trigram).

Term's Sentiment Score Measure. Sentiment score of a term measures the related volume between the term and the sentence's sentiment label. We use the following equation to compute the term's sentiment score as in [3]:

$$Score(t) = \frac{p(t|positive) - p(t|negative)}{p(t|positive) + p(t|negative)} \tag{4}$$

$p(t|positive)$ is computed by taking number of times term t appears in positive sentences then dividing it by the total number of terms in the positive sentences. $p(p|negative)$ is also computed in the similar way. The term's score $Score(t)$ ranges from -1 to 1. If $Score(t) > 0$ the sentiment orientation of the term is likely positive, and vice versa.

Language-Model-Based Feature Derivation and Coefficient Estimation.
As we mentioned in Subsect. 3.1, the simultaneous contribution of various kinds of
term to the sentence sentiment orientation is characterized by a linear combination
of their score as Eq. 2, in which each coefficient indicates how each kind of term
gives its influence on the sentence score. Therefore, identifying such influence is
equivalent to estimating such coefficient. We need to estimate coefficients in case
of assumption 1, 3.

Algorithm 1. Linear combination coefficients learning

$L = \{L_1, L_2, ..., L_m\}$ is a set of language models used.
for *each sentence s in training set* **do**
   vector := empty
   for *each $L_i \in L$* **do**
      Extracting a set of terms $L_i(s)$ in the sentence s according to L_i
      for *each term t in $L_i(s)$* **do**
         Compute $Score(t)$ by Eq. 4
      Compute score average $Score(L_i(s))$ by Eq. 1
      Append $Score(L_i(s))$ to *vector*

if *L follows assumption 1* **then**
   Train with Support Vector Machine to to identify $(w_1, w_2, ..., w_m)$
if *L follows assumption 2* **then**
   Set $w_1 = w_2 = ... = w_m = 1$
if *L follows assumption 3* **then**
   if *L1 $\subset$ L follows assumption 1* **then**
      Train with Support Vector Machine to identify coefficients
   if *L2 $\subset$ L follows assumption 2* **then**
      Set the coefficients as 1

The most likely coefficients estimation is based on the training data. Each
sentence in the training set is converted into the corresponding linear combination like Eq. 2, and then if the label of the sentence is positive the linear
combination is greater than 0, and if it is negative, the combination is smaller
than 0. For example, we assume that we convert n sentences in the training data
into a set of inequalities as the following:

$$\begin{cases} s_1 : \sum_{i=1}^{m} w_i \times Score(L_i(s_1)) < 0 \\ s_2 : \sum_{i=1}^{m} w_i \times Score(L_i(s_2)) > 0 \\ ... \\ s_n : \sum_{i=1}^{m} w_i \times Score(L_i(s_n)) > 0 \end{cases}$$

We see that determining the most likely $(w_1, w_2, ..., w_m)$ is equivalent to finding a hyperplane as a linear boundary of a data set represented by the set of
vectors $\{Score(L_1(s_k)), Score(L_2(s_k)), ..., Score(L_m(s_k))\}, k = 1, 2, ..., n$. Thus,
this problem can be solved by using Support Vector Machine (SVM) technique.
We propose algorithm 1 to for coefficients learning. In Algorithm 1, to determine which assumption L should follow, we base on assessment 2 presented in
Subsect. 4.2.

Table 2. Coefficients assumptions with groups of language models investigation

Method		MIMIC II	Movie-Review
Our method			
1	Assumption 1 with group 2	**0.823**	**0.736**
2	Assumption 1 with group 3	0.69	0.507
3	Assumption 1 with group 1 + group 2	0.799	0.747
4	Assumption 1 with group 1 + group 3	**0.827**	**0.754**
5	Assumption 1 with group 2 + group 3	0.807	0.605
6	Assumption 1 with group 1 + group 2 + group 3	0.811	0.723
7	Assumption 2 with group 2	0.817	0.732
8	Assumption 2 with group 3	0.68	0.594
9	Assumption 2 with group 1 + group 2	**0.836**	**0.756**
10	Assumption 2 with group 1 + group 3	0.823	0.738
11	Assumption 2 with group 2 + group 3	0.813	0.723
12	Assumption 2 with group 1 + group 2 + group 3	0.832	0.751
13	Assumption 3 with group 1 + group 2 + group 3 (*)	**0.836**	**0.764**
Individually sum up term's scores of each language model			
14	Terms from unigram	0.827	0.747
15	Terms from bigram	0.769	0.688
16	Terms from trigram	0.579	0.464
17	Terms from 1-skip-bigram	0.799	0.709
18	Terms from 2-skip-bigram	0.81	0.717
19	Terms from 3-skip-bigram	0.812	0.721
20	Terms from 4-skip-bigram	0.818	0.727
21	Terms from 1-skip-trigram	0.644	0.556
22	Terms from 2-skip-trigram	0.678	0.599
Bag-of-words			
23	SVM + bag-of-words	0.698	0.503

(*): The sentence's score is computed by the following equation:
$$Score(s) = \sum_{i=1}^{k} w_i \times Score(L_i(s)) + \sum_{j=1}^{h} Score(L_j(s))$$
where $L_i \in$ group 1 and group 3, $L_j \in$ group 2.

4   Experimental Evaluation

4.1   Data Preparation

In the experiment, the MIMIC II data set that contains the information of more than 32,000 patients are used for our method evaluation. 6000 sentences that are manually annotated with two labels "1" (positive) and "–1" (negative) are obtained from "NOTEEVENTS" records.

For evaluation method, the annotated data is randomly divided into 10 parts then 6 parts are used for training, and the rest for testing. This process is repeated 10 times, then we take an average of precision.

We aim to build a classifier that can work well on clinical narratives in case sentiment resources for medical domain are not available, so the classification method should not depend on a specific domain. Therefore, to investigate whether our proposed method with the derived assessments is robust and can be applied on other data set or not, we additionally use movie review data[1] for evaluation due to some fairly similar points. The text in movie review data set is also separated into sentences/snippets (short text), and also contains some kinds of sentiment shifters like the MIMIC II data set.

In case of assumption 1 and 3, we use scikit learn, a python package implementing SVM algorithm with kernel functions[2] to determine coefficients.

4.2   Experiment Results and Interpretation

Coefficient's Assumption for Language Models of Groups. The experiments aim to determine which assumption is appropriate to a given language model. In the experiments, we consider the features generated from the language models in three groups and in the combination of such groups. All sentences are represented according to the language-model-based representation method. The classification results of three assumptions with three groups are showed in Table 2.

- A comparison between group 2 and group 3

We consider language models in the same group, and make a comparison between language models in group 2 and group 3. Line 1, 2, 7, 8 in Table 2 show that the features of group 2 provide remarkably higher performance than those of group 3 with both assumption 1 and 2. To explain why there is a significant difference between the features of group 2 and group 3, we visualize the training set and testing set in Fig. 1, then observe the distribution of data points.

We observe that the language models in a same group often generate their features with similar value, so the points in Fig. 1 almost fluctuate around the bisector $y = x$ with close distance.

Figures 1a and b show a difference of the points distribution between group 2 and group 3. The data points of group 2 tend to spread along the bisector while the data points of group 3 tend to converge at the corners. The reason is that sentiment orientations of terms extracted by language models of group 3 is almost pure with very high absolute value of score because the probability of co-occurrence of three words in a sentence is very small that gives poor information for prediction. In addition, the sentences in testing set are represented through the lexicon extracted from training set, so the terms of group 3 appearing together in a training sentence have a less chance to co-occur in the testing sentence that makes the testing set significantly different from the training set. In contrast to group 3, due to the higher probability of co-occurrence of two words, features of group 2 make our method get better accuracy. We also obtain

---

[1] http://www.cs.cornell.edu/people/pabo/movie-review-data/
[2] http://scikit-learn.org/stable/modules/generated/sklearn.svm.SVC.html

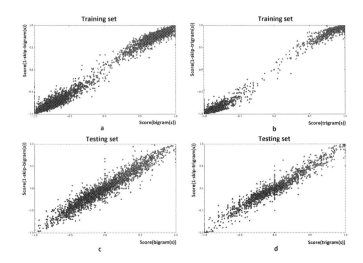

Fig. 1. A visualization of training and testing set with two features of group 2 ($Score(bigram(s))$, $Score(1 - skip - bigram(s))$) and group 3 ($Score(trigram(s))$, $Score(1 - skip - trigram(s))$). The blue, and red points indicate negative sentences, positive sentences respectively. a and c show the data points with the features of group 2, and b and d show the data points with the features of group 3. (Color figure online)

a similar result when doing classification on movie-review data set. Therefore, we have an assessment of using language models in a same group as the following:

Assessment 1: When building the feature set by language models in a same group, the language models considering the co-occurrence of two words provide better performance than ones considering the co-occurrence of three or more words.

- A comparison among different combinations of groups

Line 3, 4, 5, 6 show the accuracy when using assumption 1 with different combinations of three groups. We obtained the highest precision by incorporating language models of group 1 and group 3 (line 4), and get lower accuracy on other combinations. The quality of features depends on the linear relationship among them. If the features have a strong linear relation, there is less information to make the decision because they are considered as duplicated features, and the decision is just based on one of them. The volume of linear relationship between two features can be measured via correlation coefficient. In case the correlation coefficient is close to 1 or -1, the linear relation is strong. Table 3 shows the correlation coefficient of features generated by incorporating groups. For each group, we take a language model to generate the feature because other ones also generate the similar feature.

From Table 3, we observe that the features generated by language models of group 1, and group 3 have lowest correlation coefficient on both MIMIC and movie-review data that explains why such features get high performance of classification with assumption 1.

Table 3. Correlation coefficient among features generated by different combinations of three groups

Group pair	Correlation coefficient on MIMIC II	Correlation coefficient on movie review
Group 1 + group 2	0.901	0.893
Group 1 + group 3	0.844	0.837
Group 2 + group 3	0.977	0.983

Although the combinations of group 1 and group 2 or group 2 and group 3 do not produce the high performance with assumption 1 on the MIMIC data set and movie-review set, they get better results with assumption 2 (line 9, 11).

Through the results from line 1 to 12, we have an assessment to select the appropriate assumption for language models as the following:

Assessment 2: Assumption 1 is appropriate for language models whose generated features have a weak linear relation. In case such features have a strong relation, assumption 2 is more appropriate.

There has an interesting meaning inside this assessment. In case the features have a weak linear relationship, it will raise a conflict when aggregating, so we need a referee to judge which features are important then give such features a priority. In our method, the priority is characterized through the coefficients. Otherwise, if such features strongly linearly depend on each other, no conflict happens, so the referee is not necessary.

Line 13 shows the best result when we use assumption 3 with a combination of three groups, in which the features of group 1 and group 3 are aggregated with the different coefficients. We obtained 83.6 % on MIMIC and 76.4 % on movie-review data.

From line 14 to line 22, we show the results when using each language model to extract terms then make their score summing up. By this method, unigram has the highest performance (82.7 % on MIMIC and 74.7 % on movie-review), but it is not better than our method with assumption 3 that considers the interaction among terms extracted from different language models in voting for sentence's score.

Influence of Balance and Imbalance Training Set on Classification Performance. The experiment aims to examine the influence of balance and imbalance training data on the classification performance. A balance set contains an equal number of positive and negative sentences while a imbalance set is in contrast. The proportion between positive sentences and negative sentences impacts the term's score measure in Eq. 4. Table 4 shows how the proportion affects the classification performance.

Table 4. Influence of balance and imbalance training set on classification performance

	Method		MIMIC II	Movie-review
1	Sum up score (unigram)	B	0.827	0.747
2	Sum up score (unigram)	IB-P	0.805	0.72
3	Sum up score (unigram)	IB-N	0.813	0.726
4	Assumption 1 with group 2	B	0.823	0.731
5	Assumption 1 with group 2	IB-P	0.715	0.585
6	Assumption 1 with group 2	IB-N	0.783	0.582
7	Assumption 2 with group 1 + group 2	B	0.836	0.756
8	Assumption 2 with group 1 + group 2	IB-P	0.799	0.695
9	Assumption 2 with group 1 + group 2	IB-N	0.82	0.711

- B: Balance data set
- IB-P: Imbalance data set with greater number of positive sentences.
- IB-N: Imbalance data set with greater number of negative sentences.

Table 4 shows that imbalance sets make the accuracy reduce on both MIMIC and movie-review data set. The difference between number of positive sentences and negative sentences makes the term's score measure not fair, thus the scores are not precise.

5   Conclusion and Future Work

The paper presents our work on sentiment classification on clinical narratives. In this work, we propose a classification method to deal with two challenges of such text: the diversity of sentiment shifters, and the shortness of text. Our method uses a mixture of language models to extract terms, then estimate the sentiment score of sentences by a linear combination of such term's scores. In addition, we also derive a novel vector representation according to the language models used to extract terms that can work better on short text. Moreover, this method is flexible and independent with a specific language. The experimental results show the improvement of classification performance by using our method.

Beside the advantages, our method still has some drawbacks. The exist of sentiment shifters in training data makes the estimation of term's score sometimes is not precise. We also have to face with the problem of sparse data when using language models in group 3. Therefore, we plan to overcome these drawbacks to improve the performance of our method in the future work.

Acknowledgments. This work is partially funded by Vietnam National University at Ho Chi Minh City under the grant number B2015-42-02, and Japan Advanced Institute of Science and Technology under the Data Science Project.

References

1. Denecke, K., Deng, Y.: Sentiment analysis in medical settings: new opportunities and challenges. Artif. Intell. Med. **64**(1), 17–27 (2015)
2. Turney, P.D.: Thumbs up or thumbs down? semantic orientation applied to unsupervised classification of reviews. In: Proceedings of the 40th Annual Meeting on Association for Computational Linguistics, pp. 417–424 (2002)
3. Dave, K., Lawrence, S., Pennock, D.M.: Mining the peanut gallery: opinion extraction and semantic classification of product reviews. In: Proceedings of the 12th International Conference on World Wide Web, pp. 519–528 (2003)
4. Polanyi, L., Zaenen, A.: Contextual valence shifters. In: Shanahan, J.G., Qu, Y., Wiebe, J. (eds.) Computing Attitude and Affect in Text: Theory and Applications. The Information Retrieval Series, vol. 20, pp. 1–10. Springer, Netherlands (2006)
5. Li, J., Zhou, G., Wang, H., Zhu, Q.: Learning the scope of negation via shallow semantic parsing. In: Proceedings of the 23rd International Conference on Computational Linguistics. Association for Computational Linguistics, pp. 671–679 (2010)
6. Ikeda, D., Takamura, H., Ratinov, L.-A., Okumura, M.: Learning to shift the polarity of words for sentiment classification. In: IJCNLP, pp. 296–303 (2008)
7. Kennedy, A., Inkpen, D.: Sentiment classification of movie reviews using contextual valence shifters. Comput. Intell. **22**(2), 110–125 (2006)
8. Jia, L., Clement, Y., Meng, W.: The effect of negation on sentiment analysis and retrieval effectiveness. In: Proceedings of the 18th ACM Conference on Information and Knowledge Management, pp. 1827–1830 (2009)
9. Li, S., Lee, Sophia Yat Mei Chen, Y., Huang, C.-R., Zhou, G.: Sentiment classification and polarity shifting. In: Proceedings of the 23rd International Conference on Computational Linguistics. Association for Computational Linguistics, pp. 635–643 (2010)
10. Quan, X., Liu, G., Zhi, L., Ni, X., Wenyin, L.: Short text similarity based on probabilistic topics. Knowl. Inf. Syst. **25**(3), 473–491 (2010)
11. Song, G., Ye, Y., Du, X., Huang, X., Bie, S.: Short text classification: a survey. J. Multimedia **9**(5), 635–643 (2014)
12. Bobicev, V., Sokolova, M.: An effective and robust method for short text classification. In: Proceedings of the Twenty-Third AAAI Conference on Artificial Intelligence, pp. 1444–1445 (2008)
13. Dai, Z., Sun, A., Liu, X.-Y.: CREST: cluster-based representation enrichment for short text classification. In: Pei, J., Tseng, V.S., Cao, L., Motoda, H., Xu, G. (eds.) PAKDD 2013, Part II. LNCS, vol. 7819, pp. 256–267. Springer, Heidelberg (2013)
14. Kiritchenko, S., Zhu, X., Mohammad, S.M.: Sentiment analysis of short informal texts. J. Artif. Intell. Res. **50**, 723–762 (2014)
15. Chen, M., Jin, X., Shen, D.: Short text classification improved by learning multi-granularity topics. In: IJCAI, pp. 1776–1781 (2011)
16. Ali, T., Schramm, D., Sokolova, M., Inkpen, D.: Can I hear you? sentiment analysis on medical forums. In: Proceedings of the Sixth International Joint Conference on Natural Language Processing, pp. 667–673 (2013)
17. Deng, Y., Stoehr, M., Denecke, K.: Retrieving attitudes: sentiment analysis from clinical narratives. In: Medical Information Retrieval Workshop at SIGIR, p. 12 (2014)
18. Na, J.-C., Kyaing, W.Y.M., Khoo, C.S.G., Foo, S., Chang, Y.-K., Theng, Y.-L.: Sentiment classification of drug reviews using a rule-based linguistic approach. In: Chen, H.-H., Chowdhury, G. (eds.) ICADL 2012. LNCS, vol. 7634, pp. 189–198. Springer, Heidelberg (2012)

19. Guthrie, D., Allison, B., Liu, W., Guthrie, L., Wilks, Y.: A closer look at skip-gram modelling. In: Proceedings of the 5th International Conference on Language Resources and Evaluation (LREC-2006), pp. 1–4 (2006)
20. Chee, B.W., Berlin, R., Schatz, B.: Predicting adverse drug events from personal health messages. In: American Medical Informatics Association, pp. 217–226 (2011)
21. Madhoushi, Z., Hamdan, A.R., Zainudin, S.: Sentiment analysis techniques in recent works. In: Science and Information Conference (SAI), pp. 288–291. IEEE (2015)
22. Neethu, M.S., Rajasree, R.: Sentiment analysis in twitter using machine learning techniques. In: Computing Communications and Networking Technologies (ICC-CNT), pp. 1–5. IEEE (2013)
23. Veeraselvi, S.J., Saranya, C.: Semantic orientation approach for sentiment classification. In: Green Computing Communication and Electrical Engineering (ICGC-CEE), pp. 1–6. IEEE (2014)
24. Cambria, E.: Affective computing and sentiment analysis. IEEE Intell. Syst. **31**(2), 102–107 (2016)

Twitter Feature Selection and Classification Using Support Vector Machine for Aspect-Based Sentiment Analysis

Nurulhuda Zainuddin, Ali Selamat[✉], and Roliana Ibrahim

Faculty of Computing, Universiti Teknologi Malaysia, 81310 Johor Bahru,
Johor, Malaysia
aselamat@utm.my

Abstract. In this paper, with regards to aspect-based sentiment classification accuracy problem, we propose a Principal Component Analysis (PCA) feature selection method that can determine the most relevant set of features for aspect-based sentiment classification. Feature selection helps to reduce redundant features and remove irrelevant features which affect classifier accuracy. In this paper we present a method for feature selection for twitter aspect-based sentiment classification based on Principal Component Analysis (PCA). PCA is combined with Sentiwordnet lexicon-based method which is incorporated with Support Vector Machine (SVM) learning framework to perform the classification. Experiments on our own Hate Crime Twitter Sentiment (HCTS) and benchmark Stanford Twitter Sentiment (STS) datasets yields accuracies of 94.53 % and 97.93 % respectively. The comparisons with other statistical feature selection methods shows that our proposed approach shows promising results in improving aspect-based sentiment classification performance.

Keywords: Twitter · Aspect-based feature extraction · Aspect-based sentiment classification · Feature selection · Principal component analysis · Support vector machine

1 Introduction

Microblogging today has become a very popular communication tool among internet users. Microblogging services become valuable sources of people's opinions and sentiments. There are millions of messages appearing daily in popular microblogging websites such as Twitter, Tumblr and Facebook [1].

Twitter is a popular microblogging service where users create status messages (called tweets). Twitter messages have many unique attributes [2]. The characteristics are include the maximum length of 140 characters tweets, the magnitude of data available, the language model, domain and the diversity of contents or aspects which are not limited to any specific topics [3]. This characteristics are different from previous research, which focused on specific domains

© Springer International Publishing Switzerland 2016
H. Fujita et al. (Eds.): IEA/AIE 2016, LNAI 9799, pp. 269–279, 2016.
DOI: 10.1007/978-3-319-42007-3_23

such as movie reviews, restaurant reviews and product reviews. Twitter has emerged to become a gold mine rich with varied information that also contains opinions on current issues, complaint, and their thoughts for product they use everyday. The information also contain issues related to business, political and society.

Previous studies have approached twitter sentiment analysis problem as a tweet-level sentiment classification task that is similar to document-level sentiment classification. Tweet-level or document-level sentiment classification determines the overall sentiment orientation of a tweet. However, getting an overall positive or negative sentiment might not be useful to the organizations as it is more important to determine what exactly the opinions of their consumers. Different people or users may express their views on different aspects of the products, or services. For instance, users want to know more specific and precise information related to certain aspects or (features) of products. Even though Twitter restricts each tweet to a maximum of 140 characters, tweets can still contain mixed sentiments about a service or organization [4]. For example, customer may tweet restaurant services is poor but the price is reasonable. There are two aspects term that we can identify from the tweets; (1) services that encode negative and, (2) price that encode positive. More importantly, we can help users gain more precise information and helping them in making decisions about the target entity.

In this paper, we propose a Principal Component Analysis (PCA) feature selection method that can determine the most relevant set of features for aspect-based sentiment classification. Feature selection helps to reduce redundant features and remove irrelevant features which affect classifier accuracy. Each tweet is represented by term frequency-inverse document frequency(TF-IDF) weighting scheme. The aim is to improve classification accuracy when limited number of twitter messages use in the experiment. On the other hand, we used twitter aspect-based sentiment classification with other feature selection method as the benchmark test to see the classifier performance. The evaluation results shows that the proposed method successfully improve classifier accuracy with our own Hate Crime Twitter Sentiment (HCTS) dataset and publicly available Stanford Twitter Sentiment(STS) Dataset. In addition to that, Support Vector Machine (SVM) have been used as a learning method for aspect-based sentiment classification.

The main contributions of this paper can be described as follows:

- to introduce and implement Principal Component Analysis (PCA) feature selection method for determining the opinion words for aspect-based sentiment classification.
- to demonstrate the value of feature selection and classification method using Support Vector Machine(SVM).
- to test the accuracy of sentiment classification using our own Hate Crime Twitter Sentiment (HCTS) dataset and publicly available Stanford Twitter Sentiment (STS) dataset.

The remainder of this paper is organized as follows; Sect. 2 describes the related work for feature selection and twitter aspect-based sentiment classification. Section 3, describes the proposed method, Sect. 4 describes the results and discussion obtained from the experiments. The last section presents the conclusions and suggested future work.

2   Related Work

Sentiment analysis is important and a dynamic research field, motivated by the evolution of web social media and the chances to access the important opinions from various people on a numerous business and social issues [5]. There are five types of sentiment analysis field which are document-level, sentence-level, aspect-based, comparative and sentiment lexicon acquisition [6].

There are two widely used approaches in sentiment analysis which are machine learning approach and the semantic orientation approach [7]. Machine learning approach needs to choose extract features to do sentiment analysis. The feature which taken from a piece of text is usually converted into a feature vector that can represent the most prominent information expressed in the original text. Bag-of-words, n-grams, Part-of-Speech (POS) tags, and word position are the examples of different types of features that have been used in sentiment classification studies. Part of speech Tags (POS Tags) is frequently used in sentiment analysis because it can help to sense word uncertainty. A good Part of speech Tags (POS) features that have been widely accepted is adjectives [7] as well as other words such as nouns, verbs, and adverbs.

Generally, the performance and accuracy of machine learning techniques depend on a number of factors described by [8]. The factors include the features that will be used in sentiment analysis which influence the classifier accuracy. For example, the use of ngrams and emoticons have shown promising results by improving accuracy of classifier in research by [2,9]. Another factor is the use of feature selection method to select valuable features for classifier. This is because, not all features returned by the tokenization process can be used to determine the classifier accuracy. Furthermore, the method such as Chi-Square and Mutual Information have been used to perform this selection. Consequently, previous research shows that feature selection method has not been attempted so far especially in twitter aspect-based sentiment classification.

Aspect-based sentiment analysis is concerned with identifying the aspects of given target entities and estimating the sentiment polarity for each mentioned aspect [10]. The aspect-based sentiment analysis can be decomposed into two main tasks which are aspect-based feature extraction and aspect-based sentiment classification [11]. Aspect-based sentiment classification will determines whether the opinions on different aspects are either subjective or objectives. The main task in aspect-based sentiment classification is to extract the opinion words and also to find the polarity of opinion words towards its context. One of the method being used by researcher is to extract all the adjectives, adverbs, verbs from the tagged sentences [4, 12–14]. In this process, [15] used linguistic rules to determine the word opinion orientation. The process also extracts the set of adjectives that appeared near each aspect. Moreover, other researcher [12] also used opinion lexicon to determine the polarity of the opinion words.

3   Proposed Method

We propose a twitter aspect-based sentiment classification, which is based on Principal Component Analysis (PCA) feature selection method and Support Vector Machine (SVM). The proposed method is different compared to the previous work which highly depends on selection of tagged adjectives, adverbs, verbs and nouns words. Our work is different because we used twitter that have limitation of characters (only 140 char length) and the diversity of contents that is not specific to certain topics. Then, the combination between lexicon-based method using Sentiwordnet and PCA feature selection method can help to improve sentiment classification accuracy. Figure 1 shows the process of aspect-based sentiment classification by using feature selection method.

3.1   Data Collection and Preparation

We proposed our own Hate Crime Twitter Sentiment (HCTS) Dataset. We use Twitter Search API v1.1 for the data collection. The most common and consistent method for gathering data is to request a paged set of data for a given query.

This corpus was domain-oriented and minimal because the annotation procedure was very complex and manual. In order to build the training data, we use lexicon-based methods by exploiting information from SentiWordnet [16]. The first task is to identify objective and subjective tweets. We only considered subjective (positive or negative) tweets for the classification task. From this process, we separate facts (neutral) from opinionated text by removing all neutral tweets from the database to avoid unnecessary process.

We also used Stanford Twitter Sentiment (STS) Dataset which is publicly available to evaluate the methods. This dataset has been commonly used in the literature for evaluation. The Stanford Twitter Sentiment (STS) corpus (http://help.sentiment140.com/) was introduced by [2]. It consists of a training and testing set. But we only consider testing dataset because it contains target-dependent tweets and ignores noisy labels. The testing set consists of 173 negative, 180 positive and 139 neutral tweets which are manually classified.

Table 1 shows Stanford Twitter Sentiment(STS) Dataset and Table 2 shows Hate Crime Twitter Sentiment (HCTS)Dataset, respectively.

3.2   Preprocessing of Twitter

Twitter language has some unique attributes that may not provide relevant information in order to reduce the feature space [2]. The unique attributes that were eliminate are usernames, links and hashtags. The filtering process were done

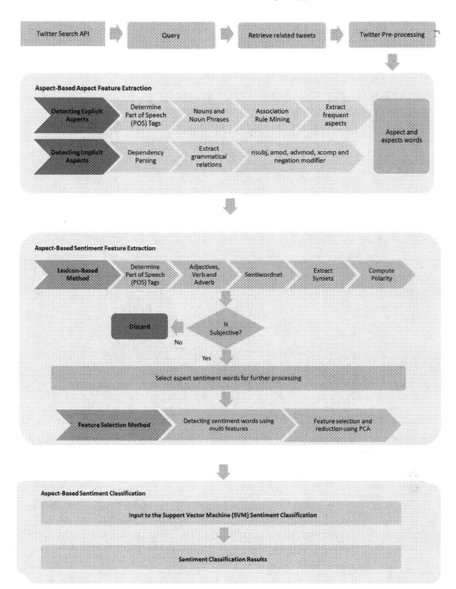

Fig. 1. Aspect-based sentiment classification using feature selection method

by removing new lines and opposite emoticons where some of the tweets contain positive and negative, repeated letters, laughter and punctuation marks. The datasets also went through the preprocessing task of tweets such as tokenization, stop word removal, lowercase conversion and stemming process.

Table 1. Stanford Twitter Sentiment (STS) dataset

Dataset	Category C_h	Sample query string	No of tweets analysed n_d
	company	aig, goodby silverstein	116
STS Datasets	event	world cup, Indian election	8
	location	North Korea, San Francisco	17
	misc	insects, exam	66
	movie	night at the museum	19
	person	obama, malcolm gladwell, cheney	64
	product	nike	63
	Total		353

Table 2. Hate Crime Twitter Sentiment (HCTS) dataset

Dataset	Category C_h	Sample query string	No of tweets analysed n_d
	race	anti-black, anti-white	255
HCTS Datasets	religion	anti-Muslim, anti-Jews	253
	sexual	anti-feminist, lesbian	214
	Total		622

4   Twitter Aspect-Based Sentiment Feature Extraction

In this paper we focus on aggregating opinion words for an aspect in tweet. There are two sections which are lexicon-based method (which has been discussed and evaluated in the previous research [17]) and feature selection method (which is our main discussion in this paper.)

4.1   Aggregating Opinions for an Aspect in Tweet

The aspect-based approach depends on opinion (or sentiment) words, which are words that express positive or negative sentiments. Words that encode a desirable state such as "love" and "good" have a positive polarity, while words that encode an undesirable state have a negative polarity such as "bad" and "hate". Although opinion polarity normally applies to adjectives and adverbs, there are verb and noun opinion words as well. From this observation, if such an adjective is found, it is considered an opinion word. The research has found that adjectives and adverbs are good indicators of subjectivity and neutral(factual) opinions [18].

Twitter Opinion Lexicon-Based Methods. The task is performed to find opinion words using lexicon-based method. We focus on finding adjectives, adverbs, verb and noun words in the tweets by using part-of-speech (POS) tagging. Besides, Sentiwordnet [19] is used to assign the polarity scores for each opinion word appear on tweets. Different polarity scores will be assigned depending on the context it appear. Hence, to depend on the word alone to determine the sentiment polarity is not enough, so we need to train a classifier using the scores provided by Sentiwordnet lexicon resources as features.

Twitter Opinion Feature Selection Methods. Principal Component Analysis(PCA) is an effective feature reduction technique which have been applied in varied fields. The reduced principle components, hence acquired from PCA is further analyzed to eliminate the least influencing attributes based on the attribute weights. Principal Component Analysis(PCA) is used to perform feature reduction method for twitter aspect-based sentiment classification. PCA has been used to reduce original feature vectors obtained from lexicon-based method. PCA has shown encouraging results in classification where it is used for feature reduction process [20, 21].

5   Twitter Aspect-Based Sentiment Classification

In this section, twitter aspect-based sentiment classification is proposed by using combination of features vectors from Principal Component Analysis (PCA) feature selection method and Sentiwordnet lexicon-based method as an input to the Support Vector Machine (SVM) for classification. The applied methods successfully improved the classification accuracy though only the small number of twitter used in the experiments.

Support Vector Machine(SVM) has been chosen for the classification in the experiments. The Support Vector Machines is a learning machine for two-group classification problems introduced by [22]. It is used to classify the texts as positives or negatives. SVM works well for text classification due to its advantages such as its potential to handle large features and its robust when there is a sparse set of examples [23]. Support Vector Machine has shown good results in previous research in sentiment analysis [9, 24].

6   Evaluation Measures

We first used accuracy to evaluate the whole classification performance of each method with binary classes, positive and negative. Standard evaluation measures of precision and recall will be employed to identify positive and negative sentiments on entities.

Four effective measures that have been used in this study are based on confusion matrix output, which are True Positive (TP), False Positive (FP), True Negative (TN), and False Negative (FN).

- Precision(P) = TP/(TP+FP)
- Recall(R) = TP/(TP+FN)
- Accuracy(A) = (TP+TN)/(TP + TN + FP + FN)
- F-Measure(Micro-averaging) = 2.(P.R)/(P+R)

7   Evaluation Results and Discussion

In this section, Support Vector Machine (SVM) is used to perform experiments on the Stanford Twitter Sentiment(STS) dataset and Hate Crime Twitter Sentiment (HCTS) dataset. STS Dataset consists of categories such as company, event, location, misc, movie, person, and product while HCTS Dataset consists of three types of hate crime category which are racial, religion and sexual category. Thus, LibSVM from WEKA with radial basis function(rbf) kernel type will be employed to perform classification.

7.1   Analysis of the Results from Lexicon-Based Method

The aim of this experiment is to provide meaningful results obtained from the proposed method for aspect-based sentiment classification. Table 3 shows the evaluation results from the lexicon-based method for HCTS Dataset. The proposed method gained polarity scores from Sentiwordnet lexicon-based method. The results is classified as hate and not hate tweets. Therefore, the sexual, racial and religion category showed tweets gained higher hate polarity scores compared to not hate polarity scores. As we can see from the table, lexicon-based Sentiwordnet has labelled tweets as hate for 77.6 % from sexual category, 77.05 % from racial category and 78.7 % from religion category. Nevertheless, the HCTS dataset labelled as not hate for 22.4 %, 22.95 % and 21.3 % tweets, respectively.

In contrast, in the results from Table 4 using lexicon-based Sentiwordnet method, STS Dataset gained higher not hate polarity scores as compared to hate polarity scores. The experimental results showed that total percentage label as hate is 71.3 % from company, 72.7 % from location, 59.6 % as misc, 11.8 % from movie, 20 % from person and 27.7 % from product category. However, we obtained total percentage label as not hate is 28.7 % from company, 100 % from event, 27.3 %, 40.4 %, 88.2 %, 80 %, and 72.3 % from other category respectively. This result will later be used as an input for SVM classifier with feature vectors from Principal Component Analysis (PCA) method.

7.2   Analysis of the Results from Principal Component Analysis (PCA) method

Tables 5 and 6 present the performance of the aspect-based classifier with lexicon-based Sentiwordnet and PCA feature selection method on HCTS and STS Datasets respectively. We measure performance using the accuracy, precision, recall and F-measure. We evaluate with five combination of different methods to incorporate with SVM classifier.

The experimental results from HCTS dataset presented in Table 5 shows that our third approach by using ABSA+Sentiwordnet+PCA method with POS Tags and unigram features giving the highest accuracy 94.53 %, 0.949 for precision, 0.945 for recall and 0.942 for F-measure, accordingly. On the other hand, by using ABSA+Sentiwordnet with POS Tags features alone, it gives only 77.51 % accuracy, compared with ABSA+Sentiwordnet+PCA with POS Tags features that

Table 3. HCTS polarity scores and total percentages of hate and not hate labels according to category

Category	Hate polarity scores	Total percentage	Not hate polarity scores	Total percentage
Sexual	21.625	77.6 %	7.5	22.4 %
Racial	43.375	77.05 %	6.75	22.95 %
Religion	18.625	78.7 %	5.625	21.3 %

Table 4. STS polarity scores and total percentages of hate and not hate labels according to category

Category	Hate polarity scores	Total percentage	Not hate polarity scores	Total percentage
Company	33.5	71.3 %	12.75	28.7 %
Event	0	0 %	4.5	100 %
Location	2.75	72.7 %	1.75	27.3 %
Misc	14.125	59.6 %	10.25	40.4 %
Movie	0.875	11.8 %	9.75	88.2 %
Person	6.5	20 %	23.75	80 %
Product	5.5	27.7 %	21	72.3 %

obtained 90.27 % accuracy and lastly, by using ABSA+Sentiwordnet+PCA and POSTags + Bigram features the accuracy achieved is 85.41 %. Furthermore, we have compared with Latent Semantic Analysis(LSA) feature selection method. The results indicated that the accuracy did not increase though we used POS Tags and unigram features. It shows only 77.51 % and F-measure indicates 0.677.

Table 6 presents the experimental results from STS Dataset, it shows that the highest accuracy achieved is 97.93 % with ABSA+Sentiwordnet+PCA method. Besides, the precision is 0.979, recall is 0.979 and F-measure is 0.979 with combination of POS Tags and unigram features. In contrast, by using ABSA+Sentiwordnet alone, the accuracy achieved only 53.44 %. Likewise using other method, the accuracy achieved is 97.24 %, and 76.55 % accordingly. The same result also achieved with STS Dataset for Latent Semantic Analysis where it only achieved only 53.45 % accuracy and F-measure shows only 0.372.

Table 5. HCTS comparison method for aspect-based sentiment classification results

Method	Features	Accuracy(%)	Precision	Recall	F-Measure
ABSA+Sentiwordnet	POS Tags	77.51	0.601	0.775	0.677
ABSA+Sentiwordnet+PCA	POS Tags	90.27	0.914	0.903	0.892
ABSA+Sentiwordnet+PCA	POS Tags+Unigram	94.53	0.949	0.945	0.942
ABSA+Sentiwordnet+PCA	POS Tags+Bigram	85.41	0.871	0.854	0.827
ABSA+Sentiwordnet+LSA	POS Tags + Unigram	77.51	0.601	0.775	0.677

Table 6. STS comparison method for aspect-based sentiment classification results

Method	Features	Accuracy(%)	Precision	Recall	F-Measure
ABSA+Sentiwordnet	POS Tags	53.44	0.286	0.534	0.372
ABSA+Sentiwordnet+PCA	POS Tags	97.24	0.973	0.972	0.972
ABSA+Sentiwordnet+PCA	POS Tags + Unigram	97.93	0.979	0.979	0.979
ABSA+Sentiwordnet+PCA	POS Tags + Bigram	76.55	0.816	0.766	0.751
ABSA+Sentiwordnet+LSA	POS Tags + Unigram	53.45	0.286	0.534	0.372

8   Conclusion and Future Work

This paper examined the twitter aspect-based sentiment analysis problem and proposed a Twitter feature selection method by using Principal Component Analysis(PCA) and aspect-based sentiment classification by Support Vector Machine (SVM). The overall results show that feature selection method based on Principal Component Analysis(PCA) helps to improve classification accuracy. In addition to that, classification by using SVM achieved higher accuracy when using small number of features. Furthermore, it also shows that using lexicon-based Sentiwordnet method is also useful for improving the performance of twitter aspect-based sentiment classification.

Acknowledgments. The authors wish to thank Universiti Teknologi Malaysia (UTM) under Research University Grant Vot- 02G31 and Ministry of Higher Education Malaysia (MOHE) under the Fundamental Research Grant Scheme (FRGS Vot-4F551) for the completion of the research.

References

1. Pak, A., Paroubek, P.: Twitter as a corpus for sentiment analysis and opinion mining. In: Proceedings of the Seventh Conference on International Language Resources and Evaluation (LREC 2010), (Valletta, Malta), European Language Resources Association (ELRA), May 2010
2. Go, A., Bhayani, R.: Twitter sentiment classificationusing distant supervision. CS224N Project Rep. Stanford **1**, 1–12 (2009)
3. Niu, Z., Yin, Z., Kong, X.: Sentiment classification formicroblog by machine learning. In: 2012 Fourth International Conference on Computational and Information Sciences (ICCIS), pp. 286–289, August 2012
4. Lek, H.H., Poo, D.: Aspect-based twitter sentimentclassification. In: 2013 IEEE 25th International Conference on Tools with Artificial Intelligence (ICTAI), pp. 366–373, November 2013
5. Ghiassi, M., Skinner, J., Zimbra, D.: Twitter brand sentiment analysis: a hybrid system using n-gram analysis and dynamic artificial neural network. Expert Syst. Appl. **40**(16), 6266–6282 (2013)
6. Feldman, R.: Techniques and applications for sentiment analysis. Commun. ACM **56**, 82–89 (2013)
7. Zhang, Y., Dang, Y., Chen, H.: Research note: examining gender emotional differences in web forum communication. Decis. Support Syst. **55**(3), 851–860 (2013)

8. Bhuta, S., Doshi, A., Doshi, U., Narvekar, M.: A review oftechniques for sentiment analysis of twitter data. In: 2014 International Conference on Issues and Challenges in Intelligent Computing Techniques (ICICT), pp. 583–591, February 2014
9. Pang, B., Lee, L., Vaithyanathan, S.: Thumbs up?: sentiment classification using machine learning techniques. In: Proceedings of the ACL-02 Conference on Empirical Methods in Natural Language Processing, EMNLP 2002, vol. 10, pp. 79–86, Association for Computational Linguistics, Stroudsburg, PA, USA (2002)
10. Brychcin, T., Konkol, M., Steinberger, J.: UWB: machine learning approach to aspect-based sentiment analysis, SemEval 2014, p. 817 (2014)
11. Liu, K.L., Li, W.J., Guo, M.: Emoticon smoothed language modelsfor twitter sentiment analysis, vol. 2, pp. 1678–1684, 2012. cited By (since 1996)
12. Kansal, H., Toshniwal, D.: Aspect based summarization of context dependent opinion words. Procedia Comput. Sci. **35**, 166–175 (2014). 2014 Proceedings of 18th Annual Conference on Knowledge-Based and Intelligent Information and amp; Engineering Systems, KES-2014 Gdynia, Poland, September
13. Zhang, W., Xu, H., Wan, W.: Weakness finder: find product weakness from chinese reviews by using aspects based sentiment analysis. Expert Syst. Appl. **39**(11), 10283–10291 (2012)
14. Jmal, J., Faiz, R.: Customer review summarization approach using twitter and sentiwordnet. In: Proceedings of the 3rd International Conference on Web Intelligence, Mining and Semantics, WIMS 2013, pp. 33:1–33:8, New York, NY, USA. ACM (2013)
15. Marrese-Taylor, E., Velsquez, J.D., Bravo-Marquez, F.: A novel deterministic approach for aspect-based opinion mining in tourism products reviews. Expert Syst. Appl. **41**(17), 7764–7775 (2014)
16. Baccianella, S., Esuli, A., Sebastiani, F.: Sentiwordnet 3.0: an enhanced lexical resource for sentiment analysis and opinion mining. In: Calzolari, N., Choukri, K., Maegaard, B., Mariani, J., Odijk, J., Piperidis, S., Rosner, M., Tapias, D. (eds.) LREC, European Language Resources Association (2010)
17. Zainuddin, N., Selamat, A., Ibrahim, R.: Improving twitter aspect-based sentiment analysis using hybrid approach. In: Nguyen, N.T., Trawiński, B., Fujita, H., Hong, T.P. (eds.) ACIIDS 2016, Part I. LNCS, vol. 9621, pp. 151–160. Springer, Heidelberg (2016)
18. Liu, B.: Sentiment Analysis and Subjectivity: Handbook of Natural Language Processing, 2nd edn. CRC Press, Boca Raton (2010)
19. Esuli, A., Sebastiani, F.: Sentiwordnet: a publicly available lexical resource for opinion mining. In: Proceedings of LREC, vol. 6, pp. 417–422 (2006)
20. Selamat, A., Omatu, S.: Web page feature selection and classification using neural networks. Inf. Sci. **158**, 69–88 (2004)
21. Vinodhini, G., Chandrasekaran, M.R.: Opinion mining using principal component analysis based ensemble model for e-commerce application. CSI Trans. ICT **2**(3), 169–179 (2014)
22. Cortes, C., Vapnik, V.: Support-vector networks. Mach. Learn. **20**(3), 273–297 (1995)
23. Joachims, T.: Text categorization with support vector machines: learning with many relevant features. In: Nédellec, C., Rouveirol, C. (eds.) ECML 1998. LNCS, vol. 1398, pp. 137–142. Springer, Heidelberg (1998)
24. Agarwal, A., Xie, B., Vovsha, I., Rambow, O., Passonneau, R.: Sentiment analysis of twitter data. In: Proceedings of the Workshop on Languages in Social Media, LSM 2011, pp. 30–38, Association for Computational Linguistics, Stroudsburg, PA, USA (2011)

Semantic Web and Social Networks

The Effectiveness of Gene Ontology in Assessing Functionally Coherent Groups of Genes: A Case Study

Nicoletta Dessì and Barbara Pes[(✉)]

Dipartimento di Matematica e Informatica,
Università degli Studi di Cagliari, Via Ospedale 72, 09124 Cagliari, Italy
{dessi,pes}@unica.it

Abstract. In recent years, ontologies have been extensively used in many biological fields to support a variety of applications. A well known example is Gene Ontology (GO) that organizes a vocabulary of terms about gene products and functions. GO offers an effective support for evaluating the similarity between two genes by measuring the distance of their respective GO terms. The advent of high-throughput technologies and the consequent production of lists of genes associated with specific conditions is stressing the need of recognizing groups of genes which cooperate within a specific biological event. This paper compares six popular similarity measures on GO in order to evaluate their effectiveness in discovering functionally coherent genes from an assigned list of genes. The aim is to discover which measure performs best. We also investigate about the potential of GO in evaluating the similarity of a set of genes according to its cardinality and the characteristics of the similarity measures. Experiments take into consideration: (a) 84 groups of genes sharing similar molecular functions through the production of enzymes within the human organism; (b) 150 groups of randomly selected genes. The paper demonstrates the efficient support of GO in detecting functionally related groups of genes, despite the GO's hierarchical structure limits the representation of richer forms of knowledge.

Keywords: Bioinformatics · Gene Ontology · Semantic similarity of genes

1 Introduction

Ontologies serve prominent roles in the life science (LS) and their formalism provides effective descriptions of biomedical knowledge. Being used only for annotation purposes, LS ontologies are controlled vocabularies which (a) provide a way to organize knowledge for subsequent retrieval, (b) make mandatory the use of authorized terms that have been preselected by expert curators.

Gene Ontology (GO) [1] is a popular ontology that provides consolidated description of gene functions for a variety of applications including disease gene prioritization, gene expression data analysis, study of protein interaction etc.

© Springer International Publishing Switzerland 2016
H. Fujita et al. (Eds.): IEA/AIE 2016, LNAI 9799, pp. 283–293, 2016.
DOI: 10.1007/978-3-319-42007-3_24

In detail, GO consists of about 30,000 terms of a controlled vocabulary with specific biological meaning. Terms are structured into a directed acyclic graph (DAG). Each term annotates a set of genes related to the event it describes.

GO consists of three sub-ontologies (i.e. three disconnected sub-graphs also called GO aspects [2]) which describe separately the biological processes (BP), the molecular functions (MF) and the cellular components (CC).

GO has drawn more and more attention from the bioinformatics researchers as a support for assessing the similarity between two genes by measuring the distance between their respective GO terms [3–5] on a sub-ontology whose choice depends on the specific interests of users. Although the abundance of tools and methods for evaluating semantic similarity measures on GO, assessing the best semantic similarity measure on GO remains a challenge for many reasons including:

(a) Different similarity measures perform differently on the same gene set. These differences depend on the GO aspect the measure takes into consideration as well as on the criteria for measuring the distance between two nodes. Recent research emphasizes difficulties that arise from the adoption of approaches that try to integrate multiple measures [6].

(b) Similarity measures evaluate the distance between two genes on a single GO sub-ontology. This procedure reflects the notion that the three GO aspects are independent, when, in reality, they are strongly correlated in all biological processes. On the other hand, it does not make sense to combine two or three similarity evaluations performed on different GO aspects [6] into a single measure.

Although a lot of research work has been done for evaluating the performance of the similarity measures on GO, there is little knowledge about the effectiveness of such measures in detecting a set of functionally coherent genes (i.e. a group of genes which cooperate to determine a single biological event [7, 8]) from a given gene list. In this case, genes that participate in the same biological process are supposed to have significant higher similarity than expected by chance in terms of GO annotations.

The advent of high-throughput technologies (such as next generation sequencing) and the consequent production of lists of genes associated with specific conditions is stressing the need of recognizing groups of functionally coherent genes in order to construct networks of genes presenting high pair-wise similarity [9] and characterize these networks with a particular transcriptional behavior [10].

Concerning the semantic similarity within sets of genes, related difficulties depend on several factors including the number of genes, the multiple functions they share, the limited amount of annotations of GO for most organisms and the narrow knowledge encoded in the GO which severely limits the representations of more complex relationships (e.g. "has-part-in", "is-a-way-of-doing" etc.).

The aim of this paper is to compare six popular similarity measures on GO in order to evaluate their effectiveness in discovering functionally coherent genes from an assigned list of genes. In more detail, our aim is twofold. First, we want to select which method or class of methods performs best. Second, we want to investigate the potential of GO ontology in evaluating the similarity of a set of genes according to the number of the considered genes and the characteristics of the similarity measure.

For this purpose, the paper presents experiments on 84 groups of genes sharing similar molecular functions through the production of enzymes. Specifically, each group is responsible for producing a single enzyme that affects a human metabolic pathway. Because genes belonging to the same group are certainly related, our basic idea is that a good similarity method should assign high scores in evaluating their similarity. At the same time, GO should offer a good support.

Our experiments compare six classical and wide used similarity measures and reveal interesting differences among their evaluations. Additionally, we demonstrate that GO is effective also in detecting small functional groups of genes, although the hierarchical structure of its catalogue limits the representation of deep knowledge about genes.

The paper is organized as follows. Focusing on the GO ontology, Sect. 2 describes and analyses the six semantic similarity measures we considered. Next, in Sect. 3, we present the dataset used in estimating the semantic similarity, as well as our experiments, and we also analyze and discuss the results. Section 4 illustrates the related work. Finally, Sect. 5 presents our conclusions and lines of our future research work.

2 Semantic Similarity Measures

GO ontology provides a way to explore the relationship of regulation or functions between gene products or genes, collectively called genes hereafter for simplicity. From a structural point of view, GO is composed by the GO graph and the GO Annotation [11].

Specifically, GO graph is a directed acyclic graph where the nodes are terms and the edges express the relationships among terms they link together. Within this graph, a GO term is a direct child of another term only if the former term is a subtype, a component or a regulator of the latter one. This means that only "*part-of*", "*is-a*" and "*regulates*" relationships are expressed within the GO graph.

GO Annotation annotates genes with the terms in the GO graph to provide a way of connecting genes to the biological process in which they are involved.

Figure 1 depicts a toy example of GO in which nodes (i.e. root, t1, t2, ..., tn) and edges represent GO terms and relationships and (g1, g2, ..., gm) are the set of genes annotated to these terms. When several genes are annotated with the same term, they are semantically similar because their potential relevance within the biological concept the term is about. As well, they are annotated with all the ancestors of that term in the GO graph.

Several semantic similarity measures have been proposed which can be broadly classified into the following categories [12, 13]:

– *Information Content (IC) measures* - These are the earlier developed methods and evaluate the semantic similarity between two genes by considering the frequencies of their annotations within GO terms and their lower common ancestor (LCA) in the GO graph.
– *Graph-based measures* - These methods compute the semantic similarity using the topology of the GO graph structure and considering the length or the types of edges by which terms are linked. As well, many strategies are based on both the terms contents and their distance in the GO graph.

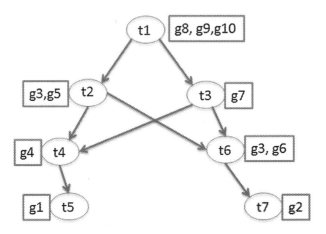

Fig. 1. An illustrative toy-example of GO

In more detail, information content measures relay on two basic concepts: the LCA and the IC of a term. Given two genes annotated by two terms t1 and t2, their LCA is the lower term they share in the GO graph. For example, in Fig. 1, the term t3 is the LCA of t5 and t7. The IC content of a term t depends on the number of the genes annotated by t and is defined as follows:

$$IC(t) = -\log(G(t)/G(r))$$

where G(t) and G(r) are the number of genes annotated to term t and the root term (including all of its descendants) respectively.

Among IC-based measures, we considered the Resnik [14] measure that expresses the similarity between two terms as the IC of their lower ancestor. Terms sharing the same LCA have the same similarity, even if they are at different levels in the GO graph. For example, in Fig. 1 the term pairs (t4, t6) and (t7, t5) have the same similarity (i.e. the IC of t3), but clearly they have different distance from t3.

Lin [15] proposed a normalized version of the Resnik method and expresses the similarity between two genes as the rate between the information content of the LCA and both terms.

In addition to Resnik and Lin measures, we considered the IC-based measure introduced by Jiang and Conrath [16]. They proposed to evaluate the distance between two terms as the difference between their IC and the IC of their LCA.

The IC methods are based on the semantic inheritance property between terms in the GO graph which ensures the transmission of information, step by step, from parents to children. Thus, the common limit of IC-based measures is that they only consider a single LCA ignoring the global structure of GO and the problem of multiple inheritance.

To take into account both the distance of terms from the root and the presence of multiple common ancestors, graph based measures were proposed.

In this paper, we consider the Schlicker measure [17] where the Lin measure is weighted by the rate between the number of genes annotated by the LCA and the number of genes annotated by the root.

In the same example in Fig. 1, the IC measures do not differentiate the similarity of the pair (t5, t6) to that of the pair (t4, t6), while the Schlicker measure is able to stress this difference even if it neglects the fact that a pair may have more than one LCA.

The Yu measure [18] takes this fact into account and considers the set of lowest common ancestors (LCAs) and the total number of gene pairs. For example, in Fig. 1, the similarity of t5 and t7 based on Yu measure considers the 45 gene pairs among the possible 10 genes and the gene pairs sharing t3 and t2.

Finally, we considered the Wang measure [19] that takes all the parents of the candidate terms into account. In the same example in Fig. 1, the similarity between t5 and t7 considers both the ancestors of t7 (i.e. t7, t6, t3, t2, t1) and the ancestors of t5 (i.e. t5, t4, t2, t1, t3).

The above measures originate evaluations of the semantic similarity that differ in their scale and distribution and solely rely on only one or few types of relationships while neglecting the others. However, the integration of outputs from different similarity measures may originate biased results. The better performance of a specific measure can be asserted only by their systematic evaluation and comparison on suitable datasets.

3   Experiments

Our experiments focus on comparing the six similarity measures presented in the previous section. As a benchmark, we consider genes involved in production of enzymes within the human model organism. Enzymes are organized in a numerical classification schema assigned by the Enzyme Commission (EC), the so called EC numbers. Specifically, an EC number identifies a group of genes (shortly denoted as proteins) which share similar molecular functions in the production of a specific enzyme. As reference ontology we considered the MF aspect of GO.

3.1   Data Preparation and Processing

First, we downloaded EC numbers and their corresponding genes from [20]. Then, we randomly selected 84 EC numbers (i.e. functional groups of genes) of different cardinality. Within these EC numbers, each gene is member of a single functional group and each group contains at least four genes. Then, we built 150 groups of random selected genes of different cardinalities. Table 1 shows how the selected groups vary in size.

Concerning the comparison of the six similarity measures, we first evaluated the pair-wise semantic similarity by all pairs of genes belonging to the same group, separately for each measure. This resulted in six sequences of values: each of them expresses the distribution of the semantic similarity evaluated by a single method on the 84 groups of enzymes. Then, we calculated the distribution of the semantic similarity evaluated by a single method on the random selected genes. Finally, we evaluated the mean and the median of each distribution.

Table 1. Selected groups and their cardinality

Cardinality	Enzymes	Random
4	6	12
5	2	4
6	6	9
7	2	4
TOTAL	84	150

3.2   Results and Discussion

Figures 2 and 3 show respectively the difference between the median and the mean of the semantic similarity estimated by each measure over the 84 enzyme groups and the 150 randomly selected groups.

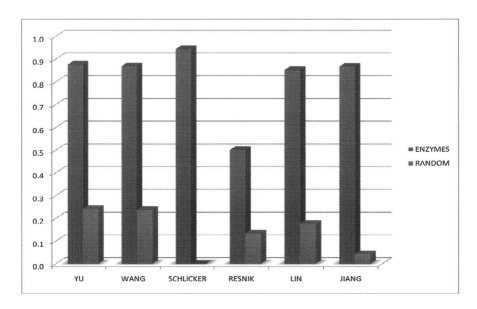

Fig. 2. Score comparison for medians evaluated by each single measure (Color figure online)

Concerning the IC-based measures, we observe that both the Lin's and Jiang's measures improve the Resnik's measure. Indeed, the former measures take into account the taxonomical depths in GO that results in better recognition of the semantic similarity of genes. Similar conclusions can be drawn for graph-based measures which obtain higher similarity scores than those based on IC. This desirable behavior depends on the circumstance that graph-based measures consider the global structure of GO and demonstrates that the GO taxonomical knowledge is properly structured and annotated.

Results also indicate that graph-based measures capture the semantics explicitly modeled in the GO taxonomy. So, this class of measures seems to be more suitable for

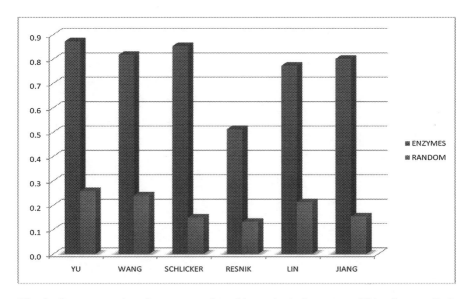

Fig. 3. Score comparison for means evaluated by each single measure (Color figure online)

devising genes that cooperate within a single biological event. Indeed, graph-based approaches seem to be able to capture more semantic evidence because they consider the taxonomical detail under each term (i.e., the topological distance and the LCA distance) in evaluating gene similarity.

Our results agree with [21] that compares Resnik, Schlicker and Wang measures on three model organisms and with [22] that shows a better performance of the Schlicker measure respect to Yu and Wang measures.

It is interesting to notice that all the methods we evaluated recognize a low similarity also between genes belonging to the random group. This is due to the large number of biological processes in which genes are involved. Indeed, genes belonging to random groups do not interact in the enzyme production, but we cannot exclude that they interact in other biological processes.

For comparing the performance of the six similarity measures, we considered only the pair-wise similarity evaluated for the 84 ECs.

Figure 4 shows a more detailed global comparison of the six methods. Specifically, top and bottom of the boxes represent 75th and 25th percentile, the division line is the median, top and bottom whiskers represent greatest and lowest values of the median.

The median, 75th and 25th percentile of Schlicker measure are 0.94, 0.99 and 0.79, significantly higher than the other evaluations. This confirms that Schlicker measure performed best among all the measures. As well, Fig. 4 better illustrates the best performance of the graph-based methods.

Finally, we investigated if and how the size of the gene groups influences the performance of the similarity measures. Towards this end, we considered the minimum of the pair-wise similarity on the 84 ECs. Due to the cooperation among genes belonging to a same EC, we should expect that this minimum scores low but, in any case, it is greater than zero, no matter the size of the group.

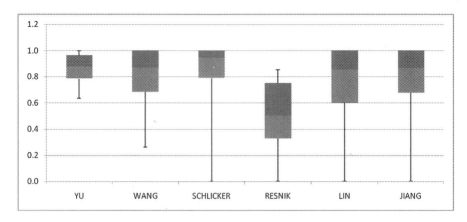

Fig. 4. A comparison of the six methods: the median

Figure 5 seems to not confirm this supposition, at least for some methods. Specifically, in Fig. 5, the x-axis depicts the group cardinality and the y-axis shows the minimum value achieved by each similarity measure.

When the size of the group is small (i.e. four and five genes per group), IC-based methods seem to fail in recognizing the similarity between genes belonging to the same group. However, this similarity exists and is well recognized by graph-based methods.

The Schlicker measure is a notable exception. Although its good performance, the minimum for this similarity measure is zero. However, because it happens for just a single couple of genes, we can consider this minimum as an outlier.

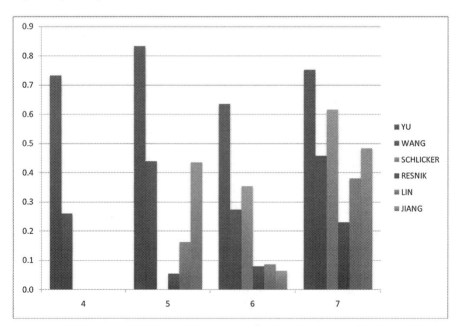

Fig. 5. Minimum values of the similarity within groups of different cardinality (Color figure online)

When the cardinality of the group increases, the IC-based measures are nonetheless able to recognize that the minimum of the pair-wise similarity is always different from zero. These results seem to confirm the best accuracy of graph-based methods, also in detecting small functionally coherent groups of genes.

4 Related Work

There are an increasing number of application scenarios where biomedical ontologies are exploited to complement and enrich knowledge acquired from experiments and domain experts. Among possible applications, the semantic similarity is a popular approach to evaluate the relatedness of two genes by considering how close two concepts are from each other in some ontology.

Tailored to the characteristics of GO, new semantic similarity measures are constantly appearing in the literature [23]. Hybrid approaches [6, 21] try to integrate state-of-art gene-to-gene similarity measures on GO and demonstrate that this integration improves the performance of single methods, almost on some datasets.

A novel approach is proposed in [22] that incorporates information from gene co-function networks in addition to using GO structure and annotation. The study demonstrates the benefits of such integration in terms of performance of semantic similarity measures. In agreement with our results, [22] also shows a better performance of the graph-based measures (i.e. Schlicker, Wang and Yu measures) against the Resnik measure. Few work [24] has been done for evaluating how a specific method performs in recognizing whether genes share or not similar functions and/or participate or not in the same biological event. Such a kind of similarity analysis would be of great importance especially in high dimensional gene lists, where the application of semantic similarity may help the selection of functionally related genes.

5 Conclusions and Future Work

Recognizing putative interactions and detecting common functions in a set of genes is a key activity for life scientists in order to assess the significance of experimentally derived gene sets and prioritizing those sets that deserve follow-up. This interest is shifting the focus of data analysis from individual genes to groups of genes that are supposed to interact each other in determining a pathological state or influencing the outcome of a single trait. The work presented in this paper is a preliminary approach to give a contribution in this direction.

We have investigated about the performance of six semantic similarity measures and our experiments demonstrate that graph-based measures provide a high degree of accuracy than the IC-based measures, especially when the target group is small.

Beside the benefits deriving from the exploitations of GO knowledge, evaluating semantic similarity on GO presents some limitations related both to completeness and coherency of its taxonomical knowledge which does not provide extensive coverage about gene functions and processes.

If some of the compared genes are not annotated by GO terms the similarity cannot be measured. The same consideration holds if GO lacks to annotate genes to some event to which genes participate.

A strategy to minimize this strict dependency consists on considering additional sources of knowledge such as several ontologies, catalogues compiled by expert curators, experimental results etc.

We are currently investigating these possibilities in order to integrate ontological knowledge with complementary information about different aspects of gene interaction.

Leveraging on our previous work [25], we are considering the exploitation of knowledge from the most important and largest collections of biomedical documents freely available on the Internet while accounting for the existence of functionally coherent groups defined by biomedical experts such as protein families, gene families etc.

References

1. Consortium GO: Gene Ontology annotations and resources. Nucleic Acids Res. **41**, D530–D535 (2013)
2. Lord, P.W., Stevens, R.D., Brass, A., Goble, C.A.: Investigating semantic similarity measures across the Gene Ontology: the relationship between sequence and annotation. Bioinformatics **19**(10), 1275–1283 (2003)
3. Lee, W.N., Shah, N., Sundlass, K., Musen, M.: Comparison of Ontology-based Semantic-Similarity Measures. AMIA Annu Symp Proceedings 2008, V2008:384-388
4. Mazandu, G.K., Mulder, N.J.: Information content-based gene ontology semantic similarity approaches: toward a unified framework theory. Biomed Res. Int. **2013**, Article ID 292063 (2013)
5. Guzzi, P.H., Mina, M., Guerra, C., Cannataro, M.: Semantic similarity analysis of protein data: assessment with biological features and issues. Briefings Bioinf. **13**(5), 569–585 (2012)
6. Peng, J., Wang, Y., Chen, J.: Towards integrative gene functional similarity measurement. BMC Bioinformatics **15**(Suppl 2), S5 (2014)
7. Richards, A.J., Muller, B., Shotwell, M., Cowart, L.A., Rohrer, B., Lu, X.: Assessing the functional coherence of gene sets with metrics based on the Gene Ontology graph. Bioinformatics **26**(12), i79–i87 (2010)
8. Kandula, S., Zeng-Treitler, Q.: Exploring relations among semantic groups: a comparison of concept co-occurrence in biomedical sources. Stud. Health Technol. Inform. **160**, 995–999 (2010)
9. Teng, Z., Guo, M., Liu, X., Dai, Q., Wang, C., Xuan, P.: Measuring gene functional similarity based on group-wise comparison of GO terms. Bioinformatics **29**, 1424–1432 (2013)
10. Soldatos, T.G., Perdigão, N., Brown, N.P., Sabir, K.S., O'Donoghue, S.I.: How to learn about gene function: text-mining or ontologies? Methods **74**, 3–15 (2015)
11. Camon, E., Magrane, M., Barrell, D., Lee, V., Dimmer, E., Maslen, J., Binns, D., Harte, N., Lopez, R., Apweiler, R.: The Gene Ontology annotation (GOA) database: sharing knowledge in Uniprot with Gene Ontology. Nucleic Acids Res. **32**, D262–D266 (2004)
12. Pesquita, C., Faria, D., Falcao, A., Lord, P., Couto, F.: Semantic similarity in biomedical ontologies. PLoS Comput. Biol. **5**(7), e1000443 (2009)

13. Pesquita, C., Faria, D., Bastos, H., Ferreira, A., Falcao, A., Couto, F.: Metrics for GO based protein semantic similarity: a systematic evaluation. BMC Bioinformatics **9**(Suppl. 5), S4 (2008)
14. Resnik, P.: Semantic similarity in a taxonomy: an information-based measure and its application to problems of ambiguity in natural language. J. Artif. Intell. Res. **11**, 95–130 (1999)
15. Lin, D.: An information-theoretic definition of similarity. In: Proceedings of the 15th International Conference on Machine Learning, pp. 296–304. Morgan Kaufmann, San Francisco (1998)
16. Jiang, J.J., Conrath, D.W.: Semantic similarity based on corpus statistics and lexical taxonomy. In: Proceedings of International Conference Research on Computational Linguistics, Taiwan, pp. 9008–9022 (1997)
17. Schlicker, A., Domingues, F., Rahnenfuhrer, J., Lengauer, T.: A new measure for functional similarity of gene products based on Gene Ontology. BMC Bioinformatics **7**, 302 (2006)
18. Yu, H., Jansen, R., Stolovitzky, G., Gerstein, M.: Total ancestry measure: quantifying the similarity in tree-like classification, with genomic applications. Bioinformatics **23**(16), 2163–2173 (2007)
19. Wang, J., Du, Z., Payattakool, R., Yu, P., Chen, C.: A new method to measure the semantic similarity of GO terms. Bioinformatics **23**(10), 1274–1281 (2007)
20. http://humancyc.org
21. Peng, J., Li, H., Jiang, Q., Wang, Y., Chen, J.: An integrative approach for measuring semantic similarities using gene ontology. BMC Syst. Biol. **8**(Suppl 5), S8 (2014)
22. Peng, J., Uygun, S., Kim, T., Wang, Y., Rhee, S.Y., Chen, J.: Measuring semantic similarities by combining gene ontology annotations and gene co-function networks. BMC Bioinformatics **16**, 44 (2015)
23. Yang, H., et al.: Improving GO semantic similarity measures using downward random walks. Bioinformatics **28**, 1383–1389 (2012)
24. Pedersen, T., Pakhomov, S.V.S., Patwardhan, S., Chute, C.G.: Measures of semantic similarity and relatedness in the biomedical domain. J. Biomed. Inform. **40**(3), 288–299 (2007)
25. Dessì, N., Pascariello, E., Pes, B.: Integrating ontological information about genes. In: 2014 IEEE 23rd International WETICE Conference, pp. 417–422. IEEE (2014)

Social Network Clustering by Using Genetic Algorithm: A Case Study

Ming-Feng Tsai[1], Chun-Yi Lu[2], Churn-Jung Liau[3], and Tuan-Fang Fan[4(✉)]

[1] Department of Computer Science and Information Engineering,
National Kaohsiung University of Applied Sciences, Kaohsiung 807, Taiwan
a5533109@yahoo.com.tw
[2] Department of Information Management,
National Penghu University of Science and Technology, Penghu 880, Taiwan
jamesleu@npu.edu.tw
[3] Institute of Information Science Academia Sinica, Taipei 115, Taiwan
liaucj@iis.sinica.edu.tw
[4] Department of Computer Science and Information Engineering,
National Penghu University of Science and Technology, Penghu 880, Taiwan
dffan@npu.edu.tw

Abstract. With the rapid growth of large-scaled social networks, the analysis of social network data has become an extremely challenging computational issue. To meet the challenge, it is possible to significantly reduce the complexity of the problem by properly clustering a large social network into groups, and then analyzing data within each group, or studying the relationship among groups. Hence, social network clustering can be regarded as one of the essential problems in social network analysis. To address the issue, we propose an evolutionary computation approach to social network clustering. We first formulate social network clustering as an optimization problem and then develop a genetic algorithm to solve the problem. We also applied the proposed approach to a case study based on data of some Facebook users.

Keywords: Social network analysis · Clustering · Evolutionary computation

1 Introduction

Humans are always connected to others in the society and cannot live alone at all times. Whether it is based on physical or psychological needs, individuals have to interact with others. Hence, social networking is a common phenomenon in human society. As small as a family, a village, to a community, a nation, and even the whole world, can be regarded as a social network of different scales. To study interaction behaviors between individuals, social network analysis has become an important tool of sociology [9,10]. In recent years, with the rapid

© Springer International Publishing Switzerland 2016
H. Fujita et al. (Eds.): IEA/AIE 2016, LNAI 9799, pp. 294–304, 2016.
DOI: 10.1007/978-3-319-42007-3_25

growth of on-line social networking sites such as Facebook, LinkedIn, MySpace, and Friendster, social network analysis has received much attention from the public.

Nowadays, there have been more than one billion social network users around the globe. While the analysis of such a large-scaled social network can induce very useful knowledge and information, it is also an extremely challenging computational issue. To meet the challenge, it is possible to significantly reduce the complexity of the problem by properly clustering a large social network into groups, and then analyzing data within each group, or studying the relationship among groups. Hence, social network clustering can be regarded as one of the essential problems in social network analysis. The paper is aimed at addressing the problem by using an evolutionary computation approach. The basic idea is that, for a fixed criterion, social network clustering can be formulated as an optimization problem that can then be solved by using genetic algorithm. In the following sections, we will explicate how such an idea can be realized.

The remainder of this paper is organized as follows. In Sect. 2, we present the basic definition of social network and its clustering problem. In Sect. 3, we propose a formulation of social network clustering as an optimization problem and its solution by using genetic algorithm. In Sect. 4, we apply the proposed algorithm to a case study based on data of some Facebook users. Finally, in Sect. 5, we present our conclusions and indicate some future research directions.

2   Social Network and Its Clustering

A social network is comprised of a finite set of individuals and a set of relationships among the individuals. In the literature of social network analysis, individuals are also called actors. Mathematically, a social network is described as a relational structure or (labeled) hypergraph $\mathfrak{N} = (U, <R_i>_{i \in I})$, where U denotes a finite set of individuals, called the universe; and $R_i \subseteq U^{k_i}$ is a k_i-ary relation on U for each $i \in I$ with I being an index set. For each R_i, the positive integer k_i is its arity. An unary relation is also called a property or a feature.

Because binary connection is the most common type of relationship in realistic social networks, we usually consider simplified social networks that contain only properties and binary relations. Thus, in this paper, a social network is defined as $\mathfrak{N} = (U, <P_i>_{i \in I}, <R_j>_{j \in J})$, where I and J are index sets, and for each $i \in i$ and $j \in j$, P_i and R_j are property and binary relation respectively. In particular, when J is a singleton, i.e., there is only one binary relation in the social network, $\mathfrak{N}$ can be rewritten as a labeled (directed) graph (V, E, L), where V is a set of vertices denoting the individuals, E is the set of edges denoting the binary relation, and $L : V \to 2^I$ is a label function such that $L(x) = \{i \in I \mid x \in P_i\}$, i.e., the set of properties possessed by x. Hence, without loss of generality, we will only consider social network in the form of labeled graph from now on.

Clustering is the process of partitioning a universe in accordance with some given criteria. In social network clustering, major criteria may be the similarity or the strength of connection between individuals. Formally, a clustering of a network (V, E, L) is a family of sets $\{C_1, C_2, \cdots, C_k\}$ such that $\bigcup_{i=1}^{k} C_i = V$ and $C_i \cap C_j = \emptyset$ for any $1 \leq i \neq j \leq k$. The criteria stipulate the conditions that should be satisfied for intra-cluster and inter-cluster elements. For *similarity-based clustering*, it is required intra-cluster elements should be similar to each other, whereas inter-cluster elements should have low similarity. On the other hand, *connection-based clustering* requires that intra-cluster elements have dense connections, whereas cross-cluster connections must be very sparse.

One well-known type of similarity-based clustering is the role assignment in social position analysis [1,4,7]. Individuals who are connected to the rest of the network in the same way are said to occupy the same position or play the same role. Different notions of *positional equivalence*, such as structural equivalence, regular equivalence, and exact equivalence, have been proposed to explain what it means that two individuals are connected to the rest of the network in the same way. Let $\mathfrak{N} = (V, E, L)$ be a social network. We define the *left and right neighborhoods* of any vertex x as $Ex = \{y | (y, x) \in E\}$ and $xE = \{y | (x, y) \in E\}$ respectively. Then, for any two vertices $x, y \in V$, x and y are *structurally equivalent* if $L(x) = L(y)$, $Ex = Ey$, and $xE = yE$. In addition, for any equivalence relation R on V and any subset $S \subset V$, we can define the set of R-equivalence classes overlapping with S as $[S]_R = \{C \mid C \cap S \neq \emptyset, C \text{ is an equivalence class of } R\}$. In an analogous way, we can define the multiset of R-equivalence classes overlapping with S as $\|S\|_R$. Then, an equivalence relation R on V is called a *regular equivalence* if for any $x, y \in V$, $(x, y) \in R$ implies $L(x) = L(y)$, $[Ex]_R = [Ey]_R$, and $[xE]_R = [yE]_R$; and we say that x and y are *regularly equivalent* if they belong to some regular equivalence relation on V. Analogously, an equivalence relation R on V is called an *exact equivalence* if for any $x, y \in V$, $(x, y) \in R$ implies $L(x) = L(y)$, $\|Ex\|_R = \|Ey\|_R$, and $\|xE\|_R = \|yE\|_R$; and we say that x and y are *exactly equivalent* if they belong to some exact equivalence relation on V.

While social position analysis provides precise criteria for similarity-based clustering, the criteria for connection-based clustering are rather vague. Despite the vagueness, the criteria of dense intra-cluster connection and sparse inter-cluster connection can be achieved by setting some thresholds. More specifically, given a network $\mathfrak{N} = (V, E, L)$, for any subset $C \subseteq V$ and $x \in V$, we define $E(x, C) = (xE \cup Ex) \cap C$, i.e., the set of vertices in C that are connected to x. Then, a clustering $\{C_1, C_2, \cdots, C_k\}$ satisfies the connection-based criterion if for any $1 \leq i \leq k$ and $x \in C_i$, both $|E(x, C_i)| \geq \alpha \cdot |C_i|$ and $|E(x, C_j)| \leq \beta \cdot |C_j| (\forall j \neq i)$ hold, where α and β are thresholds satisfying $0 \leq \beta < \frac{1}{2} < \alpha \leq 1$ and to be determined experimentally [8][1].

Although positional equivalence is a kind of crisp criterion for clustering, it is sometimes considered too restrictive because in many real applications, especially when properties of individuals are not considered and consequently

---

[1] Note that the properties of individuals do not play any role in the connection-based criterion, although it is possible to combine both connection-based and similarity-based criteria by taking such features into consideration.

the network is a simple graph (V, E), the positional equivalence relation easily degenerates into the identity or universal relation. Hence, it is desirable that the degree of similarity between individuals can be defined by relaxing some requirements of positional equivalence. For example, for the structural equivalence, we can relax the requirement of $Ex = Ey$ and $xE = yE$ to that there is a large proportion of intersection between Ex and Ey (and also between xE and yE). Thus, the degree of similarity between x and y is a monotonic function of $\frac{|Ex \cap Ey|}{|Ex \cup Ey|}$ and $\frac{|xE \cap yE|}{|xE \cup yE|}$. Then, we can induce a weighted graph from the original network that denote degrees of similarity between individuals. On the other hand, although the criterion of connection-based clustering is more flexible than positional equivalence, it may not reflect the real strength of interaction between individuals. For example, a Facebook user x may be connected to two users y and z. However, x may interact with y frequently but has rare interaction with z. To accommodate the information regarding the strength of interaction, we can attach a weight to each edge of the graph. Therefore, no matter whether the weight on an edge denotes the degree of similarity or the strength of interaction, we can unify two types of social network clusterings by using the representation of weighted graph. Consequently, the clustering of social network will become an optimization problem on a weighted graph.

3   Genetic Algorithm for Social Network Clustering

In the preceding section, we have seen that the criteria of social network clustering can be uniformly represented as a weighted graph. Thus, the purpose of the clustering process is to find the clustering that maximizes the satisfaction of these criteria. This means that social network clustering can be formulated as an optimization problem. Let $G = (V, E)$ be a weighted graph denoting the clustering criterion, where V is the set of vertices and $E : V \times V \to \Re$ is the set of weighted edges. Then, for any two vertices $u, v \in V$, we can define the outward and inward importance of the edge (u, v) as

$$I_{out}(u, v) = E(u, v) - \frac{\sum_{x \in V} E(u, x)}{|V|} \quad \text{and} \quad I_{in}(u, v) = E(u, v) - \frac{\sum_{x \in V} E(x, v)}{|V|}$$

respectively. Intuitively, $\frac{\sum_{x \in V} E(u, x)}{|V|}$ denotes the average weight of edges from u to other vertices in the graph and hence $I_{out}(u, v)$, i.e., the difference between the weight of the edge and the average weight, measures the relative importance of (u, v) among all outward edge from u. Analogously, $I_{in}(u, v)$ measures the relative importance of (u, v) among all inward edge to v. Hence, the total importance of the edge (u, v) can be defined as $I(u, v) = I_{out}(u, v) + I_{in}(u, v)$. Then, for a clustering $\mathcal{C} = \{C_1, C_2, \cdots, C_k\}$, the degree of $\mathcal{C}$ satisfying the criterion represented by G, denoted by $deg_G(\mathcal{C})$ is defined as follows:

$$deg_G(\mathcal{C}) = w_1 \cdot \sum_{1 \leq i \leq k} \sum_{u \in C_i} \sum_{v \in C_i} I(u, v) - w_2 \cdot \sum_{1 \leq i \neq j \leq k} \sum_{u \in C_i} \sum_{v \in C_j} I(u, v),$$

where $w_1, w_2 \in (0,1]$ are the parameters to be determined depending on real applications and experiments. The two summation terms of $deg_G(\mathcal{C})$ measures the intra-cluster and inter-cluster densities of the clustering respectively. Therefore, the goal is to find a clustering that can maximize the intra-cluster density and minimize the inter-cluster. Let $\mathsf{cls}(G)$ denote the set of all possible clusterings based on G. Then, the clustering problem is formulated as the following optimization problem:

$$\arg \max_{\mathcal{C} \in \mathsf{cls}(G)} deg_G(\mathcal{C}).$$

Once the optimization problem is formulated, we can solve it by using genetic algorithm (GA). GA is one of the earliest methods in evolutionary computation, which is a search heuristic inspired by the process of natural selection [6]. In the GA, an initial population of candidate solutions to the optimization problem is randomly generated and then evolved toward better solutions. Each candidate solution is encoded as a chromosome which can be altered by genetic operators such as mutation and crossover. In each generation of the GA, the fitness of every candidate solution in the population is evaluated according to the fitness function typically derived from the objective function of the problem. The more fit solutions are stochastically selected from the current population, and each solution's chromosome is modified by genetic operators to form a new generation. The new generation of candidate solutions is then used in the next iteration of the algorithm until the termination condition is meet, e.g. the maximum number of iterations has been executed or a satisfactory solution has been obtained.

The main design choices of GA depend on the genetic representation of candidate solutions and the fitness function to evaluate the solution domain. As usual, we can take the objective function of the optimization problem as the fitness function. On the other hand, while the most common encoding scheme for genetic representation is bit string, it is more convenient to encode each clustering as a string of positive integers in our applications. The length of the string is simply $|V|$ and the i-th position of the string denotes the cluster that contains vertex v_i. Hence, the encoding can be regarded as a function from $\{1, 2, \cdots, |V|\}$ to $\{1, 2, \cdots, K\}$, where K is the maximum allowed number of clusters. If there are not any constraints on the number of clusters, K is set to $|V|$. Note that we do not have to specify a fixed number of clusters in advance and the optimization

Table 1. The parameter settings of six experiments

No. of experiment	w_1	w_2	α	β	γ
Experiment 1	0.7	0.3	0.5	0.3	0.2
Experiment 2	0.7	0.3	0.2	0.3	0.5
Experiment 3	0.7	0.3	0.1	0.3	0.6
Experiment 4	0.7	0.3	0.3	0.5	0.2
Experiment 5	0.7	0.3	0.2	0.6	0.2
Experiment 6	0.7	0.3	0.2	0.7	0.1

Table 2. The results of six experiments

No. of experiment	Best fitness	Best no. of clusters
Experiment 1	70.658	6
Experiment 2	40.030	7
Experiment 3	29.583	5
Experiment 4	26.934	8
Experiment 5	28.154	7
Experiment 6	36.918	8

procedure will search through clusterings of all sizes. In addition, different selection methods and genetic operators can be chosen depending on applications.

4 A Case Study

In this section, we apply the general framework developed in the last section to clustering some users of the well-known on-line social network Facebook. These users include the first author of the paper and his 301 friends on Facebook and hence constitute a very small-scaled subgraph of the whole on-line social network. We use the number of messages posted by an user and the number of responses he received from other users during a fixed period to determine the strength of interaction between users. These numbers include

- p_j: the number of messages posted by user j,
- pl_{ij}: the number of j's posts on which i has clicked the "Like" button,

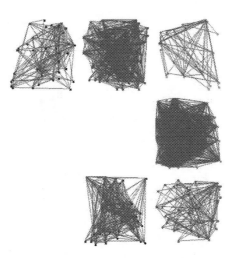

Fig. 1. The best clustering for experiment 1

– pc_{ij}: the number of j's posts on which i has commented,
– pcl_{ij}: the number of comments for j's posts on which i has clicked the "Like" button.

Then, the strength of interaction from i to j is defined as

$$\frac{\alpha \cdot pl_{ij} + \beta \cdot pc_{ij} + \gamma \cdot pcl_{ij}}{p_j},$$

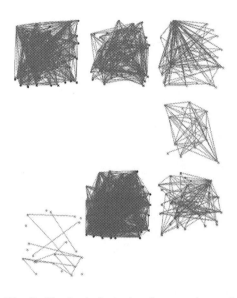

Fig. 2. The best clustering for experiment 2

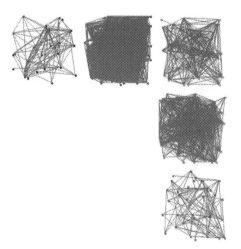

Fig. 3. The best clustering for experiment 3

where $\alpha, \beta, \gamma \in [0,1]$ satisfying $\alpha + \beta + \gamma = 1$ are used to adjust the weights of pl_{ij}, pc_{ij}, and pcl_{ij}. Therefore, we have a weighted graph $G = (V, E)$, where $|V| = 302$ and $E(i,j)$ is the normalized strength of interaction from i to j.

To apply the genetic algorithm to solve the optimization problem, we choose the tournament selection mechanism and uniform crossover and mutation operators. Tournament selection is performed by running several "tournaments" among a few chromosomes chosen at random from the population; and the winner of each tournament (the one with the maximum fitness value) is selected for crossover. For uniform crossover, a bit vector of length $|V|$ is generated randomly as indicator and the parent strings exchange their values at each position

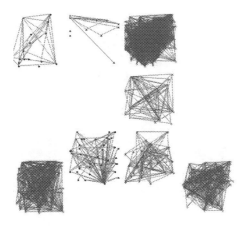

Fig. 4. The best clustering for experiment 4

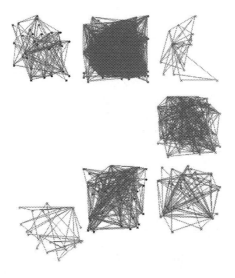

Fig. 5. The best clustering for experiment 5

with 1 in the indicator vector. Analogously, for uniform mutation, a bit vector of length $|V|$ is generated randomly as indicator and the value at each 1-position is randomly mutated.

To see how the algorithm work, we conduct six experiments with different parameter settings of w_1, w_2, α, β, and γ as shown in Table 1. For each parameter setting, the GA is run until 1000 generations are reached. Then, the best fitness value and the number of clusters of the best clustering for each experiment are shown in Table 2. The best clustering for each experiment is shown in

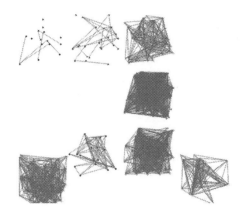

Fig. 6. The best clustering for experiment 6

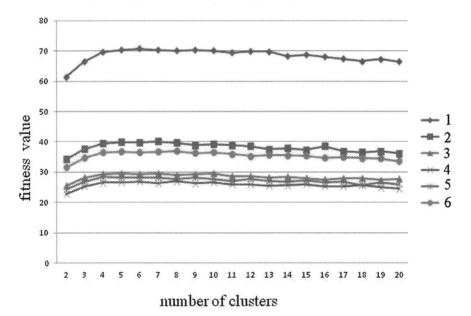

Fig. 7. The fitness values vs. numbers of clusters in different experiments

Figs. 1, 2, 3, 4, 5 and 6 respectively. In these figures, an edge between two vertices is drawn if their strength of interaction is greater than the average weight.

In addition, for each experiment, we also find the best fitness value for clusterings of different sizes and the result is plotted in Fig. 7.

5   Concluding Remarks

In this paper, we present a genetic algorithm for solving the social network clustering problem. While the clustering can be based on different criteria such as the similarity or density of connection, we show that these criteria can be uniformly represented by weighted graphs. Hence, the clustering problem can be formulated as an optimization problem to maximize the satisfaction of the clustering criterion and then GA can be naturally utilized to solving the problem. As a case study, we apply the proposed approach to a small-scaled subgraph comprised of 302 Facebook users.

In addition to GA, there exist other nature-inspired meta-heuristic methods for optimization problems, such as ant colony algorithm [2], particle swarm algorithm [3], differential evolution [5], etc. We do not see any essential difficulty in applying these methods to our general framework. Hence, in the future, it is possible to investigate and compare the effectiveness of these methods for the social network clustering problem. Furthermore, the case study only shows the application of our algorithm to a small-scaled graph. Therefore, more large-scaled experiment are needed in our future research for the practical applicability of the proposed approach.

Acknowledgment. This work was partially supported by the Ministry of Science and Technology of Taiwan under Grants MOST 103-2410-H-346-007-MY2 and MOST 104-2221-E-001-010-MY3.

References

1. Boyd, J., Everett, M.: Relations, residuals, regular interiors, and relative regular equivalence. Soc. Netw. **21**(2), 147–165 (1999)
2. Dorigo, M., Stutzle, T.: Ant Colony Optimization. MIT Press, Cambridge (2004)
3. Engelbrecht, A.: Fundamentals of Computational Swarm Intelligence. Wiley, Chichester (2005)
4. Everett, M., Borgatti, S.: Regular equivalences: general theory. J. Math. Soc. **18**(1), 29–52 (1994)
5. Feoktistov, V.: Differential Evolution: In Search of Solutions. Springer, New York (2006)
6. Goldberg, D.: Genetic Algorithms in Search, Optimization and Machine Learning. Addison-Wesley, Boston (1989)
7. Lerner, J.: Role assignments. In: Brandes, U., Erlebach, T. (eds.) Network Analysis. LNCS, vol. 3418, pp. 216–252. Springer, Heidelberg (2005)
8. Mishra, N., Schreiber, R., Stanton, I., Tarjan, R.E.: Clustering social networks. In: Bonato, A., Chung, F.R.K. (eds.) WAW 2007. LNCS, vol. 4863, pp. 56–67. Springer, Heidelberg (2007)

9. Scott, J.: Social Network Analysis: A Handbook, 2nd edn. Sage Publications, Thousand Oaks (2000)
10. Wasserman, S., Faust, K.: Social Network Analysis: Methods and Applications. Cambridge University Press, Cambridge (1994)

S-Rank: A Supervised Ranking Framework for Relationship Prediction in Heterogeneous Information Networks

Wenxin Liang[✉], Xiaosong He, Dongdong Tang, and Xianchao Zhang

School of Software, Dalian University of Technology, Dalian 116620, China
{wxliang,xczhang}@dlut.edu.cn, hxs91@126.com, tdd_ssdut@163.com

Abstract. The most crucial part for relationship prediction in heterogeneous information networks (HIN) is how to effectively represent and utilize the information hidden in the creation of relationships. There exist three kinds of information that need to be considered, namely *local structure information* (Local-info), *global structure information* (Global-info) and *attribute information* (Attr-info). They influence relationship creation in a different but complementary way: Local-info is limited to the topologies around certain nodes thus it ignores the global position of node; methods using Global-info are biased to highly visible objects; and Attr-info can capture features related to objects and relations in networks. Therefore, it is essential to combine all the three kinds of information together. However, existing approaches utilize them separately or in a partially combined way since effectively encoding all the information together is not an easy task. In this paper, a novel three-phase Supervised Ranking framework (S-Rank) is proposed to tackle this issue. To the best of our knowledge, our work is the first to completely combine Global-info, Local-info and Attr-info together. Firstly, a Supervised PageRank strategy (SPR) is proposed to capture Global-info and Attr-info. Secondly, we propose a Meta Path-based Ranking method (MPR) to obtain Local-info in HIN. Finally, they are integrated into the final ranking result. Experiments on DBLP data demonstrate that the proposed S-Rank framework can effectively take advantage of all the three kinds of information for predicting citation relation and outperforms other well-known baseline approaches.

1 Introduction

Link Prediction, introduced by Liben-Nowell and Kleinberg [10], aims at taking advantage of the "proximity" of objects to predict the emergence of links or interactions in the future. A link prediction framework or algorithm can benefit researchers and organizations in a variety of fields.

Most of the existing link prediction approaches designed for networks are under an ideal circumstance, *i.e.*, homogeneous networks, in which only one type

This work was partially supported by National Science Foundation of China (No. 61272374, No. 61300190 and No. 61572096) and 863 Project (No. 2015AA015463).

H. Fujita et al. (Eds.): IEA/AIE 2016, LNAI 9799, pp. 305–319, 2016.
DOI: 10.1007/978-3-319-42007-3_26

of object and edge can be found. However, as mentioned in [6], objects in real-world networks are interconnected, forming gigantic, informative networks with multiple types of objects and relations, and become *heterogeneous information networks* (HIN for short). Researchers [15,16] conclude that the link prediction problem could be extended to an analogous but more general problem, namely the *Relationship Prediction* problem, in the context of HIN. Due to the intrinsic complexity of HIN and the variance in modeling technology, traditional methods cannot meet all the needs in HIN. In this paper, we mainly focus on tackling the relationship prediction problem in the context of HIN. The emergence of relationship is predictable due to the assumption that there exists latent information encoded in networks, which is the main reason for why the relationship was generated. Therefore, the key issue of relationship prediction in HIN is how to effectively represent, extract and model the latent information.

1.1 Motivations

Firstly, topological features between objects can be referred as structure information. We divide the structure information into two categories by reviewing existing methods [10,15], namely *local structure information* (Local-info) and *global structure information* (Global-info). They influence the creation of relations from different aspects.

Figure 1 shows a small social network (*e.g.*, Twitter), from which we illustrate the difference between Local-info and Global-info through a simple relationship prediction problem: predicting whom s will probably create a relationship to in the future, u or v?

It is obvious that there are more paths between s and v than those between s and u. Due to the existence of more common neighbors, v is more likely to be a new friend of s in the future. This kind of information (number of paths, common neighbors between two nodes, *etc.*) is represented as Local-info. Meta path [17], connecting two types of objects through different object type composition, is one of the typical methods using the Local-info in HIN. On the other hand, u has a higher degree (indegree or outdegree) than v, which means u may have more influence and reputation in the network. In reality, u may be a famous star whom s is interested in. Hence, it is of high probability that s will follow u in the future. This kind of feature (degree, global reputation, *etc.*) is viewed as

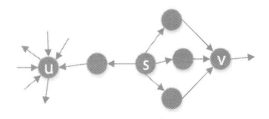

Fig. 1. A simple example of social networks

the Global-info. The most representative method using Global-info is PageRank [12] which assigns high value to nodes with high degree. Therefore, Local-info and Global-info are both useful but represent two totally different insights when solving relationship prediction problem using network topologies.

Secondly, there are many features related to edges and objects in HIN, such as the age of a person in social networks and the score of a video in movie networks. We refer such kind of features as *attribute information* (Attr-info). Let us take Facebook as an example to state how Attr-info affects the relationship emergence. There are many reasons for two students to become friends: they are in the same school and meet at the same class; they have a similar age; and they are interested in the same course, *etc.* Thus, Attr-info also has significant impact on the creation of new relationships.

In summary, Global-info, Local-info and Attr-info are all useful in predicating the relationship creation and they work in a different but complementary way. Local-info may lose the position of node in the whole network, Global-info is biased to highly visible objects and Attr-info has deep significance in modeling real-world behaviours. Therefore, it is essential to combine all these three kinds of information together. However, existing approaches utilize them separately or in a partially combined way.

1.2 Research Background and Related Work

Some methods take advantage of the information mentioned above independently. PathPredict [15] models the relationship prediction problem in a binary classification way. It proposes some measures on meta path as features and chooses the Logistic Regression as the classifier. Although PathPredict has examined the effectiveness of meta path, it only considers the topological features between objects, namely Local-info, ignoring the rich Attr-info encoded in the networks. The same situation goes to RW_ALL [9]. RW_ALL aims at tackling link prediction in a modified HIN, then a random walk with restarted method is applied to capture the Global-info. By altering an existing network into another network, the important relations in HIN can be highlighted. However, the modification process is extremely difficult since different HINs have different meta structures, which limits the applicability of RW_ALL and its similar work [3].

Other researchers partially combine the three kinds of information to improve the effectiveness of link predication. Bucket [20] has studied the citation prediction problem in heterogeneous bibliographic networks. It proposes the "Discriminative Term Bucketing" (DTM) to capture document and topic similarities without breaking possible citation relations, and then combines the meta path to predict the citation probability. Bucket takes advantage of both the Local-info and Attr-info, but this work is not a universal approach since the data structure of DTM is strongly limited to the citation prediction problem. SRW (Supervised Random Walk) [1] involves a supervised ranking method, which can effectively combine the Global-info and Attr-info together. SSP (Semi-Supervised PageRank) [5] also considers the Global-info and Attr-info to be significant and it combines them in a semi-supervised way. Contrasted with SRW, SSP regards that the functions of attributes in objects and edges are different. However, the contexts of both SRW and SSP are homogeneous networks, which cannot be directly applied to HIN.

To the best of our knowledge, there is no existing method that combines all the three kinds of information together in HIN, since it is not an easy task to effectively encode all the information in one framework. To tackle this issue, in this paper we propose a universal supervised ranking framework (S-Rank) for the relationship prediction in HIN, which completely combines Global-info, Local-info and Attr-info together.

Besides, some interesting researches have been raised in recent years. Cao et al. [2] study the collective prediction problem in HIN. Moreover, there are other topics in HIN. Shen et al. [14] propose a probabilistic model to solve linking named entities in HIN. He et al. [7] extend the similarity measurement by absorbing transitive similarity and temporal dynamics.

1.3   Contributions and Organization

Our contributions are summarized as follows.

– We propose a novel three-phase Supervised Ranking framework (S-Rank) for the relationship predication problem in HIN. To the best of our knowledge, our work is the first to completely combine Global-info, Local-info and Attr-info together.
– In S-Rank, we first propose a Supervised PageRank method (SPR) which can effectively extract the rich Attr-info and Global-info hidden in HIN. Then, we present a Meta Path-based Ranking method (MPR) to capture the Local-info. Moreover, S-Rank can overcome the lack of training example caused by class imbalance.
– Experiments on DBLP show that our framework can effectively incorporate three kinds of information, and by considering all information collectively, the performance of relationship prediction can be significantly improved compared with existing methods.

In the rest of the paper, we first introduce the preliminaries and formalize the problem in Sect. 2. S-Rank is described in Sect. 3. Experiments are presented in Sect. 4. Finally, we conclude our work in Sect. 5.

2   Preliminaries and Problem Formulation

In this section, we slightly alter the definition of **Information Network** and **Network Schema** in [17] by introducing the attribute information into them to fit our context.

Definition 1 *(Information Network). An Information Network is defined as a directed graph $G = (V, E, F)$ with an object type mapping function $\phi : V \rightarrow \mathcal{A}$, an edge type mapping function $\varphi : E \rightarrow \mathcal{R}$ and a feature type mapping* *function $\delta : F \rightarrow \mathcal{T}$, where each object $v \in V$ belongs to a particular object type $\phi(v) \in \mathcal{A}$, and each edge $e \in E$ belongs to a particular relation type $\varphi(e) \in \mathcal{R}$, and each feature $f \in F$ belongs to a particular feature type associated with an edge $\delta(f) \in \mathcal{T}$.*

When both the types of objects $|\mathcal{A}| > 1$ and the types of edges $|\mathcal{R}| > 1$, the network is called *heterogeneous information network* (we take HIN for short). Set F denotes the rich attribute information encoded in HIN, and we have $|\mathcal{R}| = |\mathcal{T}|$.

Definition 2 (Network Schema). *The network schema is a meta template of a HIN $G = (V, E, F)$ with the object type mapping $\phi : V \rightarrow \mathcal{A}$, the edge mapping $\varphi : E \rightarrow \mathcal{R}$ and the feature type mapping function $\delta : F \rightarrow \mathcal{T}$, which is a directed graph defined over object types $\mathcal{A}$, with edges as relations from $\mathcal{R}$ and denoted as $T_G = (\mathcal{A}, \mathcal{R}, \mathcal{T})$.*

Figure 2 is the network schema of DBLP[1]. The network schema serves as a template for a concrete network, and defines the rules of how entities exist, how relationship should be created and how features in both edge and node should be extracted. Based on it, we can obtain the *meta path* [17,20].

Fig. 2. Network schema of DBLP. A indicates authors, P represents papers and V denotes venues. In addition, there exist three types of edges ($|\mathcal{R}| = 3$) as well as edge features ($|\mathcal{T}| = 3$) in the networks. Both edge attributes (title similarity and author order) and object attributes (year, venue impact and author age) are encoded in F.

Definition 3 (Meta Path). *A meta path $\mathcal{P} = A_0 \xrightarrow{R_1} A_1 \xrightarrow{R_2} \dots \xrightarrow{R_l} A_l$ is a path defined on the graph of network schema $T_G = (\mathcal{A}, \mathcal{R}, \mathcal{T})$, where $A_i \in \mathcal{A}$ and $R_i \in \mathcal{R}$ for $i = 0, ..., l$, l is called the length of meta path. $A_0 = dom(R_1)$, $A_l = range(R_l)$ and $A_i = range(R_i) = dom(R_{i+1})$ for $i = 1, ..., l - 1$.*

Thus we can denote the start node type and the end node type of $\mathcal{P}$ as $dom(\mathcal{P}) = A_0$ and $range(\mathcal{P}) = A_l$ respectively.

Measures on Meta Path: Once a meta path $\mathcal{P}$ is given, some of the meta path-based measures (denoted as $\mathcal{M}_{\mathcal{P}}$) can be summarised as follows.

---

[1] http://www.informatik.uni-trier.de/~ley/db/.

– *Path Count.* Path Count, denoted as PC, simply counts the number of path instances following the given pattern $\mathcal{P}$ between two objects.
– *Random Walk score* [15]. Random Walk, denoted as RW, describes the ability one object has in visiting another object along a given meta path.
– *Symmetric Random Walk score* [15]. Compared with RW, symmetric random walk (SRW^2) considers the action started from two endpoints of a meta path.

The definition of meta path is based on the local paths between two nodes (u, v) on HIN. It is easy for us to calculate the numeric local structural features between u and v according to the measures proposed above, and these features are able to be used in MPR. It is noted that $\phi(u)$ and $\phi(v)$ can either be the same or different, depending on the kind of semantic we want to choose.

Problem Formulation: Previous studies [15,16] have defined the **target relation** as either a relation in $\mathcal{R}$ or a composite relation defined by meta path. Then the **relationship** between objects can be referred as instances of the target relation.

Generally, given a HIN $G = (V, E, F)$, a source node $s \in V$ and a set of candidate nodes $\mathbf{C} \subset V$ to which s may create relationships. A meta path $\mathcal{P}$ is utilized to represent the target relation. Then the relationship prediction task is to predict whether there will be a relationship between two nodes $s \in V$ and $v \in \mathbf{C}$ in the future, where $\phi(s) = dom(\mathcal{P})$ and $\phi(v) = range(\mathcal{P})$. We specify some details about how we select $\mathbf{C}$: $\phi(s)$ and $\phi(v)$ can either be identical or not; the self-relation is not taken into consideration, *i.e.*, $s \notin \mathbf{C}$; we are interested in predicting new relationships rather than repeated relationships, thus s never had any relationships with v.

We will address the relationship prediction problem in a ranking manner, *i.e.*, the proposed algorithm will assign higher score to nodes which s will create relationships to than to those that s will not have relationships with. Let us denote the score as $score(v)$ for $v \in \mathbf{C}$. As we will talk shortly, $score(v)$ has different representations in the two phases of S-Rank. Actually, $score(v)$ indicates to what extent source node s will probably create relationships to v.

We introduce the notation $\mathbf{L}_n$ to represent the set of labeled nodes which s ever created relation $\mathcal{P}$ to for n times. It is clear that $\mathbf{L}_0 = \mathbf{C}$, and $\forall v \in \mathbf{L}_n$, $\phi(v) = range(\mathcal{P})$. Through the labeled data, we can generate a training set of pairs: $\mathbf{T} = \{\langle u, v \rangle | u \in \mathbf{L}_i, v \in \mathbf{L}_j \ and \ i > j\}$. Hence, the potential meaning for each pair $\langle u, v \rangle \in \mathbf{T}$ is that s is more likely to have relationships with node u than v.

3   S-Rank Framework

3.1   Framework Overview

To simulate a dynamic network, we will partition two time intervals as the current network and the future network for **Training Stage** and **Testing Stage**.

---

2 Remind that SRW here is distinct from SRW method mentioned in Sect. 1.2.

In the training stage, firstly, we sample nodes that never have relationships with the source node s in interval $T_0 = [t_0, t_1]$ as candidate node set $\mathbf{C}$; then the label information from time interval $T_1 = [t_1, t_2]$ are taken out to form the supervised node pair set $\mathbf{T}$. Thus we get our training data. In the testing stage, we extract features in the time interval $T_0' = [t_0', t_1']$, and apply the learned knowledge in the training stage to predict new relationships in time interval $T_1' = [t_1', t_2']$. Then we can evaluate the prediction according to the ground truth in T_1'.

3.2   Supervised Ranking Model

Inspired by the outstanding work [1], a supervised ranking technology is proposed to learn the hidden patterns (a weight vector θ) from the historical data $\mathbf{T}$. More specifically, θ in SPR will guide the walker more likely to visit those nodes to which s will create relationships in the future; θ in MPR indicates the importance of different meta paths in the process of creating relationships. The objective of the supervised ranking model is to minimize the following function.

$$\min_{\theta} F(\theta) = ||\theta||^2 + \lambda \sum_{\langle u,v \rangle \in \mathbf{T}} l(score(v) - score(u)) \tag{1}$$

Considering the underlying meaning of pair $\langle u, v \rangle$, namely s is more likely to have relationships with node u than v, $l(.)$ will assign a non-negative penalty according to $score(v) - score(u)$. If $score(v) - score(u) >= 0$, $l(.) > 0$; otherwise $l(.) = 0$. Parameter λ controls how the fitness of the model affects the optimal value. We will use this model in both SPR and MPR.

3.3   First Phase: Supervised PageRank (SPR)

PageRank is a powerful method that can capture the Global-info in graph ranking. Its variant, the Personalized PageRank, is able to be expressed by Eq. 2, where p is the stationary distribution and Q is the probability transition matrix.

$$p^T = p^T Q \tag{2}$$

To blend both the Global-info and Attr-info, considering a HIN $G = (V, E, F)$, there is rich attribute information ($f \in F$) in it. Thus, the challenge is how can we make a connection between the weight vector θ and the feature vector f in HIN.

To achieve this goal, we extend the weight vector θ to a weight vector set W, where $W = \{w_1, w_2, ..., w_n\}$ and $n = |\mathcal{R}|$. w_i is what we utilize to weight f, namely for each edge (u, v) in G, we can calculate the capacity $c_{uv} = \pi_{w_i}(f)$ by combining w_i and $f \in F$. Function $\pi_{w_i}(f)$ takes the product of w_i and f as input. c_{uv} can indicate the ability that edge (u, v) has in guiding s to visit target node in our description above. Thus we can compute the the conductance matrix Q' by c_{uv}:

$$Q'_{uv} = \begin{cases} \frac{c_{uv}}{\sum_w c_{uw}} & if \ (u, v) \in E \\ 0 & otherwise. \end{cases} \tag{3}$$

Further, Q can be obtained through the general calculation in Eq. 4, where $0 < \alpha < 1$ is the probability of jumping (as against walking) in each step. $\alpha(v = s)$ means α is calculated only when $v = s$ and thus each row of Q sums to 1.

$$Q_{uv} = (1 - \alpha)Q'_{uv} + \alpha(v = s) \tag{4}$$

Thus, with a modification to objective function in Eq. 1, we can obtain Eq. 5, where p denotes the stationary distribution of the Personalized PageRank.

$$\min_W F(W) = \sum_{w_i \in W} ||w_i||^2 + \lambda \sum_{\langle u,v \rangle \in \mathbf{T}} l(p_v - p_u) \tag{5}$$

The gradient of $F(W)$ is derived as follows.

$$\frac{\partial F(W)}{\partial w_i} = 2w_i + \lambda \sum_{\langle u,v \rangle \in \mathbf{T}} \frac{\partial l(p_v - p_u)}{\partial (p_v - p_u)} \left(\frac{\partial p_v}{\partial w_i} - \frac{\partial p_u}{\partial w_i} \right) \tag{6}$$

Given a loss function $l(.)$, it is simple to compute the derivative $\frac{\partial l(p_v - p_u)}{\partial (p_v - p_u)}$. However, it is not an easy task to compute $\frac{\partial p_u}{\partial w_i}$, since there exists recursive relation in Eq. 2. We rewrite Eq. 2 as $p_u = \sum_j p_j Q_{ju}$ and get the recursive equation,

$$\frac{\partial p_u}{\partial w_i} = \sum_j Q_{ju} \frac{\partial p_j}{\partial w_i} + p_j \frac{\partial Q_{ju}}{\partial w_i}. \tag{7}$$

Backstrom and Leskovec [1] adopt a power-method like algorithm to compute $\frac{\partial p_u}{\partial w_i}$. It recursively applies the chain rule to Eq. 7. We extend the algorithm to fit our problem in Algorithm 1. After finishing the computation of $\frac{\partial F(W)}{\partial w_i}$, a gradient descent method can be applied to minimize $F(W)$ directly. Gradient descent may not converge to a global minimum since the optimal problem is not convex. In practice, we resolve this by randomly initializing W at several different starting points and take the best answer as the result. Note that only the value of p_v ($v \in \mathbf{C}$) is extracted from the stationary distribution of PageRank to work as $score(v)$ and form the ranking result.

3.4 Second Phase: Meta Path-Based Ranking Method (MPR)

Topologies represented by meta path have been proven to be an excellent local structure feature [15–17]. We incorporate it into a supervised ranking method.

For a candidate node $v \in \mathbf{C}$ and a source node s, we define the $score(v)$ in MPR by utilizing a linear regression model to distinguish different preferences on different meta path semantics, which is denoted as k_v:

$$k_v = \sum_{\mathcal{P} \in \mathbf{P}} g_i \times \mathcal{M}_{\mathcal{P}}, \tag{8}$$

where the length of each $\mathcal{P} \in \mathbf{P}$ is limited (e.g., 5) since a meta path will be invalid when it has a large length [17].

Algorithm 1. Iterative computation of p and all $\frac{\partial p_u}{\partial w_i}$

1: **for** each $u \in V$ **do**
2:    $p_u^{(0)} = \frac{1}{|V|}$
3: **end for**
4: $t = 1$
5: **while** *not converged* **do**
6:    **for** each $u \in V$ **do**
7:       $p_u = \sum_j p_j^{(t-1)} Q_{ju}$
8:    **end for**
9:    $t = t + 1$
10: **end while**
11: **for** $i = 1, ..., |\mathcal{R}|$ **do**
12:    Initialize w_i with zero vector
13:    $t = 1$
14:    **for** $k = 1, ..., |w_i|$ **do**
15:       **while** *not converged* **do**
16:          **for** each $u \in V$ **do**
17:             $\frac{\partial p_u}{\partial w_{ik}}^{(t)} = \sum_j Q_{ju} \frac{\partial p_j}{\partial w_{ik}}^{(t-1)} + p_j^{(t-1)} \frac{\partial Q_{ju}}{\partial w_{ik}}$
18:          **end for**
19:          $t = t + 1$
20:       **end while**
21:    **end for**
22: **end for**

Different k_v can be obtained when different measures are chosen. g_i indicates the i-th weight associated with the i-th meta path. The goal in this phase is to learn the weight vector g, which represents the importance of each meta path in determining whether two nodes will have relationships in the future. Therefore, we rewrite Eq. 1 as follows.

$$\min_g F(g) = ||g||^2 + \lambda \sum_{\langle u,v \rangle \in \mathbf{T}} l(k_v - k_u) \tag{9}$$

It is straightforward to derive Eq. 9 to obtain $\frac{\partial F(g)}{\partial(g)}$. According to the learned g, we can calculate k_v for each $v \in \mathbf{C}$ in the testing stage and obtain the ranking result.

3.5   Third Phase: Integrating Results

We formally denote the ranking result of as a mapping function: $Rank : v \rightarrow val, v \in \mathbf{C}$. val is a real number by which all nodes are sorted, and it can indicate the normalization version of $score(v)$ mentioned before. Therefore, we can record the ranking result of SPR and MPR as $Rank_1$ and $Rank_2$, respectively.

$Rank_1$ and $Rank_2$ are obtained from the same dataset and the same model (Eq. 1) but represent two different insights independently. This situation is much more similar to the search engine which must combine PageRank score and document correlation score to give a good search result, where PageRank score contains the structural information and document correlation score denotes the relevance between query and document from the aspect of textual information. Therefore, the final result $SRank$ can be calculated as

$$SRank = \Re(Rank_1, Rank_2), \tag{10}$$

where $\Re$ involves a classical issue, namely $Rank\ Aggregation$ [13]. In this paper, we mainly focus on the importance of integrating three kinds of information together instead of studying the impacts of different $\Re$. Even so, we still have conducted experiments to investigate two kinds of widely adopted implementations of $SRank$, namely Eqs. 11 and 12.

$$SRank(u) = Rank_1(u)^{\beta} \times Rank_2(u)^{1-\beta}, \beta \in (0,1) \tag{11}$$

$$SRank(u) = Rank_1(u) \times \beta + Rank_2(u) \times (1-\beta), \beta \in (0,1) \tag{12}$$

4 Experiments

4.1 Experimental Setup

The dataset used in this paper is a subset of the DBLP network [18][3], which is generated by the sampling strategy adopted in [9,19]. Its network schema can be found in Fig. 2. Just as shown in Sect. 3.1, we set $T_0 = [1991, 1998]$ and $T_1 = [1999, 2000]$. In order to comprehensively verify our framework, we randomly extract three "future" intervals for test, namely (T_0', T_1') is set to ([1991,1999], [2000,2001]), ([1991,2002], [2003,2004]) and ([1991,2004], [2005,2006]) respectively in practical. Table 1 describes the details about T_1' in three groups of data and the whole dataset.

In this work, we focus on predicting the *Citation Relation* ($\mathcal{P} = A - P \rightarrow P$). Table 2 summarizes the meta paths **P** we selected in MPR, and $S\text{-}RW$ ($\mathcal{M}_{\mathcal{P}}$) is

Table 1. Dataset statistics of DBLP

Data	#A	#V	#P	#A − P	#P → P
Group1	1008	717	3540	4576	1495
Group2	1235	852	4212	5890	2382
Group3	1461	938	5611	8046	3871
Whole	2505	3373	50910	68922	60749

Table 2. Selected Meta Paths

Number	Meta path	Length
1	$A - P$	1
2	$A - P \rightarrow P \rightarrow P$	3
3	$A - P - V - P$	3
4	$A - P - A - P$	3
5	$A - P - V - P \rightarrow P$	4
6	$A - P - A - P \rightarrow P$	4
7	$A - P - V - P - A - P$	5
8	$A - P - A - P - V - P$	5
9	$A - P - A - P - A - P$	5

---

[3] Available at http://aminer.org/billboard/DBLP_Citation.

chosen to implement MPR. In which, "→" indicates a directed edge, it means "cite" and only appears between papers in DBLP [15].

The baseline algorithms which we compare with our algorithm are: Personalized PageRank (*PPR*) and *PathPredict*. In our experiments, the LIBLINEAR [4] is adopted to implement *PathPredict*. We evaluate the method by two performance metrics: the Area under the ROC Curve (AUC) [8] and the precision at top k (*prec@k*), *i.e.*, how many of top k nodes suggested by our algorithm actually receive relationships from s. It is appropriate to choose these two measures since our dataset is *class imbalanced* [8,11] and our experiments aim at recommending the highly related papers to an author.

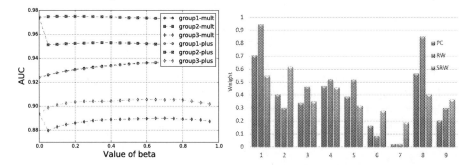

Fig. 3. Investigation for the performances of different β and *SRank* on three groups of data.

Fig. 4. Weights of different meta paths for *PC*, *RW* and *SRW*. The number in the horizontal axis represents the meta path in Table 2. (Color figure online)

For capacity function and loss function in Eqs. 5 and 9, we choose: $\pi_w(f) = \exp(f \cdot w)$ and $l(x) = \frac{1}{1+\exp(-x/b)}$. We conduct experiments to determine the function of *SRank* and the value of β. The experimental results are shown in Fig. 3, where *mult* and *plus* represent Eqs. 11 and 12, respectively. From the results we can learn that *plus* performs better than *mult*, and different β indeed affects the results. To achieve stable and overall better performance, we set β as a random float near 0.6.

Other parameters are set as follows: (1) λ is set to 1; (2) restart parameter α is set to 0.1 for S-Rank and *PPR*; (3) step of gradient descent is set initially to 0.02 and decreases by a damping factor of 0.96; (4) convergence condition ϵ is set to 10^{-9}.

4.2 Experimental Results and Analysis

We present the elaborate experiments on three groups of data in Table 3. Different $\mathcal{M}_\mathcal{P}$ is utilized to implement *PathPredict*, we denote them as *PP-PC,PP-RW,PP-SRW* and *PP-Hybrid*, respectively. To evaluate the performance, we adopt three kinds of metrics (*prec@10*, *prec@40* and *AUC*). The results for each

Table 3. Experimental results

Methods	group1			group2			group3		
	prec@10	prec@40	AUC	prec@10	prec@40	AUC	prec@10	prec@40	AUC
PPR	4.49	9.74	0.92403	7.47	15.61	0.96416	2.27	13.40	0.89354
PP-PC	3.94	4.62	0.20239	8.10	21.34	0.52881	6.34	9.81	0.33922
PP-RW	0.29	1.20	0.46884	2.47	3.67	0.39472	3.91	4.72	0.32387
PP-SRW	0.83	1.14	0.13463	3.80	4.96	0.22479	4.36	6.27	0.33760
PP-Hybrid	**5.31**	7.13	0.24720	8.35	20.41	0.51209	**8.41**	10.51	0.37133
S-Rank	5.28	**12.43**	**0.93132**	**8.91**	**22.56**	**0.97437**	5.84	**17.91**	**0.90452**

group in Table 3 are the average of multiple experiments with different source nodes. The performance of different groups differs significantly since the selected source nodes of each group are not the same.

First, S-Rank has the absolute advantage over *PathPredict* on AUC. The reason for this phenomenon is twofold. It can be seen that *PPR* has a similar but slightly poor performance on AUC compared with S-Rank, which indicates *PPR* achieves high performance in terms of AUC. As the matter of fact, *PPR* partially makes contribution to S-Rank through SPR, which results in the high AUC of S-Rank. Besides, by absorbing the information provided by Local-info (namely $\mathcal{M}_{\mathcal{P}}$) and Attr-info, S-Rank receives an even higher performance than *PPR*.

Next, from the perspective of *prec@k*, S-Rank has also been demonstrated to be effective. *prec@10* is meaningful in the situation of recommendation, which mainly concentrates on predicting the front dozens of results. In general, *PP-Hybrid* outperforms *PPR* and holds a slight advantage over S-Rank on *prec@10* in group1 and group3. However, S-Rank dominates the results of *prec@40* (improving at most 44.52 % against *PPR* and 74.64 % against *PathPredict*).

In conclusion, the experimental results indicate that the three kinds of information have their own advantages over the link prediction task. It is essential and nontrivial to combine them together. S-Rank can effectively complete the mission and significantly outperform the baseline methods on three metrics.

4.3   Additional Discussions

Weights of Different Meta Paths. As we stated before, different meta paths can capture different semantics in a HIN. Intuitively, the weight associated with each meta path is also different. S-Rank can learn the weights through MPR. Figure 4 reports the result. We can observe that the weights for three kinds of measures are unified to some extent. They all assign higher value to meta path $A - P$, $A - P - A - P$ and $A - P - A - P - V - P$ but lower value to $A - P - V - P - A - P$. Considering the semantics associated with these meta paths, the result is consistent with our knowledge in real life. $A - P$ and $A - P - A - P$ represent an author often cites papers that he or his coauthors wrote before, and this phenomenon is of common occurrence in academic world

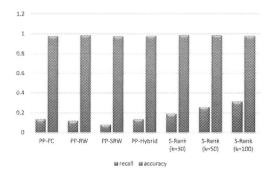

Fig. 5. Recall and accuracy of *PathPredict* and S-Rank. k means that we predict top k nodes of the ranking result will have relationship with s. (Color figure online)

since it can raise the index of author's publications; $A - P - A - P - V - P$ represents an author cites papers that have high relevance with his research interests; while paper reached through $A - P - V - P - A - P$ is too far from source author since the number of papers of one venue is too large and the other authors have generally extensive research interests.

Accuracy for Imbalanced Data. From the results shown in Table 3, we can notice the poor performance of *PathPredict* on AUC. The essence to this phenomenon is that our dataset is extremely *class imbalanced*. In our experiments, the examples with positive label (papers that source author will cite in the future) only account for 1 %–2 % of the training examples (all candidate papers). Thus, the reason why the traditional classifier works poorly is twofold [11]: (1) with fewer examples of one class, it is more difficult to infer reliable patterns; (2) it is standard to train models to optimize for 0–1 accuracy, which can be made very high by trivially predicting everything to belong to the dominant class. From Fig. 5, we can see that all methods obtain a high accuracy (around 98 %), but *PathPredict* behaves badly in contrast with the high AUC (Table 3) and high recall of S-Rank. In this case, measurement such as accuracy is untrusted. Nevertheless, by generating training data in the combination of positive examples and negative examples and absorbing Global-info provided by *PPR* and Attr-info embedded in the nodes and edges, S-Rank accomplishes a good performance over each measurement.

5 Conclusions and Future Work

In this paper, we study the relationship prediction problem in HIN. We first carefully analyze the impact of three categories of information, namely Local-info, Global-info and Attr-info, behind the creation of relationships. Then a novel supervised three-phase framework, called S-Rank, is proposed to capture all the useful information. Experimental results demonstrate that our framework can effectively integrate three kinds of information and outperforms other well-known baseline approaches in HIN. We have also conducted some extra experiments to discuss the additional issues related to our framework. For future work, the potential advantages of combining three kinds of information remains to be explored.

References

1. Backstrom, L., Leskovec, J.: Supervised random walks: predicting and recommending links in social networks. In: The Fourth ACM International Conference on Web Search and Data Mining, pp. 635–644. ACM (2011)
2. Cao, B., Kong, X., Yu, P.S.: Collective prediction of multiple types of links in heterogeneous information networks. In: ICDM 2014, pp. 50–59 (2014)
3. Deng, Z.H., Lai, B.Y., Wang, Z.H., Fang, G.D.: PAV: a novel model for ranking heterogeneous objects in bibliographic information networks. Expert Syst. Appl. **39**(10), 9788–9796 (2012)
4. Fan, R.E., Chang, K.W., Hsieh, C.J., Wang, X.R., Lin, C.J.: LIBLINEAR: a library for large linear classification. j. mach. learn. res. **9**, 1871–1874 (2008)
5. Gao, B., Liu, T., Wei, W., Wang, T., Li, H.: Semi-supervised ranking on very large graphs with rich metadata. In: SIGKDD 2011, pp. 96–104 (2011)
6. Han, J.: Mining heterogeneous information networks: the next frontier. In: SIGKDD, pp. 2–3. ACM (2012)
7. He, J., Bailey, J., Zhang, R.: Exploiting transitive similarity and temporal dynamics for similarity search in heterogeneous information networks. In: Bhowmick, S.S., Dyreson, C.E., Jensen, C.S., Lee, M.L., Muliantara, A., Thalheim, B. (eds.) DASFAA 2014, Part II. LNCS, vol. 8422, pp. 141–155. Springer, Heidelberg (2014)
8. Huang, J., Ling, C.X.: Using AUC and accuracy in evaluating learning algorithms. IEEE Trans. Knowl. Data Eng. **17**(3), 299–310 (2005)
9. Lee, J.B., Adorna, H.: Link prediction in a modified heterogeneous bibliographic network. In: ASONAM, pp. 442–449. IEEE (2012)
10. Liben-Nowell, D., Kleinberg, J.: The link-prediction problem for social networks. J. Am. Soc. Inf. Sci. Technol. **58**(7), 1019–1031 (2007)
11. Menon, A.K., Elkan, C.: Link prediction via matrix factorization. In: Gunopulos, D., Hofmann, T., Malerba, D., Vazirgiannis, M. (eds.) ECML PKDD 2011, Part II. LNCS, vol. 6912, pp. 437–452. Springer, Heidelberg (2011)
12. Page, L., Brin, S., Motwani, R., Winograd, T.: The pagerank citation ranking: bringing order to the web (1999)
13. Rajkumar, A., Agarwal, S.: A statistical convergence perspective of algorithms for rank aggregation from pairwise data. In: ICML 2014, pp. 118–126 (2014)
14. Shen, W., Han, J., Wang, J.: A probabilistic model for linking named entities in web text with heterogeneous information networks. In: SIGMOD 2014, pp. 1199–1210 (2014)
15. Sun, Y., Barber, R., Gupta, M., Aggarwal, C.C., Han, J.: Co-author relationship prediction in heterogeneous bibliographic networks. In: ASONAM, pp. 121–128. IEEE (2011)
16. Sun, Y., Han, J., Aggarwal, C.C., Chawla, N.V.: When will it happen?: relationship prediction in heterogeneous information networks. In: WSDM, pp. 663–672 (2012)
17. Sun, Y., Han, J., Yan, X., Yu, P.S., Wu, T.: Pathsim: meta path-based top-k similarity search in heterogeneous information networks. PVLDB **4**(11), 992–1003 (2011)

18. Tang, J., Zhang, J., Yao, L., Li, J., Zhang, L., Su, Z.: Arnetminer: extraction and mining of academic social networks. In: KDD, pp. 990–998 (2008)

19. Yin, Z., Gupta, M., Weninger, T., Han, J.: A unified framework for link recommendation using random walks. In: 2010 International Conference on Advances in Social Networks Analysis and Mining (ASONAM), pp. 152–159. IEEE (2010)

20. Yu, X., Gu, Q., Zhou, M., Han, J.: Citation prediction in heterogeneous bibliographic networks. In: SDM, pp. 1119–1130. SIAM (2012)

Discovering Common Semantic Trajectories from Geo-tagged Social Media

Guochen Cai, Kyungmi Lee, and Ickjai Lee(⊠)

Information Technology Academy, College of Business, Law and Governance,
James Cook University, Cairns, QLD 4870, Australia
ickjai.lee@jcu.edu.au

Abstract. Massive social media data are being created and uploaded to online nowadays. These media data associated with geographical information reflect people's footprints of movements. This study investigates into extraction of people's common semantic trajectories from geo-referenced social media data using geo-tagged images. We first convert geo-tagged photographs into semantic trajectories based on regions-of-interest, and then apply density-based clustering with a similarity measure designed for multi-dimensional semantic trajectories. Using real geo-tagged photographs, we find interesting people's common semantic mobilities. These semantic behaviors demonstrate the effectiveness of our approach.

1 Introduction

Geo-tagged social media dataset is a potential resource for understanding people's movements. One geo-tagged entity can reflect its owner's footprint that a spatial location and a time the owner visited. Therefore, a series of geo-tagged entities, connected chronologically, reflects an approximate owner's spatio-temporal movement resulting in a trajectory. Trajectory is a time ordered sequence of geographic points with time stamp. Increasing geo-tagged social data provides an information-rich environment for people's mobility mining. These rich trajectories contain some previously unknown and useful knowledge about human movement behaviors.

Extraction of people's historic common trajectories (traces) from geo-tagged images is a recent hot research topic in the data mining community [14]. A common trajectory is the movement of a group of users exhibiting similar movements. Thus, common trajectories are people's popular movements that are useful and valuable in many areas. Locational (geometric) information is of great importance in this common trajectory study, and past studies focus on the analysis of 'geometric' aspect of trajectories. However, the geometric feature of trajectory is insufficient for many applications [2]. Trajectories can be enriched with additional semantic annotations representing useful information. One important feature of trajectory is the contextual semantic information in which such a movement takes place, such as weather conditions during the travel, and the type of place of interest visited. These semantically enriched trajectories can be

© Springer International Publishing Switzerland 2016
H. Fujita et al. (Eds.): IEA/AIE 2016, LNAI 9799, pp. 320–332, 2016.
DOI: 10.1007/978-3-319-42007-3_27

used to help find much richer, detailed, novel and unknown semantic level trajectory behaviors, whose predicate bears on the specific application contextual data [12]. For instance, adding place type annotation to trajectories can help discover common movement on place type level which provides better semantic level knowledge to understanding of human mobility behaviors compared to geometric feature only trajectories. These semantic level movement behaviors are valuable to various domains, for example, understanding of tourists's mobility among place types in tourism industry.

This study mines trajectories associated with multiple semantic annotations to discover human common movement from geo-referenced social media data, in particular geo-tagged photographs. We propose a systematic framework to extract common semantic trajectories using clustering and Regions-of-Interest (RoIs) analysis. Also, we extend OPTICS [3] to handle multi-dimensional semantic annotations by proposing a new similarity measure using longest common subsequence. Experimental results demonstrate that our proposed method is able to detect various useful people movement behaviors in the semantic level, and prove the effectiveness of our proposed framework.

2   Related Work

Extracting people mobility behaviors from geo-tagged social media data is a hot research trend. [9] extracts dynamic and spatio-temporal tourist flows in urban environments using online user generated geo-referenced photographs. Using check-in data, [6] finds human mobility patterns. [5] extracts frequent trajectory patterns from geo-tagged photographs where each trajectory pattern is represented by a sequence of formed spatial RoIs with transit time. [4] adds temporal dimension into trajectories to find tourists' spatio-temporal sequential patterns for tourism science. These recent studies discover various human mobility dynamic flows and frequent sequential trajectory patterns from geotagged social media that are useful and valuable for specific applications such as in tourism.

Another popular study is mining common trajectory. This is usually achieved by grouping similar trajectories into clusters. [14] mines tourists' popular travel route from geo-tagged photographs using trajectory clustering. A common drawback of previous studies is that they focus on only geometric feature of trajectory data in the common trajectory analysis. Analyzing trajectories in geometric feature merely is insufficient for many applications which require additional information from the application context. Trajectories can be enriched with other contextual semantic annotations in the movement analysis.

Semantic trajectory mobility analysis aims to provide applications with semantic contextual information about the movement. A semantic trajectory is a trajectory that is added with annotations regarding the trajectory. Semantic trajectory mining is in its early stage, and previous research mainly focuses on the generation of semantic trajectories [1]. As Web 2.0 technologies mature such as mashup and geo-referencing, a rich set of geo-information is available that can be integrated into geometric trajectories. This integration opens a new challenge to mine semantically enriched trajectory mining that could reveal much finer, more detailed and useful patterns. This study is an attempt to fill this gap in semantically enhanced trajectory data mining.

3   Framework

This study proposes a methodological framework to discover people common semantic trajectories from geo-tagged social media data, using example of geo-tagged photograph data as shown in Fig. 1.

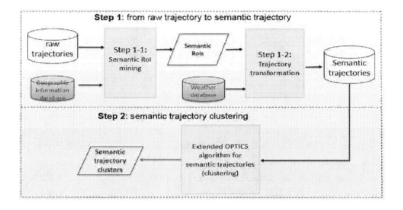

Fig. 1. Framework for common semantic trajectory mining.

3.1   Generating Semantic Trajectories from Raw Trajectories

For each photograph-taker, we firstly sort geo-tagged photographs by photo taken time and then connect them to form a raw trajectory (owner *id*, number of points, time, latitude, longitude etc.). In this study, trajectories with more than 2 points are considered useful. Semantic trajectories are then generated from raw trajectories by two sub-steps: semantic RoI mining and transformation. Spatial RoIs with place type semantics are extracted from raw trajectories during the semantic RoI mining sub-step. Based on these extracted semantic RoIs, raw trajectories are then transformed into RoI-based trajectories. During this process, place type semantic RoIs are firstly used and then further semantic features including city, temporal information and weather condition semantics are added to generate multi-dimensional semantic trajectories.

Semantic RoI Mining. The core concept behind the semantic RoI mining is that areas with place type information exhibiting a high density of moving entities are both interesting and significant. We apply the hybrid grid-based RoI mining algorithm to calculate spatial dense cell and RoI merging strategy [10]. At first, we compute dense cells using the RoI mining and then find place type semantics for those dense cells. Each dense cell has a record of all points inside it. For each spatial point, we annotate it with place name and type by geo-referencing and searching for the nearest place to the point from a background geographic information database. Once all spatial points are annotated, we calculate the most frequent annotation, the place type, which is considered as the place type of the dense cell. At last, we merge neighboring dense semantic cells to construct semantic RoIs. From this process, different place type areas can be found as several independent semantic RoIs.

Semantic Trajectory Transformation. After semantic RoIs mining, raw trajectories are then transformed into multi-dimensional semantic trajectories. Based on the semantic RoIs, a raw trajectory is transformed into a sequence of semantic RoIs called a single-dimensional semantic trajectory. These semantic RoIs contain only place type semantics. Then, RoIs are enriched with city, temporal and weather condition, and become multi-dimensional RoIs. For city name, considered as spatial semantic, we search it from geographic information database based on where RoIs located. Temporal semantic contains two features: *date type* and *day time period*. These temporal features are converted from the visit time of RoIs. In particular, *date type* is the day of week, weekday and weekend, and *day time* is the context concept which is midnight, dawn, morning, afternoon or evening. We also add weather conditions to RoIs based on the visit time and geographic location by querying from the weather observation database. At the end, RoIs contain multiple semantic annotations and semantic trajectories become multi-dimensional.

3.2 Semantic Trajectory Clustering

Investigating the common behaviors from semantic trajectories is a task of grouping into similar trajectories. We adopt and extend a density-based clustering method [7]. It is a popular method applied in previous studies on common trajectory analysis. This is because the density-based method is efficient for finding noise and detecting outliers. Also, it is able to detect clusters of arbitrary shapes. Particularly, OPTICS algorithm [3] is adopted as our clustering scheme. OPTICS algorithm does not generate object cluster explicitly. Instead, it orders objects of the dataset. The outcome of OPTICS algorithm is a reachability distance ordering of objects. In the ordering, objects which are closest become neighbors. Later, these ordered objects can be grouped into clusters based on an extra setting radius, maximum distance considered. Object clusters of varying densities can be obtained by using different radius thresholds. We also apply the EXTRACTDBCAN-Clustering method [3] to generate clusters from the ordering results. Please refer to [3] for details of OPTICS algorithm. To extend OPTICS algorithm to semantic trajectory data type, we propose a distance metric function for similarity measure between multi-dimensional semantic trajectories.

Similarity Function for Multi-dimensional Semantic Trajectories. Given two multi-dimensional sequences, in order to compute a similarity between them, we need to find how many commonalities they have according to intuitions of similarity [11]. We apply Longest Common SubSequence (LCSS) algorithm to find the common subsequence of two trajectories. Furthermore, we add a new feature of selecting dimensions as compulsory or optional. Compulsory dimensions are ones that must be matched in the comparison of two RoIs whilst optional dimensions do not have to be matched. This strategy aims to solve a problem that two multi-dimensional elements are matched where some dimensions must be matched but other dimensions with different weights are used to calculate a matching score. We also use a similar idea of weight strategy in [8] for multiple dimensions, especially for optional dimensions. Our similarity function includes two parts which find LCSS of trajectories and calculate a similarity score for two trajectories.

Calculating Match Relationship of Multi-dimensional Elements. In LCS discovering process of LCSS algorithm, finding matched RoIs is a key step. For a multi-dimensional RoI, we use a weighting strategy that assigns different weights to different dimensions. To compute the matching score, we consider the dimensions separately: compulsory dimension part and optional dimension part. For compulsory dimensions, two elements must have an exact match. Then, the other optional dimensions are assigned with weights and used to calculate matching score between two elements. When two elements have the same value on a dimension, we set '1' to the matching score of the dimension, otherwise set '0'. As shown in Eq. 1, the element matching score is the sum of every optional dimension matching score times its weight. For each optional dimension k, $match_k(mdEle1, mdEle2)$ is the matching score and $weight_k$ is its weight value. At last, a threshold is used to compare the matching score to determine whether two elements match or not.

$$MScore\,(mdEle1, mdEle2) = \sum_{k=1}^{d} (match_k\,(mdEle1, mdEle2) \times weight_k) \quad (1)$$

Calculating Similarity Score of Trajectories. LCSS algorithm finds out the longest common subsequences of two semantic trajectories. To compute the similarity between trajectories, first we need to ensure that the commonality occupies most parts in both trajectories. This is measured by checking the ratio of LCS to the length of whole trajectory respectively. When both ratios are valid, they are used to compute the similarity score between two trajectories by using the average ratio as a similarity score. Equation 2 shows the calculation of ratio of LCS to the length of one multi-dimensional trajectory. The average similarity function is in Eq. 3. Finally, the similarity score between two given semantic trajectories is obtained. Higher similarity score value means more similar. And, the distance between two trajectories is the dissimilarity score calculated by using '1' - similarity score.

$$ratio\,(mdst1) = \frac{|LCS\,(mdst1, mdst2)|}{|mdst1|}, \tag{2}$$

$$similarity(mdst1, mdst2) = \frac{ratio(mdst1) + ratio(mdst2)}{2}. \tag{3}$$

Index Structure for Semantic Trajectories. In order to facilitate efficient neighborhood queries in OPTICS algorithm, we integrate TB-tree [13], designed strictly for trajectory data, as an index structure. Each trajectory is represented as a sequence of segments, and a leaf node of TB-tree stores several segment Minimum Bounding Boxes (MBBs). Each leaf node keeps only those segments that belong to the same trajectory. Semantic trajectory is a sequence of multi-dimensional episodes. The original MBB model is not suitable for textual episode containing only texts. We modify the model of MBB of TB-tree into Minimum Term Bag (MTB) for multi-dimensional semantic RoIs. A MTB stores episodes that build minimum cover of terms for each dimension of episodes.

4   Results and Discussion

We use geo-tagged photographs from Flickr. Using Flickr APIs, we collect geo-tagged photographs taken in the Queensland area, Australia from April of 2014 to March of 2015. After pre-processing, we have 64,733 cleaned photograph records, and 1,404 valid raw trajectories including 61,322 points in total. Most trajectories lie on coastline areas where major cities of Queensland located.

To enrich additional semantic annotations to trajectories, we use two different databases which are geographic information database and weather observation database. The former database is used to query place type and city information for trajectories, and the later is used to find weather condition information. We use the Australia gazetteer data and cities 1000 dataset from GeoNames[1] as our geographic information database. In GeoNames database, all types of places are categorized into one out of nine feature classes and further subcategorized into one out of 645 feature codes. These feature codes are used as the place type in this study. And, cities 1000 database is used to query city names for RoIs. For the weather information database, we use the observation stations database and daily weather observation database from Bureau of Meteorology Australia[2]. Focusing on our study area Queensland and period from April 2014 to March 2015, we have 121 records for station list and 47,949 records for daily weather observation in our database.

---

[1] Geonames: http://www.geonames.org/.

[2] http://www.bom.gov.au/climate.

4.1 Selection of Parameters

Hybrid grid-based RoI mining algorithm and OPTICS algorithm are sensitive to parameter selections. In particular, RoI mining method relies heavily on the minimum support (*MinSup*) value for a cell to become a RoI and also on the size of cell (*CellSize*) that is used to partition the study region. It is a non-trivial problem to choose best values of parameters that produce meaningful and insightful RoIs. Thus, the approach adopted during experimentation was a systemic trial and error approach. The final parameter values chosen for this study is that a value for parameter *CellSize* is 0.003 which means 0.3 km whilst a value of for parameter MinSup is 0.002 which equals to 0.2 %.

OPTICS algorithm needs two parameters which are epsilon describing the radius of search range for querying neighbor objects and *minPts* setting the minimum number of objects a dense range needs. However, as OPTICS algorithm visits and orders closest neighbor objects, the effect of parameter epsilon has been declined. Thus, we can simply set it to the maximum possible value, 1 in this study. For parameter *minPts*, 5 is chosen as a default value.

We also need to set parameters for distance function of semantic trajectories. To measure the similarity between multi-dimensional semantic trajectories, the selection of compulsory and optional dimensions and the setting of weight values for optional dimensions are important factors. These selections are application dependent and case sensitive. There are five semantic dimensions including PLACE_TYPE, CITY, DAY_TIME, DAY_TYPE and WEATHER. Based on the preliminary experiments, we find that CITY, spatial semantic, plays an important role. As a result, in this study we present two cases: the first case is setting PLACE_TYPE only as a compulsory dimension and the other four dimensions as optional, and the second case is setting PLACE_TYPE and CITY as compulsory dimensions and the other three dimensions as optional ones. For simplicity, all optional dimensions are equally weighted (sum to 1). Another two parameters *eleMatThreshold* (validating two RoIs are matched in for the LCS finding process), and *ratioThreshod* (ensuring a valid ratio of LCS to the length of trajectory), are set to 0.3.

4.2 Common Semantic Trajectories

Using parameter values illustrated in Sect. 4.1, we are able to find interesting semantic RoIs and generate semantic trajectories at the first step. In the next step, we extract some interesting behaviors of people common semantic trajectories in two cases by applying OPTICS algorithm integrated with the defined distance function.

Semantic RoIs. Using the semantic RoI mining method, we are able to detect 763 dense spatial cells that are then enriched with PLACE_TYPE semantic annotations in the next step. After the cell merge phase, we finally find 507 RoIs with PLACE_TYPE annotations. Figure 2 shows some semantic RoIs in Cairns city area. Each arbitrary shape represents a RoI with a feature code representing

Fig. 2. Semantic RoIs in Cairns city area (*MinSup = 0.002* and *CellSize = 0.003*).

Table 1. Description of feature codes in Fig. 2.

Feature code	Description
HTL	Hotel
HSP	Hospital
DEVH	Housing development
PIER	Structure built out into navigable water providing berthing
PT	Point (a tapering piece of land projecting into a body of water)
RSTN	Railroad station
PPLX	Section of populated place

the PLACE_TYPE semantics. Table 1 lists the descriptions of feature codes from the Geonames database.

After semantic enrichment of other multiple dimensions and transformation process, there are 737 valid semantic trajectories that have at least 2 RoIs. It is about 52 % of the raw trajectories. The following example shows one 2-length semantic trajectory that begins with trajectory *id* 33 and 2 means its length.

$$33 : 2 < (HTL[midnight][weekday][Townsville][Clear])$$

$$(ISL[midnight][weekend][Tablelands][Lightrain]) >$$

Case 1: Four Optional Dimensions. In this experiment, we set PLACE_TYPE as a compulsory dimension and the other four annotations, including CITY, DAY_TIME, DAY_TYPE and WEATHER, as optional dimensions. For simplicity, we set the weight of each optional dimension to an equal value (0.25). Figure 3 is the reachability ordering plot outcome of OPTICS algorithm. We can find that many semantic trajectories reachability distances are UNDEFINED that their similarity values are too small based on our distance function.

To find final clusters of semantic trajectories, we apply ExtractDBSCAN-Clustering method with parameter epsilon value 0.5 on the ordering objects.

There are 2 clusters finally found that refer to two concave areas below 0.5 reachability distance respectively as shown in Fig. 3. The first cluster contains 141 semantic trajectories and the other contains 5 trajectories. For a sake of presentation simplicity, here we list members of the smaller cluster in Table 2. These semantic trajectories are similar in the condition that the commonality takes up at least 30 % of both trajectories and each element matching score should be no less than 0.3. In particular, the first trajectory is computed as a center object that has 4 neighbors in the searching range and the distance between each four trajectories and the first trajectory is 0.4 which means each pair trajectories has 60 % commonality.

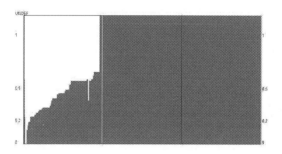

Fig. 3. Reachability ordering plot ($minPts = 5$ and epsilon $= 1$; compulsory dimension: PLACE_TYPE, optional dimensions: CITY, DAY_TIME, DAY_TYPE & WEATHER; weight of each optional dimension: 0.25).

Focusing on the behavior of people common semantic mobility, from Table 2, we can see that the common mobility is in Cairns area. People have similar moving patterns from hotels to ship berth and then go to hotels. In particular, one of the most famous travel destination in Cairns is the Great Barrier Reef which attracts millions of people to visit. This common trajectory might be one of the popular short tour routes in Cairns. This existing knowledge supports the validation of the extracted behavior of common semantic trajectory. Also we can see that similarly, they move at clear days. Though they have similar or less different in visiting time and weather condition, their trajectories are still considered to be similar. This is due to the selection of optional dimensions and the smaller RoI matching score threshold which is 0.3. When the CITY dimension values are the same, if one of the other three optional dimensions has an equal value, two visited RoIs calculated are matched. Overall, people common mobilities in semantic level have been extracted from geo-tagged photographs which is not possible in traditional geometric only trajectory mining.

Case 2: Three Optional Dimensions. To investigate the common semantic trajectories in different similarity measuring case using a different dimension combination, in this experiment, we set PLACE_TYPE and CITY (spatial semantics) as compulsory dimensions and the other three dimensions as

Table 2. Semantic trajectory cluster having 5 members.

Tra id & size	Semantic trajectory
407 : $size5$	[(HTL[weekday][Cairns][Clear][morning]), (HTL[weekday][Cairns][Clear][morning]), (PIER[weekday][Cairns][Heavy rain][morning]), (HTL[weekday][Cairns][Heavy rain][morning]), (HTL[weekday][Cairns][Clear][evening])]
49 : $size5$	[(HTL[weekday][Cairns][Light rain][morning]), (PIER[weekday][Cairns][Light rain][morning]), (HTL[weekday][Cairns][Light rain][evening]), (PPL[weekday][Cairns][Clear][dawn]), (HTL[weekday][Cairns][Clear][evening])]
646 : $size3$	[(HTL[weekday][Cairns][Clear][evening]), (HTL[weekday][Cairns][Clear][dawn]), (HTL[weekday][Cairns][Clear][midnight])]
961 : $size3$	[(HTL[weekday][Cairns][Light rain][evening]), (HTL[weekend][Cairns][Clear][dawn]), (HTL[weekday][Cairns][Clear][evening])]
704 : $size5$	[(HTL[weekday][Cairns][Clear][evening]), (PIER[weekday][Cairns][Clear][evening]), (PIER[weekday][Cairns][Clear][evening]), (PIER[weekend][Cairns][Clear][midnight]), (HTL[weekday][Cairns][Light rain][evening])]

optional. This change of dimension combination directly affects the calculation of match relationship of two multi-dimensional RoIs. The weight values of optional dimensions are that DAY_TIME and DAY_TYPE both are equal to 0.333 and WEATHER is 0.334 to make the sum equals to 1.

The reachability ordering plot is shown in Fig. 4. As expected, using a more strict similarity measure condition, more semantic trajectories have UNDEFINED reachability distance that these trajectories are not similar to others. Interestingly, setting CITY as a compulsory dimension, the shape of this plot chart is significantly different from the plot as in Fig. 3. We detect three clusters is this case. They correspond to the three concave areas below 0.5 reachability distance respectively as shown in Fig. 4. The first cluster has 53 semantic trajectories, the second cluster has 7 members and the third cluster has 9 trajectories. The total number of clustered trajectories is smaller than that of Case 1.

We select the first semantic trajectory as a representative common mobility for each cluster. Table 3 lists these representative common semantic trajectories. In cluster 1, the mobility is the route between hotels at clear days in Gold Coast. Other trajectories in this cluster have similar movement. Based on the commonality moving between hotels, these trajectories are grouped into a cluster. The second cluster presents another different common trajectory that starts from hotel at a clear morning in Brisbane, then passes an rail station in Brisbane and

Fig. 4. Reachability ordering plot ($minPts = 5$ and epsilon $= 1$; compulsory dimension 2: PLACE_TYPE & CITY, optional dimensions: DAY_TIME, DAY_TYPE & WEATHER; weight of each optional dimension: 0.333).

visits park in Logan at morning and goes back to hotel in Brisbane. Most RoIs of this mobility are in Brisbane. The other members of this cluster have similar trajectories. The third cluster shows the same common mobility as presented in Case 1. The extracted clusters show knowledge about people common trajectories at the semantic level that visits sequentially different type places with day time, day type and weather condition information.

Table 3. Representative common semantic trajectory for each cluster.

Cluster *id*	Semantic trajectory
Cluster 1	[(HTL[weekday][Clear][Gold Coast][midnight]), (HTL[weekday][Clear][Gold Coast][midnight]), (HTL[weekday][Clear][Gold Coast][dawn]), (HTL[weekend][Clear][Gold Coast][dawn])]
Cluster 2	[(HTL[weekday][Clear][Brisbane][morning]), (HTL[weekday][Light rain][Brisbane][dawn]), (RSTN[weekday][Light rain][Brisbane][midnight]), (PRK[weekday][Clear][Logan][morning]), (HTL[weekday][Clear][Brisbane][morning])]
Cluster 3	[(HTL[weekday][Light rain][Cairns][morning]), (PIER[weekday][Light rain][Cairns][morning]), (HTL[weekday][Light rain][Cairns][evening]), (PPL[weekday][Clear][Cairns][dawn]), (HTL[weekday][Clear][Cairns][evening])]

5   Conclusions

Geographical information referenced social media data becomes potential resources for the discovery of people moving behavior. We present a study of converting geo-tagged social media data, using geo-tagged photographs, into semantic trajectories for extracting knowledge about people common trajectory on the semantic level. Experimental results show that people common mobility behaviors on the semantic level have been extracted from semantically enriched trajectories. These common semantic level behaviors demonstrate that analysis of trajectories integrating with semantic annotations can generate more novel, detailed and unknown knowledge about mobility behavior on the semantic level. Geometric only trajectory mining as in [14] is about crude mobility patterns of geographical place level behaviors. However, this study enriched with semantic annotations is able to find much finer and detailed people common mobility on the semantic level, which proves the effectiveness of our proposed method.

Future work includes experiments with various real world datasets to prove the robustness of our approach. Post processing of detected semantic patterns is another interesting target to find some positive associations and cause-effect patterns.

References

1. Alvares, L.O., Bogorny, V., Kuijpers, B., de Macedo, J.A.F., Moelans, B., Vaisman, A.: A model for enriching trajectories with semantic geographical information. In: Proceedings of the 15th Annual ACM International Symposium on Advances in Geographic Information Systems, p. 22. ACM (2007)
2. Alvares, L.O., Bogorny, V., Palma, A., Kuijpers, B., Moelans, B., Macedo, J.A.F.: Towards Semantic Trajectory Knowledge Discovery. In: Technical report, Hasselt University, Belgium, October 2007
3. Ankerst, M., Breunig, M.M., Kriegel, H.P., Sander, J.: Optics: ordering points to identify the clustering structure. In: ACM Sigmod Record, vol. 28, pp. 49–60. ACM (1999)
4. Bermingham, L., Lee, I.: Spatio-temporal sequential pattern mining for tourism sciences. Procedia Comput. Sci. **29**, 379–389 (2014)
5. Cai, G., Hio, C., Bermingham, L., Lee, K., Lee, I.: Sequential pattern mining of geo-tagged photos with an arbitrary regions-of-interest detection method. Expert Syst. Appl. **41**(7), 3514–3526 (2014)
6. Cheng, Z., Caverlee, J., Lee, K., Sui, D.Z.: Exploring millions of footprints in location sharing services. In: ICWSM 2011, pp. 81–88 (2011)
7. Ester, M., Kriegel, H.P., Sander, J., Xu, X.: A density-based algorithm for discovering clusters in large spatial databases with noise. In: KDD, vol. 96, pp. 226–231 (1996)
8. Furtado, A.S., Kopanaki, D., Alvares, L.O., Bogorny, V.: Multidimensional similarity measuring for semantic trajectories. Trans. GIS **20**, 280–298 (2015)
9. Girardin, F., Fiore, F.D., Ratti, C., Blat, J.: Leveraging explicitly disclosed location information to understand tourist dynamics: a case study. J. Locat. Based Serv. **2**(1), 41–56 (2008)
10. Hio, C., Bermingham, L., Cai, G., Lee, K., Lee, I.: A hybrid grid-based method for mining arbitrary regions-of-interest from trajectories. In: Proceedings of Workshop on Machine Learning for Sensory Data Analysis, p. 5. ACM (2013)
11. Lin, D.: An information-theoretic definition of similarity. In: ICML, vol. 98, pp. 296–304 (1998)

12. Parent, C., Spaccapietra, S., Renso, C., Andrienko, G., Andrienko, N., Bogorny, V., Damiani, M.L., Gkoulalas-Divanis, A., Macedo, J., Pelekis, N., et al.: Semantic trajectories modeling and analysis. ACM Comput. Surv. (CSUR) **45**(4), 42 (2013)
13. Pfoser, D., Jensen, C.S., Theodoridis, Y., et al.: Novel approaches to the indexing of moving object trajectories. In: Proceedings of VLDB, pp. 395–406 (2000)
14. Zheng, Y.T., Zha, Z.J., Chua, T.S.: Mining travel patterns from geotagged photos. ACM Trans. Intell. Syst. Technol. (TIST) **3**(3), 56 (2012)

Analysis of Social Networks Using Pseudo Cliques and Averaging

Atsushi Tanaka[✉]

Graduate School of Science and Engineering,
Yamagata University, Yonezawa, Japan
tanaka@yamagata-u.ac.jp

Abstract. In order to analyze social networks, a improved method for Clique Percolation Method (CPM) is proposed. Using this method, which is called pseudo Alternative CPM (ACPM), network analysis of friendship networks on SNS sites for college students is carried out. As the number of lack of nodes to fuse two cliques increases, it is confirmed that small communities inside large communities can be detected. The differences between two SNS sites coming from their system, registration or invitation, are also clarified. Moreover the change of average degrees of nodes there is observed and its behavior is also discussed.

Keywords: Social network · Community analysis · Pseudo clique · Clique percolation

1 Introduction

It has been almost 20 years since the field of network science became very popular all over the world, and we study the universal characters regarding all connections among all sorts of things as network. The research field extend to Internet, ecology, physics, chemistry, sociology, epidemiology etc. [1]. Many sociologists have been interested in human related networks, and clarified its structure by some experimental studies [2–4]. However since the methods and resources for analyzing huge network were insufficient, their analyses were limited. In recent years, the rapid progress of computers and the establishment of scientific method have enabled us detailed analysis of networks. Moreover recent rapid spread of social media has also enabled us to obtain easily and cheaply connection data of human networks, streaming information in there, and whole temporal change of them. In Facebook and Twitter, more than several hundreds million people are exchanging information over the boundary of nations and the power of their influence are spreading to several kinds of directions.

Since data obtained from social media is so huge that it is not easy to clarify the whole image even in recent computer days. Hence we need to obtain sub-data by sampling based on some regulated rule and analyze them. Then we might be able to predict whole image of them. The method of predicting present or future states by analyzing temporal data are also very promising.

© Springer International Publishing Switzerland 2016
H. Fujita et al. (Eds.): IEA/AIE 2016, LNAI 9799, pp. 333–343, 2016.
DOI: 10.1007/978-3-319-42007-3_28

In this paper, from temporal analysis of data of SNS sites, one of typical social media, I clarify their dynamics and also refer to the prediction of the future.

2   Social Networks

Recent progress of social media is amazing. Facebook and Twitter especially have obtained more than several hundred million users all over the world, and also have achieved big success commercially. Many reports said many people obtained some necessary information over Twitter in the East Japan Earthquake in 2011, and it has been recognized that social media is important for the information acquisition method. Mixi, a original Japanese social media, succeeded in obtaining huge users at one time. Yuta et al. revealed its structure using network analysis [5]. Toriumi et al. also analyzed a lot of SNS sites widely and categorized them into several groups by their characteristics [6,7]. As above a lot of researches on structural analysis and modeling of inherent networks inside social media have been carried out.

Friendship network has had an interest widely since early days in the field of social science though, they could not obtain correct and huge scale data easily because of several problem like privacy. However the progress of social media enabled us to obtain human relation network data easily and minutely. You might say data on social media is virtual on Internet though, these considerable parts are formed autonomously in the real space and their growth is speculated to be fast relatively. From this feature, analyzing these data is meaningful for the analysis of distributed autonomous system.

Here let us think about the number of users in Facebook. The whole data is extremely huge and varies from hour to hour, so it is very difficult for us to grasp the whole structure. Therefore there are two kinds of analysis, sampling and prediction. Using the former, we extract and analyze some part of data by some rule, and extrapolate the results of whole data from their results. Using the latter, we predict the results in the future from the inclination of temporal data analysis. It is not certain of course that both estimation is correct or not, but it can be evaluated to some extent in the process. Hence in this paper, I focus on the estimation of the future using especially the latter method. That must be very meaningful when we concern the future trend of social media.

3   Community Analysis Methods

3.1   Existing Methods

The research of social media is suitable for studying the network dynamics. The networks in social media especially SNS strongly depend on human relationship, thus the community analysis is indispensable. The community analysis has been studied for a long time in sociology, and several kinds of methods have been proposed so far. The popular methods are the one using betweenness centrality by Girvan et al. [8], the one using Q value by Newman [9], and its developed version for larger-scale networks with more speed by Clauset et al. [10]. However, above methods are all for partitioning a large community into some distinct smaller communities. If we consider human relationship in a real world, there often happens that some people gets deeply involved in several communities at the same time. Therefore, if we consider the network in social media, it is rather natural to think that some communities overlap each other, and hence I use overlapping community analysis. Among several kinds of these techniques that have already proposed, the Clique Percolation Method (CPM) [11] is the most popular. So they have analyzed the network mainly with this method so far. For more information of community analysis, see the review article by Fortunato [12].

3.2   Improvement of CPM

In CPM, one of the cumulative algorithm, when two k degree cliques are fused to one community, $k - 1$ nodes must be shared among them. It seems to be very rigorous especially for large k, so communities are difficult to become large. Here I permit the lack σ in overlapping to fuse two communities. It means two communities are fused even if they do not share σ nodes, that is the connecting condition of cliques is alleviated. Since if I set $\sigma = 1$ in this method, I can obtain exactly the same method of original CPM, so I call it Alternative Clique Percolation Method (ACPM) [13]. The procedure of ACPM is as follows.

1. Fix the degree of clique k.
2. List k degree cliques from the network, and regard each of them as a community.
3. For all combination of two communities, if the number of shared nodes is not less than the threshold $k - \sigma$, they are fused to new one community.
4. Repeat 3 until no new fusion occurs.

Here we must note that obtained communities can be overlapped each other.

However since ACPM is based on the connection of k degree cliques, many nodes which do not form any degree cliques are all removed, thus they never become members of any community. This is a very rigorous condition. Thinking of real networks, it seems to be valid we regard a cluster with a little lack of edges as a clique, now I define the allowance parameter α as the rate of edges comparing with complete graph. For example, we consider 5-degree clique, there are 10 edges in it. If 2 edges are missing among them, the occupation rate is $8/10 = 0.8$. In the case α equals to 0.8 and below, it is considered to be a clique. I call this method pseudo ACPM.

4   Analytical Results

4.1   Target SNS Sites

In human networks, we can often observe community structures, but real relationships seldom become open, so it is often difficult to assemble the data. On the contrary, speaking of networks on social media relationships can be observed, and assembling the data is relatively easy. Thus many analyses of network structures and communities have been proceeded. In this paper, I analyze two SNS systems, tomocom.jp [14] and Fukui LEarning Community ConSortium (FLECCS) [15].

In order to investigate the network dynamics in SNS, I have constructed SNS site a.k.a. "tomocom.jp". This SNS is only for college students and completely invitation-based system. There seminar students are invited by college teachers in several areas in Japan. It possesses higher reliability thanks to its guardian system. This site was opened in 2009, and captured more than 400 users by the end of 2010. Though the interaction within their own seminars is basic, the connections between different seminars have been created by writing and browsing their blogs. Since activation event of the site over several times were held, the connections became denser. The network structure of this SNS at the end of 2010 is shown in Fig. 1(a). We have confirmed the network structure between communities has also smallworld property, that is nested structure of smallworld [16].

As another object, I analyze the SNS called FLECCS. It is a product of the cooperation of 8 colleges in Fukui prefecture in Japan and it provides not only SNS platform but also the environment of virtual college. Same as tomocom.jp, it is SNS for college students, however it adopts registration system, not invitation. Therefore the whole network is not one connected graph and there are some isolated small subnetworks (Fig. 1(b))

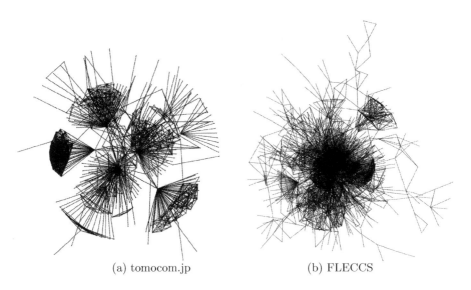

(a) tomocom.jp   (b) FLECCS

Fig. 1. Whole network figures of tomocom.jp and FLECCS using spring model [17]

4.2   Analysis Using Pseudo ACPM

I carried out community analysis on two described SNS sites using our pseudo ACPM. Results on tomocom.jp and FLECCS are shown in from Figs. 2, 3, 4 and 5. White circles, black circles, squares and gray circles in those figures mean largest, second largest, third largest community nodes and other ones respectively. Edges are restricted only on those between extracted nodes. I fix pseudo parameter $\alpha = 0.92$, and vary the parameter σ from 1 to 3. Though $\sigma = 1$ corresponds to the original CPM, They are not the same because $\alpha \neq 1$. In the case of $\sigma = 1$, only two communities are detected. On the other hand, in the

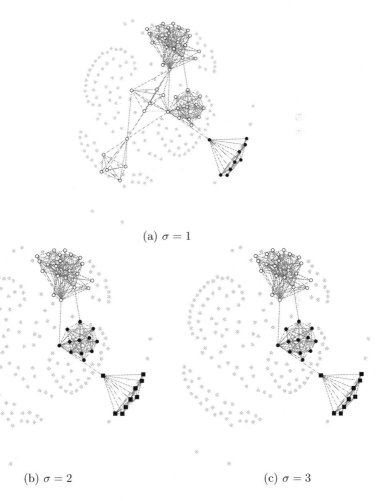

(a) $\sigma = 1$

(b) $\sigma = 2$ (c) $\sigma = 3$

Fig. 2. Communities in tomocom.jp at 30-Jun-2009. White circles, black circles, squares and gray circles represent first, second, third largest communities and other nodes respectively.

case of $\sigma = 2, 3$, A giant community is divided into middle-scale communities. Six month later, I can detect more large and dense communities in a similar way.

In FLECCS, I can also detect only two communities in the case of $\sigma = 1$ though, I can not obtain any clear and dense communities inside largest one. That is mainly because FLECCS does not have clear communities from the beginning and it is based on registration system, while tomocom.jp adopts invitation system. Six month later, such inclination does not change so much comparing with that of tomocom.jp.

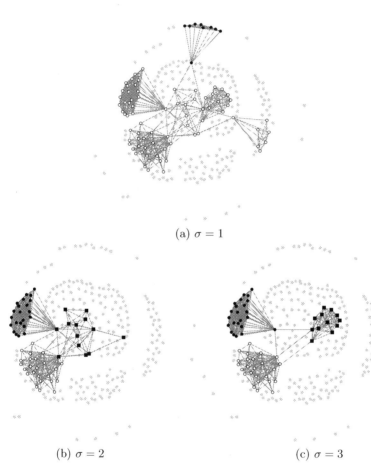

(a) $\sigma = 1$

(b) $\sigma = 2$ (c) $\sigma = 3$

Fig. 3. Communities in tomocom.jp at 31-Dec-2009. White circles, black circles, squares and gray circles represent first, second, third largest communities and other nodes respectively.

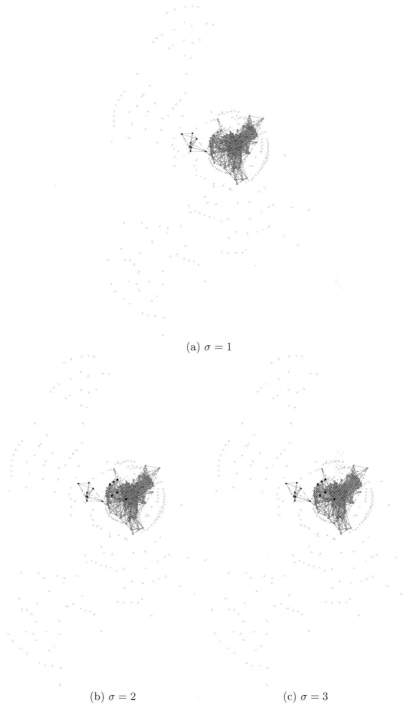

(a) $\sigma = 1$

(b) $\sigma = 2$ (c) $\sigma = 3$

Fig. 4. Communities in FLECCS at Jul-2009. White circles, black circles, squares and gray circles present first, second, third largest communities and other nodes respectively

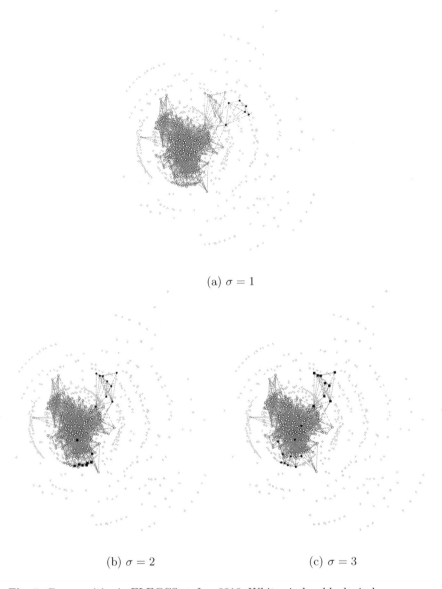

(a) $\sigma = 1$

(b) $\sigma = 2$ (c) $\sigma = 3$

Fig. 5. Communities in FLECCS at Jan-2010. White circles, black circles, squares and gray circles present first, second, third largest communities and other nodes respectively

5   Prediction of Social Networks

As mentioned above, data on social media like Facebook and Twitter is increasing daily, so it is very difficult to clarify the whole figure of it. However many people are interested in how those social media will develop.

Concerning about network model in complex network, quite many improvements have been proposed for WS model by Watts et al. [18] and BA model by Barabasi et al. [19] at the head. In social network, many models like KE and CNN have been proposed to be suitable for each social media. They are representative models with negative and positive assortativity respectively, and for details see [20,21]. One of the important factors characterizing each social network is assortativity. If some charismatic people exist, it tends to be negative, otherwise it tends to be positive. Some studies on their asymptotic behaviors using each network model. However real social network data has a large margin of error, so it is difficult to predict the future. To that problem, Hayashi proposed the method for the prediction of social media using sampling average [22], and I will conduct verification of his method using tomocom.jp and FLECCS.

At first, since real data of social media has a considerable amount of fluctuation, averaging node degree data, that is number of friends about time is carried out. I regard people who took part in this SNS in some period of time, here I select 2 days and 2 weeks, as the same group and average their degrees. I show the change of average degrees of nodes for each period of time in tomocom.jp in Fig. 6, and that in FLECCS in Fig. 7. Each line starts from the time when same group members took part in SNS.

In tomocom.jp, there convened activation events twice, so the increasing rate is not constant. On the contrary, in FLECCS the rate is relatively smooth. Since the ranges of time and degrees are narrow, it is a little difficult to specify the exponents for the verification of Hayashi's method, however their increasing curves are almost logarithmic. I only select sampling intervals 2 days and 2 weeks, so how I can select the interval is open to the future.

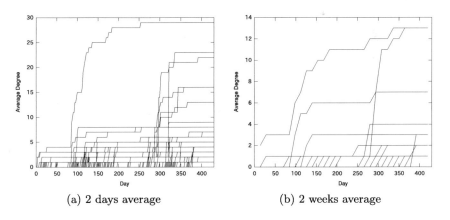

(a) 2 days average   (b) 2 weeks average

Fig. 6. Results of sampling average of degrees of nodes for each period of time tomocom.jp.

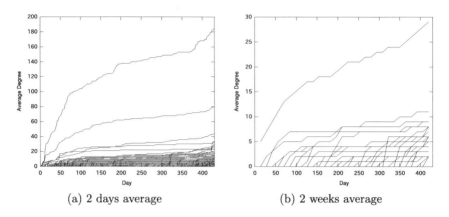

(a) 2 days average   (b) 2 weeks average

Fig. 7. Results of sampling average of degrees of nodes for each period of time in FLECCS.

6   Conclusion

In order to analyze social networks, a improved method for Clique Percolation Method (CPM), which is one of the effective community analysis with over-lapping nodes is proposed. Using this method, which is called pseudo ACPM, network analysis of friendship networks on SNS sites for college students is carried out. As the number of lack of nodes to fuse two cliques increases, it is confirmed that small communities inside large communities can be detected. The differences between two SNS sites coming from their system, registration or invitation, are also clarified. Moreover the change of average degrees of nodes there is observed and its behavior is also discussed. Their accuracy is to low because the ranges of both time and degree are low. However that can be solved by adjusting time interval of averaging, and it is our future work.

References

1. Albert, R., Barabási, A.L.: Statistical mechanics of complex networks. Rev. Mod. Phys. **74**, 47–97 (2002)
2. Milgram, S.: The small world problem. Psych. Today **1**, 61–67 (1967)
3. Travers, T., Milgram, S.: An experimental study of the small world problem. Sociometry **32**, 425–443 (1969)
4. Korte, C., Milgram, S.: Acquaintance links between White and Negro populations: application of the small world method. J. Pers. Soc. Psychol. **15**(2), 101–108 (1970)
5. Yuta, K., Ono, N., Fujiwara, Y.: Structural analysis of human network in social networking services. IPSJ J. **47**(3), 865–874 (2006)
6. Toriumi, F., Yamamoto, H., Suwa, H., Okada, I., Izumi, K., Hashimoto, Y.: Comparison analysis among large amount of SNS sites. Trans. Jpn. Soc. Artif. Intell. **25**(1), 78–89 (2010)
7. Toriumi, F., Ishii, K.: Simulation of encouragement methods for SNS based on user behavior model. In: Proceedings of 3rd World Congress on Social Simulation (2010)

8. Girvan, M., Newman, M.E.J.: Community structure in social and biological networks. Proc. Natl. Acad. Sci. U.S.A. **99**(12), 7821–7826 (2002)
9. Newman, M.E.J.: Fast algorithm for detecting community structure in networks. Phys. Rev. E **69**, 066133 (2004)
10. Clauset, A., Newman, M.E.J., Moore, C.: Finding community structure in very large networks. Phys. Rev. E **70**, 066111 (2004)
11. Palla, G., Derényi, I., Farkas, I., Vicsek, T.: Uncovering the overlapping modular structure of protein interaction networks. FEBS J. **272**(Suppl. 1), 434–434 (2005)
12. Fortunato, S.: Community detection in graphs. Phys. Rep. **486**(3–5), 75–174 (2010)
13. Tanaka, A.: Proposal of alleviative method of community analysis with overlapping nodes. In: Proceedings of the Seventh IEEE International Conference on Social Computing and Networking, C.1.30-5 (2014)
14. https://tomocom.jp
15. http://f-leccs.jp
16. Tanaka, A., Tomochi, M.: Emergence of hierarchical small-world property in SNS for college students. In: Proceedings of the International Symposium on Nonlinear Theory and Its Application, pp. 98–101 (2012)
17. Kamada, T., Kawai, S.: An algorithm for drawing general undirected graphs. Info. Process. Let. **31**(1), 7–15 (1989)
18. Watts, D.J., Strogatz, S.H.: Collective dynamics of 'small-world' networks. Nature **393**, 440–442 (1998)
19. Barabási, A.L., Albert, R.: Emergence of scaling networks. Science **286**, 509–512 (1999)
20. Klemm, K., Eguíluz, V.M.: Growing scale-free networks with small-world behavior. Phys. Rev. E **65**, 057102 (2002)
21. Vázquez, A.: Growing networks with local rules. Phys. Rev. E **67**, 056104 (2003)
22. Hayashi, Y.: Asymptotic behavior of the node degrees in the ensemble average of adjacency matrix (2015). arXiv:1512.00553

Exposing Open Street Map
in the Linked Data Cloud

Vito Walter Anelli[1], Andrea Calì[2], Tommaso Di Noia[1],
Matteo Palmonari[3], and Azzurra Ragone[3(✉)]

[1] Polytechnic University of Bari, Via Orabona, 4, 70125 Bari, Italy
`v.anelli@studenti.poliba.it, tommaso.dinoia@poliba.it`
[2] Birkbeck, University of London, Malet Street, London WC1E 7HX, UK
`andrea@dcs.bbk.ac.uk`
[3] University of Milano-Bicocca, Piazza dell'Ateneo Nuovo, 1, 20126 Milano, Italy
`{matteo.palmonari,azzurra.ragone}@unimib.it`

Abstract. After the mobile revolution, geographical knowledge has getting more and more importance in many location-aware application scenarios. Its popularity influenced also the production and publication of dedicated datasets in the Linked Data (LD) cloud. In fact, its most recent representation shows Geonames competing with DBpedia as the largest and most linked knowledge graph available in the Web. Among the various projects related to the collection and publication of geographical information, as of today, Open Street Map (OSM) is for sure one of the most complete and mature one exposing a huge amount of data which is continually updated in a crowdsourced fashion. In order to make all this knowledge available as Linked Data, we developed LOSM: a SPARQL endpoint able to query the data available in OSM by an on-line translation form syntax to a sequence of calls to the OSM overpass API. The endpoint makes also possible an on-the-fly integration among Open Street Map information and the one contained in external knowledge graphs such as DBpedia, Freebase or Wikidata.

1 Introduction

In the last years we are witnessing the spread of knowledge intensive applications that relies on the flourishing of the datasets available in the Linked Data (LD) Cloud[1]. The richness of semantic data they expose paves the way to a new generation of services and tools exploiting the ontological knowledge they encode as well as the possibility to easily mash up data coming from different sources. Among them, geographical datasets are becoming more and more important to deliver location-aware services.

The availability of a common query interface i.e. SPARQL, and the crowd-driven standardization of ontological vocabularies allows an intelligent application to grab data from diverse datasets and join them. We may imagine different

---

[1] http://lod-cloud.net.

© Springer International Publishing Switzerland 2016
H. Fujita et al. (Eds.): IEA/AIE 2016, LNAI 9799, pp. 344–355, 2016.
DOI: 10.1007/978-3-319-42007-3_29

scenarios where such a feature can be an important asset to provide a high quality service such as context-aware recommendation systems [15], on-line shopping, etc. or public services needed in situations of disaster management where the quality and timeliness of the data is of crucial importance.

In many application scenarios, geographical information is a key factor to enhance system answers to user request, e.g. in a movie recommendation scenario in order to suggests not only a movie according to user preferences, but also with reference to the closeness of the cinema.

Geographical knowledge bases and geo-spatial reasoning may play a key role also in emergency response or transportation planning [3,5,16] where it results useful to combine knowledge from different types of datasets in order to get knowledgeable information from the system. As a way of example, while organizing a trip, the user could combine information about the cultural heritage sites and means of transport with hotel accommodations, taking into account their proximity. Analogously, in emergency response situations, data can be combined to obtain a helpful and timely response. In an emergency scenario, e.g. an earthquake, the user should be able to query a knowledge base to look for collection camps, hospitals, rescue places, areas for helicopter landing as well as informal camps where the homeless have set up tent camps.

In this paper we present LOSM (Linked Open Street Map), a service that acts as a SPARQL endpoint on top of OSM. LOSM allows one to query Open Street Map data via a SPARQL query as it takes care to translate the query into a set of calls to OSM APIs. This results in exposing all the information contained in the Open Street Map geographical database as Linked Data thus making OSM a first class citizen in the Linked Data Cloud.

While LOSM supports an on-the-fly integration of OSM data with data coming from other sources, differently from similar projects (see LinkedGeoData[2] [2] for example) LOSM does not rely on a periodical dump of the data, but it always exposes fresh and up-to-date data.

Another strength of our tool is the possibility to merge different datasets of the Linked Data cloud, linking geo-spatial knowledge with the one coming from various knowledge bases, as we show in Sect. 3.1.

The remainder of this paper proceeds as follows. In Sect. 2 we start by briefly describing Open Street Map and the reason why we chose it as a provider of linked geo-spatial data, then we describe the system architecture and the query language implemented in LOSM. Then, in Sect. 3 we describe two use cases through some sample queries highlighting the capabilities of LOSM, with a particular reference to an emergency management scenario. Finally, in Sect. 4 we review related works relying on the use of geo-spatial data. Conclusion and future work close the paper.

2   LOSM: Linked Open Street Map

In the "geo-data" arena, the crowd sourced project Open Street Map [10] is currently playing a primary role due to its openness, easy of use and of integration in third party applications.

---

[2] http://linkedgeodata.org/.

OSM is a geographical database maintained by Web users containing a huge amount of data that can also be displayed on a map. Its database is updated every 15 min and as of today it contains 5,027,330,590 GPS points and 2,445,598 users who contribute to the project.

All this data is either available via weekly dumps or it is queryable through a public API. In particular, overpass[3] is a read-only API which allows the user to query Open Street Map by means of at least two different languages: XML or Overpass QL. By means of an overpass query, the API is able to retrieve nodes within an area, recognize streets or relations.

The query language is very expressive and makes possible to perform spatial reasoning by imposing constraints within the query. For instance, the user may impose relationships among nodes through filters such as `around`, `bounding box` and the `poly` function. It is easy to see that having such data available in the Linked Open Data cloud would surely enrich the amount and quality of the information available within the so called Web of Data.

This is the rationale behind the `LinkedGeoData` project [2]. It aims at triplifying Open Street Map dumps every six months by mapping OSM tags and sourceKey properties with reference to a publicly available ontology. This is a very useful resource because it makes available classes that map keys and tags used in Open Street Map nodes.

Although the big effort and work in developing and maintaining the datasets behind the project, `LinkedGeoData` suffers from the misalignment between the data available via the SPARQL endpoint (based on a dump) and the one available in Open Street Map. Indeed, the updates made by the users are available as RDF triples only when the dump is processed and loaded in the `LinkedGeoData` triple-store.

Despite the considerable effort, `LinkedGeoData` approach cannot be used for all those scenarios where timeliness and freshness of information is a must have. A flagship example is that of disaster recovery where information about collection camps, rescue places, temporary hospitals, passable roads, etc. needs to be available as soon as possible.

Starting from this observation we developed LOSM, a SPARQL endpoint that acts as a translator from a SPARQL query to a set of overpass API calls. In such a way we are sure that the data we retrieve is always fresh and up-to-date as they come directly from the OSM database, which is constantly updated by a crowd of volunteer all around the world.

The scheme in Fig. 1 shows an overview of the service architecture. In a few words, the systems is able to translate a SPARQL query to a sequence of (iterative) overpass API calls, collect the data, join and return it to the client. We currently support SPARQL queries via HTTP GET.

The **Parser** uses a scanner for the recognition of lexemes in a SPARQL query and creates the data-structures needed by the **Query Manager**. This module is in charge of breaking the query into sub-queries according to the remote functions available in the overpass API.

---

[3] http://wiki.openstreetmap.org/wiki/Overpass_API.

The **Result Manager** handles the sub-queries and the results they generate to create the final Result map. The Result Manager breaks the graph pattern described in the SPARQL query into a set of connected sub-graphs by identifying their mutual relations. Each sub-query goes through the **Translator** which is in charge of creating the overpass calls.

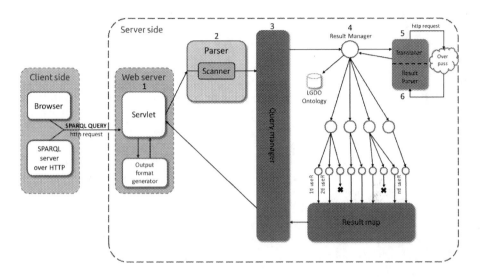

Fig. 1. Overall representation of the system architecture

The system also exposes a Web page with a query editor with autocomplete facilities with respect to the LinkedGeoData ontology classes.

LOSM is available at http://sisinflab.poliba.it/semanticweb/lod/losm/.

2.1 The SPARQL Sublanguage Implemented in LOSM

In its current version, LOSM implements a subset of the full specification of SPARQL 1.0 plus some non-standard features[4] that results very useful when querying geographical data. We currently support only the SELECT query form and the Jena Spatial[5] extension also available in GeoSPARQL [3]. We support simple graph patterns that we anyway consider representative of a large number of queries over geographical data. As for the implemented spatial functions, we may list:

- spatial:nearby (latitude longitude radius [units])[6] returns URIs nodes (Open Street Map URIs) within the radius distance of the location of the specified latitude and longitude.

---

[4] Details on the implemented subset is available at http://sisinflab.poliba.it/
   semanticweb/lod/losm/losm_grammar.html.

[5] https://jena.apache.org/documentation/query/spatial-query.html.

[6] [units] can be meters ('m' or 'M'), kilometers ('km' or 'KM') or miles ('mi' or 'MI').

- `spatial:withinCircle` (`latitude longitude radius [units]`) computes a circle centered in specified latitude and longitude and given radius and returns the OSM nodes within the circle.
- `spatial:withinBox` (`latitude_min longitude_min latitude_max longitude_max`) calculates a rectangle by specifying the list of coordinates for the edges that has to follow the order provided in the function.
- `spatial:within("POLYGON((Point1_lat Point1_lon,...,PointN_lat PointN_lon))")` calculates the polygon area expressed by Well Known Text (WKT) literals and returns OSM nodes available within it.

Regarding the URIs of classes and properties used in the graph pattern for SPARQL queries, LOSM may refer to the LinkedGeoData Ontology controlled vocabulary as well as to an ontological one which is a one-to-one mapping with OSM system of tags[7]. The rationale behind the introduction of this new vocabulary is driven by the main goal we had in mind while developing LOSM: to have a SPARQL endpoint able to timely expose all the changes that continually happen in Open Street Map even at the semantic level (represented by the tags).

Although very useful and structured, a static ontology as the one modelled within the LinkedGeoData project, cannot follow continuous data variations due to users' freedom in inserting new tags and values. In each community a lot of linguistic phenomenons happen in time that change the frequency of a term occurrence and then its importance and adoption by the community itself. To address this problem we introduced a LOSM prefix <http://sisinflab.poliba.it/semanticweb/lod/losm/ontology/> (shortened in `losm`) based on the same crowdsourcing concept. The **Parser** recognizes the use of this prefix and prepares the overpass query in the proper way. This lets the user to use any term she considers reasonable and the evaluation of the existence of the term is based only on the real data coming from Open Street Map. As an example, if the user wants to retrieve information classified with the key-value pair `key="refugee"` `value="yes"` she will refer to the corresponding property represented by the URI <http://sisinflab.poliba.it/semanticweb/lod/losm/ontology/refugee> or equivalently by the CURIE `losm:refugee` (see the example in Sect. 3.1).

Some keys are reserved to provide advanced features in LOSM that makes easy the integration with other external knowledge graphs exposing a SPARQL endpoint such as DBpedia[8] or Wikidata[9]. The `losm:dbpedia` property acts as a converter from the Wikipedia page or Wikipedia id associated to an Open Street Map node to the corresponding DBpedia resource. Analogously, `losm:wikipedia` returns the complete URI of the corresponding Wikipedia page. In both cases, the output may refer to the main English version of DBpedia/Wikipedia or to a local version depending on the value of the OSM key `wikipedia`.

---

[7] http://wiki.openstreetmap.org/wiki/Tags.

[8] http://dbpedia.org/sparql.

[9] http://query.wikidata.org.

3   Use Case

The first use case we present in this section has the only purpose to explain how LOSM works, showing the steps performed by the **Query Manager** (Fig. 1).

Suppose the following situation: *the day is over in our laboratory and the crew wants to find restaurants nearby (within 200 m) together with the cinemas which are at most one km far from each restaurant. They want to know the names of restaurants and cinemas together with the URIs of the latter.* The above use case can be modeled by the SPARQL query:

```
PREFIX rdfs: <http://www.w3.org/2000/01/rdf-schema#>
PREFIX lgdo: <http://linkedgeodata.org/ontology/>
PREFIX spatial: <http://jena.apache.org/spatial#>
PREFIX geo: <http://www.w3.org/2003/01/geo/wgs84_pos#>
SELECT ?cinema ?nameC ?nameR
WHERE   {
    ?link rdfs:label "Sisinf Lab".
    ?link geo:lat ?lat.
    ?link geo:long ?lon.
    ?object spatial:nearby(?lat ?lon 200 'm').
    ?object a lgdo:Restaurant.
    ?object rdfs:label ?nameR.
    ?object geo:lat ?lat2.
    ?object geo:long ?lon2.
    ?cinema spatial:nearby(?lat2 ?lon2 1000 'm').
    ?cinema a lgdo:Cinema.
    ?cinema rdfs:label ?nameC.
}
```

The query is processed based on a priority system relying on a weighted dependency graph. The triples composing the graph pattern are analysed and grouped into the corresponding sub-graphs by looking at their subject. The system attaches to each triple a value depending on the degree of connection with other groups measured by taking into account shared variables and predicates. Then each group is labeled with a weight proportional to its triples values. The group with the lowest value is the first sent t o the Translator component. The **Translator** converts the set of sub-queries to its overpass equivalent starting with the triple with the lowest value. Once the results from overpass API are returned, they are used to update the initial query so new weights are computed. The process iterates on triples groups until the last sub-query has been translated.

```
?link    [?link rdfs:label "Sisinf Lab" , ?link geo:lat ?lat , ?link geo:long ?lon ]
?object  [?object spatial:nearby ?lat ?lon 200 'm' , ?object a lgdo:Restaurant , ?object rdfs:label
         ?nameR , ?object geo:lat ?lat2 , ?object geo:long ?lon2 ]
?cinema  [?cinema spatial:nearby ?lat2 ?lon2 1000 'm' , ?cinema a lgdo:Cinema , ?cinema rdfs:label
         ?nameC ]
```

Based on the above grouping, the Query Manager selects first the **?link** group and generates the Overpass QL expression representing the first query to the overpass API:

```
node["name"="Sisinf Lab"];
out body;
```

Then the system executes the overpass query related to the **?object** group which is composed by taking into account the results of the previous one.

```
node(around:200,41.1095222,16.8778234)
["amenity"="restaurant"]
["name"];
out body;
```

The final sub-graph is converted into a set of overpass API calls; one for each node returned by the previous query. As an example we have:

```
node (around: 1000,41.1085645,16.8768552)
["amenity" = "cinema"]
["name"];
out body;
```

3.1 Emergency Management Scenario

We now present a use case in emergency management where the usefulness of LOSM is twofold. On the one hand, we may have access to always fresh and timely information in the context of an unpredictable disaster. On the other hand, we can exploit a third party endpoint supporting SPARQL 1.1 to perform a federated query among LOSM, DBpedia and Wikidata thus mashing up the knowledge coming from the three sources. Indeed, in such a context relevant data rapidly changes over time and the system capability of linking information from different knowledge bases is crucial.

An Italian manager is doing a business trip in the Miyagi Prefecture when an earthquake happens. The damages all around are severe and catastrophic. It is possible that aftershocks will follow and he has to find a way to rescue himself in a foreign country, plus he does not speak Japanese. In the mean time news about the event are reaching any corner of the world and mechanisms of international assistance are already on the move. Volunteers are populating Open Street Map with fresh data about collection camps, rescue places and temporary hospitals[10]. The manager has two primary needs: reaching a near refugee camp and, then, look for an airport to go back to Italy. He has a mobile phone with Gps and Internet connectivity so he tries to look for refugee camps, mapped on Open Street Map, which are located near by.

This request can be translated in the following SPARQL query[11]:

```
PREFIX rdfs: <http://www.w3.org/2000/01/rdf-schema#>
PREFIX lgdo: <http://linkedgeodata.org/ontology/>
PREFIX spatial: <http://jena.apache.org/spatial#>
PREFIX geo: <http://www.w3.org/2003/01/geo/wgs84_pos#>
SELECT ?object ?name ?lat ?long
WHERE {
?object spatial:nearby(38.2943 140.65000 5000 'm').
?object losm:refugee "yes".
?object rdfs:label ?name.
?object geo:lat ?lat.
?object geo:long ?long.
}
```

---

[10] The reader can find a real example of such scenario in [17] where it is described the Haiti-post-earthquake work done on Open Street Map: volunteers rapidly mapped the affected areas so helping the aid effort. They show the impact of a crowdsourced mapping in such emergency situation. Moreover, Open Street Map has an humanitarian team to deal with emergency situations. They keep updated a page with current and past remote mapping actions (see http://wiki.openstreetmap.org/wiki/Humanitarian_OSM_Team).

[11] It is noteworthy the use of the spatial function `spatial:nearby`, the LOSM predicate `losm:refugee` and the geo functions `geo:lat` and `geo:long`.

This query allows the user to retrieve any item containing the tag `refugee` within a circle with radius of 5000 m and returns the Open Street Map node, the name, and the GPS coordinates (lat, long). Obviously, the user does not have to write the sparql query himself, but he should rely on an end-user interface that allows him to build a sparql query without knowing the sparql language (see as a way of example the tools presented in [7,8]).

In order to show the capability of the system to link information coming from different knowledge bases, we give an example of a more complex and exhaustive queries that can be posed to the Linked Data Cloud thanks to the use of LOSM.

From the previous query, the manager has found a refugee camp whose name he cannot understand as it is returned in Japanese. Anyway, based on the result of the previous query he wants to retrieve information about the nearest cities (within 10 km) and airports to go back to Italy. He wants to retrieve info about the nearest cities in Italian and the name of the airports (together with its coordinates) in English in order to pronounce it in an understandable way.

```
PREFIX dbpedia: <http://dbpedia.org/ontology/>
PREFIX foaf: <http://xmlns.com/foaf/0.1/>
PREFIX rdfs: <http://www.w3.org/2000/01/rdf-schema#>
PREFIX lgdo: <http://linkedgeodata.org/ontology/>
PREFIX spatial: <http://jena.apache.org/spatial#>
PREFIX geo: <http://www.w3.org/2003/01/geo/wgs84_pos#>
PREFIX losm: <http://sisinflab.poliba.it/semanticweb/lod/losm/ontology/>
PREFIX schema: <http://schema.org/>
PREFIX wdt: <http://www.wikidata.org/prop/direct/>
PREFIX wd: <http://www.wikidata.org/entity/>
PREFIX dbo: <http://dbpedia.org/ontology/>
PREFIX rdf: <http://www.w3.org/1999/02/22-rdf-syntax-ns#>

SELECT  ?uriRefer ?dburi ?text ?airportname ?airportLat ?airportLong ?wdURI ?freebaseID  ?geonames
WHERE {
     SERVICE <http://sisinflab.poliba.it/semanticweb/lod/losm/sparql> {
?link losm:name " 仙台市立広陵中学校".

     ?link geo:lat ?lat .
     ?link geo:long ?lon.
     ?uriRefer spatial:nearby(?lat ?lon 10 'km') .

     ?uriRefer a lgdo:City   .
     ?uriRefer losm:dbpedia ?dburi .
     ?uriRefer losm:wikidata ?wdID .
 }
     SERVICE <http://dbpedia.org/sparql> {
     ?dburi dbpedia:abstract ?text .
     ?dburi foaf:isPrimaryTopicOf ?wikiuri .
     FILTER langMatches(lang(?text),'it').
     ?airport dbo:location ?dburi.
     ?airport rdf:type dbo:Airport.
     ?airport rdfs:label ?airportname .
     ?airport geo:lat ?airportLat .
     ?airport geo:long ?airportLong .
     filter(lang(?airportname) = 'en')

 }
  BIND(IRI(CONCAT("http://www.wikidata.org/entity/" ,str(?wdID) )) AS ?wdURI).
     SERVICE <http://query.wikidata.org/sparql> {
     SELECt ?wdURI ?freebaseID ?geonames {
     ?wdURI wdt:P646 ?freebaseID.
     ?wdURI wdt:P1566 ?geonames.
 }
 }
}
```

The previous query, by exploiting the SERVICE keyword from SPARQL 1.1, is able to combine information from different knowledge graphs with the one coming from LOSM. The first service invoked is the LOSM endpoint, the query returns cities within a radius of 10 Km from the refugee camp found in the previous query[12]. Additionally, the DBpedia resource URI and the Wikidata ID are returned. The second invoked service is the DBpedia endpoint. Here the query returns the Wikipedia URI, the Italian description of the city, the English name and the latitude and longitude of the airport. The last service is the Wikidata endpoint, form which we get the identifiers of the same city in Freebase and Geonames knowledge bases.

Summing up, the previous example shows how it is possible to get data referring to six different data sources (Open Street Map, Wikipedia, DBpedia, Wikidata, Freebase and Geonames) having only the two values of latitude and longitude available. It is worth to note that most of the data sources have a crowdsourcing nature, which usually weakens the integration between datasets because of the heterogeneity of the contributions. The problem is highly mitigated in this scenario thanks to spatial queries that can retrieve the outgoing references from points near to the starting one.

4   Related Work

In this section we first briefly describe various approaches and systems that deal with and expose geo-spatial data in a static way in the Web of Data. Then, we review some approaches that deal with and expose dynamic data sources as Linked Data.

In recent years several ontologies and languages have been proposed to model and query dataset related to geo-spatial knowledge and to extract information from these knowledge bases. The first attempts refer to Basic Geo Vocabulary and GeoOWL ontology. Basic Geo Vocabulary [9] is a simple RDF Schema vocabulary able to represent latitude, longitude and altitude information in the WGS84 reference system. The Basic Geo Vocabulary has then been extended with GeoRSS to include various geometric objects as points, lines, polygons and their associated feature descriptions [6]. A more structured and ontological representation of the GeoRSS vocabulary is available in the GeoOWL ontology[13]. Although these two projects were developed by W3C groups they never have become W3C recommendations (and they are not very used by the community).

GeoSPARQL [3] from the Open Geospatial Consortium (OGC) is a standard that has the aim to provide a way to represent and query geospatial data in the Semantic Web. GeoSPARQL addresses this task providing a small ontology to represent features and geometries and a number of SPARQL query predicates and functions. The ontology can be combined to other ontologies representing other domains, so enhancing the latter with spatial information. GeoSPARQL

---

[12] Note that the returned name is a Japanese name.
[13] http://www.w3.org/2005/Incubator/geo/XGR-geo-20071023/W3C_XGR_Geo_files/geo_2007.owl.

allows systems to infer topological information through a qualitative spatial reasoning, *e.g.*, if a monument is inside a park, and the park is in a city, then the monument is in that city [3], as well as quantitative spatial reasoning (e.g., measuring distances). A plus of GeoSPARQL is the possibility to infer qualitative knowledge starting from quantitative ones using a single languages for both types of reasoning. GeoSPARQL standards are supported by the triple-store Parliament [4] to query spatial data via RDF properties, which is able to answer queries like *"find all items located with a region X"*. Parliament does not support Basic Geo Vocabulary, differently from OWLIM-SE (now GraphDB) triple store [1] which however supports only points for storage, thus allowing queries to find points within ad-hoc polygons and circles and to compute distances between points. GraphDB data types and queries are not compliant with GeoSPARQL [16].

Strabon [12] is a semantic spatiotemporal RDF store, that can be used to store linked geospatial data and to query them using an extension of SPARQL named *stSPARQL*. *stSPARQL* can be used to query data represented in an extension of RDF called *stRDF* that model geospatial data that changes over time (e.g., the growth of a city over the years). Strabon supports spatial datatypes enabling the serialization of geometric objects in OGC standards WKT and GML, as well as a subset of GeoSPARQL.

USeekM[14] is an extension library for semantic databases that adds efficient geospatial support. The module supports OpenGIS geometry types (such as Point, Line, Polygon) and functions (such as Within, Intersects, Overlaps, Crosses) as standardized in the OGC GeoSPARQL standard.

Among database engines, Virtuoso Universal Server [18] can handle 2-dimensional points expressed with WGS84 coordinates, as well as storage of geometric shapes (lines, polygons, etc.). In order to check if two geometries are related, Virtuoso uses some built-in predicates (*e.g.* ST_contains, ST_within, ST_intersect) and supports some geometric functions (*e.g.,* ST_distance, ST_x, ST_y, ST_z). At the moment Virtuoso is not fully compliant with GeoSPARQL.

Oracle Spatial and Graph [14] supports, among others, RDF Semantic Graph data management and analysis, its applications ranging from semantic data integration to linked open data and network graphs used in transportation, utilities, energy and telcos. Oracle Spatial and Graph uses GeoSPARQL for representing and querying spatial data, even if it is not fully compliant with it.

A native RDF triple store implementation with spatial query functionality is described by Brodt et al. [5]. They model spatial features in RDF as typed complex literals and define spatial predicates as filter functions in SPARQL. However, their approach is optimized for storing and querying static RDF data with rare updates, as changes and updates in the location data can have an impact on their indexing and data processing.

Then, there are works that show how to expose dynamic data sources as Linked Data. Harth et al. [11] present an approach to expose data coming from information services as Linked Data to support their integration. Mapping is

---

[14] https://dev.opensahara.com/projects/useekm.

performed by using a tool to map RESTfull services to a reference ontology [20]. Although OSM is considered as one source, only queries based on bounding box have been supported. Thus this work does not proposes a general approach to expose OSM data as Linked Data.

Speiser et al. [19] and Norton et al. [13] present in their papers general approaches to expose data provided by services as Linked Data when invoked with a proper input, with [19] providing a more complete approach compared to [13]. Examples provided in the papers consider geospatial services like GeoNames [19] or OSM [13]. These general approaches are interesting but can hardly support the large variety of spatial queries over OSM that are supported by LOSM. In addition, vocabulary of the service is not mapped to widely adopted vocabularies as we did in LOSM.

5   Conclusion and Future Work

We presented LOSM, a service that acts as a SPARQL endpoint on top of Open Street Map data. Differently from `LinkedGeoData`, it does not work by using dumps of the OSM datasets but it queries directly the OSM database by means of a translation from SPARQL to overpass API calls. The implementation currently works on a subset of the SPARQL language plus the geographical query constructs from the Jena Spatial extension. We show how fresh and timely geographical data exposed via a SPARQL endpoint in combination with information coming from multilingual knowledge graphs can affect the search for information in a disaster recovery scenario. We are currently working to extend the expressiveness of the SPARQL sub-language supported by LOSM.

Acknowledgements. The authors acknowledge partial support of PON03PE_00136_1 Digital Services Ecosystem: DSE and Progetto Corvallis.

References

1. ADO: Graphdb (formerly owlim) triple store (2015). http://ontotext.com/products/graphdb/
2. Auer, S., Lehmann, J., Hellmann, S.: LinkedGeoData: adding a spatial dimension to the web of data. In: Bernstein, A., Karger, D.R., Heath, T., Feigenbaum, L., Maynard, D., Motta, E., Thirunarayan, K. (eds.) ISWC 2009. LNCS, vol. 5823, pp. 731–746. Springer, Heidelberg (2009)
3. Battle, R., Kolas, D.: GeoSPARQL: enabling a geospatial semantic web. Semant. Web J. **3**(4), 355–370 (2011)
4. Battle, R., Kolas, D.: Enabling the geospatial semantic web with parliament and GeoSPARQL. Semant. Web **3**(4), 355–370 (2012)
5. Brodt, A., Nicklas, D., Mitschang, B.: Deep integration of spatial query processing into native RDF triple stores. In: Proceedings of the 18th SIGSPATIAL International Conference on Advances in Geographic Information Systems, pp. 33–42. ACM (2010)

6. Consortium, O.G.: An introduction to GeoRSS: a standards based approach for geo-enabling RSS feeds. White Paper (2006). http://www.opengeospatial.org/pressroom/pressreleases/580

7. Andrawos, E., García Berrotarán, G., Carrascosa, R., Alonso i Alemany, L., Durán, H.: Quepy - transform natural language to database queries. http://quepy.machinalis.com/

8. Ferré, S.: Sparklis: a SPARQL endpoint explorer for expressive question answering. In: ISWC Posters and Demonstrations Track, pp. 45–48. CEUR (2014)

9. WSWI Group: Basic geo (wgs84 lat/long) vocabulary (2006). http://www.w3.org/2003/01/geo/

10. Haklay, M., Weber, P.: Openstreetmap: user-generated street maps. IEEE Pervasive Comput. **7**(4), 12–18 (2008)

11. Harth, A., Knoblock, C.A., Stadtmüller, S., Studer, R., Szekely, P.: On-the-fly integration of static and dynamic linked data. In: Proceedings of the Fourth International Workshop on Consuming Linked Data co-located with the 12th International Semantic Web Conference (2013). ISSN 1613-0073

12. Kyzirakos, K., Karpathiotakis, M., Koubarakis, M.: Strabon: a semantic geospatial DBMS. In: Cudré-Mauroux, P., Heflin, J., Sirin, E., Tudorache, T., Euzenat, J., Hauswirth, M., Parreira, J.X., Hendler, J., Schreiber, G., Bernstein, A., Blomqvist, E. (eds.) ISWC 2012, Part I. LNCS, vol. 7649, pp. 295–311. Springer, Heidelberg (2012)

13. Norton, B., Krummenacher, R., Marte, A., Fensel, D.: Dynamic linked data via linked open services. In: Workshop on Linked Data in the Future Internet at the Future Internet Assembly, pp. 1–10 (2010)

14. Oracle: Oracle spatial and graph. http://bit.ly/1lvCtWi

15. Ostuni, V.C., Gentile, G., Di Noia, T., Mirizzi, R., Romito, D., Di Sciascio, E.: Mobile movie recommendations with linked data. In: Cuzzocrea, A., Kittl, C., Simos, D.E., Weippl, E., Xu, L. (eds.) CD-ARES 2013. LNCS, vol. 8127, pp. 400–415. Springer, Heidelberg (2013)

16. Patroumpas, K., Giannopoulos, G., Athanasiou, S.: Towards geospatial semantic data management: strengths, weaknesses, and challenges ahead. In: Proceedings of the 22nd ACM SIGSPATIAL International Conference on Advances in Geographic Information Systems, pp. 301–310. ACM (2014)

17. Soden, R., Palen, L.: From crowdsourced mapping to community mapping: the post-earthquake work of openstreetmap haiti. In: Rossitto, C., Ciolfi, L., Martin, D., Conein, B. (eds.) COOP 2014 - Proceedings of the 11th International Conference on the Design of Cooperative Systems, pp. 311–326. Springer, Cham (2014)

18. OpenLink Software: Virtuoso universal server. http://ontotext.com/products/graphdb/

19. Speiser, S., Harth, A.: Integrating linked data and services with linked data services. In: Antoniou, G., Grobelnik, M., Simperl, E., Parsia, B., Plexousakis, D., De Leenheer, P., Pan, J. (eds.) ESWC 2011, Part I. LNCS, vol. 6643, pp. 170–184. Springer, Heidelberg (2011)

20. Taheriyan, M., Knoblock, C.A., Szekely, P., Ambite, J.L.: Rapidly integrating services into the linked data cloud. In: Cudré-Mauroux, P., et al. (eds.) ISWC 2012, Part I. LNCS, vol. 7649, pp. 559–574. Springer, Heidelberg (2012)

A MCDM Methods Based TAM for Deriving Influences of Privacy Paradox on User's Trust on Social Networks

Chi-Yo Huang[✉], Hsin-Hung Wu, and Hsueh-Hsin Lu

Department of Industrial Education,
National Taiwan Normal University, Taipei City, Taiwan
cyhuang66@ntnu.edu.tw

Abstract. Social network (SN) sites (SNSs) surged recently all over the world and have become new platforms for intimate communications. As the functionality of SNs was enhanced, users' own information can be collected, stored, and manipulated much more easily. Privacy concerns have thus become the most concerned issue by both users and SN service providers. The service providers intend to maximize the profits and need to consider how users' confidential information can be fully utilized in marketing and operations. At the same time, users usually concern over the misuse of private information by the website operations at the moment when disclosing individual details on SNSs. Apparently, a significant gap exists between the website operators' intention to fully utilize the private information as well as the users' privacy concerns about disclosing information on the SNSs. Such a cognition gap, or the "privacy paradox", influences users' trust on a specific SNS directly and further influences users' acceptance and continuous usage of the sites. In this study, the Technology Acceptance Model (TAM) was introduced as the theoretical basis by applying users' private disclosure behavior, disclosure risks perception, and the extent of privacy settings in the SNSs as the main variables. In addition, in the past works, researchers found that perceived usefulness, perceived ease of use and the interaction strength for modern technology services or products influence the use intention. So, these factors were also added as research variables in the analytic model. The Decision Making Trial and Evaluation Laboratory Based Network Process (DNP) was introduced construct the influence relationships between the variables. The weights being associated with the variables can be derived accordingly. By using the analytic model, the variables which can influence the privacy paradox on user's trust of SNs can be derived. Such variables and influence relationships can be used in developing the security policies of the SNSs.

Keywords: Social networking sites (SNS) · TAM (Technology Acceptance Model) · Privacy paradox · Decision Making Trial and Evaluation Laboratory (DEMATEL) · Decision Making Trial and Evaluation Laboratory Based Network Process (DANP)

© Springer International Publishing Switzerland 2016
H. Fujita et al. (Eds.): IEA/AIE 2016, LNAI 9799, pp. 356–363, 2016.
DOI: 10.1007/978-3-319-42007-3_30

1 Introduction

In the fast emergence process of the social networking sites (SNSs), users disclosed a huge amount of personal information and data on the sites. As more and more personal information was disclosed, information misusages, identity thefts and other privacy issues also happened. Albeit the social network (SN) service providers have developed many mechanisms to protect users' privacy, numerous users still feel insufficient privacy protections. The needs to simplify SNSs' privacy settings and users' controls over personal files or information are growing. The gap between privacy concerns and actual privacy settings of the SNS system has been described as the "privacy paradox."

Although scholars have studied the trust in the concerns of information privacy or other effects of the antecedents for trust, the research works were in short of integrating the information privacy concerns and other factors into the formation of trust in general, and the formation of trust in SNSs in special. To resolve this problem, this work aims to derive the factors influencing the trust in SNSs.

Therefore, the Technology Acceptance Model (TAM) was introduced as the theoretical basis. Further, users' private disclosure behavior, disclosure risks perception, and the extents of privacy settings in the SNSs were introduced as possible variables. In addition, the perceived usefulness, the perceived ease of use and the interaction strength for modern technology services or products were also added as the possible variables.

In order to derive the most important factors influencing users' trusts in the SNSs, the Decision Making Trial and Evaluation Laboratory (DEMATEL) was introduced to construct the influence relationships between variables. The weights being associated with each variable were further derived by using the DEMATEL Based Network Process (DNP). Thus, the variables which can influence users' future intentions to interact with the SNSs, the privacy settings, and technology acceptances were derived. These variables which influence the privacy paradox on user's trusts of SNs can be used to derive and enhance the security policies of the SNSs. An empirical study based on experts' opinions will be used to demonstrate the feasibility of the proposed analytic framework.

2 Literature Review

SNSs are Internet websites where people with similar interests and activities can establish and consolidate the friendships. Therefore, previous works define SNSs as web-based services that allow individuals to (1) construct a public or semi-public profile within a bounded system, (2) articulate a list of other users with whom they share a connection, and (3) view and traverse their list of connections and those made by others within the system [1]. People construct their own background information (profiles) and publicly disclose such profile on the SNSs. Further interpersonal relationships and businesses can be developed.

A paradox, a discrepancy between privacy concerns and actual privacy settings [], were found on SNSs and other social medias. According to Young et al. [2], the privacy paradox describes peoples' willingness to disclose personal information on social network sites despite expressing high levels of concern. Users of SNS often state that they are concerned about their privacy, yet they often disclose detailed personal information on their profiles [3].

The TAM, a theoretical basis for determinations of the external variables that affect users' internal beliefs [4], stated that a user attitude toward information system/ information technology was determined by two particular beliefs: the perceived usefulness (PU) and the perceived ease of use (PEU). The user attitude in turn leads to behavioral intention (BI) to use (accept) technology, and then generate the actual usage behavior. Venkatesh and Davis [5], and Venkatesh and Morris [6] observed that PEU and PU were influenced, to some extent, by external variables, and extensions of the TAM have been introduced by comprehensive study of the determinants of PEU. Furthermore, in distinctive research applications, or when predicting or interpreting the acceptance of technology through different theories and studies, external variables should be elaborated to expand the discussion on the degree of acceptance of technologies. The TAM is demonstrated in Fig. 1.

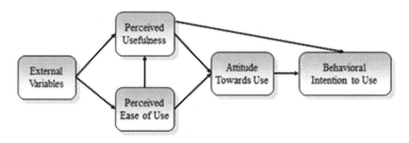

Fig. 1. Technology Acceptance Model

3   Research Methods

The DNP, a decision making process which consists of the DEMATEL as well as the network process for deriving the impact of each criterion on others as the weight, was developed by Prof. Tzeng [7]. The steps of the DNP method can be summarized as follows,

Step 1: Calculate the direct-influence matrix by scores. Based on experts' opinions, evaluations are made of the relationships among elements (or variables/attributes) of mutual influence using a scale ranging from 0 to 4, with scores representing "no influence" (0), "low influence" (1), "medium influence" (2), "high influence" (3), and "very high influence" (4). They are asked to indicate the direct effect they believe a factor i will have on factor, as indicated by d_{ij}. The matrix D of direct relations can be obtained.

Step 2: Normalize the direct-influence matrix based on the direct-influence matrix D, the normalized direct relation matrix X is acquired by using Eq. (1)

$$N = vD; v = \min\{1/\max_i \sum_{j=1}^{n} d_{ij}, 1/\max_j \sum_{i=1}^{n} d_{ij}\}, i,j \in \{1,2,\ldots,n\} \tag{1}$$

Step 3: Attaining the total-influence matrix T. Once the normalized j direct-influence matrix N is obtained, the total-influence matrix T of NRM can be obtained.

$$T = N + N^2 + \ldots + N^k = N(I - N)^{-1} \tag{2}$$

where $k \to \infty$ and T is a total influence-related matrix; N is a direct influence matrix and $N = [x_{ij}]_{n \times n}$; $\lim_{k \to \infty} (N^2 + \cdots + N^k)$ stands for a indirect influence matrix and $0 \le \sum_{j=1}^{n} x_{ij} < 1$ or $0 \le \sum_{i=1}^{n} x_{ij} < 1$, and only one $\sum_{j=1}^{n} x_{ij}$ or $\sum_{i=1}^{n} x_{ij}$ equal to 1 for $\forall i,j$. So $\lim_{k \to \infty} N^k = [0]_{n \times n}$. The (i,j) element t_{ij} of matrix T denotes the direct and indirect influences of factor i on factor j.

Step 4: Analyze the result. In this stage, the row and column sums are separately denoted as r and c within the total-relation matrix T through Eqs. (3), (4), and (5).

$$T = [t_{ij}], \ i,j \in \{1,2,\ldots,n\} \tag{3}$$

$$r = [r_i]_{n \times 1} = \left[\sum_{j=1}^{n} t_{ij}\right]_{n \times 1} \tag{4}$$

$$c = [c_j]_{1 \times n} = \left[\sum_{i=1}^{n} t_{ij}\right]_{1 \times n} \tag{5}$$

where r and c vectors denote the sums of the rows and columns, respectively. Suppose r_i denotes the row sum of the ith row of matrix T. Then, r_i is the sum of the influences dispatching from factor i to the other factors, both directly and indirectly. Suppose that c_j denotes the column sum of the jth column of matrix T. Then, c_j is the sum of the influences that factor i is receiving from the other factors. Furthermore, when $i = j$ (i.e., the sum of the row sum and the column sum) $(r_i + c_i)$ represents the index representing the strength of the influence, both dispatching and receiving), $(r_i + c_i)$ is the degree of the central role that factor i plays in the problem. If $(r_i - c_i)$ is positive, then factor i primarily is dispatching influence upon the strength of other factors; and if $(r_i - c_i)$ is negative, then factor i primarily is receiving influence from other factors [8]. Therefore, a causal graph can be achieved by mapping the dataset of $(r_i + s_i, r_i - s_i)$ providing a valuable approach for decision making [7].

Now the total-influence matrix is called as $T_C = [t_{ij}]_{nxn}$, which is obtained by criteria and $T_D = \left[t_{ij}^D\right]_{nxn}$ obtained by dimensions (clusters) from T_C. Then we normalize the ANP weights of dimensions (clusters) by using influence matrix T_D.

$$T_D = \begin{bmatrix} \begin{bmatrix} t_{11}^{D_{11}} & \cdots & t_{1j}^{D_{1j}} & \cdots & t_{1m}^{D_{1m}} \end{bmatrix} \\ \vdots & \vdots & \vdots & \vdots & \vdots \\ \begin{bmatrix} t_{i1}^{D_{i1}} & \cdots & t_{ij}^{D_{ij}} & \cdots & t_{im}^{D_{im}} \end{bmatrix} \\ \vdots & \vdots & \vdots & \vdots & \vdots \\ \begin{bmatrix} t_{m1}^{D_{m1}} & \cdots & t_{mj}^{D_{mj}} & \cdots & t_{mm}^{D_{mm}} \end{bmatrix} \end{bmatrix} \begin{matrix} \rightarrow d_1 = \sum_{j=1}^{m} t_{1j}^{D_{1j}} \\ \\ \rightarrow d_i = \sum_{j=1}^{m} t_{ij}^{D_{ij}} , d_i = \sum_{j=1}^{m} t_{ij}^{D_{ij}}, i=1,\ldots,m \\ \\ \rightarrow d_m = \sum_{j=1}^{m} t_{mj}^{D_{mj}} \end{matrix}$$

Step 5: The original supermatrix of eigenvectors is obtained from the total-influence matrix $T = [t_{ij}]$. For example, D values of the clusters in matrix T_D, as Eq. (6). Where if $t_{ij} < D$, then $t_{ij}^D = 0$ else $t_{ij}^D = t_{ij}$, and t_{ij} is in the total-influence matrix T. The total-influence matrix T_D needs to be normalized by dividing by the following formula. There, we could normalize the total-influence matrix and represent it as T_D.

$$T_D = \begin{bmatrix} t_{11}^{D_{11}}/d_1 & \cdots & t_{1j}^{D_{1j}}/d_1 & \cdots & t_{1m}^{D_{1m}}/d_1 \\ \vdots & \vdots & \vdots & \vdots & \vdots \\ t_{i1}^{D_{i1}}/d_i & \cdots & t_{ij}^{D_{ij}}/d_i & \cdots & t_{im}^{D_{im}}/d_i \\ \vdots & \vdots & \vdots & \vdots & \vdots \\ t_{m1}^{D_{m1}}/d_m & \cdots & t_{mj}^{D_{mj}}/d_m & \cdots & t_{mm}^{D_{mm}}/d_m \end{bmatrix} = \begin{bmatrix} \alpha_{11}^{D_{11}} & \cdots & \alpha_{1j}^{D_{1j}} & \cdots & \alpha_{1m}^{D_{1m}} \\ \vdots & \vdots & \vdots & \vdots & \vdots \\ \alpha_{i1}^{D_{i1}} & \cdots & \alpha_{ij}^{D_{ij}} & \cdots & \alpha_{im}^{D_{im}} \\ \vdots & \vdots & \vdots & \vdots & \vdots \\ \alpha_{m1}^{D_{m1}} & \cdots & \alpha_{mj}^{D_{mj}} & \cdots & \alpha_{mm}^{D_{mm}} \end{bmatrix} \tag{6}$$

where $\alpha_{ij}^{D_{ij}} = t_{ij}^{D_{ij}}/d_i$.

This research adopts the normalized total-influence matrix T_D (here after abbreviated to "the normalized matrix") and the unweighted supermatrix W using Eq. (7) shows theses influence level values as the basis of the normalization for determining the weighted supermatrix.

$$W^* = \begin{bmatrix} \alpha_{11}^{D_{11}} \times W_{11} & \alpha_{21}^{D_{21}} \times W_{12} & \cdots & \cdots & \alpha_{m1}^{D_{m1}} \times W_{1m} \\ \alpha_{12}^{D_{12}} \times W_{21} & \alpha_{22}^{D_{22}} \times W_{22} & \cdots & \cdots & \vdots \\ \vdots & \cdots & \alpha_{ji}^{D_{ji}} \times W_{ij} & \cdots & \alpha_{mi}^{D_{mi}} \times W_{im} \\ \vdots & \vdots & \vdots & \vdots & \vdots \\ \alpha_{1m}^{D_{1m}} \times W_{m1} & \alpha_{2m}^{D_{2m}} \times W_{m2} & \cdots & \cdots & \alpha_{mm}^{D_{mm}} \times W_{mm} \end{bmatrix} \tag{7}$$

Step 6: Limit the weighted supermatrix by raising it to a sufficiently large power k, as Eq. (8), until the supermatrix has converged and become a long-term stable supermatrix to get the global priority vectors or called ANP weights.

$$\lim_{k \to \infty} (W^*)^k \tag{8}$$

4 Empirical Study

In this Section, an empirical study will be used to demonstrate the feasibility of the proposed analytic framework. At first, literatures being published during the past decade were selected for deriving possible factors in formatting trusts in SNSs. Then, seven Taiwanese senior experts (including two engineering assistant vice president of an information company, two vice presidents from a technology company, and three directors of planning from a web design service company) with more than five years of work experiences were invited to confirm the criteria by using the modified Delphi method. Based on the experts' opinions, the criteria include: (1) privacy concerns (PC); (2) structural assurances (SA); (3) perceived ease of use (PEU); (4) perceived usefulness (PU); and (5) intends to use (IU); the determinants are introduced further as a basis of this case study.

Since the inter-relationships between the six determinants being summarized through above Delphi process seem too complicated to be analyzed, the decision problem structure will be deducted with the DEMATEL method. The major relationships were deducted by setting the threshold value as 1.94 from both the statistical and natural language aspects to derive the most important linkages between determinants (or criteria). The total relationships being derived will serve as references for calculating weights between determinants in the following ANP processes. The casual relationship map shown at Fig. 2. Then, based on the influence relationships between components in formatting trust of SNSs, and the weights being associated with each criterion are derived by the DNP and are demonstrated in Table 1.

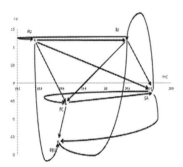

Fig. 2. The casual relationship map

Table 1. The weights

Criteria	Weight	Ranking
PC	0.204	3
SA	0.208	2
PEU	0.213	1
PU	0.183	5
IU	0.191	4
SUM	1.000	

5 Conclusions

The "privacy paradox" has influenced users' trusts on SNSs directly and have further influenced users' acceptances and continuous usages of the SNSs. Based on the theoretical framework of the TAM, the authors adopted the users' private disclosure behavior, disclosure risks perception, and the extent of privacy settings in the SNSs as main variables. A novel MCDM framework consisting of the DEMATEL technique and the DNP was proposed to address the above-mentioned problems and reached satisfactory result.

The novel MCDM model consisting of the modified Delphi, the DEMATEL and the DNP were designed to overcome (1) the privacy paradox and users' trust definition issue in the social networking site, (2) traditional MCDM approaches for resolving the privacy paradox and users' trust definition problem based on the wrong assumptions on the independences between the determinants, (3) the vague correlations between determinants and social networks as well as users' trust strategies, and (4) the lack of priorities of the social networks as well as users' trust.

To summarize the rationality of the MCDM methods being introduced into the research, apparently, the modified Delphi, DEMATEL and ANP based hybrid MCDM method can overcome following problems of a single traditional MCDM method or a combination of some methods being included: (1) the assumption of decision problem structure for the ANP without significant supports; (2) the assumption of independences between criteria; (3) a lack of an appropriate method for correlating multiple criteria with multiple strategies as a portfolio.

From the perspective of view, privacy concern will be influenced by structural assurance and perceive ease of use [9, 10], and structural assurance will influence privacy concern [11]. Our research has derived the same result as the previous study of these scholars. At last, Intend to use will be influenced by privacy concern, perceive ease of use and perceive usefulness [4, 10–12]. Based on the DNP, the most important variables that can influence users' continuous usages of the SNS were derived. The variables, which include the PEU, SA, and PC, are the most critical ones, which can be used to develop the security policies of the SNSs.

References

1. Ellison, N.B.: Social network sites: definition, history, and scholarship. J. Comput.-Mediated Commun. **13**, 210–230 (2007)
2. Young, A.L., Quan-Haase, A.: Privacy protection strategies on facebook: the internet privacy paradox revisited. Inf. Commun. Soc. **16**, 479–500 (2013)
3. Utz, S., Krämer, N.: The privacy paradox on social network sites revisited: the role of individual characteristics and group norms. Cyberpsychol.: J. Psychosoc. Res. Cyberspace **3**, 2 (2009)
4. Davis, F.D., Bagozzi, R.P., Warshaw, P.R.: User acceptance of computer technology: a comparison of two theoretical models. Manag. Sci. **35**, 982–1003 (1989)
5. Venkatesh, V., Davis, F.D.: A theoretical extension of the technology acceptance model: four longitudinal field studies. Manag. Sci. **46**, 186–204 (2000)
6. Venkatesh, V., Morris, M.G.: Why don't men ever stop to ask for directions? Gender, social influence, and their role in technology acceptance and usage behavior. MIS Q. **24**, 115–139 (2000)
7. Huang, C.-Y., Wu, M.-J., Liu, Y.-W., Tzeng, G.-H.: Using the DEMATEL based network process and structural equation modeling methods for deriving factors influencing the acceptance of smart phone operation systems. In: Jiang, H., Ding, W., Ali, M., Wu, X. (eds.) IEA/AIE 2012. LNCS, vol. 7345, pp. 731–741. Springer, Heidelberg (2012)
8. Huang, C.-Y., Shyu, J.Z., Tzeng, G.-H.: Reconfiguring the innovation policy portfolios for Taiwan's SIP mall industry. Technovation **27**, 744–765 (2007)
9. Bansal, G., Fatemeh'Mariam'Zahedi, D.G.: The role of privacy assurance mechanisms in building trust and the moderating role of privacy concern. Eur. J. Inf. Syst. **24**, 624–644 (2015)
10. Nam, C., Song, C., Lee, E., Park, C.I.: Consumers' privacy concerns and willingness to provide marketing-related personal information online. Adv. Consum. Res. **33**, 212 (2006)
11. Delvaux, E., Vanhoof, J., Tuytens, M., Vekeman, E., Devos, G., Van Petegem, P.: How may teacher evaluation have an impact on professional development? A multilevel analysis. Teach. Teach. Educ. **36**, 1–11 (2013)
12. Hsu, C.-L., Lu, H.-P.: Consumer behavior in online game communities: a motivational factor perspective. Comput. Hum. Behav. **23**, 1642–1659 (2007)

Algorithms for Quantitative-Based Possibilistic Lightweight Ontologies

Salem Benferhat[1], Khaoula Boutouhami[1,2], Faiza Khellaf[2], and Farid Nouioua[3(✉)]

[1] University of Artois, CRIL-CNRS UMR 8188, Artois Univ, 62307 Lens, France
[2] RIIMA, University of Sciences and Technology Houari Boumediene,
Bab Ezzouar, Algeria
[3] Aix-Marseille University, LSIS UMR 7296, Marseille, France
farid.nouioua@lsis.org

Abstract. This paper proposes approximate algorithms for computing inconsistency degree and answering instance checking queries in the framework of uncertain lightweight ontologies. We focus on an extension of lightweight ontologies, encoded here in DL-Lite languages, to the product-based possibility theory framework. We provide an encoding of the problem of computing inconsistency degree in product-based possibility DL-Lite as a weighted set cover problem and we use a greedy algorithm to compute an approximate value of the inconsistency degree. We also propose an approximate algorithm for answering instance checking queries in product-based possibilistic DL-Lite. Experimental studies show the good results obtained by both algorithms.

Keywords: Lightweight ontologies · Possibility theory · Product based possibility theory · Approximate algorithms · Set cover problem · Instance checking

1 Introduction

Knowledge representation for the semantic web requires an analysis of the universe of discourse in terms of concepts, definitions, objects, roles, etc., and then selecting a computer-usable version of the results. In this context, ontologies play an important role for the success of the semantic web as they provide shared vocabularies for different domains. Among the existing representation languages for ontologies, the advantage of description logics (DLs) [3] lies in their clear semantics and formal properties. Moreover, despite its syntactical restrictions, the DL-Lite family enjoys good computational properties while still offering interesting capabilities in representing terminological knowledge [1]. That is why a large amount of works has been recently dedicated to this family and this paper is a contribution to this general research line.

The information available on the web evolves continuously and comes from multiple sources that do not have the same reliability level. As a result, in real

© Springer International Publishing Switzerland 2016
H. Fujita et al. (Eds.): IEA/AIE 2016, LNAI 9799, pp. 364–372, 2016.
DOI: 10.1007/978-3-319-42007-3_31

applications, we are often confronted to uncertainties in the used information. Proposing efficient methods for handling uncertainty in DLs, and particularly in the DL-Lite family is an important research topic. Several recent works are devoted to fuzzy extensions of DLs (see e.g. [5,6,13]). Some other works propose (min-based) possibilistic extensions of DLs and focus on standard reasoning services (see e.g. [4,10,12]).

Having appropriate and efficient methods for handling uncertainty in lightweight ontologies is a key issue for many applications in a wide range of domains. Among these domains, intelligent information retrieval from large-scale ontologies (such as ontologies built on the web) is at the heart of many applications. For example, in e-business and competitive intelligence applications, it provides a valuable help in taking strategic decisions. In critical applications (eg. medicine or e-government), it is very important to know the uncertainty degrees associated with the available information. Finally, ontologies used in industrial applications often include sensor readings that usually provide uncertain data. This is also the case in multimedia processing where object recognition may be uncertain. However, in presence of uncertainty in lightweight ontologies, exact algorithms for query answering are no longer tractable and hence not useful in practice. Then, we need to propose new efficient approximate algorithms and this is the main objective of the present paper.

This paper uses a possibility theory framework [9] to represent uncertainty in lightw-eight ontologies. In out proposal, we allow ABox assertions to be uncertain. For example, in an ontology modeling the university domain, one may represent pieces of information of the form: the fact that "b is a teacher" is certain to the degree 0.85.

There are two major definitions of a possibility theory: min-based (or qualitative) possibility theory and product-based (or quantitative) possibility theory. Recently a min-based possibilitic extension of DL-Lite has been proposed in [4]. It has been shown in [4] that this extension is done without extra computational cost. Hence, query answering is tractable in min-based possibilitic DL-Lite.

In this paper, we investigate algorithms for the product-based possibilitic DL-Lite, denoted by Pb-π-DL-Lite, which has not been considered before. Unlike the min-based setting, query answering in this new setting is no longer tractable. To overcome this difficulty, we propose an approximate algorithm to compute the inconsistency degree of a Pb-π-DL-Lite KB and show how to use this degree to answer efficiently instance checking queries.

2 Product-Based Possibilistic DL-Lite

A DL-Lite knowledge base (KB) $K = \langle T, A \rangle$ consists of a set T of concept and role axioms (called TBox) and a set A of assertional facts (called ABox).

In this paper, we consider only three main members of the DL-Lite family. Namely, the DL-Lite$_{core}$ the core fragment of all DL-Lite logics, DL-Lite$_F$ and DL-Lite$_R$ that underlies OWL2-QL [7]. The syntax of the DL-Lite$_{core}$ language is defined as follows:

$$B \to A | \exists R \quad C \to B | \neg B \quad R \to P | P^- \quad E \to R | \neg R \qquad (1)$$

where A denotes atomic concepts, P atomic roles, P^- the inverse of the atomic role P, B (resp. C) are called basic (resp. complex) concepts and R (resp. E) are called basic (resp. complex) roles.

A DL-Lite$_{core}$ TBox is a finite set of inclusion axioms of the form: $B \sqsubseteq C$. An ABox is a finite set of membership assertions on atomic concepts and on atomic roles of the form: $A(a)$ and $P(a,b)$ respectively, where a and b are two individuals. The DL-Lite$_R$ language extends DL-Lite$_{core}$ with the ability of specifying in the TBox inclusion axioms between roles of the form: $R \sqsubseteq E$. The DL-Lite$_F$ language extends DL-Lite$_{core}$ with the ability of specifying functionality on basic roles of the form: $(funct\ R)$.

A Pb-π-DL-Lite KB $K = \{(\phi_i, \alpha_i) : i = 1, \ldots, n\}$ is a finite set of couples of the form $\langle \phi_i, \alpha_i \rangle$, where ϕ_i is either a DL-Lite TBox axiom or a DL-Lite ABox assertion and $\alpha_i \in]0,1]$ represents the certainty degree of ϕ_i. Only somewhat certain facts (having certainty degree > 0) are considered. We consider also that uncertainty concerns only the ABox assertions. Hence, the terminological base is assumed to be certain and if needed, only assertional facts can be removed when query answering.

Example 1. *Let us consider the Pb-π-DL-Lite KB $K = \langle T, A \rangle$ where:*

$T = \{$ $(Student \sqsubseteq \neg Teacher, 1.0),$ $(\exists teachesTo \sqsubseteq Teacher, 1.0),$
$\qquad (\exists teachesTo^- \sqsubseteq Student, 1.0)\}.$
$A = \{$ $(Teacher(Bob), 0.85),$ $(Student(Tom), 0.95),$
$\qquad (teachesTo(Tom, Bob), 0.75)\}.$

Student and Teacher are two concepts while teachesTo is a role. For instance the assertional fact $(Teacher(Bob), 0.85)$ states that Bob is a teacher with a degree of certainty of 0.85. The axiom $(Student \sqsubseteq \neg Teacher, 1.0)$ states that we are absolutely certain that a student cannot be a teacher. This KB will be used in the rest of the paper.

The semantics of a Pb-π-DL-Lite KB $K = \langle T, A \rangle$ is defined through a possibility distribution π_K which is a function that assigns to each interpretation I a possibility degree in the interval $[0,1]$ to represent the degree of compatibility of I with respect to the available information given in K. This degree is inversely proportional to the product of the weights of the assertions that I falsifies: the more certain are the formulas falsified by I, the less is the possibility degree associated with I. Formally, $\pi_K(I)$ is given by ($\models$ is the satisfaction relation between DL-Lite formulas):

Example 1 (Cont). *Consider again Example 1. Table 2 gives the possibility degrees, obtained by Eq. 2, for two interpretations over the domain $\triangle = \{Tom, Bob\}$.*

$$\pi_K(I) = \begin{cases} 1 & if\ \forall (\phi_i, \alpha_i) \in T \cup A,\ I \models \phi_i \\ *\{1 - \alpha_i : (\phi_i, \alpha_i) \in T \cup A,\ I \nvDash \phi_i\} & otherwise \end{cases} \qquad (2)$$

I	$\cdot^I$	$\pi_K(I)$
I_1	$St^I = \{Tom\}, Tch^I = \{Bob\}, Tcht^I = \{(Bob, Tom)\}$	0.25
I_2	$St^I = \{Bob\}, Tch^I = \{Tom\}, Tcht^I = \{(Tom, Bob)\}$	$0.05 * 0.15$

The interpretation I_1 does not satisfy the axiom $\langle teachesTo(Tom, Bob), 0.75 \rangle$ and I_2 does not satisfy the axioms $\langle Teacher(Bob), 0.85 \rangle$, $\langle Student(Tom), 0.95 \rangle$. None of the two interpretations is a model of K.

A Pb-π-DL-Lite KB K is said to be fully consistent if there exists an interpretation I such that $\pi_K(I) = 1$. Otherwise, K is somewhat inconsistent. In the presence of certainty degrees in K, inconsistency becomes a graduated notion. It is evaluated by the so-called inconsistency degree of K, denoted by $Inc(K)$ which is a real number between 0 and 1 defined by: $Inc(K) = 1 - \text{Max}_{I \in \Omega}(\pi_K(I))$.

Example 1 (Cont). *Consider the KB K of Example 1. $Inc(K) = 0.75$. There is no way to satisfy K with a possibility degree greater than $1 - 0.75 = 0.25$.*

3   Instance Checking in Pb-π-DL-Lite

In standard query answering, instance checking consists in deciding, given a DL-Lite KB $K = \langle T, A \rangle$ and an assertion $B(\overrightarrow{a})$ (where either B is a concept and $\overrightarrow{a}$ is a constant or B is a role and $\overrightarrow{a}$ is a couple of constants) whether $B(\overrightarrow{a})$ follows from K.

In the context of Pb-π-DL-Lite, we need first to define the necessity measure of $B(\overrightarrow{a})$ as follows: $N(B(\overrightarrow{a})) = 1 - \text{Max}\{\pi_K(I) : I \nvDash B(\overrightarrow{a})\}$.

$N(B(\overrightarrow{a}))$ represents to what extent $B(\overrightarrow{a})$ is certain given the available knowledge. In standard DL-Lite, if the KB K is inconsistent, any assertion $B(\overrightarrow{a})$ follows trivially from K. This is not the case in the context of Pb-π-DL-Lite: when π_K is sub-normalized (K is inconsistent), the deductibility of $B(\overrightarrow{a})$ from K is defined by:

$$K \models_\pi B(\overrightarrow{a}) \quad iff \quad N_\pi(B(\overrightarrow{a})) > Inc(K). \tag{3}$$

where N_π is the necessity measure induced by π_K. The following proposition shows how to use the concept of inconsistency degree to answer instance checking queries.

Proposition 1. *Let $K = \langle T, A \rangle$ be a Pb-π-DL-Lite KB, B be a concept (resp. R be a role) and a, b be two individuals. D_B (resp. D_R) is an atomic concept (resp. an atomic role) not appearing in T. Then : (i) $N(B(a)) = Inc(K_1)$ (resp. $N(R(a, b)) = Inc(K_1)$) where $K_1 = \langle T_1, A_1 \rangle$ with $T_1 = T \cup \{(D_B \sqsubseteq \neg B, 1)\}$ (resp. $T_1 = T \cup \{(D_R \sqsubseteq \neg R, 1)\}$) and $A_1 = A \cup \{(D_B(a), 1)\}$ (resp. $A_1 = A \cup \{(D_R(a, b), 1)\}$); and (ii) $B(a)$ (resp. $R(a, b)$) is a consequence of K, denoted by $K \models_\pi B(a)$ (resp. $K \models_\pi R(a, b)$) if $Inc(K_1) > Inc(K)$.*

Proposition 1 gives a characterization of instance checking by using the concept of inconsistency degree: checking whether an assertion $B(a)$ (resp. $R(a, b)$) follows from a DL-Lite KB K comes down to comparing the inconsistency degree of K and that of the augmented KB K_1 obtained from K by adding the assumption that $B(a)$ (resp. $R(a, b)$) is surely false (expressed by means of a new concept D_B (resp. role D_R)).

4 Inconsistency Degree as a Weighted Set Cover Problem

This section presents an encoding of the inconsistency degree computation problem as a weighted set cover problem (W-SCP). The W-SCP (which is NP-Hard) is defined as follows: Given a collection S of sets over a universe U such that, to each set $s \in S$ is associated a weight $w_s \geq 0$. A set cover C is a sub-collection $C \subseteq S$ such that: $\bigcup_{s \in C} = U$. The goal is to find a set cover C of minimum total weight $\sum_{s \in C} w_s$.

This encoding is motivated by the existence of a greedy algorithm for polynomial approximation of W-SCP (see [14]). It builds a cover by repeatedly choosing a set s that maximizes the ratio between the number of elements in s not yet covered by chosen sets and the weight w_s. It has been shown in [14] that the greedy algorithm returns a set cover of weight at most H_k times the minimum weight of any cover where H_k denotes the kth harmonic number given by: $H(k) = \sum_{i=1}^{k} \frac{1}{i} \leq ln(k) + 1$.

Before defining the encoding, we need first to recall that an assertional conflict is a minimal sub-base C of A which is inconsistent with A. More precisely: (1) $C \subseteq A$; (2) $\langle T, C \rangle$ is inconsistent and (3) $\forall C' \subseteq C$, $T \cup C'$ is consistent.

It has been shown in [1] that each possible conflict in a DL-Lite KB involves at most two ABox assertions and the computation of all such conflicts is achieved in a polynomial time by using the negative closure of a DL-Lite KB (see [1] for more detail).

Now the inconsistency degree computation problem is encoded as a weighted set cover Problem (W-SCP) as follows:

Let $K = \langle T, A \rangle$ be a Pb-π-DL-Lite KB. Let ζ be the set of all conflicts in A and M be a sufficiently large natural number. Let F be a scale changing function defined by: $F(x) = -10^M * (ln(1 - x))$. The W-SCP instance associated to K is denoted SCP_K and defined as follows: (1) the universe is the set of conflicts ζ; (2) to each assertional fact $(X(\overrightarrow{a}), \alpha)$ of A is associated a set denoted $X_{\overrightarrow{a}}$ having the weight $w_{X_{\overrightarrow{a}}} = F(\alpha)$ and containing the conflicts of ζ involving $X(\overrightarrow{a})$: $X_{\overrightarrow{a}} = \{C \in \zeta \mid X(\overrightarrow{a}) \in C\}$ and (3) the collection S contains all the sets constructed from all the assertions of A: $S = \{X_{\overrightarrow{a}} \mid (X(\overrightarrow{a}), \alpha) \in A\}$.

Notice that the function F transforms each weight belonging to the unit interval $[0, 1]$ into an integer in $\mathbb{N}$. The inconsistency degree of a KB K corresponds to the solution of the corresponding W-SCP.

Proposition 2. *Let K be a Pb-π-DL-Lite KB, ζ be the set of its conflicts and SCP_K be the corresponding W-SCP. $Inc(K) = \alpha$ if and only if there is a set cover C of minimum total weight $\sum_{s \in C} w_s = F(\alpha)$.*

Proposition 2 is important since it allows one to use the greedy algorithm for W-SCP as an efficient tool to compute the inconsistency degree of a Pb-π-DL-Lite KB.

Example 1 (Cont). *Consider again the KB K of Example 1. The corresponding W-SCP is defined by (we take $M = 15$): The universe is the set ζ of conflicts where $\zeta = \{C_1, C_2\}$ with $C_1 = \{(teachesTo(Tom, Bob), F(0.75)), (Student(Tom), F(0.95))\}$ and $C_2 = \{(teachesTo(Tom, Bob), F(0.75)), (Teacher(Bob), F(0.85))\}$ and the used collection of sets is: $S = \{s_1, s_2, s_3\}$ where: $s_1 = \{C_1, C_2\}$ (resp. $s_2 = \{C_1\}$, $s_3 = \{C_2\}$) is the set associated to the assertion (teaches To(Tom, Bob), F(0.75)) (resp. (Student(Tom), F(0.95)), (Teacher(Bob), F(0.85))). The weight of s_1 (resp. s_2, s_3) is $w_{s_1} = F(0.75)$ (resp. $w_{s_2} = F(0.95)$, $w_{s_3} = F(0.85)$).*

It is easy to check that $F(0.75) < F(0.95) + F(0.85)$. It follows that the solution of our W-SCP is the collection $C = \{s_1\}$ and $Inc(K) = F^{-1}(F(0.75)) = 0.75$.

Algorithm 1 computes the inconsistency degree of a Pb-π-DL-Lite KB using the greedy method for W-SCP. To assess the quality of the result given by Algorithm 1, we have conducted an experiment which compares the exact value of the inconsistency degree and its approximate value found by Algorithm 1. For each number of digits (from 1 to 10) taken after the decimal point, we computed the success rate obtained over 400 random instances of Pb-π-DL-Lite KBs. The obtained results are depicted in Fig. 1. The X-axis represents the number of digits after the decimal point and the Y-axis represents the success rate (the percentage of times where the approximate value coincides with the exact one).

Figure 1 shows that the results are very encouraging. Indeed, for one digit after the decimal point, we obtained excellent results: about 99.8 % of total success, for two digits we obtained about 94.4 % of total success. even by taking eight digits after the decimal point, the score remains relatively high (81.2 %).

Algorithm 1. *Incons_Degree*(T, A)

1: **Input:** $K = \langle T, A \rangle$: a Pb-π-DL-Lite KB
2: **Output:** Inc: approximate value of the inconsistency degree of K
3: $\zeta \leftarrow Compute\_Conflicts(T, A)$; $\triangleright$ Compute the conflicts occurring in the KB $\langle T, A \rangle$
4: $Cost \leftarrow 0$; $B \leftarrow Extract\_Assertions(\zeta)$; $\triangleright$ Extract the assertions occurring in ζ
5: **while** $(B \neq \emptyset)$ **do**
6:    $max \leftarrow 0$;
7:    **for** (every $X \in B$) **do**
8:       $Nb\_conf \leftarrow Nb\_conf(\zeta, X)$; $\triangleright$ Number of conflicts of ζ involving the assertion X
9:       **if** ($max < Nb\_conf/w_X$) **then**
10:          $max \leftarrow Nb\_conf/w_X$; $\phi \leftarrow X$;
11:    $Cost \leftarrow Cost + w_\phi$; $\triangleright$ Take the assertion ϕ which maximizes the ratio
12:    $\zeta \leftarrow Remove\_Conflicts(\zeta, \phi)$; $\triangleright$ Remove from ζ all the conflicts involving ϕ
13:    $B \leftarrow Extract\_Assertions(\zeta)$; $\triangleright$ Recompute the assertions involved in the new ζ
14: $Inc \leftarrow F^{-1}(Cost)$;
15: **return**(Inc)

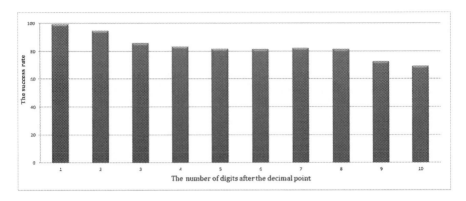

Fig. 1. Evaluation of the approximate algorithm for the inconsistency degree computation

5   Approximate Algorithm for Instance Checking

Let us now turn to the instance checking issue. Algorithm 2 decides whether an assertion $B(\overrightarrow{a})$ follows from a Pb-π-DL-Lite KB K based on Proposition 1: (1) it takes as input a Pb-π-DL-Lite K and its approximate inconsistency degree $Inc(K)$ (computed by Algorithm 1), (2) it constructs the augmented KB K_1 from K by adding the fact that $B(\overrightarrow{a})$ is false, (3) it computes the approximate inconsistency degree of K_1, $Inc(K_1)$ and (4) answers the query (yes/no) by comparing $Inc(K)$ and $Inc(K_1)$.

We have conducted a second experiment in order to (1) evaluate the results of Algorithm 2 and (2) validate our choice of using the approximate algorithm instead of the exact one to compute the consistency degree of K.

We randomly generated a large number of pairs (KB, Query). We distinguished three classes of KBs according to their size: small, medium and large. Then we computed the success rate of Algorithm 2 with respect to each class of KBs in addition to the global class containing all the generated pairs. We also compare results of Algorithm 2 with those obtained when using the exact values for the inconsistency degree of K and approximate values for the inconsistency degree of K_1. The obtained results are depicted in Fig. 2. The X-axis represents the KB size. For each of the two algorithms, the Y-axis represents the percentage of times the algorithm gives the correct answer.

Figure 2 shows that Algorithm 2 gives satisfactory results that are better than those obtained when using the exact value of the inconsistency degree of the input KB K. Notice that Algorithm 2 also distinguishes two preliminary cases where the exact answer is directly provided: If no conflict is added, then clearly $Inc(K) = Inc(K_1)$ and the algorithm answers *No* (lines 7–8) and if at least one added conflict involves the free part of K, i.e., the assertions of A that are not involved in any conflict in K, then clearly Q is a consequence of the free part of K and the algorithm answers *Yes* (lines 10–11).

Algorithm 2. Approximate algorithm for instance checking

1: **Input:** $K = \langle T, A \rangle$: a Pb-π-DL-Lite KB; Q: instance checking query; Inc: $Inc(K)$
2: **Output:** Boolean answer : YES / NO
3: $K_1 = \langle T_1, A_1 \rangle \leftarrow K \cup \neg Q;$ ▷ Constructing the augmented KB K_1 as in Proposition 1
4: $\zeta \leftarrow Compute\_Conflicts(T, A);$ $\zeta_1 \leftarrow Compute\_Conflicts(T_1, A_1);$
5: $Free \leftarrow$ *The set of assertions in A not involved in ζ;*
6: **if** $(|\zeta_1| = |\zeta|)$ **then**
7:      $Answer \leftarrow NO;$ ▷ Q is surely not a consequence of K
8: **else**
9:      **if** (ζ involves elements of $Free$) **then**
10:          $Answer \leftarrow YES;$ ▷ Q is surely a consequence of K
11:      **else**
12:          **if** $(Incons\_Degree(T_1, A_1) > Inc)$ **then**   ▷ Compare $Inc(K)$ and $Inc(K_1)$
13:              $Answer \leftarrow YES;$
14:          **else**
15:              $Answer \leftarrow NO;$
16: **return** $(Answer);$

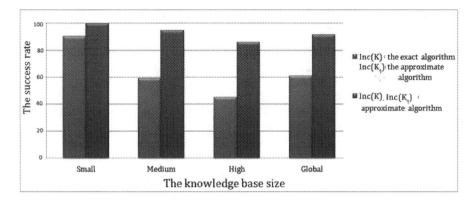

Fig. 2. Evaluation of the instance checking algorithm (Color figure online)

6   Conclusion and Future Work

In this paper, we presented two approximate and polynomial algorithms for Pb-π-DL-Lite, an extension of lightweight ontologies, encoded in DL-Lite language, to the product-based possibility theory framework. As the problem of query answering is not tractable in Pb-π-DL-Lite, the paper proposes polynomial approximate algorithms to compute the inconsistency degree of a KB and to use this degree to answer instance checking queries. The two algorithms have been validated by experimental studies. As a future work we plan to generalize our framework to general conjunctive queries and to arbitrary DL-Lite KBs where the TBox axioms are not necessarily fully certain.

Acknowledgments. This work has been supported by the european project H2020 Marie Sklodowska-Curie Actions (MSCA) research and Innovation Staff Exchange (RISE): AniAge (High Dimensional Heterogeneous Data based Animation Techniques for Southeast Asian Intangible Cultural Heritage and from ASPIQ project reference ANR-12-BS02-0003 of French National Research Agency.

References

1. Artale, A., Calvanese, D., Kontchakov, R., Zakharyaschev, M.: The DL-lite family and relations. J. Artif. Intell. Res. **36**, 1–69 (2009)
2. Baader, E., Calvanese, D., McGuinness, D.L., Nardi, D., Patel-Schneider, P.F. (eds.): The Description Logic Handbook: Theory, Implementation, and Applications. Cambridge University Press, New York (2003)
3. Baader, F., McGuinness, L., Nardi, D., Patel-Schneider, P.F.: The description logic handbook: Theory, implementation, and applications. Cambridge University Press, Cambridge (2003)
4. Benferhat, S., Bouraoui, Z.: Min-based possibilistic DL-Lite. J. Logic Comput. (2015). doi:10.1093/logcom/exv014. First published online 12 April 2015
5. Bobillo, F., Delgado, M., Gómez-Romero, J.: Delorean: a reasoner for fuzzy OWL 2. Expert Syst. Appl. **39**(1), 258–272 (2012)
6. Bobillo, F., Delgado, M., Gómez-Romero, J.: Reasoning in fuzzy OWL 2 with DeLorean. In: Bobillo, F., Costa, P.C.G., d'Amato, C., Fanizzi, N., Laskey, K.B., Laskey, K.J., Lukasiewicz, T., Nickles, M., Pool, M. (eds.) URSW 2008-2010/UniDL 2010. LNCS, vol. 7123, pp. 119–138. Springer, Heidelberg (2013)
7. Calvanese, D., De Giacomo, G., Lembo, D., Lenzerini, M., Rosati, R.: Tractable reasoning and efficient query answering in description logics: the DL-Lite family. J. Autom. Reasoning **39**(3), 385–429 (2007)
8. Darwiche, A.: Modeling and Reasoning with Bayesian Networks. Cambridge University Press, Cambridge (2009)
9. Dubois, D., Prade, H.: Possibility theory. Plenum Press, New York (1988)
10. Hollunder, B.: An alternative proof method for possibilistic logic and its application to terminological logics. Int. J. Approximate Reasoning **12**(2), 85–109 (1995)
11. Lang, J.: Possibilistic logic: complexity and algorithms. In: Gabbay, D., Smets, P. (eds.) Handbook of Defeasible Reasoning and Uncertainty Management Systems, vol. 5, pp. 1791–7220. Springer, Heidelberg (2000)
12. Qi, G., Ji, Q., Pan, J.Z., Du, J.: Extending description logics with uncertainty reasoning in possibilistic logic. Int. J. Intell. Syst. **26**(4), 353–381 (2011)
13. Straccia, U.: Foundations of Fuzzy Logic and Semantic Web Languages. CRC Press, Boca Raton (2013)
14. Young, N.E.: Greedy set-cover algorithms: 1974–1979, Chvátal, Johnson, Lovász, Stein. In: Kao, M.-Y. (ed.) Encyclopedia of Algorithms, pp. 379–381. Springer, Heidelberg (2008)

A Survey on Ontologies and Ontology Alignment Approaches in Healthcare

Vladimir Dimitrieski[1(✉)], Gajo Petrović[2], Aleksandar Kovačević[1],
Ivan Luković[1], and Hamido Fujita[2]

[1] Faculty of Technical Sciences, University of Novi Sad, Novi Sad, Serbia
{dimitrieski,kocha78,ivan}@uns.ac.rs
[2] Iwate Prefectural University, Takizawa, Iwate, Japan
gajopetrovic@gmail.com, issam@iwate-pu.ac.jp

Abstract. In the era of Internet, high connectivity and openness introduced an opportunity for a new kind of approach to healthcare information system integration. Such an approach may utilize semantic-based technologies to represent and communicate knowledge between these systems. Resource Description Framework (RDF) in conjunction with Web Ontology Language (OWL) can be considered as a de facto standard when it comes to semantic web and linked data technologies, and represents a foundation for defining healthcare ontologies. The goal of this paper is to provide an overview and critical review of existing healthcare ontologies and approaches to healthcare IS integration, focusing on OWL/RDF based solutions. With this review we want to show that although a lot work is done in this area, no universal or omnipresent solution has surfaced to allow automatic or at least semi-automatic integration of healthcare ISs. As there is a large number of established and emerging ontologies covering this subject our review will not provide an exhaustive collection of all the references in the area, but present the most notable standards, ontologies, taxonomies, and integration approaches.

Keywords: Survey · Healthcare ontology · Ontology alignment · Ontology matching · Semantic web

1 Introduction

In recent years, with informatics being omnipresent in medicine, an increased number of healthcare term representations were created. Such representations are used to systemically denote, categorize, and relate healthcare data, allowing easier handling of the data in healthcare information systems (healthcare IS). The coexistence of multiple representations introduced a major problem in healthcare IS development: the problem of healthcare IS integration. In order for an healthcare IS to fulfill its purpose of assisting medical personnel in their activities, it must be able to exchange data with other healthcare ISs. For example, patients medical history is transferred when they change a place of residence, anonymized data is provided to research facilities, and so on.

© Springer International Publishing Switzerland 2016
H. Fujita et al. (Eds.): IEA/AIE 2016, LNAI 9799, pp. 373–385, 2016.
DOI: 10.1007/978-3-319-42007-3_32

Many standards have been created to allow interchange of data and integration of such healthcare ISs, usually focusing on low level protocols and predefined message formats. In the era of Internet, high connectivity and openness introduced an opportunity for a different kind of integration approach. Such an approach may utilize semantic-based technologies to represent and communicate knowledge between the healthcare ISs. Ontologies are often used as a way of representing such knowledge. Two main reasons for using them are their ability to capture healthcare knowledge in a formal way and an easy application of reasoning processes that is performed by a medical decision support system. Resource Description Framework (RDF)[1] in conjunction with Web Ontology Language (OWL)[2] can be considered as a de facto standard when it comes to semantic web and linked data technologies, and represents a foundation for defining healthcare ontologies. Despite the popularity of OWL/RDF format, systems are often centered around traditional eXtensible Markup Language (XML) technology and relational databases, partly because of good validation tools and support from major manufacturers. Majority of traditionally used data representation languages offer some sort of input validation like parsing grammars and metamodels for domain specific languages, XML Schema (XSD) for XML, and Data Definition Language for Structured Query Language (SQL). These properties of traditional systems can be preserved while using the OWL/RDF technology as an additional layer of integration [13], in order to obtain a semantically rich representation of underlying knowledge.

OWL/RDF based integration approaches are not a new idea and are evidenced both in many research papers published in previous years and many projects and movements such as the Yosemite manifest[3], SemanticHealthNet[4], and Clinical Information Modeling Initiative[5]. The goal of this paper is to provide an overview and a critical review of existing healthcare ontologies and approaches to healthcare IS integration, with a focus on OWL/RDF based solutions. With this review we want to show that although a lot work is done in this area, no universal or omnipresent solution has surfaced to allow automatic or at least semi-automatic integration of healthcare ISs. Solutions are usually confined to a specific part of a healthcare domain and are used for specific use cases. As there is a large number of established and emerging ontologies covering this subject, our review will not provide an exhaustive collection of all the references in the area, but present the most notable standards, ontologies, taxonomies, and integration approaches.

Apart from Introduction and Conclusion, the paper is organized in three sections. In Sect. 2, we give an overview of currently used standards in healthcare domain. We also provide information about existing OWL/RDF ontologies that describe these standards. Existing approaches to integration of healthcare ontologies are described in Sect. 3. In Sect. 4, we discuss related work.

---

[1] https://www.w3.org/RDF/.

[2] https://www.w3.org/2001/sw/wiki/OWL.

[3] http://yosemitemanifesto.org/.

[4] http://www.semantichealthnet.eu/.

[5] http://www.opencimi.org/.

2 An Overview of EHR Standards and Healthcare Ontologies

Currently, there is a plethora of electronic health record (EHR) standards covering many aspects of healthcare including the management of clinical records. These standards cover exchange of messages and patient data between healthcare institutions, integration of medical devices, interfaces with clinical decision support systems, etc. Examples of such standards that are trying to prescribe common building blocks of EHRs are: Health Level Seven (HL7)[6], in particular the Clinical Document Architecture (CDA) part of the standard; CEN/ISO EN13606[7]; openEHR[8]; and The Clinical Element Model (CEM) [7]. These share some common elements and visions of how an EHR should be structured and how a system should be implemented [32]. However, as these standards are managed by different companies from different countries, the specifications diverged significantly. This makes the integration and exchange of data between systems implementing two different standards a major issue to be addressed.

Before starting with the integration process, one must gain a deeper knowledge of underlying EHR elements from the aforementioned standards. Additionally, as we plan the approach based on the OWL/RDF, it is necessary to find all ontologies implemented in these technologies for each of the standards. Furthermore, it is important to find health ontologies not implemented according to these standards as to have wider picture of the current state-of the-art in this area.

In this Section, we present standards that are directly related with information architecture for communicating patient EHRs. Where applicable, we will give an overview of OWL/RDF based ontologies related to each of the standards.

2.1 Health Level Seven (HL7) Suite

Health Level Seven International (HL7) is a non-profit organization with a goal to develop healthcare standards in order to increase interoperability of healthcare information systems. It created a set of standards for the exchange, integration, sharing, and retrieval of electronic health information. These standards aim to support clinical practice and the management, delivery and evaluation of health services. The current version of the standard suite is the HL7 Version 3 which is centered around HL7 Reference Information Model (RIM)[9]. RIM is an object model representing HL7 clinical data (domains) and identifies the life cycle that a message or groups of related messages will carry. In addition to the RIM, other HL7 standards specify elements whose data is of different types. Therefore, an additional standard named HL7 V3 Data Types (DT) is created and extensively used throughout the standard suite.

---

[6] http://www.hl7.org/.

[7] http://www.en13606.org/.

[8] http://www.openehr.org/.

[9] http://www.hl7.org/implement/standards/rim.cfm.

A part of HL7 which aims to standardize EHRs structure and semantics is Clinical Document Architecture (CDA) document markup standard. The machine derivable meaning of CDA components is defined using the HL7 RIM, HL7 DT, and by referencing a shared medical terminology such as The Systematized Nomenclature of Medicine Clinical Terms (SNOMED CT)[10]. By using such a shared terminology not only does it allow a CDA to be interoperable with other standards using the same terminology, but the SNOMED CT is also used for representing the semantics of clinical documents. The CDA uses eXtensible Markup Language (XML) for the specification of appropriate clinical information with references to SNOMED CT terms which is a way to improve interoperability with other standards using the same terminology.

Because of the benefits of OWL/RDF, in regard to universal knowledge representation and automatic inference of rules, many papers are dealing with the representation of HL7 RIM and HL7 DT in this technology [11,14,28]. Of all the papers, [14] offers the most complete and most recent implementation of HL7 RIM. In [27] an ontology for HL7v3 was developed. However, due to some reported shortcomings of HL7 from the ontological viewpoint [35,38,39] and new explanations provided by the authors of HL7 RIM in [31], we feel that the HL7 RIM OWL implementation must be further developed and refined from the ontological point of view. These RIM-based ontologies can be considered as meta-ontologies that are used as a building blocks of all other ontologies. In such a way, an ontology of a specific part of EHR can be modeled using the classes specified in RIM ontology. One of such examples is the definition of a dental ontology in [8]. Developing the CDA ontology is the important step in providing means to integrate the HL7-based systems with the systems implementing other standards. The CDA can be used in ontology based integration solutions in two ways: CDA-based XML documents may be transformed into RDF triplets (iSMART approach [19]), and developing CDA ontology with OWL/RDF tools [12].

HL7 set of standards is also being extended with the Fast Healthcare Interoperability Resources (FHIR)[11]. In the development of the FHIR currently popular web technologies are used. Therefore implementing FHIR using OWL is one of the main goals of HL7 International. This will have a positive influence on the ontology-based integration approaches as HL7 OWL/RDF implementations of FHIR concepts are added to the third-party implementations such as [2,21].

2.2   openEHR and CEN/ISO EN13606 Standards

openEHR is an open-standard set of specifications in healthcare informatics with the aim to standardize management, storage, retrieval, and exchange of health data in EHRs. It follows the dual-model approach [1] that differentiates between two levels: (i) information level represented by a reference model specifying statements that are applied to all entities of the same class and (ii) knowledge level

---

[10] https://www.nlm.nih.gov/research/umls/Snomed/snomed_main.html.
[11] https://hl7-fhir.github.io/index.html.

which is represented through archetypes that are statements about specific entities. The openEHR standard provides both functional and semantic interoperability, allowing for it to be read and processed by both humans and machines respectively.

The Electronic Health Record Communication (EN13606) is a European norm from the European Committee for Standardization (CEN) that specifies the normative for exchanging patient records between EHR systems. Although a stand-alone standard, it can be viewed as a subset of openEHR. Both EN13606 and openEHR follow the dual-model approach, however slightly different archetypes were defined. Transformations between archetypes were developed in [22] allowing future ontology-based integration approaches to consider the two standards in the following ways: (i) separately, where the ontology of each of the solutions should be developed or used if already exists, and (ii) together, using one ontology and transforming the EHRs described in one standard to the other using the predefined archetype transformations.

openEHR OWL ontology covering reference model, data types, and data structures of the openEHR was developed by Roman[12]. Also, authors of [24] have developed an OWL ontology of the archetype library by following the guidelines from official openEHR specification. Therefore, first two ontologies can be viewed as a single, complete, ontology that can be used in the integration approaches. An ontology for both EN13606 and openEHR were developed in [16] as a part of an attempt to provide semantic interoperability between the standards. However, more complete ontology for EN13606 was developed by the authors of [30] while trying to develop an architecture comprising of different EHR systems capable of inter-operating so as to offer an integrated service using interoperability patterns based on EN13606 and the semantic technologies such as OWL. A partial OWL ontology for the definition and validation of archetypes was also developed in [25].

2.3   The Clinical Element Model (CEM)

The goal behind the development of the Clinical Element Model (CEM) [7] was to provide a single, referent, architecture for representing information in EHRs. CEM comprises two models: *Abstract instance model* for representing individual instances of collected data, and *Abstract constraint model* for representing constraints on the data instances. These two models are abstract specifications and can be implemented using different programming languages. The main purpose of such an abstract implementation is to provide a way to normalize different data from EHRs.

Originally, CEM was implemented using the Clinical Element Modeling Language (CEML) [7] which has a XML-like syntax. Additional implementation was in Constraint Definition Language (CDL) [15] that extends CEML to allow specification of new constraints to the modeling language. In order for CEM to be integrated using OWL/RDF technologies, the CEM-OWL ontology is developed by the authors of [37]. They have also developed an automatic transformation

---

[12] http://trajano.us.es/~isabel/EHR/.

from CEM-XML specification to the corresponding CEM-OWL specification. This will allow future researchers to focus more on the process of integration than on data acquisition.

2.4   Other Healthcare Ontologies and Vocabularies

In addition to the previously described ontologies, that are based on widely used standards, several other healthcare ontologies were developed. These ontologies usually focus on some specific parts of EHR, but it could be beneficial, in the context of globally transferable healthcare data, to integrate systems that are using these ontologies.

Open Biomedical Ontologies (OBO) [34] is a set of ontologies developed and maintained by the scientific community with a goal to allow easier representation and integration of biomedical data. To clarify the terminology, the biomedical domain is broader than just the healthcare domain as it comprises other knowledge not only specific to patient medical care and EHRs. Disease Ontology (DO) [33] is a part of OBO repository and can be used to describe patient disease history in EHRs. A benefit of using this ontology is a fact that it is heavily referencing SNOMED and other medical thesauri. Another ontology form the OBO repository is the Gene Ontology (GO) [10] that provides structured, controlled vocabularies and classifications used in the annotation of genes, gene products, and sequences.

The Foundational Model of Anatomy Ontology (FMA)[13] is the representation of classes or types and relationships necessary for the symbolic representation of the phenotypic structure of the human body. FMA is a domain ontology that represents a coherent body of explicit declarative knowledge about human anatomy but can be also applied and extended to all other species.

Due to the lack of a common vocabulary, healthcare ontologies often reference terms from various existing vocabularies. Vocabulary standards are used to describe clinical problems, terms, categories, procedures, medications, and allergies. Various medical vocabulary standards exist and in order to implement a usable healthcare standard interoperability, these vocabularies must be also taken into the consideration. We have already described SNOMED-CT in Subsect. 2.1. Medical Subject Headings (MeSH) [18] consists of sets of terms naming descriptors in a hierarchical structure that permits searching at various levels of specificity. It is not an ontology per se, but it is referenced from a vast majority of ontologies as to provide classification and categorization of the biomedical terms. Therefore, it should be considered in all integration approaches as it provides structure and hierarchical information about the medical categories. MeSH is often used in conjunction with RxNorm [20], a pharmaceutical vocabulary used for e-prescribing, medication history, government reporting, and drug compendium mapping, and Logical Observation Identifiers Names and Codes (LOINC) [23], a database and universal standard for identifying medical laboratory observations. Another classification commonly referenced from ontologies is

---

[13] http://sig.biostr.washington.edu/projects/fm/AboutFM.html.

International Statistical Classification of Diseases and Related Health Problems (ICD) [40]. It is a medical classification provided by the World Health Organization (WHO) and contains codes for diseases, signs and symptoms, abnormal findings, complaints, social circumstances, and external causes of injury or diseases. More information about these vocabularies can be found in [29]. Another thesauri not covered by [29] is provided by the National Cancer Institute (NCIt)[14]. NCIt is a widely recognized standard for biomedical coding and reference, used by a broad variety of public and private partners both in the U.S. and internationally.

The Unified Medical Language System (UMLS) [4] aims to alleviate the problem that exists when using multiple vocabularies in a healthcare informatics system. UMLS comprises several controlled vocabularies in the biomedical sciences including SNOMED-CT, ICD, RxNorm, etc. It provides a mapping structure among these vocabularies and it may also be viewed as a comprehensive thesaurus and ontology of biomedical concepts. Therefore, it is often referred as the UMLS meta-thesaurus. Although it provides a mapping structure, it does not make semantically integrated terminology interoperable. However, it provides enough information about term relations to be used in an integration process. Additionally, UMLS provides facilities for natural language processing.

3   Approaches to the Integration of Health Ontologies

Ontology alignment in healthcare has been an issue of research for many years, with a wide variety of approaches. The main differences is the degree of automation in the process of integration, as well as the number of ontologies they cover. Ideally we would prefer systems that offer a high degree of automation, can cover multiple ontologies and even offer support for multi-domain ontology alignment. Below we give an overview of some of the different approaches and their key characteristics.

The authors of [3] have proposed a system called Artemis Message Exchange Framework (AMEF) for mediating messages between HL7 v2 and HL7 v3, which they claim might be generalized to any two different healthcare standards. The ontology alignment itself is solved manually, while the main focus of the paper is in the format conversion (HL7 V2's EDI → XML → OWL and OWL → XML → HL7 V3 message).

In paper [9] authors have attempted to integrate SNOMED-CT and disease ontologies. Relying on UMLS meta-thesaurus as a reference to disambiguate the term meaning, they have calculated the semantic similarity of concepts using Wu and Palmer's algorithm and Jiang Coranth's semantic similarity measure. This paper exemplifies usage of a reference knowledge base (the UMLS meta-thesaurus), which is a pattern followed in other papers recently.

In paper [17] authors have used HL7 and openEHR ontologies and created an ontology matching system by applying well-known tools Falcon[15] and Agreement

---

[14] https://ncit.nci.nih.gov/ncitbrowser/.

[15] http://ws.nju.edu.cn/falcon-ao/.

Maker[16] to the healthcare domain. Although Agreement Maker is primarily used for the biomedical domain, both of these systems are general purpose ontology matching tools which gives some hope that part of this problem can be generalized to different domains.

There have also been attempts to merge healthcare ontologies with ontologies from different domains. This kind of use case is interesting even for healthcare institutions that rely on a single healthcare IS provider, as it allows integration with systems that are not directly related with healthcare. One such paper [36] proposes a method to merge three multi-disciplinary ontologies related to diseases, places, and environments, relying on expert knowledge and statistical data analysis.

There are also cases of purely automated ontology matching systems. In one such example [26] authors present a machine learning based approach to ontology alignment, using the AdaBoost ensemble technique. This is done by training the system on a similarity matrix computed by one of the similarity methods (string-based, linguistic, and structural).

Even though these papers cover a majority of the ontologies mentioned in the previous section, not one of these approaches can overcome the problem of aligning ontologies implemented in different technologies. Each of these approaches works decently on its problem ontologies, but there is still a need for a universal solution. Therefore, we should utilize the advantages of these approaches in a single solution and transform different technologies into one. Currently, there are movements such as the Yosemite initiative that aim to create or at least propose a theoretical background for a unified solution, by using the OWL/RDF technology as a common ground for ontology alignment.

The Yosemite initiative suggests a two-step approach to healthcare ontology integration: (1) transforming any ontology format to OWL/RDF and (2) creating an integration algorithm for two OWL/RDF ontologies. Once the ontologies are transformed to OWL/RDF representation, in order to implement the second step, one must create an ontology alignment algorithm. Although this is the hardest and the most work-intensive part, it is always easier to preform integration and create more general solutions for a single representation technology than to create transformations on by-representation basis. Therefore the first step of the approach is the prerequisite to have such an universal technology, and due to that fact it is a subject of numerous discussions and criticism. The main issue concerns the choice of OWL/RDF for the universal representation technology for both ontologies and alignments/mapping between ontology concepts. There are several reasons of why the OWL/RDF is a suitable technology [5]:

- It is possible to map any other representation to RDF. RDF is made up of atomic statements (triplets of subject, object, and predicate). As triplets are atomic pieces of information, all other more complex information can be implemented by a set of triplets. Sometimes, this may lead to more verbose representations.

---

[16] https://github.com/AgreementMakerLight/AML-Jar

- RDF captures information, not syntax. Therefore, many different syntaxes (XML, Json, etc.) may be used to provide serialization for triplets. Therefore, usual storage mechanisms may be used for RDF-based solutions.
- RDF is self describing as it uses Uniform Resource Identifiers (URIs) as main identifiers. This reduces ambiguity and allows the creation of term definitions to be referenced by any other documents. This reduces ambiguity and allows single points of knowledge.
- OWL/RDF enables inference that derives new assertions from existing ones. This can lead to more automation of data translation processes.

In addition to these benefits, the sheer fact that the most notable healthcare ontologies and standards have their official RDF representation, or are getting RDF implementation for their next release (FHIR), is in favor to the claim that the RDF is a valid option. In the end, we feel that using the OWL/RDF can lead to the great reduction in complexity and more order in the already very confusing world of healthcare ontologies.

4   Related Work

While there have been a number of survey papers in the fields of ontology alignment as well as about healthcare ontologies in general, the number of papers focusing strictly on the state of the art healthcare ontology alignment systems is far smaller.

The authors of [29] mention five different ontologies and taxonomies. They also divide ontology alignment into three major categories by purpose: (1) mapping a global ontology view to a local ontology view in order to describe a proprietary local ontology better (2) semantic mapping between parts of the local and the target ontology and (3) mapping multiple ontologies in order to provide new knowledge not contained within the separate instances.

Paper [6] provides a slightly more detailed overview. Authors mention a wide variety of mappings between some two concrete ontology types, a lot of which base their approach on using the UMLS meta-thesaurus. They also show that it is possible to do ontology mapping by relying on First Order Logic (FOL).

Both papers deal the issue of aligning ontologies without tackling the problem of ontology format. They do not provide their thoughts on benefits and drawbacks of an approach to ontology alignment where all ontologies share a common implementation technology such as OWL/RDF.

5   Conclusion

Ontology alignment in the healthcare domain is far from solved. There is a large number of ontologies, vocabularies and taxonomies, and even though there are attempts such as HL7 to standardize and cover most of the healthcare field, it does not seem like there is going to be a single general ontology any time soon. Far more likely, we'll continue to see new ontologies appearing and some of the old ones disappearing for quite a while. Furthermore, even if there was a single healthcare ontology, alignment between multi-disciplinary ontologies remains a problem. Therefore, it's extremely important to consider creating automatic ontology alignment algorithms, as not only is the process of manual ontology matching hard, time consuming, and error prone, it's also not going to offer a complete solution simply because of the ever increasing number of different ontologies.

To better understand the current state of the art of automated algorithms in ontology alignment, an initiative (OAEI[17]) to benchmark them has been devised. Of particular interest to the healthcare field is the *largebio* challenge, featuring the alignment detection problem between three ontologies (FMA, NCIt, and SNOMED-CT) using UMLS as a meta-thesaurus. The last competition (OAEI2015) had 12 participating groups in this challenge category, and further, the benchmark data and framework provided by OAEI can also be used after the competition, as was done in [26].

The main healthcare ontologies today are defined using a number of different formats. Therefore, writing an automated tool that would work between any two ontologies is a challenging task due to the number of possible combinations. We think that in order to solve this problem a common ontology format should be used. Our stance is in accordance with the Yosemite initiative that proposes OWL/RDF as a common ontology representation format. The Yosemite initiative also suggests a two-step approach to healthcare ontology integration: (1) transforming any ontology format to OWL/RDF and (2) creating an integration algorithm for two OWL/RDF ontologies. We think that following such an approach to integration will lead to simplifying the currently complicated field of ontology alignment.

Acknowledgments. This research is supported by the Ministry of Education, Science, and Technological Development of Republic of Serbia, Grant III–47003: "Infrastructure for technology enhanced learning in Serbia" and Grant III–44010: "Intelligent Systems for Software Product Development and Business Support based on Models.".

References

1. Beale, T.: Archetypes: constraint-based domain models for future-proof information systems. In: Workshop on Behavioural Semantics, OOPSLA, vol. 105, Seattle, USA, November 2002 (2002)
2. Beredimas, N., Kilintzis, V., Chouvarda, I., Maglaveras, N.: A reusable ontology for primitive and complex HL7 FHIR data types. In: 2015 37th Annual International Conference of the IEEE Engineering in Medicine and Biology Society (EMBC), August 2015, pp. 2547–2550. IEEE (2005)
3. Bicer, V., Laleci, G.B., Dogac, A., Kabak, Y.: Artemis message exchange framework: semantic interoperability of exchanged messages in the healthcare domain. ACM Sigmod Rec. **34**(3), 71–76 (2005)

---

[17] Ontology Alignment Evaluation Initiative - http://oaei.ontologymatching.org/.

4. Bodenreider, O.: The unified medical language system (UMLS): integrating biomedical terminology. Nucleic Acids Res. **32**(1), 267–270 (2004)
5. David Booth, C., Huff, S.M., Fry, E., Dowling, C., Mandel, J.C.: RDF as a Universal Healthcare Exchange Language (2013)
6. Cardillo, E.: Mapping between international medical terminologies, May 2015
7. Coyle, J., Heras, Y., Oniki, T., Huff, S.: Clinical element model. Technical report, University of Utah (2008)
8. Feng, L., Li, Y., Liu, L., Ye, X., Wang, J., Cao, Z., Peng, X.: Towards an applied oral health ontology: a round trip between clinical data and experiential medical knowledge. In: Proceedings of 39th Annual Computer Software and Applications Conference (COMPSAC), vol. 3, pp. 288–295, July 2015. IEEE (2015)
9. Olaronke, I., Soriyan, H., Abimbola, G.P., Gambo, I.: Resolving semantic heterogeneity in healthcare: an ontology matching approach. J. Comput. Sci. Eng. **17**(2) (2013)
10. Gene Ontology Consortium: The Gene Ontology (GO) database and informatics resource. Nucleic Acids Res. **32**(1), 258–261 (2004)
11. Chen, H.: HL7 RIM V3 OWL Ontology for Semantic Web
12. Heymans, S., McKennirey, M., Phillips, J.: Semantic validation of the use of SNOMED CT in HL7 clinical documents. J. Biomed. Semant. **2**(1), 2 (2011)
13. Hoffmann, A., Kang, B.-H., Richards, D., Tsumoto, S. (eds.): PKAW 2006. LNCS (LNAI), vol. 4303. Springer, Heidelberg (2006)
14. Iqbal, A.M.: An OWL-DL ontology for the HL7 reference information model. In: Abdulrazak, B., Giroux, S., Bouchard, B., Pigot, H., Mokhtari, M. (eds.) ICOST 2011. LNCS, vol. 6719, pp. 168–175. Springer, Heidelberg (2011)
15. James, A.: Qualibria Constraint Definition Language (CDL). Technical report (2010)
16. Fernández-Breis, J.T., Vivancos-Vicente, P.J., Menárguez-Tortosa, M., Moner, D., Maldonado, J.A., Valencia-García, R., Miranda-Mena, T.G.: Using semantic technologies to promote interoperability between electronic healthcare records' information models. In: Proceedings of the 28th IEEE EMBS Annual International Conference, New York City, USA, pp. 2614–2617. IEEE (2006)
17. Khan, W.A., Khattak, A.M., Lee, S., Hussain, M., Amin, B., Latif, K.: Achieving interoperability among healthcare standards: building semantic mappings at models level. In: Proceedings of the 6th International Conference on Ubiquitous Information Management and Communication, New York, USA, p. 101. ACM (2012)
18. Lipscomb, C.E.: Lipscomb: Medical Subject Headings (MeSH). Bull. Med. Libr. Assoc. **88**(3), 265–266 (2000)
19. Liu, S., Ni, Y., Mei, J., Li, H., Xie, G., Hu, G., Liu, H., Hou, X., Pan, Y.: iSMART: ontology-based semantic query of CDA documents. AMIA Annu. Symp. Proc. **2009**, 375–379 (2009)
20. Liu, S., Ma, W., Moore, R., Ganesan, V., Nelson, S.: RxNorm: prescription for electronic drug information exchange. IT Prof. **7**(5), 17–23 (2005)
21. Luz, M.P., Rocha de Matos Nogueira, J., Cavalini, L.T., Cook, T.W.: Providing full semantic interoperability for the fast healthcare interoperability resources schemas with resource description framework. In: Proceedings of 2015 International Conference on Healthcare Informatics (ICHI), Dallas, TX, October 2015, pp. 463–466. IEEE (2015)

22. Martínez-Costa, C., Menárguez-Tortosa, M., Fernández-Breis, J.T.: An approach for the semantic interoperability of ISO EN 13606 and OpenEHR archetypes. J. Biomed. Inform. **43**(5), 736–746 (2010)
23. McDonald, C.J., Huff, S.M., Suico, J.G., Hills, G., Leavelle, D., Aller, R., Forrey, A., Mercer, K., DeMoor, G., Hook, J., et al.: LOINC, a universal standard for identifying laboratory observations: a 5-year update. Chem. Clin. **49**(4), 624–633 (2003)
24. Menárguez-Tortosa, M., Fernández-Breis, J.T.: Semantic interoperability for better health and safer healthcare: research and deployment roadmap for Europe. Semantic health report January 2009. In: Proceedings of User Centred Networked Health Care, Luxembourg. EUR-OP (2009)
25. Menárguez-Tortosa, M., Fernández-Breis, J.T.: OWL-based reasoning methods for validating archetypes. J. Biomedical Inf. **46**(2), 304–317 (2013)
26. Nejhadi, A.H., Shadgar, B., Osareh, A.: Ontology alignment using machine learning techniques. Int. J. Comput. Sci. Inf. Technol. **3**(2), 139–150 (2011)
27. Oemig, F., Blobel, B.: An ontology architecture for HL7 V3: pitfalls and outcomes. In: Dössel, O., Schlegel, W.C. (eds.) IFMBE Proceedings of the World Congress on Medical Physics and Biomedical Engineering, vol. 12, pp. 408–410. Springer, Heidelberg (2009)
28. Orgun, B., Vu, J.: HL7 ontology and mobile agents for interoperability in heterogeneous medical information systems. Comput. Biol. Med. **36**(7–8), 817–836 (2006)
29. Puri, C.A., Gomadam, K., Jain, P., Yeh, P.Z., Verma, K.: Multiple ontologies in healthcare information technology: motivations and recommendation for ontology mapping and alignment. In: International Conference on Biomedical Ontologies (ICBO) (2011)
30. Santos, M.R., Bax, M.P., Kalra, D.: Building a logical EHR architecture based on ISO 13606 standard and semantic web technologies. Stud. Health Technol. Inf. **160**(1), 161–165 (2010)
31. Schadow, G., Mead, C.N., Walker, D.M.: The HL7 reference information model under scrutiny. Stud. Health Technol. Inf. **124**, 151–156 (2006)
32. Schloeffel, P., Beale, T., Hayworth, G., Heard, S., Leslie, H.: The relationship between CEN 13606, HL7, and OpenEHR. In: Proceedings of HIC 2006 and HINZ 2006, pp. 24–28 (2006)
33. Schriml, L.M., Arze, C., Nadendla, S., Chang, Y.-W., Mazaitis, M., Felix, V., Feng, G., Kibbe, W.A.: Disease ontology: a backbone for disease semantic integration. Nucleic Acids Res. **40**(1), 940–946 (2012)
34. Smith, B., Ashburner, M., Rosse, C., Bard, J., Bug, W., Ceusters, W., Goldberg, L.J., Eilbeck, K., Ireland, A., Mungall, C.J., Leontis, N., Rocca-Serra, P., Ruttenberg, A., Sansone, S.-A., Scheuermann, R.H., Shah, N., Whetzel, P.L., Lewis, S.: The OBO Foundry: coordinated evolution of ontologies to support biomedical data integration. Nat. Biotechnol. **25**(11), 1251–1255 (2007)
35. Smith, B., Ceusters, W.: HL7 RIM: an incoherent standard. Stud. Health Technol. Inf. **124**, 133–138 (2006)
36. Sunitha, A., Babu, G.S.: Ontology-driven knowledge-based health-care system, an emerging area - challenges and opportunities - Indian Scenario. ISPRS - Int. Arch. Photogrammetry Remote Sens. Spatial Inf. Sci. **XL–8**, 239–246 (2014)

37. Tao, C., Jiang, G., Oniki, T.A., Freimuth, R.R., Zhu, Q., Sharma, D., Pathak, J., Huff, S.M., Chute, C.G.: A semantic-web oriented representation of the clinical element model for secondary use of electronic health records data. J. Am. Med. Inf. Assoc. **20**(3), 554–562 (2013)
38. Vizenor, L.: Actions in health care organizations: an ontological analysis. Stud. Health Technol. Inf. **2**(107), 1403–1407 (2004)
39. Vizenor, L., Smith, B., Ceusters, W.: Foundation for the Electronic Health Record: An Ontological Analysis of the HL7's Reference Information Model (2004)
40. World Health Organization: International Statistical Classification of Diseases and Related Health Problems. World Health Organization (2004)

Computer Vision

The Research of Chinese License Plates Recognition Based on CNN and Length_Feature

Saina He[1(✉)], Chunsheng Yang[1], and Jeng-Shyang Pan[1,2]

[1] Harbin Institute of Technology Shenzhen Graduate School, Shenzhen, China
hesainahgd@163.com
[2] College of Information Science and Engineering,
Fujian University of Technology, Fuzhou, China

Abstract. Although the license plate recognition system has been widely used, the location and recognition rate is still affected by the clarity and illumination conditions. A license plate locating (LPL) method and a license plate characters recognition (LPCR) method, respectively, based on convolution neural network (CNN) and Length_Feature (LF), are proposed in this paper. Firstly, this paper changes the activation function of CNN, and extracts local feature to train the network. Through this change, the network convergence has sped up, the location accuracy has improved, and wrong location and long time consuming, which caused by some complicated factors such as light conditions, fuzzy image, tilt, complex background and so on, have been resolved. Secondly, the LF, which is proposed in this paper, is easier to understand and has less calculation and higher speed than transform domain features, and also has higher accuracy to recognize fuzzy and sloping characters than traditional geometric features.

Keywords: License plate location · Convolution neural network · License plate characters recognition · Length_Feature

1 Introduction

As an important part of public transport management, license plates recognition (LPR) reduces the complexity of traffic management, and improves the traffic capacity and management efficiency. There are four main technologies, including LPL, characters correction, cutting and LPCR, in a practical LPR process. Up to now, many valid methods have been proposed in these four main technologies.

In LPL, there are some general methods based on different features and algorithms such as color, texture, transform domain, neural network. The color features in RGB could describe color information of license plates well, and it is enough to locate license plates in good bright images [9,16]. The result after discrete wavelet transform (DWT) is an effectual information to describe images [22]. A sliding concentric window was applied to scan images to find license plates in Giannoukos and Anagnostopoulos [7]. Some morphological methods,

© Springer International Publishing Switzerland 2016
H. Fujita et al. (Eds.): IEA/AIE 2016, LNAI 9799, pp. 389–397, 2016.
DOI: 10.1007/978-3-319-42007-3_33

such as sobel edge detection, skeleton, region growing, could combine to find license plates well [11].

In license plate character segmentation, there are the projected image analysis and connected area segmentation, etc. The simplest one is segmenting characters depending on width of characters and distance between characters, but the simplest is the most inflexible, which requires accurate plate region [18]. Analyzing projected image of license plate should find the right border of every characters [15]. A character segmentation technique based on visiting neighbor pixel algorithm, which uses the connectedness of the pixels as a property to distinguish characters, was proposed in Chitrakala et al. [5].

In LPCR, there are template matching, feature extraction, machine learning algorithm. A template matching method, calculating distance between character image and template image, has been applied to classify characters in Chen and Ding [4]. Some features of character could also be extracted to calculate the distance between templates [19] or input in machine learning algorithm [1,14]. Wan et al. [21] adopted shape context, which is proposed to match shapes and recognize objects [2] for the first time, to distinguish similar characters.

To resolve mistaken location problem and to avoid extracting bad features of license plates, this paper proposed a novel LPL method based on CNN. CNN-based LPL method is able to speed up the network convergence and avoid gradient disappearance by modifying the activation function of network. At the same time, we can improve the classification accuracy while retaining the whole color features by inputting three channels as three feature maps into the CNN. CNN can also solve the local minimum value problem, which is existed in traditional neural network. To recognize characters correctly and quickly, this paper extracts the LF of characters and inputs it into BPNN. The LF of characters is easier to figure out than transform domain features. Characters recognition based on LF, which could describe the contour of characters well, is relatively effective in recognizing fuzzy and sloping images.

The structure of this paper is as follows. In Sect. 1, the latest development and research on license plate recognition is overviewed. In Sect. 2, CNN is described. In Sect. 3, the specific work were proposed. In Sect. 4, the simulation and experiment results are presented.

2   CNN

In 2006, Hinton et al. [8] put forward the concept of deep learning which comes from the artificial neural network. CNN, which is the first real supervised deep learning multilayer algorithm, is described by Lecun as a machine learning model [6]. Comparing with traditional neural network, CNN can put images directly into the input layer without the complex feature extraction. Therefore, CNN has become the research hotspot in the field of speech analysis and image recognition [3,17]. In image processing, CNN is mostly used in handwritten character recognition and object classification recognition in original, and 3D-CNN is applied to human behavior analysis in video now [10].

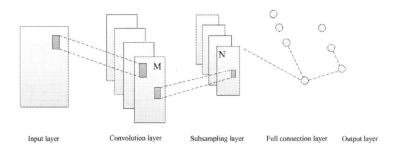

Input layer   Convolution layer   Subsampling layer   Full connection layer   Output layer

Fig. 1. The structure of CNN

In CNN, the main advantages are reducing the amount of parameters and speeding up the training time by means of the concept of receptive field and parameter sharing. In the concept of receptive field, every neurons connect with local receptive field, instead of the whole output of the former layer. The amount of parameters has been reduced greatly. Sharing the convolution kernel, which execute convolution with local receptive field, could reduce the amount of weight when extracting some kind of features. As shown in Fig. 1, the basic structure of CNN is combination of a convolution layer and a subsampling layer. In convolution layer, also called feature extraction layer, each neuron will connect with corresponding local receptive field and extract the features of this receptive field. In subsampling layer, the input data will execute convolution with a mean convolution kernel, and this convolution is called mean-pooling. Subsequently, result of mean-pooling will be mapped through Sigmoid function to ensure shift invariance. According to the local-correlation of image, subsampling layer could release calculation and preserve useful information. Through repeatedly using this basic structure, CNN simulate the iteration and abstraction of biological neural network. In CNN, iteration and abstraction are hierarchical, upgraded, and from local to global.

Different number of layers, different number of feature maps, different size of convolution kernel and different classifiers of CNN will generate different CNN structures. There are some CNN structures. LeNet-5, which has 7 layers, was used to recognize handwritten numeral on checks by many banks in America [13]. Image - Net was proposed by Alex Krizhevsky in 2012, it runs on two parallel GPU and has two full connection layers. In Image - Net, the third and the forth convolution layers have no subsampling layer [12]. Output of the third max-pooling layer and the forth convolution layer of DeepId [20], which is proposed by Sun Yi from CUHK (Chinese University of Hong Kong), input into the full connection layer. DeepId has been used in face feature learning and the accuracy achieved 97.45 %.

3   The Specific Work

3.1   License Plates Location with CNN

License Plates Location. LPL, of which result affects the subsequent operations directly, is the first part of LPR. Considering the complex illumination, background of images, and Chinese license plates have several types, traditional LPL methods are difficult to ensure high accuracy in locating all license plates without missing. Comparing with LPL methods, based on color features, texture features and traditional neural network, LPL method with CNN has higher accuracy while images are collecting in various kinds of situations.

In this paper, sliding sampling window, a set size square slightly less than the license plates, will scan images with a set step size. 3-channel images, which were covered by sampling window, were inputted into CNN to classify into license plates or not. Eliminating non-license-plates in classification results, according to the prior knowledge of license plates, could reduce mistaken locates caused by the misclassification of CNN.

Improvement of CNN. CNN, in this paper, is a simple 7-layer network with one input layer, one full connection layer, one output layer and two combinations of convolution layer and subsampling layer. Input layer has 3 feature maps which is corresponding to 3 channels of images. Two convolution layers respectively extract 6 and 12 feature maps of image with convolution kernels and the size of each kernels are both 5*5. In subsampling layer, convolution with mean convolution kernels, which size is 2*2, is also called mean-pooling. Feature maps will connect as one-dimensional vector through full connection layer, and the vector will be inputted into output layer to recognize license plates.

In this paper, CNN takes ReLu as the activation function except for Sigmoid. The formulas of Sigmoid and ReLu are presented in Eqs. (1) and (2) respectively. When Back Propagation (BP), error gradient need to be calculated. The derivation value of ReLu is either 1 or 0, so the derivation formula is simple and the amount of calculation is small. Comparing with other activation functions, ReLu simulates the sparsity of human cranial nerve network. ReLu reduces redundancy of CNN so the gradient disappearance is nonexistent in Back Propagation. But the calculation to figure out the derivative of Sigmoid is great. When Sigmoid near the saturation region, the transformation is too slow and the derivative tends to zero. For deep Neural Network, great calculation and the derivative, which close to zero, are easy to resulting in gradient disappears.

$$y = \frac{1}{1 + e^{-x}}. \tag{1}$$

$$y = \begin{cases} x & x > 0, \\ 0 & else . \end{cases} \tag{2}$$

With the same samples and parameters, the classification result of CNN, of which the activation function is Sigmoid and ReLu respectively, is shown in Table 1. The error rate in Table 1 is the rate of misclassifying an image which is covered by sampling window. Although the CNN achieve a low error rate while using Sigmoid, ReLu can performs more stable. Figure 2 shows the mean square error of CNN using Sigmoid and ReLu respectively while Alpha is 0.1. In terms of convergence, it can be seen in Fig. 2 that ReLu makes network convergence faster and more direct than Sigmoid.

Table 1. Experimental Results of Sigmoid and ReLu

Activation function	Alpha	Time	Error rate
Sigmoid	0.1	10.8 h	2.4086 %
	0.01	11.9 h	6.7797 %
ReLu	0.1	11.7 h	3.033 %
	0.01	11.05 h	3.033 %

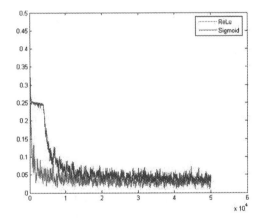

Fig. 2. The mean square error of Sigmoid and ReLu (Color figure online)

3.2 License Plates Character Recognition

In plate images, the longest line, the top and bottom border of license plate, will be found out to correct askew license plates. Then, the projection images of plate region are analyzed to remove borders and segment characters. After the above processing, characters recognition is as following. As Chinese character has a low accuracy with LF, this paper will input Chinese character images into BP network to classify. Then the features of letters and numbers can be extracted to classify.

Proposed Length-Feature. As a simple geometric feature with less calculation and good efficiency, LF is proposed to classify characters. Before extracting LF, character image would be normalized into a set size and be processed binarization. Then find the orthocenter $m\,(m_x, m_y)$ of character image as Eq. (3). The orthocenter of a binary image is the arithmetic mean of each point in the x-axis and y-axis. Thereinto, $f\,(x_i, y_j)$ is the value of point (i, j), and W, H are the width and height of binary character image.

$$
\begin{aligned}
X_m &= \sum_{i=0}^{W} \sum_{j=0}^{H} x_i \times f\,(x_i, y_j)\,, \\
Y_m &= \sum_{i=0}^{W} \sum_{j=0}^{H} y_j \times f\,(x_i, y_j)\,, \\
M &= \sum_{i=0}^{W} \sum_{j=0}^{H} f\,(x_i, y_j)\,, \\
m_x &= \frac{X_m}{M}\,, m_y = \frac{Y_m}{M}.
\end{aligned}
\tag{3}
$$

LF is divided into horizontal LF (HLF), vertical LF (VLF), right diagonal LF (RLF), and left diagonal LF (LLF) depending on the direction of scan line. In Fig. 3, vertical scan line scans the character from left to right to extract VLF. Let point vf, vl be the intersection of the scan line at the extreme end boundaries of the input character image, the VLF is the distance between vf, vl and the orthocenter m. The feature vector of character images is consisted of HLF, VLF, RLF and LLF. A character image has $6 \times H + 6 \times W - 4$ LF in total, which include $2 \times H$ HLF, $2 \times W$ VLF, $2 \times (H + W - 1)$ RLF and $2 \times (H + W - 1)$ LLF.

Substantially, LF described the character's contour information which could distinguish fuzzy sloping characters. As an extracting features method, LF is easier and quicker to be calculated, and classifies more characters.

BP Neural Network. In this paper, to classify letter, letter and numbers character separately, the LF was extracted as the input of 2 Back-Propagation neural networks (BPNN). BPNN has $6 \times H + 6 \times W - 4$ neurons to receive the LF. There are 24 kinds of letters('I', 'O' won't appear) and 10 kinds of numbers

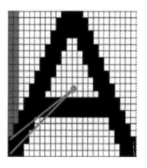

Fig. 3. Vertical LF

in Chinese license plate, therefore, the amount of output neurons are 24, 34 correspondingly. And the amount of the hidden layer neurons is the average of the amount of the input and the output layers neurons.

4   Experiments Result

11,121 images, which had been collected from the videos which are recorded by automobile data recorder, were used in the experiment. With 10,000 images, as training samples, are used for training the CNN and the rest images were test samples, the classification accuracy of CNN can reach 96.96 %. Filtrating license plate regions according to the prior knowledge, the location rate could achieve more than 99 %. The location results of few license plates are given in Fig. 4. The comparison results with various methods are shown in Table 2. In Table 2, the correct rate of Wang et al. [22] and Jiao et al. [11] are quoted from its source text. The classification accuracy is 96.96 % after CNN, and the rate of location could reach 99.3 % after filtrating according to the prior knowledge. Overall, it can be seen that the proposed method has an excellent performance.

Fig. 4. Location results of some license plates

Table 2. Comparison results of various methods

	Methods	Correct rate
[22]	With DWT	97.33 %
[11]	Edge analysis	96.8 %
Proposed	With CNN	99.3 %

LPCR has two parts, coarse classification and fine classification. As Chinese license plate consisting of one Chinese character, one letter, five letters and numbers orderly, the LF of the second and the rest characters would be inputted into 2 BPNNs respectively to classify characters coarsely. In fine classification, shape Context of characters will be extracted to recognize similar characters [21]. The results of LPCR is as Table 3. LF is useful to classify dissimilar characters in coarse classification. Similar characters, such as ('6', '8'), ('5', 'S'), ('M', 'F'), are easy to classify by extracting shape context.

Table 3. Recognition Rate

	Coarse Dissimilar	Classification Similar	Fine Classification
Letters	93.75 %	77.7 %	97.5 %
Letters and Numbers	89.7 %	66.6 %	96.5 %

5   Conclusion

The article presents a novel LPL method with CNN, which can overcome the mistaken location, caused of various and complicated factors, including fickle light, complex background. Meanwhile, the network convergence rate and the accuracy of license plate locating are both improved by optimizing the activation function of CNN. Comparing traditional features, LF, in LPCR, has obvious advantages on recognizing the fuzzy images. Actually, CNN could be used for recognizing character very well without features extraction, which is a good direction to study LPCR further.

References

1. Arora, S., Bhattacharjee, D., Nasipuri, M., Basu, D., Kundu, M.: Complementary features combined in a MLP-based system to recognize handwritten devnagari character. J. Inf. Hiding Multimedia Signal Process. **2**(1) (2011)
2. Belongie, S.J., Malik, J., Puzicha, J.: Shape matching and object recognition using shape contexts. IEEE Trans. Pattern Anal. Mach. Intell. **24**(4), 509–522 (2002)
3. Byeon, Y.H., Pan, S.B., Moh, S.M., Kwak, K.C.: A surveillance system using CNN for face recognition with object, human and face detectio. In: Kim, K.J., Joukov, N. (eds.) ICISA 2016. LNEE, vol. 376. Springer, Singapore (2016)
4. Chen, H.X., Ding, X.Y.: Research on license plate recognition based on template matching method. Appl. Mech. Mater. **668–669**, 1106–1109 (2014)
5. Chitrakala, S., Mandipati, S., Raj, S.P., Asisha, G.: An efficient character segmentation based on VNP algorithm. Res. J. Appl. Sci. Eng. Technol. **4**(24) (2012)
6. Cun, Y.L., Jackel, L.D., Boser, B., Denker, J.S., Graf, H.P., Guyon, I., Henderson, D., Howard, R.E., Hubbard, W.: Handwritten digit recognition: applications of neural net chips and automatic learning. IEEE Commun. Mag. **27**(11), 41–46 (1989)

7. Giannoukos, I., Anagnostopoulos, C.N.: Operator context scanning to support high segmentation rates for real time license plate recognition. Pattern Recognit. **43**(11), 3866–3878 (2010)

8. Hinton, G.E., Salakhutdinov, R.R.: Reducing the dimensionality of data with neural networks. Science **313**(5786), 504–507 (2006)

9. Hirose, K., Asakura, T., Aoyagi, Y.: Real-time recognition of road traffic sign in moving scene image using new image filter. In: 26th Annual Confjerence of the IEEE on Industrial Electronics Society, IECON 2000, vol. 3, pp. 2207–2212 (2000)

10. Ji, S., Xu, W., Yang, M., Yu, K.: 3d convolutional neural networks for human action recognition. IEEE Trans. Pattern Anal. Mach. Intell. **35**(1), 221–231 (2013)

11. Jiao, J., Ye, Q., Huang, Q.: A configurable method for multi-style license plate recognition. Pattern Recogn. **42**(3), 358–369 (2009)

12. Krizhevsky, A., Sutskever, I., Hinton, G.E.: Imagenet classification with deep convolutional neural networks. In: Advances in Neural Information Processing Systems, vol. 25 (2012)

13. Lecun, Y., Bottou, L., Bengio, Y., Haffner, P.: Gradient-based learning applied to document recognition. Proc. IEEE **86**(11), 2278–2324 (1998)

14. Mai, V.D., Miao, D., Wang, R., Zhang, H.: Recognition of characters and numbers in vietnam license plates based on image processing and neural network. Int. J. Hybrid Inf. Technol. **5** (2012)

15. Pan, M.S., Yan, J.B., Xiao, Z.H.: Vehicle license plate character segmentation. Int. J. Autom. **04**(4), 425–432 (2008)

16. Puranik, P., Bajaj, P., Abraham, A., Palsodkar, P., Deshmukh, A.: Human perception-based color image segmentation using comprehensive learning particle swarm optimization. In: 2nd International Conference on Emerging Trends in Engineering and Technology (ICETET 2009), pp. 630–635 (2009)

17. Sainath, T.N., Kingsbury, B., Saon, G., Soltau, H., Mohamed, A.R., Dahl, G., Ramabhadran, B.: Deep convolutional neural networks for large-scale speech tasks. Neural Netw. **64**, 39–48 (2015)

18. Sedighi, A., Vafadust, M.: A new and robust method for character segmentation and recognition in license plate images. Expert Syst. Appl. **38**(11), 13497–13504 (2011)

19. Soora, N.R., Deshpande, P.S.: Robust feature extraction technique for license plate characters recognition. IETE J. Res. **61**(1), 72–79 (2014)

20. Sun, Y., Wang, X., Tang, X.: Deep learning face representation from predicting 10,000 classes. In: IEEE Conference on Computer Vision and Pattern Recognition (CVPR 2014), pp. 1891–1898 (2014)

21. Wan, Y., Li, X.Y., Zhou, Z.G.: Character recognition of license plate image with low quality based on shape context. Comput. Appl. Softw. **30**(5), 267–270 (2013)

22. Wang, Y.R., Lin, W.H., Horng, S.J.: A sliding window technique for efficient license plate localization based on discrete wavelet transform. Expert Syst. Appl. **38**(4), 3142–3146 (2011)

View-Invariant Gait Recognition Using a Joint-DLDA Framework

Jose Portillo[1], Roberto Leyva[2], Victor Sanchez[2], Gabriel Sanchez[1],
Hector Perez-Meana[1(✉)], Jesus Olivares[1], Karina Toscano[1],
and Mariko Nakano[1]

[1] Postgraduate and Research Section ESIME Culhuacan,
Instituto Politecnico Nacional, Mexico City, Mexico
hmperezm@ipn.mx
[2] Department of Computer Science, University of Warwick, Coventry, UK

Abstract. In this paper, we propose a new view-invariant framework for gait analysis. The framework profits from the dimensionality reduction advantages of Direct Linear Discriminant Analysis (DLDA) to build a unique view-invariant model. Among these advantages is the capability to tackle the under-sampling problem (USP), which commonly occurs when the number of dimensions of the feature space is much larger than the number of training samples. Our framework employs Gait Energy Images (GEIs) as features to create a single joint model suitable for classification of various angles with high accuracy. Performance evaluations shows the advantages of our framework, in terms of computational time and recognition accuracy, as compared to state-of-the-art view-invariant methods.

Keywords: Gait recognition · View-invariant · Gait Energy Image · Direct Linear Discriminant Analysis

1 Introduction

Person identification through gait analysis using appearance-based approaches has gained considerable importance over the last few years. These approaches, which do not rely on structural models of the human walking, have been shown to attain a high recognition accuracy with low computational cost by extracting information from simple moving silhouettes [6]. However, several factors may hinder their recognition performance. Among these factors are clothes, footwear, carrying objects, walking surfaces, time elapsed, and the view angle. The later, which is defined as the angle between an optical axis and the walking direction [14], may have a big impact on the performance as a most of the appearance-based approaches rely on a fixed view angle [24].

In this paper, we present an appearance-based framework for gait recognition that tackles the challenges associated with different view angles. Our framework is based on subspace learning and employs Gait Energy Images (GEIs) as features. Specifically, it employs Direct Linear Discriminate Analysis (DLDA) to

© Springer International Publishing Switzerland 2016
H. Fujita et al. (Eds.): IEA/AIE 2016, LNAI 9799, pp. 398–408, 2016.
DOI: 10.1007/978-3-319-42007-3_34

create a single model for classification. To this end, we employ as training data GEIs computed from raw sequences captured at several view angles. We call this framework Joint-DLDA. The main novelties of Joint-DLDA are as follows:

1. No need to create independent models for classification at different view angles. This is particular useful in practical situations, where it is common that the probe data is captured at an angle not present in the gallery data. A unique model for classification of several angles can handle these situations.
2. Ability to inherently handle high-dimensional feature spaces.
3. A considerable low computation cost with a simple classification stage.

Performance evaluations on the CASIA-B gait database [24] show that our framework provides competitive results, outperforming various view-invariant approaches in terms of recognition rate and speed.

2 Related Work

Gait recognition approaches aimed at tackling the challenges associated with different view angles can be classified into three groups: visual hull-based, view transformation-based, and view-invariant approaches. Our framework is related to the later.

Visual hull-based approaches rely on 3-D gait information acquired by multiple calibrated cameras. Bodor et al. [3] apply image-based rendering on a 3-D visual hull model to automatically reconstruct gait features. Zhang et al. [25] introduce a view-independent gait recognition method based on a 3-D linear model and Bayesian rules. Zhao et al. [26] reconstruct a 3-D gait model from video sequences captured by multiple cameras. Although all of these methods are suitable for a fully controlled and cooperative multi-camera environment, they usually involve costly computations [14].

The basic idea of view transformation-based approaches is to transform features from one view to another by learning the view relationship between the two views. The transformed *virtual* features are then used for recognition [11]. Unlike the visual hull-based approaches, these approaches do not require synchronized multi-view gait data from the target subjects, thus are suitable for cases when only gallery and probe images from different views are available [11]. These approaches employ a matrix factorization process through Singular Value Decomposition (SVD) e.g. [14], or regression during the training stage [10]. The main limitation of these approaches is that the number of applicable views is limited to a discrete set of training views and consequently, recognition accuracy is degraded if the target view significantly differs from those views used for training.

View-invariant approaches can be further divided into geometry-based [8]; subspace learning-based [11]; and metric learning-based [16]. Geometry-based approaches make use of geometrical properties for feature extraction. For instance, Kale et al. propose to synthesize side-view gait images from any other arbitrary view considering that the person is represented by a planar object on a sagittal plane [9]. This method works well in cases where the angle between the sagittal plane of the person and the image plane is small; however, in cases where this angle is large, accuracy is significantly hindered [18]. Conversely, subspace and metric learning-based approaches do not rely on this angle. Metric learning-based approaches *learn* a weight vector that sets the importance of a matching score associated with each feature; they use the weight vector to calculate a final score [17]. For example, [16] applies the pairwise RankSVM algorithm [4] to improve the accuracy of gait recognition with covariate variations (view angles, clothing, and carrying objects) [16]. Subspace learning-based approaches *learn* a feature subspace using training data, and calculate view-invariant features by projecting the original features onto the learned subspace [11]. For example, Lu *et al.* [13] propose uncorrelated discriminant simplex analysis to learn the feature subspace, while Liu *et al.* [11] apply joint principal component analysis (PCA) to learn the joint subspace of gait feature pairs with different views.

Subspace learning for view-invariant gait recognition has been shown to attain high recognition accuracy. This type of dimensionality reduction can be considered as a within-class multimodality problem if the samples in a class form several separate clusters [20]. In this case, the training data tend to create clusters within the same class due to the high similarities of the view angles. To analyze the resulting subspaces after dimensionality reduction, a preprocessing step is necessary to efficiently manipulate high-dimensional data. This is particular important when using GEIs as features, since the dimensionality of the feature space is usually much larger than the size of the training set. This is known as the under-sampling problem (USP) [21] or small sample size (SSS) problem [5], which may result in a *singular* sample scatter matrix. If the USP is encountered, one common solution is to reduce the feature dimension space using principal components analysis (PCA) [21]. A potential problem is that the PCA may discard dimensions that contain important discriminating information [23]. Other approaches, e.g. Mansur *et al.* [15], propose to generate independent models for each view angle (MvDA); however, this inevitably involves more computational steps and the usage of cross-dataset information.

3   Proposed Framework

Our proposed framework is illustrated in Fig. 1. It consists of three main stages: joint model formulation, subspace learning through DLDA and k-NN classification. We describe in detail these stages next.

3.1   Joint Model

In this paper, we propose to generate a unique model for classification. We create the input feature space by joining all the training data corresponding to all samples in all available view angles. This is motivated mainly to avoid creating an independent model for each view angle. One of the advantages of using DLDA is its powerful formulation to separate low intra-variance classes. Thus, we propose to construct a projected subspace considering each subject

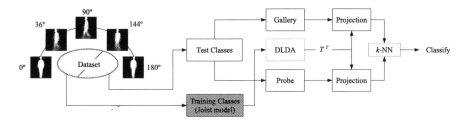

Fig. 1. Proposed framework. The dataset is first split into test and training samples. The training set is used to create a single joint model for all view angles available for training by using DLDA. The result of this is a transformation matrix T^T. Test samples are further split into galley and probe subsets; the former is used to generate a k-NN model, while the latter is used to rank the framework

with all different angles and samples as one class, as illustrated in Fig. 2. The discriminant characteristic of DLDA makes this assumption possible and helps to separate, in the projected subspace, the classes represented by the view angles well enough. This is useful to characterize query view angles not present in the training set, if several view angles are used for training.

3.2 Direct Linear Discriminant Analysis

Consider n samples stored as d-dimensional column vectors corresponding to all available view angles for all individuals in the training set. Let us define each GEI by $x_i \in \mathbb{R}^d, i = 1, 2, \ldots, n$; and the associated class labels by $y_i \in 1, 2, \ldots, c$, where c is the number of classes. The number of GEIs in class ℓ is then denoted by n_ℓ, with $\sum_{\ell=1}^{c}(n_\ell) = n$; and the matrix of all GEIs is denoted by $X \equiv (x_1|x_2|\cdots|x_n)$ (see Fig. 2). We employ DLDA to project X into a lower

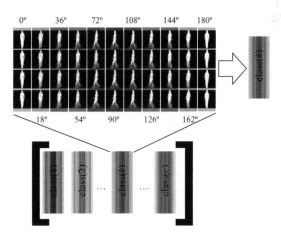

Fig. 2. Joint model: the training data involve c-classes using GEI features [27] of the CASIA-B database [24]. The class (k) consist of all different angles and samples available for the subject (k).

dimensional embedding subspace. Let $z_i \in \mathbb{R}^r : 1 \leqslant r \leqslant d$ be the low-dimensional representations of x_i, where r is the reduced dimension, i.e., the dimension of the embedding subspace. The embedded samples z_i are the given by $z_i = T^T x_i$, where T^T denotes the transpose of the transformation matrix.

DLDA aims to find a projection T that maximizes the ratio of between-class scatter matrix $S(b)$ against within-class scatter matrix $S(w)$; known as Fisher's criterion:

$$T = \arg \max_{T \in \mathbb{R}^{d \times r}} \frac{|T^T S^{(b)} T|}{|T^T S^{(w)} T|} \tag{1}$$

with:

$$S^{(b)} = \sum_{\ell=1}^{c} n_\ell \left(\mu_\ell - \mu\right) \left(\mu_\ell - \mu\right)^T \tag{2}$$

$$S^{(w)} = \sum_{\ell=1}^{c} \sum_{\forall y_i = \ell} \left(x_i - \mu_\ell\right) \left(x_i - \mu_\ell\right)^T \tag{3}$$

where μ_l is the mean of the samples in class ℓ and μ is the mean of all samples. If the number of samples is smaller than their dimension, both $S^{(b)}$ and $S^{(w)}$ become singular. Usually, in pattern recognition problems, the within-class scatter matrix $S^{(w)}$ is always singular, due to the fact that the rank of $S^{(w)}$ is at most $n - c$, while the number of features, i.e., pixels in those images is much bigger than the samples per class, for instance in face recognition, and human ID by gait. To prevent $S^{(w)}$ become singular, Belhumeur et al. [2] propose the reduction of features dimensional space through PCA, that is, pixels equal to $n - c$, and then applying standard LDA, whose formula is Fishers criterion as in Eq. 1. Maximizing Fishers criterion requires both the reduction of within-class scatter matrix $S^{(w)}$, and the increment of the between-class scatter matrix, however the reduction of the number of features through PCA, is based in of the variability in the data, and for that PCA step may discard dimensions that contain important discriminative information.

In Figs. 3 and 4, we show the projection of GEI images belonging at two-classes from CASIA-B, where one class is represented by circles and the other by crosses; furthermore, in both classes, two view angles were taken, $0°$ (thin circles and thin crosses) and $90°$ (thick circles and thick crossed), as long as these GEI are different from view angle $0°$ and $90°$, even though they might belong to the same class; the projection to be used must allow the grouping of classes regardless the view angle, since an effective projection will assemble the crosses (be it thin or thick) and at the same time separated from the circles group (where both thin and thick circles are present like a class).

In order to overcome this problem, DLDA diagonalizes the scatter matrix $S^{(b)}$. The objective is to discard the null space of $S^{(b)}$, which contains no useful information, rather than discarding the null space of $S^{(w)}$, which contains the most discriminative information [23]. DLDA is summarized in Fig. 5. By using

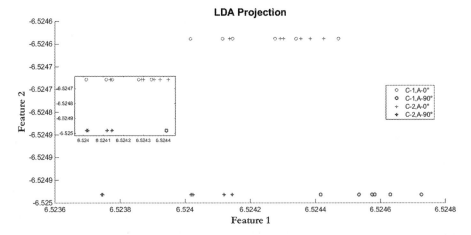

Fig. 3. A 24 GEI projection, using LDA standard, where the circles and crosses are mixed, since the class separation is not effective in this dimensional space. (Color figure online)

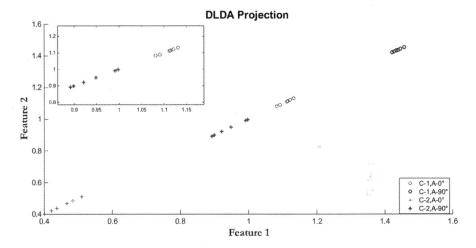

Fig. 4. A 24 GEI projection, using LDA standard, where the circles and crosses are mixed, since the class separation is not effective in this dimensional space. (Color figure online)

DLDA, we obtain a transformation matrix T that provides data projections into a low dimensional subspace and intra-class separability at the same time.

3.3 k-NN Classification

To identify the corresponding subject in the probe subset, we calculate the nearest distance relative to the projected samples. This distance must be minimal

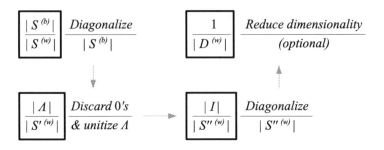

Fig. 5. Thumbnail of DLDA [23].

within the same classes, which is the final criterion for classification. In this step, we construct a 1-NN model with the transformed features obtained by Eq. 1. We then generate a class label for each subject to be identified. For each cluster, we empirically select 10 eigenvectors of the projected samples.

4 Experiments

We conduct experiments on the CASIA-B gait database [24] using the GEI features obtained by Bashir *et al.* [1], which consists of 124 subjects under 11 view angles; i.e., from $0°$ to $180°$ with a separation of $18°$. For each subject, there are 10 walking sequences, which include 6 normal walking sequences that we use for evaluation. We use the original GEI size of 240×240 pixels to create the joint model as described in Sect. 3.1.

Setup. We follow the setup proposed by Mansur *et al.* [15], which consists in selecting 62 subject as training samples and the remaining 62 as test samples. We use as the gallery subset the sequences at $90°$ and as the probe subset those at $\{0°, 18°, 36°, 54°, 72°\}$. During the training phase, we obtain one transformation matrix T^T. During the test phase, gallery and probe subsets are projected using the obtained transformation matrix. A 1-NN model is used to match the gallery and probe subsets using the Euclidean distance.

Using this setup, we perform two sets of experiments. In the first set, we do not complete the sparse samples to overcome the USP, as it is done by Mansur *et al.* [15] by using the OU-LP dataset [7]. We simply directly take the GEIs and project them using matrix T^T. In the second set of experiments, we use the idea to complete the sparse samples; however instead of using the OU-LP dataset, we use the GEIs with the reflected angles of $\{108°, 126°, 144°, 162°, 180°\}$.

Experimentally, we observe that certain combination of subjects used as training and testing achieve different performance. Thus we provide the average performance of 100 executions taking random subjects for training and testing. State-of-the-art results are shown in Table 1. The proposed Joint-DLDA framework correspond to the last two rows. Results for the first set of experiments are in row J-DLDA(1), while results for the second set of experiments are in row J-DLDA(2).[1]

---

[1] Our end-to-end implementation is available in: https://yadi.sk/d/MuEE2_tGjJxcq.

Table 1. CASIA-B recognition rate for different approaches.

Method	0°	18°	36°	54°	72°
GMLDA [19][a]	2 %	2 %	1 %	2 %	4 %
DATER [22][a]	7 %	8 %	18 %	59 %	96 %
CCA [12][a]	2 %	3 %	5 %	6 %	30 %
VTM [14][a]	17 %	30 %	46 %	63 %	83 %
MvDA [15]	17 %	27 %	36 %	64 %	95 %
J-DLDA(1)	16.29 %	21.16 %	31.69 %	49.96 %	84.37 %
J-DLDA(2)	20.20 %	24.41 %	36.86 %	58.38 %	94.33 %

[a]Reported in [15] with the available online source code.

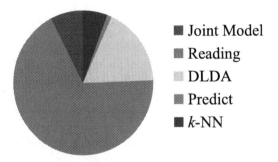

Fig. 6. Distribution of computational times of the proposed Joint-DLDA framework. Note that most of the computational time corresponds to preparing the input feature space. (Color figure online)

5 Discussion

From Table 1, we observe that the proposed framework achieves a very competitive performance. It is important to highlight the following important aspects of Joint-DLDA:

- It does not require two datasets or to modify the size of the samples in X to overcome the USP.
- It achieves the best performance for the most challenging angle, i.e. 0°.
- It provides competitive recognition accuracies with a simple 1-NN model for classification.
- It achieves up to 100 % recognition rate for the 72° angle in J-DLDA(2) by including the complementary samples available within the CASIA-B dataset.

Some existing state-of-the-art methods, e.g. VTM and MvDA, propose to build a model for each independent view angle to partly overcome the USP. This is one of their main drawbacks, because they imply pre-knowledge of how the view angles are going to be used during the evaluation. Our proposed framework does not require such pre-knowledge. Other existing approaches that also rely on a single transformation matrix formulation, require to increment the number of samples to overcome the USP. The view angles of these extra samples must be very similar to the test view angles; this inevitable hinders the ability to transfer learning across two gait datasets if the view angles are not *relatively close*.

An important advantage of our framework is the time requirements for classification. Existing methods, e.g. [14], may require up to 6 h for the training phase [10]. Joint-DLDA is an efficient framework in terms of speed; it requires as little as 25 seconds per test. Thus, a full set of experiments may be executed in approximately 40 min. The main time-consuming steps in Joint-DLDA are shown in Fig. 6, as a proportion of the total computational time. These steps correspond to reading GEI features from the dataset to create the joint model, computing matrix T, generating the k-NN model and classifying incoming samples. As we can see, the most time-consuming step is reading the dataset to create the joint model.

6   Conclusions

In this paper, we presented a framework for view-invariant gait recognition. The framework is based on the creation of a single joint model capable of classifying GEIs computed from sequences captured at different view angles, with high accuracy and low computational times. The model is based on DLDA, which helps in tackling the under-sampling problem with remarkable results. There are other common covariate factors that are daily presented in gait, for example the use of overcoats; depending on the size of such coats, the silhouette can vary significantly due to occlusion of the lower parts (e.g. using large coats or large skirts) and upper parts (e.g. hoodies) and also carrying condition of briefcases, backpacks, etc. modified the silhouettes due to changes in the appearance. Thus in both cases (clothing and carrying condition) the classification by the proposed method is affected. Further research include exploring this framework for other classification models. We also plan to explore the feasibility of transferring learning across two gait datasets.

Acknowledgments. This work has been financed by Consejo Nacional de Ciencia y Tecnologia (CONACyT), Mexico and by Secretaria de Educacion Publica, Mexico.

References

1. Bashir, K., Xiang, T., Gong, S.: Gait recognition without subject cooperation. Pattern Recogn. Lett. **31**(13), 2052–2060 (2010)
2. Belhumeur, P.N., Hespanha, J.P., Kriegman, D.J.: Eigenfaces vs. fisherfaces: recognition using class specific linear projection. IEEE Trans. Pattern Anal. Mach. Intell. **19**(7), 711–720 (1997)
3. Bodor, R., Drenner, A., Fehr, D., Masoud, O., Papanikolopoulos, N.: View-independent human motion classification using image-based reconstruction. Image Vision Comput. **27**(8), 1194–1206 (2009). http://www.sciencedirect.com/science/article/pii/S0262885608002412

4. Chapelle, O., Keerthi, S.S.: Efficient algorithms for ranking with SVMs. Inf. Retrieval **13**(3), 201–215 (2010)
5. Chen, L.F., Liao, H.Y.M., Ko, M.T., Lin, J.C., Yu, G.J.: A new lDA-based face recognition system which can solve the small sample size problem. Pattern Recogn. **33**(10), 1713–1726 (2000). http://www.sciencedirect.com/science/article/pii/S0031320399001399
6. Han, J., Bhanu, B.: Individual recognition using gait energy image. IEEE Trans. Pattern Anal. Mach. Intell. **28**(2), 316–322 (2006)
7. Iwama, H., Okumura, M., Makihara, Y., Yagi, Y.: The OU-ISIR gait database comprising the large population dataset and performance evaluation of gait recognition. IEEE Trans. Inf. Forensics Secur. **7**(5), 1511–1521 (2012)
8. Jean, F., Bergevin, R., Albu, A.B.: Computing and evaluating view-normalized body part trajectories. Image Vision Comput. **27**(9), 1272–1284 (2009). http://www.sciencedirect.com/science/article/pii/S0262885608002497
9. Kale, A., Chowdhury, A., Chellappa, R.: Towards a view invariant gait recognition algorithm. In: IEEE Conference on Proceedings of Advanced Video and Signal Based Surveillance, July 2003, pp. 143–150 (2003)
10. Kusakunniran, W., Wu, Q., Zhang, J., Li, H.: Gait recognition under various viewing angles based on correlated motion regression. IEEE Trans. Circuits Syst. Video Technol. **22**(6), 966–980 (2012)
11. Liu, N., Lu, J., Tan, Y.P.: Joint subspace learning for view-invariant gait recognition. IEEE Signal Process. Lett. **18**(7), 431–434 (2011)
12. Liu, N., Tan, Y.P.: View invariant gait recognition. In: IEEE International Conference on Acoustics Speech and Signal Processing (ICASSP), March 2010, pp. 1410–1413 (2010)
13. Lu, J., Tan, Y.P.: Uncorrelated discriminant simplex analysis for view-invariant gait signal computing. Pattern Recogn. Lett. **31**(5), 382–393 (2010). http://www.sciencedirect.com/science/article/pii/S0167865509003092
14. Makihara, Y., Sagawa, R., Mukaigawa, Y., Echigo, T., Yagi, Y.: Gait recognition using a view transformation model in the frequency domain. In: Leonardis, A., Bischof, H., Pinz, A. (eds.) ECCV 2006. LNCS, vol. 3953, pp. 151–163. Springer, Heidelberg (2006)
15. Mansur, A., Makihara, Y., Muramatsu, D., Yagi, Y.: Cross-view gait recognition using view-dependent discriminative analysis. In: 2014 IEEE International Joint Conference on Biometrics (IJCB), September 2014, pp. 1–8 (2014)
16. Martín-Félez, R., Xiang, T.: Gait recognition by ranking. In: Fitzgibbon, A., Lazebnik, S., Perona, P., Sato, Y., Schmid, C. (eds.) ECCV 2012, Part I. LNCS, vol. 7572, pp. 328–341. Springer, Heidelberg (2012). http://dx.doi.org/10.1007/978-3-642-33718-5_24
17. Muramatsu, D., Shiraishi, A., Makihara, Y., Uddin, M., Yagi, Y.: Gait-based Person recognition using arbitrary view transformation model. IEEE Trans. Image Process. **24**(1), 140–154 (2015)
18. Muramatsu, D., Shiraishi, A., Makihara, Y., Yagi, Y.: Arbitrary view transformation model for gait person authentication. In: 2012 IEEE Fifth International Conference on Biometrics: Theory, Applications and Systems (BTAS), pp. 85–90. IEEE (2012)
19. Sharma, A., Kumar, A., Daume III., H., Jacobs, D.W.: Generalized multiview analysis: a discriminative latent space. In: 2012 IEEE Conference on Computer Vision and Pattern Recognition (CVPR), pp. 2160–2167. IEEE (2012)

20. Sugiyama, M.: Dimensionality reduction of multimodal labeled data by local Fisher discriminant analysis. J. Mach. Learn. Res. **8**, 1027–1061 (2007). http://dl.acm.org/citation.cfm?id=1248659.1248694

21. Tao, D., Li, X., Wu, X., Maybank, S.: General tensor discriminant analysis and gabor features for gait recognition. IEEE Trans. Pattern Anal. Mach. Intell. **29**(10), 1700–1715 (2007)

22. Yan, S., Xu, D., Yang, Q., Zhang, L., Tang, X., Zhang, H.J.: Discriminant analysis with tensor representation. In: IEEE Computer Society Conference on Computer Vision and Pattern Recognition, CVPR 2005, vol. 1, pp. 526–532. IEEE (2005)

23. Yu, H., Yang, J.: A direct LDA algorithm for high-dimensional data with application to face recognition. Pattern Recogn. **34**(10), 2067–2070 (2001). http://www.sciencedirect.com/science/article/pii/S003132030000162X

24. Yu, S., Tan, D., Tan, T.: A framework for evaluating the effect of view angle, clothing and carrying condition on gait recognition. In: 18th International Conference on Pattern Recognition, ICPR 2006, vol. 4, pp. 441–444 (2006)

25. Zhang, Z., Troje, N.F.: View-independent person identification from human gait. Neurocomputing **69**(13), 250–256 (2005). http://www.sciencedirect.com/science/article/pii/S0925231205001797, Neural Networks in Signal Processing 2003 IEEE International Workshop on Neural Networks for Signal Processing

26. Zhao, G., Liu, G., Li, H., Pietikainen, M.: 3d gait recognition using multiple cameras. In: 7th International Conference on Automatic Face and Gesture Recognition, FGR, April 2006, pp. 529–534 (2006)

27. Zheng, W.S., Lai, J., Li, S.Z.: 1D-LDA vs. 2D-LDA: when is vector-based linear discriminant analysis better than matrix-based? Pattern Recogn. **41**(7), 2156–2172 (2008). http://www.sciencedirect.com/science/article/pii/S0031320307005274

Copyright Protection in Video Distribution Systems by Using a Fast and Robust Watermarking Scheme

Antonio Cedillo-Hernandez[1], Manuel Cedillo-Hernandez[2],
Francisco Garcia-Ugalde[1], Mariko Nakano-Miyatake[2],
and Hector Perez-Meana[2(✉)]

[1] Facultad de Ingeniería, Universidad Nacional Autónoma de México,
Mexico City, Mexico
antoniochz@hotmail.com, fgarciau@unam.mx

[2] Sección de Estudios de Posgrado E Investigación, Escuela Superior de
Ingeniería Mecánica Y Eléctrica Unidad Culhuacán, Instituto Politécnico
Nacional, Mexico City, Mexico
mcedillohdz@hotmail.com, {mnakano,hmperezm}@ipn.mx

Abstract. In this paper we propose a watermarking scheme to ensure copyright protection in video distribution systems. This research is focused on provide a low computational cost solution that allows embedding a robust and imperceptible watermarking signal in the video content, such that the original video transmission rate will not be affected. Our proposal accomplishes this challenge by introducing a fast and effective method based on spatio-temporal Human Visual System properties that allow improving imperceptibility along video sequence and at the same time, enhancing robustness. Extensive experiments were performed to prove that the proposed algorithm satisfies quickness, invisibility and robustness against several real-world video processing issues. The proposed scheme has the advantages of robustness, simplicity and flexibility, allowing it to be considered a good solution in practical situations since it can be adapted to different video distribution systems.

Keywords: Video watermarking · Human Visual System · Video distribution systems

1 Introduction

Nowadays, video experience over Internet has become a popular activity among users. Readily available, low-cost and user-friendly platforms have greatly facilitated video content consumption. These platforms usually distribute copyrighted video content with the help of trustworthy systems which control how the content can be used in a digital environment, known as Digital Rights Management (DRM) technologies. However, DRM technologies do not provide any protection once the decryption process is carried out, and thus the video can be entirely copied and distributed at a global level [1]. This illegal activity causes very important revenue loss for companies that are associated within the creation and distribution of video content [2]. In recent years,

© Springer International Publishing Switzerland 2016
H. Fujita et al. (Eds.): IEA/AIE 2016, LNAI 9799, pp. 409–421, 2016.
DOI: 10.1007/978-3-319-42007-3_35

digital watermarking has emerged as a complementary mechanism of protection that enables tracking illegal contents by putting in evidence an authorized consumer that has created an illicit copy [3]. When video watermarking techniques are used together with DRM technologies, watermarking information is embedded while the content is decoded in order to be displayed on the client's screen, and thus create an irreversible link between the watermark information and the multimedia content [4]. In this context, watermarking techniques must deal with three main requirements: low computational-cost, robustness, and imperceptibility. Low computational-cost is particularly relevant since the watermark embedding process must be performed within a specific limit of time that should not influence the original transmission rate.

In this paper we propose a watermarking scheme to provide copyright protection in video distribution systems. The research is focused on provide a low computational-cost solution that allows embedding a robust and imperceptible watermarking signal in the video content, such that the original video transmission rate will not be affected. In order to get robustness, the watermark embedding process is carried out in the frequency domain by modifying the middle frequency band of the Discrete Fourier Transform (DFT) magnitude. The main contribution of this proposal is the definition of an original, fast, and effective spatial-temporal method based on Human Visual System (HVS) properties, which improves imperceptibility along video sequence and at the same time, enhances robustness. A prototype system was implemented over an Intel® architecture computer where extensive experiments were performed to prove that the proposed algorithm satisfies quickness, invisibility and robustness against several practical video processing issues. In comparison with most video watermarking applications focused on video broadcasting that strongly depends on specific hardware and are not flexible and reusable [5], our proposed technique is a generic hardware independent solution, allowing it to be considered a good solution in practical situations since it can be adapted to different video distribution systems. The rest of the paper is organized as follows: In Sect. 2 we provide a revision of related works. Section 3 provides an explanation of the HVS-based perceptual method introduced in this proposal to get a robust and imperceptible watermark along video sequence. The proposed method, including the watermark embedding and detection processes, is described in Sect. 4. Simulation results are shown in Sect. 5 and Sect. 6 concludes this work.

2   Related Works

Several video watermarking approaches have been proposed in recent years. In general, video watermark techniques can be classified considering the domain where the watermark information is embedded, i.e. if the video frame data is directly modified then the method is referred as a spatial domain approach; on the other hand when the watermarking signal is embedding modifying the coefficients of some video frame frequency representation then is called frequency domain method. Spatial domain algorithms are often simple and fast, but they are less robust than frequency domain techniques [3]. Additionally, we can distinguish between methods that embed watermarking information within compressed video stream and those where the watermark signal is embedded into raw video data. Wang et al. [5] proposed a scalable video watermarking scheme based on a parallel MPEG-2 engine. It embeds the watermark information into DCT blocks of compressed video stream. The watermark algorithm accomplishes a good trade-off between robustness and efficiency for moderate and high-quality of MPEG-2 video. Lu et al. [6] designed a video compressed watermarking algorithm by modifying directly suitable positions in the variable length code word (VLC) over an MPEG-2 video stream. In its proposal, a good strategy against collusion and copy attacks is introduced. Zhao et al. [7] proposed a video watermarking algorithm by the least modification of the motion vectors in the compressed video stream, in which a novel fast estimation of motion vector was presented to make the algorithm effective and practical. These algorithms perform the watermark embedding process with a very low computational consumption, but they have drawbacks to extract the watermark data previously embedded after the watermarked video is subjected to aggressive video processing tasks such as low bit-rates conversion or video format conversion which result in the regeneration of whole video stream.

On the other hand, uncompressed domain proposals [8–10] can embed a larger amount of watermark information than compressed domain algorithms. Since the whole video frame data is available, several techniques can be computed to improve imperceptibility and get robustness against a wide range of attacks. However, such proposals are often computationally expensive and they cannot reach a suitable performance that not disturbs the original rate of a video transmission.

3   Low-Complexity HVS-Based Perceptual Model

Since perceptual models are considered as a suitable solution to solve the trade-off between robustness and imperceptibility [11, 12], we introduce a low-complexity HVS-based perceptual model by combining spatial and temporal analysis. This proposal has two steps: (a) Computing the region that attracts the observer's attention along video sequence and then (b) modulate the watermark energy in this region.

3.1   Visual Attention Along Video Sequence

In video analysis context, several experiments have been performed in order to determine the main factors that influence the visual attention of an observer. Motion and foreground/background separation are the factors that have been found to have strongest influence on a visual attention process [13, 14]. Based on the above, we use two HVS-based techniques in order to determine a visual attention region of an observer along a video sequence, which are explained in detail in following sections:

Location of Saliency Objects.   The first technique is calculated in the spatial domain by computing an image descriptor to distinguish saliency objects from its background. An observer is more likely to focus objects in the foreground instead of those over the background [15]. In order to locate saliency objects from an image and then isolate them from its background, we found the image signature descriptor introduced in [16] as a suitable solution to be applied in our proposal since it is simple, fast and get adequate results from different background image conditions. The reader is referred to [16] for a detailed description of the image signature descriptor algorithm. Figure 1 shows the achieved performance after this algorithm is applied to the 670th frame of the "Highway" video sequence; the saliency map m is represented as a heat map over the original frame for demonstration purposes (Fig. 1(a)). The rectangle representation of the binary map B_m represents the output of the algorithm (Fig. 1(b)).

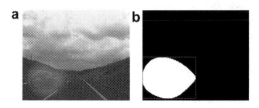

Fig. 1. An example of the foreground/background separation algorithm: (a) Original video frame with the saliency map m represented as a heat map and (b) rectangle representation of the binary map B_m which represents the output of the algorithm.

Motion Detection. It is claimed that the HVS is involuntarily attracted to areas exhibiting motion which are distinct from surrounding areas [17]. We introduce a fast and simple technique to identify those regions where the motion activity between a frame in time t and its predecessor in time $t-1$ is higher compared with the surrounding regions. Getting the luminance component of each frame I, information about the motion can be extracted from the difference between a reference frame and its predecessor $\Delta I = |I_t - I_{t-1}|$. Once the difference between frames is obtained, a binary image ΔI_{BW} is computed applying a global threshold T^Δ based on Otsu's method [18], as follows:

$$\Delta I_{BW}(x,y) = \begin{cases} 1 & \Delta I(x,y) \geq T^\Delta \\ 0 & \Delta I(x,y) < T^\Delta, \end{cases} \qquad (1)$$

Then, in order to get better performance specially for video sequences with lots of motion, the following algorithm to remove small objects is applied: (a) Determine connected components from the binary image ΔI_{BW}, (b) Remove all connected components that have fewer area than P pixels, then produce a new binary image represented as $\Delta I'_{BW}$ and (c) A motion region is defined as the rectangular area limited by the neighborhood of different binary values in $\Delta I'_{BW}$.

Figure 2 shows two consecutive frames from "Akiyo" video sequence, where the motion region is delimited by a red rectangle (Fig. 2a), and their corresponding binary frames $\Delta I'_{BW}$ representing the motion region (Fig. 2b). The visual attention region is composed by the foreground region and the scene motion region. In case regions were overlapped, visual attention will be defined by the broader region.

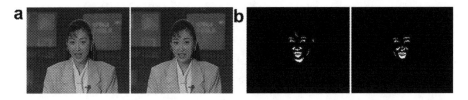

Fig. 2. Two consecutive frames of "Akiyo" video sequence: (a) Original frames with motion region represented as a red rectangle and (b) Binary frames $\Delta I'_{BW}$. (Color figure online)

3.2 Perceptual Mask

A perceptual mask based on [19] is computed over the visual attention region previously computed. This perceptual mask PM_σ is built by computing, at each frame location $I(x,y)$, the local variance over a square window S and then normalizing the frame with respect to the maximum value of the local variance over the whole image PM_{MAX}. The value of the perceptual mask PM_σ at each pixel is thus:

$$PM_\sigma(x,y) = \frac{1}{PM_{MAX}} \sum_{(i,j)\in S} [I(i,j) - \mu S(i,j)]^2 \qquad (2)$$

where $\mu S(i,j)$ is the local mean computed in the window S centered at the pixel position (x,y), and:

$$PM_{MAX} = \max_{m,n} PM_\sigma(m,n), \qquad (3)$$

In this way we built a perceptual mask PM_σ with values in the interval $[0, 1]$ which gives us a measure of the sensitivity to noise of every pixel from the original frame. Note that the proposed algorithm in [19] is based on the idea that perceptual mask will work after the watermarking embedding process is carried out, i.e. another watermark frame I'' is built by combining the original one I and its watermarked version I' as follows:

$$I'' = (1 - PM_\sigma)I + PM_\sigma I', \qquad (4)$$

In this way, the watermarking energy is higher over those areas that are less affected by common signal processing attacks which content textured, darker and brighter areas. Figure 3 shows two consecutive frames of the "Hall" video sequence, the visual attention region, and the perceptual mask computed over the visual attention region.

Please note the following strategies employed by the proposed scheme in order to reduce the computational cost:

- Computing foreground region and perceptual mask involves processing the frame in blocks rather than processing the whole frame at once, which consume large amount of computational time. In order to save computational cost, within the proposed method the above mentioned tasks are computed by performing sliding

Fig. 3. Two consecutive frames of "Hall" video sequence (a) and (b), (c) Visual attention region considering foreground/background separation (green rectangle) and scene motion region (red rectangle), and (d) perceptual mask PM_σ computed over the visual attention region. (Color figure online)

neighborhood block operations which are executed approximately 100 times faster than traditional methods. This performance is achieved by rearranging the blocks into columns and computing the core process over the resulted matrix with a single call instead of calling each block individually.

- The perceptual mask process is not calculated for the whole frame, but only on visual attention region. The rest of the frame is filling with 1's in order to avoid any change on the watermarked frame according to (4).
- According to the provisions in [17] the foreground region is calculated by resizing the original frame to a coarse 64×48 representation, that allows calculating the salient frame regions with good performance and faster.

4   Proposed Algorithm

4.1   Watermark Embedding Process

Since the security level of a watermarking scheme is proportional to the number of elements that an attacker needs to successfully estimate the secret watermark signal [20]; within the proposed algorithm three elements are introduced in order to increment its security. These elements are provided to the watermark detector as secret keys. Figure 4 shows the complete diagram of the watermark embedding process, which is described in the following 8 steps:

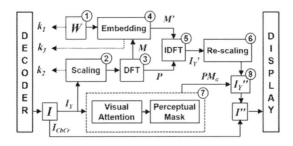

Fig. 4. Proposed watermark embedding process.

(1) Compute a 1-D zero mean binary pseudo-random pattern composed of L values generated by using the first user's secret key k_1. The watermark sequence $W = \{w_i \mid i = 1 \ldots L/2\}$ is created by dividing the above pattern into two blocks and applying the XOR operation between them. (2) Once a frame is decoded, its luminance component I_Y is isolated and rescaled to a standard size $N_1 \times N_2$ using the second user's secret key $k_2 = \{N_1, N_2\}$. (3) Compute the DFT of the rescaled version of I_Y:

$$F(u, v) = \sum_{x=1}^{N_1} \sum_{y=1}^{N_2} I_Y(x, y) e^{-j2\pi(f_1 x/N_1 + f_2 y/N_2)}, \tag{5}$$

and get the magnitude $M(u,v) = |F(u,v)|$ and phase $P(u,v)$ components. (4) Two radius r_1 and r_2 are chosen in order to be used for selecting an annular area in $F(u,v)$ forming the third user's secret key $k_3 = \{r1, r2\}$. The corresponding annular area A is defined as $A = \pi (r_2^2 - r_1^2)$. To get the best pair of radius in terms of robustness, the condition $A/4 \geq L/2$ should be satisfied, considering the four quadrants of the DFT magnitude. In order to improve robustness against video low bit rates video compression this area must cover the middle frequency components around the zero frequency term. Then, using the first and second quadrants of the upper half part of the DFT magnitude, respectively, the magnitude difference d over the annular region A is determined as $d = M_i (u_j, v_j) - M_i (-u_j, v_j)$ where $j = 1 \ldots L/2$, and denotes and index pointing to the watermark element. Introducing a watermark strength parameter α, the watermark signal is embedded in the annular region A as follows: If watermark data bit $w_i = 0$ and $d < (-\alpha)$ then $M_i (u_j, v_j)$, and $M_i (-u_j, v_j)$ remains unchanged, but if $w_i = 0$ and $d \geq (-\alpha)$ then $M_i (u_j, v_j)$, and $M_i (-u_j, v_j)$ are modified according to:

$$\begin{aligned} M_i'(u_j, v_j) &= M_i(u_j, v_j) - (\alpha + d) \\ M_i'(-u_j, v_j) &= M_i(-u_j, v_j) + (\alpha + d) \end{aligned} \tag{6}$$

On the other hand, if the watermark data bit $w_i = 1$ and $d > \alpha$ then $M_i (u_j, v_j)$, and $M_i (-u_j, v_j)$ will not be modified, but if $w_i = 1$ and $d \leq \alpha$ then $M_i (u_j, v_j)$, and $M_i (-u_j, v_j)$ are modified according to:

$$\begin{aligned} M_i'(u_j, v_j) &= M_i(u_j, v_j) + (\alpha - d) \\ M_i'(-u_j, v_j) &= M_i(-u_j, v_j) - (\alpha - d) \end{aligned} \tag{7}$$

In (6) and (7), $i = 1 \ldots L/2$, denotes an index pointing to the corresponding w_i watermark element, $M_i (u_j, v_j)$ and $M_i (-u_j, v_j)$ are the original magnitude coefficients and $M_i' (u_j, v_j)$, and $M_i' (-u_j, v_j)$ represent its watermarked versions. According to the DFT symmetrical properties in order to produce real values, after the upper half part of the DFT magnitude coefficients are modified, the lower part should be modified symmetrically. (5) A first version of the watermarked luminance component is obtained by applying the inverse DFT operation into the watermarked magnitude $M'(u,v)$ together with the corresponding original phase $P(u,v)$ component, as follows:

$$I'_Y = \text{IDFT}(F'), F' = (M', P) \tag{8}$$

(6) Watermarked luminance component I'_Y is rescaled to its original dimensions. (7) The perceptual mask PM_σ is computed over the visual attention region previously calculated by using spatial and temporal information. (8) The final watermarked luminance component I''_Y is computed by mixing the original one I_Y and its first watermarked version I'_Y according to (4). Before being displayed, the watermarked frame I'' is built by combining the watermarked luminance component and the original chrominance data. Additionally, in order to increase the security of the proposed watermark algorithm, secret keys may be renewed randomly in a desirable lapse of time, depending on the application. In this way, several video sequences may be watermarked by using different secret keys, and thus avoiding its estimation by an attacker. This security strategy would not affect the total performance of the proposed method.

4.2   Watermark Detection Process

In order to put in evidence a dishonest user that performs an illegal distribution, the video sequence is decoded and then the watermark detection process is carried out as follows (Fig. 5): (1) Isolate the luminance component from the watermarked frame and rescale it to the standard size using the second user's secret key $k_2 = \{N_1, N_2\}$. (2) Calculate the DFT transform $F' = (u,v)$ of the watermarked luminance component and get its watermarked magnitude $M' = (u,v)$. (3) Compute the annular area A using the third user's secret key $k_3 = \{r_1, r_2\}$ and then compute the subtraction operation $s_i = M_i'(u_j, v_j) - M_i'(-u_j, v_j)$ over the first and second quadrants of the upper half part of the watermarked DFT magnitude $M'(u,v)$ in the annular region A. Extract the watermark pattern by using the *sign* function as follows: if $sign(s_i)$ is a positive, or zero value then $w_i' = 1$, otherwise $w_i' = 0$, where $i = 1 \ldots L/2$. (4) Compute the original watermark sequence W using the first user's secret key k_1 according to the procedure explained above. (5) Calculate the bit error rate (BER) between the extracted watermark and the embedded one.

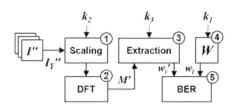

Fig. 5. The proposed watermark detection process.

5   Experimental Results

Several experiments were conducted to measure the performance of the proposed video watermarking scheme. A prototype was implemented on MATLAB© R2013a to simulate the proposed algorithm by using an Intel® Core™ i7-4770 CPU at 3.4 GHz, 16 GB DDR2 RAM and a GeForce GTX 650 graphics card. We tested five video sequences at 60 frames per second (FPS), in CIF (352 × 288) and 4CIF (704 × 576) formats, and at least 150 frames of each one. To perform the watermark embedding process, watermark code's length is defined as $L = 160$. Experimentally, the secret key values were set as $k_2 = \{512, 512\}$ and $k_3 = \{80, 81\}$. In order to refine the motion region and improve its performance, all the connected components that have a fewer area than $P = 20$ has been removed from $\Delta I'_{BW}$. Finally, the square window S employed to calculate the perceptual mask was a window of 9 × 9 pixels size.

5.1   Watermark Imperceptibility

In order to measure the visual quality distortion produced by the watermark embedding process, we use objective and subjective metrics computed between the luminance components of the watermarked I-frames and the original ones, along each video sequence. The metrics used are the well-known peak signal to noise ratio (PSNR), Structural Similarity Index (SSIM), and Video Quality Model (VQM), this one is computed by using the MSU Video Measurement Tool [21]. Figure 6 shows the average results of the three metrics applied to the first 150 frames of the five tested video sequences.

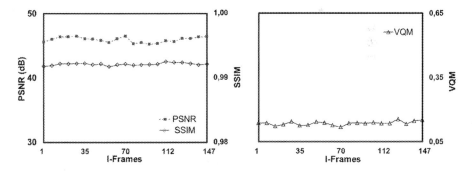

Fig. 6. Average results obtained after applying (a) PSNR, SSIM and (b) VQM metrics between original and watermarked video sequences.

PSNR values over 42 dB are considered as good as original frames. All the PSNR values obtained in our experiments are around 46 dB, which means that in terms of PSNR the video watermark embedding process does not damage the visual quality. SSIM metric was designed to improve traditional methods like PSNR and have proven to be consistent with human visual perception. The proposed scheme has an excellent performance in terms of SSIM metric obtaining an average of 0.992, remembering that a value of 1 is reached when the two frames are identical. VQM is a DCT-based video quality metric based on simplified human spatial-temporal contrast sensitivity model. Experimental results have found that VQM has a high correlation with respect to subjective video quality assessment and that's because it has been adopted by the ANSI as an objective video quality standard. Considering that VQM score decreases as the quality of the compressed video rises and it is zero for lossless compressed video, the proposed scheme get good imperceptibility by averaging VQM = 0.136. Generally speaking, with these results we can conclude that human visual perception could not recognize the difference between the original video and its watermarked version.

5.2   Watermark Robustness

To measure the robustness of the proposed scheme against video transcoding, the watermark signal is embedded in a video sequence with RGB24 uncompressed video codec with AVI container. Then, the watermarked videos are processed by changing its frame rate, resizing them to several spatial resolutions and performing video format conversion to MPEG-4. The performance achieved by the proposed technique is compared with a recently video watermarking system proposed in [22] which is one of the most robust video watermarking proposals with similar purposes than proposed. The obtained results are shown in Table 1.

Table 1. Robustness against transcoding tasks

Transcoding task		BER	
		[22]	Proposed
Format conversion	MPEG-4	0.000	0.016
Resize to	640 × 480	0.002	0.000
	320 × 240	0.004	0.001
Frame rate	FPS = 24	0.000	0.000
	FPS = 30	0.000	0.000

Changing the AVI uncompressed video to MPEG-4 video codec standard is considered as a very aggressive task since the whole video sequence information is processed and reorganized. Despite this, after the watermarked video is codified in MPEG-4 video format, watermark detection process only loses about 1 % of the embedded watermark signal. On the other hand, as we explained before the watermark embedding process keeps a standard frame resolution when the watermark signal is embedded. This feature provides a remarkable robustness respect to the change of spatial resolution. Finally, since changing original frame rate only changes the frame rate without changing the frame count, we can observe that after frame rate changing is completed, no error is returned in the extracted bits.

However, often transcoding tasks are performed at the same time in order to adapt the video characteristics and then to be displayed on different devices. Then two additional scenarios are simulated to measure the robustness of the proposed algorithm in practical situations. The watermarked video is codified by using the Advanced Simple Profile of the MPEG-4 video format standard with 30 FPS and different spatial resolutions and bit rates. The results are displayed in Table 2, where a good performance of the proposed scheme is shown, even when the video is encoded at very low bitrates.

From Table 2 we can determine that the proposed scheme can successfully detect embedded watermark signal even if aggressive transcoding tasks are performed. Finally, we measure the performance of the proposed scheme in order to determine if it can be applied in video distribution systems at the same time that the video content is streamed. To meet with this requirement, the complexity of the watermarking algorithm should be as low as possible because video frames are transmitted at a very high rate, usually 30 FPS, to get a fluent video stream. This means that video watermarking embedding process should be able to handle such rate, i.e. the process should take maximum 0.033 s. According to the elapsed time, the proposed scheme can be summarized as three main blocks: To calculate the visual attention region, compute the perceptual mask, and perform the watermark embedding process. The averaging time of each processing block is summarized in Table 3.

Table 2. Robustness in practical situations

MPEG-4@30 FPS		Spatial resolution			
		720×480	640×360	480×270	320×240
[22]	BER	0.002	0.010	0.020	0.047
	Bit rate	3.0 Mbps	2.2 Mbps	0.9 Mbps	0.4 Mbps
Proposed	BER	0.000	0.008	0.011	0.028
	Bit rate	2.7 Mbps	1.7 Mbps	869 Kbps	463 Kbps

Table 3. Average time of the watermarking embedding process

	Visual attention region		Perceptual mask	Embedding process	Total time
	Foreground region	Motion region			
Time (seconds)	0.0085	0.0004	0.0051	0.0151	0.0291

6 Conclusions

A video watermarking technique to ensure copyright protection in video distribution systems is proposed in this paper. Our main contribution consists on the introduction of a fast and effective HVS-based method which combines two complementary algorithms calculated over spatial and temporal domains given a hybrid algorithm. Combining the above techniques a visual attention region is computed and then a perceptual mask is calculated with the aim to reduce the watermark strength over those regions where the user's attention is focused. A prototype was developed and several experiments were performed through which it could be shown that the proposed scheme meets with imperceptibility, robustness and low-computational complexity requirements and it is a viable option to be implemented in practical situations.

Acknowledgements. Authors thank the Post-Doctorate Scholarships Program from DGAPA at Universidad Nacional Autónoma de México as well as Instituto Politécnico Nacional by the support provided during the realization of this research.

References

1. Zeng, W., Yu, H., Lin, C.Y. (eds.): Multimedia Security Technologies for Digital Rights Management, vol. 18. Academic Press, Cambridge (2011)
2. Ulin, J.: The Business of Media Distribution: Monetizing Film TV and Video Content in an Online World. CRC Press, Boca Raton (2013)
3. Doerr, G., Dugelay, J.L.: A guide tour of video watermarking. Sig. Process. Image Commun. **18**(4), 263–282 (2003)
4. Ku, W., Chi, C.-H.: Survey on the technological aspects of digital rights management. In: Zhang, K., Zheng, Y. (eds.) ISC 2004. LNCS, vol. 3225, pp. 391–403. Springer, Heidelberg (2004)
5. Wang, J., Liu, J.C., Masilela, M.: A real-time video watermarking system with buffer sharing for video-on-demand service. Comput. Electr. Eng. **35**(2), 395–414 (2009)
6. Lu, C.S., Chen, J.R., Fan, K.C.: Real-time frame-dependent video watermarking in VLC domain. Sig. Process. Image Commun. **20**(7), 624–642 (2005)
7. Zhao, Z., Yu, N., Li, X.: A novel video watermarking scheme in compressed domain based on fast motion estimation. In: Proceedings of International Conference on Communication Technology Proceedings, vol. 2, pp. 1878–1882 (2003)
8. Liu, Y., Zhao, J.: A new video watermarking algorithm based on 1D DFT and Radon transform. Sig. Process. **90**(2), 626–639 (2010)
9. Koz, A., Alatan, A.A.: Oblivious spatio-temporal watermarking of digital video by exploiting the human visual system. IEEE Trans. Circ. Syst. Video Technol. **18**(3), 326–337 (2008)
10. Jung, H.-S., Lee, Y.-Y., Lee, S.U.: RST-resilient video watermarking using scene-based feature extraction. EURASIP J. Appl. Sig. Process. **2004**, 2113–2131 (2004)
11. Levicky, D., Foris, P.: Implementations of HVS models in digital image watermarking. Radioengineering **16**(1), 45–50 (2007)
12. Awrangjeb, M., Kankanhalli, M.S.: Lossless watermarking considering the human visual system. In: Kalker, T., Cox, I., Ro, Y.M. (eds.) IWDW 2003. LNCS, vol. 2939, pp. 581–592. Springer, Heidelberg (2004)
13. Winkler, S.: Video quality and beyond. In: Proceedings of European Signal Processing Conference, pp. 3–7, September 2007
14. Osberger, W.M., Rohaly, A.M.: Automatic detection of regions of interest in complex video sequences. In: Photonics West 2001-Electronic Imaging, pp. 361–372. International Society for Optics and Photonics, June 2001
15. Buswell, G.T.: How People Look at Pictures. The University of Chicago Press, Chicago (1935)
16. Hou, X., Harel, J., Koch, C.: Image signature: highlighting sparse salient regions. IEEE Trans. Pattern Anal. Mach. Intell. **34**(1), 194–201 (2012)

17. Eckert, M.P., Buchsbaum, G.: The significance of eye movements and image acceleration for coding television image sequences. In: Watson, A.B. (ed.) Digital Images and Human Vision, pp. 89–98. MIT Press, Cambridge (1993)

18. Otsu, N.: A threshold selection method from gray-level histograms. IEEE Trans. Syst. Man Cybern. **9**(1), 62–66 (1975)

19. Bartolini, F., Barni, M., Cappellini, V., Piva, A.: Mask building for perceptually hiding frequency embedded watermarks. In: 1998 Proceedings in International Conference on Image Processing, ICIP 1998, vol. 1, pp. 450–454, October 1998

20. Cayre, F., Fontaine, C., Furon, T.: Watermarking security: theory and practice. IEEE Trans. Signal Process. **53**(10), 3976–3987 (2005)

21. Video Group of MSU Graphics and Media Lab. http://www.compression.ru/video/

22. Lee, M.J., Im, D.H., Lee, H.Y., Kim, K.S., Lee, H.K.: Real-time video watermarking system on the compressed domain for high-definition video contents: practical issues. Digital Signal Proc. **22**(1), 190–198 (2012)

Facial Expression Recognition Adaptive to Face Pose Using RGB-D Camera

Yuta Inoue, Shun Nishide$^{(\boxtimes)}$, and Fuji Ren

Faculty of Engineering, Tokushima University,
2-1 Minami-Josanjima-cho, Tokushima, Tokushima 770-8502, Japan
nishide@is.tokushima-u.ac.jp

Abstract. In this paper, we propose a facial expression recognition method for non-frontal faces using RGB-D camera. The method uses the depth information of the RGB-D camera to calculate the face pose, modeled using a cylinder. Feature points obtained by the RGB-D camera, modified by the face pose, are compared with Action Units of the Facial Action Coding System for recognition of facial expression. Experiments were conducted using facial images in three types of angles and four expressions: anger, sadness, happiness, and surprise. Results of the experiments have shown that the method is rather robust to roll rotations than yaw and pitch rotations.

Keywords: Facial expression recognition · Image processing · RGB-D camera

1 Introduction

Facial expression recognition is a demanding need for smooth communication between human and robot [1]. As in humans' communication, facial expressions contain various information, such as how to talk or what to say. The objective of our work is to develop a facial expression recognition method for human robot communication using camera images.

Facial expression recognition for human robot communication requires the method to be adaptable to different face poses. Most expression recognition methods use images of frontal face [2], which are not always acquirable during conversation with robot. We are focusing on creating a model capable of recognizing expressions for non-frontal faces [3]. The previous work required templates to estimate the face pose. In this paper, we propose a facial expression recognition method for non-frontal faces by applying depth information from RGB-D camera to adapt to different face poses.

The proposed method consists of two steps. In the first step, the posture of the head is modeled using a cylinder model. The rotation angle is estimated to calculate the feature points of the reconstructed frontal face. In the second step, the facial expression is recognized using Action Units (AU) of the Facial Action Coding System (FACS). Experimental results with facial images from three different angles imply the possibility of the method to adapt to different face poses for recognition of facial expression.

© Springer International Publishing Switzerland 2016
H. Fujita et al. (Eds.): IEA/AIE 2016, LNAI 9799, pp. 422–427, 2016.
DOI: 10.1007/978-3-319-42003-3_36

2   Facial Expression Recognition Using RGB-D Camera

Facial expression recognition is often conducted using appearance-based features or geometry-based features [4]. Appearance-based features are those that express the texture of the face. Geometry-based features are those that express the location and shape of the facial components. As we focus on adaptation to face poses, we use geometry-based features as they could be rotated around a certain point. In this paper, we consider recognition of four emotions (joy, surprise, sadness, and anger) as our preliminary experiment.

The proposed facial expression recognition method consists of two steps. In the first step, we calculate the geometry-based features of the frontal face from the actual image with the inclined face. A cylinder model is used to model the face pose. In the second step, the motion vectors of the features are used to calculated the facial expression. A detailed information of the methods are presented in the following subsections.

2.1   Cylinder Model for Face Modeling

We utilize a cylinder model (Fig. 1) for modeling the face pose, as used by Kumano, et al. [5]. From the camera data, the line segment connecting the two sides of the cheek is used as the diameter, and the center point of the line segment is used as the rotating point of the cylinder model. The model is rotated along the inclination of the face. The parameters are normalized using 3D affine transformation so that the diameter becomes constant and the rotating point becomes the origin.

Fig. 1. Cylinder model

Using the depth information obtained from the RGB-D camera, the rotation angle of the cylinder model is calculated using Euler angles (pitch, yaw, and roll). The feature points obtained automatically from the RGB-D camera are rotated in the opposite direction of the calculated angles to obtain the feature points of the frontal face image. An example of the feature points for the frontal face which were calculated from an inclined face is shown in Fig. 2.

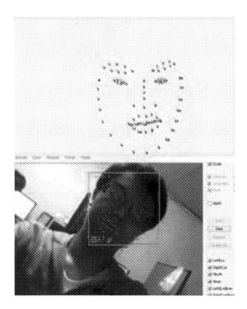

Fig. 2. Feature points of the frontal face

2.2   Facial Expression Recognition Using Action Units

In this paper, we use motion vectors of feature points for recognition of facial expression. For each feature point, motion vector is calculated by the difference of coordinates between subsequent frames as

$$\theta = tan^{-1}(\frac{\Delta x}{\Delta y}) \qquad (-\pi < \theta < \pi) \tag{1}$$

$$\sqrt{\Delta x^2 + \Delta y^2} = 1, \tag{2}$$

where θ represents the rotation angle of the motion vector along the X axis, and $(\Delta x, \Delta y)$ represents the motion vector of the feature point.

For recognition of facial expression, we utilize Action Units (AU) from Facial Action Coding System (FACS) [6]. In this paper, we use ten AUs for recognition of expressions listed in Table 1. The AUs are each used for recognizing a specific emotion, expressed as "Corresponding Emotion" in Table 1. Recognition of facial expression is done by matching motion vectors with AU. An expression is output when the matching score exceeds a threshold, determined manually in this paper.

3   Evaluation Experiment of Facial Expression Recognition

To evaluate the effectiveness of the method, we conducted a preliminary experiment for facial images of four emotions. In the first experiment, we evaluated

Table 1. Action units

Action unit number	Feature	Corresponding Emotion
AU12	Lip Corner Puller	Joy
AU5	Upper Lid Raiser	Surprise
AU27	Mouth Stretch	Surprise
AU1	Inner Brow Raiser	Sadness
AU4	Brow Lowerer	Sadness
AU15	Lip Corner Depressor	Sadness
AU2	Outer Brow Raiser	Anger
AU9	Nose Wrinkler	Anger
AU10	Upper Lip Raiser	Anger
AU17	Chin Raiser	Anger

Fig. 3. Experimental environment

the precision of the proposed method to calculate motion vectors. In the second experiment, we evaluated facial expression recognition for frontal face and six face poses.

The experiment was conducted in an indoor environment using Intel(R) RealSens(TM) 3D Camera (Fig. 3). The camera images were taken 30 fps in a resolution of 640 × 480 pixels. 78 feature points were calculated for each frame.

3.1 Precision of Motion Vector Calculation

To evaluate the precision of motion vector calculation, we calculated the motion vectors while moving the head for a normal emotionless face. As there is no facial expression change, the motion vectors should ideally be zero as the proposed method calculates motion vectors for the reconstructed frontal face. The average norm of the 78 motion vectors was calculated.

Table 2. Average norm of motion vectors

	Without proposed method	Proposed method
No motion	1.08	1.31
Horizontal motion	13.29	3.26
Vertical motion	20.97	8.50
Rotation	17.66	11.46

The evaluation was done for four cases: no motion, horizontal head motion, vertical head motion, and head rotation. The results are shown in Table 2. As shown in Table 2 the average norm decreased using the proposed method, but a better estimation of the head rotation angle is required to obtain a more promising result.

3.2 Facial Expression Recognition Results

To evaluate facial expression recognition, we conducted recognition experiments for four expressions and seven different face poses for a single person. The average recognition result for each case is shown in Table 3. The recognition rate was calculated based on 15 trials for each pose and expression.

From Table 3, it is notable that the proposed method is rather robust to Roll motions, while the performance degrades for Yaw motions. The recognition result shows better performance for Joy and Surprise than the performance for Sadness and Anger. As the recognition result for the frontal face for Joy and Surprise are better than Sadness and Anger, it could be considered that the selection of AU for Sadness and Anger should be reconsidered to achieve a better performance.

Table 3. Facial expression recognition result (in percentage)

	Frontal Face	Roll($-30°$)	Roll($30°$)
Joy	93	53	60
Surprise	93	80	73
Sadness	73	46	46
Anger	73	60	60
All	83	60	60

	Pitch($-30°$)	Pitch($30°$)	Yaw($-30°$)	Yaw($30°$)
Joy	73	53	53	53
Surprise	53	53	46	60
Sadness	26	40	26	26
Anger	46	46	26	33
All	50	48	38	43

Although the recognition performance degrades when the face pose is inclined, the experiment shows that the method is capable of compensating for the inclination to recognize facial expressions. Further evaluations and improvement of the model are left as future work.

4 Conclusions and Future Work

In this paper, we proposed a facial expression recognition method for non-frontal faces using an RGB-D camera. The proposed method uses a cylinder model to calculate the rotation angle of the face. The feature points obtained by the RGB-D camera are rotated in the opposite direction of the rotation angle. Motion vectors of the feature points are then calculated for matching with Action Units to calculate the facial expression. Preliminary experimental results imply the validity of the method to apply to non-frontal faces in the camera image.

The current work is applied for seven face poses and four emotions. For future work, we first plan to increase the number of face poses and emotions to evaluate the effectiveness of the method. Next, we plan to increase the number of AUs to achieve a higher performance of expression recognition. Further on, we plan to implement the system to an actual human robot interaction system. We believe that our work will contribute to smooth communication between human and robot.

Acknowledgments. This research was partially supported by JSPS KAKENHI Grand Number 15H01712.

References

1. Chibelushi, C.C., Bourel, F.: Facial expression recognition: a brief tutorial overview. In: CVonline: On-Line Compendium of Computer Vision (2003)
2. Shan, C., Gong, S., McOwan, P.W.: Facial expression recognition based on local binary patterns: a comprehensive study. Image Vis. Comput. **27**, 803–816 (2009)
3. Ren, F., Huang, Z.: Facial expression recognition based on AAM-SIFT and adaptive regional weighting. IEEJ Trans. Electr. Electron. Eng. **10**(6), 713–722 (2015). doi:10.1002/tee.22151
4. Tian, Y.L., Kanade, T., Cohn, J.: Facial expression analysis. In: Moeslund, T.B., Hilton, A., Krüger, V., Sigal, L. (eds.) Handbook of Face Recognition, pp. 247–275. Springer, New York (2005)
5. Kumano, S., Otsuka, K., Yamato, J., Maeda, E., Sato, Y.: Pose-Invariant facial expression recognition using variable-intensity templates. In: Yagi, Y., Kang, S.B., Kweon, I.S., Zha, H. (eds.) ACCV 2007, Part I. LNCS, vol. 4843, pp. 324–334. Springer, Heidelberg (2007)
6. Ekman, P., Friesen, W.V.: Facial Action Coding System: Investigator's Guide. Consulting Psychologists Press, Palo Alto (1978)

Visible Spectrum Eye Tracking for Safety Driving Assistance

Takashi Imabuchi[(✉)], Oky Dicky Ardiansyah Prima,
and Hisayoshi Ito

Faculty of Software and Information Science, Iwate Prefectural University,
152-52, Sugo, Takizawa, Iwate 020-0693, Japan
g236o001@s.iwate-pu.ac.jp

Abstract. Although many studies have proposed eye tracking methods to be used for driving assistance systems, these concepts have not been put into practical. The most considered issue to track drivers' eyes is the effects of ambient infrared spectrum from the sunlight, making center of pupils and the corneal reflections from the tracker's infrared light source undetectable. In this study, we propose a visible spectrum eye tracking to calculate the driver's gaze points only if the gaze detection from the infrared spectrum eye tracking failed. The proposed eye tracking uses an automated facial landmark detection to calculate the head pose which enabling head movement compensation, and a learning-based calibration model to eliminate the calibration processes.

Keywords: Visible spectrum eye tracking · Driving assistance · Head pose compensation

1 Introduction

In Japan, the number of traffic accidents shows a significant decrease for the last decade [1]. This is considerably resulted from the promotion of traffic safety measure, the comprehensive development of urban transport system, and the use of intelligent transport system. However, accidents caused by drivers with a diminished vigilance level are accounted at nearly 40 % of the total. This has been a problem of serious concern to society. Starting from 2011, the 5th advanced safety vehicle (ASV) promotion program supported by the Japan Ministry of Land, Infrastructure, Transport and Tourism (MLIT) is focusing to support the development of safety driving assistance.

Many non-intrusive techniques based on computer vision have been developed to monitor behaviors of the drivers. Typically, these techniques observe the changes of facial features to analyze the level of drowsiness and fatigue which reduce the ability of the driver to control the vehicle. More advanced techniques enable head pose and gaze estimations [2]. However, despite the wide use of eye tracking, it is not yet ready to be installed to the automobile.

The widely available eye trackers are using near-infrared (NIR) camera to detect the center of pupils and corneal reflections (glints) from its NIR illuminator. The NIR camera allows tracking to a wide range of lightning conditions. Detection of pupils can be conducted properly using bright- and dark-pupil techniques [3]. Moreover,

© Springer International Publishing Switzerland 2016
H. Fujita et al. (Eds.): IEA/AIE 2016, LNAI 9799, pp. 428–434, 2016.
DOI: 10.1007/978-3-319-42007-3_37

the relationship between the pupil center and the glints enables gaze estimation without being affected by the head pose changes. However, the detection of the pupil and glint may be affected by the presence of ambient NIR light, especially from the direct sunlight. This issue becomes more significant for users wearing glasses. Figure 1 shows two different poses of facial images captured in low and high ambient light conditions using NIR and visible spectrum cameras, respectively. The detection of pupils and glints from the NIR image in high ambient light condition failed while irises were still detectable in the same pose of the visible spectrum image.

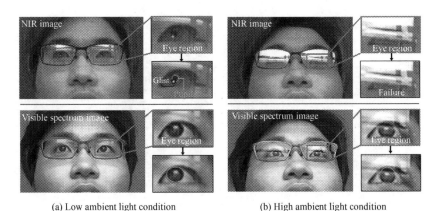

(a) Low ambient light condition   (b) High ambient light condition

Fig. 1. Eye-regions captured by NIR (upper) and visible spectrum (lower) cameras

Eye tracking based on solely NIR cameras is not adequately stable to estimate driver's gaze during the daylight. Similar problem has been reported when using a mobile eye tracker outdoor [4]. The visible spectrum camera may be effective to estimate driver's gaze along with NIR cameras. However, head movement compensation is difficult to implement because this camera cannot observe the glints.

In this paper, we propose a visible spectrum eye tracking for safety driving assistance. The proposed eye tracking includes a head movement compensation based on the head pose information derived from facial landmarks, and a calibration free mechanism by use of a learning-based calibration model.

2 Related Work

Many studies have been conducted to develop the visible spectrum eye tracking. The accuracies in terms of visual angle were approximately $2°$ for the head-mounted [5], but were less accurate for the remote (non-contact) eye tracker [6–8]. Since the scope of this study is the eye tracker for safety driving assistance, further discussion will be limited to the remote eye tracker. Iris extraction for the visible spectrum eye tracking can be done by simply thresholding the eye region, template matching, or applying starburst algorithm (Fig. 2). The iris center is calculated as the center of an ellipse or a circle fitted to the boundary of the extracted iris. The gaze direction can be estimated by calibrating the coordinates between the iris centers with some fixation points of the screen. For the eye tracker in a vehicle, the screen's coordinate can be considered as the coordinate of the view screen, the view extent of the driver represented by a scene camera.

The implementation of head movement compensation is the main issue of visible

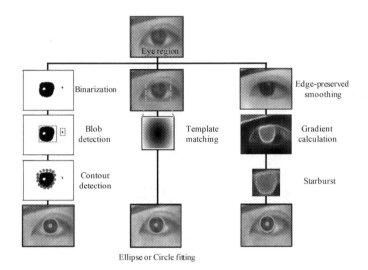

Fig. 2. Iris extraction in visible spectrum eye tracking

spectrum eye tracking because there are no reference points to estimate the relationship of the head pose changes against the view of the user. Correlating a 3D face model with the facial landmarks can be used to estimate the head pose [9]. However, the derived pose information will be less accurate if the 3D face model is greatly different with the subject' face.

3   Proposed System

To offer the visible spectrum eye tracking for safety driving assistance, the eye tracker must have a reliable head movement compensation and an automated calibration process to estimate the gaze. We design the visible spectrum eye tracking to work together with the NIR eye tracking and thus both tracker can share the information of head pose and projection matrices to estimate the gaze points.

3.1   Head Movement Compensation

To enable the head movement compensation, we implement an automated facial landmarks detection [9] to the subject's face. Perspective-n-Point camera pose determination (PnP) [10] is then applied to the derived landmarks to estimate the six degree of freedom (6-DOF) of the head pose (*x, y, z, pitch, yaw, roll*). Figure 3 shows the relationship among facial landmarks, head pose, and the view screen. Here, the initial view screen is the field of view of the subject during the calibration process. Figure 3(a) shows detected facial landmarks and the estimated head pose from those landmarks. Once the projection matrix to transform the coordinates of the iris center and to the view screen has been calculated, all gaze points will be plotted into the view screen. This view screen will be treated as an initial screen to compensate the head movement. Figure 3(b) shows the location of the new view screen after the subject's head moving toward the upper left. All gaze points fall into the new view screen can be transferred to the coordinate system of the initial system. Figure 4 shows the information resulted from the PnP to calculate the distance from the face to the view screen and to calibrate the estimated pitch and yaw angles. The roll angles estimated by the PnP can be used directly without calibration.

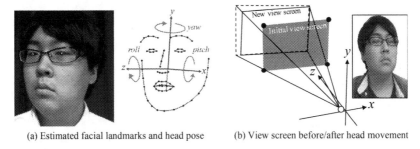

(a) Estimated facial landmarks and head pose   (b) View screen before/after head movement

Fig. 3. The relationship between facial landmarks, head pose, and the view screen

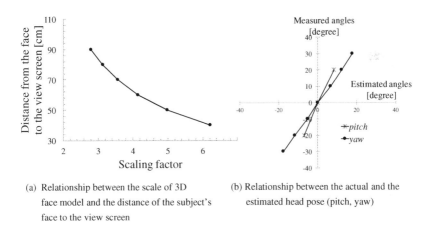

(a) Relationship between the scale of 3D face model and the distance of the subject's face to the view screen

(b) Relationship between the actual and the estimated head pose (pitch, yaw)

Fig. 4. Information resulted from the PnP

The projection matrix has to be calculated through a calibration process. However, it would be impractical for the proposed eye tracking if it requires that process to the driver. To cope with this problem, we introduce an automated calibration, described in the next section.

3.2   Automated Calibration

As the NIR eye tracking doesn't need a calibration process, we can derive the relationship between head pose and the view screen. We adapted this relationship to the visible spectrum eye tracking to enable automated calibration because the extent of the view screen derived from the NIR and visible spectrum eye tracking are considerably equal. This process needs a previously generated calibration model based on calibration data of multiple subjects. Figure 5 shows the coordinates of reference points and the corresponding points of iris center calculated from six subjects. These corresponding points are averaged (points in dots) for each reference point to calculate the projection matrix and the view screen.

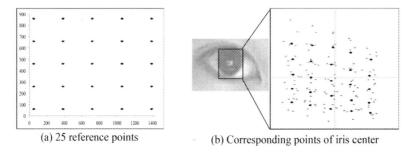

(a) 25 reference points   (b) Corresponding points of iris center

Fig. 5. Calibration model based on data learning

4   Evaluation

We conducted an experiment to assess the accuracy of the proposed visible spectrum eye tracking. Six subjects participated in the experiment. Each subject was requested to fixate nine points displayed on the screen positioned approximately 50 cm away. Figure 6 shows the distribution of gaze points of all participants. The overall accuracy

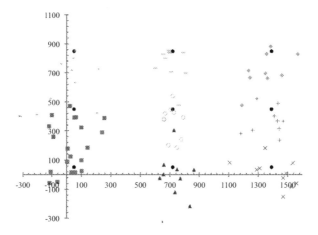

Fig. 6. Distribution of gaze points of all participants

calculated as the average error in visual angle between the reference points and the estimated gaze points was 3.45°. This accuracy is comparable with the prior research [6–8] although our experiment was done with head free and without a calibration process.

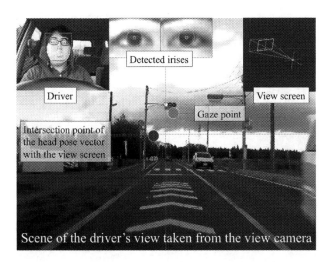

Fig. 7. Prototype system of eye tracking for safety driving assistance (Color figure online)

5   Conclusion

In this study, we have proposed a visible spectrum eye tracking for safety driving assistance. The proposed eye tracking works along with the NIR's and will calculate the driver's gaze points only if the NIR eye tracking is failed. The implementation of head movement compensation and automated calibration were proved to be effective for the visible spectrum eye tracking. Figure 7 shows our prototype system of safety driving assistance.

References

1. Cabinet Office: White paper on traffic safety in Japan. http://www8.cao.go.jp/koutu/taisaku/h25kou_haku/english/wp2013-pdf.html. Accessed 20 Dec 2015
2. Vicente, F., Huang, Z., Xiong, X., De la Torre, F., Zhang, W., Levi, D.: Driver gaze tracking and eyes off the road detection system. IEEE Trans. Intell. Transp. Syst. **16**(4), 2014–2027 (2015)
3. Ebisawa, Y.: Unconstrained pupil detection technique using two light sources and the image difference method. In: Visualization and Intelligent Design in Engineering, pp. 79–89 (1989)
4. Evans, K.M., Jacobs, R.A., Tarduno, J.A., Pelz, J.B.: Collecting and analyzing eye-tracking data in outdoor environments. J. Eye Mov. Res. **5**(2): 6, 1–19 (2012)

5. Ryan, W.J., Duchowski, A.T., Birchfield, S.T.: Limbus/pupil switching for wearable eye tracking under variable lighting conditions. In: Proceedings of the 2008 symposium on Eye Tracking Research and Applications, pp. 61–64 (2008)
6. Fukuda, T., Morimoto, K., Yamana, H.: Model-based eye-tracking method for low-resolution eye-images. In: 2nd Workshop on Eye Gaze in Intelligent Human Machine Interaction (2011)
7. Corey, H., Oleg, K.: Eye tracking on unmodified common tablets: challenges and solutions. In: Proceedings of the Symposium on Eye Tracking Research and Applications, pp. 277–280 (2012)
8. Imabuchi, T., Prima, O.D.A., Kikuchi, H., Horie, Y., Ito, H.: Visible-spectrum remote eye tracker for gaze communication. In: Proceedings of SPIE 9443, Sixth International Conference on Graphic and Image Processing (ICGIP 2014), 944333 (2015)
9. Saragih, J., Lucey, S., Cohn, J.: Deformable model fitting by regularized landmark mean-shifts. Int. J. Comput. Vision **91**(2), 200–215 (2011)
10. Wu, Y., Hu, Z.: PnP problem revisited. J. Math. Imaging Vis. **24**, 131–141 (2006)

Medical Diagnosis System and Bio-informatics

3D Protein Structure Prediction
with BSA-TS Algorithm

Yan Xu, Changjun Zhou, Qiang Zhang$^{(\boxtimes)}$, and Bin Wang

Key Laboratory of Advanced Design and Intelligent Computing,
Dalian University, Ministry of Education, Dalian, China
zhangq26@126.com

Abstract. Three-dimensional protein spatial structure prediction with the amino acid sequence can be converted to a global optimization problem of a multi-variable and multimodal function. This article uses an improved hybrid optimization algorithm named BSA-TS algorithm which combines Backtracking Search Optimization Algorithm (BSA) with Tabu Search (TS) Algorithm to predict the structure of protein based on the three-dimensional AB off-lattice model. It combines the advantage of BSA which has a simple and efficient algorithm framework, less control parameters and less sensitivity to the initial value of the control parameters and the advantage of TS which has a strong ability for the global neighborhood search, and can better overcome the short-comings of traditional algorithms which have slow convergence rate and are easy to fall into local optimum. At last we experiment in some Fibonacci sequences and real protein sequences which are widely used in protein spatial structure prediction, and the experimental results show that the hybrid algorithm has good performance and accuracy.

Keywords: Protein structure prediction · BSA · TS · 3D AB off-lattice model

1 Introduction

The study of protein structure helps us understand the function of proteins and to understand how they exercise their biological function, and understanding protein-protein interaction (or other molecules) is very important for biology, medicine and pharmacy [1]. In recent years, although the test methods for the determination of protein structures develop well, it's still time-consuming, expensive, and not applied for some proteins difficult to crystallize. Therefore, we need to develop the theoretical analysis. With the development of computer technology, it has gradually become an important tool for processing large data of protein molecule.

Theoretical Prediction of protein three-dimensional structure is mainly divided into the following three steps: Firstly, the mathematical model proposed should reflect the interaction among the amino acid residues; Secondly, we have to establish a simple calculated energy function which can also distinguish correctly between natural proteins and other proteins based on the thermodynamic hypothesis, Thirdly, we have to find the appropriate global optimization methods on the corresponding model to find the minimum of energy function. Owing to the efforts of researchers, we have had lots

© Springer International Publishing Switzerland 2016
H. Fujita et al. (Eds.): IEA/AIE 2016, LNAI 9799, pp. 437–450, 2016.
DOI: 10.1007/978-3-319-42007-3_38

of achievements about the above steps. There are two widely used models named hydrophobic-polar (HP) model [2] and AB off-lattice model [3].The HP-lattice model uses two types of residues to present the amino acid chains which are hydrophobic (H) or non-polar residues and polar (P) or hydrophilic residues, and the residues on the vertices of stack cubic lattices are linked sequentially by unit-length chemical bonds [4–6]. This model is simple but it neglects local interactions which are important in protein folding [7]. The AB off-lattice model is more accurate than HP-lattice model, because the bond angles are free-to-rotate and it takes torsional energy of each pair of bonds into account [8]. Though this model neglects the effect of side chains and provides only a coarse-grained approximation to the real proteins, its off-lattice construction still reflects some basic features of real proteins [9, 10]. Due to the complexity of protein folding, it's difficult to establish an accurate protein energy function. So many simplified energy functions are proposed as in [1, 11, 12], etc. And in this paper we use one of the most widely used energy function which is the same as that of [11] to predict the protein 3D structure. As for the global optimization methods, many algorithms have been proposed. Today, many algorithms have been proposed to predict protein structure using HP-lattice model, such as Multi-Self-Overlap Ensemble (MSOE), Ant Colony Optimization (ACO) [13], Multi-crossover and mutation Partial Swarm Optimization-Tabu Search algorithm (MCMPSO-TS) [14], etc. The MOSE algorithm uses a heuristic bias function to help the theoretical protein structure form a hydrophobic core, but it is only efficient for the proteins of simple folding [13]. The ACO algorithm's inspiration is from the observation of ants' behavior of searching for food, and the algorithm structure can be divided into three parts: construct ants solutions, update pheromone, and daemon actions [13]. The MCMPSO-TS algorithm is a hybrid search algorithm which combines the particle swarm optimizer (PSO) algorithm and tabu search (TS) algorithm to get better global optimization ability [14]. The used algorithms to predict protein structure with the AB off-lattice model are as follows, PSO, Tabu Search-Particle Swarm Optimization (TPSO) [15], Levy Flight Particle Swarm Optimizer (LPSO) [16], Genetic-Particle Swarm Optimization-Tabu Search algorithm (PGATS) [17], Artificial Bee Colony algorithm (ABC) [18], etc. The PSO algorithm is put forward on the basis of information transmission process of migrating birds that every bird can not only remember the best place to find the distance of the food but also know the optimal location found by the population of all the birds [15]. The TPSO algorithm combines PSO and TS to take advantage of each algorithm to improve the search precision and the ability to jump out local optimal [15]. The LPSO algorithm adds a random process named levy flight to PSO algorithm to improve algorithm's precision [16]. The PGATS algorithm is proposed on the basis of GA-PSO algorithm which is a hybrid algorithm using the strategy of updating parameter to guarantee the diversity of population in the late run of algorithm, while enhanced TS algorithm and particle mutation strategy are combined to jump out the local minimum [17]. The ABC algorithm is inspired from the behavior of bees, and the main feature of the algorithm is that it doesn't need to know the special information of the problem while it only needs to compare the merits of the solutions [18]. Since the algorithms aim to predict the structure of proteins, the most important criterion is the lowest energy value of corresponding protein energy function. And this paper focuses on getting lower energy value. We propose a hybrid optimization algorithm based on improved BSA and TS algorithm. It's not only less sensitivity to initial values of parameters but also more efficient than the algorithms above. Experiments using AB off-lattice model shows that it can get lower energy value than nearly all the algorithms above can get.

2 3D AB Off-Lattice Model

3D AB off-lattice model was proposed by Stillinger et al. [3] on the basis of two-dimensional AB off-lattice model, which considers that the main reason for residues chain to fold into a specific spatial structure is the hydrophobicity of amino acid residues. In this model, we divide the amino acid residues into hydrophobic (A) and hydrophilic (B) two types. So, protein sequences can be simplified as sequences consisted of only A and B [19, 20], and we can predict native structure of the proteins by the relation between bond angles and bond energy of corresponding amino acids.

In the 3D space of this model, the residues of amino acid sequences are sequentially connected into the polymer with no direction keys of unit length. We'd take the bond angles between adjacent keys and the torsional angles of the two planes constituted by the adjacent three key vectors into consideration when the model is used as 3D structure of a protein [21–23]. Thus, the 3D structure of n residues is determined by $n - 2$ bond angles $\theta_1, \theta_2 \ldots, \theta_{n-2}$ and $n - 3$ torsional angles $\beta_1, \beta_2, \ldots, \beta_{n-3}$. We set the condition $-\pi \leq \theta_i, \beta_i \leq \pi$. θ_i and β_i are positive when the angles clockwise rotate. Otherwise, the angles are negative. Figure 1 is the configuration diagram of AB off-lattice model.

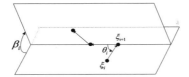

Fig. 1. Configuration diagram of AB off-lattice model

The energy function of AB off-lattice model consists of bending energy and potential energy of the Lennard-Jones type [24–27]. The energy function (E) of a sequence of n residues is expressed as Eq. (1) [28]:

$$E = \sum_{i=1}^{n-2} \frac{1}{4}(1 - \cos \theta_i) + \sum_{i=1}^{n-2} \sum_{j=i+2}^{n} 4[r_{ij}^{-12} - C(\xi_i, \xi_j)r_{ij}^{-6}] \tag{1}$$

The first part is bending potential energy which is only associated with bond angles. The second part is gravitational potential energy of any two non-adjacent residues which is not only related to polarity but also the distance of them. Here, r_{ij} represents spatial distance of non-adjacent residues, ξ_i represents the category of different residues. If the $i - th$ residue is A, $\xi_i = 1$, if it is B, $\xi_i = -1$.

Where:

$$C(\xi_i, \xi_j) = \begin{cases} +1 & \xi_i = 1, \xi_j = 1 \\ +0.5 & \xi_i = -1, \xi_j = -1 \\ -0.5 & \xi_i \neq \xi_j \end{cases} \quad (2)$$

We can see from Eq. (2) [28] that the pairs of residues AA has strong gravity, the pairs of residues BB has less gravity, and AB has weak repulsion. It reflects the true characteristic of real protein to a certain extent that hydrophobic cores form due to larger gravitational between hydrophobic residues when the protein fold into a spatial structure, at the same time, hydrophilic residues are excluded to the outside.

Therefore, the problem of protein structure prediction in the AB off-lattice model becomes the problem to find the $n - 2$ bond angles and $n - 3$ torsional angles to minimize the protein energy function E, as Eq. (3) [17]:

$$\min_{\theta_i, \beta_i \in (-\pi, \pi]} E(\theta_1, \theta_2, \ldots, \theta_{n-2}, \beta_1, \beta_2, \ldots, \beta_{n-3}) \quad (3)$$

3   BSA-TS Algorithm

3.1   BSA

Backtracking Search Optimization Algorithm (BSA) [29] was put forward by Pinar Civicioglu in 2013 for solving real-valued numerical optimization problems. Unlike many evolutionary algorithms, BSA only has one control parameter, and the initial value of the parameter has limited impact on the problem-solving performance. BSA has a simple but effective structure which enables it to easily solve multimodal problems.

3.1.1   BSA's General Algorithm Framework
The algorithm framework of BSA is analogous to differential evolution algorithms, and it can be explained by dividing its functions into five parts: initialization of population, selection-I, mutation, crossover, and selection-II [29, 30]. Figure 2 is BSA's general algorithm framework:

It has two new crossover and mutation operators to generate a trial population, and the strategy for controlling the search-space boundaries and amplitude of the search-direction matrix enables it to have very strong exploitation capabilities.

3.1.2   Improved Strategies of BSA
(1) At the end of crossover strategy, we use two randomly choices to update the individuals beyond the search-space limits. One is the standard method that we regenerate the $j - th$ element of individual M_i beyond the search-space limits using $M_{ij} = rand * (up_j - low_j) + low_j$, where up_j means the upper bound and low_j means the lower bound of the $j - th$ component of problem; another boundary control method

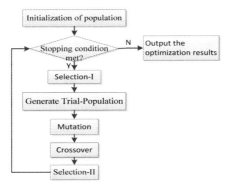

Fig. 2. BSA's general algorithm framework

is that if the value of trial population is greater than the upper bound, M_{ij} will be set as up_j, and if the value of M_{ij} is less than low_j, M_{ij} will be set as low_j.

(2) Since BSA's convergence speed is slow, this paper will use Penalty strategy to replace Selection-II strategy: First, we sort the parent population P in ascending order according to the fitness value, then we take its first half as R and combine R and the trial population T obtained after the operation of crossover as a new population Mu. Secondly, the Euclidean distance d_{ij} between the individuals of Mu is calculated, and if d_{ij} is less than D (experience value, related to the length of protein sequence), the greater fitness value between Mu_i and Mu_j will be replaced by *Penalty* (experience value, this paper take 10^{13}), and then we sort the individuals of Mu in ascending order according to the fitness value and take the first *popsize* individuals to update population P.

3.2 TS Algorithm

TS algorithm was put forward by Fred Glover in 1986 [31–33] based on the local neighborhood search algorithm. Considering that the local neighborhood search is too greedy for a local search which is easy to lead to falling into local minimum, tabu strategy is used in TS to gain more search space to jump out of a local optimal solution (not completely abandoned) and the good individuals can avoid to be missed with the application of aspiration criterion.

3.2.1 Description of TS
The process of TS can be described as follows:

Step 1: Set the algorithm parameters, initialize the current solution randomly and set tabu list blank.

Step 2: Judge whether the termination criterion is satisfied; if it is, jump to step 6.

Step 3: Generate the neighborhood solutions of the current solution, and determine the candidate solutions.

Step 4: Judge whether the aspiration criterion is satisfied; if it is, set the solution as the current solution, put its fitness into tabu list, update the optimal state, and jump to step 2.

Step 5: Judge the taboo attributes of candidate solutions, set the optimal state of candidate solutions which is not in tabu list as current solution, and jump to step 2.

Step 6: Output the optimization results.

3.2.2   Improved Strategy of TS

TS algorithm is used in the second part of BSA-TS hybrid optimization algorithm so that the algorithm can jump out of a local optimal solution. In this paper, we use a mutation operator to generate the neighborhoods of current solution. On the one hand, the mutation operator can guarantee the diversity of particles in the early stage of the algorithm; on the other hand, it produces less disruption to the current solution to ensure the global convergence of the algorithm. The strategy is as follows:

Select the $k - th$ element of individual x, then perform mutation operator, the $k - th$ element of new individual x_{new} after mutation is as follows [31]:

$$x_{new}^k = x^k + 2 * \pi * f(r) * c * rate^i \tag{4}$$

Where c is a random number between 0 and 1, $rate$ is scale factor, i is the current iteration the range of which is from 0 to K-1 (K is the size of neighbors). In this paper, we set $rate$ as 0.95 which is the same as that in reference [31]. $f(r)$ is the correlation coefficient which is defined with Eq. (5) [17]:

$$f(r) = \begin{cases} 1, r \geq 0.5 \\ -1, r < 0.5 \end{cases} \tag{5}$$

Where r is a random number between 0 and 1.

3.3   BSA-TS Algorithm

3.3.1   Parameters

In this paper, the BSA-TS algorithm is realized through MATLAB R2012a in Windows 7 system. According to the references and the experience of experiment, the parameters are set as follows: the population size $popsize$ is 200; the maximum number of iterations of the part of BSA $maxcycle$ is 3000; crossover rate $mixrate$ is 0.88; $Penalty$ is 10^{13} (appropriate adjustments are needed according to different sequence lengths). The size of neighborhood K is 800 during the part of TS algorithm; the maximum iteration of the part of TS W is 2000; tabu length $tabulength$ is 10 + N, and $rate$ is 0.95.

3.3.2   Initialization

It's need to initialize the coordinates of each individual when we calculate the corresponding fitness value. The coordinates are all calculated with Eq. (6).

$$(x_i, y_i, z_i) = \begin{cases} (0,0,0) & i = 1 \\ (0,1,0) & i = 2 \\ (\cos\theta_1, \sin\theta_1, 0) & i = 3 \\ (x_{i-1} + \cos(\theta_{i-2})\cos(\beta_{i-3}), y_{i-1} + \sin(\theta_{i-2})\cos(\beta_{i-3}), z_{i-1} + \sin(\beta_{i-3})) & 4 \le i \le n \end{cases}$$

$$(6)$$

3.3.3   Description of BSA-TS Algorithm

During the BSA-TS algorithm based on AB off-lattice model, the first step is to initialize each parameter, and generate population P and population P^{old} randomly which both contain N individuals whose range is $[-\pi, \pi]$. Then the improved BSA algorithm is used to obtain the local optimal solution. At last, the result gained by improved BSA is used as the initial solution of TS algorithm in order to avoid the low efficiency of search caused by casual initialization.

The algorithm framework of BSA-TS is as Fig. 3:

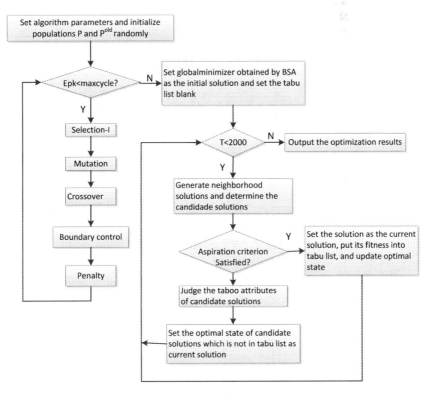

Fig. 3.   Algorithm framework of BSA-TS

As we can see from Fig. 3, the process of BSA-TS can be briefly described as follows:

Step 1: Set algorithm parameters and initialize the population P and P^{old} randomly.

Step 2: Judge the termination criterion of the part of BSA that whether $epk < maxcycle$; if it does, continue execution, else jump to step 4.

Step 3: Execute the Selection-I, mutation, crossover, boundary control, and penalty sequentially.

Step 4: Set global optimal solution obtained by BSA as the initial current solution and set the tabu list blank.

Step 5: Execute the part of improved TS, and output the optimization result when the termination criterion of TS is met.

4 Experimental Results and Discussion

4.1 Prediction with Fibonacci Sequences

Fibonacci Sequences [34] usually used in the protein structure prediction are defined as follows:

$$S_0 = A, S_1 = B, \ldots, S_{i+1} = S_{i-1} * S_i \qquad (7)$$

where '*' is a connection symbol, A denotes a hydrophilic amino acid and B denotes a hydrophobic amino acid.

Table 1 shows the trial Fibonacci Sequences. Table 2 shows the comparison of lowest energy values of the 3D structures of corresponding Fibonacci Sequences gained by PSO, TPSO, LPSO, PGATS, ABC and BSA-TS.

Table 1. Fibonacci Sequences

Length	Sequence
13	ABBABBABABBAB
21	BABABBABABBABBABABBAB
34	ABBABBABABBABBABABBABABBABBABABBAB
55	BABABBABABBABBABABBABABBABBABABBABBABABBABABBABBABABBAB

We can see from the Table 2 that the lowest energy values of the Fibonacci Sequences gained by BSA-TS are lower than those obtained by PS0, TPSO, LPSO, PGATS and ABC except the sequence length of 13. The detailed description is as

Table 2. Lowest energy values of Fibonacci Sequences

Length	E(PSO [15])	E(TPSO [15])	E(LPSO [16])	E(PGATS [17])	E(ABC [18])	E(BSA-TS)
13	−1.2329	−4.7284	−4.6159	−0.6914	−0.8875	−1.3852
21	−2.7268	−8.7379	−6.6465	−7.2423	−8.1083	−9.9878
34	−4.6756	−10.7983	−7.3375	−8.9361	−10.3953	−11.3741
55	−6.9587	−13.0352	−13.0487	−18.3413	−22.4172	−23.1124

follows. When the sequence length is 13, the lowest energy value BSA-TS get is −1.3852 while the lowest energy value TPSO gets is −4.7284 and the lowest energy value LPSO gets is −4.6159 which means the 3D structure we get is less stable than TPSO and LPSO gets but more stable than the ones other algorithms can get. Additionally, when the sequence length is 21, the lowest energy value BSA-TS gets is −9.9878 while the lowest energy value other algorithms can get is −8.7379, and when the sequence length is 34 or 55, the lowest energy values BSA-TS get are −11.3741 and −23.1124 correspondingly while the lowest energy values others get are only −10.3953 and −22.4172 correspondingly. Therefore, we can conclude that the BSA-TS algorithm can effectively predict the 3D structure of protein using Fibonacci Sequences. In addition, Table 3 lists the bond angles and torsional angles of the corresponding lowest energy configurations predicted by BSA-TS algorithm.

Table 3. The bond angles and torsional angles of the Fibonacci Sequences

Length	Bond angles						Torsional angles					
13	-0.3628	0.7875	-0.7895	0.7097	-0.6639	-0.0994	2.8530	-0.3179	-2.8831	2.8440	-1.7237	0.3422
	1.2286	-0.2492	-0.1753	-0.1750	-0.1601		-2.7739	1.9743	0.4938	2.2847		
21	1.7060	-0.2291	-1.6908	-0.1045	-1.4767	-1.5979	-2.9754	0.0819	2.4444	-2.9474	2.1367	-0.5892
	0.2081	-1.1960	0.2501	-0.7140	-1.6455	-3.0239	2.9450	-2.5439	-0.6408	-2.8518	2.9076	3.0691
	1.6622	0.5601	-2.3997	0.2236	0.5165	0.0623	-1.2841	0.1008	-2.0082	2.7850	1.2461	-3.0852
	1.3040											
34	-0.4068	0.4435	-0.0085	-0.3523	-0.1423	1.1838	2.1896	1.6225	3.1086	-1.6649	3.1415	-0.8286
	0.2099	-0.5724	-0.0215	0.1309	-0.8883	-2.1602	-0.8557	-1.4109	-0.6137	-2.8643	2.5595	2.9795
	2.4730	1.4351	2.3091	0.3918	0.6735	-0.0749	1.1119	-3.0203	1.9413	-2.8573	-1.2443	0.12009
	-1.8545	0.6571	-0.3539	1.3239	-0.8056	2.6495	-1.2598	-0.8768	-0.2696	1.9029	3.1141	-0.9851
	-0.0305	-0.0302	-1.4765	0.4551	-0.1490	-0.9108	-0.4882	2.8255	-1.7022	2.8262	1.9140	1.9823
	-0.3853	-0.3279					0.1182					
55	2.2460	1.6626	0.4867	0.0342	-2.6134	0.3197	-0.7226	-3.0247	-1.4403	-3.1272	-0.3214	1.6093
	0.0549	-0.3452	1.1502	3.0766	0.7785	0.2919	-0.1444	-2.6634	0.5139	3.0125	2.3433	2.5964
	0.8690	0.4825	-0.3523	0.5528	0.0061	0.7276	-2.7821	-0.5445	-0.5445	0.8094	2.4836	0.3339
	-0.1749	1.0492	0.0944	-1.1531	0.3756	-1.4246	-2.9688	-0.9562	-2.2109	-2.0836	0.4244	2.1284
	-0.0682	-0.8392	-0.0050	0.1595	-0.4820	-0.8138	2.3701	2.8072	-1.5283	2.9689	2.5052	1.9350
	0.3366	1.8634	-1.0231	-0.6281	-0.4771	0.8827	-2.8566	-0.7047	-0.8226	0.8132	-2.8299	0.0178
	-0.7381	-0.4107	1.9268	0.5257	1.7866	0.0016	-1.4940	-0.0431	-2.1355	0.2835	-1.5955	-2.0911
	0.5750	0.8842	0.6126	0.2646	0.6930	-2.5114	-3.0494	1.3662	1.6923	0.9561	0.3247	0.0241
	-0.5513	-0.8263	0.0431	-0.0702	0.8027		-1.1321	-1.7392	-1.5901	0.2267		

Table 4. Fibonacci Sequences

Name	Real protein sequences
2KGU	GYCAEKGIRCDDIHCCTGLKCKCNASGNCVCRKK
1CRN	TTCCPSIVARSNFNVCRLPGTPEAICATYTGCIIIPGATCPGDYAN
2KAP	KEACDAWLRATGFPQYAQLYEDFLFPIDISLVKREHDFLDRAIEALCRRLNTLNKCAVMK
1PCH	AKFSAIITDKVGLHARPASVLAKEASKFSSNITIIANEKQGNLKSIMNVMAMAIKTGIEITIQADGNDADQAIQAIKQTMIDTALIQG

4.2   Prediction with Real Protein Sequences

In this section, we use the real protein sequences which are the same as those in reference [16]. These protein sequences can be downloaded from PDB database. The real protein sequences used in this paper are as follows:

In Table 4 above, A, C, G, I, L, P, M and V are hydrophobic amino acids, and D, E, F, H, K, N, Q, R, S, T, W and Y are hydrophilic amino acids.

Table 5 shows the lowest energy values of real protein sequences.

Table 5. Lowest energy values of real protein sequences

Length	E(PSO [15])	E(TPSO [15])	E(LPSO [16])	E(PGATS [17])	E(ABC [18])	E (BSA-TS)
2KGU	−8.3635	−21.725	−20.9633	−32.2599	−31.9480	−34.2979
1CRN	−20.1826	−53.249	−28.7591	−49.6487	−52.3249	−57.6271
2KAP	−8.0448	−	−15.9988	−28.1052	−30.3643	−36.1682
1PCH	−18.4408	−	−46.4964	−49.5729	−63.4272	−64.2034

We can see from Table 5 that the lowest energy values of the real protein sequences obtained by BSA-TS are lower than all those gained by PSO, TPSO, LPSO, PGATS and ABC. When the real protein sequence is 2KGU, 1CRN, 2KAP or 1PCH, the lowest energy values BSA-TS get are −34.2979, −57.6271, −36.1682 and −64.2034 correspondingly while the lowest energy values others get are −32.2599, −52.3249, −30.3643 and −63.4272 correspondingly. So we can conclude that our algorithm is feasible and effective to predict the 3D structure of real proteins.

Tables 2 and 5 show us that BSA-TS algorithm is efficient especially in real protein 3D structure prediction. When we predict protein 3D structure using Fibonacci Sequences, our algorithm is better-behaved as the sequence length is longer.

Table 6 lists the bond angles and torsional angles of the real protein sequences.

The following Figs. 4, 5, 6 and 7 show the 3D structures of real protein sequences predicted by BSA-TS algorithm and the real 3D protein structures downloaded from PDB database. We can conclude that the 3D structures predicted by BSA-TS algorithm are similar to the native structures of real protein sequences.

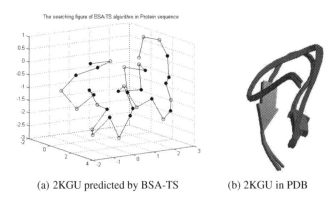

(a) 2KGU predicted by BSA-TS   (b) 2KGU in PDB

Fig. 4. Comparison of 2KGU

Table 6. The bond angles and torsional angles of the real protein sequences

Length	Bond angles						Torsional angles					
2KGU	-0.1669	-2.1412	0.5268	1.0975	0.6641	-0.6898	-0.2863	-2.4203	-0.8223	0.6329	0.3648	-0.4762
	-0.0870	-0.9542	1.2376	-0.5065	1.2986	0.1739	-2.8107	-2.2882	-2.7528	0.6952	-0.8064	1.0594
	0.9168	-0.5940	0.1924	-0.7341	-2.1755	-0.1885	3.0347	2.2126	-0.0201	3.0931	1.9193	-0.0871
	-0.2038	-1.6865	1.4663	-0.0153	1.9151	0.3536	0.2158	-1.4376	-0.4826	0.1887	-0.8352	2.9711
	-0.8108	-0.5097	-0.8404	-0.3395	-0.4083	-0.0663	2.8954	0.8762	0.4116	2.3652	2.6159	3.1284
	0.2804	-0.0695					-2.0677					
1CRN	-1.0309	1.3664	2.6891	3.1397	2.3024	-0.2724	-2.8446	-3.1320	-3.0271	1.1321	3.0215	0.1266
	3.0375	0.4026	0.7938	0.1767	-0.9387	0.5290	2.9643	3.0683	-1.6614	-0.1102	0.6941	-3.1415
	-2.5929	2.8024	1.0391	-1.3766	-1.3262	-2.1526	3.0557	2.9317	2.3983	3.0544	-3.1379	2.9053
	-1.3986	-0.9282	-0.1275	1.8422	0.4736	2.2159	1.0244	3.0738	-2.9137	-2.9375	3.0696	1.0664
	1.3804	-3.0722	-2.7121	3.1415	-1.3889	3.0151	1.3315	2.2323	-2.3403	-1.9422	-1.8556	3.0852
	-2.7292	0.7079	0.8372	2.7774	-1.0847	2.8761	-1.8929	-3.0891	-2.9064	1.8116	-2.92622	0.6080
	-1.2265	0.0101	2.9683	2.3007	1.5776	0.0646	-2.9443	-0.6404	0.7859	2.0981	1.7601	1.1664
	-0.4970	-0.2516					2.9888					
2KAP	-0.8093	-2.6397	-1.4801	2.7926	-0.9642	1.1006	2.7281	2.5670	1.2743	1.9036	3.0824	1.0521
	0.5790	-2.4767	-0.6579	-0.1268	1.5880	0.5674	2.6126	-1.1302	-1.1949	2.6947	3.1385	-2.2617
	-0.1187	-0.0890	3.1111	-1.3020	0.3872	-0.1273	-1.5645	0.2669	-0.8673	2.5987	1.5325	2.7385
	3.0466	-0.1876	-0.6100	0.6134	-0.2001	-2.2949	-0.3973	-1.7132	2.8115	-1.4590	-2.6958	-1.4388
	0.0518	0.4158	-1.0209	0.6070	-0.5793	-0.4586	2.8513	-0.5225	2.1355	2.9261	2.7558	2.1957
	-0.6472	1.5600	1.6117	-0.0652	-0.0821	1.3515	1.9057	2.8829	0.6679	-0.8247	-1.8422	3.0278
	-0.8105	0.3333	2.4782	-0.2398	1.3061	-0.2418	-2.8976	-2.1764	-0.1172	-2.8647	-1.4149	2.6605
	-2.5771	1.8776	-0.1988	-0.0387	1.1343	-1.8387	1.8947	1.5988	1.6319	1.2730	2.6464	-2.6444
	2.8144	0.8608	0.0760	-2.6360	0.1188	0.4682	-0.3847	-0.6969	-1.1354	3.1373	-2.4517	-0.9580
	-0.6844	1.6049	-1.8006	-2.8131			-3.1240	1.9528	2.4207			
1PCH	2.3837	0.1135	2.6722	1.9575	-0.5914	0.2811	1.5935	2.9751	-2.8858	1.7250	1.6557	2.7454
	-3.1011	-1.3585	2.6625	-0.7666	-0.7084	-1.7524	-2.8892	-1.1813	-0.7274	-1.7867	-2.8427	1.8842
	0.2117	-0.5654	-2.8569	2.7067	0.4575	-0.5063	-3.0660	2.1229	3.0021	0.8008	-3.1403	-2.7152
	-0.8123	-0.1471	-2.8126	2.5275	-3.1160	-2.5324	-1.6630	-2.0733	2.8908	1.3000	-3.0625	-0.9310
	0.4686	-0.4004	-0.5082	0.6098	1.8765	2.6603	-2.6517	-0.8686	-3.1177	-1.8400	-2.4446	-2.8957
	-1.5606	2.6624	1.5806	1.0635	-0.0267	3.0000	-2.8986	-3.1346	1.1236	1.6356	0.6595	1.3642
	-0.1611	1.7959	2.1088	-0.0154	2.9183	-1.1740	1.8976	-3.1415	-1.8336	3.0343	-1.8864	-1.4394
	-0.2116	-0.1058	-2.9054	3.0543	-1.1305	-0.5472	-1.8338	-1.9093	-3.1348	-2.4358	2.6950	-1.7528
	0.8016	-0.1468	0.2492	-2.1617	-1.7842	3.0632	-2.6725	1.6564	-3.0195	-2.8017	2.9032	-0.0004
	-1.5919	-2.8640	-1.4369	-0.3443	1.0223	2.9372	-2.6430	-2.4618	-0.9320	2.9721	-2.9941	-0.0785
	-0.5600	-1.0349	0.1674	-1.6850	-0.5322	-1.9706	-2.0973	-2.6544	2.7631	-1.9588	3.0461	1.2265
	-0.1369	-1.0825	0.7210	2.5296	1.5400	-1.3621	-3.0795	0.3000	3.0480	3.0234	0.9044	-2.5036
	0.2749	0.1890	0.0591	1.4680	-1.8232	2.9289	-1.2303	-1.7681	-0.9573	2.3549	2.1931	-2.9392
	3.0575	2.3969	0.9394	2.3725	1.3217	-1.5052	-3.0481	2.7907	0.4025	0.4407	-2.5874	0.1971
	0.6403	1.8016					-3.1274					

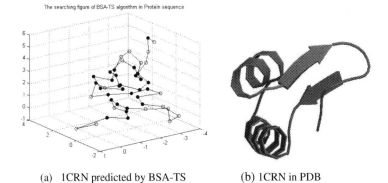

The searching figure of BSA-TS algorithm in Protein sequence

(a) 1CRN predicted by BSA-TS   (b) 1CRN in PDB

Fig. 5. Comparison of 1CRN

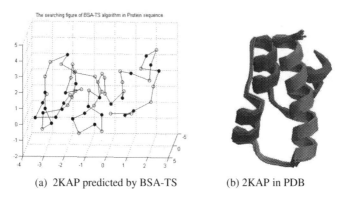

(a) 2KAP predicted by BSA-TS   (b) 2KAP in PDB

Fig. 6. Comparison of 2KAP

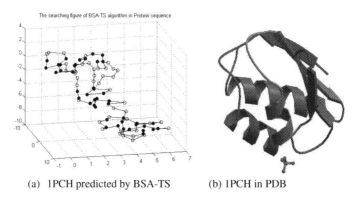

(a) 1PCH predicted by BSA-TS   (b) 1PCH in PDB

Fig. 7. Comparison of 1PCH

5   Conclusion

Due to the importance of 3D protein structure prediction, many theoretical prediction algorithms are proposed. But on account of the complexity of prediction, the present algorithms' accuracy is usually unsatisfactory. Since the most important criterion is the accuracy of algorithm, we focus on getting more stable structure with some widely used Fibonacci Sequences and real protein sequences.

This paper proposes a BSA-TS hybrid optimization algorithm which integrates the advantages of BSA that it has an efficient, fast structure, less control parameters and powerful search capability and the advantage of TS that it has strong local search capability and high precision of search. Additionally, the addition of penalty method can effectively reduce the probability of falling into local optimum caused by general BSA's Selection-II strategy. Then we use BSA-TS to predict the structure of protein based on the 3D AB off-lattice model. The experimental result shows that BSA-TS can get more stable structures which have lower energy compared with the structures gained by other algorithms. That indicates the feasibility of BSA-TS on protein structure prediction problem.

Acknowledgment. This work is supported by the National Natural Science Foundation of China (Nos. 61425002, 61402066, 61402067, 31370778, 61370005), Program for Changjiang Scholars and Innovative Research Team in University (No. IRT_15R07), the Program for Liaoning Innovative Research Team in University (No. LT2015002), the Basic Research Program of the Key Lab in Liaoning Province Educational Department (Nos. LZ2014049, LZ2015004), the Project Supported by Natural Science Foundation of Liaoning Province (No. 2014020132), and by the Project Supported by Scientific Research Fund of Liaoning Provincial Education (No. L2014499).

References

1. Lin, X.L., Zhang, X.L., Zhou, F.L.: Protein structure prediction with local adjust Tabu search algorithm. BMC Bioinformatics **15**(Suppl 15), S1 (2014)
2. Lin, C.J., Shih, C.S.: Protein 3D HP model folding simulation using a hybrid of genetic algorithm and particle swarm optimization. Int. J. Fuzzy Syst. **13**(2), 140–147 (2011)
3. Stillinger, F.H., Head-Gordon, T., Hirshfel, C.L.: Toy model for protein folding. Phys. Rev. **E48**, 1469–1477 (1993)
4. Custódio, F.L., Barbosa, H.J.C., Dardenne, L.E.: A multiple minima genetic algorithm for protein structure prediction. Appl. Soft Comput. **15**, 88–99 (2014)
5. Liu, J.F., Song, B.B., Yao, Y.L., Yu, X., Liu, W.J., Liu, Z.X.: Wang-Landau sampling in face-centered-cube hydrophobic-hydrophilic lattice model proteins. Phys. Rev. E **90**(3), 042715 (2014)
6. Liu, J.F., Li, G., Yu, J., Yao, Y.L.: Heuristic energy landscape paving for protein folding simulation in the three-dimensional HP lattice model. Comput. Biol. Chem. **38**(3), 17–26 (2012)
7. Irback, A., Peterson, C., Potthast, F., Sommelius, O.: Local interactions and protein folding: a three-dimensional off-lattice approach. J. Chem. Phys. **107**, 273–282 (1997)
8. Li, B., Chiong, R., Lin, M.: A balance-evolution artificial bee colony algorithm for protein structure optimization based on a three-dimensional AB off-lattice model. Comput. Biol. Chem. **54**, 1–12 (2015)
9. Liu, J.F., Sun, Y.Y., Li, G., Song, B.B., Huang, W.B.: Heuristic-based tabu search algorithm for folding two-dimensional AB off-lattice model proteins. Comput. Biol. Chem. **47**(3), 142–148 (2013)
10. Mansour, R.F.: Applying an evolutionary algorithm for protein structure prediction. Am. J. Bioinform. Res. **1**, 18–23 (2011)
11. Wang, W.H.: Ordering of unicyclic graphs with perfect matching by minimal energies. MATCH Commun. Math. Comput. Chem. **66**, 927–942 (2011)
12. Jin, X., Zhang, F.: The jones polynomial for polyhedral links. MATCH Commun. Math. Comput. Chem. **65**(2), 501–520 (2011)
13. Guo, H., Lv, Q., Wu, J.Z., Xu, H., Qian, P.: Solving 2D HP protein folding problem by parallel ant colonies. In: International Conference on Biomedical Engineering and Informatics, pp. 1–5 (2009)
14. Zhou, C.J., Hou, C.X., Zhang, Q., Wei, X.P.: Enhanced hybrid search algorithm for protein structure prediction using the 3D-HP lattice model. J. Mol. Model. **19**(9), 3883–3891 (2013)
15. Guo, H., Lan, R., Chen, X., Wang, Y.X.: Tabu search-particle swarm algorithm for protein folding prediction. Comput. Eng. Appl. **47**(24), 46–50 (2011)
16. Chen, X., Lv, M.W., Zhao, L.H., Zhang, X.D.: An improved particle swarm optimization for protein folding prediction. Int. J. Inf. Eng. Electron. Bus. **3**(1), 1–8 (2011)

17. Zhou, C.J., Hou, C.X., Wei, X.P., Zhang, Q.: Improved hybrid optimization algorithm for 3D protein structure prediction. J. Mol. Model. **20**(7), 1–12 (2014)
18. Li, Y.Z., Zhou, C.J., Zheng, X.D.: Artificial bee colony algorithm for the protein structure prediction based on the Toy model. Fundamenta Informaticae **136**(3), 241–252 (2015)
19. Mario, G.F., Eduardo, R.T., Gregorio, T.P.: Comparative analysis of different evaluation functions for protein structure prediction under the HP model. J. Comput. Sci. Technol. **28** (5), 868–889 (2013)
20. Zhang, X.L., Wang, T., Luo, H.P., Yang, J.Y., Deng, Y.P., Tang, J.S., Yang, M.Q.: 3D protein structure prediction with genetic tabu search algorithm. BMC Syst. Biol. **4**(Suppl 1), S6 (2010)
21. Nanda, D.J., Jaya, S.: Particle swarm optimization with backpacking in protein structure prediction problem. J. Mol. Model., pp. 734–738 (2012)
22. Gao, Y., Xie, S.L.: Particle swarm optimization algorithm based on simulated annealing. Comput. Eng. Appl. **40**(1), 47–50 (2004)
23. Van, D.B.F., Engelbrecht, A.P.: A study of particle swarm optimization particle trajectories. Inf. Sci. **176**(8), 937–971 (2006)
24. Bachmann, M., Arkin, H., Janke, W.: Multicanonical study of coarse-grained off-lattice models for folding heteropolymers. Phys. Rev. E **71**, 031906 (2005)
25. Kim, S.Y., Lee, S.B., Lee, J.: Structure optimization by conformational space annealing in an off-lattice protein model. Phys. Rev. E **72**, 011916 (2005)
26. Kim, J., Straub, J.E., Keyes, T.: Structure optimization and folding mechanisms of off-lattice protein models using statistical temperature molecular dynamics simulation: statistical temperature annealing. Phys. Rev. E **76**, 011913 (2007)
27. Liu, J.F.: Structure optimization by heuristic algorithm in a coarse-grained off-lattice model. Chin. Phys. B **18**, 2615–2621 (2009)
28. Li, W.Y., Wang, Y.: Multi-population genetic algorithm for three-dimensional protein structure prediction. Fujian Comput. **28**(11), 20–24 (2012)
29. Pinar, C.: Backtracking search optimization algorithm for numerical optimization problems. Appl. Math. Comput. **219**, 8121–8144 (2013)
30. Wang, X.J., Liu, S.Y., Tian, W.K.: Backtracking search optimization algorithm with high efficiency mutation scale factor and greedy crossover strategy. Comput. Appl. **34**(9), 2543–2546 (2014)
31. Zhang, X., Cheng, W.: An improved Tabu search algorithm for 3D protein folding problem. In: Ho, T.-B., Zhou, Z.-H. (eds.) PRICAI 2008. LNCS (LNAI), vol. 5351, pp. 1104–1109. Springer, Heidelberg (2008)
32. Glover, F.: Future paths for integer programming and links to artificial intelligence. Comput. Oper. Res. **13**, 533–549 (1986)
33. Glover, F., Kelly, J.P., Laguna, M.: Genetic algorithms and tabu search: hybrids for optimization. Comput. Oper. Res. **22**(1), 111–134 (1995)
34. Hsu, H.P., Mehra, V., Grassberger, P.: Structure optimization in an off-lattice protein model. Phys. Rev. E **68**(3), 1–4 (2003)

Training ROI Selection Based on MILBoost for Liver Cirrhosis Classification Using Ultrasound Images

Yusuke Fujita[1(✉)], Yoshihiro Mitani[2], Yoshihiko Hamamoto[1],
Makoto Segawa[3], Shuji Terai[4], and Isao Sakaida[3]

[1] Graduate School of Sciences and Technology for Innovation,
Yamaguchi University, 2-16-1 Tokiwadai, Ube, Yamaguchi 755-8611, Japan
y-fujita@yamaguchi-u.ac.jp
[2] Ube National College of Technology,
2-14-1 Tokiwadai, Ube, Yamaguchi 755-8555, Japan
[3] Yamaguchi University Graduate School of Medicine,
1-1-1 Minami-Kogushi, Ube, Yamaguchi 755-8505, Japan
[4] Graduate School of Medical and Dental Science, Niigata University,
1-757 Asahimachidori, Chuo-Ku, Niigata 951-8510, Japan

Abstract. Ultrasound images are widely used for diagnosis of liver cirrhosis. In most of liver ultrasound images analysis, regions of interest (ROIs) are selected carefully, to use for feature extraction and classification. It is difficult to select ROIs exactly for training classifiers, because of the low SN ratio of ultrasound images. In these analyses, training sample selection is important issue to improve classification performance. In this article, we have proposed training ROI selection using MILBoost for liver cirrhosis classification. In our experiments, the proposed method was evaluated using manually selected ROIs. Experimental results show that the proposed method improve classification performance, compared to previous method, when qualities of class label for training sample are lower.

Keywords: Ultrasound imaging · MILBoost · Pattern recognition · Computer-aided diagnosis (CAD) · Sample selection

1 Introduction

Computerized analysis using image processing techniques is expected to improve medical image interpretation, because it may serve as a "second opinion" in detecting lesions, assessing disease severity, and making diagnostic decisions [1]. Ultrasound imaging is a popular and non-invasive tool that is used in the diagnoses of liver disease. The purpose of our study is cirrhosis classification with high accuracy using liver ultrasound images.

However, in general, regions of interest (ROIs) on liver ultrasound images are defined manually by medical doctors. Examples of liver ultrasound images of cirrhosis and healthy cases are shown in Fig. 1. It is difficult to select the ROIs in liver ultrasound images, which are used for classifier design and classification. The classification

© Springer International Publishing Switzerland 2016
H. Fujita et al. (Eds.): IEA/AIE 2016, LNAI 9799, pp. 451–459, 2016.
DOI: 10.1007/978-3-319-42007-3_39

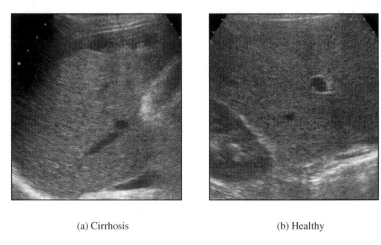

(a) Cirrhosis   (b) Healthy

Fig. 1.   Examples of liver ultrasound images.

accuracy depend on how to define ROI in the liver's tissues, because they may contain unnecessary edges or textures by interference of speckle noise. We have proposed a method of cirrhosis detection based on multiple-ROI combination [2] by the product rule [3]. Therefore, in the classification process, blood vessel exclusion and reject option were proposed to select ROIs automatically [4]. Wu et al. also used the genetic algorithm to select ROIs to improve the diagnosis accuracy [5].

In this article, training ROI selection using MILBoost [6] is introduced to improve the cirrhosis classification performance. In our experiments, the proposed method was evaluated using manually selected ROIs. Experimental results show the proposed method improve classification performance.

This article is organized as follows: Sect. 2 describes the related works. Section 3 introduces the proposed method, including feature extraction using Gabor filters, classifier design using MILBoost and multiple-ROI classification. Section 4 shows experimental results, and conclusions are described in Sect. 5.

2   Related Works

Many types of features and classification methods have been used for liver disease identification. Basset et al. presented an ultrasound interpretation system based on texture features to classify the severity of liver fibrosis [7]. Yeh et al. used the gray-level co-occurrence matrix extracted from the input ultrasound image to measure the grade of liver fibrosis [8]. The approach based on non-separable wavelet transform was also used for liver fibrosis diagnosis [9]. Zhou et al. combined features obtained from the motion curve of the liver in an M-mode ultrasound image and from the texture using a B-mode ultrasound image in the diagnosis of cirrhosis [10]. In this approach, texture features extracted from single region in the liver tissue near to the abdominal aorta. In these study, support vector machine, Fisher liner classifier, neural network or Bayes classifier are used for classification. However, these studies have focused on

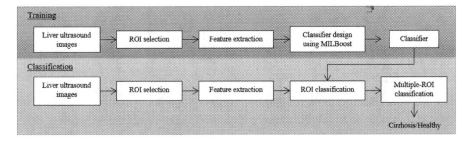

Fig. 2. System overview.

extracted features or classifiers in classification of liver ultrasound images. ROI selection remains unanswered.

3 Liver Cirrhosis Classification

In this article, we consider the two-class classification problem of identifying cirrhosis and normal using liver ultrasound images. Figure 2 shows the flow of the proposed method. In training phase, manually selected ROIs are used to design a classifier under the guidance of a medical doctor. We use Bayesian classifier based on 25 dimensional Gabor features [11, 12] extracted from each ROI. Some of training ROIs selected manually may be not valuable as training samples, because of the low SN ratio of ultrasound images and uncertainty of cirrhosis regions. Selection of training samples is important process to improve classification performance.

Regarding this issue, we applied MILBoost [6] to weight valuable ROIs in ultrasound images as training sample. In MILBoost, training samples are not singletons. Instead, they belong bags, where all of the samples in a bag share a label. In our proposed system, a cirrhosis bag means that at least one sample in the bag is cirrhosis, while a healthy bag means that all examples in the bag are healthy. In classification phase, each ROI defined on liver ultrasound images are classified into healthy or cirrhosis liver, and the final diagnosis in our method is decided by the product rule using ROI classification results.

3.1 Gabor Features

The Gabor filter is a product between a complex sinusoidal carrier and a 2D Gaussian shaped envelope. A set of Gabor filters with different frequencies and orientations is used to extract texture features from an image. These filters are defined as

$$G\left(x, y; k_x, k_y, \sigma^2\right) = \exp\left\{-\frac{x^2 + y^2}{2\sigma^2}\right\} \exp\left\{j\left(k_x x + k_y y\right)\right\} \tag{1}$$

where x and y represent coordinates in 2D space, and k_x and k_y represent the frequency coordinates. The frequency and orientation of the filter are set by k_x and k_y. σ is the

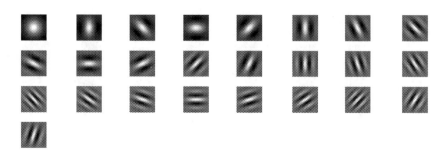

Fig. 3. Real parts of Gabor filters that have various frequency and orientation.

standard deviation of the Gaussian envelope. Figure 3 shows real parts of Gabor filters that have various frequency and orientation.

3.2 Bayesian Classifier

In our proposed system, the Bayesian classifier was used as weak classifier in MILBoost, to identify the ROI belongs cirrhosis or healthy class. The Bayesian classifier [13, 14] provides approximated posteriori probabilities of the given data based on mean vectors and covariance matrices that were estimated from the training data. It classifies the given data to the class which provided the maximum posteriori probability. The Bayesian classifier is described as

$$f_k(x) = -\frac{1}{2}(x - \mu_k)\sum{}^{-1}(x - \mu_k)^T + \frac{1}{2}\log\frac{P(\omega_k)}{|\sum_k|} \tag{2}$$

$$\mu_k = \sum_J w_{kj}x_{kj} \tag{3}$$

$$\sum{}_k = \sum_J w_{kj}(x_{kj} - \mu_k)(x_{kj} - \mu_k)^T, \tag{4}$$

where μ_k, $\sum_k$ are the mean vector and the class-conditional covariance matrices of the class ω_k, and $P(\omega_k)$ is the prior probability of the class ω_k, respectively. The function described in Eq. (2) is a logarithm approximation of the posteriori probability.

3.3 MILBoost

We construct the classifier for ROI classification by MILBoost as shown in Algorithm 1. Suppose we have a data set $X = \{x_{ij}\}$ assigned class label $Y = \{y_i\} \in \{1,0\}$, where i indexes the bag, and j indexes the sample in the ith bag. Label $y = 1$ means cirrhosis class and $y = 0$ means healthy class. In our system, training samples belonging to the cirrhosis class are not labeled individually. Instead, they reside in bags. A positive bag means that at least one sample classified as cirrhosis in the bag. On the other hand, bags of healthy class have only one sample. Thus, healthy class samples are the same that

---

Algorithm 1. Classifier design based on MILBoost.

Input: Training data set $X = \{x_{ij}\}$ assigned training labels $Y = \{y_i\} \in \{1, 0\}$, where i indexes the bag, and j indexes the sample in the bag. Initialize w_{ij} = # of class bags / # of all bags.

for $t = 1, ..., T$, **do**

    Training weak classifier $h_t(x)$ under probability condition w_{ij}.

$$h_t(x) = \frac{1}{2}\ln\frac{p(x|y=1)+\epsilon}{p(x|y=0)+\epsilon}.$$

    Update weight w_{ij} of training sample x_{ij}.

$$w_{ij} = \begin{cases} \frac{1-p_i}{p_i}p_{ij} & y_i = 1 \\ p_{ij} & y_i = 0 \end{cases}$$

$$p_i = 1 - \prod_j(1 - p_{ij})$$

$$p_{ij} = \frac{1}{1+\exp(-H_t(x))}$$

Output: Classifier $H(x)$.

$$H(x) = \sum_t^T h_t(x).$$

---

would result in a non-MILBoost framework. A set of weak classifiers $\{h_1, ..., h_T\}$ are trained by iteration of training process. We use the Bayesian classifier, which were described above, as each weak classifier on MILBoost. At each iteration of the training process, a weak classifier is designed under updated weight w_{ij}. Weights are initialized using numbers of bags, and updated by trained classifier H_t at tth step in training process. p_{ij} is likelihood of ROI x_{ij} and p_i is likelihood of ith bag.

3.4   Multiple ROI Classification

The posteriori probability of a subject belonging each class is estimated using approximated posteriori probabilities of ROIs extracted from the subject. We use the product rule [4, 13, 14] described as,

$$P(\omega_k|x_1, x_2, ..., x_n) = \prod_j^n P(\omega_k|x_j). \tag{5}$$

4   Experimental Results

In our experiments, the ultrasound images obtained from 12 cases of cirrhosis and 8 cases of healthy were used to evaluate the classification performance of the proposed method. These images were acquired at Yamaguchi University Hospital. 5 liver ultrasound images were obtained from each subject. These images have 640×480 pixels and 256 gray-level resolution. 5 ROIs, which have 32×32 pixels, have been

(a) Cirrhosis   (b) Healthy

Fig. 4. Examples of manually defined ROI images.

selected manually from each image under the guidance of a medical doctor. They were chosen to include only liver tissue, without blood vessels, acoustic shadowing, or any type of distortion. Figure 4 shows examples of ROI images defined manually. ROI images shown in Fig. 4 (a) are from a cirrhosis case and that shown in (b) are from a healthy case.

Figure 5 shows a flow of the evaluation by repeated random sub-sampling validation. Manually defined ROIs were divided into training data and test data randomly. A part of training ROIs were exchanged randomly between two classes. A Monte Carlo cross-validation method [15, 16] is used to evaluate the classification performance. In each trial, data of 7 cases of both classes are used for training, and that of 1 case of both classes are used for classification. It maintains the statistical independence between the training and the classification samples. 175 ROIs for each class are used in training. 1 case of cirrhosis and 1 case of healthy are used as test data in a trial, and the trial was repeated 800 times, independently. The accuracy rate in 800 trials was evaluated as the classification performance. The accuracy rate is defined as follows.

$$\text{Acc.} = \frac{\#\text{of test samples classified correctly}}{\#\text{of all test samples}}. \tag{6}$$

We evaluated the MILBoost classifier as the proposed method, compared to Bayesian classifier and Real AdaBoost.

Therefore, in order to evaluate effects of sample labeling qualities, classification performance were evaluated, after a certain rate r of training samples selected randomly were exchanged between cirrhosis and healthy classes. The exchange rate r were set $0, 0.01, 0.02, \ldots, 0.15$ in our evaluation. We compare the performance of classifiers designed based on MILBoost, Bayesian classifier (QDF) and Real AdaBoost.

Figure 6 shows result of the comparative performance evaluation. When all training samples were used with original label, the accuracy rate of MILBoost classifier is 0.83 ($r = 0$), that of QDF classifier is 0.81 ($r = 0$) and that of Real AdaBoost is 0.68 ($r = 0$), respectively. MILBoost classifier have higher classification performance than the others. Also, when the exchange rates r of training samples are lower than 0.10, the accuracy rates of MILBoost are over 0.80. Those are higher than those of QDF and Real AdaBoost. In the experiment, the performance of MILBoost classifier is higher than that of QDF classifier using original training data, when the exchange rate r is under 0.08. The experimental results show that MILBoost classifier has high classification performance, when the qualities of class labels set for training data are low such as using ultrasound images.

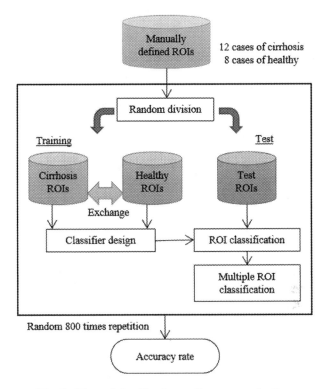

Fig. 5. Flow of classification performance evaluation.

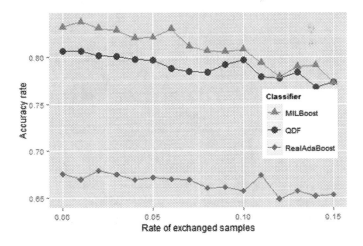

Fig. 6. Comparison of classification performance between MILBoost, QDF and Real AdaBoost.

5   Conclusions

In this article, we have proposed training ROI selection using MILBoost for liver cirrhosis classification. In our experiments, the proposed method was evaluated using manually selected ROIs. Experimental results show that the proposed method improve classification performance, compared to previous method, when qualities of class label were lower.

In the future study, the quantitative analysis is expected for early diagnostics.

Acknowledgement. We would like to thank H. Igari who carried out the preliminary experiment in his graduation thesis.

References

1. Stoitsis, J., Golemati, S., Nikita, K.S.: A modular software system to assist interpretation of medical images-application to vascular ultrasound images. IEEE Trans. Instrum. Meas. **55**(6), 1944–1952 (2006)
2. Fujita, Y., Hamamoto, Y., Segawa, M., Terai, S., Sakaida, I.: An improved method for cirrhosis detection using liver's ultrasound images. In: Proceedings of 20th International Conference on Pattern Recognition, pp. 2294–2297 (2010)
3. Kittler, J., Hatef, M., Duin, R., Matas, J.: On combining classifiers. IEEE Trans. Pattern Anal. Mach. Intell. **20**(3), 226–239 (1998)
4. Fujita, Y., Goto, T., Mitani, Y., Hamamoto, Y., Segawa, M., Terai, S., Sakaida, I.: A liver cirrhosis detection method using probabilistic ROI combination. In: Proceedings of 11th International Conference on Quality Control by Artificial Vision, pp. 81–85 (2013)
5. Wu, Y.U., Uang, H.Y., Cheng, S.C., Yang, C.K., Lin, C.L.: Evolutionary feature construction for ultrasound image processing and its application to automatic liver disease diagnosis. In: 2011 International Conference on Complex, Intelligent, and Software Intensive Systems, pp. 565–570 (2011)
6. Viola, P., Platt, J.C., Zhang, C.: Multiple instance boosting for object detection. Adv. Neural Inf. Process. Syst. **18**, 1417–1426 (2007)
7. Basset, O., Duboeuf, F., Delhay, B., Brusseau, E., Cachard, D., Tasu, J.P.: Texture analysis of ultrasound liver images with contrast agent to characterize the fibrosis stage. In: IEEE Symposium on Ultrasonics, pp, 24–27 (2008)
8. Yeh, W.C., Huang, S.W., Li, P.C.: Liver fibrosis grade classification with B-mode ultrasound. Ultrasound Med. Biol. **29**(9), 1229–1235 (2003)
9. Mojsilovic, A., Markovic, S., Popovic, M.: Characterization of visually similar diffuse diseases from B-scan liver images using nonseparable wavelet transform. IEEE Trans. Med. Imaging **17**(4), 541–549 (1998)
10. Zhou, G., Wang, Y., Wang, W., Sun, Y., Chen, Y.: Decision of cirrhosis using liver's ultrasonic images. In: Proceedings of the 2005 IEEE Engineering in Medicine and Biology 27th Annual Conference, pp. 3351–3354 (2005)
11. Idrissa, M., Acheroy, M.: Texture classification using Gabor filters. Pattern Recogn. Lett. **23**(9), 1095–1102 (2002)
12. Randen, T., Husey, J.H.: Filtering for texture classification; a comparative study. IEEE Trans. Pattern Anal. Mach. Intell. **21**(4), 291–310 (1999)

13. Fukunaga, K.: Introduction to Statistical Pattern Recognition, 2nd edn. Academic Press, Boston (1990)
14. Webb, A.R., Copsey, K.D.: Statistical Pattern Recognition, 3rd edn. Wiley, Chichester (2011)
15. Picard, R., Cook, R.: Cross-validation of regression models. J. Am. Stat. Assoc. **79**(387), 575–583 (1984)
16. Shao, J.: Linear model selection by cross-validation. J. Am. Stat. Assoc. **88**(422), 486–494 (1993)

Sleep Pattern Discovery via Visualizing Cluster Dynamics of Sound Data

Hongle Wu[1]([✉]), Takafumi Kato[2], Tomomi Yamada[2], Masayuki Numao[1], and Ken-ichi Fukui[1]

[1] The Institute of Scientific and Industrial Research, Osaka University, Suita, Japan
wu@ai.sanken.osaka-u.ac.jp
[2] Graduate School of Dentistry, Osaka University, Suita, Japan

Abstract. The quality of a good sleep is important for a healthy life. Recently, several sleep analysis products have emerged on the market; however, many of them require additional hardware or there is a lack of scientific evidence regarding their clinical efficacy. This paper proposes a novel method for discovering the sleep pattern via clustering of sound events. The sleep-related sound clips are extracted from sound recordings obtained when sleeping. Then, various self-organizing map algorithms are applied to the extracted sound data. We demonstrate the superiority of Kullback-Leibler divergence and obtain the cluster maps to visualize the distribution and changing patterns of sleep-related events during the sleep. Also, we perform a comparative interpretation between sleep stage sequences and obtained cluster maps. The proposed method requires few additional hardware, and its consistency with the medical evidence proves its reliability.

Keywords: Sleep pattern · Sound data · Self-organizing map · Pairwise F-measure · Polysomnography

1 Introduction

Sleep is an important physiological state of the human body. Almost one third of the time in a person's life is spent sleeping. The quality of sleep is very important to a person's health. However, most of us have experienced trouble sleeping at one time or another. It is normal and usually temporary, due to stress or other external factors. However, regular occurrence of such problems may be indicative of a sleep disorder. Sleep disorders cause more than just sleepiness. The lack of quality sleep can have a negative impact on our normal physical, mental, social, and emotional functioning. In Japan, data show that about 3 %– 22 % people suffer from sleep apnea, 20 % from chronic symptomatic insomnia, 2 %–12 % from periodic limb movement disorder, 0.96 %–1.80 % from restless legs syndrome, 0.16 %–0.59 % from narcolepsy, and 0.13 %–0.40 % from delayed sleep phase disorder [6]. Therefore, sleep monitoring technology has become an indispensable content in modern medical diagnosis.

© Springer International Publishing Switzerland 2016
H. Fujita et al. (Eds.): IEA/AIE 2016, LNAI 9799, pp. 460–471, 2016.
DOI: 10.1007/978-3-319-42007-3_40

The primary tool for sleep study is polysomnography (PSG) [5]. PSG monitors body functions through many methods, including electroencephalography for the brain, electrooculography for eye movements, electromyography for muscle activity, and electrocardiography for heart rhythm, and is mainly used in medical science and treatment by doctors [16,20]. Due to its professional property and financial cost, PSG usage is limited to only clinics. Currently, there are many products on the market that aim to make sleep assessment portable at a reduced cost. ZEO[1] is a popular PSG-based home sleep analysis product. Besides traditional PSG, actigraphy has also been used as an alternative tool; there are many actigraphy-based products including Beddit[2] and Fitbit[3]. One of the problems of these products is that they are invasive to users, which means that users have to wear an additional device or place a device on their bed during sleep. According to a recent survey, many people are resistant to wearing a device during sleep [4]. Even if users accept to wear the device, it is not easy to properly place the sensors in the correct position. Also, without basic medical or technical training, understanding the recorded data correctly becomes a great challenge for the users.

Moreover, additional devices add extra financial burden to the user. The efforts in the market to reduce the cost are mostly through mobile apps. Mobile apps use a smartphone's built-in sensors, and hence, users do not need to purchase additional hardware. However, according to [2], very few of the apps are based on published scientific evidence.

To solve the problems mentioned above simultaneously, and considering that many types of sleep disorder are respectively related to a distinctive type of sound, such as snoring, tooth grinding, limb movement and sleep talking, this paper proposes a method for sleep analysis based on clustering of sound data. The main features of our method are as follows:

Fine-grained sleep process visualization: We propose a novel algorithm to cluster the sleep-related events on spatio-temporal dimensions. The transition of sleep state is visualized on the cluster map, which provides a clear and easy way to understand the analysis report.

Non-invasive: The sound data can be recorded by any recording device placed near the user's bed during sleep; hence, no burden is added to the user.

No additional cost: Any off-the-shelf equipment with a microphone, including a smartphone, recording pen, or personal computer, can be used as the recording device.

Scientifically validated: The consistency of this method with the medical evidence from PSG proves its reliability.

We extracted sound clips of events from the recorded sound data, applied Fast Fourier Transform (FFT) to get the frequency spectrum as input vectors, and then applied various self-organizing map (SOM) [18] algorithms to the data to obtain cluster maps. One of these algorithms is Kulback-Leibler kernel SOM

---

[1] https://en.wikipedia.org/wiki/Zeo,_Inc.

[2] http://www.beddit.com/.

[3] https://www.fitbit.com/.

(KL-KSOM) [9], which is an extended SOM that uses KL divergence instead of Euclidean distance as a similarity measure of data points. During the study, when we compared the clustering results of the standard SOM and KL-KSOM, the latter obtained a better effect. Based on this property, we introduced KL divergence through a kernel function into the sequence-based self-organizing map (SbSOM) [11], and proposed a novel algorithm named sequenced-based kernel SOM (Sb-KSOM). The aforementioned SbSOM is another extended SOM that introduces the sequencing weight function (SWF) in SOM. By converting the spatio-temporal neighborhood into the topological neighborhood via the neighborhood function, it is able to visualize the transition of cluster dynamics. Sb-KSOM combined the advantages of KL-KSOM and SbSOM, produced a cluster map reflecting the distribution and changing of sleep-related events during the whole sleep.

To evaluate our method, we labeled the extracted events based on the synchronous PSG data scored by a medical specialist, and calculated the weighted pairwise F-measure (wPF) [10] as the validity measure of each cluster map. The total value of wPF was used to compare clustering results from different algorithms.

According to [5], the stages of sleep include rapid eye movement (REM) sleep, non-REM (NREM) sleep, and awakening; sleep proceeds in cycles of REM and NREM, usually four or five times every night. REM sleep has been discovered to be closely associated with dreaming. The American Academy of Sleep Medicine (AASM) divides NREM into three stages: NREM Stage 1 (N1), NREM Stage 2 (N2), and NREM Stage 3 (N3), the last of which is also called deep sleep or slow-wave sleep [23]. In the traditional research on sleep pattern analysis, sleep stages play a decisive role [8,25]. In order to verify the reliability of our method, we performed a comparative interpretation between the obtained cluster maps generated by Sb-KSOM and sleep stage sequences scored by medical specialists based on PSG data. The interpretation revealed that cluster distribution changes synchronously with the transferring of sleep stages. Hence, similar to sleep stage sequences, it is feasible to discover the sleep patterns through the cluster maps generated by Sb-KSOM.

2 Methodology

In this section, we introduce the key methodologies applied in this study.

2.1 Burst Extraction Algorithm

The first step to be followed after recording the sound is to determine the useful events inside an all-night-long sound recording. Manually searching the events will waste considerable time and is definitely unacceptable. In this study, we used the method in [9] to differentiate the steady noise from other types of sound events including sleep disorder symptoms, such as snoring, tooth grinding, or body movement, and environmental sound, such as air-conditioner operation or outdoor traffic. The sound events were extracted by the statistical burst extraction method [17].

By using Kleinberg's method, we no longer need to consider the size of the sliding window or amplitude threshold. Furthermore, by introducing the cost function, this method can extract an event that has been broken apart due to brief gaps during a single event; threshold methods are basically unable to perform this extraction.

We assumed the background noise during the whole night to be steady and to be generated according to a constant Gaussian distribution, and other sound events, such as snoring, to be generated by Gaussian distribution with different parameters. The burst extraction method estimates the maximum-likelihood state sequence for the sound event, and is able to extract a variable-length sound event based on the amount of activity of the signals.

2.2 SOM

The SOM is an artificial neural network and originally a model of associative memory, but has recently been widely used for visual data mining, for example, in exploratory analysis support of documents [19], for the monitoring of machinery [24], and for application to medical care or economics. In this study, we generated cluster maps by clustering algorithms based on SOM, including standard SOM, KL-KSOM, and Sb-KSOM. The generation of such a map has the following advantages:

Comprehensive evaluation: Similar sound events are assumed to have similar frequency characteristics. Therefore, a cluster of sound events corresponds to a sleep-related event type, for example, snoring. Moreover, by introducing time dimension into the clustering, the distribution of sleep disorder events transition with the sleep time elapsing can also be displayed in the cluster map.

Exploratory analysis: The user can intuitively understand the entire picture of several sleep disorder events and explore particular events or high-frequency events.

2.3 KL-KSOM

We used the frequency spectrum as input vector. The standard SOM uses Euclidean distance as a similarity measure of data points, so the distribution structure of a frequency spectrum cannot be captured since each discrete point is treated as an independent variable. The authors in [9], proposed the use of KL divergence to introduce a distribution structure into a similarity measure of frequency spectrum of acoustic emission events and obtained a good effect. To make a comparison in this study, KL-KSOM that applies the KL divergence to SOM through the kernel function [1,3] was also used to cluster the sleep-related sound events.

The KL kernel function is defined as:

$$K(x_i, x_j) = \exp\left(-\beta KL(x_i, x_j)\right), \tag{1}$$

where x_i, x_j are the input data (discrete points of frequency power spectrum), $KL(x_i, x_j)$ is the KL divergence symmetrized by Jensen-Shannon divergence, and $\beta > 0$ is a scaling parameter.

Instead of Euclidean distance, in KL-KSOM, the dissimilarity between the reference vector corresponding to the ith neuron on the cluster map and the nth input data x_n: $d_{i,n}$ is defined as follows:

$$d_{i,n} = K(x_n, x_n) - 2\gamma \sum_j h_{c(j),i} K(x_n, x_j) + \gamma^2 \sum_k \sum_l h_{c(k),i} h_{c(l),i} K(x_k, x_l), \quad (2)$$

where γ is a regularization term $\gamma = 1/\Sigma_n h_{c(n),i}$, $c(n)$ is the index of the winner neuron of the input data x_n, and $h_{c(n),i}$ is the neighborhood function defined on the topology space, where the Gaussian function is used.

2.4   Proposed Method: Sequenced-Based Kernel SOM

In order to clearly and easily understand the analysis report of a user's sleep, a fine-grained map that depicts the distribution and changing of sleep-related events is necessary. In this study, the comparison of the clustering results between standard SOM and KL-KSOM demonstrated that KL divergence exhibits better performance.

Based on the aforementioned premises, a novel algorithm, Sb-KSOM, is proposed, which is an extension of SbSOM. Different from the normal SOM that deals with static data, SbSOM introduces SWF into SOM and can visualize the transition of cluster dynamics since the spatio-temporal neighborhood is converted into the topological neighborhood by the neighborhood function. The proposed Sb-KSOM kernelized SbSOM by replacing the Euclidean distance with KL divergence to enable it to handle the frequency spectrum data.

In SbSOM, let the N input data be $x_n, (n = 1, ..., N)$; the position of M neurons in the visualization layer be $r_j = (\xi_j, \eta_j), (j = 1, ..., M)$, where ξ-direction indicates the temporal dimension; and the reference vector corresponding to the jth neuron be m_j. The nth input data are located at the ratio of n/N within the input data sequence, and the jth neuron is located at the ratio of ξ_j/ξ_M on the ξ-direction of cluster map. Let the absolute value of those differences be $\epsilon = |\xi_j/\xi_M - n/N|$. The SWF $\psi(n, \xi_j)$ is defined so as to be able to balance the spatio/temporal resolution; in case where reversal of data order is not allowed, the SWF is given as:

$$\psi(n, \xi_j) = \begin{cases} 1 & if \quad \epsilon < \frac{1}{2K} \\ \infty & otherwise \end{cases}, \quad (3)$$

where K is the number of neurons on ξ-direction. The winner neuron of the input data x_n is determined by spatial distance combined with SWF as follows:

$$c(x_n) = arg \min_j \psi(n, \xi_j) \|x_n - m_j\|. \quad (4)$$

In the proposed method, Sb-KSOM, we replaced the normal Euclidean distance calculation in Eq. (4) with the kernel function:

$$c(x_n) = arg \min_j \psi(n, \xi_j) d_{j,n}. \tag{5}$$

The dissimilarity between j^{th} reference vector and n^{th} data point: $d_{j,n}$ is calculated by Eq. (2).

2.5   wPF

The original pairwise F-measure evaluates the correlation between cluster assignment and class label. However, especially in SOM visualization, neighborhood relation is also important. Then, we employed the weighted version of the pairwise F-measure to evaluate SOM visualization comprehensively. By using events scored through PSG as ground truth labels, we applied wPF [10] to evaluate the clustering results. wPF is an extension of pairwise-based cluster validity measures [26], which introduce the likelihood function indicating a degree that a data pair belongs to the same cluster instead of the actual number of data pairs.

3   Experiment

We first applied the standard SOM and KL-KSOM to the extracted sound data, and compared the performance of these two algorithms via wPF using events scored through PSG as ground truth labels.

Then, we used Sb-KSOM on the data to obtain the spatio-temporal dimensional cluster map, and discussed the relation between the transition of sleep stages and cluster dynamics of sound events.

3.1   Experimental Setting

The data used in this study were prepared by the Graduate School of Dentistry in Osaka University. The study protocol was approved by the clinical research ethics Committee of the Osaka University Graduate School of Dentistry. Written informed consent was obtained from all subjects. All subjects were asked to sleep in a specific room from 22:30 to 8:00. The recording device included LA1250 (Ono Sokki)[4] and R-4 Pro (Roland)[5]. A microphone was placed at a distance of 50 cm from the subjects heads. The sound data were recorded on a single channel (mono) at a sampling rate of 48 kHz. In addition, all subjects were measured by PSG simultaneously.

Most of the experimental subjects are university students from Osaka University, and hence, their age was mostly around 20–24. The male to female ratio was balanced. Table 1 shows information of the subjects and the recorded sound data that were used in the experiment.

---

[4] https://www.onosokki.co.jp/English/hp_e/products/keisoku/s_v/la1200.html.
[5] http://proav.roland.com/products/r-4_pro/.

Table 1. Subject and sound data information

Subject id	Age	Gender	Recording date	Duration	Primary disorder symptoms
1	21	F	2014/05/13	08:05:22	Tooth grinding
2	22	M	2014/05/27	08:16:15	Snoring
3	22	M	2014/06/03	08:01:09	Snoring
4	23	M	2014/07/29	08:23:01	Tooth grinding, snoring
5	24	M	2014/01/20	08:17:34	Tooth grinding, snoring
6	23	F	2015/03/03	08:30:30	Tooth grinding
7	20	F	2015/06/02	07:18:30	Tooth grinding

3.2   Event Extraction

We selected seven nights of sound data. Based on the burst extraction method, we obtained a total of 6775 sound events, which included sleep disorder and other sound events such as outdoor traffic noise. FFT was applied to the extracted sound data to obtain the frequency power spectrum. From 24 Hz to 20 kHz, at intervals of 4 Hz, 4995 discretized points as an input for SOM were obtained for every sound data.

3.3   Clustering by Standard SOM and KL-KSOM

In the first part of this experiment, all the extracted data were combined into one dataset to show the differences of various disorders. The number of neurons was set to 15×15 with a two-dimensional regular grid. In general, the number of neurons is not sensitive to these results, in that SOM captures the data distribution in the feature space. A Gaussian function was used as the neighborhood function in SOM and KL-KSOM. In Fig. 1, a cell corresponds to a neuron and similar disorder events are clustered into the same or neighboring neuron. Note that the coordinates on the cluster map do not express any physical quantity other than the relative distances to the neighboring data.

Before a comparison with SOM, the scaling parameter β of KL-KSOM in Eq. (1) must be determined. Through comparison, $\beta = 0.6$ was selected since it provided the highest value of wPF. A neighborhood smoothing parameter $\sigma = 2.0$ was used in this wPF computation. This parameter was optimized with the method in [10].

The clustering results of SOM and KL-KSOM are shown in Fig. 1(a) and (b), respectively. We labeled the extracted events based on the synchronous PSG data scored by a medical specialist, and marked each neuron by the sleep disorder types that all or most of the events in the neuron belonged to. The different types of neurons are shown by different shades of gray on the map. The tooth grinding events are dispersed into five groups in Fig. 1(a) and two groups in Fig. 1(b). Also, the snoring events are dispersed into four groups in Fig. 1(a) and two groups in Fig. 1(b). By comparing the results from SOM and

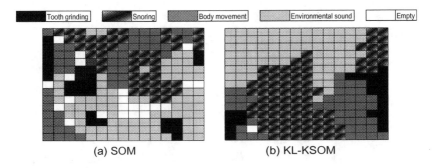

Fig. 1. Cluster maps generated by SOM and KL-KSOM

Table 2. Comparison of wPF between SOM and KL-KSOM clustering results

Subject id	SOM		KL-KSOM	
	Mean	SD	Mean	SD
1	0.537	0.033	0.604	0.037
2	0.521	0.041	0.573	0.038
3	0.506	0.031	0.551	0.031
4	0.559	0.040	0.592	0.037
5	0.602	0.039	0.629	0.039
6	0.543	0.033	0.600	0.035
7	0.483	0.042	0.523	0.047
Mean	0.535	0.037	0.581	0.037

KL-KSOM, we can intuitively realize that the same type of events was better concentrated by KL-KSOM.

In the second part of this experiment, we used the sound data from each subject as a respective dataset and compared the wPF values of SOM and KL-KSOM for each subject. In order to avoid initial value dependency, the experiments were exucuted 50 times and the average values were computed. The mean wPF values and standard deviation are shown in Table 2; the average of wPF has improved by about 10 % from the standard SOM, which indicates that KL-KSOM has a better performance.

3.4 Sleep Pattern Analysis by Sb-KSOM

In this experiment, we made a comparative analysis between cluster maps generated by Sb-KSOM and sleep stage sequences to reveal the relation between them. We analyzed all the subject respectively. Due to page limitation, only one subject is shown in this paper. Subject 4 is chosen since his tooth grinding and snoring activities are frequent, and generated more related sound events than the others. Figure 2(a) shows the result when Sb-KSOM was applied to the sound data from Subject 4; the number of neurons was set to 50×10 with a

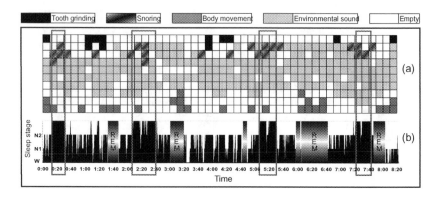

Fig. 2. Cluster map generated by Sb-KSOM on Subject 4

two-dimensional grid. Subject 4's sleep stages were scored by a medical specialist based on PSG data from the same night, with a time window size of 30 s. The sleep stage sequence is shown in Fig. 2(b), where the awakening stage is shown as "W". We defined the period that contain continuous N3 stages with intervals of other stages that less than 3 min as a deep sleep period, and periods except deep sleep periods, awakening stages and REM stages as light sleep periods. The gray box marks the deep sleep periods in the figure. The grayscale parts in Fig. 2(b) indicate REM stages. Since the REM stage is a unique phase in the sleep process, we will discuss it separately.

The sleep periods of Subject 4 were interpreted as follows:

Deep seep periods (0:13:30 - 0:31:30), (2:09:30 - 2:42:00), (5:09:30 - 05:33:00), (7:25:30 - 7:45:30): There were many snoring events during these periods, very few body movements, and no tooth grinding. We found out that a cluster center of snoring event is mostly associated with a deep sleep period.

REM stages (1:33:00 - 1:43:00), (3:02:00 - 3:19:30), (4:43:00 - 4:51:30), (6:00:00 - 6:05:30), (6:07:30 - 6:44:00), (7:50:30 - 8:07:00): Compared with other stages, REM stages have a stronger association with clusters of body movement and a weaker association with those of snoring or tooth grinding.

Light sleep periods: In each light sleep period, there were some clusters of tooth grinding and body movement event but only a few snoring events.

In this experiment, we found that the distribution of sound event clusters changed simultaneously with the sleep stage change, for not only Subject 4, but also the other subjects. Even though our analysis includes other subjects who have different primary sleep disorders and varying pattern of sleep stages, the finding led to similar conclusions. For example, on Subject 2, the deep sleep periods were also obviously associated with the clusters of snoring, the number of body movements was notably more in the light periods and REM stages than in the deep periods, and no snoring clusters were found in REM stages.

Similar discussions are found in other studies. According to [7], conventional snoring is most likely to occur during the deep sleep stage, as well as during the light sleep stage, but unlikely during the REM stage. The REM stage is always associated with dreaming [15], which triggers several body movements.

From this experiment, we found that the transition of cluster dynamics and the changing of sleep stage are related. Since the sleep stage sequence is an important tool in the study of sleep pattern, its relation provides the possibility of discovering sleep patterns based on the cluster map of sleep-related sound data from Sb-KSOM.

4 Discussion

However, currently the age range of the subjects is not general since all of subjects are university students, but with the scope of data collection enlarging, this problem will be solved. Moreover, there are lots of useless noise events in the extracted sound dataset, these noise events took up many grids in the cluster maps, impacted the accuracy of our experiments. In the following experiment, we will improve the current noise reduction algorithm, to remove more useless noise.

To further verify our work, we will apply the Sb-KSOM algorithm to the data collected over several nights with the same subject. Furthermore, although our sound-based method cannot evaluate quiet subjects, those quiet subjects might slept better than the others, and we will confirm this assumption in the future work. Also, since it is difficult to use PSG scoring in practice, we will develop an annotation tool that a user can easily use to annotate their clusters.

5 Related Work

In the academic field of sleep analysis, various studies using other methods besides PSG, such as infrared thermography [12], water filled mat [22] and Kinect [21], have been proposed. Similar to PSG, these methods require additional professional equipment to record the sleep data and specialized knowledge to use the equipment; the data collection work is limited within the scope of medical specialists. Our method, by contrast, can be applied through any off-the-shelf sound recording device including a smartphone or a personal computer, therefore greatly reducing the cost of data collection and making large-scale data collection possible.

There are several personal health products on the market that provide sleep analysis services, such as aforementioned ZEO, Beddit, Fitbit, and Silmee (Toshiba) [27]. Compared with these products, the sound data-based study that we have proposed incurs very little extra hardware cost and does not require users to have physical contact with the additional hardware.

With the popularity of smartphones, analyzing the sleep quality via smartphone applications has gradually grown. There are some academic publications regarding smartphone application for sleep analysis. Gu et al. proposed a method for scoring sleep quality by a smartphone application named Sleep Hunter [13], and Hao et al. developed an application called iSleep [14]. Gu used not only sound data but also data from the accelerometer and light sensor, which limited the range of the available equipment. Hao used only sound data; however their ground truth is another high-quality audio data, which lacks medical reliability. Currently, neither Sleep Hunter nor iSleep can be found in any application store. Moreover, we investigated two popular applications: Sleep as Android[6] and Sleep Cycle alarm clock[7]; however, no academic proof or accuracy evaluation for their outputs exists. Since our research is consistent with PSG, the application that integrates our method will be more reliable than these applications.

6   Conclusion

This paper proposes a novel approach to discover the sleep pattern through analyzing the sleep-related sound events based on clustering algorithms. With this method, we obtained a fine-grained map that depicts the distribution and changes of sleep-related events. We proved the reliability of the method, and discovered some relevant information that provides a new train of thought for studying the sleep pattern.

Acknowledgments. This research is partially supported by the Center of Innovation Program from Japan Science and Technology Agency, JST, the Grant-in-Aid for Scientific Research (B)(#25293393) from the JSPS, and Challenge to Intractable Oral Diseases from Osaka University Graduate School of Dentistry.

References

1. Andras, P.: Kernel-kohonen networks. Int. J. Neural Syst. **12**(02), 117–135 (2002)
2. Behar, J., Roebuck, A., Domingos, J.S., Gederi, E., Clifford, G.D.: A review of current sleep screening applications for smartphones. Physiol. Meas. **34**(7), R29–R46 (2013)
3. Boulet, R., Jouve, B., Rossi, F., Villa, N.: Batch kernel SOM and related laplacian methods for social network analysis. Neurocomputing **71**(7), 1257–1273 (2008)
4. Choe, E.K., Kientz, J.A., Halko, S., Fonville, A., Sakaguchi, D., Watson, N.F.: Opportunities for computing to support healthy sleep behavior. In: CHI 2010 Extended Abstracts on Human Factors in Computing Systems, pp. 3661–3666. ACM (2010)
5. Chokroverty, S.: Sleep Disorders Medicine: Basic Science, Technical Considerations, and Clinical Aspects. Butterworth-Heinemann, Boston (2013)
6. Doi, Y.: Prevalence and health impacts of sleep disorders in Japan. J. Natl. Inst. Public Health **61**, 3–10 (2012)
7. Fairbanks, D.N., Mickelson, S.A., Woodson, B.T.: Snoring and Obstructive Sleep Apnea. Lippincott Williams & Wilkins, Philadelphia (2003)
8. Feinberg, I.: Changes in sleep cycle patterns with age. J. Psychiatr. Res. **10**(3), 283–306 (1974)
9. Fukui, K., Akasaki, S., Sato, K., Mizusaki, J., Moriyama, K., Kurihara, S., Numao, M.: Visualization of damage progress in solid oxide fuel cells. J. Environ. Eng. **6**(3), 499–511 (2011)

---

[6] http://sleep.urbandroid.org/.

[7] http://www.sleepcycle.com/.

10. Fukui, K., Numao, M.: Neighborhood-based smoothing of external cluster validity measures. In: Tan, P.-N., Chawla, S., Ho, C.K., Bailey, J. (eds.) PAKDD 2012, Part I. LNCS, vol. 7301, pp. 354–365. Springer, Heidelberg (2012)

11. Fukui, K., Saito, K., Kimura, M., Numao, M.: Sequence-based SOM: visualizing transition of dynamic clusters. In: Proceedings of IEEE 8th International Conference on Computer and Information Technology (CIT 2008), pp. 47–52. IEEE (2008)

12. Fukumura, H., Okada, S., Makikawa, M.: Estimation of sleep stage using SVM from noncontact measurement of forehead and nasal skin temperature. BME **50**(1), 131–137 (2012)

13. Gu, W., Yang, Z., Shangguan, L., Sun, W., Jin, K., Liu, Y.: Intelligent sleep stage mining service with smartphones. In: Proceedings of the 2014 ACM International Joint Conference on Pervasive and Ubiquitous Computing, pp. 649–660. ACM (2014)

14. Hao, T., Xing, G., Zhou, G.: iSleep: unobtrusive sleep quality monitoring using smartphones. In: Proceedings of the 11th ACM Conference on Embedded Networked Sensor Systems, SenSys 2013, pp. 4:1–4:14. ACM (2013)

15. Hobson, J.A., Pace-Schott, E.F., Stickgold, R.: Dreaming and the brain: toward a cognitive neuroscience of conscious states. Behav. Brain Sci. **23**(06), 793–842 (2000)

16. Kato, T., Masuda, Y., Yoshida, A., Morimoto, T.: Masseter EMG activity during sleep and sleep bruxism. Archives italiennes de biologie **149**(4), 478–491 (2011)

17. Kleinberg, J.: Bursty and hierarchical structure in streams. Data Min. Knowl. Disc. **7**(4), 373–397 (2003)

18. Kohonen, T.: Self-Organisation Maps. Springer, New York (1995)

19. Kohonen, T., Kaski, S., Lagus, K., Salojärvi, J., Honkela, J., Paatero, V., Saarela, A.: Self organization of a massive document collection. IEEE Trans. Neural Netw. **11**(3), 574–585 (2000)

20. Lavigne, G., Rompre, P., Montplaisir, J.: Sleep bruxism: validity of clinical research diagnostic criteria in a controlled polysomnographic study. J. Dent. Res. **75**(1), 546–552 (1996)

21. Metsis, V., Kosmopoulos, D., Athitsos, V., Makedon, F.: Non-invasive analysis of sleep patterns via multimodal sensor input. Pers. Ubiquit. Comput. **18**(1), 19–26 (2014)

22. Noh, T., Serizawa, Y., Kimura, T., Yamazaki, K., Hayasaka, Y., Itoh, T., Izumi, S., Sasaki, T.: The assessment of sleep stage utilizing body pressure fluctuation measured by water mat sensors. J. Adv. Sci. **21**(1 and 2), 27–30 (2009)

23. Silber, M.H., Ancoli-Israel, S., Bonnet, M.H., Chokroverty, S., Grigg-Damberger, M.M., Hirshkowitz, M., Kapen, S., Keenan, S.A., Kryger, M.H., Penzel, T., et al.: The visual scoring of sleep in adults. J. Clin. Sleep Med. **3**(2), 121–131 (2007)

24. Simula, O., Kangas, J.: Process monitoring and visualization using self-organizing maps. Neural Netw. Chem. Eng. **6**, 371–384 (1995)

25. Spruyt, K., Molfese, D.L., Gozal, D.: Sleep duration, sleep regularity, body weight, and metabolic homeostasis in school-aged children. Pediatrics **127**(2), e345–e352 (2011)

26. Xu, R., Wunsch, D.C.: Cluster validity. Clustering, pp. 263–278 (2008)

27. Yamada, H., Sato, Y., Ooshima, N., Hirai, H., Suzuki, T., Minami, S.: Heterogeneous system integration pseudo-SoC technology for smart-health-care intelligent life monitor engine and eco-system (silmee). In: Proceedings of Electronic Components and Technology Conference (ECTC), pp. 1729–1734. IEEE (2014)

A Conformational Epitope Prediction System Based on Sequence and Structural Characteristics

Wan-Li Chang, Ying-Tsang Lo, and Tun-Wen Pai[✉]

Department of Computer Science and Engineering, National Taiwan
Ocean University, Keelung, Taiwan, Republic of China
twp@mail.ntou.edu.tw

Abstract. An epitope is composed of several amino acids located on structural surface of an antigen. These gathered amino acids can be specifically recognized by antibodies, B cells, or T cells through immune responses. Precise recognition of epitopes plays an important role in immunoinformatics for vaccine design applications. Conformational epitope (CE) is the major type of epitopes in a vertebrate organism, but neither regular combinatorial patterns nor fixed geometric features are known for a CE. In this paper, a novel CE prediction system was established based on physico-chemical propensities of sequence contents, spatial geometrical conformations, and surface rates of amino acids. In addition, a support vector machine technique was also applied to train appearing frequencies of combined neighboring surface residues of known CEs, and it was applied to classify the best predicted CE candidates. In order to evaluate prediction performance of the proposed system, an integrated dataset was constructed by removing redundant protein structures from current literature reports, and three testing datasets from three different systems were collected for validation and comparison. The results have shown that our proposed system improves in both specificity and accuracy measurements. The performance of average sensitivity achieves 36 %, average specificity 92 %, average accuracy 89 %, and average positive predictive value 25 %.

Keywords: Conformational epitope · Antigen-antibody complex · Sequence antigenicity · Residue surface rate · Neighboring spatial distance clustering

1 Introduction

When virus invades human body, the immune system will be induced to produce antibodies against the virus. An antibody, also called as an immunoglobulin, is a protein with Y-shape conformation produced by plasma B-cells of immune system that could identify and neutralize pathogens, such as viruses and bacteria. An antibody binds only on a specific region of antigen called as an epitope, and its complementary binding region is called a paratope of antibody. These epitopes on antigens interact with paratopes of antibodies can bring up immune responses. Successful antigen epitope recognition by bioinformatics approaches can improve and simplify immunology-related experiments, and it could be applied to various applications including disease

© Springer International Publishing Switzerland 2016
H. Fujita et al. (Eds.): IEA/AIE 2016, LNAI 9799, pp. 472–483, 2016.
DOI: 10.1007/978-3-319-42007-3_41

prevention and diagnosis, treatment, and vaccine design [1]. The epitopes, which are also called as antigenic determinants, can be divided into two categories. The first type is categorized as a linear epitope (LE) which is composed of continuous amino acids from an antigen primary sequence, and the second type is classified as a conformational epitope (CE) which is formed by contiguous surface amino-acid residues on an antigen structure but discontinuous in the protein sequence. When an antigen protein is broken down in a lysosome, it generates small peptides called LEs which lie continuously in a line and could be specifically identified by antibodies. Peptides composed of different amino acids hold different physico-chemical propensities which can be analyzed and considered as antigenicity indicators for interacting with specific antibodies. Linear B-cell epitopes appear only in a small proportion of whole epitopes (less than 10 %), nevertheless, they are significant in virology, immunology, vaccine, and biochemistry research areas. Most published LE predictors applied amino acids characteristics, such as physico-chemical properties, antigenicity, hydrophobicity, protrusion area, pocket characteristics, and surface accessibility. Using different mathematical approaches of machine learning tools to predict LE, such as PEOPLE, BcePred, LEP, LEPS, VaxiJen, BEPITOPE, etc. are all introduced in [2]. However, Blythe and Flower evaluated physico-chemical propensities of amino acids in proteins for LEs prediction, and they reported that the best performance of physico-chemical propensity scales was slightly better than a random model [3]. A CE is composed of discontinuous antigen sequences, and it is hard to design and collect peptides that could form the same conformation. Accordingly, CEs were predicted by crystallographic method instead of applying biochemical experiments. Even though, clinical and biological researchers still count on biochemical/biophysical experiments to recognize binding sites of epitopes in B-cell receptors and/or antibodies. Unfortunately, this work can be time-consuming, expensive, and sometimes unsuccessful. Furthermore, another challenge is that conformational epitopes cannot be well-formed without binding to a corresponding antibody. Thus, the information of antigen-antibody crystallographic combination is considered as an important concern for CE prediction. Up to now, several available CE prediction systems were designed based on structure-based approaches, such as the first published system of Conformational Epitope Prediction server (CEP) in 2005, which applied surface accessible characteristics [4]; ElliPro system applied residue protrusion index to predict epitopes [5]; DiscoTope 2.0 applied spatial neighboring and surface exposure measurements [6]; BEpro/PEPITO introduced the epitope residue propensity and the spatial attribute of half sphere exposure as scores [7]; PEPOP applied the segments of amino acids of accessible and contiguous sequences [8]; EPITOPIA applied a naive Bayesian classifier with physico-chemical and structural geometrical features to identify epitopes from surface structure [9]; EPCES integrated six different scores and a voting mechanism for consensus [10]; Bpredictor used random forest classifier with various attributes [11]; B-pred integrated structure quality values with solvent exposure [12]; CE-KEG used knowledge-based energy function and geometric residue characteristics [13]; SEPPA 1.0 applied the concept of unit patch of residue triangle and clustering coefficients to describe local spatial context and compactness of surface amino acids [14], and SEPPA 2.0 employed two new features including accessible surface area propensity and consolidated amino acid index [15]. In this study, a novel approach combining both sequence antigenicities and structural features, such as physico-chemical propensities, surface rates, and amino acid combinatorial patterns, was designed for predicting CEs. The proposed system was developed by firstly employing sequential antigenicity of linear peptides from LEPS [16] and then computing of surface residues from CE-KEG [13]. To achieve the goal of this study, a set of antigen-antibody complexes was initially constructed for validating and training processes, and these structures were downloaded directly from Protein Data Bank (PDB). We have observed from the collected proteins that neighboring linear peptides could be combined to form a conformational epitope. Hence, we firstly identified antigenicity fragments from an antigen protein sequence, and then selected residues on surface for the next clustering process. In order to recombine neighboring candidate epitope segments and to recover the interspersedly located residues among the candidate segments, a novel algorithm was proposed to dynamically patch the predicted CE groups. Finally, an SVM classifier was applied to remove false positive prediction cases according to statistical information obtained from the previous discovered CEs. The details of our proposed system and performance evaluation will be discussed in the following sections.

2 Material and Method

Benchmarking datasets of CE provide standards to verify whether prediction results are accurate and to evaluate performance of existing CE prediction tools. In this study, we prepared one training dataset and three testing datasets for prediction performance evaluation. In order to compare the newly proposed CE prediction tool with other systems, the testing datasets are collected from CE-KEG, SEPPA 1 and 2. These datasets are briefly described in the following sections.

2.1 ABepar Dataset

The training dataset was extracted from antibody-specified B-cell epitope prediction through association rules (ABepar dataset). The description of each complex was downloaded from PDB, and each complex contains geometrical information of antigen-antibody X-ray crystallographic molecules with resolution better than 3.0 Å. A total of 80 non-redundant protein structures were collected and all protein sequence lengths are larger than 30 amino acids. These antigen-antibody complexes are obtained and examined by four rules for removing high similarity structures. If one of the rules is satisfied, redundant complexes will be removed due to high similarity. These rules include the alignment z-score more than 4.0; root mean square deviation (RMSD) of alignment less than 3 Å; proportion of matched residues higher than eighty percent of aligned positions; proportion of overlapped epitope residues higher than eighty percent of the epitope residues of the shorter sequence.

2.2   CE-KEG Dataset

The testing dataset extracted from CE-KEG was collected by Lo et al. [13], which system was developed based on knowledge-based energy function and geometric residue characteristics. The dataset is consisted of three datasets: DiscoTope, Epitome database, and IEDB. There are a total of 163 protein structures examined by multiple structure alignment (MStA) to remove highly similar structures. We adopted the multiple structure alignment incorporate refinement system (AIR, unpublished system) developed by ourselves to perform the task. The selection criterion is according to the performance of both root mean square deviation (RMSD) and the number of aligned core residues (core size) to formulate the structural alignment score (SAS). If two structures are similar, one structure would be removed when the aligned SAS score is less than 2. Finally, a set of 28 non-redundant protein structures from CE-KEG was selected to verify our system performance.

2.3   SEPPA Datasets

Another two testing datasets containing 119 and 42 antigen structures were originally retrieved from SEPPA 1 and 2 systems. For SEPPA 1, the testing structures were prepared by removing duplicated PDB IDs from DiscoTope, Epitome, and IEDB. For SEPPA 2, testing structures obeyed the same removing method for structures from Bpredictor, EPMeta, and DiscoTope 2. Again, these testing structures were rechecked and removed through performing multiple structure alignment by AIR, and finally, we could obtain two testing datasets of 26 and 22 non-redundant protein structures from SEPPA 1 and SEPPA 2. These unique antigen structures were verified and applied for evaluating CE prediction performance and comparing with two SEPPA systems.

2.4   System Configuration

The proposed CE prediction system was designed as the following steps (Fig. 1). First, a query PDB ID is downloaded from PDB website. At the second step, protein sequences are detected by LEP for candidate epitopes. LEP was developed for finding high antigenicity of fragments in our previous work and default settings were applied. For the third step, a look-up table for integrating neighboring residues from candidate antigenic fragments is applied to form a candidate epitope region, which is a grid-based algorithm that could efficiently identify surface regions of a protein. After executing previous steps, we could obtain surface antigenic regions as anchors and further clustering and recombining operations are performed for primary candidate epitopes. The next step is to check and extend possible neighboring surface residues for constructing compact clustered groups as predicted CEs. Finally, in order to achieve a better performance, this system adopts an SVM (LIBSVM) tool to remove unsuitable CE candidates.

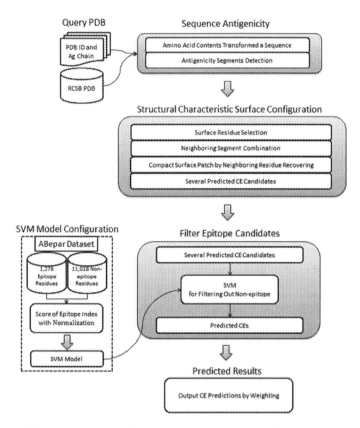

Fig. 1. A system configuration of the proposed prediction system.

2.5   Sequence Antigenicity Analysis

The developed system applied LEP to identify local continuous fragments with high antigenicity, which is designed by considering physico-chemical propensities combined with mathematical morphology approaches. Through the advantage of variable lengths of predicted linear epitopes, and different combinations of physico-chemical propensities for discovering high antigenic segments, the predicted local segments by LEP is further evaluated by combining machine learning technologies to reduce false positive prediction cases, and the sequence antigenicity analysis was adopted as the first step for CE prediction. Therefore, a query protein structure would be focused on its protein sequence contents initially, and their corresponding physico-chemical propensity scales, such as secondary structure, hydropathy, surface accessible, flexibility, and polarity would be applied to identify high antigenic fragments. Finally, the predicted local high antigenic fragments are used as initial and fundamental segments for the following CE prediction.

2.6   Structural Surface Residue Selection

Antigen-antibody interaction usually appears on surface areas. The properties of side chains of surface residues within an epitope are important factors and control protein-protein interaction mechanisms. Several literatures described the influence of side chains as key factors affecting protein binding actions. Antigen-antibody binding mechanisms may require or cause conformational changes before or after the binding events. Hence, surface amino acids possess flexible side chain characteristics may have an advantage for binding with other proteins. It is also demonstrated experimentally that nonpolar-nonpolar and polar-polar side chain interactions could stabilize protein interfaces [17]. Therefore, the characteristics of side chain atoms and their corresponding surface exposure percentage should be analyzed prior to the following prediction. Here, the use of 3D mathematical morphology operations was applied to detect the surface rate of the ith surface side-chain atom and represented as $AR_i(r)$. In other words, the surface rate of a surface residue is represented by $SR(r)$ and calculated as:

$$SR(r) = \left\{ i \in R : \frac{1}{N} \sum_{i=1}^{N} AR_i(r) \right\}$$

where i denotes the ith surface atom in the side chain of a residue, R represents all surface atoms within a residue, and N is the total number of surface atoms of residue r. Directly applying the surface rate equation for all surface residues and verified epitope residues in the ABepar training dataset. According to the statistical analysis on the ABepar training dataset, surface rate of a residue larger than 0.05 was defined as a surface residue. Therefore, the surface antigenic fragments from previously predicted LE segments could be identified according to the default settings.

2.7   Neighboring Segment Combination

From previous modules, the system provides a set of high antigenicity segments located on surface of an antigen structure. Then, compact surface patches from random distributed segments on structural surface would be analyzed. First, each segment will be initially analyzed for its corresponding radius. The radius decides the range of a disk for connecting two segments in the next step. The corresponding radius of a segment is obtained by calculating the center location of the segment, and it is obtained by searching the longest distance between two ending residues of the segment and divided by 2. Once the center location is defined, the distance of each residue of the segment and the center location will be evaluated, and a residue with the minimum distance is defined as the center residue. After obtaining all center residues of these segments, the Dijkstra's algorithm is performed for the shortest path between any two center residues, and these paths are defined as surface paths for the antigen. When the distance between two segments on the surface is less than 20 Å, these two segments will be assigned as the same cluster, and the total number of residues within a combined cluster should be less than 20 residues.

2.8   Compact Surface Patch by Neighboring Residue Recovering

According to previously detected surface antigenic segments, this module extended the surfaces neighboring residues for compact surface patch formulation. A post-processing is performed to recover residues located interspersedly within the formulated clusters. First, the maximum and the minimum distances of the two assigned segments within a cluster are applied for post recovering processes. The maximum distance represented the farthest atom of a residue to the center residue within the longer segment. An average of maximum distances of all calculated distances from all residues within a cluster is obtained and defined as a mean distance. Then, the calculated maximum and mean distances were summed and divided by 2 for representing a radius of a scanning disk for evaluating neighboring residues. When a segment containing a relatively large radius (Rmax) intersects with the other segment containing a relatively small radius (Rmin), a dynamical defined radius (Rdec) depends on the number of amino acids (#N) from the larger segment to the shorter segment in a descending trend. The concept is expressed as following equation.

$$R_{dec} = R_{max} - \left[\frac{(R_{max} - R_{min})}{\#N} \right] * N_i$$

If the surface residues located within the defined radius, they will be recovered for a compact surface patch and considered as predicted CEs. In the other way, when a segment containing a relatively small radius (Rmin) intersects with other segment containing a relatively large radius (Rmax), a dynamical defined radius (Rinc) depended on the number of amino acids (#N) from the shorter segment to the longer segment in an increasing trend. The concept is expressed as following equation.

$$R_{inc} = R_{min} + \left[\frac{(R_{max} - R_{min})}{\#N} \right] * N_i$$

If the surface residues located within the defined radius, they will be recovered for a compact surface patch and considered as predicted CEs.

2.9   SVM Features and Model Selection

The definition of surface epitopes and non-epitopes are validated as the ABepar training dataset which contain 80 protein structures. To improve prediction performance, appearing features of amino acid combinations (AASs) with lengths from two to three residues are considered. The statistical features of paired residues (AAS^2) and three combined residues (AAS^3) for both epitopes and non-epitopes are calculated. For AAS^2, a total of 210 possible combinations obtained by deleting duplicate pairs from the original 400 possible combinations. All residue pairs were analyzed for occurrence frequencies within both epitope and non-epitope groups. The minimum distance for two residues considered as neighboring pair is defined within 4 Å. Similarly, for AAS^3, 1,540 triple combinations from a total of 8,000 possible combinations were also obtained for training and evaluating occurrence frequencies. The epitope index (Epidex$_i$) of the ith pattern (AAS$_i$) was computed by calculating logarithm value of the ratio of accumulated number of AAS$_i$ among all epitopes AASs compared to the ratio in the non-epitope AASs. The index is calculated as following:

$$\text{Epidex}_i = \log\left(\frac{\frac{f_i^+}{\sum_i f_i^+}}{\frac{f_i^-}{\sum_i f_i^-}}\right)$$

where f_i^+ denoted the numbers of AAS$_i$ in the epitope dataset and f_i^- denoted the numbers of AAS$_i$ in the non-epitope dataset. $\sum_i f_i^+$ was the total number of AAS$_i$ in the corresponding epitope dataset and $\sum_i f_i^-$ was the total number of AAS$_i$ in the corresponding non-epitope dataset. Furthermore, the values of Epidex$_i$ (S) were normalized to the range of [-1, 1] to avoid extreme values of any individual Epidex$_i$ in the classifier learning processes. There are S$_{min}$ and S$_{max}$ are denoted as the minimum value of Epidex$_i$ and the maximum value of Epidex$_i$. These parameters were calculated with the following equations:

$$\text{normalization} = \frac{(S - S_{min}) * [1 - (-1)]}{S_{max} - S_{min}} + (-1)$$

Finally, the score table of AAS^2 and AAS^3 were exported to a look-up table. The ABepar training dataset was then used to construct an SVM model based on the two feature values and the target values of each epitope and non-epitope. There were four common kernel functions provided by LIBSVM, including linear, polynomial, radial basis function (RBF), and sigmoid. By trial and error, the RBF kernel function was applied to train the whole testing datasets for constructing the final SVM classifier in this study.

3   System Prediction Performance

To evaluate the performance of the proposed system, five indicators were used to measure effectiveness with default settings. These indicators include (1) sensitivity (SEN), defined as the percentage of epitopes that were correctly predicted as epitopes; (2) specificity (SP), defined as the percentage of non-epitopes that were correctly predicted as non-epitopes; (3) accuracy (ACC), defined as the proportion of correctly predicted epitopes; (4) balanced accuracy (BA), defined as the arithmetic mean value of sensitivity and specificity; (5) positive predictive value (PPV). For each predicted CE from a query protein, we could calculate the number of epitope residues correctly predicted as epitope residues (TP), the number of non-epitope residues incorrectly predicted as epitope residues (FP), the number of non-epitope predicted as not epitope residues (TN), and the number of verified epitope residues which were not correctively predicted as epitope residues by the system (FN). These parameters were calculated with the following equations:

Sensitivity (SEN) = TP/(TP + FN)
Specificity (SP) = TN/(TN + FP)
Accuracy (ACC) = (TP + TN)/(TP + TN + FN + FP)
Balanced Accuracy (BA) = (Sensitivity + Specificity)/2
Positive Predictive Value (PPV) = TP/(TP + FP)

3.1 System Performance

In this study, we applied 28 non-redundant protein structures from CE-KEG, 26 non-redundant protein structures from SEPPA 1, and 22 non-redundant protein structures from SEPPA 2, respectively. All verified CEs were received either from computational analysis or deduced from experimental observations. Moreover, all predicted CE candidates are ranked and the best three predicted CEs were evaluated for details. From Table 1, the prediction performance for three datasets were shown individually. The results showed that the proposed system achieved an average SEN of 36.11 %, an average SP of 92.34 %, an average ACC of 89.09 %, an average BA of 64.22 %, and an average PPV of 25.47 % for CE-KEG dataset; for SEPPA 1 dataset, an average SEN of 35.81 %, an average SP of 92.94 %, an average ACC of 89.78 %, an average BA of 64.38 %, and an average PPV of 25.23 %; for SEPPA 2 dataset, an average SEN of 21.13 %, an average SP of 91.90 %, an average ACC of 87.38 %, an average BA of 56.52 %, and an average PPV of 14.26 %.

Table 1. Performance of CE prediction for CE-KEG, SEPPA 1, and SEPPA 2 datasets by the proposed system.

Datasets	SEN (%)	SP (%)	ACC (%)	BA (%)	PPV (%)
CE-KEG	36.11	92.34	89.09	64.22	25.47
SEPPA 1	35.81	92.94	89.78	64.38	25.23
SEPPA 2	21.13	91.90	87.38	56.52	14.26

3.2 Comparing with Other Systems

We compared our system with SEPPA 2 by using its non-redundant 28 testing structures. The thresholding setting of SEPPA 2 was defined according to its default value of 0.1. The average CE prediction results of our proposed system against SEPPA 2 are displayed in Table 2. Due to different strategies and techniques, it can be observed that our proposed system is good at specificity while SEPPA 2 is good at sensitivity. Our system possesses higher specificity and accuracy in general compared to SEPPA 2.

Table 2. Comparison of prediction performance on our system and SEPPA 2.

System name	SEN (%)	SP (%)	ACC (%)	BA (%)	PPV (%)
Our system	36.11	**92.34**	**89.09**	64.22	**25.47**
SEPPA 2	**75.12**	61.15	63.08	**68.14**	10.97

For the second comparison scenario, we adopted the testing set of 26 protein structures from SEPPA 1. The prediction results by five systems including our system, SEPPA 1, CEP, DiscoTope, and BEpro were all evaluated simultaneously. The comparison results showed in Table 3, which can be observed that our proposed system held the best balanced accuracy of 64.38 % compared to 59.59 %, 53.22 %, 62.65 %, and 60.57 % for the other four systems. From this table, we can conclude that our proposed system based on both sequence antigenicities and surface characteristics obtained the best performance regarding the balanced accuracy measurement.

Table 3. Comparison of balanced accuracy performance for 26 testing dataset.

System name	Balanced accuracy
Our system	**64.38 %**
SEPPA 1	59.59 %
CEP	53.22 %
DiscoTope	62.65 %
BEpro	60.57 %

4 Conclusions and Discussions

Due to exponentially increased amount of resolved protein structures and powerful computational resources for immunobioinformatics, the applications of CE prediction has been widely studied by clinical immunology and biomedical scientists for vaccine design and epitope assembly. In this research, a novel approach combining both sequence antigenicities and structural features, such as surface rate, pair-pattern, and physico-chemical propensities was designed for predicting CEs. The proposed system first focused on the sequence contents retrieved from a protein structure. By adopting liner epitope prediction algorithm on the sequences, we could identify several continuous segments that possess local high antigenicities according to their corresponding physico-chemical features. With these segments as our initial candidates, we analyzed the surface rates of each identified residue within a candidate LE. Only residues located on surface will be considered for the next clustering processes. To recombine neighboring candidate epitope segments and to recover the interspersedly located residues among the neighboring segments, a novel algorithm was proposed to dynamically patch the predicted CE groups. At the final stage, an SVM classifier was applied to remove false positive prediction cases according to statistical information obtained from previous discovered CEs. All the predicted CEs would be ranked according to their measuring scores.

To evaluate the proposed system, we selected the top three predicted CEs for system performance evaluation. Three testing datasets were applied to predict CEs and compared with two other well-known systems. These datasets including 28 non-redundant protein structures filtered from 163 structures selected in CE-KEG, 26 non-redundant structures retrieved from 119 protein structures collected in SEPPA 1, and 22 non-redundant structures retrieved from 42 protein structure collected in SEPPA 2 were applied for system evaluation respectively. The outcome of the proposed system accomplished a better performance compared to previous performance. Furthermore, we adopted the balanced accuracy to evaluate prediction system performance, which is calculated by taking an average of sensitivity and specificity. To compare with SEPPA1, CEP, DiscoTope, and BEpro, we have focused on the SEPPA1's 26 testing proteins. All other systems produced averaged balanced accuracies of 59.6 %, 53.2 %, 62.7 %, and 60.6 % respectively. In comparison with our proposed system, we have achieved the best balanced accuracy values of 64.4 %. This results imply that our proposed system consider the prediction performance both in the aspects of sensitivity as well as specificity.

Acknowledgment. This work is supported by the Center of Excellence for the Oceans, National Taiwan Ocean University and Ministry of Science and Technology, Taiwan, R.O.C. (MOST 104-2627-B-019-003 and MOST 104-2218-E-019-003 to T.-W. Pai).

References

1. Sharon, J., Rynkiewicz, M.J., Lu, Z., Yang, C.Y.: Discovery of protective B-cell epitopes for development of antimicrobial vaccines and antibody therapeutics. Immunology **142**, 1–23 (2014)
2. Wang, H.W., Pai, T.W.: Machine learning-based methods for prediction of linear B-cell epitopes. Methods Mol. Biol. **1184**, 217–236 (2014)
3. Toseland, C.P., Clayton, D.J., McSparron, H., Hemsley, S.L., Blythe, M.J., Paine, K., et al.: AntiJen: a quantitative immunology database integrating functional, thermodynamic, kinetic, biophysical, and cellular data. Immunome Res. **1**, 4 (2005)
4. Kulkarni-Kale, U., Bhosle, S., Kolaskar, A.S.: CEP: a conformational epitope prediction server. Nucleic Acids Res. **33**, W168–W171 (2005)
5. Ponomarenko, J., Bui, H.H., Li, W., Fusseder, N., Bourne, P.E., Sette, A., et al.: ElliPro: a new structure-based tool for the prediction of antibody epitopes. BMC Bioinformatics **9**, 514 (2008)
6. Kringelum, J.V., Lundegaard, C., Lund, O., Nielsen, M.: Reliable B cell epitope predictions: impacts of method development and improved benchmarking. PLoS Comput. Biol. **8**, e1002829 (2012)
7. Sweredoski, M.J., Baldi, P.: PEPITO: improved discontinuous B-cell epitope prediction using multiple distance thresholds and half sphere exposure. Bioinformatics **24**, 1459–1460 (2008)
8. Moreau, V., Fleury, C., Piquer, D., Nguyen, C., Novali, N., Villard, S., et al.: PEPOP: computational design of immunogenic peptides. BMC Bioinformatics **9**, 71 (2008)
9. Rubinstein, N.D., Mayrose, I., Martz, E., Pupko, T.: Epitopia: a web-server for predicting B-cell epitopes. BMC Bioinformatics **10**, 287 (2009)
10. Liang, S., Zheng, D., Zhang, C., Zacharias, M.: Prediction of antigenic epitopes on protein surfaces by consensus scoring. BMC Bioinformatics **10**, 302 (2009)
11. Zhang, W., Xiong, Y., Zhao, M., Zou, H., Ye, X., Liu, J.: Prediction of conformational B-cell epitopes from 3D structures by random forests with a distance-based feature. BMC Bioinformatics **12**, 341 (2011)

12. Giaco, L., Amicosante, M., Fraziano, M., Gherardini, P.F., Ausiello, G., Helmer-Citterich, M., et al.: B-Pred, a structure based B-cell epitopes prediction server. Adv. Appl. Bioinform. Chem. **5**, 11–21 (2012)

13. Lo, Y.T., Pai, T.W., Wu, W.K., Chang, H.T.: Prediction of conformational epitopes with the use of a knowledge-based energy function and geometrically related neighboring residue characteristics. BMC Bioinformatics **14**(Suppl 4), S3 (2013)

14. Sun, J., Wu, D., Xu, T., Wang, X., Xu, X., Tao, L., et al.: SEPPA: a computational server for spatial epitope prediction of protein antigens. Nucleic Acids Res. **37**, W612–W616 (2009)

15. Qi, T., Qiu, T., Zhang, Q., Tang, K., Fan, Y., Qiu, J., et al.: SEPPA 2.0–more refined server to predict spatial epitope considering species of immune host and subcellular localization of protein antigen. Nucleic Acids Res. **42**, W59–W63 (2014)

16. Wang, H.W., Lin, Y.C., Pai, T.W., Chang, H.T.: Prediction of B-cell linear epitopes with a combination of support vector machine classification and amino acid propensity identification. J. Biomed. Biotechnol. **2011**, 432830 (2011)

17. Chou, W.I., Pai, T.W., Liu, S.H., Hsiung, B.K., Chang, M.D.: The family 21 carbohydrate-binding module of glucoamylase from Rhizopus Oryzae consists of two sites playing distinct roles in ligand binding. Biochem. J. **396**, 469–477 (2006)

The Factors Affecting Partnership Quality of Hospital Information Systems Outsourcing of PACS

Yi-Horng Lai[(⊠)]

Oriental Institute of Technology, New Taipei City 22061, Taiwan
FL006@mail.oit.edu.tw

Abstract. The purpose of this study is to investigate the determinants of out-sourcing partnership between clients and providers in hospital outsourcing projects, and it also explores the relationship between the quality of outsourcing partnership and the success of outsourcing. Subjects of the survey were the medical centers that involved in PACS projects in Taiwan. A total of 97 valid questionnaires were analyzed in this study by the partial least squares to test related hypotheses. The results indicated that there is a positive correlation between the following variables: (1) shared knowledge and mutual benefits (2) mutual dependency and mutual benefits (3) mutual dependency and commitment (4) mutual dependency and predisposition (5) organizational linkage and commitment (6) organizational linkage and predisposition (7) commitment and outsourcing success.

Keywords: Outsourcing · Picture archiving and communication systems (PACS) · The partial least squares (PLS)

1 Introduction

Outsourcing is the contracting out of a business process to a third-party. It usually involves transferring employees and assets from one firm to another. Outsourcing is also used to describe the practice of handing over control of public services to for-profit corporations. Outsourcing includes both foreign and domestic contracting, and sometimes includes offshoring or relocating a business function to another country. Financial savings from lower international labor rates is a big motivation for outsourcing. The opposite of outsourcing is called insourcing, which entails bringing processes handled by third-party firms in-house, and is sometimes accomplished via vertical integration. However, a business can provide a contract service to another business without necessarily insourcing that business process.

The innovative revolution that has swept through the world of information technology over the last decade has had a profound impact on radiology departments in medical facilities all across Taiwan. By adopting everything from electronic medical records (EMRs) all the way to picture archiving and communications systems (PACS), medical facilities of every description have seen the efficiency and accuracy of their radiology department's release of information (ROI) processes reach levels that were deemed impossible only a few short years ago [1].

© Springer International Publishing Switzerland 2016
H. Fujita et al. (Eds.): IEA/AIE 2016, LNAI 9799, pp. 484–492, 2016.
DOI: 10.1007/978-3-319-42007-3_42

These years, many organizations which have chosen to wait and gauge the results of these developments have also become convinced of the ability of these changes to transform their own radiology release of information procedures. For many facilities already feeling overstretched, their only question is how to transit to a viable outsourced PACS radiology release of information solution that works for them [2].

1.1 Picture Archiving and Communication System

The picture archiving and communication system (PACS) is a medical imaging technology which provides economical storage of and suitable access to images from clinic machines. Electronic images and reports are transmitted digitally through PACS; this eliminates the need to manually file, retrieve and transport films. The common format for PACS image storage and transfer is Digital Imaging and Communications in Medicine (DICOM). The PACS consists of four major components: The imaging modalities such as X-ray plain film, computed tomography (CT) and magnetic resonance imaging (MRI), workstations for interpreting and reviewing images, and archives for the storage and retrieval of images and reports. United with available and emerging web technology, PACS has the ability to deliver timely and efficient access to images, interpretations, and related data [2].

1.2 The Causal Model of an Outsourcing Partnership

One important study of partnership relationships was by Abdul-Halim et al. [3]. They aimed at examining the role of service quality in strengthening the relationship between partnership quality and human resource outsourcing success. The results of their study showed that partnership quality variables such as trust, business understanding, and communication had significant positive impact on human resource outsourcing success; whereas in general, service quality was found to partially moderate these relationships.

Teo and Bhattacherjee developed a nomological network of antecedents and outcomes of knowledge transfer and utilization in IT outsourcing relationships [4]. The result of their study showed that the characteristics of outsourcing clients, vendors, and knowledge transfer played important roles in facilitating knowledge transfer; the transferred knowledge in conjunction with the knowledge integration mechanisms affected knowledge utilization in client-firms, and that this made significant operational and strategic performance gains in information technology processes.

Lee and Kim's study premises that a proposed model consists of three major parts: behavioral variables, psychological variables, and outsourcing success [5].

To find appropriate variables to describe Lee and Kim's model, partnership-related variables that include mutual benefits, commitment, predisposition, shared knowledge, mutual dependency, and organizational linkage were first identified from the related literature in terms of the social exchange theory, which has been applied to the study of outsourcing partnerships [6].

Many researchers have considered mutual benefit [7], commitment [8], and predisposition [9] as basic variables which play important roles in intervening variables between behavioral variables and the success of outsourcing. From the previous studies, it can be found that shared knowledge [9, 10], mutual dependency [11, 12], and organizational linkage [13] have been considered important factors for the success of outsourcing. According to Park and Lee's study [12], these variables represent the behavioral constructs to create the working relationship rather than to establish the partners' belief in the partnership's sustainability. This indicates that they should be considered as antecedents of the psychological variables.

Mutual Benefits: Schwarz [7] indicates that while clients still seek financial benefits, they also seek a mutually beneficial relationship. This focuses on mutual benefits that exemplify the transformation toward the development of a partnership. The new outsourcing arrangement more closely typifies a relationship than a contractual transaction.

Commitment: The human resource practice of training outsourcing has emerged as one of the fastest growing segments of the broader business process outsourcing industry. In spite of the growing popularity in professional practice, training outsourcing continues to be subjected to critical review and ongoing debate with most attention focused on the decision to outsource or not. However, there was a shortage of research on training outsourcing as a human resource development practice and the potential relationships with desired organizational outcomes including employee commitment. Chaudhuri and Bartlett point out that those positive relationships between specific measures of employee perceptions of quality, usefulness and supervisor support for outsourced training with organizational commitment [8].

Predisposition: Francois, Isis, Sergio, and Luis integrated inter-organizational factors, such as trust, knowledge sharing, and quality of outsourcing interfaces in the model [9], and they point out that predisposition play an important role in the relationships in the information system outsourcing in the organization.

Shared Knowledge: Donga and Pourmohamadi focus on innovation matchmaking by online innovation intermediaries that operate as technology outsourcing services [10]. Francois, Isis, Sergio, and Luis integrated inter-organizational factors, such as trust, knowledge sharing, and quality of outsourcing interfaces in the model [9], and they point out that knowledge sharing play an important role in the relationships in the information system outsourcing in the organization.

Mutual Dependency: Realization of benefits may in turn inspire both the service recipient and the service provider to actively engage in further development of mutual dependency to gain further from their mutually beneficial outsourcing relationship. Therefore, Goo, Kishore, Rao, and Nam indicated that foundation characteristics of specific characteristics of service level agreements positively influence mutual dependence [11].

Organizational Linkage: Rajagopal, Zailani, and Sulaiman indicated that firms appear to confirm a positive and significant relationship between the degree of resource sharing and organizational linkage, if they see that scalable partnering efforts as hypothesized are workable [13].

2 Methodology

With the causal model of an outsourcing partnership, Fig. 1 was the summarize the research framework of this study in a model in which shared knowledge (SK), mutual dependency (MD), organizational linkage (OL), mutual benefits (MB), commitment (CM), predisposition (PD), and outsourcing success (OS). The outsourcing success variable includes business perspective, credibility, correlation, correctness, timeliness, integrity, and instant. The hypotheses in this study are:

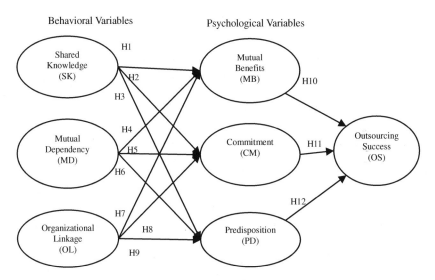

Fig. 1. Research framework.

H1:   Shared knowledge is positively associated with mutual benefits.
H2:   Shared knowledge is positively associated with commitment.
H3:   Shared knowledge is positively associated with predisposition.
H4:   Mutual dependency is positively associated with mutual benefits.
H5:   Mutual dependency is positively associated with commitment.
H6:   Mutual dependency is positively associated with predisposition.
H7:   Organizational linkage is positively associated with mutual benefits.
H8:   Organizational linkage is positively associated with commitment.
H9:   Organizational linkage is positively associated with predisposition.
H10: Mutual benefit is positively associated with outsourcing success.
H11: Commitment is positively associated with outsourcing success.
H12: Predisposition is positively associated with outsourcing success.

2.1 Research Data

The research data was obtained from Hwang's study [14] in the Survey Research Data Archive (SRDA) provided by the Academia Sinica in Taiwan. Hwang's study

"A Comparative Study of Outsourcing Partnership between Medical and Banking Industries" was finished in November 30, 2009. The purpose of Hwang's study [14] was to investigate the determinants of outsourcing partnership between clients and providers in outsourcing projects. His study focused on the organizations that partnered with healthcare organizations and banks. A total of 147 valid questionnaires were collected from healthcare organizations and 49 were from banks. This present study focuses on 97 questionnaires from healthcare organizations that adopted PACS.

2.2   Methodology

This study was developed in a way that the model constructs in the causal model of an outsourcing partnership were adapted to the context of using the picture archiving and communication system (PACS). Items on the survey include those measuring shared knowledge (SK), mutual dependency (MD), organizational linkage (OL), mutual benefits (MB), commitment (CM), predisposition (PD), outsourcing success (OS) - business perspective, outsourcing success (OS) - credibility, outsourcing success (OS) - correlation, outsourcing success (OS) - correctness, outsourcing success (OS) - timeliness, outsourcing success (OS) - integrity, and outsourcing success (OS) - accessibility. The questionnaire contains no identifying information about the individual participants. All variables exhibit a high level of reliability with the Cronbach's alpha values (Table 1) exceeding the recommended 0.6 [15].

Table 1. Scale reliability.

Scale	N	Mean	S.D.	Cornbach's α	C.R.
SK	3	3.71	.42	.912	.945
MD	2	3.86	.56	.875	.939
OL	3	3.79	.43	.970	.980
MB	2	3.85	.59	.899	.952
CM	3	3.68	.40	.746	.855
PD	3	3.70	.54	.896	.935
OS-Business Perspective	9	3.72	.56	.951	–
OS-Credibility	5	3.98	.44	.933	–
OS-Correlation	5	4.17	.50	.942	–
OS-Correctness	5	4.19	.45	.954	–
OS-Timeliness	5	4.17	.40	.932	–
OS-Integrity	6	4.14	.39	.982	–
OS-Accessibility	5	4.13	.42	.933	–

The structural model is investigated using SmartPLS 3.2.3 [16]. Path analysis was performed on the model using standardized maximum likelihood estimation. The path analytic method offers the advantage of testing the overall model fit with multiple endogenous variables as in the model as well as individual a priori hypotheses.

The result of correlation coefficient is presented in Table 2. It shows that those square of AVEs were between .814 and .971. They are greater than most of the other rations in the table.

Table 2. Results of correlation coefficient.

	CM	MB	MD	OL	OS	PD	SK
CM	.814						
MB	.367	.953					
MD	.384	.251	.941				
OL	.372	.169	.058	.971			
OS	.597	.433	.450	.482	–		
PD	.583	.356	.278	.430	.414	.910	
SK	.379	.200	.249	.510	.570	.275	.922

*: P-value < .05

3   Result

A total of 97 feedbacks from medical centers with outsourcing PACS were examined in this study. Basic demographic information indicated there were 54 males (55.67 %) and 43 females (44.33 %) in the sample population. Most of them were under 31 ∼ 40 years old (59; 60.82 %) (Table 3), and of the majority had used PACS for 4 ∼ 6 years (54; 55.67 %). Factor loadings demonstrate adequate discriminant and convergent validity as Table 4.

Table 3. Data summary.

Variable		Frequency	Percent (%)
Gender	Male	54	55.67
	Female	43	44.33
Age	∼30	22	22.68
	31 ∼ 40	59	60.82
	41 ∼ 50	13	13.40
	51∼	3	3.10
Seniority	∼5	30	30.93
	6 ∼ 10	36	37.11
	11 ∼ 15	17	17.53
	16∼	14	14.43
Experience	1 ∼ 3	25	25.77
	4 ∼ 6	54	55.67
	7 ∼ 9	12	12.37
	10∼	6	6.19
Total		97	100.00

Table 4. Factors loadings for the measurement model

	CM	MB	MD	OL	OS	PD	SK
CM1	**.827**	.274	.199	.360	.480	.598	.385
CM2	**.822**	.288	.379	.270	.403	.474	.203
CM3	**.794**	.329	.359	.278	.503	.358	.328
MB1	.378	**.948**	.268	.141	.338	.356	.194
MB2	.323	**.958**	.213	.180	.418	.322	.189
MD1	.284	.175	**.917**	-.042	.294	.198	.147
MD2	.414	.278	**.964**	.119	.472	.304	.295
OL1	.356	.136	.063	**.968**	.467	.408	.490
OL2	.372	.202	.076	**.960**	.491	.424	.488
OL3	.357	.153	.028	**.986**	.456	.419	.508
OS1	.444	.282	.402	.252	**.590**	.347	.335
OS2	.536	.396	.391	.505	**.882**	.401	.564
OS3	.515	.329	.407	.396	**.934**	.293	.444
OS4	.548	.370	.371	.434	**.923**	.302	.476
OS5	.391	.345	.400	.389	**.807**	.243	.512
OS6	.481	.379	.254	.459	**.913**	.325	.550
OS7	.462	.256	.310	.434	**.876**	.309	.433
PD1	.544	.277	.259	.405	.331	**.890**	.294
PD2	.516	.297	.195	.364	.319	**.931**	.172
PD3	.528	.387	.292	.398	.367	**.907**	.271
SK1	.335	.214	.207	.442	.536	.266	**.922**
SK2	.359	.209	.300	.486	.563	.277	**.937**
SK3	.356	.124	.174	.485	.439	.209	**.908**

This model explains .41 of the factor of outsourcing success (R square) by mutual benefits, commitment, and predisposition. This model explains .09 of the factor of mutual benefits (R square) by shared knowledge, mutual dependency, and organizational linkage. This model explains .29 of the factor of commitment (R square) by shared knowledge, mutual dependency, and organizational linkage. This model explains .25 of the factor of predisposition (R square) by shared knowledge, mutual dependency, and organizational linkage. As for the model fit, the standardized root mean square residual (SRMR) of this model is .06.

Table 5 shows that there is a positive correlation between the following variables: (1) shared knowledge and mutual benefits (2) mutual dependency and mutual benefits (3) mutual dependency and commitment (4) mutual dependency and predisposition (5) organizational linkage and commitment (6) Organizational linkage and predisposition (7) commitment and outsourcing success.

Table 5. Results of hypothesis testing.

	Path	Estimate	S.E.	t-value	P-value
H1	SK→MB	.088	.126	.701	.483
H2	SK→CM	.158	.095	1.657	.098
H3	SK→PD	-.001	.116	.010	.992
H4	MD→MB	.222	.113	1.960	.050
H4	MD→CM	.328	.093	3.510	<.001
H5	MD→PD	.253	.103	2.450	.014
H6	OL→MB	.112	.126	.885	.376
H7	OL→CM	.273	.131	2.085	.037
H8	OL→PD	.415	.119	3.498	<.001
H10	MB→OS	.216	.077	2.822	.005
H11	CM→OS	.483	.108	4.481	<.001
H12	PD→OS	.016	.111	.148	.882

*: P-value < .05

4   Conclusion

Findings of this study provide evidence that the causal model of an outsourcing partnership was an applicable model in examining partnership that influences the success of outsourcing.

In this study, it could be finding that commitment is positive with outsourcing success, and this result the same as Chaudhuri and Bartlett's study [8]. Shared knowledge is positive with mutual benefits, and this result the same as Donga and Pourmohamadi's study [10]. Mutual dependency is positive with mutual benefits, and this result same as Goo et al. study [11]. Mutual dependency is positive with commitment. Organizational linkage is positive with commitment. Mutual dependency is positive with predisposition. Organizational linkage is positive with predisposition.

Due to the high frequency of hospital information system usage, PECS has become a necessary tool in the healthcare administration. Based on the advantages of outsourcing, this study considers that it is sufficiently reliable and powerful to build PACS and improve the quality of patient service. In order to increase healthcare service quality, more and more hospitals have implemented PACS with outsourcing in Taiwan. The results of this study have shown that increasing partnership quality is a key element in the success of PACS outsourcing.

Acknowledgments. This study is based in part on data from the Survey Research Data Archive (SRDA) provided by the Academia Sinica. The interpretation and conclusions contained herein do not represent those of SRDA or Academia Sinica.

References

1. Tavakol, P., Labruto, F., Bergstrand, L., Blomqvist, L.: Effects of outsourcing magnetic resonance examinations from a public university hospital to a private agent. Acta Radiol. **52**(1), 81–85 (2011)
2. Bellon, E., Feron, M., Deprez, T., Reynders, R., Bosch, B.V.: Trends in PACS architecture. Eur. J. Radiol. **78**(2), 199–204 (2011)
3. Abdul-Halim, H., Ee, E., Ramayah, T., Ahmad, N.H.: Human resource outsourcing success leveraging on partnership and service quality. SAGE Open **4**(3), 1–14 (2014)
4. Teo, T.S.H., Bhattacherjee, A.: Knowledge transfer and utilization in IT outsourcing partnerships: a preliminary model of antecedents and outcomes. Inf. Manag. **51**(2), 177–186 (2014)
5. Lee, J.N., Kim, Y.G.: Exploring a causal model for the understanding of outsourcing partnership. In: Proceedings of the 36th Hawaii International Conference on System Sciences (2002)
6. Lee, J.N., Kim, Y.G.: Understanding outsourcing partnership: a comparison of three theoretical perspectives. IEEE Trans. Eng. Manag. **52**(1), 43–58 (2005)
7. Schwarz, C.: Toward an understanding of the nature and conceptualization of outsourcing success. Inf. Manag. **51**(1), 152–164 (2014)
8. Chaudhuri, S., Bartlett, K.R.: The relationship between training outsourcing and employee commitment to organization. Hum. Resour. Dev. Int. **17**(2), 145–163 (2014)
9. Francois, D., Isis, G.M., Sergio, P.V., Luis, F.L.R.: IT outsourcing in the public sector: a conceptual model. Transforming Gov. People Process Policy **8**(1), 8–27 (2014)
10. Dong, A., Pourmohamadi, M.: Knowledge matching in the technology outsourcing context of online innovation intermediaries. Technol. Anal. Strateg. Manag. **26**(6), 655–668 (2014)
11. Goo, J., Kishore, R., Rao, H.R., Nam, K.: The role of service level agreements in relational management of information technology outsourcing: an empirical study. MIS Q. **33**(1), 115–149 (2009)
12. Park, J.G., Lee, J.: Knowledge sharing in information systems development projects: explicating the role of dependence and trust. Int. J. Project Manag. **32**(1), 153–165 (2014)
13. Rajagopal, P., Zailani, S., Sulaiman, M.: Benchmarking on supply chain partnering effectiveness in two semiconductor companies: a case study approach. Benchmarking Int. J. **16**(5), 671–701 (2009)
14. Hwang, H.G.: A comparative study of outsourcing partnership between medical and banking industries. https://srda.sinica.edu.tw/search/gensciitem/1196. Accessed 9 Aug 2014
15. Nunnally, J.C.: Psychometric Theory. McGraw Hill, New York (1978)
16. Ringle, C.M., Wende, S., Becker, J.M.: SmartPLS 3. Boenningstedt: SmartPLS GmbH (2015). http://www.smartpls.com

A Recent Study on Hardware Accelerated Monte Carlo Modeling of Light Propagation in Biological Tissues

Jakub Mesicek[1,2], Ondrej Krejcar[1(✉)], Ali Selamat[1,3], and Kamil Kuca[1,2]

[1] Faculty of Informatics and Management, Center for Basic and Applied Research, University of Hradec Kralove, Hradec Králové, Czech Republic
{jakub.mesicek,ondrej.krejcar,ali.selamat,kamil.kuca}@uhk.cz
[2] Biomedical Research Center, University Hospital Hradec Kralove, Hradec Králové, Czech Republic
[3] Faculty of Computing, Universiti Teknologi Malaysia, Johor Bahru, Malaysia

Abstract. The Monte Carlo (MC) method is the gold standard in photon migration through 3D media with spatially varying optical properties. MC offers excellent accuracy, easy-to-program and straightforward parallelization. In this study we summarize the recent advances in accelerating simulations of light propagation in biological tissues. The systematic literature review method is involved selecting the relevant studies for the research. With this approach research questions regarding the acceleration techniques are formulated and additional selection criteria are applied. The selected studies are analyzed and the research questions are answered. We discovered that there are several possibilities for accelerating the MC code and the CUDA platform is used in more than 60 % of all studies. We also discovered that the trend in GPU acceleration with CUDA has continued in last two years.

Keywords: Monte Carlo · Turbid media · Photon migration · Parallelization · CUDA

1 Introduction

The Monte Carlo (MC) method provides the most accurate solution for modeling light propagation in tissues. The MC method describes the local rules of photon propagation as a probability distributions of step size between light-tissue interaction occurs or the angle of the scattering. However the MC method relies on calculating a large number of photon packets so computational costs are enormous [1]. There are also several alternatives how to simulate photon transport in tissues e.g. diffusion approximation (DA). While the DA of the radiative transport equation (RTE) is a fast and convinient method it fails when the bounderies appears or if the absorption is not negligible in comparison with scattering [2]. In [3] the accuracy of both the most used methods, MC and DA, was compared showing that DA diverge from MC in regions where the scattering coefficient is less then 10 times higher than the absorption coefficient.

© Springer International Publishing Switzerland 2016
H. Fujita et al. (Eds.): IEA/AIE 2016, LNAI 9799, pp. 493–502, 2016.
DOI: 10.1007/978-3-319-42007-3_43

Monte Carlo is a statistical method that relies on generating random numbers. The random numbers are used to describe the local rules of the photon behavior in scattering media [1]. In the visible and near infrared region of the light spec-trum scattering dominates over the absorption in biological tissues. A photon with the wavelength inside this "window" experience multiple scattering events before the absorption occurs. Due to the random nature of the direction change a huge number of photons has to be launched to obtain accurate results in the MC simulations [1]. Fortunately every photon propagates independently so the parallelization is straightforward.

One the most successfull MC model of steady-state light transport in multi-layered tissues (MCML) was coded 20 years ago [4]. The model assumes semi-nfinite layers where macroscopic optical properties are uniform within a layer. A huge step towards widespread of the MC method for photon propagation was done in 2006 when the first version of Compute Unified Device Architecture (CUDA) was released [5]. CUDA has allowed users parallel computing on Graphic Processing Unit (GPU). For highly parallalable tasks like the MC this presented an incredbible increase in computational efficiency. In [6] an increase in the speed of the MC simulation performed on low-cost GPU by a factor 1000 over a single standard processor was measured. In 2010 Alerstam et al. presented updated version of previous MCML code called GPU-MCML reducing execution time from 4 h when calculating on a central processing unit (CPU) to 23.2 s on a GPU [7].

These interesting results has lead us to make a research on available acceleration techniques of the MC method for light propagation simulations in biological tissues. In this study we employed the systematic literature review (SLR) method [8] to process all the available data about parallel acceleration of the MC method and select only the relevant studies. In Sect. 2 we describe the methodology of the research. The research questions (RQs) and further selection criteria are formulated here. Results of the research are described in Sect. 3 including the answers for the research questions. Section 4 contains a benchmark to compare the available models and emphasize the benefits of the parallel techniques. In the last section we summarize our findings and make the conclusion.

2   Research Method

The research was conducted using the systematic literature review (SLR) method [8]. The method consist of six phases. The first phase is focused on formulating research questions. In the second phase, search strategies such as used databases and searched terms are proposed. The data are collated in the third phase and refined in the fourth phase. In the fifth phase quality assesment criteria are applied. At the end final studies are selected for analysis and subsequent actions.

2.1 Research Questions

The Monte Carlo method is the most accurate method for modeling light propagation in tissues. However its computational costs are enormous. We were interested in the optimization of the code on many-cores architecture. Based on this interest, three RQs were formulated based of the area of our interest:

RQ1: What are the existing techniques for parallelization of the MC method ?
RQ2: What is the most used technique for accelerating the MC code ?
RQ3: What are the most cited studies in last two years ?

Since we were interested only in the visible and the near infrared regions, additional limitations regarding the wavelength of the light were added. Also papers published before 2004 were excluded from the study.

2.2 Search Strategy

When the research questions are formulated, search strings, literature resources and search process has to be designed

Search strings: Based on the formulated questions keywords and their synonyms were selected. Using boolean operators OR and AND the resulting search strings were:"Monte Carlo" AND ("Light" OR "Photon") AND ("Biological" OR "Turbid" OR "Human") AND ("Simulation" OR "Modeling" OR "Model") AND ("GPU" OR "Graphics processing unit" OR "Acceleration" OR "Cluster" OR "Multithread")
Literature resources: Scopus was employed as the main research database using our research strings. Then additional studies were looked up in Web of Science, ScienceDirect, IEEE Xplore and OSA publishing databases. The strings were founded in abstracts, keywords or titles of the journal papers, conference proceedings, symposiums and book chapters.
Search process: A comprehensive search of all sources is done in the first step. In the second step relevant studies are selected from the sources and a list of references is made. In the list another relevant studies are found.

2.3 Study Selection

The quality assessment questions are formulated in this phase. According to the answer points are given: "yes" 1 point, "partly" 0.5 point, "no" 0 point. For the purpose only studies with quality rate higher than 1.5 were selected.

QAQ1: Are the goals of the research clearly formulated ?
QAQ2: Is the proposed method clearly described ?
QAQ3: Is it clearly spoken that the study uses accelerated MC ?
QAQ4: Does the research bring something new to the community ?

3   Results

This section presents the findings and the answers to the research questions formulated in the previous section. Also short description of the applicated acceleration technologies in particular studies is mentioned.

3.1   Overview of Selected Studies

At the begining there were 263 journal papers, conference proceeding and thesis. From Scopus 42 studies proceed to the next step of the research, another 5 studies came from Web of Science and the last one from IEEE Xplore. From the references of the publications another 9 studies were found and included in the research. Refining the data and applying another filtering, final 53 studies were selected for the research. The distribution of the studies during the years is in Fig. 1.

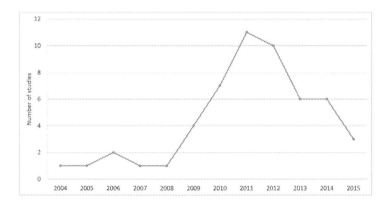

Fig. 1. Number of papers by year of publication.

3.2   Parallel Acceleration Techniques (RQ1)

The answers for RQ1 and RQ2 can be seen in Fig. 2. The most used hardware accelerated technology is the CUDA platform with 36 publications. Significantly less popular are the special instructions for modern multicore CPUs (6 publications). and Message Passing Interface (MPI) (6 publications). Three publications dealt with OpenMP acceleration of the MC code. In two older publications was used an unspecified distributed system. Also FPGA implementation was found in two cases. Intel Xeon Phi coprocessor (MIC) was used in one case. Similarly the MapReduce framework for cloud computing enviroments was reported in one case. OpenCL as an alternative technology to CUDA was also used once.

Among ten most cited works (Google Scholar, cited on 14/10/2015) five of them deal with CUDA-enabled GPU acceleration, in three works multicore

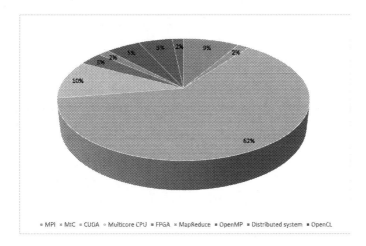

Fig. 2. The distribution of the used technologies for hardware accelerated MC. (Color figure online)

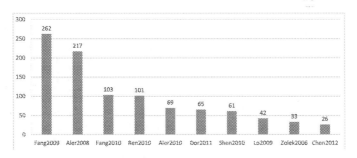

Fig. 3. Ten most cited publications involved in the research; References: Fang2009 [9], Aler2008 [6], Fang2010 [10], Ren2010 [11], Aler2010 [7], Dor2011 [12], Shen2010 [13], Lo2009 [14], Zolek2006 [15], Chen2012 [16].

CPU/OpenMP is utilized, one work involved FPGA and uses an unspecified parallel machine. The chart with ten most cited publications is in Fig. 3.

3.3   The CUDA Platform (RQ2)

CUDA has been introduced by nVidia for their GPUs in 2006 [5]. Since then it has found its applications in many scientific fields. First one to use it in the MC light transport problem was Alerstam et al. [6]. The authors rewrote the original MCML code with CUDA C programming language, refered as GPU-MCML or CUDAMCML, achieving the speedup 1000x over the classical CPU code. In 2010 several optimizations were proposed to overcome slow access to the slow GPU global memory [7]. Similar approach utilizing the original MCML and rewriting it for GPUs was done by Lo et al. [14] and Martinsen et al. [17]. Carbone et al.

extended CUDAMCML giving the possibility of including inhomogenities inside a layer [18].

Modeling time-resolved photon migration in arbitrary 3D media was accelerated with CUDA in the Monte Carlo eXtreme (MCX) project [9]. MCX also incorporated logistic-lattice (LL) based pseudorandom number generator that involves only floating-point operations making it attractive choice for GPU. With the LL algorithm 33 % acceleration was achieved in comparison with the well-known Mersenne-Twister algorithm [9]. MCX was then used to estimate the adult brain optical properties [19]. D'Alesandro et al. also developed a voxel-based MC code intended for the reconstruction of subsurface skin lesion volumes [20]. The same code was incorporated in the study of blood oxygen saturation of skin lesions [21] and the study of reconstruction the melanin and blood layer volume components [22].

In 2010, Ren et al. presented CUDA-accelerated MC with tissue surface created by triangle meshes [11]. The triangle meshes allows to model space structure and boundaries more precisely than multi-layered or voxel models [11].

In the same year, Doronin et al. started work on object-oriented MC with GPU acceleration [23]. The work continued with involving the wave nature of the light in the MC simulations [24] and releasing web-based tool [12]. The online version was then upgraded utilizing peer-to-peer (P2P) network of computers with different CUDA-enabled GPUs [25,26].

There are another studies utilizing CUDA-accelerated MC for fluorescence simulations [27], accelerating perturbation Monte Carlo [28], reconstruction in diffuse optical tomography [29], estimation of the skin optical parameters [30], or simulation of ultrasound-modulated light propagation [31]. In training of artificial neural networks for diffuse reflectance spectroscopy hardware acceleration is of significant importance [32]. Since the training sets have to contain a large amount of the optical parameters combinations (up to 10.000 combinations) producing such a database is and extraordinarily time consuming task.

3.4   Current Trend in Acceleration (RQ3)

Currently the most popular accelerated MC code was determined by number of citations of the paper where the code is described. Ten most cited papers in last two years are displayed in Fig. 4. In comparison with all-time number of citations in Fig. 3 there are two more papers dealing with CUDA and the same number of OpenMP/MulticoreCPU works. The chart indicates the trend in developing and utilizing CUDA accelerated simulations in the future.

4   Benchmark

In this section the results in terms of time consumption of a particular simulation are presented. Since the published works were presented few years ago and a lot of improvements was done we decided to test all the available and easy-to-obtain projects namely MCML, CUDAMCML, MCX, MMC and TIM-OS simulated in

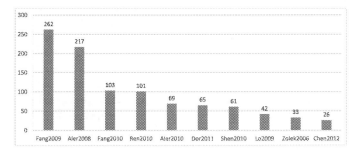

Fig. 4. Ten most cited papers in last two years; References: Fang2009 [9], Aler2008 [6], Fang2010 [10], Dor2011 [12], Aler2010 [7], Ren2010 [11], Chen2012 [16], Shen2010 [13], Dor2012 [25], Fang2012 [33].

the MMC environment. The MCML family can only calculate light propagation under assumption of perpendicular beam incident on semi-infinite layers so a benchmark have to meet these assumptions. Due to limited possibilities we performed tests in a media with a single layer with the following optical parameters: absorption coefficient $\mu_a = 0.001$, scattering coefficient $\mu_s = 1$, anisotropy parameter $g = 0.01$, refractive index $n = 1.4$. The layer thickness was 6 cm in the MCML and CUDAMCML codes, in the MCX and MMC codes $60 \times 60 \times 60$ cube was used. As was demonstrated in [10] the number of elements in a mesh does not influence the computing speed so a mesh with 135000 tetrahedrons was employed. The codes were benchmarked on a standard desktop containing Intel Core i7-4790S (8 threads) processor and GPU nVidia GeForce GTX 750 Ti (640 threads). The results of the time consumption of a single simulation are displayed in Fig. 5.

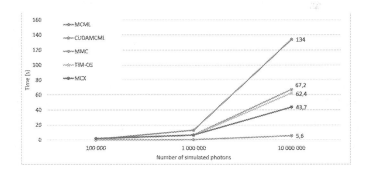

Fig. 5. The benchmark of the available MC projects. (Color figure online)

As expected for non-parallel MCML code it took the longest calculation time. On the other hand CUDA version of the exactly same code was the fastest one. However interesting results were obtained for MMC and MCX projects.

Comparing MCX and CUDAMCML it can be seen that despite running on the same GPU the measured times are very different. This can be explained by the different approach in tracing every singly photon. Since MCX has to trace every photon in every voxel it enters, find out voxel's parameters and determine the distance to the neighbors, CUDAMCML only calculates the remaining distance to the border of the semi-infinite layer. So when the assumptions of the model are met CUDAMCML is highly recommended. In the case of models capable of handling arbitrary shaped media (MCX,MMC,TIM-OS) the results do not differ so much despite the fact that MCX runs on GPUs and MMC and TIM-OS on CPU. This indicates the huge benefit from tetrahedron mesh reducing the memory consumption and the number of tracing enumeration. Unfortunately the MMC code for GPUs was not available in the time of publishing this work.

5   Conclusion

The Monte Carlo method for solving photon migration in biological tissues is a promising biomedical tool due to its accuracy and simplicity. Although its computational demands are enormous, the recent advances in parallel processing with graphics processing units make MC applicable on standard computers. In this paper we performed a research on acceleration techniques for the MC light propagation in biological tissues simulations using the systematic literature review method. We selected the most relevant studies and focused on the used technologies. As we expected the CUDA platform is the most popular hardware architecture contained in 62 % of studies. The main reason might be the availability and low prices of graphics processing units. Also nVidia offers variety of utilities making CUDA a powerful tool. Among ten most cited papers in last two years seven of them utilize the CUDA platform. On the other hand, a state-of-art tetrahedral mesh-based Monte Carlo method hasn't been rewrote into the CUDA code and this task still stays to be a challenge. So in the future one can expect spreading mesh-based Monte Carlo on graphics processing units bringing the most accurate modeling technique in complex media towards real-time simulations.

Acknowledgment. This work and the contribution were supported by project Smart Solutions for Ubiquitous Computing Environments FIM, University of Hradec Kralove, Czech Republic (under ID: UHK-FIM-SP-2016-2102). The work was also supported by project 16-13967S.

References

1. Wang, L.V., Hsin-i, W.: Biomedical Optics: Principles and Imaging. Wiley, Hoboken (2007)
2. Welch, A.J., Van Gemert, M.J.C.: Optical-Thermal Response of Laser-Irradiated Tissue. Springer, New York (2011)

3. Ancora, D., Zacharopoulos, A., Ripoll, J., Zacharakis, G.: Light propagation through weakly scattering media: a study of Monte Carlo vs. diffusion theory with application to neuroimaging. 9538(Mc), 95380G (2015)
4. Wang, L., Jacquesa, S.L., Zhengb, L.: MCML - Monte Carlo modeling of light transport in multi-layered tissues. Biomedicine 2607(713), 131–146 (1995)
5. nVidia.: Cuda C programming guide (2015)
6. Alerstam, E., Svensson, T., Andersson-Engels, S.: Parallel computing with graphics processing units for high-speed Monte Carlo simulation of photon migration. J. Biomed. Opt. 13(6), 060504 (2008)
7. Alerstam, E., Lo, W.C.Y., Han, T.D., Rose, J., Andersson-Engels, S., Lilge, L.: Next-generation acceleration and code optimization for light transport in turbid media using GPUs. Biomed. Opt. Express 1(2), 658–675 (2010)
8. Achimugu, P., Selamat, A., Ibrahim, R., Mahrin, M.N.: A systematic literature review of software requirements prioritization research. Inf. Softw. Technol. 56(6), 568–585 (2014)
9. Fang, Q., Boas, D.A.: Monte Carlo simulation of photon migration in 3D turbid media accelerated by graphics processing units. Opt. Express 17(22), 20178–20190 (2009)
10. Fang, Q.: Mesh-based Monte Carlo method using fast ray-tracing in Plücker coordinates. Biomed. Opt. Express 1(1), 165 (2010)
11. Ren, N., Liang, J., Xiaochao, Q., Li, J., Bingjia, L., Tian, J.: GPU-based Monte Carlo simulation for light propagation in complex heterogeneous tissues. Opt. Express 18(7), 6811–6823 (2010)
12. Doronin, A., Meglinski, I.: Online object oriented Monte Carlo computational tool for the needs of biomedical optics. Biomed. Opt. Express 2(9), 2461 (2011)
13. Shen, H., Wang, G.: A tetrahedron-based inhomogeneous Monte Carlo optical simulator. Phys. Med. Biol. 55(4), 947–962 (2010)
14. Lo, W.C.Y.: Hardware acceleration of a Monte Carlo simulation for photodynamic therapy treatment planning by copyright c 2009 by William Chun Yip Lo. Master's thesis. University of Toronto (2009)
15. Zołek, N.S., Liebert, A., Maniewski, R.: Optimization of the Monte Carlo code for modeling of photon migration in tissue. Comput. Methods Programs Biomed. 84(1), 50–57 (2006)
16. Chen, J., Fang, Q., Intes, X.: Mesh-based Monte Carlo method in time-domain widefield fluorescence molecular tomography. J. Biomed. Opt. 17(10), 1060091 (2012)
17. Martinsen, P., Blaschke, J., Künnemeyer, R., Jordan, R.: Accelerating Monte Carlo simulations with an NVIDIA graphics processor. Comput. Phys. Commun. 180(10), 1983–1989 (2009)
18. Carbone, N., Di Rocco, H., Iriarte, D.I., Pomarico, J.A.: Solution of the direct problem in turbid media with inclusions using Monte Carlo simulations implemented in graphics processing units: new criterion for processing transmittance data. J. Biomed. Opt. 15(3), 035002 (2010)
19. Selb, J., Zimmermann, B.B., Martino, M., Ogden, T., Boas, D.A..: Functional brain imaging with a supercontinuum time-domain NIRS system. In: SPIE BiOS, vol. 8578, no. 1, 857807–857807-9 (2013)
20. D'Alessandro, B., Dhawan, A.P.: Voxel-based, parallel simulation of light in skin tissue for the reconstruction of subsurface skin lesion volumes. In: Proceedings of the Annual International Conference of the IEEE Engineering in Medicine and Biology Society, EMBS, no. 2, pp. 8448–8451 (2011)

21. D'Alessandro, B., Dhawan, A.P.: Transillumination imaging for blood oxygen saturation estimation of skin lesions. IEEE Trans. Biomed. Eng. **59**(9), 2660–2667 (2012)
22. D'Alessandro, B., Dhawan, A.P.: 3-D volume reconstruction of skin lesions for melanin and blood volume estimation and lesion severity analysis. IEEE Trans. Med. Imaging **31**(11), 2083–2092 (2012)
23. Doronin, A., Meglinski, I.: GPU-accelerated object-oriented Monte Carlo modeling of photon migration in turbid media. In: Proceedings of SPIE 7999, Saratov Fall Meeting 2010: Optical Technologies in Biophysics and Medicine XII, vol. 7999 (2010)
24. Doronin, A., Meglinski, I.: Monte Carlo simulation of photon migration in turbid random media based on the object-oriented programming paradigm. In: Proceedings of SPIE - The International Society for Optical Engineering (2011)
25. Doronin, A., Meglinski, I.: Peer-to-peer Monte Carlo simulation of photon migration in topical applications of biomedical optics. J. Biomed. Opt. **17**(9), 0905041 (2012)
26. Doronin, A., Meglinski, I.: Using peer-to-peer network for on-line Monte Carlo computation of fluence rate distribution. In: Proceedings of SPIE 8699, Saratov Fall Meeting 2012: Optical Technologies in Biophysics and Medicine XIV and Laser Physics and Photonics XIV, vol. 8699, p. 869909 (2013)
27. Hennig, G., Stepp, H., Sroka, R., Beyer, W.: Comparison of an accelerated weighted fluorescence Monte Carlo simulation method with reference methods in multi-layered turbid media. Appl. Opt. **52**(5), 1066–1075 (2013)
28. Cai, F.: Using graphics processing units to accelerate perturbation Monte Carlo simulation in a turbid medium. J. Biomed. Opt. **17**(4), 040502 (2012)
29. Yi, X., Chen, W., Linhui, W., Zhang, W., Li, J., Wang, X., Zhang, L., Zhao, H., Gao, F.: Towards diffuse optical tomography of arbitrarily heterogeneous turbid medium using GPU-accelerated Monte-Carlo forward calculation. In: Proceedings of SPIE 8574, Multimodal Biomedical Imaging VIII, vol. 8574, p. 857400 (2013)
30. Bjorgan, A., Milanic, M., Randeberg, L.L.: Estimation of skin optical parameters for real-time hyperspectral imaging applications. J. Biomed. Opt. **19**(6), 066003 (2014)
31. Leung, T.S., Powell, S.: Fast Monte Carlo simulations of ultrasound-modulated light using a graphics processing unit. J. Biomed. Opt. **15**(5), 055007 (2014)
32. Chen, Y.-W., Tseng, S.-H.: Efficient construction of robust artificial neural networks for accurate determination of superficial sample optical properties. Biomed. Opt. Express **6**(3), 747 (2015)
33. Qianqian, F., Kaeli, D.R.: Accelerating mesh-based Monte Carlo method on modern CPU architectures. Biomed. Opt. Express **3**(12), 3223–3230 (2012)

Clustering Analysis of Vital Signs Measured During Kidney Dialysis

Kazuki Yamamoto, Yutaka Watanobe$^{(\boxtimes)}$, and Wenxi Chen

Department of Information Systems, University of Aizu,
Aizu-wakamatsu 965-8580, Japan
{m5181114,yutaka,wenxi}@u-aizu.ac.jp

Abstract. Analysis of vital data of kidney dialysis patients is presented. The analysis is based on some vital signs of pulse rate (PR), respiration rate (RR) and body movement (BM) which were obtained by a sleep monitoring system. In a series of experiments, eight patients of different genders and ages were involved. For the analysis, a hierarchical clustering method was applied with multi-dimensional dynamic time warping distance to analyze the similarity between the vital signs. The hierarchical clustering uses Ward's method to calculate the distance between two clusters. The analysis results show that daily vital sign indicates a feature related to one of the clusters and physiological rhythms based on a series of the features vary depending on the season. Based on the hypothesis, some irregular vital signs which deviate from the physiological rhythms can be detected to predict abnormal health condition and discomfort of the patients.

1 Introduction

Data science has great capability and there are a number of successful stories about big data of different areas such as web, commerce, communication and infrastructures to make industries or our daily life more efficient and productive. Biomedical data is also an useful information resource for healthcare in the medical field. In addition to the improvement of the business conditions, big data from vital data of patients can be a good basis to predict abnormal conditions and epidemics as well as to make decision for improving quality of healthcare. It is also important to consider understanding as much about a specific patient as possible to perceive warning signs at an early stage. This means that we should also consider approaches to obtain big data of vital signs from a patient which can be used for his/her own healthcare as well as for others.

In this research, the data which were obtained from patients who need to lie in beds around four hours during kidney dialysis are analyzed. The vital signs of the patients were measured by a sleep monitoring system named Umemory [1]. It can measure pressure signals through noninvasive and unconstrained ways without stimulation while patients lie in bed. From the pressure signal data, the monitoring system produces pulse rate (PR), respiration rate (RR) and body movements (BM). The vital signs from the monitoring system are the time series

© Springer International Publishing Switzerland 2016
H. Fujita et al. (Eds.): IEA/AIE 2016, LNAI 9799, pp. 503–513, 2016.
DOI: 10.1007/978-3-319-42007-3_44

data of different lengths. Many clustering algorithms need a distance function, but the time series data is difficult to be accommodated by general distance algorithms such as Euclidean distance. One of typical distances for the time series data is dynamic time warping distance (DTW) [6]. DTW is a method that calculates the optimal match between two given sequences with certain restrictions which can be used with different lengths and different cycle time series. Since three dimensional time series data related to the vital signs should be considered, an extended DTW [5] was employed for the analysis. Then, features of the vital signs were divided into several groups by a hierarchical clustering algorithm where the Ward's method was employed for the distance calculation.

In a series of experiments, eight patients were monitored for at most six months. The experimental results show that a daily vital signs indicates a feature related to one of the clusters and physiological rhythms obtained by the features of consecutive days vary depending on the season. The phenomenon also shows that a normal health condition often has regular patterns and some irregular vital signs which deviate from the physiological rhythms can be observed. So, the results can be employed to predict unusual health condition and discomfort of the corresponding patients as well as to improve some healthcare activities in medical sites.

The rest of the paper is organized as follows. In Sect. 2 some related works are surveyed. In Sect. 3 algorithms to analyze the vital signs based on the clustering are presented. Finally, Sect. 4 shows results and discuses possible future applications.

2   Related Works

Long-term pulse rate tracking during sleep has been considered in [1]. In this research, sleep condition of a patient had been tracked for about two years by using Umemory. The results of the tracking over seven successive seasons were divided into four seasons. The PR was evaluated by linear measures including time domain and frequency domain indexes. The evaluation was based on noise limit, detection rate, sample entropy and Poincare plots. In addition, noise titration algorithm was applied before the evaluation of PR. Analysis of PR dynamics shows that it began downward in winter, developed in spring and worsened seriously in summer. It was also observed that some illegal conditions show different trajectories compared with that of healthy subjects. The paper concludes that the human body presents physiological rhythms when he/she adapts to environments, and these rhythms are essential to life. The analysis also shows that once the rhythms are broken by a hostile intervention, it might lead to disease.

Polysomnography (PSG) [3] is a standard sleep test to diagnose sleep disorders. It is effective in the diagnosis of disorders as well as in the assessment of treatment effects. It can monitor brain waves, heart rate, breathing, eye movements, leg movements, throat sound, sleep stage and so on during sleep. It is based on electroencephalography, electromyography, electromyography, and electrocardiography. Status of sleep is displayed in a hypnogram where each stage of sleep is characterized by a level on the vertical axis. The main advantage of PSG is that it can directly monitor variety of vital signs at a time.

As stated above, there are several approaches to collect vital data during sleep based on monitoring systems and to analyze them with statistical procedures and the corresponding visualization techniques. On the other hand, there are some attempts to apply machine learning algorithms in medical fields. For example, clustering algorithms oriented to classification system for time series data as well as to feature detection from signal patterns have been considered [2,11]. There have also been some attempts to apply machine learning to medical documents and health records [8,10]. Although, these results can be basis to support further research in medical fields, it is still difficult to find mature analysis based on machine learning algorithms through experiments with real patients and doctors with the consideration to the situation of the necessities in the practices.

3   Algorithms to Analyze Vital Signs

The analysis includes the following steps:

1. Denoising the part of the vital data
2. Defining the distance between two vital data
3. Clustering the vital data based on the distance

3.1   Denoising

The PR and RR data which are taken from Umemory have various noises because the system finds characteristic points of pulse and respiration. These rates are calculated from distances between the characteristic points. The precision of PR and RR can be aggravated when some characteristic points are missed or extra movements are found as characteristic points. Because of this, it is difficult to use raw vital data. For example, Fig. 1 shows a raw PR signal wave with many noises which must be denoised. At the first stage of the denoising step, the two digital filtering algorithms were applied to the data. At first, the median filter [4] which can remove spike-like noises regardless of noise level was applied. Figure 2 shows the corresponding signal wave reproduced by the median filter. After the median filter, Savitzky-Golay filter [9] was applied to the data to smooth out the signals wave. The filter is not suitable for data with huge noises, but these noises should be removed by the median filter. Figure 3 demonstrates the available signal wave secondarily treated by the Savitzky-Golay filter.

The PR and RR data were processed by these two filters. On the other hand, the filtering algorithms were not applied to BM data because body movements can be reproduced easily and correctly from pressures and their transition obtained from Umemory.

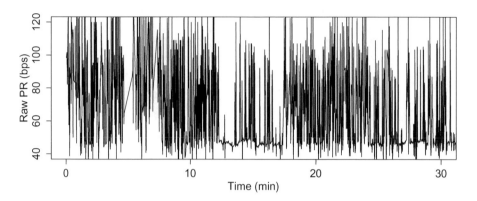

Fig. 1. A raw PR signal wave of the first 30 min from Umemory.

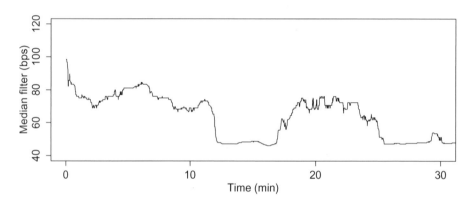

Fig. 2. A PR signal wave after the median filter is applied.

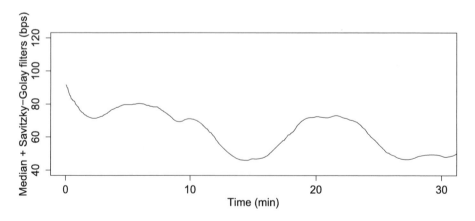

Fig. 3. A PR signal wave after both the median and Savitzky-Golay filters are applied.

3.2   Definition of Distance

The vital signs are time series data whose lengths are different from each other. It is difficult to define their distance by general methods such as Euclidean distance. One of the typical distance algorithms which can be applied to time series data is dynamic time warping (DTW) [6]. It uses the sum of the minimum distances from each point. The follow functions show the process for calculating the distance between two time series $(a_1, ..., a_n)$ and $(b_1, ..., b_n)$.

$$DTW = f(n, m)$$
$$f(0, 0) = 0$$
$$f(i, 0) = f(0, j) = \infty \tag{1}$$
$$f(i, j) = D(i, j) + \min \begin{cases} f(i - 1, j) \\ f(i, j - 1) \\ f(i - 1, j - 1) \end{cases} \tag{2}$$
$$D(i, j) = |a_i - b_j| \tag{3}$$

The distance between two time series data can be calculated by using the above functions, but it is only for one dimensional data. To analyze the data from Umemory, the distance algorithm which can deal with three or more dimensional data should be employed. So, we used multi-dimensional dynamic time warping [5] which is an algorithm for DTW on multi-dimensional time series. It considers all dimensions to find the optimum time series distance. In ordinary DTW, the distance is calculated by taking the absolute or the squared distance between feature values of each combination of points. In MD-DTW, a distance of two K-dimensional points is calculated. To combine different dimensions, it is necessary to normalize each dimension to a zero mean and unit variance. After normalizing, M by N distance matrix D is created according to

$$D(i, j) = \sum_{k=1}^{K} |A(i, k) - B(j, k)| \tag{4}$$

where A and B are two time series of length M and N in the K-dimensional space. MD-DTW uses this distance matrix to find the best synchronization within the ordinary DTW algorithm.

3.3   Clustering

We employ a hierarchical clustering method for the analysis. Advantages compared to other clustering algorithms (e.g. k-means algorithm) include:

- It is compatible with the MD-DTW algorithm.
- It is easy to hierarchically adjust the number of clusters during the analysis.
- It requires no randomized elements.

The result of the hierarchical clustering varies according to the distance method between clusters. In this research, Ward's method [12] was applied to the cluster analysis. It calculates the minimum value of the difference between sum of squares of each points and sum of squares of points in the source clusters based on the following functions.

$$d(P, Q) = E(P \cup Q) - E(P) - E(Q) \tag{5}$$

$$E(P) = \sum_{x \in P} (d(x, p))^2 \tag{6}$$

$$p = \sum_{x \in P} \frac{x}{|P|} \tag{7}$$

To implement the above mentioned clustering algorithm, Ward.D2 method [7] provided in R language was used. The method is based on Cophenetic correlations. The above functions consider x-coordinates, but the function of Ward.D2 does not need it, so the base function is defined by

$$d(i \cup j, k) = \left(\frac{n_i + n_k}{n_i + n_j + n_k} d(i, k)^2 + \frac{n_j + n_k}{n_i + n_j + n_k} d(j, k)^2 - \frac{n_k}{n_i + n_j + n_k} d(i, j)^2 \right)^{\frac{1}{2}} \tag{8}$$

where n_α is the number of elements in cluster α, $d(\alpha, \beta)$ is distance between α and β, i and j are cluster IDs for the source cluster, and k is an ID for the target cluster.

4 Analysis Results

4.1 Raw Data

The experiment was conducted under mutual cooperation between the university and a local hospital and eight patients who need kidney dialysis were involved as subjects. Tables 1 and 2 show an overview of the vital data of the patients in the hospital. These tables show averages and standard deviations of PR, RR and BM ratios obtained from Umemory. Table 1 is monthly data and Table 2 is based on patients. In the tables, the BM is calculated from a percentage of the characteristic points for each 30 s.

To conduct variety of analysis, we have collected the data from several patients in the hospital for over half a year. In this paper, because of the space limitation, we show only some analysis results considering selected patients. It is worth to mention that such experiments based on data from a specific patient is also important rather than on statistical data from many different patients for the healthcare analysis of the individual patient.

Table 1. The monthly data of the vital signs.

Term (2014)	Number of patients	Number of dates	PR (*bpm*)	RR (*bpm*)	BM (%)
March	2	22	74.19 ± 7.04	18.42 ± 3.75	11.68 ± 19.55
April	3	27	75.04 ± 8.62	18.63 ± 3.07	9.44 ± 17.05
May	2	23	77.89 ± 9.68	18.22 ± 2.22	10.66 ± 16.49
June	5	56	76.24 ± 8.51	20.02 ± 2.83	13.04 ± 19.23
July	8	96	75.40 ± 7.94	20.81 ± 2.78	19.35 ± 24.30
August	7	81	74.94 ± 8.29	20.99 ± 2.88	16.83 ± 22.43
September	5	58	76.67 ± 8.19	19.58 ± 3.04	14.84 ± 24.23
October	2	28	79.04 ± 8.86	19.43 ± 2.42	11.69 ± 18.88
All dates	8	391	75.91 ± 8.41	20.01 ± 3.03	15.07 ± 21.93

Table 2. The data of vital signs for each patient.

ID	Term	Number of dates	PR (*bpm*)	RR (*bpm*)	BM (%)
1	03/07 ∼ 09/29	82	71.62 ± 6.84	18.42 ± 2.92	18.77 ± 21.24
2	03/07 ∼ 09/29	70	75.39 ± 4.70	19.72 ± 3.64	1.81 ± 6.53
3	04/14 ∼ 10/31	85	84.21 ± 5.97	20.12 ± 1.61	6.09 ± 11.11
4	06/11 ∼ 08/25	33	71.73 ± 9.15	20.67 ± 2.21	17.28 ± 19.88
5	06/11 ∼ 10/31	47	74.90 ± 7.19	19.26 ± 2.96	21.88 ± 25.44
6	07/02 ∼ 09/29	34	78.97 ± 5.98	22.53 ± 3.09	35.68 ± 31.65
7	07/09 ∼ 08/22	20	67.61 ± 5.43	21.46 ± 2.65	15.69 ± 20.39
8	07/11 ∼ 09/10	20	71.07 ± 9.27	22.46 ± 2.67	33.88 ± 24.61

4.2 Clustering

In this paper, we show the analysis results of patient 1 among the candidates because he/she was involved for the longest term. The vital signs which include PR, RR and BM from Umemory were clustered for the analysis. The clustering results are shown in Figs. 4, 5 and 6. Figure 4 demonstrates a dendrogram of the clusters and Figs. 5 and 6 show transitions of the features within the clusters. The cluster dendrogram is one of general tree diagrams which shows a result of hierarchical clustering. The dendrogram is to show affiliations of each vital sign to one of clusters where vital signs with the similar features are aggregated. In the dendrogram, the clusters are represented by internal nodes (intersections of edges) of the tree and distance between clusters are represented by the length of the tree edges. One of advantages of the hierarchical clustering is that the number of clusters can be adjusted by height of the nodes to analyze the clustering from different perspectives and granularity even after the clustering algorithm was applied.

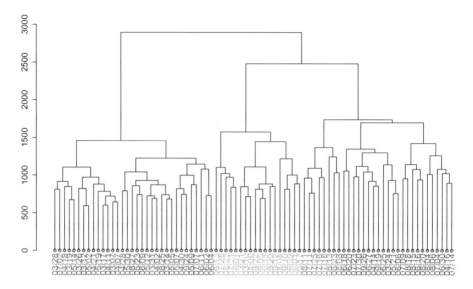

Fig. 4. Cluster dendrogram which is based on the vital signs of PR, RR and BM from March to September for patient 1.

Figures 5 and 6 show temporal transitions of the vital sign features within 3 and 4 clusters respectively. The transition shows cluster IDs for each monitoring date when the patient dialyzed. It is to show the change of clusters in consecutive monitoring dates. One of observable phenomena is that vital signs monitored in spring and autumn tend to affiliate to cluster 1 and cluster 2 respectively. On the other hand, vital signs monitored in the summer tend to affiliate to both cluster 3 and cluster 4. Besides, the clusters 3 and 4 are similar because the height of subtrees in the dendrogram between them are close (see Fig. 4). The phenomena implies that biological rhythm depends on seasons and it can be a basis to find irregular points.

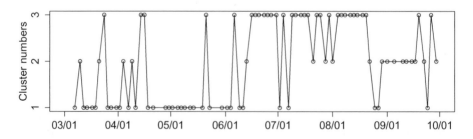

Fig. 5. Cluster transition with 3 clusters which is based on the vital signs of PR, RR and BM from March to September for patient 1.

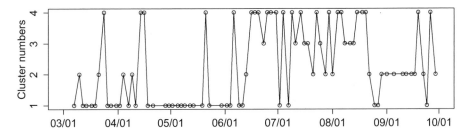

Fig. 6. Cluster transition with 4 clusters which is based on the vital signs of PR, RR and BM from March to September for patient 1.

We also analyzed the data of patient 5 as one of typical patients. However, he/she has a special condition that he/she was not dialyzed in August because of hospitalization in another facility. The analysis of such special cases can be a good basis to investigate patient's status before and after unusual events. The clustering results of the patient from June 11 to October 31 are shown in Figs. 7, 8 and 9. The results also demonstrate that there is a biological rhythm depending on seasons, even the patient was influenced by the critical events such as the hospitalization.

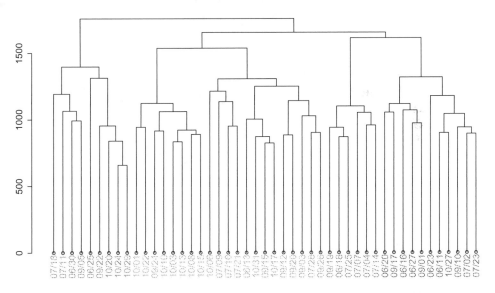

Fig. 7. Cluster dendrogram which is based on the vital signs of PR, RR and BM from June to October for patient 5.

The biological rhythm should be influenced by various factors of environment such as temperature, humidity and weather depending on seasons. In spite of the hospital efforts to maintain the better environment for the patients, the analysis

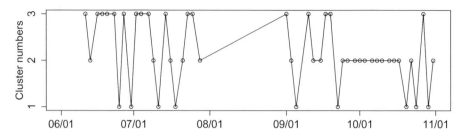

Fig. 8. Cluster transition with 3 clusters which is based on the vital signs of PR, RR and BM from June to October for patient 5.

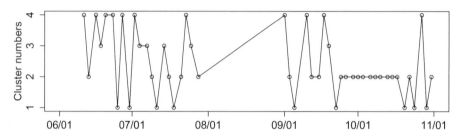

Fig. 9. Cluster transition with 4 clusters which is based on the vital signs of PR, RR and BM from June to October for patient 5.

results show the biological rhythm with a specific pattern. Although we can not speculate concrete features of each cluster medically, it can be useful to observe changes of health condition focusing on a specific patient. In addition, it is worth to mention that the clustering analysis presented can be applied to not only the experiments with the patients of kidney dialysis but also more general activities. For example, such activities include sleep monitoring for sleep disables as well as the development of pillows and bedding for improving the quality of life.

5   Conclusion

An approach to analyze health condition change of kidney dialysis patients by a clustering algorithm and it experimental results have been presented. The analysis was based on some vital signs including pulse rate, respiration rate and body movements which were obtained by a sleep monitoring system. After some filtering, a hierarchical clustering algorithm based on the MD-DTW distance was applied to the signal data. Then, temporal transitions of vital sign features within different clusters have been analyzed. The analysis results show that a daily vital signs indicates a feature related to one of the clusters and physiological rhythms based on a series of the features vary depending on the season. Based on the hypothesis, some irregular vital signs which deviate from the physiological rhythms can be detected to predict abnormal health condition and discomfort of the patients. The result also show that the presented approach can be useful for the future research and development in other fields.

References

1. Chen, Y., Chen, W.: Long-term tracking of a patient's health condition based on pulse rate dynamics during sleep. Ann. Biomed. Eng. **39**(12), 2922–2934 (2011)
2. Costa Santos, C., Bernardes, J., Vitányi, P.M., Antunes, L.H.M.: Clustering fetal heart rate tracings by compression. In: 19th IEEE International Symposium on Computer-Based Medical Systems, pp. 685–690. IEEE (2006)
3. Geyer, J.D., Talathi, S., Carney, P.R.: Introduction to Sleep and Polysomnography. Reading EEGs: A Practical Approach. Lippincott Williams & Wilkins, Philadelphia (2009)
4. Härdle, W., Steiger, W.: Optimal median smoothing (1994)
5. Holt, G.A.T., Reinders, M.J., Hendriks, E.A.: Multi-dimensional dynamic time warping for gesture recognition. In: Annual Conference of the Advanced School for Computing and Imaging (2007)
6. Keogh, E.J., Pazzani, M.J.: Scaling up dynamic time warping for datamining applications. In: ACM SIGKDD International Conference on Knowledge Discovery and Data Mining, pp. 285–289. ACM (2000)
7. Murtagh, F., Legendre, P.: Ward's hierarchical clustering method: Clustering criterion and agglomerative algorithm. ArXiv preprint (2011). arXiv:1111.6285
8. Saad, F.H., de la Iglesia, B., Bell, D.G.: A comparison of two document clustering approaches for clustering medical documents. In: 2006 International Conference on Data Mining (DMIN), pp. 425–431. Citeseer (2006)
9. Savitzky, A., Golay, M.J.: Smoothing and differentiation of data by simplified least squares procedures. Anal. Chem. **36**(8), 1627–1639 (1964)
10. Sideris, C., Shahbazi, B., Pourhomayoun, M., Alshurafa, N., Sarrafzadeh, M.: Using electronic health records to predict severity of condition for congestive heart failure patients. In: 2014 ACM International Joint Conference on Pervasive and Ubiquitous Computing: Adjunct Publication, pp. 1187–1192. ACM (2014)
11. Sugimura, H., Matsumoto, K.: Classification system for time series data based on feature pattern extraction. In: 2011 IEEE International Conference on Systems, Man and Cybernetics, pp. 1340–1345. IEEE (2011)
12. Ward, J.H.: Hierarchical grouping to optimize an objective function. J. Am. Stat. Assoc. **58**(301), 236–244 (1963)

A Smart Arduino Alarm Clock Using Hypnagogia Detection During Night

Adam Drabek[1], Ondrej Krejcar[1(✉)], Ali Selamat[1,2], and Kamil Kuca[1]

[1] Faculty of Informatics and Management,
Center for Basic and Applied Research, University of Hradec Kralove,
Rokitanskeho 62, 500 03 Hradec Kralove, Czech Republic
{adam.drabek,kamil.kuca}@uhk.cz,
Ondrej@Krejcar.org
[2] Faculty of Computing, Universiti Teknologi Malaysia,
81310 Johor Baharu, Johor, Malaysia
aselamat@utm.my

Abstract. This project describes hardware design and implementation of low-cost smart alarm clock based on Arduino platform, which uses passive infrared sensor (PIR) to detect sleep states of users. Sleep is not just a passive process. People can achieve different states during the night, which are known as Hypnagogia (state from wakefulness to sleep), NREM (non-rapid eye movement), REM (rapid eye movement), Hypnexagogium (awakening state) and dreaming. The main goal for this developed smart alarm clock is to detect these states and adjust alarm time to the best possible moment, when people are in awaking state or in light sleep. Awaking in these states is quite better and people feel much more refreshed. Hardware of this developed alarm clock is composed from LCD LED display, real-time (RTC) clock unit, temperature and humidity sensor, photosensitive module for detection of daytime, touch sensor and WiFi module for time synchronization from NTP (Network Time Protocol) servers. This Smart alarm clock could be used for better and more effective awakening for users.

Keywords: Smart alarm clock · Arduino · Sensors · Sleep · REM · NREM

1 Introduction

Sleep, which can be described as a time from falling asleep to awakening, could be divided to several phases, which may change during the night. The most common phases are REM (rapid eye movement) and NREM (non-rapid eye movement) [1].

Phase NREM is characterized by decline of brain function, physical calmness and relaxation. This phase is also divided by American Academy of Sleep Medicine (AASM) into 1–3 stages by the depth of sleep, where NREM stage 1 occurs mostly in the beginning of sleep followed by NREM stage 2 (light sleep) and NREM stage 3, which is also known as slow-wave sleep (SWS) or deep sleep phase. All these NREM stages are in common characterized by non-rapid eye movements.

Second phase REM (also known as paradoxical sleep or desynchronized sleep) is characterized by brain activity at the level of wakefulness with rapid eye movements

© Springer International Publishing Switzerland 2016
H. Fujita et al. (Eds.): IEA/AIE 2016, LNAI 9799, pp. 514–526, 2016.
DOI: 10.1007/978-3-319-42007-3_45

and loss of muscle tone for almost all the muscles controlled by our will. These stages (NREM and REM) typically alternate in cycles of four to five times per night [2].

Next common state during a sleep is dreaming, which mostly occurs during REM stage of sleep. More information about dreaming can be found in article [3]. According to some theories, dreaming has regenerative function for the body and brain reorders, sorts and recovers the stored information [9].

To complete enumeration phase, we should add phase Hypnagogia (it is the experience of the transition state from wakefulness to sleep) and Hypnexagogia (awakening, end of sleep). The best efficiency of awakening is in light sleep or in almost awake moment, when sleeping person does not have any scruples to be awakened. The indicators of this state can be sudden body movements, when sleeping person is changing his position.

There are a wide range of smart solutions that try to facilitate awakening and adapt to people. These solutions are trying to detect sleep cycles or at least they try to regulate and adjust wake-up time in the set time window.

These solutions are usually in the form of various applications for mobile phones, tablets or computers with external sensors as webcams etc. However, we can also find some hardware solutions that work mainly on Arduino, Raspberry Pi and other similar platforms on the market [11]. Most of these hardware solutions work only as alarm clocks with some enhancements (synchronization with Google calendar etc.). From this point of view, designed application will be unique at least for sleep phase detection.

Smart alarm applications are widely available on Google Play or Apple Store. However, they are also limited by their hardware specifications and limited by placement in bed, which should as close as possible to users, which can lead to equipment damage by body movements and also radiation from these devices is also not healthy [4]. Sample application can be found for example on website [6, 13].

The aim of this project is to design a functional solution on Arduino platform that will detect all body movements by motion sensor (from distance) and computing unit will use proposed algorithm to these recorded events to evaluate the best possible moment to wake up user. This solution will be also extended by external sensors as humidity and temperature sensors or photosensitive module for detection of daytime for LCD intensity adjustment.

2 Problem Definition and Related Works

All people need to get up at a certain time, which can be preset on their smart mobile devices or on classical alarm clocks. However, the time you set is static and when the time is close to minimum value for time interval the alarm easily rings. Alarm clock usually has no information about your current state, ie whether you are in deep sleep or you are already awake. There may be a problem with the alarm, which wakes you up at the wrong stage of sleep. This problem continues throughout the day and person feels tired and feels like he slept fewer hours than he sleeps usually. The question is how accurately and reliably prevents this problem? Is it possible to detect and distinguish the different sleep phases and wake user up at the right time?

In the literature we read that just before and after the REM sleeping individual moves more than at other stages. All these movement signals are easily detectable and on their basis we can assess ongoing sleep phases [19, 20].

For motion detection, we can use wide spectrum of sensors, which are divided into two general categories. In first category are sensors of electric values, where can be found sensors for scanning the brain, heart and muscle activity (EEG, ECG, EOG, EMG or ERG) [14]. These sensors are better and more accurate but they are too expensive for this solution. Second category contains sensors of non-electrical quantity, where can be found motion detectors (passive infrared sensors) and the sensors for the detection of acoustic quantities (microphones). These sensors are more affordable and practical for use.

As mentioned before, the most optimal sensor for scanning movements (also in darkness) seems infrared motion sensor (PIR), which should have at least a basic option for setting sensing range, sensitivity or the period over the sensor senses.

The advantages of these sensors are low price, high availability, low power consumption (<50 mA), high viewing angle and high scanning distance (up to 10 m).

There is also another problem, when LCD display is emitting light during a night. Light is disturbing element during sleeping and it should be suppressed. We have to add photoresistor module (also known as light-dependent resistor or photocell), which detects day/night cycles and turns off backlight of screen when the night is detected. When user needs to set backlight to high (during a night), there was added a touch sensor module to turn on the display.

The next problem may occur when people are sleeping in overheated bedrooms. Smart alarm clock has an attached temperature and humidity module, which is showing these values on display for tentative information purpose. All these external sensors with base equipment of smart alarm clock are creating the concept of smart alarm clock.

Nowadays (2015), there are several commercial or open-source solutions across different platforms, which are trying to use different sensors for finding the best possible moment, when are users in light sleep or in almost awake state. For example, we can mention a few solutions.

WakeNsmile is Windows Mobile application (for Windows Mobile 6+), which is using built-in microphone in smart devices to detect sleep phases. After analysis and evaluation of these recorded audio data is application able to set right time to wake up users. This application has also built-in algorithm for filter pathological sound manifestations of sleep like snoring or cough from recorded set of data [5].

Sleeptracker™ is brand name of wrist watches that monitoring and collecting data during sleep at night. They can evaluate right time so-called "almost awake moment" by analyze of these data to wake up user. The disadvantage of this solution is that you have to wear the watch clock during sleep and you have to tighten the strap slightly more than usual, which may not suit everyone [10].

S.M.A.R.T. Alarm Clock is solution based on Arduino platform (Arduino Yún version), which uses Google calendar and external Temboo service [12] for alarm time synchronization over ethernet port. The solution uses only the basic functionality of the Arduino platform and it does not optimize the alarm time, therefore it's not suitable for sleep cycle detection [7].

CUDA application on PC platform – These general applications on PC (also on Mac) use external webcam for movements detection. Stream processing of video data requires high computing power, so it was dropped from the use of the computer's processor to calculate movements.

All of these solutions are in some way interesting in view of the detection of sleep cycles and the approach to a solution. In the following section of this article will be described new proposed solution of smart alarm clock.

3 New Solution

The prototype of this smart alarm clock is built on an electronic prototyping platform Arduino, specific model is Arduino Mega 2560 rev. 3, which uses a microcontroller ATmega2560. Parameters are 16 MHz operating frequency for microcontroller, 256 Kbytes of flash memory (8 KB is used by boot loader), 4 KB of EEPROM and 8 KB of SRAM (Static Random Access Memory). Arduino Mega 2560 R3 has also 54 digital and 16 analog pins for use on board. The working frequency and the memory capacity have been found to be completely sufficient for this developed smart alarm clock [8].

Image output of smart alarm clock caters a two-line (16 × 2) LCD display labeled as I2C 1602. LCD displays the current time, the ambient temperature in the room and warnings like NTP synchronization or text information when the alarm rings.

LCD display 1602 has a basic setting of backlight (maximum to completely off), contrast (adjustable via a resistor - trimmer) and other software settings mediated by library directly in the source code.

Display communicates via I2C bus which allows using bidirectional two-wire data link communication between one or more devices (or integrated circuits).

The wires are known as SDA (Synchronous Data, data channel) and SCL (Synchronous Clock, clock signal), which are connected to digital pins 20 (SDA) and 21 (SCL) in Arduino Mega 2560. Signaling of alarm clock is solved by classical LEDs diodes (3 mm) connected to breadboard and by buzzers as audible alarm.

As real-time clock generator is used RTC (Real Time Clock) module with the designation DS3231, which is temperature compensated RTC. The advantage of these RTC circuits is their high accuracy at a relatively low cost price and the external supply of batteries to store the current time even after disconnecting or main power failure. Operating module handles memory chip AT24C32. For connection module to Arduino is also used I2C bus (same as connection of LCD). More information can be found on manufacturer's website [13, 15].

For sensing the temperature in the room was used DHT22 sensor which has a range 3.3 to 6 of input voltage and it uses 5 V TTL (Transistor–transistor logic) over only one wire. Sensor is very accurate (in low-cost category), the manufacturer provides for measuring the humidity (0 to 100 %) accuracy of 2–5 %. For temperature measurement in the interval −40 to 80 °C is accuracy 0.5 °C. The sensor has a maximum power consumption of 1.5 mA during measurement and is able to measure and differentiate the values every two seconds. The temperature sensor is not a key feature of a smart alarm clock, and therefore these values are still within the tolerance.

The most important sensor for smart alarm clock is passive infrared motion module (PIR). PIR is an electronic sensor which measures the infrared rays that are emitted from object in its field of vision. All objects with a temperature above absolute zero emit their heat energy as a radiation. This radiation is invisible to the human eye because it radiates at infrared wavelengths. It can be detected by electronic devices such as PIR sensors, which are designed for this a purpose [18].

As a PIR module was used HC-SR501 sensor that operates with an input voltage of 5 to 20 V. It senses 120° cone of the space to a distance of up to 7 m. This sensor has two trigger modes for scan settings, which can be changed by jumper on module board. L – it disables repeat trigger and H – it enables repeat trigger. We can measure how long each movements lasts with the time interval and respond to these events. For prolonged movement, we can say that the captured person woke up and alarm can be disabled or postponed.

For the purpose of detections sleep stages, the movements of the sleeping person occur generally just before or just after REM stage of sleep. All these events can be recorded and evaluated by proposed algorithm and alarm can be triggered in the right time of sleep (in interval of 30 min from alarm time).

User interaction with a smart alarm clock caters capacitive touch sensor W110 with a designation TTP223B. Characteristics of this sensor are particularly low power consumption in idle state and small size. External touch sensor was selected due to the absence LCD touch screen. The sensor is used to turn off the alarm and for turn on the backlight of LCD screen at night.

Since this is a prototype, setting up a new wake-up time is done through virtual serial port in format: (A day, month, year, hour, minute) where A is control character for alarm and day, month, etc are integers. This solution is new in the sense of use of the platform Arduino and it eliminates the shortcomings of previous solutions. Prototype scheme of hardware implementation can be found in (Fig. 1).

4   Implementation

The following section will describe auxiliary libraries, general structure of the program, algorithm for detecting sleep phases and instructions for use. Emphasis will be placed on the functionality and reliability of the proposed solution, but also sufficiently fast signal processing algorithms in real time [16, 17]. At first part, there is a summary of all the components of the smart alarm clock: Arduino Mega 2560 rev. 3; LCD 1602 I2C + Adapter Board w/IIC/I2C; WiFi module CC3000 with SD card reader; Motion detector HC-SR501 PIR; Capacity touch sensor W110; RTC module DS3231; Temperature and humidity sensor DHT22; Photoresistor module; Breadboard, buzzer and LED diodes for alarm signalization.

Arduino programs can be divided in three main parts: structure, values (variables and constants) and functions. In structure can be found two main methods setup() and

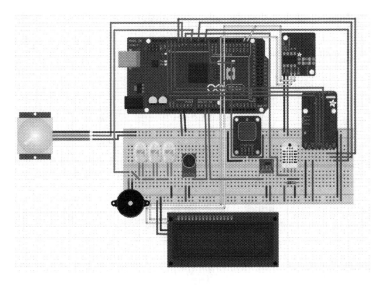

Fig. 1. Scheme of hardware implementation of smart alarm clock. Wires are color coded as blue for WiFi, orange for sensor's communication cables, red cables are for powering and black for grounding. Purple cable is for alarm signalization and white wires are SDA and SCL. (Color figure online)

loop(). Method setup() is called at the beginning of program and method loop(). The rest of control structures, variables, constants, data types and functions are almost same as in different languages.

4.1 Detection Algorithm

The most important part of implementation is proposed algorithm for sleep state detection. Detection algorithm of almost waking moment is based on theoretical findings, which were described in previous parts. Design of algorithm was partly inspired from scientific articles [21–23], that are focused to the detection of states of sleep and wakefulness by contactless biosensors (mostly accelerometers) and their results are compared with the sample sets of already measured data by polysomnography (the most accurate measurements).

These algorithms are very similar to each other. They differ mostly by different constants in proposed equations, which are usually the result of a linear combination of previously used algorithms. The goal of these algorithms is to increase the percentage chance (accuracy), that one evaluated state detected as sleep is really sleep state etc.

The scanned record is usually divided into 10–60 s intervals. In each such interval is calculated equation, into which enter as event parameters (e.g. movements) from neighboring intervals (previous and future - from pre-scanned data). Here comes the problem of the use of these equations in smart alarm clock that has information only about the current state and the previous states.

The proposed equation of almost waking moment will therefore be adapted to the possibilities of smart alarm clock. The detection rate of this evaluated time will be compared with a different solution - Android application called Sleep Cycle Alarm Clock (measuring movements by accelerometer). The proposed general equation for detecting almost waking time in the model is of the form:

$$D = P(W_i * A_i + W_{i+1} * A_{i+1} + \ldots + W_{i+n} * A_{i+n})$$

Where the $D < 1$ is for deep sleep and $D \geq 1$ is for light sleep or almost awaking moment. P is the scale (adjustment for the number of measurements in the interval) and $W_i \ldots W_{i+n}$ are weighting factors for the current interval and 9 previous intervals. Each intervals measure exactly 15 s. $A_i \ldots A_i + n$ are the total times of detected movements in individual intervals. The proposed algorithm always stores the total sum of movements in the current 15 s interval and the total sum of movements in the past 9 intervals (each separately). At each step of the calculation (after another 15 s) program

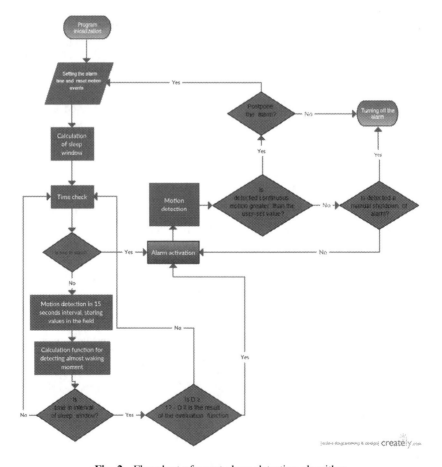

Fig. 2. Flowchart of smart alarm detection algorithm

calculates the value of D, which is also stored in the program logs. If $D \geq 1$ and current time is in the range of set time window (default time of 30 min) the alarm is triggered prematurely.

Specific equation was for testing and comparison purposes with other solution chosen in the form:

$$D = 0.25 * (1 * A_0 + 0.9 * A_1 + 0.8 * A_2 + 0.7 * A_3 + 0.6 * A_4 + 0.5 * A_5 + 0.4 * A_6 + 0.3 * A_7 + 0.2* A_8 + 0.1 * A_9).$$

The greatest weight ($W_0 = 1$) is always for movements in the current interval, followed by the next intervals with a reduced weight by 0.1 ($W_n = W_{n-1} - 0.1$) for every step. The algorithm of smart alarm clock can be described by the flowchart in (Fig. 2).

The following is a sample implementation of the algorithm within the source code of the program. Every 15 s the method *saveValues()* is called (see example below), which sorts the array. The last value is overwritten (deleted) by the previous value. Then on the first place of the array the sum of the times of movements is written (in the current 15-second intervals). Following is a calculation of the evaluation function method called *detectWakeSleep()* that determines the value of D from the values in this array and returns the result as waking time/sleep.

Based on this result it is possible to detect almost awake moment, which, if $D \geq 1$ activates an alarm early, provided that the current time is in the set time window, i.e. 30 min before the set alarm.

```
/* Method for storage of data into the array */
void saveValues() {
    for (int i = 9; i > 0; i--) { // array reordering, last value is deleted
        Movements[i] = Movements[i - 1];
    }
    Movements[0] = value; // first value in array is the actual movement
}
/* Method for detection of waking moment */
boolean detectWakeSleep() {
    float p = 0.25; // p constant
    saveValues();
    D = p * (1 * Movements[0] + 0.9 * Movements[1] + 0.8 * Movements[2] + 0.7 * Movements[3]
        + 0.6 * Movements[4] + 0.5 * Movements[5] + 0.4 * Movements[6] + 0.3 * Movements[7]
        + 0.2 * Movements[8] + 0.1 * Movements[9]);
    if (D >= 1) {
        return true; // waking moment
    } else {
        return false; // sleeping
    }
}
```

4.2   Alarm Setting and Instruction for Use

In the source code of the program, there can be defined several variables that affect the operation of the program itself. Setting up a WiFi network is defined by the following constants:

```
#define WIFI_ON 0     // 0 – disabled, 1 - enabled
#define WLAN_SSID          " SSID name"
#define WLAN_PASS          "WiFi Password"
#define WLAN_SECURITY  WLAN_SEC_WPA2
```

Setting other basic parameters for the operation of the smart alarm clock is listed in the following section:

```
#define TIME_TO_OFF 10
// The number of seconds of continuous motion to postpone the alarm
#define ALARM_WINDOW 30
// Setting the time window in minutes (0-59 minutes)
#define DELAY_ALARM 5
// Snooze time in minutes
```

```
#define LETNI_CAS 2
// The calculation for unixTime from NTP server, 2 -Summer time, 1 - Winter time
#define GMT +1
// Setting the time zone for time synchronization
#define NTP_server "time.windows.com"
```

5   Testing of Developed Application

In this part will be tested detection algorithm of almost waking moment to algorithm that is implemented in Sleep Cycle alarm clock application. This algorithm is not publicly available. Testing takes place while measuring sleep with two devices with the same settings (30 min time window) and with same alarm time.

The application (Sleep Cycle Alarm Clock) is installed on the device Xiaomi MI3, to capture movements during sleep the phone uses accelerometer labeled MPU-6050 from the manufacturer Invensense. Position of the phone has been in the top right corner of the bed approximately 15 cm from the head. In test mode the application evaluated this position as a seamless position (phone recorded even the smallest movements).

Smart alarm clock (Arduino solution) was placed on the table at a distance of approximately one meter. Position of motion detector has been chosen so that it can record the movements of the whole body (the entire length of the bed).

Measurement and analysis of movements was performed on the 9 users who don't have any sleep disorders. To analyze sleep was used the integrated function in application Sleep Cycle Alarm Clock of generating a chart that displays three states: Awake, sleep and deep sleep (NREM stage 3). To compare these results was implemented to Arduino Smart alarm clock similar function that storing temporal data and generate similar graphs.

5.1 Measurement Results

The (Table 1) below summarizes sample measured data from both solutions, which were measured simultaneously in one night. In the first column are all awake states and deep sleep states detected by Smart Alarm Clock. The second column shows same moments, which are detected by Android application Sleep Cycle Alarm Clock. As you can see, proposed solution was more accurate in detecting these states than Android application.

Table 1. Measured sleep states in the two compared solutions.

	Smart alarm clock	Sleep cycle alarm clock (Android)
Awake	1:20–1:30	1:20–1:30
	2:50–3:30	Not detected
	4:20–5:10	4:25–4:45
	6:00–6:10	Not detected
	6:20–6:40	Not detected
	7:20–7:40	Not detected
	8:50–9:05	8:00–9:05
NREM 3	1:30–2:49	1:50–2:40
	3:40–4:20	3:40–4:10
	5:10–5:49	Not detected
	6:10–6:19	Not detected
	6:40–7:09	6:40–6:55
	Not detected	7:20–7:50
	8:10–8:49	Not detected

From the measured data sets was found that the measurements from PIR sensor compared to the accelerometer sensor are more accurate and can thus record far more events during the night. On the other hand, it is possible that the graph for applications Sleep Cycle alarm clock is only indicative and some values are deliberately omitted or wrongly filtered out as false values.

Table 2. Designation of test users, their sex and age.

User	A	B	C	D	E	F	G	H	I
Gender	Male	Male	Female	Female	Male	Female	Male	Male	Male
Age	24	48	45	18	19	65	48	13	20

To compare both detection algorithms of almost waking time was compared a total of 35 measurements on the 9 test users. All users, their age and sex are shown in table (Table 2). Set of 35 measurements on all users is shown in table (Table 3).

Table 3. Alarm activation times in 24-hour clock format. Numbers 1A, 2A, etc are measurements for first user, and 8B, 9B, etc are measurements for second user. Alarm activation [1] is time for application sleep cycle alarm clock and alarm activation [2] for sleep cycle alarm clock (proposed solution). Alarm time mean the last possible time of alarm activation.

Num	1A	2A	3A	4A	5A	6A	7A	8B	9B	10B	11B
Start	23:58	23:42	23:57	23:58	0:21	0:10	1:12	23:10	22:45	0:47	23:36
Alarm time	9:30	8:30	8:30	9:30	9:30	9:30	9:30	7:30	7:15	8:30	7:45
[1]	9:03	8:02	8:04	9:02	9:01	9:01	9:04	7:06	6:49	8:05	7:18
[2]	9:03	8:02	8:04	9:02	9:01	9:01	9:05	7:07	6:49	8:05	7:19

12B	13B	14B	15C	16C	17C	18D	19D	20D	21E	22E	23E
0:12	23:58	1:31	23:17	22:51	23:56	23:11	23:49	23:08			0:16
9:00	8:15	9:00	7:00	7:00	7:00	8:00	8:00	8:00	9:15	9:30	9:00
8:36	7:45	8:31	6:42	6:32	6:36	7:33	7:38	7:31	8:48	9:02	8:36
8:36	7:45	8:32	6:42	6:33	6:36	7:33	7:38	7:31	8:48	9:03	8:36

24F	25F	26F	27G	28G	29G	30H	31H	32H	33I	34I	35I
23:13	22:47	23:09	22:35	23:32	21:58	0:56	1:14	1:01	0:32	23:41	1:11
5:30	5:30	5:30	6:30	6:30	6:30	10:00	10:00	10:00	9:20	8:15	9:30
5:09	5:02	5:06	6:08	6:10	6:02	9:32	9:39	9:31	8:57	7:49	9:16
5:09	5:02	5:06	6:08	6:11	6:03	9:32	9:39	9:31	8:57	7:49	9:16

From the data was found that the Sleep Cycle alarm clock application detects almost awake moment almost always at the beginning of sleep windows, as evidenced by the graphs generated by the application. In terms of the proposed solution it can be stated that this solution is able to detect almost awake moment, at least at the level of commercial solutions. The custom solution is also more precise, when recording sleep data from night.

Proposed application was also tested for stability and reliability. Reliability testing was realised by application run within seven days, while the application run reliably without the slightest signs of a slowdown, the shortage of memory or jam/restart alarm (assuming a constant source of alarm). The application is also able to handle all instruction up to 160.15 Hz, which is for required purposes and properties (scanning sensors in real time) better than requested value.

6 Conclusions

The result of this work is a functional prototype of smart alarm clock on the Arduino platform, which is able to detect almost awake moment during sleep. Testing and comparing the smart alarm with the commercial solution in the form of Android application Sleep Cycle alarm clock for Android showed that this application and especially the proposed algorithm is able to successfully detect the moment at the same time, maximum in the range of a few seconds from the first alarm activation.

Testing of both solutions at the same time is very difficult because the activation of the first alarm awakens the user which is followed by activation of the second alarm.

In testing part and especially in the evaluation part of sleep data was also showed that the PIR sensor is more accurate in detecting movements compared to accelerometers in mobile devices.

As regards the activation of the alarm at the right time, we cannot say with precision how accurate this result is. This impression is quite subjective, however, according to user ratings e.g. on Google Play's contribution, these types of applications are evaluated positively. This fact is evidenced by the number of downloads that are in the millions.

Acknowledgement. This work and the contribution were supported by project "SP-2102-2016 - Smart Solutions for Ubiquitous Computing Environments" Faculty of Informatics and Management, University of Hradec Kralove, Czech Republic. The Universiti Teknologi Malaysia (UTM) and Ministry of Education Malaysia under Research University grants 02G31, and 4F550 are hereby acknowledged for some of the facilities that were utilized during the course of this research work.

References

1. Botía, JA., Charitos, D.: A DIY approach to the internet of things: a smart alarm clock. In: 9th International Conference on Intelligent Environments. pp. 214–221. IOS Press (2013)
2. Polysomnography. http://www.nlm.nih.gov/medlineplus/ency/article/003932.htm. Accessed 17 Dec 2015
3. Maquet, P., et al.: Functional neuroanatomy of human rapid-eye-movement sleep and dreaming. Nature **383**(6596), 163–166 (1996)
4. Cimler, R., Matyska, J., Sobeslav, V.: Cloud based solution for mobile healthcare application. In: Proceedings of the 18th International Database Engineering and Applications Symposium, July 2014, pp. 298–301. ACM (2014)
5. Krejcar, O., Jirka, J., Janckulik, D.: Use of mobile phones as intelligent sensors for sound input analysis and sleep state detection. Sensors **11**(6), 6037–6055 (2011). doi:10.3390/s110606037
6. Smart Alarm for Android. http://sport.com/smart_alarm_android.html. Accessed 16 Dec 2015
7. S.M.A.R.T Alarm Clock. http://makezine.com/projects/s-m-a-r-t-alarm-clock/. Accessed 16 Dec 2015
8. Crick, F., Mitchinson, G.: The function of dream sleep. Nature **304**(5922), 111–114 (1983)
9. SleepTracker. http://www.sleeptracker.cz. Accessed 18 Dec 2015

10. Sleep Cycle. http://www.sleepcycle.com. Accessed 18 Dec 2015
11. Horalek, J., Sobeslav, V.: Measuring of electric energy consumption in households by means of Arduino platform. In: Sulaiman, H.A., Othman, M.A., Othman, M.F.I., Rahim, Y. A., Pee, N.C. (eds.) Advanced Computer and Communication Engineering Technology: Proceedings of ICOCOE 2015, pp. 819–830. Springer, Cham (2016)
12. Liao, W.H., et al.: iWakeUp: a video-based alarm clock for smart bedrooms. J. Chin. Inst. Eng. **33**(5), 661–668 (2010)
13. Kasik, V., Penhaker, M., Novák, V., Bridzik, R., Krawiec, J.: User interactive biomedical data web services application. In: Yonazi, J.J., Sedoyeka, E., Ariwa, E., El-Qawasmeh, E. (eds.) ICeND 2011. CCIS, vol. 171, pp. 223–237. Springer, Heidelberg (2011)
14. Augustynek, M., Pindor, J., Penhaker, M., Korpas, D., Society, I.C.: Detection of ECG significant waves for biventricular pacing treatment. In: Proceedings of the 2010 Second International Conference on Computer Engineering and Applications, ICCEA 2010, vol. 2, pp. 164–167 (2010). doi:10.1109/ICCEA.2010.186
15. Aserinsky, E., Kleitman, N.: Two types of ocular motility occurring in sleep. J. Appl. Physiol. **8**(1), 1–10 (1955)
16. Machacek, Z., Slaby, R., Vanus, J., Hercik, R., Koziorek, J.: Non-contact measurement system analysis for metallurgical slabs proportion parameters. Elektronika ir Elektrotechnika **19**(10), 58–61 (2013). ISSN 1392–1215
17. Machaj, J., Brida, P.: Optimization of rank based fingerprinting localization algorithm. In: 2012 International Conference on Indoor Positioning and Indoor Navigation (IPIN 2012), Sydney, Australia, pp. 1–7, 13–15 (2012)
18. De Chazal, P., Fox, N., O'Hare, E., Heneghan, C., Zafaroni, A., Boyle, A., Smith, S., O'Connell, C., McNicholas, W.T.: Sleep/wake measurement using a non-contact biomotion sensor. J. Sleep Res. **20**(2), 356–366 (2011). ISSN 1365-2869
19. Hoque, E., Dickerson, R.F., Stankovic, J.A.: Monitoring body positions and movements during sleep using WISPs. In: Wireless Health 2010, pp. 44–53. ACM, New York (2010). ISBN 978-1-60558-989-3
20. Automatic sleep/wake identification from wrist activity. Abstract - Europe PubMed Central. http://europepmc.org/abstract/med/1455130. Accessed 3 July 2015
21. Dolezal, R., Bodnarova, A., Cimler, R., Husakova, M., Najman, L., Racakova, V., Krenek, J., Korabecny, J., Kuca, K., Krejcar, O.: Variable elimination approaches for data-noise reduction in 3D QSAR calculations. In: Pereira, F., Machado, P., Costa, E., Cardoso, A. (eds.) EPIA 2015. LNCS, vol. 9273, pp. 313–325. Springer, Heidelberg (2015)
22. Cerny, M., Penhaker, M.: The HomeCare and circadian rhythm. Paper presented at the 5th International Conference on Information Technology and Applications in Biomedicine, ITAB 2008 in conjunction with 2nd International Symposium and Summer School on Biomedical and Health Engineering, IS3BHE 2008, Shenzhen (2008)
23. Cerny, M., Penhaker, M.: Wireless body sensor network in health maintenance systems. Elektronika Ir Elektrotechnika **115**(9), 113–116 (2011)

Flow Visualization Techniques: A Review

Yusman Azimi Yusoff$^{(\boxtimes)}$, Farhan Mohamad, Mohd Shahrizal Sunar,
and Ali Selamat

Media and Game Innovation Centre of Excellence, UTM-IRDA Digital Media Centre,
Universiti Teknologi Malaysia, Skudai 81300, Johor, Malaysia
yusman@magicx.my, {farhan,shahrizal,aselamat}@utm.my
http://magicx.utm.my/

Abstract. Flow visualization is an approach focuses on methods to get
information from the flow field datasets either in 2 or 3 dimensional.
Researchers have presented many visualization techniques with similar
goal, which is to enrich the information provided by the visualization.
The differences between each technique cause misunderstanding to new
researchers on the implementation, advantages, and their limitations.
This paper will review and discuss some of the available techniques
by classifying them based on the dimension and type of flow datasets.
The type of flow is either steady or unsteady, and the dimensions are
2, 2.5, and 3 dimensional. The classification assists readers to identify
and choose the appropriate method based on their requirement. This
paper also highlights several important information related to the history and the foundation of flow visualization before reviewing the methods in details. Discussion and tables are included to enhance the readers
understanding about the differences between reviewed methods.

Keywords: Flow visualization · Streamlines · Steady flow · Time-
dependent flow

1 Introduction

Visualization is a technique used to analyze and transform dataset to a meaningful visual representation. Back then, many researchers faced a lot of problems to
process large dataset as the completion of the visualization process in the hardware was time-consuming. The state-of-the-art of the current modern technology
has accelerated the process of understanding the large dataset tremendously [1].
In fact, high-performance hardware allows a researcher to implement the visualization algorithm to understand the information inside the dataset, especially
in flow visualization.

Flow visualization can be considered as one of the important elements when
studying fluid mechanics and should take into consideration. It could show the
flow of the fluid based on the flow dataset. It is widely used in the engineering,
meteorology, and oceanography. The dataset used to visualize the flow varies in
every research area. Thus, it requires a specific algorithm to process different
type of dataset efficiently.

© Springer International Publishing Switzerland 2016
H. Fujita et al. (Eds.): IEA/AIE 2016, LNAI 9799, pp. 527–538, 2016.
DOI: 10.1007/978-3-319-42007-3_46

Several key components are needed to visualize a flow. First of all, the main component is the dataset itself. The dataset should have flow information such as vector data either in the structure or unstructured grid. The next component is the algorithm or technique used to visualize the flow. The decision is made after determining the data structure and the information that wanted to be extracted. The common technique used to visualize 2D and 3D flow was the geometry-based technique, especially streamlines. Streamlines can be used to visualize a global and local flow pattern. An example of global flow pattern is the cloud movement during storm. Blood flow inside the arteries is one of the good examples of local flow pattern.

2   Flow Visualization Technique

As stated before, one of the ways to visualize the flow is using streamlines approach. There are other techniques that can be used in flow visualization, but depend on the desire visual output. Basic techniques for flow visualization were converting vector information in a meaningful representation such as hedgehog or glyph [2,3]. Glyph was used to show the direction of the flow by using arrow or cone glyph. The size of the glyph was adjusted based on the velocity of the flow.

Another group from geometry-based technique was n-lines. This referred to streamlines, streaklines, and pathlines. Although all the three techniques have similarity in line-based, each of the techniques has different function. Figure 1 shows the images of streamlines, streaklines, and pathlines produced by FlowVisual as presented in [4]. Streamline was primarily for steady flow whereby the flow pattern does not change. Streaklines and pathlines were suitable for unsteady flow in which the lines move with the changes of the flow pattern. It can be said that if streaklines and pathlines are implemented onto the steady flow, the result will be similar to the streamlines [5].

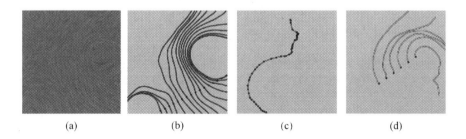

(a)	(b)	(c)	(d)

Fig. 1. Series of seed point was placed on a (a) line integral convolution (LIC) image to generate (b) streamlines, (c) streaklines, and (d) pathlines. Images were generated by FlowVisual

Certain parameters were required to initialize first before constructing and visualizing the flow. Basic parameter such as the number of lines and length of lines might help to improve the visual output. There were benefits and drawback of changing the value of these parameters. For example, increasing the number and length of streamlines will show more flow patterns. Besides, the 2D flow field does not initiate many problems when the value of the lines was increasing compared to the 3D flow field. However, when the value increased over a certain limit, the visual will become cluttered. Information that was available before might become obstructed with other lines, which lead to loss of information. Consequently, the viewer might face difficulty in visualization as it is hard to understand and confusing [6,7].

Flow visualization is used extensively in various research areas. Visualization can be used to simulate flow based on predefined parameter and post mortem analysis of a study. An example of simulation is the flow of hot and cold water in a pipe. The design of the pipe is tested to see the movement of hot and cold water and identify the distribution of the heat when the both sources meet until it moves to the pipe outlet. Simulation helped to understand the flow better without any experimental setup [1,2]. On the other hand, visualization was basically based on experimental setup. Data was collected during experiment and transformed into visual representation of the flow. The flow data cannot be manipulated since it was collected from the real experiment. Both techniques applied the same concept of flow visualization, but used at different phases. Flow visualization helped to understand the information that is not visible to the naked eye. Thus, using the correct algorithm and flow parameter was important to highlight the key information of the visualization.

When it comes to visualizing fluid, certain information need to be clarified. One of the flow parameters that differentiate between unsteady and steady flow is the velocity. Steady flow is referring to the situation where the inlet and outlet flow are consistent. In other words, there are less obstacle inside the flow path, which make the flow become steady and less turbulence. Unsteady flow occurs when there are changes in the flow pattern, making the visualization differ from time to time. The method used to visualize the unsteady flow should show the changes of the flow. This paper classified the method used to visualize the steady and unsteady flow based on the different type of flow data.

3   Flow Visualization Classification

Classification allowed a reader to focus and understand the related works with the different methods of flow visualization. Spatial resolution was the main classification followed by the type of flow. The source of visualization data was based on the spatial resolution. For example, Wind tunnel can produce 3D vector dataset while cloud movement can be transformed into 2D vector dataset. Figure 2 shows the example of 2D [8], 3D, and [9] flow visualization. Researchers will implement different algorithm and implement them to the data before producing the flow visualization. The algorithm used to generate the visualization was the main focus of this paper. This algorithm may change the way of positioning the seed point, flow presentation, and visualization focus area.

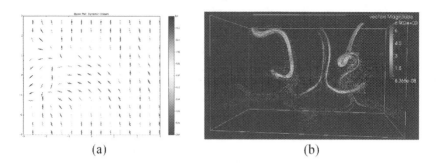

Fig. 2. (a) Flow visualization using vector plot on a 2D plane, and (b) blood flow visualization in a carotid artery (Color figure online)

3.1 2D Steady Flow

This section is separated into three sub-category in order to highlight the area of the contribution.

Streamlines Placement. An improved method for streamlines placement was proposed in this paper [10]. They used image-guided technique to place streamlines in a proper place and gap by using the low-pass filter in the energy function. This showed the differences between the current image and the target visual density. The reduction of energy improved the placement of streamlines by altering the position and length of streamlines, combining the nearly adjacent lines, and creating new streamlines to fill the large gap. The result manifested a more hand-placed appearance that does regularly or randomly placed streamlines.

In order to generate evenly spaces streamlines, an algorithm was developed and applied to visualize 2D steady flow [11]. The technical contribution of the work was computed based on a wide range of flow field sources (texture based up to hand-drawing style). The result produced clean long evenly-spaces streamlines with accurate control of visual density. The visualization become neat, which makes it easier to convey the flow information. However, this technique may have some difficulties in visualizing the small turbulence in the large flow field due to the property of the method.

Another seeding strategy [12] presented was to improve the seed placement by using the information from the flow features in the datasets. The aim of this strategy was to identify the flow pattern in the critical points even the density of the streamlines was reduced. They also placed the streamlines at the non-critical regions with varies streamlines length to make the visualization neater. The advantage of this method is the ability to visualize the flow pattern near to the critical point clearly. On the other hand, the drawback of this method is the difficulty in handling higher-order critical points, but it can be extended to position the seed correctly in the area that has higher-order critical points.

In addition, a novel algorithm that focusing on the high-quality placement to generate long streamlines was proposed according to the placement of streamlines based on the 2D steady flow vector or direction field [13]. The goal of the algorithm was to produce a high quality placement by favoring long streamlines, while retaining the uniform pattern with increasing density. This can be achieved by placing one streamlines at a time with the help of the numerical integration that started at the furthest position from all the previous streamlines placement. This technique managed to achieve the simplicity, robustness, and efficiency outcomes by applying the Delaunay triangulation to generate the streamlines.

An advance evenly-spaced streamlines placement algorithm was presented [14], which is one of the improvements from the previous research [11]. They employed a fourth-order Runge-Kutta integrator and adaptive step size error control to acquire rapid accurate streamlines advection. In order to reduce the amount of distant checking, they adopted Cubic Hermite polynomial interpolation with large sample spacing to create fewer evenly-spaces samples along the streamlines. They also proposed an ideal loop detection strategy to handle spiraling and closed streamlines. The results showed that this algorithm performed faster than the algorithm that mainly based on the order-of-magnitude [11].

Encouraged by the idea of abstract drawing, which was focusing on the explanatory quality in art, a new seeding strategy was proposed [15] to generate illustrative 2D streamlines that focus on visual clarity and evidence. The basic idea of the algorithm highlighted the flow field effectively with a minimum set of streamlines. In order to produce the result, 2D distance filed was generated to identify the distance between each grid point to the nearby streamlines. Local metric was derived to calculate the dissimilarity between the original fields to the approximate field based on the distance field. Global metric assisted the process of identifying the dissimilarity in the streamlines pattern based on the local metric result and decide whether to generate new seed point at the local point. The process was repeated until there were zero dissimilarities found in the flow field. There are 3 benefits of this technique. Firstly, the density of streamlines was related to the natural flow features of the vector field. Secondly, this technique does not consider the effect of critical point detection. Lastly, the most important point was the visual cluttering problem has been solved. Poor performance than the technique proposed previously [14] is one of the drawbacks of this technique, which was faster and do not depend on the flow field feature.

Animation. Animating the streamlines can improve the viewers understanding about the flow visualization. This concept was implemented in the previous research [16] by developing a unique technique that used to animate streamlines based on the data structure called Motion Map. This data structured valuable information such as the flow field dense representation and motion information required to animate the flow. The advantage of using this method is that computing the Motion Map does not consume more time than computing a single flow image and this step only required to execute once. The result is much better if compare to LIC-based technique. However, the only main drawback was animating the unsteady flow and added another dimension to convert into 3D flow.

The method used in animating streamlines was improved [17] by using a new approach to produce a complete cyclic and variable-speed animation for 2D steady vector fields [10,11]. The animation frames were encoded in a single image and played using color table animation technique. The cyclic effect can be produced as well and then encoded in a common animation format or used it for texture mapping on 3D objects. The advantage of this technique was that the animation produced was smoother and optimized in terms of memory requirement and the computation duration.

Visual Enhancement. A work focusing on multi-resolution flow visualization was presented [18], which was important to enrich the information for close-up view and long shot view. This paper presented a method to compute a series of streamlines-based images of a vector field with different densities, varying from sparse to texture-like representation. The streamlines position was based on the previous research [11]. It allowed a user to view a clearer flow visualization while zooming in and out in the vector field and adjusted the density for the streamlines. The density of streamlines also can be computed automatically based on the velocity and vorticity.

3.2   2D Unsteady Flow

Unsteady or time dependent flow was a challenging task when coming to visualization. Fewer works have done for 2D unsteady flow. One of the works has proposed an interesting approach of visualizing unsteady flow using streamlines with evenly-spaced streamlines placement [19]. This can be achieved by correlating the vector field instantaneously at the streamlines level. For a new frame, a feed-forward algorithm computed the set of evenly-spaced streamlines as a function of a streamlines generated for the previous frame. It can be achieved by producing a correspondence between streamlines at successive time step. The idea of highlighting the flow on the streamlines was introducing the cyclic texture that mapped onto each of the streamlines and textures of the corresponding streamline but were correlated together at a different time step. Similar work for unsteady flow visualization using evenly-spaced streamlines placement has been done with multi-resolution features [20].

3.3   Flow on Surface

Surface particle visualization was known to deliver both particle vector and magnitude as well as the flow field structure. A new method to improve the shading model for surface particle rendering was presented previously [21]. Several processes, including using Gaussian filters to prevent spatial and temporal artifacts, effective scan-conversion algorithm, occlusion handling, and lastly rendering the geometric object and surface-particles simultaneously. This method has significantly improved the image and animation, which lead to the better understanding of the flow field. One of the drawbacks of this technique was an inherent problem, which required user to choose the particle source. The process might be time-consuming and may lead to loss of information if the position does not show the interesting feature of the flow.

A new technique was proposed [8] to create a uniform distributed streamlines on a 3D parametric surface in a curvilinear grid. They extended the idea [10] by mapping the vector on a 3D surface into the computational space of the curvilinear grid. A new energy function also was designed to prevent mapping distortion caused by uneven grid by guiding the placement on streamlines in computational space with desired streamlines densities.

In order to produce a visualization of the surface in a 3D space, a new and conceptually simple method of seeding and integrating evenly-spaced streamlines for surfaces was proposed [22]. The algorithm could produce evenly-spaces streamlines for any general surface-based vector field effectively. By reversing the classic ways of producing the visualization, they could project the vector field onto the image plane, then seeding and integrating the streamlines. They also implement the idea of multi-resolution, and animating streamlines [18,19].

A dual streamlines seeding technique was proposed [7] that used greedy algorithm to solve streamlines placement problem. The algorithm was based on the dual vector fields in order to assist the seeding process. The problem solved by executing two sets of streamlines, which is primal and dual streamlines simultaneously. This technique provided several advantages such as:

(i) Only require one parameter to control the streamlines density
(ii) Naturally extended to the curved surfaces without surface parameterization

3.4   3D Steady Flow

Particle tracing can be considered as one of the common methods of visualizing 3D steady flow. One of the earliest works related to particle tracing along with simulated oil flow and shock finding technique was presented [23] to visualize 2D and 3D flow. They implemented an algorithm into an interactive graphic program or CFD flow field named PLOT3D. This method was proven to provide meaningful information about the structure of the flow. The result showed a similar structure the wind-tunnel flow visualization. This knowledge lead to a conclusion that specialized flow analysis and flow visualization were important to evaluate the data in a proper way.

Later, an advance way to process large 3D flow dataset was introduced [24] by distributing the computational process into the smaller task in the supercomputer. They developed Remote Interactive Particle-tracer (RIP) program, which use several parts from an existing program on both workstation and supercomputer, including some of the particle-tracer routines from the PLOT3D. Both machines were used for the distributed processing. RIP used supercomputer to do the time-consuming process while the workstation was used to display user interface and graphic display. The result showed that the program executed much faster and reduced memory usage because of the distributed process.

An efficient algorithm for constructing streamlines, streamribbon, and streamtube on the unstructured grid were presented [25]. A modified Rugged-Kutta method was designed to enhance the process of tracing the streamlines. The calculation was done in a canonical coordinate system to simplify the mathematical formulations and reduce the computational costs, thus increase the overall performance for large flow field exploration.

Researchers have collaborated with EXA cooperation and BMW [9] to come out with the interactive visualization of fluid dynamics simulation in locally refined Cartesian grid with a fine voxel resolution, which focusing on the interesting flow features. The works were specifically developed to be used together with PowerFLOW, a lattice-based CFD simulation tool by EXA. The visualization tool adapted the idea of virtual reality technique to enable the interactive exploration of large scalar and vector datasets. This allowed user to control the particle probe and slice probe at the interactive frame rate.

Comparison of the visualization result remained one of the biggest challenges in the research area. A tool has been developed [26] to compare the visualization for analyzing flow or vector datasets. The method allowed comparison between the individual and dense field of streamlines and streamribbons. The comparison method also can be used to study the differences in vortex cores that were represented as polylines. The tool has classified the comparison into three levels, which are image, data, and feature. Each comparison level has its own benefits and flaws. The tool was extended from UFLOW system [27] to include streamlines comparison in multiple datasets, multiple streamlines comparison, and comparison of streamlines using arc length parameterization. Researchers can use this tool to validate their technique by comparing the result with the other proved technique.

3.5   3D Unsteady Flow

A simple and intuitive method of visualizing time-dependent flow field using streaklines was presented [28]. The method was developed in a system that can process flow field in a different time frame without storing all the large data inside the memory simultaneously. In order to achieve the acceptable animation rate, all time frames were required to be stored in the memory. One of the proposed solutions (3-the virtual wind tunnel, 20-analysis, and visualization of complex unsteady three-dimensional flow) to overcome this problem was to pre-process the data so that several time steps can be visualized. The total number of time step depended on the size of physical memory. The drawback of this method was the number of time steps limitation that can be visualized if the size of physical memory is too small. By limiting the time steps with at most two in any time, it helped to remove the limitation of the number of time step that able to be visualized.

A toolkit named Unsteady Flow Analysis Toolkit (UFAT) was introduced [29] a particle tracing system that generated streaklines in unsteady flow fields. The system was visualized successfully in several 3D unsteady flow from different datasets. It also provided useful insights into the unsteady flow field. The system saved all streaklines for every time steps in a file, which allowed the system to animate the streaklines. However, this process will consume a lot of memory if use with large 3D flow datasets.

Visualizing unsteady flow was a time-consuming process. In order to improve the process, a work [30] has proposed a Tetrahedral Decomposition to optimize the process of time-dependent particle tracing. Most of the process was done in the physical space, which avoid the need of conversion and inversion. It improved the accuracy because there were no Jacobian matrix approximations. The problem of using physical space included the task of locating a particle will become harder, thus increased the computational cost. Hence, they implemented Newron-Raphson iterative method to solve this problem. They improved the performance by presenting an efficient physical space algorithm and implemented it for the interactive investigation and visualization of unsteady flow [31].

4 Discussion

Flow can be considered as an unsteady state when comes to real-world cases because of the various factors. Researchers tend to generalize the flow to steady state since it is easier to study and analyze. The amount of data produced by steady flow also is smaller compared to unsteady flow. Each case has several techniques, which highlight the importance of flow information. The dimension of the flow also plays an important role in deciding the suitable technique to visualize the flow. For example, 2D steady flow data can visualize using glyph and streamlines while Pathlines and streaklines are a proper technique used to visualize 2D unsteady flow.

Table 1 shows the classification of research focus with the flow dimension. Each dimension has its own research trend. Researchers are more focusing on enhancing the current flow visualization method for 2D flow field. On the other hand, researchers are focusing toward optimizing the process to visualize 3D flow as it is well-known with high computational cost due to the large dataset. There is also research on enhancing the visual representation of the flow. Below is the classification related to researching interest.

Table 1. Classification based on research interest with data dimension

	2D Flow		Flow on surface	3D flow	
Research interest	Steady flow	Unsteady flow	Both	Steady flow	Unsteady flow
Seeding strategy	[10–15],	[19]	[7, 8, 22]		
Computational Optimization	[16, 17]			[24, 25]	[28, 30, 31]
Enhancing visual Representation	[18]	[20]	[21]	[9]	[27, 29]
Others				[23, 26, 27]	

Our future work will focus on enhancing the current streamlines generation techniques by showing the transition between the time frames in 2D unsteady flow. The current streamlines techniques only able to visualize the flow pattern instantaneously. This will produce a different flow pattern in every time frame for unsteady flow. Identify and understand the transition between the frame assist viewers to extract the information from the visualization. Mental construction is one of the ways used to analyze and understand the flow pattern between frames, which require a lot of practice and is difficult for a beginner to master the method. Conventional visualization techniques such as pathlines and streaklines only able to show the path and position of the massless particle in a particular frame. These techniques do not depict the entire flow pattern and may lead to loss of information.

5 Conclusion

Researchers in flow visualization focused their researches on 2D flow field seeding placement to generate efficient results. These methods addressed both performance and quality of the information conveyed in the generated flow visualizations. Furthermore, conventional methods of visualizing flow are extended from streamlines to streamribbons by tracing the streamlines with a constant size vector, which provide new insights into the flow field structure.

In gaining new insight into flow above solid surface, flow visualization methods were also applied on surface of 3D model. Advancing from 2D and surface structures, 3D flow field visualization were also further explored and improved, especially in high computation visualization methods. Interactive and animated techniques were later introduced to allow focus and context information exploration by enabling users to choose selected visualization area of interests. Classifications of visualization methods are discussed in showing trends, gaps and focal areas in flow visualization studies while allowing users to select method that can visualize their flow data effectively.

References

1. Hansen, C.D., Johnson, C.R.: The Visualization Handbook. Elsevier Inc., Amsterdam (2005)
2. Yusoff, Y.A., Mohamed, F., Sunar, M.S., Chand, S.J.H.: State of the art in the 3D cardiovascular visualization. In: Lai, K.W., Dewi, D.E.O. (eds.) LNBE, pp. 143–168. Springer, Singapore (2015)
3. Borgo, R., Kehrer, J., Chung, D.H.S., Maguire, E., Laramee, R.S., Hauser, H., Ward, M., Chen, M.: Glyph-based visualization : foundations, design guidelines, techniques and applications. Eurographics State Art Rep. (2013)
4. Wang, M., Tao, J., Wang, C., Ching-kuang, S., Kim, S.H.: FlowVisual: design and evaluation of a visualization tool for teaching 2D flow field concepts flowvisual: design and evaluation of a visualization tool for teaching 2D. In: Proceedings of American Society for Engineering Education Annual Conference 2013 (2013)

5. Post, F.H., van Walsum, T.: Focus on scientific visualization. computer graphics: systems and applications. Springer, Heidelberg (1993)

6. Salzbrunn, T., Jänicke, H.: The state of the art in flow visualization: partition-based techniques. In: SimVis (2008)

7. Rosanwo, O., Petz, C., Prohaska, S., Hege, H.-C., Hotz, I.: Dual streamline seeding. In: 2009 IEEE Pacific Visualization Symposium, pp. 9–16. IEEE (April 2009)

8. Mao, X., Hatanaka, Y., Higashida, H., Imamiya, A.: Image-guided streamline placement on curvilinear grid surfaces. In: Proceedings Visualization 1998 (Cat. No. 98CB36276), vol. 98, pp. 135–142. IEEE (1998)

9. Schulz, M., Reck, F., Bertelheimer, W., Ertl. T.: Interactive visualization of fluid dynamics simulations in locally refined cartesian grids. In: Proceedings Visualization 1999 (Cat. No. 99CB37067), pp. 413–553 (1999)

10. Turk, G., Banks, D.: Image-guided streamline placement. In: Proceedings of the 23rd Annual Conference on Computer Graphics and Interactive Techniques - SIG-GRAPH 1996, pp. 453–460. ACM Press, New York (1996)

11. Jobard, B., Lefer, W.: Creating evenly-spaced streamlines of arbitrary density. In: Lefer, W., Grave, M. (eds.) Visualization in Scientific Computing 1997, pp. 43–55. Springer, Vienna (1997)

12. Verma, V., Kao, D., Pang, A.: A flow-guided streamline seeding strategy. In: Proceedings Visualization 2000, VIS 2000 (Cat. No.00CH37145), pp. 163–170. IEEE (2000)

13. Mebarki, A., Alliezy, P., Devillers, O.: Farthest point seeding for efficient placement of streamlines. In: IEEE Visualization, VIS 2005, pp. 479–486. IEEE (2005)

14. Liu, Z., Moorhead, R.J., Groner, J.: An advanced evenly-spaced streamline placement algorithm. IEEE Trans. Visual Comput. Graphics 12(5), 965–972 (2006)

15. Li, L., Hsieh, H-H., Shen, H.-W.: Illustrative streamline placement and visualization. In: 2008 IEEE Pacific Visualization Symposium, pp. 79–86. IEEE (March 2008)

16. Jobard, B., Lefer, W.: The motion map: efficient computation of steady flow animations. In: Proceedings. Visualization 1997 (Cat. No. 97CB36155), pp. 323–328. IEEE (1997)

17. Lefer, W., Jobard, B., Leduc, C.: High-quality animation of 2D steady vector fields. IEEE Trans. Visual Comput. Graphics 10(1), 2–14 (2004)

18. Jobard, B., Lefer, W.,: Multiresolution flow visualization. In: International Conference in Central Europe on Computer Graphics and Visualization (2001)

19. Jobard, B., Lefer, W.: Unsteady flow visualization by animating evenly-spaced streamlines. Comput. Graph. Forum 19(3), 31–39 (2000)

20. Ueng, S.-K., Sun, W.-Y.: Multi-resolution unsteady flow visualization. In: Third International Conference on Intelligent Information Hiding and Multimedia Signal Processing (IIH-MSP 2007), no. 2, pp. 357–360. IEEE (November 2007)

21. van Wijk, J.J.: Rendering surface-particles. In: Proceedings Visualization 1992, pp. 54–61. IEEE Comput. Soc. Press (1992)

22. Spencer, B., Laramee, R.S., Chen, G., Zhang, E.: Evenly spaced streamlines for surfaces: an image-based approach. Comput. Graph. Forum 28(6), 1618–1631 (2009)

23. Buning, P., Steger, J.: Graphics and flow visualization in computational fluid dynamics. In: 7th Computational Physics Conference, Reston, Virigina, American Institute of Aeronautics and Astronautics (July 1985)

24. Rogers, S.E., Buning, P.G., Merritt, F.J., Follin, S.E.: Distributed interactive graphics applications in computational fluid dynamics. Int. J. High Perform. Comput. Appl. 1(4), 96–105 (1987)

25. Sikorski, C., Streamline, E.: Streamribbon and streamtube constructions on unstructured grids. IEEE Trans. Vis. Comput. Graphics **2**(2), 100–110 (1996)
26. Verma, V., Pang, A., Member, S.: Comparative flow visualization. IEEE Trans. Visual Comput. Graphics **10**(6), 609–624 (2004)
27. Lodha, S.K., Alex Pang, R.E., Sheehan, C.M.: Wittenb drink uflow: visualizing uncertainty in fluid flow. In: Proceedings of Seventh Annual IEEE Visualization 1996, pp. 249–254. ACM (1996)
28. Lane, D.A.: Visualization of time-dependent flow fields. In: Proceedings Visualization 1993, pp. 32–38. IEEE Comput. Soc. Press (1993)
29. Lane, D.A.: UFAT - a particle tracer for time-dependent flow fields. In: Proceedings Visualization 1994, pp. 257–264. IEEE Comput. Soc. Press (1994)
30. Kenwright, D.N., Lane, D.A.: Optimization of time-dependent particle tracing using tetrahedral decomposition. In: Proceedings Visualization 1995, vol. 2, pp. 321–328, IEEE Comput. Soc. Press (1996)
31. Kenwright, D.N., Lane, D.A.: Interactive time-dependent particle tracing using tetrahedral decomposition. IEEE Trans. Vis. Comput. Graph. **2**(2), 120–129 (1996)

Applied Neural Networks

Hardware/Software Co-design for a Gender Recognition Embedded System

Andrew Tzer-Yeu Chen[1]([⊠]), Morteza Biglari-Abhari[1],
Kevin I-Kai Wang[1], Abdesselam Bouzerdoum[2],
and Fok Hing Chi Tivive[2]

[1] Department of Electrical and Computer Engineering,
University of Auckland, Auckland, New Zealand
{andrew.chen,m.abhari,kevin.wang}@auckland.ac.nz
[2] School of Electrical, Computer and Telecommunications Engineering,
University of Wollongong, Wollongong, Australia
{bouzer,tivive}@uow.edu.au

Abstract. Gender recognition has applications in human-computer interaction, biometric authentication, and targeted marketing. This paper presents an implementation of an algorithm for binary male/female gender recognition from face images based on a shunting inhibitory convolutional neural network, which has a reported accuracy on the FERET database of 97.2 %. The proposed hardware/software co-design approach using an ARM processor and FPGA can be used as an embedded system for a targeted marketing application to allow real-time processing. A threefold speedup is achieved in the presented approach compared to a software implementation on the ARM processor alone.

Keywords: Real-time · Embedded system · Computer vision · FPGA · Neural network · Co-design · Hardware acceleration

1 Introduction

Gender recognition has important applications for developing computer systems that are better able to identify and interact with humans, from biometric authentication to targeted marketing and advertising. However, this is a non-trivial task as there are many variations in facial features to recognise, as well as other environmental conditions that can make accurate characterisation difficult and increase the computational complexity. Even for humans, accurate gender recognition can be challenging as elements of physical appearance derived from genetic makeup, such as bone structure, may not be accurate indicators of gender or gender preference, and can lead to erroneous identification.

A targeted marketing application is our focus, where we aim to determine certain demographic characteristics of the individuals at an intersection where they can see a digital billboard, so that the digital billboard can show the most appropriate advertisement for the audience. In this application, a real-time embedded system implementation is required; images must be processed in real-time so that the right advertisement is shown to the current audience, but because it is an embedded implementation, power consumption and system cost should also be minimised.

© Springer International Publishing Switzerland 2016
H. Fujita et al. (Eds.): IEA/AIE 2016, LNAI 9799, pp. 541–552, 2016.
DOI: 10.1007/978-3-319-42007-3_47

The detection of human faces through computer vision is well established. However, we often need more information about the audience; gender recognition is one step towards more intelligent applications. There are a number of gender recognition algorithms in the literature. One of them is an algorithm developed by Tivive and Bouzerdoum [1], which uses a shunting inhibitory convolution neural network [2] to identify faces in images and classify them as male or female. The neuron model used in the neural network is based on excitatory and inhibitory weights, which is a biologically plausible explanation for how the brain processes visual images [3]. The network is also structured to allow for a certain degree of shift and distortion invariance; critical for processing real-world images that may not be as controlled as researchers would like them to be. This algorithm achieved a reported accuracy on the Facial Recognition Technology (FERET) [13] database of 97.2 %.

This paper focuses on the gender recognition part of the algorithm presented in [1], which was developed and trained in MATLAB, and investigates the feasibility and performance of implementation in an embedded system. Therefore, the first step is to implement the algorithm in C. To satisfy the real-time constraints on an embedded platform and reduce the design time, a HW/SW co-design approach has been adopted in our implementation. A Terasic DE1-SoC development board featuring an Altera Cyclone V FPGA which includes a dual-core ARM Cortex-A9 processor is used as our implementation platform.

The remainder of the paper is organised as follows; Sect. 2 discusses the motivating application in more depth and Sect. 3 investigates some of the related work in this area. Section 4 describes the steps taken to implement the algorithm and the portions of the algorithm targeted for hardware acceleration on the FPGA fabric, Sect. 5 presents the experimental results, and Sect. 6 discusses areas for future work.

2 Motivating Application

At a busy intersection, a digital billboard is mounted on one of the buildings so that it is visible to approaching vehicles. The advertisement on the billboard changes once every ten seconds; however, at this stage, the billboard cycles through a predefined set of advertisements. Decades of research have shown that there are differences in consumer behaviour and preferences between the two main gender types, male and female [4]. Marketers are therefore interested in targeting their advertisements towards specific genders; showing the wrong advertisement is a waste of time and money, and results in an inefficient advertising spend.

A camera could be placed above the billboard, scanning the faces of front-seat occupants of approaching vehicles and pedestrians. Using a real-time gender recognition algorithm, the percentage of males and females (the gender distribution) currently looking at the billboard can be determined. An appropriate advertisement can be selected that better targets that particular audience, turning the passive billboard into a more active advertising medium. This notion of a "smart billboard" is an example of intelligent systems; adding computing capabilities to an otherwise static system. Additionally, by counting the number of faces, client companies could be billed more accurately for the number of actual impressions made.

In order to achieve this, the gender distribution of the audience must be identified in real-time. This is not difficult for a single face, but more challenging when there are many faces, changing at high speed as vehicles move through the intersection. As the faces are moving, there may also be blurring effects that make some images unusable, so gender recognition may need to be performed on the same face multiple times in different positions and orientations in order to achieve accurate identification.

During peak times, assuming the vehicles are not currently stopped at a traffic light, a busy intersection could have as many as a hundred unique individuals travelling through the intersection, or in marketing terms, a hundred impressions, every ten seconds. This means that, assuming that the extraction of faces is dealt with by another processor, a gender recognition embedded system in this application should achieve at least ten successful identifications per second in order to provide an accurate gender distribution to inform advertisement selection.

3   Related Works

Ng et al. [5] presented a comprehensive survey of vision-based human gender recognition, which shows a large amount of activity in this area. They report that a human can achieve roughly 95 % accuracy in male/female binary gender recognition. In computer vision, 99.8 % accuracy has been achieved in controlled environments [6], while in uncontrolled environments the accuracy is up to 95 % [7]. Ng et al. also describe potential applications, in particular demographic classifiers for customer relationship and marketing systems, which require the ability to process 15–20 images per second.

In the embedded context, there have been a few implementations of gender recognition algorithms. Perhaps the most significant is Azarmehr et al. [8], which uses a Support Vector Machine (SVM) and Radial Basis Function (RBF) Classifier on a 1.7 GHz quad-core Snapdragon 600 SoC to characterise gender in 2.3 ms per image with 95 % accuracy. Their algorithm also detects faces and characterises age, with an average performance of 15 to 20 frames per second.

Irick et al. briefly report in [9] an SVM algorithm implemented purely on an FPGA, achieving only 88 % accuracy but processing a massive 1,100 images per second at 100 MHz. Irick et al. also reported in a separate paper [10] an artificial neural network (ANN) based system implemented on an FPGA that achieves an accuracy of 83.3 %, processing roughly 30 images per second. Ratnakar and More [11] report an FPGA-based system that achieves 78 % accuracy with a "propagation delay" of 1.9 s.

However, there are few implementations in the existing literature bringing these two paradigms together – utilising a hard processor core to better implement floating point arithmetic and maintain precision and accuracy while leveraging the FPGA fabric to improve throughput in order to meet real-time requirements. An important example is Gudis et al. [12], but in general there are gaps in the literature. There are also few implementations of gender classification using ANNs in an embedded context. While many gender recognition algorithms use SVMs for higher accuracy and modelling flexibility, the ANN can characterise multiple outputs based on multiple input factors or features, and is more suited for fixed hardware implementations that seek to avoid unused capacity or reliance on dynamic reconfiguration. To our knowledge, this paper describes the first implementation of an ANN-based gender recognition algorithm in an embedded system using both a hard core processor and FPGA fabric.

4 Algorithm Implementation

To improve the real-time performance of the software implementation of the gender recognition algorithm, one option is a hardware-only implementation using a hardware description language. However, this requires a large amount of development time and may use a lot of hardware resources for certain operations, such as floating point calculations. Considering the availability of FPGA chips which have hard core processors as well as configurable FPGA resources, HW/SW co-design can be a better approach.

The DE1-SoC development board is used as the target platform with a Cyclone V 5CSEMA5F31C6 device which has a dual-core ARM Cortex A9 (as hard processor system or HPS) and FPGA logic cells, DSP blocks, and memory resources. This allows a developer to easily segment an application, leveraging the flexibility of higher-level programming of the hard processor system as well as the reconfigurability and parallelism provided through the FPGA resources.

To compare the performance of the algorithm in an embedded context, two versions of the code are developed; one that executes purely on the ARM processor (i.e. in software), and one that executes on the ARM processor with some parts offloaded to the FPGA (i.e. software-hardware co-design). The original algorithm uses a number of built-in MATLAB functions, such as *imfilter*, which have to be rewritten from first principles in C. After the software-only implementation is complete, execution profiling is used to identify the bottlenecks, which are suitable candidates for hardware acceleration on the FPGA fabric.

The implemented algorithm has three main stages, as depicted in Fig. 1, where each stage implements one layer of the ANN, depicted in Fig. 2. The first layer (filtering) uses Gabor filters for multi-scale oriented feature extraction (the circular and regular Gabor filters have the same steps and structure but different coefficients), the NAKA-Rushton equation for contrast enhancement, and local averaging for smoothing. The hidden layer (feature detection) uses adaptive masks and activation functions based on the shunting inhibition neuron model [2], whose weights are learned from training data. The output layer (gender classification) filters the outputs of the previous stage by a set of trained weights to determine the likelihood that the face is male and the likelihood that the face is female. The outputs of the classifier are scaled to the range -1 to 1, where -1 to 0 indicates female and 0 to 1 indicates male. The classifier output can also be interpreted as probabilities that a face is male or female; for example, a gender score of 0.67 could be interpreted as a 67 % likelihood of the face being male.

As shown in Table 1, execution profiling using GNU gprof revealed that the primary bottleneck is the Gabor filters. Figure 3 shows how the Gabor filter uses a 5×5 window with real and imaginary components and therefore has a computational complexity of $\Theta(50N)$ for each pass (plus sum and absolute value operations), where N is the number of pixels in the image.

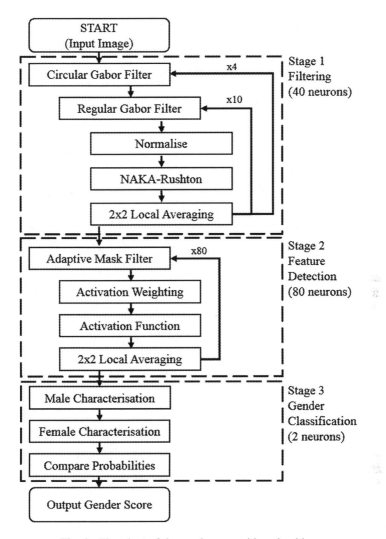

Fig. 1. Flowchart of the gender recognition algorithm

Table 1. Execution profiling of the algorithm on the ARM processor, executed over 62 iterations/images

Function	Time per call (ms)	# of calls	Total time (s)	Time (%)
Circular Gabor and Gabor filters	**2.4**	**2728**	**6.56**	**73.87**
Adaptive mask filter	0.33	4960	1.66	18.69
2 × 2 Local averaging	0.04	7440	0.29	3.27
NAKA-Rushton equation	0.06	2480	0.14	1.58
Normalisation	0.05	2480	0.12	1.35
Activation function	0.01	4960	0.06	0.68
Activation weighting	0.01	4960	0.04	0.45

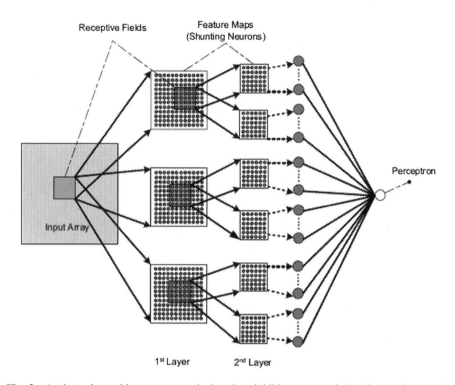

Fig. 2. A three layer binary-connected shunting inhibitory convolutional neural network (SICoNNet), from [1]

This is especially significant as the first stage of the algorithm requires 44 passes of the filter with various sets of coefficients for each image. This operation became the primary target for hardware acceleration, as the operations on the individual pixels can be executed in parallel in a single logical cycle using combinational circuits, reducing the computation time to $\Theta(N)$ (plus sum and absolute value operations). However, it is important to note that this introduces data transmission overheads between the hard core processor and the FPGA fabric, so the cost must be weighed against the benefits.

The filter described in VHDL is a kernalised correlation filter designed to complete part of the *imfilter* function from MATLAB. When passed a set of imaginary and real coefficients (which are stored in memory as fixed point numbers), the filter does the required multiplication operations, sums the products, and then calculates the absolute value by determining the magnitude of the imaginary and real sums. The filter is simulated in Modelsim and tested.

After implementing the filter in VHDL, the challenge becomes passing the data between the hard processor and the FPGA in an efficient manner. Iterating through the image is controlled by the HPS (i.e. ARM Cortex A9), with pixel and coefficient values passed to the FPGA. The HPS-FPGA bridge, which uses the AMBA AXI bus protocol, can at times be the bottleneck, as the handshaking required to retrieve data from the HPS memory and pass it to the FPGA is non-negligible. A number of steps are taken to

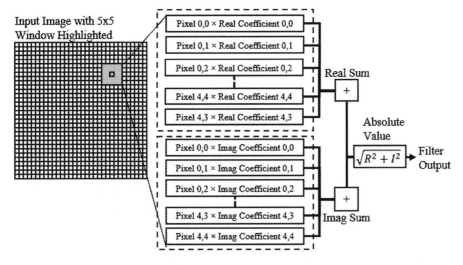

Fig. 3. Diagram of Gabor filter operation

mitigate this issue; firstly, the coefficient values and pixel values are concatenated as much as possible to use the full 32-bit bus (also known as data packing), and a shifting window (or sliding window) is used on the filter to minimise the number of data transfers required between the HPS and the FPGA fabric.

Finally, the dual-core nature of the HPS is leveraged by dividing the algorithm into two threads, each responsible for the computations of half of the processing units, running independently to ensure no race conditions. This also utilises two identical filters on the FPGA to calculate output values independently. The overall architecture is presented in Fig. 4. Using this approach, the computation time of the algorithm can be significantly reduced by using a processor with a larger number of cores in order to run more threads in parallel.

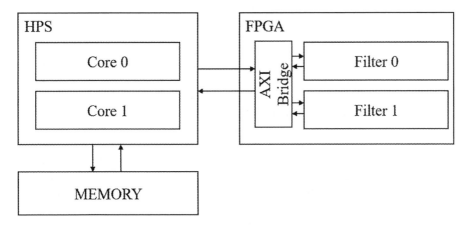

Fig. 4. Computer architecture of dual-core HPS-FPGA system

5   Results

Two test systems are set up: a desktop PC running Cygwin in Windows 7 on a 3.60 GHz i7 processor, and a DE1-SoC development board with Cyclone V FPGA (5CSEMA5F31C6) from Altera, which has a dual-core ARM Cortex-A9 processor running Linux at 400 MHz and FPGA logic running at 100 MHz.

A test set of 62 cropped images from the FERET database [13], with 30 male images and 32 female images, is used on each platform. This test set was provided in the demonstration code for [1], allowing a fair comparison of performance. The images are resized to 32×32 pixels and converted to greyscale before the processing begins, then pre-loaded into memory for the purposes of testing the algorithm speed. As shown in Table 2, the final implementation using two threads with a shifting window filter achieves a threefold speedup in comparison to the implementation that uses a single core on the HPS only. The performance of each iteration of the system is also included to show how each optimisation improves the performance. In each case the software optimisation from the gcc compiler is left at the default –O0.

Table 2. Execution times for all 62 test images, and per image

Implementation	Total execution time (s)	Execution time per image (ms)
Desktop PC – MATLAB	8.50	137.10
Desktop PC – C	0.58	9.35
ARM processor (HPS) only (single core)	**8.88**	**143.23**
Unoptimised HPS-FPGA implementation	27.03	435.97
Shifting window HPS-FPGA implementation	6.08	98.06
Dual-core shifting window HPS-FPGA implementation	**2.99**	**48.23**

The actual speed of the image processing is less important than the speed-up; images can be processed faster with higher clock frequencies or more cores, but it is important that a significant speed-up can be achieved by leveraging intelligently implemented hardware acceleration on an FPGA with relatively low development time and cost (compared to pure hardware implementation of the algorithm). With an execution time of 3 s per image, using this implementation we can process 20 faces per second, which is double the rate required for the motivating application.

As shown in Table 3, the speed-up is largely attributable to the fact that the Gabor Filter calculations are now performed in hardware. As the filter on the FPGA fabric is implemented combinationally, results are available one logical clock cycle after all inputs are provided, i.e. 10 ns in a 100 MHz system. The bottleneck becomes the data transmission between the HPS and FPGA rather than the computation itself. As mentioned previously, the overhead of the HPS-FPGA bridge is significant. This is a good place to start for future optimisations. However, it is important to note that the

Table 3. Execution times for the Gabor Filters in Stage 1 of the algorithm

Implementation	Filter total time (s)	Filter execution time (%)
Desktop PC – C	0.38	65.52
ARM processor (HPS) Only (single core)	**6.56**	**73.87**
Unoptimised HPS-FPGA implementation	24.68	91.31
Shifting window HPS-FPGA implementation	3.77	62.01
Dual-core shifting window HPS-FPGA implementation	**1.97**	**65.89**

proportion of the total execution time that is spent on the filter operation is similar for both the Desktop PC (65.52 %) and final embedded implementation (65.89 %).

Importantly, the loss of precision when moving from the MATLAB algorithm to the HPS+FPGA implementation is small and in many cases negligible. In order to save on computation resources and memory on the FPGA, some of the floating point operations were converted to fixed point.

As shown in Fig. 5, for cases where the gender is more certain (i.e. the absolute value for the score of the detected gender is larger than 0.9), the loss of precision is in the order of 1 % or less. As the gender becomes less certain, the error increases and can be as high as 10 %. Overall, the accuracy of the algorithm remains 96–97 %. Also as shown in Fig. 5, the algorithm is capable of working on a variety of image conditions, with different face orientations/poses and lighting, as well as artefacts such as glasses and beards. However, when the image is not cropped properly, and contains either part of a single face or part of more than one face, then the ability of the algorithm to make a robust characterisation of gender decreases (e.g. the last example).

6   Future Work

Future development can consider new applications as well as further improving the performance and energy efficiency of the algorithms. Further investigation should be done into improving the data transmission rate between the HPS and FPGA fabric, as this has become a significant bottleneck in the system. For example, the FPGA could be given access to the main memory as done in [12], and simply passed an address from the HPS, so that the FPGA can then retrieve 25 or 50 contiguous values directly from memory. Alternatively, a point-to-point connection could be used between the HPS and FPGA fabric.

An important application of this work may be in facial recognition systems; if an algorithm is attempting to match a face found in an image to a database of known faces, then determining the gender first as a top-level characteristic can greatly reduce the search space which may result in saving time and energy. This can be combined with other facial characteristics such as age category and hair or skin colour or tone to greatly speed up facial recognition in large databases. However, there is an important limitation; since the original neural network was trained with mostly up-right frontal face images, the gender detection algorithm may fail in situations where the camera has

	MATLAB	HPS+FPGA
	0.99973	0.99964
	-0.99524	-0.99623
	0.92512	0.89070
	-0.85450	-0.86704
	-0.70115	-0.75620
	0.58971	0.52091

Fig. 5. A sample of face images and MATLAB/Embedded gender scores – scores larger than 0 are male, scores less than 0 are female. Note that the model has low confidence where the absolute value of the gender score is lower than 0.8, such as in the last two examples. This is usually when part of the face has been obscured, or there are in fact multiple faces in the image.

an oblique or side view of the face. Since faces in the real world cannot be constrained to always be facing the camera, the algorithm should potentially be retrained to include faces in different orientations. Alternatively, multiple networks can be trained depending on the view of the face, with the weights of the neurons stored in the HPS memory. If a face detector can also determine the orientation of the face, then we can follow the same procedure as described in this paper, but loading different weights as required.

As discovered through implementing the algorithm in two threads, the algorithm is highly parallelisable as each individual neuron could be computed independently, i.e. there are no dependencies between neurons. In this paper we have not considered mapping the ANN structure directly to the FPGA fabric. More parts of the algorithm could be shifted to the FPGA or more cores could be used to leverage more parallelism. However, this should be done only if the performance gain of hardware acceleration is larger than the overhead loss of transmitting data between the HPS and FPGA.

7 Conclusion

In this paper, we have presented an embedded implementation of real-time gender recognition for a targeted marketing application, where the efficacy of a billboard can be improved by determining the gender distribution of the audience. A software-hardware co-design approach is taken to optimise the throughput of a convolutional neural network-based gender recognition algorithm while maintaining a high level of accuracy so that it can operate in real-time. After porting the algorithm from MATLAB to C, the main bottleneck is identified using execution profiling. By moving the Gabor filter into hardware on the FPGA and performing further optimisations such as data packing and using a shifting window, a threefold speedup is achieved compared to a software implementation on an ARM processor alone. This allows 20 faces to be processed per second on an embedded platform, double the throughput required in the motivating application. This implementation satisfies the embedded requirements of the target application.

References

1. Tivive, F.H.C., Bouzerdoum, A.: A gender recognition system using shunting inhibitory convolutional neural networks. In: International Joint Conference on Neural Networks, pp. 5336–5341. IEEE Press, New York (2006)
2. Tivive, F.H.C., Bouzerduom, A.: Efficient training algorithms for a class of shunting inhibitory convolutional neural networks. IEEE Trans. Neural Netw. 16(3), 541–556 (2005). IEEE Press, New York
3. Fregnac, Y., Monier, C., Chavane, F., Baudot, P., Graham, L.: Shunting inhibition, a silent step in visual computation. J. Physiol. 97, 441–451 (2003)
4. Wolin, L.: Gender issues in advertising—an oversight synthesis of research: 1970–2002. J. Advert. Res. 43, 111–129 (2003)
5. Ng, C.B., Tay, Y.H., Goi, B.M.: Recognizing human gender in computer vision: a survey. In: 12th Pacific Rim International Conference on Artificial Intelligence: Trends in Artificial Intelligence, pp. 335–346 (2012)
6. Zheng, J., Lu, B.: A support vector machine classifier with automatic confidence. Neurocomputing 74(11), 1926–1935 (2011)
7. Shan, C.: Learning local binary patterns for gender classification on real-world face images. Pattern Recogn. Lett. 4(33), 431–437 (2012)

8. Azarmehr, R., Laganiere, R., Lee, W.S., Xu, C., Laroche, D.: Real-time embedded age and gender classification in unconstrained video. In: Conference on Computer Vision and Pattern Recognition Workshops, pp. 56–64. IEEE Press, New York (2015)

9. Irick, K. M., DeBole, M., Narayanan V., Gayasen, A.: A hardware efficient support vector machine architecture for FPGA. In: International Symposium on Field-Programmable Custom Computing Machines, pp. 304–305. IEEE Press, New York (2008)

10. Irick, K., DeBole, M., Narayanan, V., Sharma, R., Moon, H., Mummareddy, S.: A unified streaming architecture for real time face detection and gender classification. In: International Conference on Field Programmable Logic and Applications, pp. 267–272. IEEE Press, New York (2007)

11. Ratnakar, A., More, G.: Real time gender recognition on FPGA. Int. J. Sci. Eng. Res. **6**(2), 19–22 (2015)

12. Gudis, E., Lu, P., Berends, D., Kaighn, K., van der Wal, G., Buchanan, G., Chai S., Piacentino, M.: An embedded vision services framework for heterogeneous accelerators. In: Conference on Computer Vision and Pattern Recognition, pp. 598–603. IEEE Press, New York (2013)

13. Phillips, P.J., Moon, H., Rauss, P.J., Rizvi, S.: The FERET evaluation methodology for face recognition algorithms. IEEE Trans. Pattern Anal. Mach. Intell. **22**(10), 1090–1104 (2000)

Style-Me – An Experimental AI Fashion Stylist

Haosha Wang[1(✉)], Joshua De Haan[2(✉)], and Khaled Rasheed[1(✉)]

[1] Institute for Artificial Intelligence, Franklin College,
The University of Georgia, Athens, GA, USA
{hswang,khaled}@uga.edu
[2] Terry College of Business, The University of Georgia,
Athens, GA, USA
jshdhn@uga.edu

Abstract. "Style endures as it is renewed and evolved" believed French fashion designer Gabrielle Chanel [1]. In this study, we propose an AI based system called "Style-Me" as our answer to the question "Can an AI machine be a fashion stylist?" Style-Me is a machine learning application that recommends fashion looks. More specifically, Style-Me learns user preferences through the use of Artificial Neural Networks (ANN). The system scores user's customized style looks based on fashion trends and users' personal style history. Although much remains to be done, our implementation shows that an AI machine can be a fashion stylist.

Keywords: Artificial Intelligence · Neutral network · Clothing styling

1 Introduction

Human creativity as one of the major challenges for the AI domain has captured the world's attention for years. Artist Harold Cohen's AI artist program, "AARON", was the first profound connection between AI and human creativity and has been in continual development since its creation in 1937 [5]. "JAPE" (Joke Analysis and Production Engine), is another example of an AI imitating human creativity. In this case, computer program generates punning riddles modeling human humor [2]. Among all of these domains of human creativity, the fashion industry's unpredictable irrationality, individual uniqueness and cultural dependence make human fashion behavior modeling one of biggest challenges in this area [3]. In a previous study [11], we compared and summarized earlier previous studies using AI techniques in the fashion domain. It provides a foundation on the design and development of our system, which we call "Style-Me". Generally speaking, a full product level system requires a large amount of data and takes significant time to build. However, the aim of this work is to present the essence of Style-Me and the major AI techniques which have been implemented.

In this study, we created a manageable database which contains 32 dresses and 20 shoes for 4 different events, encode a standard style rules engine, generate 640 looks

For further inquiry, please reach out to Haosha Wang at hswangelsa@gmail.com or Joshua De Haan at jshdhn@gmail.com as our UGA sponsored emails are deactivated after graduation.

H. Fujita et al. (Eds.): IEA/AIE 2016, LNAI 9799, pp. 553–561, 2016.
DOI: 10.1007/978-3-319-42007-3_48

and rank them by a final score in descending order. The score indicates how fashionable each look is based on users' feedback. The learning component trains an Artificial Neural Network (ANN) to learn users' personal preferences and adjust the final score. Moreover, the system provides a feature that allows users to customize a fashion look and then computer evaluates it. This feature provides a shopping guide to inform users' purchase plans. As mentioned in previous work, the Mobile Fashion Advisor (MFA) system [4] also targets assisting users shopping. The differences between MFA and Style-Me are, firstly that MFA only tells users whether this is a new item that could go with an existing item, while Style-Me provides a numeric evaluation for the users' preferences and secondly that Style-Me adapts to users' personal preferences, a feature not included in MFA. On the front-end of the Style-Me system, users initialize Style-Me by taking a fashion personality quiz and then the user has the choice to agree with the quiz result or re-do it. The User Interface (UI) design of Style-Me follows minimalistic and intuitional style, which gives users a smooth experience without instructions.

This paper presents Style-Me in five sub-sections: System Overview; Data Preparation; Experiment on Model Selection; Implementation and User Interface and Summary.

2 System Overview

Users initialize the system by taking a fashion personality quiz. The system assigns 1 of 6 standard styles as the users' initial style. Standard styles are "Classic", "Dramatic", "Gamin", "Ingénue", "Natural" and "Romantic". For each style, the system has a set of default database and styling rules for four common events. Comment events include "Cocktail", "Informal", "Formal" and "Office". For convenience in presentation the core content, we use a "Classic" style dataset in this experiment. According to the style rules, the Style Engine pairs up dress and shoes and assigns initial scores on each pair. The ANN model implemented in the Learning Components is a Multilayer Perceptron model. In this study, we experiment with various models utilizing WEKA (Waikato Environment for Knowledge Analysis), which is a Machine Learning algorithm collection and data-preprocessing tool for data mining experimental usage [12]. Details of our experiments are in Sect. 4. In Sect. 5, we demonstrate the UI design of Style-Me. One is a styling system that recommends users fashion looks and collects users' interactions. The other one is a scoring system that evaluates users' customized looks with a score to assist users' shopping plans. There is still much that remains to be done and our model is not comprehensive enough for full product level use.

3 Data Preparation

The purpose of the learning task is to predict a pair's score based on correlations between scores and items' attributes. We followed several steps preparing and pre-processing the dataset.

First of all, we collected 32 dresses and 20 shoes in typical standard style from various websites and designers' collections. Each clothes item has 12 attributes and each shoes item has 9 attributes. Then we collected 20 standard rules for the "Classic" database. Every rule interprets a relationship between pairs of attributes. The relationship is binary: "AND/OR" means "Positive" while "NOT" means "Negative". The style engine matches pairs of attributes to styling rules and counts the number of matches as follows.

$$\text{If relationship is postive, } C_{Likes} + 1; C_{Dislike} + 0;$$

$$\text{If relationship is negative, } C_{Likes} + 0; C_{Dislike} + 1; \tag{1}$$

$$Score_{pair} = \frac{C_{Likes} - C_{Dislikes}}{C_{Likes} + C_{Dislikes} + 1} \tag{2}$$

Boredom is one of the major factors in every fashion model [9]. In fashion styling for events, wearing the same looks several times is not fashionable. So, we add counts of "wear" in our model as a parameter of "boredom". Currently, we assigned 0.03 to the "wear" weight, but the optimal value for the weight of the boredom factor is an interesting topic for future study. So, the popularity for each pair in the recommender system is computed as:

$$Popularity = Score - 0.03 * C_{Wears} \tag{3}$$

The output of the Style Engine is the "pair" table that stores the pair data and pairs' score.

4   Experiments on Model Selection

We conducted several experiment in WEKA to select the best model for our dataset. WEKA is the abbreviation for Waikato Environment of Knowledge Analysis, a collection of Machine Learning algorithms and data preprocessing tools developed by the University of Waikato [12]. The learning task here is to predict the final score based on correlation with items' attributes.

4.1   Experiment 1: Model Comparison

Every machine learning application starts with choosing the right model. In the first experiment, we test five classifiers on the initial dataset with 10-fold cross validation in WEKA to find out the most suitable model for our dataset.

Here is a brief introduction of the other methods besides Linear Regression in the experiment. *SMOreg* is a regression model that implements a sequential minimal-optimization algorithm for learning [10]. There are two tree classifiers as well, *M5P*, which is a model tree learner to predict the value for a test instance and REPTree, a fast decision tree learner which builds a tree using information gain and variance and to prune the tree with reduced-error pruning [12]. Lastly, *Multilayer Perceptron (MLP)*, a feedforward Artificial Neural Network model using a backpropagation algorithm to classify instances.

The model inputs are 21 attributes, 12 from dress and 9 from shoes respectively. The output is a styling score range from -1 to 1. The initial score is initialized by style rules. There is no feature selection algorithm applied in this experiment. The experimental results as shown in Fig. 1. MLP achieves the highest correlation coefficient among all the other models.

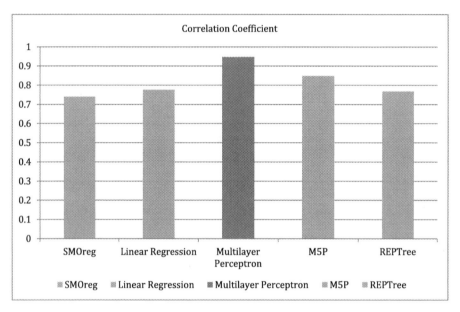

Fig. 1. Correlation Coefficient of seven different methods on initialized dataset (Color figure online)

4.2 Experiment 2: Hidden Units and Hidden Layers

We choose the MLP model that has the highest correlation coefficient in Experiment 1. The MLP with few hidden layer are proven as a universal approximator [6]. The second experiments is trying to decide the number of hidden layers and hidden units.

Firstly, we run an experiment of the number of hidden units on one hidden layer with a 0.3 Learning rate and momentum term of 0.2 (*weka.classifier.functions.MultiplayerPerceptron –L 0.3, -M 0.2, -N 500, -V 0, -S 0 –E 20 –H **). Our experiment has tested the number of hidden units from 3 to 30 (Table 1). For a single hidden layer model, 20 units achieved the highest correlation coefficient, 0.9496.

Secondly, we run an experiment on the number of hidden units in two hidden layers with a 0.3 learning rate and momentum term of 0.2 (*weka.classifier.functions.MultiplayerPerceptron –L 0.3, -M 0.2, -N 500, -V 0, -S 0 –E 20 –H **). Our experiment has tested the same hidden units in each layer (Table 2).

Table 1. Number of hidden units on one hidden layer

Number of Hidden Units	Correlation Coefficient
3	0.7578
5	0.8133
10	0.8894
20	**0.9496**
25	0.9477
30	0.9411

Table 2. Number of Hidden units on two hidden layers

Number of Hidden Units	Correlation Coefficient
3,3	0.8208
5,5	0.869
10,10	0.9169
20,20	0.9403
30,30	0.9525
35,35	**0.9619**
40,40	0.9492

This experiment shows that two hidden layers with 35 hidden units on each layer achieved the highest correlation coefficient, 0.9616. However, it takes a long time to build this model. Considering user experience, we have decided to use a single layer with 20 hidden units model in our implementation.

4.3 Experiment 3: Learning Rate

Learning rate is a decreasing function of time that determines the step size in the gradient descent search [8]. When we learn to do something new, we are very inefficient at the beginning and our efficiency gets better with more practice. Learning rate is a mathematically measure of this learning phenomenon. For a learning model, the lower learning rate the better [12].

We test the learning rate on one of the highest correlation coefficient models from the above experiment, one hidden layer with 20 hidden units and a momentum term of 0.2 (*weka.classifier.functions.MultiplayerPerceptron −L *, -M 0.2, -N 500, -V 0, -S 0 − E 20 −H 20*). The best learning rate on this data set is 0.3 (Fig 2).

4.4 Experiment 4: Momentum Term

Momentum is a technique that has been used to speed up convergence and avoid local minima. There are many ways to use it for improving the performance of the back-propagation algorithm [7]. We tested the momentum term from 0.1 to 0.9 for the model

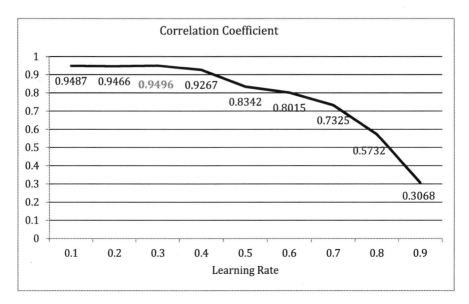

Fig. 2. Correlation Coefficient achieves the highest at Learning Rate = 0.3

of one hidden layer with 20 hidden units and 0.3 learning rate (*weka.classifier.functions.MultiplayerPerceptron –L 0.3, -M *, -N 500, -V 0, -S 0 –E 20 –H 20*). The Experiment shows that this model has the highest correlation coefficient with a momentum rate of 0.1 (Fig. 3).

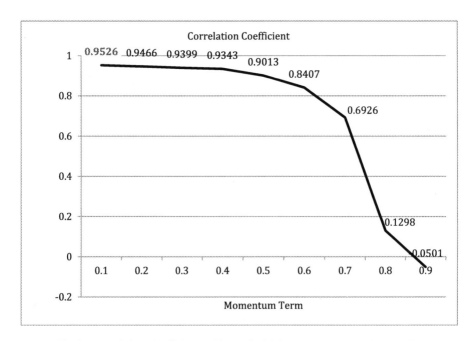

Fig. 3. Correlation Coefficient achieves the highest at Momentum Term = 0.1

5   Implementation and User Interface

We have implemented the selected MLP model in the Style-Me's demo. In the demo implementation, there are four parts: a database, a rule based recommender system, a learning component and a scoring system. The Styme-Me demo is a Java Servlet application written in Java and HTML, which uses MySQL as the back-end database and Apache Tomcat as web server.

We use the WEKA library (weka.jar) from WEKA. The UI is built with the Windowbuilder[1] from Eclipse. The design is simple and intuitive.

Fig. 4. UI of Main Program

---

[1] Windowbuilder: http://www.eclipse.org/windowbuilder/.

In the main program, there are two drop-down menus asking users' choices of "event" and "dress color" (Fig. 4). There are also two buttons in the UI. Users can click the "Style-Me" button to see recommended fashion looks and the "Reset DB" button to reset the system back to its initial state.

In the scoring program (Fig. 5), there are 14 drop-down menus for users to select garment attributes. After selecting attributes, users click the "score" button and the system will output a score to the screen. The score shows how fashionable the combination is ranging from -1 to 1. The score is computed based on users selections and users preference history from Main Program.

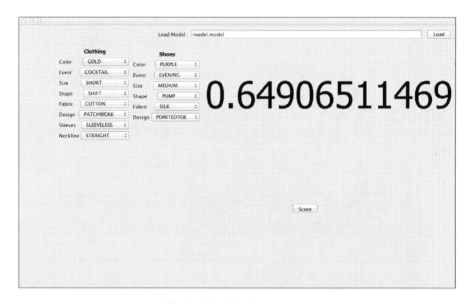

Fig. 5. UI of Scoring program

6   Summary

This study presents an experiment and implementation of AI techniques to fashion styling. In this study, we demonstrate a completed process for garment computational representation, computational styling modeling, model experiment and implementation. Our model and implementation show the capability of modeling users' preferences.

There is more work that needs to be done. The first priority is that more data and styling rules need to be collected. We would also like to conduct more experiments on modeling and implementation.

With the growing popularity of online shopping and trends in millennial consumer's shopping behavior, we believe intelligent systems such as Style-Me will become an essential form of interaction between consumers and companies, as well as a an important factor in marketing and branding efforts.

References

1. Barry, J.: Interview with Gabrielle Chanel. Interview with Chanel, 170. (G. Chanel, Interviewer) New York City, New York, United States of America: McCall Pub. Co. (1965)
2. Binsted, K.: Machine humour: an implemented model of puns. Ph.D. thesis, Univeristy of Edinburgh (1996)
3. Boden, M.A.: Creativity and artificial intelligence. Artif. Intell. **103**, 347–356 (1998)
4. Cheng, C.-I., Liu, S.-M.: Mobile fashion advisor - a nobel application in ubiquitous society. Int. J. Smart Home, **2**, 59–76 (2008)
5. Cohen, H.: The further exploits of AARON, painter. In: Franchi, S., Güzeldere, G. (eds.) Constructions of the mind. artificial intelligence and the humanities. vol. 4. Stanford Humanities Review (1995)
6. Hall, M., Frank, E., Holmes, G., Pfahringer, B., Reutemann, P., Witten, I.H.: The WEKA data mining software: an update. SIGKDD Explor. **11**(1), 10–18 (2009)
7. Hornik, K., Stinchcombe, M., White, H.: Multilayer feedforward networks are universal approximators. Neural Netw. **2**, 359–366 (1989)
8. Istook, E., Marinez, T.: Improve backpropagation learning in neural networks with windowed momentum. Int. J. Neural Syst. **12**(3&3), 303–318 (2002)
9. Sarma, A., Gollapudi, S., Panigraphy, R., Zhang, L.: Understanding cyclic trends in social choices. In: WSDM, Seattle, WA, USA (2012)
10. Shevade, S., Keerthi, S., Bhattacharyya, C., Murthy, K.: Improvements to the SMO algorithm for SVM regression. IEEE Trans. Neural Netw. **11**, 1188–1193 (2000)
11. Wang, H., Rasheed, K.: Artificial intelligence in clothing fashion. In: International Conference On Artificial Intelligence, Las Vegas (2014)
12. Witten, I., Frank, E., Hall, M.: Data Mining: Practical Machine Learning Tools and Techniques, 3rd edn. Morgan Kaufmann Publishers Inc., San Francisco (2000)

Reduction of Computational Cost Using Two-Stage Deep Neural Network for Training for Denoising and Sound Source Identification

Takayuki Morito[1(✉)], Osamu Sugiyama[1], Satoshi Uemura[1],
Ryosuke Kojima[1], and Kazuhiro Nakadai[1,2]

[1] Graduate School of Information Science and Engineering,
Tokyo Institute of Technology, 2-12-1 Ookayama, Meguro-ku, Tokyo, Japan
morito@cyb.mei.titech.ac.jp
[2] Honda Research Institute Japan Co., Ltd.,
8-1 Honcho, Wako-shi, Saitama, Japan

Abstract. This paper addresses reduction of computational cost in training of a Deep Neural Network (DNN), in particular, for sound identification using highly noise-contaminated sound recorded with a microphone array embedded in an Unmanned Aerial Vehicle (UAV), aiming at people's voice detection quickly and widely in a disastrous situation. It is known that a DNN training method called end-to-end training shows high performance, since it uses a huge neural network with high nonlinearity which is trained with a large amount of raw input signals without preprocessing. Its computational cost is, however, expensive due to the high complexity of the neural network. Therefore, we propose two-stage DNN training using two separately-trained networks; denoising of sound sources and sound source identification. Since the huge network is divided into two smaller networks, the complexity of the networks is expected to decrease and each of them can consider a specific model of denoising and identification. This results in faster convergence and computational cost reduction in DNN training. Preliminary results showed that only 71% of training time was necessary with the proposed two staged network, while maintaining the accuracy of sound source identification, compared to end-to-end training using noisy acoustic signals recorded with an 8 ch circular microphone array embedded in a UAV.

Keywords: Environment understanding · Deep learning · Sound source identification

1 Introduction

In a disastrous situation, an *Unmanned Aerial Vehicle (UAV)* is helpful to search for people since it can move quickly and widely even when traffic is cut off at places along the roads. People can be buried in the rubble, or trapped in a

© Springer International Publishing Switzerland 2016
H. Fujita et al. (Eds.): IEA/AIE 2016, LNAI 9799, pp. 562–573, 2016.
DOI: 10.1007/978-3-319-42007-3_49

collapsed building, which means that it is difficult to find them only with visual information. Thus, we are tackling to detect human utterances from an UAV with microphones although a *Signal-to-Noise Ratio (SNR)* of a target sound source is quite low due to propellers and wind noise.

To detect a sound source in such a noisy condition, Okutani et al. extended sound source localization called *MUltiple SIgnal Classification based on Generalized EigenValue Decomposition (GEVD-MUSIC)* to incrementally estimate dynamically-changing noise, which is called *incremental GEVD-MUSIC (iGEVD-MUSIC)* [8]. Since computational cost of iGEVD-MUSIC is expensive and difficult to work in real time with an embedded processor, Ohata et al. proposed *MUSIC based on incremental Generalized Singular Value Decomposition (iGSVD-MUSIC)* [7] which is much lighter than GEVD-based MUSIC as reported in [6], and they showed that a sound source 20 m away can be detected with less than 1 on the real time factor. Uemura et al. then reported sound source identification of the detected sound source, because the UAV has to distinguish a speech source from other sound sources to find the people [10]. They simply combined signal processing methods such as iGSVD-MUSIC for sound source localization and *Geometric High-order Dicorrelation-based Source Separation (GHDSS)* [5] for sound source separation with a deep learning method, *Convolutional Neural Network (CNN)*, and showed the effectiveness of CNN-based identification for the separated sound sources. However, it has many situation-dependent parameters that need to be tuned, and their exploration consumes a lot of time.

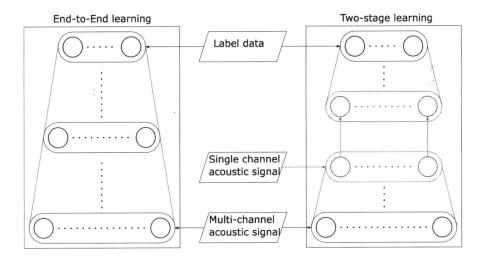

Fig. 1. Comparison of two-stage learning and end-to-end learning

Recently, many deep learning methods have been reported, and they showed high performance in various regression and identification tasks. In particular, an end-to-end training approach, which simply gives raw data as an input with the corresponding transcription to a neural network in training, is getting more common (see the left figure of Fig. 1). It basically uses a large amount of training data without assuming any preprocessing and specific model on regression/identification, and it is reported trained neural networks trained with the end-to-end approach show better performance than other methods [3]. However, it has problems as follows:

- Training is time consuming.
- It throws away application dependent knowledge.

Since a huge neural network has high non-linearity, a large amount of training data is necessary. It causes the first problem. In addition, the convergence of neural networks becomes slow, and higher computation is necessary. Also, in identification, calculations in neural networks take more time due to the use of a larger neural network, which is critical in our application in terms of real-time processing. The second one is against the concept of end-to-end training. Since end-to-end training uses raw data as an input, it does not use any knowledge depending on the application. However, we feel that such knowledge can be used effectively and efficiently. For example, in our case, after sound source detection, the use of sound source separation and sound source identification seems to be natural. When similar functions are internally trained in the end-to-end approach, it will be another choice that some knowledge on these functions is introduced in advance for more effective and efficient training.

We, therefore, propose two-stage DNN training by considering internal models to be trained for the UAV sound source identification task (see the right figure in Fig. 1). Since the input signal is highly noise-contaminated, noise suppression is necessary, and after that, sound source identification should be performed. Based on this idea, we first train a neural network for noise suppression with multi-channel acoustic signals, and then, trained another neural network for identification with a noise suppressed single-channel input. Finally these two networks were integrated into one so that it could identify sound sources with multi-channel acoustic signals. Since the number of total connections is drastically decreased, reduction in the computational cost in neural network training is expected.

The rest of this paper is organized as follows: Sect. 2 introduces our proposed method. Section 3 shows the results of evaluation experiments and discusses the effect of proposed method. The last section concludes this paper.

2   Two-Stage Training vs. End-to-End Training

Block diagrams of two-stage training (2STG) and end-to-end training (E2E) are shown in Figs. 2 and 3, respectively. Each consists of two stages; training stage and identification stage. In the training stage, the *Stacked denoising Autoencoder (SdA)* are trained using label data, single channel acoustic signals, and multi-channel acoustic signals. In the identification stage, sound source identification is performed only from multi-channel acoustic signal.

The multi-channel acoustic signals are recorded with a microphone array embedded in an UAV by emitting single-channel clean acoustic signals from a loud speaker. Label data are associated with single channel sound sources in advance. Feature extraction generates input vectors of SdA by extracting acoustic features described in Sect. 3.1. Note that we used a different dataset in training and in performance evaluation for every experiment.

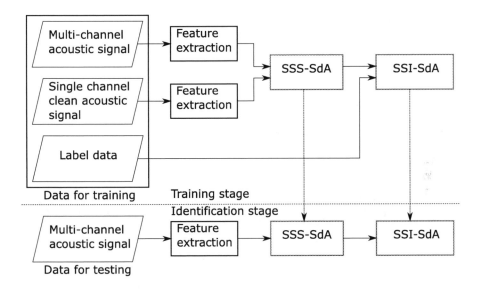

Fig. 2. A block diagram of Two-Stage training

2.1 Two-Stage Training System

In 2STG, first we trained an *SdA for Sound Source Separation (SSS-SdA)* whose input feature vector is extracted from a noise-contaminated multi-channel acoustic signal so that output can be that of the corresponding single channel acoustic signal. Secondly we trained another *SdA for Sound Source Identification (SSI-SdA)* using the output of SSS-SdA as an input. After training two SdA, we composed a sound source identification system for multi-channel signal by connecting the output of SSS-SdA with the input of SSI-SdA.

Splitting a large network into small networks is advantageous. It is well known that neural networks take a long time to converge as the number of layers is large [4]. Since this method can reduce the number of layers to be updated at once, training time of a neural network is decreased. The method fits our application. SSS-SdA is a denoising process, and thus clean single channel signals are necessary as target signals to be trained. Since the clean acoustic signals are available in advance, we can freely use them. Also the clean signals can be used as training data for SSI-SdA as additional training data, which will improve identification performance.

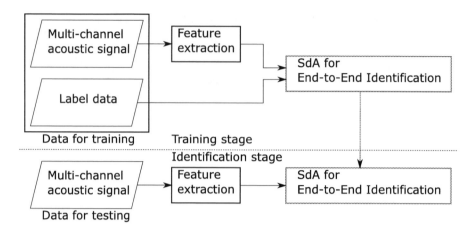

Fig. 3. A block diagram of End-to-End training

2.2 End-to-End Training System

In contrast to 2STG, E2E has only one SdA directly performing sound source identification from a feature vector of a noise-contaminated multi-channel acoustic signal.

2.3 Training with Stacked Denoising Autoencoder

Although many types of neural networks have been proposed, we selected SdA in this paper. It is known as a simple and noise-robust DNN thanks to a denoising mechanism. Training of SdA consists of two phases; pre-training and fine-tuning.

Pre-training. SdA is a multi-layer neural network with the structure of nested denoising Autoencoder (dA), which is derived from Autoencoder. We start with Autoencoder.

Autoencoder is a neural network as shown in Fig. 4, which consists of three layers, an input, output and hidden layer. The input and output layers have the same number of dimensions. The number of dimensions in the hidden layer is generally smaller than those of the other two layers. In Autoencoder, the network is trained to make the difference of the input, x, and output, z, to be smaller. Consequently, the hidden layer, h, would be a low-dimensional representation of the input layer, x.

Let the number of dimensions of the input and the output layers be d, and that of the hidden layer be d'. Let the input to the input layer, the output of the hidden layer, and the output of the output layer be $\boldsymbol{x}$, $\boldsymbol{y}$, and $\boldsymbol{z}$, respectively. $\boldsymbol{y}$ and $\boldsymbol{z}$ are calculated as:

$$y = \sigma(Wx + b), \tag{1}$$
$$z = \sigma(W^{\mathrm{T}}y + b'), \tag{2}$$

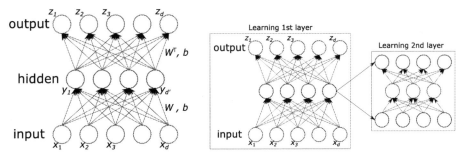

Fig. 4. Autoencoder   **Fig. 5.** Stacked denoising Autoencoder

where $\boldsymbol{\sigma}(\cdot)$, $\boldsymbol{W} \in \mathbb{R}^{d' \times d}$, $\boldsymbol{b} \in \mathbb{R}^{d}$, and $\boldsymbol{b'} \in \mathbb{R}^{d'}$ are, respectively, an element wise activation function (a sigmoid function in this paper), the weight matrix, the bias vector of the hidden layer, and that of the output layer. $\boldsymbol{W}$, $\boldsymbol{b}$ and $\boldsymbol{b'}$ are generally termed the parameters of the network. Training is performed to minimize the output of the evaluation function. In general, a mean square error function is used as the training criterion in a regression problem, while a cross entropy error function in classification problem [12]. Therefore, we adopted a mean square error function as the evaluation function for noise suppression, and a cross entropy error function for sound source identification.

Parameter updates are performed by a stochastic gradient descent [2] method, that is, parameters are updated for every N data vectors according to the following update rule:

$$\boldsymbol{W}^{new} = \boldsymbol{W}^{old} - \eta \sum_{n=1}^{N} \frac{\partial L}{\partial \boldsymbol{W}}, \tag{3}$$

$$\boldsymbol{b}^{new} = \boldsymbol{b}^{old} - \eta \sum_{n=1}^{N} \frac{\partial L}{\partial \boldsymbol{b}}, \tag{4}$$

$$\boldsymbol{b'}^{new} = \boldsymbol{b'}^{old} - \eta \sum_{n=1}^{N} \frac{\partial L}{\partial \boldsymbol{b'}}, \tag{5}$$

where η and L are, respectively, the learning rate and the evaluation function. N is referred to as batch size. An epoch is a complete pass through a given set of training data.

Denoising Autoencoder has the same structure as Autoencoder. The difference from Autoencoder is to add noise to the input during training. Since the desired output is clean data, the mapping from a noise-contaminated vector to a clean vector should be learned to make the network noise-robust. The noise was added in our experiment as follows: to train a network for noise reduction, we used multi-channel acoustic features as the input, which have already been contaminated with noise as described in Sect. 3.1 in advance. To train a network for sound source identification, we added noise with uniform distribution whose range is $\pm 20\%$ of the original value to each element of the input vector.

Stacked denoising Autoencoder is realized by replacing the hidden layer with another dA, and thus, it is a multi-layered network having a structure of nested dA shown in Fig. 5. Thanks to such a multi-layered network structure, it is possible to configure a "Deep" neural network which represents complex functions [11].

Fine-Tuning. By using the inputs and the corresponding desired outputs, i.e., training data, training the network of mapping from input to output is referred to as fine-tuning. Since all the parameters of the neural network are updated at the same time in fine-tuning, they are apt to converge to a local solution. For this reason, we first perform pre-training to set initial values to converge to a better solution.

In fine-tuning, the median layer of SdA, i.e., bottleneck layer, is regarded as the final output of the network. Usually, the numbers of dimensions in the output layer and the desired output are set to be the same in a regression problem, while in a classification problem the numbers of dimensions of the output layer and the classes to be classified are the same. In a regression problem, the output of the output layer is generally calculated with an activation function such as a sigmoid function similarly to pre-training, while a softmax function shown in Eq. (7) is mostly used as the activation function of the output layer in a classification problem [12]. An output layer with the softmax function is especially referred to as a softmax layer.

$$y = Wx + b, \tag{6}$$

$$z_i = \frac{\exp(y_i)}{\sum_{j=1}^{D} \exp(y_j)}, \tag{7}$$

where D, x, $y = [y_1, \cdots, y_D]^{\mathrm{T}}$, and $z = [z_1, \cdots, z_D]^{\mathrm{T}}$ are, respectively, the number of dimensions in the softmax layer, the output in the former layer, the input to the softmax layer, and the output in the softmax layer. According to Eq. (7), the sum of all elements of the output in the softmax layer is 1. In a classification problem, parameters are updated to maximize a likelihood of an output in the softmax layer, that is, an element of the output layer corresponds to a desired output.

The updates of the parameters are performed by stochastic gradient descent which is the same in pre-training. In this paper, we adopted the mean square error as the evaluation function of SSS-SdA and the negative log-likelihood for SSI-SdA.

3   Evaluation Experiment

We compared training time and identification accuracy for 2STG and E2E. The neural networks were implemented with Theano ver. 0.7.0, which is a framework of Python, and we used NVIDIA Tesla K20c for GPGPU in our experiments.

3.1 Conditions of Experiments

The sound sources used in our experiments are shown in Table 1. The multi-channel acoustic signals are synthesized with a numerical simulation.

Table 1. Sound sources

Categories	Classes
Voiced sounds	Male cry, female cry
Hitting sounds	China, can, pan, wooden box, bell
Instruments	Tambourine, cymbal, whistle
Electric sounds	Ringtone, ambulance, alarm clock, shaver, horn
Other sounds	Tear paper, coffee mill, cry of crow, clap, rolling stone

Numerical Simulation of Acoustics. We generated multi-channel acoustic signals with a numerical simulation. First, we calculated transfer functions between the microphone array and the sound source shown in Fig. 7. Secondly, we generated multi-channel clean acoustic signals by convoluting the transfer functions and each sound source listed in Table 1. Finally, we added the noise source to the convoluted multi-channel acoustic signals to have a desired SNR. The noise source was recorded with a microphone array mounted on the UAV bound on a tripod as shown in Fig. 6. The microphones are attached to the rim of the hull in order to avoid noise from vibration of the UAV.

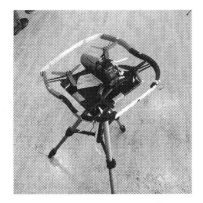

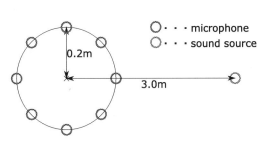

Fig. 6. UAV (Parrot Bebop Drone)   **Fig. 7.** The layout of the microphones and the sound source (Color figure online)

The SNR used in this paper is defined as Eq. (8). The SNR was standardized to be 0 dB in our experiments.

$$SNR = 20 \log_{10}(S_p/N_p), \tag{8}$$

where S_p and N_p are, respectively, the peak amplitude of the signal component and that of the noise component.

Acoustic Features. We adopted Mel filter bank features [9] as the acoustic features used as the input of the networks. We framed the acoustic signals with the frame width of 512 samples (32 ms) and the frame shift of 160 samples (10 ms). We used the complex window [1] as the window function when performing Short-Time Fourier Transform (STFT). From a complex spectrum obtained by STFT, we calculated a 27-dimentional Mel filter bank feature vector, by setting a lower and higher cut-off frequency to be 63 and 8,000 Hz, respectively.

Conditions for Training SdA. Table 2 shows the conditions used for SdA. The input vector for the two SdA is a 2,160-dimensional acoustic feature vector obtained as 10 frames × 8 channels × 27-dimentional Mel filter bank feature vector. The output of SSS-SdA, and the input of SSI-SdA are the same, that is, 270-dimensional features defined as 10 frames × 1 channel × 27-dimensional feature vectors. Since the number of sound source categories is 20, SSI-SdA has 20-dimensional outputs. For comparison, the number of dimensions in the hidden layers for 2STG and E2E are the same.

Table 2. Structure of SdA

	Input	Hidden layers	Output
Suppression	2160	[1620, 1080, 540]	270
Identification	270	[100, 50]	20
End-to-End	2160	[1620, 1080, 540, 270, 100, 50]	20

The training conditions are as follows: The number of training epochs is 20 in pre-training, 500 in fine-tuning for SSS-SdA in 2STG and 1000 in fine-tuning for SSI-SdA in 2STG and E2E at maximum. Note that after the 100th epoch, parameter updating stops when the loss of the cost function(L) only changes within a certain threshold (early-stopping). The batch size and the learning rate in pre-training, respectively, were set to 10, and 0.01. The learning rate for fine-tuning was set to 1.0 for noise suppression, and 0.1 for sound source identification.

3.2   Results

Table 3 shows the results of the experiments. Figure 8 shows the comparison of 2STG and E2E on a dataset. Training time in Fig. 8 does not start from 0 due to pre-training for E2E and the training of SSS-SdA for 2STG.

We generated 30 pieces of 3-s wave files for each of the 20 types of sound sources shown in Table 1 and then obtained about 172,000 sets of data vectors as described in Sect. 3.1. We divided these data vectors into 5 groups and evaluated the performance of the networks with 5-fold cross validation. The identification

Table 3. Performance of Sound Source Identification

	Error rate [%]		Time [min]		Fine-tuning epochs		
	2STG	E2E	2STG	E2E	Suppressor	Identifier	E2E
1	**23.41**	25.39	**684**	952	500	1000	1000
2	**24.42**	26.14	**675**	961	500	1000	1000
3	**27.54**	28.84	**680**	947	500	1000	1000
4	**25.98**	27.62	**680**	978	500	1000	1000
5	**29.41**	30.69	**682**	965	500	1000	1000
Ave.	**26.15**	27.74	**680**	961	500	1000	1000

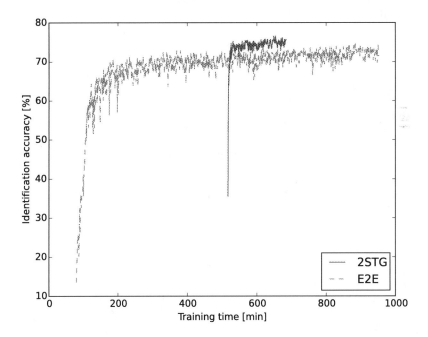

Fig. 8. Training time and identification accuracy of 2STG and E2E (Color figure online)

accuracy was measured for every data vector. Note that training time for 2STG is the sum of ones for noise suppression and for sound source identification.

According to Table 3, 2STG outperforms E2E both in accuracy and training time. The p-value of t-test was 0.00028 ($p < 0.01$), which means that these two results have a statistically significant difference. We guess that this is caused by two reasons. One is the number of layers and parameters to be updated at once is reduced with 2STG as we expected. The other is that 2STG was able to use the clean acoustic signals effectively in training, and it reduce the variance of the input to SSI-SdA, which resulted in the two SdA converging faster.

4   Conclusion

This paper presented a technique to reduce computational cost of neural network training by proposing two stage training, i.e., dividing a neural network into smaller ones, and then integrating them into one. It was applied to sound source identification of a multi-channel and noise-contaminated acoustic signal. We showed the proposed method could reduce the training time by approximately 29 % compared to end-to-end training, which trains a neural network to identify sound sources directly from a multi-channel acoustic signal. We also showed that the proposed method has better performance in terms of identification accuracy. Our future work will inspect the effectiveness of our method in a real environment.

Acknowledgment. This work was supported by KAKENHI No. 24220006 and ImPACT Tough Robotics Challenge.

References

1. Mainpage - hark wiki. http://www.hark.jp/. Accessed 21 Dec 2015
2. Bottou, L.: Online learning and stochastic approximations. On-line Learn. Neural Netw. **17**(9), 25 (1998)
3. Hannun, A., Case, C., Casper, J., Catanzaro, B., Diamos, G., Elsen, E., Prenger, R., Satheesh, S., Sengupta, S., Coates, A., et al.: Deepspeech: Scaling up end-to-end speech recognition. arXiv preprint (2014). arXiv:1412.5567
4. Lawrence, S., Giles, C.L., Tsoi, A.C.: What size neural network gives optimal generalization? convergence properties of backpropagation. Technical report (1996)
5. Nakajima, H., Nakadai, K., Hasegawa, Y., Tsujino, H.: Correlation matrix estimation by an optimally controlled recursive average method and its application to blind source separation. Acoust. Sci. Technol. **31**(3), 205–212 (2010)
6. Nakamura, K., Nakadai, K., Okuno, H.G.: A real-time super-resolution robot audition system that improves the robustness of simultaneous speech recognition. Adv. Robot. **27**(12), 933–945 (2013)
7. Ohata, T., Nakamura, K., Mizumoto, T., Taiki, T., Nakadai, K.: Improvement in outdoor sound source detection using a quadrotor-embedded microphone array. In: 2014 IEEE/RSJ International Conference on Intelligent Robots and Systems (IROS 2014), pp. 1902–1907 (2014)

8. Okutani, K., Yoshida, T., Nakamura, K., Nakadai, K.: Outdoor auditory scene analysis using a moving microphone array embedded in a quadrocopter. In: Proceedings of 2012 IEEE/RSJ International Conference on Intelligent Robots and Systems (IROS 2012), pp. 3288–3293. IEEE (2012)

9. Stevens, S., Volkmann, J., Newman, E.: A scale for the measurement of the psychological magnitude pitch. J. Acoust. Soc. Am. **8**(3), 185–190 (1937)

10. Uemura, S., Sugiyama, O., Kojima, R., Nakadai, K.: Outdoor acoustic event identification using sound source separation and deep learning with a quadrotor-embedded microphone array. In: The 6th International Conference on Advanced Mechatronics (2015)

11. Vincent, P., Larochelle, H., Lajoie, I., Bengio, Y., Manzagol, P.-A.: Stacked denoising autoencoders. J. Mach. Learn. Res. **11**, 3371–3408 (2010)

12. Yu, D., Deng, L.: Automatic Speech Recognition - A Deep Learning Approach. Springer, London (2014)

KANSEI (Emotional) Information Classifications of Music Scores Using Self Organizing Map

Satoshi Kawamura[1]([⊠]) and Hitoaki Yoshida[2]

[1] Iwate University Super Computing
and Information Sciences Center, Morioko, Iwate 020-8550, Japan
kawamura@iwate-u.ac.jp
[2] Faculty of Education, Iwate University, Morioko, Iwate 020-8550, Japan
hitoaki@iwate-u.ac.jp

Abstract. We classified KANSEI (emotional) information for musical compositions by using only the notes in the music score. This is in contrast to the classification of music by using audio files, which are taken from a performance with the emotional information processed by the instrumentalists. The first is classification into one of two classes, duple meter or irregular meter. The second is classification into one of the two classes, slow vs. fast (threshold tempo: $\downarrow$ = 110). The classification of the musical meter is based on identifying the meter indicated in the score. For tempo classification, we generally used the tempo indication in the score, but we evaluate classification that includes tempo revisions through a subject's emotions to be accurate. We performed classification for both the meter and tempo evaluations with a recognition rate above 70 % by using self-organizing maps for unsupervised online training. Particularly, in the tempo classification, a computer successfully processed the emotional information directed.

Keywords: KANSEI information · Emotional information · Music score classification · Self-organizing · Feature map

1 Introduction

Most musical research of an information technology nature is based on audio files that are the result of a person performing a music score, and the research involves carrying out classification after obtaining characteristics such as frequency [1–5]. Such studies do not include processing of human KANSEI, which is described as sensibility or emotion. Instead, these studies consider musics as information that has already been processed by human emotion. Studies based on the score itself have been conducted [6–8], but virtually no studies have covered the processing of emotional information.

The score does not include all of the information about a music [9–12]. The lines of notes printed in the score provide a framework, but in order to transform them into musical phrases that do not sound unnatural or mechanical, it is necessary for the instrumentalist to perform (transform) the music while relying on information not written in the score [9, 10]. Note that a phrase consists of a certain number of notes that are grouped together, including melody, bass, etc.

© Springer International Publishing Switzerland 2016
H. Fujita et al. (Eds.): IEA/AIE 2016, LNAI 9799, pp. 574–586, 2016.
DOI: 10.1007/978-3-319-42007-3_50

1.1 Information Indicated in the Score and Information Processing Using Emotions

Included in the score are the notes to be played at various times by various parts and instruments, along with symbols related to expression. If the notes are played with "mathematical correctness" (e.g., by a synthesizer), the performance sounds extremely artificial [9]. Various reasons for this can be surmised. One reason is that in the case of a person's performance, the instrumentalist supplements the information in the score and infuses the music with his or her own interpretations.

When performing a music, an instrumentalist takes the information indicated in the score and plays it while interpreting information not indicated in the score, for example, the composer's intent, the historical background, the phrase structure, and the part and phase roles of the music [9–12]. The instrumentalist performs some but not all of the attributes described below [9–12]. Every one of the attributes involves the instrumentalist reading the lines of notes printed in the score and interpreting them. This is because almost all scores include only the following general indications, and almost all other musical elements, including KANSEI information, must be generated by the instrumentalists.

- Tempo and rhythm: These are significantly influenced by variables such as the part of a music or the location of a phrase. It is rare for tempos and rhythms to be given rigid mechanical treatment. That is, they need to be handled flexibly. The values of notes are also not absolute designations, and it is common for the length and modification to differ according to variables such as the music's character or the instrumentalist. A single phrase denoted at a single tempo might be played at that same tempo, but it also might be played differently.
- Articulation: Articulation is adding connections notes, dynamics, and expressions in units shorter than phrases. The treatment of the beginnings and endings of notes, expression of notes, rest values, joining and separation of notes, places to breathe, and types of breathing (e.g., circular breathing) must all be considered to give the performance meaning. Since articulation is also influenced by expressions and the character of music (e.g., syncopation, swing), it is very rare that even the rest values are handled as written in the score. In general, the effects of changes in the note values of continuous notes, changes in playing methods, fluctuations from standard beats and tempos, etc., differ according to the instrumentalist, even in the same music.
- Overall image of the music: This is related to other elements in the performance of the score. A person does not perform mechanically exactly as written in the score, because emotional information processing in the broad human sense is indispensable. For the emotional information processing to take place, the instrumentalist needs to have an image of the music. Such an image requires not only feeling the emotions from the music, but also performing it to express what is being felt by the person at that time. This is difficult even for a human instrumentalist and is an attribute that is not considered possible in the case of an automatic performance by a computer. Furthermore, since information regarding emotions is virtually never indicated in the score, the instrumentalist has to conjure up an image of the music. Naturally, interpretations differ from instrumentalist to instrumentalist.

In these ways, merely performing the score as written is extremely insufficient. The instrumentalist processes information on a number of levels to produce the kind of performance that we are used to hearing; that is, we hear the end result of the emotional information processed by the instrumentalist. We need to emphasize that decisions based on the instrumentalist's individual experience and sensibilities are included in his or her processing of emotional information.

1.2   Classification Problem of Musical Compositions (Songs)

As mentioned in Sect. 1.1, the score includes mainly information on which notes to play. Not only is information on expression incomplete, but information on emotions is largely or completely absent [9–12]. In the case of certain lines of notes, the human instrumentalist carries out processing based on that individual's sensibility to address questions, such as which notes should be played as a single phrase, how each note should be accented/stressed or separated, what the volume level should be (low, normal, or loud), and what kind of mood (stiff, flexible, stirring, sad, etc.) is set by changing the tone of the musical instrument (straight, vibrate, shrill, mellow, etc.). Anyone who has ever performed in a chorus or played an instrument answers these questions from experience and the answers are self-evident to the instrumentalist [9–12].

The research to date on the classification of music has been almost exclusively carried out by using audio files of music played by human instrumentalists [1–5]. Although several studies [6–8] have been based on scores, these studies have primarily covered the question of the recognition of score markings for the image recognition of music notation using modern musical symbols, and nearly all of the studies focus on specific applications, such as the extraction of characteristics from a music score in a particular genre. Studies on processing emotional information in music in the true sense are virtually nonexistent.

In this study, we considered two problems: meter classification and tempo classification. In the meter classification problem, meter information is clearly indicated (written) in the score, so it is a self-evident problem. In the tempo classification problem, tempo involves an emotional decision by the instrumentalist, who generates emotional information, KANSEI, because no emotional information is given in the score [9–12]. Thus, tempo is not a self-evident problem. For the tempo problem, we tested whether a computer could simulate an instrumentalist's personal emotional information processing.

2   Converting a Score into Numerical Data

To a certain extent, a set method can specify the musical notation in a score. However, differences do exist, depending on factors such as the historical period when the score was written and the type of instrument(s) to be played. In addition, some unique notation methods depend on the composer or arranger, as well as differences among publishers, types of performance, and full scores and parts of scores [9, 10]. In some cases, differences in clef are found even for the same instrument. However, one common element is that musical notes printed in a score use the 12-tone chromatic scale to represent the relative duration and pitch of a sound.

In this section, we describe the method employed for producing numerical data from musical notes. Music notes has a musical note (scale) and a note length. In musicology, the note length is called the note value.

2.1 Numerical Conversion of Musical Notes

In Fig. 1(a), the note "A", which is located in the space between the second and third lines of the five-line staff marked with a G (treble) clef, is referred to as "A4" in scientific pitch notation. The "4" after the note "A" (La) represents octave 4. The pitch A4 is defined as 440 Hz under ISO16 [13], and it is known as the standard tuning pitch.

Since the fourth octave is used most often in common music scores, we set the values as shown in Fig. 1(b) relative to the fourth octave: C4 = 0.0, C2 = −1.0, and C6 = 1.0. For each musical scale, we assigned values by setting the distance of one octave to 0.5 and dividing this into 12 equal parts. For example, the difference of one musical scale becomes 0.5/12. This was done to make the range of notes that appear frequently correspond to the input-output characteristics of a sigmoid-like activation function in an artificial neuron model [14–16]. In addition, we assigned a value of −2.0 to a rest value, which indicates a silence. Musical notes lower than C0 do not exist in music.

(a) A4

(b) Convesion table of musical note

C4 (musical note name = C, octave = 4) = 0.0, C2 (musical note name = C, octave = 2) = −1.0, C6 (musical note name = C, octave = 6) = 1.0. The value of each musical note is the value of 1 octave (0.5) equally divided by 12. The scale is 12 notes of chromatic scale (in C).

Scale	Octave						
	O1	O2	O3	O4	O5	O6	O7
C	−1.500	−1.000	−0.500	0.000	0.500	1.000	1.500
C♯, D♭	−1.458	−0.958	−0.458	0.042	0.542	1.042	1.542
D	−1.417	−0.917	−0.417	0.083	0.583	1.083	1.583
D♯, E♭	−1.375	−0.875	−0.375	0.125	0.625	1.125	1.625
E	−1.333	−0.833	−0.333	0.167	0.667	1.167	1.667
F	−1.292	−0.792	−0.292	0.208	0.708	1.208	1.708
F♯, G♭	−1.250	−0.750	−0.250	0.250	0.750	1.250	1.750
G	−1.208	−0.708	−0.208	0.292	0.792	1.292	1.792
G♯, A♭	−1.167	−0.667	−0.167	0.333	0.833	1.333	1.833
A	−1.125	−0.625	−0.125	0.375	0.875	1.375	1.875
A♯, B♭	−1.083	−0.583	−0.083	0.417	0.917	1.417	1.917
B	−1.042	−0.542	−0.042	0.458	0.958	1.458	1.958

Musical Note Names (the 12 notes of a chromatic scale built in C)

Fig. 1. Conversion table for musical notes. Rest notes does not match the musical scale. So, the value of a rest note is −2.0. −2.0 is a musical scale not to exist.

2.2   Numerical Conversion of the Note Value (Note Lengths)

In musicology, the note length is called the note value. Note values include additional symbols, such as the quarter note (crotchet) and the eighth note (quaver). Two additional symbols are related to note value. One symbol represents dotted notation, which includes single-dotted notes (1.5 times as long) that often appear in jazz and swing, double-dotted notes $(1 + 0.5 + 0.25 = 1.75$ times as long), and triple-dotted notes $(1 + 0.5 + 0.25 + 0.125 = 1.875$ times as long). The other symbol represents a tie, where two notes with the same pitch are connected by a tie (curved line) with the intent of having them played as a single note. An infinite number of patterns exists due to the fact that multiple notes connected by a tie are handled as a single note. As a result, it is necessary to devise unified rules for the conversion in order to generate numerical values for all note lengths.

In this study, we conceived a scale that uses two whole notes (semibreves) as the "standard" note that the length is some fraction of the standard (double whole note (breve) = whole note (semibreve) $\times$ 2 = quarter note $\times$ 8), and enables notes converted into a desired range of numerical values to be obtained as the reciprocal of the log Eq. (1). This allows us to handle not only note values as short as the sixteenth notes (semiquaver) and thirty-second notes (demisemiquaver), but also note values obtained through division into non-integral values, such as multiplets. Multiplets are notes obtained by dividing the beat into a different number of equal subdivisions than those usually permitted by the time signature (e.g., triplets, duplets, etc. [11, 12]). Furthermore, the results of Eq. (1), shown in Table 1, are converted into the desired range of positive values. As an exception, when a longer note exceeds the double whole note (breve), we set the value of the longer note that is longer than two whole notes to 4.0.

$$\text{Conversion value of the note value} = \frac{-1}{\log(X)} \tag{1}$$

Where, X = note value in the score/double whole note (breve).

2.3   Extraction of Phrases from the Score

The score contains all sources of information about the music, but the constituents of the music are variable. The music may have only a single melody, such as an ABA form where the middle (B) section differs, or an ABA'C form that ends (C) with a different melody (BA') after the original melody (A) is repeated. In many cases, the A, B, C, or A' parts differ to the extent that when each is omitted, the melody is recognized as a different composition; that is, all the parts do not sound like they are from the same composition. A large-scale and easy-to-grasp example of a variable score is a symphony score, where the length of range of one phrase is not clearly specified [9, 11, 12].

In this study, since we are considering classification based on human emotional information, KANSEI, as it pertains to music, it is necessary to extract meaningful chunks that have identical phrases and convert them into data. The problem here is determining which part of a music to convert into data, because it involves processing

Table 1. Conversion table of note values. The left column are note values, and the most right column are conversion results. Here, X represents the ratio of a note's length to the length of the double whole note (breve). We assigned 4.0 to notes longer than a double whole note (breve). The value of each rest note (silence) was given the negative value of a note with the same length.

Note Value	Note Velue Name	X = (Note Vlue / 8)	X	$\dfrac{-1}{\ln X}$
	double whole note (breve)	8/8	1.00000	**4.00000**
	whole note (semibreve)	4/8	0.50000	1.44270
	dotted half note	3/8	0.37500	1.01955
	half note (minim)	2/8	0.25000	0.72135
	dotted quarter note	1.5/8	0.18750	0.59738
	quarter note (crotchet)	1/8	0.12500	0.48090
	dotted eighth note	0.75/8	0.09375	0.42245
	eighth note (quaver)	0.5/8	0.06250	0.36067
	triple quarter note	(1/3)/8	0.04167	0.31466
	dotted sixteenth note	(1/4+(0.5/4))/8	0.04688	0.32677
	quintuplet quarter note	(1/5)/8	0.02500	0.27109
	sixteenth note (semiquaver)	(1/4)/8	0.03125	0.28854
	dotted thirty-second note	((1/8)+(0.5)/8))/8	0.02344	0.26642
	thirty-second note (demisemiquaver)	(1/8)/8	0.01563	0.24045
	sixty-fourth note (semidemisemiquaver)	(1/16)/8	0.00781	0.20610

that originates from the "way of feeling," which is a human sensibility. For example, parts A and B may refer to the interpretation of the internal structure (e.g., the tempo changes or the mood of the piece differs partway through). Furthermore, since the beginnings and endings of phrases are not clearly indicated in the score, their interpretation differs depending on the score reading by the instrumentalist.

We used a melody that comprises phrases and the accompanying bass as the data for classification. To create the data, we carried out identical processing on the melody and the bass to make a single set of "melody + bass" data.

Depending on the music, various phrase lengths and numbers of notes occur with-in the phrases. Therefore, to obtain pieces of data of the same length, we set the number of notes to 70. If we had 70 notes, the phrase was converted to data as-is. If a phrase had less than 70 notes, we made up the difference by padding, that is, by adding a repeat of melody material until the number of notes reached 70 (Fig. 2). As is customary, the number of notes in the bass is not the same as that in the melody, so we padded the bass in the same way as we did the melody to make its number of notes also equal to 70. We used material from the bass part for the bass padding.

Fig. 2. Padding to make all phrase data have the same length. In this score, 14 are insufficient to assume the number of notes 70 (as for the number of notes 56). It filled 14 insufficient notes sequentially from the top.

3   Results of Computer Experiment

SOMs that performed unsupervised online training in two stages of the classification of music [17–19]. Using SOMs, we can project high-dimensional data nonlinearly onto a two-dimensional map (called competitive layer or output layer) and carry out pattern classification of the data. The SOM projected data, which were phrases consisting of 280 numbers (information related to notes), onto a competitive layer and performed pattern classification. In a two-stage SOM, changing the range of the winner neuron neighbors in the 1st-stage training and 2nd-stage training is easier. Therefore, we set the grid size to be the initial value for the 1st-stage range, and set the initial value for the 2nd-stage range to 3. In addition, we used a linear field for initialization of the SOM weighting matrix, based on the direction of the first two principle components. We performed data classification by drawing a separating line in the competitive layer. We made a separating line that could best divide the training data and used it to judge the class to which the evaluation data corresponded. In other words, we conducted the classification while considering the data sets projected on the two-dimensional competitive layer as linearly separable.

The experimental setup is shown in Table 2. We obtained the optimal value for each parameter from the results of a preliminary experiment. The shaded areas in the table are explained in detail in the test results. We ran two types of experiments: one for the time (meter) signature (duple meter vs. irregular/mixed meter), and one for the tempo (beats per minute (BPM)) including KANSEI (fast tempo vs. slow tempo, threshold of tempo: ♩ = 110). In the preliminary tests, SOMs were not used to classify training data and test data; instead, we randomly introduced data belonging to different classes. Therefore, in the classification experiment, we presented Class-1 data first and then Class-2 data. We sampled the weights of the neurons (input layer to competitive layer) evenly from the subspace spanned by the two largest principal component eigenvectors. We initialized the weighting matrix as a linear field based on the direction of the first two principle components and obtained equivalent experimental values. We verified our procedure by running multiple experiments under identical conditions.

Table 2. Details of the computer experiment. We set the number of training steps to two [14–19]. A traing data set size is 80 pieces. The 1st stage is two times larger than the training data set (160 data pieces), and the 2nd stage is 10 times larger than the training data set (800 data pieces).

(a) SOM Parameters. R package som

	Arguments	Value
grid size	Specifying size of competitive layer.	10x10, 20x20, 30x30, 40x40
initializing	Specifying the initializing method. Uses the linear grids upon the first two principle components directin.	Linear
alpha	Initial learning rate parameter. Decreases linearly to zero during traininig (inv.alp.c). In this study, 2nd alpha is half of 1st alpha.	Variable 2nd alpha is half of 1st alpha
alphatype	Specifying learning rate funciton type.	Linear function (linear decrease)
neigh	Specifying the neighborhood function type.	Bubble function, Gaussian function
topol	Specifying the topology type of competitive layer (measuring distance).	Hexagonal, Rectangle
radius	A vector of initial radius of the training area in som-algorithm for the training.	1st : grid size 2nd :3
rlen	A vector of running length (number of steps) in the two training phases.	1st : 160 2nd : 800
inv.alp.c	The constant C in the inverse learning rate function: alpha0 * C / (C + t)	1st : 1.6 2nd : 8.0

(b) Experimental Settings. BPM means beats per minute (Tempo).

Data type		Time (meter) Signature	Tempo (BPM)
Music for the classification	Class-1	Duple meter (only 4/4)	Slow : ♩<110
	Class-2	Irregular meter or Mixing meter	Fast : ♩≧110
Training Data	Class-1	Duple meter : 40	Slow : 40
	Class-2	Irregular meter : 40	Fast : 40
Evaluation Data	Class-1	Duple meter : 10	Slow : 10
	Class-2	Irregular meter : 10	Fast : 10
Total		100	100

(c) Structure of a Data Set

♩ = 2 elements : {Musical Note, Note Value}

Data = {Melody Part, Bass Part}

Total Data Set

♩ : 70 (Melody) + 70 (Bass) = 140

140 X 2 elements (Note Value, Musical Note) = 280

(d) Standing Data Sets

Class-1 Class-1 ⋯ Class-1 Class-2 Class-2 ⋯ Class-2

(e) Software Environment

R-3.2.2 x86-64 (Microsoft Windows 8.1, 64 bit)

som: Self-Organizing Map Version: 0.3-5 (R Package)

4 Experimental Results and Consideration

4.1 Experiment of Time (Meter) Signatures

In the experiment to classify meters from notes in the SOM training data, we presented 40 pieces of duple meter data (Class-1), and then 40 pieces of irregular meter data (Class-2). As the results according to order of data presentation, the recognition rate for classification was lower when we presented Class-1 after Class-2 in preliminary experiments. In addition, since the recognition rate decreased when we used a bubble function as the neighborhood function for the competitive layer's reinforcement training, we used a Gaussian function. Similarly, the recognition rate was highest when we used a rectangular topology and a grid size of 40 × 40 for the competitive layer. Therefore, we ran the experiments by using a Gaussian neighborhood function, a rectangular grid topology, and a grid size of 40 × 40.

Among the parameters in Table 2(a), alpha (initial training parameter) had a significant impact. Therefore, we ran tests with increments of 0.1, from alpha = 0.1 to alpha = 1.0. The results are shown in Table 3. For alpha = 0.5, the recognition rate for both training and evaluation data was 70 %. For alpha = 0.3, the recognition rate for evaluation data was 90 %, but the recognition rate for training data was 65 %. Figure 3 shows the case for the highest recognition rate in (a) (alpha = 0.5), and the case in which a singular classification could not be performed (b) (alpha = 0.7). In the case of impossible of a classification (alpha = 0.7), two classes overlap in the competing layer. As a result, the class classification is not possible. This cause is unclear.

Table 3 shows that in almost all cases, the recognition rate for the irregular meter was lower (incorrect recognitions were made) than that for the duple meter. This result coincides with the degree of difficulty in training music and is of great interest.

Table 3. Influence of initial training rate alpha.

Initial Training Rate Alpha	Training data			Evaluation Data			Recognition Rate (All)
	Duple Meter	Irregular Meter	Average	Duple Meter	Irregular Meter	Average	
0.1	77.5%	57.5%	67.5%	90.0%	60.0%	75.0%	69.0%
0.2	72.5%	57.5%	65.0%	90.0%	70.0%	80.0%	68.0%
0.3	67.5%	62.5%	65.0%	90.0%	90.0%	90.0%	70.0%
0.4	70.0%	62.5%	66.3%	70.0%	30.0%	50.0%	63.0%
0.5	82.5%	57.5%	70.0%	90.0%	50.0%	70.0%	70.0%
0.6	72.5%	62.5%	67.5%	80.0%	60.0%	70.0%	68.0%
0.7	unclassifiable	unclassifiable	unclassifiable	unclassifiable	unclassifiable	unclassifiable	unclassifiable
0.8	65.0%	60.0%	62.5%	90.0%	60.0%	75.0%	65.0%
0.9	70.0%	57.5%	63.8%	90.0%	60.0%	75.0%	66.0%
1.0	75.0%	62.5%	68.8%	90.0%	60.0%	75.0%	70.0%

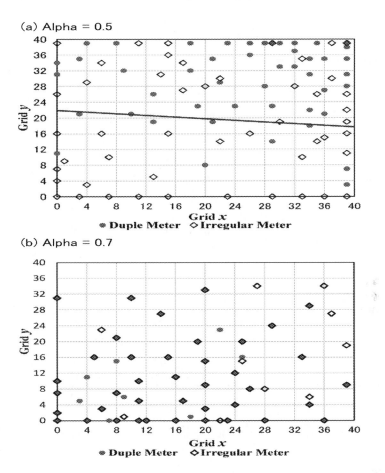

Fig. 3. Classification of time (meter) signatures. Neighborhood function: Gaussian, Topology: rectangular.

4.2   Experiment of Tempo (BPM)

In the tests for classifying the tempo by processing the notes with the author's emotional information, we used the same procedure as that for the meter classification problem. After presenting Class-1, which was the slow tempo data group, we presented Class-2, which was the fast tempo data group. The preliminary experimental results differed from those for the meter experiment in that no significant impact of the initial training parameter alpha was found. Therefore, we ran the experiment with the focus on the effects of the competitive layer topology type and grid size, and the type of neighborhood function.

Table 4 shows the results of the recognition experiment using the parameters of the competitive layer neighborhood function, the topology type, and the grid size. Depending on the competitive layer combination, 10 % or more differences occurred in the recognition rate, and the recognition rate was higher for evaluation data than for

Table 4. Fluctuations in recognition rate due to differences in competitive layer parameters. In the following table, the highest recognition rate (all) are 71 %.

Competivive Layer			Training Data			Evaluation Data			Recognition
Neighbourhood	Topology	Size	Slow Tempo	Fast Tempo	Average	Slow Tempo	Fast Tempo	Average	Rate (All)
Bubble Function	Hexagonal	10x10	62.5%	67.5%	65.0%	70.0%	70.0%	70.0%	66.0%
		20x20	65.0%	67.5%	66.3%	70.0%	80.0%	75.0%	68.0%
		30x30	65.0%	67.5%	66.3%	80.0%	80.0%	80.0%	69.0%
		40x40	67.5%	65.0%	66.3%	70.0%	70.0%	70.0%	67.0%
	Rectangle	10x10	65.0%	67.5%	66.3%	70.0%	70.0%	70.0%	67.0%
		20x20	67.5%	67.5%	67.5%	80.0%	80.0%	80.0%	70.0%
		30x30	62.5%	62.5%	62.5%	70.0%	80.0%	75.0%	65.0%
		40x40	67.5%	67.5%	67.5%	70.0%	70.0%	70.0%	68.0%
Gaussian Function	Hexagonal	10x10	67.5%	55.0%	61.3%	70.0%	50.0%	60.0%	61.0%
		20x20	65.0%	67.5%	66.3%	70.0%	80.0%	75.0%	68.0%
		30x30	60.0%	70.0%	65.0%	70.0%	80.0%	75.0%	67.0%
		40x40	65.0%	67.5%	66.3%	80.0%	70.0%	75.0%	68.0%
	Rectangle	10x10	72.5%	67.5%	70.0%	70.0%	70.0%	70.0%	70.0%
		20x20	70.0%	67.5%	68.8%	70.0%	70.0%	70.0%	69.0%
		30x30	72.5%	67.5%	70.0%	80.0%	70.0%	75.0%	71.0%
		40x40	72.5%	67.5%	70.0%	70.0%	70.0%	70.0%	70.0%

training data. In Table 4, the highest recognition rate was 71 % for a Gaussian neighborhood function, hexagonal and rectangular topologies, and a size of 30 × 30.

4.3   Analysis of the Data Which Failed in Recognition

From the standpoint of playing a musical instrument, we verified the scores for the data that failed to result in recognition in Sects. 4.1 and 4.2. As further clarification, a real example is one of the authors (Kawamura) who plays the trumpet. Because he takes into consideration the reading of music scores and then plays them with the trumpet, the author is carrying out sensibility processing (decision-making). Readers should note that this is a common practice in music [11, 12].

In Sect. 4.1, the recognition rate was lower in the case of the irregular meter. This also corresponds to the degree of difficulty of the performance (understandability). An irregular meter is not used for primary education in instrumental performance, but tends to be used in highly difficult scores. This is consistent with the fact that it is necessary to interpret odd meters such as 7/8 time signatures as 3/8 + 2/8 + 2/8 or 2/8 + 2/8 + 3/8, and the fact that the instrumentalist must decide which division is suitable depending on the score, phrases, etc. [9–12]. It is conceivable that a decision must be made based on the instrumentalist's sensibility, and that focusing too much on a portion of a phrase could cause the instrumentalist to mistake it for another meter (for example, 2/8 + 2/8 = 4/8 = 2/4 is included in the time signature 7/8).

In Sect. 4.2, the tendency was to fail in the classification of scores that reflect characteristics that are often seen in music with slow or fast tempos. The scores of music with slow tempos commonly indicate a relaxed mood and do not use many notes of short value (eighth and sixteenth notes) or tuplets. Generally, music with a fast tempo uses many eighth and sixteenth notes. Thus, a tendency to fail occurred in the classification of scores of slow pieces that had many short notes and tuplets, or scores of fast pieces that had many long notes.

5   Conclusion

In this paper, classification of KANSEI information (emotional information) for musical compositions by using only the notes in the music score is described. This type of classification does not include processing of emotional information or sensitivity (KANSEI) by the instrumentalist, or any hint information. We extracted phrases from the score, converted them into numerical values according to a set of rules, and built data sets for training and evaluation. We performed classifications on the meter and the tempo, which includes KANSEI information in the SOM that we used for evaluation. We found that meter classification for a duple meter (a normal meter) is easier than that for an irregular meter, and that tempo classification tends to fail in cases where the score is structured in a way that differs from the tempo designation, such as in slow pieces that have many short notes and tuplets, or scores of fast pieces that have many long notes. In both the meter and tempo evaluations the recognition rate was above 70 %. Particularly, in the problem of tempo classification, the fact that a computer was able to process the emotional information of a specific instrumentalist is of great interest in terms of emotional information processing directed at personalization.

Challenges for the future include carrying out experiments by using a batch-type SOM rather than an online SOM, expanding the data set, and considering whether the same results can be obtained even with data sets for different individuals.

Acknowledgements. We thanks to Kawamura's students, Mika Watanabe and Soh Sato for their supports who participated in the experiment.

References

1. Bill, M., Juan, R., Penousal, M., Dwight, K., Timothy, H., Walter, P., Robert, B.D.: Zipf's law, music classification, and aesthetics. Comput. Music J. **29**(1), 55–69 (2005)
2. Saadia, Z., Fawad, H., Muhammad, R., Muhammad, H.Y., Hafiz, A.H.: Optimized audio classification and segmentation algorithm by using ensemble methods. Math. Probl. Eng. **2015**, 1–10 (2015)
3. Srimani, P.K., Parimala, Y.G.: Artificial neural network approach to develop unique classification and raga identification tools for pattern recognition in carnatic music. In: AIP Conference Proceedings, vol. 1414, pp. 227–231 (2011)
4. Juhász, Z., Sipos, J.: A comparative analysis of Eurasian folksong corpora, using self organising maps. J. Interdisc. Music Stud. **4**(11), 1–16 (2010)

5. Andreas, R., Elias, P., Dieter, M., Andreas, R., Elias, P., Dieter, M.: The SOM-enhanced JukeBox: organization and visualization of music collections based on perceptual models. J. New Music Res. **32**(2), 193–210 (2013)
6. Ofer, D., Yoram, R.: An evaluation of musical score characteristics for automatic classification of composers. Comput. Music J. **35**(3), 86–97 (2011)
7. van Peter, K.: A comparison between global and local features for computational classification of folk song melodies. J. New Music Res. **42**(1), 1–18 (2013)
8. Moise A., Constantin A., Bucur G.: Musical notes recognition using artificial neural networks. In: Annals of DAAAM & Proceedings, pp. 1159–1160 (2009)
9. Gerhard, M.: Interpretation Vom Text zum Klang. Schott Music GmbH & Co. KG, Mainz (2006)
10. Oshima F.: "Wie man richtig die Noten list" Historische Musikpraxis von Bach bis Schubert NOtizen nach Vortra gen von Ingomar Rainer. Gendai Guitar Co. ltd., Tokyo (2009)
11. Yasushi, A.: Basics of Music. Iwanami Shoten Publishers, Tokyo (1971)
12. Carl, H.: The Piano Handbook: A Complete Guide for Mastering Piano. Backbeat Books, Milwaukee (2002)
13. ISO 16:1975, Acoustics– Standard tuning frequency (Standard musical pitch), ISO (1975)
14. Kohonen, T.: Self-Organizing Maps, 3rd edn. Springer, Heidelberg (2001)
15. Sandhya, S.: Neural Networks for Applied Sciences and Engineering: From Fundamentals to Complex Pattern Recognition. Auerbach Publications, New York (2006)
16. Pavel, S., Orga, K.: Investigation on training parameters of self-organizing maps. Baltic J. Mod. Comput. **2**(2), 45–55 (2014)
17. Ron, W., Lutgarde, M.C.B.: Self-and Super-organizing maps in R: the kohonen Package. J. Stat. Softw. **21**(5), 1–19 (2007)
18. The R Project for Statistical Computing. https://www.r-project.org/
19. Yan, J.: SOM Self-Organizing Map. R package version 0.3-4. http://CRAN.R-project.org/

Origin of Randomness on Chaos Neural Network

Hitoaki Yoshida[1]($\boxtimes$), Takeshi Murakami[2], Taiki Inao[1], and Satoshi Kawamura[3]

[1] Faculty of Education, Iwate University, Ueda, Morioka, Iwate 020-8550, Japan
{hitoaki, e0112007}@iwate-u.ac.jp
[2] Technical Division, Iwate University, Morioka, Japan
mtakeshi@iwate-u.ac.jp
[3] Super-Computing and Information Sciences Center, Iwate University, Morioka, Japan
kawamura@iwate-u.ac.jp

Abstract. We have proposed a hypothesis on the origin of randomness in the chaos time series of a chaos neural network (CNN) according to empirical results. An improved pseudo-random number generator (PRNG) has been proposed on the basis of the hypothesis and contamination mechanisms. PRNG has been implemented also with the fixed-point arithmetic (Q5.26). The result is expected to apply to embedded systems; for example the application of protecting personal information in smartphone and other mobile devices.

Keywords: Chaos · Pseudo-random number · Chaos neural network · Cipher

1 Introduction

Cipher for consumer use has been important as the Internet community has developed; for protection of personal information and privacy, for prevention against information leakage and so on.

We have continuously studied on the chaos neural network (CNN) that consists of conventional artificial neurons and generates chaotic outputs [1–3]. Recently, we have reported a high-speed (more than 300 Gbps with GPGPU) and highly secure novel pseudo-random number generator based on CNN. In particular, the period of the pseudo-random number (PRN) series is more than 10^{9726} ($\approx 2^{32308}$) [4]. A period of theoretical chaos is generally infinite, but it is not true in computer generated chaos. The time series from CNN is chaotic but also *eventually periodic*. The CNN pseudo-random number generator (CNN-PRNG) is expected to apply to a high-speed and highly secure stream cipher [4, 5].

In this paper, we have investigated properties of chaos time series and corresponding PRNs, and proposed a mechanism of random number generation, and then a hypothesis on the origin of randomness. Improved methods of random number generation also have been proposed, according to the hypothesis.

© Springer International Publishing Switzerland 2016
H. Fujita et al. (Eds.): IEA/AIE 2016, LNAI 9799, pp. 587–598, 2016.
DOI: 10.1007/978-3-319-42007-3_51

2   Deterministic Chaos as Pseudo-random Number Generator

Pseudo-random number generators (PRNGs) based on chaotic time series have been reported so far. Chaotic time series, however, do not always show uniform distribution, and adjacent PRNs show a strong correlation (or determinism). As for the time series from a logistic map some improved methods have proposed as follows:

  (i) direct transformation of the series to uniform PRN series [6, 7]
  (ii) using a threshold the series transform to uniform 1 bit binary number [8, 9]
  (iii) a part of a mantissa is extracted from the series as uniform PRN [10].

The correlation should be reduced by some other method as to (i), (ii). While, (iii) is also useful method of reducing the correlation [4, 5, 10]. These methods (i–iii), however, the origin of randomness is still unclear. Randomness is only discussed as empirical facts.

If the origin of randomness becomes clear, the statistical properties of PRNs will be controlled and excellent huge scale PRNGs will be designed. In this work, we have proposed a hypothesis on the origin of random numbers in the chaos outputs of CNN, and also have proposed improved methods of random number generation.

3   Chaos Neural Network as Pseudo-random Number Generator (CNN-PRNG)

CNN that composed of 4 neurons in discrete-time system has been used for a chaos generator (Fig. 1).

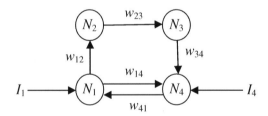

Fig. 1. CNN having cyclic structure (C-4nn).

A total value of inputs and a threshold of jth neuron at time t are defined as

$$u_j(t) = \sum_{i=1}^{n} w_{ij}x_i(t) + I_j \tag{1}$$

$$x_j(t+1) = f(u_j(t)) \tag{2}$$

where w_{ij} is a synaptic weight from ith neuron to jth neuron, x_i is an input from ith neuron, I_j is an external input of jth neuron. An output from jth neuron at time $t + 1$ is defined as Eq. (2) with the asymmetric piecewise-linear-function (APLF) f (Fig. 2).

APLF is useful for avoiding the periodic window corresponds to a non-chaotic periodic orbit, and 7 independent parameters of APLF can be used as secret keys in a cipher system [4, 11, 12]. A function f_1 is used unless otherwise stated, but for periodic chaos f_2 is adopted.

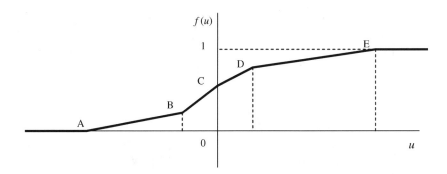

Function 1 (f_1)	A	B	C	D	E
x-coordinate value	-51.0001	-1.980101	0.0	1.980101	64.0001
y-coordinate value	0.0	0.09899	0.499012	0.8891	1.0
Function 2 (f_2)	A	B	C	D	E
x-coordinate value	-51.0001	-1.980101	0.0	1.980101	64.0001
y-coordinate value	0.0	0.08899	0.498012	0.8891	1.0

Fig. 2. Asymmetric piecewise-linear-function (APLF).

The iteration of CNN is computed by double precision arithmetic, and a pseudo-random number is extracted from a chaos output of CNN by the method shown in Fig. 3. The lowest 3 bits of the mantissa are discarded because it contains statistical deviation. The lower 24 bits in the remaining mantissa part are extracted as a pseudo-random number (Fig. 3).

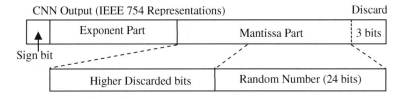

Fig. 3. The method for extraction of a random number from a CNN output [5].

The number of the extracted bits is optimized with NIST SP800-22 tests [13–15]. In the following study, 49 bits of the mantissa part except lowest 3 bits are used for analysis.

4 Component of Chaos Time Series and Hypothesis on Origin of Randomness

4.1 Experimental Condition and Component of Chaos Time Series

In order to extract pure and unadulterated outputs from CNN, the time series (x_1) from N_1 is separated 2 independent subseries; α series and β series (Eqs. (3), (4)). The 2 time series do not mix to each other. Following tests have been performed the 2 subseries, respectively.

$$\alpha(k) = \{x_1(t) \,|\, t = 2k + 1, k = 0, 1, 2, \ldots\} \tag{3}$$

$$\beta(k) = \{x_1(t) \,|\, t = 2k, k = 0, 1, 2, \ldots\} \tag{4}$$

We have reported the result of fractal analysis, chaos time series analysis and NIST800-22 tests on the time series of CNN outputs [2–4, 15, 16]. The results of multi-fractal analysis and $f(\alpha)$ spectrum analysis suggest that the higher 7–8 bits have fractality (Fig. 4). The results of recurrence plots suggest that the higher 7 bits also have determinism [17]. The results of NIST SP800-22 tests suggest that the lower 24 bit has the appropriate randomness for cryptography [13, 14].

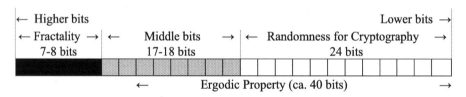

Fig. 4. The composition of the mantissa of the chaos output from CNN. 49 bits of the mantissa part except lowest 3 bits are used for analysis.

In this work, the accuracy of analysis has improved prior to further experiments. At first, NIST SP800-22 tests must be the latest version sts-2.1.2 which corrects problems of a non-overlapping template matching test and an overlapping template matching test [14, 19]. Maurer's "universal statistical" test has been improved according to Coron's approximation [18]. A test data ($10^6 \times 1000$ bits) is extracted from a chaos time series after idle time (10^6 iterations) as shown in Fig. 3, and all the NIST tests have been executed by 1000 times to study the averaged ratio of failed tests. In this paper, only the selected results of the NIST tests are shown in Tables for simplicity.

4.2 Results of NIST SP800-22 Tests

Results of the selected NIST tests on α series from CNN are shown in Table 1 and that of β series is omitted because it is nearly the same tendency. If a test is designed appropriately and a tested PRN series has statistically acceptable properties, the ratio of failed tests should be less than 1 % level [15]. The ratio of UN, however, is still slightly larger. Coron reported the inaccuracy due to the heuristic estimation can make the test 2.67 times more permissive than what is theoretically admitted [18]. Further modification of UN test is not adapted, because it is beyond the NIST test suite. The results are sufficient to use cipher applications, although a few data shows slightly over 1 % level. In the next, we have investigated the origin of randomness for further improvement.

Table 1. Results of selected NIST SP 800-22 tests on the lower-24-bit of α series.

I	FR[a,b]	RU[a,b]	OT[a,b]	UN[a,b]	LC[a,b]
0.31	0.0	0.0	0.5	1.0	0.1
0.32	0.6	1.4	0.6	0.5	0.1
0.33	0.4	0.9	0.5	1.2	0.1
0.34	0.1	0.4	0.2	0.7	0.5
0.35	0.0	0.5	0.8	0.7	0.0
0.45	0.0	0.2	0.8	0.8	0.2
0.50	0.0	1.1	1.0	1.6	0.4
0.55	0.1	0.3	0.1	1.1	0.2

[a]The averaged ratio of failed tests (%). [b]Abbreviations of test names: FR: frequency test, RU: runs test, OT: overlapping template matching test, UN: Maurer's "universal statistical" test, LC: linear complexity test.

The lower 24 bits in Fig. 4 have ergodic property through the experimental result of the NIST tests, because Maurer's "universal statistical" test (UN) is designed to be able to detect any one of the very general class of statistical defects that can be modeled by an ergodic stationary source [14]. There is no contradiction on ergodicity of chaotic maps [20]. The middle bits in Fig. 4 are also tested by the NIST tests. Results are partly shown in Table 2. It often failed in FR or OT test, yet the result of UN is not so worse. It suggests that the ergodic property exists in lower ca.40 bits of the mantissa.

4.3 Hypothesis on Origin of Randomness

Because fractality and determinism exist just in upper 7–8 bits and ergodic property exists in lower ca. 40 bits, ergodic bits are shuffled and mixed during repeated iterations of CNN (Fig. 5). A hypothesis on the origin of randomness in a lower-24-bit is following: the repeated shuffle of the ergodic bits improves the randomness and diminishes the correlation. Middle bits, however, is often contaminated with the upper deterministic and fractal bits. In the next section, the possible mechanism to diminish randomness is discussed.

Table 2. Results of selected NIST SP 800-22 tests on the middle bits of α series.

I	FR[a]	OT[a]	UN[a]	LC[a]
0.31	100.0	0.8	0.6	0.1
0.32	100.0	62.6	0.7	0.2
0.33	49.0	99.5	0.7	0.2
0.34	100.0	0.7	2.0	0.4
0.35	100.0	1.5	1.1	0.4
0.45	0.4	3.1	0.4	0.4
0.50	1.3	1.2	0.6	0.2
0.55	0.2	0.2	0.0	0.2

[a]The averaged ratio of failed tests (%). [b]Result of RU is omitted. When FR is failed, RU should not have been run. If the test is not applicable, then the P-value is set to 0.0000 [14].

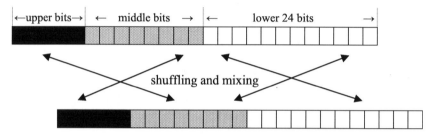

Fig. 5. Process of shuffling and mixing chaos outputs.

5   Factors to Affect Randomness

Possible factors to affect randomness are enumerated as follows: (i) period of time series, (ii) dynamical properties of time series, (iii) contamination by a constant value.

The factor (i) is detectable by NIST test, but it is not sensitive to longer period. Because 10^6 bits data is used for each NIST test, a period of a 24-bit-pattern is not detectable beyond $10^6/24$. The period of the sub-series from CNN is more than 10^8, and consequently the period of CNN cannot be detectable in NIST tests. As to the factor (ii) if the system is periodic vibration or quasi-periodic vibration, the randomness is changed drastically. Otherwise the property of chaos may affect randomness (*vide infra*). The factor (iii) can probably be a main cause of contamination. For example, when the external input $I = 0.2$, does it affect lower bits? At first, we will discuss this problem in the next subsection.

5.1   Contamination by Constant Value

Because all calculations in this work execute with a digital computer, binary codes should be considered. For example, decimal fraction 0.2 becomes circulating binary fraction $0.00110011001100110011\ldots = 0.00\dot{1}\dot{1}$. The patterns can contaminate the lower-24-bit even for 0.2, 0.3 and 0.4. To confirm this effect the point C in APLF (Fig. 4) is fixed at C (0.0, 0.5) and effect on external input I is studied (Table 3). When $I = 0.3$, the dynamics of the time series changes to periodic vibration. The other time series corresponding to $I = 0.2, 0.4, 0.5$ shows nearly the same properties of chaos [21], results of the NIST tests of the lower-24-bit, however, are quite different. The results suggested that the lower-24-bit is contaminated by the circulating pattern.

Table 3. Results of the selected NIST tests using I which involves circulating pattern.[a]

I	Circulating pattern	FR[b]	OT[b]	UN[b]	LC[b]	Dynamics
0.2	1100	3.2	0.5	1.6	0.3	chaos
0.3	1001	100.0	100.0	100.0	100.0	period 5
0.4	1100	99.9	27.6	18.5	0.6	chaos
0.5	–	0.0	0.6	0.8	0.2	chaos

[a]The point C of APLF is fixed at C (0.0, 0.5), and the other points are the same as f_1. [b]The averaged ratio of failed tests (%) on the lower-24-bit of α series.

The effect on the external input I is weakened by using f_1. The results corresponding to $I = 0.23$ are worse although the circulating pattern is longer (Table 4). While, $I = 0.23012345$ which have no circulating pattern within a mantissa, the results have considerably improved. The property of chaos time series are nearly the same for every Is in Table 4, but the randomness of the lower bits changed drastically.

Table 4. Results of comparable tests on some characteristic external inputs (I).[a]

I	Series	Circulating pattern	FR[b]	OT[b]	UN[b]	LC[b]
0.23	α	111000010100011111010	89.2	4.9	0.6	0.1
0.23012345	α	–	0.0	0.6	1.4	0.2
0.23	β	111000010100011111010	88.9	5.0	1.0	0.1
0.23012345	β	–	0.1	0.2	1.8	0.5

[a]f_1 is used as APLF. [b]The averaged ratio of failed tests (%) on the lower-24-bit.

5.2 Behavior of Periodic Chaos

In this system *periodic chaos* is rarely emerged. An example of bifurcation diagram of periodic chaos is shown in Fig. 6b when the function f_2 is used. $I > $ ca.0.4 the attractor is separated to 2 islands. That of normal chaos is also shown in Fig. 6a for comparison. The results of the NIST tests for periodic chaos are shown in Table 5; the OT test becomes slightly worse. Periodic chaos is known to be also ergodic but not mixing [22]. The cause of the test result is now under investigation, but it should not be used when the best security is required. The correlation coefficient between α and β series also is shown in Table 5, where the series means whole bits not lower 24 bits. The correlation coefficient between α and β series is normally negligible (< 0.01), because the 2 series is independent. Periodic chaos, however, has only 2 modes, (i) α and β

series simultaneously visit the same island (plus correlation), (ii) both series singly visit the other island (negative correlation). In this system, the periodic chaos can be numerically detected by correlation coefficient, so that the periodic chaos is removed easily.

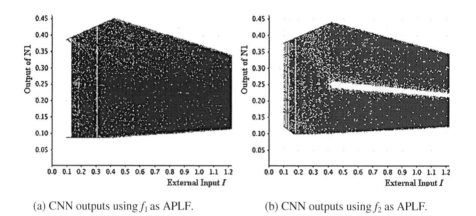

(a) CNN outputs using f_1 as APLF.   (b) CNN outputs using f_2 as APLF.

Fig. 6. Bifurcation diagram of CNN outputs.

Table 5. Results of selected NIST tests on α and β series from the periodic chaos.[a]

I	Series	Correlation coefficient [b]	FR[c]	RU[c]	OT[c]	UN[c]	LC[c]
0.45	α	0.82982731	0.2	0.3	1.0	0.7	0.1
0.45	β		0.0	0.4	0.2	0.4	0.0
0.50	α	-0.831300471	0.2	0.2	1.1	0.5	0.0
0.50	β		0.4	0.4	1.2	0.2	0.0
0.55	α	-0.835412049	0.0	0.3	1.5	0.8	0.4
0.55	β		0.0	0.7	2.0	0.9	0.4

[a]f_2 is used as APLF. [b]Correlation between whole bits of α and β series. [c]The averaged ratio of failed tests (%) on lower-24-bit of α and β series.

6   Improved Method of Pseudo-random Number Generation

According to the above-mentioned results and hypothesis, we have proposed an improved method. To prevent from contamination of regularity, an external input I is determined by a following equation:

$$I = 0.23000 + (\text{random number below } 0.00001) \tag{5}$$

The mean value of the averaged ratio over 15 Is which are determined by improved method is FR: 0.24 %, RU: 0.15 %, OT: 0.625 %, UN: 0.944 %, LC: 0.169 % (typical results are shown in Table 6). The results are considerably improved compared with $I = 0.23$ in Table 4.

Table 6. Results of selected NIST tests on α series by the improved method.[a]

I	Hexadecimal notation[b]	FR[c]	RU[c]	OT[c]	UN[c]	LC[c]
0.2300004270391	0 × 3FCD70A76C197601	0.0	1.2	0.7	1.3	0.1
0.2300011948672	0 × 3FCD70ADDCFF7277	0.1	0.0	0.4	0.4	0.1
0.2300009896898	0 × 3FCD70AC2461F0E6	0.0	0.0	0.3	0.5	0.2
0.2300010027085	0 × 3FCD70AC40570CC3	0.0	0.0	0.0	0.2	0.2
0.2300009996263	0 × 3FCD70AC39B8A093	0.0	0.0	0.8	1.3	0.3

[a]f_1 is used as APLF. [b]Hexadecimal notation of I as IEEE 754 representations. [c]The averaged ratio of failed tests (%) on lower-24-bit of α series.

7 Implementation of CNN with Fixed-Point Arithmetic

If we consider on the application of a smartphone or other mobile devices on embedded systems, it will be important to implement CNN with single precision floating-point arithmetic or fixed-point arithmetic. A mantissa of the single precision floating-point type data is only 23 bits and the significant decimal digit is only 6–7. If the hypothesis is true, the bit length is not sufficient for random number generation. Actually we have never succeeded so far.

Next, we have tried to implement CNN with 32-bit-fixed-point arithmetic operation (Q5.26). A period of CNN is expected to be shorter because the period of chaos time series is depend on the number of a significant digit of data [23]. In this case the significant decimal digit is 9 at most. Periods of most time series in a preliminary experiment are less than 10^4 as expected, which are not long enough for cipher. We have proposed new method for practical use in the following.

Internal state of CNN is defined as $X(t) = \{x_1(t), x_2(t), x_3(t), x_4(t)\}$, and the iteration is described with the map F which is corresponding to the whole CNN operation.

$$X(t+1) = F[X(t)] \tag{6}$$

Here, we introduce a simple map, bit pattern rotation of 32 bits data. That is, 7-bit-rotate-left instruction $Rot7$ is executed before the iteration (Eqs. (7), (8)). $Rot7\{X(t)\}$ no longer exists in the same chaos orbit that belongs before. In other words, $X(t)$ wanders over various chaos orbits during iterations.

$$Rot7\{X(t)\} = \{Rot7(x_1), Rot7(x_2), Rot7(x_3), Rot7(x_4)\} \tag{7}$$

$$X(t+1) = F[Rot7\{X(t)\}] \tag{8}$$

It has prolonged the period of CNN with the fixed-point arithmetic (Q5.26) to 10^{14}–10^{18}, which is comparable to the period of double precision floating point. If we chose the value of external inputs, the period can keep 10^{16}–10^{18}, which surpasses expectations (Fig. 7).

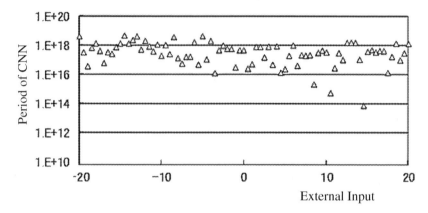

Fig. 7. Periods of time series with 32-bit-fix-point arithmetic operation.

The bifurcation diagram of the new time series from CNN shows no periodic window so that all initial values are available for chaos generation (Fig. 8a). The attractor of CNN (embedded in 3D phase space) does not have clear fractal structure (Fig. 8b). The results of chaos time series analysis suggest that the new time series is chaotic and deterministic, but not fractal. Probably because the 7-bit-rotate-left

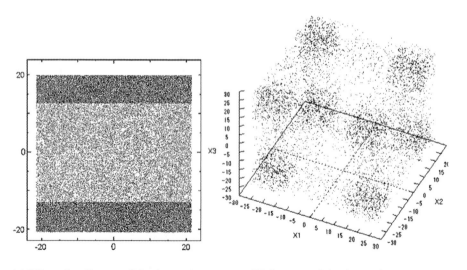

(a) Bifurcation diagram of the time series.   (b) Attractor of the time series.

Fig. 8. The bifurcation diagram and the attractor of time series with fixed-point arithmetic.

instruction moves the middle bits to the upper position, the fractal property is diluted. It suggests that during an idle time (10^6) of CNN iterations the whole bits are mixing and lowering the fractality and afford the attractor which has not clear fractal structure (Fig. 8b). The study of this attractive new chaotic time series just has begun, the precise properties as chaos will be reported in further publication.

The results of NIST tests also have been fine, if RN is extracted from the lower-16-bit of an output of N_1 (Table 7). While RN is extracted from the lower-24-bit, RU test is failed 100 %.

In the next place, CNN has been implemented with CUDA 5 on PC mounted with GPU. The rate of RNG has reached 190 Gbps with Tesla C2070 (448 CUDA cores), and 226 Gbps with Tesla K10 (1536 CUDA cores). The detail condition of the experiment is same as reference [4].

Table 7. Results of selected NIST tests on α series with the fixed-point arithmetic.

I	Lower 16 bits					Lower 24 bits				
	FR[a]	RU[a]	OT[a]	UN[a]	LC[a]	FR[a]	RU[a]	OT[a]	UN[a]	LC[a]
−20	0.0	0.0	0.7	0.8	0.1	1.8	100.0	0.0	1.0	0.1
−10	0.1	0.0	0.7	0.9	0.0	0.1	100.0	0.4	1.1	0.3
10	1.0	0.2	0.7	1.5	0.0	1.5	100.0	0.6	1.3	0.2
20	0.0	0.0	0.9	0.1	0.3	0.0	100.0	0.2	1.4	0.1

[a]The averaged ratio of failed tests (%) on the lower-24-bit.

8   Conclusion

We have proposed the hypothesis on the origin of randomness in the chaos time series of CNN, according to the results of time series analysis and NIST SP800-22 tests.

The improved PRNG method has been proposed on the basis of the hypothesis and the contamination mechanisms. CNN-PRNG has been also implemented with the fixed-point arithmetic (Q5.26). As future work, we will apply CNN-PRNG with the fixed-point arithmetic to a cipher application on embedded systems; for example the system with Cortex (ARM architecture) to protect personal information in smartphone and other mobile devices.

Acknowledgements. The calculations in this study have performed with the SGI UV-100 and the GPGPU computers in Iwate University Super-Computing and Information Sciences Center (ISIC). Special thanks to Mr. Mitsuaki SASAKI for help with experiments.

References

1. Yoshida, H., Yoneki, K., Tsunekawa, Y., Miura, M.: Chaos neural network. In: Proceedings of Papers, ISPACS 1996, vol. 1 of 3, pp. 16.1.1–16.1.5 (1996)
2. Yoshida, H., Nihei, Y., Nakanishi, T.: Comparative study on structurally different chaos neural networks. In: Proceedings of Papers, ISITA 2004, pp. 1046–1050 (2004)

3. Nihei, Y., Nakanishi, T., Yoshida, H.: Time series analysis for chaos neural network. Tech. Rep. IEICE **104**, 7–10 (2004)

4. Yoshida, H., Murakami, T., Liu, Z.: High-speed and highly secure pseudo-random number generator based on chaos neural network. In: Proceedings of Papers, ICSSE 2015, pp. 224–237 (2015)

5. Kawamura, S., Yoshida, H., Miura, M., Abe, M.: Implementation of uniform pseudo random number generator and application to stream cipher based on chaos neural network. In: Proceedings of Papers, ICFS 2002, R-18, pp. 4–9 (2002)

6. Ulam, S.M., von Neumann, J.: On combination of stochastic deterministic processes. Bull. AMS **53**, 1120 (1947)

7. Phatak, S.C., Rao, S.S.: Logistic map: a possible random-number generator. Phys. Rev. E **51** (4), 3670–3678 (1995)

8. Kohda, T., Ogata, E.: Bernoulli trials and chaotic trajectories in the logistic map. IEICE Trans. Fundam. **J68-A**(2), 146–152 (1985)

9. Kozak, J.J., Musho, M.K., Hartlee, M.D.: Chaos, periodic chaos, and the random-walk problem. Phys. Rev. Lett. **49**, 1801–1804 (1982)

10. Watanabe, H., Kanada, Y.: Pseudorandom Numbers generator using logistic map. In: Proceedings of the 53th National Convention of IPSJ, pp. 65–66 (1996)

11. Yoshida, H., Murakami, T.: Japan patent JP5504501B (2014)

12. Komori, T., Yi, H., Nakanishi, T., Murakami, T., Yoshida, H.: Behavior analysis of chaos neural network with simplification of nonlinear function. Tech. Rep. IEICE **109**, 53–58 (2009)

13. Soto, J., Bassham, L.: Randomness Testing of the Advanced Encryption Standard Finalist Candidates. National Institute of Standards and Technology (NIST) (2000)

14. Rukhin, A., Soto, J., Nechvatal, J., Smid, M., Barker, E., Leigh, S., Levenson, M., Vangel, M., Banks, D., Heckert, A., Dray, J., Vo, S.: A statistical test suite for random and pseudorandom number generators for cryptographic applications, NIST SP800-22 rev.1a, Accessed July 2015 (sts-2.1.1). Lawrence E. Bassham III (2015)

15. Yoshida, H., Murakami, T., Kawamura, S.: Study on testing for randomness of pseudo-random number sequence with NIST SP800-22 rev. la. Tech. Rep. IEICE **110**, 13–18 (2012)

16. Yoshida, H., Ohira, O., Taira, H., Nakanishi, T.: Fractal analysis of chaos neural network outputs in transient state and steady state. In: Proceedings of Papers, NOLTA 2006, pp. 103–106 (2006)

17. Horai, S., Yamada, T., Aihara, K.: Determinism analysis with Iso-directional recurrence plots. IEEE Trans. – Inst. Electr. Eng. Jpn. **C122**, 141–147 (2002)

18. Coron, J.-S., Naccache, D.: An accurate evaluation of Maurer's Universal test. In: Tavares, S., Meijer, H. (eds.) SAC 1998. LNCS, vol. 1556, pp. 57–71. Springer, Heidelberg (1999)

19. Okutomi, H., Nakamura, K., Aihara, K.: A study on rational judgement method of randomness property using NIST randomness test (NIST SP.800-22). IEICE Trans. Fundam. **J93-A**(1), 11–22 (2010)

20. Lasota, A., Mackey, M.C.: Probabilistic Properties of Deterministic Systems. Cambridge University Press, Cambridge (1985)

21. Aihara, K. (ed.): Basics and Application of Chaos Time Series Analysis. Sangyo Tosho, Tokyo (2000)

22. Thompson, J.M.T., Stewart, H.B.: Nonlinear Dynamics and Chaos: Geometrical Methods for Engineers and Scientists. Wiley, Chichester (1986)

23. Murakami, T., Kawamura, S., Yoshida, H.: Prediction of periods on chaos time series: dependence on precision of chaos neural network outputs. Tech. Rep. IEICE **107**, 21–26 (2008)

Artificial Neural Network Application for Parameter Prediction of Heat Induced Distortion

Cesar Pinzon[✉], Kazuhiko Hasewaga, and Hidekazu Murakawa

Department of Naval Architecture and Ocean Engineering,
Osaka University, Suita, Japan
Cesar_pinzon@naoe.eng.osaka-u.ac.jp

Abstract. Heat induced distortion has been widely studied over the years, in order to provide reliable results, thermal elastic-plastic FEM analysis have been used to estimate the distortions produced by the heat source. However this type of analysis often involves long computational time and requires high degree of technical knowledge by the user, moreover it's mainly performed to specific regions that limit the scope of the analysis. In order to provide a tool for the prediction of the line heating phenomena, an artificial neural network (ANN) is used. ANN is a powerful tool to predict complex phenomena, and in addition, it is very attractive because of the relatively modest hardware requirements and fast computational time. In this paper, parameter prediction for the heat induced distortion as an inverse problem is performed by ANN, using, the inherent deformation from a gas heating FEM analysis and their heating conditions as the training data. Exploratory analysis of the data and the model were performed to accurately predict the heating conditions. The prediction of the necessary heating conditions to generate an arbitrary deformation in the plate is a step forward in the automation of the line heating forming process. The possibility of predicting arbitrary heat induced distortion problem by an ANN model is shown.

Keywords: Inverse problem · Neural networks · Heat induced distortion · Line heating

1 Introduction

For many decades line heating has been used as a method for forming three dimensional surfaces in the shipbuilding industry [1], however the procedure is often done manually depending on the skill of experienced workers. Therefore this dependency results in a low productivity rate for shipyards which leads to high cost. The prediction of the distortion produced by line heating have been studied for several years, however, it is a highly nonlinear problem consisting of many factors such as the amount of heat, the plate thickness, the speed of the heating source and secondary factors like the cooling method, residual stress, etc. [2, 3]. In order to analyze this phenomena it is required to have an appropriate mathematical tool dealing with all the variables involved in this problem. Finite element method (FEM), through a three dimensional

© Springer International Publishing Switzerland 2016
H. Fujita et al. (Eds.): IEA/AIE 2016, LNAI 9799, pp. 599–608, 2016.
DOI: 10.1007/978-3-319-42007-3_52

thermal elastic plastic analysis, is the tool for it, but due to the complexity of the problem, it takes long computational time [4]. Final plate deformation have been successfully predicted by using an ANN model [5], however this ANN model is only able to predict the final deformation of a flat plate due to a line heating process. By this limitation, the ANN model cannot be used to predict the required heating conditions to produce an arbitrary deformation in the plate, which is often requested, in the ship-building industry. In this study, in order to improve this problem, another ANN model is proposed and discussed.

2   Gas Heating FEM Analysis

First of all, to understand heat induced distortion, thermal-elastic-plastic FEM is perform. In this code proposed by Osawa et al. [2], gas torch is regarded as the heating source. The process of forming a plate for a given heating conditions can be viewed as the process to deform a plate into a desired shape by giving the heating conditions, using the shrinkage and the angular distortion induced by the heating and cooling process [6]. Plate deformation by line heating can be described by four deformation components as is shown in Fig. 1. These deformation components (inherent deformation) are inherent longitudinal shrinkage, inherent transverse shrinkage, inherent longitudinal bending and inherent transverse bending respectively.

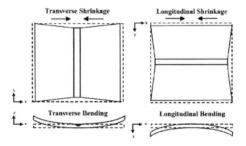

Fig. 1. Line heating distortion components

In this study the attention is focused in the process of deforming thick plates, where herein the term thick plate refers to those plates with thickness equal or greater than 10 mm, because it is more difficult to deform them by using the line heating technique. Therefore for this analysis, the plate thickness range from 10 to 50 mm and the heating speed range from 4 to 40 mm/s as shown in Table 1.

Table 1. Heating Conditions

Heating conditions	
Plate thickness [mm]	10, 20, 30, 40 & 50
Heating speed [mm/s]	4, 8, 12, 16, 20, 24, 28, 32, 36, 40

For better prediction of the inherent distortion, squared plate of 1200 mm (length) by 1200 mm (width) is selected as shown in the Fig. 2. Furthermore the effect of the plate thickness and the heating speed are evaluated. Once the model and the heating condition are defined, the thermal-elastic-plastic analysis is carried out. The analysis can be divided in two steps: the thermal analysis, where the heat source is applied to the model, and then the mechanical analysis where the deformation of the plate it's obtained. Later these results are used to train the ANN model to estimate the heating speed.

Fig. 2. Subject flat plate and its FEM model

From the thermal analysis the temperature distribution over the plate is obtained. Figure 3, show the maximum temperature in the plate with different heating speed. As a common practice in line heating process, the maximum temperature of the plate should be kept below 800 °C [7], in order to avoid material degradation,. Therefore the heating speed is restricted to a range over 12 mm/s.

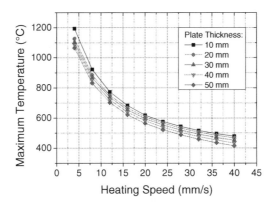

Fig. 3. Maximum temperature with heating speed and plate thickness (thermal analysis)

The transient temperature distribution obtained by the thermal analysis is employed as a thermal load in the subsequent mechanical analysis and finally the inherent distortion is obtained as shown in Fig. 4.

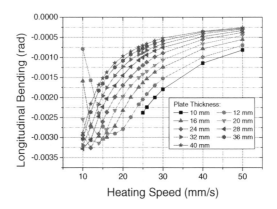

Fig. 4. Longitudinal bending distortion with plate thickness and heating speed (mechanical analysis)

Then the gas heating to a plate is simulated by FEM analysis with the given transient temperature distribution. The result is shown in Fig. 4. It can be noticed clear continuous relation between the plate thickness, the heating speed and the four inherent distortion components. On the other hand in the case of thinner plates with low heating speed, there exists slightly different behavior especially for the case of the longitudinal bending. This behavior is mainly because the heat distribution in the direction of the plate depth. At low speed the temperature at the top and the bottom of the plate are nearly similar.

2.1   Generation of the Training Data

In the case of the thermal-elastic-plastic FEM analysis, the model is able to predict the inherent distortion for a given plate thickness and heating speed. This type of problems, where the cause is given and the effect is determined is known as forward problem as it is presented in Fig. 5. In the other hand, the prediction of the heating speed by using the actual deformation of the plate and its thickness is an inverse problem where the effect is given and the cause is estimated.

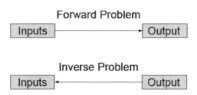

Fig. 5. Forward and inverse problems

Inverse problem is generally difficult, because it may not have a unique solution [8], furthermore inverse problem has a deductive nature rather than the inductive nature of forward problem. To overcome this situation ANN is used based on its ability to approximate unknown input-output mapping.

To perform the inverse problem analysis, the ANN model use the inherent distortion predicted in the forward problem as shown in the Fig. 6. From the figure, it can be seen that the plate thickness is taken as an input parameter since the plate thickness is known beforehand as well as the inherent deformation, then the ANN model will be able to predict the heating to deform the plate into the desired condition.

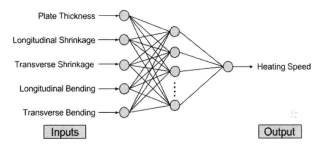

Fig. 6. Neural network model topology

The training data plays a key role in the development of the model. During the training stage of the ANN model, it was noticed that not only the size but also the quality of the data have a positive impact in the performance of the model.

3   Neural Network Development

In general terms, ANN can be seen as a tool that is designed to mimic the way in which the brains perform a specific task or function [9]. They are characterized by: the architecture, the training method and its activation function [10].

In this study, the development of the ANN model were performed using the neural network toolbox of MATLAB [11].

3.1   Neural Network Topology

Major advantages of ANN are that it is not necessary to presume any predetermined model nor coefficients and it copes with the cases not prepared as teaching data. In order to develop this feature, the topology of the network is quite important. The performance of the ANN model will depend on the chosen topology.

For simplicity of the ANN model, one hidden is selected as it is shown in Fig. 6, then the number of neurons in the hidden layer is evaluated so as to get the best performance of the ANN model, as shown in Fig. 7. From the figure it can be noticed that as the number of neurons in the hidden layer is increased, the MSE of the ANN

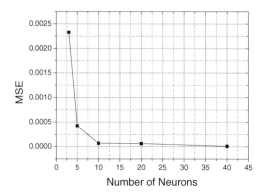

Fig. 7. Influence of number of hidden layers neurons in terms of MSE

model is improved. However, five neurons is selected for the hidden layer because further increase in the number of neurons does not improve the response of the ANN model drastically. Since the ANN model is used to perform an approximation task, a hyperbolic tangent sigmoid transfer function is selected for the hidden layer while a linear transfer function is used for the output layer.

The Levenberg-Marquardt optimization is chosen as the training algorithm for the backpropagation ANN model, as the ANN model present few hundred weights [12], while the Mean Squared Error (MSE) is used as the performance function.

Early stopping technique is used while monitoring the behaviour of the network, dividing the data into three subsets: 70 % for training, 15 % for test and the other 15 %, for validation, where the training data consist in 120 different cases taken from the gas torch FEM analysis. Furthermore another set of data is prepared for the purpose of testing the tolerance of the model, i.e. this set is not used to train the network, but to confirm the performance of the untaught cases.

4 Results

Once the training stage is finished, the performance of the ANN model during the training is evaluated as it is shown in Fig. 8. As different solutions were obtained even using the same model, the performance of each candidate was evaluated in order to select the best ANN model. From the figure, can be noticed that beyond 150 epochs the MSE is almost constant, then further training of the network will not improve the performance of the ANN model.

After the ANN model is selected, the response of the ANN model is tested with the training data and simultaneously is compared with the FEM results as shown in Fig. 9. In the figure, the relationship of the longitudinal bending and the heating speed for a plate of 10 mm thickness is presented, where the ANN model predicted the FEM results quite well, therefore the heating speed is accurately predicted by the ANN model.

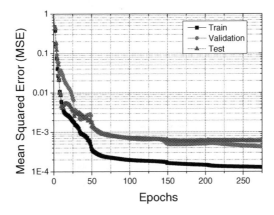

Fig. 8. ANN model training performance

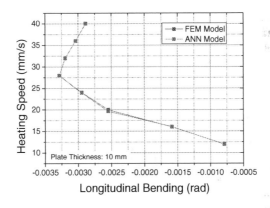

Fig. 9. FEM & ANN model comparison (Color figure online)

In order to have a wider view of the response of the ANN model, the error generated by predicting the training data is measured and plotted as shown in Fig. 10. From the figure is noticed that the error in the prediction is almost below 2 %, thus it proves the good performance for almost the given conditions.

Besides the response of the ANN model for data used during the training stage, the response of the ANN model for unseen data must be tested, as the prediction of the heating speed is performed using the plate thickness and the inherent distortion, they were taken in consideration for testing the response of the ANN model.

In Fig. 11, the relationship of the heating speed and the longitudinal bending for a 10 mm thickness plate is presented again, but the model is tested with conditions that were not used during the training stage. The response of the ANN model correctly matches with the FEM analysis.

Again in Fig. 12, the relationship of the heating speed with the longitudinal bending is presented but for an 11 mm plate thickness. Since the training data do not have any information for an 11 mm thickness plate, it can be noticed the good

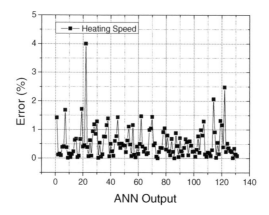

Fig. 10. ANN model response

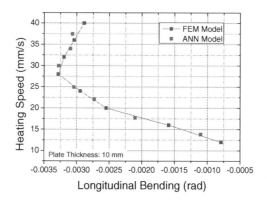

Fig. 11. ANN model prediction for longitudinal bending intermediate points

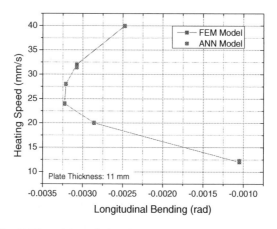

Fig. 12. ANN model prediction for plate thickness intermediate points

estimation of the ANN model for intermediate values of plate thickness. To confirm this results, several cases (blue square symbols) is calculated by the FEM model for an 11 mm thickness plate. The ANN model has a good capability to predict the heating speed even for untrained cases.

5   Conclusion

The heating speed for a given plate thickness with certain conditions by line heating process is accurately predicted using an ANN model. This model is capable to predict the heating speed by using the plate thickness and the four components of inherent distortion measured at the middle of the plate. Although this model is limited to a gas torch heating under the condition that the maximum temperature of the plate is kept below 800 °C to avoid material degradation, the following conclusions are drawn.

1. ANN models can successfully solve line heating distortion as an inverse problem.
2. The topology and the training data used in the ANN model have greatly affect the performance of the model, so they should be chose carefully.
3. The developed ANN model can successfully predict the heating speed to obtain certain distortion for a given steel plate even for conditions outside the training data making the ANN model an alternative tool for the heat induced distortion problem prediction.

References

1. Tango, Y., Ishiyama, M., Suzuki, H.: IHIMU a fully automated steel plate bending system for shipbuilding. IHI Eng. Rev. **44**(1), 6–11 (2011)
2. Osawa, N., Hashimoto, K., Sawamura, J., Kikuchi, J., Deguchi, Y., Yamaura, T.: Development of heat input estimation technique for simulation of shell forming by line-heating. J. Comput. Model. Eng. Sci. (CMES) **20**(1), 45–53 (2007)
3. Tango, Y., Ishiyama, M., Nagahara, S., Nagashima, T., Kobayashi, J.: Automated line heating for plate forming by IHI-ALPHA system and its application to construction of actual vessels-system outline and application record to date. J. Soc. Naval Archit. Jpn. **193**, 85–95 (2003)
4. Vega, A.: Development of inherent deformation database for automatic forming of thick steel plates by line heating considering complex heating patterns, Doctoral Dissertation Thesis, Osaka University (2009)
5. Pinzon, C., Plazaola, C., Banfield, I., Fong, A., Vega, A.: Development of a neural network model to predict distortion during the metal forming process by line heating, ship science & technology, vol 6, no. 12 (2012)
6. Rashwan, H.: Computer aided planning system for plate bending by line heating. Doctoral Dissertation Thesis, Osaka University (1994)
7. Blandon, J., Osawa, N., Masanori, S., Rashed, S. and Murakawa, H.: Numerical study on heat straightening process for welding distortion of a stiffened panel structure. In: Proceedings of the Twenty-Fifth International Offshore and Polar Engineering Conference, Kona, Hawaii (2015)

8. Tarantola, A.: Inverse problem theory and methods for model parameter estimation. In: Society of Industrial and Applied Mathematics (2005)
9. Haykin, S.: Neural Networks and Learning Machines. Prentice-Hall Inc., Upper Saddle River (1999)
10. Fausett, L.: Fundamentals of Neural Networks Architectures: Algorithms and Applications. Prentice-Hall Inc., Upper Saddle River (1994)
11. Beale, M., Hagan, M., Demuth, H.: Neural Network Toolbox. The Math Works Inc., Natick (2015)
12. Hagan, M.T., Menhaj, M.: Training feed-forward networks with the Marquardt algorithm. IEEE Trans. Neural Netw. 5(6), 989–993 (1994)

The Optimization of a Lathing Process Based on Neural Network and Factorial Design Method

Karin Kandananond[✉]

Valaya Alongkorn Rajabhat University, Prathumthani, Thailand
kandananond@hotmail.com

Abstract. The capability to optimize the surface roughness is critical to the surface quality of manufactured work pieces. If the performance of the available CNC machine is correctly characterized or the relationship between inputs and output is clearly identified, the operators on the shop floor will be able to operate their machine at the highest efficiency. In order to achieve the desired objective, this research is based on the empirical study which is conducted in such a way that the optimization method is utilized to analyze the empirical data. The focused process in this study is the lathing process with three input factors, spindle speed, feed rate and depth of cut while the corresponding output is surface roughness. Two methods, namely artificial neural network (ANN) and 2^k factorial design, are used to construct mathematical models exploring the relationship between inputs and output. The performance of each method is compared by considering the forecasting errors after fitting the model to the empirical data. The results according to this study signify that there is no significant difference between the performance of these two optimization methods.

Keywords: Artificial neural network · Factorial design · Lathing · Surface roughness

1 Introduction

A large number of work pieces are manufactured by a lathing process, and the surface quality of these work pieces is known to depend on the irregularities of materials resulting from machining operations or surface roughness, which is a critical quality characteristic. There are many factors known to have influence on the surface roughness. However, selecting the appropriate method to explore the relationship between turning process factors and surface quality is still a challenging issue. In this study, the performance of a machine learning method, ANN, and an empirical study method, factorial design, are quantified in order to come up with the best method in this case.

2 Literature Review

Among the most basic operations performed by machine tools are drilling, milling, grinding and turning or lathing. The lathing process is a machining method that removes material from the surface using a rotating cutting tool that moves to a work

© Springer International Publishing Switzerland 2016
H. Fujita et al. (Eds.): IEA/AIE 2016, LNAI 9799, pp. 609–619, 2016.
DOI: 10.1007/978-3-319-42007-3_53

piece. The surface quality, which is measured in terms of surface roughness, is utilized to evaluate the performance of the turning operation. The surface roughness is known to be significantly affected by different cutting parameters, i.e., the depth of cut, spindle speed and feed rate.

Therefore, the surface roughness will be minimized if the appropriate cutting conditions are selected. Experimental design methods, such as the two-levels (2^k) factorial design, are frequently utilized to model the surface roughness, so the desired levels of machining parameters are achieved. There are numerous works reporting the success of implementing factorial design to study the relationship between machining factors and surface roughness. Among these works, Choudhury and El-Baradie [1] utilized a factorial design technique to study the effect of cutting speed, feed rate and depth of cut on surface roughness. The experimental study was conducted on a turning machine equipped with uncoated carbide inserts. The workpiece material used was EN24T steel. Wang and Feng [2] utilized a factorial design to develop an empirical model for suface quality in turning processes. The predicting model are based on workpiece hardness; feed rate; cutting tool point angle; depth of cut; spindle speed and cutting time. Arbizu and Perez [3] deployed a 2^3 factorial design to construct a first order model to predict the surface roughness in a turning process of testpieces which followed ISO 4287 norm. Benga and Abrao [4] investigated the machining properties of hardened 100Cr6 bearing steel under continuous dry turning using mixed alumina, whisker reinforced alumina and polycrystalline cubic boron nitride (PCBN) inserts. A full factorial experimental design was used to determine the effects of feed rate and cutting speed on surface finish. Ozel et al. [5] studied the effects of workpiece hardness, feed rate, cutting speed and cutting edge geometry on multi-responses, surface roughness and resultant forces, in the finish hard turning of AISI H13 steel. The experiments were conducted using two-level fractional factorial experiments while the statistical analysis was concluded in the form of analysis of variance (ANOVA).

Other experimental design approaches commonly utilized for modeling responses are the Taguchi technique and response surface methodology (RSM). Davim [6] studied the influence of velocity, feed rate and depth of cut on the surface roughness using Taguchi design. The material used in this turning process was free machining steel, 9SMnPb28k (DIN). The model for predicting the surface roughness was developed in order to optimize the cutting conditions. Sahin and Motorcu [7] utilized RSM to construct a surface roughness model for the turning process of AISI 1040 mild steel coated with TiN. Three machining parameters, depth of cut, cutting speed and feed rate, were included in the predicted model.

Another approach to model and optimize surface roughness has been accomplished through the utilization of machine learning method such as artificial neural network. For example, ANN is deployed by Pontes et al. [8] to predict the surface roughness of test specimen operated from a turning process while radial basis function was used to train neural networks. ANN was also deployed by Natarajan et al. [9] to optimize surface roughness in the turning process of another type of metal, brass. Input parameters are cutting speed, feed rate and depth of cut while response is the surface roughness.

According to the literature, both experimental design and artificial neural network method are utilized to characterize and optimize the surface roughness of work pieces manufactured by lathing process. However, research works focusing on the comparison of these two methods is still rarely available. Therefore, it is interesting to compare the performance of both methods based on the standardized process so the guidelines for implementing each optimization method will be available for practitioners.

3 Method

3.1 Factorial Design Method

The factorial design method is a combination of statistical and mathematical techniques to analyse and model processes. The purpose of this method is to establish the unknown relationship between the independent variables (input factors) and the process responses. The experiments are performed to fit a first order model or linear function to the input variables in order to predict the output response. The efficiency of the factorial design is significantly influenced by selecting the proper choice of experimental designs. Factorial designs are the experiment in which all possible combinations of the levels of the factors are investigated. This design is one of the mostly used types of experiment involving the study of the effects of two or more factors. As experimental results, the effect of primary factor or main effect is defined to be the change in response caused by a change in the level of the factor. In some experiments, when the difference in response between the levels of one factor is not the same at all levels of the other factor, there is an interaction between the factors. The most important case of factorial design is the design for k factors, when the experiment is conducted at two levels for each factor, the high and low levels of a factor. In this case, a complete replicate of such a design requires 2^k observations or 2^k factorial design. The relationship between inputs and output is illustrated in (1).

$$\hat{y} = b_0 + \sum_{i=1}^{k} b_i x_i + \sum \sum_{i<j} b_{ij} x_i x_j \tag{1}$$

As shown in Fig. 1, all treatment combinations of the factorial design can display geometrically as a cube.

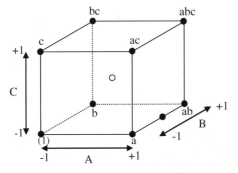

Fig. 1. Geometric view of 2^3 factorial design.

3.2 Artificial Neural Network (ANN)

The objective of ANN models is to explore the relationship between input variables and output variables. Basically, the neural architecture consists of three or more layers, i.e., input layer, output layer and hidden layer as shown in Fig. 2. The function of this network can be described as follows:

$$y_j = f\left(\sum_i w_{ij} x_{ij}\right), \tag{2}$$

where y_j is the output of node j, f (.) is the transfer function, w_{ij} the connection weight between node j and node i in the lower layer and x_{ij} is the input signal from the node i in the lower layer to node j.

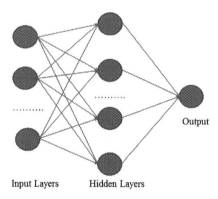

Input Layers   Hidden Layers

Output

Fig. 2. ANN Structure.

One of the algorithms used in ANN is the Levenberg–Marquardt algorithm or LMA. The main purpose of this algorithm is to solve the non-linear problems based on the least square method. When y_i and x_i are the inputs and outputs, β is the regression parameter and $f(x,\beta)$ is the curve fitting function, the LMA is based on the idea that the sum of the squares of the deviations $S(\beta)$ as shown in (3) is minimized.

$$S(\beta) = \sum_{i=1}^{n} [y - f(x_i, \beta)]^2 \tag{3}$$

4 Lathing Process

In this study, there are three input factors, spindle speed, feed rate and depth of cut, which are controlled to achieve the desired surface finish. Each factor is set at the low and high level in order to facilitate the 2^k factorial design of experiment. The quality of surface finish is measured in term of surface roughness which is the response or output of the process. Surface roughness is the index which quantifies how smooth the surface of the specimen is. As a result, the ouput of this study is surface roughness while the inputs are spindle speed, feed rate and depth of cut. The performance index of the output is surface roughness which is measured in terms of the average surface roughness (R_a). According to (4), R_a is the average deviation of the profile from the centerline along the sampling length (L):

$$R_a = \frac{1}{L} \int_0^L |Y(x)|dx, \tag{4}$$

where x is the profile direction and Y is the ordinate of the profile curve.

The range of each input is shown in Table 1. According to Table 1, the range of spindle speed is between 950 and 1100 Ft/min and feed rate is set at 0.008 and 0.009 inch per round. The depth of cut is between 0.005 and 0.01 in. The standard work piece is illustrated in Figs. 3 and 4. Work pieces in this study are manufactured by following British standard, BS 465638: 1995, which addresses the accuracy of machine tools and methods of test and specification for surface finish of test pieces. The focused tool path where the measurement of surface roughness is conducted are also shown in Fig. 3. The make of CNC machine used is KIA lathe model KT20 controlled by Fanuc system. The cutting tool in this study is the Carbide insert grade C5 with nose radius of 0.0156 in. and side cutting edge angle of 5° while the material used to manufacture work pieces is Aluminium Al-T3.

Table 1. Ouput and inputs.

Type of variable	Variable	Range		Unit
		Low	High	
Output	Surface roughness	–	–	Microinch
Input	Spindle speed	950	1100	Ft/min
Input	Feed rate	0.008	0.009	Inch per round
Input	Depth of cut	0.005	0.01	Inch

5 Empirical Results

The experiment was conducted by varying three inputs at the different levels. All inputs, spindle speed, feed rate and depth of cut, is graphically illustrated in Figs. 4, 5 and 6 consecutively. The work pieces are manufactured every minute and total number of work pieces are 334 pieces. The corresponding output in term of surface roughness is shown in Fig. 7. Due to Fig. 4, the spindle speed is set at 950 Ft/min from piece 1–40, 81–120, 181–220 and 261–300. From Figs. 4, 5, 6 and 7, the x-axis represents the time which each work piece is manufactured.

According to Fig. 5, feed rate is set at 0.008 in. per round from piece 1–80 and 181–260. On the other hand, the depth of cut is set at 0.005 in. from piece 1–180 while the rest (piece 181–340) is manufactured at 0.01 in. The surface roughness (R_a) corresponded to the above three factors are measured and shown in Fig. 7 which indicates

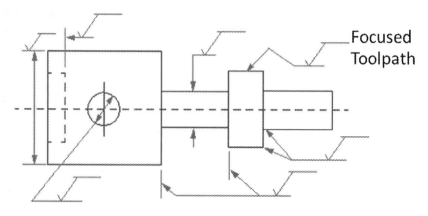

Fig. 3. Manufactured work pieces.

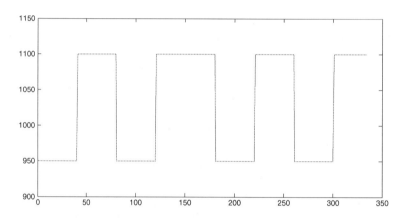

Fig. 4. Spindle speed.

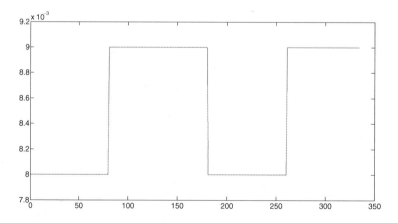

Fig. 5. Feed rate.

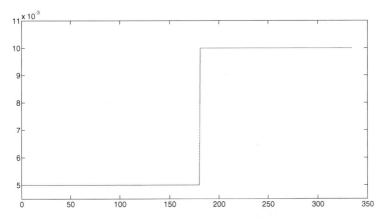

Fig. 6. Depth of cut.

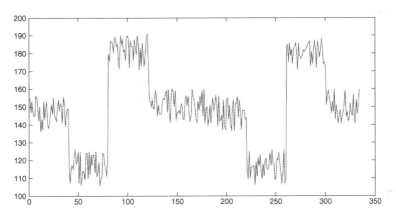

Fig. 7. Surface roughness.

that surface roughness moves like a wave form at seven levels. At each level, the surface roughness is fluctuated around the fixed value like a random jitter.

6   Analysis

Toolboxes equipped in MATLAB software package are utilized to conduct the data analysis of experimental data. Since there are two optimization methods used in this research study, the results are categorized into two separating reports according to the deployed method.

6.1 Artificial Neural Network

To construct a model to predict surface roughness by ANN, there are three inputs (spindle speed, feed rate and depth of cut) and the output is surface roughness while the number of hidden layers are 20 as concluded in Fig. 8. The network is trained by using Levenberg–Marquardt back-propagation algorithm. The total number of 234 samples are used to train the neural network while 100 samples are holded for the purposes of validation and testing (50 samples for each purpose).

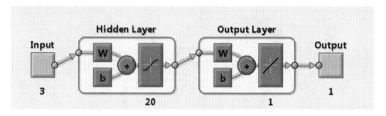

Fig. 8. Inputs, layers and output of neural network.

Due to Fig. 9, the mean squared errors (MSE) of the training, validation and testing are as follows: 34.6357, 34.85068 and 32.60469. Moreover, the R^2 of each model is between 96 to 97 % which indicates that ANN is sufficient enough to explain the relationship between inputs and output. The regression plot of the model derived from each stage is shown in Fig. 10.

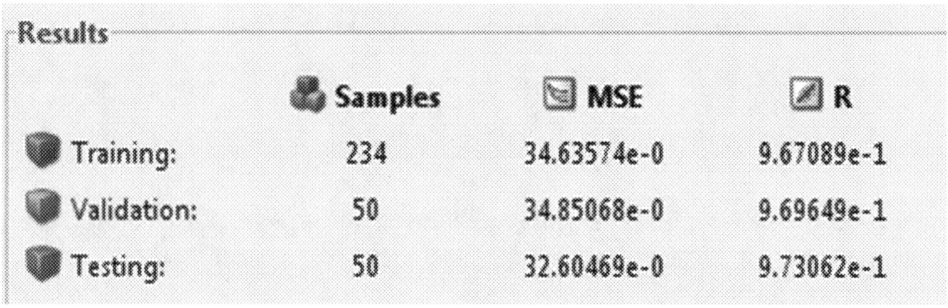

Results	🐌 Samples	✉ MSE	☑ R
🔴 Training:	234	34.63574e-0	9.67089e-1
🔴 Validation:	50	34.85068e-0	9.69649e-1
🔴 Testing:	50	32.60469e-0	9.73062e-1

Fig. 9. The summary of ANN model.

6.2 Factorial Design

The factorial design 2^k is utilized to analyze the relationship between inputs and output. The main effect and the interaction between factors are explored and shown in Fig. 11. According to Fig. 11, x_1 is the spindle speed while x_2 is the feed rate and depth of cut is x_3. When spindle speed increases, the surface roughness tends to reduce significantly. The same tendency is also found in the case of depth of cut but the surface roughness seems to decrease with a slow rate. On the other hand, the surface roughness seems to

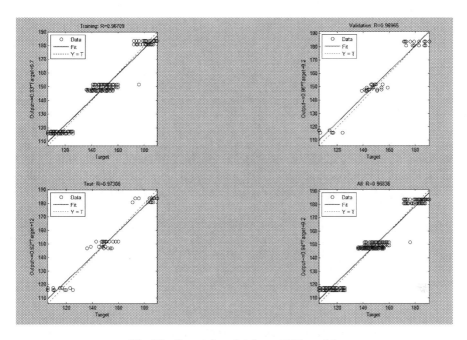

Fig. 10. Regression plot from ANN model.

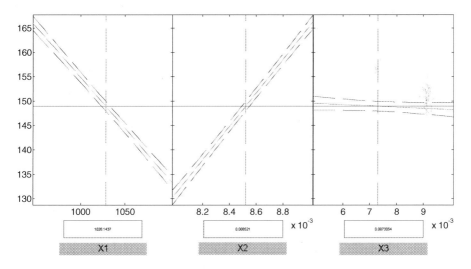

Fig. 11. Main effect plot.

increase dramatically when feed rate is set at the high level. The mathematical model obtained from this method is shown in (5) which indicates that there is no interaction among each input. In this case, the forecasting error (MSE) is as low as 34.32788.

$$\text{Surface roughness} = 71.16 - 0.21 * \text{Spindle speed} + 34389.79 * \text{Feed rate} \\ - 260.55 * \text{Depth of cut} \tag{5}$$

7   Conclusions

Two optimization methods, neural network and factorial design, are deployed to model and forecast the surface roughness of work pieces based on three machining inputs, i.e., spindle speed, feed rate and nose radius. Both methods are capable to construct models which are able to efficiently predict the surface roughness since the value of R^2 for these methods is higher than 96 %. In term of forecasting errors, the MSE of factorial design is not much different from the MSE of ANN. Therefore, for this case study, there is a reason to believe that the performance of these two methods is not significantly different.

8   Discussions

According to the results, the forecasting errors obtained from neural network and factorial design models are not significantly different. The possible reason is that empirical experiment is conducted at the fixed levels of inputs, i.e., each input is set at two different levels and there are only eight combinations of the experimental levels because it is the limitation of the manufacturer. Since the number of data for the same treatment is high, it is possible that the capability of training the networks of ANN is not used at its full potential. As a result, the knowledge achieved from the experiment is limited to only the eight different conditions. Therefore, the factorial design which is repeated the experiment for the same treatment has about the same performance as the neural network.

References

1. Choudhury, I.A., El-Baradie, M.A.: Tool-life prediction model by design of experiments for turning high strength steel (290 BHN). J. Mater. Process. Tech. **77**, 319–326 (1998)
2. Wang, X., Feng, C.X.: Development of empirical models for surface roughness prediction in finish turning. Int. J. Adv. Manuf. Tech. **20**(5), 348–356 (2002)
3. Arbizu, I.P., Perez, C.J.L.: Surface roughness prediction by factorial design of experiments in turning processes. J. Mater. Process. Tech. **143**, 390–396 (2003)
4. Benga, G.C., Abrao, A.M.: Turning of hardened 100Cr6 bearing steel with ceramic and PCBN cutting tools. J. Mater. Process. Tech. **143–144**, 237–241 (2003)
5. Ozel, T., Hsu, T.-K., Zeren, E.: Effects of cutting edge geometry, workpiece hardness, feed rate and cutting speed on surface roughness and forces in finish turning of hardened AISI H13 steel. Int. J. Adv. Manuf. Tech. **25**(3–4), 262–269 (2005)

6. Davim, J.P.: A note on the determination of optimal cutting conditions for surface finish obtained in turning using design of experiments. J. Mater. Process. Tech. **116**(2–3), 305–308 (2001)
7. Sahin, Y., Motorcu, A.R.: Surface roughness model for machining mild steel with coated carbide tool. Mater. Design. **26**, 321–326 (2005)
8. Pontes, F.J., De Paiva, A.P., Balestrassi, P.P., Ferreira, J.R., Da Silva, M.B.: Optimization of radial basis function neural network employed for prediction of surface roughness in hard turning process using Taguchi's orthogonal array. Expert Syst. Appl. **39**, 7776–7787 (2012)
9. Natarajan, C., Muthu, S., Karuppuswamy, P.: Investigation of cutting parameters of surface roughness for brass using artificial neural networks in computer numerical control turning. Aust. J. Mech. Eng. **9**, 35–45 (2012)

FPGA Implementation of Neuron Model Using Piecewise Nonlinear Function on Double-Precision Floating-Point Format

Satoshi Kawamura[1(✉)], Masato Saito[2], and Hitoaki Yoshida[3]

[1] Iwate University Super Computing and Information Sciences Center,
Ueda, Morioka, Iwate 020-8550, Japan
kawamura@iwate-u.ac.jp
[2] P&A Technologies Inc., Ueda, Morioka, Iwate 020-0834, Japan
m_saito@pa-tec.com
[3] Faculty of Education, Iwate University, Ueda, Morioka, Iwate 020-8550, Japan
hitoaki@iwate-u.ac.jp

Abstract. The artificial neurons model has been implemented in a field programmable gate array (FPGA). The neuron model can be applied to learning,
training of neural networks; all data types are 64 bits, and first and second-order
functions is employed to approximate the sigmoid function. The constant values
of the model are tuned to provide a sigmoid-like approximate function which is
both continuous and continuously differentiable. All data types of the neuron are
corresponding to double precision in C language. The neuron implementation is
expressed in 48-stage pipeline. Assessment with an Altera Cyclone IV predicts
an operating speed of 85 MHz. Simulation of 4 neurons neural network on
FPGA obtained chaotic behavior. An FPGA output chaos influenced by calculation precision and characteristics of the output function. The circuit is the
estimation that above 1,000 neurons can implement in Altera Cyclone IV. It
shows the effectiveness of this FPGA model to have obtained the chaotic
behavior where nonlinearity infuences greatly. Therefore, this model shows
wide applied possibility.

Keywords: Artificial neurons model · Field programmable gate array
(FPGA) · Sigmoid function · Chaotic behavior · Piecewise nonlinear function

1 Introduction

Neural networks (NNs) are used for analysis of problems that defy computer processing, such as pattern recognition and some kinds of time series analyses [1–3].
Generally, recent applications of NNs employ large numbers of neurons. For some
problems, the neurons perform numerous operations (multiplications or additions), but
for others, they perform calculations of nonlinear functions. Software can accomplish
such tasks with adequate precision, but can be too slow. This is the reason research into
NN processing has used specialized hardware.

When a neuron model is implemented in hardware, it is critical to consider the
number of bits in the internal state of the neuron and the method for realizing the

© Springer International Publishing Switzerland 2016
H. Fujita et al. (Eds.): IEA/AIE 2016, LNAI 9799, pp. 620–629, 2016.
DOI: 10.1007/978-3-319-42007-3_54

nonlinear output function. Systems using 16-bit fixed point numbers in the internal state and table lookup or piecewise linear output functions have been reported [4, 5], but it is not enough to compare the number of bits in the internal state with the number of bits in the software solver. Other research has attempted to provide a more precise approximation of a sigmoid function using functions based on a continued fractional expansion, but increases in precision require expansion of the circuit scale, slowing the clock speed [6]. Also, since that author used a nonlinear output function which differed considerably from the sigmoid function, the behavior of the model was expected to differ considerably from that of the software representation. Thus, the hardware implementation of NNs is still running quite far behind software implementation. The article that uses Look-up Table (LUT) for implementation of the sigmoidal activation function for the implementation of the nonlinear function [7, 8]. A differential coefficient is necessary for NNs learning, however, these models are not enough for precision as an output function of a neuron. By and large, there are not most of hardware models who balanced the precision of data (data type) with a function of the precision that learning requires. Furthermore, there is not a differential coefficient in most hardware models.

This study employs floating-point expressions in the internal state to express neuron hardware and considered an implementation of a nonlinear output function [9]. The synaptic weight (weight coefficient) and internal state are double-precision floating-point format, according to IEEE 754 standard. Instead of a sigmoid function, which has difficulties with exponential functions and division, first and second-order functions are employed in piecewise nonlinear approximation functions to comprise the model. The differential coefficients of the functions are set so as to provide approximate continuity and differentiability of the boundary. As for this model, precision of the number representation is a double-type, and a differential coefficient is present. The differential coefficient is necessary for a learning process of neural networks [1–3]. Therefore, this model is useful for general neural network applications widely. This eliminated the discrepancies between software-derived solutions and hardware-derived solutions and expanded the scope of problems that can be addressed with hardware.

Verilog HDL, a hardware description language, is used to write this neuron model and the modeling is carried out on an Altera Cyclone IV field programmable gate array (FPGA). As a result of using the pipeline architecture with this model, a clock speed of 85 MHz has been obtained. To verify the behavior of this neuron model, an FPGA is estimated by implementation of 4 neurons neural network which output chaos. Chaotic behavior is strongly influenced by calculation precision and characteristics of the function. As a result of simulation, chaotic output are obtained from an FPGA.

2 Neurons Model and Neural Network Model

Figure 1(a) shows the neuron model employed in this study and Fig. 1(b) shows an example of a neural network model which provides a chaotic response [9, 10]. This neuron model is commonly used in NN studies [1–3]. We use NN of Fig. 1(b) to verify this NN model. This model outputs chaos with 4 normal neurons [10, 11]. We called a chaos neural network (CNN) [10, 11]. Chaotic behavior is sensitive to output functions and the representation of data type and accuracy of data type. CNN is used for the

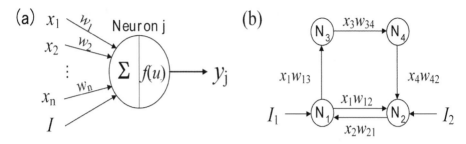

Fig. 1. Artificial Neurons model and a Chaos Neural Network (CNN) model. (a) Artificial neurons model [1–3, 10, 11]. (b) A Chaos Neural Network model [10, 11], which is composed of 4 neurons (4CNN). Where, N_j: neuron j, x_j: input of neuron j, y_j: output of neuron j, w_j: weight coefficient of input of neuron j, I_j: External input of neuron j, u: internal state (summation).

verification of the circuit, the chaos which nonlinearity influences shows the effectiveness of the proposal method. CNN output the chaos which nonlinearity influences uses for the verification of the circuit.

It obtains its internal state u, multiplying the inputs from other neurons x_i by their respective weight coefficients w_i and external input I (see Eq. (1)). The internal state u is entered into the nonlinear output function f (a sigmoid function) shown in Eq. (2) to obtain the neuron output $y_j = f(u)$. In this study, each element of the neuron model (inputs x_i from other neurons, weight coefficients w_i, external input I, neuron internal state u, neuron output y_j) are expressed in accordance with IEEE 754's standard for double-precision floating-point format.

$$y_j = f(u), \quad u = \sum_{i=1}^{n} w_i x_i \tag{1}$$

$$f(u) = \frac{1}{1 + \exp(-u)} \tag{2}$$

Let us consider neuron operations in three ways: (i) Multiplying the inputs from other neurons x_i by their weight coefficients w_i; (ii) Calculating the total sum of the products and the external input to obtain u, the internal state of the neuron; and (iii) applying the nonlinear approximation function f to the internal state to find the neuron's output $y_j = f(u)$ (Eq. (1)). In contrast, the nonlinear output function for point (iii) in Eq. (2) (the sigmoid function) includes division and the exponential function, so is difficult to express with hardware as is [6, 8]. Even if a table lookup is used, the scale of the circuit becomes impractically large for longer data types [4–6].

The behavior of NNs can be affected by the lower precision of hardware in approximations when handling the sigmoid function [9]. For example, the principle of learning through back propagation requires an inverse for the output function, but up to the present, this has not been practical because it has been difficult to obtain derivatives for the functions implemented by circuits [6, 8]. A second-order function-based piecewise nonlinear approximation function was used in this study to ease implementation in a circuit [9] (Eq. (3)). This can be implemented using just multiplication and addition. One consideration in setting the locations for the boundary in the piecewise function is that the sigmoid function has abrupt changes in output values in the domain – 12.0 < u < 12.0. Therefore, this domain is divided into four zones and the approximation function is constructed while imposing constraint conditions at the boundaries between the partial domains for each second-order function. The constants are determined by forcing the first derivatives of adjacent functions to be equal at the boundaries. The output value of the sigmoid function converges to zero in the negative portion of the domain, where its values are low, and to one in the positive portion, so it is set to constant values of 0 in u less than -12.0 and 1 in u more than 12.0. The behavior of the neuron with the approximation function has been confirmed by simulation experiment using C-language. Statistical properties of random numbers extracted from outputs of 4CNN have also confirmed by NIST SP800-22 test suite [12].

$$\begin{cases} f(u)=1.0 & ,12.0<u \\ f(u)=-0.000337592(u-12.0)^2+1.0 & ,3.75<u\leq12.0 \\ f(u)=-0.0324362(u-3.83587)^2+0.977262 & ,0.0<u\leq3.75 \\ f(u)=0.0324362(u+3.83587)^2-0.022738 & ,-3.75<u\leq0.0 \\ f(u)=0.000337592(u+12.0)^2+0.0 & ,-12.0<u\leq-3.75 \\ f(u)=0.0 & ,u\leq-12.0 \end{cases} \quad (3)$$

3   Implementation of the Neuron Model Using FPGA

Figure 2(a) is a block diagram of the circuit realized using FPGA. The implementation for numerical values in neuron inputs and outputs conformed to the IEEE 754 standard for double data types (64-bit) in this circuit. The neuron operations explained in Sect. 2 are used, i.e., the internal state u is calculated as Eq. (1). Next, the output function f is applied to u find $y_i = f(u)$. Pipeline architecture is used to speed this procedure in the computational unit and each block. Figure 2(a) and (b) shows latencies which each blocks and processing required. The total latency is 48. This model is composed of a 48-stage pipeline. Equation (3) is implemented in the form $f(u) = A(u + B)^2 + C$, rather than $f(u) = Au^2 + Bu + C$. This allows the computation to proceed with one second-degree calculation, one multiplication and two additions (including subtraction), while the latter case would have required one second-degree calculation, two multiplications and two additions, i.e., one extra operation. Constant values of A, B and C satisfies these requirements. Figure 2(c) shows constant values of the approximate function of Eq. (3).

Individual neurons are used in a time-sharing architecture in this NN model. In other words, the maximum number of neurons that can be implemented is equal to the number of steps in the pipeline. In the case of 48-stage pipeline, 48 neurons work by time sharing. Nevertheless, in order to simplify the calculations for neuron internal state u, a neural network consisting of four neurons is considered, as shown in Fig. 1(b). The inputs into a neuron at each moment are implemented by the controller, which selects and produces a combination of the inputs with the past outputs in the memory block and their weight coefficients. In this model, the input from other neurons is coded only

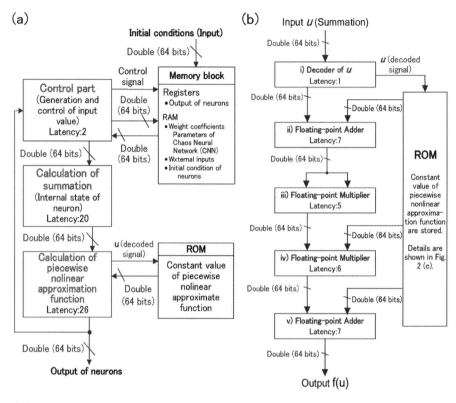

Fig. 2. Neuron model implemented on an FPGA (it also shows latency, as it was used for pipeline processing). (a) Block diagram of neuron model on FPGA. Total latency is 48. (b) Computational unit for piecewise nonlinear approximation function. The input values for u were found using the value of u, resulting in output of corresponding constants. (c) The constants of the approximation function stored in a ROM.

with two neurons. Therefore, an internal state of one neuron is up to three inputs, other 2 neurons and an external input.

4 Results of Evaluation by FPGA

Figure 3 shows a simulation result in FPGA. A development environment of a compilation and the simulation describes Fig. 3(a). In Fig. 3(b), compilation results are shown.

(a)

Software	Quartus II 64–Bit Version 14.0.0 Build 200 06/17/2014 SJ Full Version
Target FPGA	Cyclone IV
Target Device	EP4CGX150DF31C7
Frequency (fmax)	85.46MHz

(b)

Resource	Usage of this model	Amount of the whole FPGA	Usage rate (%)
Total logic elements	12,612	149,760	8.4%
Total combinational functions	11,486	149,760	7.7%
Dedicated logic registers	5,979	149,760	4.0%
Total pins	288	508	56.7%
Total memory bits	23,564	6,635,520	0.4%
Embedded Multiplier 9–bit elements	72	720	10.0%

(c)

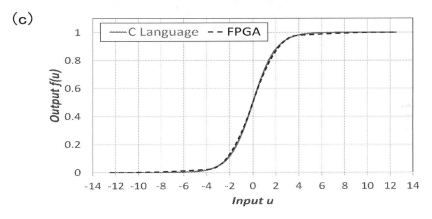

Fig. 3. Results of comparison with an FPGA. (a) Compiling conditions (software, settings, target FPGA). The model operated at a maximum of 85 MHz. (b) Compilation result. This shows the amount of the resources which this model uses on this FPGA. (c) Output of C language's sigmoid function vs. FPGA's piecewise nonlinear approximation function (this model)

Figure 3(c) provides the results when the input and output characteristics of this model are simulated with the Quartus II; these are similar to the characteristics resulting when the sigmoid function is used. The results confirm the effectiveness of this model and it's piecewise nonlinear functions. The joint of the function is smooth, and the output of sigmoid function correspond to the output of the approximate function. From this result, this model verified that it functioned expectedly.

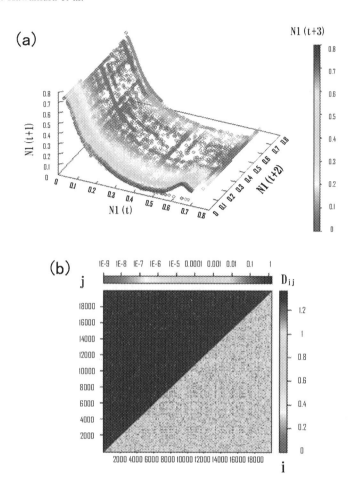

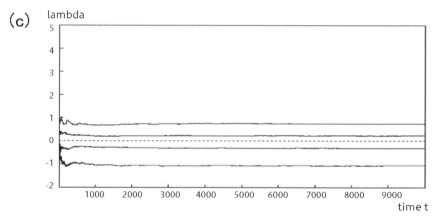

Fig. 4. A chaotic output created by the FPGA. The system consisting of 4 neurons produced chaos [10, 11]. (a) Strange Attractor. (b) Recurrence Plot. (c) Lyapunov spectrum.

4CNN is simulated on FPGA to verify the effectiveness of this model. 4CNN is a neural network system for constituting chaos (Fig. 1(b)) [9–11]. In addition, the chaos where nonlinearity influences shows the effectiveness of the proposal method.

The outputs of 4CNN are analyzed with chaos time series analysis [11, 13, 14]. The results are shown in Fig. 4(a) a reconstructed attractor in 4D phase space, (b) a recurrence plot, and (c) a Lyapunov spectrum. The results suggest that the 4CNN constructed in the FPGA reproduces same chaotic behavior as 4CNN implemented by C-languages [9].

Finally the realizability of the proposal method is estimated. When it is current commercially available FPGA, the number to be equipped with of neurons was calculated. Table 1 shows estimation of numbers of neurons that it is implemented the same model in the largest same FPGA series. By the estimation, this FPGA can implement 1104 neurons. In this manner, as for this model, FPGA can implement many neurons.

Table 1. The number of neurons on FPGA implimentation.

	Number of logic elements	The number of neurons
Size of 1 Neuron	12616	48 48–stage pipeline, time sharing
Cyclone IV (Used in this study)	149760	$11 \times 48 = 528$ Calculated value, 48–stage pipeline, time sharing
Cyclone V (Maximum circuit scale)	301000	$23 \times 48 = 1104$ Calculated value, 48–stage pipeline, time sharing

From these results, it is considered that this model is the performance that is equal to the neuron model that C language described. And this model uses the approximate function with the differential coefficient. Therefore, this model is usable by general applications.

5   Conclusion

A model of a hardware neuron has been implemented in a simulated FPGA. The neuron internal state is expressed in double-precision floating-point format so that it is expected to apply to a general NN, and the model has approximated a sigmoid function with piecewise second-degree functions; no exponential functions or no division operations are used, since these are difficult to realize in circuit form with sigmoid functions.

The components of the approximation function are also controlled to obtain identical first derivatives of the function boundaries. In order to fast operation of the computational units, pipeline processing is used in the design of the circuits. The model is assessed using Altera Cyclone IV, and thus indicated that a hardware model can be built to attain a speed of 85 MHz. A 4CNN composed of this hardware model was created and showed similar results to those found in the software version. It shows the effectiveness of this FPGA model to have obtained the chaotic behavior where non-linearity influences greatly. In this manner, this model shows wide applied possibility. Furthermore, it is suggested that commercial FPGA is equipped with above 1,000 neurons. As for this model, NN applications by the learning need a nonlinearity and a differential coefficient is expected.

Acknowledgments. This study was supported by a grant from the Yume-Kendo Dream-Land Iwate Foundation of the Iwate Prefecture. We thanks to Dr. Takeshi Murakami for his support who participated in the experiment. The calculations in this study have partly performed with the SGI UV-100 in Iwate University Super-Computing and Information Sciences Center (ISIC). Special thanks to the stuff members of ISIC.

References

1. Russell, B., Tom, B.: Neural Computing: An Introduction. Adam Hilger, Bristol (1990)
2. Fredric, M.H., Ivica, K.: Principles of Neurocomputing for Science & Engineering. McGraw-Hill, Singapore (2000)
3. Sandhya, S.: Neural Networks for Applied Sciences and engineering. Auerbach Publications, New York (2006)
4. Rafael G., Joaquin C., Joaquin C., Francisco B., Francisco B., Antonio M.S., Antonio M.S.: Artificial neural network implementation on a single FPGA of a pipelined on-line backpropagation. In: Proceedings of the 13th International Symposium on System Synthesis, pp. 225–230 (2000)
5. Kwan, H.K., Tang, C.Z.: Multiplierless multilayer feedforward neural network design using quantised neurons. Electron. Lett. **38**(13), 645–646 (2002)
6. Basterretxea, K., Tarela, J.M., del Campo, I.: Digital design of sigmoid approximator for artificial neural networks. Electron. Lett. **38**(1), 35–37 (2002)
7. Sledevic, T., Navakauskas, D.: The lattice-ladder neuron and its training circuit implementation in FPGA. In: 2014 IEEE 2nd Workshop on Advances in Information, Electronic and Electrical Engineering (AIEEE), pp. 1–4 (2014)
8. Abrol, S., Mahajan, R.: Implementation of single artificial neuron using various activation functions and XOR Gate on FPGA chip. In: 2015 Second International Conference on, Advances in Computing and Communication Engineering (ICACCE), pp. 118–123 (2015)
9. Kawamura, S., Nakanishi, T., Yoshida, H., Ozeki, K., Fujimaki, K., Gotoh, R.: Studies on the accuracy of numerical operations with embedded CPUs. IEICE Electron. Express **3**(8), 149–155 (2006)
10. Kawamura, S., Yoshida, H., Miura, M.: Minimum constituents of chaos neural network composed of conventional neurons. Electron. Commun. Jpn. Part III **86**(7), 62–71 (2003)
11. Yoshida, H., Murakami, T., Zhongda, L.: High-speed and highly secure pseudo-random number generator based on chaos neural network. Front. Artif. Intell. Appl. **276**, 224–237 (2015)

12. Rukhin, A., Soto, J., Nechvatal, J., Smid, M., Barker, E., Leigh, S., Levenson, M., Vangel, M., Banks, D., Heckert, A., Dray, J., Vo, S.: A Statistical Test Suite for Random and Pseudorandom Number Generators for Cryptographic Applications, NIST SP800-22 rev.1a, Revised: July 2015 (sts-2.1.1). Lawrence E. Bassham III (2015)
13. Mario, M.: Introduction to Discrete Dynamical Systems and Chaos. Wiley InterScience, New York (1999)
14. Henry, D.I.A.: Analysis of Observed Chaotic Data. Springer, New York (1997)

Innovations in Intelligent Systems and Applications

Video Inpainting in Spatial-Temporal Domain Based on Adaptive Background and Color Variance

Hui-Yu Huang[(⊠)] and Chih-Hung Lin

Department of Computer Science and Information Engineering,
National Formosa University, Yunlin 632, Taiwan
hyhuang@nfu.edu.tw

Abstract. Video inpainting is repairing the damage regions. Nowadays, video camera is usually used to record the visual memory in our life. When people recorded a video, some scenes (or some objects) which unwanted are presented in video sometimes, but it doesn't record repeatedly based on some reasons. In order to solve this problem, in this paper, we propose a video inpainting method to effectively repair the damage regions based on the relationships of frames in temporal sequence and color variability in spatial domain. The procedures of the proposed method include adaptive background construction, removing the unwanted objects, and repairing the damage regions in temporal and spatial domains. Experimental results verify that our proposed method can obtain the good structure property and extremely reduce the computational time in inpainting.

Keywords: Video inpainting · Repair · Removal object · Color variance

1 Introduction

Video inpainting is restored the damaged region which removed the foreground objects (motion objects) in video. In order to obtain the foreground object from background, the Gaussian mixture model (GMM) [1] or background subtraction method [2, 3] is usually adopted. Although these methods can remove the foreground objects, it still exists some noise data around the boundary for those objects which cannot be completely removed from the background. Hence, in this paper, we will establish an adaptive background to obtain the foreground regions more contact, thus, it will benefit to repair the damaged regions after object removal in the background.

As for inpainting literature, there are many existing inpainting technologies have been published [2–6]. Koochari and Soryanni [2] used the exemplar-based inpainting method to fill the damaged regions based on large patches. The moving objects separated from the background and the missing areas of the background were adopted exemplar-based method to repair. Ghanbari and Soryani [3] proposed a video inpainting used patch-based method with the help of contour-based method and large patches to fill the holes. Criminisi et al. [4] proposed an exemplar-based method to fill the removing regions from an image which mainly computed the priority values with

© Springer International Publishing Switzerland 2016
H. Fujita et al. (Eds.): IEA/AIE 2016, LNAI 9799, pp. 633–644, 2016.
DOI: 10.1007/978-3-319-42007-3_55

confidence term and data term given a patch. The higher the priority value at the point in the target region, the earlier the repairing of the point. The filling strategy is region-based filling. The best matching patch in source region is most similar to the patch with the highest priority and then to fill it with data extracted from this best patch. Zarif et al. [5] proposed a recovery method for video based on frame local similarity. Authors used the color or texture in neighboring block to estimate the local symmetrical similarity for horizontal or vertical axis in Fourier transform of any frame. The missing region is filled by using this local symmetrical similarity, it can further maintain the spatial and temporal consistency. Ling et al. [6] proposed an object-based and patch-based video inpainting mechanism to repair occluded object using posture sequence matching technique. This method first samples a 3-D volume of the video into directional spatio-temporal slices, and then used patch-based image inpainting to repair the partially damaged object trajectories in the 2-D slices.

In this paper, we propose a novel video inpainting method in spatial-temporal domain. In temporal property, the repairing is based on the correlation between frames. In spatial domain, the repairing is to modify the data term with color variability in exemplar-based inpainting method.

The remainder of this paper is organized as follows. Review of exemplar-based inpainting method is presented in Sect. 2. Section 3 presents the proposed method. Experimental results are presented in Sect. 4. Finally, Sect. 5 concludes this paper.

2   Review of Exemplar-Based Inpainting Algorithm

In this section, we will briefly describe the exemplar-based inpainting method proposed by [4]. The core of exemplar-based method is an isophote-driven image process. This method is capable of propagating both linear structure and texture information into the target region.

First, given an input image I, the target region Ω by user defined is filled by selecting the best patch from the source region Φ. After computing the filling priority of all the pixels along the boundary of target region, the pixel p with the highest priority is used as the center pixel to choose the target patch Ψ_p (9×9 size) to be filled. The priority of a pixel p is calculated as follows

$$P(p) = C(p)D(p), \tag{1}$$

where $C(p)$ and $D(p)$ are confidence and data terms, respectively. The confidence $C(p)$ a measure of the amount of reliable information surrounding the pixel p, $D(p)$ is data term which includes the structure information the fill front, and they are defined as follows

$$C(p) = \frac{\sum_{q \in \Psi_p \cap \Phi} C(q)}{|\Psi_p|}, \quad 0 \le C(p) \le 1, \tag{2}$$

$$D(p) = \frac{|\nabla I_p^\perp \cdot n_p|}{\alpha}, \quad 0 \le D(p) \le 1, \tag{3}$$

where $|\Psi_P|$ is the area of Ψ_P, n_p denotes the unit vector normalized the boundary of the target region at pixel p, $\nabla I_p^\perp$ denotes the unit vector which is orthogonal to the image gradient, signal $\perp$ denotes the orthogonal operator, and α is the normalization factor (e.g. $\alpha = 255$ for grayscale image).

Finding the patch $\Psi_{\hat{p}}$ with the highest priority, the method begins to search a best match patch $\Psi_{\hat{q}}$ from the source region.

$$\Psi_{\hat{q}} = \arg \min_{\Psi_q \in \Phi} d\left(\Psi_{\hat{p}}, \Psi_q\right), \tag{4}$$

where the distance $d\left(\Psi_{\hat{p}}, \Psi_q\right)$ calculates the sum of squared differences (SSD) of the known pixels between the patch $\Psi_{\hat{p}}$ and the patch Ψ_q.

Copying the known pixels of patch $\Psi_{\hat{q}}$, each of unknown pixels of patch $\Psi_{\hat{p}}$ corresponding to the positions will be filled. After filled with new pixel values for patch $\Psi_{\hat{p}}$, the confidence value is updated

$$C(q) = C(\hat{p}), \quad \forall q \in \Psi_{\hat{p}} \cap \Omega. \tag{5}$$

The method is executed iteratively until the target region is filled. The details of the exemplar-based procedures can survey [4].

3 Proposed Approach

In this section, we will describe our method which includes adaptive background, temporal domain, and spatial domain inpainting.

3.1 Adaptive Background

According to background subtraction method, the foreground objects (motion objects) obtained by computing the difference between input data and pre-built background [7] can be extracted and removed from each frame in video sequence. But the removed regions are usually not compactly close to the real foreground objects, it is because the threshold decided the foreground region is used a fixed value. In order to solve this problem, a gradient degree characteristic derived from pixels in building background model is used to define the adaptive threshold and then to construct the adaptive background. The construction of adaptive background describes in the following subsection.

(1) **Background model**

First, a progress background is firstly constructed by [7]. According to the property of progress background, each of frames in video can generate the belonging partial background image, and a complete background is obtained with increasing input frame, as shown in Fig. 1. In this paper, we take the background generated by the last frame as background model (*BG*) which is used to obtain the adaptive threshold and adaptive background.

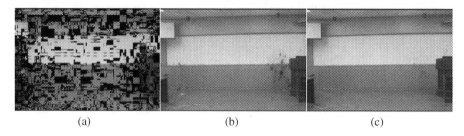

(a)   (b)   (c)

Fig. 1. The partial background image. (a) no. 1 frame, (b) no. 30 frame, and (c) no. 50 frame.

(2) **Adaptive background construction**

Based on progress background, it can obtain the gradient degree value corresponding to each pixel position for each frame in a video sequence, and then we compute the number of each of degree values for each pixel position. The gradient degree value $g_i(t, a)$ is gray-level value. Afterwards, we will further divide this histogram of gradient degree into some grounds. The difference between two adjacent gradient degree values is less and equal to 10, thus it divides the same group. Then we can obtain the minimum and the maximum values of each group. Figure 2 shows an example.

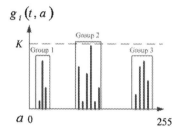

Fig. 2. An example for groups of the gradient degree value for the ith pixel point of frame at t time.

According to the above result, we also take the gradient degree value obtained from the last frame to decide the adaptive threshold, each of pixel positions can be given an adaptive threshold by

$$Adaptive\_th(x, y) = \max(|BG(x, y) - G_{n\_\min}|, |BG(x, y) - G_{n\_\max}|) + d, \quad (6)$$

where $Adaptive\_th(x, y)$ and $BG(x, y)$ denote the adaptive threshold and background model at the coordinates (x, y), respectively. $G_{n\_\min}$ and $G_{n\_\max}$ are the maximum and minimum values belonging to the nth group for the last frame, respectively. d is an offset value and here we set 5. By Eqs. (7) and (8), the foreground objects can be more compactly removal from the background and an adaptive background can be obtained. The foreground object and adaptive background are given by

$$binaryB_i(x, y) = \begin{cases} 255, & \text{if } |S_i(x, y) - BG(x, y)| > Adaptive\_th(x, y), \\ 0, & \text{otherwise,} \end{cases} \quad (7)$$

$$AdaptiveB_i(x, y) = \begin{cases} 0, & \text{if } binaryB_i(x, y) = 255, \\ S_i(x, y), & \text{otherwise,} \end{cases} \quad (8)$$

where $binaryB_i(x, y)$ and $S_i(x, y)$ denote the binary image and the original one at the coordinates (x, y) for the ith frame, respectively. However, this $binaryB_i(x, y)$ may have some noise data around the boundary of foreground objects. In order to solve this problem, we will further take this binary image to do morphologic process. And the final results are shown in Fig. 3.

(a)   (b)   (c)

Fig. 3. Result of the removal object for an image. (a) no. 10 frame, (b) no. 50 frame, and (c) no. 90 frame.

3.2 Temporal Domain Inpainting

In our inpainting strategy, it divided into two parts: temporal domain inpainting and spatial domain inpainting. In temporal domain, we use the characteristic for high relationship within frames in temporal sequence to refill the damaged areas. The inpainting process is pixel-based process. First, we search and take the pixel values for non-damaged area within some forward and backward frames corresponding to the position of the damaged areas in current frame, and compute the average for these pixel values in each position. Then, the current damaged areas can be filled by using the corresponding average pixel value for each position. If the damaged area in current frame is presented in all of reference frames (forward and backward frames), then this position within the damaged area doesn't repair in temporal domain. It will repair in spatial domain. In our experiments, we set three forward and three backward reference frames with four intervals based on Chen [8]. Figure 4 shows the profile for temporal domain inpainting. F_1, F_C, and F_N denote the first frame, current frame and the end one, respectively. Interval is the frame interval and set four. Red color denotes the damaged region for each frame in temporal sequence.

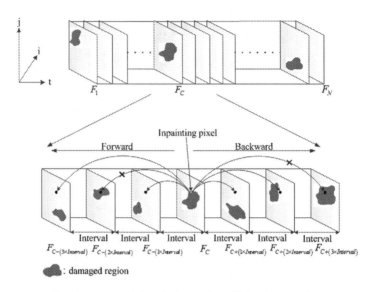

Fig. 4. Temporal domain inpainting. (Color figure online)

The inpainting in temporal domain is defined as:

$$CurFrame(x,y) = \frac{Sum(x,y)}{Count(x,y)},\qquad(9)$$

$$Sum(x,y) = \begin{cases} \sum\limits_{\substack{t = F_C - (R \times Interval), \\ R = -3 \sim 3, \neq 0}} F(x,y,t), & \text{if } F(x,y,t) \neq \text{Target region}, \\ \text{Skip}, & \text{otherwise}, \end{cases}\qquad(10)$$

$$Count(x,y) = \begin{cases} \sum\limits_{\substack{t = F_C - (R \times Interval), \\ R = -3 \sim 3, \neq 0}} Count(x,y), & \text{if } F(x,y,t) \neq \text{Target region}, \\ \text{Skip}, & \text{otherwise}, \end{cases}\qquad(11)$$

where the initial $Count(x,y)$ and $Sum(x,y)$ set 0. $F(x,y,t)$ denotes the pixel value at t time in temporal sequence. Figure 5 shows an example results for video inpainting. The black regions in Fig. 5 cannot fill in this stage. It is because all of reference frames are the damaged regions. That is, there is not any data that can be used to repair. Hence, we will repair the rest of damaged regions in the spatial domain.

3.3   Spatial Domain Inpainting

In spatial domain inpainting, we use exemplar-based method to repair the rest of the damaged regions. In exemplar-based method, we further modify the data term which involved with color variance within the target region and then use this to decide which the damage regions have high priority for inpainting. On the other hand, local searching

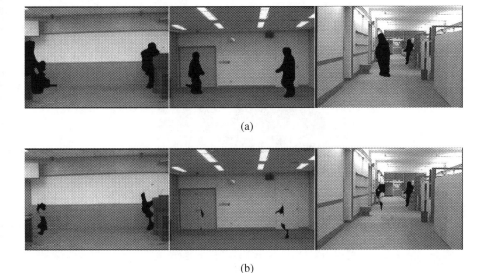

Fig. 5. Results of the temporal domain inpainting. (a) Removal objects. (b) The results.

strategy instead to global searching is used to find the best matching block which can extremely reduce the cost of inpainting time. Finally, the target regions have been replaced with visually plausible backgrounds after inpainting. According to our strategy and improved [4] method, the priority is rewritten as:

$$P_V(p) = C(p)D_V(p), \tag{12}$$

where $D_V(p)$ denotes the modified data term and is defined as:

$$D_V(p) = D(p)H(p), \tag{13}$$

where $H(p)$ denotes color variance in the patch Ψ_p for hue component in HSV space. The default patch size is of 9×9 pixels. The bigger the $H(p)$, the richer the color in the patch. First, we assume that the weight value closed to the point p in the patch is higher than other positions, as shown in Fig. 6. The weight value sets $\omega_{L1} = 0.4$, $\omega_{L2} = 0.3$, $\omega_{L3} = 0.2$, and $\omega_{L4} = 0.1$. Then, we compute and record the hue value with the highest number in the patch. Next, according to the positions of this hue value and its corresponding weight value in the patch, a color weight for hue component is obtained by

$$Color_{weight} = \begin{cases} \sum_{y=0}^{8} \sum_{x=0}^{8} \omega(x,y), & \text{if } \omega(x,y) \neq \text{Target region,} \\ \text{Skip,} & \text{otherwise.} \end{cases} \tag{14}$$

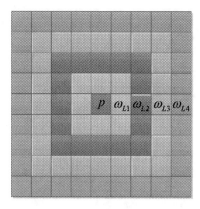

Fig. 6. Assignment of the weight value in the patch Ψ_p. (Color figure online)

In addition, we normalize this $Color_{weight}$ by

$$Color_{normal} = \frac{Color_{weight}}{N}, \quad 0 \leq Color_{normal} \leq 1, \tag{15}$$

where N is normalization factor of weight. Color variance is defined as follows:

$$H(p) = 1 - Color_{normal}. \tag{16}$$

The general searching strategy is usually used the global search by means of the sum of squared differences (SSD) method. In order to reduce the cost of inpainting time, we adopt the local search instead of global search. In addition, in order to obtain more advantage information, we further extend the interval of local search for 4 times of a patch. That is, the search region is a size of 81 × 81 pixels.

4   Experimental Results

Here, we used three test data to verify our proposed method. The first two test data are a resolution of 352 × 240 with a frame rate 29.97 fps and 119 frames from [6, 9], and the last one is a resolution of 360 × 240 with a frame rate 29.97 fps and 109 frames from [10]. We worked a 3.10 GHz Intel ® Core(TM) i5-2400 CPU with 4 GB RAM PC and C ++ language.

According to the experimental results, the inpainting in temporal domain can reduce target regions by the correlation between frames. Then, the remained target regions can also be repaired in spatial domain by means of extending the local searching interval and modifying the data term. The data term uses the characteristic of color variance in target region, it can decide which one of pixel points in target regions has high priority in inpainting. So the inpainting results are clearly improved the details of the target regions.

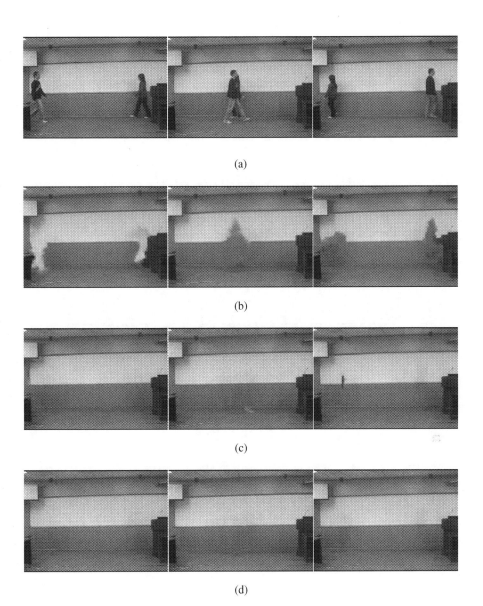

Fig. 7. Some of inpainting results in test data 1. (a) Original, (b) Das and R's method [11], (c) Chen's method [8], and (d) our proposed method.

Figures 7, 8 and 9 show a part of the experimental results in test data. Figures 7, 8 and 9(b) obtained [11] have the serious errors for inpainting. From these results, it is clear that our inpainting results are superior to [8, 11].

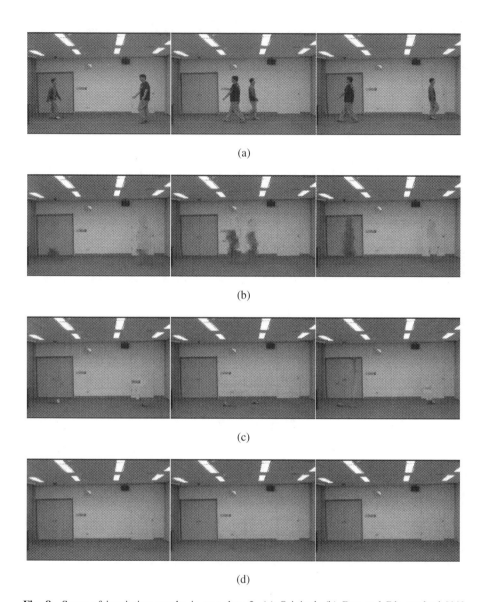

Fig. 8. Some of inpainting results in test data 2. (a) Original, (b) Das and R's method [11], (c) Chen's method [8], and (d) our proposed method.

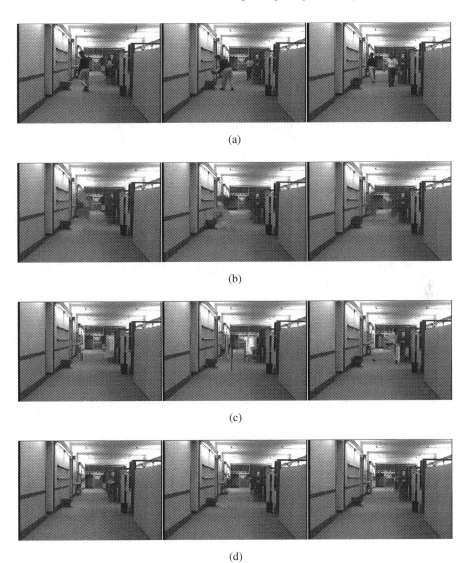

(a)

(b)

(c)

(d)

Fig. 9. Some of inpainting results in test data 3. (a) Original, (b) Das and R's method [11], (c) Chen's method [8], and (d) our proposed method.

5   Conclusions

In this paper, we proposed a video inpainting technology which consisted of temporal domain inpainting by the correlation characteristics for frames and spatial domain inpainting by the modifying exemplar-based inpainting method. Consequently, the experimental results show that the proposed method has a good inpainting result compared with [8, 11] methods.

References

1. Xia, A., Gui, Y., Yao, L., Ma, L., Lin, X.: Exemplar-based object removal in video using GMM. In: International Conference on Multimedia and Signal Processing (CMSP), pp. 366–370 (2011)
2. Koochari, A., Soryani, M.: Exemplar-based video inpainting with large patches. J. Zhejiang Univ. Sci. Comput. Electron. **11**, 270–277 (2010)
3. Ghanbari, A., Soryani, M.: Contour-based video inpainting. In: 7th Iranian Machine Vision and Image Processing (MVIP), pp. 1–5 (2011)
4. Criminisi, A., Perez, P., Toyama, K.: Region filling and object removal by exemplar-based image inpainting. IEEE Trans. Image Process. **13**, 1200–1212 (2004)
5. Zarif, S., Faye, I., Rohaya, D.: Static object removal from video scene using local similarity. In: 9th IEEE International Colloquium on Signal Processing and its Applications (CSPA), pp. 54–57 (2013)
6. Ling, C.H., Lin, C.W., Su, C.H., Mark Liao, H.Y., Chen, Y.S.: Video object inpainting using posture mapping. In: 16th IEEE International Conference on Image Processing (ICIP), pp. 2785–2788 (2009)
7. Chung, Y.C., Wang, J.M., Chen, S.W.: Progressive background images generation. In: 15th IPPR Conference on Computer Vision, Graphics and Image Processing, pp. 858–865 (2002)
8. Chen, Y.R.: Multi-moving object removal and repair in video, Master thesis. National Formosa University. Taiwan (2013)
9. Video object inpainting using posture mapping (test sequence). http://www.ee.nthu.edu.tw/cwlin/inpainting/object_inpainting/object_inpainting.htm
10. Yang, Y., Wang, Y.: Automatic video object segmentation algorithm based on background reconstruction. In: International Conference on Multimedia Technology, pp. 235–241 (2011)
11. Das, S., Reeba, R.: Robust exemplar based object removal in video. Int. J. Sci. Eng. Res. (IJSER) **1**, 65–69 (2013)

Multi-core Accelerated Discriminant Feature Selection for Real-Time Bearing Fault Diagnosis

Md. Rashedul Islam[1], Md. Sharif Uddin[1], Sheraz Khan[1],
Jong-Myon Kim[1(✉)], and Cheol-Hong Kim[2]

[1] Department of Electrical, Electronics, and Computer Engineering,
University of Ulsan, Ulsan, South Korea
rashed.cse@gmail.com, sharifruet@gmail.com,
sherazalik@gmail.com, jongmyon.kim@gmail.com
[2] School of Electronics and Computer Engineering,
Chonnam National University, Gwangju, South Korea
cheolhong@gmail.com

Abstract. This paper presents a real-time and reliable bearing fault diagnosis scheme for induction motors with optimal fault feature distribution analysis based discriminant feature selection. The sequential forward selection (SFS) with the proposed feature evaluation function is used to select the discriminative feature vector. Then, the k-nearest neighbor (k-NN) is employed to diagnose unknown fault signals and validate the effectiveness of the proposed feature selection and fault diagnosis model. However, the process of feature vector evaluation for feature selection is computationally expensive. This paper presents a parallel implementation of feature selection with a feature evaluation algorithm on a multi-core architecture to accelerate the algorithm. The optimal organization of processing elements (PE) and the proper distribution of feature data into memory of each PE improve diagnosis performance and reduce computational time to meet real-time fault diagnosis.

Keywords: Fault diagnosis · Feature selection · Class compactness · Class separability · Multi-core architecture

1 Introduction

Induction motors are widely used as a low-speed rotating machine to support steady rotational speed with heavy loads in industries [1]. In rotating machines, bearings are the most significant element which supports stationary rotational speed with heavy load. In a bearing, due to variable-speed, improper loading, and rapid rising of voltage pulse, faults are generated, which results in an unexpected shutdown of manufacturing process [2, 3]. Therefore, real-time and reliable bearing fault diagnosis is needed to avoid these unexpected failures.

An effective data-driven bearing fault diagnosis model involves three basic steps: (1) data acquisition from bearings during operation, (2) feature extraction for identifying specific fault signatures, and (3) classification of different faults generated from

© Springer International Publishing Switzerland 2016
H. Fujita et al. (Eds.): IEA/AIE 2016, LNAI 9799, pp. 645–656, 2016.
DOI: 10.1007/978-3-319-42007-3_56

the bearing. Many signal processing techniques have been developed in time and frequency domains to extract significant fault information from incoming signals of defective bearings [4–6]. However, these methods are only suitable in different situations depending on signal characteristics [7]. Thus, hybrid feature extraction has been widely used for a fault diagnosis system. Hybrid feature selection can explore maximum possible fault signatures by extracting features using different feature extraction paradigms. The resultant high-dimensional feature vector, however, may have redundant information, which can lead to performance degradation of the diagnostic system. The feature selection process can reduce the redundant feature for a reliable fault diagnosis.

A feature selection method selects the best feature subset from the original high dimensional feature vector based on feature evaluation criteria or rules from a classifier [8, 9]. The feature searching process includes the complete search, sequential search, or heuristic search. The complete search ensures a subset with high quality features, but it is costly in terms of its computational time. If there are n number of feature elements in the original feature vector, then the complete search evaluates 2^n feature subsets to find an optimal feature set. Thus, the complete search process is computationally inefficient for a high-dimensional feature vector. On the other hand, the heuristic search like a genetic algorithm (GA) uses several thousands of generations, and hundreds of iterations of crossover and mutation are included in each generation to find an optimal chromosome that represents the optimal feature set. In contrast to these, the sequential search requires less computation time to find an optimal feature set by using an appropriate feature subset evaluation technique. Thus, the sequential forward search (SFS) is employed in this study to find the optimal feature set.

In these approaches, there is a common task to evaluate feature subsets based on some evaluation criteria. Several feature evaluation methods have been proposed depending on the classification accuracy or Euclidean distance based feature distribution. Kanan et al. proposed an evaluation method based on average classification accuracy to select optimal features [10]. In this study, the selected features can ensure better classification performance, but calculating the classification accuracy in the feature selection process requires high computation time. Kang et al. proposed a feature subset evaluation method based on average Euclidian distance by employing within class distance and between class distances [11]. These average distance based feature evaluation techniques do not consider the proper fault data distribution for evaluation.

To overcome the limitations of conventional average distance based evaluation methods, this paper proposes an evaluation criterion based on optimal class distribution analysis using Euclidean distance. The proposed feature selection method is applied to hybrid feature vectors, which are extracted from AE signals. Then, a k-NN classifier is used to classify each fault condition using the optimal feature set. However, the complexity of the feature evaluation and feature selection processes, and the large amount of analysis data demand very high computation time. To accelerate this method, an optimal parallel implementation of the sequential forward feature selection with the proposed optimal class distribution analysis based feature evaluation on a multi-core architecture is presented in this paper.

The rest of the paper is organized as follows. Section 2 describes the AE signal acquisition process. Section 3 presents the proposed optimal class distribution analysis based feature evaluation technique with sequential forward selection. Section 4 illustrates a parallel implementation of the proposed fault diagnosis model. Section 5 shows the experimental results. Finally, Sect. 6 concludes this paper.

2 AE Fault Signal Acquisition

Vibration, current, voltage, thermal based sensors have been used to collect motor's fault signals. In contrast to these sensors, acoustic emission (AE) based technique can detect the fault propagation before the vibration level [14], which makes it suitable for incipient fault detection in bearings. Thus, this study employs the AE sensor. In the experimental setup, as depicted in Fig. 1, an induction motor is connected to the drive-end bearing house, which in turn is connected to the non-drive-end bearing house via a gearbox with 1.52:1 gear reduction factor. Cylindrical rolling element bearings are used and the physical cracks on bearing surface are seeded using a diamond cutter bit. The AE sensor is attached on the top of the non-drive-end bearing house. The sampling rate for both the normal and faulty signals is set to 250 kHz.

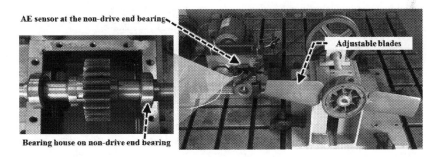

Fig. 1. Self-designed experimental data acquisition setup.

In this paper, eight different types of signals are acquired (i.e., one from a fault-free bearing and seven from faulty bearings with different seeded defects as shown in Fig. 2). In addition, AE signals are collected under five different shaft speeds (300 rpm, 350 rpm, 400 rpm, 450 rpm, and 500 rpm). Thus, the data for this study is divided into five datasets (i.e., one dataset for each shaft speed). Each dataset contains 720 AE signals of five seconds duration each, which is 90 signals for each of the 8 different fault types considered in this study.

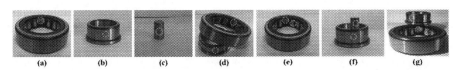

Fig. 2. Different types of faulty bearing, where (a), (b), (c) are inner raceway (BCI), outer raceway (BCO), roller (BCR), respectively, and (d), (e), (f), and (g) are outer & inner raceway (BCIO), outer raceway & roller (BCOR), inner raceway & roller (BCIR), Inner & outer raceway & roller (BCIOR), respectively.

3   The Online Fault Diagnosis Model

The proposed online fault diagnosis model consists of two processes, one for analysis and the other for evaluation, as illustrated in Fig. 3. In the analysis process, initially (at time T_0), hybrid features are extracted from known AE signals and the optimal feature selection process selects the optimized feature vector, which can be used for real-time fault diagnosis. In the evaluation process (at T_1 time), only optimal features are extracted from unknown bearing fault signals, which are then used as input to the k-NN classifier for classification. Finally, the classified test signals are added to the training set and the analysis process is repeated to determine the optimal features and training set for future analysis (at time T_2). The proposed model is discussed in detail in the following sections.

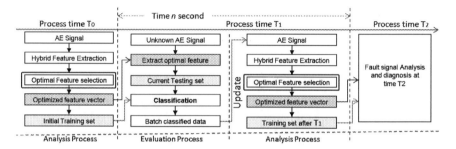

Fig. 3.   The online bearing fault diagnosis model.

3.1   Hybrid Feature Extraction

Data-driven fault diagnosis approaches generally employ time-domain, frequency-domain or time-frequency domain techniques for extracting features [14]. However, a hybrid feature extraction technique is effective to extract most possible fault signatures. Thus, this study uses hybrid features, which include 10 statistical features from time-domain signal, 3 statistical features from frequency domain signal, and 9 RMS frequency features from three defect frequency ranges up-to 3rd harmonics of envelope power spectrum of AE signal.

The time domain statistical features include f1: RMS (root mean square), f2: SRA (square root of the amplitude), f3: KV (kurtosis value), f4: SV (skewness value), f5: PPV (peak-to-peak value), f6: CF (crest factor), f7: (IF) impulse factor, f8: MF (margin factor), f9: SF (shape factor) and f10: KF (kurtosis factor), and the frequency domain statistical features are f11: FC (frequency center), f12: RMSF (RMS frequency), and f13: RVF (root variance frequency). The remaining 9 RMS features are extracted from the defect frequency ranges of inner, outer, and roller faults. Figure 4 represents the defect frequency ranges of different bearing faults. The frequency ranges of each defect frequency and its harmonics are shown in Fig. 5, where N_{OFreqR}, N_{IFreqR}, and N_{RFreqR} are the ranges of outer defect frequency (BPFO), inner defect frequency (BPFI), and roller defect frequency ($2 \times$ BFS), respectively, based on the operating frequency (F_{shaft}), the case frequency (FTF), and random variation of frequency (RV_{order}).

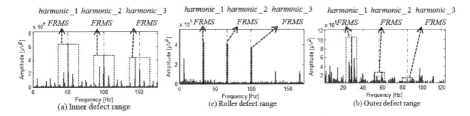

Fig. 4. Defect frequency ranges of each fault and its harmonics: (a) inner raceway fault, (b) outer raceway fault, and (c) roller fault

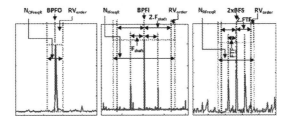

Fig. 5. Frequency range calculation for (a) BPFO, (b) BPFI, and (c) 2 × BFS

3.2 Proposed Optimal Feature Evaluation Technique Using SFS

3.2.1 Sequential Forward Selection (SFS)

The SFS is a simple yet effective feature selection algorithm, which selects an optimal feature set X_k from the original set of features F. The set F consists of N feature variables (i.e., $f_1, f_2, \ldots, f_N$). Initially, the optimal feature set X_k is empty (i.e., the feature index $k = 0$). The value of the objective function V_{obj}, is zero for the empty feature set. Next, a feature x^+ is selected from the set, F, and the objective function $J[X_k + x^+]$ is evaluated for the existing optimal feature set and x^+. If $J[X_k + x^+] > J[X_k]$, then x^+ is included in the optimal feature set, otherwise it is ignored. This process is repeated iteratively until all the features in the original feature set F are exhausted. The result is an optimal feature set X_k, with features that were sequentially found to improve the objective function value, V_{obj}.

3.2.2 Optimal Class Distribution Analysis Based Feature Evaluation

In SFS, feature subsets are evaluated using an objective function. This paper proposes an optimal feature distribution analysis based objective function, which is given in Eq. (1). The proposed objective function is a ratio of the separability between classes and their compactness. The calculation of these measures is shown in Fig. 6. The within-class compactness is the Euclidean distance from the class median point to the most distant point within that class. The between-class distance is calculated by first calculating distances from all the samples of one class to all the samples of other classes and then finding the minimum distance from the boundary point of one class to the nearest boundary point of the nearest class. The main goal of the discriminant feature distribution analysis is to find the maximum objective value (i.e., minimize the

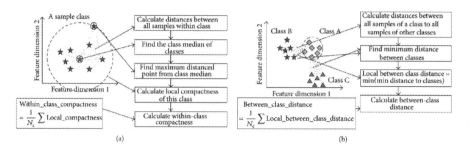

Fig. 6. (a) Within class compactness value calculation, (b) between class separability value calculation.

within-class compactness and maximize the between-class distance). The objective value of the optimal feature set ensures a good distribution of samples of different fault classes, which can increase the classification performance.

$$objective\_value = \frac{between\_class\_separability}{within\_class\_compactness} \tag{1}$$

3.3 Fault Classification

The proposed fault diagnosis model employs the non-parametric k-NN classifier to identify unknown fault signals using discriminant fault signatures. The k-NN is a popular and computationally efficient classification algorithm, which classifies an unknown test data point by finding its k nearest neighbors in the training dataset, using some distance measure; the test data point is classified based on the class of the majority of its k neighbors [15]. In this study the Euclidean distance is used as distance criterion, which can be calculated as follows.

$$dist(x_i, x_j) = \sqrt{\sum_{d=1}^{D} (x_{i,d} - x_{j,d})^2}, \tag{2}$$

where $dist(x_i, x_j)$ represents the Euclidean distance between two data points x_i and x_j, whereas each data point has d feature dimensions.

4   Parallel Implementation of Proposed Feature Evaluation Method on Multi-core Architecture

In feature selection, the objective function is evaluated for each feature subset and this process is repeated until the original feature set is exhausted. The objective function involves the calculation of distances between many data points, which is a time-consuming process. Hence, the feature subset evaluation method is implemented on the multi-core architecture to support real-time bearing fault diagnosis.

4.1   Overview of Multi-core Architecture

Figure 7 presents an overview of the multi-core architecture along with the interconnection between its processing elements (PE).

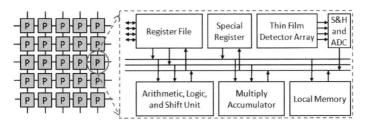

Fig. 7. Multi-core architecture along with the interconnection between processing elements.

This multi-core architecture is a SIMD (single instruction multiple data) based parallel architecture. It consists of a 2-D array of processing elements (PE), where each PE has a local memory, an arithmetic and logical unit (ALU), a barrel shifter, a multiply accumulate (MACC) unit, a sleep unit, a north–east–west–south (NEWS) network and a serial input/output unit [12, 13]. All the PEs are interconnected through a north–east–west–south (NEWS) network and concurrently execute the same instruction but operate on different data. An array control unit (ACU) fetches a vector instruction from instruction memory and dispatches it to the PEs for execution.

4.2   Parallel Implementation of Feature Evaluation Method

The feature subset evaluation process involves plenty of sequential calculations on $N_{class} \times N_{sample} \times N_{feature}$ fault signatures, which is time consuming. To improve performance, this section explains an optimal parallel implementation of the feature evaluation algorithm using an $N \times N$ array of PEs. There are several calculations involved in the algorithm such as calculating Euclidean distance between samples, finding the class median, finding the maximum and minimum distance values of samples, calculating within-class compactness and between-class distance. The following sub-sections describe the parallel implementation of these calculations.

4.2.1   Euclidean Distance Calculation

In the proposed feature subset evaluation algorithm, Euclidean distance is calculated between all the samples. The $N \times N$ array of PEs can calculate the Euclidean distance between N samples of a class in a single iteration. Figure 8 shows an array of 40×40 PEs, where each PE calculates the distance between respective samples depending on its position in the array. For example, the 40th PE, which is located in the 2nd row and 1st column, calculates distance between the 1st and 2nd samples by loading all feature values of these samples from the local memory to its register file. The calculated distances are stored in the register file of the corresponding PE.

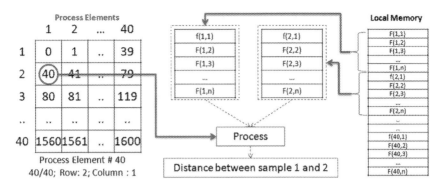

Fig. 8. Euclidean distance calculation using $N \times N$ number of PEs.

4.2.2   Calculation of Within-Class Compactness

The within-class compactness is the distance between the class median and the point, which is most distant from it. Therefore, its calculation involves two steps; the class median is determined, and then, the point, which is at maximum distance from the class median, is calculated. Class median represents a data point, which is located approximately in the middle of the class (i.e., it is at a minimum average Euclidian distance from other samples of the class). Figure 9 shows the sequential algorithm and its parallel implementation. From our discussion in the last subsection, we know that all calculated distances are stored in the general register of the corresponding PE. To calculate the class median, the average distance of each sample with other samples is calculated and then the minimum average distance is determined.

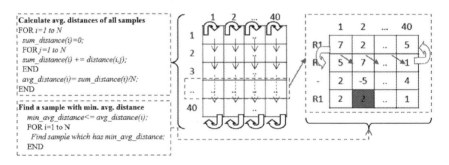

Fig. 9. Class median calculation using $N \times N$ number of PE.

In order to calculate the sum of all distances, inter PE communication and addition operations are used. In the first cycle, the distance values of PEs in each row are shifted to south. In the next cycle, each PE adds its own distance value with the one shifted into its memory in the last cycle. This process is applied for N number of cycles to calculate the sum of all distances for each sample, which is then divided by N to get the average distance. In step 2, the minimum average distance and its index is determined. This is done by shifting the average distance from one PE to the next and comparing the two. Each PE retains the smaller of the two values and hence after the first round all PEs hold the minimum value as compared to their neighbors. This process is repeated N times to get the minimum value among all the PEs. The index (sample number) of the minimum value is selected as the class median.

Finally, the compactness of a class is calculated, which is the distance between the class median and the point at maximum distance from it. This is done by first calculating the distance of each point from the class median and then searching for the maximum distance value, the same way we did for the minimum distance.

4.2.3   Between-Class Distance Calculation

The between-class distance is calculated by first finding the distances from all samples of a class to all samples of other classes and then searching for the minimum distance value. This process is repeated for all the classes. The minimum of all the distance values represents the between-class distance.

5   Experimental Result and Analysis

To validate the proposed algorithm, signals are acquired from a healthy bearing and bearings with seven different faults. In this study, 90 different signals are recorded for each fault type; out of which 40 become part of the analysis dataset ($N_{AnalysisData}$) and the remaining 50 signals are used as evaluation dataset ($N_{EvaluationData}$). In the analysis process, a hybrid feature vector, with 22 features, is evaluated for each signal.

The main objective of the proposed optimal feature distribution analysis based feature selection is to select the minimum number of discriminative features. Figure 10 shows the distribution of samples from different classes (samples from each class are shown in a different color) and the corresponding objective function values. We can observe that the samples from different classes are very well separated for higher values of the objective function, whereas they overlap for lower values of the objective function. Thus, the proposed objective function for SFS ensures that only those features are selected, which result in the maximal separation of all the classes. The SFS selects optimal features based on the proposed feature evaluation function for different datasets. The analysis process is executed in different time windows. Table 1 presents the

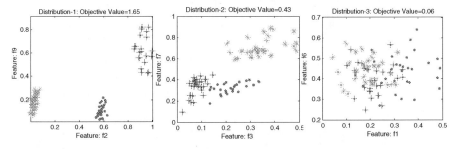

Fig. 10. Class distribution and their corresponding calculated objective value. (Color figure online)

Table 1. Final optimal feature sets of the different datasets.

Execution time window	Dataset 1 300 rpm	Dataset 2 350 rpm	Dataset 3 400 rpm	Dataset 4 450 rpm	Dataset 5 500 rpm
Time T_0	f_2, f_9	f_2, f_3, f_9	$f_2, f_{10}, f_{11}, f_{13}$	f_2, f_9, f_{11}	f_2, f_3, f_9
Time T_1	f_2, f_9, f_{13}	f_2, f_3, f_9	f_2, f_{11}, f_{13}	f_2, f_9, f_{13}	f_2, f_3, f_9

selected features for time windows T_o and T_1 for different datasets. In subsequent time windows (e.g., T_2, T_3, T_4…), the analysis process is repeated, which can result in the selection of a different set of optimal features for each dataset.

5.1   Classification Accuracy

In the proposed online fault diagnosis model, the optimal features obtained through the analysis process are used as training data in the evaluation process in the next time window. In the evaluation process, the optimal features are extracted from unknown signals of the evaluation dataset. Finally, k-NN classification algorithm is used to classify the unknown fault signals. To evaluate the performance of the feature selection method, we compare the classification accuracies of the proposed diagnosis model and a model with no feature selection. In this validation process, k-cross validation ($k_{cv} = 3$) is applied and the final classification accuracy is calculated by averaging the accuracy of k_{cv}. Figure 11 shows the classification performance of the two models for different datasets. It is clearly evident that the proposed feature selection scheme improves the average classification accuracy by 2–6 % across different datasets. This enables the proposed fault diagnosis model to achieve average classification accuracy of around 99 %.

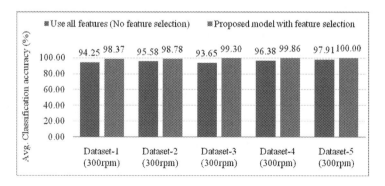

Fig. 11. Classification performance of the proposed model. (Color figure online)

5.2   Time Efficiency of Feature Selection on Multi-core Architecture

To validate the performance of the multi-core architecture at the application level, an instruction-level multi-core simulator is utilized to simulate different execution kernels of the proposed algorithm. On the other hand, the sequential version of the proposed algorithm is implemented using Matlab in a CPU environment to compare the time efficiency of this algorithm. Figure 12 shows the overall execution time of the proposed feature selection algorithm for time T_0 and T_1 on both the CPU and multi-core systems. These results show that the multi-core implementation is 40× to 60× faster than the CPU based implementation.

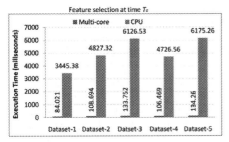
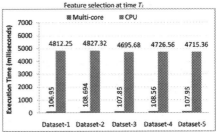

Fig. 12. Execution time of multi-core and CPU-based implementations. (Color figure online)

6   Conclusions

This paper proposed an online bearing fault diagnosis model with discriminant feature selection. In the feature selection process, optimal fault class distribution was considered by calculating the ratio of minimum within-class compactness and maximum between-class separability. The feature selection process starts with a hybrid feature vector that includes 22 features. The SFS based feature selection, which uses the proposed feature subset evaluation algorithm, was applied to extract discriminant features. However, the feature selection process is computationally expensive. It was accelerated using an efficient parallel implementation on a multi-core system. In the evaluation process, optimal features of the analysis dataset were used to train the k-NN classifier, which is then used to classify the unknown fault signals and validate the proposed model in terms of average classification accuracy. The parallel implementation of the proposed algorithm on a multi-core architecture enables the proposed fault diagnosis system to operate in real time.

Acknowledgements. This work was supported by the National Research Foundation of Korea (NRF) grant funded by the Korea government (MSIP) (Nos. NRF-2015K2A1A2070866 and NRF-2013R1A2A2A05004566).

References

1. Widodo, A., Kim, E.Y., Son, J.-D., Yang, B.-S., Tan, A.C.C., Gu, D.-S., Choi, B.-K., Mathew, J.: Fault diagnosis of low speed bearing based on relevance vector machine and support vector machine. J. Expert Syst. Appl. **36**(3), 7252–7261 (2009). Part 2

2. Zhao, M., Jin, X., Zhang, Z., Li, B.: Fault diagnosis of rolling element bearings via discriminative subspace learning: visualization and classification. Expert Syst. Appl. **41**(7), 3391–3401 (2014)
3. Uddin, J., Islam, R., Kim, J.: Texture feature extraction techniques for fault diagnosis of induction motors. J. Convergence **5**(2), 15–20 (2014)
4. Prieto, M.D., Cirrincione, G.A., Espinosa, G., Ortega, J.A., Henao, H.: Bearing fault detection by a novel condition-monitoring scheme based on statistical-time features and neural networks. IEEE Trans. Ind. Electron. **30**(8), 3398–3407 (2013)
5. Yu, J.: Local and nonlocal preserving projection for bearing defect classification and performance assessment. IEEE Trans. Ind. Electron. **59**(5), 2363–2376 (2012)
6. Bediaga, I., Mendizabal, X., Arnaiz, A., Munoa, J.: Ball bearing damage detection using traditional signal processing algorithms. IEEE Instrum. Meas. Magz. **16**(2), 20–25 (2013)
7. Namsrai, E., Munkhdalai, T., Li, M., Shin, J., Namsrai, O., Ryu, K.H.: A feature selection-based ensemble method for arrhythmia classification. J. Inf. Process. Syst. **9**(1), 31–40 (2013)
8. Mahrooghy, M., Nicolas, H.Y.: On the use of the genetic algorithm filter-based feature selection technique for satellite precipitation estimation. IEEE Geosci. Remote Sens. Lett. **9** (5), 963–967 (2012)
9. Rauber, T.W., de Assis Boldt, F., Flavio, M.V.: Heterogeneous feature models and feature selection applied to bearing fault diagnosis. IEEE Trans. Ind. Electron. **62**(1), 637–646 (2015)
10. Kanan, H.R., Faez, K.: GA-based optimal selection of PZMI features for face recognition. Appl. Math. Comput. **205**(2), 706–715 (2008)
11. Kang, M., Kim, J., Kim, J.-M.: reliable fault diagnosis for incipient low-speed bearings using fault feature analysis based on a binary bat algorithm. Inf. Sci. **294**, 423–438 (2015)
12. Kang, M., Kim, J., Kim, J.-M.: An FPGA-based multicore system for real-time bearing fault diagnosis using ultrasampling rate AE signals. IEEE Trans. Ind. Electron. **62**(4), 2319–2329 (2015)
13. Seo, J., Kang, M., Kim, C.-H., Kim, J.-M.: An optimal many-core model based supercomputing for accelerating video-equipped fire detection. J. Supercomput. **71**(6), 2275–2308 (2015)
14. Rashedul Islam, M., Khan, S.A., Kim, J.-M.: Maximum class separability-based discriminant feature selection using a GA for reliable fault diagnosis of induction motors. In: Huang, D.-S., Han, K. (eds.) ICIC 2015. LNCS, vol. 9227, pp. 526–537. Springer, Heidelberg (2015)
15. Yigit, H.: A weighting approach for KNN classifier. In: Proceedings of International Conference on Electronics, Computer and Computation, pp. 228–231 (2013)

QUasi-Affine TRansformation Evolution (QUATRE) Algorithm: A New Simple and Accurate Structure for Global Optimization

Jeng-Shyang Pan[1,2], Zhenyu Meng[1(✉)], Huarong Xu[3], and Xiaoqing Li[4]

[1] Department of Computer Science and Technology,
Innovative Information Industry Research Center,
Harbin Institute of Technology Shenzhen Graduate School, Shenzhen, China
`jengshyangpan@gmail.com`, `mzy1314@gmail.com`
[2] College of Information Science and Engineering,
Fujian University of Technology, Fuzhou, China
[3] Department of Computer Science and Technology,
Xiamen University of Technology, Xiamen, China
[4] Shenzhen Institute of Advanced Technology,
Chinese Academy of Sciences, Shenzhen, China
`http://www.springer.com`

Abstract. QUasi-Affine TRansformation Evolution (QUATRE) algorithm is a simple but powerful structure for global optimization. Six different evolution schemes derived from this structure will be discussed in this paper. There is a close relationship between our proposed structure and Different Evolution (DE) structure, and DE can be considered as a special case of the proposed QUATRE algorithm. The performance of DE is usually dependent on parameter control and mutation strategy. There are 3 control parameters and several mutation strategies in DE, and this makes it a little complicated. Our proposed QUATRE is simpler than DE algorithm as it has only one control parameter and it is logically powerful from mathematical perspective of view. We also use COCO framework under BBOB benchmarks and CEC Competition benchmarks for the verification of the proposed QUATRE algorithm. Experiment results show that though QUATRE algorithm is simpler than DE algorithm, it is more powerful not only on unimodal optimization but also on multimodal optimization problem.

Keywords: Benchmark function · Differential Evolution · Global optimization · QUATRE algorithm

1 Introduction

Optimization demands exist everywhere not only throughout the scientific community but also in our daily lives. Generally, the task is to find out the optimal

© Springer International Publishing Switzerland 2016
H. Fujita et al. (Eds.): IEA/AIE 2016, LNAI 9799, pp. 657–667, 2016.
DOI: 10.1007/978-3-319-42007-3_57

value of a certain parameter, and the standard approach to tackle these problems often begins by evaluating a certain target function which models the problems' objectives while incorporating any constraints [1]. Optimization problems, such as flight vehicle design, circuit design, path planning, electricity supply, time scheduling, function parameter tuning, and so on, often have different kinds of object functions, even some of them do not have a specific object function but operate with so-called regions of acceptability [2,3]. A better designed optimization algorithm usually can solve a wider scope of all these discuss optimization problems in different areas. According to the different characteristic of target problems, optimization algorithms can be divided into different categories, such as deterministic optimization, probabilistic optimization, continuous optimization, combinatorial optimization, and they also have other names of these different optimization field, stochastic optimization, global optimization, discrete optimization, numerical optimization, etc.

With the development of computer technology, computational intelligence shows new computational techniques and methodologies to tackle kinds of complex optimization problem in the real world. There are many branches in the field of computational intelligence, such as fuzzy logic, neural networks, evolutionary computation, memetic computing, etc., and there arise many simple but powerful optimization algorithms in them. Particle Swarm Optimization (PSO) [4] is a powerful algorithm inspired by bird flock. Differential Evolution (DE) [1] is originated with Genetic Annealing algorithm, an algorithm hybrids Genetic algorithm and Simulated Annealing [5]. Ant colony optimization [6] is inspired by the foraging behavior of ant colony. Article Bee Colony (ABC) [7] Optimization mimics the behavior of bees. Harmony search [8] is derived from the improving process of composing a piece of music. Bat algorithm [9] is inspired by bat, and it extends the harmony search algorithm and PSO algorithm. Monkey King Evolution [10] algorithm is a powerful algorithm inspired by behaviors of the character Monkey King in the novel "Journey to the West". These are all powerful algorithms and have disadvantages and advantages of their own.

There are mainly two direction of the development in optimization algorithm, one is improvement that enhancing an existing algorithm, and the other is new powerful algorithm with totally new evolution scheme. Some of the newly proposed algorithms usually use some old canonical algorithms for comparison, and these often may be of less use because big improvement has been made since the inception of the canonical algorithm. So contrasts made with some state-of-the-art algorithms are highly recommended for these comparison. In this paper, we first discuss the new QUATRE algorithm, show six different evolution schemes of it. Then we discuss the relationship between QUATRE algorithm and DE algorithm, show some mathematical analysis of the differences between the two algorithms. The analysis explains why the QUATRE algorithm is logically better from mathematical perspective of view than the canonical DE algorithm. Some state-of-the-art DE variants are also used in the comparison with QUATRE algorithm, COCO framework and CEC Competition benchmarks are also used to verify the new algorithm. The rest of the paper is organized as follows,

Sect. 2 gives the details of QUATRE algorithm, Sect. 3 shows experiment test beds with benchmark functions, Sect. 4 shows the experiment analysis, Sect. 5 gives the final conclusion.

2 The QUasi-Affine TRansformation Evolution (QUATRE) Algorithm

The newly proposed algorithm in this paper is named QUasi-Affine TRansformation Evolution (QUATRE) algorithm, as the evolution equation is in an affine-transformation-like form. Affine transformation is a term in geometry, it describes a process of transformation from one affine space to another. The transformation function $(f : X \to Y)$ is usually written in the following form: $X \mapsto MX + B$. We extend the form $(X \mapsto MX + B)$ in this paper to $X \mapsto MX + \overline{M}B$, and we use such a similar evolution equation in our proposed QUATRE algorithm. The exact evolution equation used in QUATRE is shown in Eq. 1.

$$X \mapsto M \bigotimes X + \overline{M} \bigotimes B \qquad (1)$$

$\bigotimes$ means component-wise multiplication, same as the ".∗" operation in Matlab. X denotes coordinate matrix of the particle population, if the coordinate of the ith particle is $X_i = [x_1, x_2, \ldots, x_D]$, then X can be written in equation $X = [X_1, X_2, \ldots, X_{ps}]^T$. Matrix B has the same number (ps) of row vector components as matrix X, and it usually has several schemes for the calculation of matrix B. Table 1 shows the six evolution schemes of matrix B calculation in QUATRE algorithm. $X_{r1,G}$, $X_{r2,G}$, $X_{r3,G}$, $X_{r4,G}$ and $X_{r5,G}$ all denote random matrices which are generated by randomly permutating row vectors of coordinate matrix (X) of the Gth generation. $X_{gbest,G}$ is a row-vector-duplicated matrix with each row equaling to the Gth global best particle. Equation 2 shows the definition.

$$X = \begin{bmatrix} X_1 \\ X_2 \\ \ldots \\ X_{ps} \end{bmatrix} X_{gbest,G} = \begin{bmatrix} X_{gbest} \\ X_{gbest} \\ \ldots \\ X_{gbest} \end{bmatrix} \qquad (2)$$

Table 1. The six schemes of matrix B calculation in QUATRE algorithm.

No.	QUATRE/B	Equation
1	QUATRE/target/1	$B = X + c * (X_{r1,G} - X_{r2,G})$
2	QUATRE/rand/1	$B = X_{r1,G} + c * (X_{r2,G} - X_{r3,G})$
3	QUATRE/best/1	$B = X_{gbest,G} + c * (X_{r2} - X_{r3})$
4	QUATRE/target/2	$B = X + c * (X_{r1,G} - X_{r2,G}) + c * (X_{r3,G} - X_{r4,G})$
5	QUATRE/rand/2	$B = X_{r1,G} + c * (X_{r2,G} - X_{r3,G}) + c * (X_{r4,G} - X_{r5,G})$
6	QUATRE/best/2	$B = X_{gbest,G} + c * (X_{r1,G} - X_{r2,G}) + c * (X_{r3,G} - X_{r4,G})$

M in QUATRE can be considered as evolution matrix/selection matrix, and $\overline{M}$ means a binary inverse operation of the matrix elements of M, the inverse values of non-zero elements are zeros, and inverse values of zero elements are ones. An example of binary inverse operation is given in Eq. 3.

$$M = \begin{bmatrix} 1 & & & \\ 1 & 1 & & \\ & & \cdots & \\ 1 & 1 & \cdots & 1 \end{bmatrix}, \overline{M} = \begin{bmatrix} 0 & 1 & \cdots & 1 \\ 0 & 0 & \cdots & 1 \\ & & \cdots & 1 \\ 0 & 0 & \cdots & 0 \end{bmatrix} \tag{3}$$

As to the evolution matrix, M is transformed from a lower triangular matrix M_{tmp}. The transformation is shown in Eq. 4 with $ps = D$. There are two steps for the transformation, the first step is to randomly permute the elements of each D-dimension row vector in M_{tmp}, and the second step is to randomly permute the row vectors with the elements of each row vector unchanged, so we can get M.

$$M_{tmp} = \begin{bmatrix} 1 & & & \\ 1 & 1 & & \\ & & \cdots & \\ 1 & 1 & \cdots & 1 \end{bmatrix} \sim \begin{bmatrix} 1 & 1 & & \\ & \cdots & & \\ 1 & 1 & \cdots & 1 \\ & & 1 & \end{bmatrix} = M \tag{4}$$

Usually, the size of the particle population is larger than particle coordinate dimension, matrix M_{tmp} needs to be extended according to population size ps. We take $ps = 2D + 2$ for example, the extension is described in Eq. 5. Generally, when $ps\%D = k$, the first k rows of the $D \times D$ lower triangular matrix are included in M_{tmp}, and M is adaptively changed according to M_{tmp}.

$$M_{tmp} = \begin{bmatrix} 1 & & & \\ 1 & 1 & & \\ & & \cdots & \\ 1 & 1 & \cdots & 1 \\ 1 & & & \\ 1 & 1 & & \\ & & \cdots & \\ 1 & 1 & \cdots & 1 \\ & & \vdots & \\ 1 & & & \\ 1 & 1 & & \end{bmatrix} \sim \begin{bmatrix} 1 & 1 & & \\ & \cdots & & \\ 1 & 1 & \cdots & 1 \\ & & 1 & \\ & \cdots & & \\ 1 & \cdots & 1 & \\ & & 1 & \\ 1 & & \cdots & 1 \\ & & \vdots & \\ & & 1 & 1 \\ 1 & & \cdots & 1 \end{bmatrix} = M \tag{5}$$

c is the coefficient factor/step size of differential matrix (the result of $X_{r1} - X_{r2}$ is considered as differential matrix). X_{r1} and X_{r2} are generated by randomly permuting the row vectors of matrix X.

3 Benchmark Function for Optimization

COmparing Continuous Optimizers (COCO) framework as a tool for benchmarking algorithms for black-box optimization is used herein the paper for the verification of QUATRE algorithm. This framework is easy to use because it provides an experimental framework for algorithm testing. Benchmark functions in Congress of Evolutionary Computation (CEC) Competition 2013 are also used for algorithm validation, and the benchmark equations and domains are listed in Tables 2 and 3 respectively. These benchmarks can be divided into different categories, the unimodal function, the basic multimodal function, and even some composition functions. They all reflect kinds of real world complex optimization problems. COCO framework under BBOB2009 testbeds uses fix-target measurement while CEC competition benchmarks uses fix-target measurement. Both of these two measurements are tested in the paper to make a thorough verification of QUATRE algorithm.

4 Experiment Analysis of QUATRE Algorithm

4.1 Relationship Between de and QUATRE

DE is a powerful algorithm, and all the variants of DE can be notated in the form: $DE/x/y/z$ [1]. x denotes the vector to be mutated, y denotes the number of difference vectors. x and y together determine the mutation vector V_m. z denotes the crossover scheme, and it can be regarded as the movement from the current position to a certain location in a higher dimensional cube with its body diagonal formed by the two point X and V_m. The DE variants with different V_m vectors are shown in Table 4. There are several crossover schemes for DE algorithm, including One-Point Crossover, N-Point Crossover, Exponential Crossover, Uniform (Binomial) Crossover, etc. Uniform (Binomial) Crossover is the commonly used and also the recommended crossover schemes in classical DE algorithm. In order to analysis the relationship between DE and QUATRE algorithm, we rewrite DE algorithm ($DE/best/1/bin$) in a matrix evolution style, Eq. 6 shows the evolution equation, and $\widehat{I}$ is the extended identity matrix. The probability of vertex selection (vertexes are selected in the hypercube formed by X and V_m, two vertexes of the body diagonal in the hypercube) is not equal when using evolution matrix M in DE variants, and it is relative equal in QUATRE algorithm. This is why QUATRE performs better especially in multimodal optimization problems.

$$\begin{cases} V_m = X_{gbest,G} + F * (X_{r1,G} - X_{r2,G}) \\ X \mapsto \overline{M}_{DE} \bigotimes X + M_{DE} \bigotimes V_m \\ M_{DE} = M_{rand \leq Cr} | \widehat{I} \end{cases} \qquad (6)$$

Table 2. CEC2013 test suite for real-parameter optimization benchmarks

No.	Name	Benchmark function
1	Sphere Function	$f_1(x) = \sum_{i=1}^{D} z_i^2 + f_1^*, Z = X - O$
2	Rotated High Conditioned Elliptic Function	$f_2(x) = \sum_{i=1}^{D} (10^6)^{\frac{i-1}{D-1}} z_i^2 + f_2^*, Z = T_{osz}(M_1(X - O))$
3	Rotated Bent Cigar Function	$f_3(x) = z_1^2 + 10^6 \sum_{i=2}^{D} z_i^2 + f_3^*, Z = M_2 T_{asy}^{0.5}(M_1(X - O))$
4	Rotated Discuss Function	$f_4(x) = 10^6 z_1^2 + \sum_{i=2}^{D} z_i^2 + f_4^*, Z = T_{osz}(M_1(X - O))$
5	Different Powers Function	$f_5(x) = \sqrt{\sum_{i=1}^{D} |z_i|^{2+4\frac{i-1}{D-1}}} + f_5^*, Z = X - O$
6	Rotated Rosenbrock's Function	$f_6(x) = \sum_{i=1}^{D-1} (100(z_i^2 - z_{i+1})^2 + (z_i - 1)^2) + f_6^*, Z = M_1(\frac{2.048(X-O)}{100}) + 1$
7	Rotated Schaffers F7 Function	$f_7(x) = (\frac{1}{D-1} \sum_{i=1}^{D-1} (\sqrt{z_i} + \sqrt{z_i} sin^2(50 z_i^{0.2})))^2 + f_7^*,$ $z_i = \sqrt{y_i^2 + y_{i+1}^2}, Y = \Lambda^{10} M_2 T_{asy}^{0.5}(M_1(X - O))$
8	Rotated Ackley's Function	$f_8(x) = -20 exp(-0.2\sqrt{\frac{1}{D}\sum_{i=1}^{D} z_i^2}) - exp(\frac{1}{D}\sum_{i=1}^{D} D cos(2\pi z_i)) + 20 + e + f_8^*$ $Z = \Lambda^{10} M_2 T_{asy}^{0.5}(M_1(X - O))$
9	Rotated Weierstrass Function	$f_9(x) = \sum_{i=1}^{D} (\sum_{k=0}^{kmax} [a^k cos(2\pi b^k(z_i + 0.5))]) - D \sum_{k=0}^{kmax} [a^k cos(2\pi b^k 0.5)] + f_9^*$ $a = 0.5, b = 3, kmax = 20, Z = \Lambda^{10} M_2 T_{asy}^{0.5}(M_1 \frac{0.5(X-O)}{100})$
10	Rotated Griewank's Function	$f_{10}(x) = \sum_{i=1}^{D} \frac{z_i^2}{4000} - \prod_{i=1}^{D} cos(\frac{z_i}{\sqrt{i}}) + 1 + f_{10}^*,$ $Z = \Lambda^{100} M_1 \frac{600(X-O)}{100}$
11	Rastrigin's Function	$f_{11}(x) = \sum_{i=1}^{D} (z_i^2 - 10 cos(2\pi z_i) + 10) + f_{11}^*,$ $Z = \Lambda^{10} T_{asy}^{0.2}(T_{osz}(\frac{5.12(X-O)}{100}))$
12	Rotated Rastrigin's Function	$f_{12}(x) = \sum_{i=1}^{D} (z_i^2 - 10 cos(2\pi z_i) + 10) + f_{12}^*,$ $Z = M_1 \Lambda^{10} M_2 T_{asy}^{0.2}(T_{osz}(M_1 \frac{5.12(X-O)}{100}))$
13	Non-continuous Rotated Rastrigin's Function	$f_{13}(x) = \sum_{i=1}^{D} (z_i^2 - 10 cos(2\pi z_i) + 10) + f_{13}^*, Z = M_1 \Lambda^{10} M_2 T_{asy}^{0.2}(T_{osz}(Y))$ $\hat{x} = M_1 \frac{5.12(X-O)}{100}, y_i = \begin{cases} \hat{x}_i, & if \; |\hat{x}_i| \leq 0.5 \\ round(2\hat{x}_i)/2, & if \; |\hat{x}_i| > 0.5 \end{cases}$
14	Schwefel's Function	$f_{14}(Z) = 418.9829 * D - \sum_{i=1}^{D} g(z_i) + f_{14}^*,$ $Z = \Lambda^{10}(\frac{100(X-O)}{100}) + 4.209687462275036e + 002$
15	Rotated Schwefel's Function	$f_{15}(Z) = 418.9829 * D - \sum_{i=1}^{D} g(z_i) + f_{15}^*,$ $Z = \Lambda^{10} M_1(\frac{100(X-O)}{100}) + 4.209687462275036e + 002$
16	Rotated Katsuura Function	$f_{16}(x) = \frac{10}{D^2} \prod_{i=1}^{D} (1 + i \sum j = 132 \frac{|2^j z_i - round(2^j z_i)|}{2^j})^{\frac{10}{D^{1.2}}} - \frac{10}{D^2} + f_{16}^*$ $Z = M_2 \Lambda^{100}(M_1(\frac{5(X-O)}{100}))$
17	Lunacek Bi-Rastrigin Function	$f_{17}(x) = min(\sum_{i=1}^{D} y_0^2, dD + s \sum_{i=1}^{D} y_1^2) + 10(D - \sum_{i=1}^{D} cos(2\pi \hat{z}_i)) + f_{17}^*$ $y_0 = (\hat{x}_i - \mu_0), y_1 = (\hat{x}_i - \mu_1), z = \Lambda^{100}(\hat{x} - \mu_0)$
18	Rotated Lunacek Bi-Rastrigin Function	$f_{18}(x) = min(\sum_{i=1}^{D} y_0^2, dD + s \sum_{i=1}^{D} y_1^2) + 10(D - \sum_{i=1}^{D} cos(2\pi \hat{z}_i)) + f_{18}^*$ $y_0 = (\hat{x}_i - \mu_0), y_1 = (\hat{x}_i - \mu_1), z = M_2 \Lambda^{100}(M_1(\hat{x} - \mu_0))$
19	Expanded Griewank's plus Rosenbrock's Function	$f_{19}(x) = g_1(g_2(z_1, z_2)) + g_1(g_2(z_2, z_3)) + ... + g_1(g_2(z_D, z_1)) + f_{19}^*$ $g_1(x) = \sum_{i=1}^{D} \frac{x_i^2}{4000} - \prod_{i=1}^{D} cos(\frac{x_i}{\sqrt{i}}) + 1, z = M_1(\frac{5(X-o)}{100}) + 1$

(continued)

Table 2. (*continued*)

No.	Name	Benchmark function
20	Expanded Scaffer's F6 Function	$f_{20}(x) = g(z_1, z_2) + g(z_2, z_3) + \ldots + g(z_D, z_1) + f_{20}^*$ $$g(x,y) = 0.5 + \frac{sin^2(\sqrt{x^2 + y^2}) - 0.5}{(1 + 0.001(x^2 + y^2))^2}, \, Z = M_2 T_{asy}^{0.5}(M_1(X - O))$$
21	Composition Function 1	$$f(x) = \sum_{i=1}^{n} \omega_i * [\lambda_i g_i(x) + bias_i] + f^*$$ $f_i' = f_i - f_i^*, g_i = f_6', g_2 = f_5', g_3 = f_3', g_4 = f_4', g_5 = f_1'$
22	Composition Function 2	$$f(x) = \sum_{i=1}^{n} \omega_i * [\lambda_i g_i(x) + bias_i] + f^*$$ $f_i' = f_i - f_i^*, g_{1-3} = f_{14}'$
23	Composition Function 3	$$f(x) = \sum_{i=1}^{n} \omega_i * [\lambda_i g_i(x) + bias_i] + f^*$$ $f_i' = f_i - f_i^*, g_{1-3} = f_{15}'$
24	Composition Function 4	$$f(x) = \sum_{i=1}^{n} \omega_i * [\lambda_i g_i(x) + bias_i] + f^*$$ $f_i' = f_i - f_i^*, g_1 = f_{15}', g_2 = f_{12}', g_3 = f_9', \sigma = [20, 20, 20]$
25	Composition Function 5	$$f(x) = \sum_{i=1}^{n} \omega_i * [\lambda_i g_i(x) + bias_i] + f^*$$ $f_i' = f_i - f_i^*, g_1 = f_{15}', g_2 = f_{12}', g_3 = f_9', \sigma = [10, 30, 50]$
26	Composition Function 6	$$f(x) = \sum_{i=1}^{n} \omega_i * [\lambda_i g_i(x) + bias_i] + f^*$$ $f_i' = f_i - f_i^*, g_1 = f_{15}', g_2 = f_{12}', g_3 = f_2', g_4 = f_9', g_5 = f_{10}'$
27	Composition Function 7	$$f(x) = \sum_{i=1}^{n} \omega_i * [\lambda_i g_i(x) + bias_i] + f^*$$ $f_i' = f_i - f_i^*, g_1 = f_{10}', g_2 = f_{12}', g_3 = f_{15}', g_4 = f_9', g_5 = f_1'$
28	Composition Function 8	$$f(x) = \sum_{i=1}^{n} \omega_i * [\lambda_i g_i(x) + bias_i] + f^*$$ $f_i' = f_i - f_i^*, g_1 = f_{19}', g_2 = f_7', g_3 = f_{15}', g_4 = f_{20}', g_5 = f_1'$

4.2 Verification on CEC2013 Benchmarks and COCO Framework

In order to validate and evaluate the proposed algorithm, comparisons are made under CEC competition benchmarks and COCO framework. Several state-of-the-art DE variants are also contrasted with the QUATRE algorithm. Table 5 shows the contrasts between DE algorithm and QUATRE algorithm under CEC2013 benchmarks for real-parameter optimization. From the table, we can see that QUATRE algorithm is more likely to find the better solution of an optimization, and it wins 20/28 of the global best comparison, wins 19/28 on mean performance comparison. Some state-of-the-art DE variants are also contrasted with QUATRE algorithm on COCO framework under BBOB2009 benchmarks, the compared variants include canonical DE [1], Opposition-based DE (ODE) [11], Self-adaptive control parameter DE [12], JADE [13]. All these comparison use the recommend parameter settings, and the comparison results are shown in Fig. 1.

Table 3. Search domain and minimum of CEC2013 benchmark functions

No.	Name	Minimum value
1	Sphere Function	$f(o_1, o_2, \ldots, o_d) = -1400$
2	Rotated High Conditioned Elliptic Function	$f(o_1, o_2, \ldots, o_d) = -1300$
3	Rotated Bent Cigar Function	$f(o_1, o_2, \ldots, o_d) = -1200$
4	Rotated Discuss Function	$f(o_1, o_2, \ldots, o_d) = -1100$
5	Different Powers Function	$f(o_1, o_2, \ldots, o_d) = -1000$
6	Rotated Rosenbrock's Function	$f(o_1, o_2, \ldots, o_d) = -900$
7	Rotated Schaffers F7 Function	$f(o_1, o_2, \ldots, o_d) = -800$
8	Rotated Ackley's Function	$f(o_1, o_2, \ldots, o_d) = -700$
9	Rotated Weierstrass Function	$f(o_1, o_2, \ldots, o_d) = -600$
10	Rotated Griewank's Function	$f(o_1, o_2, \ldots, o_d) = -500$
11	Rastrigin's Function	$f(o_1, o_2, \ldots, o_d) = -400$
12	Rotated Rastrigin's Function	$f(o_1, o_2, \ldots, o_d) = -300$
13	Non-continuous Rotated Rastrigin's Function	$f(o_1, o_2, \ldots, o_d) = -200$
14	Schwefel's Function	$f(o_1, o_2, \ldots, o_d) = -100$
15	Rotated Schwefel's Function	$f(o_1, o_2, \ldots, o_d) = 100$
16	Rotated Katsuura Function	$f(o_1, o_2, \ldots, o_d) = 200$
17	Lunacek Bi-Rastrigin Function	$f(o_1, o_2, \ldots, o_d) = 300$
18	Rotated Lunacek Bi-Rastrigin Function	$f(o_1, o_2, \ldots, o_d) = 400$
19	Expanded Griewank's plus Rosenbrock's Function	$f(o_1, o_2, \ldots, o_d) = 500$
20	Expanded Scaffer's F6 Function	$f(o_1, o_2, \ldots, o_d) = 600$
21	Composition Function1 (n = 5, Rotated)	$f(o_1, o_2, \ldots, o_d) = 700$
22	Composition Function2 (n = 3, Unrotated)	$f(o_1, o_2, \ldots, o_d) = 800$
23	Composition Function3 (n = 3, Rotated)	$f(o_1, o_2, \ldots, o_d) = 900$
24	Composition Function4 (n = 3, Rotated)	$f(o_1, o_2, \ldots, o_d) = 1000$
25	Composition Function5 (n = 3, Rotated)	$f(o_1, o_2, \ldots, o_d) = 1100$
26	Composition Function6 (n = 5, Rotated)	$f(o_1, o_2, \ldots, o_d) = 1200$
27	Composition Function7 (n = 5, Rotated)	$f(o_1, o_2, \ldots, o_d) = 1300$
28	Composition Function8 (n = 5, Rotated)	$f(o_1, o_2, \ldots, o_d) = 1400$
All	Search Domain	$[-100, 100]^D$

We can see that the proposed MKE algorithm outperforms the compared DE variants. So we can say that the overall performance of QUATRE algorithm is excellent.

Table 4. The corresponding V_m values with different DE variants

No.	$DE/x/y/z$	Equation
1	$DE/target-to-best/1/z$	$V_m = X + F * (X_{gbest,G} - X) + F * (X_{r2,G} - X_{r3,G})$
2	$DE/rand/1/z$	$V_m = X_{r1,G} + F * (X_{r2,G} - X_{r3,G})$
3	$DE/best/1/z$	$V_m = X_{gbest,G} + F * (X_{r2,G} - X_{r3,G})$
4	$DE/target-to-rand/1/z$	$V_m = X + F * (X_{r1,G} - X) + F * (X_{r2,G} - X_{r3,G})$
5	$DE/rand/2/z$	$V_m = X_{r1,G} + F * (X_{r2,G} - X_{r3,G}) + F * (X_{r4,G} - X_{r5,G})$
6	$DE/best/2/z$	$V_m = X_{gbest,G} + F * (X_{r1,G} - X_{r2,G}) + F * (X_{r3,G} - X_{r4,G})$

Table 5. Best value (minimum), mean and standard deviation of 20−run fitness error comparisons between DE variants and QUATRE algorithm. The population size $ps = 1000$ and generation of particles $gen = 5000$ are the same ($100000 * 50$ NFEs). The best results of the comparisons are emphasized in **BOLDFACE**.

50D	DE/best/1/bin			QUATRE/best/1		
No.	Best	Mean	Std	Best	Mean	Std
f_1	2.2737E−13	2.2737E−13	**0**	**0**	1.0232E−13	1.1606E−13
f_2	4.1844E+07	5.7181E+07	1.4067E+07	**4.2211E+05**	**7.6320E+05**	**2.5036E+05**
f_3	6.0496E+08	1.2512E+09	5.2202E+08	**1.0591E+05**	**8.2695E+06**	**1.1699E+07**
f_4	2.5173E+04	2.9266E+04	2.5478E+03	**1.1439E+02**	**2.4178E+02**	**9.9987E+01**
f_5	*1.1369E−13*	2.0464E−13	4.6656E−14	*1.1369E−13*	**1.1369E−13**	**0**
f_6	*4.3447E+01*	**4.3447E+01**	**2.8360E−06**	*4.3447E+01*	4.3731E+01	1.2701E+00
f_7	4.5521E+01	6.2100E+01	7.0529E+00	**3.5657E+00**	**1.8333E+01**	**9.3588E+00**
f_8	2.1038E+01	**2.1070E+01**	**1.7406E−02**	**2.0980E+01**	2.1071E+01	4.1055E−02
f_9	4.9120E+01	5.5174E+01	**2.7377E+00**	**1.2047E+01**	**2.2417E+01**	6.1640E+00
f_{10}	2.0323E+00	3.4720E+00	1.1064E+00	**5.6843E−14**	**1.9336E−02**	**1.0332E−02**
f_{11}	**5.6843E−14**	**5.6843E−14**	**0**	1.9899E+01	2.8307E+01	5.5489E+00
f_{12}	2.6164E+02	2.9096E+02	**1.7201E+01**	**7.3627E+01**	**1.0657E+02**	2.3831E+01
f_{13}	2.6290E+02	3.1499E+02	**2.2954E+01**	**7.7258E+01**	**2.0640E+02**	4.8023E+01
f_{14}	**1.0453E+00**	**5.5185E+00**	**3.4508E+00**	3.1598E+02	8.1165E+02	2.9142E+02
f_{15}	1.2132E+04	1.2640E+04	**2.8700E+02**	**5.6966E+03**	**6.3868E+03**	8.0317E+02
f_{16}	2.3790E+00	2.7649E+00	2.0284E−01	**6.8763E−01**	**1.7048E+00**	**5.6878E−01**
f_{17}	**5.0786E+01**	**5.0827E+01**	**8.8978E−02**	6.1534E+01	6.7517E+01	4.2382E+00
f_{18}	3.5033E+02	3.9295E+02	1.4678E+01	**1.2309E+02**	**1.7282E+02**	2.5671E+01
f_{19}	7.6386E+00	8.4198E+00	4.7192E−01	**2.9139E+00**	**4.2863E+00**	**7.3743E−01**
f_{20}	2.1397E+01	2.1826E+01	2.5408E−01	**1.9045E+01**	**2.0078E+01**	**7.0689E−01**
f_{21}	*2.0000E+02*	4.9419E+02	4.1873E+02	*2.0000E+02*	6.8188E+02	4.1891E+02
f_{22}	**1.8817E+01**	**1.3045E+02**	**2.5502E+02**	5.1095E+02	9.6705E+02	2.8410E+02
f_{23}	1.0693E+04	1.2691E+04	6.3703E+02	**5.8164E+03**	**7.0678E+03**	**7.0716E+02**
f_{24}	3.0196E+02	3.1605E+02	8.7894E+00	**2.0728E+02**	**2.5584E+02**	1.6771E+01
f_{25}	3.3992E+02	3.5693E+02	**8.2616E+00**	**2.6748E+02**	**2.7878E+02**	1.2327E+01
f_{26}	2.0351E+01	**2.0546E+02**	1.0825E+00	**2.0003E+02**	2.7489E+02	7.7796E+01
f_{27}	1.5326E+03	1.6556E+03	6.3619E+01	**6.6279E+02**	**8.6198E+02**	**1.3027E+02**
f_{28}	*4.0000E+02*	5.5489E+02	9.6087E+02	*4.0000E+02*	1.1338E+03	1.3040E+03
Win	4	9	22	20	19	6
Draw	4	0	0	4	0	0

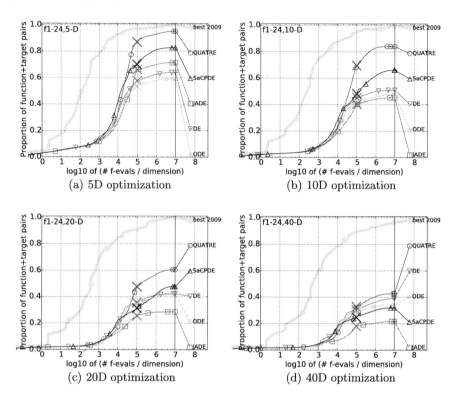

Fig. 1. COCO framework under BBOB2009 testbeds over 24 functions for DE variants group.

5 Conclusion

In this paper, we propose QUasi-Affine TRansformation Evolution (QUATRE) algorithm. QUATRE algorithm is a simple but powerful structure for global optimization. Six different evolution schemes derived from this structure is shown in the paper and the relationship between QUATRE algorithm and DE algorithm is also discussed. Verification is also made both on COCO framework and on CEC competition benchmarks. Experiment results show that the proposed QUATRE algorithm performs very well not only on unimodal optimization but also on multimodal optimization problem.

Acknowledgement. This work was supported by National Natural Science Foundation of China (61273290).

References

1. Storn, R., Price, K.: Differential evolution - a simple and efficient heuristic for global optimization over continuous spaces. J. Glob. Optim. **11**(4), 341–359 (1997)
2. Meng, Z., Pan, J.-S.: A simple and accurate global optimizer for continuous spaces optimization. In: Sun, H., Yang, C.-Y., Lin, C.-W., Lin, J.-S., Snasel, V., Abraham, A. (eds.) Genetic and Evolutionary Computing. AISC, vol. 329, pp. 121–129. Springer, Switzerland (2015)
3. Meng, Z., Pan, J.-S., Alelaiwi, A.: A new meta-heuristic ebb-tide-fish-inspired algorithm for traffic navigation. Telecommun. Syst. **62**, 403–415 (2016)
4. Kennedy, J., Eberhart, R.: Particle swarm optimization. In: Proceedings of IEEE International Conference on Neural Networks, 1995, vol. 4. IEEE (1995)
5. Kirkpatrick, S., Vecchi, M.P.: Optimization by simmulated annealing. Science **220**(4598), 671–680 (1983)
6. Dorigo, M., Maniezzo, V., Colorni, A.: Ant system: optimization by a colony of cooperating agents. IEEE Trans. Syst. Man Cybern. Part B Cybern. **26**(1), 29–41 (1996)
7. Karaboga, D., Basturk, B.: A powerful and efficient algorithm for numerical function optimization: artificial bee colony (ABC) algorithm. J. Glob. Optim. **39**(3), 459–471 (2007)
8. Yang, X.-S.: Harmony search as a metaheuristic algorithm. In: Geem, Z.W. (ed.) Music-Inspired Harmony Search Algorithm. SCI, vol. 191, pp. 1–14. Springer, Heidelberg (2009)
9. Yang, X.-S.: A new metaheuristic bat-inspired algorithm. In: González, J.R., Pelta, D.A., Cruz, C., Terrazas, G., Krasnogor, N. (eds.) NICSO 2010. SCI, vol. 284, pp. 65–74. Springer, Heidelberg (2010)
10. Meng, Z., Pan, J.-S.: Monkey king evolution: a new memetic evolutionary algorithm and its application in vehicle fuel consumption optimization. Knowl.-Based Syst. **97**, 144–157 (2016)
11. Rahnamayan, S., Tizhoosh, H.R., Salama, M.: Opposition-based differential evolution. IEEE Trans. Evol. Comput. **12**(1), 64–79 (2008)
12. Brest, J., et al.: Self-adapting control parameters in differential evolution: a comparative study on numerical benchmark problems. IEEE Trans. Evol. Comput. **10**(6), 646–657 (2006)
13. Zhang, J., Sanderson, A.C.: JADE: adaptive differential evolution with optional external archive. IEEE Trans. Evol. Comput. **13**(5), 945–958 (2009)

Decision Support Systems

Mobile Gaming Trends and Revenue Models

Khaled Mohammad Alomari[1]([⊠]) [iD], Tariq Rahim Soomro[2],
and Khaled Shaalan[3]

[1] Faculty of Arts and Sciences, Abu Dhabi University, Abu Dhabi, UAE
khaled.alomari@adu.ac.ae
[2] Faculty of Computing and Engineering Sciences,
SZABIST Dubai Campus, Dubai, UAE
tariq@szabist.ac.ae
[3] Faculty of Engineering and IT, British University in Dubai, Dubai, UAE
khaled.shaalan@buid.ac.ae

Abstract. The study tries to find out the most important features in building games based on the grossing. The study is limited to fifty iPhone games that have achieved top grossing in the USA. The game features were extracted from a previous study [1] and classified through ARM funnel into five groups ("A", "R", "M", "AR", and "RM"). The paper follows CRISP-DM approach under SPSS Modeler through business and data understanding, Data preparation, model building and evaluation. The researcher uses Decision Tree model since the features have closed value i.e. (Yes/No) on the grossing weight. The study reached to the most important 10 features out of 31. These features are important to build successful mobile games. The study emphasizes on the availability of (Acquisition, Retention and Monetization) elements on every successful game and if any is missed, will lead to the failure of the game.

Keywords: Mobile games · Mobile game trends · Mobile games revenue models

1 Introduction

According to the UN International Telecommunications Union, the of end 2014 there were 7 billion subscriptions in the world for mobile phone - compared to 2.2 billion in 2005 and 719 million in 2000 [2]. Mobile phones are used extensively for entertainments, communications and it became a part of personal accessories. The new trends and applications developments in the mobile phone open the door for developers to offer their expertise in the development of better to mobile phone applications [3]. Currently, mobile phones support a wide range of connectivity features like Bluetooth, NFC, Wi-Fi and Data Mobile. Today, mobile data become more prevalent. According to the UNInternational Telecommunications Union there 2.3 billion mobile-broadband subscriptions globally [2] and 352 operators globally. The 4G mobile technology was launched in December 2009 and until the end of January 2015, more than a third of the global users were using 4G mobile data [4]. The technology is not going to stop here and 5G was predicted at 360-europe event held in Brussels last year [5].

© Springer International Publishing Switzerland 2016
H. Fujita et al. (Eds.): IEA/AIE 2016, LNAI 9799, pp. 671–683, 2016.
DOI: 10.1007/978-3-319-42007-3_58

As far as games are concerned on mobile devices, "Snake" was the first game installed on Nokia mobile phone in 1997 – the now-ubiquitous Snake – in one of its models [6]. As compared to game console and PC games using mobile devices to play games are quite easier [7, 8]. Today mobile game becomes more important issue in technical development of Smartphone and/or tablets device [9] and mobile games become the prime component of the video game industry [6]. Mobile games part is one of the fastest growing fields in the mobile market [9] and increasing by around 20 % annually [10].

Mobile Games is not limited to entertainment only, but entered in the fields of learning and treatment. In the field of Learning and Education, it is observed that learning potentials for players Games higher than non-players [9]. It is noted in the literature that serious games can make a positive contribution [11]. Mobile game-based learning (mGBL) use gameplay to strengthen the motivation to learn, engage in the acquisition of knowledge and to improve the effectiveness of learning content transfer and should mix between education and purpose entertainment to make mGBL successful [3]. The games promote skills of thinking and planning more than assigned to content knowledge [12] also promote problem-solving and collaboration [13]. According to [3] it was found that the students preferred mobile phone for learning rather than other devices. There are several examples in the field of treatment too. "Lumosity" is a mobile game to brain training based on the latest discoveries in neuroscience and the goal of this game is to make Human Cognition more rapidly and efficiently [14]. Another example is, e-Health game "Re-Mission" to improve behavioral outcomes for young patients with cancer in a multisite randomized controlled trial [15]. There is also SPARX program that uses a platform of a fantasy game to teach CBT skills for adolescents with depression [16]. According to the Apple app store and Google play, the popular mobile game markets, the mobile games can be classified into 18 classes [17, 18], as shown in Table 1 below.

Table 1. Mobile game categories [17, 18]

Action	Adventure	Arcade	Board	Music
Casino	Dice	Educational	Family	Sports
Puzzle	Racing	Role playing	Simulation	Strategy
Trivia	Word	Card		

Few years ago, most video games industry moves their development towards mobile games and today reap billions of dollars. This paper will discuss important features to build mobile games and will also cover how to build successful game development environment and will study top grossing games and analyze data through CRISP-DM approach where this approach plays a big role in building models, utilizing the logical sequence of construction steps which applied in this study, such as business and data understanding, Data preparation, model building and evaluation.

This paper is organized as follows; Sect. 2 will discover trends and revenue models in mobile gaming; Sect. 3 will discuss Materials and methods used in this study; Sect. 4 will represent analysis and findings; Sect. 5 will conclude this study.

2   Trends and Revenue in Mobile Gaming

In the year 2014, there were more than 850,000 apps in Apple App Store and 700,000 apps in Google Play store [1]. The statistic shows in Feb 2014 that the game applications get on 41.2 % of all applications category in Google Play store [19] and it shows in Sep 2014 that the game applications were the 1st popular category, with a share of 21.14 % of all applications in Apple app store [20]. The year 2014 has been a jammy year for the mobile games industry. United States get on the highest earnings occurred, where games accounted for nearly all revenue of the platform (Leonov 2014). That mobile games remained the largest segment in terms of revenue, accounting for around a quarter of the total market [10].

Most games are based on "Free to play" model it is the fastest growing during the past 15 years [21]. Following are few important models described:

- **Free to Play (F2P):** These games are free to download and play, but contain micro transactions [22].
- **Pay to play (P2P) or Premium:** In these types of models the users required to pay before playing [1].
- **Freemium:** These are part of F2P [23], and are based on the combination between "free" and "Premium". Freemium become the popular business model for smartphone app developers today [24]. In the year 2014, there are 69 % of gross revenue from IOS and 75 % from Android devices coming from the freemium model. Freemium is achieving monetization from two ways direct and indirect [1].
  - **Direct monetization:** In-App Purchases (IAP) for example, it sales Virtual Goods inside games, best example, Hay Day and clash of clan. In the year 2014, they were able to generate more than 2 million US$ a day from these two titles [25].
  - **Indirect monetization:** Coming from advertising and the best example is Fruits Ninja of Half Brick, this game monthly generates 400000 US$ from Ads only.
- **Paynium:** This model is like a freemium, but the difference is that it requires a payment on an initial purchase, then playing is completely free [26].

This study focuses on the latest trends of the mobile game market and explores its revenue model in general and freemium revenue model in particular. Also, what is the importance of acquisition, retention and monetization for F2P games? Mobile Games that depend on freemium revenue models are analyzed along with the relationship between revenue and games features are explained. Other important relationships are also analyzed, for example, the relationship between revenue, category, developer, users daily play, users daily install and others, for the future researchers and developers. Features in this study are extracted from Askelöf's [27] and Moreira et al. [1] they used ARM (Acquisition, Retention and Monetization) funnel, as shown in Fig. 1, to chosen a group of features related to F2P Model. Total 31 features were used in this study. Figure 5 later in the paper shows the features and classification.

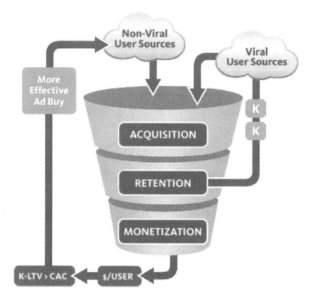

Fig. 1. ARM funnel

3   Materials and Methods

Applying the experimental method to extract, transform, and analyze data, the researchers followed CRISP-DM (Cross Industry Standard Process for Data Mining) approach to handling the problem. Researcher this approach adopted for its success in providing scientific contributions in a previous study [1] in the same field. Where presented, what the features are needs to build a successful game.

3.1   Data

Step-1: Researcher extracted data from "thinkgaming" [28]. Apple app store's Top 50 Grossing iPhone Games were investigated. Out of 50, two paid games were removed, as study focus on a freemium model; all data inserted in an MS Excel spreadsheet, with attributes, for example, Game Name, Revenue, New Installs, Daily Active User, Category and Developer Name for each game to facilitate data analysis.

Step-2: Researchers create another MS Excel spreadsheet with attributes Game Name in Columns and game features in rows. Out of 48 further 2 games were removed because of missing data.

Step-3: After understanding the phases of business and data, data preparation was initiated to use it in the modeling tool.

3.2 Transformation

Total 1824 entries collected through two ways. First install top grossing games in iPhone and extracted 31 features for each game; second extracted other data from thinkgaming for further analysis. The data was processed and converted in such a way to the values to fit with statistical operations in new MS Excel spreadsheet. This prepared data now can use in software, such as Excel, SpSS statistical & SpSS Modeler, to get the best result.

3.3 Normalization

In order to get the easiest value in the columns (Revenue, New Installs and Daily Active User) data was normalization to get value between (0,1) by using following formula:[1]

$$value = \frac{(xi - Min(range\ xi))}{(Max(range\ xi) - Min(range\ xi))}$$

To measure performance evaluation used ROC Curve test, where this test just accepts values between (0–1) i.e. Candy Crush Saga revenue is 1005806\$ after normalization Candy Crush Saga revenue the result becomes (0.590196166).

3.4 Model

In this study, three step processes have been used to generate a model. In the first step, a "Decision Tree" is used to extract knowledge about the problem domain. In the second step, a "Regression Analysis" used to support the most significance between variables. Finally, the third step measures "Performance Evaluation" to ensure the validity of results that get from two previous steps.

Decision Tree. In the finding of this study, a decision tree is the best way for modeling data, as data contents features have close values i.e. (yes/no) were used to find the features are available in some particular game or not. The data representation by decision tree allows advantage compared with other approaches and easy to interpret. The goal of a data model is to create classification model to predict target attributes using SpSS modeler software. The value was used in this model was: (label: Game-Name, values: All Features, weight: Revenue). Figure 2 below depicts the strong relationship between revenue and the features through decisions tree model was generated and it was found that there are total 10 features, which are most important among all 31 features. These 10 features model was generated and find that there is a realistic choice of features to achieve higher profits, as shown below in Fig. 3.

---

[1] "Xi" is value for each game under columns (Revenue, New Installs and Daily Active User) and "range Xi" mean all games values in the same columns.

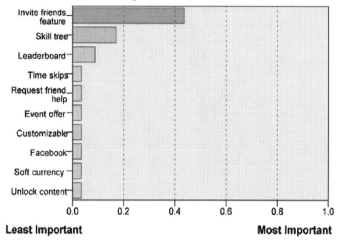

Fig. 2. Features through decision tree

Fig. 3. Decision tree rules for 10 Features

Regression Analysis. To extract the relationship between attributes in the database, linear regression analysis has been used through SpSS statistical software. Before analysis, two variable dependent and independent variable were chosen. All features that got from Decision Tree after applying analysis through revenue attributes were used as dependent variable and feature factors were used as an independent variable as shown in Table 2. From Table 2, it is depicted that the statistical value T-Test of (2.176) at the 0.05 level. This value is statistically significant, which indicates the existence of a relationship between features factor and revenue.

Table 2. Linear regression between features factor and revenue

Coefficients[a]							
Model		Un-standardized coefficients		Standardized coefficients		t	Sig.
		B	Std. error	Beta			
1	(Constant)	−104368.711	130133.878			−.802	0.427
	Features.fac	45103.448	20727.667	.305		2.176	0.035

[a]Dependent Variable: Revenue

Performance Evaluation. ROC Curve is the fundamental tool for diagnostic test evaluation. The ROC Curve test was used for this step. ROC Curve is binary classifier system for illustrates performance. ROC Curve was acceptable if the result is between 1.0 and 0.5 and if it is less than 0.5 is not acceptable. The SpSS statistical software was used to apply ROC Curve through variable "Revenue" and as a test variable all features were extracted from Decision Tree, as shown in Fig. 4 below and result of test variable as shown in Table 3.

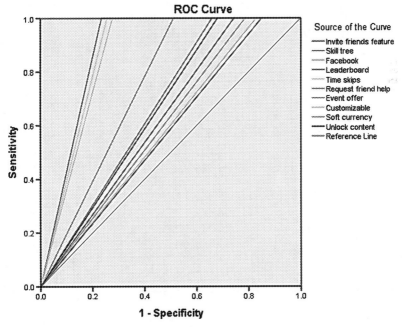

Diagonal segments are produced by ties.

Fig. 4. ROC Curve (Color figure online)

Table 3. Result of test ROC curve variable

Test result variable (s)	Area
Invite friends feature	.628
Skill tree	.883
Facebook	.585
Leaderboard	.574
Time skips	.872
Request friend help	.670
Event offer	.745
Customizable	.862
Soft currency	.606
Unlock content	.660

4 Analysis and Findings

This part will discuss the approach to game features and its relationship of revenue with the category, developers, other features, for example, the number of users' daily play, a number of users' daily install.

4.1 Category

Study finds 14 categories out of 18 categories of 48 top grossing games (as in 2 games researchers were not able to find all related data), as shown in Table 4:

Table 4. Category-wise

Category	Trivia	Card	Arcade	Strategy	Casual	Casino	Family	Action	Role Playing	Sports	Simulation	Adventure	Puzzle	racing	Total
Frequency	1	1	1	6	7	13	2	2	2	2	3	2	4	2	48
Percent	2.1	2.1	2.1	12.5	14.6	27.1	4.2	4.2	4.2	4.2	6.3	4.2	8.3	4.2	100.0
Gross $	83007	37537	32345	3281339	1813050	1165649	224565	172057	165401	175827	233260	131956	218554	65134	7799681

As shown in Table 4, the casino games listed the highest 13 games out of 48 top grossing games in the USA on iPhone devices. This is due to culture and lifestyle. The casual and strategy games came next at 7 and 6 respectively.

4.2 Developers

Study finds 8 developers out of 31 Developers of 48 top grossing games (as in 2 games researchers were not able to find all related data), as shown in Table 5:

Table 5. Developers-wise

Developers	KING.COM	ELECTRONIC ARTS	ZYNGA INC	SUPERCELL	PLAYTIKA LTD	BIG FISH GAMES INC	SGN	FUNZIO INC
Frequency	5	4	4	3	3	2	2	2
Percent	10.4	8.3	8.3	6.3	6.3	4.2	4.2	4.2
Gross $	1716487	341659	140003	2199992	322896	318159	124277	72314
Percent	22.01%	4.38%	1.79%	28.2 1%	4.14%	4.08%	1.59%	0.93%

As shown in Table 5, King.com had 5 games out of 48 games. The net gross exceeded $ 1 m and 700 thousand per day. Whereas Supercell had 3 games out of 48 but it reached nearly 2 m and 200 thousand per day. These results were measured on iPhone devices in the USA.

There are many developers in the field of games. However, a successful game developer might attract the majority of users if one of their successful games was the most popular and used among others.

4.3 Features Analysis

Study used 31 features and divided it into five groups, such as "A", "R", "M", "AR", and "RM" of ARM Funnel approach (Fig. 1 above) and the results are shown in Fig. 5. Through review the 10 features, as shown in Table 6 below, which are important to build the successful game with the ARM Funnel Column finding the three elements (Acquisition, Retention and Monetization) in the F2P games are very important for the success of the game and a failure to one side leads to the failure of the game, where two features have classified under acquisition and four features have classified under each retention and monetization.

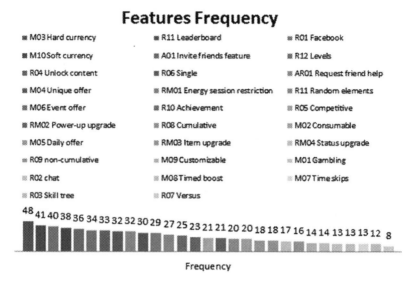

Fig. 5. Feature frequency of 48 top grossing mobile game and groups (Color figure online)

Table 6. Ten important features with ARM funnel classification

Nodes	Importance	ARM funnel
Invite friends feature	0.4356	A01
Skill tree	0.1708	R03
Leaderboard	0.0869	R11
Unlock content	0.0341	R04
Soft currency	0.0341	M10
Facebook	0.0341	R01
Customizable	0.0341	M09
Event offer	0.0341	M06
Request friend help	0.0341	AR01
Time skips	0.0341	M07

4.4   Other Relationships

Study finds other relationships belonging to problem and findings were summaries in Table 7 using linear regression.

Table 7. Linear regression between other relationships belonging to problem and findings

Dependent	Independent	Sig
Unique offer	Daily Active User	.014
Daily offer	Revenue	.002
	Daily Active User	.027
Event offer	Revenue	.010
Time skips	Revenue	.011
Timed boost	Revenue	.018
Soft currency	Revenue	.019
	Daily Active User	.001
Chat	Daily New Installs	.027
	Revenue	.008
	Daily Active User	.009
Competitive	Revenue	.001
	Daily Active User	.006
Single	Revenue	.024
	Daily Active User	.046
Cumulative	Revenue	.011
	Daily Active User	.043
Item upgrade	Revenue	.010
	Daily Active User	.031
Status upgrade	Revenue	.020
	Daily Active User	.135
M.Factor	Revenue	.001
	Daily Active User	.008
R.Factor	Revenue	.004
RM.Factor	Revenue	.037
SUM.ALL.factor	Revenue	.000
	Daily Active User	.014
Developer	Daily Active User	.049
Daily Active User	Revenue	.000
	Daily New Installs	.000
Daily New Installs	Revenue	.029
	Daily Active User	.000

Table 7 depicted that the statistical value T-Test at the 0.05 level. This value is statistically significant, which indicates the existence of a relationship between Dependent and Independent variables. As far as there are daily active users of the games, there will surely be revenue and new installations of games, where a regression linear analysis (Table 7) shows significant equal 0.0 between the elements above. This proves a strong relationship between them.

5 Conclusions

This paper presents the development of previous research regard building a successful game. The 31 features were extracted from the decision tree. The study found that there are 10 most important features to build a successful game. The result is useful for future researchers and developers who want to develop and implement a successful game. Other features were also analyzed in Table 7 and their significant values were presented with the relationship among revenue and other features the strongest relation affecting the revenue was the Daily Active User. So the developer should strengthen the retention element in building games. It is also believed that the other features will also play important roles in successful games, such as culture, lifestyle and loyalty to brand names.

Acknowledgement. Special thanks to ÁtilaValgueiro Malta Moreira who provides some data, which help this research to be completed (Moreira et al. 2014).

References

1. Moreira, Á.V.M., Filho, V.V., Ramalho, G.L.: Understanding mobile game success: a study of features related to acquisition, retention and monetization. SBC J. Interact. Syst. **5**, 2–13 (2014)
2. ITU: ITU releases 2014 ICT figures. Mobile-broadband penetration approaching 32 per Mobile-broadband penetration approaching 32 per cent Three billion Internet users by end of this year, Geneva (2014)
3. Shiratuddin, N., Zaibon, S.B.: Local content game: the preferred choice for mobile learning space. In: ASCILITE, Auckland (2009)
4. GSMA: 4G Networks to Cover More Than a Third of the Global Population This Year, According to New GSMA Intelligence Data. Number of 4G Operators Worldwide Surpasses 350; 4G Forecast to Account for More Than 30 Per Cent of Mobile Connections Globally by 2020, London (2015)
5. GSMA: GSMA publishes new report outlining 5G future, Brussels (2014)
6. EAS: Mobile Games, Washington (2014)
7. Weber, R.: 7 Predictions for the Mobile Gaming Market (2013). http://www.forbes.com/sites/sungardas/2015/03/05/the-1-thing-employees-do-that-compromises-their-own-companys-cyber-security/
8. Wentling, I., Dutton, C.J.: Mobile game application. 20. Google Patents (2014). https://www.google.com/patents/US20140256389
9. Ganguin, S., Hoblitz, A.: Mobile worlds: mobile gaming and learning? In: Göbel, S., Müller, W., Urban, B., Wiemeyer, J. (eds.) GameDays 2012 and Edutainment 2012. LNCS, vol. 7516, pp. 111–120. Springer, Heidelberg (2012)
10. GmbH & Co. KG: Global Mobile Gaming Market 2014 - Mobile Games Remained the Largest Segment in Terms of Revenue. M2 PressWIRE (2014)
11. Connolly, T.M., Boyle, E.A., MacArthur, E., Hainey, T., Boyle, J.M.: A systematic literature review of empirical evidence on computer games and serious games. Comput. Educ. **59**, 661–686 (2012)

12. Ke, F.: A qualitative meta-analysis of computer games as learning tools. In: Ferdig, R.E. (ed.) Handbook of Research on Effective Electronic Gaming in Education, pp. 1–32. IGI Global, Hershey (2009)
13. McClarty, K.L., Orr, A., Frey, P.M., Dolan, R.P., Vassileva, V., McVay, A.: A Literature Review of Gaming in Education (2012)
14. Lumosity: Lumosity's Big Data Provides New Approach to Understanding Human Cognition. Business Wire (2013). (English)
15. Kato, P.M., Cole, S.W., Bradlyn, A.S., Pollock, B.H.: A video game improves behavioral outcomes in adolescents and young adults with cancer: a randomized trial. Pediatrics **122**, e305–e317 (2008)
16. Stasiak, K., Merry, S.: e-Therapy using computer and mobile technologies in treatment. In: Rey, J.M. (ed.) IACAPAP e-Textbook of Child and Adolescent Mental Health. IACAPAP, Geneva (2013)
17. Google: Game Categories. https://play.google.com/store/apps/category/GAME
18. Apple: Game Categories. https://itunes.apple.com/en/genre/ios/id36?mt=8
19. Statista.com: Most popular Google Play app categories, February 2014. By device installs, http://www.statista.com/statistics/279286/distribution-of-worldwide-google-play-app-downloads-by-category/
20. Statista.com: Most popular Apple App Store categories, January 2015. By share of available apps, http://www.statista.com/statistics/270291/popular-categories-in-the-app-store/
21. Sheppard, A.: F2P the "most democratic form of development" – Kabam. http://www.gamesindustry.biz/articles/2014-03-05-f2p-the-most-democratic-form-of-development-kabam
22. Hahl, K.: The Success of Free to Play Games and the Possibilities of Audio Monetization, Tampere, Finland (2014)
23. Alha, K., Koskinen, E., Paavilainen, J., Hamari, J., Kinnunen, J.: Free-to-Play Games: Professionals' Perspectives. In: Proceedings of the 2014 International DiGRA Nordic Conference, DiGRA Nordic 2014. DiGRA, Sweden (2014)
24. Kumar, V.: Making "Freemium" Work. https://hbr.org/2014/05/making-freemium-work/ar/1
25. MobileGameArch: The Mobile Game Arch Roadmap for the Future of the Mobile Game Ecosystem, http://www.mobilegamearch.eu/wp-content/uploads/2013/09/Mobile-Game-Arch_D33_13September_2013_v20.pdf
26. Counsell, D.: Paid, Paymium or Freemium. http://dancounsell.com/articles/paid-paymium-or-freemium
27. Askelöf, P.: Monetization of Social Network Games in Japan and the West. Lund, Sweden (2013)
28. thinkgaming: Top Grossing iPhone Games. https://thinkgaming.com/app-sales-data/

Decision Making Based on Different Dimension Direction

Yung-Lin Chen[1,2(✉)]

[1] Business Office of Salomon Island, Kaohsiung, Taiwan
michaelchen19730318@gmail.com
[2] Department of Administrative Management, College of Public Administration,
Huazhong University of Science and Technology, Wuhan, China

Abstract. "Dimension direction" means the evaluation direction in decision making problem. Decision-making sometimes should consider multi-criteria, multi-period and multi-place. The criteria, period and place are different dimension direction for making decision. In special situation, decision maker must think over more than three dimension directions in the decision making problem. However, past research literatures at most consider two dimension directions. The goal of this research is to develop the execution process to deal with the decision making problem under multi dimension direction. Proposed execution process can flexibly apply in various kinds of multi dimension direction problem. A numerical example will be implemented. Some analysis and comparison will be executed. Conclusion and future research will be discussed as ending.

Keywords: MCDM · TOPSIS · Entropy · Decision making

1 Introduction

Decision-Making is the procedure to find the best action among a set of feasible actions [1, 2]. Decision-making must consider multi-dimension such as period dimension, space dimension and criterion dimension. Traditional multi criteria decision making (MCDM) is the decision-making based on different criteria. Multi period decision making (MPDM) is another kind of decision making problem which will occur in the product utility evaluation. For example, the product will possess different utility for customer in different period (the utility of new car is better than the utility of old car; the overall utility of the car is the integration utility of the car under different period). The technology product will possess different sale volume in different periods (the sale volume of technology product in maturity period is higher than it in the introduction period, growth period and decline period) [3]. Multi space decision making (MSDM) is the decision-making based on different space. Multi space decision making is another kind of decision making problem which will happen in marketing problem. For instance, the product will possess different customer demand in different place (the sale volume of air conditioning is usually different in different place because the climate condition is different in different place).

© Springer International Publishing Switzerland 2016
H. Fujita et al. (Eds.): IEA/AIE 2016, LNAI 9799, pp. 684–700, 2016.
DOI: 10.1007/978-3-319-42007-3_59

"Dimension direction" means the evaluation direction in decision making problem. For example, the performance of alternative in different periods is one of dimension direction. The performance of alternative in different place is another dimension direction. The performance of alternative in different criteria is the other dimension direction. The performance of alternative for different user group is another dimension direction (Some alternatives such as public facilities should consider the usefulness to different group). The different dimension directions are independent and can be integrated to analyze the alternative in detail.

Period, space and criteria can be said as different dimension direction for making decision. In some situation, decision maker maybe can limit dimension direction in special decision situation for making decision. However, making decision based on multi dimension direction can select or evaluate alternative more exhaustively. Decision making can be at least separated as three dimension directions and can be combined as seven decision making processes (Refer to Table 1). So, the goal of this research is to introduce the concept of the dimension direction in making decision and introduce decision process for making decision based on the combination of different decision dimension.

Table 1. Decision making dimension direction

Volume of dimension direction	Decision Making Method
Single dimension direction	Multi Criteria Decision Making (MCDM)
	Multi Period Decision Making (MPDM)
	Multi Space Decision Making (MSDM)
Two dimension direction	Multi Period Multi Criteria Decision Making (MPCDM)
	Multi Space Multi Criteria Decision Making (MSCDM)
	Multi Period Multi Space Decision Making (MPSDM)
Three dimension direction	Multi Period Multi Space Multi Criteria Decision Making (MCPSCDM)

2 Decision Making Model

2.1 Decision Making Based on Single Dimension Direction

Decision maker only needs to consider one dimension direction when the decision situation focuses on special condition. Multi criteria decision making (MCDM), multi period decision making (MPDM), multi space decision making (MSDM) are three kinds of decision making based on single dimension direction. The detail of MCDM, MPDM and MSDM are as follows:

2.1.1 Multi Criteria Decision Making Method (MCDM)
Multi criteria decision making is one of the most widely used decision methodologies in the sciences, business, government and engineering worlds [2, 4]. MCDM methods can help to improve the quality of decisions by making the decision-making process

more explicit, rational, and efficient [4]. Every kind of MCDM approach includes analytic hierarchy process (AHP) [5], analytic network process (ANP) [6], ELECTRE [7], Fuzzy Integral [8], preference ranking organization method for enrichment evaluation (PROMETHEE) [9], rough set [10], technique for order preference by similarity to ideal solution (TOPSIS) [11], utility function [12] has its suitable scope and limitation.

The notation of MCDM can be illustrated as follows.

(1) A set of alternatives $A = \{A_1, A_2, \ldots, A_m\}$ where m means volume of alternatives.
(2) A set of criteria $C = \{C_1, C_2, \ldots, C_n\}$ where n means volume of criteria.
(3) A set of criteria weight $W = \{W_1, W_2, \ldots, W_n\}$ where n means volume of criteria weight.
(4) A set of performance ratings of each alternative with respect to each criterion $X = [x_{ij}]_{m*n}$.

According to above discussion, there are various kinds of MCDM method. This research uses TOPSIS [11] and entropy method [13] to implement the process of MCDM technique. The execution process is as follows:

Step 1. Decision maker collects information from the society according to the problem which decision maker wants to handle.

Step 2. Normalize the information according to below formula

$$s_{ij} = (x_{ij} - \min_i x_{ij})/(\max_i x_{ij} - \min_i x_{ij}), \text{ if } C_j \text{ is a benefit criterion} \qquad (1)$$

$$s_{ij} = 1 - (x_{ij} - \min_i x_{ij})/(\max_i x_{ij} - \min_i x_{ij}), \text{ if } C_j \text{ is a cost criterion} \qquad (2)$$

where s_{ij} means the normalized value of alternative i respect to criterion j.

Step 3. Calculate the weight of each criterion according to entropy method as $£_j = 1/\ln(n) \sum_{i=1}^{n} s_{ij} * \ln(1/s_{ij})$, where $£_j$ means the entropy value of j-th criterion.

The information in the criterion is low when the entropy value is high. So, the information performance value of j-th criterion is $\Psi_j = 1 - £_j$. The weight formula about j-th criterion is as $w_j = \Psi_j / \left(\sum_{z=1}^{n} \Psi_z\right)$.

Step 4. Calculate the weighted matrix as $V = [v_{ij}]_{m*n}$, where $v_{ij} = s_{ij} * w_j$.

Step 5. Calculate the positive ideal solution as formula $F_j^+ = \max_i v_{ij}$, where F_j^+ means the positive ideal value respect to criterion j.

Step 6. Calculate the negative ideal solution as formula $F_j^- = \min_i v_{ij}$, where F_j^- means the negative ideal value respect to criterion j.

Step 7. Calculating the distance between the alternative and the positive ideal solution is as $d_i^+ = (\sum_{j=1}^{n} (F_j^+ - v_{ij})^2)^{0.5}$.

Step 8. Calculating the distance between the alternative and negative ideal solution can be calculated as $d_i^- = (\sum_{j=1}^{n} (v_{ij} - F_j^-)^2)^{0.5}$.

Step 9. The closeness coefficient of each alternative can be computed as $CC_i = d_i^-/(d_i^+ + d_i^-)$ $i = 1, 2, \ldots, m$.

The ranking order of alternatives can be determined in accordance with the closeness coefficient. If $CC_y > CC_z$, then alternative A_y is better than alternative A_z.

2.1.2 Multi Period Decision Making Method (MPDM)

Multi period decision making method is evaluating and selecting the alternative based on the utility of alternative in each period which can be occurred in the overall production utility analysis, health insurance analysis, enterprise operation overall performance analysis, BOT project operation overall performance analysis (Build–Operate–Transfer, BOT). All of above problem should consider the performance of alternative in each period. So, measuring the weight (importance) of each period for evaluating performance of alternative is an important issue.

The notation of MPDM can be illustrated as follows:

(1) A set of alternatives $A = \{A_1, A_2, \ldots, A_m\}$ where m means volume of alternatives.
(2) A set of periods $T = \{T_1, T_2, \ldots, T_n\}$ where n means the volume of periods.
(3) A set of period weight $W = \{W_1, W_2, \ldots, W_n\}$ where n means volume of period weight.
(4) A set of performance ratings of alternatives with respect to periods $X = \left[x_{ij}\right]_{m*n}$.

In MPDM problem, the period weight evaluation is an important issue because the importance of each period is usually related. Arithmetic series based method, geometric series based method and normal distribution based method are three kinds of period weight decision method. The detail of above three methods is as follows:

In arithmetic series based method, the weight difference between each adjacent period is λ. When λ is positive, it means that the high period is more important to evaluate the alternative; On the other hand, it means that the few period is more important to evaluate the alternative when λ is negative.

The formula of arithmetic series based method is as follows [14]:

$$W_j = \Gamma + (j - 1) \in / \sum_{i=1}^{n} \Gamma + (j - 1) \in \qquad (3)$$

where Γ means the importance of first period, $\in$ means the amount of change between adjacent period.

The formula of geometric series based method is as follows [14]:

$$W_j = \Gamma * \in^{j-1} / (\sum_{i=1}^{n} \Gamma * \in^{j-1}) \qquad (4)$$

where Γ means the importance of first period, $\in$ means the degree of change between adjacent period.

The formula of normal distribution based method is as follows [15]:

$$W_j = e^{\frac{(z-u_n)^2}{2\sigma_n^2}} / \left(\sum\nolimits_{z=1}^n e^{\frac{(z-u_n)^2}{2\sigma_n^2}} \right) \tag{5}$$

where u_n means the mean of period volume and $u_n = (1+n)/2$, σ_n means the standard deviation of period volume and $\sigma_n = (\frac{1}{n} \sum_{i=1}^n (i - u_n)^2)^{0.5}$.

This research uses TOPSIS [11] and one kind of period weight evaluation technique to implement the process of MPDM approach. The execution process is as follows:

Step 1. Decision maker collects information from the society according to the problem which decision maker wants to handle.

Step 2. Normalize the information as $s_{ij} = (x_{ij} - \min_i x_{ij})/(\max_i x_{ij} - \min_i x_{ij})$ where s_{ij} means the normalized value of alternative i in the j-th period.

Step 3. Calculate the weight of each period according to period weight evaluation technique which decision maker choose (The period weight evaluation technique can refer to formulas 3, 4 or 5).

Step 4. Calculate the weighted matrix as $V = [v_{ij}]_{m*n}$, where $v_{ij} = s_{ij} * w_j$.

Step 5. Calculate the positive ideal solution as $F_j^+ = \max_i v_{ij}$, where F_j^+ means the positive ideal value in the j-th period.

Step 6. Calculate the negative ideal solution as $F_j^- = \min_i v_{ij}$, where F_j^- means the negative ideal value in the j-th period.

Step 7. Calculating the distance between the alternative and positive ideal solution is as $d_i^+ = (\sum_{j=1}^n \left(F_j^+ - v_{ij} \right)^2)^{0.5}$.

Step 8. Calculating the distance between the alternative and negative ideal solution is as $d_i^- = (\sum_{j=1}^n \left(v_{ij} - F_j^- \right)^2)^{0.5}$.

Step 9. The closeness coefficient of each alternative can be computed as $CC_i = d_i^- / (d_i^+ + d_i^-)$. The ranking order of alternatives can be determined in accordance with the closeness coefficient.

2.1.3 Multi Space Decision Making Method (MSDM)

The notation and execution process of MSDM is the same with MCDM, the only difference between MCDM and MSDM is that the evaluation direction in MCDM is using different criteria to evaluate alternative and the evaluation direction in MSDM is the utility of alternative in different space.

The notation of MSDM can be illustrated as follows.

(1) A set of alternatives $A = \{A_1, A_2, \ldots, A_m\}$ where m means volume of alternatives.

(2) A set of spaces $S = \{S_1, S_2, \ldots, S_n\}$ where n means volume of spaces.

(3) A set of space weight $W = \{W_1, W_2, \ldots, W_n\}$ where n means volume of space weight.

(4) A set of performance ratings of alternatives with respect to each space $X = [x_{ij}]_{m*n}$.

According to above discussion, there are various kinds of MCDM method. This research uses TOPSIS [11] and entropy method [13] to implement the process of MSDM approach. The execution process is as follows:

Step 1. Decision maker collects information from the society according to the problem which decision maker wants to handle.

Step 2. Normalize the information according to formula $s_{ij} = (x_{ij} - min_i x_{ij})/(max_i x_{ij} - min_i x_{ij})$ where s_{ij} means the normalized value of alternative i respect to space j.

Step 3. Calculate the weight of each space according to entropy method by formula $£_j = (1/\ln(n)) \sum_{i=1}^{n} s_{ij} * \ln(1/s_{ij})$ where $£_j$ means the entropy value of j-th space. The information in the space is low when the entropy value is high. So, The information performance value of j-th space is $\Psi_j = 1 - £_j$, The formula about j-th space is as $w_j = \Psi_j/(\sum_{z=1}^{n} \Psi_z)$.

Step 4. Calculate the weighted matrix as $V = [v_{ij}]_{m*n}$, where $v_{ij} = s_{ij} * w_j$.

Step 5. Calculate the positive ideal solution as $F_j^+ = max_i v_{ij}$ where F_j^+ means the positive ideal value respect to space j.

Step 6. Calculate the negative ideal solution as $F_j^- = min_i v_{ij}$ where F_j^- means the negative ideal value respect to space j.

Step 7. Calculating the distance between the alternative and positive ideal solution as $d_i^+ = (\sum_{j=1}^{n} (F_j^+ - v_{ij})^2)^{0.5}$.

Step 8. Calculating the distance between the alternative and negative ideal solution as $d_i^- = (\sum_{j=1}^{n} (v_{ij} - F_j^-)^2)^{0.5}$.

Step 9. The closeness coefficient of each alternative can be computed as $CC_i = d_i^-/(d_i^+ + d_i^-)$. The ranking order of alternatives can be determined in accordance with the closeness coefficient.

2.2   Decision Making Based on Two Dimension Direction

The decision making based on two dimension directions will occur in every situation, if decision maker wants to evaluate and select alternative more detail.

2.2.1   Multi Period Multi Criteria Decision Making Method (MPCDM)

The notation of MPCDM is as follows:

(1) A set of alternatives $A = \{A_1, A_2, ..., A_m\}$ where m means volume of alternatives.
(2) A set of criteria $C = \{C_1, C_2, ..., C_n\}$ where n means volume of criteria.
(3) A set of period $T = \{T_1, T_2, ..., T_o\}$ where o means volume of periods.
(4) A set of criteria weight $W^C = \{W_1^C, W_2^C, ..., W_n^C\}$ where n means volume of criteria weight.

(5) A set of period weight $W^T = \{W_1^T, W_2^T, \ldots, W_o^T\}$ where o means volume of period weight.
(6) A set of performance ratings of each alternative with respect to each criterion in each period, called $X = \left[x_{ij}^k\right]$ i = 1, 2, ... m; j = 1, 2, ... , n; k = 1, 2, ... , o.

The execution process of MPCDM is as follows:

Step 1. Decision maker collects information $X = \left[x_{ij}^k\right]$ from the society according to the MPCDM problem which decision maker wants to handle.
Step 2. Normalize the information according to below formula.

$$s_{ij}^k = \left(x_{ij}^k - \min_i x_{ij}^k\right)\Big/\left(\max_i x_{ij}^k - \min_i x_{ij}^k\right), \text{if } C_j \text{ is a benefit criterion} \qquad (6)$$

$$s_{ij}^k = 1 - \left(x_{ij}^k - \min_i x_{ij}^k\right)\Big/\left(\max_i x_{ij}^k - \min_i x_{ij}^k\right), \text{if } C_j \text{ is a cost criterion} \qquad (7)$$

where s_{ij}^k means normalized value of alternative i respect to criterion j in k-th period.

Step 3. Calculate the weight of each period according to period weight evaluation technique which decision maker chooses (The period weight evaluation technique can refer to formulas 3, 4 or 5).
Step 4. Calculate the weight of each criterion according to entropy method as $£_j = \frac{1}{\ln(m*o)}\sum_{i=1}^m \sum_{k=1}^o s_{ij}^k * \ln(\frac{1}{s_{ij}^k})$, where $£_j$ means the entropy value of j-th criterion. The information in the criterion is low when the entropy value is high. So, The information performance value of j-th criterion is $\Psi_j = 1 - £_j$, The weight formula about j-th criterion is $w_j^c = \Psi_j/(\sum_{z=1}^n \Psi_z)$.
Step 5. Calculate the weighted matrix as $V = \left[v_{ij}^k\right]$, where $v_{ij}^k = s_{ij}^k * w_j^C * w_k^T$, i = 1, 2, ... , m; j = 1, 2, ... , n; k = 1, 2, ... , o.
Step 6. Calculate the positive ideal solution as $F_{jk}^+ = \max_i v_{ij}^k$ where F_{jk}^+ means the positive ideal value respect to j-th criterion in the k-th period.
Step 7. Calculate the negative ideal solution as $F_{jk}^- = \min_i v_{ij}^k$ where F_{jk}^- means the negative ideal value respect to j-th criterion in the k-th period.
Step 8. Calculating the distance between the alternative and positive ideal solution is as $d_i^+ = (\sum_{j=1}^n \sum_{k=1}^o \left(F_{jk}^+ - v_{ij}^k\right)^2)^{0.5}$.
Step 9. Calculating the distance between the alternative and negative ideal solution is as $d_i^- = (\sum_{j=1}^n \sum_{k=1}^o \left(v_{ij}^k - F_{jk}^-\right)^2)^{0.5}$.
Step 10. The closeness coefficient of each alternative can be computed as $CC_i = d_i^+/(d_i^+ + d_i^-)$. The ranking order of alternatives can be determined in accordance with the closeness coefficient.

2.2.2   Multi Criteria Multi Space Decision Making Method (MSCDM)

The notation of MSCDM is as follows:

(1) A set of alternatives $A = \{A_1, A_2, \ldots, A_m\}$ where m means volume of alternatives.
(2) A set of criteria $C = \{C_1, C_2, \ldots, C_n\}$ where n means volume of criteria.
(3) A set of spaces $S = \{S_1, S_2, \ldots, S_o\}$ where o means volume of spaces.
(4) A set of criteria weight $W^C = \{W_1^C, W_2^C, \ldots, W_n^C\}$ where n means volume of criteria weight.
(5) A set of space weight $W^S = \{W_1^S, W_2^S, \ldots, W_o^S\}$ where o means volume of space weight.
(6) A set of performance ratings of each alternative with respect to each criterion in each space, called $X = \left[x_{ij}^k\right]$ i = 1, 2, ... m; j = 1, 2, ... , n; k = 1, 2, ... , o.

The execution process of MSCDM is as follows:

Step 1. Decision maker collects information $X = \left[x_{ij}^k\right]$ from the society according to the MSCDM problem which decision maker wants to handle.

Step 2. Normalize the information according to below formula.

$$s_{ij}^k = \left(x_{ij}^k - min_i x_{ij}^k\right) \Big/ \left(max_i x_{ij}^k - min_i x_{ij}^k\right), \text{ if } C_j \text{ is a benefit criterion} \qquad (8)$$

$$s_{ij}^k = 1 - \left(x_{ij}^k - min_i x_{ij}^k\right) \Big/ \left(max_i x_{ij}^k - min_i x_{ij}^k\right), \text{ if } C_j \text{ is a cost criterion} \qquad (9)$$

where s_{ij}^k means normalized value of alternative i respect to criterion j in k-th period.

Step 3. Calculate the weight of each criterion according to entropy method as $£_j^C = \frac{1}{\ln(m*o)} \sum_{i=1}^m \sum_{k=1}^o s_{ij}^k * \ln(\frac{1}{s_{ij}^k})$, where $£_j^C$ means the entropy value of j-th criterion. The information in the criterion is low when the entropy value is high. So, The information performance value of j-th criterion is $\Psi_j^c = 1 - £_j^c$, The weight formula about j-th criterion is $w_j^c = \Psi_j^c / (\sum_{z=1}^n \Psi_z^c)$.

Step 4. Calculate the weight of each space according to entropy method as $£_k^S = \frac{1}{\ln(m*n)} \sum_{i=1}^m \sum_{j=1}^n s_{ij}^k * \ln(\frac{1}{s_{ij}^k})$, where $£_k^S$ means the entropy value of the k-th space. The information in the space is low when the entropy value is high. So, The information performance value of the k-th space is $\Psi_k^S = 1 - £_k^S$, The weight formula about k-th space is $w_k^S = \Psi_k^S / (\sum_{z=1}^o \Psi_z^S)$.

Step 5. Calculate the weighted matrix as $V = \left[v_{ij}^k\right]$, where $v_{ij}^k = s_{ij}^k * w_j^C * w_k^S$, i = 1, 2, ... , m; j = 1, 2, ... , n; k = 1, 2, ... , o.

Step 6. Calculate the positive ideal solution as $F_{jk}^+ = max_i v_{ij}^k$ where F_{jk}^+ means the positive ideal value respect to j-th criterion in the k-th space.

Step 7. Calculate the negative ideal solution as $F_{jk}^- = min_i v_{ij}^k$ where F_{jk}^- means the negative ideal value respect to j-th criterion in the k-th space.

Step 8. Calculating the distance between the alternative and positive ideal solution is as $d_i^+ = (\sum_{j=1}^{n} \sum_{k=1}^{o} (F_{jk}^+ - v_{ij}^k)^2)^{0.5}$.

Step 9. Calculating the distance between the alternative and negative ideal solution is as $d_i^- = (\sum_{j=1}^{n} \sum_{k=1}^{o} (v_{ij}^k - F_{jk}^-)^2)^{0.5}$.

Step 10. The closeness coefficient of each alternative can be computed as $CC_i = d_i^- / (d_i^+ + d_i^-)$. The ranking order of alternatives can be determined in accordance with the closeness coefficient.

2.2.3   Multi Period Multi Space Decision Making Method (MPSDM)

The notation of MPSDM is as follows:

(1) A set of alternatives $A = \{A_1, A_2, \ldots, A_m\}$ where m means volume of alternatives.

(2) A set of spaces $S = \{S_1, S_2, \ldots, S_n\}$ where n means volume of spaces.

(3) A set of periods $T = \{T_1, T_2, \ldots, T_o\}$ where o means volume of periods.

(4) A set of criteria weight $W^S = \{W_1^S, W_2^S, \ldots, W_n^S\}$ where n means volume of space weight.

(5) A set of period weight $W^T = \{W_1^T, W_2^T, \ldots, W_o^T\}$ where o means volume of period weight.

(6) A set of performance ratings of each alternative with respect to each space in each period $X = \left[x_{ij}^k\right]$ i = 1, 2, ... m; j = 1, 2, ... , n; k = 1, 2, ... , o.

The execution process of MPSDM is as follows:

Step 1. Decision maker collects information $X = \left[x_{ij}^k\right]$ from the society according to the MPSDM problem which decision maker wants to handle.

Step 2. Normalize the information according to below formula.

$$s_{ij}^k = \left(x_{ij}^k - \min_i x_{ij}^k\right) \Big/ \left(\max_i x_{ij}^k - \min_i x_{ij}^k\right) \tag{10}$$

where s_{ij}^k means the normalized value of alternative i respect to space j in k-th period.

Step 3. Calculate the weight of each period according to period weight evaluation technique which decision maker chooses (The period weight evaluation technique can refer to formulas 3, 4 or 5).

Step 4. Calculate the weight of each space according to entropy method as $£_j = \frac{1}{\ln(m*o)} \sum_{i=1}^{m} \sum_{k=1}^{o} s_{ij}^k * \ln(\frac{1}{s_{ij}^k})$, where $£_j$ means the entropy value of j-th space. The information in the space is low when the entropy value is high. So, The information performance value of j-th space is $\Psi_j = 1 - £_j$, The weight formula about j-th space is $w_j^S = \Psi_j / (\sum_{z=1}^{n} \Psi_z)$.

Step 5. Calculate the weighted matrix as $V = \left[v_{ij}^k \right]$ where $v_{ij}^k = s_{ij}^k * w_j^S * w_k^T$, $i = 1$, $2, \ldots, m$; $j = 1, 2, \ldots, n$; $k = 1, 2, \ldots, o$.

Step 6. Calculate the positive ideal solution as $F_{jk}^+ = \max_i v_{ij}^k$ where F_{jk}^+ means the positive ideal value respect to j-th space in the k-th period.

Step 7. Calculate the negative ideal solution as $F_{jk}^- = \min_i v_{ij}^k$ where F_{jk}^- means the negative ideal value respect to j-th space in the k-th period.

Step 8. Calculating the distance between the alternative and positive ideal solution is as $d_i^+ = (\sum_{j=1}^n \sum_{k=1}^o \left(F_{jk}^+ - v_{ij}^k \right)^2)^{0.5}$.

Step 9. Calculating the distance between the alternative and negative ideal solution is as $d_i^- = (\sum_{j=1}^n \sum_{k=1}^o \left(v_{ij}^k - F_{jk}^- \right)^2)^{0.5}$.

Step 10. The closeness coefficient of each alternative can be computed as $CC_i = d_i^- / (d_i^+ + d_i^-)$. The ranking order of alternatives can be determined in accordance with the closeness coefficient.

2.3 Decision Making Based on Three Dimension Direction

The notation of multi periods, spaces and criteria decision making method (MPSCDM) is as follows:

(1) A set of alternatives $A = \{A_1, A_2, \ldots, A_m\}$ where m means volume of alternatives.

(2) A set of periods $T = \{T_1, T_2, \ldots, T_n\}$ where n means volume of periods.

(3) A set of spaces $S = \{S_1, S_2, \ldots, S_o\}$ where o means volume of spaces.

(4) A set of criteria $C = \{C_1, C_2, \ldots, C_p\}$ where p means the volume of criteria.

(5) A set of period weight $W^T = \{W_1^T, W_2^T, \ldots, W_n^T\}$ where n means volume of period weight.

(6) A set of space weight $W^S = \{W_1^S, W_2^S, \ldots, W_o^S\}$ where o means volume of space weight.

(7) A set of criteria weight $W^S = \{W_1^C, W_2^C, \ldots, W_p^C\}$ where p means volume of criteria weight.

(8) A set of performance ratings of each alternative with respect to each criteria in each space and period, called $X = \left[x_{ijk}^l \right]$ $i = 1, 2, \ldots m$; $j = 1, 2, \ldots, p$; $k = 1$, $2, \ldots, o$; $l = 1, 2, \ldots, n$.

The execution process of MPSCDM is as follows:

Step 1. Decision maker collects information from the society according to the problem which decision maker wants to handle.

Step 2. Normalize the information according to below formula.

$$s_{ijk}^l = (x_{ijk}^l - \min_i x_{ijk}^l)/(\max_i x_{ijk}^l - \min_i x_{ijk}^l), \text{ if } C_j \text{ is a benefit criterion} \tag{11}$$

$$s_{ijk}^l = 1 - (x_{ijk}^l - \min_i x_{ijk}^l)/(\max_i x_{ijk}^l - \min_i x_{ijk}^l), \text{ if } C_j \text{ is a cost criterion} \tag{12}$$

where s_{ijk}^l means normalized value of alternative i respect to criterion j in k-th space in l-th period.

Step 3. Calculate the weight of each period according to period weight evaluation technique which decision maker chooses (The period weight evaluation technique can refer to formulas 3, 4 or 5).

Step 4. Calculate the weight of each criterion according to entropy method as $\pounds_j^C = \frac{1}{\ln(m*n*o)} \sum_{i=1}^m \sum_{k=1}^o \sum_{l=1}^n s_{ijk}^l * \ln(\frac{1}{s_{ijk}^l})$, where $\pounds_j^C$ means the entropy value of j-th criterion. The information respect to this criterion is low when the entropy value is high. So, The information performance value of j-th criteria is $\Psi_j^C = 1 - \pounds_j^C$, The weight formula about j-th criterion is $w_j^C = \Psi_j^C/(\sum_{z=1}^n \Psi_z^C)$.

Step 5. Calculate the weight of each space according to entropy method as $\pounds_k^S = \frac{1}{\ln(m*o*p)} \sum_{i=1}^m \sum_{j=1}^p \sum_{l=1}^n s_{ijk}^l * \ln(\frac{1}{s_{ijk}^l})$, where $\pounds_k^S$ means the entropy value of k-th space. The information in the k-th space is low when the entropy value is high. So, The information performance value of k-th space is $\Psi_k^S = 1 - \pounds_k^S$, The weight formula about k-th space is $w_k^S = \Psi_k^S/(\sum_{y=1}^n \Psi_y^S)$.

Step 6. Calculate the weighted matrix as $V = \left[v_{ijk}^l \right]$ where $v_{ijk}^l = s_{ijk}^l * w_l^T * w_k^S * w_j^C$, i = 1, 2, ... , m; j = 1, 2, ... , p; k = 1, 2, ... , o; l = 1, 2, ... , n.

Step 7. Calculate the positive ideal solution as $F_{jkl}^+ = \max_i v_{ijk}^l$ where F_{jkl}^+ means positive ideal value respect to j-th criteria in k-th space in l-th period.

Step 8. Calculate the negative ideal solution as $F_{jkl}^- = \min_i v_{ijk}^l$ where F_{jkl}^- means negative ideal value respect to j-th criteria in k-th space in l-th period.

Step 9. Calculating the distance between the alternative and positive ideal solution is as $d_i^+ = (\sum_{j=1}^p \sum_{k=1}^o \sum_{l=1}^n \left(F_{jkl}^+ - v_{ijk}^l \right)^2)^{0.5}$.

Step 10. Calculating the distance between the alternative and negative ideal solution is as $d_i^- = (\sum_{j=1}^p \sum_{k=1}^o \sum_{l=1}^n \left(v_{ijk}^l - F_{jkl}^- \right)^2)^{0.5}$.

Step 11. The closeness coefficient of each alternative can be computed as $CC_i = d_i^-/(d_i^+ + d_i^-)$. The ranking order of alternatives can be determined in accordance with the closeness coefficient.

2.4 The Execution Process of Decision Making Based on Multi Dimension Direction

In decision making problem, its dimension direction can include "criteria", "time", "space", "the characteristic of user who use the alternative" etc. For example, The new car manufacture project selection maybe can include four decision dimension (1) consumer relative criteria dimension (such as maintain cost, useful life, additional outfit purchase consume volume), (2) time dimension (the performance of the new car in consumer relative criteria in each period), (3) space dimension (the performance of the new car in consumer relative criteria in each place), (4) demographics dimension (the performance of the new car in consumer relative criteria in the demographics characteristics). If there are x_1 criteria in consumer relative criteria dimension, x_2 periods in time dimension, x_3 place in space dimension and x_4 characteristics in demographics dimension. The decision making space of this new car manufacture project selection is $x_1 * x_2 * x_3 * x_4$. So, it needs an execution process to make decision based on multi dimension directions. The execution process of decision making based on multi dimension directions is as follows:

At first, decision maker should decide the alternatives and decision dimensions, the content in each decision dimension. And then, decision maker must collect relative information according to the alternatives and the content in each decision dimension. The relative information is about the performance of alternatives respect to the criteria in different kinds of dimension direction. After that, the information will be normalized according to the attribute of criteria (benefit criteria or cost criteria). Afterward, the weight of each criterion in each dimension direction is calculated by entropy method except the time dimension. The weight of each period in time dimension is calculated by period weight evaluation technique. Finally, the execution process of TOPSIS includes the weighted matrix, the positive ideal solution, the negative ideal solution, the distance between the alternative and the positive ideal solution, the distance between the alternative and the negative ideal solution, the closeness coefficient can be calculated. The rank of alternatives can be acquired according to closeness coefficient of each alternative (Refer to Fig. 1).

3   Numerical Example

In order to let reader understand the above method, this research decides a numerical example to implement. Suppose that an air condition manufacture enterprise wants to produce a new air condition product to sell in the market. So, the enterprise investigates the different kinds of air condition information under different time and space. There are three dimension directions (criteria, time and space) in this problem. In the criteria dimension, the criteria includes four criteria: Air Condition Sell Volume (C_1), Air Condition Unit Maintain Cost (C_2), Air Condition Customer Satisfaction Ratio (C_3) and Air Condition Unit Sell profit (C_4). There are five years (period) in the time dimension ($T_1, T_2, \ldots, T_5$). In the space dimension, the space includes five places: Center Area (S_1)), East Area (S_2)), West Area (S_3), North Area (S_4) and South Area (S_5). According to enterprise's past manufacture and sell information, there are four kinds of air conditions which enterprise has ever been manufactured. Wall-Mounted Air Condition (A_1), Floor-Standing Air Condition (A_2), 4-Way Cassette Air Condition (A_3) and Compact Floor Air Condition (A_4).

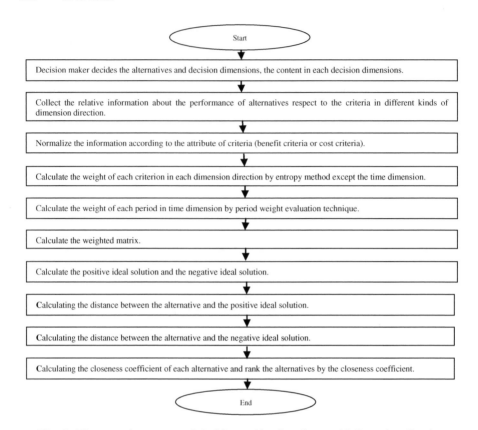

Fig. 1. The execution process of decision making based on multi dimension direction

The execution process of MPSCDM is as follows:

Step 1. The air condition enterprise collects information about the sell volume, unit maintain cost, customer satisfaction ratio of air condition and unit sell profit of four kinds of air condition under different periods and places (Refer to Table 2).

Step 2. Enterprise decision maker normalizes the information about the sell volume, unit maintain cost, customer satisfaction ratio of air condition and unit sell profit of four kinds of air condition under different periods and places.

Step 3. Enterprise decision maker decides to use geometric series based method to make decision about importance of each period. He/she decides importance of first period $\Gamma = 1$ and degree of change between adjacent period $\in = 1.1$. So, importance of period (year) is $W_1^T = 0.1638$, $W_2^T = 0.1802$, $W_3^T = 0.1982$, $W_4^T = 0.2180$ and $W_5^T = 0.2398$.

Step 4. Enterprise decision maker calculates the weight of each criterion according to entropy method. So, the importance of each criterion is $W_1^C = 0.2501$, $W_2^C = 0.2510$, $W_3^C = 0.2497$ and $W_4^C = 0.2493$

Table 2. Raw data of air condition under the combination dimension based on criteria, periods and spaces

		S_1				S_2				S_3			
		C_1	C_2	C_3	C_4	C_1	C_2	C_3	C_4	C_1	C_2	C_3	C_4
T_1	A_1	680	1504	0.8263	2630	517	1747	0.7634	2789	397	1562	0.8005	2865
	A_2	681	1668	0.8987	2736	484	166	0.8536	2631	420	1733	0.8605	2862
	A_3	636	1589	0.8938	2616	506	1927	0.7449	2672	448	1230	0.7660	3044
	A_4	610	1669	0.8415	2522	449	1701	0.8260	2910	354	1730	0.8902	2989
T_2	A_1	575	1965	0.7088	2561	529	1773	0.7246	2804	388	1499	0.8767	2638
	A_2	622	1558	0.7742	2548	436	1972	0.8160	2694	434	1685	0.7732	2414
	A_3	574	1595	0.8746	2444	423	1896	0.8366	2926	362	1309	0.8766	2956
	A_4	535	1504	0.8576	2580	548	1814	0.7720	2999	362	1541	0.7973	2341
T_3	A_1	638	1667	0.8755	2571	409	1814	0.7905	2621	393	1561	0.7763	2942
	A_2	640	1472	0.8536	2493	523	1804	0.7578	3028	449	1411	0.8577	2878
	A_3	631	1940	0.8449	2388	505	1744	0.8203	2964	368	1321	0.7912	2665
	A_4	564	1701	0.8718	2532	545	1716	0.7121	2729	389	1384	0.8634	3044
T_4	A_1	649	1803	0.8537	2442	516	1978	0.7319	2745	391	1779	0.7633	2519
	A_2	538	1690	0.7785	2630	556	1791	0.7020	2634	439	1202	0.8188	2932
	A_3	657	1642	0.7617	2372	406	1811	0.8254	2840	436	1442	0.8260	2324
	A_4	600	1927	0.8087	2427	548	1775	0.7445	2983	381	1502	0.8737	2495
T_5	A_1	687	1796	0.8181	2412	453	1966	0.8718	2862	424	1292	0.8608	2655
	A_2	580	1501	0.7845	2838	526	1747	0.7934	2635	441	1633	0.8041	2322
	A_3	525	1890	0.7680	2597	410	1826	0.7522	2882	362	1281	0.7513	2650
	A_4	589	1567	0.7350	2387	448	1761	0.8385	2845	386	1407	0.8124	2759

		S_4				S_5			
		C_1	C_2	C_3	C_4	C_1	C_2	C_3	C_4
T_1	A_1	674	2069	0.6537	1821	535	1560	0.7907	2375
	A_2	659	1980	0.7410	2149	596	1568	0.7546	2431
	A_3	703	1970	0.7646	2021	645	1666	0.7866	2167
	A_4	780	1960	0.7830	2037	564	1799	0.8173	2246
T_2	A_1	742	1902	0.8775	2065	602	1456	0.8024	2282
	A_2	719	1960	0.8637	2126	597	1626	0.8057	2150
	A_3	728	1903	0.6538	2039	557	1790	0.8218	2409
	A_4	726	2007	0.7992	2021	573	1591	0.7654	2247
T_3	A_1	720	2094	0.6889	1925	610	1587	0.8402	2253
	A_2	791	2011	0.6921	1953	681	1721	0.8453	2349
	A_3	763	1989	0.7048	1807	506	1499	0.8225	2409
	A_4	691	2100	0.7245	1854	591	1646	0.8054	2351
T_4	A_1	757	2000	0.8155	1864	455	1452	0.8477	2445
	A_2	771	1916	0.7624	2232	636	1768	0.7874	2224
	A_3	703	2014	0.8811	1982	661	1730	0.7870	2225
	A_4	787	1909	0.8996	2165	637	1635	0.8259	2218
T_5	A_1	775	1952	0.7033	1898	670	1669	0.7528	2189
	A_2	792	2054	0.7330	2120	585	1653	0.8493	2340
	A_3	722	1952	0.7035	2093	501	1573	0.7517	2191
	A_4	650	1984	0.8374	2202	502	1750	0.7879	2207

Step 5. Enterprise decision maker calculates the weight of each space according to entropy method. So, the importance of each space is $W_1^S = 0.1998$, $W_2^S = 0.2001$, $W_3^S = 0.1996$, $W_4^S = 0.2007$ and $W_5^S = 0.1998$.

Step 6, 7 and 8. Weighted matrix, positive ideal solution and negative ideal solution is calculated.

Step 9. The distance between the air condition manufacture project and positive ideal solution is calculated as Table 3.

Step 10. The distance between the air condition manufacture project and negative ideal solution is calculated as Table 3.

Step 11. The closeness coefficient (CC) of each air condition manufacture project is calculated as Table 4. After calculating the closeness coefficient, the ranking order of air condition manufacture project is $A_2(CC = 0.55057) > A_4(CC = 0.50260) > A_1(CC = 0.49640) > A_3(CC = 0.48581)$.

Table 3. The distance between air condition manufacture project and PIS, The distance between air condition manufacture project and NIS, closeness coefficient

	A_1	A_2	A_3	A_4
Distance between air condition manufacture project and positive ideal solution	0.0649	0.0586	0.0674	0.0636
Distance between air condition manufacture project and negative ideal solution	0.0640	0.0718	0.0637	0.0643
Closeness coefficient	0.4964	0.5505	0.4858	0.5026

Table 4. The comparison of different kinds of decision making model

	First rank	Second rank	Third rank	Fourth rank
Multi Periods, Spaces and Criteria Decision Making (MPSCDM)	A_2(CC = 0.5505)	A_4(CC = 0.5026)	A_1(CC = 0.4964)	A_3(CC = 0.4858)
Multi Period Multi Criteria Decision Making (MPCDM)	A_2(CC = 0.6741)	A_4(CC = 0.5371)	A_1(CC = 0.4746)	A_3(CC = 0.3349)
Multi Space Multi Criteria Decision Making (MSCDM)	A_2(CC = 0.5658)	A_3(CC = 0.4391)	A_4(CC = 0.4389)	A_1(CC = 0.4135)
Multi Criteria Decision Making (MCDM)	A_2(CC = 0.5003)	A_4(CC = 0.3377)	A_3(CC = 0.1709)	A_1(CC = 0.1321)

4   Analysis and Comparison of Different Kinds of Decision Making Model

For comparing and analyzing the difference of decision making based on different kinds of dimension direction. This research uses the same data to execute multi criteria decision making method, multi space multi criteria decision making method, multi period multi criteria decision making method and multi period multi space multi criteria decision making method (Refer to Table 4).

According to experiment result, the rank order of air condition manufacture project which is handled by MPSCDM method is the same with it which is copied with by MPCDM method. It can conclude that decision make analyze the row data roughly by MCDM method. By using MPCDM method or MSCDM method, decision maker will handle and analyze the raw data more detail. But, decision maker will only carefully and completely cope with and analyze the raw data by MPSCDM method.

Although, the first rank of air condition manufacture project is the same in MCDM, MPCDM, MSCDM and MPSCDM. But, only the rank order of air condition manufacture projects which is executed by MPSCDM method can be the same with it which is copied with by MPCDM method. So, the raw data analyze roughly or detail will influence the ranking order of alternatives in the decision problem. But, decision maker will acquire relatively stable experiment result (ranking order of alternatives) if he/she can choose decision making method as much dimension direction as possible.

In decision making problem, decision maker uses the same raw data to analyze and compare the performance of each alternative by different kind of decision making method (such as MCDM, MPCDM, MSCDM and MPSCDM) is appropriate in the short term. In the long term, the benefit and the cost should be discounted for real financial environment.

5   Conclusion and Future Research

In this research, we introduce the concept of "decision dimension". The practical applicant field of decision making based multi dimension is discussed in the study. Seven kinds of decision making method are introduced. We also illustrate the execution process of decision making method if decision maker wants to make decision under multi dimension direction. A numerical example is implemented by multi period multi space multi criteria decision making method. We also justify that the rank order of alternatives will be more stable if the more direction dimensions is used to make decision.

In the future, the core decision making method used in proposed seven kinds of decision making method can be replace from TOPSIS to VIKOR, ELECTRE or PROMETHEE etc. And then, the difference of such methods can be analyzed. Decision making based on different dimension direction is a relatively noble concept to deal with decision making problem. So, all of the application field can use this concept to design decision making execution process for making decision.

References

1. Figueira, J., Greco, S., Ehrgott, M.: Multiple Criteria Decision Analysis: State of the Art Surveys. Springer, New York (2005)
2. Pai, P.F., Chen, C.T., Hung, W.Z.: Applying linguistic information and intersection concept to improve effectiveness of multi-criteria decision analysis technology. Int. J. Inf. Technol. Decis. **13**, 291–315 (2014)
3. Stark, J.: Product Lifecycle Management, pp. 1–16. Springer, London (2011)
4. Wang, X., Triantaphyllou, E.: Ranking irregularities when evaluating alternatives by using some ELECTRE methods. Omega **36**, 45–63 (2008)
5. Wei, C.C., Chien, C.F., Wang, M.J.J.: An AHP-based approach to ERP system selection. Int. J. Prod. Econ. **96**, 47–62 (2005)

6. Yüksel, İ., Dagdeviren, M.: Using the analytic network process (ANP) in a SWOT analysis–a case study for a textile firm. Inform. Sci. **177**, 3364–3382 (2007)
7. Mousseau, V., Slowinski, R., Zielniewicz, P.: A user-oriented implementation of the ELECTRE-TRI method integrating preference elicitation support. Comput. Oper. Res. **27**, 757–777 (2000)
8. Chen, Y.W., Tzeng, G.H.: Using fuzzy integral for evaluating subjectively perceived travel costs in a traffic assignment model. Eur. J. Oper. Res. **130**, 653–664 (2001)
9. Behzadian, M., Kazemzadeh, R.B., Albadvi, A., Aghdasi, M.: PROMETHEE: a comprehensive literature review on methodologies and applications. Eur. J. Oper. Res. **200**, 198–215 (2010)
10. Swiniarski, R.W., Skowron, A.: Rough set methods in feature selection and recognition. Pattern Recogn. Lett. **24**, 833–849 (2003)
11. Chen, C.T.: Extensions of the TOPSIS for group decision-making under fuzzy environment. Fuzzy Set Syst. **114**, 1–9 (2000)
12. Lahdelma, R., Salminen, P.: Pseudo-criteria versus linear utility function in stochastic multi-criteria acceptability analysis. Eur. J. Oper. Res. **141**, 454–469 (2002)
13. Zou, Z.H., Yi, Y., Sun, J.N.: Entropy method for determination of weight of evaluating indicators in fuzzy synthetic evaluation for water quality assessment. J. Environ. Sci. **18**, 1020–1023 (2006)
14. Xu, Z.: On multi-period multi-attribute decision making. Knowl. Based Syst. **21**, 164–171 (2008)
15. Xu, Z., Da, Q.L.: An overview of operators for aggregating information. Int. J. Intell. Syst. **18**, 953–969 (2003)

Non-Conformity Detection in High-Dimensional Time Series of Stock Market Data

Akira Kasuga[1(✉)], Yukio Ohsawa[1], Takaaki Yoshino[2], and Shunichi Ashida[2]

[1] Department of System Innovation, The University of Tokyo, Hongo 7-3-1,
Bunkyo-ku, Tokyo, Japan
me7te7or.sai.dsw@gmail.com, ohsawa@sys.t.u-tokyo.ac.jp
[2] Daiwa Securities Co. Ltd., GranTokyo North Tower, 9-1, Marunouchi 1-chome,
Chiyoda-ku, Tokyo, Japan
{takaaki.yoshino,shunichi.ashida}@daiwa.co.jp

Abstract. It is desired to extract important information on-line from high-dimensional time series because of difficulty in reflecting the data fully in decision making. In econometric techniques, previous works primarily focus on prediction of price. To make decision in business practice, it is important to focus on human-machine interaction based on chance discovery. Here we propose Non-Conformity Detection as a method for aiding to humans to discover chances. Non-Conformity Detection is designed to detect a noteworthy point that behaves exceptionally compared to surrounding points in time series. In the experiment, the method of Non-Conformity Detection is applied to the time series of 29 stocks return in the electrical machine industry. As the result, four dates among the detected top five non-conformity points coincide with the important dates that professional analysts judged for making investment decision. These results suggest Non-Conformity Detection support the discovery of chances for decision making.

Keywords: Non-Conformity Detection · Chance discovery · High-dimension · Time series · Stock return · Decision making · On-line

1 Introduction and Motivation of Research

In recent years, it has been enabled to obtain a large quantity and high-dimensional data, which here means data including several dozens or hundreds variables for each time, called data-rich environment. It is desired to extract important information from high-dimensional time series because it is difficult to reflect the data fully in decision making. Dynamic factor model has been proposed for predicting future price from high-dimensional data on economic time series. In the recent trends on machine learning, deep learning is applied to prediction of stock price. However, the prediction accuracy hardly exceeds 60 %. For the time, if a stock investor seeks to obtain much confidence in investment, one cannot completely rely on automated reasoning by these algorithms in business practice. Therefore, the interaction between data mining system and human

© Springer International Publishing Switzerland 2016
H. Fujita et al. (Eds.): IEA/AIE 2016, LNAI 9799, pp. 701–712, 2016.
DOI: 10.1007/978-3-319-42007-3_60

intelligence is important than relying on only automated quantitative analysis or machine learning in order to make decision.

In this paper, we propose Non-Conformity Detection in high-dimensional time series as a method for aiding to humans to discovery chances through human-machine interaction. We define a non-conformity point as a data point that behaves exceptionally compared to surrounding data points in time series. The reason we focus on non-conformity is based on chance discovery. An event as a non-conformity point tends to be overlooked due to contextual gap from trendy events that most people are conscious of. In contrast, the overlooked events, if not highlighted, may worth attention if it has latent influence to events in the future. In business practice, it is an essential requirement to just observe events with highlighting non-conformity points in time series. Non-Conformity Detection is designed to detect a data point that has specificity on-line so as to aid humans to perceive when a noteworthy event happened. This paper attempts to integrate chance discovery and analysis of high-dimensional time series and to provide insight into these research areas.

In the following sections, the second section reviews previous works related to our research, the third section describes the proposed Non-Conformity Detection in details, the forth section shows the experiment that Non-Conformity Detection is applied to the time series of 29 stocks return in the electrical machine industry, the fifth section explains the evaluation of the utility of the proposed method according to nine professional analysts. Finally in the last section, we conclude the whole research with discussion about future research.

2   Related Works

In this section, we highlight some previous works related to our method. This introduction mentions two categories: chance discovery and high-dimensional time series analysis.

2.1   Chance Discovery

Chance discovery is defined as to become aware of a chance, defined as a significant event for decision making, and to explain its significance, especially if the chance is rare and its significance is unnoticed [1]. Three keys for the progress have been extracted from fundamental discussion on how to realize chance discovery: communication, imagination, and data mining. Then, the double helical model that means decision process modeling across the internet and the real world has been proposed [2]: An automated data mining system finds candidates of chance, humans realize the meaning of the chance, and finally they create solutions and make their decision. Tools for data mining with visualization function are formalized as tools aiding chance discoveries on this basic idea of chance discovery [3,4]. Furthermore, researchers on chance discovery invented Innovators' Marketplace, two games - Innovators' Market Game and Analogy

Game that accelerate the spiral of innovation with visualizing data on the connectivity of pieces of existing knowledge using methods for data visualization [5]. Regarding an approach of data utilization and application, Innovators Marketplace on Data Jackets has been proposed so as to design a market of data [6]. In this market, participants communicate to externalize and share the value of datasets so that one can buy or sell it in a reasonable condition, e.g., for a reasonable price. These researches aim to make a social environment where analysts and decision makers in active businesses can create innovative solutions with using data they have or provided data by others.

2.2 High-Dimensional Time Series Analysis

Bernanke and Boivin define rich data as the data contains much more information than can be extracted from a relatively small set of macroeconomic time series [7]. Analysis of high-dimensional time series is challenging because nontrivial data mining and indexing algorithms tend to degrade exponentially with dimensionality [8]. In econometric techniques, the traditional statistical methods where only a few variables are considered are not always applicable when analyzing a dataset of several dozens or hundreds variables, and so new methods are required [9]. Stock and Watson proposed the dynamic factor model using a small number of indexes constructed by principal component analysis [10] and a method of estimating turning points using large data sets [11]. Moreover, Chao proposed the use of deep belief network (DBN) to tackle the exchange rate forecasting problem as an approach in Machine Learning [12]. As a result, the maximum prediction accuracy is below 60 %. In terms of stock return analysis, Ludvigson and Ng proposed dynamic factor analysis for large data sets, to summarize a large amount of economic information by few estimated factors, and find that three new factors - termed volatility, risk premium, and real factors [13]. These previous works primarily focus on prediction of stock price.

3 Non-Conformity Detection

In this section, we propose Non-Conformity Detection that aids in humans' discovery of chances, defined in Sect. 2.1, in high-dimensional time series. We define a non-conformity point as a data point that behaves exceptionally compared to surrounding data points in time series. Non-Conformity Detection is designed to detect a data point that has specificity on-line so as to aid humans to perceive when an important event that may have latent influence happened.

3.1 Concept of Non-Conformity Detection

The concept of Non-Conformity Detection is based on chance discovery, i.e., the research for discovering an event or a situation that has significant impact on decision making. The proposed method attempts to combine methods for chance discovery and methods for analyzing high-dimensional time series. An event as

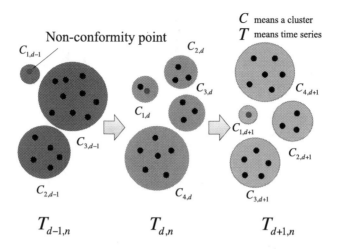

Fig. 1. Concept of Non-Conformity Detection

a non-conformity point tends to be overlooked due to temporarily exceptional movement from trendy events which most people follow. In contrast, the overlooked events, if not highlighted, may worth attention if it has latent influence to events in the future. It is desired in business practice to just observe events with highlighting non-conformity points that may be a risk or an opportunity in time series.

The developed method consists of subsequence time series clustering and scoring the specificity for quantification of non-conformity degree. This process is related to a method for detecting changes of clustering structures [14], however our method is fundamentally different from the method in terms of chance discovery where we focus on the interaction between data analysis with computer and human intelligence. After we divide high-dimensional time series into segments using a sliding window, each segment shall be clustered by Affinity Propagation [15] in order to optimize the number of clusters. By use of sliding window in time series, the method allows us to detect a specific point on-line and take contextual relation into consideration. In each subsequence, a data point classified in a smaller cluster is heavily weighted and a data point classified in a larger cluster is lightly weighted. Non-conformity degree of each data point is defined as the accumulation of weighting in each subsequence. The non-conformity degree as weighting shows specificity and is used for choosing candidates of chance as an important event to make decision (shown in Fig. 1).

3.2 Algorithm of Non-Conformity Detection

The algorithm consists of 2 steps: clustering by Affinity Propagation and scoring by focusing on a data point in a small cluster. In previous researches, the problem that the k-means cluster centers in sliding window become a sine wave [16,17] has been pointed out. Therefore, the method of Affinity Propagation is adopted

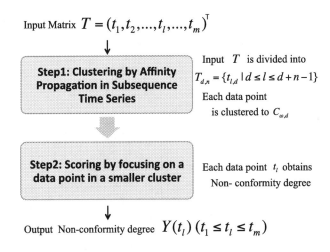

Fig. 2. Flow chart of Non-Conformity Detection

into our method in step 1. The non-conformity degree is calculated based on density of data points in a cluster in step 2. The algorithm overview is shown in Fig. 2.

Step 1. At first, there is the matrix whose row means time, T of length m : $T = (t_1, t_2, ..., t_m)^T$. A subsequence of length n in time series is $T_{d,n}$, where $1 \leq d \leq m-n+1, n \leq m$. A data point corresponding to each time in $T_{d,n}$ is clustered by Affinity Propagation.

In Affinity Propagation, two messages are exchanged with each data point so as to find a subset of exemplar points that best represent the data (shown in Fig. 3). Input data are similarities $s(i, j)$ between data points, where $i, j \in \{d, d + n - 1\}$. The similarity is called preference, measured by the negative Euclidean distance. Affinity Propagation determines an exemplar that is a data point representing a cluster. Between data points, they exchanges responsibility and availability till these messages from and to all data points are converged to stable values. As a result, the best exemplars and the other data points are determined uniquely. Responsibility $r(i, j)$ is defined as appropriateness that data point j is an exemplar representing the cluster including data point i. Data point i sends this appropriateness to data point j as a message. Availability $a(i, j)$ is defined as appropriateness that data point i is assigned to the cluster represented by data point j. Data point j sends this appropriateness to data point i as well. In order to converge $r(i, j)$ and $a(i, j)$, these values are calculated recursively as the update rules Eqs. (1), (2) and (3) [15,18]. In first step, these values are initialized to zero. Then, the responsibility $r(i, j)$ are computed using the update rules and availability $a(i, j)$ is computed as well.

$$r(i, j) \leftarrow (1 - \lambda)\rho(i, j) + \lambda r(i, j)$$
$$a(i, j) \leftarrow (1 - \lambda)\alpha(i, j) + \lambda a(i, j)$$

(1)

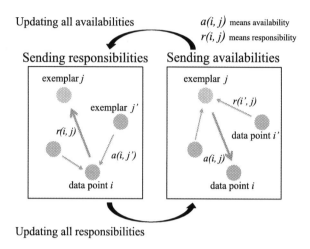

Fig. 3. Algorithm of affinity propagation [15]

In (1), λ is a damping factor to prevent the oscillation of iterating calculation. We can adopt the convergence criteria with setting a damping factor and number of iterations. $\rho(i,j)$ is a propagation value in the calculation of responsibility and $\alpha(i,j)$ is a propagation value in the calculation of availability. If $i \neq j$, $\rho(i,j)$ and $\alpha(i,j)$ are calculated as update rules Eq. (2).

$$\rho(i,j) \leftarrow s(i,j) - \max_{j's.t.j'\neq j}\left[\alpha(i,j') + s(i,j')\right]$$

$$\alpha(i,j) \leftarrow \min\left[0, r(j,j) + \sum_{i's.t.i'\notin\{i,j\}} \max\left[0, r(i',j)\right]\right] \tag{2}$$

if $i = j$, they are calculated as update rules Eq. (3)

$$\rho(i,j) \leftarrow s(i,j) - \max_{j's.t.j'\neq j}\left[s(i,j')\right]$$

$$\alpha(i,j) \leftarrow \sum_{i's.t.i'\neq i} \max\left[0, r(i',j)\right] \tag{3}$$

After $\rho(i,j)$ and $\alpha(i,j)$ are converged, an appropriate exemplar of data point i is determined as Eq. (4).

$$\arg\max_i\left[r(i,j) + a(i,j) : i,j \in \{d, d+n-1\}\right] \tag{4}$$

Step 2. The calculation of non-conformity degree is based on density of data points in the cluster determined in step1. Here in the subsequent $T_{d,n}$, each data point is $\{t_{l,d} | d \leq l \leq d+n-1, 1 \leq d \leq m-n+1\}$ as follows. Each data point $t_{l,d}$ is clustered to $C_{\omega,d} = \{C_{1,d}, C_{2,d}, \cdots, C_{K,d}\}(1 \leq \omega \leq K)$, where K is the number of exemplars determined by Affinity Propagation. Let $N_{\omega,d}(1\leq\omega\leq K)$ mean the number of data points in cluster $C_{\omega,d}$. If data point $t_{l,d}$ is assigned to cluster $C_{\omega,d}$, $t_{l,d}$ obtains $W_{l,d}$ that means the specificity in the cluster (shown in Eq. (5)). $W_{l,d}$ is defined as inverse number of density, the ratio of the number of data points in a cluster to the number of all data points in subsequence. Each data point $t_{l,d}$ obtains specificity $W_{l,d}$ in each subsequence, thereby t_l accumulatively obtains non-conformity degree $Y(t_l)$ $(t_1\leq t_l\leq t_m)$ (shown in Eq. (6)). Finally, we can quantify non-conformity degree of a data point in high-dimensional time series.

$$W_{l,d} = \left(\frac{N_{\omega,d}}{\sum_{\omega=1}^{K} N_{\omega,d}}\right)^{-1} : t_{l,d}\in C_{\omega,d}, \text{for the subsequence of} T_{d,n} \qquad (5)$$

$$Y(t_l) = \sum_{d=l-n+1}^{l} W_{l,d} \qquad (6)$$

the detail algorithm of Non-Conformity Detection is shown in Algorithm 1.

Algorithm 1. Non-Conformity Detection

Input: T=matrix $(t_1, t_2, \cdots, t_m)^{\mathrm{T}}$, length of subsequence: n, Max Iteration: MI
Output: Non-conformity degree $Y(t_l)$ $(t_1\leq t_l\leq t_m)$
  for $d = 1$ to $d = m - n + 1$ **do**
    (Step1)
    $T_{d,n} = \{t_{l,d}|d\leq l\leq d+n-1\}$
    $s(i,j)$= negative Euclidean distance(t_i, t_j), where $i, j \in \{d, d+n-1\}$
    $r(i,j)$ and $a(i,j)$ are initialized to 0
    for $iterations = 1$ to MI **do**
      for $i, j \in \{d, d+n-1\}$ **do**
        $r(i,j)$ are updated by $a(i,j)$ as Eq.(1),(2), and (3)
        $a(i,j)$ are updated by $r(i,j)$ as Eq.(1),(2), and (3)
        if $r(i,j)$ and $a(i,j)$ is converged **then**
          Break;
        end if
      end for
    end for
    $C_{\omega,d}=\{C_{1,d}, C_{2,d}, \cdots, C_{K,d}\}(1\leq\omega\leq K)$ are determined by exemplars as Eq.(4)
    (Step2)
    for $l = d$ to $d+n-1$ **do**
      $W_{l,d}$ is weighted as Eq.(5) and $t_{l,d}$ obtains $W_{l,d}$
      $Y(t_l)+ = W_{l,d}$
    end for
  end for

Table 1. Stock return in electrical machine industry on a daily basis

Date	4902	6479	6501	⋯	7751	7752	8035
1/5/2015	–0.0121	0.0294	0.0006	⋯	–0.0139	–0.0024	–0.006
1/6/2015	–0.0428	-0.0307	–0.0315	⋯	–0.0255	–0.0294	–0.048
1/7/2015	0.0072	0.0044	–0.0001	⋯	–0.0038	–0.0092	–0.0274
⋮	⋮	⋮	⋮	⋮	⋮	⋮	⋮
9/30/2015	0.0245	0.0211	0.015	⋯	0.0144	0.0186	0.015
10/1/2015	–0.004	0.0285	0.0459	⋯	0.0237	0.0137	0.0495
10/2/2015	0.012	–0.0023	–0.0091	⋯	0.0037	0.0188	0.0087

4 Experiment on Stock Market Data and the Result

In this section, we explain our demonstration experiments to verify the effectiveness of the proposed method. Non-Conformity Detection is here applied to the data of stock return consisting of the Nikkei 225 components on a daily basis, which Daiwa Securities Co. Ltd. provided as research data. The stock return on a daily basis is calculated as the ratio of the closing price to closing price on the previous day. Although the stock return includes a dividend, the ratio is adjusted to the actual dividend till the next business day if the dividend is determined. We conducted the experiment about the 29 stocks in the category of electrical machine industry with proposed method, which means we dealt 29 dimensional time series here. However, this analysis is a starting step before expanding to higher dimensional data, i.e., all stocks exceeding hundreds. We have the hypothesis that the method may identify a company risk, e.g. a scandal or an accident that had significant impact on decision making. In this experiment, the stocks in electrical machine industry have been put into focus since especially essential events happened in the industry in 2015. The matrix whose rows m are 185 business dates in 2015 and columns are the 29 brand codes of stocks is constructed from the provided data between January 5, 2015 and October 2, 2015 (shown in Table 1). In order to take the influence of whole market on an independent brand into consideration, we subtract the average of each row from the stock return in each row.

T is the input matrix that, the preset length n of subsequence is 20, the damping factor λ is 0.9 [18] and max iteration MI is 200 in this experiment. The reason we determine 20 as length n is that we compare each data point in about one month according to the hypothesis. The length n could be changeable based on user's hypothesis. As the execution result of the proposed algorithm, we obtained non-conformity degree in the time series as output (shown in Fig. 4). The axis of ordinate means the values of non-conformity degree and the axis of abscissa means the business date in 2015. Referring to Fig. 4, May 11, 2015 marks the highest value 400 of non-conformity degree, i.e., this result suggests that May 11, 2015 was the most non-conformable point and an important event for

Table 2. Detected top five non-conformity points and non-conformity degree

Date	Non-conformity degree
May 11, 2015	400.00
February 5, 2015	386.67
February 2, 2015	382.86
May 1, 2015	371.43
July 29, 2015	363.50

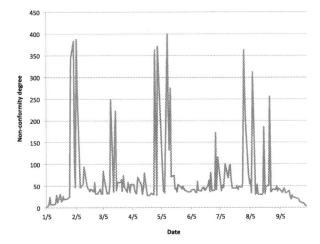

Fig. 4. Output of the proposed algorithm applied to the stock return in electrical machine industry on a daily basis

decision making during this experiment period. Top five non-conformity points are extracted in the quantification result as Table 2.

5   Validation of the Result and Implications

In the last section, we presented the quantification of non-conformity degree using the proposed method. In this section, we explain the evaluation of the obtained result according to the actual judgment of professional analysts.

We conducted questionnaire so as to collect actual judgments: what time is the important date for professional analyst to make investment decision. In terms of investment decision in electrical machine industry during the experiment period, nine professional analysts evaluated 16 dates that were listed by the experiment organizers, including results of our Non-Conformity Detection and dummy dates, by five grades, 5: There was especially noteworthy movement, 4: There was noteworthy movement, 3: I cannot judge, 2: There was not much noteworthy movement, and 1: There was not noteworthy movement. The listed 16

Table 3. Top five non-conformity points

$n = 20$	$n = 10$
May 11, 2015	May 11, 2015
February 5, 2015	July 29, 2015
February 2, 2015	March 13, 2015
May 1, 2015	March 18, 2015
July 29, 2015	August 7, 2015

Table 4. Important date evaluated by professional analysts

Actual judgments of professional analysts
February 5, 2015
March 13, 2015
April 28, 2015
May 1, 2015
May 11, 2015
July 29, 2015

business dates are extracted as top five data points, whose features three different methods quantify in high-dimensional time series: Non-Conformity Detection ($n = 10, 20$) (shown in Table 3), Principle Component Analysis [10], and Subspace Outlier Detection [19]. Principal Component Analysis and Subspace Outlier Detection are essentially different from Non-Conformity Detection, hence that are not accuracy comparison but just reference in order to set dummy dates in questionnaire. Nine professional analysts who have been working as a analyst for more than ten years answered these question. After collecting that questionnaire, we compared top five non-conformity points with the actual judgment of professional analysts. The six important dates by their actual judgment are extracted as the dates that obtain more than 3 on the average of nine professional analysts' evaluation in listed dates (shown in Table 4).

As a result, four dates in top five non-conformity points ($n = 20$) coincide with the important dates that professional analysts judged for making investment decision. On May 11, 2015, the stock of TOSHIBA Corp. hit limit down due to the dishonesty of the treasurer and the stock of Sharp Corp. fell hard due to drastic capital reduction. These accidents caused much uncertain in market and influenced on decision making according to the professional analysts. On the other hand, the stock of Sony Corp. hit limit up in about five years because of shifting to account surplus on February 5, 2015. This event attracted a lot of attention of investors in market. As May 11 and February 5, 2015, there were the important events that have significant influence on May 1 and July 29, 2015. Although stock return of two brands fluctuated widely on February 2, 2015, there was not the clear fact related to the movement. Accordingly, the professional analyst didn't pay attention to the date for decision making. In case of $n = 10$, three dates in top five non-conformity points ($n = 10$) coincide with the important dates by professional's judgment. If n is changed, the extracted dates are changed. n can control whether you want to detect non-conformity points in terms of short or long period. Although we conducted the experiment about past data, the proposed method can detect non-conformity points on-line, because the proposed methods involve just available neighboring points of each data point for estimating the non-conformity degree.

With this evidence, the proposed method can detect important dates when significant events happened for making investment decision. It is difficult to reflect the data fully in decision making from high-dimensional stock return, and so it is expected to discover chances that may influence analysts to make their decision by the aiding effect of the proposed method.

6   Conclusion and Future Works

In this paper, we proposed a method for Non-Conformity Detection so as to aid humans to discover chances in high-dimensional time series. The proposed method attempts to integrate chance discovery and analysis of high-dimensional time series and to provide insight into these research areas. It can quantify non-conformity degree with clustering by Affinity Propagation in subsequence time series and scoring specificity by focusing on a data point in a small cluster. Then, we applied the proposed method to the time series of 29 stocks return in the electrical machine industry. As the result, four dates in top five non-conformity points coincide with important dates that professional analysts judged significantly for making investment decision. This result suggests that extraction of important information by Non-Conformity Detection is serviceable for making investment decision. Besides, our method provides insight that significant but overlooked events can be detected on-line in high-dimensional time series into decision support in chance discovery and analysis of high-dimensional time series.

There are some future works toward expansion of our method. In the first, visualization method would be developed. It is better to present not only non-conformity degree but also the brand codes related to important date. To visualize this kind of related information contributes decision making of analysts much more. In the second, we will be enable users, i.e., stock analysts and investors here, to detect an important event as non-conformity point on-line, observing all stock brands as higher-dimensional time series in market. Consequently, future works will widen application range of Non-Conformity Detection.

Acknowledgments. This research was partially supported by JST-CREST.

References

1. Ohsawa, Y.: Chance discoveries for making decisions in complex real world. New Gener. Comput. **20**(2), 143–163 (2002)
2. Ohsawa, Y., Nara, Y.: Decision process modeling across internet and real world by double helical model of chance discovery. New Gener. Comput. **21**(2), 109–121 (2003)
3. Maeno, Y., Ohsawa, Y., Ito, T.: Catalyst personality for fostering communication among groups with opposing preference. In: Okuno, H.G., Ali, M. (eds.) IEA/AIE 2007. LNCS (LNAI), vol. 4570, pp. 806–812. Springer, Heidelberg (2007)

4. Chiu, T.-F., Hong, C.-F., Chiu, Y.-T.: Visualization of financial trends using chance discovery methods. In: Nguyen, N.T., Borzemski, L., Grzech, A., Ali, M. (eds.) IEA/AIE 2008. LNCS (LNAI), vol. 5027, pp. 708–717. Springer, Heidelberg (2008)
5. Ohsawa, Y., Nishihara, Y.: Innovators' Marketplace: Using Games to Activate and Train Innovators. Understanding Innovation. Springer, Heidelberg (2012)
6. Ohsawa, Y., Liu, C., Suda, Y., Kido, H.: Innovators marketplace on data jackets for externalizing the value of data via stakeholders requirement communication. In: 2014 AAAI Spring Symposium Series, pp. 45–50 (2014)
7. Bernanke, B.S., Boivin, J.: Monetary policy in a data-rich environment. J. Monetary Econ. **50**(3), 525–546 (2003)
8. Lin, J., Keogh, E., Lonardi, S., Chiu, B.: A symbolic representation of time series, with implications for streaming algorithms. In: Proceedings of the 8th ACM SIGMOD Workshop on Research Issues in Data Mining and Knowledge Discovery, pp. 2–11. ACM (2003)
9. Hayakawa, K.: Recent development of high-dimensional time series analysis. J. Jpn. Stat. Soc. Series J **43**(2), 275–292 (2014)
10. Stock, J.H., Watson, M.W.: Macroeconomic forecasting using diffusion indexes. J. Bus. Econ. Stat. **20**(2), 147–162 (2002)
11. Stock, J.H., Watson, M.W.: Estimating turning points using large data sets. J. Econometrics **178**, 368–381 (2014)
12. Chao, J., Shen, F., Zhao, J.: Forecasting exchange rate with deep belief networks. In: The 2011 International Joint Conference on Neural Networks, pp. 1259–1266 (2011)
13. Ludvigson, S.C., Ng, S.: The empirical riskreturn relation: a factor analysis approach. J. Financ. Econ. **83**(1), 171–222 (2007)
14. Hirai, S., Yamanishi, K.: Detecting changes of clustering structures using normalized maximum likelihood coding. In: Proceedings of the 18th ACM SIGKDD International Conference on Knowledge Discovery and Data Mining, pp. 343–351. ACM (2012)
15. Frey, B.J., Dueck, D.: Clustering by passing messages between data points. Science **315**(5814), 972–976 (2007)
16. Chen, J.R.: Making subsequence time series clustering meaningful. In: Fifth IEEE International Conference on Data Mining, p. 8. IEEE (2005)
17. Idé, T.: Why does subsequence time-series clustering produce sine waves? In: Fürnkranz, J., Scheffer, T., Spiliopoulou, M. (eds.) PKDD 2006. LNCS (LNAI), vol. 4213, pp. 211–222. Springer, Heidelberg (2006)
18. Wang, K., Zhang, J., Li, D., Zhang, X., Guo, T.: Adaptive affinity propagation clustering. Acta Automatica Sinica **33**(12), 1242–1246 (2008)
19. Kriegel, H.-P., Kröger, P., Schubert, E., Zimek, A.: Outlier detection in axis-parallel subspaces of high dimensional data. In: Theeramunkong, T., Kijsirikul, B., Cercone, N., Ho, T.-B. (eds.) PAKDD 2009. LNCS, vol. 5476, pp. 831–838. Springer, Heidelberg (2009)

Recent Study on the Application of Hybrid Rough Set and Soft Set Theories in Decision Analysis Process

Masurah Mohamad[1] and Ali Selamat[2(✉)]

[1] UTM-IRDA Digital Media Centre,
Universiti Teknologi Malaysia, 81310 Johor Bahru, Johor, Malaysia
masur480@perak.uitm.edu.my
[2] Faculty of Computing, Universiti Teknologi Malaysia,
81310 Johor Bahru, Johor, Malaysia
aselamat@utm.my

Abstract. Many approaches and methods have been proposed and applied in decision analysis process. One of the most popular approaches that has always been investigated is parameterization method. This method helps decision makers to simplify a complex data set. The purpose of this study was to highlight the roles and the implementations of hybrid rough set and soft set theories in decision-making especially in parameter reduction process. Rough set and soft set theories are the two powerful mathematical tools that have been successfully proven by many research works as a good parameterization method. Both of the theories have the capability of handling data uncertainties and data complexity problems. Recent studies have also shown that both rough set and soft set theories can be integrated together in solving different problems by producing a variety of algorithms and formulations. However, most of the existing works only did the performance validity test with a small volume of data set. In order to prove the hypothesis, which is the hybridization of rough set and soft set theories could help to produce a good result in the classification process, a new alternative to hybrid parameterization method is proposed as the outcome of this study. The results showed that the proposed method managed to achieve significant performance in solving the classification problem compared to other existing hybrid parameter reduction methods.

Keywords: Hybrid · Rough set · Soft set · Parameter reduction · Medical and big data

1 Introduction

Decision analysis always deals with complex data that have different characteristics and structures. The most problematic data is categorized as uncertain data whereas the criteria value is practically difficult to be determined. This problem will remain unsolved if the decision-making process is involved with a complex data or more specifically, if it deals with a big data set [31]. A very powerful method needs to be selected in order to avoid the problem of ineffectiveness in the computational work and also to produce the best computational results. Based on the literature, one of the most popular approaches that has always been investigated is parameterization method.

© Springer International Publishing Switzerland 2016
H. Fujita et al. (Eds.): IEA/AIE 2016, LNAI 9799, pp. 713–724, 2016.
DOI: 10.1007/978-3-319-42007-3_61

This method helps decision makers to simplify a complex data set. Even though the best method has been selected to handle this kind of problem, it still has some disadvantages. For instance, the classical rough set theory which is a well-known method that is capable of handling complex problems [17] still needs assistance from other methods to deal with parameterization problem. Thus, various theories and concepts such as granular computing, deep learning, mathematical theories, artificial intelligent (AI), and hybrid approaches have been proposed in order to overcome these problems.

The purposes of this study are: (i) to highlight the existing works of hybrid rough set and soft set theories in decision analysis especially in parameter reduction (PR) process; (ii) to propose a new hybrid rough set and soft set theories as a parameter reduction method; and (iii) to evaluate the performance of the proposed hybrid rough set and soft set parameter reduction method towards medical data sets. Several factors have inspired this study to integrate the rough set and soft set theories as a parameter reduction method. Firstly, based on the literature study conducted, soft set theory and other existing hybrid methods have not been tested with a complex medical data set. Secondly, the AI techniques such as rough set theory and fuzzy c-means are claimed as the best techniques to handle complex multi-criteria decision problems as stated by [23]. Thirdly, the soft set theory is one of the emerging mathematical tools which has been proven to be a good parameterization tool by many researchers [10, 19]. Therefore, it is selected to be integrated with the rough set theory. Finally, the medical data set is selected because of its complexity and vagueness structure which make it suitable to test the uncertainty and complexity issues. It is also believed that a new medical knowledge will be discovered by studying the relationships and the patterns of the medical data. Furthermore, the analysis of the medical data usually involves the study of improving the incomplete information as uncertain information handling as well as organizing different level of data representation [15].

This paper is divided into 4 sections. Section 1 introduces the purposes of the study and Sect. 2 briefly explains the basic concept of parameter reduction process, rough set theory, soft set theory, and the existing hybrid methods of these two theories. Section 3 provides the experimental work and its results to benchmark the existing rough set and soft set hybrid parameterization methods. Lastly, Sect. 4 concludes the overall works.

2   Implementation of Hybrid Rough Set and Soft Set Theories in Decision-Making

This section discusses the related theories and the existing hybrid rough set and soft set methods in handling medical data. It begins with the basic explanation of the parameter reduction process, followed by rough set theory, soft set theory, and in-depth discussion of hybrid rough set and soft set theories in decision-making process. Parameter reduction is one of the important processes that could be applied in decision-making. It is applied in the pre-processing phase in which it helps to reduce the volume of the data set before other tasks—such as classification or ranking task—is executed. It is also used to eliminate less important attributes and uncertainty values. Reduction process can be divided into two parts; parameter reduction and parameter value reduction. Most of the publications deal with parameter reduction instead of parameter value reduction [19].

2.1 Rough Set Theory

Rough set theory (RST) deals with approximation concept. It has been proven by many researchers that rough set theory has the strong capability of handling big and uncertain data [21]. According to the publication written by [22], the philosophies of rough set theory are to minimize the data size and to deal with uncertainty data. RST will construct a set of rules that provides useful information from the uncertain and inconsistent data [14]. The capability of rough set theory in handling uncertainty and ambiguity data has been successfully tested by many researchers either by extending the rough set theory or by combining it with other theories.

RST can also be applied as a parameter reduction technique to remove unnecessary attributes by preserving the original information. RST has also been proven as a good parameter reduction technique by many researchers and has been extensively used in many areas such as medical diagnosis, decision-making, image processing, as well as economic and data analysis. RST is also recommended to be a tool that can effectively reduce unnecessary attributes when it is integrated with other techniques in decision-making process. Recently, various parameter reduction techniques which are based on rough set theory have been proposed such as dominance-based rough set approach (DRSA), variable consistency dominance-based rough set approach (VC-DRSA), and variable precision dominance-based rough set approach (VP-DRSA) [12]. Each of these techniques has their own ability and limitation.

2.2 Soft Set Theory

Soft set theory is a theory that utilizes the advantages of rough set theory in handling imprecise and vague data [6, 18]. Soft set theory allows the object to be defined without any restricted rules. In other words, to identify the membership function, adequate parameters are needed [13]. It is a mathematical tool that has been proposed by Molodtsov and it is independent of any insufficient parameterization tools that are inherited by several approaches such as rough set and fuzzy set theories [31]. Guan et al. stated that soft set is a set of data that comprises of a record set, a set of parameters, and a mapping set of selected parameters from a power set of universe [9].

As stated by [13], recently, the theory of soft set has attracted many researchers to further improve the theory or apply in various areas such as operational research, medical research, and decision-making; especially in an environment with uncertain information. Soft set has emerged recently due to its functionality and ability in handling uncertainties. Molodtsov has claimed that soft set is better than fuzzy set and rough set in decision-making process. Besides, it does not need any parameterization tools [30]. Consequently, motivated by this theory, researchers have done many works related to the decision-making area. Most of the works published were investigating and proving the ability of soft set theory to assist decision makers in making a good decision.

2.3 Hybrid Methods

Method hybridization is a process of integration between one method and other methods. Hybridization presents the alternative to solve the limitation of single method in a particular process. For example, fuzzy set relies on the expert's knowledge, rough set suffers from nondeterministic polynomial-time hard (NP-hard) problem in attribute reduction and optimal rule discovery, and genetic algorithm (GA) might face convergence problems [3]. Recently, many researchers have proposed several enhancements and one of them is the integration of artificial intelligent (AI) methods in order to maximize the functionality of the methods and to minimize the shortcoming of the original method. For instance, interval valued fuzzy ANP (IVF-ANP) was developed to solve the multiple attribute problems by determining the weights of each defined criteria [28] and rough AHP was proposed to measure the system performance [2]. The followings are some of the hybrid methods that integrate either soft set theory or rough set theory in their proposed works.

(1) Fuzzy soft sets: Fuzzy soft set is another soft set approach which has been introduced in 2001 to solve many problems including uncertainties. It is an extension to the classical soft set approach which was proposed to solve decision-making problems in real world situation. Many publications have contributed to fuzzy soft sets. One of the publications was written by Agarwal et al. [1] in which they proposed an expert system that generalizes the intuitionistic fuzzy soft set approach to solve medical diagnosis problem. Meanwhile, Xiao et al. [29] initiated an optimization approach based on interval-valued fuzzy soft set in solving multi-attribute group decision-making problems under uncertain environment. Geng et al. [7] proposed a model that provides an approximate description of objects in an intuitionistic fuzzy environment and also with the additional information of weight attributes in solving multi-attribute decision-making problems in 2011. Besides, in the same year, [8] also had proposed a method that considers multiple parameters group decision-making by implementing the interval-valued intuitionistic fuzzy soft set approach. Furthermore, some of the publications have considered to apply the fuzzy soft set in certain particular application problems to accomplish different objectives. For example, these research works were conducted in order to reduce the chances of piracy in image transmission [25], to rank the technical attributes in quality function deployment [30], to apply new soft information order algorithm to solve problems [9], and to solve ranking problems by using the concept of intuitionistic multi-fuzzy soft set [4].

(2) Soft fuzzy rough set and soft rough fuzzy set: Soft fuzzy rough set is a combination of three mathematical tools; soft set theory, fuzzy set theory, and rough set theory. These tools are almost related when dealing with uncertainties and vagueness problems. It was introduced by Feng et al. [5] who investigated the problem and the consequences of integrating these three theories in which it was inspired by Dubois and Prades research work named rough fuzzy sets. Later in 2011, Meng et al. [20] redefined the concept proposed by Feng et al. [5] and introduced a new soft approximation space by considering several issues arose from the previous research works. Then, still in 2011, another definition for the soft fuzzy rough set was presented by Sun et al. [27]. They proposed a new concept of soft fuzzy rough set by integrating soft set, rough set, and fuzzy set with traditional fuzzy rough set. Later, in 2012, an enhancement of intuitionistic fuzzy soft set approach with rough set theory was proposed by Zhang et al. [31]. A new alternative related to the intuitionistic fuzzy soft sets problems was

suggested for decision makers in making a scientific and appropriate decision. He successfully proofs his work was more suitable than the other works in dealing with the crisp soft set decision-making problems.

(3) Rough set, modified soft rough set and rough soft set: Rough soft set and soft rough set were introduced by Feng et al. with the conjunction of integrating the three approaches—rough set, soft set, and fuzzy set—to produce one hybrid model named soft rough fuzzy set [24]. As pointed out by Feng et al. in [6], rough soft set is the approximation of soft set in rough approximation space which was introduced by Pawlak meanwhile soft rough set is grounded by soft rough approximation in soft approximation space. According to [16], soft rough set which was introduced by Feng et al. was a generalization of the rough set model over the soft set model which promotes in providing a better approximation in certain cases compared to Pawlaks work. Based on the findings, the researchers are recommended to make enhancement and modification to their work by exploring the associations between other rough set models and soft rough sets [6]. In 2013, Shabir et al. made an improvement to the soft rough set approximation theory which is called modified soft rough set (MSR sets) [24]. This work claims that the proposed approach is more robust and the granules of information produced are finer than the original soft rough sets.

3   Experimental Work and Results

Based on the literature works done on rough set and soft set hybrid methods, most of the hybrid research works have only provided the theorems and algorithms without any experimental works to proof the proposed theory either it can be applied to real data or not. Therefore, in this section, an experimental work was conducted by using real data sets with a new proposed hybrid framework. The proposed hybrid framework integrated the rough set and soft set theories in the parameter reduction process to help the classifier deals with complex medical data as mentioned in Sect. 1. Four types of medical data sets were applied to the experimental task and it can be downloaded at UCI Machine Learning Repository, http://archive.ics.uci.edu/ml/. The data sets used were breast cancer Wincostin, BUPA liver disorders, heart disease, and diabetes Pima Indian. The characteristics of the data sets are depicted in Table 1.

Table 1. Data sets description.

Data sets	Number of attributes	Number of instances	Attribute characteristics	Missing values
Heart disease	14	294	Categorical, integer, real	Yes
Breast cancer	10	699	Integer	Yes
Diabetes pima Indian	9	768	Integer, real	Yes
Bupa liver disorder	7	345	Categorical, integer, real	No

Figure 1 presents the framework of the proposed work which has four phases; (1) data cleaning and formatting; (2) parameter reduction process; (3) classification process; and (4) performance evaluation. The software that was used in the experimental work were MATLAB version 2014a, Rough Set Exploration System (RSES) and WEKA 3.6. MATLAB was used to execute the soft set parameter reduction algorithm and neural network classification process, RSES which can be downloaded from http://www.mimuw.edu.pl/~ szczuka/rses/about.html was used to execute the rough set reduction process. Meanwhile, WEKA 3.6 was used to execute other hybrid parameter reduction methods listed in Table 3.

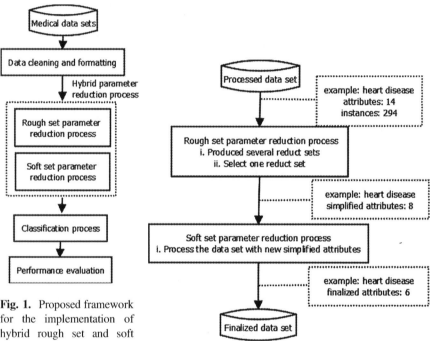

Fig. 1. Proposed framework for the implementation of hybrid rough set and soft set parameter reduction method in the classification process

Fig. 2. Hybrid rough set and soft set parameter reduction process

3.1 Experimental Work

In phase 1, the raw data sets were cleaned up and formatted according to the certain format based on the software requirements. Basically, the attributes and the number of instances are defined during the pre-processing task. The data sets then went through the hybrid parameter reduction process in order to simplify the uncertain and complex issues of the data in phase 2. During this phase, two level of parameter reduction processes were executed. The first step was implementing the rough set parameter reduction process and the second step was applying the soft set parameter reduction process. Basically, all the attributes that contain missing values and ranked as less important based on the specified formulations are removed. For rough set parameter reduction process, several algorithms have been provided by the RSES in which it can be selected to execute the reduction process. After several tests with available algorithms had been performed, the exhaustive algorithm was used to generate the reduct sets because it helps the classifier in generating a good classification result. The reduct sets contain several sets of attributes which can be used in the next processing task. One of the reduct sets generated by the rough set parameter reduction process was used as an input for the soft set parameter reduction process. The soft set parameter reduction process applied the parameter reduction algorithm introduced by [10, 11]. It was used to determine the best attributes that need to be used in the classification task. Both parameter reduction processes identified the important attributes to be used in the classification task. Only one reduct set for each parameter reduction level was used in the classification task. Figure 2 demonstrates the flow of the two-level reduction process in detail with a simple example. Meanwhile, Fig. 3 provides the example of the data set after it went through the hybrid parameter reduction process. Data set heart disease which has 14 attributes was reduced to 6 attributes after it when through the hybrid PR process. Most of the uncertain attributes and missing values were eliminated after the parameter reduction process was executed.

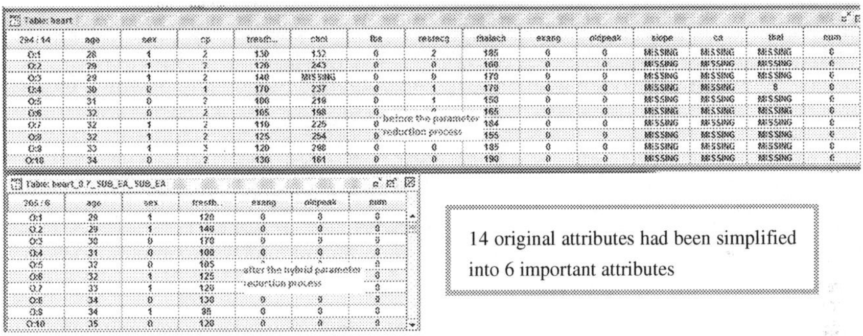

Fig. 3. Example of the obtained result from the hybrid rough set and soft set parameter reduction process for heart disease data set

After the new simplified data set had been generated through the hybrid parameter reduction process, the classification task was executed in phase 3. At this stage, neural network was used as a classifier to train and test the performance of the proposed hybrid parameter reduction method. Before the classification process was executed, the processed data sets were randomly divided into three groups, 70 % for training process, 15 % for testing process, and another 15 % for validation process. Training process was done several times until the best classification result was obtained. Finally, in phase 4, the obtained classification results for each data set were evaluated by using six standard performance measures; accuracy (ACC), sensitivity (SENS), specificity (SPEC), positive predictive value (PPV), negative predictive value (NPV), and receiver operating characteristic (ROC) curves [26]. The classification results are presented in Tables 2, 3, 4 and Figs. 4, 5.

Table 2. Overall performance before implementation of hybrid parameter reduction method.

Data sets	Overall performance (%)				
	ACC	SENS	SPEC	PPV	NPV
Breast cancer	91	90.8	91.3	95.2	83.9
Liver disorder	75.7	84	64.1	76.4	74.4
Heart disease	64.3	100.0	0.9	64.2	100.0
Diabetes	65.1	99.8	0.4	65.1	50

Table 3. Overall performance after implementation of hybrid parameter reduction method.

Data sets	Overall performance (%)				
	ACC	SENS	SPEC	PPV	NPV
Breast cancer	97.1	97.4	97.6	98.2	95.1
Liver disorder	66.4	84	42.1	66.7	65.6
Heart disease	81.6	93.6	60.4	80.7	84.2
Diabetes	66.1	98.8	5.2	66.0	70.0

Table 4. Performance of existing and proposed hybrid parameter reduction methods.

Methods	Accuracy rate (%)			
	Breast cancer	Liver disorder	Heart disease	Diabetes
Bijective soft set [15]	98.0	80.0	NA	96.0
Rough set and decision tree approaches [26]	NA	NA	97.5	NA
Information gain and soft set	81.8	72.5	76.5	65.8
Principle component and rough set	95.3	64.6	66.0	70.0
Proposed work	97.1	66.4	81.6	66.1

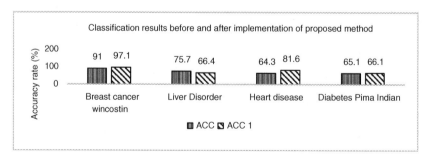

Fig. 4. Overall classification results

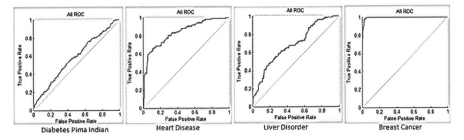

Fig. 5. Receiver operating characteristic (ROC) plot for each data sets

3.2 Results and Discussion

Based on the accuracy rates in Tables 2 and 3, all the data sets delivered satisfactory results after the implementation of hybrid PR in the classification task. Three data sets (breast cancer, heart disease, and diabetes) showed a good increment of the classification accuracy rate after the implementation of the proposed hybrid parameter reduction method. Heart disease showed the most significant result compared to the other three data sets in which the accuracy rate improved by 17.3 %, from 64.3 % to 81.6 %. However, the accuracy rate for BUPA liver disorder data set decreased by 9.3 %, from 75.7 % to 66.4 %, after the implementation of the parameter reduction process. The differences of the performance accuracy are presented in Fig. 4 which are denoted by the accuracy rate before the hybrid PR is implemented (ACC) and the accuracy rate after the hybrid PR is implemented (ACC1). The performance of each data set is presented in Fig. 5. via receiver operating characteristic (ROC) curve plots. Among all the ROC plots, breast cancer showed the nearest curve to the upper-left corner in which the classification accuracy was 97.1 %, that is near to 100 %. It can be concluded that the neural network classifier performed well with the breast cancer data set.

To demonstrate the significant of this study, the obtained results were compared to the results of selected existing works that applied similar or nearly similar data sets. The results from two existing works by [15, 26] were used as the comparison and benchmarking since many existing works did not use real data set to test their proposed work. However, not all data sets were tested by the selected existing works. In this case, it is represented by 'NA' or non-available symbol. Unfortunately, these two existing works did not apply the same research method as the proposed work. Therefore, two additional hybrid parameter reduction methods were applied to all data sets as a comparison to the proposed work. The comparison results are presented in Table 4.

The proposed hybrid parameter reduction method produced significant results for two data sets, breast cancer and heart disease, when it is compared to the other two hybrid PR methods (information gain with soft set and principle component with rough set). Meanwhile, by referring to the existing works, there are some possibilities that make the classification results better than the proposed work. Firstly, the objective of the work proposed by [15] was to test the ability of the bijective soft set to generate the classification rules instead of using it as a parameter reduction method. Secondly, the data set used by [26] in their work were from their own collection in which there were no missing values and other limitations such as small number of instances. Among the four data sets used, it can be concluded that breast cancer is one of the best data set that can be applied in any decision analysis problem since it helps in producing good results as shown in Table 4. A number of possibilities that might cause low accuracy rate of the data sets such as the characteristic of the data set, data size, and the chosen classifier should be further investigated. Since the work proposed by [15] produced good classification results, it might be considered to be implemented in the proposed framework in order to increase the classification performance.

4 Conclusion

To ensure a good decision is made, one of the most important approaches that researchers should apply in the decision analysis task is parameter reduction process. This process will eliminate complex data such as uncertainty values and simplify data volume into an acceptable data format. Recently, many research works have proposed different hybrid parameter reduction methods to overcome the limitation of a single parameter reduction method and to increase the performance of the previous works. Driven by the existing works highlighted in the previous sections, this study proposed a new framework of hybrid parameter reduction method which manipulates the advantages of rough set and soft set theories in dealing with complex data problem. Consequently, this study analysed the existing hybrid works and also investigated the ability and the performance of the proposed hybrid parameter reduction method in processing complex medical data for classification problem.

Most of the publications preferred to explore and apply the fuzzy concept into the soft set theory instead of integrating the rough set theory with the soft set theory in solving multi-criteria decision-making problems. Each of these methods has proven that a good solution could be obtained based on the given numerical examples without conducting real experimental works with real data sets. Evidently, most of the beneficial works have provided a simple and small data set in the validation test as stated in [6] which provided 'life expectancy' example that consists of six people with four decision parameters. It is difficult to prove whether the proposed hybrid methods are really efficient in producing the best solution without facing any computational problem. Thus, this study tested the performance of the proposed hybrid method with a large data set that consists of more than 100 instances as described above.

The outcome proved that parameter reduction method is needed when the data used are complex and contain uncertain or missing values. It helps in reducing the complexity without changing the structures and meaning of the data. As a conclusion, the hybrid rough set and soft set parameter reduction method have a great potential for researchers to further their research directions towards this area especially in solving big data phenomena, uncertainties, and data complexity problems. It is beneficial if all the proposed hybrid methods can be applied to any application areas such as medical science, social science, and economy.

Acknowledgements. The authors would like to thank anonymous reviewers for their constructive comments and valuable suggestions. The authors wish to thank Universiti Teknologi Malaysia (UTM) under Research University Grant Vot-02G31 and Ministry of Higher Education Malaysia (MOHE) under the Fundamental Research Grant Scheme (FRGS Vot-4F551) for completion of the research.

References

1. Agarwal, M., Hanmandlu, M., Biswas, K.K.: Generalized intuitionistic fuzzy soft set and its application in practical medical diagnosis problem. IEEE Int. Conf. Fuzzy Syst. **3**, 2972–2978 (2011)
2. Aydogan, E.K.: Performance measurement model for Turkish aviation firms using the rough-AHP and TOPSIS methods under fuzzy environment. Expert Syst. Appl. **38**, 3992–3998 (2011)
3. Bello, R., Verdegay, J.L.: Rough sets in the soft computing environment. Inf. Sci. **212**, 1–14 (2012)
4. Das, S., Kar, S.: Intuitionistic multi fuzzy soft set and its application in decision making. In: Maji, P., Ghosh, A., Murty, M., Ghosh, K., Pal, S.K. (eds.) PReMI 2013. LNCS, vol. 8251, pp. 587–592. Springer, Heidelberg (2013)
5. Feng, F., Li, C., Davvaz, B., Ali, M.I.: Soft sets combined with fuzzy sets and rough sets: a tentative approach. Soft. Comput. **14**(9), 899–911 (2010)
6. Feng, F., Liu, X., Leoreanu-Fotea, V., Jun, Y.B.: Soft sets and soft rough sets. Inf. Sci. **181**(6), 1125–1137 (2011)
7. Geng, S., Li, Y., Feng, F., Wang, X.: Generalized intuitionistic fuzzy soft sets and multiattribute decision making. In: Proceedings of 2011 4th International Conference on Biomedical Engineering and Informatics, BMEI 2011, vol. 4, pp. 2206–2211 (2011)
8. Gong, Z.T., Xie, T., Shi, Z.H., Pan, W.Q.: A multiparameter group decision making method based on the interval-valued intuitionistic fuzzy soft sets. In: Proceedings of the 2011 International Conference on Machine Learning and Cybernetics, pp. 10–13 (2011)
9. Guan, X., Li, Y., Feng, F.: A new order relation on fuzzy soft sets and its application. Soft. Comput. **17**(1), 63–70 (2013)
10. Herawan, T., Deris, M.M.: Soft decision making for patients suspected influenza. In: Taniar, D., Gervasi, O., Murgante, B., Pardede, E., Apduhan, B.O. (eds.) ICCSA 2010, Part III. LNCS, vol. 6018, pp. 405–418. Springer, Heidelberg (2010)
11. Herawan, T., Deris, M.M., Abawajy, J.H.: Matrices representation of multi soft-sets and its application. In: Taniar, D., Gervasi, O., Murgante, B., Pardede, E., Apduhan, B.O. (eds.) ICCSA 2010, Part III. LNCS, vol. 6018, pp. 201–214. Springer, Heidelberg (2010)
12. Inuiguchi, M., Yoshioka, Y., Kusunoki, Y.: Variable-precision dominance-based rough set approach and attribute reduction. Int. J. Approx. Reasoning **50**(8), 1199–1214 (2009)
13. Ali, M.I.: A note on soft sets, rough soft sets and fuzzy soft sets. Appl. Soft Comput. J. **11**(4), 3329–3332 (2011)
14. Karami, J., Ali Mohammadi, A., Seifouri, T.: Water quality analysis using a variable consistency dominance-based rough set approach. Comput. Environ. Urban Syst. **43**, 25–33 (2014)
15. Kumar, S.U., Inbarani, H.H., Kumar, S.S.: Bijective soft set based classification of medical data. In: Proceedings of the 2013 International Conference on Pattern Recognition, Informatics and Mobile Engineering, PRIME 2013, pp. 517–521 (2013)

16. Li, Z., Liang, P., Avgeriou, P., Guelfi, N.: A systematic mapping study on technical debt and its management. J. Syst. Softw. **101**, 193–220 (2015)
17. Liou, J.J.H.: Variable consistency dominance-based rough set approach to formulate airline service strategies. Appl. Soft Comput. J. **11**(5), 4011–4020 (2011)
18. Ma, X., Wang, G.: An extended soft set model: type-2 fuzzy soft sets. In: Proceedings of IEEE International Conference on Cloud Computing and Intelligence Systems, pp. 128–133 (2011)
19. Ma, X., Sulaiman, N., Qin, H.: Parameterization value reduction of soft sets and its algorithm. IEEE Colloquium on Humanities, Science and Engineering, pp. 261–264 (2011)
20. Meng, D., Zhang, X., Qin, K.: Soft rough fuzzy sets and soft fuzzy rough sets. Comput. Math Appl. **62**(12), 4635–4645 (2011)
21. Mohamad, M., Selamat, A., Krejcar, O., Kuca, K.: A recent study on the rough set theory in multi-criteria decision analysis problems. Comput. Collective Intel. **2**, 265–274 (2015)
22. Nguyen, H.S., Skowron, A.: Rough sets: From rudiments to challenges. Intel. Syst. Ref. Libr. **42**, 75–173 (2013)
23. Omurca, S.I.: An intelligent supplier evaluation, selection and development system. Appl. Soft Comput. J. **13**(1), 690–697 (2013)
24. Shabir, M., Ali, M.I., Shaheen, T.: Another approach to soft rough sets. Knowl.-Based Syst. **40**, 72–80 (2013)
25. Shah, T., Medhit, S., Farooq, G.: Intuitionistic fuzzy soft set decision criterion for selecting appropriate block cipher. 3D Res. **6**(3), 32 (2015)
26. Son, C.S., Kim, Y.N., Kim, H.S., Park, H.S., Kim, M.S.: Decision-making model for early diagnosis of congestive heart failure using rough set and decision tree approaches. J. Biomed. Inf. **45**(5), 999–1008 (2012)
27. Sun, B., Ma, W.: Soft fuzzy rough sets and its application in decision making. Artif. Intel. Rev. **41**(1), 67–80 (2011)
28. Vahdani, B., Hadipour, H., Tavakkoli, M.R.: Soft computing based on interval valued fuzzy ANP-A novel methodology. J. Intell. Manuf. **23**, 1529–1544 (2012)
29. Xiao, Z., Chen, W., Li, L.: A method based on interval-valued fuzzy soft set for multi-attribute group decision-making problems under uncertain environment. Knowl. Inf. Syst. **34**(3), 653–669 (2013)
30. Yang, Z., Chen, Y.: Fuzzy soft set-based approach to prioritizing technical attributes in quality function deployment. Neural Comput. Appl. **23**(78), 2493–2500 (2013)
31. Zhang, Z.: A rough set approach to intuitionistic fuzzy soft set based decision making. Appl. Math. Model. **36**(10), 4605–4633 (2012)

Multivariate Higher Order Information for Emergency Management Based on Tourism Trajectory Datasets

Ye Wang, Kyungmi Lee, and Ickjai Lee[✉]

Information Technology Academy, College of Business, Law and Governance,
James Cook University, Cairns, QLD 4870, Australia
`ickjai.lee@jcu.edu.au`

Abstract. Higher order information of trajectory dataset for "what-if" analysis is getting more popular for providing better informed decision making. However, existing research had been studied trajectory higher order information based on univariate dataset only. There is yet no research to study higher order information of multivariate dataset for trajectory analysis. This paper will introduce a unified data structure which supports multivariate datasets for trajectory analysis and approaches to analyse information based on multivariate higher order information for making decisions related to emergency management. Interactive visualisation tools are implemented to facilitate the analysis. Tourism trajectory dataset is used to demonstrate the proposed approach.

1 Introduction

Emergency/disaster management is a discipline related to avoiding and dealing both nature and man-made disasters. Decision-making can be a crucial difficulty throughout unexpected emergencies. Improving operational efficiency by allocating resources and efforts strategically is one of the important issues in emergency management, especially when there are multiple operations with different emergency situations. How many people will be needed to provide relief support in a disaster region? Where, what and how many emergency supplies such as food, water, and sanitation kits should be pre-positioned in a regional depot? How much financial support is required from government or humanitarian organisations for the long time recovery of the region? These are some common questions facing by humanitarian operations [2]. These questions often require immediate decisions in order to quickly respond to the emergency situations. To make informed and quick decisions to balance the needs and resources in different regions as well as to improve the operational efficiency of humanitarian organisation, an efficient and effective data analytics system is required. In particular, spatio-temporal data analytics with trajectory data is of great relevancy for answering these kind of location-based and time-based resource allocation questions.

Trajectory data are spatio-temporal data capturing valuable sources of information for trails of object dynamics. They include spatial and temporal aspects

© Springer International Publishing Switzerland 2016
H. Fujita et al. (Eds.): IEA/AIE 2016, LNAI 9799, pp. 725–736, 2016.
DOI: 10.1007/978-3-319-42007-3_62

of moving entities and additional quantitative and qualitative attributes about the movement as well as the environment or context in which the movement takes place. Trajectory of moving objects such as cars, humans, cows, or other objects is useful in decision making processes in many application domains such as surveillance analysis [9], animal movement [4], transport analysis [1,15], sport team strategies [11], supermarket shopping paths [12], and emergency management [7,8].

Trajectory data plays an important role in the increasing number of emergency management applications. The rise of GPS-empowered mobile devices, advance of internet technologies, and affordable data storage mechanisms allowed collection and distribution of huge amounts of user-generated trajectory data. It has presented new opportunities for spatial computing applications and geo-informatics technologies with regard to emergency management. Trajectory analysis is uncertain and complex with spatio-temporal dynamicity in nature. Higher Order Information (HOI) for "what-if" analysis allows users to deal with the complexity of trajectory analysis. However, HOI for trajectory analysis has received little attention in the literature [5,16]. HOI includes k-Nearest Neighbor (kNN) information and k-Order Region (kOR) information that are of great significancy when the first order or lower order information is not functioning properly. For instance, the second nearest evacuation centre information is required when the first nearest evacuation centre is full or closed in emergencies. Analysing trajectory dataset and HOI could provide great benefits for emergency management. While existing research has proposed different ways to analyse HOI for trajectory analysis, they can only deal with univariate Point of Interests (POIs). As our society moves into a more data-rich, information-rich, but knowledge-poor environment, there are many different datasets available to support decision making. To analyse multiple datasets along with the trajectory data, multivariate HOI analysis approach is needed.

This study introduces a new Multivariate-HOI (Mul-HOI) data structure for support multivariate generators (targets); presents a new visualisation approach by Parallel Coordinates for analysing large trajectory datasets with Mul-HOI and introduces a new measure called Parallel Coordinates Multivariate Measure (PCMM) to quantify qualitative and quantitative multivariate HOI with regards to kNN information. These three proposed approaches provide new ways to analyse and compare trajectory datasets information based on Mul-HOI. We provide a case study that demonstrates the usefulness and applicability of our proposed approaches.

2   Preliminaries

kNN information and kOR information are the two most often used HOI for analysing trajectory data. Given a point set S, a query point q and a positive integer k, the kNN query gives a point set kNNS(q) which is a subset of S such that $d(q, r) \leq d(q, p)$ holds for any point $r \in k$NNS(q) and for any point $p \in (S - k$NNS$(q))$.

kNNQ is supplemented by the k-th Order Region Query (kORQ) which returns a region R for a certain subset $P_i^{(k)}$ in P where the set of kNN for any location l in R is the same as $P_i^{(k)}$. kORQ can also be represented by a kth-Nearest Point Diagram. kORQ associated with a given site is the set of points in the plane such that the site ranks number k in the ordering of the sites by distance from the point, and is a collection of one or more unconnected cells [10]. It is useful to establish a boundary on the number of cells, as their complexity can affect the accuracy of the final output the k-th order Voronoi diagram. kNNQ simply returns kth nearest neighbor, and typically returns quantitative geometrical and topological information. kNN has been widely used for classification, regression, "what-if" analysis. While kOR returns kth nearest neighbor Voronoi regions. Typically, it returns qualitative geometrical and topological information. kOR has been extensively used for location optimisation, catchment analysis, market analysis, disaster management and "what-if" analysis [6]. Build on top of kNN and kOR, this study introduces a new approach to analyse Mul-HOI of trajectory data. Both kNNQ and kOR are modeled by order-k Voronoi diagram families which are generalisations of the Ordinary Voronoi Diagram (OVD).

Existing research of HOI mainly focuses on geometrical information with limited attention on topological information [14] and directional information [13] and yet no research has been conducted in multivariate data analysis. This study attempts to fill part of this research gap; and hence focuses on relationship information between multivariate datasets and trajectory datasets.

Consider Fig. 1 with three sets of POIs data representing 8 tourist attractions (red point) and 8 information centres (blue point) and 6 emergency centres (magenta point) and 1 trajectory denoting tourist movements. The trajectory has 10 timestamps (nodes) representing geo-tagged photos taken and uploaded to a photo-sharing website. The geometrical HOI based on nearest neighbours as the trajectory travels along the study region with regards to the generators. A sequence $(A_4 - A_3 - A_6)$ is shown in Fig. 1(a). Whilst trajectory geometrical HOI with regards to generators B sequence $(B_2 - B_1 - B_7 - B_6)$ is shown in Fig. 1(b). At the same time, Fig. 1(c) shows generator C sequence $(C_1 - C_2 - C_4 - C_5)$ based on trajectory geometrical HOI. Considering building three sets of POIs together, it can get more detailed information from these three generator groups. Figure 1(d), shows these three sets of POIs (generators) with geometrical trajectory. This mixed sequence not only provides the combined HOI for the three sets of POIs but also provides a way to analyse and build the Mul-HOI data structure. This Mul-HOI, mix of three sets of POIs, reveals interesting multivariate HOI patterns with trajectory, but traditional approaches fail to address this.

3   Computing Multivariate Higher Order Information

3.1   Unified Data Structure

This paper is based on the unified Delaunay triangle based data structure which consists of a complete set of Order-k Delaunay triangles (from Order-0 to

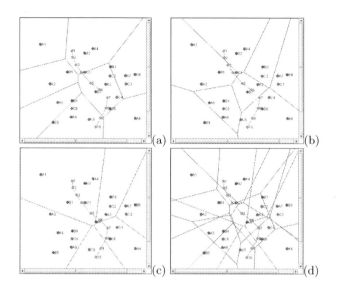

Fig. 1. Multivariate POIs with trajectory dataset T_1: (a) O1VD of generator $P_{A1} = \{p_1, ..., p_8\}$; (b) O1VD of generator $P_{B1} = \{p_1, ..., p_8\}$; (c) O1VD of generator $P_{C1} = \{p_1, ..., p_6\}$; (d) O1VD of 3 generators $P_{A1} + P_{B1} + P_{C1}$. (Color figure online)

Table 1. The relationship between OkVDs and the unified data structure.

Order-k	OkVD
1	Order-0 triangle
2	Order-0 & Order-1 triangle
3	Order-1 & Order-2 triangle
4	Order-2 & Order-3 triangle
...	...
k	Order-(k-2) & Order-(k-1) triangle

Order-(k-1)) [7]. A complete set of OkVD could be drawn from this data structure, and the relationship between the data structure is shown in Table 1. For example, O2VD can be obtained by a combination of Order-0 triangle and Order-1 triangle. However, for multivariate OkVD data structure, we need to present to each generator Order-k Delaunay triangles and combining all generators together to one generator group.

The Delaunay triangulation is a dual graph of the OVD with edges connecting neighboring points. It can be constructed by linking two adjacent Voronoi generators if they share a Voronoi edge together. For Order-0 triangles, no generator is on the circumcircle of each Delaunay triangle in the triangulation. However, there could be triangles whose circumcircles include a number of generators within them. Order-1 triangles are those triangles whose corresponding circumcircles include only one generator in it. Therefore, Order-k triangles are those ones whose corresponding circumcircles include k generators in it. Please refer to [7] for more details.

3.2   Algorithm

Figure 2 displays screenshots of the visualisation tools developed for this study with three generators P_{A2}, P_{B2} and P_{C2}, and trajectory datasets T_1. The Mul-HOI implementation enables users to change various k values to get different types of Mul-HOI from trajectory data. Since different OkVDs are visualised from the same dataset P_{A2}, P_{B2} and P_{C2}, users could interact with the program to retrieve different dataset points for any user-interested location within the study query points. P_{A2} and the corresponding Voronoi diagrams are represented in red. P_{B2} and the corresponding Voronoi diagrams are represented in blue. P_{C2} and the corresponding Voronoi diagrams based on geometrical HOI are represented in magenta. Geometrical HOI is computed based on the Euclidean distance of the trajectory data points to a set of nearest generators. O1VD and O5VD of the trajectory data are shown in Fig. 2(a) and (c) in green.

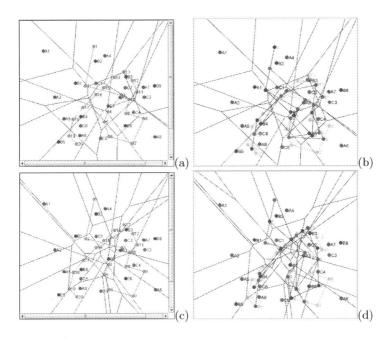

Fig. 2. OkVDs: (a) O1VD; (b) O5VD with generators P_{A2} in red, P_{B2} in blue, P_{C2} in magenta and trajectory data T_2 in green. Topologically simplified trajectory dataset for each generators: (c) O1VD; (d) O5VD with generators, P_{A2} for topologically simplified trajectory T_{A2} in red, P_{B2} for topologically simplified trajectory T_{B2} in blue and P_{C2} for topologically simplified trajectory T_{C2} in magenta. (Color figure online)

Figure 2(b) and (d) display topologically simplified trajectories with multivariate generators. For topologically simplified trajectories, the visualisation tools will try to go through every single data point in the trajectory to perform a classification in relation to P_{A2}, P_{B2} and P_{C2}. Topologically simplified trajectory for P_{A2} is presented in red. Topologically simplified trajectory for P_{B2} is presented in blue. Topologically simplified trajectory for P_{C2} is in magenta.

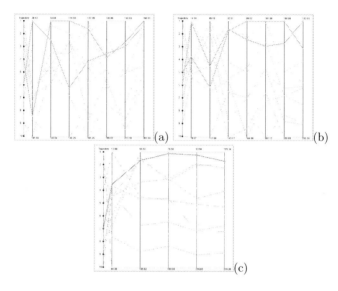

Fig. 3. PCMMs: (a) Parallel Coordinates - P_{A2}; (b) Parallel Coordinates - P_{B2}; (c) Parallel Coordinates - P_{C2} based on trajectory data T_2. Quantitative PCMM is shown in orange. Qualitative PCMM is shown in cyan. Once quantitative and qualitative PCMM are the same, it is shown in green. (Color figure online)

Based on HOI trajectory, we can define Parallel Coordinates Multivariate Measure (PCMM) for mining trajectory exhibiting relationship between different generators as shown in Fig. 3. For each single data point of the HOI trajectory (T_2) of three POIs (P_{A2}, P_{B2}, P_{C2}), we calculate distance between each single POI and trajectory to draw Parallel Coordinates. As each layer from left to right, it shows the Order-1 to Order-k information. Each trajectory data goes through each layer by calculating distance to POIs, then each trajectory data draws the distance to each k layer. Whilst we use v_{Short} as shown in Eq. 1 to determine qualitative PCMM. For each data point in the HOI trajectory of three POIs groups, distance of the data point to the 1NN to kNN HOI is used to calculate v_{Short}. If the data point has the shortest distance from HOI in P_{A2}, this trajectory point is calculated and determines quantitative PCMM for P_{A2}. Quantitative PCMM is shown in orange, qualitative PCMM is shown in cyan. For the same quantitative and qualitative PCMM, it is shown in green.

$$v_{Short} = \frac{v - Min}{Max - Min}(newMax - newMin) + newMin. \qquad (1)$$

Thus, with this approach, the best PCMM of P_{A2} or P_{B2} or P_{C2} is the same of quantitative and qualitative PCMM. The runtime of this approach is linear. This approach is faster and can provide more details compared to other data mining approaches. Example applications of Mul-HOI are shown in the next section.

4   Applications

Emergency management is composed of four basic phases: mitigation, preparedness, response, and recovery [3]. This section demonstrates how the trajectory with Mul-HOI can be used for each stage of emergency management. We explain this with a study using geo-tagged photos from Flickr in Queensland, Australia, in the years of 2010 and 2011 and tourism trajectory datasets. For the Flickr dataset, there are 17,066 records preprocessed to produce the dataset used in this study. Figure 4(a) shows the distribution of the dataset using Google Earth 6.0 (earth.google.com).

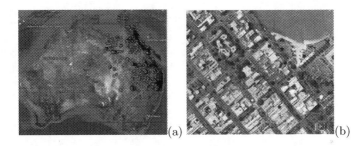

Fig. 4. (a) Geo-tagged photos from Flickr; (b) Distribution of clusters in CBD of Cairns (P_{A3} in red), and information office in CBD of Cairns (P_{B3} in blue), and emergency office in CBD of Cairns (P_{B3} in magenta). (Color figure online)

Most of these records appear along the coastline of Queensland, where the major cities are located. This application will take a subset of the dataset with data points from Cairns, a regional capital city in Queensland, Australia, to demonstrate the usefulness and applicability of our approaches. Figure 4(b) shows a subset P_{A3} of clusters (red) within the CBD of Cairns. These locations can be set up as potential tourist centres related to emergency management. Blue datasets P_{B3} are current information offices within the CBD of Cairns. Magenta datasets P_{C3} are current emergency offices within Cairns city.

4.1   Mitigation

Mitigation is a realisation that eliminates or decreases the causes of risk/disaster, and is a factor that provides prevention of (removing, eliminating, reducing) the effects of inescapable disaster.

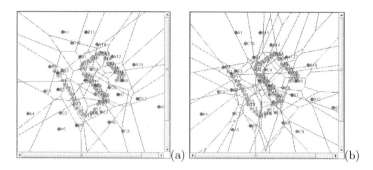

Fig. 5. (a) O1VD; (b) O5VD trajectory T_3 with Mul-HOI P_{A3}, P_{B3}, P_{C3}. (Color figure online)

Figure 5 shows different values of k with Order-k Voronoi diagrams. A trajectory dataset is collected every 30 s for 25 min from a tourist group which walked around the CBD of Cairns. The data collected from tourist T_2 is represented in green.

If we assume that P_{A3} represents tourist attractions, P_{B3} represents information centres, P_{C3} represents emergency centres, Fig. 5(a) shows O1VD, Fig. 5(b) displays O5VD with only T_3 data information. Considering the three POIs datasets are different, k is different for each. Based on Mul-HOI, it can provide useful information to indicate where are the nearest centres they can seek help from along the path of the trajectory, assuming that they can seek help from the emergency centres to come to the tourist attractions or the information centres.

4.2   Preparedness

In the preparedness phase, governments, organisations, and emergency centers involve in making plans to save lives and minimise disaster damage such as having emergency personnel on standby or determining the nearest evacuation places with dynamic (moving/changing over time) data. In other words, preparedness also explores ways to improve damage response operations.

Figure 6 shows different values of k with 3 new topologically simplified trajectories datasets. T_{A3} is represented in red, T_{B3} is represented in blue, T_{C3} is represented in magenta. The government can make decisions about the best location for setting up emergency tables or information tables for tourists before a disaster arrives. For the regions, the nearest neighbour emergency units give suggestions about areas which have more activities. This provides information for tourists to make back up plans when the nearest information centre is not available or functional. With the additional information from the topologically simplified visual analytic trajectory, governments can make better informed decisions for planning emergency support before damage from the disaster occurs.

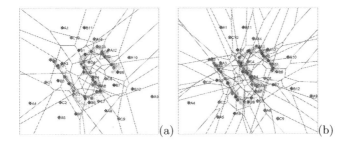

Fig. 6. (a) O1VD; (b) O5VD with 3 new topologically simplified trajectories dataset. T_{A3} in red, T_{B3} in blue, T_{C3} in magenta. (Color figure online)

4.3   Response

Response consists of providing activities following the emergency situation to assist disaster victims. It also explores how to stabilise the situation and to prevent the probability of secondary disaster and to give suggestions for recovery. Typically, the nearest neighbor emergency unit is required to respond to emergencies quickly. When emergencies occur, the nearest neighbor emergency units are required to cooperate and collaborate in order to decrease the damage from the disaster.

Figure 7 shows different values of k with Order-k information, it displays information to determine the relationship of the trajectory based on generator P_{A3} and P_{B3} and P_{C3}. It also highlights the quantitative PCMM and qualitative PCMM information between Mul-HOI trajectory T_3 and generators P_{A3} and P_{B3} and P_{C3}. The government focuses on these top priority regions where the emergency centre and the information centre has the shortest distance to the trajectory, based on quantitative and qualitative PCMM. PCMM provides two ways to compare a set of POIs to trajectory Mul-HOI. It allows governments to make better decisions on where (location) and which (centre) begins to provide the emergency response.

4.4   Recovery

Recovery happens when the disaster is over. It is a set of activities that returns all systems back to normal and also prevents secondary damage. These include two phases, short term and long term. From Fig. 8, we can see that although cluster numbers of topologically simplified trajectories are different, HOI is very similar. Figure 8(a) shows the common PCMM in red based on P_{A3}. Figure 8(b) shows the common PCMM in blue based on P_{B3}. Figure 8(a) shows the common PCMM in magenta based on P_{C3}. In the short term of emergency recovery, we can choose $T_{24} - T_{28}$ as recovery plans based on common quantitative and qualitative PCMM. For long term, recovery plans can be introduced based on Parallel Coordinates pattern k of higher order information. Interestingly, PCMM, based on those datasets, are better PCMM after $k = 5$, either trajectory with P_{A3} or

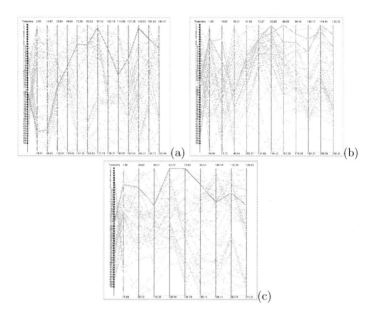

Fig. 7. Response PCBCs: (a) generator P_{A3}'s PCBC for T_{A3}; (b) generator P_{B3}'s PCBC for T_{B3}; (c) generator P_{C3}'s PCBC for T_{C3}.

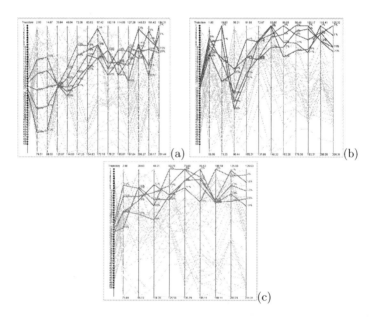

Fig. 8. Recovery PCBCs: (a) generator P_{A3}'s PCBC for T_{A3}; (b) generator P_{B3}'s PCBC for T_{B3}; (c) generator P_{C3}'s PCBC for T_{C3}. (Color figure online)

P_{B3} or P_{C3}. Therefore, for a long term recovery plan, k after 5 of these datasets of HOI are the best reference for decision making.

4.5   Discussion

This study proposes a new approach to analyse trajectory dataset for emergency management. The unified data structure gives general suggestion or information before making plans in mitigation procedures. Data structure can support with multivariate data information and can prepare information for future analysis. Topologically simplified trajectory dataset can filter and well organise with different groups of people, to quickly make decisions in preparedness steps. People can logically get different information based on different POIs. In response steps, organisations or governments can well support and control tourist attractions or information centres or emergency centres. Which, what and how many items should be considered can be prepared for the next recovery step. To quickly make decisions for short term recovery is important after disaster. This approach introduces short term and long term suggestions. Organisations can make decisions which cover multivariate centres.

5   Conclusion

Disaster and emergencies can lead to various forms of financial, structural, and environmental damage. Even though it is almost impossible to avoid occurrences of disasters, effective prediction and preparedness along with a post-emergency management program can mitigate the risk and damage. Trajectory analysis has emerged in the last decade as a very active research domain with many potential applications for emergency management. While previous research mainly focused on univariate dataset relationships with trajectories, this research focuses more on analysis of HOI with multivariate datasets associated trajectory data.

This paper proposes a unified data structure for multivariate datasets that can support with Mul-HOI; a visual analytical approach for topologically simplified HOI of trajectory datasets; and PCMM for quantified multivariate HOI with regards to k-nearest neighbour information. Those three approaches enable users to analyse large volume of trajectories with different POIs, related to the trajectories to understand the relationship of POIs and trajectory HOI. We test our proposed approaches with a case study explaining the usefulness and applicability of our approaches.

References

1. Chu, D., Sheets, D., Zhao, Y., Wu, Y., Yang, J., Zheng, M., Chen, G., et al.: Visualizing hidden themes of taxi movement with semantic transformation. In: 2014 IEEE Pacific Visualization Symposium (PacificVis), pp. 137–144. IEEE (2014)
2. Gonçalves, P.: Balancing provision of relief and recovery with capacity building in humanitarian operations. Oper. Manag. Res. 4(1–2), 39–50 (2011)

3. Haddow, G., Bullock, J., Coppola, D.P.: Introduction to Emergency Management. Butterworth-Heinemann, Boston (2013)
4. Handcock, R.N., Swain, D.L., Bishop-Hurley, G.J., Patison, K.P., Wark, T., Valencia, P., Corke, P., ONeill, C.J.: Monitoring animal behaviour and environmental interactions using wireless sensor networks, GPS collars and satellite remote sensing. Sensors 9(5), 3586–3603 (2009)
5. Beni, H., Mostafavi, M.A., Pouliot, J., Gavrilova, M.: Toward 3D spatial dynamic field simulation within GIS using kinetic Voronoi diagram and delaunay tetrahedralization. Int. J. Geogr. Inf. Sci. 25(1), 25–50 (2011)
6. Keim, D.A., Andrienko, G., Fekete, J.-D., Görg, C., Kohlhammer, J., Melançon, G.: Visual analytics: definition, process, and challenges. In: Kerren, A., Stasko, J.T., Fekete, J.-D., North, C. (eds.) Information Visualization. LNCS, vol. 4950, pp. 154–175. Springer, Heidelberg (2008)
7. Lee, I., Lee, K.: A generic triangle-based data structure of the complete set of higher order Voronoi diagrams for emergency management. Comput. Environ. Urban Syst. 33(2), 90–99 (2009)
8. Lee, I., Pershouse, R., Phillips, P., Christensen, C.: What-if emergency management system: a generalized Voronoi diagram approach. In: Yang, C.C., Zeng, D., Chau, M., Chang, K., Yang, Q., Cheng, X., Wang, J., Wang, F.-Y., Chen, H. (eds.) PAISI 2007. LNCS, vol. 4430, pp. 58–69. Springer, Heidelberg (2007)
9. Lin, L., Lu, Y., Pan, Y., Chen, X.: Integrating graph partitioning and matching for trajectory analysis in video surveillance. IEEE Trans. Image Process. 21(12), 4844–4857 (2012)
10. Okabe, A., Boots, B., Sugihara, K., Chiu, S.N.: Spatial Tessellations: Concepts and Applications of Voronoi Diagrams, vol. 501. Wiley, New York (2009)
11. Pingali, G., Opalach, A., Jean, Y., Carlbom, I.: Visualization of sports using motion trajectories: providing insights into performance, style, and strategy. In: Proceedings of the Conference on Visualization 2001, pp. 75–82. IEEE Computer Society (2001)
12. Popa, M.C., Rothkrantz, L.J., Shan, C., Gritti, T., Wiggers, P.: Semantic assessment of shopping behavior using trajectories, shopping related actions, and context information. Pattern Recogn. Lett. 34(7), 809–819 (2013)
13. Wang, Y., Lee, K., Lee, I.: Directional higher order information for spatio-temporal trajectory dataset. In: 2014 IEEE International Conference on Data Mining Workshop (ICDMW), pp. 35–42. IEEE (2014)
14. Wang, Y., Lee, K., Lee, I.: Visual analytics of topological higher order information for emergency management based on tourism trajectory datasets. Procedia Comput. Sci. 29, 683–691 (2014)
15. Wang, Z., Lu, M., Yuan, X., Zhang, J., Van De Wetering, H.: Visual traffic jam analysis based on trajectory data. IEEE Trans. Vis. Comput. Graph. 19(12), 2159–2168 (2013)
16. Xiong, X., Mokbel, M.F., Aref, W.G.: SEA-CNN: Scalable processing of continuous k-nearest neighbor queries in spatio-temporal databases. In: 2005 Proceedings of 21st International Conference on Data Engineering, ICDE 2005, pp. 643–654. IEEE (2005)

Hourly Solar Radiation Forecasting Through Model Averaged Neural Networks and Alternating Model Trees

Cameron R. Hamilton[✉], Frederick Maier, and Walter D. Potter

Institute for Artificial Intelligence, 30602 Athens, Georgia
cameron@cs.uga.edu, {fmaier,potter}@uga.edu

Abstract. The objective of the current study was to develop a solar radiation forecasting model capable of determining the specific times during a given day that solar panels could be relied upon to produce energy in sufficient quantities to meet the demand of the energy provider, Southern Company. Model averaged neural networks (MANN) and alternating model trees (AMT) were constructed to forecast solar radiation an hour into the future, given 2003–2012 solar radiation data from the Griffin, GA weather station for training and 2013 data for testing. Generalized linear models (GLM), random forests, and multilayer perceptron (MLP) were developed, in order to assess the relative performance improvement attained by the MANN and AMT models. In addition, a literature review of the most prominent hourly solar radiation models was performed and normalized root mean square error was calculated for each, for comparison with the MANN and AMT models. The results demonstrate that MANN and AMT models outperform or parallel the highest performing forecasting models within the literature. MANN and AMT are thus promising time series forecasting models that may be further improved by combining these models into an ensemble.

Keywords: Solar radiation · Time series forecasting · Artificial neural networks · Decision trees

With the world population projected to increase to 9.6 billion by 2050 [33], there is an urgent need to utilize renewable energy sources. The necessity of harnessing renewable energy is more apparent when one considers that the average farm uses 3 kcal of fossil fuel energy to produce 1 kcal of food energy before that food is even processed or transported to the market [12]. In order to ensure that the energy needs of a population are met, it is vital that there are systems in place for forecasting how much energy can be anticipated in the future, such that usage of non-renewable energy resources can be scaled back accordingly. Thus, a highly accurate model of solar radiation prediction is imperative as such predictions can inform the expected yield from crops in a given year as well as the amount of energy that will be produced from a solar panel at a given location and time [12, 16]. The present study proposes a model for hourly solar radiation prediction whose accuracy is competitive with the leading models of solar radiation (compare with [3, 29, 31]).

With the ability to predict the amount of solar energy that will be produced by a set of solar panels, assuming a constant efficiency of the panels, it is possible for energy

H. Fujita et al. (Eds.): IEA/AIE 2016, LNAI 9799, pp. 737–750, 2016.
DOI: 10.1007/978-3-319-42007-3_63

providers to reduce their usage of non-renewable energy resources. The forecasting model utilized measurements of weather conditions collected at the Georgia Automated Environmental Monitoring Network's (GAEMN) Griffin weather station as inputs. As shown in Figs. 1 and 2, the overall trend of increasing solar radiation through the spring and summer, and the decrease starting in the fall, is present in the solar radiation data from year to year. However, the amount of solar radiation occurring on a given day at a given time can vary drastically between years, which prevents linear approximation techniques from yielding accurate predictions. In order to model the non-linearity present within the data, a number of machine learning models were implemented including least median squares, random forest, alternating model trees [6], and artificial neural networks.

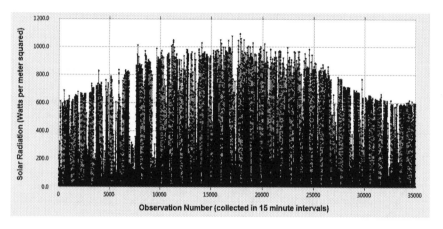

Fig. 1. Solar radiation data collected in 15 min intervals from the Griffin, Georgia weather station in 2003.

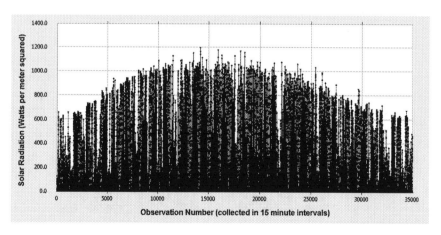

Fig. 2. Solar radiation data collected in 15 min intervals from the Griffin, Georgia weather station in 2013.

To further bolster prediction performance, model averaged neural networks were constructed from combinations of single networks with the same architecture, but with their respective weight vectors initialized randomly to distinct sets of values. It was conjectured that since the final weight vector for a network is deterministic, given its initial weight vector, the training set, and the learning algorithm, a greater extent of the search space for the weight vector could be explored by creating multiple copies of the same network, but with different initial weights. It was hypothesized that the average of the outputs from these networks would be closer to the global optimum, as more of the weight space would be explored by initializing each network with a distinct weight vector. This is to say that by initializing the networks to different weights, training each network on the same time series, and averaging their output would help reduce the bias inherent in any network's initial weights. As a consequence, the model averaged neural network should have a smaller mean square error for its predictions in comparison to a single neural network [32]. Thus, the model averaged neural network appears to be an effective approach for improving the accuracy of continuous value predictions, such as the amount of solar radiation to occur, when compared to a single neural network.

Alternating model trees (AMTs) were also implemented, due to their demonstrated efficacy in time series domains [6]. Two nodes comprise AMTs: splitter nodes, where numeric attributes are split at the median value of the attribute, and predictor nodes, which utilize linear regression to predict the numeric output at that node. In addition, an AMT is grown via forward stage-wise additive modeling, where the residual errors made by the current AMT are fitted to a base learner (e.g. a decision stump or a linear regression model), after which the fitted base learner is added into the regression predictions made by the AMT. The AMT is tuned by specifying the number of iterations to grow the tree for (i) as well as the shrinkage (λ), which dampens the predictions of each base learner within the additive model towards predicting the mean of the target series. Thus, an AMT was deemed appropriate for the domain of solar radiation, as the replacement of constant values with linear regression models at the leaf nodes of the decision tree allow AMTs to model non-linear curves in a piece-wise fashion.

1 Literature Review

Elizondo et al. [4] developed an artificial neural network that provided daily solar radiation predictions with an MSE between 0.0077–0.0121 ($2.92-3.64$ MJ m^{-2}) for four sites: Tifton, GA, Clayton, NC, Gainesville, FL, and Quincy, FL. Sfetsos and Coonick [29] evaluated several model types, including a linear regression model, Elman networks (ELM), radial basis function networks (RBF), adaptive neuro-fuzzy inference systems (ANFIS), and neural networks trained using backpropagation (BP) and the Levenberg-Marquardt (LM) algorithm. Of these algorithms, the LM algorithm attained the smallest error, likely due to the fact that it combines gradient descent (backpropagation) and Newton's method, and thus has the strengths of both [8, 19, 21]. The LM algorithm is described by Eq. (1):

$$W_{min} = W_0 - (H + \lambda I)^{-1}g \tag{1}$$

Within Eq. (1), W_0 is the weight matrix of the neural network, H is the Hessian matrix, λ is the damping factor, I is the identity matrix, and g represents the gradients of the neural network. Their study indicated that a neural network trained with the LM algorithm provided superior predictions to the other models, with an RMSE of $27.58\ \text{Wm}^{-2}$.

Mellit and colleagues [22] implemented an artificial neural network and Markov transition matrix (MTM)[1] hybrid model to achieve daily solar radiation predictions with an mean absolute percentage error (MAPE) not exceeding eight percent. Spokas and Forcella's [30] model predicted hourly solar radiation as the sum of direct beam radiation and diffuse solar radiation; these values were calculated based upon the angle between the sun's location and the earth's surface (i.e. the zenith angle), the seasonal variability of the Earth's tilt (i.e. declination angle), the percentage of direct radiation passes through the atmosphere without being reflected (i.e. atmospheric transmittance), and the optical air mass number. Spokas and Forcella's model provided estimations of hourly solar radiation with an RMSE ranging from $77-167\ \text{Wm}^{-2}$ ($M = 112\ \text{Wm}^{-2}$) and a mean absolute error from $36-92\ \text{Wm}^{-2}$ ($M = 57\ \text{Wm}^{-2}$) across 18 sites between the years 1996–2005, with a maximum observed solar radiation of $1100\ \text{Wm}^{-2}$. Hocaoglu et al. [9] derived a unique representation of their solar radiation data by converting the 1D time series into a 2D image signal. The value for each pixel (x, y) within the image was determined as the solar radiation that occurred on the x hour of the yth day of the year. The solar radiation values were scaled within the range 0–255, such that the corresponding pixel had a value of 0 if no solar radiation occurred and a value of 255 if the maximum amount of solar radiation within the dataset occurred. A 3-3-1 neural network was then trained on the resulting input representation. Hocaoglu et al. confirmed that the LM algorithm also yielded the most accurate predictions of all the training algorithms applied, with a resulting RMSE of $34.57\ \text{Wm}^{-2}$, given a maximum observed solar radiation of $600\ \text{Wm}^{-2}$. Thus, the 2-D representation of the input data was a significant improvement upon the 1-D representation, which achieved a RMSE of $43.73\ \text{Wm}^{-2}$ with the same network architecture and training algorithm. More recently, the resilient propagation algorithm has been shown to outperform LM on some solar radiation data [7].

Cao and Lin [3] implemented a diagonal recurrent wavelet neural network (DRWNN), which is a recurrent neural network that uses wavelet bases as its activation functions. The DRWNN was trained on a year of solar radiation data and attained an RMSE of $13.2\ \text{Wm}^{-2}$ when tested on a month of data. Ji and Chee [18] implemented a novel hybrid model that aggregated the outputs of an autoregressive moving average (ARMA) model and a time delay neural network (TDNN) in order to determine its prediction of hourly solar radiation.

---

[1] A Markov transition matrix is a matrix which characterizes the transitions of a Markov chain [1]. For a given element i, j describes the probability of moving from state i to state j in one time step. It is also known as a probability matrix, substitution matrix, or a stochastic matrix.

One of the reoccurring problems within the literature on solar radiation forecasting is the lack of unity in performance metrics for comparing models. As Hoff et al. [11] note, while RMSE and mean absolute error (MAE) are calculated with the same equation across studies, their calculations as percentages are not. The most apparent case is with calculations of normalized RMSE (NRMSE). Equations (2) and (3) demonstrate the two most common formulations of NRMSE:

$$NRMSE = \frac{RMSE}{y_{max} - y_{min}} = \frac{\sqrt{\frac{\sum_{i=1}^{n} \left(\widehat{y_i} - y_i\right)^2}{n}}}{y_{max} - y_{min}} \tag{2}$$

$$NRMSE = \frac{RMSE}{\bar{y}} = \frac{\sqrt{\frac{\sum_{i=1}^{n} \left(\widehat{y_i} - y_i\right)^2}{n}}}{\bar{y}} \tag{3}$$

It is necessary to calculate a normalized RMSE when comparing models, as RMSE itself is not calculated with respect to scale of the predicted value across models. For instance, suppose model A achieves a RMSE of $15 \, \mathrm{W/m^2}$ on a dataset with an $\bar{y} = 250$, $y_{min} = 0 \, \mathrm{W/m^2}$ and $y_{max} = 400 \, \mathrm{W/m^2}$. In contrast, model B achieves a RMSE of $30 \, \mathrm{W/m^2}$ on a dataset with an $\bar{y} = 500 \, \mathrm{W/m^2}$, $y_{min} = 0 \, \mathrm{W/m^2}$ and $y_{max} = 1000 \, \mathrm{W/m^2}$. If Eq. (2) is used to calculate NRMSE for models A and B, then model B will be characterized as the more accurate model as 15/400-0 = 0.0375 and 30/1000-0 = 0.03, respectively. However, if Eq. (3) is used to calculate NRMSE (often referred to as coefficient of variation of the RMSE), models A and B can be said to have equal performance as 15/250 = 30/500 = 0.06. Thus, a single equation should be used to calculate NRMSE for the models to be assessed, so that the comparison between them is fair. Within the current study, Eq. (2) is used to calculate NRMSE as the maximum and minimum values for solar radiation are more readily available within the literature than average solar radiation. A comparison between some of the most prominent models for hourly solar radiation forecasting is provided in Table 1.

Before interpreting the results of these forecasting studies, it is important to note several assumptions and caveats associated with the comparison of these studies/models. First, a number of studies either did not report RMSE explicitly or provided measurements in units other than $\mathrm{W/m^2}$. Models that did not have MSE/RMSE calculations within the respective article were excluded, as there was generally not enough additional information within these articles to calculate RMSE manually. Metrics such as mean absolute percentage error (MAPE), mean bias error (MBE), mean absolute error (MAE), and relative absolute error (RAE) occurred within some publications, but without enough of a consensus to warrant their use for comparing models. In addition, many articles did not specify explicit values for overall RMSE on the test set, but instead displayed error trend plots. In these cases, the error trend was averaged over time in order to determine a scalar value for RMSE for the model. Furthermore, some studies reported RMSE on a seasonal or monthly basis. The assumption was made that there was likely a uniform distribution of instances within

Table 1. A performance comparison between hourly solar radiation forecasting models prominent within the literature, measured as normalized root mean square error (NRMSE).

Authors	Model type	Max W/m² observed	RMSE W/m²	NRMSE (%)
Sfetsos and Coonick [29]	ANN trained with LM	1000	27.6	2.8
Spokas and Forcella [30]	Clear Sky/Direct Normal Irradiance model	1100	112	10.2
Cao and Lin [3]	Diagonal recurrent wavelet neural network (DRWNN)	558	13.2	2.4
Hocaoglu et al. [9]	ANN trained with 2D visual representation of time series and LM	600	34.57	5.8
Reikard [26]	ARIMA	1100	322.3	29.3
Perez et al. [25]	Cloud motion vectors derived from difference in consecutive GHI Index grids	1000	87.57	8.8
Marquez [20]	ANN w/input selection via genetic algorithm	1000	72	7.2
Izgi et al. [17]	ANN w/air temp., cell temp., and solar radiation as input	400	55	13.8
Pedro and Coimbra [24]	ANN w/parameter optimization via genetic algorithm	1000	131	13.1
Wang et al. [31]	ANN w/statistical feature parameters (ANN-SFP)	1200	63.5	5.3
Huang et al. [13]	Coupled autoregressive dynamical systems (CARDS)	1146	80.6	7
Benmouiza and Cheknane [2]	Hybrid k-means and NARX neural network model	950	64.3	6.8
Fidan et al. [5]	ANN w/periodic Fourier series coefficients as input	400	64.3	16.1

these subsets, such that the overall RMSE of the model could be calculated as the model's average RMSE on the respective subsets.

The central difficulty in calculating NRMSE for these models is the rarity in which authors explicitly provided values for y_{min} and y_{max}. However, y_{min} was assumed to be equal to zero, while the remaining y_{max} values for these respective models was either extracted from figures within the given publication or from other metrics therein provided. In addition, some publications reported NRMSE, but did not report how normalization was performed. Future publications could be greatly improved not only by providing RMSE calculations, but also the range of values for the target variable (i.e. y_{min} and y_{max}), the average target value ($\bar{y}$), and the normalization method/formula used, in addition to other performance metrics.

Table 1 appears to demonstrate that Sfetsos and Coonick's [29] feedforward neural network trained with the LM algorithm and Cao and Lin's [3] diagonal recurrent wavelet neural network are the top performing models, as they have achieved the lowest NRMSE, in comparison to the other models accounted for within the solar radiation forecasting literature. However, these models also highlight a further problem with comparing models between studies: there are sometimes significant differences in the size of the training and testing sets between models. Both Sfetsos & Coonick and Cao & Lin's models were tested on a dataset spanning a month or less, which suggests it is inappropriate to assume these models accurately predict solar radiation for the remaining months of the year, without further evaluation. In a similar vein, there is no sense in which solar radiation datasets across distinct studies can be treated as equivalent, as one data set may contain a significant amount of noise, due to recording equipment error or other factors, in comparison to another dataset. Therefore, in order to further ensure the validity of model comparison between studies, standard baseline models such as a persistence model, where y_{t+1} is predicted as equal to y_t, should be used for determining how the proposed model improves upon the baseline model. In sum, despite the aforementioned limitations, these models serve as a basis of comparison to illustrate the efficacy and precision of models developed within the current study.

2   Materials

Data from the Griffin, Georgia weather station from 2003–2014 were used to build the observations for the input layer to the neural network. Observations were collected at 15 min intervals over the duration of each year for a total of 35,040 observations per year (350,400 observations total). Forty-three data fields were observed, though only a subset of these fields was used for solar radiation prediction: year, day of year, time of day, air temperature (°C), humidity (%), dew point (°C), vapor pressure (kPa), barometric pressure (kPa), wind speed (m/s), solar radiation (W/m^2), total solar radiation (KJ/m^2), photosynthetically active radiation ($umole/m^2s$), and rainfall (mm). As subsequent tests of the model demonstrated that these fields did not bolster forecasting performance, only solar radiation and its respective lag were used as inputs within the models of this study.

3   Methods

The fields used for prediction of future solar radiation values were first extracted from the raw measurement files from the Griffin station. For each value of the extracted fields at time step t (with the exception of year, day, and time), values from the n previous time steps $(t_{-1}, t_{-2}, t_{-3}, \ldots, t_{-n})$ were added to the observation file that would serve as input into the input layer of the neural network. This is known as the sliding window technique and has been shown to significantly increase the accuracy of time series predictions with neural networks [23].

Trials to determine the optimal network architecture, input fields, and hyper-parameter configuration began with a standard multi-layer perceptron (MLP) network with 57 hidden nodes and 1 output node that provided the prediction of solar radiation one hour into the future. The number of nodes in the input layer of each network was determined by Eq. (4):

$$\# \text{ input nodes} = n + 1 + \text{the number of non-solar radiation input fields used} \quad (4)$$

In subsequent trials, the number of hidden nodes was adjusted within a range of 17–257 nodes. The initial trials used air temperature, humidity, dew point, barometric pressure, wind speed, solar radiation, total solar radiation, photosynthetically active radiation, and rainfall as the input fields into the neural network. Combinations of these fields were then implemented, in order to determine which fields consistently provided for predictions with the lowest MSE. In later trials, only solar radiation at t and prior solar radiation values from t_{-1} to t_{-n} were used as input, as other input fields did not appear to positively influence the models' performance. The activation function for this network type is formalized as shown in (5):

$$a_j^l = \sigma\left(\sum_k w_{jk}^l a_k^{l-1} + b_j^l\right) \quad (5)$$

In Eq. (5), a_j^l is the activation of the jth neuron in the lth layer, as determined by activation of neurons within the $(l_{-1})^{\text{th}}$ layer [5]. The sum $\sum_k w_{jk}^l a_k^{l-1}$ is the total activation of each neuron k in the $(l_{-1})^{\text{th}}$ multiplied by their respective weighted connections to each neuron j in the lth layer. The term b_j^l is the bias value of neuron j in the lth layer. For each training instance d within the training set D_{train}, where $d \in D_{train}$, the weights of the network were updated through the backpropagation algorithm as shown in Eq. (6) [8].

$$\Delta w_{(t)} = \in \frac{\partial E}{\partial w_{(t)}} + \alpha \Delta w_{(t-1)} \quad (6)$$

Thus, the change of a given weight at iteration t $(\Delta w_{(t)})$ is equal to the product of the learning rate $(\in)$ and the gradient $\left(\frac{\partial E}{\partial w_{(t)}}\right)$, in addition to the product of momentum (α) and the change of that weight at the previous time step $(\Delta w_{(t-1)})$.

In subsequent trials, the backpropagation algorithm was replaced with the resilient propagation algorithm (iRPROP+) [14, 15]. The RPROP and backpropagation algorithms are similar in that gradients must be calculated for each weight, however the gradient used in RPROP is the inverse of the gradient used in backpropagation and the RPROP gradients are utilized such that specifying a learning rate and momentum is not required [8, 27]. First, the gradient of the current iteration is compared with the gradient of the previous iteration. The calculation of sign change is shown in Eq. (7):

$$c = \frac{\partial E^{(t)}}{\partial w_{ij}} \cdot \frac{\partial E^{(t-1)}}{\partial w_{ij}} \tag{7}$$

The value of c is then used to determine the weight change, such that c is greater than 0, than the weight change is equal to the negative of the weight update value and positive if c is less than 0. Otherwise, no change is made to the weight. The update value for weight w_{ij} is shown in Eq. (8):

$$\Delta_{ij}^{(t)} = \begin{cases} \eta^+ \cdot \Delta_{ij}^{(t-1)}, & \text{if } c > 0 \\ \eta^- \cdot \Delta_{ij}^{(t-1)}, & \text{if } c < 0 \\ \Delta_{ij}^{(t-1)}, & \text{otherwise} \end{cases} \tag{8}$$

Within Eq. (8), η^- and η^+ are constant parameters specified prior to training, typically with the values 0.5 and 1.2, respectively. As only the sign of the gradient is used to determine the weight update value, rather than considering the gradient itself as in backpropagation, RPROP is able to train much faster than backpropagation.

A model averaged neural network (MANN) was constructed by initializing a number of MLPs with identical architectures, but distinct weight vectors, and training these networks in parallel using the resilient propagation algorithm. The MANN was first formalized by Ripley [28]. In effect, a MANN is a voting ensemble of MLPs. The implementation of the algorithm is straight-forward: first, N neural networks are initialized with the same architecture, and for each weight vector w_{ij} between layers i and j of network n, the weights are randomized using a different seed than what was used to initialize the weight vectors of the other networks. Each of the networks is trained in parallel until the minimum gradient for each network is reached. Within the present study, training was configured to continue while 1. The training error was greater than the specified maximum acceptable error, 2. The number of epochs was less than the specified maximum number of epochs, and 3. The testing error continued to decrease from the previous testing epoch. A testing iteration was performed every 100 training iterations to ensure the network did not overfit. If any of these three conditions were violated, training of that network was halted. Once the training of each network was complete, the networks were tested as a single MANN. The error for testing iterations was calculated by averaging the predictions of each network and then finding the difference between this average predicted value and the actual value of the output for the given testing instance. After the training of each network is complete, predictions from the MANN were simply calculated as the average of the predicted output from each network for a given instance in the data.

It is critical to note that weights of the networks comprising the MANN implemented in the current study are not updated based on the error of the MANN's predicted output value, but rather based on each network's individual error. Were the weights of each network updated based on the error of the MANN's predicted output value, the weight updates of each network would likely not shift the error function of the MANN's predictions towards a global or local minima, as the error of the MANN may differ substantially from an individual network comprising it. For instance, one of the networks comprising the MANN may be predicting values lower than the actual values, while another network within the MANN may be providing higher than the actual values. Though this example is somewhat of a simplification, it demonstrates the need to update networks within the MANN based on their individual error.

In addition to MANN, generalized linear models (GLM), least median squares (LMS), random forests (RF), and alternating model trees (AMT) were implemented. Like the single MLP and MANN implemented, variations in the lag input into the respective models was varied to minimize prediction error. To further optimize model performance, the number of trees used in the RF models, and the shrinkage (λ) and the number of iterations (i) used in the AMTs were varied.

4 Results

The MANN architecture consistently outperformed single neural network models in predicting hourly solar radiation, as shown in Table 1. However, the number of networks within the MANN did not appear to have a significant effect upon the RMSE of predictions when forecasts of two or more networks were averaged. In fact, prediction accuracy marginally decreased when the number of networks comprising the MANN exceeded five. One possible explanation for this phenomenon is that larger MANN may overfit the training data, such that their performance on the testing data is not improved in comparison to MANN comprised of fewer networks. This explanation is supported by observations of MANN comprised of six or more networks forecasting with a scaled MSE less than 0.0018 (< 50.91 W/m^2) on training data, while MANN comprised of 5 or less networks forecasting with a scaled MSE greater than 0.0020 (> 53.67 W/m^2) on training data (Fig. 3).

Two unexpected observations can be made about the non-neural network models: First, the generalized linear model (GLM) significantly outperformed the MLP with same amount of lag, despite the non-linear nature of the data (see Table 2). To understand this discrepancy in expectation, it is important to note that the GLM implemented ridge regression according to Eq. (9):

$$\beta_\gamma = \left(Z^T Z + \gamma I_p\right)^{-1} Z^T y \qquad (9)$$

Within Eq. (9) $\beta\gamma$ is the vector of coefficients of the model (i.e. the weight vector), Z is the standardized (i.e. zero mean, unit variance) training data input matrix, y is the centered training data output vector, and γ is the tuning parameter [10]. The tuning parameter thus regularizes $\beta\gamma$ (i.e. prevents $\beta\gamma$ from growing too large), such that the variance of $\beta\gamma$ is reduced while some bias is introduced, therefore reducing overall prediction error.

The second observation is the high prediction accuracy of the AMT models in comparison to both the neural network and non-neural network models. The AMT with lag = 64, the number of boosting iterations (i) = 20, and shrinkage (λ) = 1.0 attained a marginally lower RMSE $(62.7\,\text{W/m}^2)$ than the MLP and MANN models, which

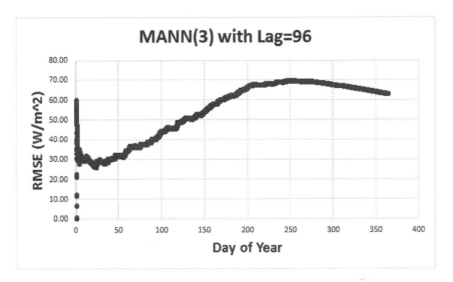

Fig. 3. RMSE (Wm^{-2}) of the MANN(3) model with a lag of 96 on the Griffin 2013 data set. The model achieves an RMSE of 26 Wm^{-2} on Day 24, which equates to an NRMSE of 2.2 %. In addition, the average RMSE for Janurary is 29.8 Wm^{-2}, which equates to an NRMSE of 2.5 %. The performance of the MANN is thus competive with the top models reviewed: Sfetsos and Coonick [29] and Cao and Lin [3].

Table 2. A comparison of the best performance of single neural networks (MLP) and model averaged neural networks (MANN) with varied amounts of lag with respect to solar radiation. The results shown are the observations taken from a given model for its best performance over five trials.

Model	Lag	Scaled MSE	RMSE (W/m^2)	NRMSE (%)
MLP	16	0.00434	79.05	6.59
MLP	96	0.00280	63.50	5.29
MANN(2)	96	0.00274	62.81	5.23
MANN(3)	96	0.00274	62.81	5.23
MANN(5)	96	0.00274	62.81	5.23
MANN(8)	96	0.00275	62.93	5.24
MANN(10)	96	0.00275	62.93	5.24

utilized a lag of up to 96 time steps. As AMT also tend to increase in prediction accuracy with a greater amount of lag used, it is possible that a lower RMSE could be attained by further exploring the parameter space defined by i and λ (Table 3).

Table 3. A comparison of the performance of generalized linear model (GLM), least median squares (LMS), random forest (RF), and alternating model trees (AMT) across varied amounts of lag. The bolded values indicate the best performance for a given error measure.

Model	Lag	MAE	RRSE	DA	RAE	MSE	RMSE	NRMSE
LMS	12	31.6	105.7	41	103.4	5561.3	74.6	6.22
GLM	16	34.4	94.9	42.5	112.6	4482.3	66.9	5.58
RF 100 trees	16	29.9	92.9	63.3	98	4296.1	65.5	5.46
AMT $i = 10$ $\lambda = 1.0$	12	27.6	92	67.2	90.4	4209.9	64.9	5.41
AMT $i = 10$ $\lambda = 1.0$	16	26.6	91	60.6	87.1	4118.8	64.2	5.35
AMT $i = 20$ $\lambda = 1.0$	16	**25.9**	89.8	60.3	**84.9**	4014.4	63.4	5.28
AMT $i = 10$ $\lambda = 1.0$	24	28.1	90.5	54.2	92.1	4082	63.9	5.33
AMT $i = 20$ $\lambda = 1.0$	32	26.4	89.1	**70.1**	86.5	3955.5	62.9	5.24
AMT $i = 30$ $\lambda = 0.1$	32	33.6	92.7	49.4	110	4281.4	65.4	5.45
AMT $i = 30$ $\lambda = 1.0$	32	26	**88.8**	69.6	85.2	3929.5	**62.7**	**5.23**
AMT $i = 20$ $\lambda = 1.0$	64	26.2	**88.8**	51	85.6	3928.4	**62.7**	**5.23**

5   Conclusion

The model averaged neural network (MANN) and the alternating model tree (AMT) models were shown to produce the most accurate hourly predictions of solar radiation of the models assessed within the current study. Furthermore, MANN and AMT outperform or are competitive with the leading models of hourly solar radiation forecasting within the literature. The increase in performance of the MANN in comparison to single neural networks is likely due to the capacity for a MANN to explore more of the weight space for units within its hidden layers, as it is composed of multiple networks with different initial weights but identical architectures. However, aggregating the predictions of more than two networks did not appear to improve prediction performance (i.e. reduce prediction error), though it is possible these networks were overfit due to their significantly lower prediction error on the training data. In order to further bolster prediction accuracy in future studies, a more rigorous stopping strategy should be implemented, where a separate weight vector of a given network is saved for t time steps, such that if training is permitted to continue for t time steps after the first instance of the testing error increasing, the weights of the network can be reverted to their values t time steps back from the current time step. Prediction accuracy may be further improved by using a greater lag for both MANN and AMT models, as well as by aggregating these models into an ensemble model. As decision trees and neural networks have radically different methods for attaining their model, it is likely the aggregation of their predictions will help to alleviate the shortcomings of both model types [32].

References

1. Asmussen, S.R.: Markov chains. In: Applied Probability and Queues. Stochastic Modelling and Applied Probability, vol. 51, pp. 3–8 (2003)
2. Benmouiza, K., Cheknane, A.: Forecasting hourly global solar radiation using hybrid k-means and nonlinear autoregressive neural network models. Energy Convers. Manag. **75**, 561–569 (2013)
3. Cao, J., Lin, X.: Study of hourly and daily solar irradiation forecast using diagonal recurrent wavelet neural networks. Energy Convers. Manag. **49**(6), 1396–1406 (2008)
4. Elizondo, D., Hoogenboom, G., McClendon, R.W.: Development of a neural network model to predict daily solar radiation. Agric. For. Meteorol. **71**(1), 115–132 (1994)
5. Fidan, M., Hocaoğlu, F.O., Gerek, Ö.N.: Harmonic analysis based hourly solar radiation forecasting model. IET Renew. Power Gener. **9**(3), 218–227 (2014)
6. Frank, E., Mayo, M., Kramer, S.: Alternating model trees. In: Proceedings of the 30th Annual ACM Symposium on Applied Computing, pp. 871–878 (2015)
7. Hamilton, C., Potter, W., Hoogenboom, G., McClendon, R., Hobbs, W.: Solar radiation time series prediction. Int. J. Comput. Control Quant. Inf. Eng. **9**(5), 656–661 (2015)
8. Heaton, J.: Introduction to the math of neural networks. Heaton Research, Inc. (2011)
9. Hocaoğlu, F.O., Gerek, Ö.N., Kurban, M.: Hourly solar radiation forecasting using optimal coefficient 2-D linear filters and feed-forward neural networks. Solar Energy **82**(8), 714–726 (2008)
10. Hoerl, A.E., Kennard, R.W.: Ridge regression: biased estimation for nonorthogonal problems. Technometrics **12**(1), 55–67 (1970)
11. Hoff, T.E., Perez, R., Kleissl, J., Renne, D., Stein, J.: Reporting of irradiance modeling relative prediction errors. Prog. Photovoltaics: Res. Appl. **21**(7), 1514–1519 (2013)
12. Horrigan, L., Lawrence, R.S., Walker, P.: How sustainable agriculture can address the environmental and human health harms of industrial agriculture. Environ. Health Perspect. **110**, 445–456 (2002)
13. Huang, J., Korolkiewicz, M., Agrawal, M., Boland, J.: Forecasting solar radiation on an hourly time scale using a Coupled AutoRegressive and Dynamical System (CARDS) model. Solar Energy **87**, 136–149 (2013)
14. Igel, C., Hüsken, M.: Improving the Rprop learning algorithm. In: Second International Symposium on Neural Computation (NC 2000), pp. 115–121 (2000)
15. Igel, C., Hüsken, M.: Empirical evaluation of the improved Rprop learning algorithm. Neurocomputing **50**, 105–123 (2003)

16. Intergovernmental Panel on Climate Change. Climate Change 2014: Impacts, Adaptation, and Vulnerability. Geneva, Switzerland (2015)

17. Izgi, E., Öztopal, A., Yerli, B., Kaymak, M.K., Şahin, A.D.: Short–mid-term solar power prediction by using artificial neural networks. Solar Energy 86(2), 725–733 (2012)

18. Ji, W., Chee, K.C.: Prediction of hourly solar radiation using a novel hybrid model of ARMA and TDNN. Solar Energy 85(5), 808–817 (2011)

19. Levenberg, K.: A method for the solution of certain non–linear problems in least squares. Q. Appl. Math. 2, 164–168 (1944)

20. Marquez, R., Coimbra, C.F.: Forecasting of global and direct solar irradiance using stochastic learning methods, ground experiments and the NWS database. Solar Energy 85(5), 746–756 (2011)

21. Marquardt, D.W.: An algorithm for least-squares estimation of nonlinear parameters. J. Soc. Ind. Appl. Math. 11(2), 431–441 (1963)

22. Mellit, A., Menghanem, M., Bendekhis, M.: Artificial neural network model for prediction solar radiation data: application for sizing stand-alone photovoltaic power system. In: Power Engineering Society General Meeting, pp. 40–44, IEEE, June 2005

23. Paoli, C., Voyant, C., Muselli, M., Nivet, M.L.: Forecasting of preprocessed daily solar radiation time series using neural networks. Solar Energy 84(12), 2146–2160 (2010)

24. Pedro, H.T., Coimbra, C.F.: Assessment of forecasting techniques for solar power production with no exogenous inputs. Solar Energy 86(7), 2017–2028 (2012)

25. Perez, R., Kivalov, S., Schlemmer, J., Hemker, K., Renné, D., Hoff, T.E.: Validation of short and medium term operational solar radiation forecasts in the US. Solar Energy 84(12), 2161–2172 (2010)

26. Reikard, G.: Predicting solar radiation at high resolutions: a comparison of time series forecasts. Solar Energy 83(3), 342–349 (2009)

27. Riedmiller, M.: Rprop-Description and implementation details: technical report. Inst. f. Logik, Komplexität u. Deduktionssysteme (1994)

28. Ripley, B.: Pattern Recognition and Neural Networks. Cambridge University Press, New York (1996)

29. Sfetsos, A., Coonick, A.H.: Univariate and multivariate forecasting of hourly solar radiation with artificial intelligence techniques. Solar Energy 68(2), 169–178 (2000)

30. Spokas, K., Forcella, F.: Estimating hourly incoming solar radiation from limited meteorological data. Weed Sci. 54(1), 182–189 (2006)

31. Wang, F., Mi, Z., Su, S., Zhao, H.: Short-term solar irradiance forecasting model based on artificial neural network using statistical feature parameters. Energies 5(5), 1355–1370 (2012)

32. Witten, I., Frank, E., Hall, M.: Data Mining: Practical Machine Learning Tools and Techniques. Morgan Kaufmann, Los Altos (2011)

33. World Population Prospects, the 2012 Revision (2012). http://esa.un.org/wpp/. Accessed 5 April 2015

Adaptive Control

Fully Automated Learning for Position and Contact Force of Manipulated Object with Wired Flexible Finger Joints

Kanta Watanabe[1], Shun Nishide[2], Manabu Gouko[3], and Chyon Hae Kim[1(✉)]

[1] Faculty of Engineering, Department of Electrical Engineering
and Computer Science, Iwate University, Iwate, Japan
`tenkai@iwate-u.ac.jp`

[2] Institute of Technology and Science, Tokushima University, Tokushima, Japan
`nishide@is.tokushima-u.ac.jp`

[3] Faculty of Engineering, Department of Mechanical Engineering
and Intelligent Systems, Tohoku Gakuin University, Sendai, Japan
`gouko@mail.tohoku-gakuin.ac.jp`

Abstract. We discuss about the modeling technology in the object manipulation of the robot arm that is equipped with flexible finger joints. In recent years, flexible robot fingers are getting attention because of their handling capability and safety. However, the position and contact force of manipulated object take much non-linear uncertainty from the flexibility. In this paper, we propose the modeling framework of the position and contact force of the manipulated object. The proposed framework is an online learning method that is composed of motor babbling, dynamics learning tree (DLT), and ϵ-greedy method. In the experiments, the effectiveness of DLT was compared with neural network (NN), the effectiveness of the proposed framework was validated using a drawing task of a humanoid robot that equipped with flexible finger joints. The proposed framework was able to realize a fully automated incremental-manipulation-learning.

Keywords: Learning · Robot arm · Manipulation

1 Introduction

For the robots that work nearby humans, 'soft' functions are important [1]. One of the means of 'soft' functions is softness of the body structures that do not injure the people around a robot. Another is software that gives dexterous manipulation.

After Sugano *et al.* presented a keyboard performance robot with wired flexible fingers [1], many types of flexible fingers have been developed. For example, Kajikawa *et al.* developed a flexible finger joint of human care robot [2]. Yoshimi *et al.* developed a two-fingered parallel soft gripper [3]. However, formulating the control rules of the manipulated object is more difficult than the case of rigid fingers, because of the uncertainty in the soft mechanism.

© Springer International Publishing Switzerland 2016
H. Fujita et al. (Eds.): IEA/AIE 2016, LNAI 9799, pp. 753–767, 2016.
DOI: 10.1007/978-3-319-42007-3_64

Sugaiwa *et al.* proposed a heuristic methodology for picking up objects using a robot hand with flexible finger joints, soft skin, and tactile sensors. However the applicability of this method depends on the shapes of manipulated objects [4].

Formulating the deformation specifications of soft material and flexible mechanizm mathematically may be a way to design a commonly usable manipulation method. Doulgeri *et al.* proposed an asymptotic stability controller that minimizes force and position errors in order to handle the uncertainty around soft finger tip [5]. Kim *et al.* formulated the dynamic deformation of soft fingertips [6]. Yoshida *et al.* formulated the dynamics of pinch motion of a pair of 2–3 degrees of freedom (DOF) fingers [7]. However, obtaining mathematical equations for the other than round finger tips has been difficult.

Learning technique has much potential to solve the uncertainty around the soft material and flexible mechanizm. Funabashi *et al.* demonstrated in-hand-object-manipulations using a deep neural network with the training data that was captured from a data glove [8]. However, this learning requires much effort of manual settings. The first is the trajectory generation using a data glove. The developer must make much training data through the control of the robot fingers using a data glove. The second is the management of the training data. The developer must estimate the sufficiency of the amount of training data. If the estimation is not correct and the learning is done by less amount of data, another learning must be rerun using larger number of data. When the amount of training data is large, this rerunning cost so much time. The third is the management of the learning parameters of neural network. In order to assure correct learning, the developer needs to find a good learning parameters (number of nodes, learning rate, and so on) by his/her intuition. If the parameters are not correct, the learning is easily collapsed. We need a fully automated system that learns the uncertainty around flexible fingers.

In this paper, we propose a learning framework that is composed of motor babbling, dynamics learning tree (DLT), and ϵ-greedy method. Motor babbling is an automated data sampling method that is on the basis of infant's learning. DLT is an online learning system that was developed by the authors. Online learning realizes a learning that is free from manual decision of sampling data amount. The parameters of the framework are decidable quite easily by a common sense. For example, the depth of DLT is decidable with the required resolution of the input data, we can choose any value of ϵ because it does not collapse the learning. This framework increases its knowledge about the uncertainty without any help of humans after starting obtaining training data. ϵ-greedy method is optionally used in order to assure the efficiency of this framework. It realized efficient learning with smaller number of training data.

2   Proposed Framework

We propose a learning framework in Fig. 1 against the previous framework in Fig. 2. The proposed framework is composed of motor babbling, DLT, and ϵ-greedy method. Motor babbling is employed for the learning through random

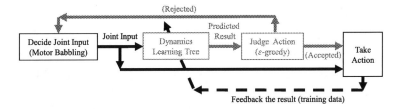

Fig. 1. Proposed framework. In proposed framework, learning is done simultaneously with obtaining training data. "Generation of joint input" is done by using motor babbling. In order to improve the efficiency, the action result is predicted using DLT before taking action. When the prediction satisfies the constraint that is defined in the judgment process, the action is taken. After taking the action, the learning system is updated with an online process.

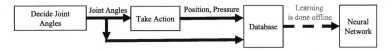

Fig. 2. Previous framework. In previous work [8,16], joint angles are decided by a data grove and motor babbling respectively. The action results of the joint angles are learned by the offline learning process of neural network. This framework suffers from the above mentioned three problems in Introduction.

joint input. We developed DLT in order to implement the learning system of Fig. 1. ϵ-greedy method is used for the judgment of action in Fig. 1.

2.1 Motor Babbling

In recent years, motor babbling is actively researched in the field of constructive research for the cognitive development, which focuses on the of the recognition and behavior architectures in humans [9]. Motor babbling is the motions through that infants learn their body dynamics models. In that, infants try to sample the relationship between their joint input and resulting motion. Schillaci et al. modeled the dynamics of a robot arm [10]. Saegusa et al. developed a humanoid robot that learns reaching motions through motor babbling [11]. Dawood et al. and Grimes et al. developed robots that imitate the motion of others through motor babbling [12,13]. Especially, Mochizuki et al. developed a humanoid robot that learns the dynamics of its arm and a grabbed pen using Tani's neural network model [14,15]. The performance of this model was improved by applying stopping/pausing sequences [16]. This theory is supported by several researches [19,20].

In this paper, motor babbling was applied to the object manipulation task using flexible finger joints as in [16].

2.2 Dynamics Learning Tree (DLT)

For the proposed framework, we developed an online learning algorithm named dynamics learning tree[1]. Neural network is not suitable to make a fully automated learning system, because of the problems of the amount of training data and parameter settings.

Also, extension of neural network to online learning includes the following problem. Neural network with three or more layered structure must use back propagation method in order to avoid catastrophic forgetting problem [18]. Back propagation method assures the avoidance only when it is done by offline learning. On the other hand, neural network with less than two layers is not able to learn arbitrary I/O map. This limits the learning capability and causes unstable learning.

DLT is able to learn new training data with the same learning weight as that of previously learned data one by one. Its online incremental learning is supported by the theory of statistics. Thus, it is not suffered from catastrophic forgetting.

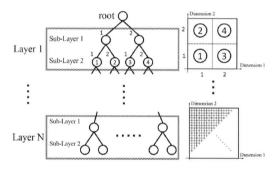

Fig. 3. DLT (N Layer 2 Dim. 2-ary). The tree type data structure of DLT (left) represents the division of input space (right). In the Layer 1, assigned numbers 1–4 in the tree correspond the numbers 1–4 of the divisions of the input space. In this figure, all of the branches are illustrated, although these branches are incrementally created in the online learning process.

DLT is a tree typed multi-layered learning system. Figure 3 shows the example of DLT with N layer 2 sub-layers (dimensions) 2-ary tree. DLT's root node corresponds to the whole region of n dimensional input space. Each main layer has n sub-layers (dimensions) with d-ary. The leaf nodes represent the divided sub-space as numbers 1–4 in Fig. 3.

DLT learns the relationship of continuous I/O functions. The example of its learning is shown in Fig. 4. When input data is given as the circle, a sub-space is created according to the input, so that the input data is in the sub-space.

---

[1] The same algorithm is used in another proposition [17] that is presented in this conference.

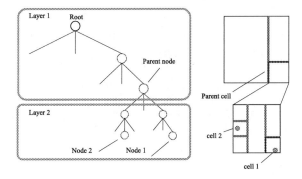

Fig. 4. Input space and DLT

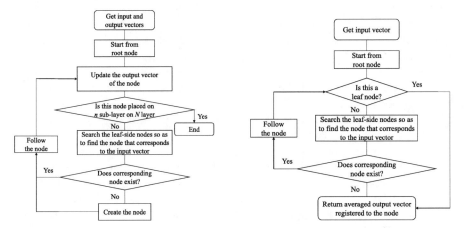

Fig. 5. Learning flow of DLT   **Fig. 6.** Prediction flow of DLT

According to the position-on-the-tree of the node that represents the sub-space, a sequence of nodes are created from Root. Resulting tree of DLT is shown in Fig. 4 upper left. In these figures Cells 1 and 2 are corresponding to Nodes 1 and 2 respectively. In every node of DLT, an average output vector was retained using following update functions:

$$\hat{O}_{Cell\ n} \leftarrow (N_{Cell\ n} \times \hat{O}_{Cell\ n} + O)/(N_{Cell\ n} + 1) \tag{1}$$

$$N_{Cell\ n} \leftarrow N_{Cell\ n} + 1 \tag{2}$$

where $N_{Cell\ n}$ is the learned number of output data in Cell n, $\hat{O}_{Cell\ n}$ is the average output vector learned by Cell n, O is training output vector. DLT cancels Gaussian noise around the training output vectors by the averaging process. These equations are characterized by online incremental update processes statistically. Also, its averaging process assures the same weight of learning between previously and currently learned training data. This update function is applied to a sequence of nodes from Root to the leaf node that corresponds to training output vector O. Thus, all of the nodes retain the average output vectors.

In the prediction process of DLT, DLT returns the averaged output vector of the node that represents the partition of the input space in that input data is placed. Thus, when the number of learned training input data is sparse around the new input data, DLT's output is calculated using shallow nodes in the tree. On the other hand, when it is dense, deep nodes are used. In intuitive understanding, DLT is controlling its complementation method against a new input according to the density of the training input data. The flows of the learning and prediction processes are shown in Figs. 5 and 6.

2.3 ϵ-Greedy Method

In order to learn object manipulation tasks effectively, the constraint of manipulated object is important. We implemented the constraint as an initial setting of the learning system using ϵ-greedy method. Implementing the constraint with motor babbling has been difficult, because of the randomness of the joint angle generation and the uncertainty of the state of the manipulated object. Even if joint angles are decided to fixed values, the object may be out of the constraint because of the uncertainty of the flexible fingers. To overcome this problem, we applied ϵ-greedy method.

In the field of reinforcement learning, exploration and exploitation are frequently discussed [22]. However, its implementation and effectiveness in the online learning with motor babbling have not been researched.

In the proposed framework, we assign exploration and exploitation with probability ϵ. In Exploration mode ($\epsilon = 1$), the proposed framework accepts all of the joint angles that are generated from the motor babbling process. This results in the free exploration in the input space. In Exploitation mode ($\epsilon = 0$), the joint angle where the learning system predicted an inhibited state of the manipulated object was rejected. This function realizes efficient and safe data sampling, because the manipulator is able to avoid not required sampling and risky state in that the object may collide to another object intensively. At a glance, using only Exploitation mode may look like better than Exploration mode. However, it makes sticky actions around predictable object states, because it rejects the input that does not result in a good prediction. Thus, we combined Exploration and Exploitation modes with a probability.

3   Experiment

We conducted three experiments. In Experiment 1, the learning capability of DLT was compared with NN for the validation. In Experiment 2, the three modes (Exploration, Exploitation, and ϵ-greedy modes) are compared in order to validate the effectiveness of the proposed framework. In Experiment 3, using trained DLT, drawing task was performed.

3.1 Settings

Experiment 1. We validated the learning capability of DLT with simple-harmonic-motion (SHM) learning task. For SHM, we used the following equation.

$$\ddot{q} = -q \tag{3}$$

DLT with 6 layer 2 dim. 3-ary was employed. NN with three layers and back propagation method was employed for the comparison. For both learning systems, their input and output are set at $(q, \dot{q})$ and $\ddot{q}$. For NN, the learning rate and the number of middle layer nodes are preliminary optimized before the comparison. We selected the best of them according to the error of 1000000-step learning.

The learning of DLT was done by online incremental learning, in which training data is leaned one by one. The learning of NN was done by batch learning, in which 2000 training data was simultaneously learned in a learning step.

Experiment 2. We validated the proposed babbling methods using 5 joints of the right arm of a humanoid robot NAO (Fig. 7) in a drawing task. The hand of NAO was composed of wired flexible finger joints with 1 DOF (open and close). Using a pen tablet, we obtained the position and pressure of the pen that was grabbed by NAO.

We obtained pen pressure (force) values from the pen tablet in the range of $[0, 1]$ (0: without pressure). On the pen tablet, 1 pixel is about 0.25 [mm].

DLT with 6 layer 5 dim. 3-ary was employed. The input and output of DLT were set at 5 joint angles and pen tip information respectively. The pen tip information is composed of x, y coordinate and pen tip pressure. We made two categories for the pen tip data in order to implement the judgment function of the proposed framework. The first was effective data that was sampled when pen tip was on the tablet. The second was ineffective data that was sampled out of the tablet. When 'pen tip position was $y > 500$' or 'pen tip pressure was 0 or 1', the input was rejected. The data with 1 of pen pressure was rejected

Fig. 7. Humanoid robot NAO and pen tablet

exceptionally, because pen was too strongly fitting on the tablet in this state. Also, for the initialization of DLT of the proposed framework, 100 data was sampled using previous exploration mode. We set $\epsilon = 0.5$.

Experiment 3. In order to compare the three modes. We tested the drawing capability of NAO.

The original figure was given by the experimenters. We generated the data of the original figure using the technique of stopping/pausing sequences [16]. The predicted figures were given by the rehearsal of DLT that was trained in Experiment 2. The procedure of this rehearsal is as follows:

1. DLT got a target pen tip position from the original data.
2. DLT took 1000 sets of random joint angles in order to predict resulting pen tip state.
3. A set of joint angles that gives the most close pen tip position to the target was selected from the 1000 sets of predicted results.
4. Go to (1) again.

Figures were obtained from the actual drawing of NAO. In the drawing, the joint angles that were calculated by the rehearsal were used as target joint angles of the robot. In the actual implementation, the rehearsal and drawing are done simultaneously. Thus, the prediction is included in the online control process of the robot.

3.2   Results

Experiment 1. The learning results of DLT are shown in Fig. 8(1) and (2). After 100 data was learned (Fig. 8(1)), the answer and prediction were not close still. After 5000 data was learned (Fig. 8(2)), they were almost the same. Prediction errors of DLT and NN are plotted in Fig. 9. DLT converged its error much faster than NN.

In Table 1, the learning results of NN are listed. The best result of NN was compared with DLT in Table 2. From the result, DLT converged to 2.23 times smaller error and about 500 times smaller computational time than NN.

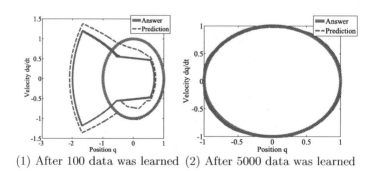

(1) After 100 data was learned (2) After 5000 data was learned

Fig. 8. Learning results of SHM (Color figure online)

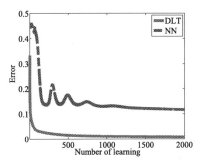

Fig. 9. Number of learning and prediction error (Color figure online)

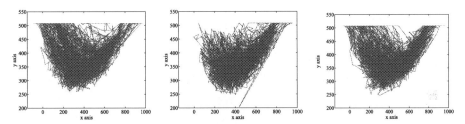

Fig. 10. Sampled pen tip positions of Exploration mode

Fig. 11. Sampled pen tip positions of Exploitation mode

Fig. 12. Sampled pen tip positions of ϵ-greedy mode

Experiment 2. The numbers of sampled effective and ineffective data are listed in Table 3. Exploitation mode and ϵ-greedy mode sampled a large number of effective data compared with Exploration mode.

The pen tip positions that were used for the sampling are plotted in Figs. 10, 11 and 12[2]. Exploitation mode made a bias about the sampling positions. The region of data distribution in Fig. 11 lacks upper left part compared with the others.

By using 1268 test data sampled by Exploration mode, the prediction errors of the three modes are validated. For the test data, the pen tip information with pen tip pressure in the range of (0,1) was used. The prediction errors are plotted in Figs. 13, 14 and 15. In this validation, ϵ-greedy mode converged fastest.

Experiment 3. Figure 16(1)–(3) show the errors between original-prediction, prediction-drawn, and drawn-original of the drawn figures of the three modes. Figures 16(4)–(6) show the original, predicted, and actually drawn figures of

---

[2] The humanoid robot NAO was placed on the top of the figures. It moved a pen using its right hand. The result of Exploitation mode has a bias compared with the others. This might be because Exploitation mode searches pen tip positions conservatively by exploiting the knowledge of DLT.

Table 1. Error and computing time of NN. The results show the values after each NN was trained with 2000 data of SHM while 1000000 steps. The best one is selected on Table 2, according to the errors.

Learning rate	Number of middle nodes	Average error	Computational time [ms]
0.0002	5	0.063083	630613
	10	0.10998	1094978
	20	0.109632	2026054
0.0005	5	0.0438948	652052
	10	0.105286	1124058
	20	0.109136	2079837
0.001	5	0.0420112	604970
	10	0.0439611	1125948
	20	0.108682	2042482
0.01	2	0.0429104	371751
	3	0.0182211	463264
	5	0.0572928	636039
0.02	3	0.322405	462179
	5	0.302984	630325
	10	0.315269	1096139

Table 2. Comparison of the best NN and DLT. The best NN of Table 1 was compared with DLT. The result of DLT is one step online learning for 2000 data. The result of NN is 1000000 steps offline learning for 2000 data.

	Average error	Computational time [ms]
DLT	0.0081701	921
Best NN	0.0182211	463264

Table 3. Rate of effective data (ED)

Mode	Sampled data	Effective data (ED)	Rate of ED [%]
Exploration mode	5000	4192	84
Exploitation mode	5000	4614	92
ϵ-greedy mode	5000	4574	91

them. Table 4 shows the average prediction errors that were calculated from the data of Figs. 16(1)–(3).

From Figs. 16(1)–(3), the error between original-prediction was not large, but the others. Even if the prediction was on the original figures, their actual results had large errors in some cases. Among the three modes, ϵ-greedy mode

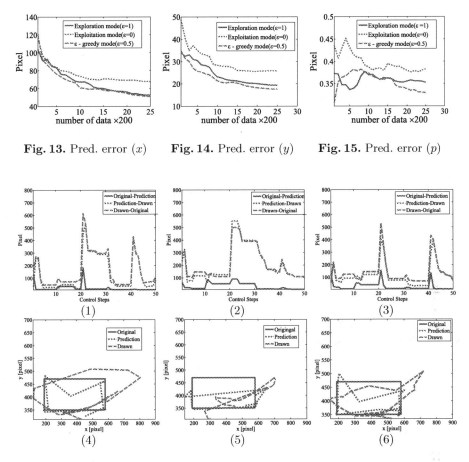

Fig. 13. Pred. error (x) **Fig. 14.** Pred. error (y) **Fig. 15.** Pred. error (p)

Fig. 16. Results of drawing. (1)–(3) show the errors of Exploration mode, Exploitation mode, and ϵ-greedy mode, respectively. (3)–(6) show the trajectories of these modes on the tablet.

Table 4. Average errors of drawing task

Mode	Original-prediction	Prediction-drawn	Drawn-original
Exploration	19.49	131.52	157.63
Exploitation	31.61	210.58	212.70
ϵ-greedy	23.00	114.19	137.01

showed smallest error between drawn-original. This means that the prediction of ϵ-greedy mode was more trustable than the others. As in Figs. 16(1)–(3), the drawn lines of ϵ-greedy mode was closest to the original.

4   Discussion

4.1   Required Manual Settings

The required manual settings of the proposed framework were much reduced from the previous framework. Compared with NN, DLT has less parameters while maintaining high learning capability. DLT requires only the setting of maximum depth (number of layers) that is related to the resolution of the input space, because the number of sub-layers is decided by the dimension of the input, the number of ary is fixable to 3 (our recommendation) without losing its capability. The online learning capability of DLT helps the management of training data too. The users of the proposed framework do not need to estimate the required amount of training data. Also, rerun of learning process is not required because DLT is able to learn new data incrementally while combining the data with previously learned data with the same learning weight.

4.2   Validation of DLT

From Experiments 1 and 2, DLT is applicable to the mapping of continuous input and output of dynamical systems. Also, DLT realized 2.23 times smaller error and about 1/500 calculation effort than NN. The reason of this result might be because DLT adaptively tunes its generalization according to given data samples using an adaptive tree. In the learning of NN, the volume of the input space region where a set of training I/O affects to the corresponding output is not governed with a theoretical back ground. On the other hand, DLT adaptively tunes the affection according to the density of data.

4.3   Validation of Exploitation Mode

From Table 3, Exploitation mode sampled the largest number of the effective data in that pen tip position is on the tablet. The number of ineffective data was about the half of that of Exploration mode. Exploitation mode successfully learned how to keep the constraint that was defined by the judgment process simultaneously with maintaining motor babbling.

4.4   Validation of ϵ-Greedy Mode

From the results of Experiments 2 and 3, the learning performance of ϵ-greedy mode always exceeded the other modes. Exploration mode does not sample data efficiently, because it does not keep constraint. Thus, it waists the number of sampling trials to sample not required data that is out of constraint. Exploitation mode samples data on the constraint. However, it sticks around the confidential pen tip positions that were already used for its training. ϵ-greedy mode relaxes the sticky tendency. When Exploitation mode is performed, pen tip position is almost on the constraint. Small exploration around the position results in a good exploration that almost keeps the constraint. Thus, in ϵ-greedy mode, explored region does not expand too far from the constraint like Exploration mode, and constraint does not too strong like Exploitation mode. This results in a good sampling for the new data that is sufficiently different from already learned one.

4.5   Planning Algorithm

The proposed framework might improve sampling based motion planning (SBMP, such as [23–26]). SBMP has a problem of simulation time, because it requires many number of simulations. The proposed framework, which is able to simulate dynamics using random access, reduces the simulation time drastically.

5   Conclusion and Future Work

In this paper, we proposed the framework of manipulation learning with a robot arm equipped with flexible fingers. From the validation results, DLT realized 2.23 times smaller error and about $1/500$ calculation effort than NN. Proposed framework with Exploitation mode successfully learned how to keep the constraint that was defined by the judgment process simultaneously with maintaining motor babbling. The learning performance of ϵ-greedy mode of the proposed framework always exceeded the other modes. Using ϵ-greedy mode, the drawing skill was improved more than the others.

The framework has much potential to be applied to a variety of tasks and robots. We will confirm the limitation of its applicability in the future.

Acknowledgement. This work was supported by JSPS KAKENHI Grant Numbers, 15K00363, 15K20850, 24119003, 25730159, and SCAT Technology Research Foundation.

References

1. Sugano, S., Kato, I.: WABOT-2: autonomous robot with dexterous finger-arm - finger-arm coordination control in keyboard performance. In: IEEE International Conference on Robotics and Automation (1987)
2. Kajikawa, S., Kon, H., Kikuchi, K., Satoh, S.: Development of a soft finger-joint for human-care robot. In: IEEE International Conference on Mechatronics and Automation (2005)
3. Yoshimi, T., Iwata, N., Mizukawa, M., Ando, Y.: Picking up operation of thin objects by robot arm with two-fingered parallel soft gripper. In: IEEE International Workshop on Advanced Robotics and its Social Impacts (2012)
4. Sugaiwa, T., Fujii, G., Iwata, H., Sugano, S.: A methodology for setting grasping force for picking up an object with unknown weight, friction, and stiffness. In: IEEE/RAS International Conference on Humanoid Robots (2010)
5. Doulgeri, Z., Simeonidis, A., Arimoto, S.: A position/force control for a soft tip robot finger under kinematic uncertainties. In: IEEE International Conference on Robotics and Automation (2000)

6. Kim, B.H., Hirai, S.: Characterizing the dynamics deformation of soft fingertips and analyzing its significance for multi-fingered operations. In: IEEE International Conference on Robotics and Automation (2004)

7. Yoshida, M., Arimoto, S., Bae, J.H.: Blind grasp and manipulation of a rigid object by a pair of robot fingers with soft tips. In: IEEE International Conference on Robotics and Automation (2007)

8. Funabashi, S., Schmitz, A., Sato, T., Somlor, S., Sugano, S.: Robust in-hand manipulation of variously sized and shaped objects. In: IEEE/RSJ International Conference on Intelligent Robots and Systems (2015)

9. Asada, M., Dorman, K.M., Ishiguro, H., Kuniyoshi, Y.: Cognitive developmental robotics as a new paradigm for the design of humanoid robots. Robot. Auton. Syst. **33**, 185–193 (2001)

10. Schillaci, G., Hafner, V.V.: Random movement strategies in self-exploration for a humanoid robot. In: IEEE/ACM International Conference on Human Robot Interaction, pp. 245–246 (2011)

11. Saegusa, R., Metta, G., Sandini, G.: Active learning for multiple sensorimotor coordination based on state confidenc. In: IEEE/RSJ International Conference on Robotics and Systems, pp. 2598–2603 (2009)

12. Dawood, F., Loo, C.K.: Humanoid behavior learning through visuomotor association by self-imitation. In: SCIS and ISIS, pp. 922–929 (2014)

13. Grimes, D.B., Rao, R.P.N.: Learning actions through imitation and exploration: towards humanoid robots that learn from humans. In: Sendhoff, B., Körner, E., Sporns, O., Ritter, H., Doya, K. (eds.) Creating Brain-Like Intelligence. LNCS, vol. 5436, pp. 103–138. Springer, Heidelberg (2009)

14. Mochizuki, K., Nishide, S., Okuno, H., Ogata, T.: Developmental imitation learning of robot on drawing with neuro dynamical mode. In: Annual Conference of IPSJ (2013). (In Japanese)

15. Yamashita, Y., Tani, J.: Emergence of functional hierarchy in a multiple timescale neural network model: a humanoid robot experiment. PLoS Comput. Biol. **4**(11), e1000220 (2008)

16. Nishide, S., Mochizuki, K., Okuno, H.G., Ogata, T.: Insertion of pause in drawing from babbling for robot's developmental imitation learning. In: IEEE International Conference on Robotics and Automation, pp. 4785–4791 (2014)

17. Eto, K., Kobayashi, Y., Kim, C.H.: Vehicle dynamics modeling using FDA learning. In: International Conference on Industrial, Engineering and Other Applications of Applied Intelligent Systems (2016, in press)

18. French, R.M.: Catastrophic forgetting in connectionist networks. Trends Cogn. Sci. **3**(4), 128–135 (1999)

19. Nagai, Y., Asada, M., Hosoda, K.: Learning for joint attention helped by functional development. Adv. Robot. **20**(10), 1165–1181 (2006)

20. Nakaoka, S., Nakazawa, A. Yokoi, K., Hirukawa, H., Ikeuchi, K.: Generating whole body motions for a biped humanoid robot from captured human dances. In: IEEE International Conference on Robotics and Automation, pp. 905–3910 (2003)

21. Takahashi, K., Ogata, T., Yamada, H., Tjandra, H., Sugano, S.: Effective motion learning for a flexible-joint robot using motor babbling. In: IEEE International Conference on Robotics and Systems (2015)

22. Sutton, R.S., Barto, A.G.: Reinforcement Learning: An Introduction. MIT Press, Cambridge (1998)

23. Kim, C.H., Sugano, S.: Tree based trajectory optimization based on local linearity of continuous non-linear dynamics. IEEE Trans. Autom. Control **61**(9) (2016, in press)

24. Kim, C.H., Sugano, S.: Closed loop trajectory optimization based on reverse time tree. Int. J. Control Autom. Syst. (in press)
25. Kim, C.H., Yamazaki, S., Sugano, S.: Motion optimization using state-dispersion based phase space partitions. Multibody Syst. Dyn. **32**(2), 159–173 (2013)
26. Kim, C.H., Sugano, S.: A GPU parallel computing method for LPUSS. Adv. Robot. **27**(15), 1199–1207 (2013)

Vehicle Dynamics Modeling Using FAD Learning

Keigo Eto[1], Yuichi Kobayashi[2], and Chyon Hae Kim[1(✉)]

[1] Department of Electrical Engineering and Computer Science, Iwate University,
203 Room, 4 Bldg. East, 4-3-5, Ueda, Morioka-shi, Iwate, Japan
tenkai@iwate-u.ac.jp
[2] Department of Mechanical Engineering, Shizuoka University, Shizuoka, Japan
kobayashi.yuichi@shizuoka.ac.jp

Abstract. Highly precise vehicle dynamics modeling is indispensable for self-driving technology. We propose a model learning framework, which utilizes FAD (The abbreviation of the capital letters of free dynamics, actuator, and disturbance.) learning, motor babbling, and dynamics learning tree. In the proposed framework, modeling error was decreased compared with conventional neural network approach. Also, this framework is applicable to online learning. In experiments, FAD learning and dynamics learning tree decreased learning error. The dynamics of a simulated car was learned using motor babbling. The proposed framework is applicable to a variety of mechanical systems.

Keywords: Learning · Dynamics · Automobile

1 Introduction

Self-driving technology is getting attention, because it might suppress many types of accidents caused by human error (e.g. inadequate distance between vehicles, traffic jam caused by insufficient driving skill, increase of environmental load caused by unnecessary acceleration and deceleration, loose driving and inattentive driving). In order to realize safe self-driving vehicle, precise modeling is important.

In previous research, Lee *et al.* proposed a path planning algorithm with a novel path representation for self-driving vehicle [1]. The algorithm was successfully implemented to KAIST Self-Driving Car. Li *et al.* proposed short term trajectory generation algorithm [2]. Kim *et al.* proposed parallel scheduling method for the physical attributes of a vehicle [3]. Kim *et al.* proposed linear prediction based uniform state sampling (LPUSS) that effectively find almost the optimal trajectory based on a vehicle model [4–7]. In order to use these trajectory planning algorithms, precise model of a vehicle is required. However, there is much error between the trajectory calculated by the used model and that of real machine in many cases.

In recent years, Bang developed a vehicle dynamics model by focusing on spring mass forces and moments [8]. Satar proposed a five degree of freedom (DOF) longitudinal model [9]. Setiawan proposed 14 DOF vehicle model [10].

© Springer International Publishing Switzerland 2016
H. Fujita et al. (Eds.): IEA/AIE 2016, LNAI 9799, pp. 768–781, 2016.
DOI: 10.1007/978-3-319-42007-3_65

However, these mathematical models have also a gap from a real vehicle. Aerodynamic affection, friction around wheels, specification of engine, and so on, are difficult to be determined precisely using mathematical equations.

Using real data for the modeling is a key technology to overcome the difficulty. Kabiraj *et al.* applied nonlinear tire data to their model [11]. Yim *et al.* developed a learning system on the basis of neural network, fuzzy logic, and evolutionally algorithm [12]. Tanizoe *et al.* realized the learning for the parking skill of driver [13]. In these methods, the vehicle models were learned from real data. However, the learning capability of the used algorithms has rarely discussed.

Even if real data was used for the learning, the precision of the model may not be improved sufficiently. The learning algorithm must have sufficient learning capability. This fact limits the applicability of the methods and scenario. In [11], real data was applied only to the limited dynamics of a vehicle (tires). In [13], the targeted dynamics is limited to low speed. The scenario is limited to parking. In [12], learnable vehicle dynamics is limited, because the used learning system has theoretical limitation about its representation capability. The employed online learning neural network has two layers, but neural network requires more than three layers in order to assure arbitrary mapping between the input and output. Increasing the number of layers is not easy, because of catastrophic forgetting problem [14]. In previous research, the learned dynamics are limited to a part of the dynamics of a vehicle. The applied learning algorithm may not sufficient to represent the whole dynamics of a vehicle.

In this paper, we propose a new learning framework on the basis of the following concept:

1. Extendable to whole dynamics learning
2. Assuring arbitrary nonlinear I/O map learning

For the proposed framework, FAD learning is formulated by considering the characteristic of vehicle dynamics. FAD learning decreases the amount of required learning data. FAD learning is applicable to many types of vehicle dynamics. A specific motor babbling method is developed in order to assure the learning. This motor babbling effectively samples trajectory data from a vehicle. Dynamics learning tree is developed in order to enlarge the online learning capability than that of neural network. Dynamics learning tree is able to represent arbitrary nonlinear I/O map.

In the experiment, dynamics learning tree showed better precision and learning time than neural network, motor babbling realized the learning of simulated vehicle dynamics, FAD learning required less training data than direct acceleration learning.

2   Proposed Framework

We propose a vehicle modeling framework that is composed of FAD learning, motor babbling, and dynamics learning tree.

2.1 FAD Learning

The acceleration around a vehicle is classified to three types. Free dynamics acceleration a_F includes inertia, gravity, friction, aerodynamic acceleration (without wind), and so on. These factors are always affecting the vehicle. Actuator acceleration a_A is generated from the power of engine and brake. Disturbance acceleration a_D is generated mainly from the wind, condition of the road, and contact force. Total acceleration a of a vehicle is defined as $a = a_F + a_A + a_D$.

The modeling target (whole dynamics of a vehicle) is considered as a_F and a_A, because a_D is an unpredictable term. In order to model a_F and a_A, we developed FAD learning in that a_F is modeled using the training data and equation $a_F = a$. While the instant when $a_A = a_D = 0$, a_A is modeled using equation $a_A = a - \hat{a}_F$. While the instant that $a_D = 0$, $\hat{a}_F$, which is the prediction of a_F, is obtained from the output of a dynamics learning tree that was trained the above process.

Modeling a_F using mathematical equations is difficult because its function changes according to the mass balance of equipment and aero-dynamical force which is a function of the form of a vehicle. With mathematical equations, a_F and a_A are defined as $a_F(q, \dot{q}) := -B(q, \dot{q})$ and $a_A(q, Q) := A^{-1}(q)Q$ from the following general motion equation of rigid body systems.

$$a = \ddot{q} := B(q, \dot{q}) + A^{-1}(q)Q + a_D, \tag{1}$$

where q is position (posture) of the rigid body system, Q is generalized joint force/torque, A is inertia matrix, B is bias vector. In many vehicles, modeling A and B is difficult because they are much complex nonlinear functions. In recent learning technology, A and B are approximated by function approximators like neural network (e.g. [12]).

2.2 Motor Babbling

We introduce motor babbling in order to sample data from a vehicle. In recent years, the development of body dynamics models through motor babbling is actively researched in the field of constructive research for the cognitive development, which focuses on the mechanism of the recognition and behavior architectures of humans [15]. The methodology of motor babbling is applicable to many types of dynamics and effective to sample training data that is related to whole dynamics of the targeted body.

Motor babbling is the random motions through which infants learn their body dynamics models. Infants learn and obtain the relationship between the input of their body joints and their body motions that occur as a result of the input. Schillaci *et al.* modeled the dynamics of a robot arm [16]. Saegusa *et al.* developed a humanoid robot that learns reaching motions through motor babbling [17]. Dawood *et al.* and Grimes *et al.* developed robots that imitate the motion of others through motor babbling [18,19]. Especially, Mochizuki *et al.* developed a humanoid robot that learns the dynamics of its arm and a grabbed pen using Tani's neural network model [20,21]. The performance of this model was

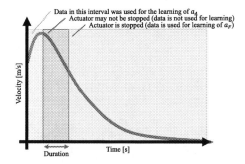

Fig. 1. Developed motor babbling method. For real machine, we may need an allowance before sampling free dynamics acceleration a_F, in order to stop the motors completely. (Color figure online)

improved by applying stopping/pausing sequences [22]. This theory is supported by some research [23,24].

We developed a new motor babbling method in order to effectively sample the data of a_F and a_A as in Fig. 1. In the typical motor babbling, random motor input was used for sampling the states of the body. However, it is not efficient for sampling a_F, because a_F is observable only when a_A and a_D equal to zero. In the motor babbling method in Fig. 1, a vehicle is accelerated with a random motor input at first, then the acceleration is stopped. After that, the vehicle is slowing down its velocity, and finally stops its motion. When we perform this kind of process under no disturbance ($a_D = 0$), we can sample two types of data. The first type of data is $d_1 := a_F + a_A$. That is observable while the acceleration is performed (the yellow region of Fig. 1). The second type of data is $d_2 := a_F$. That is observable while no acceleration is performed (the green region of Fig. 1). After motor command is stopped, there may be a small time delay before the actuator stops completely. In this interval, observed data is not just a_F. Also, actual motor input that causes a_A is uncertain in this interval (the red region of Fig. 1). Thus, we do not use the data that is sampled in this interval for the learning.

In order to get the data of a_A, we use the prediction of a learning system. When a_F is learned sufficiently, we can estimate a_A like $d_1 - \hat{a}_F$, where $\hat{a}_F$ is predicted a_F using the learning system.

2.3 Dynamics Learning Tree

In order to approximate a_F and a_A from sampled data, we developed dynamics learning tree (DLT)[1]. DLT is able to learn arbitrary I/O map using a tree structure.

---

[1] The same algorithm is used in another proposition [25] that is presented in this conference.

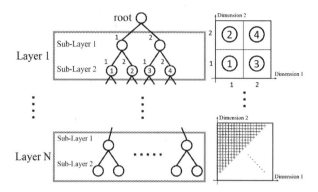

Fig. 2. Dynamics learning tree with N Layer 2 Dimension (Sub-layer) 2-ary tree. The tree structure of DLT (left) and the regions of the input space corresponded by the nodes at every layer (right). The numbers of leaf nodes at the first layer correspond to the numbers of the regions of the input space. This figure shows the tree structure of DLT when all of the branches are created, although the branches can be created one by one with an online-incremental learning process.

In typical vehicle dynamics learning, the input of DLT is $(q, \dot{q}, \tau)$, the output is a. When we apply FAD learning, the input and output of the first DLT are $(q, \dot{q})$ and a_F respectively. Those of the second DLT are (q, τ) and a_A respectively. DLT is a multi-layered learning system implemented by a tree structure. DLT's root node represents n dimensional input space. Every main layer of DLT is composed of n dimensions (sub-layers) with d-ary tree. At the first main layer, the leaf nodes represent the subspace of the input space of the root node. The size of the subspace is $1/d^n$ of the input space. DLT with N layer n dimension (sub-layer) d-ary tree is obtained by constructing N times of n dimension d-ary tree from every leaf node of main layer. Figure 2 shows the example of DLT with N layer 2 dimension 2-ary tree.

Figure 3 shows the example of the learning process. DLT with N layer n dimension d-ary tree divides the input space according to the input data, when it obtains a pair of input and output data. Figure 3 shows the example of 2 layer 2 dimension 3-ary tree, which maps the input data to the input space. When the input data illustrated by the circle in Fig. 3 (right) is obtained, all of the cells that include the input data are divided. As a result, DLT obtains the tree structure as in Fig. 3 (left). Cell 1 of Fig. 3 (right), which is the smallest cell that includes the input data, is represented by the Node 1 of Fig. 3 (left). Every node of DLT has an averaged output vector that was calculated from the output data. DLT updates the averaged output vector using following equations when new pair of input and output data is given.

$$\hat{O}_{Cell\ n} \leftarrow (N_{Cell\ n} \times \hat{O}_{Cell\ n} + O)/(N_{Cell\ n} + 1) \tag{2}$$
$$N_{Cell\ n} \leftarrow N_{Cell\ n} + 1, \tag{3}$$

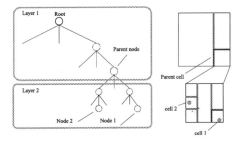

Fig. 3. Input space and DLT. The input space of the learning system is represented by a tree. The tree structure is incrementally constructed when new training I/O data is given. According to the density of data, DLT tunes the discretization of the input space.

where $N_{Cell\ n}$ is the number of output data that is learned by Cell n, $\hat{O}_{Cell\ n}$ is the average vector that is learned by Cell n, O is given output vector. DLT hypothesizes the noise with a Gauss distribution for the set of output data vectors obtained inside a cell. The median of the Gauss distribution is estimated by the average of the output data vectors. We can update the averaged output data when single output vector is given one by one. Thus, DLT is able to learn with online and incremental process.

DLT applies this update to multiple nodes when a pair of input and output data that should be learned is obtained. Actually, as in Fig. 3, when data is given in Cell 1, this update is applied to all of the nodes from Node 1 to Root. As same, when data is given in Cell 2, this update is applied to all of the nodes from Node 2 to Root.

When DLT predicts the corresponding output vector from an input vector of a continuous system. DLT searches on the tree from Root using the input vector as the key in order to find the bottom node that represents the sub-input-space that includes the input vector. DLT returns averaged output vector $\hat{O}_{Cell\ n}$ of this bottom node. Using this search process, DLT precisely predicts the output of a system around the sub-input-space where the density of learned data is large. Around the sub-input-space where the density of learned data is small, DLT generalizes the input-output pairs and predicts the output using the generalization.

The response of the prediction process of DLT is quite high, because it requires only the calculation cost for the tree search of d-ary tree. DLT realizes robust prediction for noisy data by using the nodes in the upper layer nodes instead of the bottom nodes, although this method is not applied to the experiments in this paper. The flows of the learning and prediction processes are shown in Fig. 4(1) and (2).

3   Experiment

We conducted several experiments in order to confirm the effectiveness of the framework.

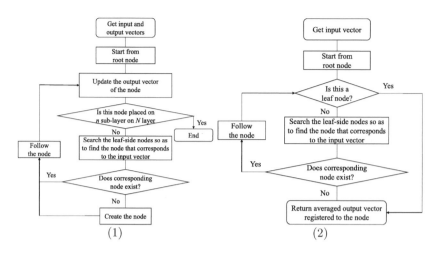

Fig. 4. (1) Learning flow and (2) prediction flow

3.1 Settings

Experiment 1. We validated the learning capability of DLT with simple-harmonic-motion (SHM) learning task. For SHM, we used the following equation.

$$\ddot{q} = -q \tag{4}$$

DLT with 6 layer 2 dim. 3-ary was employed. NN with three layers and back propagation method was employed for the comparison. For both learning systems, their input and output are set at $(q, \dot{q})$ and $\ddot{q}$. For NN, the learning rate and the number of middle layer nodes are preliminary optimized before the comparison. We selected the best of them according to the error of 1000000-step learning.

The learning of DLT was done by online incremental learning, in which training data is leaned one by one. The learning of NN was done by batch learning, in which 2000 training data was simultaneously learned in a learning step.

Experiment 2. To validate the learning capability of DLT, the vehicle model in a simulator was learned by DLT. We conducted vehicle simulation using Open Dynamics Engine [27]. The appearance of the vehicle model is in Fig. 5(1).

In this experiment, DLT learned a_0 using the developed body babbling. The input data of DLT is composed of forward direction velocity $\dot{x}$, side direction velocity $\dot{y}$, angular velocity $\dot{\theta}$, steering angle ϕ. The output of DLT is composed of $\ddot{x}$, $\ddot{y}$, and $\ddot{\theta}$. In order to generate training data by motor babbling, ϕ and x direction force were given at random. These values were recalculated and shifted when the velocity of the vehicle is almost 0. Each data was obtained by 0.01[s] in simulation time. DLT learned 2 million training data. Every 10000 data, the

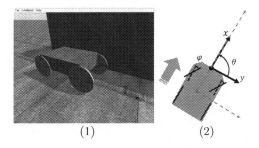

(1)   (2)

Fig. 5. (1) Simulated vehicle and (2) coordinate system

precision of the learning of DLT was validated using 0.1 million test data that
was not learned by DLT. The training data was prepared before the experiment
using the same motor babbling process. The error between the acceleration and
predicted acceleration are calculated as follows:

$$E_x = \frac{1}{N} \sum_{i=1}^{N} (|\ddot{x}_{0_i} - \hat{\ddot{x}}_{0_{i-1}}|) \tag{5}$$

$$E_\theta = \frac{1}{N} \sum_{i=1}^{N} (|\ddot{\theta}_{0_i} - \hat{\ddot{\theta}}_{0_{i-1}}|), \tag{6}$$

where E is error, N is number of data, $\circ_0$ is measured value, $\hat{\circ}_0$ is predicted
value.

Experiment 3. Using the trained dynamics learning tree of Experiment 3,
the trajectory of a vehicle while $a_A = a_D = 0$ was predicted from given initial
state $(\dot{x}, \dot{y}, \dot{\theta}, \phi)$. For the initial state, we prepared two states. One of them is
the state in training data. The other one is the state in test data. The farmer
is an already-learned state. The latter is a not-learned state. We validated the
precision against both of them while 5 [s] in simulation time.

Experiment 4. In order to validate the effectiveness of FAD learning, we com-
pared it with the direct learning of the map from $(\dot{x}, \dot{y}, \dot{\theta}, \phi, F)$ to a. Both learning
was done by using dynamics learning tree. While motor babbling, ϕ was selected
from the range of -0.7–0.7 [rad] with 15 levels, F was selected in the range
of -100–100 [N] with 200 levels. The learning systems were trained with 50000
data. We used 10000 test data that was captured by motor babbling. It was not
learned by the learning systems. Comparing with the test data and prediction
of the learning systems, we calculated the errors.

3.2   Results

Experiment 1. The learning results of DLT are shown in Fig. 6(1) and (2). After 100 data was learned (Fig. 6(1)), the answer and prediction were not close still. After 5000 data was learned (Fig. 6(2)), they were almost the same. Prediction errors of DLT and NN are plotted in Fig. 7. DLT converged its error much faster than NN.

In Table 1, the learning results of NN are listed. The best result of NN was compared with DLT in Table 2. From the result, DLT converged to 2.23 times smaller error and about 500 times smaller computational time than NN.

Experiment 2. Figures 8 and 9 show the convergence of the error. E_x and E_θ converged to 0.06 and 0.02 [m/s^2] respectively. The convergence was done almost 50000 (500 [s] in simulation time).

Experiment 3. Figures 10 and 11 show the resulted trajectory. These trajectories were quite close to those of the original simulator. Also, from the state that was not included in the training data, DLT predicted sufficiently close trajectories (Figs. 12 and 13).

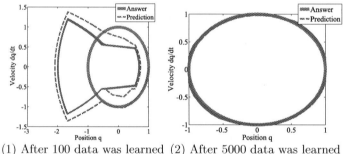

(1) After 100 data was learned (2) After 5000 data was learned

Fig. 6. Learning results of SHM (Color figure online)

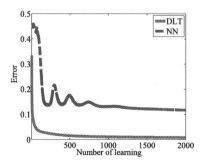

Fig. 7. Number of learning and prediction error (Color figure online)

Table 1. Error and computing time of NN. The results show the values after each NN was trained with 2000 data of SHM while 1000000 steps. The best one is selected on Table 2, according to the errors.

Learning rate	Number of middle nodes	Average error	Computational time [ms]
0.0002	5	0.063083	630613
	10	0.10998	1094978
	20	0.109632	2026054
0.0005	5	0.0438948	652052
	10	0.105286	1124058
	20	0.109136	2079837
0.001	5	0.0420112	604970
	10	0.0439611	1125948
	20	0.108682	2042482
0.01	2	0.0429104	371751
	3	0.0182211	463264
	5	0.0572928	636039
0.02	3	0.322405	462179
	5	0.302984	630325
	10	0.315269	1096139

Table 2. Comparison of the best NN and DLT. The best NN of Table 1 was compared with DLT. The result of DLT is one step online learning for 2000 data. The result of NN is 1000000 steps offline learning for 2000 data.

	Average error	Computational time [ms]
DLT	0.0081701	921
Best NN	0.0182211	463264

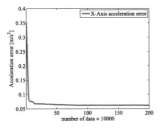

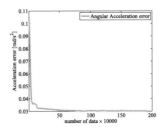

Fig. 8. E_x against learned number of data

Fig. 9. E_θ against learned number of data

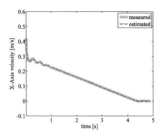

Fig. 10. Prediction result against training data (x)

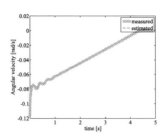

Fig. 11. Prediction result against training data (θ)

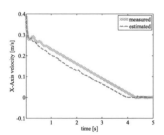

Fig. 12. Prediction result from not-trained state (x)

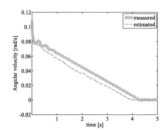

Fig. 13. Prediction result from not-trained state (θ)

Experiment 4. Figures 14 and 15 show the results. Before the learning of a_A was started, FAD learning is not predict $a = a_F + a_A$ precisely, because it only was learning a_F in this period. After shifting the learning to a_F FAD learning drastically improved its prediction. The learning of FAD learning was faster than the direct learning against a.

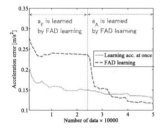

Fig. 14. Comparison of FAD and direct learning (x)

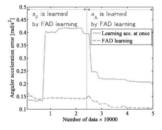

Fig. 15. Comparison of FAD and direct learning (θ)

4 Discussion

From Experiments 1 and 2, we could confirm the online learning capability of DLT. From Experiments 3 and 4, DLT successfully modeled a_F with quite small error. From Experiment 5, the effectiveness of FAD learning was confirmed. FAD learning realized lower error about the learning of a than direct learning. This means that FAD learning requires less number of data than direct learning to assure the same amount of error in a. From the fact that a was successfully learned against the input $(\dot{x}, \dot{y}, \dot{\theta}, \phi)$, we could confirm that the whole dynamics of plainer vehicle model was learned by the proposed framework. This framework has much possibility to be extended to the whole dynamics learning of a vehicle, because the input of DLT is easily extendable without any change of the theory.

5 Conclusion and Future Work

In this paper, we proposed a vehicle modeling framework that is extendable to whole body dynamics learning of a vehicle. The proposed framework is composed of FAD learning, motor babbling, and dynamics learning tree. The theories of FAD learning and motor babbling are on the basis of the whole body dynamics. Dynamics learning tree is applicable to arbitrary I/O map learning including the mapping between the state (and action) and acceleration of a vehicle. In the experiments, FAD learning successfully modeled the plainer vehicle dynamics of a simulator. Also, FAD learning showed smaller error than that of direct learning of acceleration. For the future work, we are going to apply the proposed framework to the whole dynamics of a vehicle including roll and pitch directions.

Acknowledgement. A part of this work was supported by JSPS KAKENHI Grant Number 15K20850.

References

1. Lee, U., Yoon, S., Shim, H.C., Vasseur, P., Demonceaux, C.: Local path planning in a complex environment for self-driving car. In: Annual IEEE International Conference on Cyber Technology in Automation, Control and Intelligent Systems (2014)
2. Li, C., Wang, J., Wang, X.: A model based path planning algorithm for self-driving cars in dynamic environment. In: Chinese Automation Congress (2015)
3. Kim, J., Kim, H., Lakshmanan, K., Rajkumar, R.: Parallel scheduling for cyber-physical systems: analysis and case study on a self-driving car. In: ACM/IEEE International Confernce on Cyber-Physical Systems (2013)
4. Kim, C.H., Sugano, S.: Tree based trajectory optimization based on local linearity of continuous non-linear dynamics. IEEE Trans. Autom. Control **61**(9) (2016)
5. Kim, C.H., Sugano, S.: Closed loop trajectory optimization based on reverse time tree. Int. J. Control Autom. Syst. (in press)
6. Kim, C.H., Yamazaki, S., Sugano, S.: Motion optimization using state-dispersion based phase space partitions. Multibody Sys. Dyn. **32**(2), 159–173 (2013)

7. Kim, C.H., Sugano, S.: A GPU parallel computing method for LPUS. Adv. Robot. **27**(15), 1199–1207 (2013)

8. Bang, K.: Development of dynamics modeling in the vehicle simulator for road safety analysis. In: SICE Annual Conference (2007)

9. Satar, M.I.: Modeling and verification of 5 degree of freedom vehicle longitudinal model. In: Asian Control Conference (2015)

10. Setiawan, J.D., Safarudin, M., Singh, A.: Modeling, simulation and validation of 14 DOF full vehicle model. In: International Conference on Instrumentation, Communications, Information Technology and Biomedical Engineering (2009)

11. Kabiraj, K.: Mathematical modeling for vehicle dynamics, steady state cornering performance prediction using nonlinear tire data employing a four wheel model. In: IEEE International Conference on Current Trends in Engineering and Technology (2014)

12. Yim, Y.U., Oh, S.Y.: Modeling of vehicle dynamics from real vehicle measurements using a neural network with two-stage hybrid learning for accurate long-term prediction. IEEE Trans. Veh. Technol. **53**(4), 1076–1084 (2004)

13. Tanizoe, T., Kawamoto, M., Muto, D., Yokoi, H.: A system to park automatically in various cases. Technical note of the Institute of Electronics, Information and Communication Engineering, vol. 105, no. 419 (2005). (In Japanese)

14. French, R.M.: Catastrophic forgetting in connectionist networks. Trends Cogn. Sci. **3**(4), 128–135 (1999)

15. Asada, M., Dorman, K.M., Ishiguro, H., Kuniyoshi, Y.: Cognitive developmental robotics as a new paradigm for the design of humanoid robots. Robot. Auton. Syst. **33**, 185–193 (2001)

16. Schillaci, G., Hafner, V.V.: Random movement strategies in self-exploration for a humanoid robot. In: IEEE/ACM International Conference on Human Robot Interaction, pp. 245–246 (2011)

17. Saegusa, R., Metta, G., Sandini, G.: Active learning for multiple sensorimotor coordination based on state confidence. In: IEEE/RSJ International Conference on Robotics and Systems, pp. 2598–2603 (2009)

18. Dawood, F., Loo, C.K.: Humanoid behavior learning through visuomotor association by self-imitation. In: SCIS and ISIS, pp. 922–929 (2014)

19. Grimes, D.B., Rao, R.P.N.: Learning Actions through Imitation and Exploration: Towards Humanoid Robots That Learn from Humans. In: Sendhoff, B., Körner, E., Sporns, O., Ritter, H., Doya, K. (eds.) Creating Brain-Like Intelligence. LNCS, vol. 5436, pp. 103–138. Springer, Heidelberg (2009)

20. Mochizuki, K., Nishide, S., Okuno, H., Ogata, T.: Developmental imitation learning of robot on drawing with neuro dynamical mode. In: Annual Conference of IPSJ (2013). (In Japanese)

21. Yamashita, Y., Tani, J.: Emergence of functional hierarchy in a multiple timescale neural network model: a humanoid robot experiment. PLoS Comput. Biol. **4**(11), e1000220 (2008)

22. Nishide, S., Mochizuki, K., Okuno, H.G., Ogata, T.: Insertion of pause in drawing from babbling for robot's developmental imitation learning. In: IEEE International Conference on Robotics and Automation, pp. 4785–4791 (2014)

23. Nagai, Y., Asada, M., Hosoda, K.: Learning for joint attention helped by functional development. Adv. Robot. **20**(10), 1165–1181 (2006)

24. Nakaoka, S., Nakazawa, A., Yokoi, K., Hirukawa, H., Ikeuchi, K.: Generating whole body motions for a biped humanoid robot from captured human dances. In: IEEE International Conference on Robotics and Automation, pp. 905–3910 (2003)

25. Watanabe, K., Numakura, A., Nishide, S., Gouko, M., Kim, C.H.: Fully automated learning for position and contact force of manipulated object with wired flexible finger joints. In: International Conference on Industrial, Engineering and Other Applications of Applied Intelligent Systems (2016, in press)
26. Takahashi, K., Ogata, T., Yamada, H., Tjandra, H., Sugano, S.: Effective motion learning for a flexible-jointrobot using motor babbling. In: IEEE International Conference on Robotics and Systems (2015)
27. Smith, R.L.: Open Dynamics Engine. http://ode.org

Adaptive Model for Traffic Congestion Prediction

Pankaj Mishra[✉], Rafik Hadfi, and Takayuki Ito

Department of Computer Science and Engineering, Nagoya Institute of Technology,
Gokiso, Showa-ku, Nagoya 466-8555, Japan
{pankaj.mishra,rafik}@itolab.nitech.ac.jp, ito.takayuki@nitech.ac.jp

Abstract. Traffic congestion is influenced by various factors like the weather, the physical conditions of the road, as well as the traffic routing. Since such factors vary depending on the type of road network, restricting the traffic prediction model to pre-decided congestion factors could compromise the prediction accuracy. In this paper, we propose a traffic prediction model that could adapt to the road network and appropriately consider the contribution of each congestion causing or reflecting factors. Basically our model is based on the multiple symbol Hidden Markov Model, wherein correlation among all the symbols (congestion causing factors) are build using the bivariate analysis. The traffic congestion state is finally deduced on the basis of influence from all the factors. Our prediction model was evaluated by comparing two different cases of traffic flow. We compared the models built for uninterrupted (without traffic signal) and interrupted (with traffic signal) traffic flow. The resulting prediction accuracy is of *79 %* and *88 %* for uninterrupted and interrupted traffic flow respectively.

1 Introduction

In recent years, there has been lot of research in the field of traffic congestion estimation in road networks. Traffic congestion is caused by multiple factors such as the road physical condition, the traffic flow parameters, etc. Most of the work in literature focused on pre-decided and fixed number of traffic flow parameters such as average speed, density, queue length, flow rate, etc., for estimating the traffic congestion. However, the influence or contribution of these factors in traffic congestion varies from one location to another. That is, in some locations, the low average speed of all the vehicles reflects the presence of congestion may be at some other locations congestion is not reflected by the recorded average speed of all the vehicles. Similarly at some locations, the weather factors also need to be considered. Therefore, we propose an adaptive model that can effectively incorporate different pairs of traffic factors based on road network.

From the literature, the existing methodologies to predict the traffic congestion were based on Hidden Markov Model (HMM) [4], Artificial Neural Network (ANN) [5] or on plain mathematical model [2]. For instance, traffic congestion estimation based on probabilistic model relying on video data [8,12] or plain

© Springer International Publishing Switzerland 2016
H. Fujita et al. (Eds.): IEA/AIE 2016, LNAI 9799, pp. 782–793, 2016.
DOI: 10.1007/978-3-319-42007-3_66

probabilistic models [9] considered few parameters to build the prediction model. Additionally from the work [4,9] HMM is proved to be one of the best suited model for modelling a traffic prediction model. Since HMM [10] architecture can very well reflect the problem of traffic congestion prediction, wherein traffic congestion state represents the hidden state and the set of congestion influencing factors $(\chi)^1$ being the observed symbols. But native discrete HMM architecture can incorporate only single observed symbols. Therefore discrete multiple symbol HMM (MS-HMM) [11] was implemented which can very well represent all the values in χ.

Moreover, factors in χ vary with type of road considered [6] namely, freeways, uninterrupted roads, interrupted roads, etc. Other than this, contribution of each factor in traffic congestion vary with considered road network. Therefore an adaptive model which can efficiently incorporate any number of factors and adapt to any road network is needed, but the current state of traffic prediction models works on pre decided fixed number of congestion causing or reflecting factors[2]. Due to these limitations in the current state of traffic prediction models, we propose a model for traffic prediction that can easily be extensible (add or remove) to congestion factors and adapt to any road network. Wherein adaptiveness in model is realised by calculating impact factor *(I)* from correlation parameters of all factors in χ as discussed in the later sections. In short we propose an adaptive traffic congestion prediction model based on MS-HMM which can easily incorporate any number of factors effectively. Further, this adaptive model can be extended to many application such as dynamic routing, dynamic traffic signal scheduling, etc. In this paper, our two main contributions are, firstly an extensible model that can efficiently consider the contribution of different congestion reflecting factors based on the correlation parameters calculation and impact factor calculation. Other contribution is the usage of MS-HMM to build the prediction model based on the multiple congestion reflecting factors.

The paper is organised as follows, in Sect. 2 we discuss the system overview by briefly describing each step in building the prediction model. In Sect. 3, we discuss the prediction algorithm implemented in our system. In Sect. 4 experimental results are discussed and the proposed model is validated by building an baseline system. Finally, we conclude and outline the future work.

2   System Overview

The primary aim of our work is to build an adaptive traffic prediction model. Such adaptive model would predict the traffic congestion considering the congestion causing factors relevant to considered road network, and not based on the pre defined factor or factors. Moreover consider the appropriate contribution of each considered factor, whereas data is recorded by loop detectors installed on the road network. In this paper, we propose a novel method according to

---

[1] χ is a set of congestion reflecting factors that may consist of factors such as occupancy, speed, waiting time, etc.

[2] Congestion reflecting factors and congestion causing factors are used interchangeably.

best of our knowledge which can adapt to the network, and reflect the appropriate effect of each congestion causing factor. The brief description of the model building steps are discussed in the next subsection, where whole model building is divided into 3 main steps (i) Adaptive Modelling, (ii) Data Discretisation and (iii) Prediction Model Building.

2.1 Adaptive Modeling

As discussed in the previous section, an adaptive model is the primary aim of our work. Wherein adaptability is realised by calculating the correlation parameters and later the impact factor. Basically adaptability is achieved in two step named as (i) Correlation parameter calculation and (ii) Traffic impact factor calculation. The correlation parameters are based on bivariate analysis of different factors with respect to single factor, using a characteristics graph[3]. Later from the characteristic graph of a particular factor two values, namely strength value *(c)* and relationship value *(r)* are calculated. These two parameters represent the contribution of individual factors, where c quantify the strength of correlation between different factors, which is calculated from the Eq. 1; which depicts c for factor f. Whereas, r denotes the degree at which one parameter changes with the change in other parameter, basically it is represented by the degree of curve in the characteristic graph.

$$c_f = \frac{covar(f^*, f)}{\sqrt[2]{var(f) \times var(f^*)}} \qquad (1)$$

Where f is the congestion reflecting factor and f^* is common congestion causing factors, with respect to which correlation parameters of all the other factors are calculated. Whereas, n is number of total instance of factors or size of χ, and $var(f^*)$ and $var(f)$ are variance of common factor f^* and any other factor f respectively, $covar(f^*, f)$ is covariance value and c_f is strength value of factor f. Once we have pairwise correlation parameters for all the factors f with respect to a common factor f^*, we calculate the traffic impact factor I_t for every time step t; calculated from Eq. 2. Where, 2 represents the collective effect from all the factors by summing linear combination of correlation parameters of different factors for a particular time instance. Where n being the size of χ or number of factor considered and *val(f)* being the recorded value of respective factors.

The intuition behind the Eq. 2 is that each factor's involvement is defined by their correlation parameters, and also the correlation parameters will change with locations. So first we calculate the correlation parameters for each factor with respect to a single factor, so as to examine the influence of each factor from a common reference point. Later we calculate the resultant impact for every time step t, based on the recorded magnitude $(val(f_t))$ of factor f at time t

---

[3] The curve representing the change in one factor with change in other factor with time.

and its respective correlation parameters (c_f and r_f). Wherein c normalises the magnitude of all the factors and r denotes the degree at which factor f changes with respect to change in factor f^*; f^* can be any one of the congestion influential factor among the chosen set of factors. Finally, the summation of contribution by all the factors in the considered road network gives the resultant influence for traffic congestion. Moreover, this makes the model easily extensible to any other factors defining weather conditions, physical condition of road etc.

$$I_t = \sum_{f=1}^{n} (val(f_t) \times c_f)^{r_f} \tag{2}$$

2.2 Data Discretisation

The magnitude of congestion causing factors are recorded for every instance of time, similarly impact factor (I) is calculated for corresponding time stamp. Therefore congestion causing factors and impact factor are continuous in nature. However, our system is based on discrete MS-HMM, therefore we discretise the recorded data and I. We adopted a discretisation method based on equal width K-means clustering algorithm. Wherein, the continuous data is clustered into different classes on the basis of neighbourhood and these class represents a label. After discretisation of all the n factors and impact factor, we get n + 1 different set of labels. Where, n set of labels denotes the set of observation symbol (O) for n congestion causing factors and the other 1 set of labels represent the set of hidden states (Q) in MS-HMM. The K-means clustering algorithm is depicted in the Algorithm 1.

Algorithm 1. Factor Labelling

1: **procedure** CLUSTER
2:     Input: L is number of labels
3:     $X \rightarrow \{x_1, x_2, x_3, \ldots, x_n\}$ is set of data points
4:     Output: Indices of cluster
5:     Start
6:     $C \rightarrow \{c_1, c_2, c_3, \ldots, c_L\}$ is set of centre of cluster
7:     **for** all X $i \rightarrow 1$ to n **do**
8:         **for** all C $j \rightarrow 1$ to L **do**
9:             $d_{ij} \rightarrow |c_i - x_j|$
10:    **for** all d $i \rightarrow 1$ to k **do**
11:        (It gives value of i to which j belongs)
12:        $cluster_j \rightarrow$ value_of_i$(min(d_{ji}))$
13:    **Recalculating the new cluster centre**
14:    $v_i \rightarrow \left(\frac{1}{c_i}\right) * \sum_{i=1}^{c_i}(x_i)$
15:    (where, c_i is number of points in i^{th} cluster)
16:    **Repeat 5 to 16 until cluster remain same**
17:    **End**

2.3   Prediction Model Building

Our Prediction model based on discrete MS-HMM, which is built using discretised data obtained from the previous section. The MS-HMM [11] is mathematically represented as $\lambda = (T, E, \pi)$, producing the sequence of observation symbols O. Where $T = P(Q_{n+1}|P(Q_n)$ represents transition probability, $\pi = P(Q_{t=1})$ represents the initial probability and $E = P(O|\lambda)$ represents emission probability [11]. The graphical representation of the MS-HMM for 4 symbols model is depicted in the Fig. 1, where s_t, wt_t, ql_t, and oc_t are labels of average speed, waiting time, queue length and occupancy [6] respectively at time t. Wherein, the average speed (s) is the average speed of all the vehicles at particular time instance, queue length (ql) is the average length of queue of the vehicle waiting at the cross section, waiting time (wt) is the average waiting time of the vehicles and occupancy (oc) is the percentage of road occupied by the vehicles.

Further, training and building of the prediction model λ is done according to the steps depicted in Fig. 2. Wherein, a model in the new learning environment is first initialised to random traffic data or traffic data of previous road network if any, then this initialised model is trained to adapt to this new road network. Further, sequence segmentation is used to segment the training and test data in equal sequence length for prediction. Training is done on the training set of the previously recorded data. Finally we get model λ, which has been adapted to the considered road network based on the recorded data.

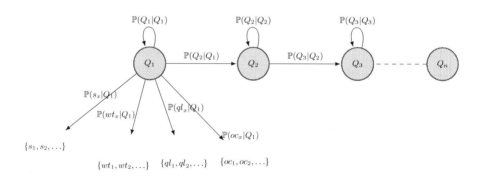

Fig. 1. MS-HMM architecture

3   Prediction Algorithm

After building a trained prediction model λ, the model λ is used for prediction of traffic state. In this section, we discuss the extension of the trained model λ to predict current state using model hidden state predictor (HSP) and future state using the model future state predictor (FSP). The trained model is used

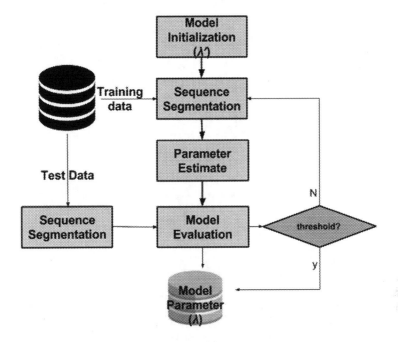

Fig. 2. Model training

for the traffic state prediction. Prediction of sequence of hidden states from the sequence of observation symbols is done by implementing the Viterbi algorithm [10]. Wherein, sequence of hidden states are predicted from the past observed symbols. The Eq. 3 represents the hidden state predictor on the recorded stream of observation symbols. Where, 3 represents the conditional probability of occurrence of a hidden states in the set Q, on the given sequence of observation symbols s, wt, ql, and oc at time $t = n$. Furthermore, the HSP module output is used by FSP for further state prediction.

Our next interest is to predict the future state of the road network from the given past sequence of data. The FSP is a modified backward algorithm [10] to fit into MS-HMM. The *FSP* algorithm is depicted in Fig. 3, where *FSP* algorithm takes input from *HSP* along with learnt model λ. It can be seen in the Fig. 3, each *FSP* block gives the next hidden state label. In order to maintain the accuracy at each step, the model λ is trained with online data with the building step shown in Fig. 2. Then the modified model λ is given to next *FSP* block to predict the next future state. The validation of proposed methodology is discussed in the next section.

$$P(Q_{t=n}|(s, wt, ql, oc)_{t=n}) \approx P(Q_{t=n}|s_{t=n}) \times P(Q_{t=n}|wt_{t=n})$$
$$\times P(Q_{t=n}|qt_{t=n}) \times P(Q_{t=n}|oc_{t=n}) \quad (3)$$

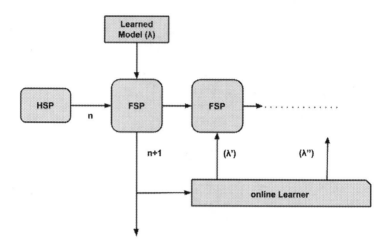

Fig. 3. FSP

4   Experimental Results

The tools used in the experiments were WEKA [13] for clustering and labelling of the data, SUMO [3] for running the simulation, and MATLAB toolbox [7] for building the prediction model. In order to study the working of the proposed model, which should be able to reflect the adaptability to different types of road networks and their respective traffic causing factors, we considered 4 values in χ; namely, s, oc, wt and ql to build the baseline prediction model as per the proposed method. The model is based on MS-HMM, built using the steps discussed in the above sections. Wherein building and validation of the model was done by simulating a traffic of a road network. The main objective of the experiment was to showcase the capability of the model to adapt to different road types and predict the traffic congestion for the same. So, in order to examine the adaptability of the model, we calculate the correlation parameters for two different road network setting fed with the same traffic flow data. Later, correlation parameters for both the road setting is analysed and traffic congestion is predicted. We used $SUMO$ to simulate traffic of few roads from Enfield area in London along with its traffic counts data of the heterogeneous traffic, available at [1]. From which we record the values for each χ in two different settings, namely, interrupted traffic flow (with traffic signal) and uninterrupted traffic flow (without traffic signal).

Table 1. Uninterrupted traffic flow

	Avg. speed	Wt. time	Q. length
r	1.2	2	2.5
c	1	0.80	0.60

Table 2. Interrupted traffic flow

	Avg. speed	Wt. time	Q. length
r	0.5	1.5	3
c	0.4	0.65	0.96

The prediction models for both the scenarios, interrupted traffic flow *(inp)* and uninterrupted traffic flow *(unp)* is represented as λ_{inp} and λ_{unp} respectively. The experimental results of each step are mentioned in the next section.

4.1 Correlation Parameter

As discussed above, 4 parameters were extracted from the simulation for both the scenarios, *inp* and *unp*. Later, the correlation parameters for each factor in the respective scenario is calculated. Figures 4 and 5 depicts the characteristics graphs for scenario *unp* and *inp* respectively. Further, from these characteristics graph, two correlation parameters c and r with respect to occupancy *(oc)* are calculated for both the scenarios. These correlation parameters for *unp* and *inp* scenarios are given in Tables 1 and 2 respectively.

As discussed before, the state of traffic is defined by the labels of the I obtained after discretisation. Whereas, labels of each factor in χ represents the sequence of observed symbols. In order to evaluate both the trained models λ_{unp} and λ_{inp}, we compared the predicted traffic states and the actual traffic states. The actual traffic state was calculated manually using Eq. 2. Now, in order to test the prediction model, we consider a total of 100 instances of data from the recorded data, where each instance is of length 5 min. Moreover, for prediction of traffic state, 15 latest known observation sequence, and the updated λ from the online learning module is fed to the *FSP* algorithm. Further, we repeated the prediction process 3 times, to test if the result is replicable.

4.2 Model Evaluation

The prediction results of both the models λ_{unp} and λ_{inp} are compared with their corresponding actual data, which is depicted in Fig. 6. Where *y-axis* represents the traffic state based on value of I, the *x-axis* represents the time series starting from the 16th time step. Now, from the prediction characteristic graphs (Fig. 6) of both the models, it can be said that the predicted state curve accurately follows the actual state curve, and also accuracy % for the models λ_{unp} and λ_{inp} were *79 %* and *88 %* respectively. The accuracy was calculated using the Eq. 4, which represents the ratio of correctly predicted data and total data to be predicted.

$$\%error = \frac{Correctly\ labeled\ stream\ of\ data}{Total\ length\ of\ stream} \times 100 \qquad (4)$$

4.3 Discussion

In order to validate the adaptiveness and prediction of the proposed traffic congestion model, we compare the correlation parameters and the results obtained from both the models λ_{unp} and λ_{inp}. Firstly, by comparing their characteristic graphs for the average speed from Figs. 4c and 5c, of λ_{unp} and λ_{inp} respectively, It can be seen in Fig. 4c that the occupancy is inversely correlated to the average

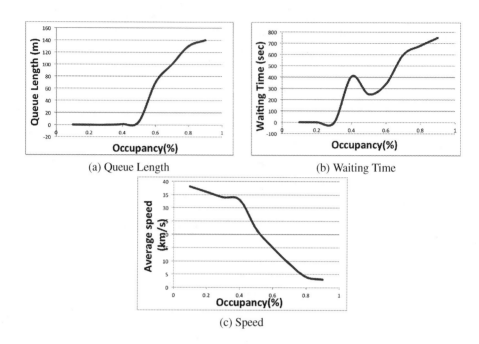

(a) Queue Length   (b) Waiting Time

(c) Speed

Fig. 4. Characteristics graph for *unp*

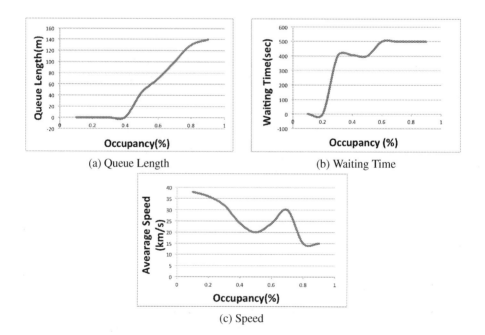

(a) Queue Length   (b) Waiting Time

(c) Speed

Fig. 5. Characteristics graph for *inp*

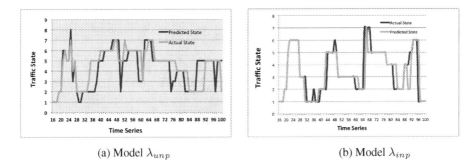

(a) Model λ_{unp} (b) Model λ_{inp}

Fig. 6. Evaluation of the model

speed. But such correlation cannot be seen from Fig. 5c. Such change in correlation with the change in model can be interpreted as, the average speed decreases with increase in the occupancy in the uninterrupted traffic flow. But this is not true in the interrupted traffic flow. Because in the interrupted traffic flow, the average speed can also decrease because of interruption[4], which is the red traffic signal in our case. Therefore contribution of average speed for causing traffic congestion, in the uninterrupted traffic flow is more as compared to contribution in the interrupted traffic flow. Furthermore, such change in contribution is reflected from the values of their correlation parameters of the average speed in Tables 1 and 2. Similarly contribution of other factors, change with the change in traffic flow environment. Thus, on basis of correlation parameters values, one can decide the significance of a factor in traffic congestion. So, if the values of correlation parameters for any factor are negligible, then that factor can be ignored, thus removing the insignificant factors. Other than this, introduction of impact factor calculation makes the model extensible, and also incorporate the factors based on their respective contribution. This gives the flexibility to add or remove factors from the model, based on the considered road network.

Thus by comparing the correlation parameters for different road setting on the same traffic dataset, it can be said that the correlation among the congestion reflecting factors changes with the road traffic settings. Since our prediction model is based on the values of these correlation parameters, therefore it can be said that model prediction is adapted to the learning environment. Moreover, the prediction accuracy of the model is not compromised over extensibility, thus giving satisfying prediction accuracy.

5   Conclusion

We proposed an adaptive mathematical model for traffic prediction based on MS-HMM. The adaptiveness of the model was achieved by impact factor calculation. Our primary aim was to build an extensible model, and incorporate

---

[4] Interruption: Hindrance in normal flow of traffic because of traffic signal, cross sectional traffic flow, speed limit zone, etc.

792 P. Mishra et al.

different traffic congestion factors. The model should appropriately reflect the varying contribution of each factor with the road networks. Therefore correlation parameters were calculated for each factor. As a baseline system, we built our prediction model based on 4 factors: occupancy, average speed, queue length and waiting time. To validate the adaptability of the model to a road network, we compared the correlation parameters in uninterrupted and interrupted traffic flow environment. Finally, the prediction accuracy for the model in uninterrupted and interrupted environment was *79 %* and *88 %* respectively. Thus our proposed model has the ability to adapt to road network, and also give satisfying prediction of traffic state. Other parameters such as road, weather conditions and other sensor data can be incorporated for the better prediction of the congestion based on the considered road networks. Otherwise any factor can be removed having negligible correlation parameters.

As a future work, we intend to focus on decision making problem by integrating a prediction model with dynamic routing and dynamic traffic signals. Thus building a smart recommendation system for the drivers in the sense that it can provide the shortest and the less congested paths. Integrating the current model with the extended application can solve real life problems and contribute to a smarter traffic management system.

Acknowledgement. The research results have been achieved by Congestion Management based on Multiagent Future Traffic Prediction, Researches and Developments for utilizations and platforms of social big data, the Commissioned Research of National Institute of Information and Communications Technology (NICT).

References

1. Road traffic dataset. http://www.dft.gov.uk/traffic-counts/index.php
2. Antoniou, C., Koutsopoulos, H.N., Yannis, G.: Traffic state prediction using Markov chain models. In: Proceedings of the European Control Conference (2007)
3. Behrisch, M., Bieker, L., Erdmann, J., Krajzewicz, D.: Sumo-simulation of urban mobility. In: The Third International Conference on Advances in System Simulation (SIMUL 2011), Barcelona, Spain (2011)
4. Geroliminis, N., Skabardonis, A.: Prediction of arrival profiles and queue lengths along signalized arterials by using a Markov decision process. Transp. Res. Rec. J. Transp. Res. Board **1934**, 116–124 (2005)
5. Li, X., Parizeau, M., Plamondon, R.: Training hidden Markov models with multiple observations - a combinatorial method. IEEE Trans. Pattern Anal. Mach. Intell. **22**(4), 371–377 (2000)
6. Manual, H.C.: Highway capacity manual, Washington, D.C. (2000)
7. Murphy, K.: Hidden Markov model (HMM) toolbox for matlab (1998). http://www.ai.mit.edu/~murphyk/Software/HMM/hmm.html
8. Porikli, F., Li, X.: Traffic congestion estimation using HMM models without vehicle tracking. In: IEEE Intelligent Vehicles Symposium, 2004, pp. 188–193. IEEE (2004)
9. Qi, Y.: Probabilistic models for short term traffic conditions prediction. Ph.D. thesis, Louisiana State University (2010)

10. Rabiner, L.R.: A tutorial on hidden Markov models and selected applications in speech recognition. Proc. IEEE **77**(2), 257–286 (1989)
11. Schenk, J., Schwärzler, S., Rigoll, G.: Discrete single vs. multiple stream HMMs: a comparative evaluation of their use in on-line handwriting recognition of white-board notes. In: Proceedings of the International Conference on Frontiers in Handwriting Recognition, pp. 550–555 (2008)
12. Sen, R., Cross, A., Vashistha, A., Padmanabhan, V.N., Cutrell, E., Thies, W.: Accurate speed and density measurement for road traffic in India. In: Proceedings of the 3rd ACM Symposium on Computing for Development, p. 14. ACM (2013)
13. Witten, I.H., Frank, E., Trigg, L.E., Hall, M.A., Holmes, G., Cunningham, S.J.: Weka: practical machine learning tools and techniques with Java implementations (1999)

Soft Computing and Multi-agent Systems

Intellectual Processing of Human-Computer Interruptions in Solving the Project Tasks

P. Sosnin[(⊠)]

Ulyanovsk State Technical University,
Severny Venetc str. 32, 432027 Ulyanovsk, Russia
sosnin@ulstu.ru

Abstract. The paper focuses on an approach that facilitates improving the management processes in solving the project tasks. The improvement can be achieved if Agile managerial mechanisms will be combined with the managing of human-computer interruptions when tasks are reflected on programmable queues. The offered combining opens the possibility for planned and situational interactions of the designer with queues of tasks in points of interruptions taking into account the evolving multitasking. Furthermore, the approach supports intellectual processing the reasons of interruptions for their more effective using.

Keywords: Conceptual experimenting · Human-computer interruption · Multitasking · A queue of project tasks · Semantic memory · Question-answering · Workflow

1 Introduction

The question "How will project success be affected by changing the way of working?" occupies the central place in the search for innovations in SEMAT (Software Engineering Methods And Theory) [1] aimed at reshaping the software engineering. In normative documents of SEMAT, a way of working used by a team of designers is defined as "the tailored set of practices and tools used by the team to guide and support their work."

This definition indicates some directions of possible changing among which we can mark the search of innovations in managing of personal and collective human-computer activity. One of these directions is an innovative development of agile approaches in the project management.

Below we present combining of Agile managerial mechanisms with the managing of human-computer interruptions. The choice of such combining is caused by inevitable multitasking in a personal behavior of designers with multitasking in their collective activity. Moreover, the main feature of the offered approach is the use of intellectual processing the creative situations that often arise as the reasons for human-computer interruptions or unplanned events during their processing. The suggested solutions facilitate increasing the adequateness in specifying the work of the designer and open the possibility for the use of experimenting in processing the reasons of interruptions.

© Springer International Publishing Switzerland 2016
H. Fujita et al. (Eds.): IEA/AIE 2016, LNAI 9799, pp. 797–807, 2016.
DOI: 10.1007/978-3-319-42007-3_67

The offered approach is based on the use of reflections of the operational space of designing on the semantic memory of the toolkit WIQA (Working In Questions and Answers) that has been created for conceptual designing of Software Intensive Systems (SISs).

We start the rest of this paper with highlighting the features of human-computer interruptions in setting the multitasking. The third section introduces related works used as a source of prompts for the offered approach. The fourth section opens the instrumental environment WIQA with the accent on its semantic memory. The fifth section presents basic solutions of the offered approach and its application. The aim of the sixth section is the comparative discussion.

2 Preliminary Bases

Interruptions inevitably accompany any human activity. They can be caused by different reasons and can lead to negative or positive effects. For example, they can be caused by made errors and can lead to other mistakes or they can be bound with a discovery of a more effective idea for the task being solved.

Taking into account the inevitability and importance of interruptions, reactions on them should be managed by the type of the human activity, reasons for interruptions and aims of their processing. The offered approach is developed for human-computer interruptions in the human-computer activity in setting of multitasking.

The typical understanding of multitasking is the "ability to handle the demands of multiple tasks simultaneously" [2] while a useful definition of the interruption is "the process of coordinating abrupt changes in people's activities" [3]. It should be noted; that multiple tasks can be simultaneously implemented by a group of designers each of which can operate with a set of multiple tasks switching among them in planned or situational conditions. This indicates that the multitasking behavior of the designer is very complicated kind of the interrupted activity which additionally has a creative character. It is this kind of the human activity occupies the central place in the approach described below. We shall describe the offered approach from the viewpoint of its use by a single designer.

Nowadays, agile methods are widely used for managing the collaborative work of designers with multiple tasks. These methods suggest a reflection of the activity process and its state on their descriptions, estimations of which are applied to managerial aims. For example, Kanban-method orients the designer on the use of a set of visualized cards registering the information about the tasks being solved.

These cards are prepared for interactions with them any of which fulfills the role of the corresponding interruption. Thus, any access to the chosen card or a queue of cards can be interpreted as interrupting the previous item (task) of the designer's behavior for the planned or situational processing in accordance with the definite Kanban-rule. It should be noted that Kanban-rules are specified and being implemented under following basic restrictions:

1. On any card presenting the corresponding task, intermediate states of its solution are not reflected.

2. Arising the new tasks (creatively generated by the designer in the real time) is out the responsibility of Kanban mechanisms.

However, in the reality of designing, an interruption can be happened at an intermediate step (and state) of the task solution, and creative generating the new tasks is the very important reason for self-interruptions initiated by the designer in a real-time solving of project tasks.

Except the above, there are other reasons for interruptions of human-computer activity that are taking place but not managed with using of Kanban rules. All of these notes prompt combining of the managerial potential of interruptions with Kanban mechanisms.

The creation of the offered version of combining had two stages the first of which has dealt with developing a kernel of version. This kernel includes a set of agile means that was extended by the use of programmable queues of tasks. Features of our solutions embedded in the kernel were described in the paper [4] where Fig. 1 presented the workspace of the offered version of combining.

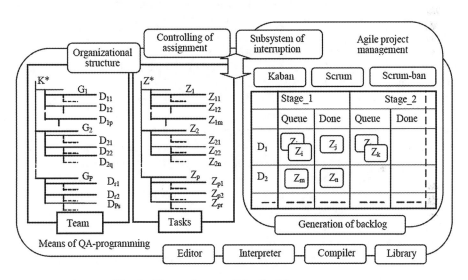

Fig. 1. The workspace of the flexible management

The scheme of the workspace consists of following components:

1. The organizational structure that reflects relations among team K^*, its groups $\{G_p(\{D_{ps}\})\}$, their members $\{D_{ps}\}$ and project tasks $\{Z_i\}$ in real-time.
2. A number of means that can be fitted on agile project management in Kanban or Scrum or Scrum-ban version.
3. A number of means, including Editor, Interpreter, Compiler and Library, which support conceptually algorithmic programming the work with queues of tasks.
4. Subsystem "Controlling of assignment" that allows appointing the estimated characteristics of time for planned work with each task Z_i included to the front of tasks (backlog of Kanban or Scrum or Scrum-ban).

5. The subsystem of interruption that supports the work with tasks in parallel and in pseudo-parallel mode.

Four positions of this list were presented in our previous publications [4, 5] where the interruption subsystem was only mentioned. At that moment, the necessity of this component was envisaged, but it was specified in general without deep delving and without the possibility of using in practice. Developing the second stage of our approach is aimed at an extension of the kernel by instruments supporting of the personal multitasking that is implemented in coordination with the agile management by taking into account the human-computer interruptions. Moreover, the main attention is paid on self-interruptions that can be inevitable without interactions with natural experience and its models.

3   Related Works

The paper [3] of McFarlane and Latorella occupies the first place among related works. This work includes the thorough analysis of human-computer interruption as a phenomenon and specifies the number of viewpoints on it. The special attention is devoted to factors of human interruptions and examples of their values. The paper also discusses the use of program agents as an important source of interruptions. It also presents a number of methods for coordinating interruptions.

An important group of related work includes reviews of the domain "Interruption management." In this group, the paper [2] discusses a number of viewpoints on this domain in attempts to find their integrity description, the review [6] describes effects of interruptions as the outcome of a complex set of variables and the publication [7] compares ways of switching ad interrupting in different multitasking environments.

The next group combines the papers presents self-interruptions in the human-computer activity. In this group, we mark the paper [8] that develops a typology of self-interruptions based on the Multitasking integration of Flow Theory and Self-regulation Theory. The typology is based on negative (Frustration, exhaustion, and obstruction) and positive (stimulation, reorganization, and exploration) triggers. The very useful paper [9] also suggest a typology of self-interruptions including seven basic types ("adjustment", "break", "inquiry", "recollection", "trigger" and "wait") each of which generates for increasing the effectiveness of multitasking.

One more group of related tasks deals with the explicit and implicit programming of a human behavior caused by an interaction potential embedded in screen interfaces. Registering the human activity in program forms has been offered and specified constructively for Human Model Processor (MH-processor) in the paper [10]. The good example of implicit programming of the human behavior is described in [11] where multitasking and interruptions are estimated for a web-oriented activity. Both these examples indicate the usefulness of implementing the human behavior in program forms.

Moreover, finally, we mark the group of related works, which are bound to the domain "Concept Development and Experimentation (CD&E)." In [12] CD&E is defined as "A method which allows us to explore and predict, by way of experimentation, whether new concepts that may impact people, organization, process and systems will contribute to transformation objectives and will fit in a larger context." This document specifies the processes in this subject area where the role and place of conceptual experimentation in military applications are indicated in details. The following publication [13] defines the version of the occupational maturity of the CD&E-process. The publication also demonstrates some specialized solutions that are focused on a behavioral side of experimenting.

It should be noted, all papers indicated in this section were used as sources of requirements in developing the set of instrumental means provided the offered version of the interruption management. Any of these papers concerns only a part of the offered approach to interrupting in conditions of multitasking.

4   Reflection on the Semantic Memory

The features of our solutions are caused by applying the artificial semantic memory the use of which orients on reflections of the human behavior based on interactions with experience and its models. This memory supports controlled intertwining of doing and thinking both of which are implemented by the designer, first of all, in the process of self-interruptions.

In the investigated case, automated doing includes a part that automates a part of used thinking with conceptual artifacts. Moreover, this part of doing simulates actions in thought experimenting. The controlled intertwining of doing and thinking is coordinated with an activity of consciousness that has a dialogue nature. That is why the semantic memory has a question-answer type (QA-memory) oriented on supporting the workflows "Interactions with Experience" [4] in conceptual designing of SISs.

In the toolkit WIQA, the semantic memory and its instrumental surrounding are specified and implemented so that they allow conceptually modeling and simulating any component and any relation among components of an operational space where the designer solves delegated (appointed) tasks. What is particularly important, they help to build and apply experience models combined in an "Experience Base" with an embedded "Ontology." All of the indicated reflections constitute a conceptual space (generally presented in Fig. 2) where the designer can work with conceptual artifacts of tasks being solved.

The scheme indicates that the operational space S includes a conceptual space (C-space) as an activity area where designers fulfill the automated part of own conceptual thinking when they work with conceptual artifacts. It should be noted, the C-space is a kind of the actuality, that inherits definite regularity from the operational space, and it has owned regularities expressed the behavioral nature of the human activity.

In solutions of project tasks $\{Z_j\}$, any result of the reflection (or shortly R^{QA}) is accessible for any designer as an interactive object (QA-object). Conceptually thinking, designers work in the C-space. In such actions, they interact with necessary concepts, units of the Experience Base and other components of the C-space. Thus, the C-space is the system of conceptual artifacts that are created and used by designers in processes of conceptual thinking when they conceptually solve the project tasks. In this kind of the

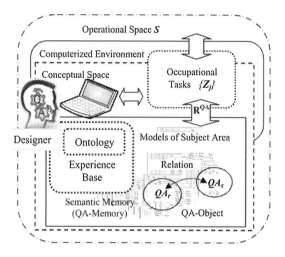

Fig. 2. Operational space

activity, the C-space models entities and processes of designing, and such modeling is oriented on their reflections in the consciousness of designers. For example, reasoning, the designer combines corresponding dialog processes in consciousness with their models registered in cells of the QA-memory.

Such potential of the QA-memory is provided by features of the memory cells intended for storing the interactive objects of the QA-type. Any cell is a kernel of the corresponding QA-object, and this kernel can be expanded by components that are shown in Fig. 3.

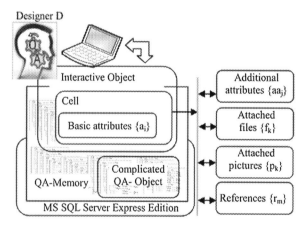

Fig. 3. Structure of QA-objects

To the kernel with its basic attributes, the designer can attach additional attributes, files, pictures and references to external informational sources if they will be useful for the work with the corresponding QA-object. Thus, QA-memory with its cells is intended for registering the conceptual content of units (reflected from the operational space) in view of the semantics of their textual descriptions. Moreover, for the definite QA-object, the accessible semantics can include information about relations with other objects.

The QA-memory has an algorithmic potential because cells of this memory can be adjusted on emulating of any type of data and coding of any operator an interpretation of which will implement the necessary processing of corresponding data. This potential has been investigated and tested in the work with some practical tasks. The mastered experience has allowed creating the original pseudo-code language L^{WIQA}, built-in the toolkit WIQA. This language opens the possibility for programming in the operational space and C-space where the designer can build the complicated QA-object of declarative and procedural types. This language also helps to include conceptually algorithmic reasoning in conceptual thinking and experimenting with conceptual artifacts.

5   Features of Intellectual Processing the Self-interruptions

The offered approach helps to manage the human-computer interruptions in any of their cases, but the more interesting class of these cases is self-interruptions when the designer discovers the necessity to include a new task in the implemented multitasking work. The typical scheme of processing for this type of self-interruptions is presented in Fig. 4.

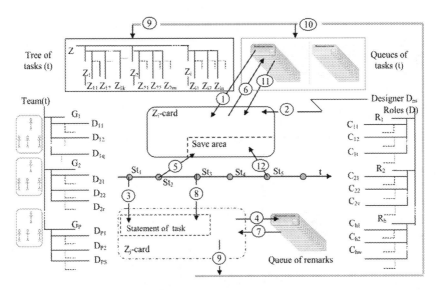

Fig. 4. Steps of processing the self-interruption

This scheme indicates that the designer D_{ps} (as a member of a definite group G_p of the designers' team) implements the delegated task Z_i that has been chosen, for example, from the corresponding queue of the Kanban board. (label one on the scheme). In accordance with Kanban-rules, any task of any queue has been extracted from the task tree of the corresponding project In a management process; the necessary information describes the task Z_i in the corresponding visualized card (Z_i-card) that is mapped in the QA-memory. Such cards have a normative structure and can be used as units of data in pseudocode programming of interactions with any card and/or queues of cards. Such possibility is useful for an automation of behavioral actions not only in the agile management but also in the interruption management in conditions of multitasking.

When the designer implements a current task Z_i, a reason of a self-interruption (label 2) unpredictably appears in brain structures where this reason must get its expression in the consciousness of the designer for explicit initiating the process of the corresponding interruption. Let us assume that such reason potentially leads to the new task Z_j. In this case, the first step St_1 of such initiating should aim at registering the arisen reason outside of brains with using the reason description (for example in the form of a number of words and phrase as keys for remembering the reason) that must be sufficient for formulating an initial statement of the task Z_j. Without such registering, there is a danger to "lose" the reason or incorrectly recover it. In the described research, the special utility activated by the designer creates the card (for the potential task Z_j) where it registers the keys in the field "statement of task" (label 3). After registering, the Z_j-card relocates to the queue of temporal remarks (label 4).

Only after that, it is better to start the second step St_2 of preparing the task Z_i for the future resumption. Actions of this step are similar to actions of the computer interruptions when the system of interruption registers the necessary information in the save area (label 5) that extends the informational structure of the card used in the Kanban management. It should be noted, the point between the first and the second steps can be the point of the next self-interruption if it is necessary or useful for the designer. Therefore, for ordering, the Z_i-card relocates to the corresponding queue of tasks (label 6) after the second step.

In the offered approach, at the third step St_3, extracting the Z_j-card from the queue of temporal remarks (label 7), the designer defines the task Z_j in a form that adequately reflects the interruption reason for its successful processing. Such work is creative, and it requires interactions with accessible experience or, by other words it requires intellectual actions of the designer. That is why the title of this paper includes the phrase "intellectual processing." More definitely, the designer creates the statement of the task (label 8) by interacting with the Experience Base and Ontology as its part.

In this step, the task Z_j is included in the tree of project tasks (label 9) and in the set of queues of tasks (label 10) that are solved by the designer. The choice of the appropriate queue is caused by the priority of the new task Z_j and its type. If the priority is low then the task Z_i then the task Z_j embeds in a backlog of the used agile managing. In another case, the task Z_j is located in the definite queue of tasks for its solving in multitasking mode.

The fourth step St_4 suggests processing of the reason when the designer creatively solves the task Z_j. This step will begin and fulfill by normative rules of the described management that are applied to an ongoing set of tasks. Solving the tasks corresponding the interruption reasons, the designer can use conceptual experiments any of which helps to understand the reason, outline the way of its processing and check the preliminary solution of the task Z_j. All of these actions include using the interactions with the experience. After conceptual experimenting, the task Z_j will be a member of a multitasking set, and it will be managed similarly as other members of this set.

At the fifth step St_5, after activating Z_i-card (label 11), actions of the designer provide the recovery of the task Z_i from the point of its interruption. Actions of this step use data from the corresponding save area (label 12). In the described approach, these tasks are reflected in a set of queues oriented on types of tasks and roles played by the designer in solving the tasks. The hierarchical structure of roles and competencies is presented in Fig. 4. The use of types, roles, and competencies open additional opportunities for effective managing the multitasking work of the designer. In concluding this section we note, the offered approach is realized in the instrumental environment WIQA, which supports conceptually algorithmic programming of designer's actions. In this case, solutions of tasks are described by behavioral programs that are executed by the designer who fulfills the role of the (intellectual) processor. Thus, the activity of the designer divided on operators that facilitate their interruptions. Moreover, the designer has the opportunity of using the interpreter of behavioral programs and this interpreter provides step-by-step executing the operators of such programs. Therefore, ways of computer interruptions can be used as a source of inheritance for the interruption management of behavioral programs. This similarity has been used in the offered approach.

6 Comparative Estimation

Focusing on SEMAT documents, underlining the important role of way-of-working in software engineering, we have studied the new version of the work management based on the controlled use of human-computer interruptions.

Called documents pay special attention to applying the Agile management means in the real-time work of designers. Solutions described in these documents are based on visualized cards registering the current states of tasks being solved in predefined points of their life cycles. It was assumed that designers use appropriate possibilities and means for (human-computer) interacting with these cards. The lack of SEMAT recommendations for useful forms of human-computer interactions with the cards has led us to the intention of developing our approach to the Agile management.

Expectations of novelty and effects from the development of this approach were caused by features of the toolkit WIQA that provides the experiential kind of human-computer interaction with the process of designing. Moreover, these expectations were met.

The implementation of the approach included two steps. The first step has led to the management version that combines Agile means with programmable queues of tasks [4]. Basic components of this version are presented above (Sect. 2).

In this version, the designer can work with cards and queues of cards only in their predefined states that correspond to the normative division of the life-cycle of the task on stages. Sets of these states and stages were depended on the model of the life cycle used by designers. Technologically, in the WIQA-environment, any normative stage is defined by the corresponding practice previously programmed in the pseudocode language for its implementation by the designer. This feature leads to following useful possibilities:

1. Tasks solved by the team at the definite stage are reflected on their queue that uses a priority for each task. The programmable access to all queues of tasks helps to build the personal queue in interacting with which the designer uses priorities.
2. Additionally, the designer can interrupt the own work with the active practice in any point of its (pseudocode) program. The basic aim of interruptions is to conduct the useful conceptual experiments [5].

In the first version, interruptions were used, but the interruption management was absent. This managerial function is included in the second version supporting the multitasking mode that simulates mechanisms of computer interruptions. The scenario simulation is described in the previous section. This scenario provides controlled intertwining offered managerial functions. First of all, such intertwining is achieved by the use of the special area on the visualized card for saving information that provides the resumption of the interrupted task. So, in the definite moment of time, life-cycles of tasks can be crossed in any their points (not only in predefined points). The additional feature of the second version is bound with the offered way of processing the reason of the self-interruption caused by the appearance of the new project task. The above, we described this reason and its processing without details.

It should be noted that both versions of offered means are realized and successively applied (the first for two years, the second with last October) in the project organization (about 2,000 employees). This fact indicates that offered approach was tested in the real designing.

7   Conclusion

The following conclusions can be drawn from the present study of human-computer interruptions in the multitasking mode. The use of programmable queues of tasks for combining the agile management with interruption management facilitates increasing the level of automation in the behavioral activity of designers' team. In conditions of pseudocode programming the designer' behavior, such way-of-working helps to organize and control the multitasking work of any designer by analogies with computer interruptions and the use of visualized cards for registering the recovery information. The offered mechanisms open the important possibilities for self-interruptions, and especially for their class that is bound with appearing the new tasks.

In this case, the designer can creatively build the owned behavior that can include formulating the statement of the new task and experimenting with its conceptually algorithmic solution. Useful intellectual and practical effects are achieved, first of all, by using the interactions of the designer with the Experience Base and Ontology located in the environment of the toolkit WIQA. Offered combining is possible for any version of agile management that uses the board of visualized cards.

References

1. Jacobson, I., Ng, P.-W., McMahon, P., Spence, I., Lidman, S.: The essence of software engineering: the SEMAT kernel. Queue **10**(10), 1–12 (2012)
2. Janssen, C.P., Gould, S.J.J., Li, S.Y.W., Brumby, D.P., Cox, A.L.: Integrating knowledge of multitasking and interruptions across different perspectives and research methods. Int. J. Hum. Comput. Stud. **79**(C), 1–5 (2015)
3. McFarlane, D.C., Latorella, K.A.: The scope and importance of human interruption in human-computer interaction design. Hum. Comput. Interact. **17**, 1–61 (2002)
4. Sosnin, P.: Combining of Kanban and Scrum means with programmable queues in designing of software intensive systems. In: Fujita, H., Guizzi, G. (eds.) SoMeT 2015. CCIS, vol. 532, pp. 367–377. Springer, Heidelberg (2015)
5. Sosnin, P., Lapshov, Y., Svyatov, K.: Programmable managing of workflows in development of software-intensive systems. In: Ali, M., Pan, J.-S., Chen, S.-M., Horng, M.-F. (eds.) IEA/AIE 2014, Part I. LNCS, vol. 8481, pp. 138–147. Springer, Heidelberg (2014)
6. Ratwani, R.M., Andrews, A.E., Sousk, J.D., et al.: The effect of interruption modality on primary task resumption. In: Proceedings of the Human Factors and Ergonomics Society 52nd Annual Meeting, pp. 393–397. Human Factors and Ergonomics Society, Santa Monica (2008)
7. Darmoul, S., Ahmad, A., Ghaleb, M., Alkahtani, M.: Interruption management in human multitasking environment. In: 15th IFAC Symposium on Information Control Problems in Manufacturing, Ottawa, Canada, 11–13 May 2015, vol. 48, issue 3, pp. 1179–1185 (2015, preprints). IFAC-Papers Online
8. Adler, R.F., Benbunan-Fich, R.: Self-interruptions in discretionary multitasking. Comput. Hum. Behav. **29**(4), 1441–1449 (2013)
9. Jin, J., Dabbish, L.A.: Self-interruption on the computer: a typology of discretionary task interleaving. In: Proceedings of the SIGCHI Conference on Human Factors in Computing Systems (CHI 2009), pp. 1799–1808. ACM, New York (2009)
10. Card, S.K., Thomas, T.P., Newell, A.: The Psychology of Human-Computer Interaction. Lawrence Erlbaum Associates, London (1983)
11. Leiva, L.A.: MouseHints: easing task switching in parallel browsing. In: Extended Abstracts on Human Factors in Computing Systems, CHI 2011, pp. 1957–1962. ACM, New York (2011)
12. MCM-0056-2010: NATO Concept Development and Experimentation (CD&E) Process. NATO HQ, Brussels (2010)
13. Wiel, W.M., Hasberg, M.P., Weima, I., Huiskamp, W.: Concept maturity levels bringing structure to the CD&E process. In: Proceedings of Interservice Industry Training, Simulation and Education, I/ITSEC 2010 Conference, Orlando, pp. 2547–2555 (2010)

How the Strategy Continuity Influences the Evolution of Cooperation in Spatial Prisoner's Dilemma Game with Interaction Stochasticity

Xiaowei Zhao[1], Xiujuan Xu[1], Wangpeng Liu[1], Yixuan Huang[2],
and Zhenzhen Xu[1](✉)

[1] School of Software Technology,
Dalian University of Technology, Dalian 116024, China
{xiaowei.zhao,xjxu,xzz}@dlut.edu.cn, 459267269@qq.com
[2] School of Software and Microelectronics at Wuxi,
Peking University, Beijing 214125, China
orangehix@outlook.com

Abstract. The evolution of cooperation among selfish individuals is a fundamental issue in artificial intelligence. Recent work has revealed that interaction stochasticity can promote cooperation in evolutionary spatial prisoner's dilemma games. Considering the players' strategies in previous works are discrete (either cooperation or defection), we focus on the evolutionary spatial prisoner's dilemma game with continuous strategy based on interaction stochasticity mechanism. In this paper, we find that strategy continuity do not enhance the cooperation level. The simulation results show that the cooperation level is lower if the strategies are continuous when the interaction rate is low. With higher interaction rate, the cooperation levels of continuous-strategy system and the discrete-strategy system are very close. The reason behind the phenomena is also given. Our results may shed some light on the role of continuous strategy and interaction stochasticity in the emergence and persistence of cooperation in spatial network.

Keywords: Evolution of cooperation · Spatial altruism · Prisoner's dilemma · Interaction stochasticity · Strategy continuity

1 Introduction

Cooperation among agents is critical for agents' artificial intelligence (AI) and multi-agent system (MAS). Example applications include cooperative target observation [1], foraging [2], and peer-to-peer systems [3]. Coordination is important to improve the performance of the entire system; and cooperation is the first step of coordination [4]. Agent can decide whether to cooperate or defect with another agent based on their past interactions. Among previous researches, the Prisoner's Dilemma (PD) game has been adopted widely in the field of evolution of cooperation [5]. Under the setting of prisoner's dilemma, two players decide to choose one decisions from the

© Springer International Publishing Switzerland 2016
H. Fujita et al. (Eds.): IEA/AIE 2016, LNAI 9799, pp. 808–817, 2016.
DOI: 10.1007/978-3-319-42007-3_68

two alternative decisions (cooperation, C or defection, D) at the same time. If two players take cooperation strategy, their benefits will both be R. If both players take defection strategy, they will both get P. If one player take cooperation strategy and another player take defection strategy, the cooperator will get S and the defector will obtain T. In the PD, it is assumed that $T>R>P>S$ and $T+S<2R$. Obviously, if the PD game only happens for one single time, defection will be the best choice, and this cause both of players only get P. Thus, if players make decision based on their own maximization of fitness, it will result in low individual benefits.

In order to explain the phenomenon of evolution of cooperation in social and biological systems, a series of cooperation mechanisms has been proposed, including kin selection, direct reciprocity, indirect reciprocity, graph selection and group selection [6]. Graph selection (or spatial reciprocity) has become one of the most important mechanisms, and has gained wide attention in recent years [7–13]. The graph theory provides a very natural and convenient model to describe the spatial structure of cooperation evolutionary groups. Each node of the graph represents an individual participating in the game, and the link between the nodes is used to represent the adjacent links between individuals. At the same time, individuals constantly change (or maintain) their relationship with the neighbors according to the gains from the previous game. Based on such research framework, the evolution of the PD game in the past decade has been extended to various network models. It has been found that the structure of individuals' interaction plays an important role in the evolution of cooperation.

Most of the researches on spatial PD game are based on a simplified assumption that each individual will interact with all his neighbors. That is to say, as long as two individuals are "neighbors", they will play games definitely. However, in real life, whether to join a game with neighbors, the decision is not necessarily to be 100 %, because it is likely to be changed based on the individual's environment or risk propensity. Traulsen et al. [14] put forward that the interaction between individuals is Stochasticity for the first time. Chen et al. proposed a model that individuals will interact with neighbors in a random rate in the spatial PD game [15]. Simulation results show that interaction intensity can promote the cooperation level maximization. Li et al. extended the research to the next step [16]. They proposed an adaptive interaction mechanism in which an individual can adjust their interaction rate with neighbors according to their incomes. If an individual's income is higher than before, he will strengthen interaction rate with neighbors; otherwise he will reduce the interaction rate. They found that if the individual can adjust their interaction rate within a certain range, the cooperation level will be improved effectively. In addition, the individuals with low interaction rate usually occupy the cooperation cluster's edge. Thus, the level of overall cooperation system will be improved by decreasing in the interaction rate between cooperators and defectors. It is also proved that the spatial structure can effectively reduce the invasion of defectors.

Previous researches about "interaction stochasticity" are all assume that the players' strategies are "either all or nothing", that is to say, the player of the game is either complete cooperator or defector. Numerous evidences in nature and human society show that strategy may be continuous, in other words, the player of the game is neither complete cooperator nor complete defector. In recent years, the scholars have carried out some researches on the strategy continuity in spatial games. Zhong et al. [17] found that in the population with a spatial structure, the equilibrium point of the continuous strategy and the discrete strategy are different. Especially the type of the game is more close to the "Chicken game", the system of continuous strategy will show a phenomenon of "internal equilibrium", which leads to the cooperation level in continuous-strategy system higher than in discrete-strategy system.

Based on the above consideration, the influence of strategy continuity on the evolution of cooperation in spatial PD game with interaction stochasticity is worth to research. According to our known, this problem has not been studied yet. Therefore, in this paper, we study the system in which the individuals play the PD game on a network with interaction stochasticity, and we focus on the influence of the continuity of strategies on the evolution of cooperation. We use the method of computer simulation, and compare the system with continuous strategy and the system with discrete strategy in the same initial settings. To explain conveniently, the system with players adopting continuous strategies is called continuous-strategy system (CSS) in the rest of this paper; and the system with players adopting discrete strategies is called discrete-strategy system (DSS).

The structure of this paper is as follows. In the second chapter, we give an introduction of the simulation model. The third part is the result and analysis of the imitation. Finally, the conclusion is given in the fourth part.

2 Model Description

In this paper, we adopt the matrix of PD game as Tanimoto and Sagara's [18] work. We use $Dr = P-S$ to represent the games close to "stag-hunt" game and use $Dg = T-R$ to represent the games close to like "chicken" game. We also set $P = 0$ and $R = 1$, so the pay-off matrix can be shown as formula (1).

$$M = \begin{pmatrix} R & S \\ T & P \end{pmatrix} = \begin{pmatrix} 1 & -Dr \\ 1+Dg & 0 \end{pmatrix} \text{ and } 0 \leq Dr, Dg \leq 1 \qquad (1)$$

We use the real number $si (si \in [0, 1])$ to represent continuous strategy. The strategy is fully cooperation strategy when $si = 1$ and the strategy is fully defection strategy when $si = 0$. When the value of si is between 0 and 1, the players adopt partial cooperation strategy. Suppose that a player has adopted strategy p in the game and his opponent uses strategy q, its income can be calculated by formula (2) according to Zhong et al. [17].

$$\pi(p, q) = (R - S - T + P)pq + (S - P)p + (T - P)q + P \qquad (2)$$

When $p = 0$ or 1 and $q = 0$ or 1, $\pi (D, C) = T$, $\pi (D, D) = P$, $\pi (C, C) = R$, $\pi (C, D) = S$. Thus, the formula (2) can also be used to calculate the income of discrete strategy. Because $Dr = P-S$, $Dg = T-R$, $P = 0$ and $R = 1$, we can get formula (3) from formula (2).

$$\begin{aligned}
\pi(p, q) &= (Dr - Dg)pq - Drp + (1 + Dg)q + P \\
&= (Dr - Dg)pq + (-Dr)p + (1 + Dg)
\end{aligned} \tag{3}$$

In this paper, the population is placed on a square-lattice network and the length of the side is L ($L = 100$). Each individual sits in a single grid with 4 neighbors and can interact with its every single neighbor. For the CSS, at the initial stage, the individual strategy is a random real number between [0, 1] which is evenly distributed on the network. For the DSS, at the initial stage, half individuals choose cooperation strategy and the others choose defection strategy. We adopt the interaction stochasticity model as Chen [15], using the variable ω ($\omega \in [0, 1]$) to represent the interaction rate between each individual. If ω increases, it represents the individual has more possibility to participating in games, otherwise the possibility of interaction will be lower. When $\omega \rightarrow 1$, the possibility of individual interact with its neighbors will be almost 100 %. When $\omega \rightarrow 0$, the possibility of interact will be almost 0 %, in another words, the interaction is frozen.

After the evolution starts, individuals play games neighbors with according to ω, and then update strategies by comparing the benefits with his neighbors. Based on the literature [19], individual's fitness can be improved if individual has good performance when it collects neighbor's strategies. We adopt the strategy updating mechanism as Chen [15]. The probability of the individual x adopting the strategy of neighbor y will be $Px \rightarrow y$.

$$P_{x \rightarrow y} = \frac{P_y}{\sum_{z \in \Omega_x} P_z} \tag{4}$$

In the formula (4), Py represents the neighbor y's benefit. $\sum_{z \in \Omega_x} P_z$ represents the sum of all neighbors' benefits. After calculating each neighbor' income respectively, individual x will learn a neighbor's strategy by probability which is proportional to each neighbor benefit. Based on formula (4), a neighbor will have a higher probability to be learned with more income. If $\sum_{z \in \Omega_x} P_z$ is zero, x will choose a neighbor randomly and adopt the neighbor' strategy with the probability according to the Fermi rule shown in formula (5).

$$f(P_y - P_x) = \frac{1}{1 + \exp[-(P_y - P_x)/K]'} \tag{5}$$

Px is the total income of x with all its neighbor. K represents the impact of noise. When $K \rightarrow 0$, x imitates y deterministically. When $K \rightarrow 1$, x adopts y's strategy randomly. In order to focus on the core of the research, we set $K = 0.1$ in this paper.

3   Result and Analysis

Based on the model introduced in part two, we use numerical simulations to test how the strategy continuity influences on the evolution of cooperation in lattice population under different rates of player's interaction. We examine two categories of parameters:

(1) The interaction rate ω;

(2) The PD game's type which is classified by the space of Dr and Dg.

According to Tanimoto [18], there are three sub-classes of PD which are illustrated in Fig. 1. Based on $Dr \in [0, 1]$ and $Dg \in [0, 1]$, Dr and Dg increase from 0.0 to 1.0 by 0.1, so the game space is represented by $11 \times 11 = 121$ points (the whole space of PD game, AllPD). The first sub-class of PD game is Donor & Recipient Game (DRG) which is the average of 11 points featured by $Dr = Dg$. The second sub-class is the boundary games between the PD and Chicken games (BCH) which is the average of 11 points featured by $Dr = 0$. The third sub-class is the boundary games between the PD and Stag–Hunt games (BSH) which is another average of 11 points featured by $Dg = 0$.

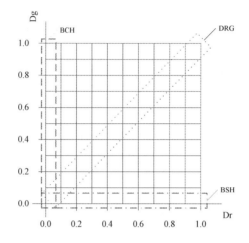

Fig. 1. Three sub-classes of PD, BCH, BSH and DRG.

In the following simulation experiment, all the statistical results are reached an equilibrium in system (arrives at quasi-equilibrium). Cooperation rate (the fraction of cooperation) is an important indicator to describe system average cooperation level. It is defined as $fc = \sum_{i \in N} S_i / N$. All the results are average values after 50 independent runs.

Firstly we research the dependence of fc on the interaction rate ω in CSS and DSS respectively. As shown in Fig. 2, both continuous and discrete strategies on the whole value of interval have a great influence on cooperation evolution. Especially in the interval when $\omega < 0.4$, the difference is particularly large. For both DSS and CSS, ω have a larger impact on fc. According to its general trend, fc will increase at the beginning and then drop in the DSS; and fc in the CSS is down to the bottom and increase after that in ALLPD, BCH and DRG games. From Fig. 2a we can observe that in AllPD game, when $\omega > 0.3$, the fcs in the DSS and CSS are no difference, but when $\omega < 0.3$, the fc in CSS drops fast at the beginning which makes the gap of fc between DSS and CSS biggest. From Fig. 2b we can observe that in BCH game, the difference of CSS and DSS is relatively large. In the CSS, fc drops to the bottom firstly ($\omega = 0.1$)

and then rises. In the DSS, fc rise firstly ($\omega = 0.1$) and then decreases. In all, when $\omega < 0.4$, fc in DSS is higher than that in CSS; and when $\omega > 0.4$, fc in CSS is higher than that in DSS. From Fig. 2c we can observe that in the BSH game, the difference of fc between CSS and DSS is very small when $\omega > 0.3$. When $\omega < 0.3$, fc in the DSS is higher than that in the CSS; especially when $0.1 < \omega < 0.2$, the gap of fc between DSS and CSS is biggest. From Fig. 2d we can observe that the trend of DRG game is like BCH game, the difference of fc in CSS and DSS reaches the maximum value at $\omega = 0.1$.

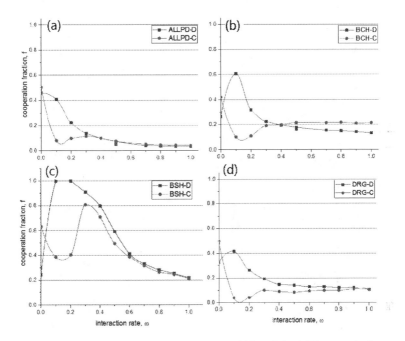

Fig. 2. Correlations between ω and fc in the CSS and the DSS respectively.

From Fig. 2, we can draw the basic conclusion that the overall performance of the CSS is worse than that of DSS based on interaction stochasticity mechanism. The difference is prominent when interaction rate is small. In order to further analyze this phenomenon, we choose four specific values $\omega = 0.1, 0.3, 0.5, 0.9$ and compare the specific difference between the CSS and DSS. Experimental results are shown in Fig. 3.

From Fig. 3a, c, e, the region in the space $Dr\text{-}Dg$ that is close to BCH, with the increase of Dg we observe the obvious "phase transition" from cooperation to defection. When $\omega = 0.1$, transition point is $Dg = 0.5$. At this point, a rapid falling of fc to 0.0 with a slight increase of Dg. On the other hand in the CSS, we see different situations, as shown in Fig. 3b, d, f. In the CSS, in the region in the $Dr < 0.4$ and $Dg > 0.5$ (the square A), fc is always greater than 0, especially $fc = 0.1$ when $\omega = 0.1$ (Fig. 3b). With the increase of ω, in square A gradually declines to 0.05 when

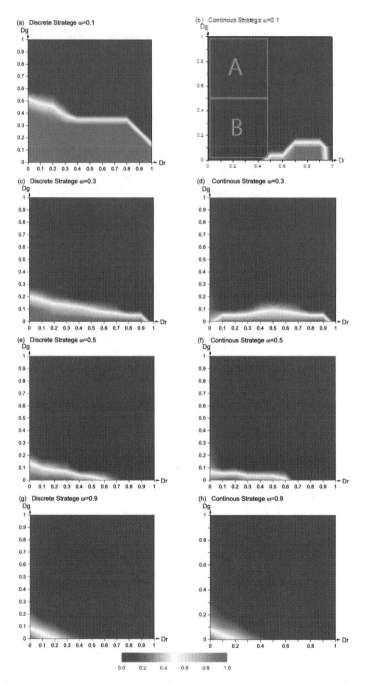

Fig. 3. The fc as a function of Dr and Dg in the CSS and the DSS under different ω.

ω reaches 1.0. In all, when $Dr < 0.4$ and $Dg > 0.5$, fc in the CSS is slightly greater than that in the DSS. The advantage of continuous strategy in this region on cooperation gradually diminishes with the increase of ω. In the CSS, in the region in the $Dr < 0.4$ and $Dg < 0.5$ (the square B), the influence of strategy continuity on cooperation shows complex changing phenomenon. In this region, fc in CSS is always lower than that in DSS, especially $fc = 0.1$ when $\omega = 0.1$ (Fig. 3b). With the increase of ω, fc in square B gradually increases closely to fc in DSS when ω reaches 1.0. Besides the regions of square A and B, the difference of fc in CSS and DSS is insignificant.

From the experimental results of Figs. 2 and 3, we can see that in the full Dr-Dg parameter space, interaction stochasticity and strategy continuity influences the evolution of cooperation together. It is at a relatively low level of ω, the slightly positive influence of the strategy continuity on the fraction of cooperation is clear in the region where Dr is relatively low and Dg is relatively high. On the other hand, the massive negative influence of the strategy continuity on the fraction of cooperation is obvious in the region where Dr is relatively low and Dg is relatively low. This explains the remarkably lower fraction of cooperation in CSS than in the DSS in Fig. 2 when $\omega < 0.4$. When ω increases, the difference between the CSS and DSS becomes smaller which implies the influence of strategy continuity reduces with greater interaction rate. This explains the results shown in Fig. 2 $\omega > 0.4$ and also explains the significant difference of CSS and DSS in DRG.

From the above simulations, we can draw the conclusion that, strategy continuity and interaction stochasticity are two different types of dynamics in promoting cooperation. The reason of the strategy continuity can promote the cooperation level of the system is the evolution of strategies instinctively tends toward the internal equilibrium which leads higher fc in the CSS than DSS when the game is closer to BCH game. The reason that the interaction stochasticity can promote cooperation level is that it weakens the interaction strength between individuals, equivalent to reducing the individual effective number of neighbors, especially in the type of game closer to BSH game. Simulation experiment shows that strategy continuity and interaction stochasticity cannot promote cooperation in the same time, because their deep mechanisms are incompatible, especially when the type of game is closer to the DSG game.

4   Conclusions and Future Work

In this paper, we have studied the influence of strategy continuity on evolution of cooperation in lattice population with interaction stochasticity. From the simulation results, we find that the strategy continuity has a great effect on cooperation level in spatial PD game with interaction stochasticity. According to our research, when the interaction rate is low, the cooperation level is reduced if the strategy is continuous in the system. When the interaction rate is high, the cooperation levels of continuous-strategy system and the discrete-strategy system are close. The reason behind this is strategy continuity and interaction stochasticity are two different types of dynamics in promoting cooperation, but their deep mechanisms are incompatible. When interaction rate is low, the incompatible feature is prominent, especially when the type of game is closer to DSG game.

For future work: our existing approach lacks an analysis of the interaction between the microscopic individual and we need to study the mechanism behind the incompatibility phenomenon further.

Acknowledgement. This work is partly supported by National Natural Science Foundation of China under grant No. 61300087.

References

1. Luke, S., Sullivan, K., Balan, G., Panait, L.: Tunably decentralized algorithms for cooperative target observation. In: Proceedings of the Fourth International Joint Conference on Autonomous Agents and Multiagent Systems, Utrecht Netherlands, July 2005
2. Gheorghe, M., Holcomb, M., Kefalas, P.: Computational models of collective foraging. Biosystems 61(2–3), 133–141 (2001)
3. Camorlinga, S., Barker, K., Anderson, J.: Multiagent systems for resource allocation in peer-to-peer systems. In: Proceedings of International Symposium on Information Communication and Technologies, pp. 1–6 (2004)
4. Ciacarini, P., Omicini, A., Zambonelli, F.: Multiagent system engineering: the coordination viewpoint. In: Jennings, N.R., Lespérance, Y. (eds.) ATAL 1999. LNCS (LNAI), vol. 1757, pp. 250–259. Springer, Heidelberg (2000)
5. Alexrod, R.: The Evolution of Cooperation. Basic Books, New York (1984)
6. Nowak, M.A.: Five rules for the evolution of cooperation. Science 314(5805), 1560–1563 (2006)
7. Suri, S., Watts, D.J.: Cooperation and contagion in web-based, networked public goods experiments. PLoS ONE 6, e16836 (2011)
8. Gracia-Lázaro, C., Cuesta, J.A., Sánchez, A., Moreno, Y.: Human behavior in prisoner's dilemma experiments suppresses network reciprocity. Sci. Rep. 2, 325 (2012)
9. Skyrms, B., Pemantle, R.: A dynamic model for social network formation. Proc. Natl. Acad. Sci. USA 97, 9340–9346 (2000)
10. Zimmermann, M.G., Eguíluz, V.M., Miguel, M.S.: Coevolution of dynamical states and interactions in dynamic networks. Phys. Rev. E 69, 065102(R) (2004)
11. Santos, F.C., Pacheco, J.M., Lenaerts, T.: Cooperation prevails when individuals adjust their social ties. PLOS Comput. Biol. 2, 1284–1291 (2006)
12. Tanimoto, J.: Dilemma solving by the coevolution of networks and strategy in a 262 game. Phys. Rev. E 76, 021126 (2007)
13. Pestelacci, E., Tomassini, M., Luthi, L.: Evolution of cooperation and coordination in a dynamically networked society. J. Biol. Theory 3, 139–153 (2008)
14. Traulsen, A., Nowak, M.A., Pacheco, J.M.: J. Theor. Biol. 244, 349 (2007)
15. Chen, X.: Interaction stochasticity supports cooperation in spatial Prisoner's dilemma. Phys. Rev. E 78, 051120 (2008)
16. Li, Z.: Evolution of cooperation in lattice population with adaptive interaction intensity. Physica A 392, 2046–2051 (2013)
17. Zhong, W., Kokubo, S., Tanimoto, J.: How is the equilibrium of continuous strategy game different from that of discrete strategy game? Biosystems 107(2), 88–94 (2012)
18. Tanimoto, J.: Difference in dynamic between discrete strategies and continuous strategies in a multi-player game with a linear payoff structure. BioSystem 90, 568–572 (2007)

19. Guan, J.-Y., Wu, Z.-X., Huang, Z.-G., Xu, X.-J., Wang, Y.-H.: Europhys. Lett. **76**, 1214 (2006)
20. Zhao, X., Xia, H., Yu, H., Tian, L.: Agents' cooperation based on long-term reciprocal altruism. In: Jiang, H., Ding, W., Ali, M., Wu, X. (eds.) IEA/AIE 2012. LNCS, vol. 7345, pp. 689–698. Springer, Heidelberg (2012)
21. Zhao, X., Yu, H., Xu, Z., Tian, T., Xu, X.: The role of probability of learning and reconnecting in the evolution of cooperation. In: 2014 IEEE 12th International Conference on Dependable, Autonomic and Secure Computing, 22–24 August 2014, pp. 377–381 (2014)
22. Zhao, X., Xu, Z., Yu, H., Tian, T., Xu, X.: Evolution of mixed strategies on cooperative. In: 2015 IEEE Conference on Collaboration and Internet Computing, 28–30 October 2015

π-SROIQ(D): Possibilistic Description Logic for Uncertain Geographic Information

Safia Bal-Bourai[1,2(✉)] and Aicha Mokhtari[2]

[1] Ecole nationale Suprieure d'Informatique,
BP 68M, Oued-Smar, 16309 Algiers, Algeria
s_bourai@yahoo.fr
[2] RIIMA, Computer Science Department, USTHB, Algiers, Algeria
http://www.esi.dz
http://www.usthb.dz

Abstract. The central question addressed in this paper is to determine how to represent uncertain knowledges present in many applications in geographic domain. The resulting proposition is a possibilistic extension of the very expressive Description Logic $\mathcal{SROIQ(D)}$, the underlying logic of the OWL2, called π-$\mathcal{SROIQ(D)}$. It is a solution for handling uncertainty and for dealing with inconsistency in geographic applications. Both syntax and semantics of the proposed description logic are considered. In addition we provide a tableau algorithm for reasoning with the possibilistic DL π-$\mathcal{SROIQ(D)}$.

Keywords: Description logics · Geographic information · Uncertainty · Possibilistic logic

1 Introduction

The spatial data are an approximation of reality. They can be localized in space according to their different degrees of precision and dated with more or less precision and different levels of temporal granularity. This leads us to ask about the structure of the spatial data in their different dimensions, namely: spatial and temporal attributes under uncertainty. Which is the inability to accurately specify something. The need to model and reason with uncertainty has been found in many contexts specially in geographical applications such as environmental, archaeological or touristic ones. For example in archaeological applications, one might want to express the certainty of position of objects or the accessibility of a road in a given period. GIS (geographic information system) can organize objects and phenomena in space, but it cannot, for example, determine whether a statement is true or false or determines at what degree of certitude the statement is true or infer new informations. Description logics (DLs) constitute a knowledge representation framework equipped with tools to automatically infer implicit information present in a knowledge base. Classical description logics are not appropriate to deal with uncertain information. To overcome this limit,

© Springer International Publishing Switzerland 2016
H. Fujita et al. (Eds.): IEA/AIE 2016, LNAI 9799, pp. 818–829, 2016.
DOI: 10.1007/978-3-319-42007-3_69

a number of approaches have been proposed on extending DLs with uncertainty: probabilistic [10,14,17], fuzzy [5,22,24] and possibilistic [3,6,19] extensions that have fundamental differences in terms of semantics and thus in the types of knowledge they can represent.

Recently, possibilistic Description Logics have been proposed as formalisms capable of representing and reasoning with uncertain informations [3,6,19]. However, no framework for reasoning with possibilistic extensions of SROIQ(D) exist today. Recall that this logic is particularly interesting for geographic information. For this reason, we proposed an extension of SROIQ(D)[1] named Poss-SROIQ(D) [1]. In the latest version, the syntax and the semantics of Poss-SROIQ(D) (π-SROIQ(D)) have been developed. In this new version the semantics has been refined, accurately, the proposed weighted concept (for more details see the semantics of $(A, \alpha)^I(x)$ in Table 1 relatively to the semantics of this constructor in [1]). But the main contribution, on this paper, is the proposition of a reasoning procedure for uncertainty geographical knowledge expressed in π-SROIQ(D). The paper is organized as follows: Sect. 2 overviews the background and the related work. Section 3 presents the description logic π-$\mathcal{SROIQ}(\mathcal{D})$. Section 4 describes the reasoning algorithm for the possibilistic DL π-SROIQ(D). Finally, Sect. 5 concludes the paper and presents some future works.

2   Background and Related Work

In this section, we first introduce the context of our approach, then, we will review related work.

2.1   Background

The development of π-SROIQ(D) was motivated by the existence of the uncertainty in most geographical applications. But such knowledge cannot be expressed nor reasoned with the DL SROIQ(D) [13]. We have favored the expressiveness of the DL SROIQ(D), the subjacent DL of OWL2 [15] as our main formalism, because this logic is better suited to represent the geographic knowledge, i.e., the spatio-temporal knowledge. Naturally, geographic knowledges appear uncertain and qualitative. For this reason we focus on these properties by representing their uncertainty.

SROIQ(D) is equipped with concrete datatypes and named individuals which are particularly useful for many geographic applications. Concrete datatypes can be used to represent thematic, spatial and temporal properties of geographic objects. For example, we can represent completed time of a new hotel: To do so, we may define a concrete role "completed time" and use it in the concept ≺ Hotel ⊓ ∃ completed time.string ≻. Indeed, SROIQ(D) specifies nominals which are an

---

[1] S used for ALC extended with transitive roles (R+), R for limited complex role inclusion axioms, role disjointness, O for nominals, I for inverse roles, Q for qualified number restrictions and (D) for data values or data types.

important feature, used to represent a concept by enumerating its elements. For example, we can define the concept of "Rainy Week", relating to the weather prevision as follows: rainy week starts from February 12 to 18 is represented by the concept: RainyWeek $\equiv \{(Feb : 12), (Feb : 13), (Feb : 14), (Feb : 15), (Feb : 16), (Feb : 17), (Feb : 18)\}$. Where RainyWeek is a concept, Feb:12, ..., Feb:18 are individuals. From the above examples, we cannot exactly assert the completed time of a new hotel or the rainy days, due to incomplete knowledges, but we can estimate to which degree of uncertainty these statements are possible. In addition, roles can be declared to be (Symmetric, Asymmetric, Inverse, Transitive, Complex, ...). These properties are important for us. We can, for example, represent the fact that the roles relative to spatial relations between two spatial areas: EC (Externally Connected with), DC (Disconnected from), PO (Partially Overlaps) and EQ (EQual to) are symmetric.

2.2   Related Work

A few approaches for managing uncertainty in geographic information have been proposed recently. Shu et al. [23] suggest an uncertainty model including uncertain spatio-temporal data types and spatio-temporal relationships. Li et al. [11] describe a framework, based on fuzzy approach, to handle both uncertainty and time in spatial domain. In [21], spatial objects are modeled as a set of points and fuzziness modeled by means of memberships. Pfoser et al. [18] present probabilistic models to model position and attributes errors. These proposals are theoretical and not followed by implementation. de Saint-Cyr and Prade [9] develops logical framework in a possibilistic approach for handling uncertain spatial information called attributive formula and merging it when it comes from multiple sources. The notion of spatial relationships is not considered. In addition, several general approaches have been proposed to extend description logics with probabilistic [10,14,17], fuzzy [5,22,24] and possibilistic logic [3,6,16,19]. The fuzzy description logic handles vagueness information and can model imprecise knowledge, whereas the probabilistic and the possibilistic deals with information certainty and can model uncertain knowledge. The probabilistic logic is based on the classical probability theory. It relies the use of a single measure refers to the probability that the proposition is true. The possibilistic approach, based on possibility theory, allows both certainty and possibility be handled in the same formalism. It is developed in two directions: qualitative and quantitative [7]. It provides an efficient solution for handling uncertainty or prioritized formulas and coping with inconsistency. Contrary to the probabilistic logic, the possibilistic logic provides a good representation of partial or total ignorance. As for the main differences between probability and possibility theory, the probability of an event is the sum of the probabilities of all worlds that satisfy this event, which increases errors, whereas the possibility of an event is the maximum of the possibilities of all worlds that satisfy the event.

The relevant works related to our solution are [4–6,19,22]. Giorgos and Stamou [22] presents an algorithm for reasoning with the fuzzy DLs f-SHOIQ and f-SROIQ. Bobillo and Straccia [4,5] propose a fuzzy version of $\mathcal{SROIQ}$ and provide a reasoning capabilities of fuzzy $\mathcal{SROIQ}$. This work is more adapted for fuzzy information than uncertain one. Qi et al. [19] proposes a possibilistic description logic as an extension of the description logic ALC applying the classic tableau algorithm. The syntax of description language is the same as the standard DL and the interpretation is based on possibility theory with a thorough study of its semantics. Couchariere et al. [6] propose a direct extension of the tableau algorithm ALC. They introduce extensions of the clash definition and completion rules to handle necessity values. These solutions are then not adapted for managing the various forms of geographic uncertainty.

3   The Description Logic π-SROIQ(D)

In this section, we describe the DL π-SROIQ(D). This includes the definition of the syntax and the semantics.

- **Syntax**
  The π-SROIQ(D) DL is a combination of both DL SROIQ(D) and possibilistic logic. Alphabet of symbols is proposed: possibilistic concepts, possibilistic roles and possibilistic individuals. Possibilistic concepts denote possibilistic sets of individuals and possibilistic roles denote possibilistic relationships. Let C, D are (possibly complex) concepts, A is an atomic concept, R is a (possibly complex) abstract role, S is a simple role, T is a concrete roles, a, b are abstract individuals, d is a concrete predicate, v is a concrete individual, n a natural numbers with $n \geq 0$ and $\alpha \in (0,1]$.
  The concepts can be built using the following rule:

$$C, D \rightarrow \top \mid \bot \mid A \mid C \sqcap D \mid C \sqcup D \mid \neg C \mid$$
$$\forall R.C \mid \exists R.C \mid \forall T.d \mid \exists T.d \mid$$
$$\geq nS.C \mid \leq nS.C \mid \geq nT.d \mid \leq nT.d \mid$$
$$\exists S.Self \mid \{(o_i, \alpha_i)\} \mid (A, \alpha).$$

The roles can be built from atomic role (R_A), inverse role (R^-), universal role U according to the following rule:

$$R \rightarrow R_A \mid R^- \mid U.$$

The main difference with the non-possibilistic case is the presence of weighted nominals and weighted concepts [1].
A Possibilistic knowledge base Σ is a finite set of weighted axioms in the form (φ, α). The axioms consist of a possibilistic TBox denoted by πTBox $\mathcal{T}$ (axioms about concepts and roles) and possibilistic ABox denoted by πABox $\mathcal{A}$ (axioms about individuals)[2].

---

[2] A πTBox consists of a finite set of possibilistic GCIs (General Concept Inclusions) and a finite set of possibilistic role axioms RIAs (Role Inclusion Axioms). A πABox is a finite set of possibilsitic concepts and role assertions axioms expressed in the following forms: $(C(a), \alpha)$, $(R(a,b), \alpha)$ or $(\neg R(a,b), \alpha)$, $T((a, v), \alpha)$ or $\neg T((a, v), \alpha)$.

- **Semantics**

  Concepts and roles are viewed as the open first order formula. We are inspired by the semantics given in [8,25]. Which defines conjunction as minimum, disjunction as maximum and the negation function as $1 - \alpha$, universal quantifier as infinum and existential quantifier as supremum. The semantic level is based on the notion of a possibility distribution, which is a mapping from a set of interpretations to the interval $[0,1]$. The possibility distribution of an interpretation I, denoted $\pi(I)$, is the degree of compatibility of interpretation I with available beliefs. From a possibility distribution, two important measures can be processed:

  – The possibility degree of a formula φ, defined as $\Pi(\alpha) = max\{\pi(I), I \models \varphi\}$
  – The certainty degree of a formula φ, defined as $N(\varphi) = 1 - \Pi(\neg\varphi)$.

  A possibilistic interpretation $I = (\triangle^I, \cdot^I)$ with respect to the possibilistic concrete domain $\triangle_D$, consists of a non empty set $\triangle^I$, disjoint with $\triangle_D$ called the domain of I, and a valuation $\cdot^I$ which associates, with each concept C a function $C^I : \triangle^I \to [0,1]$, with each abstract role R, a function $R^I : \triangle^I \times \triangle^I \to [0,1]$, with each concrete role T a function $T^I : \triangle^I \times \triangle_D \to [0,1]$, with each individual a, an element in $\triangle^I$, with each concrete individual v, an element in $\triangle_D$, and to each n-ary concrete predicate d the interpretation $d^D : \triangle_D^n \to [0,1]$. Tables 1 and 2 show respectively the semantics of concepts and roles and axioms of π-SROIQ(D).

Table 1. Semantics of π-SROIQ(D) concepts and roles

Constructor	Semantics
$\top^I(x)$	1
$\bot^I(x)$	0
$(C \sqcap D)^I(x)$	$min(C^I(x), D^I(x))$
$(C \sqcup D)^I(x)$	$max(C^I(x), D^I(x))$
$(\neg C)^I(x)$	$1 - C^I(x)$
$(\forall R.C)^I(x)$	$inf_{y \in \triangle^I} \{max(1 - R^I(x,y), C^I(y)\}$
$(\exists R.C)^I(x)$	$sup_{y \in \triangle^I} \{min(R^I(x,y), C^I(y)\}$
$(\forall T.d)^I(x)$	$inf_{v \in \triangle_D}\{max(1 - T^I(x,v), d^D(v)\}$
$(\exists T.d)^I(x)$	$sup_{v \in \triangle_D}\{min(T^I(x,v), d^D(v)\}$
$(\geq nS.C)^I(x)$	$sup_{y_i \in \triangle^I} \, min_{i=1}^n\{min(S^I(x,y_i), C^I(y_i))\}$
$(\leq nS.C)^I(x)$	$inf_{y_i \in \triangle^I} \, max_{i=1}^{n+1}\{max(1 - S^I(x,y_i), 1 - C^I(y_i))\}$
$(\geq nT.d)^I(x)$	$sup_{v_i \in \triangle_D} \, min_{i=1}^n\{min(T^I(x,v_i), d^D(v_i))\}$
$(\leq nT.d)^I(x)$	$inf_{v_i \in \triangle_D} \, max_{i=1}^{n+1}\{max(1 - T^I(x,v_i), 1 - d^D(v_i))\}$
$(\exists S.Self)^I(x)$	$S^I(x,x)$
$(\{(o_1, \alpha_1), ..., (o_n, \alpha_n)\})^I(x)$	$sup\{\alpha_i/x = o_i^I\}$
$(A, \alpha)^I(x)$	$A^I(x)$ if $A^I(x) \geq \alpha$, otherwise $A^I(x) = 0$
$(R_A)^I(x,y)$	$R_A^I(x,y)$
$(R^-)^I(x,y)$	$R^I(y,x)$
$U^I(x,y)$	1

Table 2. Semantics of π-SROIQ(D) axioms

Table 2. Semantics of π-SROIQ(D) axioms

Axiom	Semantics
$(C \sqsubseteq D)^I$	$inf_{x \in \Delta}^I \{C^I(x) \Rightarrow D^I(x)\}$
$(R_1, ... R_n \sqsubseteq R)^I$	$inf_{x_1, x_{n+1} \in \Delta}^I \{sup\{inf(R_1^I(x_1, x_2), ..., R_n^I(x_n, x_{n+1})) \Rightarrow R_n^I(x_1, x_{n+1})\}\}$
$(T_1 \sqsubseteq T_2)^I$	$inf_{x \in \Delta, v \in \Delta D}^I \{T_1^I(x, v) \Rightarrow T_2^I(x, v)\}$
$C^I(a)$	$C^I(a^I)$
$R^I(a, b)$	$R^I(a^I, b^I)$
$\neg R^I(a, b)$	$1 - R^I(a^I, b^I)$
$T^I(a, v)$	$T^I(a^I, v_D)$
$\neg T^I(a, v)$	$1 - T^I(a^I, v_D)$

4 Reasoning in π-SROIQ(D)

In order to decide the consistency of a possibilistic KB $\Sigma = (\pi\text{TBox}, \pi\text{ABox})$ a tableau T for Σ has to be constructed. The proposed reasoning algorithm is based on π-SROIQ(D), which extends the DL SROIQ(D) [12,13] with uncertainty. We introduce extensions of the clash definition and completion rules to take into account the uncertainty. Inspired by the approaches proposed in [6,20] which incorporate uncertainty in the standards DL language ALC, the interesting feature of our approach covers new uncertainty parameters that are associated with the axioms and assertions in the π-SROIQ(D) KB. So different type of geographic uncertainty can be modeled and reasoned with the proposed algorithm. In addition, similar to Couchariere [6], the extension computes an inconsistency degree that quantifies the level of inconsistency of the possibilistic knowledge base.

Given $\Sigma = (\mathcal{T}, \mathcal{A})(\mathcal{T}:\pi\text{TBox}, \mathcal{A}:\pi\text{ABox})$. To show that a knowledge base Σ is satisfiable, a tableau algorithm constructs a sequence of ABoxes A0, A1, ... An where A0 = $\mathcal{A}$ and each Ai is obtained from Ai-1 by an application of one completion rule (see paragraph 4.2).

In addition:

– Each concept description is transformed into its negation normal form (NNF),
– each axiom of the form $(C \sqsubseteq D, \alpha)$ is replaced with $(\top \sqsubseteq \neg C \sqcup D, \alpha)$,
– for each individual "a" in $\mathcal{A}$ and each axiom $(\top \sqsubseteq \neg C \sqcup D, \alpha)$ in $\mathcal{T}$ add $(a:\neg C \sqcup D, \alpha)$ to $\mathcal{A}$.

Let $\Sigma = (\mathcal{T}, \mathcal{A} \sqcup \{((\neg C \sqcup D)(a), \alpha)\})$ be an π-SROIQ(D) KB.

- R (R_A: Abstract role together with their inverses, R_D: Concrete role) be the set of roles occurring in Σ.
- I_Σ be the set of individuals appearing in Σ (either in assertions or in nominals concepts).
- For a concept D, cl(D) is the smallest set that contains D and is closed under sub-concepts and $\sim$ (negation).

- if ∀R.C ∈ cl(D), P ⊑* R (a role P is sub-role of a role R), then ∀ P.C ∈ cl(D). Finally, cl(Σ) = $\bigcup_{(a:D,\alpha)\in A}$cl(D)$\bigcup_{(C\sqsubseteq D,\alpha)}$cl(C)∪cl(D).

Definition. Let T = $(\mathbf{S}, \mathcal{L}, \mathcal{E}_A, \mathcal{E}_D, \mathcal{J})$ is the tableau for Σ such that:
- **S** is a non-empty set of elements,
- $\mathcal{L}$: **S** x cl(Σ) → [0, 1]: is a function that maps each pair of elements of **S** and cl(Σ) to an uncertainty degree (the degree of element being an instance of the concept).
- $\mathcal{E}_A$: R_A x (**SxS**)→ [0, 1]: is a function that maps each abstract role in R_A and pair of elements of **S** to an uncertainty degree.
- $\mathcal{E}_D$: R_D x (**Sx**$\triangle_D$) → [0, 1]: is a function that maps each concrete role in R_D to pair set of pairs of elements of **S** and concrete values to an uncertainty degree.
- $\mathcal{J}$: I_Σ → **S** maps individuals occurring in Σ to element of S.

For all s, t ∈ **S**, C, C1, C2 ∈ cl(Σ), R, S ∈ R_A, T, T′ ∈ R_D, α ∈ [0, 1], the tableau T satisfies the following conditions. Inspired with [6], we extend the satisfying conditions of [12,13] with uncertainty by taking into account the necessity degree (certainty value).

(C1) if ($\mathcal{L}(s, C1⊓C2)$, α), then ($\mathcal{L}(s, C1)$, α) and ($\mathcal{L}(s, C2)$, α),
(C2) if ($\mathcal{L}(s, C1⊔C2)$, α), then ($\mathcal{L}(s, C1)$, α) or ($\mathcal{L}(s, C2)$, α),
(C3) if ($\mathcal{L}(s, ∀S.C)$, α) ∈ $\mathcal{L}(x)$ and (S(s, t), β) ∈ $\mathcal{E}_A$ (s, t) then ($\mathcal{L}(t,C)$, γ) and γ ≥ min((α, β),
(C4) if ($\mathcal{L}(s, ∃S.C)$, α) ∈ $\mathcal{L}(x)$ and (S(s, t), β) ∈ $\mathcal{E}_A$(s, t) then ($\mathcal{L}(t,C)$, γ) and γ ≥ min((α, β),
(C5) $\mathcal{L}(s,\top) = 1$ and $\mathcal{L}(s, \bot) = 0$ for all s ∈ **S**,
(C6) if (C ∈ $\mathcal{L}(s)$, α) then (¬C ∉ $\mathcal{L}(s)$, 1 − α) (C atomic or ∃ R.Self),
(C7) if ((s,t), α) ∈ $\mathcal{E}_A$ (R) iff ((t,s), α) ∈ $\mathcal{E}_A$ (Inv(R)),
(C8) if ((s,t), α) ∈ $\mathcal{E}_A$ (R) and R ⊑* S then ((s,t), α) ∈ $\mathcal{E}_A$ (S),
if ((s,t), α) ∈ $\mathcal{E}_D$ (T) and T ⊑* T′ then ((s,t), α) ∈ $\mathcal{E}_D$(T′),
(C9) if (∀S.C, α) ∈ $\mathcal{L}(s)$ and ((s,t), α) ∈ $\mathcal{E}_A$(R) for some R ⊑* S with Trans(R), then (∀R.C, α) ∈ $\mathcal{L}(t)$ or ((s,t), 1 − α) ∈ $\mathcal{E}_A$ (R),
if (∀T.d, α) ∈ $\mathcal{L}(s)$ and ((s,t), α) ∈ $\mathcal{E}_D$(T), then (T ∈ d^D, α),
(C10) if ((≥nS.C), α) ∈ $\mathcal{L}(s)$ then ♯ S^T((s, C), α) ≥ n,
(C11) if ((≤nS.C), α) ∈ $\mathcal{L}(s)$ then ♯ S^T((s, C), 1 − α) ≤ n,
(C12) if ((≤nS.C), α) ∈ $\mathcal{L}(s)$ then ♯ S^T((s,C), 1 − α) ≥ n then (C ∈ $\mathcal{L}(t)$, α) or ¬ C ∈ ($\mathcal{L}(t)$, 1 − α),
(C13) if {O,α} ∈ $\mathcal{L}(s)∩\mathcal{L}(t)$ then s = t,
(C14) if ((C ⊑ D), α) ∈ $\mathcal{T}$ then (¬C ∈ $\mathcal{L}(s)$, 1 − α) or (D ∈ $\mathcal{L}(s)$, α).

4.1 π-SROIQ(D) Tableau Algorithm

The proposed algorithm works on completion-forest for Σ. Unlike in the algorithm presented in [12,13] where nodes correspond to objects and edges to relations that connect two nodes:

- Each our node x is labeled with a set $\mathcal{L}(x)$ containing π-SROIQ(D) concepts (C, α). If (C, α) $\in \mathcal{L}(x)$ implies that x is an instance of C at least with the degree α.
- and each edge $\prec$ x, y $\succ$ is labeled with a set $\mathcal{L}(x, y)$ of roles (R, α). If (R, α) $\in \mathcal{L}(x, y)$ implies that α represents the degree of uncertainty of being $\prec$ x, y $\succ$ an instance of R.

The tableau algorithm evolves a completion forest according to completion rules applied in arbitrary order. The algorithm stops until a clash or when no more rules can be applied.

The notion of a clash is used in order to denote that a contradiction has occurred in the completion forest and for detecting possible inconsistencies in the possibilistic knowledge base.

$\mathcal{L}(x)$ is said to contain a clash if there are nodes x and y such that:

1. a pair $\prec \bot, \alpha \succ$, with $\alpha > 0$ or a pair $\prec \top, \alpha \succ$ with $\alpha < 0$, or
2. for some concept name A, $\{(A, \alpha), (\neg A, \beta)\} \subseteq \mathcal{L}(x)$ where $\alpha, \beta > 0$, or
3. x is an S-neighbour of x and $\neg \exists S.\text{Self} \in \mathcal{L}(x)$, or
4. for some $o \in I$, $x \neq y$ and $o \in \mathcal{L}(x) \cap \mathcal{L}(y)$ or
5. there is some $\text{Dis}(R, S) \in R_A$ and y is an R- and an S-neighbour of x, or
6. there is some $\text{Asy}(R) \in R_A$ and y is an R-neighbour of x and x is an R-neighbour of y, or
7. there is some concept $(\leq nS.C) \in \mathcal{L}(x)$ and x has $n + 1$ s-neighbours y0, ..., yn with yi $\neq$ yj for all $0 \leq i < j \leq n$.

If the algorithm stops and all of the forest i ($i = 0, ..., n$) contain an inconsistency degree α_i, then the inconsistency degree is $\min(\alpha_1, ..., \alpha_n)$. Otherwise, if the algorithm stops and there is not inconsistency, then the knowledge base is consistent.

4.2   Possibilistic Completion Rules

The proposed completion rules are extensions of the rules presented in [12,13] where necessity degrees are taken into account.

The negation rule
if $(\neg C, \alpha) \in \mathcal{L}(x)$, x is not blocked[3], and (C(x), α) not in $\mathcal{L}(x)$
then $\mathcal{L}(x) \rightarrow \mathcal{L}(x) \cup \{C, 1 - \alpha)\}$
The $\rightarrow \sqcap$-rule
if $(C_1 \sqcap C_2, \alpha) \in \mathcal{L}(x)$, x is not indirectly blocked,
and $\{(C_1, \alpha), (C_2, \alpha)\} \nsubseteq \mathcal{L}(x)$,
then $\mathcal{L}(x) \rightarrow \mathcal{L}(x) \cup \{(C_1, \alpha), (C_2, \alpha)\}$.

---

[3] Blocked Node: A node x is blocked iff it is not a root node and it is either directly or indirectly blocked or if there is an ancestor z of x such that z is blocked. If a node x is blocked and none of its ancestors is blocked, then we say that x is directly blocked.

For example. We suppose the assertion $\prec$dangerous $\sqcap$ accessible (R10), O.6 $\succ$ then we infer that dangerous(R10), 0.6$\succ$ and dangerous(R10), 0.6$\succ$.

The $\rightarrow \sqcup$-rule

if $(C_1 \sqcup C_2, \alpha) \in \mathcal{L}(x)$, x is not indirectly blocked,
and $\{(C_1, \alpha), (C_2, \alpha)\} \cup \mathcal{L}(x) = \emptyset$
then $\mathcal{L}(x) \rightarrow \mathcal{L}(x) \cup \{C\}$ for C $\in \{(C_1, \alpha), (C_2, \alpha)\}$.

The $\rightarrow \exists$-rule

if $(\exists\, S.C, \alpha) \in \mathcal{L}(x)$, x is not blocked,
and x has no S-neighbour y with $(C, \beta) \in \mathcal{L}(y)$ and $\beta \geq \alpha$
then create a new node y with $\mathcal{L}(x, y) := (S, \alpha)\}$ and $\mathcal{L}(y) := (C, \alpha)\}$.

The $\rightarrow \exists$D-rule

if $(\exists\, T.d, \alpha) \in \mathcal{L}(x)$, x is not blocked,
with $(d, \beta) \in \mathcal{L}(y)$
then create a new node y with $\mathcal{L}(x, y) := (T, \alpha)\}$ and $\mathcal{L}(y) := (d, \alpha)\}$
For example. We suppose the assertion $\prec(\exists$ haslocalisation.String)(area1), $0.8 \succ$ we generate a new individual $\prec$ haslocation(area1, (x, y)), $0.8 \succ$ and (String(x, y), 0.8).

The $\rightarrow$ Self rule

if $(\exists S.Self, \alpha) \in \mathcal{L}(x)$, x is not blocked and $(S, \beta) \notin \mathcal{L}(x, x)$ with $\beta \geq \alpha$
then Add the edge (x, x), if it does not yet exist,
and set $\mathcal{L}(x, x) \rightarrow \mathcal{L}(x, x) \cup \{S, \alpha\}$.

The $\rightarrow \forall$-rule

if $(\forall R.C, \alpha) \in \mathcal{L}(x)$, x is not indirectly blocked,
and there is y such that $(R, \beta) \in \mathcal{L}(x, y)$ and $(C(y), \gamma) \notin \mathcal{L}(y)$ and $\gamma \geq \min(\alpha, \beta)$.
then $\mathcal{L}(y) \rightarrow \mathcal{L}(y) \cup \{(C(y), \gamma)\}$ with $\gamma = \min(\alpha, \beta)$.

The $\rightarrow \forall$ D-rule

if $(\forall\, T.d, \alpha) \in \mathcal{L}(x)$, x is not indirectly blocked, and there is y such that $(T, \beta) \in \mathcal{L}(x, y)$ and $(d, \gamma) \notin \mathcal{L}(y)$ and $\gamma \geq \min(\alpha, \beta)$
then $\mathcal{L}(y) \rightarrow \mathcal{L}(y) \cup \{(d, \gamma)\}$ with $\alpha = \min(\gamma, \beta)$.
For example. If we have the assertion $\prec \forall$haslocalisation.String, $0.8 \succ$ and $\prec$ haslocation(area1, (x, y)),0.7 $\succ$ we infer that $\prec$ String(x, y), $\beta \succ$ with the constraint: $\alpha = \min(\gamma, \beta)$.

The $\rightarrow \forall$+ rule

if $(\forall\, S.C, \alpha) \in \mathcal{L}(x)$, x is not indirectly blocked,
and there is some R with Trans(R) and R $\sqsubseteq^* S$ (S has a reflexive and transitive subrole R),
and there is an R-neighbour y of x with $(\forall R.C, \alpha) \notin \mathcal{L}(y)$,
then $\mathcal{L}(y) \rightarrow \mathcal{L}(y) \cup \{ (\forall R.C, \alpha)\}$

The choose-rule

if $(\leq n\, S.C, \alpha) \in \mathcal{L}(x)$ or $(\geq n\, S.C, \alpha) \in \mathcal{L}(x)$, x is not indirectly blocked, and there is an S-neighbour y of x with $\{(C, \alpha), (\neg C, 1 - \alpha)\} \cap \mathcal{L}(y) = \phi$
then $\mathcal{L}(y) \longrightarrow \mathcal{L}(y) \cup \{E\}$ for some E $\in \{(C, \alpha), (\neg C, 1 - \alpha)\}$

The $\rightarrow \geq$-rule

if $(\geq n\ S.C, \alpha) \in \mathcal{L}(x)$, x is not blocked, there are no n S-neighbours y1,..yn of x
with $(C, \alpha) \in L(yi)$ and yi $\neq$ yj for $1 \leq i < j \leq n$
then create n new nodes y1,..yn with $\mathcal{L}(\prec x,yi\succ) = \{(S,\alpha)\}$, $L(y_i) = \{(C, \alpha)\}$ and
yi $\neq$ yj for $1 \leq i < j \leq n$.

The $\rightarrow \leq$-rule

if $(\leq n\ S.C, \alpha) \in \mathcal{L}(z)$, z is not indirectly blocked,
and there are more than n S-neighbours of z and there are two of them x, y with
$(C, \alpha) \in \mathcal{L}(x) \sqcap L(y)$ and not x $\neq$ y.
Then:

1. if x is a nominal then Merge $(y, x)^4$
2. else if y is a nominal or an ancestor of x, then Merge (x, y)
3. else Merge (y, x)

The $\rightarrow$ O-rule

if for some $O \in I_{\Sigma}$ there are two nodes x, y with $(\{o\}, \alpha) \in \mathcal{L}(x) \cap L(y)$ and not
x$\neq$y
then Merge(x, y).
For example, we consider the rainy days starts from February 12 to 18 with
degrees uncertainty for each day: 0.9, 0.6, 0.7, 0.3, 0.4, 0.2, 0.2 are represented
as follows:

1. $L(x) := \{(Feb.12, 0.9), (Feb.13, 0.6), (Feb.14, 0.7), (Feb.15, 0.3), (Feb.16, 0.4)\}$
2. $L(y) := \{(Feb.15, 0.3), (Feb.16, 0.4), (Feb.17, 0.2), (Feb.18, 0.2)\}$.
3. We have $L(x) \cap L(y) := \{(Feb.15, 0.3), (Feb.16, 0.4)\}$ then
4. Merge (x, y)

The $\rightarrow$ NN-rule

if 1. $(\leq n\ S.C, \alpha) \in \mathcal{L}(x)$ x is a nominal node, and there is a blockable S-neighbour
y of x such that $(C, \alpha) \in \mathcal{L}(x)$ and x is a successor of y,
2. there is no m such that $1 \leq m \leq n$, $(\leq m\ S.C, \alpha) \in \mathcal{L}(x)$, and there exist m
nominal S-neighbours z1, ..., zm of x with $(C, \alpha) \in \mathcal{L}(zi)$, and zi $\neq$ zj, for all
$1 \leq i < j \leq m$ then

1. guess m, with $1 \leq m \leq n$ and set $\mathcal{L}(x) := \mathcal{L}(x) \cup \{(\leq mS.C, \alpha)\}$
2. create m new nodes y1, ..., ym with $\mathcal{L}(x, yi) := \{S, \alpha\}$, $L(yi) = \{(oi, \alpha), (C, \alpha)\}$
   for each oi $\in I_{\Sigma}$ new in G and yi $\neq$ yj for $1 \leq i < j \leq m$.

---

[4] Merging a node y into a node x, means that we add L(y) to L(x), all edges leading
to y so that they lead to x and add all the edges leading from y to nominal nodes
(contain a nominal) so that they lead from x to the same nominal nodes; then if y is
not a root node we remove y (and blockable sub-trees below y) from the completion-
graph, otherwise we set L(y) to the empty set and assert that x = y.

5   Conclusion

In this paper, we proposed a possibilistic extension of the $\mathcal{SROIQ}(\mathcal{D})$ description logic called π-$\mathcal{SROIQ}(\mathcal{D})$ by incorporating certainty level for different elements of the $\mathcal{SROIQ}(\mathcal{D})$ DL. This allows us to consider real objects, described by, the uncertainty of their concepts, individuals, attributes and relationships. We first introduced the syntax and the semantics of the possibilistic DL π-$\mathcal{SROIQ}(\mathcal{D})$. Then we proposed a possibilistic extension of the $\mathcal{SROIQ}(\mathcal{D})$ tableau algorithm. Different notions of uncertainty can be modeled and reasoned with, to reduce the gap between the real knowledge and those represented in the knowledge base. However, currently we can not compare the performance of our proposition to others because of the non existence of similar approaches. Major topics for future direction are indeed the extension of the OWL 2, based on π-$\mathcal{SROIQ}(\mathcal{D})$ to support possibilistic ontology and study the feasibility and evaluate the complexity of our proposition over real geographic ontologies. Examples of application taken from [2] illustrating the uncertainty reasoning are given. Experimental study related to the urban domain is underway.

References

1. Bourai, S.B., Mokhtari, A., Khellaf, F.: Poss–SROIQ(D): possibilistic description logic extension toward an uncertain geographic ontology. In: Catania, B., Cerquitelli, T., Chiusano, S., Guerrini, G., Kämpf, M., Kemper, A., Novikov, B., Palpanas, T., Pokorny, J., Vakali, A. (eds.) New Trends in Databases and Information Systems. AISC, vol. 241, pp. 277–286. Springer, Heidelberg (2014)
2. Bourai, S., R202015: Une logique de description possibiliste pour des connaissances géographiques: application au domaine touristique. Rapport interne, RIIMA, Computer Science Department, USTHB
3. Benferhat, S., Bouraoui, Z.: Possibilistic DL-Lite. In: Liu, W., Subrahmanian, V.S., Wijsen, J. (eds.) SUM 2013. LNCS, vol. 8078, pp. 346–359. Springer, Heidelberg (2013)
4. Bobillo, F., Straccia, U.: Reasoning with the finitely many-valued Lukasiewicz fuzzy Description Logic SROIQ. Inf. Sci. **181**, 758–778 (2011)
5. Bobillo, F., Straccia, U.: Fuzzy ontology representation using OWL 2. Int. J. Approximate Reasoning **52**, 1073–1094 (2011)
6. Couchariere, O., Lesot, M.-J., Bouchon-Meunier, B.: Consistency checking for extended description logics. In: Proceedings of the 21st International Workshop on Description Logics, DL 2008. Description Logics, vol. 9, pp. 602–607. CEURWS.org/CEUR Workshop Proceedings (2008)
7. Dubois, D., Lang, J., Prade, H.: Possibilistic logic. In: Gabbay, D.M., Hogger, C.J., Robinson, J.A., Siekmann, J.H. (eds.) Handbook of Logic in Artificial Intelligence and Logic Programming, pp. 439–513. Oxford University Press, Oxford (1994)
8. Dubois, D., Mengin, J., Prade, H.: Possibilistic uncertainty and fuzzy features in description logic: a preliminary discussion. In: Sanchez, E. (ed.) Capturing Intelligence: Fuzzy Logic and the Semantic Web, pp. 101–113. Elsevier, Amsterdam (2006)

9. de Saint-Cyr, F.D., Prade, H.: Logical handling of uncertain, ontology-based, spatial information. Fuzzy Sets Syst. (Science Direct) **159**, 1515–1534 (2008)

10. Heinsohn, J.: Probabilistic description logics. In: Proceedings UAI 1994, pp. 311–318. Morgan Kaufmann (1994)

11. Li, J., Simon, W., Huang, G.: Handling temporal uncertainty in GIS domain: a fuzzy approach. In: Symposium on Geospatial Theory, Ottawa (2002)

12. Horrocks, I., Sattler, U.: Ontology reasoning in the SHOQ(D) description logic. In: Proceedings of the Seventeenth International Joint Conference on Artificial Intelligence (2001)

13. Horrocks, I., Kutz, O., Sattler, U.: The even more irresistible SROIQ. In: Doherty, P., Mylopoulos, J., Welty, C.A. (eds.) Proceedings of the 10th International Conference of Knowledge Representation and Reasoning (KR 2006), Lake District, UK (2006)

14. Jaeger, M.: Probabilistic role models and the guarded fragment. In: Proceedings IPMU 2004, pp. 235–242 (2004). Extended version in Int. J. Uncertain. Fuzz. Knowl.-Based Syst. **14**(1), 43–60 (2006)

15. Krotzsch, M., et al.: OWL 2 web ontology language primer. Technical report, W3C October (2009)

16. Liau, C.-J., Yao, Y.Y.: Information retrieval by possibilistic reasoning. In: Mayr, H.C., Lazanský, J., Quirchmayr, G., Vogel, P. (eds.) DEXA 2001. LNCS, vol. 2113, pp. 52–61. Springer, Heidelberg (2001)

17. Lukasiewicz, T.: Expressive probabilistic description logics. Artif. Intell. **172**(67), 852–883 (2008)

18. Pfoser, D., Tryfona, N., Jensen, C.S.: Indeterminacy and spatiotemporal data: basic definitions and case study. GeoInformatica **9**(3), 211–236 (2005)

19. Qi, G., Pan, J., Ji, Q.: Possibilistic description logics extension. In: Proceedings of the International Workshop on Description Logics (DL 2007), pp. 435–442 (2007)

20. Qi, G., Ji, Q., Pan, J.Z., Du, J.: PossDL — a possibilistic DL reasoner for uncertainty reasoning and inconsistency handling. In: Aroyo, L., Antoniou, G., Hyvönen, E., Teije, A., Stuckenschmidt, H., Cabral, L., Tudorache, T. (eds.) ESWC 2010, Part II. LNCS, vol. 6089, pp. 416–420. Springer, Heidelberg (2010)

21. Schneider, M.: Uncertainty management for spatial data in databases: fuzzy spatial data types. In: Güting, R.H., Papadias, D., Lochovsky, F.H. (eds.) SSD 1999. LNCS, vol. 1651, pp. 330–351. Springer, Heidelberg (1999)

22. Giorgos, G., Stamou, G.: Reasoning with Fuzzy Extensions of OWL and OWL 2 (2013)

23. Shu, H., Spaccapietra, S., Parent, C., Sedas, D.Q.: Uncertainty of geographic information and its support in MADS. In: ISSDQ03 Proceedings (2003)

24. Straccia, U.: Reasoning within fuzzy description logics. J. Artif. Intell. Res. **14**, 137–166 (2001)

25. Zadeh, L.A.: Fuzzy sets as a basis for a theory of possibility. Fuzzy Sets Syst. **1**(1), 3–28 (1978)

Smart Solution of Alternative Energy Source for Smart Houses

Jakub Vit and Ondrej Krejcar[(⊠)]

Faculty of Informatics and Management,
Center for Basic and Applied Research, University of Hradec Kralove,
Rokitanskeho 62, 500 03 Hradec Kralove, Czech Republic
jakub.vit@seznam.cz, Ondrej@Krejcar.org

Abstract. This paper describes the design and implementation of smart photovoltaic power source. It describes the principle of solar irradiation to energy transformation and influences on its effectiveness, as well as ways to achieve higher energy gain. Based on the operation of the control algorithm the microcontroller-controlled electronics is capable of providing optimal adjustment of the front surface of the panel towards the Sun. It is also capable of measuring the energy balance of the whole device. All individual electronic devices that was built as a products of this project are directly using or are based on Arduino development kit. This work also designs and implements solutions that provide visualization and storage of measured data. Evaluation of the benefit of smart power source in practical operation is performed on the basis of comparison with the measured data obtained from the photovoltaic panel with fixed position.

Keywords: Arduino · Photovoltaics · Renewable power source · Sun tracker

1 Introduction

As a result of long-term technological development we live in a world where we are increasingly dependent on a variety of technologies. This our dependence has been increasing in response to declining prices and expansion of the possibilities of electronic equipment. Therefore, we increasingly encounter with their use in such areas of human life, seemed to be a utopia or a matter for the distant future.

Energy can be considered as a very important commodity [3], which is necessary to run complex intelligent home. It is therefore necessary to ensure its supply even at times when traditional way is not possible or is limited. The aim of this work is to design and implement a device that would be able to act as an independent source of energy within the smart home, or some of its parts.

At present time, we can definitely say that sunlight represents a virtually inexhaustible source of energy for our civilisation. Emitted intensity of solar radiation is equal to approximately 64000 kW/m^2. Hence, use of this energy for a smart power source is very suitable.

Converting sunlight directly into electricity is carried out by photovoltaic cells, which operate on the principle of photovoltaic effect. This occurs when electrons are

© Springer International Publishing Switzerland 2016
H. Fujita et al. (Eds.): IEA/AIE 2016, LNAI 9799, pp. 830–840, 2016.
DOI: 10.1007/978-3-319-42007-3_70

excited from the material due to incident radiation. Free movement of electrons emerging in semiconductor material then forms a voltage potential and direct electrical current. There are many types of photovoltaic cells, which are currently used. The most commonly used cells are based on silicon, which can be divided according to the type of production to monocrystalline, polycrystalline and amorphous cells. Every cell type has its own advantages and disadvantages and it should be considered for every particular application. These kinds of cells differ from each other in the way of its production, price, sensitivity to direct and diffuse sunlight, and output power [1, 3, 5].

The aim of this work is to design and build a microcontroller-controlled smart power source that will gather energy through photovoltaic panel [10]. To be able to talk about this power source as a smart device, it must be capable of changing of the orientation of the photovoltaic panel to be set always towards the Sun. It also must be able to measure amount of consumed energy for its operation and amount of energy made by photovoltaic cell [1].

2 Problem Definition

There is also a large number of influences that affect the efficiency of converting of the sunlight into electrical energy. Above all, it is the latitude, time of year, atmospherical conditions and the orientation of the panel towards the Sun that have the major influence. For example gas and particles in the atmosphere can disperse, reflect and absorb the incident sunlight. It results in reducing overall energy of the sunlight and strengthening its diffuse component. The amount of energy produced is also influenced by the ambient temperature. With each degree of temperature reduced, the rated output of polycrystalline and monocrystalline panels drops by about 5 % [4].

When solar panel with fixed position is used, its orientation towards the Sun must be optimally set. It minimizes the reflections of radiation and therefore increases its efficiency. Orientation of the front side of the panel must be selected to the South. Its inclination towards the ground must be then set accordingly to the latitude so that at noon the Sun's rays fell on the panel vertically. The ideal slope of the panel should be in the range of 35°–45°, depending on the season.

An increase of the efficiency of sunlight conversion into electrical energy can be made by using an appropriate mechanical design of the device. This design must ensure changes of the orientation of the panel in accordance with the movement of the Sun across the sky. A device that performs such position corrections is called tracker. Many types of such devices can be used. These types mainly differ in the way they evaluate position of the source of radiation, in their mechanical and electronical complexity of construction and also in the amount of degrees of freedom. These devices set position of the solar panel in one or two axes. Single-axis systems can be divided into polar and azimuthal depending on the axis, by which the solar panel is rotated [14]. Panel inclination towards the surface of the ground is then fixed, depending on the current latitude. On the other hand, two-axis trackers are capable of adjusting the position of the photovoltaic panel in both axes depending on the ambient conditions [1, 2, 4].

According to [6], a solar tracker [12, 14] can be considered beneficial if the system, taking into account its own consumption, produces at least twenty percent more power than panel with fixed position. Contribution of tracking the Sun can also be found in the results presented in [7] where at least 24 % benefit was evident when tracking devices are used instead of panels with fixed position. It is also interesting to compare the difference in the benefits of single and two-axis trackers on the amount of produced energy shown in Fig. 1.

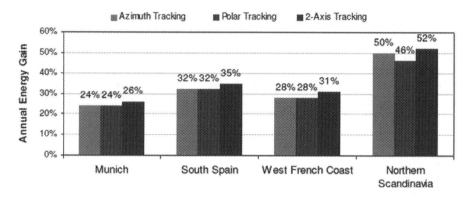

Fig. 1. Energy gain when tracking devices are used [7]. (Color figure online)

2.1 Passive Tracking

Tracking devices that use passive tracking as shown in Fig. 2 are based on a simple, but not very accurate principle. They may use, for example, thermal expansion of low boiling point liquids like Freons. Mechanical construction of the solar panel is placed in an equilibrium position on the axis of rotation. On opposite sides of the panel there are

Fig. 2. Passive solar tracker mechanism [8].

connected tanks filled with Freon. Due to the movement of the sun and exposure to solar radiation, the tanks are unevenly heated. It causes evaporation of Freon and its transition into a second tank, which is colder. Condensation of Freon in the second tank changes center of gravity of the entire mechanism and it therefore spontaneously inclines towards the Sun [1].

2.2 Active Tracking

For active tracking of the Sun, it is necessary to use electronic control unit, which provides evaluation of the position of the solar panel. That's why active trackers are more complex and more expensive than passive tracking systems. The current position of the Sun can be evaluated in several ways. It is of course possible to combine both following methods.

The first method uses time-based positioning. The position of the Sun is calculated according to the current time and date by special algorithms. Panel is then optimally adjusted towards it. This method of adjustment is not dependent on actual ambient conditions. It is therefore not affected by the amount of incident radiation, heavy clouds and so on.

Another way how to evaluate the position of the Sun is to use optical sensors for searching for the source of radiation. Many components can be used for this purpose. For example photoresistors, phototransistors, photodiodes and photovoltaic cells. Generally, it is necessary to use at least a pair of sensors for each axis. Output signals of sensors are in equilibrium at the optimum orientation of the panel towards the Sun. By constant measurement of their output signals, it can be then decided which sensor receives more solar radiation. The control electronics then sets the panel in such a manner that output signals of the sensors are equal again [1, 6, 11].

3 Implementation

Designed and developed device is an active two-axis solar tracker that can be seen in the Fig. 3. It uses a 40 Wp monocrystalline solar panel for producing of electricity. Its mechanical construction is composed of iron profiles, ensuring sufficient durability of the entire device. The bottom part of the device is a stand with motors and transmissions that ensures the azimuthal rotation of the output shaft [7]. On the shaft, the upper part of the device is mounted. Both parts can be separated because of easier transport. This part of the mechanism ensures inclination adjustment of the photovoltaic panel. The controlling electronics in waterproof boxes as well as the voltage regulator and sun sensor are also attached to the upper part of the device [11]. Linking of the stand and upper module and all accessories is secured by the wiring harness through which the battery is also connected [9]. Angle of movement of the upper part is limited by limit switches in range of 180°.

The main component of controlling electronics is a microcontroller ATmega$328P that ensures the execution of all actions of the assembled device. Microcontroller is also equipped by a number of other support circuits and sensors. The amount of analog

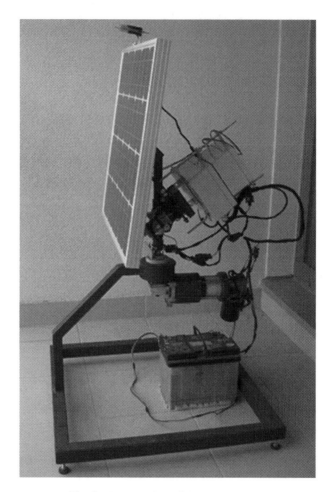

Fig. 3. An overview of the smart source.

inputs is extended by a multiplexer, which provides for example information from sun sensor and limit switches. For measuring of an amount of produced and consumed energy via shunt resistors the INA219 circuits are used. Temperature is then measured by digital thermometer NCT75 Motors for adjusting of the solar panel are controlled by dual h-bridge L298N. All activities of the controlling algorithm are strictly dependent on the current date and time. That's why a real time clock circuit DS1307 has also an important role for the entire system.

The key part for the Sun position evaluation is the sun sensor. It is assembled by four photoresistors separated by a shade which are connected as two voltage dividers. Each photoresistor then receives an irradiation from a certain part of the sky only. Both output signals of voltage dividers are regularly evaluated by a microcontroller. Adjustment of the photovoltaic panel is then performed so that both output signals are in an equilibrium state i.e. all photosensitive elements are receiving the same amount of radiation (Fig. 4).

Fig. 4. Smart power source and panel with fixed position during testing.

Smart source uses a lead-acid battery as energy storage element. Its charging is ensured by constant-voltage produced by voltage regulator LM2596-adj which is also equipped with a fan to ensure sufficient cooling.

4 Controlling Algorithm

The main program regularly evaluates the position of the Sun, adjusts a photovoltaic panel, measures energy flow and communicates with a web interface. It works according to the state diagram shown in Fig. 5. The program is executed in an endless loop. After each iteration microcontroller goes into sleep mode. Important parts of the main algorithm are described in the following paragraphs.

4.1 System Monitoring

After every wakeup of the microcontroller, the status check of the system is performed. It provides a simple diagnostics of whole device and capturing of fast events. For example, it checks multiplexer operation, the battery voltage and the status of real-time clock [8, 9]. It also checks the state of limit switches and even whether they are connected to the device or not. If any problem occurs, the system halts all its activities.

4.2 Panel Adjustment

Validation of solar panel position is performed in regular intervals in predefined time period of the day based on information from real-time clock. According to the data from the sun sensor is then decided about the need to make any adjustments. When position of the panel is changed, its new setting is checked for predefined period and adjusted again if necessary. This ensures the accuracy of the newly set position. If the device is properly positioned, the panel is in the evening adjusted to its initial position heading towards the East.

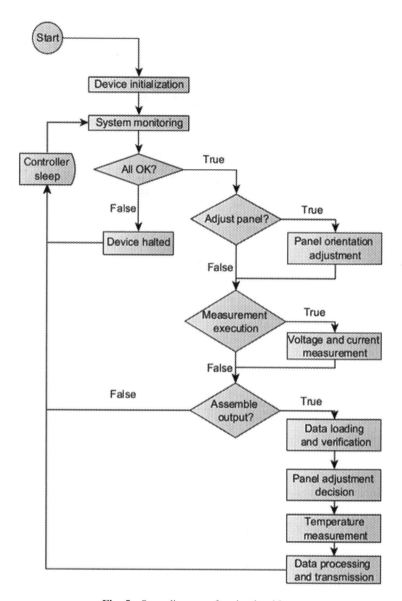

Fig. 5. State diagram of main algorithm.

4.3 Energy Measurement

Amount of consumed and produced energy by the smart source is measured regularly via reading out of an appropriate sensors. The amount of consumed energy includes self-consumption of the smart source as well as the consumption of optionally connected devices. All measured values are reset at midnight.

4.4   Data Processing and Transmission

Smart source periodically sends data to a web server by a serial bus. Transmitted string of characters contains information about current date, time, energy balance and the measured values of voltage and current.

An Arduino Mega2560 based device equipped with Ethernet module Wizznet 5100 is used for processing and displaying all data received from the smart power source. It is capable of creating a decent web server that interprets important data through a web page [6]. It can be seen in Fig. 6 [13]. Hence, it is possible to check device status and monitor amounts of produced and consumed energy at any time. All received data are also stored onto a memory card for further usage.

5   Testing of Constructed Device

Smart source was tested under various real-life conditions in Czech Republic nearby the city of Hradec Kralove (Fig. 4). The amount of energy it produced was compared with the amount of energy that was produced by photovoltaic panel with fixed position. The example of such comparison is shown in Fig. 7. The performed measurements

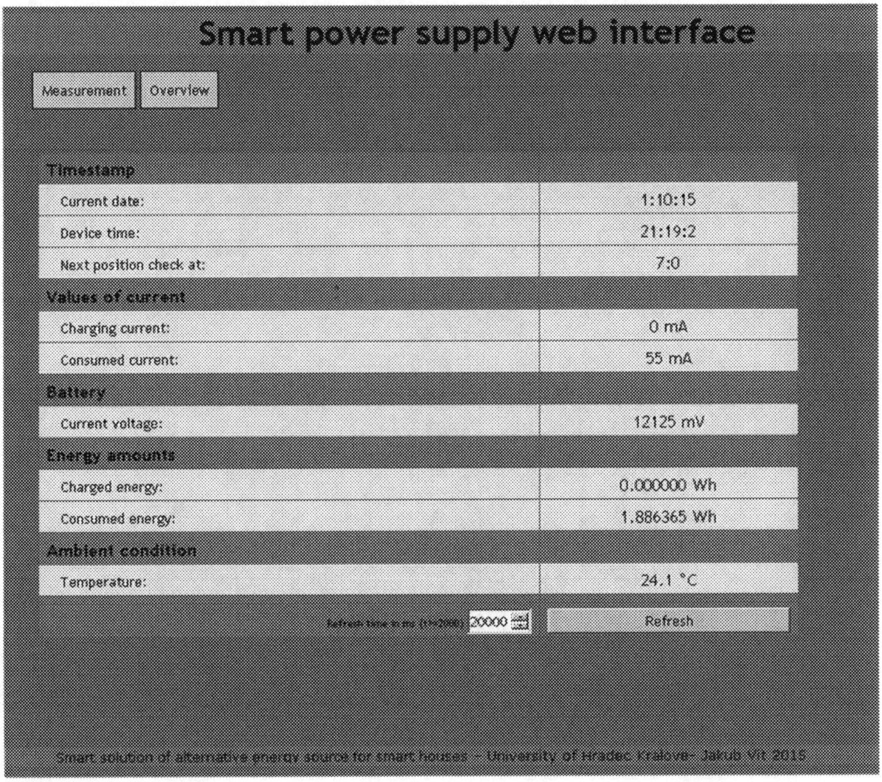

Fig. 6. Web page for system monitoring.

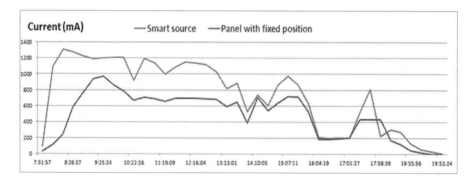

Fig. 7. Comparison of charging currents during one day of measurement. (Color figure online)

produced some interesting results. At first, active tracking of the panel towards the Sun is the most beneficial while device is placed in a location with great view of the Sun throughout all day. Amount of produced energy also reaches its maximum at a time when there are suitable atmospheric conditions. Benefits of the proposed smart source

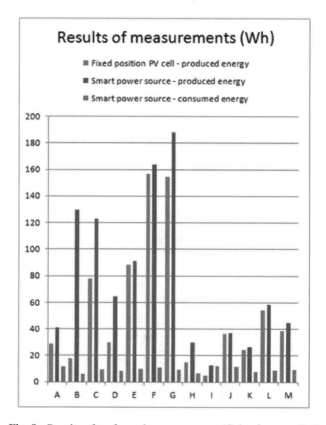

Fig. 8. Results of performed measurements. (Color figure online)

were, however, minimal in the case when the direct view of the Sun was limited during measurement, e.g. by surrounding buildings and atmospheric conditions like heavy clouds. In that case the smart source has produced a roughly the same amount of energy as a panel with fixed position. In some cases this also meant that the energy produced was just enough sufficient for the operation of the device itself. Individual measurements and their results are shown in the Fig. 8.

6 Conclusion

Subject of this work was to create a smart power source for use in smart houses or its parts. All demands for this device were defined and implemented in the final prototype. All components of this two-axis solar tracker device were designed, developed and also tested in real conditions. The overall cost of assembly of the device was only 300 USD. It includes costs of all electronic components of the smart source and web server as well as costs of accumulator and photovoltaic panel.

The web server for data processing and displaying was made as well as the electronics for data acquisition of the solar panel with fixed position. There were also made a few measurements that describe power gain of smart source in comparison with photovoltaic panel with fixed position. Developed device is special because it is capable of adjusting of orientation of photovoltaic panel towards the Sun and measuring consumed and produced amounts of energy. All data are also transmitted via serial bus for further processing. The device in this configuration is not very powerful yet it still could be easily used as a power source for some autonomous subsystems of the smart house.

Acknowledgment. This work and the contribution were supported by project "SP/2016/2102 - Smart Solutions for Ubiquitous Computing Environments" from FIM, University of Hradec Kralove.

References

1. Krejcar, O., Mahdal, M.: Optimized solar energy power supply for remote wireless sensors based on IEEE 802.15.4 Standard. Int. J. Photoenergy **2012**, 9 (2012). doi:10.1155/2012/305102. Article ID: 305102
2. Tutsch, M. Vojcinak, P., Koziorek, J., Skrepek, M.: Using automated evaluation of efficiency for photovoltaic power plant. In: Proceedings of the IEEE 16th Conference on Emerging Technologies and Factory Automation (ETFA 2011), Toulouse, France, pp. 1–4 (2011)
3. Machacek, Z., Ozana, S., Pies, M., Nevriva, P.: Mathematical modeling of turbine as a part of flexible energy system. In: Sambath, S., Zhu, E. (eds.) Frontiers in Computer Education. AISC, vol. 133, pp. 465–472. Springer, Heidelberg (2012). ISSN: 1867-5662
4. Tofighi, A., Kalantar, M.: Power management of PV/battery hybrid power source via passivity-based control. Renewable Energy **36**(9), 2440–2450 (2011). doi:10.1016/j.renene.2011.01.029. ISSN: 0960-1481

5. Krejcar, O., Spicka, I., Frischer, R.: Implementation of full-featured PID regulator in microcontrollers. Electron. Electr. Eng. **7**(113), 77–82 (2011). doi:10.5755/j01.eee.113.7. 617. ISSN: 1392-1215

6. Kasik, V., Penhaker, M., Novák, V., Bridzik, R., Krawiec, J.: User interactive biomedical data web services application. In: Yonazi, J.J., Sedoyeka, E., Ariwa, E., El-Qawasmeh, E. (eds.) ICeND 2011. CCIS, vol. 171, pp. 223–237. Springer, Heidelberg (2011). ISBN: 978-3-642-22728-8, ISSN: 1865-0929

7. David, J., Jancikova, Z., Frischer, R., Vroyina, M.: Crystallizer's desks surface diagnostics with usage of robotic system. Arch. Metallurgz Mater. **58**(3), 907–910 (2013)

8. Krejcar, O., Frischer, R.: Real time voltage and current phase shift analyzer for power saving applications. Sensors **12**(8), 11391–11405 (2012). ISSN: 1424-8220

9. Krejcar, O., Frischer, R.: Batteryless powering of remote sensors with reversed peltier power source for ubiquitous environments. Int. J. Distrib. Sens. Netw. **2013**, Article Id 789405, 9 (2013). ISSN: 1550–1329

10. Tutsch, M., Machacek, Z., Krejcar, O., Konarik, P.: Development methods for low cost industrial control by WinPAC controller and measurement cards in Matlab Simulink. In: Proceedings of Second International Conference on Computer Engineering and Applications, ICCEA 2010, Bali Island, Indonesia, 19–21 March 2010, vol. 2NJ, pp. 444–448. IEEE Conference Publishing Services (2010). doi:10.1109/ICCEA.2010.235, ISBN: 978-0-7695-3982-9

11. Dvorak, J., Berger, O., Krejcar, O.: Universal central control of home appliances as an expanding element of the smart home concepts - case study on low cost smart solution. In: Saeed, K., Snášel, V. (eds.) CISIM 2014. LNCS, vol. 8838, pp. 479–488. Springer, Heidelberg (2014). doi:10.1007/978-3-662-45237-0_44

12. Alexandru, C., Pozna., C.: Simulation of dual-axis solar tracker for improving the performance of a photovoltaic panel. J. Power Energy **224**(6), 797–811 (2015). doi:10.1243/09576509jpe871, http://pia.sagepub.com/content/224/6/797. Accessed 21 Aug 2015

13. Cimler, R., Matyska, J., Balik, L., Horalek, J., Sobeslav, V.: Security issues of mobile application using cloud computing. In: Abraham, A., Krömer, P., Snasel, V. (eds.) AECIA 2014. AISC, vol. 334, pp. 347–358. Springer, Heidelberg (2015). doi:10.1007/978-3-319-13572-4_29

14. Huld, T., Dunlop, E.: Optimal mounting strategy for single-axis tracking non-concentrating PV in Europe. In: 23rd European Photovoltaic Solar Energy Conference (PVSEC), Valencia, Spain (2008)

An Assembly Sequence Planning Approach with a Multi-state Particle Swarm Optimization

Ismail Ibrahim[1], Zuwairie Ibrahim[1(✉)], Hamzah Ahmad[1],
and Zulkifli Md. Yusof[2]

[1] Faculty of Electrical and Electronics Engineering,
Universiti Malaysia Pahang, 26600 Pekan, Pahang, Malaysia
peel2001@stdmail.ump.edu.my,
{zuwairie,hamzah}@ump.edu.my
[2] Faculty of Manufacturing Engineering,
Universiti Malaysia Pahang, 26600 Pekan, Pahang, Malaysia
zmdyusof@ump.edu.my

Abstract. Assembly sequence planning (ASP) becomes one of the major challenges in the product design and manufacturing. A good assembly sequence leads in reducing the cost and time of the manufacturing process. However, assembly sequence planning is known as a classical hard combinatorial optimization problem. Assembly sequence planning with more product components becomes more difficult to be solved. In this paper, an approach based on a new variant of Particle Swarm Optimization Algorithm (PSO) called the multi-state of Particle Swarm Optimization (MSPSO) is used to solve the assembly sequence planning problem. As in of Particle Swarm Optimization Algorithm, MSPSO incorporates the swarming behaviour of animals and human social behaviour, the best previous experience of each individual member of swarm, the best previous experience of all other members of swarm, and a rule which makes each assembly component of each individual solution of each individual member is occurred once based on precedence constraints and the best feasible sequence of assembly is then can be determined. To verify the feasibility and performance of the proposed approach, a case study has been performed and comparison has been conducted against other three approaches based on Simulated Annealing (SA), Genetic Algorithm (GA), and Binary Particle Swarm Optimization (BPSO). The experimental results show that the proposed approach has achieved significant improvement.

Keywords: Combinatorial optimization problem · Assembly sequence planning · Meta-heuristic · Multi-state particle swarm optimization algorithm

1 Introduction

The cost of assembly processes are determined by assembly plans. Assembly sequence planning, which is an important part of assembly process planning, plays an essential role in the manufacturing industry. Given the product-assembly model of an assembly

© Springer International Publishing Switzerland 2016
H. Fujita et al. (Eds.): IEA/AIE 2016, LNAI 9799, pp. 841–852, 2016.
DOI: 10.1007/978-3-319-42007-3_71

sequence planning (ASP), the shorter assembly time or reducing cost can be found after determining the sequences of components. This problem is regarded as a large-scale, highly constrained combinatorial optimization problem, and it is nearly impossible to generate and then evaluate all the assembly sequences in order to obtain the optimal one, either with human's interaction or through computer programs.

Historically, the typical combinatorial explosion problem needs experienced assembly technicians to determine assembly plans. Nonetheless, this manual assembly planning approach involves more time and makes quantitative assembly costs of the assembly solution cannot be achieved. Thus, many studies in the last two decades have intensely done based on geometric reasoning capability and full automatism to find more efficient algorithms for the automated ASP. Approaches used for representation of assembly sequence planning can be categorized into four groups. These groups are:

1. **Graph based representation.** In this representation, the source of data acquires from user or a CAD system. By using these approaches, most particulars of assembly analysis can be carried out. Mello and Sanderson [1] and Zhang [2] proposed graph based representation methods based on AND/OR and directed graph. Lee and Shin [3], Moore et al. [4], and Zha [5] proposed graph based representation methods based on PETRI nets.
2. **Lingual representation.** The representation involves the usage of a special language for representing subassemblies, parts and relations between them. A few approaches include, but is not restricted to Part and Assembly Description Language (PADL), Automated Parts Assembly System (AUTOPASS) and Geometric Design Processor (GDP) [1].
3. **An ordered list representation**. This kind of representation can be categorized as an ordered list of task representation, binary vectors, partitions of the set of parts and connections. Garrod and Everett [6] represented each assembly sequence in form of a set of list.
4. **Meta-heuristics based representation**. Those meta-heuristic approaches include, but is not restricted to rule-based [7], heuristic search [8], neural network based [8–10], genetic algorithm based [11–17], simulated annealing based [18, 19], ant colony optimization based [20], memetic algorithm based [21], particle swarm optimization [22–24], and hybrid based [25, 26].

The implementation of meta-heuristics in solving discrete optimization problems, particularly in the ASP problem leads to significant reduction of computation times, which by its nature sacrifices the guarantee of finding exact optimal solutions [27, 28]. However, these approaches are permitted to obtain acceptable performance at acceptable costs in a large number of possible assembly sequences. In other words, these approaches have a capacity to find good solution on large-size problem instances.

Over a span of two decades, there has been numerous studies inspired by the swarming behaviour of animals and human social behaviour called Particle Swarm Optimization introduced by Kennedy and Eberhart [29]. Researchers found that the synchrony of animals' behaviour was presented through constantly conserving optimal distances between individual members of swarm and their neighbours. The conventional PSO was originally designed to solve problems in continuous-valued space. Later, Kennedy and Eberhart [30] developed a reworked of the conventional PSO known as binary particle swarm optimization algorithm (BPSO) to allow PSO to operate in discrete binary variables.

In this paper, the research of assembly sequence planning using multi-state particle swarm optimization developed by Ibrahim *et al.* [31] is introduced and investigated. The MSPSO is applied to generate and optimize assembly sequences of mechanical products. The purpose is to investigate the applicability of an alternative intelligent approach to ASP.

The organization of this paper is as follows. Section 2 explains the original PSO. Section 3 describes the concept of the MSPSO. Section 4 presents the description of the problem. In Sect. 5 implementation of the proposed ASP approach using the MSPSO is described. Section 6 provides the experimental results of the MSPSO compared with SA, GA, and BPSO. Section 7 finally offers a general conclusion.

2 Particle Swarm Optimization (PSO)

In PSO algorithm, an optimal or good enough solution is found by simulating social behavior of bird flocking. The PSO algorithm consist of a group of individuals named particles which encode the possible solutions to the optimization problem using their positions. The group can attain the solution effectively by using the common neighboring particles. Using this information, each particle compares its current position with the best position found by its neighbours so far.

Figure 1 presents the general principle of the original PSO. Consider the following minimization problem: there are I-particles flying around in a D-dimensional search space, where their position, $s_i(d)(i = 1, 2, \ldots, I; d = 1, 2, \ldots, D)$, represent the possible solutions. In the initialization stage, all particles are randomly positioned in the search space. All particles are then assigned with random velocity $v_i(k, d)$, where k represents the iteration number. Next, the objective fitness $F_i(k)$, for each particle is evaluated by calculating the objective functions with respect to $s_i(k)$. Each particle's best position is $pbest_i(k)$ then initialized to its current position. The global best among the all $pbest_i(k)$ is called $gbest(k)$, is chosen as the swarm's best position as:

$$
\begin{aligned}
gbest &= \{pbest_i \in S | f(pbest_i) \\
&= \min f(\forall pbest_i \in S)\}
\end{aligned}
\tag{1}
$$

where S is the swarm of particles.

Subsequently, the algorithm iterates until the stopping condition is met, either the maximum number of iteration is reached or a particular amount of error is obtained.

Each particle updates its velocity and position as:

$$
\begin{aligned}
v_i(k+1, d) &= \omega v_i(k, d) + c_1 r_1 (pbest_i(k, d) - s_i(k, d)) \\
&+ c_2 r_2 (gbest(k, d) - s_i(k, d))
\end{aligned}
\tag{2}
$$

$$
s_i(k+1, d) = s_i(k, d) + v_i(k+1, d)
\tag{3}
$$

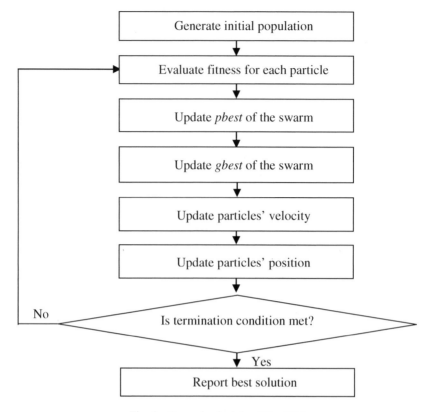

Fig. 1. General principle of the PSO

where c_1 and c_2 are the cognitive and social coefficients, respectively. r_1 and r_2 are random number uniformly distributed between 0 and 1, and ω is called inertia weight, which used to control the impact of the previous history of velocities on the current velocity of each particle. After updating the velocity and position, $F_i(k)$ for each particle is calculated again. $pbest_i(k)$ is then updated by a more optimal, obtained either from the new position of the ith particle or $pbest_i(k)$. The $gbest(k)$ is also updated by the most optimal $pbest_i(k)$ of all the particles. Finally, the optimum solution of the problem represented by $gbest(k)$ is yielded when the stopping condition is met.

3   Multi-state Particle Swarm Optimization (MSPSO)

The MSPSO is explained in this section. The MSPSO follows similar general principle of original PSO with two modifications; updating particle's velocity and position. For simplicity, the term state will be used to describe the position's representation in the MSPSO.

Each particle's vector in the MSPSO is represented by state; neither continuous nor discrete value. To elaborate state representation, Burma14 benchmark instance of

Travelling Salesman Problem (TSP) is used as an example as shown in Fig. 2. All the cities in Burma14 can be represented as a collective of states, in which the states are represented by the small black circle as presented in Fig. 3. A centroid of the circle shows the current state. Radius of the circle represents velocity value possessed by the current state. These three elements occur in each dimension for each particle. The process of updating velocity and state of each state in the MSPSO are executed after all particles' *pbest* and *gbest* are updated.

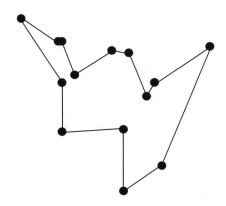

Fig. 2. Burma14 benchmark instance of Travelling Salesman Problem (TSP)

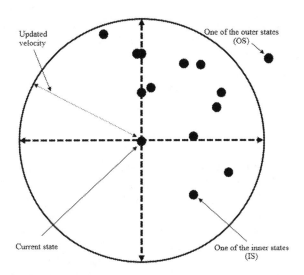

Fig. 3. The illustration of the multi-state representation in MSPSO for Burma14 benchmark instance of TSP. Each particle's vector applies similar representation

The way to calculate velocity value in the MSPSO algorithm is different compared with PSO algorithm due to $pbest_i(k)$, $gbest(k,d)$, and $s_i(k,d)$ in the form of state.

Referring to PSO algorithm, a particle has three movement components; the inertia, cognitive, and social component. The effect of the first, second, and third component are the particle bias to follow in its own way, to go back to its best previous position, and to go towards the global best particle, respectively. However, in the MSPSO algorithm, the velocity value is the summation of previous velocity, cost function ($C(.)$) subjected to particle's best position and current particle's position multiplies with a cognitive coefficient and a uniform random value, and cost function ($C(.)$) subjected to global best particle and current particle's position multiplies with a social coefficient and a uniform random value. The velocity equation is derived as follow:

$$v_i(k+1,d) = \omega v_i(k,d) + c_1 r_1 C(pbest_i(k,d), s_i(k,d))$$
$$+ c_2 r_2 C(gbest(k,d), s_i(k,d)) \tag{4}$$

Cost function can be used to find the difference of distance or time between two points or activities, such as in Assembly Sequence Planning (ASP) and Traveling Salesman Problem (TSP), respectively [23, 31]. The cost between two states in ASP is in time unit. Meanwhile, it is in distance for TSP. In addition, the cost between two states is a positive number given by $C(s_j(t,d), s_i(t,d))$.

In the MSPSO, once the velocity is updated, the process of updating current to next state for each dimension of each particle is executed. We define the current state as a centroid and the updated velocity value as a radius, a circle is built. Any state that is located in the area of the circle is defined as a member of inner states (IS) group. Given a set of j IS members $IN_i(t,d) = (IN_{i_1}((t,d), \ldots, IN_{i_j}(t,d)))$. Meanwhile, any state that is located outside of the area of the circle is then defined as a member of outer states (OS) group. Given a set of l OS group members $OUT_i(t,d) = (OUT_{i_1}(t,d), \ldots, OUT_{i_l}(t,d))$.

Based on the current state and the updated velocity of the current state, a next state can be selected by referring just to the information of inner states. The formulation can be derived as:

$$s_i(t+1,d) = random(IN_{i_1}((t,d), \ldots, IN_{i_j}(t,d)) \tag{5}$$

In order to update state in the MSPSO algorithm, a random function as shown above is applied. The equation may lead to the existence of repeated state in an updated solution. Let us consider a solution of a particle at a particular iteration consisting 14-dimensional vector $\{s_5, s_3, s_{14}, s_{11}, s_2, s_8, s_9, s_{13}, s_{12}, s_{10}, s_1, s_4, s_6, s_7\}$. Note that this solution has no repeated state. This solution is then subjected to dimension-by-dimension updates. After updating each state in the solution, the updated solution is a 14-dimensional vector $\{s_4, s_7, s_8, s_{11}, s_{13}, s_8, s_5, s_{12}, s_{13}, s_{10}, s_1, s_8, s_{12}, s_9\}$. Obviously, the state in the 3rd, 5th, 6th, 8th, 9th, 12th, and 13th dimension occurs more than once, making the updated solution infeasible for combinatorial optimization problems. For example, the updated state in the 5th and 9th dimension of the updated solution is s_{13}, which are identical.

In order to overcome the limitation of the MSPSO, an additional procedure is implemented after updating velocity and state to obtain unrepeated states for each particle. The additional procedure is discussed in [31] for details.

4 Assembly Sequence Planning

The main objective of ASP is to generate a feasible assembly sequence in which it will take less time to assemble, thereby reducing assembly costs. The most important factor in reducing assembly time and costs are setup time, including transfer time, number of tool changes, proper fixture selection, and so on.

In this paper, assumptions for ASP are as follows;

1. Setup time and the actual assembly time for each part and component are given.
2. Transfer time between workstations is included in set up time.
3. Downtime of machines and workstations is omitted.

A precedence matrix (PM) is used to show the relation using the precedence constraints between components in assembly. The relationship includes the nature of the connection (free to assembly components) and the relative assembly precedence between two components. For this purpose, Table 1 can be build according to precedence constraints between two components (i.e., a and b). If component a must be assembled after component b, $PM(P_a, P_b) = 1$, otherwise $PM(P_a, P_b) = \emptyset$ where (P_a, P_b) is a pair of components that has a geometric information in which P_a must be assembled without interfering with P_b. To decide which pair is feasible, precedence constraints for a product should be described using PM. Given ψ as the set of the components have been assembled before component a and the union of PM is a feasible assembly sequence FAS (P_a, P_b) with constraints, therefore:

$$FAS \ (P_a, P_b) = \cup \, PM(P_a, P_b), P_b \in \psi \tag{6}$$

The generation of feasible sequences is discussed in [23] for details.

Table 1. Best results and their associated assembly sequences of the proposed approach based on the MSGSA and the approach based on SA, GA, and BPSO.

Approach based	Total assembly time	Assembly sequence
MSPSO	511.5	15-1-2-4-12-9-3-13-5-18-6-16-11-7-8-14-10-17-19
SA	528.7	2-1-4-9-3-12-13-16-5-15-18-6-11-7-8-10-14-17-19
GA	524.1	2-18-3-12-1-13-16-5-11-15-4-6-9-7-8-10-14-17-19
BPSO	514.4	16-2-13-4-1-15-11-9-6-5-18-7-8-14-12-10-3-17-19

5 Solving Assembly Sequence Planning Using Multi-state PSO

To search an optimal solution, each particle must be evaluated to measure its fitness value. The evaluation of fitness is performed after the initial population is generated and the PM, coefficient table and actual assembly times are loaded. The *pbest* and the *gbest* are then updated. Next, the velocity and position for each particle is updated. The updated assembly sequence of each particle is then evolved to feasible assembly sequence. Occasionally, some respective assembly components cannot be integrated into a feasible assembly sequence. The determination of the assembly components that do not correspond to a feasible assembly sequence is achieved by satisfying all PM constraints between the components in the assembly, which are determined earlier, either from CAD or a disassembly analysis [3]. As a result, each particle produces a feasible assembly sequence. The optimum sequence is then selected from the feasible assembly sequences by evaluating the fitness of each particle. The best of the population is the sequence that is more optimal up until the stopping condition is met. After the stopping condition is met, the performance of the proposed approach based on the MSPSO can be investigated.

The assembly of a hypothetical product with 19 components is considered according to [12, 19] and its associated coefficient table is outlined in [23]. Figure 4 shows the precedence diagram. In this diagram, the components that are free to be assembled are the components that can be placed regardless of any part of a sequence.

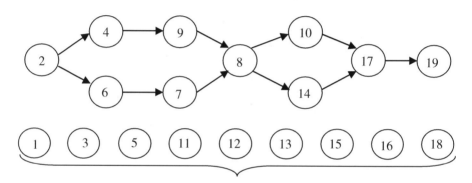

Free to be assembled components

Fig. 4. The assembly precedence diagram for the case study

During initialization, the initial population is randomly generated. Second, the generated initial population is subjected to a validity check using the PM described in the previous section to ensure the initial population is valid and feasible. The vector representation corresponding to the assembly components of a particle is represented in Fig. 5; in this case, the sequence is 2-3-1-5-4. The length of a string depends on the total number of components used in the assembly process.

2	3	1	5	4

Fig. 5. An example of assembly sequence represented by a particle

The evaluation procedure is performed to improve the objective value. Hence, for each iteration, each particle represented by a feasible assembly sequence is evaluated. The particle with the highest objective value for current iteration is then compared with the particle with the highest objective value for previous iterations. If the particle with the highest objective value for the current iteration is better than the particle with the highest objective value for previous iterations, the particle with the highest objective value for the current iteration is updated to be the particle with the highest objective value for all iterations.

To evaluate the fitness for each particle, the total assembly time should be found. The total assembly time is the combination of the setup time and the actual assembly time. It is assumed that regardless of the assembly sequence, the actual assembly time is constant, and a proper tool and setup for each component to be assembled is required. These two items depend on the geometry of the component itself and the components assembled up to that point. The setup time for a component can be calculated using:

$$Time_{Setup}(a) = p_{a0} + \sum_{b=1}^{e} p_{ab}q_{ab} \qquad (7)$$

where (a) is the component to be assembled, p_{a0} is the setup time for product (a) being the first component, p_{ab} is the contribution to the setup time due to the presence of part (b) when entering part (a), and $q_{ab} = 1$ if component (b) has already been assembled. Otherwise, $q_{ab} = 0$ for a = 1, 2 ,..., e. e is the number of components that forms each sequence in the ASP.

The total assembly time is the summation of setup time and actual assembly time. Hence, the objective function for minimizing the assembly time should be calculated by:

$$Min\ Time_{Assembly} = \sum_{b=1}^{e} (Time_{Setup}(a) + A_a) \qquad (8)$$

where A_a is the assembly time for component a. The calculation of time is in time units.

To confirm the production of the feasible assembly sequence of each particle, the updated assembly sequence of each particle produced by the updating position process is evolved to feasible assembly sequence. To assemble the components of the product in valid manner, only the feasible assembly sequences should be used to perform that task. The feasibility of the sequences can be determined by referring to the PM.

The PM gives information which position of each sequence should be swapped randomly (an infeasible component and a feasible component). The swapping process ends when each component occurs in an assembly sequence in which the sequence is now feasible.

6   Experimental Results

Table 1 demonstrates the best results and theirs associated assembly sequences of the proposed approach based on the MSPSO and the approach based on SA [19], GA [12], and BPSO [23]. To simplify the understanding of this work, fitness or objective value is now called total assembly time and feasible assembly sequence is the solution.

The success of the MSPSO is heavily depend on setting of control parameters namely; inertia weight ω, coefficient factors c_1 and c_2, number of particles, and number of iteration T. These control parameters should be carefully selected when using the MSPSO in order to know the best parameters, so a successful implementation of the algorithm can be achieved. A series of experiments are carried out to tune the MSPSO best parameters for the assembly sequence planning problem. It is clear from results shown in Table 1 that the best parameters for inertia weight ω, coefficient factors c_1 and c_2, number of particles, and number of iteration T are $\omega = 0.9 - 0.4$, c_1 and $c_2 = 2$, number of particles $NOP = 25$, $T = 500$ respectively. The best objective value obtained for these parameters is 511.5. The assembly sequence generated for the best objective value using these parameters is 15-1-2-4-12-9-3-13-5-18-6-16-11-7-8-14-10-17-19.

The result clearly shows that the MSPSO successfully provides the best assembly sequence compared to SA, GA, and BPSO. Convergence pattern of the best assembly sequence obtained by the MSPSO is then portrayed in Fig. 6. It seems that the MSPSO converges fast at iteration 245.

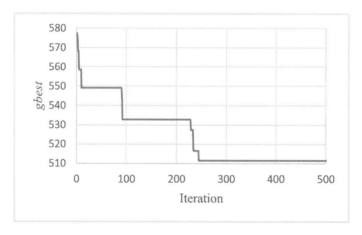

Fig. 6. Convergence pattern of the best assembly sequence for the MSPSO in solving the ASP

7   Conclusions

ASP is a combinatorial problem with a large-scale solution. The main goal of ASP is to seek for the assembly sequences with the minimum assembly cost while satisfying the assembly process constraints. This paper presents an approach based on the MSPSO to solve the assembly sequence planning problem. Two new modifications (i.e., updating the velocity and updating the position) in the MSPSO fit efficiently together with the rest of other mechanism of original PSO to find the optimal or nearly optimal assembly sequence of a mechanical product. To show the relations between components in the ASP, the PM assembly has been used. To evaluate the performance of the proposed approach, a case study of the ASP consisting of 19 components is examined, and the performance of the proposed approach based on the MSPSO is evaluated with other three different approaches that use SA, GA, and BPSO. The experimental results obtained shows that the performance of the proposed approach is better than the three other approaches in searching the best assembly sequence. In the future, we may examine the performance of this approach with other constraints in assembly sequence planning, including assembly stability, machine and workstation assignment, and workload.

Acknowledgement. This work is financially supported by the Ministry of Higher Education Malaysia through the Fundamental Research Grant Scheme (FRGS) VOT RDU140114 granted to Universiti Malaysia Pahang.

References

1. De Mello, L.S.H., Arthur, C.D.: A correct and complete algorithm for the generation of mechanical assembly sequences. IEEE Trans. Robot. Autom. **7**, 228–240 (1991)
2. Zhang, W.: Representation of assembly and automatic robot planning by petri net. IEEE Trans. Syst. Man. Cybern. **19**, 418–422 (1989)
3. Lee, S., Shin, Y.G.: Assembly planning based on geometric reasoning. Comput. Graph. **14**, 237–250 (1990)
4. Moore, E.K., Aşkıner, G., Surendra, M.G.: Petri net approach to disassembly process planning for products with complex and/or precedence relations. Eur. J. Oper. Res. **135**, 428–449 (2001)
5. Zha, X.F.: An object-oriented knowledge based petri net approach to intelligent integration of design and assembly planning. Artif. Intell. Eng. **14**, 83–112 (2000)
6. Garrod, W., Everett, L.J.: Automated sequential assembly planner. In: ASME International Computers in Engineering Conference, pp. 139–150. Texas A&M University, Boston (1990)
7. Chakrabarty, S., Wolter, J.: A structure-oriented approach to assembly sequence planning. IEEE Trans. Robot. Autom. **13**, 14–29 (1997)
8. Hong, D.S., Cho, H.S.: A neural network based computation scheme for generating optimized robotic assembly sequences. Eng. Appl. Artif. Intell. **8**, 129–145 (1995)
9. Chen, W.C., Tai, P.H., Deng, W.J., Hsieh, L.F.: A three-stage integrated approach for assembly sequence planning using neural networks. Expert Syst. Appl. **34**, 1777–1786 (2008)
10. Huang, H.H., Wang, M.H., Johnson, M.R.: Disassembly sequence generation using a neural network approach. J. Manuf. Syst. **19**, 73–82 (2000)
11. Bonneville, F., Perrard, C., Henrioud, J.M.: A genetic algorithm to generate and evaluate assembly plans. In: IEEE Symposium on Emerging Technologies and Factory Automation, pp. 231–239. IEEE Press, Paris (1995)
12. Choi, Y.K., Lee, D.M., Cho, Y.B.: An approach to multi-criteria assembly sequence planning using genetic algorithms. Int. J. Adv. Manuf. Technol. **42**, 180–188 (2008)
13. De, L., Latinne, P., Rekiek, B.: Assembly planning with an ordering genetic algorithm. Int. J. Prod. Res. **39**, 3623–3640 (2001)
14. Lu, C., Wong, Y.S., Fuh, J.Y.H.: An enhanced assembly planning approach using a multi-objective genetic algorithm. Proc. Inst. Mech. Eng. B J. Eng. Manuf. **220**, 255–272 (2006)

15. Marian, R.M., Luong, L.H.S., Abhary, K.: Assembly sequence planning and optimization using genetic algorithms: part I. automatic generation of feasible assembly sequences. Appl. Soft Comput. **2**, 223–253 (2003)
16. Tseng, Y.J., Yu, F.Y., Huang, F.Y.: A multi-plant assembly sequence planning model with integrated assembly sequence planning and plant assignment using GA. Int. J. Adv. Manuf. Technol. **48**, 333–345 (2010)
17. Zhou, W., Zheng, J.R., Yan, J.J., Wang, J.F.: A novel hybrid algorithm for assembly sequence planning combining bacterial chemotaxis with genetic algorithm. Int. J. Adv. Manuf. Technol. **52**, 715–724 (2011)
18. Milner, J.M., Graves, S.C., Whitney, D.E.: Using simulated annealing to select least-cost assembly sequences. In: IEEE International Conference on Robotics and Automation, pp. 2058–2063. IEEE Press, California (1994)
19. Motavalli, S., Islam, A.: Multi-criteria assembly sequencing. Comput. Ind. Eng. **32**, 743–751 (1997)
20. Wang, J.F., Liu, J.H., Zhong, Y.F.: A novel ant colony algorithm for assembly sequence planning. Int. J. Adv. Manuf. Technol. **25**, 1137–1143 (2005)
21. Gao, L., Qian, W.R., Li, X.Y., Wang, J.F.: Application of memetic algorithm in assembly sequence planning. Int. J. Adv. Manuf. Technol. **49**, 1175–1184 (2010)
22. Guo, Y.W., Li, W.D., Mileham, A.R., Owen, G.W.: Applications of particle swarm optimization in integrated process planning and scheduling. Robot. Comput. Integr. Manuf. **25**, 280–288 (2009)
23. Mukred, J.A.A., Ibrahim, Z., Ibrahim, I., Adam, A., Wan, K., Yusof, Z.M., Mokhtar, N.: A binary particle swarm optimization approach to optimize assembly sequence planning. Adv. Sci. Lett. **13**, 732–738 (2012)
24. Tseng, Y.J., Yu, F.Y., Huang, F.Y.: A green assembly sequence planning model with a closed-loop assembly and disassembly sequence planning using a particle swarm optimization method. Int. J. Adv. Manuf. Technol. **57**, 1183–1197 (2011)
25. Cheng, H., Li, Y., Zhang, K.F.: Efficient method of assembly sequence planning based on GA and optimizing by assembly path feedback for complex product. Int. J. Adv. Manuf. Technol. **42**, 1187–1204 (2009)
26. Li, M., Wu, B., Hu, Y., Jin, C., Shi, T.: A hybrid assembly sequence planning approach based on discrete particle swarm optimization and evolutionary direction operation. Int. J. Adv. Manuf. Technol. **68**, 617–630 (2013)
27. Blum, C., Roli, A.: Metaheuristics in combinatorial optimization: overview and conceptual comparison. ACM Comput. Surv. **35**, 268–308 (2003)
28. Talbi, E.G.: Metaheuristics: From Design to Implementation. Wiley, Hoboken (2009)
29. Kennedy, J., Eberhart, R.: Particle swarm optimization. In: IEEE International Conference on Neural Networks, pp. 1942–1948. IEEE Press, Perth (1995)
30. Kennedy, J., Eberhart, R.C.: A discrete binary version of the particle swarm algorithm. In: IEEE International Conference on Systems, Man, and Cybernetics, pp. 4104–4108. IEEE Press, Florida (1997)
31. Ibrahim, I., Ahmad, H., Ibrahim, Z., Jusoh, M.F.M., Yusof, Z.M., Nawawi, S.W., Khalil, K., Rahim, M.A.A.: Multi-state particle swarm optimization for discrete combinatorial optimization problem. Int. J. Simulat. Sys. Sci. Technol. **15**, 15–25 (2014)

Evolutionary Algorithms
and Heuristic Search

A Black Hole Algorithm for Solving the Set Covering Problem

Ricardo Soto[1]([✉]), Broderick Crawford[1], Ignacio Figueroa[1],
Stefanie Niklander[2,3,4], and Eduardo Olguín[5]

[1] Pontificia Universidad Católica de Valparaíso, Valparaíso, Chile
{ricardo.soto,broderick.crawford}@ucv.cl, ignacio.figueroa.b@mail.pucv.cl
[2] Universidad Adolfo Ibañez, Viña del Mar, Chile
stefanie.niklander@uai.cl
[3] Universidad Autónoma de Chile, Santiago, Chile
[4] Universidad Cientifica del Sur, Lima, Peru
[5] Universidad San Sebastián, Santiago, Chile
eduardo.olguin@uss.cl

Abstract. The set covering problem is a classical optimization benchmark with many industrial applications such as production planning, assembly line balancing, and crew scheduling among several others. In this work, we solve such a problem by employing a recent nature-inspired metaheuristic based on the black hole phenomena. The core of such a metaheuristic is enhanced with the incorporation of transfer functions and discretization methods to handle the binary nature of the problem. We illustrate encouraging experimental results, where the proposed approach is capable to reach various global optimums for a well-known instance set from the Beasley's OR-Library.

Keywords: Meta-heuristics · Soft computing · Black Hole Algorithm · Set covering problem

1 Introduction

The Set Covering Problem (SCP) is a classic benchmark in the subject of combinatorial optimization that belongs to the NP-complete class of problems [19]. The purpose of the SCP is to find a set of solutions that cover a range of needs at the lowest possible cost. The SCP can be applied to many real-world problems, such are airline crew scheduling [14], network discovery [10], plant location [12], and service allocation [6] among others. Different algorithms have been developed to solve the classic SCP, ranging from classic exact methods to more recent bio-inspired metaheuristics. Exact methods can be applied to solve SCPs [1,2], the main problem is when the instance size increases the algorithm is commonly unable to reach a solution in a reasonable amount of time. Approximate methods such as the well-known metaheuristics tackle this concern, being capable to generally provide good enough local optimums in a limited time interval.

© Springer International Publishing Switzerland 2016
H. Fujita et al. (Eds.): IEA/AIE 2016, LNAI 9799, pp. 855–861, 2016.
DOI: 10.1007/978-3-319-42007-3_72

In this context a large list of metaheuristics have been proposed to solve the SCP [4,5,8,9,17,20].

In this paper, a new approach for SCPs based on the black hole algorithm is presented. The Black Hole Algorithm (BHA) is a population-based metaheuristic based on the gravitational force that has a black hole to attract everything that is around it. The core of the BHA is enhanced with the incorporation of binarization through transfer and discretization functions in order to handle the binary nature of the SCP. Repairing operators are also employed to rapidly discard the unfeasible solutions and as a consequence to alleviate the search. We present promising results on 40 well-known pre-processed instances from the Beasley's OR-Library, where a considerable amount of global optimums are reached.

This paper is organized as follows: In Sect. 2, we describe the SCP. Next section presents the BHA including binarization and repairing. Section 4 provides the experimental results, followed by conclusions and future work.

2   The Set Covering Problem

The Set Covering Problem consists in finding a set of solutions at the lowest possible cost to cover a set of needs. Formally, we define the problem as follows: Let $A = (a_{ij})$ be a binary matrix with $m-rows \times n-columns$, and let $C = (c_j)$ be a vector representing the cost of each column j, assuming that $c_j > 0$ for $(j \in N)$. So we can say that column $(j \in N)$ cover a row i that exists in M if $a_{ij} = 1$. The mathematical model is as follows:

$$min\,(z) = \sum_{j=1}^{n} c_j X_j$$

Subject to:

$$\sum_{j=1}^{n} a_{ij}x_j \geq 1 \quad \forall i \in M \qquad x_j = \begin{cases} 1\,j \in S \\ 0\;if\;not \end{cases} \forall\,j \in N$$

3   The Black Hole Algorithm

A black hole is a region of space that has so much mass concentrated in it that there is no way for a nearby object to escape its gravitational pull [15]. Anything falling into a black hole, including light, cannot escape. The BHA is inspired on this phenomena [11].

Similar to other population-based metaheuristics, The BHA begins by randomly generating a population of candidate solutions, called stars, which are placed in the search space of some problem or function. After initialization, the fitness values of the population are evaluated and the best candidate, which has the best fitness values is introduced as black hole and the other solutions are selected as normal stars. Then, all the stars commence moving towards the black hole due of the power absorbing of the black hole.

The absorption of stars by the black hole is formulated as follows: $x_i(t+1) = x_i + rand * (x_{bh} - x_i(t)) \geq 1$ for $i = \{1, 2, 3, \ldots, N\}$. Where $x_i(t + 1)$ is the location of the i_{th} star at the iteration t+1, $Rand$ is a random number between zero and one, x_{bh} is the location of the black hole in the search space, and N is the number of solutions (stars). In addition, there is a distance between stars and black hole, the stars that crosses the event horizon of the black hole will be absorbed by the black hole, in carrying out this event another candidate solution (star) is born and distributed randomly in the search space and starts a new iteration, this is known as probability of crossing the event horizon. This is done to keep the number of candidate solutions constant.

The radius of the event horizon in the black hole algorithm is calculated by using the following equation: $E = f_{BH} / \sum_{i=1}^{N} f_i$. Where f_{bh} is the fitness value of the black hole, f_i is the fitness value of the i_{th} star, and N is the number of candidate solutions (stars). When the distance from the black hole with the star is less than the radio, or in other words when the difference in fitness between the black hole and the star is less than the radio, that star is swallowed by the black hole.

3.1 Binarization

When the star moves toward the black hole, the algorithm generates a real number which must be transformed to a binary domain due to the nature of the problem treated. To this end, we firstly employ a transfer function, which map a real value to a $[0, 1]$ real interval. As transfer function we employ the V-shaped-V4 (Eq. 1), which is was the best-performing one among the 8 tested transfer functions (4 S-shaped and 4 V-shaped) [8,13]. Then, the resulting value from the transfer function is discretized via the half method depicted in Eq. 2 in order to obtain a binary value.

$$T(x) = \left| \frac{2}{\pi} arctan \left(\frac{\pi}{2}x \right) \right| \tag{1}$$

$$x_i(t + 1) = \begin{cases} 1 \text{ if } rand > 0.5 \\ \\ 0 \text{ otherwise} \end{cases} \tag{2}$$

3.2 Repairing

The BHA, as most metaheuristics do, generates a random population with solutions that violate the constraints, i.e., solutions holding uncovered rows. Repairing operators are responsible for turning unfeasible solutions on feasible ones. To this end, we incorporate a heuristic operator that achieves the generation of feasible solutions, and additionally eliminates column redundancy [3].

To make all feasible solutions we compute a ratio based on the sum of all the constraint matrix rows covered by a column c_j/N_{uc}, where N_{uc} is the amount of uncovered columns. The unfeasible solution are repaired by covering the columns of the solution that had the lower ratio. After this, a local optimization step is applied, where column redundancy is eliminated. A column is redundant when it can be deleted and the feasibility of the solution is not affected.

4   Experiment Results

The performance of the proposed black hole algorithm was experimentally evaluated by using 40 preprocessed instances of the SCP from the Beasley's OR-Library[1]. This algorithm has been implemented in Java and the experiments have been launched on a 2.3 Ghz Intel Core i3 with 4 GB RAM machine running Windows 7. We employ an initial population of 20 stars, 4000 iterations and 20 executions per instance. The results are given in Table 1 where column 1

Table 1. Results obtained by BHA for the tested SCP instances.

Instance	Opt	Best	Avg	RPD	Instance	Opt	Best	Avg	RPD
4.1	429	430	430.25	0.23	6.1	138	140	142.95	1.45
4.2	512	512	512	0.0	6.2	146	147	149.1	0.68
4.3	516	516	517.2	0.0	6.3	145	145	147.7	0.0
4.4	494	495	495.25	0.20	6.4	131	131	131	0.0
4.5	512	514	514.9	0.39	6.5	161	161	163.5	0.0
4.6	560	560	560.9	0.0	A.1	253	253	255.5	0.0
4.7	430	430	431.1	0.0	A.2	252	253	257.35	0.39
4.8	492	493	496.3	0.20	A.3	232	233	235.65	0.43
4.9	641	644	648.05	0.46	A.4	234	234	234.95	0.0
4.10	514	514	515.05	0.0	A.5	236	236	236.7	0.0
5.1	253	253	255.6	0.0	B.1	69	69	70.3	0.0
5.2	302	305	306.2	0.99	B.2	76	76	77.6	0.0
5.3	226	228	228	0.88	B.3	80	80	80.9	0.0
5.4	242	242	242.25	0.0	B.4	79	79	80.1	0.0
5.5	211	211	211.4	0.0	B.5	72	72	72.3	0.0
5.6	213	213	213.15	0.0	C.1	227	229	231.25	0.88
5.7	293	293	295.05	0.0	C.2	219	219	221.4	0.0
5.8	288	288	289	0.0	C.3	243	245	250.7	0.82
5.9	279	279	282.35	0.0	C.4	219	219	222.7	0.0
5.10	265	265	265.1	0.0	C.5	215	215	216.6	0.0

[1] Available at http://www.brunel.ac.uk/~mastjjb/jeb/info.html.

shows the SCP instance, column 2 depicts the best known optimum for the instance, column 3 provides the best optimal value found by the algorithm, while columns 4 and 5 show average of results and the relative percentage deviation, respectively. The relative percentage deviation (RPD) is computed as follows: $RDP = (Z - Z_{opt})/Z_{opt} \times 100$, where Z is the best optimum value found by the metaheuristic and Z_{opt} depicts the best known optimum value for the instance.

In Table 2, the proposed approach is compared with three recently reported metaheuristics for the SCP, namely, shuffled frog leaping algorithm (SFLA) [7], XOR-based artificial bee colony (xABC) [16], and a binary firefly algorithm (BFF) [8]. Table 3 depicts the amount of global optimums reached by each algorithm. BHA is able to reach 27 global optimums, while the results for the remaining 13 instances stay very near to the global optimum (RPDs around 1 %). The

Table 2. Results obtained using BHA for instances SCP

Instance	Opt	BHA	SFLA	xABC	BFF	Instance	Opt	BHA	SFLA	xABC	BFF
4.1	429	430	430	430	429	6.1	138	140	140	142	138
4.2	512	512	513	512	517	6.2	146	147	147	147	147
4.3	516	516	519	519	519	6.3	145	145	147	148	147
4.4	494	495	501	495	495	6.4	131	131	131	131	131
4.5	512	514	514	514	514	6.5	161	161	166	165	164
4.6	560	560	563	561	563	A.1	253	253	255	254	255
4.7	430	430	431	431	430	A.2	252	253	160	257	259
4.8	492	493	497	493	497	A.3	232	233	237	235	238
4.9	641	644	656	649	655	A.4	234	234	235	236	235
4.10	514	514	518	517	519	A.5	236	236	236	236	236
5.1	253	253	254	254	257	B.1	69	69	70	70	71
5.2	302	305	307	309	309	B.2	76	76	76	78	78
5.3	226	228	228	229	229	B.3	80	80	80	80	80
5.4	242	242	242	242	242	B.4	79	79	79	80	79
5.5	211	211	211	211	211	B.5	72	72	72	72	72
5.6	213	213	213	214	213	C.1	227	229	229	231	230
5.7	293	293	297	298	298	C.2	219	219	223	222	223
5.8	288	288	291	289	291	C.3	243	245	253	254	253
5.9	279	279	281	280	284	C.4	219	219	227	231	225
5.10	265	265	265	267	265	C.5	215	215	217	216	217

Table 3. Optimums reached for the 40 instances

	BHA	SFLA	xABC	MBFF
Opt. reached	**27/40**	10/40	7/40	11/40

results also illustrate that BHA greatly outperforms its competitors, which were unable to reach more than 12 optimum values from the 40 tested instances. Let us also note the robustness of the proposed BHA, whose averages for 20 executions remain very close to the best optimum value found.

5   Conclusions

In this paper we have presented a new approach for solving SCPs based on the black hole algorithm. We have incorporated a transfer function and a discretization method in order to handle the binary nature of the problem. Repairing operators are also employed to avoid unfeasible solutions and column redundancy. We have tested 40 non-unicost instances from the Beasley's OR-Library where the quality of results clearly outperform very recent reported metaheuristics for the SCP. The proposed approach is also robust able to provide averages very near to global optimums. As future work, we plan to test larger instances of the SCP as well as to incorporate adaptive capabilities to the BHA for performing parameter tuning during solving as exhibited in [18].

References

1. Balas, E., Carrera, M.C.: A dynamic subgradient-based branch-and-bound procedure for set covering. Locat. Sci. **5**(3), 203–203 (1997)
2. Beasley, J.E.: An algorithm for set covering problem. Eur. J. Oper. Res. **31**(1), 85–93 (1987)
3. Beasley, J.E., Chu, P.C.: A genetic algorithm for the set covering problem. Eur. J. Oper. Res. **94**(2), 392–404 (1996)
4. Brusco, M.J., Jacobs, L.W., Thompson, G.M.: A morphing procedure to supplement a simulated annealing heuristic for cost and coveragecorrelated set covering problems. Ann. Oper. Res. **86**, 611–627 (1999)
5. Caserta, M.: Tabu search-based metaheuristic algorithm for large-scale set covering problems. In: Doerner, K.F., Gendreau, M., Greistorfer, P., Gutjahr, W., Hartl, R.F., Reimann, M. (eds.) Operations Research/Computer Science Interfaces Series, vol. 39, pp. 43–63. Springer, New York (2007)
6. Ceria, S., Nobili, P., Sassano, A.: Annotated Bibliographies in Combinatorial Optimization. Wiley, Chichester (1997)
7. Crawford, B., Soto, R., Peña, C., Palma, W., Johnson, F., Paredes, F.: Solving the set covering problem with a shuffled frog leaping algorithm. In: Nguyen, N.T., Trawiński, B., Kosala, R. (eds.) ACIIDS 2015. LNCS, vol. 9012, pp. 41–50. Springer, Heidelberg (2015)
8. Crawford, B., Soto, R., Riquelme-Leiva, M., Peña, C., Torres-Rojas, C., Johnson, F., Paredes, F.: Modified binary firefly algorithms with different transfer functions for solving set covering problems. In: Silhavy, R., Senkerik, R., Oplatkova, Z.K., Prokopova, Z., Silhavy, P. (eds.) Software Engineering in Intelligent Systems. AISC, vol. 349, pp. 307–315. Springer, Heidelberg (2015)
9. Cuesta, R., Crawford, B., Soto, R., Paredes, F.: An artificial bee colony algorithm for the set covering problem. In: Silhavy, R., Senkerik, R., Oplatkova, Z.K., Silhavy, P., Prokopova, Z. (eds.) Modern Trends and Techniques in Computer Science. AISC, vol. 285, pp. 53–63. Springer, Switzerland (2014)

10. Grossman, T., Wool, A.: Computational experience with approximation algorithms for the set covering problem. Eur. J. Oper. Res. **101**(1), 81–92 (1997)
11. Hatamlou, A.: Black hole: a new heuristic optimization approach for data clustering. Inf. Sci. **222**, 175–184 (2013)
12. Krarup, J., Bilde, O.: Plant location, set covering and economic lot size: an 0 (mn)-algorithm for structured problems. In: Collatz, L., Meinardus, G., Wetterling, W. (eds.) Numerische Methoden bei Optimierungsaufgaben. International Series of Numerical Mathematics, vol. 36, pp. 155–180. Birkhuser, Basel (1977)
13. Mirjalili, S., Hashim, S., Taherzadeh, G., Mirjalili, S., Salehi, S.: A Study of Different Transfer Functions for Binary Version of Particle Swarm Optimization. CSREA Press, Las Vegas (2011)
14. Mirjalili, S., Lewis, A.: S-shaped versus V-shaped transfer functions for binary particle swarm optimization. Swarm Evol. Comput. **9**, 1–14 (2013)
15. Ruffini, R., Wheeler, J.A.: Introducing the black hole. Phys. Today **24**(1), 30 (1971)
16. Soto, R., Crawford, B., Lizama, S., Johnson, F., Paredes, F.: A XOR-based ABC algorithm for solving set covering problems. In: Gaber, T., Hassanien, A.E., El-Bendary, N., Dey, N. (eds.) Proceedings of the 1st International Conference on Advanced Intelligent System and Informatics (AISI). AISC, vol. 407, pp. 208–218. Springer, Switzerland (2016)
17. Soto, R., Crawford, B., Muñoz, A., Johnson, F., Paredes, F.: Pre-processing, repairing and transfer functions can help binary electromagnetism-like algorithms. In: Silhavy, R., Senkerik, R., Oplatkova, Z.K., Prokopova, Z., Silhavy, P. (eds.) Artificial Intelligence Perspectives and Applications. AISC, vol. 347, pp. 89–97. Springer, Heidelberg (2015)
18. Soto, R., Crawford, B., Palma, W., Galleguillos, K., Castro, C., Monfroy, E., Johnson, F., Paredes, F.: Boosting autonomous search for CSPs via skylines. Inf. Sci. **308**, 38–48 (2015)
19. Ullman, J.D.: Np-complete scheduling problems. J. Comput. Syst. Sci. **10**(3), 384–393 (1975)
20. Valenzuela, C., Crawford, B., Soto, R., Monfroy, E., Paredes, F.: A 2-level meta-heuristic for the set covering problem. Int. J. Comput. Commun. Control **7**(2), 377 (2014)

Challenging Established Move Ordering Strategies with Adaptive Data Structures

Spencer Polk$^{(\boxtimes)}$ and B. John Oommen

School of Computer Science, Carleton University, Ottawa, Canada
andrewpolk@cmail.carleton.ca, oommen@scs.carleton.ca

Abstract. The field of game playing is a particularly well-studied area within the context of AI, leading to the development of powerful techniques, such as the alpha-beta search, capable of achieving competitive game play against an intelligent opponent. It is well known that tree pruning strategies, such as alpha-beta, benefit strongly from proper move ordering, that is, searching the best element first. Inspired by the formerly unrelated field of Adaptive Data Structures (ADSs), we have previously introduced the History-ADS technique, which employs an adaptive list to achieve effective and dynamic move ordering, in a domain independent fashion, and found that it performs well in a wide range of cases. However, previous work did not compare the performance of the History-ADS heuristic to any established move ordering strategy. In an attempt to address this problem, we present here a comparison to two well-known, acclaimed strategies, which operate on a similar philosophy to the History-ADS, the History Heuristic, and the Killer Moves technique. We find that, in a wide range of two-player and multi-player games, at various points in the game's progression, the History-ADS performs at least as well as these strategies, and, in fact, outperforms them in the majority of cases.

1 Introduction

Achieving competitive play in a strategic board game, against one or more intelligent opponents, is a canonical problem within the field of AI. From the inception of the field to the present, a broad corpus of literature has been published on this topic, introducing a wide range of strategies to achieve effective game play, in a wide range of board games [1–3]. In particular, one of the most studied and acclaimed techniques is the alpha-beta search, which is capable of achieving a much greater look-ahead, or search depth, in game trees, by pruning large sections of the search space [4,5]. It is furthermore well-established that the efficiency of alpha-beta pruning is highly dependent on proper move ordering, that is, searching the strongest moves at each level of the tree first, and a range of move ordering heuristics have been developed to achieve this [2,6]. These

---

Chancellor's Professor; _Fellow: IEEE_ and _Fellow: IAPR_. The second author is also an _Adjunct Professor_ with the Department of ICT, University of Agder, Grimstad, Norway.

H. Fujita et al. (Eds.): IEA/AIE 2016, LNAI 9799, pp. 862–872, 2016.
DOI: 10.1007/978-3-319-42007-3_73

include the highly regarded and well-studied History Heuristic, and the Killer Moves strategy [6,7].

Although a broad range of techniques for achieving competitive game play have been introduced, the majority of the literature focuses on Two-Player games, such as Chess and Go, with substantially less emphasis placed on Multi-Player (MP) games [2,8]. Indeed, it is well known that most of the techniques for MP games are extensions of corresponding two-player strategies, and often have trouble performing on the level of their counterparts, for a variety of reasons [9–12]. In an attempt to improve MP strategies, in earlier papers, we derived techniques from the formerly unrelated field of Adaptive Data Structures (ADSs), a field concerned with reorganizing a data structure dynamically, to match query frequencies [13,14]. A technique to improve move ordering in a state-of-the-art MP algorithm, the Threat-ADS heuristic, was introduced in [15], and expanded upon in [16][1]. Based on this success, we later generalized ADS-based strategies to both two-player and MP environments, introducing the History-ADS heuristic [17]. The History-ADS heuristic operates by ranking potential moves, based on their previous performance within the tree, using a list-based ADS.

The History-ADS heuristic demonstrated an ability to produce substantial improvements in terms of tree pruning in a wide range of cases, and without a substantial investment in terms of computational resources [17,18]. However, while the History-ADS heuristic has produced known benefits, it has not been directly compared to known previously-reported move ordering strategies. Given its conceptual similarities to the History Heuristic and the Killer Moves strategy, in this work we present a comparison of its performance to these two well-known, highly regarded techniques, and demonstrate that it is capable of performing on their level, and in fact, outperforming them in some cases.

The rest of the paper is laid out as follows. Section 2 discusses in detail the motivation behind our work in this paper, and Sect. 3 describes the Threat-ADS and History-ADS techniques in depth. Section 4 describes our experimental design, and the game models we will be employing in our work. Section 5 presents our results for both two-player and MP games, and Sect. 6 provides our discussion and analysis of these results. Lastly, Sect. 7 concludes the paper.

2 Motivation

Our previous work in [17,18] demonstrated the potential benefits of the History-ADS heuristic in a wide range of environments, and explored a large number of possible configurations within these environments. Unlike the Threat-ADS heuristic, however, which pioneered an entirely new concept, i.e., that of using opponent threats to achieve move ordering, the History-ADS heuristic achieves move ordering through move history, a known metric. While large reductions, of over 75 %, for a relatively lightweight technique, clearly demonstrate the success

---

[1] The latter paper won the *Best Paper Award* of the IEA/AIE conference in 2015.

of the History-ADS, we cannot be sure of its actual relative benefits, unless it is compared to established, well-known techniques of a similar nature.

Given the similarities in principle behind the History-ADS heuristic, and both the History Heuristic and Killer Moves, we have elected to present a comparison between its performance and these two well-known methods, under a similar testing domain to that employed in our earlier work. By providing this comparison, we believe that we will be able to place the History-ADS heuristic's achievements in the proper context and perspective. Furthermore, it may actually be the case that the History-ADS heuristic is able to outperform one or both of them, which would be a very valuable result, given its inexpensive cost.

In fact, we have reason to suspect that the History-ADS heuristic may be capable of outperforming the History Heuristic in at least some domains. This is based on the premise of the results presented in [18]. The History Heuristic employs a relatively complex mechanism to rank moves, compared to the History-ADS heuristic. It is, however, less sensitive to change than the History-ADS heuristic, when employing a "Move-to-Front" adaptive list, which was found to perform better than strategies that are more conservative in their structural changes in our previous work. It may prove to be the case that in at least some domains, the extreme adaptability of the Move-to-Front rule will outperform even the elaborate History Heuristic. This begs investigation.

The Killer Moves heuristic is already conceptually very similar to the History-ADS operating with a "multi-level" ADS, introduced in [18] (this concept will be described in detail in the next section). In that paper, we found that a single ADS generally outperformed the multi-level variant, despite some potential drawbacks of applying the same list at all levels of the tree. Given that the Killer Moves strategy typically retains fewer moves, at each level of the tree, than any of the multi-level approaches we have previously explored, we highly suspect that the History-ADS, employing a single list, will be able to outperform it.

The potential to achieving performance on the level of the well-regarded History Heuristic and Killer Moves, using an ADS-based strategy, motivates the work presented in this paper.

3   Previous Work

ADSs were, as mentioned earlier, were originally designed to reorganize their structure, in response to queries over time, to better match access frequencies [13,14]. An example of a specific ADS *update mechanism*, for adaptive lists, is the Move-to-Front rule, where the accessed element is moved to the head of the list, and thus, will tend to remain close to the front if it is frequently accessed. The reader will observe, however, that this organization also provides an intuitive mechanism by which the elements of the data structure could be ranked. Our previous work is based on harnessing the lightweight mechanics of ADSs to serve as a ranking mechanism for elements of a game, such as players, moves, or board positions, and leveraging this ranking to achieve improvements in performance. Currently, we have focused on achieving better move ordering, and thus tree pruning, although this strategy may have broader applications.

Threat-ADS: Our first attempt to apply ADS-based techniques to game playing, was motivated by the desire to improve performance in the under-studied field of MP games. In a MP environment, as opposed to the two-player case, there are many opponents to consider, instead of just one. Intuitively, each of these opponents may threaten the player to a different extent. We focused on the Best-Reply Search (BRS), a recent, powerful MP strategy, which seeks to manage the complex case of multiple opponents by unifying them into a single "super-opponent" in its search, which minimizes the player [10]. We observed that the BRS did not unify the moves for each opponent, at each MIN level of the tree, in a specific manner, and thus introduced the Threat-ADS, which uses a small adaptive list containing opponents, to dynamically determine the best method to do this. An example of the Threat-ADS heuristic in action is shown in Fig. 1. The Threat-ADS heuristic was found to produce statistically significant results in a wide range of cases, considering different update mechanisms, ply depths, and games [16,19].

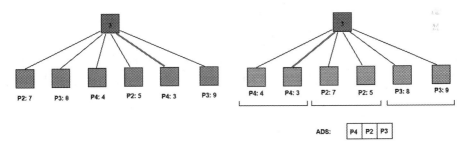

Fig. 1. The BRS without Threat-ADS on the left, and with it on the right. Notice how a cut is made faster in the second case.

History-ADS: Based on the success of the ADS-based Threat-ADS heuristic, we sought to generalize its specific, MP approach, to be applicable to both MP and two-player games. Rather than seeking to rank opponents, which only has applicability in the MP space, we drew inspiration from the well-known History Heuristic, and employed a list-based ADS to rank moves. The History-ADS heuristic operates in the context of the alpha-beta search, maintaining a list of possible moves. When a move produces a cut, the ADS is "queried" with the identity of that move, and it is moved towards the head of the list, according to the ADS' update mechanism. When moves are explored at a new level of the tree, this is done in the order dictated by the ADS, if applicable, similar to exploring the killer moves first. An example of the History-ADS heuristic in action is provided in Fig. 2.

Results from previous work demonstrated clearly that the History-ADS heuristic was capable of obtaining very large reductions in the tree size, through improved pruning, in a wide range of cases [17]. It was furthermore shown to

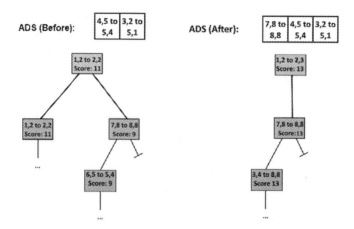

Fig. 2. A demonstration of how an ADS can be used to manage move history over time. The move (7,8) to (8,8) produces a cut, and so it is moved to the head of the list, and informs the search later.

perform best using the Move-to-Front update mechanism, relative to less sensitive strategies, and that it could keep most of its performance with a strong restriction on the length of the list [18].

4   Game Models and Experimental Setup

Given that our work in this paper is a logical "next step" of our previously published work, it is sensible for us to employ an analogous set of experiments, and employ a similar software framework, to that which was employed in [17]. We are interested in the improvement in tree pruning, when employing the History-ADS heuristic, relative to the established History Heuristic and Killer Moves techniques. We accomplish this measurement by recording an aggregate of the Node Count (NC) over several turns of the game. The NC is defined as the number of nodes that are expanded during the search, i.e. excluding those generated but then pruned before being visited. Historically, this metric has been shown to be highly correlated to runtime, while also being platform-agnostic [6]. For a variety of MP and two-player games, we average this value over fifty trials, for each technique in question.

As in our previous work, we will employ the known MP games Focus, and Chinese Checkers, as well as a territory control game of our own devising, which we have named the Virus Game, the rules for which are described in [15]. We will employ the same two-player games as well, including the two-player variant of Focus, the game Othello, and the very well-known Checkers, or Draughts. However, the requirement in Checkers that forces jumps when possible, often leads to a game with a very small branching factor, with a highly variable number of moves available at the midgame state. Thus, for our experiments, we choose to relax this rule, and do not require that a player must necessarily make an available jump. We shall refer to this game as "Relaxed Checkers". While Checkers has been solved, nevertheless serves as a useful testing environment for the general applicability of a domain-independent strategy, given how well-known and documented the game is in the literature [20].

Rather than simply test the performance of the History-ADS and other techniques from the initial board state, which can be relatively unchanging between games, we also provide results for the midgame case, as we did previously in [16]. In order to generate reasonable midgame states, we have intelligent agents play the game for a number of turns, before measurements take place, in each trial. The details of how midgame states are generated is described in greater detail in [16]. The number of turns the games are advanced for has been refined through observation, and the values we use in this work are 15 for the Virus Game, 5 for Focus (both MP and two-player versions) and Relaxed Checkers, and 10 for Othello and Chinese Checkers.

The number of turns we aggregate the NC over, for both initial and midgame cases, is 5 for Relaxed Checkers, Othello, and Chinese Checkers, 3 for Focus (given its short duration), and 10 for the Virus Game. To determine statistical significance, we employ the Mann-Whitney test due to a lack of guaranteed normalcy in the data. Lastly, we provide the Effect Sizes, which serve as an easily readable indication of the degree of the savings in terms of tree pruning. We employ a domain-independent version of the History Heuristic, based on its original specification in [6].

Our results are presented in the following section.

5   Results

The following sections present our results, as well as our statistical analysis, of the History-ADS heuristic in comparison to the History Heuristic, and the Killer Moves technique, in both two-player and multi-player contexts.

5.1   Results for Two-Player Games

Our results for Othello are presented in Table 1. We observe that in both the initial board position and midgame cases, the History-ADS heuristic outperformed both the History Heuristic, and Killer Moves, which behaved very similarly to each other. For example, in the midgame case, the History-ADS represented a 14 % improvement over the History Heuristic and Killer Moves techniques.

Table 2 showcases our results for Relaxed Checkers. Again, the History-ADS performed best in all situations, however the Killer Moves technique outperformed the History Heuristic, in this game. The History-ADS did 25 % better than the History Heuristic, when measurements were taken from the initial board position.

Table 1. Results comparing the History-ADS, History Heuristic, and Killer Moves in Othello.

Midgame?	Technique	Avg. NC	Std. Dev	P-Value	Effect Size
No	None	5,061	2,385	-	-
No	History-ADS	3,727	1,552	1.7×10^{-3}	0.56
No	History Heuristic	4,136	1,711	0.071	0.37
No	Killer Moves	4,013	1,720	0.015	0.44
Yes	None	20,100	9,899	-	-
Yes	History-ADS	13,300	6,916	7.0×10^{-5}	0.69
Yes	History Heuristic	15,500	6,939	9.4×10^{-3}	0.47
Yes	Killer Moves	15,500	6,696	0.015	0.47

Table 2. Results comparing the History-ADS, History Heuristic, and Killer Moves in Relaxed Checkers.

Midgame?	Technique	Avg. NC	Std. Dev	P-Value	Effect Size
No	None	78,600	10,600	-	-
No	History-ADS	41,000	5,588	$< 1.0 \times 10^{-5}$	3.55
No	History Heuristic	54,800	9,018	$< 1.0 \times 10^{-5}$	2.25
No	Killer Moves	52,200	6,723	$< 1.0 \times 10^{-5}$	2.50
Yes	None	64,000	25,700	-	-
Yes	History-ADS	34,400	12,400	$< 1.0 \times 10^{-5}$	1.15
Yes	History Heuristic	42,800	14,500	$< 1.0 \times 10^{-5}$	0.83
Yes	Killer Moves	39,100	14,700	$< 1.0 \times 10^{-5}$	0.97

In Table 3, we present our results for two-player focus. As is consistent with previous work, all techniques led to a drastic reduction in NC, although again, History-ADS did best of all, and Killer Moves outperformed the History Heuristic.

5.2   Results for Multi-player Games

Table 4 holds our results for the Virus Game, where the same patterns as previously were observed. The History Heuristic performed particularly poorly here, with History-ADS representing 28 % improvement over it in the midgame situation.

In Table 5, we show our results for MP Focus, which again, follow the established pattern. In this case, the difference between History-ADS and Killer Moves was negligible, although History-ADS outperformed it in both cases.

Finally, Table 6 presents our results for Chinese Checkers. We see the pattern observed with the other games repeated again, although Killer Moves was very close to History-ADS in this case, being approximately equivalent in the midgame case, and the History Heuristic did noticeably worse. In the midgame case, the History-ADS heuristic did 37 % better than the History Heuristic.

Table 3. Results comparing the History-ADS, History Heuristic, and Killer Moves in two-player focus.

Midgame?	Technique	Avg. NC	Std. Dev	P-Value	Effect Size
No	None	5,250,000	381,000	-	-
No	History-ADS	1,260,000	90,900	$< 1.0 \times 10^{-5}$	10.46
No	History Heuristic	1,980,000	221,000	$< 1.0 \times 10^{-5}$	8.59
No	Killer Moves	1,420,000	105,100	$< 1.0 \times 10^{-5}$	10.04
Yes	None	10,600,000	3,460,000	-	-
Yes	History-ADS	2,390,000	631,000	$< 1.0 \times 10^{-5}$	2.37
Yes	History Heuristic	3,500,000	1,040,000	$< 1.0 \times 10^{-5}$	2.05
Yes	Killer Moves	2,680,000	648,000	$< 1.0 \times 10^{-5}$	2.29

Table 4. Results comparing the History-ADS, History Heuristic, and Killer Moves in the Virus Game.

Midgame?	Technique	Avg. NC	Std. Dev	P-Value	Effect Size
No	None	10,500,000	1,260,000	-	-
No	History-ADS	4,650,000	767,000	$< 1.0 \times 10^{-5}$	4.60
No	History Heuristic	6,860,000	1,080,000	$< 1.0 \times 10^{-5}$	2.86
No	Killer Moves	5,210,000	858,000	$< 1.0 \times 10^{-5}$	4.16
Yes	None	12,800,000	1,950,000	-	-
Yes	History-ADS	5,870,000	863,000	$< 1.0 \times 10^{-5}$	3.55
Yes	History Heuristic	8,190,000	1,080,000	$< 1.0 \times 10^{-5}$	2.36
Yes	Killer Moves	6,380,000	991,000	$< 1.0 \times 10^{-5}$	3.29

Table 5. Results comparing the History-ADS, History Heuristic, and Killer Moves in Multi-player focus.

Midgame?	Technique	Avg. NC	Std. Dev	P-Value	Effect Size
No	None	6,970,000	981,000	-	-
No	History-ADS	2,150,000	165,000	$< 1.0 \times 10^{-5}$	4.92
No	History Heuristic	3,360,000	351,000	$< 1.0 \times 10^{-5}$	3.69
No	Killer Moves	2,220,000	175,000	$< 1.0 \times 10^{-5}$	4.84
Yes	None	14,200,000	8,400,000	-	-
Yes	History-ADS	3,160,000	1,700,000	$< 1.0 \times 10^{-5}$	1.31
Yes	History Heuristic	5,050,000	3,010,000	$< 1.0 \times 10^{-5}$	1.09
Yes	Killer Moves	3,260,000	1,530,000	$< 1.0 \times 10^{-5}$	1.30

Table 6. Results comparing the History-ADS, History Heuristic, and Killer Moves in Chinese Checkers.

Midgame?	Technique	Avg. NC	Std. Dev	P-Value	Effect Size
No	None	3,370,000	1,100,000	-	-
No	History-ADS	1,280,000	368,000	$< 1.0 \times 10^{-5}$	1.90
No	History Heuristic	1,550,000	445,000	$< 1.0 \times 10^{-5}$	1.66
No	Killer Moves	1,310,000	341,000	$< 1.0 \times 10^{-5}$	1.88
Yes	None	8,260,000	1,950,000	-	-
Yes	History-ADS	3,200,000	863,000	$< 1.0 \times 10^{-5}$	1.92
Yes	History Heuristic	5,050,000	1,090,000	$< 1.0 \times 10^{-5}$	1.64
Yes	Killer Moves	3,200,000	799,000	$< 1.0 \times 10^{-5}$	1.92

6   Discussion

Our results clearly demonstrate the power of the History-ADS heuristic, even when compared to the established, highly-regarded techniques, specifically, the Killer Moves strategy, and the History Heuristic. We found that in nearly every case examined, the History-ADS heuristic outperformed both of these established techniques, or performed on a level comparable to them.

Although we were somewhat surprised that the History Heuristic was outperformed by the History-ADS heuristic in every case examined, reviewing our findings from [17] with this knowledge in mind, such an outcome is rather predictable. We had earlier observed that, in the context of the History-ADS heuristic, the Move-to-Front rule consistently outperformed the less sensitive Transposition rule. The strategy of the History Heuristic, however, is even less sensitive to change than the Transposition rule. This is because, according to its specification, the ranking of the moves will only change when one move's counter exceeds another. As opposed to this, use of the Transposition rule leads to some change in structure every time it is queried, even if it is slight.

Our results suggest that the History Heuristic allows very strong moves to gain a substantial lead over all others. Indeed, when viewing the History Heuristic's internal updates as the search proceeded, in both the Virus Game and Othello, a single move would quickly gain a nearly insurmountable lead. This is the likely reason for the History Heuristic's poor performance, compared to the History-ADS heuristic, and is consistent with our previous observations. Our results strongly suggest that the History-ADS heuristic outperforms the History Heuristic under a broad set of board games, and we hypothesize, based on these results, that it is likely to do so in others as well.

The fact that the single ADS, Move-to-Front History-ADS heuristic outperforms the Killer Moves strategy is not at all surprising, considering our previous observations from [18]. Indeed, we had earlier determined that a single ADS would generally outperform a multi-level ADS, and that a multi-level ADS with a restriction on its length would have its performance hampered even further.

As the Killer Moves technique functionally identical to the History-ADS with a multi-level ADS, and with a limit of two on the length, we would expect it to be outmatched by the single, unbounded ADS. Our results, clearly, support that.

The degree by which the Killer Moves technique was outperformed varied between the various game models, with it doing best in Chinese Checkers, and worst in Relaxed Checkers. Given that it can only maintain a very small number of moves, this suggests that storing more information achieves a superior move ordering in the case of Relaxed Checkers, but it is not so critical in the case of Chinese Checkers, with the other games falling between these extreme cases.

7   Conclusions

Our results reinforce our previous findings, that the History-ADS heuristic is able to produce strong gains in terms of tree pruning. Additionally, we have also clearly demonstrated that the History-ADS heuristic is capable of outperforming the established Killer Moves technique and History Heuristic in a wide range of game models and configurations, in some cases by a substantial margin. This is a particularly strong result, which serves to justify its usage in game playing engines, particularly given its lightweight qualities, and the fact that it does not need any additional sorting.

Our results further reinforce the idea that, in the context of the History-ADS heuristic, the most basic configuration tends to perform best. We confirm this because the single, unbound, Move-to-Front implementation of the History-ADS outperformed both established heuristics. This suggests that within the perspective of move ordering, that is based on a move history criterion, the adage "simpler is better" holds true.

References

1. Rimmel, A., Teytaud, O., Lee, C., Yen, S., Wang, M., Tsai, S.: Current frontiers in computer go. IEEE Trans. Comput. Intell. Artif. Intell. Games **2**(4), 229–238 (2010)
2. Russell, S.J., Norvig, P.: Aritificial Intelligence: A Modern Approach, 3rd edn. Prentice-Hall Inc., Upper Saddle River (2009)
3. Shannon, C.E.: Programming a computer for playing Chess. Phil. Mag. **41**, 256–275 (1950)
4. Baudet, G.M.: An analysis of the full alpha-beta pruning algorithm. In: Proceedings of the Tenth Annual ACM Symposium on Theory of Computing, pp. 296–313 (1978)
5. Knuth, D.E., Moore, R.W.: An analysis of alpha-beta pruning. Artif. Intell. **6**, 293–326 (1975)
6. Schaeffer, J.: The history heuristic and alpha-beta search enhancements in practice. IEEE Trans. Pattern Anal. Mach. Intell. **11**, 1203–1212 (1989)
7. Akl, S., Newborn, M.: The principal continuation and the killer heuristic. In: Proceedings of ACM 1977 the 1977 Annual Conference, pp. 466–473 (1977)

8. Sturtevant, N., Games, M.-P.: Algorithms and approaches. Ph.D. thesis, University of California (2003)
9. Luckhardt, C., Irani, K.: An algorithmic solution of n-person games. In: Proceedings of the AAAI 1986, pp. 158–162 (1986)
10. Schadd, M.P.D., Winands, M.H.M.: Best Reply Search for multiplayer games. IEEE Trans. Comput. Intell. AI Games **3**, 57–66 (2011)
11. Sturtevant, N., Bowling, M.: Robust game play against unknown opponents. In: Proceedings of AAMAS 2006, The 2006 International Joint Conference on Autonomous Agents and Multiagent Systems, pp. 713–719 (2006)
12. Sturtevant, N., Zinkevich, M., Bowling, M.,:Prob-Maxn: playing n-player games with opponent models. In: Proceedings of AAAI 2006, 2006 National Conference on Artificial Intelligence, pp. 1057–1063 (2006)
13. Gonnet, G.H., Munro, J.I., Suwanda, H.: Towards self-organizing linear search. In: Proceedings of FOCS 1979, The 1979 Annual Symposium on Foundations of Computer Science, pp. 169–171 (1979)
14. Hester, J.H., Hirschberg, D.S.: Self-organizing linear search. ACM Comput. Surv. **17**, 285–311 (1985)
15. Polk, S., Oommen, B.J.: On applying adaptive data structures to multi-player game playing. In: Proceedings of AI 2013, The Thirty-Third SGAI Conference on Artificial Intelligence, pp. 125–138 (2013)
16. Polk, S., Oommen, B.J.: Novel AI strategies for multi-player games at intermediate board states. In: Proceedings of IEA/AIE 2015, The Twenty-Eighth International Conference on Industrial, Engineering, and Other Applications of Applied Intelligent Systems, pp. 33–42 (2015)
17. Polk, S., Oommen, B.J.: Enhancing history-based move ordering in game playing using adaptive data structures. In: Núñez, M., et al. (eds.) ICCCI 2015. LNCS, vol. 9329, pp. 225–235. Springer, Heidelberg (2015). doi:10.1007/978-3-319-24069-5_21
18. Polk, S., Oommen, B.J.: Space and depth-related enhancements of the history-ads strategy in game playing. In: Proceedings of CIG 2015, The 2015 IEEE Conference on Computational Intelligence and Games, pp. 322–327 (2015)
19. Polk, S., Oommen, B.J.: On enhancing recent multi-player game playing strategies using a spectrum of adaptive data structures. In: Proceedings of TAAI 2013, The 2013 Conference on Technologies and Applications of Artificial Intelligence (2013)
20. Schaeffer, J., Burch, N., Bjornsson, Y., Kishimoto, A., Muller, M., Lake, R., Lu, P., Sutphen, S.: Checkers is solved. Science **14**, 1518–1522 (2007)

An Binary Black Hole Algorithm
to Solve Set Covering Problem

Álvaro Gómez Rubio[1(✉)], Broderick Crawford[1,4(✉)],
Ricardo Soto[1,2,3(✉)], Adrián Jaramillo[1(✉)], Sebastián Mansilla Villablanca[1(✉)],
Juan Salas[1(✉)], and Eduardo Olguín[4(✉)]

[1] Pontificia Universidad Católica de Valparaíso, Santiago de Chile, Chile
{alvaro.gomez.r,adrian.jaramillo.s,
sebastian.mansilla.v,juan.salas.f}@mail.pucv.cl,
{broderick.crawford,ricardo.soto}@ucv.cl
[2] Universidad Autónoma de Chile, Santiago de Chile, Chile
[3] Universidad Científica del Sur, Lima, Peru
[4] Universidad San Sebastián, Santiago de Chile, Chile
eduardo.olguin@uss.cl

Abstract. The set covering problem (SCP) is one of the most representative combinatorial optimization problems and it has multiple applications in different situations of engineering, sciences and some other disciplines. It aims to find a set of solutions that meet the needs defined in the constraints having lowest possible cost. In this paper we used an existing binary algorithm inspired by Binary Black Holes (BBH), to solve multiple instances of the problem with known benchmarks obtained from the OR-library. The presented method emulates the behavior of these celestial bodies using a rotation operator to bring good solutions. After tray this algorithm, we implemented some improvements in certain operators, as well as added others also inspired by black holes physical behavior, to optimize the search and exploration to improving the results.

Keywords: Set covering problem · Binary black hole · Methaheuristics · Combinatorial optimization problem

1 Introduction

The SCP is one of 21 NP-Hard problems, representing a variety of optimization strategies in various fields and realities. Since its formulation in the 1970s has been used, for example, in minimization of loss of materials for metallurgical industry [1], preparing crews for urban transportation planning [2], safety and robustness of data networks [3], focus of public policies [4], construction structural calculations [5].

Considering a binary numbers array A, of m rows and n columns (a_{ij}), and a C vector (c_j) of n columns containing the costs assigned to each one, then we can then define the SCP such as:

© Springer International Publishing Switzerland 2016
H. Fujita et al. (Eds.): IEA/AIE 2016, LNAI 9799, pp. 873–883, 2016.
DOI: 10.1007/978-3-319-42007-3_74

$$\text{Minimize} \sum_{j=1}^{n} c_j x_j \tag{1}$$

where a:

$$\sum_{j=1}^{n} a_{ij} x_j \geq 1 \ \forall \ i \in \{1, ..., n\}$$

$$x_j \in \{0, 1\}; \ j \in \{1, ..., n\}$$

This problem was introduced in 1972 by Karp [6] and it is used to optimize problems of elements locations that provide spatial coverage, such as community services, telecommunications antennas and others.

The present work applied a strategy based on a binary algorithm inspired by black holes to solve the SCP, developing some operators that allow to implement an analog version of some characteristics of these celestial bodies to support the behavior of the algorithm and improve the processes of searching for the optimum. This type of algorithm was presented for the first time by Abdolreza Hatamlou in September 2012 [7], registering some later publications dealing with some applications and improvements. In this paper it will be detailed methodology, developed operators, experimental results and execution parameters and handed out some brief conclusions about them, the original version for both the proposed improvements.

2   Black Holes

Black holes are the result of the collapse of a big star's mass that after passing through several intermediate stages is transformed in a so massively dense body that manages to bend the surrounding space because of its immense gravity. They are called "black holes" due to even light does not escape their attraction and therefore is undetectable in the visible spectrum, knowing also by "singularities", since inside traditional physics loses meaning. Because of its immense gravity, they tend to be orbited by other stars in binary or multiple systems consuming a little mass of bodies in its orbit [8]. When a star or any other body is approaching the black hole through what is called "event horizon", collapses in its interior and is completely absorbed without any possibility to escape, since all its mass and energy become part of singularity (Fig. 1). This is because at that point the exhaust speed is the light one [8].

On the other hand, black holes also generate a type of radiation called "Hawking radiation", in honor of its discoverer. This radiation have a quantum origin and implies transfer of energy from the event horizon of the black hole to its immediate surroundings, causing a slight loss of mass of the dark body and an emission of additional energy to the nearby objects [9].

3   Algorithm

The algorithm presented in September 2012 by Hatamlou [7] faces the problem of determination of solutions through the development of a set of stars called

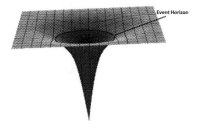

Fig. 1. Event horizon in a black hole

"universe", using an algorithm type population similar to those used by genetic techniques or particles swarm. It proposes the rotation of the universe around the star that has the best fitness, i.e., which has the lowest value of a defined function, called "objective function". This rotation is applied by an operator of rotation that moves all stars in each iteration of the algorithm and determines in each cycle if there is a new black hole, that it will replace the previous one. This operation is repeated until it find the detention criteria, being the last of the black holes founded the proposed solution. Eventually, a star can ever exceed the defined by the radius of the event horizon [8]. In this case, the star collapses into the black hole and is removed from the whole universe being taken instead by a new star. Thus, stimulates the exploration of the space of solutions. The following is the proposed flow and the corresponding operators according to the initial version of the method (Fig. 2).

3.1 Big Bang

It consists the creation of the initial random universe for the algorithm. The number of stars generated will remain fixed during the iterations, notwithstanding that many of the vectors (or stars) are replaced. The mechanism of creation of vectors is as follows and shall also apply in the intermediate steps that require the generation of new stars:

Algorithm 1. Random initial generation of stars

1: $n \leftarrow Cols\ quantity$
2: **for** $row = 1$ to $Stars\ quantity$ **do**
3:     $Star_{row} = \text{StarGeneration}(n)$
4: **end for**

Where StarGeneration is the creation of a binary vector of n elements that comply with the restrictions of A matrix.

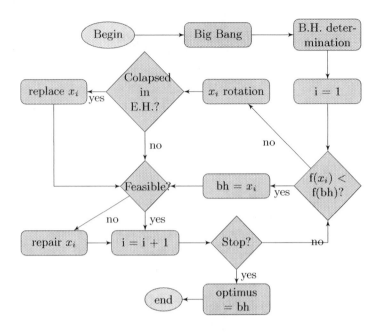

Fig. 2. Original black hole algorithm

3.2   Fitness Evaluation

The each star x_i fitness is calculated by evaluating the objective function, according to the initial definition of the problem. In algorithmic terms described in the following way:

Algorithm 2. Fitness evaluation

1: $Fitness \leftarrow 0$
2: **for** $X_j = 1$ to $Cols\ quantity$ **do**
3:     $Fitness \leftarrow Fitness + x_j c_j$
4: **end for**

It should be remembered that c_j corresponds to the cost of that column in the matrix of costs. In other words, the fitness of a star is the sum of the product of the value of each column covered with a star in particular, multiplied by the corresponding cost. The black hole will be those who have minor fitness among all existing stars at the time of the evaluation.

3.3   Rotation Operator

The rotation operation occurs above all the universe of x_i^d stars of t iteration, with the exception of the black hole, which is fixed in its position. The operation sets the new t+1 position follows:

$$X_i^d(t+1) = X_i(t) + random(X_{BH} - X_i(t)), where\ i = 1, 2, ..., N \qquad (2)$$

Where random $\in \{0,1\}$ and change in each iteration, $x_i(t)$ and $x_i(t+1)$ are the positions of the star x_i at t and t+1 iterations respectively, x_{BH} is the black hole location in the search space, random is a random number in the range [0,1] and N is the number of stars that make up the universe (candidate solution). It should be noted that the only exception in the rotation is designated as black hole star, which retains the position.

3.4 Collapse into the Black Hole

When a star is approaching a black hole at a distance called event horizon is captured and permanently absorbed by the black hole, being replaced by a new randomly generated one. In other words, it is considered when the collapse of a star exceeds the radius of Schawarzchild (R) defined as:

$$R = \frac{f_{BH}}{\sum_{i=1}^{n} f_i} \qquad (3)$$

where f_{BH} is the value of the fitness of the black hole and f_i is the ith star fitness. N is the number of stars in the universe.

3.5 Algorithm Implementation

The algorithm implementation was carried out with a I-CASE tool, generating Java programs and using a relational database as a repository of the entry information and gathered during executions. The parameters that will be presented are the result both of the needs of the original design of the algorithm improvements made product of the tests performed. In particular, attempted to improve the capacity of exploration of the metha heuristics. Is contrast findings with tables of known optimal values [10, 11], in order to quantitatively estimate the degree of effectiveness of the presented metha heuristics. We found no publications that present information regarding algorithmic specifications or the original version of the algorithm implementation, so there are various aspects to which the authors do not point solution, but we can speculate that they are similar to other metaheuristic algorithms that is does have information from other authors:

The process begins with the random generation of a population of binary vectors (Star) in a step that we will call "big bang". With a universe of m star formed by binary vectors of n digits, you must identify that with better fitness, i.e. one that is evaluated with the objective function release the lowest value among all those that make up the universe generated. The next step is to rotate the other stars around the black hole detected until some other presents a better fitness and take its place.

The number of star generated will remain fixed during the iterations, notwithstanding that many vectors (or star) will be replaced by one of the operators.

3.6   Transfer Functions and Binarization

The transfer functions aims to take values from the domain of the real to the range [0..1]. For this, many functions was tested, been the inverse exponential function used for the definitive benchmarks.

$$\frac{1}{1 + (e^{-x/3})} \tag{4}$$

In addition, the binarization function is aimed at conveying the value obtained in the previous transformation in a binary digit. Therefore be tested the following routines, where random is a random value between 0 and 1 inclusive.

Algorithm 3. Standard binarization

1: **if** random $\leq$ value **then**
2:    Digit = 1
3: **else**
4:    Digit = 0
5: **end if**

The binarization best results have been achieved with that was the standard to be applied in the subsequent benchmarks.

3.7   Feasibility and Repair Operators

The feasibility of a star is given by the condition if it meets each of the constraints defined in the matrix A. In those cases which unfeasibility was detected, opted for repair of the vector to make it comply with the constraints. We implemented a repair function in two phases, ADD and DROP, as way to optimize the vector in terms of coverage and costs. The first phase changes the vector in the column that provides the coverage at the lowest cost, while the second one removes those columns which only added cost and do not provide coverage.

3.8   Collapse into the Black Hole

One of the main problems for the implementation of this operator is that the authors refer to vectorial distances determinations or some other method. However, in a 2015 publication, Farahmandian, and Hatamlouy [12] intend to determine the distance of a star x_i to the radius R as:

$$|f(x_{BH}) - f(x_i)| \tag{5}$$

I.e. a star x_i will collapse if the absolute value of the black hole and his fitness subtraction is less than the value of the radius R:

$$|f(x_{BH}) - f(x_i)| < R \tag{6}$$

Table 1. Execution parameters

Parameter	Best value
Universe size in stars	50
Iterations max	20.000
Transference function	$\frac{1}{1+(e^{-x/3})}$

Table 2. Experimental results

Instance	Z_{BKS}	Z_{min}	Z_{max}	Z_{avg}	RPD	Instance	Z_{BKS}	Z_{min}	Z_{max}	Z_{avg}	RPD
4.1	429	455	603	529,00	6,06	C.1	227	252	287	269,5	9,92
4.2	512	544	633	588,50	6,25	C.2	219	245	289	267	10,61
4.3	516	551	696	623,50	6,78	C.3	243	266	399	332,5	8,65
4.4	494	527	749	638,00	6,68	C.4	219	252	301	276,5	13,10
4.5	512	448	730	639,00	7,03	C.5	215	247	295	271	12,96
4.6	560	601	674	637,50	7,32	D.1	60	71	146	108,5	15,49
4.7	430	461	514	487,50	7,21	D.2	66	73	177	125	9,59
4.8	492	528	613	570,50	7,32	D.3	72	81	120	100,5	11,11
4.9	641	688	767	727,50	7,33	D.4	62	70	135	102,5	11,43
4.10	514	547	660	603,50	6,42	D.5	61	72	208	140	15,28
5.1	253	269	398	333,50	6,32	E.1	5	9	53	31	44,44
5.2	302	322	430	376,00	6,62	E.2	5	12	61	36,5	58,33
5.3	226	246	275	281,50	8,85	E.3	5	10	112	61	50,00
5.4	242	261	287	268,00	7,85	E.4	5	11	76	43,5	54,55
5.5	211	228	258	243,00	8,06	E.5	5	13	71	42	61,54
5.6	213	230	359	294,50	7,98	NRE1	29	81	169	125	64,20
5.7	293	322	372	347,00	9,90	NRE2	30	44	152	98	31,82
5.8	288	308	459	383,50	6,94	NRE3	27	435	522	478,5	93,79
5.9	279	296	449	372,50	6,09	NRE4	28	44	62	53	36,36
5.10	265	283	412	347,50	6,79	NRE5	28	213	346	279,5	86,85
6.1	138	151	201	176,00	9,42	NRF1	14	658	711	684,5	97,87
6.2	146	157	281	219,00	7,53	NRF2	15	18	163	90,5	16,67
6.3	145	153	195	175,50	7,59	NRF3	14	69	116	92,5	79,71
6.4	131	144	233	188,50	9,92	NRF4	14	45	147	96	68,89
6.5	161	177	258	217,50	9,94	NRF5	13	222	362	292	94,14
A.1	253	298	414	356,00	17,79	NRG1	176*	770	797	783,5	77,14
A.2	252	301	430	365,50	19,44	NRG2	151*	876	1006	941	82,76
A.3	232	256	390	323,00	10,34	NRG3	166*	1012	1046	1029	83,60
A.4	234	268	316	292	14,53	NRG4	168*	289	398	343,5	41,87
A.5	236	266	369	317,50	12,71	NRG5	168*	1211	1339	1275	620,83
B.1	69	82	149	115,50	18,84	NRH1	63*	2143	2242	2192,5	3301,59
B.2	76	99	184	133,50	30,26	NRH2	63*	701	810	755,5	1012,70
B.3	80	89	145	117	11,25	NRH3	59*	893	915	904	1413,56
B4	79	88	104	96	11,39	NRH4	59*	329	464	396,5	457,63
B.5	72	88	119	99,50	22,22	NRH5	55*	715	845	780	1200

* = Best results found in literature [15]

3.9   Parameters

For the purpose of implementing all the features and operators that are detailed in this document in multiple configurations, a table of parameters was built for the algorithm. Below are parameters values used and which values are those that gave the best results (Table 1).

The first parameter refers to the fixed number of stars that comprise the full universe to be processed, while the second specifies the maximum number of iterations to perform in total. The third and fourth parameters refer to the functions that will be used in the transfer and binarization of the variables in each iteration.

4   Experimental Results

The original algorithm was subjected to a test by running the benchmark 4, 5, 6, A, B, C, D, NRE, NRF, NRG and NRH from OR library [13]. Each of these data sets ran 30 times with same parameters [14], presenting the following results (Table 2).

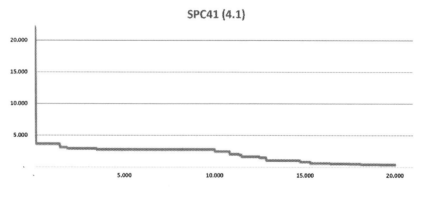

Fig. 3. SPC41 results

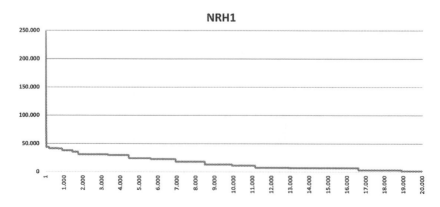

Fig. 4. NRH1 results

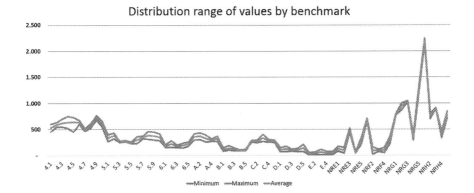

Fig. 5. Evolution of maxs and mins (Color figure online)

5   Experimental Analisis and Conclusions

Comparing the results of experiments with optimal ones reported in the literature, we can warn that the results obtained are acceptably close to the best known optimum for benchmarks 4, 5 and 6 and very far away from them in the case of the final ones (A, B, C, D, E and NR). In the case of the first ones are deviations between 6,06 % and 61,54 %, while in the case of the latter ones reach 3.301,59 % of deviation. Both cases despite the execution of the algorithm a lot of times. The rapid initial convergence is achieved, striking finding very significant improvements in the first iteration, finding very significant improvements in early iterations, being much more gradual subsequent and requiring the execution of those operators that stimulate exploration, such as collapse and Hawking radiation. This suggests that the algorithm has a tendency to fall in optimal locations, where cannot leave without the help of scanning components. In order to illustrate these trends, some graphics performance benchmarks are presented (Figs. 3 and 4).

While in the benchmark results which threw poor results have significant percentages of deviation from the known optimal, in absolute terms the differences are low considering the values from which it departed iterating algorithm. It is probably that these tests require greater amount of iterations to improve its results, since the values clearly indicate a consistent downward trend, the number of variables is higher and the difference between the optimum and the start values is broader. An interesting analysis element is that the gap between the best and the worst outcome is small and relatively constant in practically all benchmarks, indicating the algorithm tends continuously towards an improvement of results and the minimums are not just a product of suitable random values. The following chart explains this element (Fig. 5).

On the other hand, it is also important to note that in those initial tests in which the stochastic component was greater than that has been postulated as the optimal, the algorithm presented lower performance, determining optimal much higher probably by the inability to exploit areas with better potential solutions. All this is what it can be noted that the associated parameters to define the roulette for decision-making are quite small ranges in order that the random component be moderate. Other notorious elements are the large differences in results obtained with different methods of transfer and binarization, some ones simply conspired against acceptable results. Various possibilities already exposed to find a satisfactory combination were explored. Some investigation lines that can be interesting approach for a possible improvement of results may be designed to develop a better way to determine the concept of distance, with better tailored criteria to the nature of the algorithm, as well as a more sophisticated method of mutation for those stars subjected to Hawking radiation. Additionally, some authors treat the rotation operator by adding additional elements such as mass and electric charge of the hole negros [16], what was not considered in this work because the little existing documentation.

Acknowledgements. Álvaro Gómez is supported by Postgraduate Grant Pontificia Universidad Católica de Valparaíso 2015 (INF-PUCV 2015). Broderick Crawford is supported by Grant CONICYT/FONDECYT/REGULAR/1140897. Ricardo Soto is supported by Grant CONICYT/FONDECYT/INICIACION/11130459. Adrián Jaramillo is supported by Postgraduate Grant Pontificia Universidad Católica de Valparaíso 2015 (INF-PUCV 2015). Sebastián Mansilla is supported by Postgraduate Grant Pontificia Universidad Católica de Valparaíso 2015 (INF-PUCV 2015). Juan Salas is supported by Postgraduate Grant Pontificia Universidad Católica de Valparaíso 2015 (INF-PUCV 2015).

References

1. Vasko, F., Wolf, F., Stott, K.: Optimal selection of ingot sizes via set covering. Oper. Res. **35**(3), 346–353 (1987)
2. Desrochers, M., Soumis, F.: A column generation approach to the urban transit crew scheduling problem. Transp. Sci. **23**(1), 1–13 (1989)
3. Bellmore, M., Ratliff, H.D.: Optimal defense of multi-commodity networks. Manage. Sci. **18**(4–part–i), B-174 (1971)
4. Garfinkel, R.S., Nemhauser, G.L.: Optimal political districting by implicit enumeration techniques. Manage. Sci. **16**(8), B–495 (1970)
5. Amini, F., Ghaderi, P.: Hybridization of harmony search and ant colony optimization for optimal locating of structural dampers. Appl. Soft Comput. **13**(5), 2272–2280 (2013)
6. Karp, R.: Reducibility among combinatorial problems (1972). http://www.cs.berkeley.edu/~luca/cs172/karp.pdf
7. Hatamlou, A.: Black hole: a new heuristic optimization approach for data clustering. Inf. Sci. **222**, 175–184 (2013)
8. Hawking, S.: Agujeros negros y pequeños universos. Planeta, Buenos Aires (1994)
9. Hawking, S., Jackson, M.: A Brief History of Time. Dove Audio, Beverly Hills (1993)
10. Ataim, P.: Resolución del problema de set-covering usando un algoritmo genético (2005)

11. Crawford, B., Soto, R., Olivares-Suarez, M., Palma, W., Paredes, F., Olguín, E., Norero, E.: A binary coded firefly algorithm that solves the set covering problem. Sci. Technol. **17**(3), 252–264 (2014)
12. Farahmandian, M., Hatamlou, A.: Solving optimization problems using black hole algorithm. J. Adv. Comput. Sci. Technol. **4**(1), 68–74 (2015)
13. Beasley, J.: Or-library (1990). http://people.brunel.ac.uk/~mastjjb/jeb/orlib/scpinfo.html
14. Beasley, J.E.: An algorithm for set covering problem. Eur. J. Oper. Res. **31**(1), 85–93 (1987)
15. Gervasi, O., Murgante, B., Misra, S., Gavrilova, M.L., Rocha, A.M.A.C., Torre, C., Taniar, D., Apduhan, B.O.: Computational Science and Its Applications-ICCSA 2015, vol. 9157. Springer, Cham (2015)
16. Nemati, M., Salimi, R., Bazrkar, N.: Black holes algorithm: a swarm algorithm inspired of black holes for optimization problems. IAES Int. J. Artif. Intell. (IJ-AI) **2**(3), 143–150 (2013)

Solving the Set Covering Problem
with the Soccer League Competition Algorithm

Adrián Jaramillo[1(✉)], Broderick Crawford[1,4], Ricardo Soto[1,2,3],
Sebastián Mansilla Villablanca[1], Álvaro Gómez Rubio[1], Juan Salas[1],
and Eduardo Olguín[4]

[1] Pontificia Universidad Católica de Valparaíso, Valparaíso, Chile
{adrian.jaramillo.s,sebastian.mansilla.v,alvaro.gomez.r,
juan.salas.f}@mail.pucv.cl,
{broderick.crawford,ricardo.soto}@ucv.cl
[2] Universidad Autónoma de Chile, Santiago, Chile
[3] Universidad Científica del Sur, Lima, Peru
[4] Universidad San Sebastian, Santiago, Chile
eduardo.olgin@uss.cl

Abstract. The Soccer League Competition (SLC) algorithm is a new
metaheuristic approach intendended to solve complex optimization problems. It is based in the interaction model present in soccer teams and the
goal to win every match, becoming the best team and league of players.
This paper presents adaptations to the initial mode of SLC for the purpose of being applied to the Set Covering Problem (SCP) with a Python
implementation.

Keywords: Soccer League Competition · Combinatorial · Optimization · Set Covering Problem · Constraint satisfaction

1 Introduction

The Set Covering Problem (SCP) is a widely studied optimization problem
present in many real-life scenarios like operations research, machine learning,
planning, data mining, data quality and information retrieval. Its main goal in
general terms is to find the smallest subcollection of items from a given universe
so that its union is that universe and a set of constraints are met. In the last
decades, several techniques have been proposed to find the best solutions for
complex scenarios of SCP as is discussed in [3–6,11,12].

Soccer League Competition (SCL) is a newly metaheuristic approach based
on the soccer competitions, as discussed in [8–10]. This model considers feasible
solutions as soccer players, sets of them as soccer teams, and certain movement
operators applied on players, emulating the dynamic generated in the competition process to reach the team's victory and become the best soccer player of the
seasson. Mapped to a mathematical model, the goal is to find the best feasible
solution that achieves the best value of an evaluation function defined for the
problem, using specific explotation and exploration movement operators.

© Springer International Publishing Switzerland 2016
H. Fujita et al. (Eds.): IEA/AIE 2016, LNAI 9799, pp. 884–891, 2016.
DOI: 10.1007/978-3-319-42007-3_75

We propose a derived model from SLC to solve optimization problems inside a binary search space, specially applied to solve the Set Covering Problem. This derived model is built in Python programming language and it can lead to a wide variety of benchmarks.

2   The Set Covering Problem

Set Covering Problem, defined as NP-hard problem, establishes a set of m constraints over n decision variables in the $\{0,1\}$ domain. Its formulation is as follow:

$$\min \quad C = \sum_{j=1}^{n} c_j x_j \tag{1}$$

s.a.:

$$\sum_{j=1}^{n} a_{ij} x_j \geq 1 \quad \forall i \in I = \{1, 2, ..., m\} \tag{2}$$

$$x_j \in \{0, 1\} \quad \forall j \in J = \{1, 2, ..., n\} \tag{3}$$

The main idea is to find the solution vector $\mathbf{X} \in \{0,1\}^n$ such every constraint is verified and $C : R^n \to R$ is minimum.

3   Soccer League Competition

SLC intends to find the solution vector $\mathbf{X}$ for which a function $F(\mathbf{X})$ has the best minimun (or maximum) value, providing a displacement mechanism within a space of solutions. Identifying and defining some model elements are relevant to understand the relation between reality and the underlying mathematical model of SLC. Allow to define $N_{players}$ solution vectors $\mathbf{X} = (x^1, x^2, ..., x^d)$ as n-dimensional vectors inside a solution space $\mathcal{S}$.

$$\mathbf{X}_i = (x_i^1, x_i^2, ..., x_i^d) \mid \mathbf{X}_i \in \mathcal{S} \\ i \in I = \{1, 2, ..., N_{players}\} \tag{4}$$

Each $\mathbf{X}_i$, defined as a player, can be a *fixed* or *substitute* player, $\mathbf{F}_i$ and $\mathbf{S}_i$, respectively, according to certain ranking definition to be explained in the next sections. Associated to each player $\mathbf{X}_i$ we will find an indicator scalar value called *power player* (pp_i). We will define $PP : \mathbb{R}^n \to \mathbb{R}$ as the power player function:

$$pp_i = PP(\mathbf{X}_i) \tag{5}$$

It allows to compare performance among players. If two vectors $\mathbf{X}_j$ and $\mathbf{X}_k$ verifies $PP(\mathbf{X}_j) > PP(\mathbf{X}_k)$ then we say $\mathbf{X}_j$ has a better performance than $\mathbf{X}_k$ and it is a better known solution in relation to $\mathbf{X}_k$ for the problem we are solving. For example, in a optimizacion problem looking for minimization fitness, we can define:

$$PP(\mathbf{X}) = \frac{1}{C(\mathbf{X})} \tag{6}$$

Each team consists in a set of N_f fixed players and N_s substitute players. In a scenario with N_{teams} teams, the total number of players will be given by:

$$N_{players} = N_{teams} * (N_f + N_s) \tag{7}$$

The Star Player (SP_k) of a team $\mathbf{T}_k$ will be the player with the highest player power. In a similar way, the Super Star Player will be the player with the highest player power considering all the teams. We will define the team power TP_i for the team $\mathcal{T}_i$ as the average value of power player regarding all its players (fixed and substitutes), i.e.:

$$TP_i = \sum_{X_k \in \mathcal{T}_i} \frac{PP(\mathbf{X}_k)}{N_f + N_s} \qquad i \in M = \{1, 2, ..., N_{teams}\} \tag{8}$$

Given two teams $\mathcal{T}_j$ and $\mathcal{T}_k$, SLC defines a single winner always. The winner team is unknown until the match ends. The victory chance of $\mathcal{T}_k$ is directly proportional to its team performance and inversely proportional to opposing team performance. Given TP_j and TP_k as the team power for $\mathcal{T}_j$ and $\mathcal{T}_k$, respectively, the probability of victory of $\mathcal{T}_j$ defined as PV_j, and $\mathcal{T}_k$ defined as PV_k, is given as:

$$PV_j = \frac{TP_j}{TP_j + TP_k} \tag{9}$$

$$PV_k = \frac{TP_k}{TP_j + TP_k} \tag{10}$$

It is clear from (9) and (10) that:

$$PV_j + PV_k = 1 \tag{11}$$

In a scenario for $\binom{N_{teams}}{2}$ matches, where all teams confront each other without repetition, and define winners and losers, the matches end with a single winner team, which include the best player of the league. This best player will correspond to the **SSP** associated with the best solution vector for the problem.

Movement Operators. SLC defines [8] four movement operator types: *imitation*, *provocation*, *mutation* and *substitution*. In the *imitation operator* each fixed player in the winner team tries to improve their performance in order to become like team's Super Player or league's Super Star Player. As indicated in [8], we define new candidates $\mathbf{F}_{new_1}$ and $\mathbf{F}_{new_2}$, being $\mathbf{F_k}$ and $\mathbf{SP_k}$ a fixed player and Star Player from the winner team, respectively:

$$\mathbf{F}_{new_1} = \mu_1 \mathbf{F_k} + \tau_1(\mathbf{SSP} - \mathbf{F_k}) + \tau_2(\mathbf{SP_k} - \mathbf{F_k}) \tag{12}$$

$$\mathbf{F}_{new_2} = \mu_2 \mathbf{F_k} + \tau_1(\mathbf{SSP} - \mathbf{F_k}) + \tau_2(\mathbf{SP_k} - \mathbf{F_k}) \tag{13}$$

where $\mu_1 \sim U(\theta, \beta)$, $\mu_2 \sim U(0, \theta)$, $\theta \in [0, 1]$, $\beta \in [1, 2]$ and $\tau_1, \tau_2 \sim (0, 2)$ are random numbers with uniform distribution. If $\mathbf{F}_{new_1}$ improve player power of $\mathbf{F_k}$ then it is replaced by $\mathbf{F}_{new_1}$. In the other case, if $\mathbf{F}_{new_2}$ improve the player power of $\mathbf{F_k}$ then it is replaced by $\mathbf{F}_{new_2}$. If none of both improve power player, then $\mathbf{F_k}$ remains without change. In the *provocation operator* each substitute player tries to become a fixed player in the winner team. Given a fixed player centroid $\mathbf{G}_k$ of the winner team, we calculate two new candidates for the substitute player $\mathbf{S_k}$ in the winner team:

$$\mathbf{S}_{new_1} = \mathbf{G}_k + \chi_1(\mathbf{G}_k - \mathbf{S_k}) \tag{14}$$

$$\mathbf{S}_{new_1} = \mathbf{G}_k + \chi_2(\mathbf{S_k} - \mathbf{G}_k) \tag{15}$$

where $\chi_1 \sim U(0.9, 1)$, $\chi_2 \sim U(0.4, 0.6)$ are random numbers with uniform distribution. We define the d-dimension of $\mathbf{G}_k$ for the winner team $\mathcal{T}_k$ as follow:

$$\mathbf{G}_k^d = \frac{\sum\limits_{F_k \in \mathcal{T}_k} F_k^d}{N_f} \tag{16}$$

According (14) and (15), if $\mathbf{S}_{new_1}$ is a better player power value than $\mathbf{S}_k$ then it is replaced by $\mathbf{S}_{new_1}$. In the other case, if $\mathbf{S}_{new_2}$ is a better player power value than $\mathbf{S}_k$ then it is replaced by $\mathbf{S}_{new_2}$. If none of them improve their player power, then $\mathbf{S}_k$ is replaced by a new random generated solution vector.

Fixed players in the looser team try to apply small changes to avoid repeating the previous failures. In this scenario, we will apply some mutation operator like Genetic Algorithm (GA). Some substitute players in the looser team are replaced by young talents. In SLC it is achieved by generating new substitutes vector as follow. We choose two (any) substitute vectors $\mathbf{S}_k$ and $\mathbf{S}_l$ in the looser team and calculate two new solutions vector for each one:

$$\mathbf{S}_{new_k} = \alpha \times \mathbf{S}_k + (\mathbf{1} - \alpha) \times \mathbf{S}_l \tag{17}$$

$$\mathbf{S}_{new_l} = \alpha \times \mathbf{S}_l + (\mathbf{1} - \alpha) \times \mathbf{S}_k \tag{18}$$

where $\alpha \in R^n$ is a vector with values in $\{0, 1\}$ defined randomly with uniform distribution, and $\mathbf{1}$ is the unitary vector in R^n. If $\mathbf{S}_{new_k}$ is a better player power value than $\mathbf{S}_k$ then this is replaced for the new vector $\mathbf{S}_{new_k}$. In the same way, if $\mathbf{S}_{new_l}$ is a better power player value than $\mathbf{S}_l$ then it is replaced by the new vector $\mathbf{S}_{new_l}$.

Iteration Process and Convergence. The iteration starts defining $N_{teams} *$ $(N_f + N_s)$ random solutions vectors (or players) and ranking them by player power from the highest to lowest value. First $N_f + N_s$ players at the top ranking corresponds to the first team, next $N_f + N_s$ players corresponds to team two and so on. In each team, first N_f players in the ranking corresponds to the team's fixed players, and the next N_f players corresponds to substitutes. First player at the top ranking corresponds to the SSP. First player at each team ranking corresponds to its SP.

Two teams faced define a winner and a loser. Each winner team fixed player try to imitate SSP and/or SP to improve his performance applying *imitation operator*. Each winner team substitute player try to improve their performance to be closer to the average power of the fixed players by *provocation operator*. The looser team apply changes to the player configuration by *mutation* and *substitution*. Next two teams are faced and so on. As a result of this dynamic, player power of each player could be changed and the raking in the same way. The best players (solution vectors with highest player player) will be ranked at the top of the raking. SSP at the end of the seasons will correspond to the best solution to the problem.

4   Solving SCP by Using SLC and Python Implementation

SLC is intended to solve optimization problems inside continuous search spaces and to apply SLC solving SCP means to adopt a binary search space.

We can define the player power function $PP(X)$ from the SCP cost function in (1) when the problem is looking for a maximum, or its inverse in case of minimum. The SCP constraint statements (2) and (3) can be build as a test-feasibility function with the capability to repair an unfeasible solution if necessary under an ADD/REMOVE approach.

The *imitation operator* in (12) and (13) in a binary space could be achieved by reducing Hamming distance instead vector arithmetic. We consider the follow statement to get two new positions for **F** considering a Hamming approach:

$$\mathbf{F}^d_{new_1} = \begin{cases} \mathbf{SSP}^d & \text{if } rand() \leq p_{imitation} \\ \mathbf{F}^d & \text{other case} \end{cases} \tag{19}$$

$$\mathbf{F}^d_{new_2} = \begin{cases} \mathbf{SP}^d & \text{if } rand() \leq p_{imitation} \\ \mathbf{F}^d & \text{other case} \end{cases} \tag{20}$$

where $rand() \sim U(0,1)$ is a random generated value with uniform distribution and $p_{imitation}$ is a probability of imitation defined as parameter as part of model. In the *provocation operator* on a binary solution space a new approach is presented to obtain a centroid vector based on (16) and the probability for a dimension d to get 1 or 0 value.

$$\mathbf{BG}^d_k = \begin{cases} 1 & \text{if } \mathbf{G}^d_k \geq 0.5 \\ 0 & \text{other case} \end{cases} \tag{21}$$

In the *provocation operator* to calc new positions as defined in (14) and (15) will generate a not feasible solution vectors because each x^d dimension value will be in the domain R instead $\{0, 1\}$. It requires a binarization operator to transform not binary solutions vectors into binary solutions. We could define a binarization function for a x^d dimension as follows:

$$B(x^d) = \begin{cases} 1 & \text{if } rand() \leq T(x(t)^d) \\ 0 & \text{other case} \end{cases} \tag{22}$$

being $T(x^d)$ the transfer function as defined in [7] as follows:

$$T(x^d) = \left| \frac{2}{\pi} arctan(\frac{\pi}{2} x^d) \right| \tag{23}$$

A mutation operator for fixed players could be considered as follows:

$$\mathbf{F}^d_{new} = \begin{cases} \mathbf{F}^d_{new} & \text{if } rand() \leq p_{mutation} \\ {}^\sim\mathbf{F}^d_{new} & \text{other case} \end{cases} \tag{24}$$

where $rand() \sim U(0,1)$ is a random generated value with uniform distribution and $p_{mutation}$ is a probability of mutation defined as parameter part of the model.

Table 1. Experimental results of SLC resolving SCP benchmarks (4, 5, 6, A sets) from OR-Library.

Instance	Z_{BKS}	Z_{min}	Z_{MAX}	Z_{AVG}	RPD
4.1	429	431	461	444.5	0.47
4.2	512	519	570	544.1	1.37
4.3	516	520	549	535.0	0.78
4.4	494	503	549	525.8	1.82
4.5	512	518	550	531.4	1.17
4.6	560	566	640	585.3	1.07
4.7	430	435	464	447.4	1.16
4.8	492	499	541	518.2	1.42
4.9	641	678	709	689.5	5.77
4.10	514	524	575	548.4	1.95
5.1	253	254	13893	722.1	0.40
5.2	302	311	17408	1404.2	2.98
5.3	226	229	43344	1958.6	1.33
5.4	242	242	250	245.8	0.00
5.5	211	212	227	219.0	0.47
5.6	213	217	230	221.8	1.88
5.7	293	301	315	309.7	2.73
5.8	288	294	315	302.0	2.08
5.9	279	292	294	292.7	4.66
5.10	265	269	13,601	942.5	0.02
6.1	138	144	153	147.0	4.35
6.2	146	149	162	154.0	2.05
6.3	145	150	157	152.3	3.45
6.4	131	131	135	132.7	0.00
6.5	161	171	176	174.0	6.21
A.1	253	29,235	44,037	34,892.7	11,455.34
A.2	252	32,524	41,013	35,855.0	12,806.35
A.3	232	30,864	44,422	35,501.3	13,203.45
A.4	234	29,311	42,167	35,287.7	12,426.07
A.5	236	29814	41348	34462.7	12,533.05

Python 3.5.1 was used for coding internal data representation, imitation, provocation, mutation, substitution operators, test-feasibility and repair functions. For data generated by the process, a relational database MySQL was used to store, summarize and do analysis. pypyodbc lib [2] served as middle-ware between Python and MySQL. From OR-Library [1], first 30 SCP benchmarks where tested on the Python implementation, the rest of SCP benchmarks test data was discarded in this stage work by a low and poor convergence velocity of the Python implementation. The results obtained on each benchmark is summarized in the Table 1 above.

Results. Benchmarks 4.1 to 4.10, involving 200 constraints and 1000 decision-variables, was tested 31 instances each one, using 7 teams, 11 fixed players and 5 substitute players, matching in 5 seasons, showing good RPD values regarding Z_{BKS} known optimum. Benchmarks 5.1 to 5.10, involving 200 constraints and 2000 decision-variables, was tested 31 instances each one, using 7 teams, 11 fixed players and 5 substitute players, matching in 5 seasons, showing good RPD values regarding Z_{BKS} known optimum. Benchmarks 6.1 to 6.5, involving 200 constraints and 1000 decision-variables was tested 31 instances each one, using 7 teams, 11 fixed players and 5 substitute players matching in 5 seasons showing good RPD values regarding Z_{BKS} known optimum. Benchmarks A1 to A5, involving 300 constraints and 3000 decision-variables, it was not possible to obtain final evidence about convergence to the optimum Z_{BKS}, but first iterations showed a poor and extremely slow convergence to the Z_{BKS} known optimum.

5   Conclusions

A Python implementation of SLC algorithm has been presented in this work to solve SCP. It has also been applied in 30 OR-Library SCP benchmark sets to test the convergence capability of the implementation obtaining acceptable convergence up to 200 constraints and 2000 decision-variables.

Benchmark test was performed in several servers and PC's with different process computing capability, but in I7 Intel processor with 16GB in RAM it was possible to run simultaneously up to 10 instances of the algorithm without overload CPU (closer to 87 % CPU workload).

Time consumption is an important issue to face for benchmarks with a big decision-variables size and parallel-processing scenarios is a good alternative, but it is necessary to modify the SLC model to take the most to work in this kind of scenario and get capability convergence using complex benchmark like A1 to A5 OR-Library data sets.

Acknowledgements. Broderick Crawford is supported by Grant CONICYT/ FONDECYT/REGULAR/1140897. Ricardo Soto is supported by Grant CONI- CYT/FONDECYT/REGULAR/1160455. Sebastián Mansilla, Álvaro Gómez and Juan Salas are supported by Postgraduate Grant Pontificia Universidad Católica de Valparaiso 2015 (INF-PUCV 2015).

References

1. OR-Library a collection of test data sets for a variety ofoperations research (or) problems. http://people.brunel.ac.uk/mastjjb/jeb/orlib/scpinfo.html. Accessed 30 Mar 2015
2. PyPyODBC a pure Python ODBC module by ctypes. https://pypi.python.org/pypi/pypyodbc. Accessed 20 Jan 2015
3. Crawford, B., Soto, R., Cuesta, R., Paredes, F.: Using the bee colony optimization method to solve the weighted set covering problem. In: Stephanidis, C. (ed.) HCI 2014, Part I. CCIS, vol. 434, pp. 493–497. Springer, Heidelberg (2014)
4. Crawford, B., Soto, R., Peña, C., Palma, W., Johnson, F., Paredes, F.: Solving the set covering problem with a shuffled frog leaping algorithm. In: Nguyen, N.T., Trawiński, B., Kosala, R. (eds.) ACIIDS 2015. LNCS, vol. 9012, pp. 41–50. Springer, Heidelberg (2015)
5. Crawford, B., Soto, R., Peña, C., Riquelme-Leiva, M., Torres-Rojas, C., Johnson, F., Paredes, F.: Binarization methods for shuffled frog leaping algorithms that solveset covering problems. In: Silhavy, R., Senkerik, R., Oplatkova, Z.K., Prokopova, Z., Silhavy, P. (eds.) CSOC 2015. AISC, vol. 349, pp. 317–326. Springer, Cham (2015)
6. Crawford, B., Soto, R., Riquelme-Leiva, M., Peña, C., Torres-Rojas, C., Johnson, F., Paredes, F.: Modified binary firefly algorithms with different transfer functions for solving set covering problems. In: Silhavy, R., Senkerik, R., Oplatkova, Z.K., Prokopova, Z., Silhavy, P. (eds.) CSOC 2015. AISC, vol. 349, pp. 307–315. Springer, Cham (2015)
7. Mirjalili, S., Lewis, A.: S-shaped versus V-shaped transfer functions for binary particle swarm optimization. Swarm Evol. Comput. 9, 1–14 (2013)
8. Moosavian, N.: Soccer league competition algorithm, a new method for solving systems of nonlinear equations. Sci. Res. 4, 7–16 (2014)
9. Moosavian, N.: Soccer league competition algorithm for solving knapsack problems. Swarm Evol. Comput. 20, 14–22 (2015)
10. Moosavian, N., Roodsari, B.K.: Soccer league competition algorithm: a novel meta-heuristic algorithm for optimal design of water distribution networks. Swarm Evol. Comput. 17, 14–24 (2014)
11. Soto, R., Crawford, B., Olivares, R., Barraza, J., Johnson, F., Paredes, F.: A binary cuckoo search algorithm for solving the set covering problem. In: Vicente, J.M.F., Álvarez-Sánchez, J.R., López, F.P., Toledo-Moreo, F.J., Adeli, H. (eds.) Bioinspired Computation in Artificial Systems. LNCS, vol. 9108, pp. 88–97. Springer, Heidelberg (2015)
12. Soto, R., Crawford, B., Vilches, J., Johnson, F., Paredes, F.: Heuristic feasibility and preprocessing for a set covering solver based on firefly optimization. In: Silhavy, R., Senkerik, R., Oplatkova, Z.K., Prokopova, Z., Silhavy, P. (eds.) Artificial Intelligence Perspectives and Applications. AISC, vol. 347, pp. 99–108. Springer, Heidelberg (2015)

An Artificial Fish Swarm Optimization Algorithm to Solve Set Covering Problem

Broderick Crawford[1,2(✉)], Ricardo Soto[1,3,4], Eduardo Olguín[2],
Sebastián Mansilla Villablanca[1], Álvaro Gómez Rubio[1],
Adrián Jaramillo[1], and Juan Salas[1]

[1] Pontificia Universidad Católica de Valparaíso, Valparaíso, Chile
{broderick.crawford,ricardo.soto}@ucv.cl,
{sebastian.mansilla.v,alvaro.gomez.r,
adrian.jaramillo.s,juan.salas.f}@mail.pucv.cl
[2] Universidad San Sebastián, Santiago, Chile
eduardo.olguin@uss.cl
[3] Universidad Autónoma de Chile, Temuco, Chile
[4] Universidad Científica del Sur, Lima, Peru

Abstract. The Set Covering Problem (SCP) consists in finding a set of solutions that allow to cover a set of necessities with the minor possible cost. There are many applications of this problem such as rolling production lines or installation of certain services like hospitals. SCP has been solved before with different algorithms like genetic algorithm, cultural algorithm or firefly algorithm among others. The objective of this paper is to show the performance of an Artificial Fish Swarm Algorithm (AFSA) in order to solve SCP. This algorithm, simulates the behavior of a fish shoal inside water and it uses a population of points in space to represent the position of a fish in the shoal. Here we show a study of its simplified version of AFSA in a binary domain with its modifications applied to SCP. This method was tested on SCP benchmark instances from OR-Library website.

Keywords: Set Covering Problem · Artificial fish swarm optimization algorithm · Metaheuristics · Combinatorial optimization

1 Introduction

SCP is a classical combinatorial optimization problem which has many practical applications in the world such as deciding construction of firemen stations in different places or installing of cell phone networks in order to obtain the maximum coverage with a minimal possible cost. The SCP can be formulated as follows [1]:

$$\text{minimize} \quad Z = \sum_{j=1}^{n} c_j x_j \tag{1}$$

© Springer International Publishing Switzerland 2016
H. Fujita et al. (Eds.): IEA/AIE 2016, LNAI 9799, pp. 892–903, 2016.
DOI: 10.1007/978-3-319-42007-3_76

Subject to:

$$\sum_{j=1}^{n} a_{ij} x_j \geq 1 \quad \forall i \in I \tag{2}$$

$$x_j \in \{0,1\} \quad \forall j \in J \tag{3}$$

Let $A = (a_{ij})$ be a $m \times n$ 0–1 matrix with $I = \{1, \ldots, m\}$ and $J = \{1, \ldots, n\}$ be the row and column sets respectively. Column j can cover a row i if $a_{ij} = 1$. Where c_j is a nonnegative value that represents the cost of selecting the column j and x_j is a decision variable, it can be 1 if column j is selected or 0 otherwise. The objective is to find a minimum cost subset $S \subseteq J$, such that each row $i \in I$ is covered by at least one column $j \in S$. The SCP was also successfully solved with meta-heuristics such as artificial bee colony [2,3], cultural algorithm [4], swarm optimization particles [5], ant colony optimization [6], firefly algorithm [7,8], shuffled frog leaping algorithm [9] or genetic algorithm [10].

2   Artificial Fish Swarm Algorithm

According to [11] AFSA simulates the behavior of a fish swarm inside the water and this algorithm was proposed and applied in order to solve problems of optimization in an engineering context. Moreover, this method was applied in global optimization problems and bound constrained global optimization problems. In optimization problems a fish represents a point in a population and the swarm movements are randomly. However, they are synchronized. A group of fish stay close to the swarm in order to obtain protection from predators, find food and avoid collisions with other members of the swarm. Therefore, this behavioral model is used to solve optimization problems in an efficient way and it seeks to imitate as well as to make variations on the swarm behavior in nature, and to create new types of abstract movements. The fish swarm behavior is summarized as follows [11]:

1. **Random Behavior:** Fish swims randomly inside water in order to find companion and food.
2. **Chasing Behavior:** When food is discovered by fish, the others in the neighborhood find the food dangling quickly after it.
3. **Swarming Behavior:** Fish assembles in groups which is a living habit in order to guarantee the existence of the swarm and avoid dangers from predators.
4. **Searching Behavior:** When a region is discovered with more food by fish, it could be by vision or sense, it goes directly and quickly to that region.
5. **Leaping Behavior:** When fish stagnates in a region, it leaps to look for food in other regions.

In accordance with [11], the five behaviors mentioned above are simulations and interpretations from an artificial fish which is a fictitious entity of a real fish. The environment in which the artificial fish moves, searching for the optimum, is the feasible search space of the problem.

Another description of AFSA is proposed in [12] where the aim is to optimize problems of type $minimize_{x \in \Omega} f(x^i)$ Here $f : \mathbb{R}^n \rightarrow \mathbb{R}$ is a nonlinear function to be minimized and $\Omega = \{x^i \in \mathbb{R} : l_j \leq x_j \leq uj, j = 1, 2, ..., n\}$ is the search space. l_j and u_j are the lower and upper bounds of x_j, respectively, and n is the number of variables of the optimization problem.

According to [12] N points x^i are the population of AFSA, where $i \in \{1, 2, ..., N\}$ is to identify better regions looking for a global solution. x^i is a floating-point encoding that covers the entire search space Ω. The crucial issue of AFSA is the "visual scope" of each point x^i. This represents a closed neighborhood of x^i with a radius equal to a positive quantity ν defined by [12]:

$$\nu = \delta \ max(u_j - l_j) \ j \in \{1, 2, ..., n\} \tag{4}$$

where $\delta \in (0, 1)$ is a positive visual parameter. This parameter may be reduced along the iterative process. Let $\mathbf{I}^i$ be the set of indices of the points inside the "visual scope" of point x^i, where $i \notin \mathbf{I}^i$ and $\mathbf{I}^i \subset \{1, 2, ..., N\}$, and let np^i be the number of points in its "visual scope". Depending on the relative positions of the points in the population, three possible situations may occur [12]:

(a) When $np^i = 0$, the "visual scope" is empty, and the point x^i, with no other points in its neighborhood, moves randomly looking for a better region.
(b) When the "visual scope" is not crowded, the point x^i is able either to chase moving towards the best point inside the "visual scope", or, if this best point does not improve the objective function value corresponding to x^i, moving swarm towards the central point of the "visual scope".
(c) When the "visual scope" is crowded, the point x^i has some difficulty in following any particular point, and searches for a better region by choosing randomly another point (from the "visual scope") and moving towards it.

The condition that decides when the "visual scope" of x^i is not crowded is [12]:

$$C_f \equiv \frac{np^i}{N} \leq \theta, \tag{5}$$

where C_f is the crowding factor and $\theta \in (0, 1)$ is the crowd parameter. In this situation, the point x^i has the ability to swarm or to chase. The swarming behavior is characterized by a movement towards the central point inside the "visual scope" of x^i defined by [12]:

$$\bar{x} = \frac{\sum_{l \in \mathbf{I}^i} x^l}{np^i} \tag{6}$$

3   Artificial Fish Swarm Algorithm and Its Simplified Binary Version

First of all, it will be showed the pseudocode of the proposed method for solving the SCP. It is shown in Algorithm 1. Then it will be explained the main steps of AFSA to solve SCP.

Algorithm 1. AFSA applied to SCP

Require: T_{max} and z_{opt} and other values of parameters
1: Set $t = 1$ Initialize population $x^{i,t}$, $i = 1, 2, ..., N$
2: Perform the repair function in order to evaluate the population,
      identify x^{min} and z_{min}
3: **while** 'termination conditions are not met' **do**
4:     **if** $t\%R = 0$ **then**
5:         Reinitialize 50% of the population, keeping x^{min} and z_{min}
6:         Perform the repair function in order to evaluate population,
           identify x^{min} and z_{min}
7:     **end if**
8:     **for** $i = 1$ to N **do**
9:         **if** $i = x^{min}$ **then**
10:            Perform 4 flip-bit mutation to create trial point $y^{i,t}$
11:         **else**
12:            **if** $rand(0, 1) \leq \tau_1$ **then**
13:                Perform random behavior to create trial point $y^{i,t}$
14:            **else if** $rand(0, 1) \geq \tau_2$ **then**
15:                Perform chasing behavior to create trial point $y^{i,t}$
16:            **else**
17:                Perform searching behavior to create trial point $y^{i,t}$
18:            **end if**
19:         **end if**
20:     **end for**
21:     Perform the repair function in order to evaluate and get $y^{i,t}$, $i = 1, 2, ..., N$ and evaluate them
22:     Select new population $x^{i,t+1}$, $i = 1, 2, ..., N$
23:     Perform local search
24:     Identify x^{min} and z_{min}
25:     Set $t = t + 1$
26: **end while**
27: **return** x^{min} and z_{min}

3.1 Proposed Algorithm

In [12], each trial point is created from the current one by using the concept of "visual scope" of a current point for identifying the points inside the "visual scope" of each current point and the Hamming distance is used. For points of equal bits length, this distance is the number of positions at which the corresponding bits are different. The computational requirement of this procedure grows rapidly with problem's dimension, according to [13]. Therefore, in some cases, the population stagnates and the algorithm converges to a non-optimal solution. To deal with these problems, a new version of AFSA was proposed in [13] to solve knapsack problem, with other properties and they were modified in order to solve SCP in this paper.

According to [13], in this simplified version of AFSA the "visual scope" concept was discarded, the selection of the behavior depends on two new probabilities, τ_1 and τ_2, warming behavior is never performed, among other modifications that are explained in [13].

Then, with the modifications explained above and doing the necessary changes to apply on SCP, it means minimization instead of maximization. Moreover, it was introduced a proper function in order to obtain good solutions to solve SCP.

Also, in order to increase efficiency and to improve the quality of the solutions, for solving large 0–1 problems. Some other modifications were introduced [13]: an effect-based crossover is used instead of an uniform crossover and a simple local search with two steps is implemented. First, a flip-bit mutation is operated on a pre-defined number of points randomly selected from the population, with a pre-specified probability; second, at the end of the selection procedure, the best point is refined using a flip-bit mutation on a pre-defined number of positions.

Next, it will be explained the steps of AFSA in order to obtain SCP results.

3.2   Initialization

As well as in [13], N current points are randomly generated, x^i, where $i \in \{1, 2, ..., N\}$ each one represented by a binary string of $0/1$ of length n.

3.3   Generating Trial Points in AFSA

In accordance with [13], in order to create trial points, y^i, at each iteration and based on behaviors of random, chasing, and searching is necessary utilize crossover, and mutation after initializing the N current points. Probabilities of $0 \leq \tau_1 \leq \tau_2 \leq 1$ are the responsible to reach this objective and the main behaviors are [13]:

Random Behavior: If a fish does not have companion in its neighborhood, then it moves randomly looking for food in another region. This behavior is used when $rand(0, 1) \leq \tau_1$. Here, y^i is created in random way, setting $0/1$ bits of length n.

Chasing Behavior: When a fish, or a group of fish in the swarm, discover food, and the others find the food dangling quickly after it. This behavior is implemented when $rand(0, 1) \geq \tau_2$ and it is related to the movement towards the best point found so far in the population, x^{min}. Here, y^i is created using an effect-based crossover (see Algorithm 1) between x^i and x^{min}.

Searching Behavior: When fish discovers a region with more food, by vision or sense, it goes directly and quickly to that region. This behavior is related to the movement towards a point x^{rand} where "rand" is an index randomly chosen from the set $\{i = 1, 2, ..., N\}$. When $\tau_1 < rand(0, 1) < \tau_2$ it is implemented. An effect-based crossover (see Algorithm 1) between x^{rand} and x^i is performed to create the trial point y^i.

Trial Point Corresponding to the Best Point: These 3 behaviors explained above are implemented to create $N-1$ trial points; the best point x^{min} is treated separately. A 4 flip-bit mutation is performed on the point x^{min} to create the corresponding y^i. In this operation 4 positions are randomly selected, and the bits of the corresponding positions are changed from 0 to 1 or vice versa.

3.4 The Effect-Based Crossover

In [13], an effect-based crossover is used in chasing and searching behaviors in order to create the trial points. Here, each bit of the trial point is created by copying the corresponding bit from one or the other current point based on the *effect ratio*. To compute the *effect ratio* ER_{u,x^i} of u on the current point x^i, where [13]:

(1) $u = x^{min}$ when chasing is performed.
(2) $u = x^{rand}$ when searching is performed.

It is used:

$$ER_{u,x^i} = \frac{q(u)}{q(u) + q(x^i)} \tag{7}$$

where:

$$q(x^i) = \exp[\frac{-(z(x^{min}) - z(x^i))}{(z(x^{min}) - z(x^{max}))}] \tag{8}$$

x^{max} is the worst point of the population and z is the objective function value. The effect-based crossover to compute the trial point y^i is displayed in Algorithm 2.

3.5 Dealing with SCP Constraints

In accordance with [12,13], AFSA was utilized for the 0–1 multidimensional knapsack problems and it is showed an algorithm for constraints handling. Therefore, it is going to be showed a different and proper repair function for constraints handling to solve SCP in the following algorithm.

According to [15], Algorithm 3 shows a repair method where all rows not covered are identified and the columns required are added. So, in this way all the constraints will be covered. The search of these columns are based in the relationship showed in the next equation.

$$\frac{cost\ of\ one\ column}{amount\ of\ columns\ not\ covered} \tag{9}$$

Once the columns are added and the solution is feasible, a method is applied to remove redundant columns of the solution. A redundant column are those

---

Algorithm 2. Effect-Based Crossover SCP

Require: current point x^i, u and $ER_{u,\,x^i}$
1: for $j = 1$ to n do
2:     if $rand(0,1) < ER_{u,\,x^i}$ then
3:         $y^i_j = u_j$
4:     else
5:         $y^i_j = x_j$
6:     end if
7: end for
8:     return trial point y^i

Algorithm 3. Repair Operator To Dealing With SCP Constraints

1: $w_i \leftarrow |S \cap J_i| \; \forall i \in I$;
2: $U \leftarrow \{i \mid w_i = 0\}, \forall i \in I$;
3: **for** $i \in U$ **do**
4:     find the first column j in J_i that minimize $\frac{c_j}{|U \cap I_j|} S \leftarrow S \cap j$;
5:         $w_i \leftarrow w_i + 1, \forall i \in I_j$;
6:         $U \leftarrow U - I_j$;
7: **end for**
8: **for** $j \in S$ **do**
9:     **if** $w_i \geq 2, \forall i \in I_j$ **then**
10:             $S \leftarrow S - j$;
11:             $w_i \leftarrow w_i - 1, \forall i \in I_j$;
12:     **end if**
13: **end for**

that are removed, the solution remains a feasible solution. The algorithm of this repair method is detailed in the Algorithm 3. Where:

(a) I is the set of all rows
(b) J is the set of all columns
(c) J_i is the set of columns that cover the row $i, i \in I$
(d) I_j is the set of rows covered by the column $j, j \in J$
(e) S is the set of columns of the solution
(f) U is the set of columns not covered
(g) w_i is the number of columns that cover the row $i, \forall i \in I$ in S

3.6   Selection of a New Population

As well as in [13], each trial point y^i competes with the current x^i, in order to decide which one should become a member of the new population in the next iteration. Hence, if $z(y^i) \leq z(x^i)$, then the trial point becomes a member of the new population in the next iteration; otherwise, the current point is maintained to the next iteration.

3.7   Reinitialization of the Population

In accordance with [13] in other versions of AFSA, the points in a population converge to a non-optimal solution, this could be considered region with food shortage for all the swarm and it is necessary looking for a better region. In nature, this may be occurring only in a seasonal way. So, it was introduced a reinitialization of part of the population only at every R iterations where R gives the seasonal time period. As well as in [13], to diversify the search and look for a promising region, a randomly reinitialization of 50 % of the population is implemented, guaranteeing that the best solution found so far is maintained, is implemented. In practical terms, this technique has greatly improved the quality of the solutions.

3.8 Local Search

According to [13], in many cases the exploitation is good to find high-quality solutions around a particular good region. When a better solution is found then it replaces the last best solution of the current set. In this context, local search may be interpreted by the swarm as a procedure that allows first of all a small percentage of the swarm to progress towards food and then confers to the best one further progress into a promising region. Therefore, the local search is based on a flip-bit mutation that is operated on N_{loc} points selected randomly from the population, where $N_{loc} = \tau_3 N$ where $\tau_3 \in (0,1)$. This flip-bit mutation that operates on a point changes the value of a 0 bit to 1 and vice versa according to a probability p_m. After the flip-bit operation, the new points are made feasible by using the repair function in order to solve SCP. Then they become members of the population, if they improve the objective function value with respect to the corresponding current points. This flip-bit operation is repeated L times in order to find good solutions. Then, the best point of the population is identified and a flip-bit mutation is operated on N_{ref}, with $N_{ref} = \tau_3 n$, randomly selected positions of the point. Each time, a new point is created, the repair function is implemented to make the point feasible. Then this new point will replace the best point if it improves the objective function value with respect to the current best point.

3.9 Termination Conditions

In accordance with [13], AFSA terminates when the known optimal solution is reached or a maximum number of iterations, T_{max}, is exceeded.

$$t > T_{max} \ or \ z_{min} \leq z_{opt} \tag{10}$$

where z_{min} is the best objective function value reached at iteration t and z_{opt} is the known optimal value available in the literature.

4 Experimental Results

In this section, it is going to be showed the results obtained after performing AFSA to solve SCP. In the appendix section it is possible to find the Table 1, which shows the results of SCP with more details.

The algorithm proposed was run 30 times for majority of instances, specifically for the fifty first files and 20 times for the twenty last instances which have a great deal of rows and columns because our algorithm needs a big quantity of hours to evaluate the last 20 files. Thus, this algorithm tested the 70 data files from the OR-Library website, 25 of them are the instance sets 4,5,6 was originally from Balas and Ho [14], the others 25, the sets A, B, C, D, E from Beasley [15] and 20 of these data files are the test problem sets E, F, G, H from Beasley [16]. These 70 files are formatted as: number of rows n, number of columns m, the cost of each column $c_j, j \in \{1, \ldots, n\}$, and for each row

$i, i \in \{1, ..., m\}$ the number of columns which cover row i followed by a list of the columns which cover rows i.

This algorithm was implemented in Java programming language, using Eclipse IDE in two computers with the following hardware, Intel core i5 dual core 2.60 GHz processor, 8 GB RAM and it was run under OSX Yosemite and Intel core i5 2.30 GHz processor, 4 GB RAM and it was run under OSX Snow Leopard.

Finally, the algorithm was executed with a population of $N = 20$ fish, probability $\tau_1 = 0.1$, probability $\tau_2 = 0.9$, probability $\tau_3 = 0.1$, probability $p_m = 0.1$, $L = 50$, reinitialization of population $R = 10$ and each time was run 1000 iterations.

The Table 1 shows the results obtained where the first column is the number of experiment of each file, the second column *Instance* indicates each benchmark evaluated, Z_{opt} shows the best known solution value of each instance. The next columns $Z_{min}, Z_{max}, Z_{avg}$ represents the minimum, maximum among minimums, and average of minimums solutions obtained. The last column reports the relative percentage deviation RPD which represents the deviation of the best known solution f_{opt} from f_{min} which is the minimum value obtained for each instance. RPD was calculated as follows:

$$\text{RPD} = \frac{100(f_{min} - f_{opt})}{f_{opt}} \tag{11}$$

5   Conclusions

This paper was a study of AFSA in its simplified binary version which imitates the behavior of a fish in a shoal inside the water, in order to solve SCP. As it has seen in Table 1, the algorithm converges to very good solutions. Also, it has seen during experiments that it converges quickly to the solutions. In some cases it obtained optimal solutions, but it requires time of processing, many hours in some cases. Moreover, it has seen that it has a great variability in its results. However, These results mean that $AFSA$ is an algorithm that could be used to obtain global optimums, because it has obtained results with a smaller percentage than 8 % in RPD column and only one instance had a percentage greater than 10 %, $NRH.3$ instance, but smaller than 12 %. In other cases, it has obtained the optimal result or almost the optimal result in the group of instances such as $4, 5, 6$ and E instances where it has obtained the optimal result in all cases. With these results, it is possible to say that it could be possible that it is necessary more than 1000 iterations or search for a better configuration of its parameters in order to obtain the optimal results in all of its instances or in the majority of the 70 instances.

Acknowledgements. Broderick Crawford is supported by Grant CONICYT/ FONDECYT/REGULAR/1140897. Ricardo Soto is supported by Grant CON-ICYT/FONDECYT/REGULAR/1160455. Sebastián Mansilla Villablanca, Álvaro Gómez, Adrián Jaramillo and Juan Salas are supported by Postgraduate Grant Pontificia Universidad Católica de Valparaíso 2016 (INF-PUCV 2016).

A Appendix

Table 1. Experimental results of SCP benchmarks (4, 5, 6, A, B, C, D, E, NRE, NRF, NRG and NRH sets)

Number	Instance	Z_{opt}	Z_{min}	Z_{max}	Z_{avg}	RPD
1	4.1	429	430	445	437,4	0,23
2	4.2	512	515	546	530,83	0,59
3	4.3	516	519	543	528,27	0,58
4	4.4	494	495	532	514,83	0,20
5	4.5	512	514	536	521,73	0,39
6	4.6	560	565	597	580,9	0,89
7	4.7	430	432	447	437,37	0,47
8	4.8	492	492	514	501,73	0,0
9	4.9	641	658	688	669,8	2,65
10	4.10	514	525	559	539,6	2,14
11	5.1	253	254	271	263,03	0,40
12	5.2	302	310	318	314,27	2,65
13	5.3	226	228	244	232,77	0,88
14	5.4	242	242	247	244,77	0,0
15	5.5	211	212	215	212,6	0,47
16	5.6	213	214	242	227,77	0,47
17	5.7	293	299	315	307,9	2,05
18	5.8	288	291	313	298,97	1,04
19	5.9	279	279	296	285,73	0,0
20	5.10	265	266	276	272,07	0,38
21	6.1	138	138	153	146,37	0,0
22	6.2	146	149	156	151,97	2,05
23	6.3	145	145	161	149,63	0,0
24	6.4	131	131	137	134,17	0,0
25	6.5	161	164	181	172,67	1,86
26	A.1	253	256	270	259,6	1,19
27	A.2	252	258	276	264,4	2,38
28	A.3	232	235	255	246,2	1,29
29	A.4	234	243	266	252,25	3,85
30	A.5	236	237	259	244,9	0,42
31	B.1	69	72	88	78,3	4,35
32	B.2	76	79	94	84,67	3,95
33	B.3	80	82	89	85,6	2,5

(Continued)

Table 1. *(Continued)*

Number	Instance	Z_{opt}	Z_{min}	Z_{max}	Z_{avg}	RPD
34	B.4	79	82	96	86,45	3,80
35	B.5	72	72	89	79,5	0,0
36	C.1	227	231	252	238,85	1,76
37	C.2	219	227	254	236,5	3,65
38	C.3	243	251	274	263,15	3,29
39	C.4	219	223	253	240,1	1,,83
40	C.5	215	217	250	228,3	0,93
41	D.1	60	60	81	66,6	0,0
42	D.2	66	69	83	73,35	4,54
43	D.3	72	76	87	82,4	5,56
44	D.4	62	64	76	69,05	3,23
45	D.5	61	64	78	68,9	4,92
46	E.1	5	5	6	5,87	0,0
47	E.2	5	5	6	5,5	0,0
48	E.3	5	5	6	5,2	0,0
49	E.4	5	5	6	5,7	0,0
50	E.5	5	5	6	5,57	0,0
51	NRE.1	29	29	39	32,1	0,0
52	NRE.2	30	32	40	32,25	6,67
52	NRE.3	27	28	35	32,1	3,70
54	NRE.4	28	30	38	33,05	7,14
55	NRE.5	28	30	35	31,95	7,14
56	NRF.1	14	15	18	16,75	7,14
57	NRF.2	15	16	18	17,05	6,67
58	NRF.3	14	15	20	17,25	7,14
59	NRF.4	14	15	19	16,45	7,14
60	NRF.5	13	14	18	15,75	7,69
61	NRG.1	176	184	249	194,05	4,54
62	NRG.2	151	162	170	166,5	7,82
63	NRG.3	166	174	268	184,7	4,82
64	NRG.4	168	178	284	190,55	5,95
65	NRG.5	168	178	344	193,6	5,95
66	NRH.1	63	66	100	72,15	4,76
67	NRH.2	63	66	129	72,0	4,76
68	NRH.3	59	66	79	68,7	11,86
69	NRH.4	59	63	123	70,5	6,78
70	NRH.5	55	58	71	60,7	5,45

References

1. Garey, M.R., Johnson, D.S.: Computers and Intractability: A Guide to the Theory of NP-Completeness. W. H. Freeman & Co., New York (1990)
2. Crawford, B., Soto, R., Aguilar, R.C., Paredes, F.: A new artificial bee colony algorithm for set covering problems. Electr. Eng. Inf. Technol. **63**, 31 (2014)
3. Crawford, B., Soto, R., Aguilar, R.C., Paredes, F.: Application of the artificial bee colony algorithm for solving the set covering problem. Sci. World J. **2014**, 1–8 (2014)
4. Crawford, B., Soto, R., Monfroy, E.: Cultural algorithms for the set covering problem. In: Tan, Y., Shi, Y., Mo, H. (eds.) ICSI 2013, Part II. LNCS, vol. 7929, pp. 27–34. Springer, Heidelberg (2013)
5. Crawford, B., Soto, R., Monfroy, E., Palma, W., Castro, C., Paredes, F.: Parameter tuning of a choice-function based hyperheuristic using Particle Swarm Optimization. Expert Syst. Appl. **40**(5), 1690–1695 (2013)
6. Crawford, B., Soto, R., Monfroy, E., Paredes, F., Palma, W.: A hybrid Ant algorithm for the set covering problem (2014)
7. Crawford, B., Soto, R., Olivares-Suárez, M., Paredes, F.: A binary firefly algorithm for the set covering problem. Modern Trends Tech. Comput. Sci. **285**, 65–73 (2014)
8. Crawford, B., Soto, R., Riquelme-Leiva, M., Peña, C., Torres-Rojas, C., Johnson, F., Paredes, F.: Modified binary firefly algorithms with different transfer functions for solving set covering problems. In: Silhavy, R., Senkerik, R., Oplatkova, Z.K., Prokopova, Z., Silhavy, P. (eds.) CSOC 2015. AISC, vol. 349, pp. 307–315. Springer, Cham (2015)
9. Crawford, B., Soto, R., Peña, C., Palma, W., Johnson, F., Paredes, F.: Solving the set covering problem with a shuffled frog leaping algorithm. In: Nguyen, N.T., Trawiński, B., Kosala, R. (eds.) ACIIDS 2015. LNCS, vol. 9012, pp. 41–50. Springer, Heidelberg (2015)
10. Michalewicz, Z.: Genetic Algorithms + Data Structures = Evolution Programs, 3rd edn. Springer, Heidelberg (1996)
11. Azad, M.A.K., Rocha, A.M.A.C., Fernandes, E.M.G.P.: Solving multidimensional 0–1 knapsack problem with an artificial fish swarm algorithm. In: Murgante, B., Gervasi, O., Misra, S., Nedjah, N., Rocha, A.M.A.C., Taniar, D., Apduhan, B.O. (eds.) ICCSA 2012, Part III. LNCS, vol. 7335, pp. 72–86. Springer, Heidelberg (2012)
12. Azad, M.A.K., Rocha, A.M.A., Fernandes, E.M.: Improved binary artificial fish swarm algorithm for the 0–1 multidimensional knapsack problems. Swarm Evol. Comput. **14**, 66–75 (2014)
13. Azad, M.A.K., Rocha, A.M.A., Fernandes, E.M.: Solving large 0–1 multidimensional knapsack problems by a new simplified binary artificial fish swarm algorithm. J. Math. Model. Algorithms Oper. Res. **14**, 313–330 (2015)
14. Balas, E., Ho, A.: Set covering algorithms using cutting planes, heuristics, and subgradient optimization: a computational study. In: Padberg, M.W. (ed.) Combinatorial Optimization, pp. 37–60. Springer, Heidelberg (1980)
15. Beasley, J.E.: An algorithm for set covering problem. Eur. J. Oper. Res. **31**(1), 85–93 (1987)
16. Beasley, J.E.: A Lagrangian heuristic for set-covering problems. Naval Res. Logist. (NRL) **37**(1), 151–164 (1990)

The Impact of Using Different Choice Functions When Solving CSPs with Autonomous Search

Ricardo Soto[1(✉)], Broderick Crawford[1], Rodrigo Olivares[1,2],
Stefanie Niklander[3,4,5], and Eduardo Olguín[6]

[1] Pontificia Universidad Católica de Valparaíso, Valparaíso, Chile
{ricardo.soto,broderick.crawford}@ucv.cl, rodrigo.olivares@uv.cl
[2] Universidad de Valparaíso, Valparaíso, Chile
[3] Universidad Adolfo Ibañez, Viña del Mar, Chile
stefanie.niklander@uai.cl
[4] Universidad Autónoma de Chile, Santiago, Chile
[5] Universidad Cientifica del Sur, Lima, Peru
[6] Universidad San Sebastián, Santiago, Chile
eduardo.olguin@uss.cl

Abstract. Constraint programming is a powerful technology for the efficient solving of optimization and constraint satisfaction problems (CSPs). A main concern of this technology is that the efficient problem resolution usually relies on the employed solving strategy. Unfortunately, selecting the proper one is known to be complex as the behavior of strategies is commonly unpredictable. Recently, Autonomous Search appeared as a new technique to tackle this concern. The idea is to let the solver adapt its strategy during solving time in order to improve performance. This task is controlled by a choice function which decides, based on performance information, how the strategy must be updated. However, choice functions can be constructed in several manners variating the information used to take decisions. Such variations may certainly conduct to very different resolution processes. In this paper, we study the impact on the solving phase of 16 different carefully constructed choice functions. We employ as test bed a set of well-known benchmarks that collect general features present on most CSPs. Interesting experimental results are obtained in order to provide the best-performing choice functions for solving CSPs.

Keywords: Autonomous Search · Constraint Programming · Constraint satisfaction · Optimization · Choice functions

1 Introduction

During the last years, Constraint Programming (CP) has widely been used to solve different constraint satisfaction and optimization problems in multiple application domains, such as: computer graphics, engineering design, database systems, electrical engineering, molecular biology, manufacturing, scheduling,

© Springer International Publishing Switzerland 2016
H. Fujita et al. (Eds.): IEA/AIE 2016, LNAI 9799, pp. 904–916, 2016.
DOI: 10.1007/978-3-319-42007-3_77

among other [6]. In CP, problems are modeled by using variables, domains, and constraints. Variables are the unknowns of the problem and hold a domain of possible values, while constraints are relations among these variables that limits the values that variables can take [8]. The resolution of a CSP is performed by a solving engine, commonly called solver. The solver employs a tree data structure that holds the potential solutions. A backtracking-based procedure is used to explore the tree by interleaving enumeration and propagation phases. Enumeration is responsible for instantiating the variables in order to create the branches of the tree, while propagation tries to prune the tree by filtering from domains the values that do not conduct to a feasible solution. Enumeration and propagation are controlled by the so-called solving strategy.

A main problem of this technology is that the efficient problem resolution depends on the use of a proper solving strategy. In fact, the use of the correct one can dramatically improve the performance of the search process [1,5]. Unfortunately, selecting the right strategy is known to be a hard task as the behavior of solving strategies is commonly unpredictable and certainly depends on the problem at hand. Recently, Autonomous Search (AS) [7] appeared as a new technique to tackle this concern. The idea is to let the solver to autonomously adapt its solving strategy during solving time in order to improve performance. The idea is employ various strategies that are interleaved along the process instead of using a single static one. In this way, when bad performing strategies are detected they are autonomously replaced by more promising ones. Such replacement process is controlled by a choice function which decides based on a set of performance indicators which are the strategies that must act. However, choice functions can be constructed in several manners variating the information used to take decisions. Such variations may certainly conduct to very different resolution processes.

In this paper, we study the impact on the solving phase of 16 different carefully constructed choice functions. Those choice functions are responsible for selecting the best strategy for each part of the search tree. We incorporate those choice functions to the AS framework reported in [11], which is able to operate with 24 solving strategies. We solve a set of well-known benchmarks that collect general features that may be present on most constraint satisfaction and optimization problems. Particularly, we use the N-Queens problem, the Sudoku puzzle, the Magic & Latin Square, the Knight Tour as well as the Langford and Quasigroup problem. Interesting experimental results are obtained in order to provide the best-performing choice functions for solving CSPs. As far as we know, the knowledge provided in this work as not been reported yet in the literature.

This paper is organized as follows: In Sect. 2, basic notions of constraint solving are introduced. In Sect. 3, the AS framework employed is described. Finally, experimentation results and conclusions are presented in Sects. 4 and 5, respectively.

2   Constraint Solving

As previously mentioned, Constraint Programming (CP) is a powerful technology for the efficient solving of optimization and constraint satisfaction problems (CSPs). In this context, problems are modeled as a sequence of variables and a set of constraints. The variables have a non-empty domain of candidate values and constraints restrict the values that variables can adopt. Formally, a CSP $\mathcal{P}$ is defined by a triple $\mathcal{P} = \langle \mathcal{X}, \mathcal{D}, \mathcal{C} \rangle$ where $\mathcal{X}$ is an n-tuple of variables $\mathcal{X} = \langle x_1, x_2, \ldots, x_n \rangle$, $\mathcal{D}$ is a corresponding n-tuple of domains $\mathcal{D} = \langle D_1, D_2, \ldots, D_n \rangle$ such that $x_i \in D_i$, and D_i is a set of values, for $i = 1, \ldots, n$. Finally, $\mathcal{C}$ is an m-tuple of constraints $\mathcal{C} = \langle C_1, C_2, \ldots, C_m \rangle$, and a constraint C_j is defined as a subset of the Cartesian product of domains $D_{j_1} \times \cdots \times D_{j_{n_j}}$, for $j = 1, \ldots, m$.

A solution to a CSP is an assignment $\{x_1 \rightarrow a_1, \ldots, x_n \rightarrow a_n\}$ such that $a_i \in D_i$ for $i = 1, \ldots, n$ and $(a_{j_1}, \ldots, a_{j_{n_j}}) \in C_j$, for $j = 1, \ldots, m$.

Algorithm 1 represents a general procedure for solving CSPs.

Algorithm 1. A general procedure for solving CSPs

Require: Variable set & variable domains.
Ensure: Assignments of type $x_i \rightarrow a_i$ that solve the CSP.
 1: $load\_CSP()$
 2: **while not** $all\_variables\_fixed$ **or** $failure$ **do**
 3:     $heuristic\_variable\_selection()$
 4:     $heuristic\_value\_selection()$
 5:     $propagate()$
 6:     **if** $empty\_domain\_in\_future\_variable()$ **then**
 7:         $shallow\_backtrack()$
 8:     **end if**
 9:     **if** $empty\_domain\_in\_current\_variable()$ **then**
10:         $backtrack()$
11:     **end if**
12: **end while**

---

3   Autonomous Search and Bat Optimizer

As previously presented, AS aims at providing self-tuning capabilities to the solver. In this context, the idea is to autonomously control which strategy is applied to each part of the search tree during resolution. The exchange of these strategies is carried out according to a quality ranking which is prepared by a choice function (CF). A CF is mainly composed of performance indicators (see indicators employed in Table 1) of the search process and weights that controls its relevance within the equation. In general, a choice function is defined as fallows: $CF_t(S_j) = w_1 a_{1t}(S_j) + \ldots + w_{IN} a_{INt}(S_j)$, where IN corresponds to the indicator set, w_i is a weight that controls the relevance of the ith-indicator within the CF and $a_{it}(S_j)$ is the score of the ith-indicator for the strategy S_j in time t. Under this scenario, the main component of a CF are the weights, that need to be finely tuned by an optimizer. This is done by carrying out a sampling phase where the problem is partially solved to a given cutoff. In each iteration, the solver attempts to determine the most successful weight set for the CF by using as input data the performance information gathered via the indicators.

Let us remark that this tuning process is dramatically important, as the correct configuration of the CF may have essential effects on the ability of the solver to properly solve specific problems. Parameter (weights) tuning is hard to achieve as parameters are problem-dependent and their best configuration is not stable along the search [9]. Additional and detailed information about this framework can be seen in [2,10].

Table 1. Indicators used during the search process.

Name	Description
VFP	Number of variables fixed by propagation
VFE	Number of variables fixed by enumeration
Step	Number of steps or decision points (n increments each time a variable is fixed during enumeration)
$T_n(S_j)$	Number of steps since the last time that an enumeration strategy S_j was used until step n^{th}
SB	Number of Shallow Backtracks [3]
B	Number of Backtracks
d_{max}	Maximum depth in the search tree
MDV	Represents a Variation of the Maximum Depth. It is calculated as: $CurrentDepth_{Maximum} - PreviousDepth_{Maximum}$
DV	Calculated as: $CurrentDepth - PreviousDepth$. A positive value means that the current node is deeper than the one explored at the previous step
SSR	Search Space Reduction. It is calculated as: $(PreviousSearchSpace - SearchSpace) / PreviousSearchSpace$
TR	The solving process alternates enumerations and backtracks on a few variables without succeeding in having a strong orientation. It is calculated as: $d_{t-1} - VFP_{t-1}$

3.1 Bat Optimizer

As previously mentioned, to determine the most successful weight set for performance indicators, the choice function must be finely tuned by an optimizer. To this end, is that also we propose the use of Bat Algorithm (BA), which is a recent metaheuristic inspired on the echolocation behavior of bats [12]. Bats and particularly micro-bats are able to identify objects in their surrounding areas by emitting pulses of sound and retrieving the corresponding produced echoes. This advanced capability allows bats even to distinguish obstacles from preys, being able to hunt in complete darkness.

The bat algorithm as been developed following three rules:

(1) It is assumed that all bats use echolocation to determine distances, and all of them are able to distinguish food, prey, and background barriers.

(2) A bat b_i searches for a prey with a position x_i and a velocity v_i. The pulses of sound emitted have the following features: a frequency f_{min}, (or varying λ), a varying wavelength λ (or frequency f), loudness A_0, and a rate of pulse emission $r \in [0,1]$. All sound features can be automatically adjusted depending on their target proximity.

(3) Although the loudness can vary in many ways, it is assumed that the loudness varies from a large (positive) A_0 to a minimum constant value A_{min}.

Algorithm 2. Bat algorithm

1: Objective function $f(x), x = (x_1, \ldots, x_d)$.
2: Initialize the bat population x_i and velocity v_i, $i = 1, 2, \ldots, m$.
3: Define pulse frequency f_i at x_i, $i = 1, 2, \ldots, m$.
4: Initialize pulse rates r_i and the loudness A_i, $i = 1, 2, \ldots, m$.
5: **while** The termination conditions are not met. **do**
6:     **for all** b_i **do**
7:         Generate new solutions through the equations (1), (2) and (3).
8:         **if** $rand > r_i$ **then**
9:             Select a solution among the best solutions.
10:            Generate a local solution around the best solution.
11:        **end if**
12:        **if** $rand < A_i$ and $f(x_i) < f(\hat{x})$ **then**
13:            Accept the new solutions.
14:            Increase r_i and reduce A_i.
15:        **end if**
16:    **end procedure**
17: **end while**
18: Rank the solutions and find the current best.

At the beginning, a population of n bats is initialized with position x_i and velocity v_i. Then, the frequency f_i at position x_i is set followed by pulse rates and loudness. Then, a while loop encloses a set of actions to be performed until the termination condition is reached. The first action to be done between the loop is the movement of bats according to Eqs. 1, 2 and 3.

$$f_i = f_{min} + (f_{max} - f_{min})\beta \tag{1}$$

$$v_i^j(t+1) = v_i^j(t) + [(\hat{x})^j - x_i^j(t)]f_i \tag{2}$$

$$x_i^j(t) = x_i^j(t-1) + v_i^j(t) \tag{3}$$

Equation 1 is used to control the pace and range of bats movements, where β is a randomly generated number drawn from a uniform distribution within the interval $[0,1]$. Equation 2 defines the velocity of decision variable j held by bat i in time t, where $(\hat{x})^j$ represents the current global best position encountered from the m bats for decision variable j; and finally Eq. 3 defines the position of decision variable j held by bat i in time t. Then, at line 8, a condition handles the variability of the possible solutions. Firstly, a solution is selected among the current best solutions, and a new solution is generated via random walks as proposed in [13]. Next, at line 12, a second condition is responsible for accepting the new best solution and for updating r_i and A_i according to Eqs. 4 and 5, where α and γ are ad-hoc constants between with $0 < \alpha < 1$ and $\gamma > 0$. Finally, the bats are ranked in order to find $\hat{x}$.

$$A_i(t+1) = \alpha A_i(t) \tag{4}$$

$$r_i(t+1) = r_i(0)[1 - exp(-\gamma t)], \tag{5}$$

For the experiments, we employ the following bat configuration as suggested in [14]: $\alpha = \gamma = 0.9$; $f_{min} = 0.75$; and $f_{max} = 1.25$. At the beginning, $r_i(0)$ and $A_i(0)$ are selected randomly, with $A_i(0) \in [1, 2]$ and $r_i(0) \in [0, 1]$.

4   Experimental Results

We have performed an experimental evaluation of the proposed approach on different instances of classic and known-well constraint satisfaction problems: N-queens with N = $\{10, 12, 15, 20, 50, 75\}$, Sudoku puzzle $\{1, 2, 5, 7, 9\}$, Knight's Tour with N=$\{5, 6\}$, Magic Square with N=$\{3, 4, 5\}$, Latin Square with N=$\{4, 5, 6\}$, Quasigroup with N=$\{5, 6, 7\}$, and Langford problems with $M = 2$ and N=$\{12, 16, 20, 23\}$.

Table 2. CFs employed for the experiments.

Choice functions	
CF_1:	$w_1 B + w_2 Step + w_3 T_n(S_j)$
CF_2:	$w_1 SB + w_2 MDV + w_3 DV$
CF_3:	$w_1 B + w_2 Step + w_3 MDV + w_4 DV$
CF_4:	$w_1 VF + w_2 dmax + w_3 DB$
CF_5:	$w_1 VFP + w_2 SSR - w_3 B$
CF_6:	$w_1 VFP + w_2 SSR - w_3 SB$
CF_7:	$w_1 VFP + w_2 SSR - w_3 SB - w_4 B$
CF_8:	$w_1 VFE + w_2 SSR - w_3 B$
CF_9:	$w_1 VFE + w_2 SSR - w_3 SB$
CF_{10}:	$w_1 VFE + w_2 SSR - w_3 SB - w_4 B$
CF_{11}:	$w_1 VFP + w_2 VFE - w_3 SSR - w_4 B$
CF_{12}:	$w_1 VFP + w_2 VFE - w_3 SSR - w_4 SB$
CF_{13}:	$w_1 VFP + w_2 VFE - w_3 SSR - w_4 SB - w_5 B$
CF_{14}:	$w_1 VFP + w_2 SSR - w_3 TR$
CF_{15}:	$w_1 VFE + w_2 SSR - w_3 TR$
CF_{16}:	$w_1 VFP + w_2 VFE - w_3 SSR - w_4 TR$

Table 3. Portfolio of the enumerations strategies used

Variable ordering	Heuristic description
Input Order	The first entry in the list is chosen
First Fail	The entry with the smallest domain size is chosen
Anti First Fail	The entry with the largest domain size is chosen
Occurrence	The entry with the largest number of attached constraints is chosen
Smallest	The entry with the smallest value in the domain is chosen
Largest	The entry with the largest value in the domain is chosen
Most Constrained	The entry with the smallest domain size is chosen
Max Regret	The entry with the largest difference between the smallest and second smallest value in the domain is chosen
Value ordering	Heuristic description
Min	Values are tried in increasing order.
Mid	Values are tried beginning from the middle of the domain.
Max	Values are tried in decreasing order.
Enumeration strategies	S_j
$S_1 = Input\ Order + Min$	$S_{13} = Smallest + Mid$
$S_2 = First\ Fail + Min$	$S_{14} = Largest + Mid$
$S_3 = Anti\ First\ Fail + Min$	$S_{15} = Most\ Constrained + Mid$
$S_4 = Occurrence + Min$	$S_{16} = Max\ Regret + Mid$
$S_5 = Smallest + Min$	$S_{17} = Input\ Order + Max$
$S_6 = Largest + Min$	$S_{18} = First\ Fail + Max$
$S_7 = Most\ Constrained + Min$	$S_{19} = Anti\ First\ Fail + Max$
$S_8 = Max\ Regret + Min$	$S_{20} = Occurrence + Max$
$S_9 = Input\ Order + Mid$	$S_{21} = Smallest + Max$
$S_{10} = First\ Fail + Mid$	$S_{22} = Largest + Max$
$S_{11} = Anti\ FirstFail + Mid$	$S_{23} = Most\ Constrained + Max$
$S_{12} = Occurrence + Mid$	$S_{24} = Max\ Regret + Max$

The adaptive enumeration component has been implemented on the Ecl^ips^e Constraint logic Programming Solver v6.10, and the bat-optimizer has been developed in Java. The experiments have been launched on a 3.30 GHz Intel Core $i3 - 2120$ with 4 Gb RAM running Windows 7 Professional 32*bits*. The instances are solved to a maximum number of 65535 steps as equally done in previous work [4]. If no solution is found at this point the problem is set to *t.o.* (time-out).

Table 4. Backtracks required for N-Queens and Sudoku puzzles.

CF_j	Problems NQ						$\sum$	RPD %	Problems SK					$\sum$	RPD %
	10	12	15	20	50	75			1	2	5	7	9		
CF_1	1	7	1	8	3	818	838	48.294	0	1	1	90	0	92	0.095
CF_2	4	0	1	11	11	8	35	1.059	0	2	32	50	0	84	0
CF_3	2	4	1	11	1	0	19	0.118	0	2	308	78	0	388	3.619
CF_4	4	1	1	11	146	0	163	8.588	0	2	85	114	0	201	1.393
CF_5	2	1	1	12	3	0	19	0.118	0	2	58	46	0	106	0.262
CF_6	1	2	0	37	0	4	44	1.588	0	2	51	47	0	100	0.19
CF_7	4	2	1	14	2	0	23	0.353	0	2	36	156	0	194	1.31
CF_8	2	3	2	16	1	0	24	0.412	0	0	8	121	0	129	0.536
CF_9	4	0	1	11	2	7	25	0.471	0	3	372	118	0	493	4.87
CF_{10}	4	3	1	14	1	4	27	0.588	0	8	69	46	0	123	0.464
CF_{11}	1	2	1	8	0	128	140	7.235	0	2	38	52	0	92	0.095
CF_{12}	3	3	1	15	1	2	25	0.471	0	0	261	65	0	326	2.881
CF_{13}	3	3	2	11	1	1	21	0.235	0	8	221	71	0	300	2.571
CF_{14}	2	2	1	12	0	0	17	0	0	3	212	45	0	260	2.096
CF_{15}	4	2	1	9	1	1	18	0.059	0	2	411	77	0	490	4.833
CF_{16}	1	0	1	2	4	18	26	0.529	0	2	475	58	0	535	5.369

Table 2 shows the CFs that we employed for the experiments. These CFs were the best performing ones after the corresponding training phase of the algorithm. Finally, we have analyzed the choice function impact on 8 variable selection heuristics and 3 value selection heuristics, which when combined, provide a portfolio of 24 strategies (see Table 3).

Results are evaluated in terms of the minimum called backtracks and solving time (in ms), which are the most employed indicators to evaluate performance in constraint programming. For this comparison, we used the relative percentage deviation (RPD). RPD value quantifies the deviation of the accumulated value $\sum_i^n CF_i$ from $min(\sum_i^n CF_i)$, where n corresponds to the number of instances, and it is calculated as follows:

$$RPD = \left(\frac{\sum_i^n CF_i - min(\sum_i^n CF_i)}{min(\sum_i^n CF_i)} \right) \times 100, \ \forall i \in \{1, \ldots, 16\} \tag{6}$$

Tables 4, 6 and 8 illustrate the performance in terms of called backtrack and Tables 5, 7 and 9 expose the efficiency in terms of the solving time required to find a solution for each choice function and the proposed optimizer. Results demonstrate the impact to use a choice function to correctly select the strategy to each part of the search tree. For instance, taking into account the total number

Table 5. Solving time for N-Queens and Sudoku puzzles.

CF$_j$	Problems NQ						$\sum$	RPD %	Problems SK					$\sum$	RPD %
	10	12	15	20	50	75			1	2	5	7	9		
CF$_1$	610	760	1025	1570	8107	21747	**33819**	0.069	635	565	640	595	632	**3067**	0.069
CF$_2$	627	760	1030	1569	8004	19787	**31777**	0.004	514	616	604	603	532	**2869**	0
CF$_3$	620	770	1030	1580	7930	19725	**31655**	0.007	520	620	610	630	580	**2960**	0.032
CF$_4$	620	760	1035	1635	8230	19775	**32055**	0.013	670	625	615	605	560	**3075**	0.072
CF$_5$	622	774	1014	1592	8169	21060	**33231**	0.05	585	610	610	595	515	**2915**	0.016
CF$_6$	610	755	1020	1570	7945	19735	**31635**	0	620	580	610	595	520	**2925**	0.02
CF$_7$	620	760	1020	1570	8040	19775	**31785**	0.005	585	650	620	615	515	**2985**	0.04
CF$_8$	620	780	1025	1660	8195	19725	**32005**	0.012	625	605	605	600	510	**2945**	0.026
CF$_9$	620	765	1015	1585	7832	19915	**31732**	0.003	630	650	610	600	505	**2995**	0.044
CF$_{10}$	615	775	1025	1575	8040	19905	**31935**	0.009	580	645	615	605	520	**2965**	0.033
CF$_{11}$	620	770	1065	1575	7995	20865	**32890**	0.04	610	600	595	610	525	**2940**	0.025
CF$_{12}$	620	765	1030	1555	7920	20760	**32650**	0.032	560	610	610	600	515	**2895**	0.009
CF$_{13}$	610	760	1055	1570	8040	19845	**31880**	0.008	545	640	615	605	515	**2920**	0.018
CF$_{14}$	615	765	1025	1585	8105	19740	**31835**	0.006	640	665	605	600	520	**3030**	0.056
CF$_{15}$	615	765	1020	1580	7985	19810	**31775**	0.004	585	595	615	600	515	**2910**	0.014
CF$_{16}$	610	755	1020	1570	8065	20295	**32315**	0.021	635	605	615	590	499	**2944**	0.026

Table 6. Backtracks required for Knight's Tour, Magic and Latin Square.

CF$_j$	Problems KT		$\sum$	RPD %	Problems MS			$\sum$	RPD %	Problems LS			$\sum$	RPD %
	5	6			3	4	5			4	5	6		
CF$_1$	767	13456	**14223**	0.196	0	0	26	**26**	0.857	0	0	0	**0**	0
CF$_2$	767	14998	**15765**	0.326	0	2	185	**187**	12.357	0	0	0	**0**	0
CF$_3$	767	14998	**15765**	0.326	0	0	102	**102**	6.286	0	0	0	**0**	0
CF$_4$	767	14998	**15765**	0.326	0	1	185	**186**	12.286	0	0	0	**0**	0
CF$_5$	454	11557	**12011**	0.01	0	0	153	**153**	9.929	0	0	0	**0**	0
CF$_6$	454	19504	**19958**	0.678	0	1	53	**54**	2.857	0	0	0	**0**	0
CF$_7$	454	13501	**13955**	0.173	0	1	15	**16**	0.142	0	0	0	**0**	0
CF$_8$	454	11557	**12011**	0.01	0	3	81	**84**	5	0	0	0	**0**	0
CF$_9$	454	11557	**12011**	0.01	0	1	139	**140**	9	0	0	0	**0**	0
CF$_{10}$	741	11557	**12298**	0.034	0	0	96	**96**	5.857	0	0	0	**0**	0
CF$_{11}$	454	11557	**12011**	0.01	0	0	146	**146**	9.429	0	0	0	**0**	0
CF$_{12}$	336	11557	**11893**	0	0	2	12	**14**	0	0	0	0	**0**	0
CF$_{13}$	454	11557	**12011**	0.01	0	0	91	**91**	5.5	0	0	0	**0**	0
CF$_{14}$	336	11557	**11893**	0	0	0	154	**154**	10	0	0	0	**0**	0
CF$_{15}$	454	11557	**12011**	0.01	0	3	18	**21**	0.5	0	0	0	**0**	0
CF$_{16}$	336	11557	**11893**	0	0	3	148	**151**	9.786	0	0	0	**0**	0

Table 7. Solving time for Knight's Tour, Magic and Latin Square.

CF$_j$	Problems KT		Σ	RPD %	Problems MS			Σ	RPD %	Problems LS			Σ	RPD %
	5	6			3	4	5			4	5	6		
CF$_1$	3265	4555	7820	0.014	620	950	750	2320	0.097	350	400	465	1215	0.061
CF$_2$	3247	4537	7784	0.009	629	928	760	2317	0.096	340	395	440	1175	0.026
CF$_3$	3250	4560	7810	0.013	620	905	770	2295	0.085	340	395	455	1190	0.039
CF$_4$	3270	4493	7763	0.007	640	1005	780	2425	0.147	345	390	440	1175	0.026
CF$_5$	3185	4675	7860	0.019	546	845	775	2166	0.024	335	385	445	1165	0.017
CF$_6$	3250	4780	8030	0.041	545	860	780	2185	0.033	340	385	440	1165	0.017
CF$_7$	3240	4585	7825	0.015	545	820	750	2115	0	340	390	445	1175	0.026
CF$_8$	3240	4525	7765	0.007	545	900	775	2220	0.05	335	380	430	1145	0
CF$_9$	3245	4580	7825	0.015	545	865	775	2185	0.033	335	385	440	1160	0.013
CF$_{10}$	3250	4461	7711	0	550	920	790	2260	0.069	345	390	435	1170	0.022
CF$_{11}$	3245	4705	7950	0.031	545	980	765	2290	0.083	340	385	440	1165	0.017
CF$_{12}$	3275	4530	7805	0.012	540	930	780	2250	0.064	330	385	440	1155	0.009
CF$_{13}$	3265	4500	7765	0.007	575	940	775	2290	0.083	335	390	445	1170	0.022
CF$_{14}$	3270	4800	8070	0.047	545	805	790	2140	0.012	340	395	445	1180	0.031
CF$_{15}$	3260	4590	7850	0.018	550	835	755	2140	0.012	340	390	450	1180	0.031
CF$_{16}$	3285	4545	7830	0.015	565	810	775	2150	0.017	330	385	440	1155	0.009

Table 8. Bactracks required for Quasigroup and Langford problems.

CF$_j$	Problems QG					Σ	RPD %	Problems LF$_{m=2}$				Σ	RPD %
	1	3	5	6	7			12	16	20	23		
CF$_1$	0	0	0	0	1	1	1	1	0	0	0	1	0
CF$_2$	0	0	0	0	0	0	0	1	0	1	0	2	1
CF$_3$	0	0	0	0	0	0	0	1	0	0	0	1	0
CF$_4$	0	0	0	0	1	1	1	1	0	1	0	2	1
CF$_5$	0	0	0	0	1	1	1	1	0	0	0	1	0
CF$_6$	0	0	0	0	1	1	1	1	0	0	0	1	0
CF$_7$	0	0	0	0	0	0	0	1	0	0	1	2	1
CF$_8$	0	0	0	0	0	0	0	1	0	0	0	1	0
CF$_9$	0	0	0	0	0	0	0	1	0	0	0	1	0
CF$_{10}$	0	0	0	0	0	0	0	1	0	0	0	1	0
CF$_{11}$	0	0	0	0	0	0	0	1	0	1	1	3	2
CF$_{12}$	0	0	0	0	0	0	0	1	0	0	0	1	0
CF$_{13}$	0	0	0	0	0	0	0	1	0	0	0	1	0
CF$_{14}$	0	0	0	0	0	0	0	1	0	0	1	2	1
CF$_{15}$	0	0	0	0	0	0	0	1	0	0	0	1	0
CF$_{16}$	0	0	0	0	0	0	0	1	0	0	0	1	0

Table 9. Solving time for Quasigroup and Langford problems.

CF_j	Problems QG					$\sum$	RPD %	Problems $LF_{m=2}$				$\sum$	RPD %
	1	3	5	6	7			12	16	20	23		
CF_1	291	225	695	670	546	**2427**	0.134	543	671	765	869	**2848**	0.032
CF_2	239	211	655	642	432	**2179**	0.018	546	661	759	837	**2803**	0.016
CF_3	260	215	665	660	455	**2255**	0.054	565	660	770	850	**2845**	0.031
CF_4	275	220	660	640	480	**2275**	0.063	570	675	730	845	**2820**	0.022
CF_5	260	215	690	635	430	**2230**	0.042	530	645	765	820	**2760**	0
CF_6	240	210	655	665	425	**2195**	0.026	555	675	760	865	**2855**	0.034
CF_7	240	220	660	645	445	**2210**	0.033	550	670	785	835	**2840**	0.029
CF_8	245	215	670	615	430	**2175**	0.016	540	660	735	830	**2765**	0.002
CF_9	245	215	683	620	435	**2198**	0.027	530	670	775	845	**2820**	0.022
CF_{10}	255	210	685	635	425	**2210**	0.033	545	670	775	830	**2820**	0.022
CF_{11}	245	215	650	630	435	**2175**	0.016	545	650	760	855	**2810**	0.018
CF_{12}	245	215	645	625	430	**2160**	0.009	560	670	765	835	**2830**	0.025
CF13	275	210	675	620	440	**2220**	0.037	535	660	750	855	**2800**	0.014
CF_{14}	245	215	690	640	425	**2215**	0.035	540	665	775	855	**2835**	0.027
CF_{15}	265	215	640	650	425	**2195**	0.026	540	675	760	865	**2840**	0.029
CF_{16}	240	210	645	625	420	**2140**	0	535	675	770	855	**2835**	0.027

Table 10. Summary of computational results in backtracks

CF_j	CF_1	CF_2	CF_3	CF_4	CF_5	CF_6	CF_7	CF_8
$\sum$ (*Backtraks*)	1085	1149	1163	1167	879	1440	1014	876
$\sum$ (*Solving time*)	53500	50944	51075	51618	52372	51050	50985	51055
CF_j	CF_9	CF_{10}	CF_{11}	CF_{12}	CF_{13}	CF_{14}	CF_{15}	CF_{16}
$\sum$ (*Backtraks*)	906	897	886	876	888	881	896	901
$\sum$ (*Solving time*)	50975	51136	52280	51800	51095	51355	50935	51409

of backtracks needed for solving the CSPs (Table 10), the best choice functions are CF_8 and CF_{12}, where both selection techniques required only 876 backtracks.

Now, evaluating the choice functions in terms of the solving time needed to reach a solution, we can see that the best choice function is CF_{15} required 50935 ms, fallowed from close by CF_2, CF_9 and CF_7, respectively.

5   Conclusions

Autonomous search is an interesting approach to provide more capabilities to the solver in order to improve the search process based on some performance indicators and self-tuning. In this paper, we have focused in the impact to use choice functions for correct selection the strategy and improving as a consequence the performance of the solver during the resolution process. To this end, we have presented a performance evaluation of 16 different carefully constructed choice functions tuned by the bat algorithm. The experimental results have demonstrated which the CF_8 and CF_{12} are better than others in terms of the backtracks, and CF_{15} converges faster than its closest competitors: CF_2, CF_9 and CF_7, respectively. As future work, we plan to test new modern metaheuristics for supporting CFs for AS. The incorporation of additional strategies to the portfolio would be an interesting research direction to follow as well.

Acknowledgments. Ricardo Soto is supported by Grant CONICYT/FONDECYT/ REGULAR/1160455, Broderick Crawford is supported by Grant CONICYT/ FONDECYT/REGULAR/1140897 and Rodrigo Olivares is supported by Postgraduate Grant Pontificia Universidad Católica de Valparaíso 2016.

References

1. Crawford, B., Soto, R., Castro, C., Monfroy, E.: A hyperheuristic approach for dynamic enumeration strategy selection in constraint satisfaction. In: Ferrández, J.M., Álvarez Sánchez, J.R., de la Paz, F., Toledo, F.J. (eds.) IWINAC 2011, Part II. LNCS, vol. 6687, pp. 295–304. Springer, Heidelberg (2011)
2. Crawford, B., Soto, R., Montecinos, M., Castro, C., Monfroy, E.: A framework for autonomous search in the Eclipse solver. In: Mehrotra, K.G., Mohan, C.K., Oh, J.C., Varshney, P.K., Ali, M. (eds.) IEA/AIE 2011, Part I. LNCS, vol. 6703, pp. 79–84. Springer, Heidelberg (2011)
3. Barták, R., Rudová, H.: Limited assignments: a new cutoff strategy for incomplete depth-firstsearch. In: Proceedings of the 20th ACM Symposium on Applied Computing (SAC), pp. 388–392 (2005)
4. Crawford, B., Castro, C., Monfroy, E., Soto, R., Palma, W., Paredes, F.: Dynamic selection of enumeration strategies for solving constraint satisfaction problems. Rom. J. Inf. Sci. Tech. **15**(2), 106–128 (2012)
5. Crawford, B., Soto, R., Castro, C., Monfroy, E., Paredes, F.: An extensible autonomous search framework for constraint programming. Int. J. Phys. Sci. **6**(14), 3369–3376 (2011)
6. Rossi, F.: Handbook of Constraint Programming. Elsevier, Amsterdam (2006)
7. Hamadi, Y., Monfroy, E., Saubion, F., Optimization, H.: What is autonomous search? In: van Hentenryck, P., Milano, M. (eds.) The Ten Years of CPAIOR. Springer, New York (2011)
8. Apt, K.R.: Principles of Constraint Programming. Cambridge Press, Cambridge (2003)
9. Maturana, J., Saubion, F.: A compass to guide genetic algorithms. In: Rudolph, G., Jansen, T., Lucas, S., Poloni, C., Beume, N. (eds.) PPSN 2008. LNCS, vol. 5199, pp. 256–265. Springer, Heidelberg (2008)
10. Soto, R., Crawford, B., Monfroy, E., Bustos, V.: Using autonomous search for generating good enumeration strategy blends in constraint programming. In: Murgante, B., Gervasi, O., Misra, S., Nedjah, N., Rocha, A.M.A.C., Taniar, D., Apduhan, B.O. (eds.) ICCSA 2012, Part III. LNCS, vol. 7335, pp. 607–617. Springer, Heidelberg (2012)

11. Soto, R., Crawford, B., Misra, S., Palma, W., Monfroy, E., Castro, C., Paredes, F.: Choice functions for autonomous search in constraint programming: GA vs PSO. Tech. Gaz. **20**(4), 621–629 (2013)
12. Yang, X.-S.: A new metaheuristic bat-inspired algorithm. In: González, J.R., Pelta, D.A., Cruz, C., Terrazas, G., Krasnogor, N. (eds.) NICSO 2010. SCI, vol. 284, pp. 65–74. Springer, Heidelberg (2010)
13. Yang, X.-S.: Bat algorithm for multi-objective optimisation. IJBIC **3**(5), 267–274 (2011)
14. Yang, X.-S., He, X.: Bat algorithm: literature review and applications. IJBIC **5**(3), 141–149 (2013)

Binary Harmony Search Algorithm for Solving Set-Covering Problem

Juan Salas[1]([✉]), Broderick Crawford[1,4,5], Ricardo Soto[1,2,3],
Álvaro Gómez Rubio[1], Adrián Jaramillo[1], Sebastián Mansilla Villablanca[1],
and Eduardo Olguín[5]

[1] Pontificia Universidad Católica de Valparaíso, Santiago, Chile
{juan.salas.f,alvaro.gomez.r,adrian.jaramillo.s,
sebastian.mansilla.v}@mail.pucv.cl,
{broderick.crawford,ricardo.soto}@ucv.cl
[2] Universidad Autónoma de Chile, Santiago, Chile
[3] Universidad Científica del Sur, Lima, Peru
[4] Universidad Central de Chile, Santiago, Chile
[5] Facultad de Ingeniería y Tecnología, Universidad San Sebastián,
Bellavista 7, 8420524 Santiago, Chile
eduardo.olguin@uss.cl

Abstract. This paper is intended to generate solutions to Set Covering Problem (SCP) through the use of a metaheuristic. The results were obtained using a variation of Harmony Search called Binary Global-Best Harmony Search Algorithm. To measure the effectiveness of the technique against other metaheuristics, Weasly benchmark was used.

Keywords: Binary Harmony Search · Set Covering Problem · Metaheuristics

1 Introduction

1.1 SCP Formulation

The Set Covering Problem (SCP) is a well-known mathematical problem, which tries to cover a set of needs at the lowest possible cost. The SCP was included in the list of 21 $\mathcal{NP}$-*complet* problems of Karp [2]. There are many practical uses for this problem, such as: crew scheduling [3,4], location of emergency facilities [5,6], production planning in industry [7–9], vehicle routing [10,11], ship scheduling [12,13], network attack or defense [14], assembly line balancing [15,16], traffic assignment in satellite communication systems [17,18], simplifying boolean expressions [19], the calculation of bounds in integer programs [20], information retrieval [21], political districting [22], crew scheduling problems in airlines [23], among others. The SCP can be formulated as follows:

$$\text{Minimize} \quad Z = \sum_{j=1}^{n} c_j x_j \tag{1}$$

© Springer International Publishing Switzerland 2016
H. Fujita et al. (Eds.): IEA/AIE 2016, LNAI 9799, pp. 917–930, 2016.
DOI: 10.1007/978-3-319-42007-3_78

Subject to:

$$\sum_{j=1}^{n} a_{ij}x_j \geq 1 \quad \forall i \in I \tag{2}$$

$$x_j \in \{0,1\} \quad \forall j \in J \tag{3}$$

Let $A = (a_{ij})$ be a $m \times n$ 0-1 matrix with $I = \{1,\dots,m\}$ and $J = \{1,\dots,n\}$ be the row and column sets respectively. We say that column j can be cover a row i if $a_{ij} = 1$. Where c_j is a nonnegative value that represents the cost of selecting the column j and x_j is a decision variable, it can be 1 if column j is selected or 0 otherwise. The objective is to find a minimum cost subset $S \subseteq J$, such that each row $i \in I$ is covered by at least one column $j \in S$.

In the following section, we present a simple way to understand the SCP, through an example:

1.2   SCP Sample Solution

Imagine that an ambulance station can meet the needs of an geographic zone. Similarly the ambulance station can cover all the needs of the nearby areas. For example, if a station is built in Zone 1 (Fig. 1) ambulance station can meet the needs of neighboring areas, that is, it could also cover: Zone 1, Zone 2, Zone 3 and Zone 4. This can be appreciated in Eq. (5).

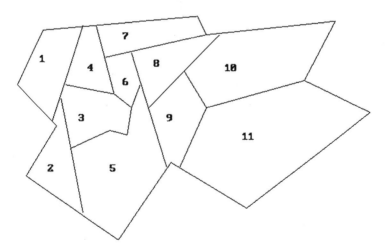

Fig. 1. Set Covering Problem example.

In this example, we must fulfill the need to cover the geographical areas defined in accordance with the restrictions.

The restriction of this case is that all areas must be covered by at least one ambulance station and the goal is to minimize the number of stations built, the cost of building a station is the same for all areas. The x_j variable represents the area j which is 1 if the ambulance station is built, and will be 0 if not. As above, it can be formulated as follows:

$$\text{Min}\quad c_1x_1+c_2x_2+c_3x_3+c_4x_4+c_5x_5+c_6x_6+c_7x_7+c_8x_8+c_9x_9+c_{10}x_{10}+c_{11}x_{11} \tag{4}$$

Subject to:

$$
\begin{aligned}
x_1+ x_2+ x_3+ x_4 &\geq 1 \tag{5}\\
x_1+ x_2+ x_3\phantom{{}+{}} + x_5 &\geq 1 \tag{6}\\
x_1+ x_2+ x_3+ x_4+ x_5+ x_6 &\geq 1 \tag{7}\\
x_1\phantom{{}+{}} + x_3+ x_4\phantom{{}+{}} + x_6+ x_7 &\geq 1 \tag{8}\\
x_2+ x_3\phantom{{}+{}} + x_5+ x_6\phantom{{}+{}} + x_8+ x_9 &\geq 1 \tag{9}\\
x_3+ x_4+ x_5+ x_6+ x_7+ x_8 &\geq 1 \tag{10}\\
x_4\phantom{{}+{}} + x_6+ x_7+ x_8 &\geq 1 \tag{11}\\
x_5+ x_6+ x_7+ x_8+ x_9+ x_{10} &\geq 1 \tag{12}\\
x_5\phantom{{}+{}} + x_8+ x_9+ x_{10}+ x_{11} &\geq 1 \tag{13}\\
x_8+ x_9+ x_{10}+ x_{11} &\geq 1 \tag{14}\\
x_9+ x_{10}+ x_{11} &\geq 1 \tag{15}
\end{aligned}
$$

The first constraint (5) indicates that to cover zone 1, it is possible to locate a station in the same area or in the border. The following restriction is for zone 2 and so on. One possible optimal solution for this problem is to locate ambulance stations in zones 3, 8 and 9. That is, $x_3 = x_8 = x_9 = 1$ y $x_1 = x_2 = x_4 = x_5 = x_6 = x_7 = x_{10} = x_{11} = 0$. As shown in (Fig. 2).

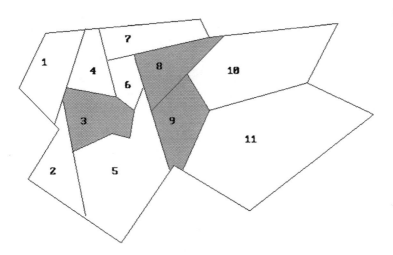

Fig. 2. Set Covering Problem solution.

In this document, we propose solve the SCP, with metaheuristic Harmony Search (HS) to obtain satisfactory solutions within a reasonable time. HS mimics the process of musical improvisation, where musicians make adjustments in tone to achieve aesthetic harmony.

1.3 Main Goal

Solving the SCP using the HS metaheuristic algorithm.

1.4 Specific Objectives

- Solve the SCP with the HS metaheuristic and compare the results with benchmark of Beasley [26]
- Introduce new operators that make the results more close to the global optimum.
- Search nearest optimal solutions to problems in the SCP.
- Compare and analyze the results using different resolution strategies.

2 Harmony Search in Detail

First of all, for the purpose of properly understand what is looking like a good solution in this metaheuristic, we must know the meaning of Aesthetic Quality of a Harmony (AQH), which is set forth in the following paragraph.

The AQH instrument it is essentially determined by its pitch (or frequency), sound quality, and amplitude (or loudness). The sound quality is largely determined by the harmonic content that is in turn determined by the waveforms or modulations of the sound signal. However, the harmonics that it can generate will largely depend on the pitch or frequency range of the particular instrument [27].

Different notes have different frequencies. For example, the note A above middle C has a fundamental frequency of $f_0 = 440\,\mathrm{Hz}$ (Table 4).

Given the above, it is established that a good harmony has good AQH.

HS is a population-based metaheuristic algorithm inspired from the musical process of searching for a perfect state of Harmony or AQH. The HS was proposed by Geem [28].

The pitch of each musical instrument determines the AQH, just as the fitness function values determines the quality of the decision variables.

In the music improvisation process, all musicians sound pitches within possible range together to make one harmony.

If all pitches make a good harmony, each musician stores in his memory that experience and possibility of making a good harmony in increased next time. The same thing in optimization, the initial solution is generated randomly from decision variables within the possible range.

If the objective function values of these decision variables is good to make promising solution, then the possibility to make a good solution is increased next time.

In this document, we focus on studying the classical SCP problem (Eq. 1), where we want minimize the cost. The problem is to achieve find the variable that when activated to achieve the previously stated goal. The decision version of set covering is *NP-complete*, and the optimization version of set cover is *NP-hard* [29].

2.1 HS Operation in Depth

In general, the procedure for HS metaheuristic, consists of the following four steps. *Step 1: Initialization parameters* Initialize the control parameters and a Harmony Memory (HM). At this step, an initial HM is filled with a population of Harmony Memory Size (HMS) harmonies generated randomly. In addition, the parameters of HS, that is, Harmony Memory Consideration Rate (HMCR) and Pitch Adjusting Rate PAR are given when the metaheuristic begins. *Step 2: New Harmony* Improvise a new harmony from the current HM. *Step 3: Replace worst Harmony in HM* If the new generated harmony is better than the worst one in HM, then replace the worst harmony with the new one; otherwise, go to the next step. *Step 4: Check the stop criteria* If a stopping criterion is not satisfied, go to Step 2.

Table 1. HS components - Musician

	Each musician represents a decision variable, according to the example shown, there would be 11 musicians, since there are 11 decision variables $x_1 \ldots x_{11}$.

Table 2. HS components - Pitch range

	The pitch range of the instrument represents the range of values that can take a decision variable. Given the nature of the problem, the possible values are $\{0, 1\}$

Table 3. HS components - Solution

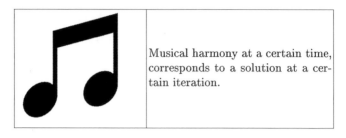	Musical harmony at a certain time, corresponds to a solution at a certain iteration.

Table 4. HS components - Aesthetics audience

$f = 440 \times 2^{(p_n - 69)/12}$	Aesthetics audience, judges whether harmony is good or not. In the problem it refers to the objective function

3   Global-Best Harmony Search

To further improve the convergence performance of Harmony Search (HS) and overcome some shortcomings of HS, a new variant of HS, called GHS, was proposed by Omran and Mahdavi [30]. First, the GHS dynamically updates parameter PAR according to Eq. (16):

$$PAR(t) = PAR_{min} + \frac{PAR_{max} - PAR_{min}}{NI} t \qquad (16)$$

where $PAR(t)$ represents the pitch adjusting rate at generation t, PAR_{min} and PAR_{max} are the minimum and maximum adjusting rate, respectively. The parameter t is the iterative variable, and parameter NI is the number of improvisations.

4   Binary Global-Best Harmony Search Algorithm

The HS is good at identifying the high performance regions of the solution space in a reasonable time, but poor at performing local search [31]. Namely, there is imbalance between the exploration and the exploitation of HS. Furthermore, HS designed for continuous space cannot be directly used to solve discrete combinatorial optimization problems.

In order to overcome the drawbacks of HS, a novel binary global-best harmony search (BGHS) is designed for binary optimization problems.

Owing to better performance of GHS, some modifications to GHS are introduced to further enhance the convergence performance of GHS. Then a novel binary coded GHS, a two-phase repair operator, and a greedy selection mechanism are integrated into the BGHS [31]. And they are described in detail as follows.

4.1 Initialization in BGHS

The initial population in BGHS is generated randomly using a Bernoulli process, which is a finite or infinite sequence of binary random variables, so it is a discrete-time stochastic process that takes only two values, canonically 0 and 1. A Bernoulli process is a repeated coin flipping.

Depending probability variable which is a real number that lies between 0 and 1. While the value of the variable probability p is closer to 1, then there is greater chance of success, this means that the variable has a higher chance of being one [32].

In addition, another harmony vector $\overrightarrow{x_0}$ is generated based on a greedy operation. The greedy operation is based on the idea that the item with higher profit density ratio should be selected first. And the profit density ratio can be calculated by the following equation:

$$\mu_j = \frac{1}{c_j} \tag{17}$$

The first way is to sort the items by μ_j. Then we select the items with higher value of μ_j while the solution remains feasible. This way, we can get a harmony vector $\overrightarrow{x_0}$. And if $\overrightarrow{x_0}$ is better than the worst one of previously initialized HM, then substitute the worst one with $\overrightarrow{x_0}$.

4.2 Dynamically Updating of the Parameters

The control parameters HMCR and PAR play an important role in standard HS. More specifically, the parameters HMCR and PAR set to constants may have some adverse effects on the performance of HS, which is the idea behind designing the varying parameters. For HS with the guidance of the best harmony, the parameter HMCR with a larger value can be helpful to accelerate the convergence speed of HS variants with the guidance of the best harmony (individual) at the beginning of search, while the parameter HMCR with a smaller value can help the corresponding HS variant to get out of local minima at the end of search. Like the parameter HMCR, a varying parameter PAR is also considered in this paper in order to further balance the exploration and exploitation of HS variants.

$$HMCR(t) = HMCR_{max} - \frac{HMCR_{max} - HMCR_{min}}{NI}t \tag{18}$$

$$PAR(t) = PAR_{max} - \frac{PAR_{max} - PAR_{min}}{NI}t \tag{19}$$

Where $HMCR_{max}$ and $HMCR_{min}$ represent the lower and upper bounds of $HMCR$, respectively. And the other parameters are the same as those in (16).

4.3 Improvising a New Harmony

The musicians generally choose a perfect state of a harmony from their memory or harmony memory (HM) during the process of improvising a new harmony

($\overrightarrow{x_0}$). Next, they may select a pitch from the current harmony memory randomly and then they would perform a fine tune operation, that is, pitch adjustment, for the chosen pitch to improve the effectiveness of music. For SCP, the states of a pitch just include zero and one 0,1; that is, any state of a pitch is ranging in. So discrete genetic mutation [33] used is suitable for pitch adjustment.

4.4   Two-Phase Repair Operator: ADD and DROP

In this section, a repair operator is introduced. For BGHS, new generated harmony vector needs to be repaired under two cases. One is that the harmony vector violates the constraints. The other corresponds to inactivate the most expensive columns, maintaining the feasibility of the solution. Hence, the repair operator consists of two phases. The first phase, called ADD, is responsible for repairing a harmony vector violating the constraint. The second phase, named DROP, is applied to remove any redundant column such that by removing it from the solution, the solution still remains feasible.

4.5   Selection Mechanism

In order to avoid being clustered in the best harmony, its employ a selection mechanism, in which a new generated harmony vector $\overrightarrow{x'}$ is compared with $\overrightarrow{x}_{best}$ first, and if $\overrightarrow{x'}$ is better than $\overrightarrow{x}_{best}$, replace $\overrightarrow{x}_{best}$ with $\overrightarrow{x'}$; otherwise, is compared with the worst harmony $\overrightarrow{x}_{worst}$ in HM again, and a greedy selection is applied between $\{\overrightarrow{x'}, \overrightarrow{x}_{worst}\}$. In this way, the number of harmony around the best harmony would be small at the early stage of evolution so that the diversity of harmony memory would be kept better. As a consequence, BGHS not only speeds up the convergence speed but also avoids being trapped in a local optimum.

4.6   The Proposed Algorithm

According to the analysis and modifications mentioned above, an initialization of HM based on greedy operation, a novel scheme of improvising a new harmony with the direction information of the best harmony, and a repair operator with greedy strategy make up the proposed BGHS designed for SCP. The Python code of BGHS is given in Algorithm 1.

Algorithm 1 Pseudocode Binary Harmony Search

```
1    FUNCTION HS_execution(inputfile):
2            GLOBAL HARMONY_MEMORY
3            GLOBAL FILE
4            FILE <- inputfile
5            parse([FILE])
6            HS_initiation()
7
8            i <- 0
9
10           FOR harmony_in_HM in HARMONY_MEMORY:
11                   harmony_repaired <- repair_of_harmony(harmony_in_HM)
12                   HARMONY_MEMORY[i] <- harmony_repaired
13                    i = i + 1
14           ENDFOR
15
16           i <- 1
17
18           WHILE i < MAX_improvisations:
19                   storesBetterAndWorseHarmony()
20                   new_harmony_vector <- createNewHarmony(i)
21                   new_harmony_vector <- repair_of_harmony(new_harmony_vector)
22
23                   IF betterOrEqualToBest(new_harmony_vector):
24                           replaceBest(new_harmony_vector)
25                   ELSEIF betterOrEqualToWorst(new_harmony_vector):
26                           replaceWorst(new_harmony_vector)
27                   ENDIF
28
29                   insert_best_and_worst()
30                   # Improvisation increases in one
31                   i = i + 1
```

5   Results and Analysis

The Table 5 show the detailed results obtained by BGHS. The second column Z_{BKS} details the best known solution value of each instance evaluated. The next columns Z_{MIN}, Z_{MAX}, and Z_{AVG} represents the minimum, maximum, and average objective function value respectively. The last column reports the relative percentage deviation (RPD) value that represents the deviation of the objective value Z (fitness) from Z_{OPT} which in our case is the minimum value obtained for each instance.

$$\mathrm{RPD} = \frac{100(Z_{MIN} - Z_{BKS})}{Z_{BKS}} \qquad (20)$$

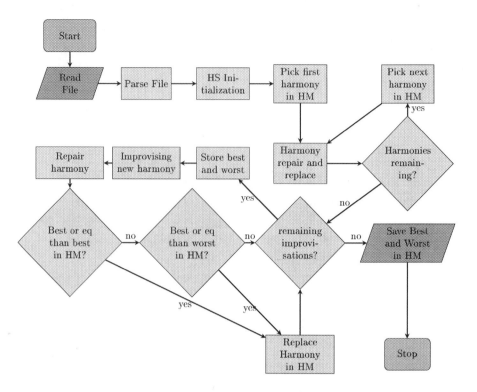

Fig. 3. Flowchart Binary Harmony Search

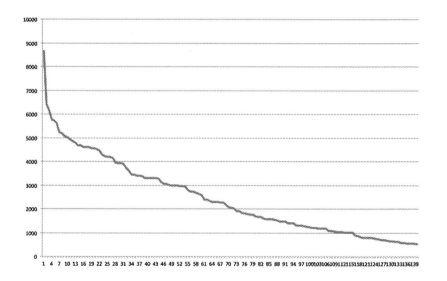

Fig. 4. SCP41

Table 5. Experimental results of SCP benchmarks (4, 5, 6, A, B and C sets).

Instance	Z_{BKS}	Z_{MIN}	Z_{MAX}	Z_{AVG}	RPD
4.1	429	429	435	432	0
4.2	512	514	520	517	0,39
4.3	516	516	537	526,5	0
4.4	494	494	532	513	0
4.5	512	512	520	516	0
4.6	560	560	583	571,5	0
4.7	430	430	440	435	0
4.8	492	492	512	502	0
4.9	641	641	952	796,5	0
4.10	514	514	809	661,5	0
5.1	253	253	280	266,5	0
5.2	302	302	393	347,5	0
5.3	226	226	239	232,5	0
5.4	242	242	255	248,5	0
5.5	211	211	220	215,5	0
5.6	213	213	215	214	0
5.7	293	293	311	302	0
5.8	288	288	299	293,5	0
5.9	279	279	295	287	0
5.10	265	265	280	272,5	0
6.1	138	138	272	205	0
6.2	146	146	226	186	0
6.3	145	145	248	196,5	0
6.4	131	131	153	142	0
6.5	161	161	172	166,5	0
A.1	253	256	280	268	1,19
A.2	252	270	300	285	7,14
A.3	232	243	302	272,5	4,74
A.4	234	234	329	281,5	0
A.5	236	250	258	254	5,93
B.1	69	69	90	79,5	0
B.2	76	76	120	98	0
B.3	80	80	97	88,5	0
B.4	79	79	84	81,5	0
B.5	72	72	92	82	0
C.1	227	281	400	340,5	23,79
C.2	219	220	420	320	0,46
C.3	243	254	399	326,5	4,53
C.4	219	230	528	379	5,02
C.5	215	215	623	419	0
Total**					31

** = Total optimal results.

6 Conclusion and Future Work

In this document, we have worked on the basis of test instances Beasley [26]. On those instances, repeatedly it ran the BGHS metaheuristic, getting the results displayed in Sect. 5.

In all experiments, the BGHS was executed 30 times over each SCP instance. The results for the instances 4 (as it is shown in the graph of convergence Fig. 4), 5 and 6 are good, achieving a ratio 0 or close to zero. However for higher instances A, B and C, the results are not good, given that a comprehensive approach to global optimum is not achieved.

Discovered the above, continuously making use of the meta-heuristic parameters as a "key tool" for experiments. It was checked by manual adjustment of parameters and laboratory tests: one configuration does not cover all benchmark, so we worked on the setting of parameters for each instance.

It is intended to work on improving parameters for each instance or group of instances to improve the RPD. We will continue conducting experiments including improved parameters and include evolutionary architecture to maintain knowledge of diversity and fitness learned over each generation during the search process [34].

Likewise, a statistical evaluation of the results is proposed compared with other metaheuristics, accomplishing more in-depth understanding a simple ratio.

Acknowledgements. Broderick Crawford is supported by Grant CONICYT/ FONDECYT/REGULAR/1140897. Ricardo Soto is supported by Grant CONI-CYT/FONDECYT/REGULAR/1160455. Juan Salas, Sebastian Mansilla, Alvaro Gomez and Adrian Jaramillo are supported by Postgraduate Grant Universidad Catolica de Valparaiso 2016 (INF-PUCV 2016).

References

1. Crawford, B., Soto, R., Guzmán, N., Johnson, F., Paredes, F.: Recent harmony search algorithms for 0–1 optimization problems. In: Stephanidis, C., Tino, A. (eds.) HCII 2015 Posters. CCIS, vol. 528, pp. 567–572. Springer, Heidelberg (2015). doi:10.1007/978-3-319-21380-4_96

2. Karp, R.M.: Reducibility among combinatorial problems. In: 50 Years of Integer Programming 1958–2008 - From the Early Years to the State-of-the-Art, pp. 219–241 (2010)

3. Ali, A.I., Thiagarajan, H.: A network relaxation based enumeration algorithm for set partitioning. Eur. J. Oper. Res. **38**(1), 76–85 (1989)

4. Bartholdi, J.J.: A guaranteed-accuracy round-off algorithm for cyclic scheduling and set covering. Oper. Res. **29**(3), 501–510 (1981)

5. Walker, W.: Using the set-covering problem to assign fire companies to fire houses. Oper. Res. **22**, 275–277 (1974)

6. Vasko, F.J., Wilson, G.R.: Using a facility location algorithm to solve large set covering problems. Oper. Res. Lett. **3**(2), 85–90 (1984)

7. Vasko, F.J., Wolf, F.E., Stott, K.L.: Optimal selection of ingot sizes via set covering. Oper. Res. **35**, 346–353 (1987)

8. Vasko, F.J., Wolf, F.E., Stott, K.L.: A set covering approach to metallurgical grade assignment. Eur. J. Oper. Res. **38**(1), 27–34 (1989)
9. Vasko, F.J., Wolf, F.E., Stott, K.L., Scheirer, J.W.: Selecting optimal ingot sizes for bethlehem steel. Interfaces **19**(1), 68–84 (1989)
10. Balinski, M.L., Quandt, R.E.: On an integer program for a delivery problem. Oper. Res. **12**(2), 300–304 (1964)
11. Foster, B.A., Ryan, D.M.: An integer programming approach to the vehicle scheduling problem. Oper. Res. **27**, 367–384 (1976)
12. Fisher, M.L., Rosenwein, M.B.: An interactive optimization system for bulk-cargo ship scheduling. Nav. Res. Logist. **36**(1), 27–42 (1989)
13. Bellmore, M., Geenberg, H.J., Jarvis, J.J.: Multi-commodity disconnecting sets. Manage. Sci. **16**(6), B427–B433 (1970)
14. Bellmore, M., Ratliff, H.D.: Optimal defense of multi-commodity networks. Manage. Sci. **18**(4-part-I), B174–B185 (1971)
15. Freeman, B.A., Jucker, J.V.: The line balancing problem. J. Ind. Eng. **18**, 361–364 (1967)
16. Salveson, M.E.: The assembly line balancing problem. J. Ind. Eng. **6**, 18–25 (1955)
17. Ribeiro, C.C., Minoux, M., Penna, M.C.: An optimal column-generation-with-ranking algorithm for very large scale set partitioning problems in traffic assignment. Eur. J. Oper. Res. **41**(2), 232–239 (1989)
18. Ceria, S., Nobili, P., Sassano, A.: A lagrangian-based heuristic for large-scale set covering problems. Math. Program. **81**(2), 215–228 (1998)
19. Breuer, M.A.: Simplification of the covering problem with application to boolean expressions. J. Assoc. Comput. Mach. **17**, 166–181 (1970)
20. Christofides, N.: Zero-one programming using non-binary tree-search. Comput. J. **14**(4), 418–421 (1971)
21. Day, R.H.: Letter to the editor—on optimal extracting from a multiple file data storage system: an application of integer programming. Oper. Res. **13**(3), 482–494 (1965)
22. Garfinkel, R.S., Nemhauser, G.L.: Optimal political districting by implicit enumeration techniques. Manage. Sci. **16**(8), B495–B508 (1970)
23. Housos, E., Elmroth, T.: Automatic optimization of subproblems in scheduling airline crews. Interfaces **27**(5), 68–77 (1997)
24. Musliu, N.: Local search algorithm for unicost set covering problem. In: Ali, M., Dapoigny, R. (eds.) IEA/AIE 2006. LNCS (LNAI), vol. 4031, pp. 302–311. Springer, Heidelberg (2006)
25. Azimi, Z.N., Toth, P., Galli, L.: An electromagnetism metaheuristic for the unicost set covering problem. Eur. J. Oper. Res. **205**(2), 290–300 (2010)
26. Beasley, J.E.: OR-library: distributing test problems by electronic mail. J. Oper. Res. Soc. **41**(11), 1069–1072 (1990)
27. Geem, Z.W.: Music-Inspired Harmony Search Algorithm: Theory and Applications, 1st edn. Springer, Heidelberg (2009)
28. Geem, Z.W., Kim, J., Loganathan, G.V.: A new heuristic optimization algorithm: harmony search. Simulation **76**(2), 60–68 (2001)
29. Cormen, T.H., Leiserson, C.E., Rivest, R.L., Stein, C.: Introduction to Algorithms, 2nd edn. The MIT Press and McGraw-Hill Book Company, Cambridge (2001)
30. Omran, M.G.H., Mahdavi, M.: Global-best harmony search. Appl. Math. Comput. **198**(2), 643–656 (2008)
31. Xiang, W., An, M., Li, Y., He, R., Zhang, J.: An improved global-best harmony search algorithm for faster optimization. Expert Syst. Appl. **41**(13), 5788–5803 (2014)

32. Klenke, A.: Probability Theory: A Comprehensive Course. Springer, London (2008)
33. Srinivas, M., Patnaik, L.: Adaptive probabilities of crossover and mutation in genetic algorithms. IEEE Trans. Syst. Man Cybern. **24**, 656–667 (1994)
34. Crawford, B., Soto, R., Monfroy, E.: Cultural algorithms for the set covering problem. In: Tan, Y., Shi, Y., Mo, H. (eds.) ICSI 2013, Part II. LNCS, vol. 7929, pp. 27–34. Springer, Heidelberg (2013)

Differential Evolution for Multi-objective Robust Engineering Design

Andrew Linton and Babak Forouraghi[(⊠)]

Computer Science Department,
Saint Joseph's University, Philadelphia, PA, USA
{al573005,Babak.Forouraghi}@sju.edu

Abstract. There have been various algorithms based on Differential Evolution (DE) that have been demonstrated to optimize various engineering problems. While there has been work on using variations of DE for solving multi-objective problems, there has not been much work on applying this methodology to tolerance design. The ability to handle tolerances is extremely important in robust design, where a product of process is to be optimized while minimizing its responses to uncontrollable manufacturing and/or operational variations. This paper will demonstrate how Multi-Objective Differential Evolution can be applied to the area of multiple objective tolerance design. The algorithm's utility is highlighted with two canonical engineering design optimization problems.

Keywords: Multi-objective optimization · Pareto front · Differential Evolution · Tolerance design

1 Introduction

There are numerous engineering design problems that have multiple and often-conflicting objectives that need to be satisfied simultaneously [13]. Due to presence of multiple conflicting objectives, there is not just one ideal solution but rather a collection of solutions, which are considered optimal [3]. This set of optimal solutions is known as the Pareto front [12]. The discovered Pareto solutions are non-dominated with respect to each other, i.e., for each Pareto solution there is always some sacrifice in one objective for gain in another [10]. Without the addition of any further information, one solution cannot be declared superior to another, so this demands that an algorithm find as many Pareto-optimal solutions as possible [6].

Various types of Evolutionary Algorithms (EA) have been shown to be effective at finding these sets of solutions. EA algorithms typically take a random population of possible solutions and evolve them through an iterative process [5]. During each iteration (generation), the population of solutions are combined through various mutation and mating methods. This allows the algorithm to find multiple members of the Pareto front in a single run instead of having to run a series of separate runs [3]. These algorithms have two primary goals. The first goal is to discover solutions along the Pareto front or as close to it as possible. The other important goal is to find a diverse set of these Pareto optimal solutions [5].

H. Fujita et al. (Eds.): IEA/AIE 2016, LNAI 9799, pp. 931–943, 2016.
DOI: 10.1007/978-3-319-42007-3_79

In recent years, there have been various evolutionary algorithms introduced to solve Multi-Objective problems [5]. One such algorithm in particular, Differential Evolution (DE), is a simple but powerful algorithm that has been shown to quickly converge solutions to the Pareto front [14]. DE mutates its population based on the distribution of the current solutions in the population. This is unlike the other traditional EAs, which use probability density functions to adjust their parameters. Also, unlike genetic algorithms, it does not use binary encoding. Since its first introduction, numerous researchers have tried to adapt DE to Multi-Objective problems. Abbass did the first of these works when the Pareto Differential Algorithm (PDE) was presented [5]. Robič and Filipič introduced a new way to extend DE to MO problems, called the Differential Evolution for Multi-Objective Optimization (DEMO) [14]. DEMO maintains a single population, and it allows new solutions to immediately partake in the creation of new candidates. This allows DEMO to quickly converge to the Pareto front [22].

While there have been many algorithms designed to achieve these goals of finding a set of solutions, another important aspect of engineering design is tolerance design. The main goal of robust design is create stable products, which have minimized sensitivity to uncontrollable manufacturing errors [7, 8]. For instance, in the area of manufacturing, it is important to understand the dimension tolerances for manufactured parts since a higher precision typically leads to higher manufacturing costs [11, 15]. Traditionally, finding the objectives and tolerances are two separate processes, but by leveraging the power of DE, it can be shown that both the Pareto front and tolerance design can occur simultaneously. Many quality-engineering approaches have an issue that they cannot handle multiple conflicting objectives and constraints [7]. Therefore, evolutionary algorithms can improve two important aspects of quality control, tolerance design (determining tolerances around a nominal values of design variables) and parameter optimization (reducing the sensitivity of objectives to disturbing factors) [7].

This present work will discuss how a Multi-Objective version of Differential Evolution based on DEMO has been developed to concurrently handle multiple objective optimization and tolerance design. To approach the tolerance design, the algorithm represents each design variable as a tolerance range instead of a single value. Since each design variable will be a range, each solution will be a region in the design space instead of a single point. Therefore, the algorithm evolves these regions to reduce their sensitivity to uncontrollable variances. Further, the algorithm uses the average from of full-factorial experiments [7, 8] for Pareto ranking in the selection process.

The remainder of this paper is organized as follows. Section 2 will provide background information on Differential Evolution. Section 3 will discuss how the newly proposed differential evolution algorithm, called HDE, can be used for multi-objective robust engineering design. In Sects. 4 and 5, the design of a Welded Beam and the I-beam problems will be presented to demonstrate the HDE algorithm's performance. Finally, Sect. 6 will present the summary and conclusions.

2 Differential Evolution

Differential Evolution (DE) is a simple and powerful evolutionary algorithm (EA) for optimizing problems over continuous domains [2, 3, 14]. DE was developed to solve real-valued numerical optimization problems [4]. It was first developed by Price and Storn in the mid 1990's and has been shown to be successful in many optimization problems [3]. While similar to traditional EAs, there are multiple differences between the two algorithms. DE does not use binary encoding for its design variables but instead uses real values. Also, DE performs mutations based on the distribution of the current generation, unlike EAs, which use a probability density function [3]. DE uses three parents for the crossover operation instead of just two parents. The final difference is that a candidate solution replaces its parent only if it has a better fitness. There are several variations of DE, the most popular of which is referred to as "DE/rand/1/bin" [3]. Where "DE" is Differential Evolution, "rand" means the parents are randomly chosen, "1" represents the number of pair picked and finally "bin" stands for binomial recombination is used.

The pseudo-code for the DE/rand/1/bin algorithm is shown in Fig. 1. Each P_i is a vector of D design variables. *MAX_GEN* is the total number of generations to run the algorithm. There are three algorithmic control variables: *popSize*, *CR*, and *P*. The parameter *popSize* is the number of solutions in the population. The parameter, $CR \in [0, 1]$, controls the crossover frequency; the higher its value, the less influence its parent has on the new solution. $F \in [0, 2]$ is the scaling factor that is used to control the amount of change that the difference of the two parents $P_{2,j,G}$, $P_{3,j,G}$ contribute. The random value x (see Fig. 1) ensures that at least one decision variable is passed on from the P_i solution. During the crossover process a new solution may fall within an invalid range. There are various studies conducted on different methods to repair invalid solutions. These repair strategies include re-initialization, reflection, projection,

```
1     Create the initial random population P = {P₁, ... , P_popSize}
2     For G = 1 to MAX_GEN Do
3        For I = 1 to popSize Do
4           Select randomly 3 parents such that P₁ ≠ P₂ ≠ P₃ ≠ Pᵢ
5           x = randInt(1,D)
6           For j=1 to D Do
7              If rand[0,1) < CR or j = x Then
8                 U_{i,j,G+1} = P_{1,j,G} + F(P_{2,j,G} − P_{3,j,G})
9              Else
10                U_{i,j,G+1} = P_{i,j,G}
11             End For
12          If f(U_{i,G+1}) ≤ f(P_{i,G}) Then
13             P_{i,G+1} = U_{i,G+1}
14          Else
15             P_{i,G+1} = P_{i,G}
16          End For
17    End For
```

Fig. 1. The DE/rand/1/bin algorithm

wrapping, and resampling [2]. Once the new solution $U_{i,G+1}$ is created, it is compared with the original parent $P_{i,G}$. If the new solution is better than the original, then it replaces $P_{i,G}$ in the next generation.

3   Hyper Differential Evolution

To apply DE to multiple-objective problems, there have been several variations created. The first was the Pareto Differential Evolution (PDE) algorithm developed by Abbass [1]. The multiple-objective DE algorithm, which the proposed Hyper Differential Evolution (HDE) is based on, is called Differential Evolution Multiple Objective (DEMO) algorithm and it is used for applying DE to MO problems [14]. The DEMO algorithm has a different approach to the selection of solutions for the next generation. The selection of the new solution vs. the original parent is based on dominance. If the new solution dominates the parent then it replaces it, and if the parent dominates the new solution, then the parent remains. However, if neither solution dominates the other, then the new solution is added to the population. Therefore, during the creation of the next generation, the population can temporarily grow up to twice the original population size. To truncate the population back down to the correct size, the algorithm ranks the new population and then keeps the top solutions.

The HDE algorithm developed in this work utilizes the concept of hyper solutions for tolerance design. Instead of using single points to represent decision and objective variables, tolerances are used instead. Thus, a solution with three decision variables x_1, x_2, and x_3 would be represented as $[(x_1^{lo}, x_1^{hi}), (x_2^{lo}, x_2^{hi}), (x_3^{lo}, x_3^{hi})]^T$. The fitness evolution of the solutions is determined by full factorial experiments. The selection of the fittest solutions is done via the Pareto Ranking Method based on each solution's objectives' average values and tolerances [9].

The outline of the HDE algorithm is shown in Fig. 2, where the population P is an array of *popSize* solutions. Each solution is a vector of D decision intervals, where for a solution P_i, values $P_{i,x}^{lo}$ and $P_{i,x}^{hi}$, represent the lower and upper bounds for the random decision variable x, respectively.

The fitness for each solution is determined by using the quality-control notion of factorial experiments. Since each decision variable is actually a range of values, different combinations of values are evaluated [13]. A variable called the experiment level determines the number of equally distant values between the lower and upper bounds of each design variable. For example if there are three design variables, each having three levels (Low, Mid, Upper), then there will be 3^3 combinations of design values to be evaluated. This method creates a grid of test points within the hyper area of each solution. The objective values gained from these experiments are then averaged and the variance is determined. Both of these values are used for determining the dominance of the solution. A solution is dominant only if both its objective values and tolerances are less than another solution. A solution is only considered valid if a set percentage of experiments are valid. When a solution is considered invalid then it is removed, and the original parent is kept for the next generation.

1	Create the initial random population $P = \{P_i, \ldots, P_{popSize}\}$
2	**For** G = 1 to MAX_GEN **Do**
3	**For** I = 1 to popSize **Do**
4	Select randomly 3 parents such that $P_1 \neq P_2 \neq P_3 \neq P_i$
5	x = randInt(1,D)
6	**For** j=1 to D **Do**
7	**If** rand[0,1) < CR **or** j = x **Then**
8	$U_{i,j}^{Lower}{}_{,G+1} = P_{1,j}^{Lower}{}_{,G} + F(P_{2,j}^{Lower}{}_{,G} - P_{3,j}^{Lower}{}_{,G})$
9	**Else**
10	$U_{i,j}^{Lower}{}_{,G+1} = P_{i,j}^{Lower}{}_{,G}$
11	**If** rand[0,1) < CR **or** j = x **Then**
12	$U_{i,j}^{Upper}{}_{,G+1} = P_{1,j}^{Upper}{}_{,G} + F(P_{2,j}^{Upper}{}_{,G} - P_{3,j}^{Upper}{}_{,G})$
13	**Else**
14	$U_{i,j}^{Upper}{}_{,G+1} = P_{i,j}^{Upper}{}_{,G}$
15	**End For**
16	Calculate Fitness for $U_{i,G+1}$
17	**If** $U_{i,G+1}$ is valid
18	**If** $U_{i,G+1}$ dominates $P_{i,G}$ **Then**
19	$P_{i,G+1} = U_{i,G+1}$
20	**Else If** $P_{i,G}$ dominates $U_{i,G+1}$ **Then**
21	$P_{i,G+1} = P_{i,G}$
22	**Else**
23	$P_{popSize+1,G+1} = U_{i,G+1}$
24	**Else**
25	$P_{i,G+1} = P_{i,G}$
26	**End For**
28	Truncate P to popSize individuals
29	**End For**

Fig. 2. Outline of the proposed Hyper DE algorithm

4 Welded Beam Problem

This engineering problem's goal is to design a beam that is welded to a support structure [5]. This beam must be able to support a downward force (P) of 6,000 lbs (Fig. 3).

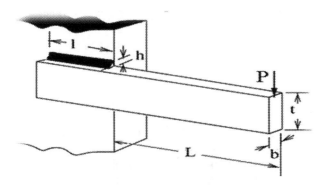

Fig. 3. Welded beam problem

The welded beam design problem has two objectives, minimize the fabrication costs and minimize the end deflection of the welded beam. The problem has 4 design parameters, $X = [h, l, t, b]$, where the design variable b is the thickness of the beam, and t is the width of the beam. The length and thickness of the weld are defined by l and h, respectively. The problem requires identifying $X*$ which minimizes $F(X*) = [f_1(X*), f_2(X*)]^T$ where:

$$f_1(X) = 1.10471h^2\ell + 0.04811tb(14.0 + \ell) \tag{1}$$

$$f_2(X) = \frac{2.1952}{t^3b} \tag{2}$$

Subject to the following constraints:

$$g_1(X) = 13,600 - \tau(x) \geq 0 \tag{3}$$

$$g_2(X) = 30,000 - \sigma(x) \geq 0 \tag{4}$$

$$g_3(X) = b - h \geq 0 \tag{5}$$

$$g_4(X) = P_c(X) - 6,000 \geq 0 \tag{6}$$

Where,

$$\tau(X) = \sqrt{(\tau')^2 + (\tau'')^2 + (\ell\tau'\tau'')/\sqrt{0.25(\ell^2 + (h+t)^2)}} \tag{7}$$

$$\tau' = \frac{6,000}{\sqrt{2h\ell}} \tag{8}$$

$$\tau'' = \frac{6,000(14 + 0.5\ell)\sqrt{0.25(\ell^2 + (h+t)^2)}}{2(0.707h\ell(\ell^2/12 + 0.25(h+t)^2))} \tag{9}$$

$$\sigma(X) = \frac{504,000}{t^3b} \tag{10}$$

$$P_c(X) = 64,746.022(1 - 0.0282346t)tb^3 \tag{11}$$

With the design boundaries of $0.125 \leq h, b \leq 5.0$ and $0.1 \leq \ell, t \leq 10.0$.

The cost and deflection objectives are conflicting goals. To minimize the cost of the beam, the smallest size is needed; however, to minimize the deflection, a larger beam is required. To optimize this problem, HDE was run with a population of 250 solutions over 30 generations with $CR = 0.6$ and $F = 0.25$. The factorial experiment level is set to four levels (256 experiments per a single solution). The results shown in Figs. 4 and 5 depict how the population converges along the Pareto front. The black points represent solution averages while the gray points are the partial experiment values.

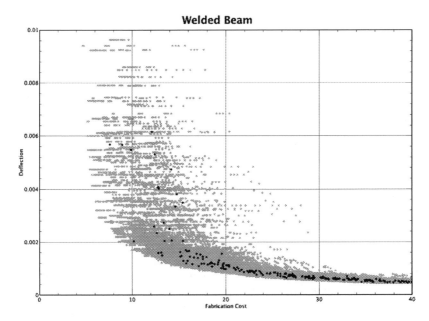

Fig. 4. Generation 15 of the welded beam design problem

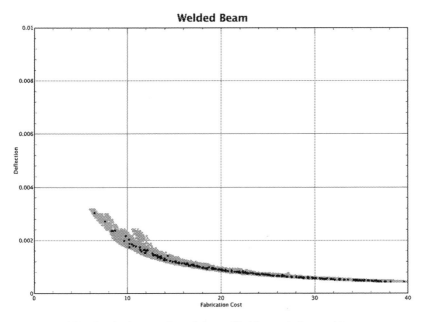

Fig. 5. Final generation of the welded beam design problem

The results clearly indicate that the algorithm correctly converges to the Pareto front with a good spread of solutions. Table 1 shows how HDE compares to similar algorithms producing solutions for the Welded Beam Problem.

Table 1. Comparison of solutions to the welded beam design problem

Method	(h, l, t, b)	$F^*(X^*) = (f_1(X), f_2(X))$
PSO [12]	(0.2, 6.52, 0.25,10)	(2.79, 0.009)
MOGA [6]	(0.63, 1.64, 9.99, 0.66)	(5.70, 0.0033)
HDE	([0.36, 0.43], [0.57, 0.64], [3.39,3.77], [10,10])	(5.683 ± 0.231, 0.00366 ± 0.00015)

As shown in Table 1, HDE is capable of identifying tolerances for design parameters, for which, the resulting response surfaces exhibit small operational variances. This is of prime importance in the field of robust engineering design where large functional variations in design of products and processes can potentially lead into loss of quality and variable resources.

5   Design of an I-Beam

This section presents the problem of multi-objective design of an I-beam which has previously been approached using classical vector optimization techniques [12, 13] (Fig. 6).

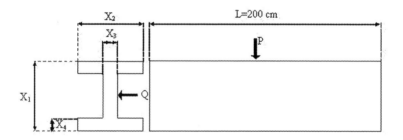

Fig. 6. The frontal and side views of an I-beam

Assuming that the I-beam is subject to maximal bending forces of $P = 600$ kN and $Q = 50$ kN at the midspan, the objective of the design is to find the optimum dimensions of the beam ($X^* = [x_1, x_2, x_3, x_4]^T$) such that the cross section area (f_1 in cm^2) and static deflection of the beam (f_2 in cm) are both minimized subject to the constraint that the beam's bending stress (f_3) does not exceed 16 kN/cm^2.

The optimization of the beam can mathematically be stated as follows. Find X^* which minimizes $F(X) = [f_1(X), f_2(X)]^T$ where:

$$f_1(X) = 2x_2x_4 + x_3(x_1 - 2x_4) \tag{12}$$

$$f_2(X) = \frac{60,000}{x_3(x_1 - 2x_4)^3 + 2x_2x_4\left[4x_4^2 + 3x_1(x_1 - 2x_4)\right]} \tag{13}$$

Subject to the bending stress constraint:

$$f_3(X) = \frac{180,000x_1}{x_3(x_1 - 2x_4)^3 + 2x_2x_4\left[4x_4^2 + 3x_1(x_1 - 2x_4)\right]}$$
$$+ \frac{15,000x_2}{(x_1 - 2x_4)x_3^3 + 2x_4x_2^3} \leq 16 \tag{14}$$

and the geometric side constraints: $10 \leq x_1 \leq 80$, $10 \leq x_2 \leq 50$, and $0.9 \leq x_3$, $x_4 \leq 5.0$. Given the boundaries of the feasible design region, the computed ranges of responses for the two objective functions reveal that f_1 is in conflict with f_2 and that the ideal solution $f^{id} = [25.38, 0.0059]^T$, where the two objectives are simultaneously minimized, can never be attained.

To account for any statistical fluctuations that could potentially produce misleading results the proposed HDE algorithm was ran over three statistically independent runs, each time with a population of 150 solutions over eighteen generations. The CR value was set to 0.45 and the F parameter was fixed to 0.6 for each of the runs. The experiment level was set to five levels; therefore, a total of 625 experiments were performed to determine the fitness of each solution. The progress of the HDE algorithm through eighteen generations for the I-Beam problem is shown in Fig. 8 through Fig. 10.

The black points in the graphs are the average of the 625 experiments for a single solution. The grey points are each of 625 experiments for each solution. In Fig. 7, the randomly created first generation is shown. It can be seen in this figure how the random population's design variables ranges are large. After ten generations (Fig. 8), the tolerances are greatly decreased and the averages are starting to align along the Pareto front. In generation 18 (Figs. 9 and 10), the Pareto front can be seen. These graphs show how DE and the tolerance ranges quickly converge along the Pareto front.

Table 2 shows results from the Hyperparticle swarm optimization (HPSO) algorithm which also incorporates tolerance design similarly to this proposed method. The HDE results shown in the table demonstrates the spread of solutions that the algorithm generates. This proposed algorithm share similar results as those listed here.

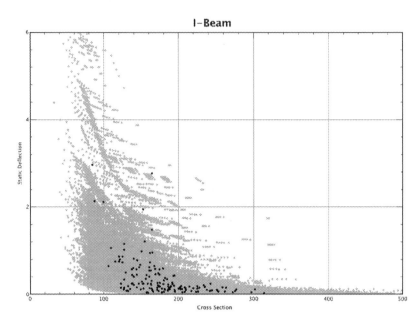

Fig. 7. Initial Population (Black = Solution averages; Gray = Solution experiments)

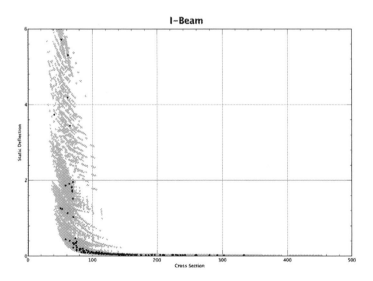

Fig. 8. Generation 10 (Black = Solution averages; Gray = Solution experiments)

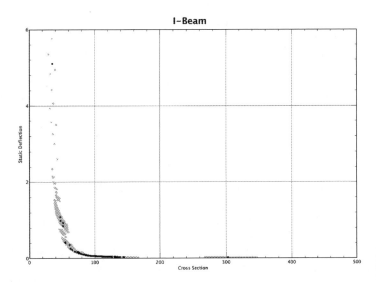

Fig. 9. Generation 18 (Black = Solution averages; Gray = Solution experiments)

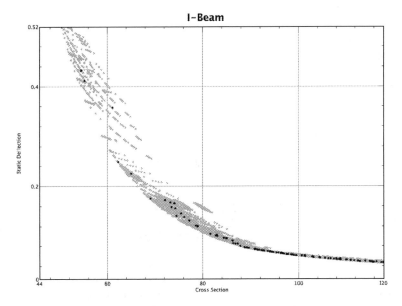

Fig. 10. Generation 18 - the final pareto front

Table 2. Comparing optimal solutions from two MO algorithms

Method	$X^* = (x_1, x_2, x_3, x_4)$	$F^*(X^*) = (f_1(X^*), f_2(X^*))$
HPSO [13]	([79.17, 80.00], [45.30, 50.00], [0.900, 0.903], [0.90 1.64])	(176.31 ± 0.265, 0.025 ± 4.020E-5)
HDE	([79.91, 80.00], [10.93, 15.88], [0.900, 0.901], [1.30, 2.04])	(111.72 ± 8.37, 0.051 ± 0.006)
	([80.00, 80.00], [13.35, 13.51], [1.37, 1.96], [2.90,3.09])	(200.64 ± 15.20, 0.029 ± 0.001)

6 Conclusions

This work demonstrates how Multi-Objective Differential Evolution can be applied to robust engineering design. By altering the evolutionary algorithm to use tolerances for each design variable, the modified algorithm not only discovers valid solutions along the Pareto front, but it also finds the design tolerances around those solutions. The proposed algorithm determines the average objective values and tolerances by using full-factorial experiments. Dominance is determined by the average value of each solution and its tolerances. Using these values, it performs the selection of the next generation using Pareto ranking and truncates the population to the specified size. The two design examples presented in this paper demonstrate that the proposed methodology can not only find the Pareto front quickly but it can concurrently perform tolerance design.

References

1. Abbass, H.A.: The self-adaptive Pareto differential evolution algorithm. In: Proceedings of the 2002 Congress on Evolutionary Computation, vol. 1, pp. 831–836 (2002)
2. Arabas, J., Szczepankiewicz, A., Wroniak, T.: Experimental comparison of methods to handle boundary constraints in differential evolution. In: Schaefer, R., Cotta, C., Kołodziej, J., Rudolph, G. (eds.) PPSN XI. LNCS, vol. 6239, pp. 411–420. Springer, Heidelberg (2010)
3. Coello Coello, C.A.: A short tutorial on evolutionary multiobjective optimization. In: Zitzler, E., Deb, K., Thiele, L., Coello Coello, C.A., Corne, D.W. (eds.) EMO 2001. LNCS, vol. 1993, pp. 21–40. Springer, Heidelberg (2001)
4. Coello Coello, C., Lamont, G.B., Van Veldhuizen, D.A.: Evolutionary Algorithms for Solving Multi-objective Problems, 2nd edn. Springer, New York (2007)
5. Deb, K.: Multi-objective Optimization Using Evolutionary Algorithms. Wiley, Chichester (2001)
6. Deb, K., Pratap, A., Agarwal, S., Meyarivan, T.A.M.T.: A fast and elitist multi-objective genetic algorithm: NSGA-II. IEEE Trans. Evol. Comput. 6(2), 182–197 (2002)
7. Forouraghi, B.: A genetic algorithm for multiobjective robust design. Appl. Intell. 12(3), 151–161 (2000)
8. Forouraghi, B.: Optimal tolerance allocation using a multiobjective particle swarm optimizer. Int. J. Adv. Manuf. Technol. 44(7–8), 710–724 (2009)

9. Gen, M., Cheng, R.: Genetic Algorithms and Engineering Optimization, vol. 7. Wiley, New York (2000)
10. Konak, A., Coit, D.W., Smith, A.E.: Multi-objective optimization using genetic algorithms: a tutorial. Reliab. Eng. Syst. Saf. **91**(9), 992–1007 (2006)
11. Kumar, M.S., Kannan, S.M.: Optimum manufacturing tolerance to selective assembly technique for different assembly specifications by using genetic algorithm. Int. J. Adv. Manuf. Technol. **32**(5–6), 591–598 (2007)
12. Ma, L., Forouraghi, B.: A modified particle swarm optimizer for engineering design. In: Jiang, H., Ding, W., Ali, M., Wu, X. (eds.) IEA/AIE 2012. LNCS, vol. 7345, pp. 187–196. Springer, Heidelberg (2012)
13. Ochlak, E., Forouraghi, B.: A particle swarm algorithm for multiobjective design optimization. In: The 18th IEEE International Conference on Tools with Artificial Intelligence, pp. 765–772 (2006)
14. Robič, T., Filipič, B.: DEMO: differential evolution for multiobjective optimization. In: Coello Coello, C.A., Hernández Aguirre, A., Zitzler, E. (eds.) EMO 2005. LNCS, vol. 3410, pp. 520–533. Springer, Heidelberg (2005)
15. Sivakumar, K., Balamurugan, C., Ramabalan, S.: Concurrent multi-objective tolerance allocation of mechanical assemblies considering alternative manufacturing process selection. Int. J. Adv. Manuf. Technol. **53**(5–8), 711–732 (2011)

Multiple Objectives Reconfiguration in Distribution System Using Non-Dominated Sorting Charged System Search

Cheng-Chieh Chu[(⊠)] and Men-Shen Tsai

Graduate Institute of Automation Technology,
National Taipei University of Technology, Taipei, Taiwan
Java.Sunday@gmail.com, mstsai@mail.ntut.edu.tw

Abstract. Distribution system reconfiguration is achieved by changing the statuses of the switches. A number of targets of distribution system operation can be achieved after feeder reconfiguration operation. In general, while constructing the feeder reconfiguration, some factors need be considered. For example, primary feeder losses minimization, the number of switch actions reduction and voltage profile. Weighted sum method is used when multiple objective problems have to be solved. However, through the use of the weighted sum method, only one solution can be found. This is not preferred by distribution systems operators. So as to provide multiple compromise solutions, the multi-objective approach is one of the methods. In order to provide operators with different compromise solutions, A Non-Dominated Sorting Charged System Search (NDSCSS) is proposed to solve the multi-objective problems of distribution systems. Because the values of different factors are made using diverse topologies, these topologies can find different solutions. In order to generate a legal topology, the Zone Real Number Strings (ZRNS) encoding/decoding scheme is used. The 33-bus is implemented. The performance of Non-Dominated Sorting Evolutionary Programming (NSEP), Multi-Objective Particle Swarm Optimization (MOPSO) and Non-Dominated Sorting Charged System search (NDSCSS) are compared. The results indicate that NDSCSS can search for the best solutions among the three considered algorithms for distribution system reconfiguration problems.

Keywords: Non-dominated · Multi-objectives · Charged system search · Distribution system

1 Introduction

Distribution systems is comprised of feeders, switches and zones. In order to find a valid distribution systems topology, the statuses of switches in a distribution systems are crucial. The topology of a distribution systems may be changed by altering the statuses of the switches. Primary feeder losses is kept as low as possible during normal operations. When an emergency happens, the outage zone can be kept to a minimum. In reality, both the line currents and the feeder voltage profile within the allowed range are considered. In addition, the topology of distribution systems remains in a radial

© Springer International Publishing Switzerland 2016
H. Fujita et al. (Eds.): IEA/AIE 2016, LNAI 9799, pp. 944–955, 2016.
DOI: 10.1007/978-3-319-42007-3_80

structure. Therefore, the distribution system reconfiguration problem is defined as a constraint combinatorial optimization problem. Traditional optimization methods are used to solve most problems. The methods cannot be used to solve distribution system reconfiguration related problems effectively because of the distribution system reconfiguration properties of the distribution system.

Many researchers use different soft computation methods to solve the distribution system reconfiguration problems. The general formulation and solution method is proposed for the primary feeder loss reduction problem [2]. In order to find the topology and minimize the primary feeder loss, the following traditional methods are used. [5, 9, 10, 25]. A number of soft computing methods such as Artificial neural network (ANN) [19], simulated annealing (SA), genetic algorithm (GA) [23] and evolutionary programming (EP) [12, 13], have been used to handle the distribution system reconfiguration problem. Recently, some swarm intelligence algorithms, such as Ant Colony Optimization [20, 25, 26], Particle Swarm Optimization [7, 17, 18] and Artificial Bee Colony algorithm [3] have been applied to solve the distribution system reconfiguration problems. The new solutions, which are produced through the operation of generation of algorithms, may be invalid. Thus, the validation of all the solutions will take time. When validation processes are integrated with these algorithms, the algorithms' performance will be affected. A proper encoding/decoding scheme which is to reduce the validation time is needed. The application of these algorithms is vital to solving the distribution system reconfiguration problems and to decreasing the validation time between solutions space and encoding space.

In the past, a number of different encoding/decoding schemes had been developed. For instance, the binary and the integer encoding schemes are commonly used to represent a solution [8, 30]. Lately, various intelligence algorithms have been developed based on natural environmental concepts to solve problems with continuous variables such as GA, PSO and ABC. Since the distribution system reconfiguration problem is a discrete combination problem, these algorithms cannot be used directly without modifications. Hence, a proper encoding/decoding scheme is developed for decreasing the optimal solution time. For identifying the solution and reducing the validation time, ZRNS representation of the solution for the distribution systems is proposed to improve the search performance. A novel encoding/decoding scheme can produce a legal distribution systems topology from a solution in encoding space.

On the other hand, the common objectives of distribution systems problems include: primary feeder loss minimization and the number of switch actions reduction [24, 27]. These objectives are evaluated individually. In reality, the distribution systems operators simultaneously considers some objectives. Under normal operations, these objectives may conflict with each other. In the past, objectives were integrated using the weighted sum method to find a solution for multiple objective problems [24, 25]. In these papers, the single objective optimization method was applied to obtain the optimal solution. However, there is only one solution identified. In order to provide a suitable decision strategy for system operators, multiple solutions may be preferred. Distribution system reconfiguration solutions are treated as the "Pareto optimal solution set" in this paper. In 1971, the concept of "Pareto optimization solution set" is different from the weighted-sum approach. The solutions in Schwier are compromised solutions, which are special cases in the "Pareto optimization solution set" found through the weighted-sum method. Several algorithms have been proposed [6, 22, 28] to identify the solution set. For solving multiple objective problems, Non-dominated Sorting Genetic Algorithm-II (NSGA-II) [6] is one of the most widely used methods among these algorithms. In [1, 10], NSGA-II has been applied to power system problems. The solution through use of reproduction procedures of NSGA-II is not suitable for the distribution system reconfiguration problems.

In order to improve the problems of NSGA-II, CSS is considered for handling the multiple objective problems of distribution system reconfiguration. CSS is an optimization algorithm based on rules including the governing law of motion from Newtonian mechanics and the Coulomb and Gauss laws from electrical physics [21]. The Non-Dominated Sorting Charged System Search (NDSCSS) is proposed to handle distribution system reconfiguration problems. A 1-feeder, 33-zone and 37-switch distribution systems which is used to verify the usefulness and effectiveness of the NDSCSS. Primary feeder loss minimization and the number of switch actions reduction are used for verifying efficiency of NSEP [13], MOPSO [14] and NDSCSS. The numeric results are discussed in different scenarios.

2   Problem Formulation

By changing the statuses of switches, distribution system reconfiguration is accomplished. These switches can be categorized as "Sectionalized Switches" and "Tie Switches". The different combinations represent different statuses of the switches. Every combination represents a different topology. Different topologies are different solutions to the distribution systems problem. During algorithm operation, topologies are produced by using encoding/decoding schemes in advance, the values of factors will be decided. In order to find an effective solution and increase algorithm efficiency, an apt encoding/decoding scheme is needed. In the past, some researchers needed to consider system constraints and algorithms characteristics while designing different encoding/decoding schemes. For instance, binary string representation of the statuses of the switches [29], binary string representation of the identities of the tie switch [30], integer string representation of the identities of the tie switch [29] and integer string representation of the identities of the source zone [8] were proposed in the past. The mentioned representation has been integrated with various algorithms in order to handle the distribution system reconfiguration problem; however the validation of the solutions is needed due to distribution systems properties. In 2002, the topology validation can be saved by using the Irving's approach [16]. This paper proposes the Zone Real Number String encoding/decoding (ZRNS) scheme. The topology of distribution systems can be represented through the use of the ZRNS. The valid topology can then be found after the decoding procedure.

2.1   Zone Real Number Strings Encoding/Decoding Scheme

Since the distribution system reconfiguration is accomplished by changing the statuses of the switches, some algorithms have applied different encoding/decoding schemes to create valid or non-valid solutions. Some algorithms based on natural inspirations were developed with continuous variables to deal with different problems. The forms of solutions are discrete because of the distribution systems characteristics. In order to use the discrete variable in these algorithms, the modification of the algorithm and valid solutions are needed. For the distribution system reconfiguration problem, an integer number string representation have been discussed [16]. The proposed encoding/decoding scheme cannot only improve the search efficiency, but also generate valid solutions for the distribution system reconfiguration problems. In 2013, the switch real number strings were used to solve the distribution systems problems, and the results were better than with other algorithms [4]. Nevertheless, the proposed method has some drawbacks. For example, the switch real number strings need to be sorted in advance, and the tie switches are always located at the end positions of the switch real number strings. These parts will increase execution time of algorithms and decrease search efficiencies. This paper represents different topologies through the use of a ZRNS encoding/decoding scheme to avoid the drawbacks of the switch real number strings decoding method. The best solution can be identified without modifying reproduction operations of the algorithms.

The topology representation is made by reproduction operation during CSS operation. The reproduction will produce a number of ZRNSs. Different ZRNSs represent the different topologies of distribution systems. In order to find a valid topology, a robust decoding method is developed so that the topology can be used directly without any validation. This paper provide a decoding method for finding the valid topology. The decoding process is explained as follows:

- Initialization:

  1. All switches statuses are "Open".
  2. Each zone has a real number obtained from the real number string which is between [0, 1] in turn.
  3. Each feeder has a real number from 1 to the number of distribution systems feeders.
  4. To determine the operator sequence without any sequence from the ZRNS.

- Decoding process:

  1. Replace the index of the switch adjacent to a feeder with the feeder index.
  2. The operator zone (OZ), which is obtained from a real number string, is decided through the operator's sequence.
  3. Determine the index of adjacent OZs individually based on the sequences determined in step 1.
  4. MZ represents the neighboring zone of OZ with the largest index. MZI and OZI represent the index of MZ and index of OZ. The rules that determine indices of the zones and the statuses of switches between two adjacent zones are associated with each switch.
  5. Find the largest index from all the neighboring zones of OZs. The OZI is replaced with MZI only if the MZI is larger than OZI. The status of the switch which is connected with the OZ and MZ is changed to 'CLOSED'.
  6. If the OZI is larger than MZI, then the MZI is replaced by the OZI. The switch status between the MZ and the OZ is changed to 'CLOSED'.

7. No operation is needed when the OZI is equal to the index of a feeder.
8. For every zone in the system, if the index of zones is equal to OZI, the index of the zone is replaced by the MZI.
9. The tie switches are determined when the decoding scheme operation is finished.

For multiple objectives distribution system reconfiguration problems, a combination of switch statuses is a solution. Each solution indicates a topology of the distribution systems. The value of different objectives is calculated in the solution set. This paper considers two objectives. The primary feeder loss minimization is one of the objectives. The 3-phase power flow calculation with the constant PQ load model is applied. In order to find the primary feeder loss, the following equation is used:

$$y_1 = P_{losses} = P_{real\ power\ delivery} - \sum\nolimits_{x=1}^{n} PL_a \qquad (1)$$

P_{losses} and $P_{real\ power\ delivery}$ are the primary feeder losses and the real power delivery from the substation; $\sum_{a=1}^{n} PL_a$ represents the total real power demand of the distribution systems. a is the total number of zones of the distribution systems. The second objective is the number of switch actions reduction. The number of switch actions is obtained by using Eq. (2).

$$y_2 = \begin{cases} 1, & if\ S_h(t^+) \neq S_h(t^-) \\ 0, & if\ S_h(t^+) = S_h(t^-) \end{cases}. \qquad (2)$$

In this paper, three considerations are as follows:

- The conductors' current cannot exceed their thermal limits.
- The feeders' voltage profile cannot surpass 5 %.
- The system must be kept in a radial structure.

3   Proposed Optimization Method

This paper is to handle the distribution system reconfiguration problem's multiple objectives. When the topology is changed, the values of factors, such as primary feeder loss, voltage profile, and the number of switch actions will be different. In order to find the proper topology of distribution systems, many different soft computing methods are used for identifying the optimal solutions. In this paper, the NDSCSS is proposed to guide the search process. The Non-Dominated Sorting, Crowding Distance and CSS are explained as follows:

3.1   Non-Dominated Sorting and Crowding Distance

Generally, the operator will change the topology of distribution systems in order to improve the efficiency of distribution systems. During normal operations, the operator needs to consider the primary feeder loss, voltage profile, and the number of switch actions reduction. These factors can be in conflicted. For example, when the primary feeder loss needs to be minimized, the number of switch actions may need to be increased. Hence, the distribution system reconfiguration problem is a multiple objective problem. In the past, a number of algorithms were designed to solve the single-objective problem. The CSS is one of these algorithms. In order to solve the distribution systems' multi-objective problem, the CSS needs to be modified. Because the particles of CSS represent different solutions, the particles will be initialized in advance and sorted based on non-domination of each front. The first front being a non-dominant set in the present particles, the individual particles in the front only dominate a second front and the front continues. Each particle in each front is assigned rank values or values based on the front where they belong to. Each particle in the front is set a fitness value which is 1. The individuals in the second front are set fitness values of 2 and so on.

A new parameter, which is called "Crowding Distance", is calculated through the use of fitness values for each particle. The concept of crowding distance is to find the distance between all particles in a front based on their M objectives in the M dimensional hyperspace. The particles in the boundary are selected because they have infinite distance assignments. The goal of crowding distance is to measure the distance of an in particle to its neighbors. The results will be more diverse when the crowding distance is large for all particles. Based on the rank and crowding distance, the best particles are selected from all particles through the use of the binary tournament selection process. If crowding distance is greater than the others or the rank is less than the others, a particle is selected. The new position and the new velocity of each particle will be generated from the proposed algorithms operation. This will be discussed in detail in a later section. Based on non-domination, all the particles are sorted again.

3.2 Charged System Search

CSS is an efficient search algorithm based on the movement of charged particles (CPs) in n-dimensional space. The standard CSS algorithm process is discussed as follows:

Step 1: Initialization: Different random number is assigned to each CP's position. The fitness of each CP is calculated. Equation (3) represents the magnitude of each CP:

$$q_i = \frac{fit(i) - fit(i)_{worst}}{fit(i)_{best} - fit(i)_{worst}}, \quad i = 1, 2, 3. \ldots x. \tag{3}$$

$fit(i)_{best}$ and $fit(i)_{worst}$ are the best fitness and the worst fitness of particles. $fit(i)$ is the fitness of the i-th CP. x is the total number of CPs. The distance r_{ij} between two CPs is calculated by using Eq. (4).

$$r_{ij} = \frac{\|x_i - x_j\|}{\left\|\frac{x_i - x_j}{2} - x_{best}\right\| + \varepsilon}. \tag{4}$$

X_i and X_j represent the positions of the i-th and the j-th CPs. X_{best} is the position of the best CP. ε is a small real number.

Step 2: Charged Memory Creation: A part of the CPs have better fitness which are filled in the CM.

Step 3: Moving Probability Calculation: For each CP, Eq. (5) is used to calculate the moving probability.

$$P_{ij} = \begin{cases} 1, \dfrac{fit(i) - fit(i)_{best}}{fit(j) - fit(i)} > r \text{ and } \vee fit(j) > fit(i) \\ \\ 0, otherwise \end{cases} \tag{5}$$

Step 4: Resultant Force Determination: The resultant force of the i-th CP is calculated.

$$F_{ij} = q_i \sum_{i, i \neq j} \left(\frac{q_i}{a^3} r_{ij} . i_1 + \frac{q_i}{r_{ij}^2} . i_2 \right) P_{ij}, \ i = 1, 2, , \ n, \tag{6}$$

$$i_1 = 1, \ i_2 = 0 \text{ when } r_{ij} < a, \ i_1 = 0, \ i_2 \text{ when } r_{ij} \geq a. \tag{7}$$

Equation (8) is used to calculate the value of a.

$$a = 0.10 \times \max \left(\{ X_{i,max} - X_{i,min} | i =, \ 2, \ldots, \ n \} \right) \tag{8}$$

Step 5: New Position $(X_{i,new})$ and Velocity $(V_{i,new})$ Determination of Each CP:

$$X_{i,new} = r_1 \cdot k_a \cdot \frac{F_i}{q_i} \cdot \Delta t^2 + r_2 \cdot k_v \cdot V_{i,old} + X_{i,old}, \tag{9}$$

$$V_{i,new} = \frac{X_{i,new} - X_{i,old}}{\Delta t}. \tag{10}$$

r_1, r_2 are two different random numbers between [0, 1]. k_a and k_v represent the acceleration and velocity coefficients. The value of Δt is 1.

Step 6: Position Evaluation and Correction of Each CP: If the new location of a CP surpasses the predefined boundaries, the new CP will be replaced through the harmony search method.

Step 7: CM updating: The best CP substitutes for the worst CP in the CM.

Step 8: Terminating Condition: Repeat steps 3–7 until the pre-determined iteration is accomplish. In this paper, when the number of iterations is the same as the pre-determined number of iterations, the algorithms is finished.

3.3   Non-Dominated Sorting CSS (NDSCSS)

This section presents a brief overview of the Non-dominated Sorting Charged System Search (NDSCSS). NDSCSS spreads the basic procedure of CSS by making good use of the particles' best offspring for effective non-domination comparisons. In place of only comparing between a particle's position and velocity, NDSCSS compares all particles' positions and velocities in all the particles. All the particles are stored into various non-dominated levels. The dominance comparisons are not entirely utilized in the process of updating the positions and the velocities of each particle. In order to overcome this problem, raise the sharing level between particles. NDSCSS creates a form, which is a temporary population of 2 N particles. The domination comparisons of 2 N individuals are then carried out. This method can ensure more non-dominated solutions and allow the sorting of the entire population into different non-domination levels. The goals of the multi-objective optimization are to acquire a set of non-dominated solutions as close as possible to the true Pareto front, and to maintain a well-distributed solution along the Pareto front. With the purpose of satisfying the second goal, the crowding distance concept is used. When the non-dominated sorting and crowding distance are integrated using CSS, the distribution systems' multiple objective problem can be solved effectively, as shown in Fig. 1, which is a flow chart of proposed NDSCSS.

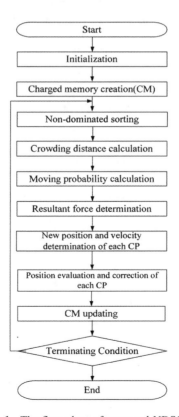

Fig. 1. The flow chart of proposed NDSCSS

4 Simulation Results

In order to evaluate the performance of a proposed method, a 33-bus distribution systems. The performances of NSEP, MOPSO and NDSCSS are compared in this paper. The parameters used in these algorithms are described in Table 1.

Table 1. Parameter setting of each algorithm.

Algorithms	Parameters
NSEP	generation = 1500, Population size = 25, Reproduction rate = 80 %
MOPSO	generation = 1500, number of particles = 25
NDSCSS	generation = 1500, CPs = 25, ε = 0.001

A 33-bus distribution systems is used to test the performance of NSEP, MOPSO and NDSCSS. This systems is comprised of a 1-feeder, 37-switch and 33-zone. The initial loss is 202.65 Kw. In order to calculate the average efficiency, each algorithm was performed 600 times. All algorithms can find the primary feeder losses and the number of switch actions reduction as shown in Table 2. All algorithms can obtain the primary feeder loss of 202.65 Kw when the number of switch actions is 0. When the number of switch actions is 4, the primary feeder are 163.27 and 165.27 through the use of MOPSO and NSEP. NDSCSS can find the minimum primary feeder loss of 132.44 Kw as shown in Table 2. The Pareto front can be shown for each algorithm in Fig. 2.

Table 2. Primary feeder loss with the number of switch actions for the 33-bus system

NDSCSS		MOPSO		NSEP	
Primary feeder loss (Kw)	Switch actions (times)	Primary feeder loss (Kw)	Switch actions (times)	Primary feeder loss (Kw)	Switch actions (times)
202.65	0	202.65	0	202.65	0
132.44	4	163.27	4	165.27	4
137.29	3	177.29	3	180.29	3
139.77	2	179.16	5	184.77	2
147.37	1	182.77	2	185.89	5
		194.37	1	197.37	1

The ZRNS encoding/decoding scheme is used to represent the topologies of a distribution systems. The new solution for each algorithm can be used directly without modifications. In order to identify the performance, a 33-bus system is used, NESP, MOPSO and NDSCSS can all deal with the multiple objectives as shown in Table 2. Compared with the performance of the NESP, MOPSO and NDSCSS, the proposed algorithms can find the best non-dominated sorting solution.

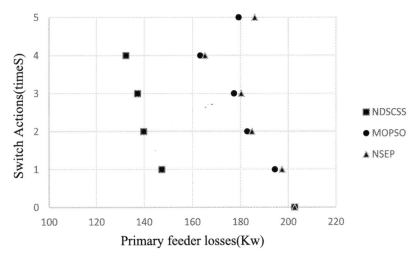

Fig. 2. The praetor front can be shown for each algorithms

5 Conclusions

This paper proposes a novel non-dominated sorting charged systems search for solving distribution systems' multiple objectives problems. In the past, MOPSO and NSEP had been used in different fields to solve the multiple objectives problems. The effectiveness of MOPSO or NESP has been confirmed using many study results. Because of the distribution systems' topological constraints, the ZRNS encoding/decoding scheme was added to NDSCSS. This scheme found the distribution systems topology without validation. The numeric results indicated that the NDSCSS can find optimal Pareto front solutions among MOPSO, NESP and NDSCSS. Hence, NDSCSS can more effectively deal with distribution systems problems. The system operators can provide proper decisions through use of the proposed method.

Acknowledgment. This work was supported by the National Science Council of Republic of China under Contract MOST 105-3113-E-006-007

References

1. Abido, M.A.: Multi-objective optimal power flow using strength Pareto evolutionary algorithm. In: Proceedings of 39th International Universities Power Engineering Conference, Bristol, UK, 8 September 2004, vol. 1, pp. 457–461 (2004)
2. Baran, M.E., Wu, F.F.: Network reconfiguration in distribution systems for loss reduction and load balancing. IEEE Trans. Power Delivery **4**(2), 1401–1407 (1989)
3. Basturk, B., Karaboga, D.: An Artificial Bee Colony (ABC) algorithm for numeric function optimization. In: Proceedings of the IEEE Swarm Intelligence Symposium, Indianapolis, Indiana, USA (2006)

4. Chu, C.-C., Tsai, M.-S.: Application of novel charged system search with real number string for distribution systems loss minimization. IEEE Trans. Power Syst. **28**(4), 3600–3609 (2013)
5. DcDermott, T.E., Drezga, I., Broadwater, R.P.: A heuristic nonlinear constructive method for distribution systems reconfiguration. IEEE Trans. Power Syst. **14**(2), 478–483 (1999)
6. Deb, K., Pratap, A., Agarwal, S., Meyarivan, T.: A fast and elitist multiobjective genetic algorithm: NSGA-II. IEEE Trans. Evol. Comput. **6**(2), 182–197 (2002)
7. Esmin, A.A.A., Lambert-Torres, G., de Souza, A.C.Z.: A hybrid particle swarm optimization applied to loss power minimization. IEEE Trans. Power Syst. **20**(2), 859–866 (2005)
8. Fukuyama, Y., Endo, H., Nakanishi, Y.: A hybrid system for service restoration using expert system and genetic algorithm. In: Proceedings of International Conference on Intelligent Systems Applications to Power Systems, Orlando, FL, 28 January–2 February 1996, pp. 394–398 (1996)
9. Goswami, S.K., Basu, S.K.: A new algorithm for the reconfiguration of distribution feeders for loss minimization. IEEE Trans. Power Delivery **7**(3), 1484–1491 (1992)
10. Gomes, F.V., Carneiro Jr., S., Pereira, J.L.R., Vinagre, M.P., Garica, P.A.N., Araujo, L.R.: A new heuristic reconfiguration algorithm for large distribution systems. IEEE Trans. Power Syst. **20**(3), 1373–1378 (2005)
11. Gomes, A., Antunes, C.H., Martins, A.G.: Improving the responsiveness of NSGA-II using an adaptive mutation operator: a case study. Int. J. Adv. Intell. Paradigms **2**(1), 4–18 (2010)
12. Hsiao, Y.-T.: Multi-objective evolution programming method for feeder reconfiguration. IEEE Trans. Power Syst. **19**(1), 594–599 (2004)
13. Hsu, F.-Y., Tsai, M.-S.: A multi-objective evolution programming method for feeder reconfiguration of power distribution system. In: Proceedings of the 13th International Conference on Intelligent Systems Application to Power Systems, Avlington, VA, 6–10 November 2005, pp. 55–60 (2005)
14. Margarita, R.S., Coello Coello, C.A.: A multi-objective particle swarm optimizers: a survey of the state-of-the-art. Comput. Intell. Res. Int. J. Program. **2**(2), 287–308 (2006)
15. Hsiao, Y.-T.: Multi-objectives evolution programming method for feeder reconfiguration. IEEE Trans. Power Syst. **19**(1), 594–599 (2003)
16. Irving, M.R., Luan, W.P., Daniel, J.S.: Supply restoration in distribution network using a genetic algorithm. Int. J. Electr. Power Energ. Syst. **24**(6), 447–457 (2002)
17. Kennedy, J.: The particle swarm: social adaptation of knowledge. In Proceedings of IEEE International Conference on Evolutionary Computation Indianapolis, IN, 13–16 April 1997, pp. 303–308 (1997)
18. Karaboga, D., Basturk, B.: A powerful and efficient algorithm for numerical function optimization Artificial Bee Colony (ABC) algorithm. J. Global Optim. **39**(3), 459–471 (2007)
19. Kim, H., Ko, Y., Jung, K.-H.: Artificial neural networks based feeder reconfiguration for loss reduction in distribution systems. IEEE Trans. Power Delivery **8**(3), 1356–1366 (1993)
20. Khoa, T.Q.D., Phan, B.T.T.: Ant colony search based loss minimum for reconfiguration of distribution systems. In: Proceedings of IEEE Power India Conference, New Delhi, India (2006)
21. Kaveh, A., Talatahari, S.: A novel heuristic optimization method: charged system search. Acta Mech. **213**(3–4), 267–289 (2010)
22. Murata, T., Ishibuchi, H.: MOGA: Multi-Objective Genetic Algorithms. In Proceedings of IEEE International Conference on Evolutionary Computation, Perth, WA, Australia, 29 November–1 December 1995, Vol: 1, pp. 289–294

23. Nara, K., Shiose, A., Kitagawa, M., Ishihara, T.: Implementation of genetic algorithm for distribution systems loss minimum re-configuration. IEEE Trans. Power Syst. **7**(3), 1044–1051 (1992)
24. Nara, K., Mishima, Y., Satoh, T.: Network reconfiguration for loss minimization and load balancing. In: Proceedings of IEEE Power Engineering Society General Meeting, Ibaraki University, Japan (2003)
25. Shirmonhammadi, D., Hong, H.W.: Reconfiguration of Electric Distribution Networks for Resistive Line Losses Reduction. Power Delivery, IEEE Transaction on **4**(2), 1492–1498 (1989)
26. Teng, J.-H., Liu, Y.-H.: A novel ACS-based optimum switches relocation method. IEEE Trans. Power Syst. **18**(1), 113–120 (2003)
27. Tsai, M.-S., Hsu, F.-Y.: Application of grey correlation analysis in evolutionary programming for distribution system feeder reconfiguration. IEEE Trans. Power Syst. **25**(2), 1126–1133 (2009)
28. Veldhuizen, D.A.V., Lamont, G.B.: Evolutionary computation and convergence to a Pareto front. In: Proceedings of Late Breaking Papers at the Genetic Programming Conference, Madison, Wisconsin, USA (1998)
29. Wu, W.-C., Tsai, M.-S.: Application of enhanced integer coded particle swarm optimization for distribution system feeder reconfiguration. IEEE Trans. Power Syst. **26**(3), 1591–1599 (2011)
30. Zhu, J.Z.: Optimal reconfiguration of electrical distribution network using the refined genetic algorithm. Electr. Power Syst. Res. **62**(1), 37–42 (2002)

System Integration for Real-Life Applications

Design of a Communication System that Can Predict Situations of an Absentee Using Its Behavior Log

Hironori Hiraishi[(⌧)]

Department of Electrical and Computer Engineering,
National Institute of Technology, Akita College, Akita, Japan
hiraishi@akita-nct.ac.jp

Abstract. We designed a communication system that can predict the current situation of an absentee on the basis of its behavior log and provide some communication tools suitable for the situation to a visitor. An operative evaluation revealed that the proposed system could predict user situations with high accuracy. Further, it can output the expected values in short terms by calculating the feedback values to the possibility table with our Bayesian network. Furthermore, the proposed system has realized the communication suitable for a situation and for information sharing including prediction by using the current information and communication technology.

Keywords: Behavior log · Probability table · Bayesian network · Information and communication technology

1 Introduction

When we leave our desk or our room for some time, we often use an indicator about a destination or a situation, such as meeting or lunch. This indicator tells a visitor how long he/she should wait or where he/she needs to go to meet us.

In particular, in a higher education institution such as our college, which includes classes from high school to junior college, we have to provide a wide range of education and instruction, such as learning instruction, club activities, internship, vocational guidance, and even life guidance. Therefore, the relationship between students and teachers is very important. An indicator such as that shown in Fig. 1 is used as a useful tool for smooth communication among us. In general, we need to provide information voluntarily. Therefore, there are some problems: We sometimes forget to operate the indicator, or we have to go back to operate it. If the indicator is displaying a wrong place, we do not have a clue as to where to search for an absentee.

Therefore, we have designed a system that can automatically provide some information to visitors, by predicting situations (or places) of an absentee on the basis of a user's behavior logs. The proposed system records the behavior logs and creates a probability table for every period. Further, the system displays situations as probability values. Here, when a new behavior log is entered into the system, the system does not change a probability value simply as the frequency increased. Therefore, we have

© Springer International Publishing Switzerland 2016
H. Fujita et al. (Eds.): IEA/AIE 2016, LNAI 9799, pp. 959–970, 2016.
DOI: 10.1007/978-3-319-42007-3_81

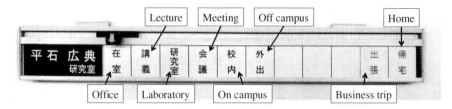

Fig. 1. Example of the indicator in front of our room

adopted a Bayesian network [5, 9], which represents the relationships between user operations and situations. It allows the system to adjust to new surroundings at an earlier stage.

When a person is absent, we can contact the person by phone or by e-mail. However, if the person is attending a meeting or giving a lecture, it is difficult for him/her to answer the phone. Moreover, the ringtone itself may disturb the meeting or the lecture. Therefore, we have designed the proposed system not only for predicting but also for the communication. The proposed system has the functions of a message board, e-mail, and TV phone, and it limits the use of tools and facilitates appropriate communication suitable for the absentee situations.

The rest of this paper is organized as follows: Sect. 2 explains the proposed communication system and discusses the user interfaces and system architecture. Section 3 describes the proposed prediction method based on the behavior log. Some extensions using physical computing [2], cloud computing [8], smartphones, and the tablet terminal of the proposed system are introduced in Sect. 4. Finally, Sect. 5 summarizes this paper.

2 Proposed Communication System

Figure 2 shows an example of an actual use case and the user interfaces of the proposed system. We designed the system as a Kiosk terminal and placed it in front of a personal room in a university. The probability table calculated using the behavior log is at the bottom of the menu window. Unlike the general indicator shown in Fig. 1, the proposed system can suggest several situations as the probability values to visitors. If an absentee is not in a place of the highest possibility, a visitor can go and check the places of the next highest possibility. In consideration of user visibility, the colors of each column are gradually thinned from a high probability value to a low probability value. Further, the red double circle emphasizes the column of the highest probability value.

At the top of the menu window, there are three buttons to start each communication tool. Visitors can leave a message by using the message board tool, send an e-mail by using the e-mail tool, and communicate directly with an absentee through the TV phone.

Here, we set up the limitations of the use of each tool on the basis of the situation predicted as the highest probability. For example, a TV phone is not suitable for the situation of a lecture or a meeting and the ringtone itself may disturb these events.

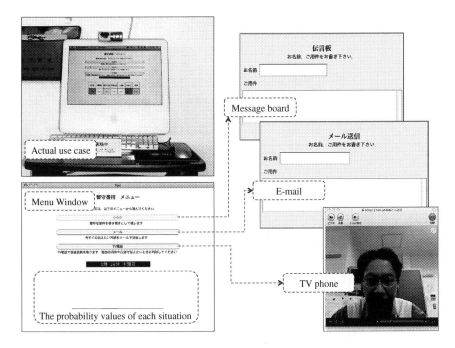

Fig. 2. Actual use case and user interfaces of proposed communication system (The proposed system is very light and is compatible even with an old machine such as Apple iMac PowerPC G5 1.8 GHz, Memory 1 GB)

Table 1 summarizes the available tools for each situation. The limitations are implemented by making the corresponding buttons inactive. Figure 2 shows the case in which the probability of the person being in a laboratory is the highest. The corresponding button on the message board becomes inactive. This implies that the absentee would like a visitor to contact him/her by e-mail or TV phone, because he/she is relatively free in the laboratory.

Further, in the proposed system, we can select each position manually by clicking each column such as "Lecture" or "Laboratory." This operation is recorded in the behavior log, which is fed back to the probability table. Therefore, the proposed system can adapt to the user behavior gradually and display the user's situations automatically, in spite of the fact that it requires user operations similar to those of an ordinary indicator in an early stage. When clicking a column, it is certain that a user go to the indicated place. Therefore, the proposed system temporarily displays 100 % in this column, and changes this value back to the value in the probability table gradually.[1]

Figure 3 shows the architecture of the proposed system. This system is implemented and integrated in the Java language. The user operations are stored as the behavior log. The Bayesian network, which represents the relationships between user operations and situations, is used to feed back the behavior log to the probability table.

---

[1] The column returns to the original probability value in 1 h.

Table 1. Limitations of the use of each tool

Tools \ Situations	Office	Lecture	Laboratory	Meeting	On campus	Off campus
Message board	NG	OK	NG	OK	OK	OK
E-mail	NG	NG	OK	NG	OK	OK
TV phone	NG	NG	OK	NG	OK	OK

We have adopted JavaBayes[2] as the tool of the Bayesian network and have used the JavaMail API to implement the e-mail function. We have selected Skype for the TV phone. The proposed system can access Skype directly by using the Skype API, which allows us to contact an absentee through Skype.

3   Prediction Based on Behavior Log

The proposed system creates a probability table for each period from the behavior log and represents the situations of an absentee as probability values. We have considered six user situations (Office, Lecture, Laboratory, Meeting, On campus, and Off campus), as shown in Table 1. Further, we have divided the time into 10 periods (1, 2, 3, 4, Lunch, 5, 6, 7, 8, and After school) according to the timetable of our college.

In general, the class schedules of universities are divided into several periods and the working schedules of companies are divided into some periods by some rest time and lunchtime and user situations (or places) are associated with each of these periods. Further, class schedules or regular meetings are usually set for every day of the week, and the schedule is the same for the same day of the week. Therefore, the proposed system creates a probability table for every day of the week.

If the user of the proposed system finds that the system indicates a wrong situation when the user goes out of the room, the user sets the appropriate situation manually. The operation is stored in the behavior log. The probability table can be created from the behavior log by using the frequency of the situations. For example, if Lecture was selected thrice and Meeting was selected twice, the probability value of Lecture would be 60 % and that of Meeting would be 40 %. However, in order to achieve certain accuracy, this method requires a considerable amount of data and we have to spend a significant amount of time to operate the system. Furthermore, the change in probability brought about by one data item decreases with an increase in the log data. Therefore, recent operations tend to be reflected less than the past operations. Hence, it is difficult to cope with a schedule change such as a change in the lecture or meeting time.

Comparing a situation predicted by the system and a situation selected by the user, the proposed system judges whether the operation is for a schedule change or for a temporary change. For example, if a user selects Lecture when the system outputs

---

[2] http://www.cs.cmu.edu/~javabayes/.

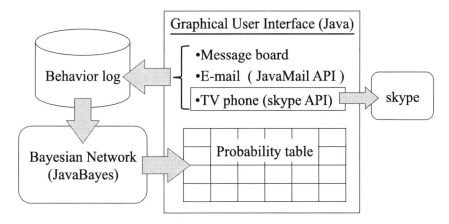

Fig. 3. System architecture

Office, we can assume that the operation implies a schedule change, because a new lecture may have been inserted at this time. Further, if the user selects Off campus when the system outputs Meeting, we can assume that the operation denotes a temporary change, such as a case wherein a meeting ended earlier than planned.

Figure 4 shows the basic policy of changing the possibility values of each situation. The situations on the left side are those with a high possibility of a schedule change, and those on the right side are related to a temporary change. The left value of each situation denotes the possibility value that is added to the situation when a user selects it in the proposed system. On the other hand, the right value represents the possibility value that is added to a situation that the user selects, when he/she changes each situation that is displayed in the proposed system to another situation. Therefore, the sum of these two values is fed back to the possibility of the situation that the user selects. For example, if a user changes Office to Lecture, the left value 30 % of Lecture, as shown in Fig. 4, and the right value 10 % of Office are added; then, the sum 40 % is fed back to the possibility of Lecture. Considering the degree of change or temporariness of each situation, the values shown in Fig. 4 have been decided through our heuristics, which will be described in detail in the following section. In order to infer the feedback value based on our policy, we have constructed the Bayesian network shown in Fig. 5.

The nodes of this network are as follows:

- **System output** (left 6 nodes)
  Office (s-office), Lecture (s-lecture), Laboratory (s-lab),
  Meeting (s-meeting), On campus (s-on), Off campus (s-off)
- **User selection** (right 6 nodes)
  Office (office), Lecture (lecture), laboratory (lab),
  Meeting (meeting), On campus (on), Off campus (off)
- **Type of change** (2 nodes)
  Schedule change (schedule), temporary change (temporary)
- **Feedback value** (4 nodes)
  40 %, 30 %, 20 %, 10 %

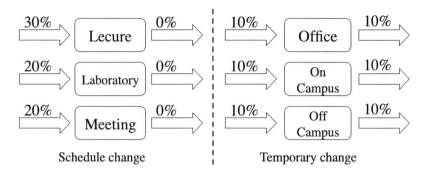

Fig. 4. Basic policy of changing possibility values

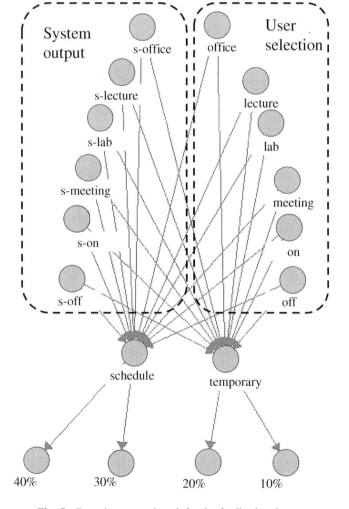

Fig. 5. Bayesian network to infer the feedback value

The probability values from the system are input to the left nodes. Further, a user selection is input to a right node; here, the value of 100 % is set in the selected node and 0 % is set in the other nodes. Then, the judgment of whether a change is a schedule change or a temporary change is made, and finally, the feedback value is inferred.

We have coordinated the Bayesian network that outputs feedback values that are the same as those shown in Fig. 4, when the system outputs only one situation whose probability value is 100 %. If the system outputs several situations, the Bayesian network outputs the probability values of each feedback value. Thus, the actual feedback value is calculated by using the probability values. If 40 % is 0.5 and 30 % is 0.5, the feedback value becomes 40 * 0.5 + 30 * 0.5 = 35 %. The feedback value is added to the probability of the situation until the upper limit of 100 % is reached. The added value is equally divided into other situations and is subtracted from them until the lower limit of 0 % is reached.

3.1 Consideration of Feedback Values

In the proposed system, the feedback value becomes 40 % or 30 %, if the system judges the user selection as a schedule change. On the other hand, the feedback value becomes 20 % or 10 %, if the system judges the user selection to be a temporary change.

Here, take Lecture that has a high possibility of a schedule change for example. When a user selects Lecture in the situation of Office 100 %, 40 % feedback is provided. Thus, the value for Office becomes 60 % and that for Lecture becomes 40 %. Therefore, Office still has the highest possibility. If the user selects Lecture again, Lecture is provided nearly 40 % feedback, and its value becomes about 80 %. Therefore, the possibility of Lecture becomes the highest when it is selected twice. A lecture will sometimes be changed as a business trip or something. Since it is rare that a lecture is delivered in a different period for two weeks, we can assume that the feedback values for the schedule change are appropriate.

As for the temporary change, the maximum feedback value is about 20 %. A situation needs to be selected thrice at least for it to have the highest possibility (more than 50 %). Although Office or On campus tend to be selected many times, they are actually temporary, and going out (Off campus) during office hours is rare. Therefore, the feedback values are set up for situations of temporary change so as to not result in a relatively high possibility even if these situations are selected several times. Thus, we can regard the feedback values for a temporary change as appropriate, too.

3.2 Operative Evaluation of Proposed System

We recorded the behavior log of a professor in our college for six weeks. We operated the proposed system according to the behavior log. We started from the situation of Office 100 % in every period. Further, considering the real operations, we decided that we would click on a situation only when the highest probability situation of the proposed system was different from the situation of the behavior log.

Here, in order to evaluate the proposed method, we assume the frequency of the six-week data to be the correct probability. We explain our evaluation by using the behavior log shown in Table 2. First, the correct probability for the evaluation is calculated by the frequency of the six-week log. Since there are two instances of Office and four of Lecture, the probability of Office is $2/6 \times 100 = 33.33$ % and that of Lecture is 66.67 %. Further, we operate the proposed system from the first week to the sixth week as described above and then, check the probability values that the system predicts. In this case, the probability of Office is 35 % and that of Lecture is 65 %. We have summarized the probability values of every period in Table 3.

Table 2. Example of the behavior log of a period

1st week	2nd week	3rd week	4th week	5th week	6th week
Office	Lecture	Lecture	Office	Lecture	Lecture

Table 3. Example of the probability values of a period

System output [%]					
Office	Lecture	Laboratory	Meeting	On campus	Off campus
35	65	0	0	0	0

Frequency of behavior log [%]					
Office	Lecture	Laboratory	Meeting	On campus	Off campus
33.33	66.67	0	0	0	0

Next, the error rate is calculated from Table 3. The error rate is the absolute value of the difference between the system output and the frequency. Therefore, the error rate of both Office and Lecture is 1.67 %. We have calculated the error rates for every period along with their average. However, to calculate the average, we did not use the situations whose probability was 0 % in both the system output and the frequency. In such situations, the error rate was 0 % even when the user did not operate the system at all. Therefore, we did not consider such situations in order to now lower the whole error rate to a very low value.

All the system outputs and the frequencies of every period after six weeks are given in the appendix. Further, the average of all the error rates was about 7.8 %. This implies that we can manage the proposed system with 92.2 % accuracy after six weeks from the condition indicating only Office 100 %.

Table 4 shows the change in the accuracy of each week. The accuracy reached more than 80 % in the second week. As described in the previous section, the proposed system can output the expected values in short terms.

Thus, the possibility of the selected situation does not reach 100 %. However, we can select a situation several times at once, or we can set the feedback value as 20 % to

Table 4. Change in the accuracy of each week [%]

1st week	2nd week	3rd week	4th week	5th week	6th week
73.7	84.4	87.4	90.4	91.1	92.2

50 %. However, note that there are some cases in which the error rates will increase as a result with an increase in the change in the possibility of one operation.[3]

4   Extensions of Proposed Communication System

Since the proposed system needs user operations to collect the behavior log, particularly in the early stage, more natural input methods for users are required for this system. Therefore, we can adopt a physical user interface [4] that can be operated physically, such as the indicator shown in Fig. 1.

Figure 6 shows an example of our physical user interface. The user can select each situation by pushing the corresponding button. A change in the probability values is represented by the gradation of the LED. Red light indicates 100 %, and then, the light gradually changes to blue to denote 0 %. Therefore, the user can do basic operations of the proposed system by using only this device and without using the mouse, keyboard, or the computer itself. This device is implemented as a USB device [7].

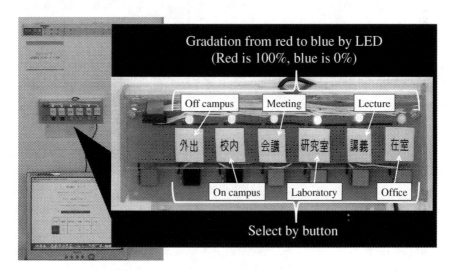

Fig. 6. Example of our physical user interface

Figure 7 shows an example of the use of Android terminals and cloud computing. An Android terminal[4] is a type of high-performance telephone, which is placed on a desk in the office environment. There are some situations such as Working, Meeting, Break, Lunch, and Going out. Therefore, the proposed system can be applied to the office environment. We adopted cloud computing [3], which allows us to access the proposed system anywhere by using a smartphone, tablet, or web browser. Therefore, the proposed system can be used as groupware to share schedules, including predictions from the behavior log [6].

---

[3] In the case of 20 % to 50 %, the error rate increases to more than 10 % after six weeks.

[4] GRANYC produced by NAKAYO Inc.

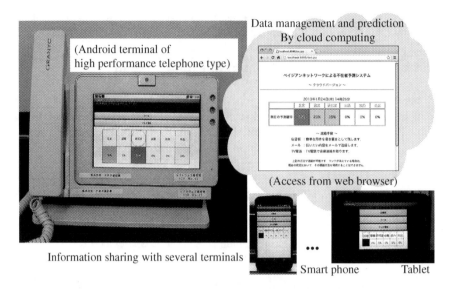

Fig. 7. Example of using Android terminals and cloud computing

5   Conclusions

In this study, we designed a communication system that can predict the current situation of an absentee on the basis of its behavior log and provide some communication tools suitable for the situation to a visitor.

Through our operative evaluation, we have clarified that the proposed system can predict user situations with high accuracy. Further, it has been shown that it can output the expected values in short terms by calculating the feedback values for the possibility table with our Bayesian network.

Unlike the ordinary indicators, the proposed system can automatically provide visitors several candidate situations (or locations) of an absentee. Furthermore, the proposed system can realized communication suitable for a situation and for information sharing, including prediction sharing [6]. This implies that we can successfully generate a new value of the indicator by using the current inforation and communication technology.

There are some cases in which users (not visitors) do not want to inform others of their own situation, for example, when they want to concentrate on something. Therefore, 100 % accuracy might not always be necessary. Further, the proposed system does not always perform with 100 % accuracy and predicts several candidate situations by probability. Therefore, we can consider the prediction of the situations of an absentee as one of the suitable applications for a life log [1] or a Bayesian network [9].

A prediction based on the behavior log does not require special devices such as GPS or sensors. Therefore, the application of the proposed system to problems in which the use of such special devices is difficult, such as the problem of the loitering of an elderly person, can be expected.

Appendix: Results of the operative evaluation

		System output [%]						Frequency of behavior log [%]						The average of the error
		Office	Lecture	Labo.	Meeting	On	Off	Office	Lecture	Labo.	Meeting	On	Off	
Mon.	1	23	77	0	0	0	0	16.67	83.33	0	0	0	0	6.33
	2	35	65	0	0	0	0	33.33	66.67	0	0	0	0	1.67
	3	17	0	83	0	0	0	16.67	0	83.33	0	0	0	0.33
	4	17	0	83	0	0	0	33.33	0	66.67	0	0	0	16.33
	Lunch	100	0	0	0	0	0	100	0	0	0	0	0	0.00
	5	23	77	0	0	0	0	0	100	0	0	0	0	23.00
	6	21	62	17	0	0	0	0	80	20	0	0	0	14.00
	7	11	64	18	0	0	7	0	40	40	0	0	20	17.50
	8	62	0	0	0	0	38	60	0	0	0	0	40	2.00
	After	62	0	0	0	0	38	60	0	0	0	0	40	2.00
		Office	Lecture	Labo.	Meeting	On	Off	Office	Lecture	Labo.	Meeting	On	Off	
Tue.	1	100	0	0	0	0	0	100	0	0	0	0	0	0.00
	2	80	0	0	0	20	0	75	0	0	0	25	0	5.00
	3	100	0	0	0	0	0	100	0	0	0	0	0	0.00
	4	80	0	0	0	20	0	75	0	0	0	25	0	5.00
	Lunch	100	0	0	0	0	0	100	0	0	0	0	0	0.00
	5	47	0	20	33	0	0	50	0	25	25	0	0	5.33
	6	51	0	49	0	0	0	50	0	50	0	0	0	1.00
	7	51	0	49	0	0	0	50	0	50	0	0	0	1.00
	8	44	0	42	14	0	0	25	0	50	25	0	0	12.67
	After	70	0	0	30	0	0	75	0	0	25	0	0	5.00
		Office	Lecture	Labo.	Meeting	On	Off	Office	Lecture	Labo.	Meeting	On	Off	
Wed.	1	70	0	0	30	0	0	80	0	0	20	0	0	10.00
	2	100	0	0	0	0	0	100	0	0	0	0	0	0.00
	3	100	0	0	0	0	0	100	0	0	0	0	0	0.00
	4	100	0	0	0	0	0	100	0	0	0	0	0	0.00
	Lunch	100	0	0	0	0	0	100	0	0	0	0	0	0.00
	5	17	0	66	0	11	6	0	0	40	0	40	20	21.50
	6	25	0	60	0	6	9	20	0	40	0	20	20	12.50
	7	38	0	0	0	49	13	20	0	0	0	60	20	12.00
	8	36	0	0	27	27	10	20	0	0	20	40	20	11.50
	After	46	0	0	27	16	11	40	0	0	20	20	20	6.50
		Office	Lecture	Labo.	Meeting	On	Off	Office	Lecture	Labo.	Meeting	On	Off	
Thu.	1	23	77	0	0	0	0	0	100	0	0	0	0	23.00
	2	23	77	0	0	0	0	0	100	0	0	0	0	23.00
	3	44	0	0	0	56	0	40	0	0	0	60	0	4.00
	4	58	0	25	0	17	0	60	0	20	0	20	0	3.33
	Lunch	100	0	0	0	0	0	100	0	0	0	0	0	0.00
	5	33	0	54	0	13	0	0	0	80	0	20	0	22.00
	6	33	0	54	0	13	0	0	0	80	0	20	0	22.00
	7	70	0	30	0	0	0	80	0	20	0	0	0	10.00
	8	70	0	30	0	0	0	80	0	20	0	0	0	10.00
	After	37	0	0	63	0	0	60	0	0	40	0	0	23.00
		Office	Lecture	Labo.	Meeting	On	Off	Office	Lecture	Labo.	Meeting	On	Off	
Fri.	1	80	0	0	0	20	0	83.33	0	0	0	16.67	0	3.33
	2	80	0	0	0	0	20	83.33	0	0	0	0	16.67	3.33
	3	35	23	0	0	13	29	33.33	16.67	0	0	16.67	33.33	4.00
	4	41	27	0	0	0	32	50	16.67	0	0	0	33.33	6.89
	Lunch	80	0	0	0	0	20	83.33	0	0	0	0	16.67	3.33
	5	55	0	31	0	0	14	66.67	0	16.67	0	0	16.67	9.56
	6	39	0	17	0	13	31	33.33	0	16.67	0	16.67	33.33	3.00
	7	49	0	20	0	17	14	50	0	16.67	0	16.67	16.67	1.83
	8	49	0	19	0	0	32	33.33	0	16.67	0	0	50	12.00
	After	62	0	0	0	0	38	50	0	0	0	0	50	12.00
											The average of all the error			7.84

References

1. Sellen, A.J., Whittaker, S.: Beyond total capture: a constructive critique of lifelogging. Commun. ACM **53**(5), 70–77 (2010)
2. O'Sullivan, D., Igoe, T.: Physical Computing: Sensing and Controlling the Physical World with Computers. Thomson Course Technology, Boston (2004)
3. Sanderson, D.: Programming Google App Engine with Java: Build & Run Scalable Java Applications on Google's Infrastructure. O'Reilly Media, Sebastopol (2015)
4. Ishii, H.: Tangible bits: beyond pixels. In: Proceedings of the 2nd International Conference on Tangible and Embedded Interaction, pp. xv–xxv (2008)
5. Iwasaki, H., Sega, S., Hiraishi, H., Mizoguchi, F.: Design and evaluation of the user-adapted program scheduling system based on Bayesian network and constraint satisfaction. Trans. Jpn. Soc. Artif. Intell. **23**(4), 268–280 (2008)
6. Rawassizadeh, R.: Towards sharing life-log information with society. Behav. Inf. Technol. **31**(11), 1057–1067 (2012)
7. Sasaki, T., Hiraishi, H.: Design and evaluation of user interface for whereabout forecast. In: The 75th National Convention of IPSJ, pp. 307–308 (2013)
8. Erl, T., Puttini, R., Mahmood, Z.: Cloud Computing: Concepts, Technology & Architecture. Prentice Hall, Englewood Cliffs (2013)
9. Fenton, N., Neil, M.: Risk Assessment and Decision Analysis with Bayesian Networks. CRC Press, Boca Raton (2012)

Autonomic Smart Home Operations Management Using CWMP: A Task-Centric View

Chun-Feng Liao$^{(\boxtimes)}$, Shih-Ting Huang, and Yi-Ching Wang

Department of Computer Science and Program in Digital Content and Technologies,
National Chengchi University, Taipei, Taiwan
{cfliao,101703006,103753028}@nccu.edu.tw

Abstract. Despite the well-development of smart living space research field, Smart Home is still more like a luxury product than a daily necessity for most families. Operations management issues are essential for a new technology to be accepted by the mass consumer market. However, only few attempts have been made toward this direction. In this paper, we present the design and implementation of a CWMP-based platform that supports autonomic operations management. The experiment results show that the proposed approach is stable and is able to drive the operations tasks smoothly. We also demonstrate the feasibility of the platform by realizing two application scenarios supported by the prototype of the proposed approach.

Keywords: CWMP · TR-069 · Operations management · Smart Home

1 Introduction

The concept of Smart Home was envisioned twenty years ago [1]. Although the essential technology for constructing smart living spaces has also been an object of study for more than two decades [2] and the costs of embedded computers, sensors, and home appliances are much lower than before in the last few years, Smart Home is still more like a luxury product than a daily necessity for most families. As pointed out in recent researches, central to this issue is the problem of autonomic operations management, namely, the ability of a Smart Home system to be self-deployable [3], self-diagnosable [4,5], and self-configurable [6,7]. The above-mentioned self-* properties of systems have been proposed by the researchers in the field of "autonomic computing" [8].

Despite the importance of operations management in Smart Home, only few attempts have been made toward this direction. Rachidi and Karmouch's work [9] is a pioneer study in this issue. Based on the MAPE-K (Monitor, Analyze, Plan, and Execute using Knowledge) model of autonomic computing, they present an approach that facilitates self-configuration for home gateway based on CWMP (CPE WAN Management Protocol, where CPE abbreviates Consumer Premises

© Springer International Publishing Switzerland 2016
H. Fujita et al. (Eds.): IEA/AIE 2016, LNAI 9799, pp. 971–982, 2016.
DOI: 10.1007/978-3-319-42007-3_82

Equipments) [10]. CWMP, propose by Broadband Forum and also know as TR-069, is a SOAP-based [11] application layer protocol for remote management and configuration of CPEs, where CPEs are usually home gateways in real-world applications. As CWMP has been implemented on more than 250 million devices world-wide [12], it is apparently a good basis for designing the Smart Home operations managing mechanisms.

Figure 1 is an UML deployment diagram that illustrates a typical application architecture of autonomic operations management using CWMP. The service provider's server hosts an MAPE-K Manager module to proactively analyze, plan, and determine the strategies of management. After a decision is made, several commands are executed by an ACS (Auto-Configuration Server). In a typical scenario, an ACS is responsible for performing administrative operations of home gateways and a CPE is hosted by a home gateway. Generally, a home gateway is usually equipped with one or more protocol gateway modules to HAN (Home Area Network). For instance, to enforce the management strategy remotely in an UPnP-based home network, an UPnP Control Point module has to be implemented and deployed in the home gateway. Figure 1 also reveals that CWMP's scope is limited to communication among an ACS and CPEs.

In fact, the CWMP specification only defines signatures of remote procedures (listed in Table 1) and their SOAP representations, figuring out how to compose these remote procedures so that certain operations management tasks can be carried out is a burden of developers. In other words, CWMP specification [10] does not define how to orchestrate these methods to perform operations management tasks.

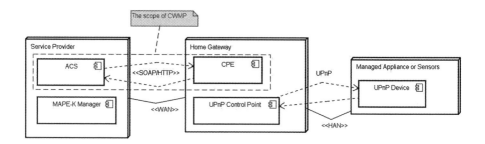

Fig. 1. A typical autonomic operations management architecture using CWMP

Inspired by Rachidi and Karmouch's work [9] on CWMP-based self-configuration, this paper aims to design and to implement other two essential self-* mechanisms, namely self-deployment and self-diagnosis, that supports MAPE-K-based autonomic operations management using CWMP. To relief the burden of developers, this work takes a task-centric approach. Specifically, we design the system by first identifying the core tasks that is critical to a Smart Home's daily operations management. Then, we examined and identified the call sequence of CPE/ACS operations that collaboratively accomplish these tasks. To

Table 1. Core methods specified in CWMP

Subject	Method name	Subject	Method name
CPE	GetRPCMethods	ACS	GetRPCMethods
CPE	SetParameterValues	ACS	Inform
CPE	GetParameterValues	ACS	TransferComplete
CPE	GetParameterNames	ACS	AutonomousTransferComplete
CPE	SetParameterAttributes	CPE	GetParameterAttributes
CPE	AddObject	CPE	DeleteObject
CPE	Reboot	CPE	Download

evaluate the proposed call sequences, this paper demonstrates the feasibility of the proposed approaches by implementing a prototype and by conducting experiments on the prototype. We hope that our work is helpful for the developers of infrastructure for Smart Home service providers when developing operations management tasks.

2   Related Work

The advent of Smart Home brings about the issues of operations management. Operations management means the configuration, deployment, upgrade, and monitoring of devices or systems in Smart Home. Obviously, the best way to address this issue would be to reuse an existing mature standards or technologies. In network equipment industry, several standards have been proposed for operations management such as NETCONF [13], MUWS (Management for Using Web Service) [14], WS-Management [15] and CWMP [10]. As mentioned, CWMP has been widely deployed in home gateway, and therefore are considered the most competitive among the standards [9].

As a result, several works have been done for developing management functions based on CWMP. Nikolaidis et al. [6] proposed an MIB-based (Management Information Base) framework and a graphical development environment for control devices in a home network. Other works focused on providing the tools to assist service engineers to diagnose hazards in Smart Home system [16,17]. There are relatively fewer studies aim to realize the operations management in a Smart Home. Rachidi and Karmouch [9] are one of the pioneering works toward this direction. Based on CWMP, they presented an approach for realizing self-configuration using MAPE-K model of autonomic computing. This work differs from Rachidi and Karmouch's work in that we take a task-centric approach and focus on different operations management issues, namely self-deployment and self-diagnosis.

3   Design

In this section, we shall present the approaches for supporting self-deployment and self-diagnosis using CWMP. Before turning to the details of design, the essential tasks of deployment and diagnosis must be clarified first.

- **Deployment tasks:** without self-deployment, a service engineer is required to perform on-site setup and configuration tasks for newly installed system. Moreover, when users want to buy a new service or to upgrade an existing service, a service engineer also must be present. Thus, the deployment tasks are labor intensive. In this work we focus on the following deployment tasks: (1) setup a newly installed system; (2) download and install a new software module, which is called a Deployment Unit (DU), in CWMP; (3) update an existing DU.
- **Diagnosis tasks:** It is widely recognized that robustness is a paramount concern of Smart Home users [18,19], since most of the domestic technologies are expected to work 24-7. The occupants of Smart Homes are usually non-technical users, so that Smart Home is in lack of professional system administrator. Since the consumers would be unable to pinpoint the source of failures [20], the Smart Home system must be highly reliable and be able to detect and to recover from failures autonomously.

Having considered the essential tasks of self-deployment and self-diagnosis, let us now turn to detailed mechanisms for accomplishing these tasks, which are presented in the following sub-sections.

3.1   Supporting Self-deployment Tasks

Ease of installation is a key factor that determines if a new technology can be accepted by mass consumer market. As mentioned, installation tasks including setting up a new system, a new deployment unit and upgrade an existing deployment. These tasks can be automated by a defined sequence of CWMP interactions between CPE and ACS, as shown below.

Setup a Newly Installed System: As depicted in Fig. 2a, when a CPE is installed and boots for the first time, it establishes a connection to ACS using the factory-default IP address. CPE then sends an **Inform 0 BOOTSTRAPE** to ACS, which is an event notification indicates that the CPE boots for the first time. Meanwhile, CPE also register its identity in ACS's database (ACSDB) for further management. After that, CPE sends an **Inform 4 VALUE CHANGE** to ACS, and then ACS calls CPE's **setParameter** method to configure the CPE. Finally, CPE sends **Inform M VALUE CHANGED**, which means auto-configuration of a newly installed system has been finished.

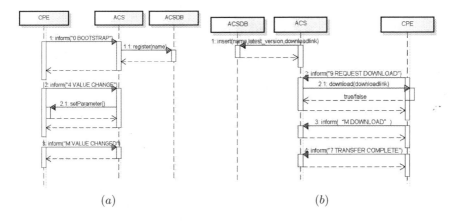

Fig. 2. Supporting self-deployment tasks: (a) setup a newly installed system (b) install a new deployment unit

Install a New Deployment Unit (DU): In this task, we assume that the user buys a new DU and users CPE is responsible for downloading and installing the DU. After the DU is ready for deployment, ACS inserts the new DUs name, latest version, and download link into a table in ACSDB. Then, CPE sends **inform 9 REQUEST DOWNLOAD** to ACS and ACS calls **download** method of CPE and pass the download link and access token as arguments. After the download process is complete, CPE sends **inform 7 TRANSFER COMPLETE** to ACS. Finally, ACS updates the current DU version number in ACSDB. The overall process is indicated in Fig. 2b.

Check and Upgrade a Deployment Unit: After a DU is installed, CPE periodically sends **inform 2 PERIODIC** to ACS to check the version of DU. If there is a new version available, CPE sends **inform 9 REQUEST DOWNLOAD** to ACS, and then ACS calls CPE **download** and pass the download link and access token as arguments. After finishing the download, CPE sends **inform 7 TRANSFER COMPLETE** to ACS. Again, the last step is to update the current DU version number in ACSDB. The overall process is indicated in Fig. 3.

3.2   Supporting Self-diagnosis Tasks

As the consumers are usually unable to pinpoint the source of failures, it is desirable that a Smart Home system being able to detect and to recover from failures autonomously. In the following, we present our design of realizing self-diagnosis tasks for CPE and DU using CWMP.

Diagnosing CPE: CWMP defines a set of CPE methods to monitor the status of CPE. CPE can periodically send **inform 2 PERIODIC** to ACS to indicate the "liveness" of itself. This mechanism is usually called "heartbeat". Once ACS

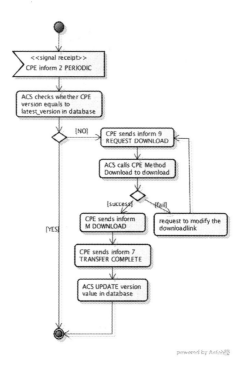

Fig. 3. Supporting self-diagnosing tasks: check and upgrade a deployment unit

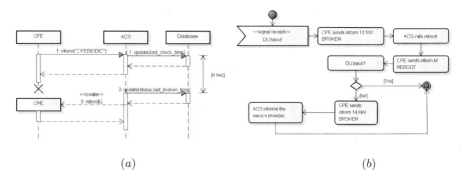

(a) (b)

Fig. 4. Supporting self-diagnosing tasks: (a) diagnose CPE (b) diagnose and recover DU

does not receive the heartbeat from CPE, it first tries to reboot CPE by calling the **Reboot** method. If CPE does not respond, it assumes that there is something wrong with CPE and then contacts the service provider automatically. The overall procedure is shown in Fig. 4a.

Diagnosing Deployment Units: Figure 4b depicts the process of diagnosing and recovering of DU. There are typically several DUs being installed on a CPE. As the original CWMP does not define event types that indicate software of hardware failures, we defined two CWMP inform extension using the **inform X** prefix, namely **inform X HW BROKEN** and **inform X SW BROKEN**. When CPE finds a DU fails, the CPE sends **inform X SW BROKEN** to ACS, and then ACS calls the CPE **Reboot** method (try to recover DU). After the CPE is rebooted, it sends **inform M REBOOT** to ACS. If the problem still exists, CPE sends **inform X HW BROKEN** to ACS and then ACS contacts the service provider automatically.

4   Implementation

This section discuss the implementation issues of our design. Currently, we implement a working prototype using JAX-WS (Java API for XML Web Services) [21]. JAX-WS is a Java API for creating web services. It is mainly used to build web services and corresponding clients that communicate using XML-based remote procedure calls, which implemented based on SOAP [11]. The detailed implementation mechanisms for accomplishing the self-deployment and self-diagnosis tasks are presented in the following sub-sections.

4.1   Self-deployment Tasks

As mentioned, deployment tasks are (1) setup a newly installed system, (2) install a new deployment unit, and (3) check and upgrade a deployment unit. The detailed steps of implementing setup a newly installed system task is listed below.

```
Step1: CPE boots
Step2: CPE sends "inform 0 BOOTSTRAP"
Step3: ACS process inform
Step4: CPE sends "inform 4 VALUE CHANGE"
Step5: ACS calls the "setParameterValue" method
Step6: CPE sends "inform M VALUE CHANGED"
```

To implement the setup a newly installed system task, the CPE Inform and ACS setPrarmeterValue methods are being used. According to the CWMP specification, argument type of the CPE Inform method is *EventStruct*, which is used to carry required information for the notification. When CPE boots, CPE sends **inform** to ACS to notify CPE has booted and then ACS uses **setParameterValue** method to set values which need to be initialized. To implement the install a new DU and upgrade DU task, we use the JDK **TimerTask** class to make CPE send inform to ACS periodically. Then, getParameterValue is called to get the version number of a DU. Upon a new install or an upgrade is required, the **downlaod** method of CPE is called. The following procedure indicates steps of DU download and upgrade.

```
Step1: CPE regularly (e.g. every n seconds) sends "inform 2 PERIODIC"
Step2: ACS calls the CPE "getParameterValue" to get CPE version number
Step3: ACS compares the version number with latest_version value in ACSDB
Step3: If version number < latest_version, CPE sends "inform 9 REQUEST DOWNLOAD"
Step4: ACS calls the CPE "download"
Step5: CPE sends "inform M download"
Step6: CPE sends "inform 7 TRANSFER COMPLETE"
Step7: Back to Step1
```

4.2   Self-diagnosis Tasks

As mentioned, deployment tasks are (1) diagnose CPE and (2) diagnose DU. Let us first take a look at the detailed steps of diagnosing CPE.

```
Step1: CPE regularly (e.g. n seconds) sends "inform 2 PERIODIC"
Step2: ACS receives heartbeats from CPE
Step3: ACS does not receive the inform from CPE for a period of time
Step4: ACS calls the CPE "reboot" method
Step5: CPE sends inform "M reboot"
Step6: ACS does not receive the inform from CPE for a period of time
Step7: ACS notifies service provider that the CPE has hardware failure
```

Similar to the check and upgrade a deployment unit task, CPE regularly sends heartbeat to ACS as heartbeat messages. When ACS does not receive heartbeat from CPE, the ACS first tries to reboot the CPE and if it still not receive the heartbeat form CPE, the ACS notifies service provider that the CPE has hardware failure. The diagnose DU task starts when the CPE finds that DU fails. Generally, this is reasonable as CPE and DU located in the same host. CPE then sends inform to ACS to notify that DU is down. Again, ACS first tries to restart the failed DU and then if it is still not recovered then ACS notifies service provider that the CPE has hardware failure. The overall process is listed below.

```
Step1: CPE finds that the DU fails
Step2: CPE sends "inform 13 SW BROKEN"
Step3: ACS calls the CPE "reboot" method
Step4: CPE sends "inform M REBOOT"
Step5: If DU still not available
Step6: CPE sends "inform 14 HW BROKEN"
Step7: ACS notifies service provider that the DU has unrecoverable failure
```

5   Evaluation

To evaluate the proposed approach, we first build two application scenarios to verify the feasibility. Then, we conducted experiments to test the performance of performing deployment and diagnosis tasks using the proposed approach.

5.1 Feasibility

We constructed application scenarios based on the prototype detailed in Sect. 4 to show the feasibility. In these scenarios, it is assumed that the user buys the TV media service, and that the service contains two DUs: TV controller and a media server. The CPE prototype is implemented and deployed on a Raspberry Pi Model B+ model and the ACS prototype is deployed on a normal PC.

Deploying the TV Media Service. First, CPE downloads the media server DU. Then, the DU is downloaded and installed on CPE. From CWMP's perspective, the objective is to install a new DU so that the operations include to check the CPE version number and sends an **inform 9 REQUEST DOWN-LOAD**. After the ACS is ready, it calls **download** of CPE. Finally, CPE replies **inform 7 TRANSFER COMPLETE** to finish this transaction. After that, the user can enjoy the TV media service, which is provided by connecting TV to the media server.

Diagnosing the TV Media Service. As the objective is to diagnose the DU and recover the DU from failures if necessary. In this scenario, the media server DU does not respond **inform 2 PERIODIC** for a period of time. CPE detects the problem and informs ACS. Then, ACS first reboots the broken DU by calling **reboot** method of CPE, then CPE responds **inform M reboot**. Unfortunately, ACS finds that the DU still not responding. Thus, CPE informs ACS about this failure via email service and the service provider sends a service engineer to the user's home to repair the service.

5.2 Performance

This section reports the experiment results that are performed to evaluate the required time to download a DU file with various sizes and to study how the quantity of DUs impacts the performance of diagnosing.

Performance of Deployment Tasks. This experiment runs deployment tasks and measure the total time required starting from CPE downloading to the completion of installation. We respectively test the file sizes ranges from 10 KB to 10 MB. Figure 5 shows the experiment results. The experiment is performed repeatedly using 10, 20, 40 and 60 CPEs. The turnaround time appears to increase linearly as the DU size increases. However, there is a steeply change after the DU size is larger than 1 MB. However, as the general size of DUs typically ranges from 100K to 200K, which, according to the results, can be downloaded within 10 s. Also, the turnaround time also increases slightly as the number of CPEs increases. The increasing trend of turnaround time is consistent among experiments with different sizes of CPEs.

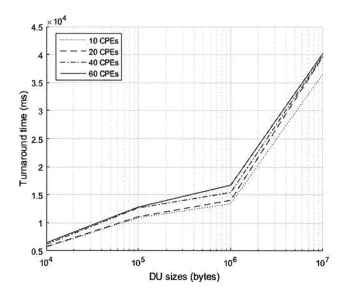

Fig. 5. Performance evaluation of DU deployment

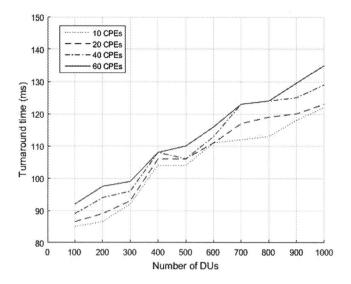

Fig. 6. Performance evaluation of DU diagnosis

Performance of Diagnosis Tasks. In this experiment, we test the impact to performance of DU monitoring when number of DU increases. The test is performed when the number of DUs is 100, 200, 300, 400, 500, 600, 700, 800, 900, and 1000. The experiment is performed repeatedly using 10, 20, 40 and 60 CPEs. The results are depicted in Fig. 6, which shows that the increase of DU number leads to the linear growth of monitor time. Based on this result, the monitor can be finished within 10 s with a reasonable number of DUs (100–200). Also,the turnaround time also increases slightly as the number of CPEs increases. Again, the increasing trend of turnaround time is also consistent among experiments with different sizes of CPEs.

6   Conclusion

This paper present approaches for supporting MAPE-K control loop between the service provider and the gateway of a Smart Home. Specifically, these approaches enables self-deployment and self-diagnosis of a Smart Home system based on CWMP. To relief the burden of developers, we take a task-centric approach, that is, we first identify the essential tasks for operations management in Smart Home and then define the interactions among entities using CWMP methods. The results show that both deployment and diagnosis have good performance when the DU size and DU number is reasonable. This paper present our fist step toward the vision of autonomic operations management. In the future, we are going to investigate how to transparently integrate the smart home network such as UPnP with the CWMP so that the devices in the home area network become also manageable.

Acknowledgements. This work is sponsored by Ministry of Science and Technology, Taiwan, under grant 104-2221-E-004-001 and 104-2815-C-004-008VE.

References

1. Gates, B., Myhrvold, N., Rinearson, P.: The Road Ahead. Wheeler Hardcover. Wheeler Pub. (1996). ISBN:9781568953069
2. Caceres, R., Friday, A.: Ubicomp systems at 20: Progress, opportunities, and challenges. IEEE Pervasive Comput. **1**, 14–21 (2011)
3. Hnat, T.W., Srinivasan, V., Lu, J., Sookoor, T.I., Dawson, R., Stankovic, J., Whitehouse, K.: The hitchhiker's guide to successful residential sensing deployments. In: Proceedings of the 9th ACM Conference on Embedded Networked Sensor Systems, pp. 232–245. ACM (2011)
4. Mennicken, S., Vermeulen, J., Huang, E.M.: From today's augmented houses to tomorrow's smart homes: new directions for home automation research. In: Proceedings of the 2014 ACM International Joint Conference on Pervasive and Ubiquitous Computing, pp. 105–115. ACM (2014)
5. Cetkovic, M., Nemet, N., Samardzic, T., Teslic, N.: Auto-configuration server architecture with device cloud cache. In: 2014 IEEE Fourth International Conference on Consumer Electronics Berlin (ICCE-Berlin), pp. 296–298. IEEE (2014)

6. Nikolaidis, A.E., Papastefanos, S.S., Stassinopoulos, G., Drakos, M.P.K., Doumenis, G., et al.: Automating remote configuration mechanisms for home devices. IEEE Trans. Consum. Electron. **52**(2), 407–413 (2006)
7. Feminella, J., Pisharoty, D., Whitehouse, K.: Piloteur: a lightweight platform for pilot studies of smart homes. In: Proceedings of the 1st ACM Conference on Embedded Systems for Energy-Efficient Buildings, pp. 110–119. ACM (2014)
8. Kephart, J.O., Chess, D.M.: The vision of autonomic computing. Computer **36**(1), 41–50 (2003)
9. Rachidi, H., Karmouch, A.: A framework for self-configuring devices using TR-069. In: 2011 International Conference on Multimedia Computing and Systems (ICMCS), pp. 1–6. IEEE (2011)
10. Bernstein, J., Spets, T.: CPE WAN management protocol. In: Technical report TR-069, Absorption in the Earth's Atmosphere, DSL Forum (2004)
11. Box, D., Ehnebuske, D., Kakivaya, G., Layman, A., Mendelsohn, N., Nielsen, H.F., Thatte, S., Winer, D.: Simple object access protocol (SOAP) 1.1 (2000)
12. Zapata, J., Fernández-Luque, F.J., Ruiz, R.: Wireless sensor network for ambient assisted living. In: Wireless Sensor Networks: Application-Centric Design, pp. 127–146. InTech (2010)
13. Enns, R., Bjorklund, M., Schoenwaelder, J.: Netconf configuration protocol. In: Network (2011)
14. Murray, B., Wilson, K., Ellison, M.: Web services distributed management: MUWS primer. OASIS WSDM Committee Draft (2006)
15. Arora, A., Cohen, J., Davis, J., Golovinsky, E., He, J., Hines, D., McCollum, R., Milenkovic, M., Montgomery, P., Schlimmer, J., et al.: Web services for management (WS-management). In: Distributed Management Task Force (DMTF) (2004)
16. Bjelica, M.Z., Golan, G., Radovanovic, S., Papp, I., Velikic, G.: Adaptive device cloud for internet of things applications. In: 2014 IEEE International Conference on Consumer Electronics, China, pp. 1–3. IEEE (2014)
17. Nemet, N., Radovanovic, S., Cetkovic, M., Ikonic, N., Bjelica, M.Z.: User self-help module for a device management cloud based on the TR-069 protocol. In: 2014 IEEE Fourth International Conference on Consumer Electronics Berlin (ICCE-Berlin), pp. 199–201. IEEE (2014)
18. Edwards, W.K., Grinter, R.E.: At home with ubiquitous computing: seven challenges. In: Proceedings of the 3rd International Conference on Ubiquitous Computing (UbiComp 2001), pp. 256–272 (2001)
19. Grimm, R., Davis, J., Hendrickson, B., Lemar, E., MacBeth, A., Swanson, S., Anderson, T., Bershad, B., Borriello, G., Gribble, S., Wetherall, D.: Systems directions for pervasive computing. In: Proceedings of the 8th Workshop on Hot Topics in Operating Systems (2001)
20. Dixit, S., Prasad, R.: Technologies for Home Networking. Wiley-Inderscience (2008). ISBN:9780470196526
21. Kohlert, D., Gupta, A.: The Java Api for XML-based web services (JAX-WS) 2.1 (2007)

An Event-Driven Adaptive Cruise Controller

Jessica Jreijiry and Mohamad Khaldi$^{(\boxtimes)}$

Department of Electrical Engineering, Faculty of Engineering,
University of Balamand, P.O. Box 100, Tripoli, Lebanon
mohamad.khaldi@balamand.edu.lb

Abstract. The proposed Event-Driven Adaptive Cruise Controller (EDACC) serves as longitudinal driver assistant by accelerating, decelerating, and stopping the host vehicle given the readings of various sensors. EDACC uses a simplified non-linear longitudinal vehicle model and a hierarchical control structure of PI and PID controllers integrated with an embedded specific logic. In addition to the adaptability of the ACC, events can be added such as host vehicle entering speed-limit zone and/or having punctured tire. The proposed EDACC is designed and implemented using Matlab and Simulink to simulate multiple scenarios.

Keywords: Adaptive Cruise Control · Event-driven systems · Longitudinal vehicle dynamics

1 Introduction

Adaptive Cruise Control (ACC) has the purpose of maintaining the speed of the host vehicle at a speed set by the driver as a conventional cruise control system would. ACC also has the ability to detect a slower leading vehicle and follow it at constant speed equal to that of the leading vehicle while maintaining a constant relative time headway policy.

The proposed Event-Driven ACC (EDACC) serves as longitudinal control driver assistant by accelerating, decelerating, and stopping the host vehicle given the measurements of the range sensors that measure the relative speed and distance of the leading vehicle [1]. Pressure sensors that can detect the presence of a tire puncture or leak and pre-entered data combined with GPS technology that transfers the speed limit zone to the EDACC if and when the vehicle enters a speed limit zone [2]. EDACC does not attempt any lateral control, such as lane change. The speed of the host vehicle is specified by the controller depending on the specified mode of operation by the logic incorporated in the Stateflow Chart. If the logic leads to conventional cruise mode, the controller will only attempt to achieve a constant vehicle speed equal to that set by the driver. If the logic leads to vehicle following mode, the controller will attempt to maintain a speed equal to the leading vehicle, i.e. a null relative speed, while reserving an inter-vehicle distance with constant time headway policy, where the constant time headway is specified by the driver. The controller must achieve the desired distance and speed, without colliding with the leading vehicle and without using excessive acceleration and deceleration that may discomfort or endanger passengers. Accelerating and

© Springer International Publishing Switzerland 2016
H. Fujita et al. (Eds.): IEA/AIE 2016, LNAI 9799, pp. 983–994, 2016.
DOI: 10.1007/978-3-319-42007-3_83

decelerating the vehicle may lead to discomfort and probably to safety endangerment. The controllers manipulate the throttle and brake and hard braking may lead to swaying and in some cases sliding [3, 4].

There are many controllers which have been proposed for ACC [5–7]. From Proportional-Integral (PI), Proportional-Integral-Derivative (PID), and Constant Time Gap (CTG) controllers in [6], to Sliding Mode Control in [7] and Model-Predictive Control (MPC) in [5]. There are also different inter-vehicle distance policies, such as constant spacing, constant time headway policy, and variable time gap policy [6–8]. Constant spacing policies are deemed to be unsuitable for autonomous control applications [9], since they do not ensure string stability. The main purpose of this study is the introduction of discrete-events to the conventional ACC model.

2 Vehicle Model

The vehicle model, introduced in this study, is constituted of several subsystems, divided into two large groups, longitudinal vehicle dynamics and driveline dynamics. Longitudinal vehicle dynamics present all external forces acting on the vehicle; rolling resistance, gravitational, aerodynamic drag, and longitudinal tire forces, while driveline dynamics present the internal forces; engine, transmission and wheel dynamics.

2.1 Longitudinal Vehicle Dynamics

The application of Newton's law presents the longitudinal dynamics equation [6],

$$m\ddot{x} = F_x - F_{ad} - R_x - F_g \qquad (1)$$

where F_x is the longitudinal tire force, F_{ad} is the equivalent longitudinal aerodynamic drag force, R_x is the force due to the rolling resistance at the tires, F_g is the gravitational force given by $mg \sin \theta$, where m. is the mass of the vehicle, g is the acceleration due to gravity, and θ is the road inclination angle. The angle θ is defined to be positive clockwise when the longitudinal direction of motion x is towards the left. It is defined to be positive counter clockwise when the longitudinal direction of motion x is towards the right.

The aerodynamic drag force can be determined through the following function [6],

$$F_{ad} = \frac{1}{2}\rho A_F C_d (V_x + V_{wind})^2 = C_{ad}(V_x + V_{wind})^2 \qquad (2)$$

where ρ is the mass air density, C_d is the aerodynamic drag coefficient, A_F is the frontal area of the vehicle, which is the projected area of the vehicle in the direction of travel, V_x is the longitudinal vehicle velocity, V_{wind} is the wind velocity, in this study wind velocity is considered negligible in comparison to vehicle velocity and will be - omitted when considering F_{ad}.

Wong developed the following relation between frontal area and car mass [10],

$$A_F = 1.6 + 0.00056(m - 765) \tag{3}$$

where, $800\,\text{kg} \le m \le 2000\,\text{kg}$, and the drag coefficient C_d for automobiles ranges between 0.3 and 0.4. The rolling resistance can be obtained from [11],

$$R_x = C_r mg \cos \theta. \tag{4}$$

Where C_r is the rolling resistance coefficient.

F_x is a function of the slip ratio and the friction coefficient between tire and road [6]. But in this paper, the slip between tires and road is assumed to be zero, i.e. wheel velocity will be considered equal to vehicle velocity; and the longitudinal tire force expression will be extracted from the wheel dynamics.

2.2 Driveline Dynamics

The driveline is composed of engine, transmission, wheel, torque converter and final drive [6]. The engine dynamics equation is,

$$I_e \dot{\omega}_e = T_e - T_p \tag{5}$$

where, T_e is the net engine torque, T_p is the pump torque, I_e is the engine's moment of inertia, and ω_e is the engine angular velocity. In the simulation the net engine torque is obtained through a look up table to form the engine map that determines this torque as a function of throttle opening percentage and engine rotational speed.

The torque converter is considered to be locked, and that there is ideal fluid coupling, which yields the following equation [6],

$$T_t = T_p = -6.1 \times 10^{-6} \omega_e^2 \tag{6}$$

where, T_t is the turbine torque. And the transmission dynamics equation is [6],

$$I_t \dot{\omega}_t = T_t - RT_w \tag{7}$$

where T_w is the wheel torque, I_t is the transmission's moment of inertia, ω_t is the transmission angular velocity, and R the total gear speed transmission ratio.

As for the wheel dynamics equation it is given by [6],

$$I_w \dot{\omega}_w = T_w - T_{br} - r_{eff} F_x \tag{8}$$

where T_{br} is the brake torque, I_w is the wheel's moment of inertia, ω_w is the wheel angular velocity, and r_{eff} is the wheel's effective radius. Given the previous assumptions that there is zero slip and the transmission is locked; i.e. $T_t = T_p$, $\omega_w = R\omega_e$ and $\omega_t = \omega_e$ leads to the following expression of F_x,

$$F_x = \frac{-6.1 \times 10^{-6}\omega_e^2 - \dot{\omega}_e[I_t + I_w R^2] - T_{br}R}{r_{eff}R} \quad (9)$$

And the longitudinal vehicle dynamic Eq. (1) becomes;

$$\ddot{x} = \frac{1}{m}\left[\frac{\dot{\omega}_e[I_t + I_w R_T^2] + 6.1 \times 10^{-6}\omega_e^2 - T_{br}R_T}{r_{eff}R_T} - C_{ad}V_x^2 - mg(C_r\cos\theta - \sin\theta)\right]$$

$$(10)$$

3   ACC Controllers

The ACC itself is composed of the Lower Level Controller (LLC) and Upper Level Controller (ULC) [5]. The ULC role is to decide on whether the vehicle should be in cruise control mode or in vehicle following mode, and then compute the desired acceleration, a_d. The desired acceleration is then transferred to the LLC that uses host vehicle dynamics to determine the throttle position or brake input torque, T_{br} [6].

3.1   System Limitations & Switching Logic

The ACC is restricted by the following physical constraints:

1. An ACC vehicle cannot have a negative velocity during a transitional maneuver [1].
2. For vehicle drag force limitation [6], driving comfort issues [4–6], and minimum jerking purposes [5], an ACC vehicle's acceleration is limited between –4.9 m/s^2 and 2.45 m/s^2 [12].
3. The inter-vehicular distance between host and leading vehicles must always remain bigger than a minimum safety distance d_{min} that is needed to avoid a collision if the vehicles applied brakes to the fullest at the same time, and smaller than a distance that can allow another car to cut-in between the two vehicles [6].

The ACC needs to follow a switching strategy that determines whether the ACC should be in regular CC mode or if it should be trailing a vehicle and enter the ACC mode. The following approach is based on the logical operation algorithm suggested in [13], which considers that an ACC vehicle does not need to switch modes, unless the leading vehicle has a lower velocity than that of the host vehicle, and is at distance less or equal to the change distance. Table 1 identifies the ACC's mode of operation, function of the set vehicle speed v_{set}, leading vehicle speed v_l, inter-vehicular distance d, and the change distance d_c.

For a constant headway spacing policy, d_c has the following relationship [13];

$$d_c = l + d_0 + t_h v_x \quad (11)$$

Table 1. Switching logic between CC and ACC

	$v_l < v_{set}$	$v_l \geq v_{set}$
$d < d_c$	ACC	CC
$d \geq d_c$	CC	CC

Where, l is the vehicle's length, d_0 is an additional safety distance needed to keep the vehicles from colliding, t_h is the time headway constant and v_x is the vehicle's longitudinal speed.

3.2   Upper Level Controller

The ULC is responsible for calculating the acceleration at which the vehicle must travel in order to attain the desired speeds and spacing or headway from other vehicles. This study will be based on the simple ACC model, where the control input is a first-order lag [6]. The model for this controller is,

$$\ddot{x}_{des} = \frac{1}{\tau s + 1} u \tag{12}$$

where $\ddot{x}_{des}$ is the desired acceleration in order to meet the system's requirements, τ is the time lag, and will be considered equal to 0.5 s in this study as analytical and experimental studies show that this is its corresponding value, and u is the system's control input.

Cruise Mode PI Controller. A typical controller for the cruise control mode is the Proportional-Integral controller [6],

$$u(t) = -k_p(v_x - v_s) - k_I \int_0^t (v_x - v_s)dt \tag{13}$$

where v_s is the driver's set vehicle speed, $k_p = 0.473$ is the proportional gain, and $k_I = 6.782$ is the integrator's proportional gain, these values were computed using Matlab optimal design.

3.3   Lower Level Controller

The LLC is responsible for interpreting the calculated acceleration value in the upper controller to yield the throttle position and the torque brake inputs. Here, all assumptions that took part of vehicle modeling will be adopted, and the longitudinal vehicle speed it is approximated to be [6],

$$\dot{x} \cong r_{eff}\omega_w. \tag{14}$$

Hence, the longitudinal acceleration can be expressed as,

$$\ddot{x}_{des} = r_{eff} R \dot{\omega}_e .$$ (15)

Under these assumptions the dynamics relating ω_e, T_e, T_{br} can be linked with the expression [6],

$$\dot{\omega}_e = \frac{T_e - C_{ad} R^3 r_{eff}^3 \omega_e^2 - R\left(r_{eff} R_x + r_{eff} mg \sin \theta + T_{br}\right)}{J_e}$$ (16)

where, $J_e = I_e + I_t + R^2 I_w + m R^2 r_{eff}^2$.

Since using brake and throttle pedals at the same time will not be allowed, T_{br} will be zero during throttle control and T_e can be computed using the following relationship,

$$T_e = \frac{J_e}{R r_{eff}} \ddot{x}_{des} + \left(C_{ad} R^3 r_{eff}^3 \omega_e^2 + R\left(r_{eff} R_x + r_{eff} mg \sin \theta\right)\right).$$ (17)

Then using an inverse engine map, having obtained T_e, and deriving ω_e from the expression of $\ddot{x}_{des}$, the percentage of throttle input is obtained [14].

If the ULC decides that braking is needed then the LLC will use the actual engine torque, which is obtained from the vehicle itself, will be used, and T_{br} will be obtained using the following expression,

$$T_{br} = \frac{T_{e_{act}} - \dfrac{J_e}{R r_{eff}} \ddot{x}_{des} + \left(C_a R^3 r_{eff}^3 \omega_e^2 + R\left(r_{eff} R_x + r_{eff} mg \sin \theta\right)\right)}{R}$$ (18)

Connecting both the ULC and LLC will give the overall ACC controller. The inputs to the ULC are the host vehicle speed, v_x, the leading vehicle speed, v_l, desired vehicle speed or the set speed, v_s, and the measured distance, d. the ULC computes the desired acceleration, a_d, for the LLC. The LLC also takes in from the vehicle model the host vehicle speed, v_x, the engine angular speed, Ne, and the engine torque, Te. The LLC can also consider the Road Gradient in determining the throttle position in percentage and the brake torque, T_{br}, which will be returned to the vehicle, and the overall model of interaction between ACC and vehicle is shown in Fig. 1. for the purpose of simplifying the simulation, the range sensor is considered to give only the leading vehicle's speed, even though its real function is to give the relative velocity, but since the leading vehicle's velocity is equal to the relative velocity added to the host vehicle velocity, this simplification can be made. Also the relative position can be obtained through the integration of the difference between the two vehicles velocities.

4   Event Driven ACC

A Discrete Event System (DES) is a dynamic system that deals with specific events that may occur abruptly at an unknown instant. In order to relate DES to ACC, one can consider possible events that can occur while driving [15]. The ACC must deal with a

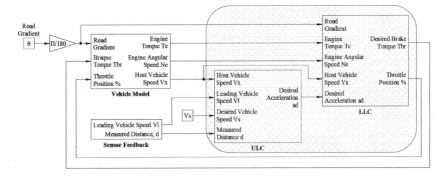

Fig. 1. Host vehicle with Adaptive Cruise Control that includes both the upper and lower level controllers

car that cuts in, or a leading car that slows down abruptly, a road block, a car that fully stopped at a red light, etc. A StateFlow Chart (SFC) can represent graphically a Finite State Machine (FSM) which simplifies and facilitates the application of DESs and the inclusion of special events [14]. The use of a SFC to represent transitions between modes facilitates the modeling of an Event-Driven System (EDS). In another words, EDS gives the ACC the ability to adapt and include new conditions to improve its capability of assisting the driver in case of emergencies and abrupt incidents that could lead to an accident, as well as being the milestone that leads to complete vehicle autonomy.

The use of a stateflow chart to represent transitions between modes facilitates the modeling of event-driven systems. In other words, the event-driven system gives the ACC system the ability to adapt and include new conditions to improve its ability to assist the driver in case of emergencies and abrupt incidents that could lead to an accident and as well as being the milestone of the road that leads to complete vehicle autonomy.

The ULC relies on the stateflow chart, Fig. 2, to determine the appropriate mode of operation [15].

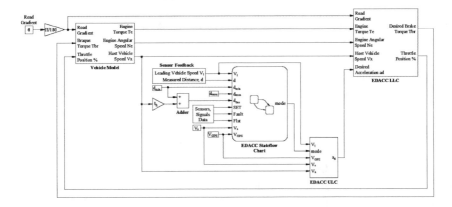

Fig. 2. The Event-Driven ACC

In order to give the driver full control over the activity of the ACC, the first two evident states of an ACC are whether it is ON or OFF, these states are conveyed to the controller through the SET signal, coming from the ACC push button present on the car's dashboard. After activation, a conventional ACC, goes through four different states [15]:

1. When there are no close vehicles travelling at a speed lower than the driver's set speed v_s, the ACC's goal is to get the host vehicle to travel at this exact set speed, this state shall be known as: Cruise.
2. When a preceding vehicle is detected, the controller's first goal is to reach the desired inter-vehicle pre-set distance d_{des}, it must start slowing down the car after it reaches the maximum following headway distance d_{max}, and then after reaching the desired distance it will move on to the next state, this state shall be known as: Gap_Reduction.
3. After the inter-vehicular distance is achieved, the controller's next goal is to achieve the leading vehicle's speed v_l in order to maintain this distance, this state shall be known as: Adaptive_Speed.
4. If the controller was not able to slow down the vehicle fast enough to keep the vehicle within the desired headway, then a braking operation must occur before the car reaches an unsafe distance d_{min}, i.e. a distance smaller than that needed to avoid a collision if the two vehicles were to apply maximum braking at the same time, this state shall be known as: Brake.

In order to simplify and lessen the states, the ACC controller can be used to merge the Gap_Reduction state and the Adaptive_Speed state, into one state that achieves both their goals, this state shall be known as: Trail.

In this paper, an EDACC is introduced, and presents two new events that have been added to the conventional functions of an ACC. The first is the ability to detect a puncture or leak in a tire, Flat_Tire, through a pressure sensor, and the second uses pre-entered data that uses GPS technology to determine speed limit zones, Speed_Limit, and transmits it to the ACC in order to calculate the needed throttle or brake actuators position for the vehicle to travel at the speed limit v_{GPS}, while also notifying the driver that he has entered a speed limit zone with the help of GPS technology [2]. A flat tire requires a special treatment, as a first step the throttle input must be maintained momentarily, and then slowly and gradually must be decreased, and hard braking is disallowed before the car comes to a safe stop [16]. As for the speed limit v_{GPS}, it can be treated the same as the driver's set speed. The details of the stateflow chart shown in Fig. 2 are exemplified in Fig. 3.

Table 2 shows explicitly how the system transitions from one state to the other.

5 Simulation

To simulate the adaptability of the proposed ACC, the desired host vehicle speed, v_s, is set to 60 km/h. Figure 4 shows the host vehicle speed, v_x (with CC dashed line and with ACC solid line) against the assumed leading vehicle speed, v_l (dotted line) with a headway of 8 m. At zero simulation time the host vehicle is traveling at 36 km/h where

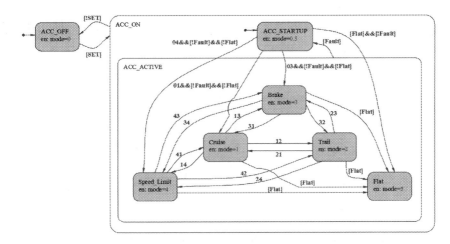

Fig. 3. Stateflow chart of the introduced Event-Driven ACC

Table 2. EDACC state transitions

From	To	Condition	State
OFF	ON	SET = 1	[SET]
ON	OFF	SET = 0	[!SET]
Active	Standby	Fault = 1	[Fault]
Standby	Cruise	$v_{GPS} \geq v_s$ and $d \geq d_{max}$	01&&[Flat]&& [!Fault]
Standby	Brake	$d \leq d_{min}$	03&&[Flat]&& [!Fault]
Standby	Speed_Limit	$v_{GPS} < v_s$ and $d \geq d_{max}$	04&&[Flat]&& [!Fault]
Standby	Flat	Flat = 1 and Fault = 0	[Flat]&&[!Fault]
Cruise	Trail	$d < d_{max}$ and $v_l < vs$ and $d > d_{des}$	12
Cruise	Brake	$d < d_{min}$	13
Cruise	Speed_Limit	$(v_{GPS} < v_s)$ and $(d \geq d_{max})$ or $(v_s \leq v_l$ and $d > d_{min})$	14
Cruise	Flat	Flat = 1	[Flat]
Trail	Cruise	$(d \geq d_{max})$ or $(v_s \leq v_l$ and $d > d_{min})$	21
Trail	Brake	$d < d_{min}$	23
Trail	Speed_Limit	$(d \geq d_{max})$ or $(v_{GPS} \leq v_l$ and $d > d_{min})$	24
Trail	Flat	Flat = 1	[Flat]
Brake	Cruise	$(d \geq d_{max})$ or $(v_s \leq v_l$ and $d > d_{min})$	31
Brake	Trail	$d_{des} < d$ and $v_l < v_s$ and $d > d_{min}$	32
Brake	Speed_Limit	$(d \geq d_{max})$ or $(v_{GPS} \leq v_l$ and $d > d_{min})$	34
Brake	Flat	Flat = 1	[Flat]
Speed_Limit	Cruise	$(v_{GPS} \geq v_s)$ and $(d \geq d_{max})$ or $(v_s \leq v_l$ and $d > d_{min})$	41
Speed_Limit	Trail	$d < d_{max}$ and $v_l < v_s$ and $d > d_{des}$	42
Speed_Limit	Brake	$d < d_{min}$	43
Speed_Limit	Flat	Flat = 1	[Flat]

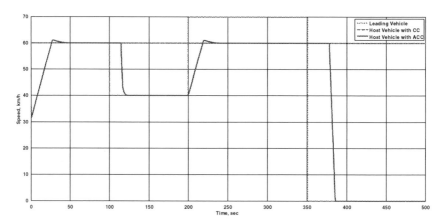

Fig. 4. Comparative performance between two host vehicles one with CC and another with ACC with desired speed was set to 60 km/h

the ACC system was activated at this instant and the driver sets the speed to 60 km/h. Note that when the leading vehicle speed, $v_l = 40$ km/h at $100 < t < 200$, the host vehicle follows the leading vehicle. When the leading vehicle speed, $v_l = 70$ km/h at $200 < t < 350$, the host vehicle speed remains at the set speed of 60 km/h. And eventually, when the leading vehicle stops, the host vehicle simply follows and stops for $t \geq 350$. Note that a host vehicle with a regular CC, its speed is equal to the set speed of 60 km/h at all times regardless what the leading vehicle speed is.

The EDACC model was tested through the following scenario, Fig. 5, with a headway of 8 m between the host and leading vehicles: at zero simulation time the host vehicle is traveling at 36 km/h where the EDACC system was activated at this instant and the driver sets the speed to 60 km/h.

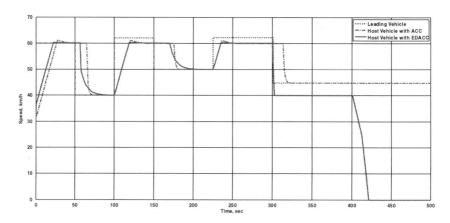

Fig. 5. Comparative performance between two host vehicles one with ACC and another with EDACC with desired speed was set to 60 km/h

A leading vehicle is moving at a speed, $v_l = 60$ km/h at $t \geq 0$. At $50 < t < 100$, the leading vehicle slows down to 40 km/h. Consequently, the host vehicle follows and its speed drops to 40 km/h. When the leading vehicle speed, $v_l = 65$ km/h at $100 < t < 150$, the host vehicle speed remains at the set speed of 60 km/h. Then at $t = 150$, a new vehicle traveling at 50 km/h cuts in and then changes lane at $t = 225$. The host vehicle adjusts its speed to 50 km/h until the intruder vehicle changes lane at $t = 225$ and then picks up its speed to the set speed of 60 km/h since the leading vehicle speed is above 60 km/h between $225 < t < 300$. Until now the performance of two host vehicles one with ACC and the other with EDACC is the same. However, at $t = 300$, the EDACC slows the host vehicle down to 40 km/h because it enters a 40 km/h speed limit zone whereas the host vehicle with ACC kept following the leading vehicle and it did not honor the speed limit. Finally, At $t = 400$, the host vehicle with EDACC stops because of a warning message indicating that one of the tires' air pressure is low while the host vehicle with a regular ACC kept going and did not stop for this emergency.

6   Conclusion

In this paper, a simplified longitudinal vehicle model was derived. The Adaptive Cruise Controller (ACC) composed of the upper and lower level controllers was also designed and simulated. Also, the importance of Discrete Event System (DES) and how state-flow charts can be used in order to facilitate the design of an Event-Driven Adaptive Cruise Controller (EDACC) was introduced. Two new modes of operation, namely speed-limit zone and flat tire occurrence, have been added to the conventional ACC model to yield the EDACC. Future work can be done on improving the system response in going from one state to another. This work also suggests that well designed EDACC can ultimately lead to an autonomous vehicle.

References

1. Ali, Z.: Transitional controller design for adaptive cruise control systems. Ph.D. thesis, University of Nottingham, UK (2011)
2. Labbani, O., Rutten, É., Dekeyser, J.L.: Safe design methodology for an Intelligent cruise control system with GPS. In: 64th IEEE Vehicular Technology Conference, Montréal, Québec, Canada, September 2006 (2006)
3. Hiraoka, T., Kunimatsu, T., Nishihara, O., Kumamoto, H.: Modeling of driver following behavior based on minimum-jerk theory. In: 12th World Congress on ITS, San Francisco, USA, November 2005, vol. 113, no. 4, pp. 27–30 (2005)
4. Martinez, J.J., Canudas-de-Wit, C.: A safe longitudinal control for adaptive cruise control and stop-and-go scenarios. IEEE Trans. Control Syst. Technol. **15**(2), 246–258 (2007)
5. Luo, L., Liu, H., Li, P., Wang, H.: Model predictive control for adaptive cruise control with multi-objectives: comfort, fuel-economy, safety and car-following. J. Zhejiang Univ. Sci. A **I**(3), 191–201 (2010)
6. Rajamani, R.: Vehicle Dynamics and Control. Mechanical Engineering Series. Springer, New York (2006)

7. Hedrick, J.K., Gerdes, J.C., Maciuca, D.B., Swaroop, D.: Brake system modeling, control and integrated brake/throttle switching phase I. California PATH Research report, University of California, Berkeley, California (1997)

8. Wang, J., Rajamani, R.: Should adaptive cruise control systems be designed to maintain a constant time gap between vehicles? IEEE Trans. Veh. Technol. **53**(5), 1480–1490 (2004)

9. Wang, J., Rajamani, R.: Adaptive cruise control system design and its impact on traffic flow. In: Proceedings of the American Control Conference, Anchorage, Alaska, May 2002 (2002)

10. Wong, J.Y.: Theory of Ground Vehicles, 4th edn. Wiley, New York (2008)

11. Shakouri, P., Ordys, A., Askari, M., Laila, D.S.: Longitudinal vehicle dynamics using Simulink/Matlab. In: UKACC International conference on CONTROL, Coventry, 6–10 September 2010 (2010)

12. Kulkarni, M., Shim, T., Zhang, Y.: Shift dynamics and control of dual-clutch transmissions. Mech. Mach. Theor. **42**(2), 168–182 (2007)

13. Shakouri, P., Czeczot, J., Ordys, A.: Adaptive cruise control system using balance-based adaptive control techniques. In: 17th International Conference on Methods and Models in Automation and Robotics (MMAR), Miedzyzdrojie, Poland, August 2012, pp. 510–515 (2012)

14. Güvenç, B.A., Kural, E.: Adaptive cruise control simulator: a low-cost, multiple-driver-in-the-loop simulator. IEEE Control Syst. Mag. **26**(3), 42–55 (2006)

15. Breimer, B.: Design of adaptive cruise control model for hybrid systems fault diagnosis. MSc thesis, Department of Computing and Software, McMaster University (2013)

16. How to control your car during a Puncture or Tyre Blowout, June 2014. roaddriver.co.uk, online driving community. Article No. 37 RoadDriver 2010

Design and Implementation
of a Smartphone-Based Positioning System

Chun-Chao Yeh[✉], Yu-Ching Lo, and Chin-Chun Chang

Department of Computer Science,
National Taiwan Ocean University, Keelung, Taiwan
{cceyh,10057052,cvml}@mail.ntou.edu.tw

Abstract. In this research study, we investigate feasibility of a smart
phone based positioning system. The system allows users to get their
current location information via their smartphones. The client software
required for user smartphones can be distributed and installed easily
through standard mobile Apps. Once the mobile App is ready in user
smartphones, they just need to take one or two pictures at target objects
around, and upload the pictures to the positioning system through Internet connection. The positioning system will identify the location, based
on the upload pictures and other related information. We conducted a
field test to identify locations among a building complex in a campus. We
collected thousands of images taken from outward appearance of several
buildings in a campus at different days. The images were classified into
17 classes, based on the location the picture images were taken. From the
experiment results, we found that under well control of image quality for
both training and testing images the correct classification rate can be as
high as 98.3 %. Even under the cases of large scope-of-view mismatching
between the raining images and the tested images, the proposed scheme
can still generate good correct classification rate (86.7 % and 77.3 % for
both covering and covered cases respectively) compared with random
guest ($1/17 = 5.88$ %).

Keywords: Feature extraction · Bag of word · Classifier

1 Introduction

When people visit a place they do not familiar with, it would be much helpful if they can easily get the location information about where their are. GPS
(Global Positioning System) can provide such a functionality. Combining with
geographic information and GPS sensors, positioning service (for example Google
Map service)have been one of most popular services of smart phones especially
when people are traveling outdoors. However, due to the LOS (line of sight)
constraint of satellite signals, GPS services can not be applied to indoor environment. Aside from providing useful geographical information to users, indoor
positioning information can be applied to a variety of location-based services
(LBS). Consequently, how to effectively provide indoor positioning information

© Springer International Publishing Switzerland 2016
H. Fujita et al. (Eds.): IEA/AIE 2016, LNAI 9799, pp. 995–1006, 2016.
DOI: 10.1007/978-3-319-42007-3_84

attracts researchers to investigate various feasible schemes with a variety of sensor devices under different application scenarios. These approaches are such as WiFi [1,2], Zigbee [3], bluetooth [4], RFID [5], to name a few. A good survey of recent indoor positioning development can be found in [6].

Most of proposed indoor positioning systems (IPS) depend on interactions of well-deployed active sensor nodes (e.g. WiFi Access Point(AP), Zigbee node, and RFID readers/transmitter) and corresponding receiver/tag devices carried by users. For example, most WiFi-based IPS use the fingerprints of receiving signal strength indication (RSSI) and/or the node id information detected by the user WiFi device from all the well deployed WiFi APs. comparing the fingerprint the user detected at current location with a well-established WiFi AP RSSI fingerprint database, the IPS system predicts a most likely location the user located. Similar mechanisms can be applied to the cases of Zigbee and RFID.

Different from many others, in this research study we propose a scheme without the need to deploy active sensor nodes precisely. It is not only because of avoiding high deployment cost but also because it is possible that we can not do it due to time and other constraints. Moreover, availability of user devices should be taken into account. It would be more easy to use the devices most of users available in hands. For example, among the four sensor devices aforementioned: WiFi, Bluetooth, Zigbee, and RFID. We think the cases for WiFi and Bluetooth are more feasible as the two sensors (WiFi and Bluetooth) are widely available in smartphones. In contrast, the cases for Zigbee and RFID are more restricted since most of people do not have a zigbee receiver or the specified RFID tag in hands at any time. While it is possible to enable the services if the IPS system providers offer the zigbee receiver/RFID tag to users, it introduces user device management and accessibility overhead.

In this paper, we present an universal position identification scheme for both indoor and outdoor environment. The proposed scheme utilizes visual and orientation sensor information to detect user location indoors (and outdoors). We designed and implemented a smartphone based positioning system which allows users to get their current location information via their smartphones. The client software for user smartphones can be distributed and installed easily through standard mobile Apps. Once the mobile App is ready in user smartphones, they just need to take a picture at target objects, and upload the picture to the remote positioning detection servers. The servers identify the location, based on the upload picture and other related information.

The remainder of the paper is organized as follows. In Sect. 2, we present the proposed smartphone-based positioning system. Section 3 describes detailed design and implementation of the proposed system. In Sect. 4, we present the results of performance evaluation of the proposed system. Finally, we describe future work in Sect. 5.

2   Smartphone-Based Positioning Service System

Our intention for this research study is to develop a positioning system to help internet users to acquire position information of the location where the users currently are. For example, when we visit an organization/institute such as schools, hospitals, museums, and industrial plants, we might get lost in the direction to the target spot and wonder where we are now in the building complex. While it might be helpful to visitors if the organization/institute can offer digital map and/or floor plan information via internet, it would be much better if the system can provide visitor position information to let the visitors know where they are right now. Moreover, with enabling the position information, many location based services can be offered. To provide a cost-effective positioning system, we consider following system demands.

- High accessibility. Ideally, we hope users are able to use the positioning information service with a device they are available in hands. No extra access devices are needed as the premise for users to use the positioning services.
- High Usability. The system is easy to use. No long learning curve is expected.
- Low deployment cost. Here the cost includes time and TOC (total cost of ownership, including for example Capex (Capital expenditures) and Opex (Operational expenditure)). We hope the TOC is as low as possible and the time to deploy such a system is as short as possible.
- Low deployment limits. We hope the deployment can be easily done without much of construction/engineering work. For example, if the deployment involves to install massive active sensors to cover all the building complex (both internal and external), both sensor device installation and related wiring (for data and power transmission) issues would be a big concern.
- High system performance. This includes, for example, firs system response and high correction rate of user position detection.

With such design considerations and application scenarios in mind, we proposed a smartphone-based position detection system as shown in Fig. 1. On the users side, we assume users have a smartphone in hands and they are able to connect to internet via mobile telecommunication networks (e.g. 3G/4G mobile networks) or WiFi wireless LAN. Considering high penetration rate of mobile phone usage and popularity of smartphone devices, the assumptions of user smartphone availability and internet connection are valid at most of the time. Meanwhile, we assume the users have installed the mobile App for the positioning service provided by the organization/institute. This installation procedure is same as for many others mobile Apps, and should be no difficulty for most of internet users. After properly install the positioning service App on their smartphones, visitors can access the positioning service via the following three steps as shown in Fig. 1.

1. When the visitor would like to know the position where he/she located, he/she actives the position service App and finds a scene around him/her with rich texture, then takes a picture at the target objects.
2. Follow the instruction of the position service App to upload the picture and other related information to the remote server.
3. The remote server uses the user upload picture and the related information to generate a fingerprint for the user location. Based on the fingerprint, the

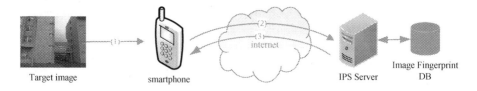

Target image   smartphone   IPS Server   Image Fingerprint
                                                                                                    DB

Fig. 1. Architecture concept of the proposed smartphone-based position detection system.

server find a best match location corresponding to the user upload data, and send the result to the user. For example, a map/floor plan corresponding to the user location is sent to the smartphone of the request user, with a mark on the map/floor plan image to indicate user current location.

3   Position Detection System Design

Due to characteristics of image signal presented in a picture, it is a non-trivial task to find a best match between the picture uploaded by a user and candidate pictures stored in the database. Since user may take a picture at different location and/or with different angle and scaling, it would be hard to guarantee there always exists a picture in the database with high similarity to the picture uploaded by users. More likely, the scene of the user upload picture is overlapped with multiple pictures in the database. Each of the overlapped pictures cover parts of the scene of the uploaded picture. In addition, for each overlapped scene between the database picture and the uploaded picture, they are more likely with different scaling size, viewing angle, and lighting effect. And, parts of the scenes included in the uploaded picture might not be covered in any pictures in the database. For example, a moving object (such as a bird, a car or a person) might run into the scene while users take the picture to upload. Even two pictures with same focused objects, there might exist big differences in term of visual signal between the two pictures, due to the different effects and parameters such as shooting angles, depth of field, focus length, lighting, exposure, and aperture, to name a few. How to do the image matching is the key design problem in the proposed positioning system.

We use both of image and orientation information uploaded by users. The orientation information is obtained by smartphone's orientation sensor sampled when the picture to upload is taken. We use the orientation information associated with the image to narrow down our search space to focus on those images in the database associated with similar orientation angles. The main source information to find out user current location is based on the picture uploaded by users. Instead of directly comparing two images pixel-by-pixel, we adopt indirect matching scheme based on machine learning strategies. Figure 2 shows our image matching scheme, which is inspired by [7]. In the following subsections we describe how we do the feature extraction and classification.

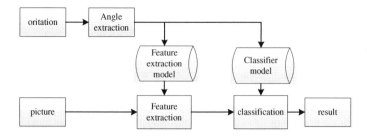

Fig. 2. Block diagram of the proposed position detection system.

3.1 Image Feature Extraction

We use SURF (Speeded Up Robust Features) algorithm [9] to extract image features. SURF is a popular image feature extraction scheme in computer vision and image classification. SURF is improved from SIFT(Scale-invariant feature transform) algorithm [8]. One of main reasons for us to use SURF algorithm is the image features extracted by SURF/SIFT is scale-invariant, which is important to our system as the images taken by users and images collected for comparison are very likely with different scale factors even both of the two pictures focus on the same objects.

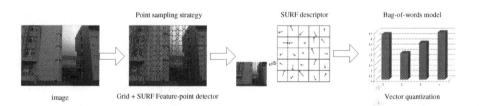

Fig. 3. Image feature extraction steps.

Figure 3 shows the feature extraction scheme. For a image I_i, first we apply SURF point sampling strategy to get some SURF key points. Then, these SURF key points are further manipulated to get a set of SURF key point descriptors $s_i = \{s_{i,j} | 1 \leq j \leq N_i\}$, where N_i is the number of SURF key points in the image I_i. Each of the $s_{i,j}$ is a D dimension vector of real number, where the number D is a predefine integer (=128 in current implementation) representing the dimension of a SURF key point descriptor. These N_i SURF key point descriptors ($s_{i,j}, 1 \leq j \leq N_i$) represent the visual information corresponding to the N_i SURF key points in the image I_i. To predict which geographical location is most likely to be the place the picture was taken, we apply bag-of-words (BOW) voting strategy to transform SURF key point descriptors of an image to a feature vector with fixed dimension required for following classification procedure. The bag-of-words voting strategy is a popular machine learning feature extraction scheme to transform a set of observations obtained from an unclassified object instance to an unified measurement vector which represents a fingerprint to the unclassified object instance. The fingerprint then can serve as a feature vector for classification.

3.2   SURF Key Point Selection

As image features are derived from SURF point descriptors in the proposed scheme, how to choose SURF points is a critical design issue. On the one hand, we would like to select top-k SURF key points among all possible candidate points in the image, which can be done by applying the SURF algorithm with whole image as search space. These SURF key points are with better resistance against image scaling. However, the question is non-uniform distribution of these key points in the whole image. This might cause troubles when the user uploaded image covers only those areas with few SURF key points. To alleviate such a non-uniform distribution problem, we adapted local SURF key point search as well. For each image I_i we partition the whole image into $12 \times 12 = 144$ grids. For each grid, we apply the SURF algorithm to find out one SURF key point in the grid. To sum up, we use a hybrid approaches to generate SURF key points. We equally partition a image into 144 grids and generate one key point per grid. Also, we enlarge the search space to whole image and have the SURF algorithm to find up to 32 best scale-invariant SURF key points. The actual number of SURF key points can be found depends on the image. Consequently, for each image I_i, the number of SURF key points we generate, N_i, is not a fixed number. It is between 144 and $176(144 + 32)$.

We did not use the SURF key point descriptors as the fingerprint for classification directly. Instead, we apply bag-of-word procedure to transform the SURF key point descriptors extracted from the image to a real-number vector corresponding to the correlation response of the input image to each groups of images belonging to a specified location label. There are two reasons why we did not use the SURF key point descriptors of an image as the classification feature. First, we would like the feature vector for classification is with fixed dimension, which is required for most of classification algorithms. As aforementioned, feature vector based on the SURF key point descriptors extracted from each image is with a variable dimension range from 144 up to 176. Second, instead of comparing user uploaded image with each images in the database directly, we would like to compare the uploaded image with each groups of images represented a special location in the database. Consequently, we need one more step to fuse all SURF key point descriptors contributed by all the images associated with same location tag. Accordingly, we use bag-of-word(BOW) voting mechanism to transform $s_i = \{s_{i,j} | 1 \leq j \leq N_i\}$, the SURF key point descriptors of image I_i, to a fixed-dimension feature vector $f_i = <f_{i,1}, f_{i,2}, ..., f_{i,B}>$ for classification. In the following two sections we describe how to generate the dictionary for the BOW voting and how to do the voting.

3.3   Bag-of-Word Voting and Classification Feature Extraction

In this section, we present how to generate BOW dictionary for BOW voting. For each location label, we generate a set of pseudo SURF key point descriptors as the "words" in the BOW dictionary. These "words" play a role as pivot points in the feature space. Assume I^c is the set consists of all the images in the training data set labeled with class c. We employed clustering methods to generate a set of B^c pseudo SURF key point descriptors as the pivot points in the feature space for location label c.

Bag-of-Word Dictionary Generation. For each location labeled with class c, we collect a set of image I^c to cover the locations. All the images in the set is labeled with class c. Similar procedures are applied to all the location in the positioning service space. We assume there are C distinct locations to be labeled. Without loss of generality, we assume the locations are labeled with integer number from 1 to C. For each of location labeled with class c, we generate B^c pivot words for the location, from all the images in the database labeled with location c. For each image I_i^c in the images database labeled with location c, we extract N_i^c SURF key point descriptors from the image according to the SURF key point selection procedure mentioned in previous sections. All the SURF key point descriptors for location c are collected to a set s^c. Then, we apply Kmeans clustering algorithm to cluster all the data points in s^c into B^c clusters. The pseudo center points of each clusters are chosen as the pivot points for the location class.

BOW Voting and Feature Extraction. Given a BOW dictionary, we then calculate BOW profile of an input image to the BOW dictionary. Conceptually, each SURF key point descriptor can be treated as a "word" in the BOW voting procedure. The N_i SURF key point descriptors extracted from image I_i can be treated as N_i "words" associated with the image. Given a BOW dictionary, the BOW profile of image I_i is defined to be the word occurrence pattern of each word in the BOW dictionary. Algorithm 1 shows detail steps of the BOW voting procedure. The feature vector associated with the image for classification is then derived from the voting results, as shown in the algorithm.

3.4   Prototype System Implementation

We have implement the proposed positioning system. The prototype system consists of three major components: mobile App software to serve as user interface, web server for user web connection, and the back-end server for image processing. The mobile App was developed on Eclipse for Android-based smartphones. When users active the mobile App, the App instructs users to take a picture to the scenes around them, and then requests the user to upload the picture to a web server. We setup the web server page with PHP web programming language to handle user picture upload and send back the results. The back-end server is

---

Algorithm 1. Classification_Feature_Extraction

1: **procedure** CLASSFEATUREEXT
2: **input:**
3:    $m = < m_1, m_2, ..., m_B >$; //a sequent of B pivot words, $B = \sum_c B^c$
4:    I_i; //the input image.
5: **output:**
6:    $f_i = < f_{i,1}, f_{i,2}, ..., f_{i,B}]$; //classification feature of the input image.
7: **begin:**
8:        $v[b] = 0, \forall 1 \leq b \leq B$; // reset the voting count for each pivot words.
9:        //get the SURF key point descriptors for the input image
10:       s_i=get_SURF_key_point_descriptors(I_i);
11:       **for** each SURF key point $s_{i,j} \in s_i$ **do**
12:            // find the index of the most similar pivot words in m
13:            $b = \arg_k \min\{dist(s_{i,j}, m_k)|1 \leq k \leq B\}$;
14:            v[b]++;
15:       **end for**
16:       $c = \sum_b v[b]$; //total vote count
17:       $f_i[b] = v[b]/c$; // normalize the vote count vector v
18:       return f_i;
19: **end procedure**

---

a Intel Core i5 PC. We implemented all the image processing and classification code with MS Visual Studio 2010 and OpenCV 2.4 Library [10]. The classifier we used in the prototype system is SVM (Support Vector Machine).

4   Performance Evaluation

To evaluate system performance of the proposed mechanism, intensive experiments had been done. We evaluated classification performance with different classifiers. The images for the experiments are from a set of outward appearance of several buildings (in a campus). The reason we used outward appearance of buildings instead of indoor scene is due to the limitation of a proper indoor scene which should be large enough we can have at the time. Figure 4 shows the building complex for the positioning system performance evaluation. We took picture at the building complex alone the surrounding roads of the buildings. The images were classified into 17 classes, based on the location the images belonging to, as labeled (1 to 17) in the figure. The roads are indicated (A, B, C, D) in the figure. We took pictures at about every five meters alone each roads. The location spots we took the pictures are marked as a dot along each road in the figure, with a label combining the road and a sequence number. For example, location label "12A" is the 12-th location spot alone road A. For each location spots we took about 150–180 pictures cover different shooting angles ($+/-30°$ perpendicular to the building) at different days. Taking pictures at different shooting angles and different days is to emulate user possible behaviors to take the pictures under different outdoor conditions. Meanwhile, each of the images are associated with its device azimuth (got from smartphone's orientation sensor) and true location label observed when the image was taken.

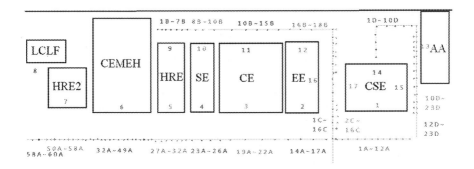

Fig. 4. The building complex for the positioning system performance evaluation.

We evaluated our proposed scheme under different machine learning classifiers (ANN, Bayesian, Decision Tree, and SVM) and different degrees of azimuth angle partitions. The azimuth angle is obtained from the smartphone device (orientation-senor) while the picture was taken. We use the azimuth angle information to filter out possible image candidates to be compared. Assume the system partition the azimuth angle into γ non-overlapped regions. Each of the regions would cover $\delta = \frac{360}{\gamma}$ degree. Consequently, k-th region would includes images associated with azimuth angles in the range $[k * (\delta - 1), k * \delta)$. A special case is to set $delta = 1$ (1 angle region partition), which means no partition. We also evaluated system performance under small image size. As shown in Fig. 5, we artificially created a small image size pictures (320×240) from original 640×480 resolution pictures. Hereafter, we refer the original image resolution as full-sized image, and the small-sized image as quarter-sized image.

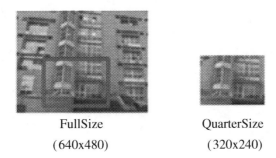

FullSize   QuarterSize

(640x480)   (320x240)

Fig. 5. Examples of image size tested in the experiment.

Figure 6(a) show the experiment results for the cases of full-sized images. Among the four classifiers (Ann, Bayesian, Decision Tree, and SVM) used in the experiments, SVM out performs the other three classifiers. With SVM classifier, our proposed scheme can achieve as high as 98.3 % correction rate (under 12 angle region partition) in the experiment. Similar results are found for the cases of quarter-sized image (Fig. 6(b)), in which SVM performs best followed by Bayesian classifier. The best correction rate under the case of quarter-sized images is about 91.8 % achieved by SVM (under 12 angle region partition) as well. Meanwhile, as shown in Fig. 6(a) and (b), the results indicate that angle region partition can improve correction rate of location positioning to some degrees depending on the classifiers being used. The results show the correction rate under 12-partition-region always outperforms that under 1-partition-region (that is, no region partition), no matter which of the four classifiers is applied. However, we also found that increasing partition regions did not always result in higher correction rate. For example, all the schemes under 48-partition-region result in lower correction rate, as shown in the figure. We guest possible reasons are due to insufficiency of training samples in the cases of finer region partition. In the experiments, when region partition scheme is applied, all the training images we collected are assigned to different partition regions according to the azimuth angle associated with each of the images. Each partition regions are treated independently. When the number of partition regions gets higher, the number of training images assigned to a partition region become lower. In our experiments, we found some of partition regions are with less than 10 training sample images.

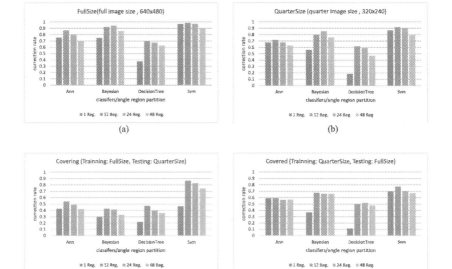

Fig. 6. Correction rate of location classification under different classifiers and number of region partitions. (a) Full image size (640×480), (b) Quarter image size (320×240), (c) Covering, (d) Covered. (Color figure online)

Meanwhile, we also evaluated the effects of scope-of-view mismatching between the training images and test images. Here scope-of-view means the the covering of objects (scenes) appeared in the image. For example, in Fig. 5, the objects (scenes) shown in the image on the left covers all the objects (scenes) shown in the image on the right. The situation could happen as users might take a picture with wide or narrow scope-of-view which is not consistent with the images we used to build the classifier model (as training image data). We simulated the effects by using the full-sized images and their corresponding quarter-sized images at the same time. One of them serves as training data set, and the other serves as testing data set. The results are show in Fig. 6(c) and (d). Again, SVM performs best. For the case of "covering" where we used full-sized images as training data and tested by quarter-sized images, the best correction rate is 86.7 % (SVM with 12-partition-region). The case for "covered" where we used quarter-sized images as training data and tested by full-sized images, the best correction rate is 77.3 % (SVM with 12-partition-region). It is worthy of noting that The scope-of-view of full-sized images is four times larger than that in the corresponding quarter-sized images. The results show even with such large difference in scope-of-view, the proposed scheme still can achieve over 77 % correction rate which is significantly better than random guest (that is $1/17 = 5.88\%$).

5   Concluding Remarks

In this paper, we present an universal position identification scheme for both indoor and outdoor environment. The proposed scheme utilizes visual and orientation sensor information to detect user location indoors (and outdoors). We designed and implemented a smartphone based positioning system which allows users to get their current location information via their smartphones. The client software for user smartphones can be distributed and installed easily through standard mobile Apps. Once the mobile App is ready in user smartphones, they just need to take a picture at target objects, and upload the picture to the remote positioning detection servers. The servers identify the location, based on the upload picture and other related information. We implemented the proposed positioning system for Android-based smartphones. We setup the web server page with PHP web programming language to handle user picture upload and send back the results. The back-end server was implemented with MS Visual Studio 2010 and OpenCV 2.4 Library.

Meanwhile, we evaluated our proposed scheme under different machine learning classifiers (ANN, Bayesian, Decision Tree, and SVM) and different degrees of azimuth angle partitions. Our experiment results indicate that angle region partition can improve correction rate of location positioning to some degrees depending on the classifiers being used. Among all the possible system configuration considered in the experiments, the method combining SVM classifier with 12-angle-region partition scheme can achieve best correction rate for most of the cases. With such a system configuration, the proposed positioning scheme can achieve 98.3 % and 91.8 % correction rate for full-sized images (640×480)

and quarter-sized image (320×240) respectively. We also evaluated the effects of scope-of-view mismatching between the training images and test images. For the case of "covering" where we used full-sized images as training data and tested by quarter-sized images, the best correction rate is 86.7 %. The case for "covered" where we used quarter-sized images as training data and tested by full-sized images, the best correction rate is 77.3 %. The results show even with such large difference in scope-of-view, the proposed scheme still can achieve over 77 % correction rate which is significantly better than random guest (that is $1/17 = 5.88\%$).

References

1. Leu, J.-S., Tzeng, H.-J.: Received signal strength fingerprint and footprint assisted indoor positioning based on ambient Wi-Fi signals. In: 75th IEEE Vehicular Technology Conference (VTC Spring), pp. 1–5. IEEE Press, New York (2012)
2. Chen, F., Au, W.S.A., Valaee, S., Tan, Z.: Received-signal-strength-based indoor positioning using compressive sensing. IEEE Trans. Mob. Comput. **11**(12), 1983–1993 (2012)
3. Chai, J.: Patient positioning system in hospital based on Zigbee. In: International Conference on Intelligent Computation and Bio-Medical Instrumentation (ICBMI), pp. 159–162. IEEE Press, New York (2011)
4. Zhou, S., Pollard, J.K.: Position measurement using bluetooth. IEEE Trans. Consum. Electron. **52**(2), 555–558 (2006)
5. Ni, L.M., Zhang, D., Souryal, M.R.: RFID based localization and tracking technologies. IEEE Wirel. Commun. **18**(2), 45–51 (2011)
6. Deng, Z., Yu, Y., Yuan, X., Wan, N., Yang, L.: Situation and development tendency of indoor positioning. China Commun. **10**(3), 42–55 (2013)
7. Van de Sande, K.E.A., Gevers, T., Snoek, C.G.M.: Evaluating color descriptors for object and scene recognition. IEEE Trans. Pattern Anal. Mach. Intell. **32**(9), 1582–1596 (2010)
8. Lowe, D.G.: Object recognition from local scale-invariant features. In: 7th IEEE International Conference on Computer Vision, pp. 1150–1157. IEEE Press, New York (1999)
9. Bay, H., Ess, A., Tuytelaars, T., Van Gool, L.: SURF: speeded up robust features. Comput. Vis. Image Underst. **110**(3), 346–359 (2008)
10. OpenCV. http://opencv.org

Prototypical Design and Implementation of an Intelligent Network Data Analysis Tool Collaborating with Active Information Resource

Kazuto Sasai$^{(\boxtimes)}$, Hideyuki Takahashi, Gen Kitagata, and Tetsuo Kinoshita

Research Institute of Electrical Communication, Tohoku University, Sendai, Japan
kazuto@riec.tohoku.ac.jp

Abstract. The methodologies of data analysis such as data mining, statistical analysis and machine learning are important notion in the network management in order to deal with complex network structure and growing network threats. Although a number of analytics tools are available in present day, it is not easy to apply the tools for network management because the ordinal administrators do not have the professional knowledge about mathematics and statistics. To improve the knowledge problem of network management, in this paper, we propose an agent-based analysis support system for network data and its collaboration mechanism with autonomic network management system. The evaluation experiment using prototypical system shows that the system can reduce intellectual load of the analysis task of the users.

Keywords: AIR-NMS · Data analysis · Multiagent system · Network data

1 Introduction

Data analysis is one of the important element of recent network and systems management. For the cloud resource management, data mining is used to predict the future demand of computing resources [1], and for the traffic anomaly detection, model analysis is applied to huge amount of traffic data which is obtained by log-term observation [2]. In the field of data center management, correlation analysis is used to detect anomalies in the sensors which is allocated in the large data centers [3,4]. On the other hand, there is a study about the tools to support data analysis processes using rich computing resource in the cloud foundation [5,6]. Actually, some cloud services to support data analysis have been recently released for consumer users [7–9]. Although the tools and the services provide many support functions for data analysis processes, it is not enough to solve network management problems because it remains some

This work was supported by Council for Science, Technology and Innovation (CSTI), Cross-ministerial Strategic Innovation Promotion Program (SIP), Enhancement of societal resiliency against natural disasters (Funding agency: JST).

© Springer International Publishing Switzerland 2016
H. Fujita et al. (Eds.): IEA/AIE 2016, LNAI 9799, pp. 1007–1018, 2016.
DOI: 10.1007/978-3-319-42007-3_85

domain dependent matters how to collect the data, how to format the data and how to extract knowledge from the data. Therefore, new data analysis support foundation which is easily collaborate with network data collection systems such as network management system and which is easy to reconstruct the structure according to the requirement.

To reduce complexity of network and systems management, there is a challenge to assemble the autonomic features of biological systems to the network devices, which is called autonomic network management [10]. Conventionally, autonomic actions in biological systems is represented with the sensory motor coupling [11], which indicates that a biological system has several sensors and complex actuators, and cognitive and/or decision-making functions bridge the two elements. Some practical implementation of decision-making functions of network systems is demonstrated [12,13]. However, since the studies depend on the predefined action and decision rules, the rules have to be updated according to the environmental changes. Further, although the researchers continuously provide new methodologies to understand the data collected from the network systems, it is difficult to deploy new methodologies to the autonomic systems. Therefore, a new kind of autonomic network management foundation on which human developers and maintainers easily build, reuse, and exploit knowledge resources to the network management systems.

Recently, Kephart [14] asserted that a view of decision-making systems should include both of machines and humans, and they collaborate each other to adapt knowledge resources to target problems, and they called that symbiotic cognitive computing. In this paper, as a foundation for symbiotic cognitive computing, we discuss about a system to support network data analysis which connect to our data-centric network management system which is called Active Information Resource-based Network Management System (AIR-NMS) [15]. AIR-NMS is an information system for network and systems management which consists of individual information resource including the autonomic features to realize self-management [16]. We introduce an agent-based design of network data analysis tools that is autonomously organized by according to the request of users. It is expected that highly reusable and deployable symbiotic cognitive systems can be established by connecting agent-based network data analysis tools and AIR-NMS.

In Sect. 2, we briefly introduce about AIR-NMS and agent-based analysis tools for network and systems management. Section 3 includes the practical design scheme and a prototypical implementation. Section 4 shows some experiments which were conducted to evaluate the prototypical system. Section 5 shows the conclusion of this paper.

2   Agent-Based Network Data Analysis Tools

Agent-based network management is a concept that the agents perform the tasks of network management instead of the human administrators, and the complexity of the network management tasks is reduced by the collaboration between the

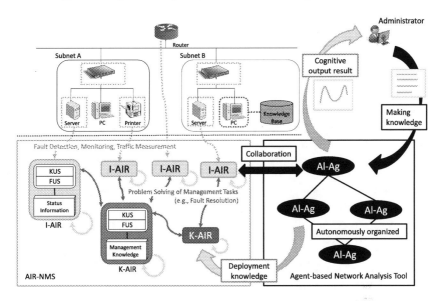

Fig. 1. A schematic diagram of the proposed analytics tool and of the AIR-NMS structure proposed in [15].

agents [17,18]. On the other hand, the software agent technologies are applied to data analysis to reduce the complexity of data analysis process by using self-organization as conceptually introduced in [19]. However, since the source of raw data such as sensors and these analysis tools are separated and distributed, it is difficult to deploy new analysis methodologies into on-going analysis systems. Thus, we propose an agent-based network data analysis support system that is connected to AIR-NMS.

2.1 Active Information Resource-Based Network Management System (AIR-NMS)

AIR-NMS [15] is a network management system including various Active Information Resources (AIRs) about network management. An AIR is structured by the information resource itself, the functions to access and process, and the knowledge to understand meanings and to link the other AIRs. To deal with the information resources about network management, we categorized the AIRs into two types, I-AIR and K-AIR as shown in Fig. 1. I-AIR is categorized that includes status data collected from servers, network equipment, and other devices. K-AIR includes the knowledge that recognize what happen in the target network, and that execute some recovery action to the network system. The I-AIRs and the K-AIRs in an AIR-NMS are always linking each other, and autonomously exchange messages. An example of prototypical implementation is presented in [15]. The knowledge for the fault detection and recovery such as symptoms, causes, and counter measures are stored in the K-AIRs, and they can apply knowledge by collecting specific information of the current status and the configurations from the I-AIRs.

2.2   Analysis Agent

As shown in Fig. 1, we define an analysis agent (Al-Ag) is an agent that performs the analysis tasks that are the component of a complete analysis process, which includes collecting, preprocessing, analyzing, and visualizing objective data. Al-Ags are designed to collaborate each other in order to connect the output of an agent to the input of another agent. Since the Al-Ags have the capability of self-organization as below, the analysis process is composed on demand, and the administrator can easily achieve the analysis results without considering the compatibility of the components, which mostly requires professional knowledge. In addition to Al-Ags, we design the collaboration scheme between an organization of Al-Ags and AIR-NMS. It supports the task of data collection using the function of I-AIRs, which can respond to the request messages of data retrieve. Further, by extracting and forming knowledge from the analysis results, it is possible to return the exploitable knowledge that can be used as the management knowledge included in K-AIRs.

In this paper, we design Al-Ags according to the development framework of software agents, ADIPS/DASH [20,21], which provides a production system for knowledge utilization and a FIPA compatible message exchange mechanism among DASH agents. The DASH framework has an agent repository where the agents are active and ready to construct multiagent organizations according to the users' requests. A structured organization can be allocated to the distributed workplaces. Here, we design and implement the prototypical Al-Ags, which includes experimental and simplified templates of the three agent types, Analyzer, Preprocessor, and Visualizer.

Analyzer has a function to perform processing an analysis method to objective data, and also has the knowledge to support efficient processing which consists of required preprocessing, possible visualization methods and some result meaning support information of the result.

Preprocessor has a function to transform a data to the appropriate format for the successive agents. For instance, in the case of data mining, preprocessing means preparation of target data set such as case balancing, probabilistic sampling and so on. Since the role of Preprocessor is important to improve performance and accuracy of analysis, it should have knowledge to control parameters for the transformation.

Visualizer receives processed results from Analyzer. If the data format satisfies the requirements to plot format, Visualizer displays plotting results in the user interface, else if it is not, sending request message to Preprocessor to convert data format, else if there is no conversion method, it come back to waiting state.

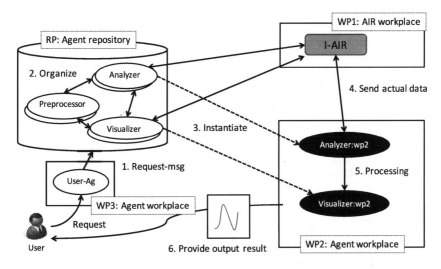

Fig. 2. A schematic diagram of organization procedure of analysis agents and schema of collaboration with AIR-NMS.

3   Design and Implementation of Prototype System

In this section, we explain how to design the collaboration scheme between the agents, and also between the agents and AIR-NMS. Figure 2 shows a schematic diagram of the prototypical system. The processing flow of the prototypical system is as follows:

1. When the user tells a request to user agent (User-Ag), the User-Ag sends a request message to the agents in the agent repository.
2. When an agent receives the request message, a self-organization process is executed, and an organization is constructed as a result. In this process, the message is also send to AIR-NMS, and an I-AIR which has the required data is selected.
3. After the self-organization process, the agents in the organization are instantiated (activated) in the target agent workplace, which is depicted as WP2 in the Fig. 2.
4. The I-AIR start to send the required data to the prescribed agent.
5. Then the agents start to execute their functions.
6. The selected output agent (probably a visualizer in usual) show the output to the user.

3.1   Self-organization of Agents

In this section, we explain the detailed process of self-organization in the agent repository. When the user agent sends user's request to the Analyzers in the agent repository, and the Analyzers check the request whether it can be accepted or not. After that, if an analyzer accepted the request, it sends messages to the other Al-Ags necessary to make the analysis process, and simultaneously send broadcast message to ask the required data to AIR-NMS. If all conditions are satisfied, the Analyzer sends messages to the members of the organization which involves the AIRs and the Al-Ags. When the agents in the organization are instantiated, they check the individual destination, which indicates the subsequent agent, and then they can establish the link states. Finally, the I-AIR start to send data to the next Al-Ags as mentioned above.

3.2   Collaboration with AIR-NMS

The AIR-NMS is also designed and implemented according to the agent framework ADIPS/DASH. In this paper, we focus on the I-AIR which stores various status information about the target network system and design the message exchange scheme between the I-AIRs and the Al-Sgs. The detail and the other part of AIR-NMS are explained in [15]. In the self-organization process of the Al-Ags, the agents should check whether the data required for processing exists or not. The data retrieve process is realized by message exchange between the Al-Ag and the I-AIRs. The Al-Ag send the data retrieve message into the I-AIRs by using the addressing service of agent repository, which manages the name of the connected workplaces and the agents instantiated on the workplaces, and it is a function of ADIPS/DASH runtime environment. Additionally, we redesign the response action of the I-AIRs to send the return message to the Al-Ag. Since I-AIRs are originally designed to manage the stored data autonomously, from the perspective of the Al-Ags, they can retrieve the data without considering about the data sources. Consequently, we have established the interoperability between Al-Ags and I-AIRs, and the analysis support system consists of the Al-Ags can effectively utilize the function of the I-AIRs.

3.3   Implementation

In this section, in order to develop the prototypical system to support analysis task of network data, we consider two cases of the practical analysis of network data, a performance evaluation of a server and a time series analysis of a network traffic data. The performance evaluation is a task to calculate the load of the target server, and the time series analysis is a task to apply Fourier transformation to the time series of network traffic [22]. According to the aforementioned design, we implement eight agents including two Analyzers, two Preprocessors, three Visualizers and User-agent, and their agents run on the ADIPS/DASH environment. Figure 3 is a display of IDEA, which is the development environment of ADIPS/DASH systems. The oval shapes represent the agents instantiated in the repository and the workplaces. Figure 4(a) shows the user interface provided by User-Ag in Fig. 2, and the user can input the four entities, analysis method, object, target server and interval. "Analysis method" represents the method that the user will use. The user can select the method by drop down box. "Object" indicates what type of data is used for the analysis. The user

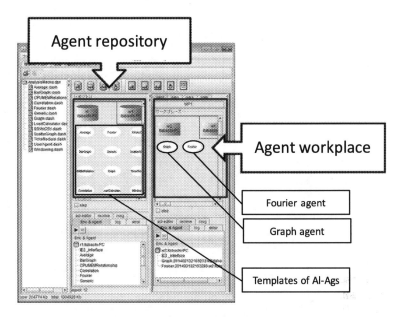

Fig. 3. The screenshot of IDEA, the development environment of ADIPS/DASH. The IDEA shows the agents in the repository and the workplace by representing the oval shapes.

can also select the types by drop down box. "Target server" represents the host to which analysis is applied. "Interval" indicates the time interval of the data. Figure 4(b) shows the window generated by visualizer. Here, the result of Fourier transformation of the network traffic data is displayed. The user can recognize the result of analysis by using individual visualization method of the agents.

4   Experiment

In order to evaluate the prototypical implementation as an analysis support system, we conducted an experiment with the human participants. In the experiment, the participants are asked to deal with some tasks of analyzing network data. We compare the usability of our prototypical system with manual strategy that the user can only use existing tools with some user-manuals. In this paper, we employed the Fourier analysis of the network traffic data for this experiment. The task was defined by investigating the researches about analysis of network data. Figure 5 shows the experimental environment. The experimental system consists of two computers, Node 1 and Node 2, connecting to the same LAN. Node 1 is used as a console that the participants use. Node 2 is a server which observes network traffic and stores the data. The participant in the picture (the left bottom side of Fig. 5) is performing the analysis task, and we record and

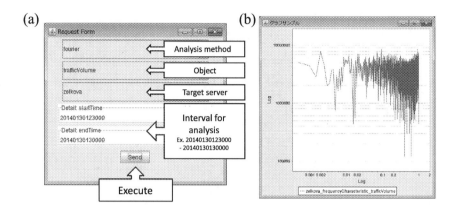

Fig. 4. The screenshots of (a) the user interface of User-Ag, and (b) an example of displayed result output, which is from Fourier-Ag.

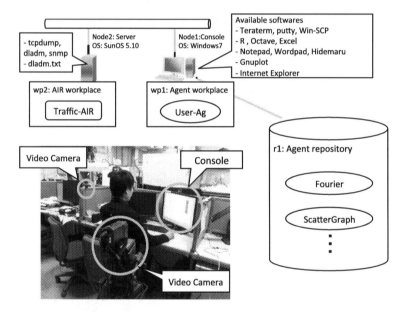

Fig. 5. Experiment environment for usability evaluation. There are two nodes that are in the same LAN. Node 1 is a console PC where a participant does various analysis tasks, and Node 2 is the target server in which analyzing data was captured. The DASH runtime environments are installed on each node. We putted two video cameras for side and back view of the participant and analyzed the records after the experiments.

monitor the behavior of the participant by the cameras and screen capturing. After the experiment, we classified the behaviors of the participants to some categories of actions, e.g., run software, search on a browser, etc.

4.1 Configuration

In an experiment, the participant can use any software on the Node 1 including search engines and analysis tools except asking to experimenter. For the case of manual strategy, some particular values such as IP address are preliminarily given by the instruction paper, which supposes a manual. The values and instructions are listed below:

1. *Objective Summery* gives meanings of Fourier analysis for traffic amount data, that are to detect malicious activities on the node and to characterize statistical behavior of clients. Further, we give an additional information that usual server traffic data shows simple white noise or sometimes 1/f noise [23].
2. *Analysis Steps (manual)* are presented as,
   (a) Retrieve data file from Node 2,
   (b) Extract traffic amount per 1 min,
   (c) Apply Fourier transformation to the target data,
   (d) Make a graph from the result of power spectrum,
   (e) Estimate meaning of the result.
3. *Analysis Steps (proposal)* are presented as,
   (a) Execute IDEA (with introduction to DASH/IDEA),
   (b) Instructions for UI (shown in Fig. 4(b)),
   (c) Estimate meaning of the result.
4. *Available Tools* installed in Node 2 are listed.
5. *Additional Information* are given as follows:

Table 1. Comparison of required knowledge.

Manual		Proposal	
Subject	Knowledge	Subject	Knowledge
Human	Hostname and IP address	Human	Name of data
Human	Data location	System	Hostname and IP address
Human	Name of target data	System	Data location
Human	Move data location	System	Data format
Human	Schema of data	System	Calculate traffic amount
Human	Value format of data	System	Modify data format
Human	Calculate traffic amount	System	Move data location
Human	Modify data format	System	Apply Fourier transformation
Human	Apply Fourier transformation	System	Suitable visualization by graph
Human	Visualize output by graph	System	Schema of outputted data
Human	Schema of outputted data	System	Selection of graph type
Human	Selection of graph type	Human	Understand results
Human	Understand results		
Required human knowledge: 13		Required human knowledge: 2	

(a) The IP address and hostname of Node 2.

(b) Traffic data is stored on Node 2 in the form of single file. The directory where the file is located is given in this part.

When the participants use our prototypical system, the experimenter tells how to use the system at the beginning of the experiment, though the instruction is just how to run the system.

4.2   Results

From the classified actions, we extracted knowledge required for each action. Table 1 shows the comparison of knowledge between proposal and manual. In the manual case, 13 categories of knowledge are required to link the network data and particular analysis method. On the other hand, the knowledge required for our proposal are only 2 categories, and it implies that our prototypical system can reduce the intellectual load of the users.

Table 2 shows average value of consumption time, steps and their reduction rate of the participants. For all participants, the consumption time and steps are

Table 2. Reduction of burden for data analysis.

Value	Manual	Proposal	Reduction rate [%]
Time consumption [min]	67	2	96.7
Number of Steps	72	7	90.2

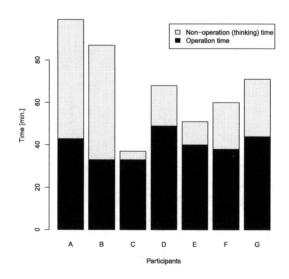

Fig. 6. Comparison of operating time with thinking time in the time consumption. Although the operating time is same among participants, the thinking time is different. It implies the time in thinking is susceptible to the variation of the skill levels.

drastically reduced over 90 %. From this result, we can confirm our prototypical system effectively supported the participants' tasks. Further, we analyzed the support function of our prototypical system in terms of the types of knowledge. Figure 6 shows the consumption time of manual case according to the individual participant by separating with operational steps and the other intellectual steps, where the reason why intellectual is that most of the steps except for operations are spent for understanding or investigating next operation. From the result, it is found that the time spent for operations are almost same length among the participants. It implies the deference of consumption time and steps for manual task depends on the skills, experiences and knowledge for network data analysis of individual participant. Thus, the prototypical system supported intellectual load rather than the operations, in other words, the effectiveness of the proposal is not only the automation of the task but also intellectual support.

5   Conclusion

In this paper, we propose an agent-based network data analysis support by collaboration with AIR-NMS. The Analysis Agents (AL-Ags) are categorized to three templates, Analyzer, Preprocessor, and Visualizer, and the templates can be organized according to the requested analysis pattern. The collaboration with AIR-NMS provides the function to retrieve appropriate network status data by using functions in I-AIRs. To evaluate the prototypical implementation, we conducted an experiment to compare with and without the proposed tool in an analysis task. As a result, we conclude the prototypical analysis support system can reduce the burden of the users because the raw data is automatically retrieved and seamlessly processed in the system. For the future work, it needs to evaluate reusability and scalability of the organized analysis agents, and some anomaly detection and recovery situation is preferable to test effectiveness of the proposed analysis tool.

References

1. Huang, Q., Shuang, K., Xu, P., Li, J., Liu, X., Su, S.: Prediction-based dynamic resource scheduling for virtualized cloud systems. J. Netw. **9**(2), 375–383 (2014)
2. Scherrer, A., Larrieu, N., Owezarski, P., Borgnat, P., Abry, P.: Non-Gaussian and long memory statistical characterizations for internet traffic with anomalies. IEEE Trans. Dependable Secur. Comput. **4**(1), 56–70 (2007)
3. Ma, X., Hu, C., Chen, K.: Error tolerant address configuration for data center networks with malfunctioning devices. In: 32nd IEEE International Conference on Distributed Computing Systems, pp. 708–717 (2012)
4. El-Sayed, N., Stedanovici, I.A., Amvrosiadis, G., Hwang, A.A., Schroeder, B.: Temperature management in data centers: why some (might) like it hot. In: the 12th ACM SIGMETRICS/PERFORMANCE Joint International Conference on Measurement and Modeling of Computer Systems, pp. 163–174 (2012)
5. Demirkan, H., Delen, D.: Leveraging the capabilities of service-oriented decision support systems: putting analytics and big data in cloud. Decis. Support Syst. **55**(1), 412–421 (2013)

6. Chong, D., Shi, H.: Big data analytics: a literature review. J. Manag. Anal. **2**(3), 175–201 (2015)
7. Google Analytics. https://www.google.com/analytics/
8. Microsoft Azure. https://azure.microsoft.com/en-us/
9. IBM Watson Analytics. http://www.ibm.com/analytics/watson-analytics/
10. Samaan, N., Karmouch, A.: Towards autonomic network management: an analysis of current and future research directions. IEEE Commun. Surv. Tutor. **11**(3), 22–36 (2009)
11. Flanders, M.: What is the biological basis of sensorimotor integration? Biol. Cybern. **104**, 1–8 (2011)
12. Maggio, M., Hoffmann, H., Papadopulos, A.V., Panerati, J., Santambrogio, M.D., Agarwal, A., Leva, A.: Comparison of decision-making strategies for self-optimization in autonomic computing systems. ACM Trans. Auton. Adapt. Syst. **7**(4), 36: 1–36: 32 (2012)
13. Paton, N., de Aragão, M.A.T., Lee, K., Fernandes, A.A.A., Sakellariou, R.: Optimizing utility in cloud computing through autonomic workload execution. Bull. Techn. Committee Data Eng. **32**(1), 51–58 (2009)
14. Kephart, J.O., Lechner, J.: A symbiotic cognitive computing perspective on autonomic computing. In: IEEE 12th International Conference on Autonomic Computing, pp. 109–114 (2015)
15. Sasai, K., Sveholm, J., Kitagata, G., Kinoshita, T.: A practical design and implementation of active information resource based network management system. Int. J. Energy Inf. Commun. **2**(4), 67–86 (2011)
16. Kinoshita, T., Kitagata, G., Takahashi, H., Sasai, K., Kalegele, K.: An agent-based network management system using active information resources. Int. J. Adv. Smart Convergence **2**(2), 10–15 (2013)
17. Beszczad, A., Pagurek, B., White, T.: Mobile agents for network management. IEEE Commun. Surv. **1**(1), 2–9 (1998)
18. Terauchi, A., Akashi, O., Maruyama, M., Sugawara, T., Fukuda, K., Hirotsu, T., Kurihara, S., Koyanagi, K.: Agent organization system for multi-agent based network management. Trans. Jpn. Soc. Artif. Intell. **22**(5), 482–492 (2007)
19. Kalegele, K., Sasai, K., Takahashi, H., Kitagata, G., Kinoshita, T.: Four decades of data mining in network and systems management. IEEE Trans. Knowledge Data Eng. **27**(10), 2700–2716 (2015)
20. Kinoshita, T., Sugawara, K.: ADIPS framework for flexible distributed systems. In: Ishida, T. (ed.) PRIMA 1998. LNCS (LNAI), vol. 1599, p. 18. Springer, Heidelberg (1999)
21. Uchiya, T., Maemura, T., Li, X., Kinoshita, T.: Design and implementation of interactive design environment of agent system. In: Okuno, H.G., Ali, M. (eds.) IEA/AIE 2007. LNCS (LNAI), vol. 4570, pp. 1088–1097. Springer, Heidelberg (2007)
22. Shevtekar, A., Anantharam, K., Ansari, N.: Low rate TCP denial-of-service attack detection at edge routers. IEEE Commun. Lett. **9**(4), 363–365 (2005)
23. Csabai, I.: 1/f noise in computer network traffic. J. Phys. A Math. Gen. **27**(12), L417 (1994)

Author Index

Printed in the United States
By Bookmasters